O

D0400639

A Dictionary of

# Science

 SEE WEB LINKS

Many entries in this dictionary have recommended web
links. When you see the above symbol at the end of an entry
go to the dictionary's web page at http://www.oup.com/
uk/reference/resources/science, click on **Web links** in the
Resources section and locate the entry in the alphabetical
list, then click straight through to the relevant websites.

**Oxford Paperback Reference**

The most authoritative and up-to-date reference books for both students and the general reader.

Many of these titles are also available online at www.oxfordreference.com

# A Dictionary of
# Science

## SIXTH EDITION

# OXFORD

UNIVERSITY PRESS

Great Clarendon Street, Oxford OX2 6DP

Oxford University Press is a department of the University of Oxford.
It furthers the University's objective of excellence in research, scholarship,
and education by publishing worldwide in

Oxford New York

Auckland Cape Town Dar es Salaam Hong Kong Karachi
Kuala Lumpur Madrid Melbourne Mexico City Nairobi
New Delhi Shanghai Taipei Toronto

With offices in

Argentina Austria Brazil Chile Czech Republic France Greece
Guatemala Hungary Italy Japan Poland Portugal Singapore
South Korea Switzerland Thailand Turkey Ukraine Vietnam

Oxford is a registered trade mark of Oxford University Press
in the UK and in certain other countries

Published in the United States
by Oxford University Press Inc., New York

First published 1984 as *Concise Science Dictionary*
Second edition 1991
Third edition 1996
Fourth edition 1999
Fifth edition 2005
Sixth edition 2010

British Library Cataloguing in Publication Data
Data available

Library of Congress Cataloging in Publication Data
Data available

Typeset by Market House Books Ltd.
Printed in Great Britain by
Clays Ltd, St Ives plc

ISBN 978–0–19–956146–9

9

REF
503
DIC
2010

# Contents

# Preface

This sixth edition of *A Dictionary of Science*, like its predecessors, aims to provide school and first-year university students with accurate explanations of any unfamiliar words they might come across in the course of their studies, in their own or adjacent disciplines. For example, students of the physical sciences will find all they are likely to need to know about the life sciences, and vice versa. The dictionary is also designed to provide non-scientists with a useful reference source to explain the scientific terms that they may encounter in their work or in their general reading.

At this level the dictionary provides full coverage of terms, concepts, and laws relating to physics, chemistry, biology, biochemistry, palaeontology, and the earth sciences. There is also coverage of key terms in astronomy, cosmology, mathematics, biotechnology, and computer technology. In addition, the dictionary includes:

- over 160 short biographical entries on the most important scientists in the history of the subject

- ten features (each of one or two pages) on concepts of special significance in modern science

- ten chronologies showing the development of selected concepts, fields of study, and industries

- eight Appendices, including the periodic table, tables of SI units and conversion tables to and from other systems of units, summary classifications of the plant and animal kingdoms, and useful websites.

For this sixth edition nearly 700 new entries have been added to the text, incorporating recent advances in all the major fields and increased coverage of astronomy, forensic chemistry, and computing.

In compiling the dictionary, the contributors and editors have made every effort to make the entries as concise and comprehensible as possible, always bearing in mind the needs of the readers. Particular features of the book are its lack of unnecessary scientific jargon and its extensive network of cross-references. An asterisk placed before a word used in an entry indicates that this word can be looked up in the dictionary and will provide further explanation or clarification. However, not every word that is defined in the dictionary has an asterisk placed before it when it is used in an entry. Some entries simply refer the reader to another entry, indicating either that they are synonyms or abbreviations or that they are most conveniently explained in one of the dictionary's longer articles. Synonyms and abbreviations are usually placed within brackets immediately after the headword. Terms that are explained within an entry are highlighted by being printed in boldface type. Where appropriate, the entries have been supplemented by fully labelled line-drawings or tables *in situ*.

JD
EM
2009

# Credits

## Editors

John Daintith BSc, PhD
Elizabeth Martin MA

## Advisers

B. S. Beckett BSc, BPhil, MA(Ed)
R. A. Hands BSc
Michael Lewis MA

## Contributors

Richard Batley PhD, MInstP
Tim Beardsley BA
Lionel Bender BSc
John Clark BSc
H. M. Clarke MA, MSc
W. M. Clarke BSc
Derek Cooper PhD, FRIC
John Cullerne DPhil
R. Cutler BSc
E. K. Daintith BSc
D. E. Edwards BSc, MSc
A. J. H. Goddard PhD, MIMechE,
  FInstNucE, FInstP, MSRP
William Gould BA

Malcolm Hart BSc, MIBiol
Robert S. Hine BSc, MSc
Elaine Holmes BSc, PhD
Valerie Illingworth BSc, MPhil
Anne Lockwood BSc
J. Valerie Neal BSc, PhD
R. A. Prince MA
Richard Rennie BSc, MSc, PhD
Michael Ruse BSc, PhD
Jackie Smith BA
Brian Stratton BSc, MSc
Elizabeth Tootill BSc, MSc
David Eric Ward BSc, MSc, PhD
Edmund Wright DPhil

**aa** *See* LAVA.

**AAR** *See* AMINO ACID RACEMIZATION.

**AAS** *See* ATOMIC ABSORPTION SPECTROSCOPY.

**ab-** A prefix attached to the name of a practical electrical unit to provide a name for a unit in the electromagnetic system of units (*see* ELECTROMAGNETIC UNITS), e.g. abampere, abcoulomb, abvolt. The prefix is an abbreviation of the word 'absolute' as this system is also known as the **absolute system**. *Compare* STAT-. In modern practice both absolute and electrostatic units have been replaced by *SI units.

**abdomen** The posterior region of the body trunk of animals. In vertebrates it contains the stomach and intestines and the organs of excretion and reproduction. It is particularly well defined in mammals, being separated from the *thorax by the *diaphragm. In many arthropods, such as insects and spiders, it may be segmented.

**Abelian group** *See* GROUP.

**aberration** **1.** (in optics) A defect in the image formed by a lens or curved mirror. In **chromatic aberration** the image formed by a lens (but not a mirror) has coloured fringes as a result of the different extent to which light of different colours is refracted by glass. It is corrected by using an *achromatic lens. In **spherical aberration**, the rays from the object come to a focus in slightly different positions as a result of the curvature of the lens or mirror. For a mirror receiving light strictly parallel with its axis, this can be corrected by using a parabolic surface rather than a spherical surface. Spherical aberration in lenses is minimized by making both surfaces contribute equally to the ray deviations, and can be lessened (though with reduced image brightness) by the use of diaphragms to let light pass only through the centre part of the lens. *See also* ASTIGMATISM; COMA. **2.** (in astronomy) The apparent displacement in the position of a star as a result of the earth's motion round the sun. Light appears to come from a point that is slightly displaced in the direction of the earth's mo-

tion. The angular displacement $\alpha = v/c$, where $v$ is the earth's orbital velocity and $c$ is the speed of light.

**abiogenesis** The origin of living from nonliving matter, as by *biopoiesis. *See also* SPONTANEOUS GENERATION.

**abiotic factor** Any of the nonliving factors that make up the **abiotic environment** in which living organisms occur. They include all the aspects of climate, geology, and atmosphere that may affect the biotic environment. *Compare* BIOTIC FACTOR.

**abomasum** The fourth and final chamber of the stomach of ruminants. It leads from the *omasum and empties into the small intestine. The abomasum is referred to as the 'true stomach' as it is in this chamber that protein digestion occurs, in acidic conditions. *See* RUMINANTIA.

**ABO system** One of the most important human *blood group systems. The system is based on the presence or absence of *antigens A and B on the surface of red blood cells and of *antibodies against these in blood serum. A person whose blood contains either or both these antibodies cannot receive a transfusion of blood containing the corresponding antigens as this would cause the red cells to clump (*see* AGGLUTINATION). The table overleaf illustrates the basis of the system: people of blood group O are described as 'universal donors' as they can give blood to those of any of the other groups. *See also* IMMUNE RESPONSE.

**abscisic acid** A naturally occurring plant hormone that appears to be involved primarily in seed maturation, stress responses (e.g. to heat and waterlogging), and in regulating closure of leaf pores (stomata). In seeds, it promotes the synthesis of storage protein and prevents premature germination. In leaves, abscisic acid is produced in large amounts when the plant lacks sufficient water, promoting closure of stomata and hence reducing further water losses. It was formerly believed to play a central role in *abscission, hence the name.

| Group | Antigens on red cell surface | Antibodies in serum | Blood group of people donor can receive blood from | Blood group of people donor can give blood to |
|---|---|---|---|---|
| A | A | anti-B | A, O | A, AB |
| B | B | anti-A | B, O | B, AB |
| AB | A and B | none | A, B, AB, O | AB |
| O | neither A nor B | anti-A and anti-B | O | A, B, AB, O |

**ABO system.**

**abscissa** *See* CARTESIAN COORDINATES.

**abscission** The separation of a leaf, fruit, or other part from the body of a plant. It involves the formation of an **abscission zone**, at the base of the part, within which a layer of cells (**abscission layer**) breaks down. This process is suppressed so long as sufficient amounts of *auxin, a plant hormone, flow from the part through the abscission zone. However, if the auxin flow declines, for example due to injury or ageing, abscission is activated. Ethylene (*ethene) acts as the primary trigger for abscission, inducing cells in the abscission zone to produce cellulase enzymes that degrade cell walls.

**absolute** 1. Not dependent on or relative to anything else, e.g. *absolute zero. 2. Denoting a temperature measured on an **absolute scale**, a scale of temperature based on absolute zero. The usual absolute scale now is that of thermodynamic *temperature; its unit, the kelvin, was formerly called the degree absolute (°A) and is the same size as the degree Celsius. In British engineering practice an absolute scale with Fahrenheit-size degrees has been used: this is the Rankine scale.

**absolute alcohol** *See* ETHANOL.

**absolute configuration** A way of denoting the absolute structure of an optical isomer (*see* OPTICAL ACTIVITY). Two conventions are in use: The D–L convention relates the structure of the molecule to some reference molecule. In the case of sugars and similar compounds, the dextrorotatory form of glyceraldehyde (HOCH$_2$CH(OH)CHO), 2,3-dihydroxypropanal) was used. The rule is as follows. Write the structure of this molecule down with the asymmetric carbon in the centre, the –CHO group at the top, the –OH on the right, the –CH$_2$OH at the bottom, and the –H on the left. Now imagine that the central carbon atom is at the centre of a tetrahedron with the four groups at the corners and that the –H and –OH come out of the paper and the –CHO and –CH$_2$OH groups go into the paper. The resulting three-dimensional structure was taken to be that of *d*-glyceraldehyde and called D-glyceraldehyde. Any compound that contains an asymmetric carbon atom having this configuration belongs to the D-series. One having the opposite configuration belongs to the L-series. It is important to note that the prefixes D- and L- do not stand for dextrorotatory and laevorotatory (they are not the same as *d*- and *l*-). In fact the arbitrary configuration assigned to D-glyceraldehyde is now known to be the correct one for the dextrorotatory form, although this was not known at the time. However, all D-compounds are not dextrorotatory. For instance, the acid obtained by oxidizing the –CHO group of glyceraldehyde is glyceric acid (1,2-dihydroxypropanoic acid). By convention, this belongs to the D-series, but it is in fact laevorotatory; i.e. its name can be written as D-glyceric acid or *l*-glyceric acid. To avoid confusion it is better to use + (for dextrorotatory) and – (for laevorotatory), as in D-(+)-glyceraldehyde and D-(–)-glyceric acid.

The D–L convention can also be used with alpha amino acids (compounds with the –NH$_2$ group on the same carbon as the –COOH group). In this case the molecule is imagined as being viewed along the H–C bond between the hydrogen and the asymmetric carbon atom. If the clockwise order of the other three groups is –COOH, –R, –NH$_2$, the amino acid belongs to the D-series; otherwise it belongs to the L-series. This is known as the **CORN rule**.

The R–S convention is a convention based on priority of groups attached to the chiral carbon atom. The order of priority is I, Br, Cl, SO$_3$H, OCOCH$_3$, OCH$_3$, OH, NO$_2$, NH$_2$,

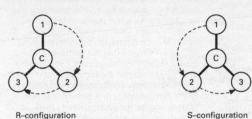

planar formula      structure in 3 dimensions      Fischer projection

D-(+)-glyceraldehyde (2,3-dihydroxypropanal)

D-alanine (R is $CH_2$ in the CORN rule). The molecule is viewed with H on top

R–configuration          S–configuration

R–S system. The lowest priority group is behind the chiral carbon atom

### Absolute configuration.

$COOCH_3$, $CONH_2$, $COCH_3$, CHO, $CH_2OH$, $C_6H_5$, $C_2H_5$, $CH_3$, H, with hydrogen lowest. The molecule is viewed with the group of lowest priority behind the chiral atom. If the clockwise arrangement of the other three groups is in descending priority, the compound belongs to the R-series; if the descending order is anticlockwise it is in the S-series. D-(+)-glyceraldehyde is R-(+)-glyceraldehyde. See illustration.

**⊕ SEE WEB LINKS**

• Information about IUPAC nomenclature for the R–S system

**absolute expansivity** See EXPANSIVITY.

**absolute humidity** See HUMIDITY.

**absolute permittivity** See PERMITTIVITY.

**absolute pitch (perfect pitch)** The ability of a person to identify and reproduce a note without reference to a tuned musical instrument.

**absolute space** Space that exists as a background to events and processes and is not affected by objects or other entities in the universe. The idea underpins Newtonian physics, although many physicists have always regarded absolute space as an undesirable concept and suggested, as in *Mach's principle, that fundamental physics should be described by *relational theories.

**absolute temperature** See ABSOLUTE; TEMPERATURE.

**absolute time** Time that exists independently of any events or processes in the universe. Like *absolute space, absolute time is a basic concept in Newtonian physics.

**absolute value (modulus)** The square root of the sum of the squares of the real numbers in a *complex number, i.e. the absolute value of the complex number $z = x + iy$ is $|z| = \sqrt{(x^2 + y^2)}$.

**absolute zero** Zero of thermodynamic *temperature (0 kelvin) and the lowest temperature theoretically attainable. It is the temperature at which the kinetic energy of atoms and molecules is minimal. It is equivalent to $-273.15°C$ or $-459.67°F$. *See also* ZERO-POINT ENERGY; CRYOGENICS.

**absorbed dose** *See* DOSE.

**absorptance** Symbol $\alpha$. The ratio of the radiant or luminous flux absorbed by a body to the flux falling on it. Formerly called **absorptivity**, the absorptance of a *black body is by definition 1.

**absorption** 1. (in chemistry) The take up of a gas by a solid or liquid, or the take up of a liquid by a solid. Absorption differs from *adsorption in that the absorbed substance permeates the bulk of the absorbing substance. 2. (in physics) The conversion of the energy of electromagnetic radiation, sound, streams of particles, etc., into other forms of energy on passing through a medium. A beam of light, for instance, passing through a medium, may lose intensity because of two effects: scattering of light out of the beam, and absorption of photons by atoms or molecules in the medium. When a photon is absorbed, there is a transition to an excited state. 3. (in biology) The movement of fluid or a dissolved substance across a plasma membrane. In many animals, for example, soluble food material is absorbed into cells lining the alimentary canal and thence into the blood. In plants, water and mineral salts are absorbed from the soil by the *roots. *See* OSMOSIS; TRANSPORT PROTEIN.

**absorption coefficient** 1. (in physics) *See* LAMBERT'S LAWS. 2. (in chemistry) The volume of a given gas, measured at standard temperature and pressure, that will dissolve in unit volume of a given liquid.

**absorption indicator** *See* ADSORPTION INDICATOR.

**absorption spectrum** *See* SPECTRUM.

**absorptivity** *See* ABSORPTANCE.

**ABS plastic** Any of a class of plastics based on acrylonitrile–butadiene–styrene copolymers.

**abundance** 1. The ratio of the total mass of a specified element in the earth's crust to the total mass of the earth's crust, often expressed as a percentage. For example, the abundance of aluminium in the earth's crust is about 8%. 2. The ratio of the number of atoms of a particular isotope of an element to the total number of atoms of all the isotopes present, often expressed as a percentage. For example, the abundance of uranium–235 in natural uranium is 0.71%. This is the **natural abundance**, i.e. the abundance as found in nature before any enrichment has taken place.

**abyssal zone** The lower depths of the ocean (below approximately 2000 metres), where there is effectively no light penetration. Abyssal organisms are adapted for living under high pressures in cold dark conditions. *See also* APHOTIC ZONE.

**a.c.** *See* ALTERNATING CURRENT.

**acac** The symbol for the *acetylacetonato ligand, used in formulae.

**accelerant** A flammable material used to start and spread a fire in cases of arson. Petrol and paraffin are the substances commonly used. Traces of accelerant are detectable by gas chromatography in forensic work.

**acceleration** Symbol $a$. The rate of increase of speed or velocity. It is measured in m s$^{-2}$. For a body moving linearly with constant acceleration $a$ from a speed $u$ to a speed $v$,

$$a = (v - u)/t = (v^2 - u^2)/2s$$

where $t$ is the time taken and $s$ the distance covered.

If the acceleration is not constant it is given by $dv/dt = d^2s/dt^2$. If the motion is not linear the vector character of displacement, velocity, and acceleration must be considered. *See also* ROTATIONAL MOTION.

**acceleration of free fall** Symbol $g$. The acceleration experienced by any massive object falling freely in the earth's gravitational field. Experimentally this is almost constant for all positions near the earth's surface, independent of the nature of the falling body (provided air resistance is eliminated). This is taken to indicate the strict proportionality of *weight (the force causing the acceleration) and inertial *mass, on the basis of Newton's second law of motion (*see* NEWTON'S LAWS OF MOTION). There is some variation of $g$ with latitude, because of the earth's rotation and because the earth is not completely spherical. The standard value is taken as 9.806 65 m s$^{-2}$. The acceleration of free fall is also called the **acceleration due to gravity**.

**accelerator** 1. (in physics) An apparatus for increasing the kinetic energies of charged particles, used for research in nuclear and particle physics. *See* CYCLOTRON; LINEAR ACCELERATOR; SYNCHROCYCLOTRON; SYNCHROTRON. 2. (in chemistry) A substance that increases the rate of a chemical reaction, i.e. a catalyst.

**⊕ SEE WEB LINKS**

• A list of world particle accelerators and accelerator laboratories, with links, from the ELSA website at the University of Bonn

**acceptor** 1. (in chemistry and biochemistry) A compound, molecule, ion, etc., to which electrons are donated in the formation of a coordinate bond. 2. (in biochemistry) A *receptor that binds a hormone without any apparent biological response. 3. (in physics) A substance that is added as an impurity to a *semiconductor because of its ability to accept electrons from the valence bands, causing *p*-type conduction by the mobile positive holes left. *Compare* DONOR.

**acceptor levels** Energy levels of an acceptor atom in a *semiconductor, such as aluminium, in silicon. These energy levels are very near the top of the valence band, and therefore cause *p*-type conduction. *See also* ENERGY BANDS

**access point** (wireless access point) A device that acts as the core of a wireless network. It communicates with wireless nodes within its range and provides the necessary facilities for them to network successfully. Access points commonly also manage a link to a wired network, allowing their nodes to link with corporate networks, the Internet, etc.

**acclimation** The physiological changes occurring in an organism in response to a change in a particular environmental factor (e.g. temperature), especially under laboratory conditions. Thermal acclimation studies reveal how such properties as metabolic rate, muscle contractility, nerve conduction, and heart rate differ between cold- and warm-acclimated members of the same species. These changes occur naturally during *acclimatization and equip the organism for living in, say, cold or warm conditions.

**acclimatization** 1. The progressive adaptation of an organism to any change in its natural environment that subjects it to physiological stress. 2. The overall sum of processes by which an organism attempts to compensate for conditions that would substantially reduce the amount of oxygen delivered to its cells. *Compare* ACCLIMATION.

**accommodation** 1. (in animal physiology) Focusing: the process by which the focal length of the *lens of the eye is changed so that clear images of objects at a range of distances are displayed on the retina. In humans and some other mammals accommodation is achieved by reflex adjustments in the shape of the lens brought about by relaxation and contraction of muscles within the *ciliary body. 2. (in animal behaviour) Adjustments made by an animal's nervous or sensory systems in response to continuously changing environmental conditions.

**accretion** The way in which collisions with relatively slow-moving smaller objects add to the mass of a larger celestial object. The process accelerates as the increased mass strengthens the gravitational field of the larger object. For example, the planets are thought to have formed by the accretion of dust particles onto *planetesimals. Other accreting objects probably include black holes and protostars.

**accretion disc** A disc-shaped rotating mass formed by gravitational attraction. *See* BLACK HOLE; NEUTRON STAR; WHITE DWARF.

**accumulator** (secondary cell; storage battery) A type of *voltaic cell or battery that can be recharged by passing a current through it from an external d.c. supply. The charging current, which is passed in the opposite direction to that in which the cell supplies current, reverses the chemical reactions in the cell. The common types are the *lead–acid accumulator and the *nickel–iron and nickel–cadmium accumulators. *See also* SODIUM–SULPHUR CELL.

**acellular** Describing tissues or organisms that are not made up of separate cells but often have more than one nucleus (*see* SYNCYTIUM). Examples of acellular structures are muscle fibres. *Compare* UNICELLULAR.

**acentric** Describing an aberrant chromosome fragment that lacks a centromere. Such fragments are normally lost because they are unable to orientate properly during cell division.

**acetaldehyde** *See* ETHANAL.

**acetaldol** *See* ALDOL REACTION.

**acetals** Organic compounds formed by addition of alcohol molecules to aldehyde molecules. If one molecule of aldehyde (RCHO) reacts with one molecule of alcohol (R'OH) a **hemiacetal** is formed (RCH(OH)OR'). The rings of aldose sugars are hemiacetals. Further reaction with a second alcohol molecule produces a full acetal (RCH(OR')$_2$). It is common to refer to both types of compounds simply as 'acetals'. The formation of acetals is reversible; acetals can be hydrolysed back to aldehydes in acidic solutions. In synthetic organic chemistry aldehyde groups are often converted into acetal groups to protect them before performing other reactions on different groups in the molecule. *See also* KETALS.

**(⊕) SEE WEB LINKS**
• Information about IUPAC nomenclature

**Formation of acetals.**

**acetamide** *See* ETHANAMIDE.

**acetanilide** A white crystalline primary amide of ethanoic acid, $CH_3CONHC_6H_5$; r.d. 1.2; m.p. 114.3°C; b.p. 304°C. It is made by reacting phenylamine (aniline) with excess ethanoic acid or ethanoic anhydride and is used in the manufacture of dyestuffs and rubber. The full systematic name is N-**phenylethanamide**.

**acetate** *See* ETHANOATE.

**acetate process** *See* RAYON.

**acetic acid** *See* ETHANOIC ACID.

**acetoacetic acid** *See* 3-OXOBUTANOIC ACID.

**acetoacetic ester** *See* ETHYL 3-OXOBU-TANOATE.

**acetone** *See* PROPANONE; KETONE BODY.

**acetone–chlor–haemin test (Wagenaar test)** A *presumptive test for blood in which a small amount of acetone (propenal) is added to the bloodstain, followed by a drop of hydrochloric acid. Haemoglobin produces derivatives such as haematin and haemin, forming small characteristic crystals that can be identified under a microscope.

**acetylacetonato** The ion ($CH_3COCHCOCH_3$)$^-$, functioning as a bidentate ligand coordinating through the two oxygen atoms. In formulae, the symbol **acac** is used.

**acetylation** *See* ACYLATION.

**acetyl chloride** *See* ETHANOYL CHLORIDE.

**acetylcholine (ACh)** One of the main *neurotransmitters of the vertebrate nervous system. It is released at some (**cholinergic**) nerve endings and may be excitatory or inhibitory; it initiates muscular contraction at *neuromuscular junctions. Once acetylcholine has been released it has only a transitory effect because it is rapidly broken down by the enzyme *cholinesterase.

**acetylcholinesterase** *See* CHOLINESTERASE.

**acetyl coenzyme A (acetyl CoA)** A compound formed in the mitochondria when an acetyl group ($CH_3CO-$), derived from the breakdown of fats, proteins, or carbohydrates (via *glycolysis), combines with the thiol group (–SH) of *coenzyme A. Acetyl CoA feeds into the energy generating *Krebs cycle and also plays a role in the synthesis and oxidation of fatty acids.

**acetylene** *See* ETHYNE.

**acetylenes** *See* ALKYNES.

**acetyl group** *See* ETHANOYL GROUP.

**acetylide** *See* CARBIDE.

**achene** A dry indehiscent fruit formed from a single carpel and containing a single seed. An example is the feathery achene of clematis. Variants of the achene include the *caryopsis, *cypsela, *nut, and *samara. *See also* ETAERIO.

**Acheson process** An industrial process for the manufacture of graphite by heating coke mixed with clay. The reaction involves the production of silicon carbide, which loses silicon at 4150°C to leave graphite. The process was patented in 1896 by the US inventor Edward Goodrich Acheson (1856–1931).

**achiral** Describing a molecule that does not contain a *chirality element.

**achondrite** A stony meteorite that has no spherical silicate particles (chondrules) found in the meteorites called chondrites. Achondrites do not contain iron or nickel and have a coarser crystal structure than chondrites.

**achromatic lens** A lens that corrects for chromatic *aberration by using a combination of two lenses, made of different kinds of glass, such that their *dispersions neutralize each other although their *refractions do not. The aberration can be reduced further by using an **apochromatic lens**, which consists of three or more different kinds of glass.

**aciclovir (acyclovir; acycloguanosine)** A drug used to treat cold sores, shingles, genital blisters, or other lesions caused by herpesvirus infection. It is an analogue of the base guanine and acts by interfering with DNA replication of the virus.

**acid** **1.** A type of compound that contains hydrogen and dissociates in water to produce positive hydrogen ions. The reaction, for an acid HX, is commonly written:

$$HX \rightleftharpoons H^+ + X^-$$

In fact, the hydrogen ion (the proton) is solvated, and the complete reaction is:

$$HX + H_2O \rightleftharpoons H_3O^+ + X^-$$

The ion $H_3O^+$ is the **oxonium ion** (or **hydroxonium ion** or **hydronium ion**). This definition of acids comes from the **Arrhenius theory**. Such acids tend to be corrosive substances with a sharp taste, which turn litmus red and give colour changes with other *indicators. They are referred to as **protonic acids** and are classified into **strong acids**, which are almost completely dissociated in water (e.g. sulphuric acid and hydrochloric acid), and **weak acids**, which are only partially dissociated (e.g. ethanoic acid and hydrogen sulphide). The strength of an acid depends on the extent to which it dissociates, and is measured by its *dissociation constant. *See also* BASE.

**2.** In the **Lowry–Brønsted theory** of acids and bases (1923), the definition was extended to one in which an acid is a proton donor, and a base is a proton acceptor. For example, in

$$HCN + H_2O \rightleftharpoons H_3O^+ + CN^-$$

the HCN is an acid, in that it donates a proton to $H_2O$. The $H_2O$ is acting as a base in ac-

cepting a proton. Similarly, in the reverse reaction $H_3O^+$ is an acid and $CN^-$ a base. In such reactions, two species related by loss or gain of a proton are said to be **conjugate**. Thus, in the reaction above HCN is the **conjugate acid** of the base $CN^-$, and $CN^-$ is the **conjugate base** of the acid HCN. Similarly, $H_3O^+$ is the conjugate acid of the base $H_2O$. An equilibrium, such as that above, is a competition for protons between an acid and its conjugate base. A strong acid has a weak conjugate base, and vice versa. Under this definition water can act as both acid and base. Thus in

$$NH_3 + H_2O \rightleftharpoons NH_4^+ + OH^-$$

the $H_2O$ is the conjugate acid of $OH^-$. The definition also extends the idea of acid–base reaction to solvents other than water. For instance, liquid ammonia, like water, has a high dielectric constant and is a good ionizing solvent. Equilibria of the type

$$NH_3 + Na^+Cl^- \rightleftharpoons Na^+NH_2^- + HCl$$

can be studied, in which $NH_3$ and HCl are acids and $NH_2^-$ and $Cl^-$ are their conjugate bases.

**3.** A further extension of the idea of acids and bases was made in the **Lewis theory** (G. N. Lewis, 1923). In this, a **Lewis acid** is a compound or atom that can accept a pair of electrons and a **Lewis base** is one that can donate an electron pair. This definition encompasses 'traditional' acid–base reactions. In

$$HCl + NaOH \rightarrow NaCl + H_2O$$

the reaction is essentially

$$H^+ + {:}OH^- \rightarrow H{:}OH$$

i.e. donation of an electron pair by $OH^-$. But it also includes reactions that do not involve ions, e.g.

$$H_3N{:} + BCl_3 \rightarrow H_3NBCl_3$$

in which $NH_3$ is the base (donor) and $BCl_3$ the acid (acceptor). The Lewis theory establishes a relationship between acid–base reactions and *oxidation–reduction reactions. *See* HSAB PRINCIPLE. *See also* AQUA ACID; HYDROXOACID; OXOACID.

**acid anhydrides (acyl anhydrides)** Compounds that react with water to form an acid. For example, carbon dioxide reacts with water to give carbonic acid:

$$CO_2(g) + H_2O(aq) \rightleftharpoons H_2CO_3(aq)$$

A particular group of acid anhydrides are anhydrides of carboxylic acids. They have a general formula of the type R.CO.O.CO.R′, carboxylic acids          acid anhydride

**Acid anhydride.** Formation of a carboxylic acid anhydride.

where R and R′ are alkyl or aryl groups. For example, the compound ethanoic anhydride ($CH_3.CO.O.CO.CH_3$) is the acid anhydride of ethanoic (acetic) acid. Organic acid anhydrides can be produced by dehydrating acids (or mixtures of acids). They are usually made by reacting an acyl halide with the sodium salt of the acid. They react readily with water, alcohols, phenols, and amines and are used in *acylation reactions.

**(∰) SEE WEB LINKS**
• Information about IUPAC nomenclature

**acid–base balance** The regulation of the concentrations of acids and bases in blood and other body fluids so that the pH remains within a physiologically acceptable range. This is achieved by the presence of natural *buffer systems, such as the haemoglobin, hydrogencarbonate ions, and carbonic acid in mammalian blood. By acting in conjunction, these effectively mop up excess acids and bases and therefore prevent any large shifts in blood pH. The acid–base balance is also influenced by the selective removal of certain ions by the kidneys and the rate of removal of carbon dioxide from the lungs.

**acid–base indicator** See INDICATOR.

**acid dissociation constant** See DISSOCIATION.

**acid dye** See DYES.

**acid halides** See ACYL HALIDES.

**acidic** 1. Describing a compound that is an acid. 2. Describing a solution that has an excess of hydrogen ions. 3. Describing a compound that forms an acid when dissolved in water. Carbon dioxide, for example, is an acidic oxide.

**acidic hydrogen (acid hydrogen)** A hydrogen atom in an *acid that forms a positive ion when the acid dissociates. For instance, in methanoic acid

$$HCOOH \rightleftharpoons H^+ + HCOO^-$$

the hydrogen atom on the carboxylate group is the acidic hydrogen (the one bound di-rectly to the carbon atom does not dissociate).

**acidic stains** See STAINING.

**acidimetry** Volumetric analysis using standard solutions of acids to determine the amount of base present.

**acidity constant** See DISSOCIATION.

**acid rain** Precipitation having a pH value of less than about 5.0, which has adverse effects on the fauna and flora on which it falls. Rainwater typically has a pH value of 5.6, due to the presence of dissolved carbon dioxide (forming carbonic acid). Acid rain results from the emission into the atmosphere of various pollutant gases, in particular sulphur dioxide and various oxides of nitrogen, which originate from the burning of fossil fuels and from car exhaust fumes, respectively. These gases dissolve in atmospheric water to form sulphuric and nitric acids in rain, snow, or hail (**wet deposition**). Alternatively, the pollutants are deposited as gases or minute particles (**dry deposition**). Both types of acid deposition affect plant growth – by damaging the leaves and impairing photosynthesis and by increasing the acidity of the soil, which results in the leaching of essential nutrients. This acid pollution of the soil also leads to acidification of water draining from the soil into lakes and rivers, which become unable to support fish life. Lichens are particularly sensitive to changes in pH and can be used as indicators of acid pollution (*see* INDICATOR SPECIES).

**acid rock** A low-density igneous rock containing a preponderance (more than 65%) of light-coloured *silicate minerals. Examples include granite and rhyolite.

**acid salt** A salt of a polybasic acid (i.e. an acid having two or more acidic hydrogens) in which not all the hydrogen atoms have been replaced by positive ions. For example, the dibasic acid carbonic acid ($H_2CO_3$) forms acid salts (hydrogencarbonates) containing the ion $HCO_3^-$. Some salts of monobasic acids are also known as acid salts. For instance, the compound potassium hydrogendifluoride, $KHF_2$, contains the ion $[F...H–F]^-$, in which there is hydrogen bonding between the fluoride ion $F^-$ and a hydrogen fluoride molecule.

**acid value** A measure of the amount of free acid present in a fat, equal to the number of milligrams of potassium hydroxide needed to neutralize this acid. Fresh fats contain glycerides of fatty acids and very little free acid, but the glycerides decompose slowly with time and the acid value increases.

**acinus** The smallest unit of a multilobular gland, such as the pancreas. Each acinus in the pancreas is made up of a hollow cluster of **acinar cells**, which produce the digestive enzymes secreted in pancreatic juice. Minute ducts from the pancreatic acini eventually drain into the pancreatic duct.

**acoustics 1.** The study of sound and sound waves. **2.** The characteristics of a building, especially an auditorium, with regard to its ability to enable speech and music to be heard clearly within it. For this purpose there should be no obtrusive echoes or resonances and the reverberation time should be near the optimum for the hall. Echoes are reduced by avoiding sweeping curved surfaces that could focus the sound and by breaking up large plane surfaces or covering them with sound-absorbing materials. Resonance is avoided by avoiding simple ratios for the main dimensions of the room, so that no one wavelength of sound is a factor of more than one of them. If the reverberation time is too long, speech will sound indistinct and music will be badly articulated, with one note persisting during the next. However, if it is too short, music sounds dead. It is long in a bare room with hard walls, and can be deliberately reduced by carpets, soft furnishings and sound-absorbent ('acoustic') felt. Reverberation times tend to be reduced by the presence of an audience and this must be taken into account in the design of the building.

**(((•))) SEE WEB LINKS**
- The website of the Acoustical Society of America
- The website of the UK Institute of Acoustics

**acoustoelectronic devices (electroacoustic devices)** Devices in which electronic signals are converted into acoustic waves. Acoustoelectronic devices are used in constructing *delay lines and also in converting digital data from computers for transmission by telephone lines.

**acquired characteristics** Features that are developed during the lifetime of an individual, e.g. the enlarged arm muscles of a tennis player. Such characteristics are not genetically controlled and cannot be passed on to the next generation. *See also* LAMARCKISM; NEO-LAMARCKISM.

**acquired immune deficiency syndrome** *See* AIDS.

**Acrilan** A trade name for a synthetic fibre. *See* ACRYLIC RESINS.

**acrolein** *See* PROPENAL.

**acromegaly** A chronic condition developing in adulthood due to overproduction of (or oversensitivity to) *growth hormone, usually caused by a tumour in the pituitary gland. This leads to a gradual enlargement of the bones, causing characteristic coarsening of the facial features and large hands and feet.

**acrosome** *See* SPERMATOZOON.

**acrylamide** An inert gel (polyacrylamide) employed as a medium in *electrophoresis. It is used particularly in the separation of macromolecules, such as nucleic acids and proteins.

**acrylate** *See* PROPENOATE.

**acrylic acid** *See* PROPENOIC ACID.

**acrylic resins** Synthetic resins made by polymerizing esters or other derivatives of acrylic acid (propenoic acid). Examples are poly(propenonitrile) (e.g. **Acrilan**), and poly(methyl 2-methylpropenoate) (polymethyl methacrylate, e.g. **Perspex**).

**acrylonitrile** *See* PROPENONITRILE.

**ACTH (adrenocorticotrophic hormone; corticotrophin)** A hormone, produced by the anterior *pituitary gland, that controls secretion of certain hormones (the *corticosteroids) by the adrenal glands. Its secretion, which is controlled by corticotrophin-releasing hormone and occurs in short bursts every few hours, is increased by stress.

**actin** A contractile protein found in muscle tissue, in which it occurs in the form of filaments (called thin filaments). Each thin filament consists of two chains of globular actin molecules, around which is twisted a strand of *tropomyosin and interspersed *troponin. Units of muscle fibre (*see* SARCOMERE) consist of actin and *myosin filaments, which interact to bring about muscle contraction. Actin is also found in the microfilaments that form part of the *cytoskeleton of all cells.

**actinic radiation** Electromagnetic radiation that is capable of initiating a chemical reaction. The term is used especially of ultra-violet radiation and also to denote radiation that will affect a photographic emulsion.

**actinides** See ACTINOIDS.

**actinium** Symbol Ac. A radioactive metallic element belonging to group 3 of the periodic table; a.n. 89; mass number of most stable isotope 227 (half-life 21.7 years); m.p. $1050 \pm 50°C$; b.p. 3200°C (estimated). Actinium–227 occurs in natural uranium to an extent of about 0.715%. Actinium–228 (half-life 6.13 hours) also occurs in nature. There are 22 other artificial isotopes, all radioactive and all with very short half-lives. Its main use is as a source of alpha particles. The element was discovered by A. Debierne in 1899.

**⊕ SEE WEB LINKS**
• Information from the WebElements site

**actinium series** See RADIOACTIVE SERIES.

**Actinobacteria (Actinomycetes; Actino-mycota)** A phylum of Gram-positive mostly anaerobic nonmotile bacteria. Many species are fungus-like, with filamentous cells producing reproductive spores on aerial branches similar to the spores of certain moulds. The phylum includes bacteria of the genera *Actinomyces*, some species of which cause disease in animals (including humans); and *Streptomyces*, which are a source of many important antibiotics (including streptomycin).

**actinoid contraction** A smooth decrease in atomic or ionic radius with increasing proton number found in the *actinoids.

**actinoids (actinides)** A series of elements in the *periodic table, generally considered to range in atomic number from thorium (90) to lawrencium (103) inclusive. The actinoids all have two outer *s*-electrons (a $7s^2$ configuration), follow actinium, and are classified together by the fact that increasing proton number corresponds to filling of the $5f$ level. In fact, because the $5f$ and $6d$ levels are close in energy the filling of the $5f$ orbitals is not smooth. The outer electron configurations are as follows:

89 actinium (Ac) $6d^1 7s^2$
90 thorium (Th) $6d^2 7s^2$
91 protactinium (Pa) $5f^2 6d^1 7s^2$
92 uranium (Ur) $5f^3 6d^1 7s^2$
93 neptunium (Np) $5f^5 7s^2$ (or $5f^4 6d^1 7s^2$)
94 plutonium (Pu) $5f^6 7s^2$
95 americium (Am) $5f^7 7s^2$

96 curium (Cm) $5f^7 6d^1 s^2$
97 berkelium (Bk) $5f^8 6d7s^2$ (or $5f^9 7s^2$)
98 californium (Cf) $5f^{10} 7s^2$
99 einsteinium (Es) $5f^{11} 7s^2$
100 fermium (Fm) $5f^{12} 7s^2$
101 mendelevium (Md) $5f^{13} 7s^2$
102 nobelium (Nb) $5f^{14} 7s^2$
103 lawrencium (Lw) $5f^{14} 6d^1 s^2$

The first four members (Ac to Ur) occur naturally. All are radioactive and this makes investigation difficult because of self-heating, short lifetimes, safety precautions, etc. Like the *lanthanoids, the actinoids show a smooth decrease in atomic and ionic radius with increasing proton number. The lighter members of the series (up to americium) have *f*-electrons that can participate in bonding, unlike the lanthanoids. Consequently, these elements resemble the transition metals in forming coordination complexes and displaying variable valency. As a result of increased nuclear charge, the heavier members (curium to lawrencium) tend not to use their inner *f*-electrons in forming bonds and resemble the lanthanoids in forming compounds containing the $M^{3+}$ ion. The reason for this is pulling of these inner electrons towards the centre of the atom by the increased nuclear charge. Note that actinium itself does not have a $5f$ electron, but it is usually classified with the actinoids because of its chemical similarities. See also TRANSITION ELEMENTS.

**actinometer** Any of various instruments for measuring the intensity of electromagnetic radiation. Recent actinometers use the *photoelectric effect but earlier instruments depended either on the fluorescence produced by the radiation on a screen or on the amount of chemical change induced in some suitable substance.

**actinomorphy** See RADIAL SYMMETRY.

**Actinomycetes** See ACTINOBACTERIA.

**action at a distance** The direct interaction between bodies that are not in physical contact with each other. The concept involves the assumption that the interactions are instantaneous. This assumption is not consistent with the special theory of *relativity, which states that nothing (including interactions) can travel through space faster than the *speed of light in a vacuum. For this reason it is more logical to describe interactions between bodies by *quantum field theories or by the exchange of virtual particles (*see* VIRTUAL STATE) rather than theories based on action at a distance.

**action potential** The change in electrical potential that occurs across a plasma membrane during the passage of a nerve *impulse. As an impulse travels in a wavelike manner along the *axon of a nerve, it causes a localized and transient switch in electric potential across the membrane from −60 mV (millivolts; the *resting potential) to +45 mV. The change in electric potential is caused by an influx of sodium ions. Nervous stimulation of a muscle fibre has a similar effect.

**action spectrum** A graphical plot of the efficiency of electromagnetic radiation in producing a photochemical reaction against the wavelength of the radiation used. For example, the action spectrum for photosynthesis using light shows a peak in the region 670–700 nm. This corresponds to a maximum absorption in the absorption *spectrum of chlorophylls in this region.

**activated adsorption** Adsorption that involves an activation energy. This occurs in certain cases of chemisorption.

**activated alumina** *See* ALUMINIUM HYDROXIDE.

**activated charcoal** *See* CHARCOAL.

**activated complex** The association of atoms of highest energy formed in the *transition state of a chemical reaction.

**activation analysis** An analytical technique that can be used to detect most elements when present in a sample in milligram quantities (or less). In **neutron activation analysis** the sample is exposed to a flux of thermal neutrons in a nuclear reactor. Some of these neutrons are captured by nuclides in the sample to form nuclides of the same atomic number but a higher mass number. These newly formed nuclides emit gamma radiation, which can be used to identify the element present by means of a gamma-ray spectrometer. Activation analysis has also been employed using high-energy charged particles, such as protons or alpha particles.

**⊕ SEE WEB LINKS**
- An overview of neutron activation analysis from the Archaeometry Laboratory, Missouri Research Reactor
- A description of neutron activation analysis from the University of Wisconsin

**activation energy** Symbol $E_a$. The minimum energy required for a chemical reaction to take place. In a reaction, the reactant molecules come together and chemical bonds are stretched, broken, and formed in producing the products. During this process the energy of the system increases to a maximum, then decreases to the energy of the products (see illustration). The activation energy is the difference between the maximum energy and the energy of the reactants; i.e. it is the energy barrier that has to be overcome for the reaction to proceed. The activation energy determines the way in which the rate of the reaction varies with temperature (*see* ARRHENIUS EQUATION). It is usual to express activation energies in joules per mole of reactants. An activation energy greater than 200 kJ mol⁻¹ suggests that a bond has been completely broken in forming the transition state (as in the $S_N1$ reaction). A lower figure suggests incomplete breakage (as in the $S_N2$ reaction).

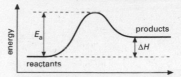

**Activation energy.** Reaction profile (for an endothermic reaction).

**activator** 1. A type of *transcription factor involved in assembling proteins to form an initiation complex at the *promoter of a gene in readiness for transcription. *Compare* REPRESSOR. 2. A substance that – by binding to an allosteric site on an enzyme (*see* INHIBITION) – enables the active site of the enzyme to bind to the substrate. 3. Any compound that potentiates the activity of a drug or other foreign substance in the body.

**active device** 1. An electronic component, such as a transistor, that is capable of amplification. 2. An artificial *satellite that receives information and retransmits it after amplification. 3. A radar device that emits microwave radiation and provides information about a distant body by receiving a reflection of this radiation. *Compare* PASSIVE DEVICE.

**active galactic nucleus** *See* ACTIVE GALAXY.

**active galaxy** A galaxy that contains an

**active galactic nucleus (AGN)**, i.e. a central region that gives off a great deal of electromagnetic radiation. This emission is thought to be due to the accretion of matter into a supermassive *black hole, at the centre of the galaxy. *See also* RELATIVISTIC JETS.

**active immunity** *Immunity acquired due to the body's response to a foreign antigen.

**active mass** *See* MASS ACTION.

**active site (active centre)** **1.** A site on the surface of a catalyst at which activity occurs. **2.** The site on the surface of an *enzyme molecule that binds and acts on the substrate molecule. The properties of an active site are determined by the three-dimensional arrangement of the polypeptide chains of the enzyme and their constituent amino acids. These govern the nature of the interaction that takes place and hence the degree of substrate specificity and susceptibility to *inhibition.

**active transport** The movement of substances through membranes in living cells, often against a *concentration gradient: a process requiring metabolic energy. Organic molecules and inorganic ions are transported into and out of both cells and their organelles. The substance binds to a *transport protein embedded in the membrane, which carries it through the membrane and releases it on the opposite side. Active transport serves chiefly to maintain the normal balance of ions in cells, especially the concentration gradients of sodium and potassium ions crucial to the activity of nerve and muscle cells. *Compare* FACILITATED DIFFUSION.

**activity** **1.** Symbol $a$. A thermodynamic function used in place of concentration in equilibrium constants for reactions involving nonideal gases and solutions. For example, in a reaction

$$A \rightleftharpoons B + C$$

the true equilibrium constant is given by

$$K = a_B a_C / a_A$$

where $a_A$, $a_B$, and $a_C$ are the activities of the components, which function as concentrations (or pressures) corrected for nonideal behaviour. **Activity coefficients** (symbol $\gamma$) are defined for gases by $\gamma = a/p$ (where $p$ is pressure) and for solutions by $\gamma = aX$ (where $X$ is the mole fraction). Thus, the equilibrium constant of a gas reaction has the form

$$K_p = \gamma_B p_B \gamma_C p_C / \gamma_A p_A$$

The equilibrium constant of a reaction in solution is

$$K_c = \gamma_B X_B \gamma_C X_C / \gamma_A X_A$$

The activity coefficients thus act as correction factors for the pressures or concentrations. The activity is given by an equation

$$\mu = \mu^{\ominus} + R T \ln a$$

where $\mu$ is chemical potential. *See also* FUGACITY.
**2.** Symbol $A$. The number of atoms of a radioactive substance that disintegrate per unit time. The **specific activity** ($a$) is the activity per unit mass of a pure radioisotope. *See* RADIATION UNITS.

**activity series** *See* ELECTROMOTIVE SERIES.

**acyclic** Describing a compound that does not have a ring in its molecules.

**acyclovir (acycloguanosine)** *See* ACICLOVIR.

**acyl anhydrides** *See* ACID ANHYDRIDES.

**acylation** The process of introducing an acyl group (RCO–) into a compound. The usual method is to react an alcohol with an acyl halide or a carboxylic acid anhydride; e.g.

$$RCOCl + R'OH \rightarrow RCOOR' + HCl$$

The introduction of an acetyl group ($CH_3CO$–) is **acetylation**, a process used for protecting –OH groups in organic synthesis.

**acyl fission** The breaking of the carbon–oxygen bond in an acyl group. It occurs in the hydrolysis of an *ester to produce an alcohol and a carboxylic acid.

**acylglycerol** *See* GLYCERIDE.

**acyl group** A group of the type RCO–, where R is an organic group. An example is the acetyl group $CH_3CO$–.

**acyl halides (acid halides)** Organic compounds containing the group –CO.X, where X is a halogen atom (see formula). Acyl chlorides, for instance, have the general formula RCOCl. The group RCO– is the **acyl group**. In systematic chemical nomenclature acyl-halide names end in the suffix *-oyl*; for example, ethanoyl chloride, $CH_3COCl$. Acyl

**Acyl halide.** X is a halogen atom.

halides react readily with water, alcohols, phenols, and amines, and are used in *acylation reactions. They are made by replacing the –OH group in a carboxylic acid by a halogen using a halogenating agent such as $PCl_5$.

**(⊕) SEE WEB LINKS**
• Information about IUPAC nomenclature

**Ada** A high-level computer programming language developed in the late 1970s for the US military. It was originally employed in missile control systems and is now used in various other real-time applications. Ada was named after Augusta Ada Lovelace (1815–52), the mathematician daughter of Lord Byron, who worked with Charles *Babbage on his mechanical computer, the 'analytical engine'.

**(⊕) SEE WEB LINKS**
• The Ada 2005 Language Reference Manual

**adamantane** A colourless crystalline hydrocarbon, $C_{10}H_{16}$; m.p. 269°C. It is found in certain petroleum fractions. The structure contains three symmetrically fused cyclohexane rings.

**Adams, John Couch** (1819–92) British astronomer who became professor of astronomy and geometry at Cambridge University in 1858. He is best known for his prediction (1845) of the existence and position of the planet *Neptune, worked out independently the following year by Urbain Leverrier (1811–77). The planet was discovered in 1846 by Johann Galle (1812–1910), using Leverrier's figures. Adams's priority was not acknowledged.

**Adams catalyst** A dark brown powder, a hydrated form of platinum(IV) oxide ($PtO_2$), produced by heating chloroplatinic acid ($H_2PtCl_6$) with sodium nitrate ($NaNO_3$). Platinum nitrate is produced, and this decomposes to platinum(IV) oxide with evolution of $NO_2$ and oxygen. It is used in hydrogenation of alkenes to alkanes, nitro compounds to aminos, and ketones to alcohols. The actual catalyst is not the oxide but finely divided *platinum black, which forms during the hydrogenation reaction.

**adaptation** 1. (in evolution) Any change in the structure or functioning of successive generations of a population that makes it better suited to its environment. *Natural selection of heritable adaptations ultimately leads to the development of new species. Increasing adaptation of a species to a particular environment tends to diminish its ability to adapt to any sudden change in that environment. **2.** (in physiology) The alteration in the degree of sensitivity (either an increase or a decrease) of a sense organ to suit conditions more extreme than normally encountered. An example is the adjustment of the eye to vision in very bright or very dim light.

**adaptive optics (AO)** Computer-aided techniques for making virtually instantaneous corrective adjustments to the shape of a deformable mirror in a ground-based optical or near-infrared telescope in order to reduce or eliminate distortions of an image caused by dynamic turbulence in the earth's atmosphere. Adaptive optics also has applications outside astronomy. For example, in ophthalmology it can be used to measure aberration within the human eye and help to provide sharp images of the retina at the cellular level.

**adaptive radiation (divergent evolution)** The evolution from one species of animals or plants of a number of different forms. As the original population increases in size it spreads out from its centre of origin to exploit new habitats and food sources. In time this results in a number of populations each adapted to its particular habitat: eventually these populations will differ from each other sufficiently to become new species. A good example of this process is the evolution of the Australian marsupials into species adapted as carnivores, herbivores, burrowers, fliers, etc. On a smaller scale, the adaptive radiation of the Galapagos finches provided Darwin with crucial evidence for his theory of evolution (*see* DARWIN'S FINCHES).

**addition polymerization** *See* POLYMERIZATION.

**addition reaction** A chemical reaction in which one molecule adds to another. Addition reactions occur with unsaturated compounds containing double or triple bonds, and may be *electrophilic or *nucleophilic. An example of electrophilic addition is the reaction of hydrogen chloride with an alkene, e.g.

$$HCl + CH_2{:}CH_2 \rightarrow CH_3CH_2Cl$$

An example of nucleophilic addition is the addition of hydrogen cyanide across the carbonyl bond in aldehydes to form *cyanohydrins. **Addition–elimination** reactions are ones in which the addition is followed by

a

elimination of another molecule (*see* CONDENSATION REACTION).

**additive** A substance added to another substance or material to improve its properties in some way. Additives are often present in small amounts and are used for a variety of purposes, as in preventing corrosion, stabilizing polymers, etc. **Food additives** are used to enhance the taste and colour of foods and improve their texture and keeping qualities. *See* FOOD PRESERVATION.

**additive process** *See* COLOUR.

**adduct** A compound formed by an addition reaction. The term is used particularly for compounds formed by coordination between a Lewis acid (acceptor) and a Lewis base (donor). *See* ACID.

**adenine** A *purine derivative. It is one of the major component bases of *nucleotides and the nucleic acids *DNA and *RNA.

**adenosine** A nucleoside comprising one adenine molecule linked to a D-ribose sugar molecule. The phosphate-ester derivatives of adenosine, AMP, ADP, and *ATP, are of fundamental biological importance as carriers of chemical energy.

**adenosine diphosphate (ADP)** *See* ATP.

**adenosine monophosphate (AMP)** *See* ATP; CYCLIC AMP.

**adenosine triphosphate** *See* ATP; CYCLIC AMP.

**adenovirus** One of a group of DNA-containing viruses found in rodents, fowl, cattle, monkeys, and humans. In humans they produce acute respiratory-tract infections with symptoms resembling those of the common cold. They are also implicated in the formation of tumours (*see* ONCOGENIC).

**adenylate cyclase** The enzyme that catalyses the formation of *cyclic AMP. It is bound to the inner surface of the plasma membrane. Many hormones and other chemical messengers exert their physiological effects by increased synthesis of cyclic AMP through the activation of adenylate cyclase. The hormone binds to a receptor on the outer surface of the plasma membrane, which then activates adenylate cyclase on the inner surface via a *G protein.

**ADH** *See* ANTIDIURETIC HORMONE.

**adhesive** A substance used for joining surfaces together. Adhesives are generally colloidal solutions, which set to gels. There are many types including animal glues (based on collagen), vegetable mucilages, and synthetic resins (e.g. *epoxy resins).

**adiabatic approximation** An approximation used in *quantum mechanics when the time dependence of parameters such as the inter-nuclear distance between atoms in a molecule is slowly varying. This approximation means that the solution of the *Schrödinger equation at one time goes continuously over to the solution at a later time. This approximation was formulated by Max Born and the Soviet physicist Vladimir Alexandrovich Fock (1898–1974) in 1928. The *Born–Oppenheimer approximation is an example of the adiabatic approximation.

**adiabatic demagnetization** A technique for cooling a paramagnetic salt, such as potassium chrome alum, to a temperature near *absolute zero. The salt is placed between the poles of an electromagnet and the heat produced during magnetization is removed by liquid helium. The salt is then isolated thermally from the surroundings and the field is switched off; the salt is demagnetized adiabatically and its temperature falls. This is because the demagnetized state, being less ordered, involves more energy than the magnetized state. The extra energy can come only from the internal, or thermal, energy of the substance. It is possible to obtain temperatures as low as 0.005 K in this way.

**adiabatic process** Any process that occurs without heat entering or leaving a system. In general, an adiabatic change involves a fall or rise in temperature of the system. For example, if a gas expands under adiabatic conditions, its temperature falls (work is done against the retreating walls of the container). The **adiabatic equation** describes the relationship between the pressure ($p$) of an ideal gas and its volume ($V$), i.e. $pV^\gamma = K$, where $\gamma$ is the ratio of the principal specific *heat capacities of the gas and $K$ is a constant.

**adipic acid** *See* HEXANEDIOIC ACID.

**adipose tissue** A body tissue comprising cells containing *fat and oil. It is found chiefly below the skin (*see* SUBCUTANEOUS TISSUE) and around major organs (such as the kidneys and heart), acting as an energy reserve, providing insulation and protection, and generating heat. Secretion of the hormone *leptin by adipose tissue regulates the amount of adipose tissue and adjusts the body's energy balance. *See* BROWN FAT; THERMOGENESIS.

**admittance** Symbol $Y$. The reciprocal of *impedance. It is measured in siemens.

**adolescence** The period in human development that occurs during the teenage years, between the end of childhood and the start of adulthood, and is characterized by various physical and emotional changes associated with development of the reproductive system. It starts at **puberty**, when the reproductive organs begin to function, and is marked by the start of menstruation (*see* MENSTRUAL CYCLE) in females and the appearance of the *secondary sexual characteristics in both sexes. In males the secondary sexual characteristics are controlled by the hormone testosterone and include deepening of the voice due to larynx enlargement, the appearance of facial and pubic hair, rapid growth of the skeleton and muscle, and an increase in *sebaceous gland secretions. In females the secondary sexual characteristics are controlled by oestrogens and include growth of the breasts, broadening of the pelvis, redistribution of fat in the body, and appearance of pubic hair.

**ADP** *See* ATP.

**adrenal cortex** The outer layer of the *adrenal gland, in which several steroid hormones, the *corticosteroids, are produced.

**adrenal glands** A pair of endocrine glands situated immediately above the kidneys (hence they are also known as the **suprarenal glands**). The inner portion of the adrenals, the **medulla**, secretes the hormones *adrenaline and *noradrenaline; the outer **cortex** secretes small amounts of sex hormones (*androgens and *oestrogens) and various *corticosteroids, which have a wide range of effects on the body. *See also* ACTH.

**adrenaline (epinephrine)** A hormone, produced by the medulla of the *adrenal glands, that increases heart activity, improves the power and prolongs the action of muscles, and increases the rate and depth of breathing to prepare the body for 'fright, flight, or fight'. At the same time it inhibits digestion and excretion and stimulates mobilization of body fat (*lipolysis) and energy metabolism (*glycolysis). Similar effects are produced by stimulation of the *sympathetic nervous system. Adrenaline causes these

effects by binding to *adrenoceptors on target cells. It can be administered by injection to relieve bronchial asthma and reduce blood loss during surgery by constricting blood vessels.

**adrenal medulla** The inner part of the *adrenal gland, in which *adrenaline is produced.

**adrenergic** 1. Describing a cell (especially a neuron) or a cell receptor that is stimulated by *adrenaline, *noradrenaline, or related substances. *See* ADRENOCEPTOR. 2. Describing a nerve fibre or neuron that releases adrenaline or noradrenaline when stimulated. *Compare* CHOLINERGIC.

**adrenoceptor (adrenoreceptor; adrenergic receptor)** Any cell receptor that binds and is activated by the catecholamines adrenaline or noradrenaline. Adrenoceptors are therefore crucial in mediating the effects of catecholamines as neurotransmitters or hormones. There are two principal types of adrenoceptor, alpha ($\alpha$) and beta ($\beta$). The **alpha adrenoceptors** fall into two main subtypes: $\alpha_1$-adrenoceptors, which mediate the contraction of smooth muscle and hence cause constriction of blood vessels; and $\alpha_2$-adrenoceptors, which occur, for example, in presynaptic neurons at certain nerve synapses, where they inhibit release of noradrenaline from the neuron. The **beta adrenoceptors** also have two main subtypes: $\beta_1$-adrenoceptors, which stimulate cardiac muscle causing a faster and stronger heartbeat; and $\beta_2$-adrenoceptors, which mediate relaxation of smooth muscle in blood vessels, bronchi, the uterus, the bladder, and other organs. Activation of $\beta_2$-adrenoceptors thus causes widening of the airways (bronchodilation) and blood vessels (vasodilation). *See also* BETA BLOCKER.

**adrenocorticotrophic hormone** *See* ACTH.

**Adrian, Edgar Douglas, Baron** (1889–1977) British neurophysiologist, who became a professor at Cambridge in 1937, where he remained until his retirement. He is best known for his work on nerve impulses, establishing that messages are conveyed by changes in the frequency of the impulses. He shared the 1932 Nobel Prize for physiology or medicine with Sir Charles *Sherrington for this work.

**ADSL (asymmetric digital subscriber line)** A mechanism by which *broadband communication via the Internet can be made available via pre-existing telephone lines, while allowing simultaneous use of the line for normal telephone calls. Data communication via ADSL is asymmetric in that upstream (transmitting) communication is slower than downstream (receiving) communication, typically half as fast. Commonly available downstream data rates in the UK are 512 Kbps, 1 Mbps, and 2 Mbps. Faster rates are available in other countries. ADSL coexists with standard telephone operation on the same line by the use of band separation filters at each telephone socket.

**adsorbate** A substance that is adsorbed on a surface.

**adsorbent** A substance on the surface of which a substance is adsorbed.

**adsorption** The formation of a layer of gas, liquid, or solid on the surface of a solid or, less frequently, of a liquid. There are two types depending on the nature of the forces involved. In **chemisorption** a single layer of molecules, atoms, or ions is attached to the adsorbent surface by chemical bonds. In **physisorption** adsorbed molecules are held by the weaker *van der Waals' forces. Adsorption is an important feature of surface reactions, such as corrosion, and heterogeneous catalysis. The property is also utilized in adsorption *chromatography.

**adsorption indicator (absorption indicator)** A type of indicator used in reactions that involve precipitation. The yellow dye fluorescein is a common example, used for the reaction

$$NaCl(aq) + AgNO_3(aq) \rightarrow AgCl(s) + NaNO_3(aq)$$

As silver nitrate solution is added to the sodium chloride, silver chloride precipitates. As long as $Cl^-$ ions are in excess, they adsorb on the precipitate particles. At the end point, no $Cl^-$ ions are left in solution and negative fluorescein ions are then adsorbed, giving a pink colour to the precipitate. The technique is sometimes known as **Fajans' method**.

**adsorption isotherm** An equation that describes how the amount of a substance adsorbed onto a surface depends on its pressure (if a gas) or its concentration (if in a solution), at a constant temperature. Several adsorption isotherms are used in surface chemistry including the *BET isotherm and the *Langmuir adsorption isotherm. The different isotherms correspond to different assumptions about the surface and the adsorbed molecules.

**adulterant** See CUTTING AGENT.

**advanced gas-cooled reactor (AGR)** See NUCLEAR REACTOR.

**adventitious** Describing organs or other structures that arise in unusual positions. For example, ivy has adventitious roots growing from its stems.

**aerial (antenna)** The part of a radio or television system from which radio waves are transmitted into the atmosphere or space (**transmitting aerial**) or by which they are received (**receiving aerial**). A **directional** or **directive aerial** is one in which energy is transmitted or received more effectively from some directions than others, whereas an **omnidirectional aerial** transmits and receives equally well in all directions.

**aerobe** See AEROBIC RESPIRATION.

**aerobic respiration** A type of *respiration in which foodstuffs (usually carbohydrates) are completely oxidized to carbon dioxide and water, with the release of chemical energy, in a process requiring atmospheric oxygen. The reaction can be summarized by the equation:

$$C_6H_{12}O_6 + 6O_2 \rightarrow 6CO_2 + 6H_2O + energy$$

The chemical energy released is stored mainly in the form of *ATP. The first stage of aerobic respiration is *glycolysis, which takes place in the cytosol of cells and also occurs in fermentations and other forms of *anaerobic respiration. Further oxidation in the presence of oxygen is via the *Krebs cycle and *electron transport chain, enzymes for which are located in the *mitochondria of eukaryote cells. Most organisms have aerobic respiration (i.e. they are **aerobes**); exceptions include certain bacteria and yeasts.

**aerodynamics** The study of the motion of gases (particularly air) and the motion of solid bodies in air. Aerodynamics is particularly concerned with the motion and stability of aircraft. Another application of aerodynamics is to the *flight of birds and insects. The branch of aerodynamics concerned with the flow of gases through compressors, ducts, fans, orifices, etc., is called **internal aerodynamics**.

    **Aerodynamic drag** is the force that opposes the motion of a body moving relative to a gas and is a function of the density of the gas, the square of the relative velocity, the

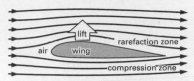

section through an aircraft wing

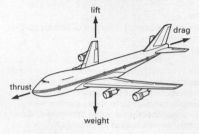

forces on an aircraft

**Aerodynamics.**

surface area of the body, and a quantity called the **drag coefficient**, which is a function of the *Reynolds number. **Aerodynamic lift** is an upward force experienced by a body moving through a gas and is a function of the same variables as aerodynamic drag.

**aerogel** A low-density porous transparent material that consists of more than 90% air. Usually based on metal oxides or silica, aerogels are used as drying agents and insulators.

**aerogenerator** *See* WIND POWER.

**aerosol** A colloidal dispersion of a solid or liquid in a gas. The commonly used aerosol sprays contain an inert propellant liquefied under pressure. Halogenated alkanes containing chlorine and fluorine (*chlorofluorocarbons, or CFCs) were formerly used in aerosol cans. They have now largely been replaced by volatile hydrocarbons because of their effect on the *ozone layer.

**aerospace** The earth's atmosphere and the space beyond it.

**aerotaxis** *See* TAXIS.

**aestivation 1.** (in zoology) A state of inactivity occurring in some animals, notably lungfish, during prolonged periods of drought or heat. Feeding, respiration, movement, and other bodily activities are considerably slowed down. *See also* DORMANCY.

*Compare* HIBERNATION. **2.** (in botany) The arrangement of the parts of a flower bud, especially of the sepals and petals.

**aetiology** The study of causes, especially the causes of medical conditions.

**afferent** Carrying (nerve impulses, blood, etc.) from the outer regions of a body or organ towards its centre. The term is usually applied to types of nerve fibres or blood vessels. *Compare* EFFERENT.

**aflatoxin** Any of four related toxic compounds produced by the mould *Aspergillus flavus*. Aflatoxins bind to DNA and prevent replication and transcription. They can cause acute liver damage and cancers: humans may be poisoned by eating stored peanuts and cereals contaminated with the mould.

**AFM** *See* ATOMIC FORCE MICROSCOPY.

**afterbirth** The *placenta, *umbilical cord, and *extraembryonic membranes, which are expelled from the womb after a mammalian fetus is born. In most nonhuman mammals the afterbirth, which contains nutrients and might otherwise attract predators, is eaten by the female.

**after-heat** Heat produced by a nuclear reactor after it has been shut down. The afterheat is generated by radioactive substances formed in the fuel elements.

**agamospermy** *See* APOMIXIS.

**agar** An extract of certain species of red seaweeds that is used as a gelling agent in microbiological *culture media, foodstuffs, medicines, and cosmetic creams and jellies. **Nutrient agar** consists of a broth made from beef extract or blood that is gelled with agar and used for the cultivation of bacteria, fungi, and some algae.

**agate** A variety of *chalcedony that forms in rock cavities and has a pattern of concentrically arranged bands or layers that lie parallel to the cavity walls. These layers are frequently alternating tones of brownish-red. **Moss agate** does not show the same banding and is a milky chalcedony containing mosslike or dendritic patterns formed by inclusions of manganese and iron oxides. Agates are used in jewellery and for ornamental purposes.

**ageing** *See* SENESCENCE.

**age of the earth** The time since the earth

emerged as a planet of the sun, estimated by *dating techniques to be about $4.6 \times 10^9$ years. The oldest known rocks on earth are estimated by their *radioactive age to be about $3.5 \times 10^9$ years old. The earth is older than this because of the long time it took to cool. An estimate for the cooling time is included in the estimate for the age of the earth.

**age of the universe** A time determined by the reciprocal of the value of the *Hubble constant to be about 13.7 billion years. The calculation of the Hubble constant, and hence the age of the universe, depends on which theory of *cosmology is used. Usually, the age of the universe is calculated by assuming that the *expansion of the universe can be described by the *big-bang theory.

**agglutination** The clumping together by antibodies of microscopic foreign particles, such as red blood cells or bacteria, so that they form a visible pellet-like precipitate. Agglutination is a specific reaction, i.e. a particular antigen will only clump in the presence of its specific antibody; it therefore provides a means of identifying unknown bacteria and determining *blood group. When blood of incompatible blood groups (e.g. group A and group B – *see* ABO SYSTEM) is mixed together agglutination of the red cells occurs (**haemagglutination**). This is due to the reaction between antibodies in the plasma (**agglutinins**) and *agglutinogens (antigens) on the surface of the red cells.

**agglutinogen** Any of the antigens that are present on the outer surface of red blood cells (erythrocytes). There are more than 100 different agglutinogens and they form the basis for identifying the different *blood groups. Antibodies in the plasma, known as **agglutinins**, react with the agglutinogens in blood of an incompatible blood group (*see* AGGLUTINATION).

**aggression** Behaviour aimed at intimidating or injuring another animal of the same or a competing species. Aggression between individuals of the same species often starts with a series of ritualized displays or contests that can end at any stage if one of the combatants withdraws, leaving the victor with access to a disputed resource (e.g. food, a mate, or *territory) or with increased social dominance (*see* DOMINANT). It is also often seen in *courtship. Aggression or threat displays usually appear to exaggerate the performer's size or strength; for example, many

fish erect their fins and mammals and birds may erect hairs or feathers. Special markings may be prominently exhibited, and **intention movements** may be made: dogs bare their teeth, for example. Some animals have evolved special structures for use in aggressive interactions (e.g. antlers in deer) but these are seldom used to cause actual injury; the opponent usually flees first or adopts *appeasement postures. Fights 'to the death' are comparatively rare. *See* AGONISTIC BEHAVIOUR; DISPLAY BEHAVIOUR; RITUALIZATION.

**AGN** Active galactic nucleus. *See* ACTIVE GALAXY.

**agnathan** Any jawless craniate animal. Agnathans were formerly classified in the subphylum (or superclass) Agnatha, with the living representatives – lampreys and hagfishes – constituting the class Cyclostomata. However, it is now accepted that the lampreys are more closely related to the jawed vertebrates than to the hagfishes, and the agnathans are now placed in two distinct clades of the subphylum *Craniata, the hagfishes forming the class Myxini (Hyperotreti) and the lampreys the Hyperoartia. The closest living relatives of the latter are the jawed vertebrates (Gnathostomata), which with the lampreys constitute the superclass Vertebrata. Fossil agnathans, covered in an armour of bony plates, are the oldest known fossil vertebrates. They have been dated from the Silurian and Devonian periods, 440–345 million years ago.

**agonist** A drug, hormone, neurotransmitter, or other signal molecule that forms a complex with a *receptor site, thereby triggering an active response from a cell. *Compare* ANTAGONIST.

**agonistic behaviour** Any form of behaviour associated with *aggression, including threat, attack, *appeasement, or flight. It is often associated with defence of a territory; for example, a threat display by the defending individual is often met with an appeasement display from the intruder, thus avoiding harmful conflict.

**AGR** Advanced gas-cooled reactor. *See* NUCLEAR REACTOR.

**agranulocyte** Any white blood cell (*see* LEUCOCYTE) with a nongranular cytoplasm and a large spherical nucleus; *lymphocytes and *monocytes are examples. Agranulocytes are produced either in the lymphatic system or in the bone marrow and account for 30% of all leucocytes. *Compare* GRANULO-CYTE.

**agriculture** The study and practice of cultivating land for the growing of crops and the rearing of livestock. The increasing demands for food production since the mid-20th century have seen many developments in agricultural technology and practices that have greatly increased crop and livestock production. However, these advances in modern **intensive farming** techniques have had their impact on the environment, particularly with increased use of *fertilizers and *pesticides. The now widespread practice of crop **monoculture** (in which one crop is grown densely over an extensive area) has required an increase in the use of *pesticides, as monoculture provides an ideal opportunity for crop pests. Monoculture also requires vast areas of land, which has meant that natural habitats have been destroyed. *Deforestation has resulted from the clearing of forests for crop production and cattle rearing. Advances in technology have included ploughing machines with hydraulic devices that can control the depth to which the soil is ploughed, and seed drills that automatically implant seeds in the soil so that ploughing is not necessary. Food supply in many less-developed countries relies on **subsistence farming**, in which the crops and livestock produced are used solely to feed the farmer and his family. In such countries a system known as **slash and burn** is common, in which the vegetation in an area is cut down and then burnt, thus returning the minerals to the soil. The area can then be used for crop cultivation until the soil fertility drops, at which point it is then abandoned for a number of years and another site is cultivated.

The selective *breeding of crop plants and farm animals has had an enormous impact on productivity in agriculture. Modern varieties of crop plants have increased nutritional value and greater resistance to disease, while animals have been selectively bred to enhance their yields of milk, meat, and other products. Developments in genetic engineering have enabled the introduction to commercial cultivation of genetically modified crop plants, such as tomatoes and soya, which contain foreign genes to enhance crop growth, nutritional properties, or storage characteristics. Ge-

netic modification can also confer resistance to herbicides, thereby allowing more effective weed control, as well as improved resistance to insects and other pests and to diseases. The application of similar technology to animal production is being researched. *See also* GENETICALLY MODIFIED ORGANISMS (Feature).

***Agrobacterium tumefaciens*** A Gram-negative soil bacterium that infects a wide range of plants and causes *galls, especially at the root/stem junction (crown gall). It is of interest because the bacterial cells contain a *plasmid, the **Ti plasmid** (tumour-inducing plasmid), a segment of which is transferred to cells of the plant host. This T-DNA (transfer DNA) segment, which comprises the genes responsible for the gall, becomes integrated into the genome of infected plant cells. Possession of the Ti plasmid has made *A. tumefaciens* an important tool in genetic engineering for the introduction of foreign genes into plant tissue. The tumour-inducing genes are usually replaced with the gene of interest, and a marker gene (e.g. the antibiotic resistance gene) is added to enable selection of transformed cells. *See* GENETICALLY MODIFIED ORGANISMS (Feature).

**AI 1.** *See* ARTIFICIAL INTELLIGENCE. **2.** *See* ARTIFICIAL INSEMINATION.

**AIDS (acquired immune deficiency syndrome)** A disease of humans characterized by defective cell-mediated *immunity and increased susceptibility to infections. It is caused by the retrovirus *HIV (human immunodeficiency virus). This infects and destroys helper *T cells, which are essential for combating infections. HIV is transmitted in blood, semen, and vaginal fluid; the major routes of infection are unprotected vaginal and anal intercourse, intravenous drug abuse, and the administration of contaminated blood and blood products. A person infected with HIV is described as **HIV-positive**; after the initial infection the virus can remain dormant for up to ten years before AIDS develops. *Antiviral drugs can delay the development of full-blown AIDS, in some cases for many years.

**air** *See* EARTH'S ATMOSPHERE.

**air bladder** *See* SWIM BLADDER.

**air mass** (in meteorology) An area of the atmosphere that in the horizontal field possesses more or less uniform properties, especially temperature and humidity, and extends for hundreds of kilometres. The transition zone at which one air mass meets another is known as a *front. Air masses develop over extensive areas of the earth's surface, known as source regions, where conditions are sufficiently uniform to impart similar characteristics to the overlying air. These areas are chiefly areas of high pressure. As an air mass moves away from its source region it undergoes modification.

**air pollution (atmospheric pollution)** The release into the atmosphere of substances that cause a variety of harmful effects to the natural environment. Most air pollutants are gases that are released into the troposphere, which extends about 8 km above the surface of the earth. The burning of *fossil fuels, for example in power stations, is a major source of air pollution as this process produces such gases as sulphur dioxide and carbon dioxide. Released into the atmosphere, carbon dioxide is the major contributor to the *greenhouse effect; methane, derived from livestock and rice cultivation, is another significant greenhouse gas. Sulphur dioxide and nitrogen oxides, released in car exhaust fumes, are air pollutants that are responsible for the formation of *acid rain; nitrogen oxides also contribute to the formation of *photochemical smog. *See also* OZONE LAYER; POLLUTION.

**air sac 1.** Any one of a series of thin-walled sacs in birds that are connected to the lungs and increase the efficiency of ventilation. Some of the air sacs penetrate the internal cavities of bones. **2.** A structural extension to the *trachea in insects, which increases the surface area available for the exchange of oxygen and carbon dioxide in respiration.

**alabaster** *See* GYPSUM.

**alanine** *See* AMINO ACID.

**albedo 1.** The ratio of the radiant flux reflected by a surface to that falling on it. **2.** The probability that a neutron entering a body of material will be reflected back through the same surface as it entered.

**albinism** Hereditary lack of pigmentation (*see* MELANIN) in an organism. Albino animals and human beings have no colour in their skin, hair, or eyes (the irises appear pink from underlying blood vessels). The *allele responsible is *recessive to the allele for normal pigmentation.

**albumen** *See* ALBUMIN.

**albumin** One of a group of globular proteins that are soluble in water but form insoluble coagulants when heated. Albumins occur in egg white (the protein component of which is known as **albumen**), blood, milk, and plants. Serum albumins, which constitute about 55% of blood plasma protein, help regulate the osmotic pressure and hence plasma volume. They also bind and transport fatty acids. $\alpha$-lactalbumin is a protein in milk.

**albuminous cell** *See* COMPANION CELL.

**alburnum** *See* SAPWOOD.

**alcoholic fermentation** *See* FERMENTATION.

**alcohols** Organic compounds that contain the –OH group. In systematic chemical nomenclature alcohol names end in the suffix *-ol*. Examples are methanol, $CH_3OH$, and ethanol, $C_2H_5OH$. **Primary alcohols** have two hydrogen atoms on the carbon joined to the –OH group (i.e. they contain the group $-CH_2-OH$); **secondary alcohols** have one hydrogen on this carbon (the other two bonds being to carbon atoms, as in $(CH_3)_2CHOH$); **tertiary alcohols** have no hydrogen on this carbon (as in $(CH_3)_3COH$): see formulae. The different types of alcohols may differ in the way they react chemically. For example, with potassium dichromate(VI) in sulphuric acid the following reactions occur

primary alcohol → aldehyde → carboxylic acid

secondary alcohol → ketone

tertiary alcohol – no reaction

Other characteristics of alcohols are reaction with acids to give *esters and dehydration to give *alkenes or *ethers. Alcohols that have two –OH groups in their molecules are

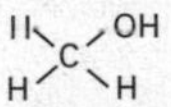

primary alcohol (methanol)

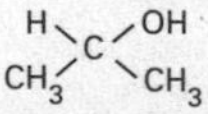

secondary alcohol (propan-2-ol)

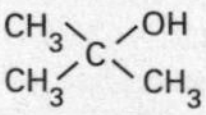

tertiary alcohol (2-methylpropan-2-ol)

**Examples of alcohols.**

**diols** (or **dihydric alcohols**), those with three are **triols** (or **trihydric alcohols**), etc.

((())) SEE WEB LINKS
• Information about IUPAC nomenclature

**aldehydes** Organic compounds that contain the group –CHO (the **aldehyde group**; i.e. a carbonyl group (C=O) with a hydrogen atom bound to the carbon atom). In systematic chemical nomenclature, aldehyde names end with the suffix -*al*. Examples of aldehydes are methanal (formaldehyde), HCOH, and ethanal (acetaldehyde), $CH_3CHO$. Aldehydes are formed by oxidation of primary *alcohols; further oxidation yields carboxylic acids. They are reducing agents and tests for aldehydes include *Fehling's test and *Tollens reagent. Aldehydes have certain characteristic addition and condensation reactions. With sodium hydrogensulphate(IV) they form addition compounds of the type $[RCOH(SO_3)H]^-$ $Na^+$. Formerly these were known as **bisulphite addition compounds.** They also form addition compounds with hydrogen cyanide to give *cyanohydrins and with alcohols to give *acetals and undergo condensation reactions to yield *oximes, *hydrazones, and *semicarbazones. Aldehydes readily polymerize. *See also* KETONES.

((())) SEE WEB LINKS
• Information about IUPAC nomenclature

**Aldehyde structure.**

**aldohexose** *See* MONOSACCHARIDE.

**aldol** *See* ALDOL REACTION.

**aldol reaction** A reaction of aldehydes of the type

$$2RCH_2CHO \rightleftharpoons RCH_2CH(OH)CHRCHO$$

where R is a hydrocarbon group. The resulting compound is a hydroxy-aldehyde, i.e. an aldehyde–alcohol or **aldol**, containing alcohol (–OH) and aldehyde (–CHO) groups on adjacent carbon atoms. The reaction is base-catalysed, the first step being the formation of a carbanion of the type $RHC^-CHO$, which adds to the carbonyl group of the other aldehyde molecule. For the carbanion to form, the aldehyde must have a hydrogen atom on the carbon next to the carbonyl group.

Aldols can be further converted to other products; in particular, they are a source of unsaturated aldehydes. For example, the reaction of ethanal gives 3-hydroxybutenal (**acetaldol**):

$$2CH_3CHO \rightleftharpoons CH_3CH(OH)CH_2CHO$$

This can be further dehydrated to 2-butenal (**crotonaldehyde**):

$$CH_3CH(OH)CH_2CHO \rightarrow H_2O +$$
$$CH_3CH{:}CHCHO$$

**aldose** *See* MONOSACCHARIDE.

**aldosterone** A hormone produced by the adrenal glands (*see* CORTICOSTEROID) that controls excretion of sodium by the kidneys and thereby maintains the balance of salt and water in the body fluids. *See also* AN-GIOTENSIN.

**algae** Any of various unrelated simple organisms that contain chlorophyll (and can therefore carry out photosynthesis) and live in aquatic habitats and in moist situations on land. The algal body may be unicellular or multicellular (filamentous, ribbon-like, or platelike). Molecular studies have confirmed that red and green algae are related to modern plants, whereas other algal groups, such as brown algae, are not. Algae are assigned to separate phyla based primarily on the composition of the cell wall, the nature of the stored food reserves, and the other photosynthetic pigments present. *See* BACILLARIOPHYTA; CHLOROPHYTA; PHAEOPHYTA; RHODOPHYTA.

The organisms formerly known as blue-green algae are now classified as bacteria (*see* CYANOBACTERIA).

**algal bloom (bloom)** The rapid increase in populations of algae and other phytoplankton, in particular the *Cyanobacteria, that occurs in inland water systems, such as lakes. The density of the organisms may be such that it may prevent light from passing to lower depths in the water system. Blooms are caused by an increase in levels of nitrate, a mineral ion essential for algal and bacterial growth. The source of increased nitrate may be from agricultural *fertilizers, which are leached into water systems from the land, or *sewage effluent. Blooms contribute to the eutrophication of water systems. *See also* EUTROPHIC.

**algebra** The branch of mathematics in which variable quantities and numbers are represented by symbols. Statements are usually made in the form of equations, which are manipulated into convenient forms and solved according to a set of strictly logical rules.

**algebraic sum** The total of a set of quantities paying due regard to sign, e.g. the algebraic sum of 3 and –4 is –1.

**Algol** An early high-level block-structured computer programming language. Algol 60 dates from about 1960. Algol 68, of 1968, is a more powerful abstract language. The final version, Algol W, was the precursor of Pascal. The name is short for *algo*rithmic *l*anguage (*see* ALGORITHM).

**algorithm** A method of solving a problem, involving a finite series of steps. In computing practice the algorithm denotes the expression on paper of the proposed computing process (often by means of a flowchart) prior to the preparation of the program. If no algorithm is possible a *heuristic solution has to be sought.

**alicyclic compound** A compound that contains a ring of atoms and is aliphatic. Cyclohexane, $C_6H_{12}$, is an example.

**alignment** (in bioinformatics) The process of matching up base sequences (e.g. of genes) or amino acid sequences (of proteins) to reveal similarities and differences between them. It enables researchers to compare, for example, a newly sequenced gene or protein fragment with well-characterized sequences and is a key step in identifying the nature, possible function, and evolutionary relationships of novel genes and proteins. Alignment is performed by any of various computer programs and makes use of the vast amount of sequence data stored on public databases, which can be accessed via the Internet. In a pairwise alignment, just two sequences are compared, whereas multiple sequence alignment compares three or more. The program compares the sequences and computes the best alignment(s), allowing for gaps and mismatches.

**alimentary canal (digestive tract; gut)** A tubular organ in animals that is divided into a series of zones specialized for the ingestion, *digestion, and *absorption of food and for the elimination of indigestible material (see illustration overleaf). In most animals the canal has two openings, the mouth (for the intake of food) and the *anus (for the elimination of waste). Simple animals, such as cnidarians (e.g. *Hydra* and jellyfish) and flatworms, have only one opening to their alimentary canal, which must serve both functions.

**aliphatic compounds** Organic compounds that are *alkanes, *alkenes, or *alkynes or their derivatives. The term is used to denote compounds that do not have the special stability of *aromatic compounds. All noncyclic organic compounds are aliphatic. Cyclic aliphatic compounds are said to be **alicyclic**.

**alizarin** An orange-red dye, $C_{14}H_8O_4$, which forms *lakes when heavy metal salts are added to its alkaline solutions. It occurs naturally in madder, but is generally synthesized from *anthraquinone.

**alkali** A *base that dissolves in water to give hydroxide ions.

**alkali metals (group 1 elements)** The elements of group 1 (formerly IA) of the *periodic table: lithium (Li), sodium (Na), potassium (K), rubidium (Rb), caesium (Cs), and francium (Fr). All have a characteristic electron configuration that is a noble gas structure with one outer *s*-electron. They are typical metals (in the chemical sense) and readily lose their outer electron to form stable $M^+$ ions with noble-gas configurations. All are highly reactive, with the reactivity (i.e. metallic character) increasing down the group. There is a decrease in ionization energy from lithium (520 kJ $mol^{-1}$) to caesium (380 kJ $mol^{-1}$). The second ionization energies are much higher and divalent ions are not formed. Other properties also change down the group. Thus, there is an increase in atomic and ionic radius, an increase in density, and a decrease in melting and boiling point. The standard electrode potentials are low and negative, although they do not show a regular trend because they depend both on ionization energy (which decreases down the group) and the hydration energy of the ions (which increases).

All the elements react with water (lithium slowly; the others violently) and tarnish rapidly in air. They can all be made to react with chlorine, bromine, sulphur, and hydrogen. The hydroxides of the alkali metals are strongly alkaline (hence the name) and do not decompose on heating. The salts are generally soluble. The carbonates do not decompose on heating, except at very high temperatures. The nitrates (except for lithium) decompose to give the nitrite and oxygen:

$$2MNO_3(s) \rightarrow 2MNO_2(s) + O_2(g)$$

Lithium nitrate decomposes to the oxide. In fact lithium shows a number of dissimilarities to the other members of group 1 and in many ways resembles magnesium (*see* DIAGONAL RELATIONSHIP). In general, the stability of salts of oxo acids increases down the group (i.e. with increasing size of the $M^+$ ion). This trend occurs because the smaller cations (at the top of the group) tend to polarize the oxo anion more effectively than the larger cations at the bottom of the group.

**alkalimetry** Volumetric analysis using standard solutions of alkali to determine the amount of acid present.

**alkaline** 1. Describing an alkali. 2. Describing a solution that has an excess of hydroxide ions (i.e. a pH greater than 7).

**alkaline-earth metals (group 2 elements)** The elements of group 2 (formerly IIA) of the *periodic table: beryllium (Be), magnesium (Mg), calcium (Ca), strontium (Sr), and barium (Ba). The elements are sometimes referred to as the 'alkaline earths', although strictly the 'earths' are the oxides of the elements. All have a characteristic electron configuration that is a noble-gas structure with two outer *s*-electrons. They are typical metals (in the chemical sense) and readily lose both outer electrons

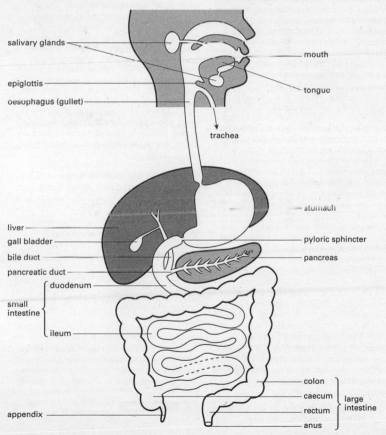

salivary glands
mouth
epiglottis
tongue
oesophagus (gullet)
trachea
stomach
liver
gall bladder
pyloric sphincter
bile duct
pancreas
pancreatic duct
duodenum
small intestine
ileum
colon
caecum
large intestine
appendix
rectum
anus

**The human alimentary canal.**

to form stable $M^{2+}$ ions; i.e. they are strong reducing agents. All are reactive, with the reactivity increasing down the group. There is a decrease in both first and second ionization energies down the group. Although there is a significant difference between the first and second ionization energies of each element, compounds containing univalent ions are not known. This is because the divalent ions have a smaller size and larger charge, leading to higher hydration energies (in solution) or lattice energies (in solids). Consequently, the overall energy change favours the formation of divalent compounds. The third ionization energies are much higher than the second ionization energies, and trivalent compounds (containing $M^{3+}$) are unknown.

Beryllium, the first member of the group, has anomalous properties because of the small size of the ion; its atomic radius (0.112 nm) is much less than that of magnesium (0.16 nm). From magnesium to radium there is a fairly regular increase in atomic and ionic radius. Other regular changes take place in moving down the group from magnesium. Thus, the density and melting and boiling points all increase. Beryllium, on the other hand, has higher boiling and melting points than calcium and its density lies between those of calcium and strontium. The standard electrode potentials are negative and show a regular small decrease from magnesium to barium. In some ways beryllium resembles aluminium (see DIAGONAL RELATIONSHIP).

All the metals are rather less reactive than the alkali metals. They react with water and oxygen (beryllium and magnesium form a protective surface film) and can be made to react with chlorine, bromine, sulphur, and hydrogen. The oxides and hydroxides of the metals show the increasing ionic character in moving down the group: beryllium hydroxide is amphoteric, magnesium hydroxide is only very slightly soluble in water and is weakly basic, calcium hydroxide is sparingly soluble and distinctly basic, strontium and barium hydroxides are quite soluble and basic. The hydroxides decompose on heating to give the oxide and water:

$$M(OH)_2(s) \rightarrow MO(s) + H_2O(g)$$

The carbonates also decompose on heating to the oxide and carbon dioxide:

$$MCO_3(s) \rightarrow MO(s) + CO_2(g)$$

The nitrates decompose to give the oxide:

$$2M(NO_3)_2(s) \rightarrow 2MO(s) + 4NO_2(g) + O_2(g)$$

As with the *alkali metals, the stability of salts of oxo acids increases down the group. In general, salts of the alkaline-earth elements are soluble if the anion has a single charge (e.g. nitrates, chlorides). Most salts with a doubly charged anion (e.g. carbonates, sulphates) are insoluble. The solubilities of salts of a particular acid tend to decrease down the group. (Solubilities of hydroxides increase for larger cations.)

**alkaloid** One of a group of nitrogenous organic compounds, mostly derived from plants, and having diverse pharmacological properties. They are biosynthesized from amino acids and classified according to some structural feature. A simple classification is into:

the pyridine group (e.g. coniine, nicotine)

the tropine group (e.g. atropine, cocaine)

the quinoline group (e.g. quinine, strychnine, brucine)

the isoquinoline group (e.g. morphine, codeine)

the phenylethylamine group (e.g. methamphetamine, mescaline, ephedrine)

the indole group (e.g. tryptamine, lysergic acid)

the purine group (e.g. caffeine, theobromine, theophylline)

**alkanal** An aliphatic aldehyde.

**alkanes (paraffins)** Saturated hydrocarbons with the general formula $C_nH_{2n+2}$. In systematic chemical nomenclature alkane names end in the suffix -*ane*. They form a *homologous series (the **alkane series**) methane ($CH_4$), ethane ($C_2H_6$), propane ($C_3H_8$), butane ($C_4H_{10}$), pentane ($C_5H_{12}$), etc. The lower members of the series are gases; the high-molecular weight alkanes are waxy solids. Alkanes are present in natural gas and petroleum. They can be made by heating the sodium salt of a carboxylic acid with soda lime:

$$RCOO^-Na^+ + Na^+OH^- \rightarrow Na_2CO_3 + RH$$

Other methods include the *Wurtz reaction and *Kolbe's method. Generally the alkanes are fairly unreactive. They form haloalkanes with halogens when irradiated with ultraviolet radiation.

**(((⊕))) SEE WEB LINKS**

• Information about IUPAC nomenclature
• Further details about nomenclature

**alkanol** An aliphatic alcohol.

**alkenes (olefines; olefins)** Unsaturated hydrocarbons that contain one or more double carbon–carbon bonds in their molecules. In systematic chemical nomenclature alkene names end in the suffix *-ene*. Alkenes that have only one double bond form a homologous series (the **alkene series**) starting ethene (ethylene), $CH_2:CH_2$, propene, $CH_3CH:CH_2$, etc. The general formula is $C_nH_{2n}$. Higher members of the series show isomerism depending on position of the double bond; for example, butene ($C_4H_8$) has two isomers, which are (1) but-1-ene ($C_2H_5CH:CH_2$) and (2) but-2-ene ($CH_3CH:CHCH_3$): see formulae.

Alkenes can be made by dehydration of alcohols (passing the vapour over hot pumice):

$$RCH_2CH_2OH - H_2O \rightarrow RCH:CH_2$$

An alternative method is the removal of a hydrogen atom and halogen atom from a haloalkane by potassium hydroxide in hot alcoholic solution:

$$RCH_2CH_2Cl + KOH \rightarrow KCl + H_2O + RCH:CH_2$$

Alkenes typically undergo *addition reactions to the double bond. They can be tested for by the *Baeyer test. *See also* HYDROGENATION; OXO PROCESS; OZONOLYSIS; ZIEGLER PROCESS.

**((⊕)) SEE WEB LINKS**
• Information about IUPAC nomenclature

but-1-ene

but-2-ene

**Alkenes.** Butene isomers.

**alkoxides** Compounds formed by reaction of alcohols with sodium or potassium metal. Alkoxides are saltlike compounds containing the ion $R-O^-$.

**alkyd resin** A type of *polyester resin used in paints and other surface coating. The original alkyd resins were made by copolymerizing phthalic anhydride with glycerol, to give a brittle cross-linked polymer. The properties of such resins can be modified by adding monobasic acids or alcohols during the polymerization.

**alkylation** A chemical reaction that introduces an *alkyl group into an organic molecule. The *Friedel–Crafts reaction results in alkylation of aromatic compounds.

**alkylbenzenes** Organic compounds that have an alkyl group bound to a benzene ring. The simplest example is methylbenzene (toluene), $CH_3C_6H_5$. Alkyl benzenes can be made by the *Friedel–Crafts reaction.

**alkyl group** A group obtained by removing a hydrogen atom from an alkane, e.g. methyl group, $CH_3-$, derived from methane.

**alkyl halides** *See* HALOALKANES.

**alkynes (acetylenes)** Unsaturated hydrocarbons that contain one or more triple carbon–carbon bonds in their molecules. In systematic chemical nomenclature alkyne names end in the suffix *-yne*. Alkynes that have only one triple bond form a *homologous series: ethyne (acetylene), $CH\equiv CH$, propyne, $CH_3CH\equiv CH$, etc. They can be made by the action of potassium hydroxide in alcohol solution on haloalkanes containing halogen atoms on adjacent carbon atoms; for example:

$$RCHClCH_2Cl + 2KOH \rightarrow 2KCl + 2H_2O + RCH\equiv CH$$

Like *alkenes, alkynes undergo addition reactions.

**((⊕)) SEE WEB LINKS**
• Information about IUPAC nomenclature

**allantois** One of the membranes that develops in embryonic reptiles, birds, and mammals as a growth from the hindgut. It acts as a urinary bladder for the storage of waste excretory products in the egg (in reptiles and birds) and as a means of providing the embryo with oxygen (in reptiles, birds, and mammals) and food (in mammals; *see* PLACENTA). *See also* EXTRAEMBRYONIC MEMBRANES.

**allele (allelomorph)** One of the alternative forms of a gene. In a diploid cell there are usually two alleles of any one gene (one from each parent), which occupy the same relative position (*locus) on *homologous chromosomes. These alleles may be the same, or one allele may be *dominant to the other (known as the *recessive), i.e. it determines which aspects of a particular characteristic the organism will display. Within a population there may be many alleles of a gene; each has a unique nucleotide sequence.

**allelomorph** *See* ALLELE.

**allelopathy** The secretion by plants of chemicals, such as phenolic and terpenoid compounds, that inhibit the growth or germination of other plants, with which they are competing. For example, the aromatic oils released by certain shrubs of the Californian chaparral pass into the soil and inhibit the growth of herbaceous species nearby. Some plants produce chemicals that are toxic to grazing herbivorous animals.

**allenes** Compounds that contain the group >C=C=C<, in which three carbon atoms are linked by two adjacent double bonds. The outer carbon atoms are each linked to two other atoms or groups by single bonds. The simplest example is 1,2-propadiene, $CH_2CCH_2$. Allenes are *dienes with typical reactions of alkenes. Under basic conditions, they often convert to alkynes. In an allene, the two double bonds lie in planes that are perpendicular to each other. Consequently, in an allene of the type $R_1R_2C:C:CR_3R_4$, the groups $R_1$ and $R_2$ lie in a plane perpendicular to the plane containing $R_3$ and $R_4$. Under these circumstances, the molecule is chiral and can show optical activity.

**allergen** An antigen that provokes an abnormal *immune response. Common allergens include pollen and dust (*see* ALLERGY).

**allergy** A condition in which the body produces an abnormal *immune response to certain *antigens (called **allergens**), which include dust, pollen, certain foods and drugs, or fur. In allergic individuals these substances, which in a normal person would be destroyed by antibodies, react with preexisting antibodies or trigger responses in primed T cells. A common mechanism is binding of allergen to IgE on *mast cells. This causes the latter to secrete *histamine and other vasoactive agents, leading to inflammation and other characteristic symptoms of the allergy (e.g. asthma or hay fever). *See also* ANAPHYLAXIS; MAST CELL.

**allogamy** Cross-fertilization in plants. *See* FERTILIZATION.

**allograft** *See* GRAFT.

**allometric growth** The regular and systematic pattern of growth such that the mass or size of any organ or part of a body can be expressed in relation to the total mass or size of the entire organism according to the allometric equation: $Y = bx^\alpha$, where $Y$ = mass of the organ, $x$ = mass of the organism, $\alpha$ = growth coefficient of the organ, and $b$ = a constant.

**allopatric** Describing or relating to groups of similar organisms that could interbreed but do not because they are geographically separated. *Compare* SYMPATRIC. *See* SPECIATION.

**allopolyploid** A *polyploid organism, usually a plant, that contains multiple sets of chromosomes derived from different species. Hybrids are usually sterile, because they do not have sets of *homologous chromosomes and therefore *pairing cannot take place. However, if doubling of the chromosome number occurs in a hybrid derived from two diploid ($2n$) species, the resulting tetraploid ($4n$) is a fertile plant. This type of tetraploid is known as an **allotetraploid**; as it contains two sets of homologous chromosomes, pairing and crossing over are now possible. Allopolyploids are of great importance to plant breeders as advantages possessed by different species can be combined. The species of wheat, *Triticum aestivum*, used to make bread is an **allohexaploid** ($6n$), possessing 42 chromosomes, which is six times the original haploid number ($n$) of 7. *Compare* AUTOPOLYPLOID.

**all-or-none response** A type of response that may be either complete and of full intensity or totally absent, depending on the strength of the stimulus; there is no partial response. For example, a nerve cell is either stimulated to transmit a complete nervous impulse or else it remains in its resting state; a stinging *thread cell of a cnidarian is either completely discharged or it is not.

**allosteric enzyme** An enzyme that has two structurally distinct forms, one of which is active and the other inactive. In the active form, the quaternary structure (*see* PROTEIN) of the enzyme is such that a substrate can interact with the enzyme at the active site (*see* ENZYME–SUBSTRATE COMPLEX). The conformation of the substrate-binding site becomes altered in the inactive form and interaction with the substrate is not possible. Allosteric enzymes tend to catalyse the initial step in a pathway leading to the synthesis of molecules. The end product of this synthesis can act as a feedback inhibitor (*see* INHIBITION) and the enzyme is converted to the inactive form, thereby controlling the amount of product synthesized.

**allotropy** The existence of elements in two or more different forms (**allotropes**). In the case of oxygen, there are two forms: 'normal' dioxygen ($O_2$) and ozone, or trioxygen ($O_3$). These two allotropes have different molecular configurations. More commonly, allotropy occurs because of different crystal structures in the solid, and is particularly prevalent in groups 14, 15, and 16 of the periodic table. In some cases, the allotropes are stable over a temperature range, with a definite transition point at which one changes into the other. For instance, tin has two allotropes: white (metallic) tin stable above 13.2°C and grey (nonmetallic) tin stable below 13.2°C. This form of allotropy is called **enantiotropy**. Carbon also has two allotropes – diamond and graphite – although graphite is the stable form at all temperatures. This form of allotropy, in which there is no transition temperature at which the two are in equilibrium, is called **monotropy**. *See also* POLYMORPHISM.

**allowed bands** *See* ENERGY BANDS.

**allowed transitions** *See* SELECTION RULES.

**alloy** A material consisting of two or more metals (e.g. brass is an alloy of copper and zinc) or a metal and a nonmetal (e.g. steel is an alloy of iron and carbon, sometimes with other metals included). Alloys may be compounds, *solid solutions, or mixtures of the components.

**alloy steels** *See* STEEL.

**alluvial deposits** Sediments deposited in a river, which range in particle size from fine silts to coarse gravels.

**allyl group** *See* PROPENYL GROUP.

**Alnico** A trade name for a series of alloys, containing iron, aluminium, nickel, cobalt, and copper, used to make permanent magnets.

**alpha-iron** *See* IRON.

**alphamethyltryptamine (AMT)** A synthetic derivative of *tryptamine with stimulant and hallucinogenic properties. It is used illegally as a club drug.

**alpha-naphthol test** A biochemical test to detect the presence of carbohydrates in solution, also known as **Molisch's test** (after

the Austrian chemist H. Molisch (1856–1937), who devised it). A small amount of alcoholic alpha-naphthol is mixed with the test solution and concentrated sulphuric acid is poured slowly down the side of the test tube. A positive reaction is indicated by the formation of a violet ring at the junction of the liquids.

**alpha particle** A helium–4 nucleus emitted by a larger nucleus during the course of the type of radioactive decay known as **alpha decay**. As a helium–4 nucleus consists of two protons and two neutrons bound together as a stable entity the loss of an alpha particle involves a decrease in *nucleon number of 4 and decrease of 2 in the *atomic number, e.g. the decay of a uranium–238 nucleus into a thorium–234 nucleus. A stream of alpha particles is known as an **alpha-ray** or **alpha-radiation**.

**alternating current (a.c.)** An electric current that reverses its direction with a constant *frequency (*f*). If a graph of the current against time has the form of a *sine wave, the current is said to be **sinusoidal**. Alternating current, unlike direct current, is therefore continuously varying and its magnitude is either given as its peak value ($I_0$) or its *root-mean-square value ($I_0/\sqrt{2}$ for a sinusoidal current). This r.m.s. value is more useful as it is comparable to a d.c. value in being a measure of the ability of the current to transmit power. The instantaneous value of a sinusoidal current (*I*) is given by $I = I_0\sin2\pi ft$.

If a direct current is supplied to a circuit the only opposition it encounters is the circuit's *resistance. However, an alternating current is opposed not only by the resistance of the circuit but also by its *reactance. This reactance is caused by *capacitance and *inductance in the circuit. In a circuit consisting of a resistance (*R*), an inductance (*L*), and a capacitance (*C*) all in series, the reactance (*X*) is equal to $(2\pi fL) - (1/2\pi fC)$. The total opposition to the current, called the *impedance (*Z*), is then equal to the ratio of the r.m.s. applied p.d. to the r.m.s. current and is given by $\sqrt{(R^2 + X^2)}$.

**alternation of generations** The occurrence within the *life cycle of an organism of two or more distinct forms (generations), which differ from each other in appearance, habit, and method of reproduction. The phenomenon occurs in some protists and other simple multicellular organisms, certain lower animals (e.g. cnidarians and parasitic flatworms), and in plants. The malaria parasite (*Plasmodium*), for example, has a complex life cycle involving the alternation of sexually and asexually reproducing generations. In plants the generation with sexual reproduction is called the *gametophyte and the asexual generation is the *sporophyte, either of which may dominate the life cycle, and there is also alternation of the haploid and diploid states. Thus in vascular plants the dominant plant is the diploid sporophyte; it produces spores that germinate into small haploid gametophytes. In mosses the gametophyte is the dominant plant and the sporophyte is the spore-bearing capsule.

**alternative splicing** The splicing of the RNA transcript of a gene in different ways by the cellular machinery to create distinct mature messenger RNAs (mRNAs) encoding different proteins. Hence, the same gene can give rise to related but variant forms of a protein at different stages of development or in different tissues. Alternative splicing accounts for the disparity between the relatively small number of human genes identified so far, roughly 22 000, and the 35 000 or so different gene transcripts discovered.

**alternator** An *alternating-current generator consisting of a coil or coils that rotate in the magnetic field produced by one or more permanent magnets or electromagnets. The electromagnets are supplied by an independent direct-current source. The frequency of the alternating current produced depends on the speed at which the coil rotates and the number of pairs of magnetic poles. In the large alternators of power stations the electromagnets rotate inside fixed coils; many bicycle dynamos are alternators with rotating permanent magnets inside fixed coils.

**altimeter** A device used to measure height above sea level. It usually consists of an aneroid *barometer measuring atmospheric pressure. Aircraft are fitted with altimeters, which are set to the atmospheric pressure at a convenient level, usually sea level, before take off. The height of the aircraft can then be read off the instrument as the aircraft climbs and the pressure falls.

**altitude** In horizontal coordinate systems, the distance of a celestial object above or below the observer's horizon, expressed as an angle i.e. its elevation. *Compare* AZIMUTH.

**altruism** Behaviour by an animal that de-

creases its chances of survival or reproduction while increasing those of another member of the same species. For example, a lapwing puts itself at risk by luring a predator away from the nest through feigning injury, but by so doing saves its offspring. Altruism in its biological sense does not imply any conscious benevolence on the part of the performer. Altruism can evolve through *kin selection, if the recipients of altruistic acts tend on average to be more closely related to the altruist than the population as a whole. Some animals perform altruistic acts in the expectation that the 'favour' will be returned in the future. This **reciprocal altruism** is practised most notably by humans. *See also* INCLUSIVE FITNESS.

**ALU (arithmetic/logic unit)** The part of the central processor of a *computer in which simple arithmetic and logical operations are performed electronically. For example, the ALU can add, subtract, multiply, or compare two numbers, or negate a number.

**alum** *See* ALUMINIUM POTASSIUM SULPHATE; ALUMS.

**alumina** *See* ALUMINIUM OXIDE; ALUMINIUM HYDROXIDE.

**aluminate** A salt formed when aluminium hydroxide or γ-alumina is dissolved in solutions of strong bases, such as sodium hydroxide. Aluminates exist in solutions containing the aluminate ion, commonly written $[Al(OH)_4]^-$. In fact the ion probably is a complex hydrated ion and can be regarded as formed from a hydrated $Al^{3+}$ ion by removal of four hydrogen ions:

$$[Al(H_2O)_6]^{3+} + 4OH^- \rightarrow 4H_2O + [Al(OH)_4(H_2O)_2]^-$$

Other aluminates and polyaluminates, such as $[Al(OH)_6]^{3-}$ and $[(HO)_3AlOAl(OH)_3]^{2-}$, are also present. *See also* ALUMINIUM HYDROXIDE.

**aluminium** Symbol Al. A silvery-white lustrous metallic element belonging to *group 13 (formerly IIIB) of the periodic table; a.n. 13; r.a.m. 26.98; r.d. 2.7; m.p. 660°C; b.p. 2467°C. The metal itself is highly reactive but is protected by a thin transparent layer of the oxide, which forms quickly in air. Aluminium and its oxide are amphoteric. The metal is extracted from purified bauxite ($Al_2O_3$) by electrolysis; the main process uses a *Hall–Heroult cell but other electrolytic methods are under development, including conversion of bauxite with chlorine and electrolysis of the molten chloride. Pure aluminium is soft and ductile but its strength can be increased by work-hardening. A large number of alloys are manufactured; alloying elements include copper, manganese, silicon, zinc, and magnesium. Its lightness, strength (when alloyed), corrosion resistance, and electrical conductivity (62% of that of copper) make it suitable for a variety of uses, including vehicle and aircraft construction, building (window and door frames), and overhead power cables. Although it is the third most abundant element in the earth's crust (8.1% by weight) it was not isolated until 1825 by H. C. Oersted (1777–1851).

**SEE WEB LINKS**
• Information from the WebElements site

**aluminium acetate**  *See* ALUMINIUM ETHANOATE.

**aluminium chloride** A whitish solid, $AlCl_3$, which fumes in moist air and reacts violently with water (to give hydrogen chloride). It is known as the anhydrous salt (hexagonal; r.d. 2.44 (fused solid); m.p. 190°C (2.5 atm.); sublimes at 178°C) or the hexahydrate $AlCl_3.6H_2O$ (rhombic; r.d. 2.398; loses water at 100°C), both of which are deliquescent. Aluminium chloride may be prepared by passing hydrogen chloride or chlorine over hot aluminium or (industrially) by passing chlorine over heated aluminium oxide and carbon. The chloride ion is polarized by the small positive aluminium ion and the bonding in the solid is intermediate between covalent and ionic. In the liquid and vapour phases dimer molecules exist, $Al_2Cl_6$, in which there are chlorine bridges making coordinate bonds to aluminium atoms (see formula). The $AlCl_3$ molecule can also form compounds with other molecules that donate pairs of electrons (e.g. amines or hydrogen sulphide); i.e. it acts as a Lewis *acid. At high temperatures the $Al_2Cl_6$ molecules in the vapour dissociate to (planar) $AlCl_3$ molecules. Aluminium chloride is used commercially as a catalyst in the cracking of oils. It is also a catalyst in certain other organic reactions, especially the Friedel–Crafts reaction.

**Aluminium chloride.** Structure of aluminium trichloride dimer.

**aluminium ethanoate (aluminium acetate)** A white solid, $Al(OOCCH_3)_3$, which decomposes on heating, is very slightly soluble in cold water, and decomposes in warm water. The normal salt, $Al(OOCCH_3)_3$, can only be made in the absence of water (e.g. ethanoic anhydride and aluminium chloride at 180°C); in water it forms the basic salts $Al(OH)(OOCCH_3)_2$ and $Al_2(OH)_2(OOCCH_3)_4$. The reaction of aluminium hydroxide with ethanoic acid gives these basic salts directly. The compound is used extensively in dyeing as a mordant, particularly in combination with aluminium sulphate (known as **red liquor**); in the paper and board industry for sizing and hardening; and in tanning. It was previously used as an antiseptic and astringent.

**aluminium hydroxide** A white crystalline compound, $Al(OH)_3$; r.d. 2.42–2.52. The compound occurs naturally as the mineral **gibbsite** (monoclinic). In the laboratory it can be prepared by precipitation from solutions of aluminium salts. Such solutions contain the hexaquoaluminium(III) ion with six water molecules coordinated, $[Al(H_2O)_6]^{3+}$. In neutral solution this ionizes:

$$[Al(H_2O)_6]^{3+} \rightleftharpoons H^+ + [Al(H_2O)_5OH]^{2+}$$

The presence of a weak base such as $S^{2-}$ or $CO_3^{2-}$ (by bubbling hydrogen sulphide or carbon dioxide through the solution) causes further ionization with precipitation of aluminium hydroxide

$$[Al(H_2O)_6]^{3+}(aq) \rightarrow Al(H_2O)_3(OH)_3(s) + 3H^+(aq)$$

The substance contains coordinated water molecules and is more correctly termed **hydrated aluminium hydroxide**. In addition, the precipitate has water molecules trapped in it and has a characteristic gelatinous form. The substance is amphoteric. In strong bases the *aluminate ion is produced by loss of a further proton:

$$Al(H_2O)_3(OH)_3(s) + OH^-(aq) \rightleftharpoons$$
$$[Al(H_2O)_2(OH)_4]^-(aq) + H_2O(l)$$

On heating, the hydroxide transforms to a mixed oxide hydroxide, AlO.OH (rhombic; r.d. 3.01). This substance occurs naturally as **diaspore** and **boehmite**. Above 450°C it transforms to γ-alumina.

In practice various substances can be produced that are mixed crystalline forms of $Al(OH)_3$, AlO.OH, and aluminium oxide ($Al_2O_3$) with water molecules. These are known as **hydrated alumina**. Heating the hydrated hydroxide causes loss of water, and produces various **activated aluminas**, which differ in porosity, number of remaining –OH groups, and particle size. These are used as catalysts (particularly for organic dehydration reactions), as catalyst supports, and in chromatography. Gelatinous freshly precipitated aluminium hydroxide was formerly widely used as a mordant for dyeing and calico printing because of its ability to form insoluble coloured *lakes with vegetable dyes. See also ALUMINIUM OXIDE.

**aluminium oxide (alumina)** A white or colourless oxide of aluminium occurring in two main forms. The stable form α-alumina (r.d. 3.97; m.p. 2015°C; b.p. 2980 ± 60°C) has colourless hexagonal or rhombic crystals; γ-alumina (r.d. 3.5–3.9) transforms to the α-form on heating and is a white microcrystalline solid. The compound occurs naturally as **corundum** or **emery** in the α-form with a hexagonal-close-packed structure of oxide ions with aluminium ions in the octahedral interstices. The gemstones ruby and sapphire are aluminium oxide coloured by minute traces of chromium and cobalt respectively. A number of other forms of aluminium oxide have been described (β-, δ-, and ζ-alumina) but these contain alkali-metal ions. There is also a short-lived spectroscopic suboxide AlO. The highly protective film of oxide formed on the surface of aluminium metal is yet another structural variation, being a defective rock-salt form (every third Al missing).

Pure aluminium oxide is obtained by dissolving the ore bauxite in sodium hydroxide solution; impurities such as iron oxides remain insoluble because they are not amphoteric. The hydrated oxide is precipitated by seeding with material from a previous batch and this is then roasted at 1150–1200°C to give pure α-alumina, or at 500–800°C to give γ-alumina. The bonding in aluminium hydroxide is not purely ionic due to polarization of the oxide ion. Although the compound might be expected to be amphoteric, α-alumina is weakly acidic, dissolving in alkalis to give solutions containing aluminate ions; it is resistant to acid attack. In contrast γ-alumina is typically amphoteric dissolving both in acids to give aluminium salts and in bases to give aluminates. α-alumina is one of the hardest materials known (silicon carbide and diamond are harder) and is widely used as an abrasive in both natural (corundum) and synthetic forms. Its refractory nature makes alumina

brick an ideal material for furnace linings and alumina is also used in cements for high-temperature conditions. See also ALUMINIUM HYDROXIDE.

**aluminium potassium sulphate (potash alum; alum)** A white or colourless crystalline compound, $Al_2(SO_4)_3.K_2SO_4.24H_2O$; r.d. 1.757; loses $18H_2O$ at 92.5°C; becomes anhydrous at 200°C. It forms cubic or octahedral crystals that are soluble in cold water, very soluble in hot water, and insoluble in ethanol and acetone. The compound occurs naturally as the mineral **kalinite**. It is a double salt and can be prepared by recrystallization from a solution containing equimolar quantities of potassium sulphate and aluminium sulphate. It is used as a mordant for dyeing and in the tanning and finishing of leather goods (for white leather). See also ALUMS.

**aluminium sulphate** A white or colourless crystalline compound, $Al_2(SO_4)_3$, known as the anhydrous compound (r.d. 2.71; decomposes at 770°C) or as the hydrate $Al_2(SO)_3.18H_2O$ (monoclinic; r.d. 1.69; loses water at 86.5°C). The anhydrous salt is soluble in water and slightly soluble in ethanol; the hydrate is very soluble in water and insoluble in ethanol. The compound occurs naturally in the rare mineral **alunogenite** $(Al_2(SO)_3.18H_2O)$. It may be prepared by dissolving aluminium hydroxide or china clays (aluminosilicates) in sulphuric acid. It decomposes on heating to sulphur dioxide, sulphur trioxide, and aluminium oxide. Its solutions are acidic because of hydrolysis.

Aluminium sulphate is commercially one of the most important aluminium compounds; it is used in sewage treatment (as a flocculating agent) and in the purification of drinking water, the paper industry, and in the preparation of mordants. It is also a fireproofing agent. Aluminium sulphate is often wrongly called **alum** in these industries.

**aluminium trimethyl** See TRIMETHYLALUMINIUM.

**aluminosilicate** The chief rock-forming mineral in, for example, some clays, feldspar, mica, and zeolite. Aluminosilicates are also key constituents of china, glass, and cement. Most have a tetrahedral silicate structure with aluminium atoms replacing some of the silicon.

**alums** A group of double salts with the formula $A_2SO_4.B_2(SO_4)_3.24H_2O$, where A is a monovalent metal and B a trivalent metal. The original example contains potassium and aluminium (called **potash alum** or simply **alum**); its formula is often written $AlK(SO_4)_2.12H_2O$ (aluminium potassium sulphate-12-water). **Ammonium alum** is $AlNH_4(SO_4)_2.12H_2O$, **chrome alum** is $KCr(SO_4)_2.12H_2O$ (*see* POTASSIUM CHROMIUM SULPHATE), etc. The alums are isomorphous and can be made by dissolving equivalent amounts of the two salts in water and recrystallizing. *See also* ALUMINIUM SULPHATE.

**alunogenite** A mineral form of hydrated *aluminium sulphate, $Al_2(SO_4)_3.18H_2O$.

**Alvarez, Luis Walter** (1911–88) US physicist most of whose working life was spent at the University of California, Berkeley. After working on radar and the atomic bomb during World War II, he concentrated on particle physics. In 1959 he built the first large *bubble chamber and developed the technique for using it to study charged particles, for which he was awarded the 1968 Nobel Prize for physics.

**Alvarez event** The collision of a giant meteorite with the earth 65 million years ago that caused catastrophic changes to the earth's climate and environment and a *mass extinction of species, including the dinosaurs. This hypothesis was advanced in 1980 by Luis Walter *Alvarez and his geologist son Walter Jr, based on the unusually high concentration of the element iridium in a thin layer of clay deposited at the end of the Cretaceous (*see* IRIDIUM ANOMALY). This clay marks the boundary between the Cretaceous period and the more recent Tertiary (the so-called **K–T boundary**). Subsequently, geologists discovered a possible impact crater, roughly 160 km in diameter, along the coast of eastern Mexico, and other evidence has tended to support the hypothesis. Such a collision would have produced a massive tidal wave and fireball and sent a vast cloud of rock and other debris into the atmosphere. The resulting upheaval in the climate is estimated to have caused the extinction of some 75% of all species.

**alveolus** (*pl.* **alveoli**) **1.** The tiny air sac in the *lung of mammals and reptiles at the end of each *bronchiole. It is lined by a delicate moist membrane, has many blood capillaries, and is the site of exchange of respiratory gases (carbon dioxide and oxygen). **2.** The socket in the jawbone in which a tooth is

rooted by means of the *periodontal membrane.

**Alzheimer's disease** A neurological disease characterized by progressive loss of intellectual ability. The disease, which is named after German physician Alois Alzheimer (1864–1915), is associated with general shrinkage of the brain tissue – with deposits of β-amyloid peptide (a glycoprotein) and abnormal filaments of tau protein associated with *microtubules – and changes in the neurotransmitter systems within the brain that include a loss in the activity of *cholinergic neurons.

**AM (amplitude modulation)** *See* MODULATION.

**amalgam** An alloy of mercury with one or more other metals. Most metals form amalgams (iron and platinum are exceptions), which may be liquid or solid. Some are simple solutions; others contain definite intermetallic compounds, such as $NaHg_2$.

**amatol** A high explosive consisting of a mixture of ammonium nitrate and trinitrotoluene.

**amber** A yellow or reddish-brown fossil resin. The resin was exuded by certain trees and other plants and often contains preserved insects, flowers, or leaves that were trapped by its sticky surface before the resin hardened. Amber is used for jewellery and ornaments. It also has the property of acquiring an electrical charge when rubbed (the term electricity is derived from *electron*, the Greek name for amber). It occurs throughout the world in rock strata from the Cretaceous to the Pleistocene, but most commonly in Cretaceous and Tertiary rocks.

**ambidentate** Describing a ligand that can coordinate at two different sites. For example, the $NO_2$ molecule can coordinate through the N atom (the **nitro ligand**) or through an O atom (the **nitrido ligand**). Complexes that differ only in the way the ligand coordinates display **linkage isomerism**.

**ambient** Denoting the immediate surroundings or environment. **Ambient light** is light generated by outside sources, such as the sun, in relation to the environment of a specific optical system. **Ambient noise** is the background noise in relation to a particular sound, such as music. **Ambient pressure** and **ambient temperature** are the pressure and temperature of the surroundings, e.g. of the atmosphere, in relation to a specific object or system.

**ambient computing** See UBIQUITOUS COMPUTING.

**americium** Symbol Am. A radioactive metallic transuranic element belonging to the *actinoids; a.n. 95; mass number of most stable isotope 243 (half-life $7.95 \times 10^3$ years); r.d. 13.67 (20°C); m.p. $994 \pm 4$°C; b.p. 2607°C. Ten isotopes are known. The element was discovered by G. T. Seaborg and associates in 1945, who obtained it by bombarding uranium–238 with alpha particles.

**(⊕) SEE WEB LINKS**
• Information from the WebElements site

**Ames test (*Salmonella* mutagenesis test)** A test to determine the effects of a chemical on the rate of mutation in bacterial cells, and hence its likely potential for causing cancer in other organisms, including humans. Devised by US biologist Bruce Ames (1928– ), it is widely used in screening chemicals occurring in the environment for possible carcinogenic activity. The chemical is applied to plates inoculated with a special mutant strain of bacteria, usually *Salmonella typhimurium*, and cells that mutate back to the wild type are detected by the occurrence of colonies able to grow on the medium.

**amethyst** The purple variety of the mineral *quartz. It is found chiefly in Brazil, the Urals (Russia), Arizona (USA), and Uruguay. The colour is due to impurities, especially iron oxide. It is used as a gemstone.

**amides** 1. Organic compounds containing the group $-CO.NH_2$ (the **amide group**). Compounds containing this group are **primary amides**. **Secondary** and **tertiary amides** can also exist, in which the hydrogen atoms on the nitrogen are replaced by one or two other organic groups respectively. Simple examples of primary amides are ethanamide, $CH_3CONH_2$, and propanamide, $C_2H_5CONH_2$. They are made by heating the ammonium salt of the corresponding carboxylic acid. Amides can also be made by reaction of ammonia (or an amine) with an acyl halide. 2. Inorganic compounds containing the ion $NH_2^-$, e.g. $KNH_2$ and $Cd(NH_2)_2$. They are formed by the reaction of ammonia with electropositive metals.

**(⊕) SEE WEB LINKS**
• Information about IUPAC nomenclature

**amination** A chemical reaction in which an amino group ($-NH_2$) is introduced into a molecule. Examples of amination reaction include the reaction of halogenated hydrocarbons with ammonia (high pressure and temperature) and the reduction of nitro compounds and nitriles.

**amines** Organic compounds derived by replacing one or more of the hydrogen atoms in ammonia by organic groups (see illustration). **Primary amines** have one hydrogen replaced, e.g. methylamine, $CH_3NH_2$. They contain the functional group $-NH_2$ (the **amino group**). **Secondary amines** have two hydrogens replaced, e.g. methylethylamine, $CH_3(C_2H_5)NH$. **Tertiary amines** have all three hydrogens replaced, e.g. trimethylamine, $(CH_3)_3N$. Amines are produced by the decomposition of organic matter. They can be made by reducing nitro compounds or amides. See also IMINES.

**(⊕) SEE WEB LINKS**
• Information about IUPAC nomenclature of primary amines
• Information about IUPAC nomenclature of secondary and tertiary amines

**amine salts** Salts similar to ammonium salts in which the hydrogen atoms attached to the nitrogen are replaced by one or more organic groups. Amines readily form salts by reaction with acids, gaining a proton to form a positive ammonium ion, They are named as if they were substituted derivatives of ammonium compounds; for example, dimethylamine $((CH_3)_2NH)$ will react with hydrogen chloride to give dimethylammonium chloride, which is an ionic compound $[(CH_3)_2NH_2]^+Cl^-$. When the amine has a common nonsystematic name the suffix -*ium* can be used; for example, phenylamine (aniline) would give $[C_6H_5NH_3]^+Cl^-$, known as anilinium chloride. Formerly, such compounds were sometimes called **hydrochlorides**, e.g. aniline hydrochloride with the formula $C_6H_5NH_2.HCl$.

Salts formed by amines are crystalline substances that are readily soluble in water. Many insoluble *alkaloids (e.g. quinine and atropine) are used medicinally in the form of

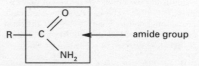

**Amide structure.**

R—C ⚌ O, —NH₂ — amide group

**Examples of amines.**

soluble salts ('hydrochlorides'). If alkali (sodium hydroxide) is added to solutions of such salts the free amine is liberated.

If all four hydrogen atoms of an ammonium salt are replaced by organic groups a **quaternary ammonium compound** is formed. Such compounds are made by reacting tertiary amines with halogen compounds; for example, trimethylamine $((CH_3)_3N)$ with chloromethane $(CH_3Cl)$ gives tetramethylammonium chloride, $(CH_3)_4N^+Cl^-$. Salts of this type do not liberate the free amine when alkali is added, and quaternary hydroxides (such as $(CH_3)_4N^+OH^-$) can be isolated. Such compounds are strong alkalis, comparable to sodium hydroxide.

**amino acid** Any of a group of water-soluble organic compounds that possess both a carboxyl (–COOH) and an amino (–NH₂) group attached to the same carbon atom, called the α-carbon atom. Amino acids can be represented by the general formula R–CH(NH₂)COOH. R may be hydrogen or an organic group and determines the properties of any particular amino acid. Through the formation of peptide bonds, amino acids join together to form short chains (*peptides) or much longer chains (*polypeptides). Proteins are composed of various proportions of about 20 commonly occurring amino acids (see table overleaf). The sequence of these amino acids in the protein polypeptides determines the shape, properties, and hence biological role of the protein. Some amino acids that never occur in proteins are nevertheless important, e.g. *ornithine and citrulline, which are intermediates in the urea cycle.

Plants and many microorganisms can synthesize amino acids from simple inorganic compounds, but animals rely on adequate supplies in their diet. The *essential amino acids must be present in the diet whereas others can be manufactured from them.

**SEE WEB LINKS**
- Interactively depicts molecular structures of all the amino acids using Jmol
- Information about IUPAC nomenclature

**amino acid racemization (AAR)** A dating technique used in archaeology based on the relative amounts of the optical isomers of an amino acid in a sample. In most organisms, the L-isomer of the amino acid is the one produced by metabolism. When the organism dies, this isomer slowly converts into the D-form, and eventually an equilibrium is reached in which the two forms are present in equal amounts. Measuring the proportions of the L- and D-forms in a sample can, in principle, give an estimate of the time since death. Not all amino acids racemize at the same rate, and the rate of the process depends on other factors, such as moisture and temperature. Most work has been done using leucine or aspartic acid.

A particular application in forensic science involves measuring the D/L ratio of aspartic acid in the dentine of teeth. Once a tooth has fully formed, the dentine is isolated by the enamel and then racemization takes place in the living subject at a fairly constant temperature and moisture level. Measuring the ratio gives a fairly good estimate of the age of the subject (rather than the time since death).

**aminobenzene** See PHENYLAMINE.

**amino group** See AMINES.

a

| amino acid | abbreviation 3-letter | 1-letter | formula |
|---|---|---|---|
| alanine | Ala | A | $CH_3 - \overset{\overset{H}{\vert}}{\underset{\underset{NH_2}{\vert}}{C}} - COOH$ |
| arginine | Arg | R | $H_2N - \overset{}{\underset{\underset{NH}{\parallel}}{C}} - NH - CH_2 - CH_2 - CH_2 - \overset{\overset{H}{\vert}}{\underset{\underset{NH_2}{\vert}}{C}} - COOH$ |
| asparagine | Asn | N | $H_2N - \overset{}{\underset{\underset{O}{\parallel}}{C}} - CH_2 - \overset{\overset{H}{\vert}}{\underset{\underset{NH_2}{\vert}}{C}} - COOH$ |
| aspartic acid | Asp | D | $HOOC - CH_2 - \overset{\overset{H}{\vert}}{\underset{\underset{NH_2}{\vert}}{C}} - COOH$ |
| cysteine | Cys | C | $HS - CH_2 - \overset{\overset{H}{\vert}}{\underset{\underset{NH_2}{\vert}}{C}} - COOH$ |
| glutamic acid | Glu | E | $HOOC - CH_2 - CH_2 - \overset{\overset{H}{\vert}}{\underset{\underset{NH_2}{\vert}}{C}} - COOH$ |
| glutamine | Gln | Q | $\underset{O}{\overset{H_2N}{\diagdown}} C - CH_2 - CH_2 - \overset{\overset{H}{\vert}}{\underset{\underset{NH_2}{\vert}}{C}} - COOH$ |
| glycine | Gly | G | $H - \overset{\overset{H}{\vert}}{\underset{\underset{NH_2}{\vert}}{C}} - COOH$ |
| *histidine | His | H | $HC = \overset{}{\underset{\underset{N}{\vert}}{C}} - CH_2 - \overset{\overset{H}{\vert}}{\underset{\underset{NH_2}{\vert}}{C}} - COOH$ with imidazole ring $N$ $NH$ $\underset{H}{C}$ |
| *isoleucine | Ile | I | $CH_3 - CH_2 - \overset{}{\underset{\underset{CH_3}{\vert}}{CH}} - \overset{\overset{H}{\vert}}{\underset{\underset{NH_2}{\vert}}{C}} - COOH$ |
| *leucine | Leu | L | $\underset{H_3C}{\overset{H_3C}{\diagup}} CH - CH_2 - \overset{\overset{H}{\vert}}{\underset{\underset{NH_2}{\vert}}{C}} - COOH$ |
| *lysine | Lys | K | $H_2N - CH_2 - CH_2 - CH_2 - CH_2 - \overset{\overset{H}{\vert}}{\underset{\underset{NH_2}{\vert}}{C}} - COOH$ |

| | | | |
|---|---|---|---|
| *methionine | Met | M | $CH_3-S-CH_2-CH_2-\overset{\overset{\displaystyle H}{\mid}}{\underset{\underset{\displaystyle NH_2}{\mid}}{C}}-COOH$ |
| *phenylalanine | Phe | F | ⬡$-CH_2-\overset{\overset{\displaystyle H}{\mid}}{\underset{\underset{\displaystyle NH_2}{\mid}}{C}}-COOH$ |
| proline | Pro | P | proline → 4–hydroxyproline |
| serine | Ser | S | $HO-CH_2-\overset{\overset{\displaystyle H}{\mid}}{\underset{\underset{\displaystyle NH_2}{\mid}}{C}}-COOH$ |
| *threonine | Thr | T | $CH_3-\underset{\underset{\displaystyle OH}{\mid}}{CH}-\overset{\overset{\displaystyle H}{\mid}}{\underset{\underset{\displaystyle NH_2}{\mid}}{C}}-COOH$ |
| *tryptophan | Trp | W | $C-CH_2-\overset{\overset{\displaystyle H}{\mid}}{\underset{\underset{\displaystyle NH_2}{\mid}}{C}}-COOH$ |
| *tyrosine | Tyr | Y | $HO-$⬡$-CH_2-\overset{\overset{\displaystyle H}{\mid}}{\underset{\underset{\displaystyle NH_2}{\mid}}{C}}-COOH$ |
| *valine | Val | V | $\underset{H_3C}{\overset{H_3C}{\diagup}}CH-\overset{\overset{\displaystyle H}{\mid}}{\underset{\underset{\displaystyle NH_2}{\mid}}{C}}-COOH$ |

*an essential amino acid

**Amino acid.** The amino acids occurring in proteins.

**aminopeptidase** Any enzyme that cleaves amino acids from the N-terminus of peptides or polypeptides. For example, membrane-bound aminopeptidases in the small intestine break down peptides and dipeptides into amino acids.

**amino sugar** Any sugar containing an amino group in place of a hydroxyl group. The **hexosamines** are amino derivatives of hexose sugars and include **glucosamine** (based on glucose) and **galactosamine** (based on galactose). The former is a constituent of *chitin and the latter occurs in cartilage.

*α*-**aminotoluene** *See* BENZYLAMINE.

**ammeter** An instrument that measures electric current. The main types are the **moving-coil** ammeter, the **moving-iron** ammeter, and the **thermoammeter**. The moving-coil instrument is a moving-coil *galvanometer fitted with a *shunt to reduce its sensitivity. It can only be used for d.c., but can be adapted for a.c. by using a *rectifier. In moving-iron instruments, a piece of soft iron moves in the magnetic field created when the current to be measured flows through a fixed coil. They can be used with a.c. or d.c. but are less accurate (though more robust) than the moving-coil instruments. In thermoammeters, which can also be used with a.c. or d.c., the current is passed through a resistor, which heats up as the current passes. This is in contact with a thermocouple, which is connected to a galvanometer. This indirect system is mainly used for measuring high frequency a.c. In the **hot-wire** instrument the wire is clamped at its ends and its elongation as it is heated causes a pointer to move over a scale.

**ammine** A coordination *complex in which the ligands are ammonia molecules. An example of an ammine is the tetraamminecopper(II) ion $[Cu(NH_3)_4]^{2+}$.

**ammonia** A colourless gas, $NH_3$, with a strong pungent odour; r.d. 0.59 (relative to air); m.p. −77.7°C; b.p. −33.35°C. It is very soluble in water and soluble in alcohol. The compound may be prepared in the laboratory by reacting ammonium salts with bases such as calcium hydroxide, or by the hydrolysis of a nitride. Industrially it is made by the *Haber process and over 80 million tonnes per year are used either directly or in combination. Major uses are the manufacture of nitric acid, ammonium nitrate, ammonium phosphate, and urea (the last three as fertilizers), explosives, dyestuffs, and resins.

Liquid ammonia has some similarity to water as it is hydrogen bonded and has a moderate dielectric constant, which permits it to act as an ionizing solvent. It is weakly self-ionized to give ammonium ions, $NH_4^+$ and amide ions, $NH_2^-$. It also dissolves electropositive metals to give blue solutions, which are believed to contain solvated electrons. Ammonia is extremely soluble in water giving basic solutions that contain solvated $NH_3$ molecules and small amounts of the ions $NH_4^+$ and $OH^-$. The combustion of ammonia in air yields nitrogen and water. In the presence of catalysts NO, $NO_2$, and water are formed; this last reaction is the basis for the industrial production of nitric acid. Ammonia is a good proton acceptor (i.e. it is a base) and gives rise to a series of ammonium salts, e.g. $NH_3 + HCl \rightarrow NH_4^+ + Cl^-$. It is also a reducing agent.

The participation of ammonia in the *nitrogen cycle is a most important natural process. Nitrogen-fixing bacteria are able to achieve similar reactions to those of the Haber process, but under normal conditions of temperature and pressure. These release ammonium ions, which are converted by nitrifying bacteria into nitrite and nitrate ions.

**ammoniacal** Describing a solution in which the solvent is aqueous ammonia.

**ammonia clock** A form of atomic clock in which the frequency of a quartz oscillator is controlled by the vibrations of excited ammonia molecules (*see* EXCITATION). The ammonia molecule ($NH_3$) consists of a pyramid with a nitrogen atom at the apex and one hydrogen atom at each corner of the triangular base. When the molecule is excited, once every 20.9 microseconds the nitrogen atom passes through the base and forms a pyramid the other side: 20.9 microseconds later it returns to its original position. This vibration back and forth has a frequency of 23 870 hertz and ammonia gas will only absorb excitation energy at exactly this frequency. By using a *crystal oscillator to feed energy to the gas and a suitable feedback mechanism, the oscillator can be locked to exactly this frequency.

**ammonia–soda process** *See* SOLVAY PROCESS.

**ammonite** An extinct aquatic mollusc of the class *Cephalopoda. Ammonites were abundant in the Mesozoic era (225–65 million years ago) and are commonly found as fossils in rock strata of that time, being used as *index fossils for the Jurassic period. They were characterized by a coiled shell divided into many chambers, which acted as a buoyancy aid. The external suture lines on these shells increased in complexity with the advance of the group.

**ammonium alum** *See* ALUMS.

**ammonium carbonate** A colourless or white crystalline solid, $(NH_4)_2CO_3$, usually encountered as the monohydrate. It is very soluble in cold water. The compound decomposes slowly to give ammonia, water, and carbon dioxide. Commercial 'ammonium carbonate' is a double salt of ammonium hydrogencarbonate and ammonium aminomethanoate (carbamate), $NH_4HCO_3.NH_2COONH_4$. This material is manufactured by heating a mixture of ammonium chloride and calcium carbonate and recovering the product as a sublimed solid. It readily releases ammonia and is the basis of sal volatile. It is also used in dyeing and wool preparation and in baking powders.

**ammonium chloride (sal ammoniac)** A white or colourless cubic solid, $NH_4Cl$; r.d. 1.53; sublimes at 340°C. It is very soluble in water and slightly soluble in ethanol but insoluble in ether. It may be prepared by fractional crystallization from a solution containing ammonium sulphate and sodium chloride or ammonium carbonate and calcium chloride. Pure samples may be made directly by the gas-phase reaction of ammonia and hydrogen chloride. Because of its ease of preparation it can be manufactured industrially alongside any plant that uses or produces ammonia. The compound is used in dry cells, metal finishing, and in the preparation of cotton for dyeing and printing.

**ammonium ion** The monovalent cation $NH_4^+$. It may be regarded as the product of the reaction of ammonia (a Lewis base) with a hydrogen ion. The ion has tetrahedral symmetry. The chemical properties of ammonium salts are frequently very similar to those of equivalent alkali-metal salts.

**ammonium nitrate** A colourless crystalline solid, $NH_4NO_3$; r.d. 1.72; m.p. 169.6°C; b.p. 210°C. It is very soluble in water and soluble in ethanol. The crystals are rhombic when obtained below 32°C and monoclinic above 32°C. It may be readily prepared in the laboratory by the reaction of nitric acid with aqueous ammonia. Industrially, it is manufactured by the same reaction using ammonia gas. Vast quantities of ammonium nitrate are used as fertilizers (over 20 million tonnes per year) and it is also a component of some explosives.

**ammonium sulphate** A white rhombic solid, $(NH_4)_2SO_4$; r.d. 1.77; decomposes at 235°C. It is very soluble in water and insoluble in ethanol. It occurs naturally as the mineral **mascagnite**. Ammonium sulphate was formerly manufactured from the 'ammoniacal liquors' produced during coal-gas manufacture but is now produced by the direct reaction between ammonia gas and sulphuric acid. It is decomposed by heating to release ammonia (and ammonium hydrogensulphate) and eventually water, sulphur dioxide, and ammonia. Vast quantities of ammonium sulphate are used as fertilizers.

**amniocentesis** The taking of a sample of amniotic fluid from a pregnant woman to determine the condition of an unborn baby. A hollow needle is inserted through the woman's abdomen and wall of the uterus and the fluid drawn off. Chemical and microscopical examination of cells shed from the embryo's skin into the fluid are used to detect spina bifida, *Down's syndrome, or other serious biochemical or chromosomal abnormalities.

**amnion** A membrane that encloses the embryo of reptiles, birds, and mammals within the **amniotic cavity.** This cavity is filled with **amniotic fluid,** in which the embryo is protected from desiccation and from external pressure. *See also* EXTRAEMBRYONIC MEMBRANES.

**amniote** A vertebrate whose embryos are totally enclosed in a fluid-filled sac – the *amnion. The evolution of the amnion provided the necessary fluid environment for the developing embryo and therefore allowed animals to breed away from water. Amniotes comprise the reptiles, birds, and mammals. *Compare* ANAMNIOTE.

*Amoeba* A genus of protists formerly placed in the phylum Rhizopoda but now, on the basis of molecular systematics, classified as *amoebozoans. *Amoeba* spp. have temporary body projections called *pseudopodia. These are used for locomotion and feeding and result in a constantly changing body shape (*see* AMOEBOID MOVEMENT). Most species are free-living in soil, mud, or water, where they feed on smaller protoctists and other single-celled organisms, but a few are parasitic. The best known species is the much studied *A. proteus.*

**amoebocyte** An animal cell whose location is not fixed and is therefore able to wander through the body tissues. Amoebocytes are named after their resemblance, especially in their movement, to *Amoeba* (*see* AMOEBOID MOVEMENT) and they feed on foreign particles (including invading bacteria). They occur, for example, in sponges and mammalian blood (e.g. some *leucocytes).

**amoeboid movement** The mechanism of movement demonstrated by *Amoeba* and other cells that are capable of changing their shape (e.g. *phagocytes). The cytoplasm of *Amoeba* consists of a central fluid endoplasm surrounded by a more viscous ectoplasm. The fluid endoplasm slides towards the front of the cell, forming a *pseudo-podium and propelling the cell forward. On reaching the tip of the pseudopodium, the endoplasm is converted into ectoplasm; at the same time the ectoplasm at the rear of the cell is converted into endoplasm and streams forward, thus maintaining continuous movement.

**amoebozoans** An assemblage of eukaryotic protists, based largely on molecular systematics, that includes the plasmodial (or true) *slime moulds, the dictyostelid cellular slime moulds, the lobose amoebas (e.g. *Amoeba* species), and archamoebas, such as the pelobionts (which lack mitochondria and many other features of eukaryotic cells apart from a membrane-bounded nucleus).

**amorphous** Describing a solid that is not crystalline; i.e. one that has no long-range order in its lattice. Many powders that are described as 'amorphous' in fact are composed of microscopic crystals, as can be demonstrated by X-ray diffraction. *Glasses are examples of true amorphous solids.

**amount concentration** *See* CONCENTRATION.

**amount of substance** Symbol $n$. A measure of the number of entities present in a substance. The specified entity may be an atom, molecule, ion, electron, photon, etc., or any specified group of such entities. The amount of substance of an element, for example, is proportional to the number of atoms present. For all entities, the constant of proportionality is the *Avogadro constant. The SI unit of amount of substance is the *mole.

**AMP** *See* ATP; CYCLIC AMP.

**ampere** Symbol A. The SI unit of electric current. The constant current that, maintained in two straight parallel infinite conductors of negligible cross section placed one metre apart in a vacuum, would produce a force between the conductors of $2 \times 10^{-7}$ N m$^{-1}$. This definition replaced the earlier international ampere defined as the current required to deposit 0.001 118 00 gram of silver from a solution of silver nitrate in one second. The unit is named after A. M. Ampère.

**Ampère, André Marie** (1775–1836) French physicist who from 1809 taught at the Ecole Polytechnique in Paris. He is best known for putting electromagnetism (which

he called 'electrodynamics') on a mathematical basis. In 1825 he formulated *Ampère's law. The *ampere is named after him.

**ampere-hour** A practical unit of electric charge equal to the charge flowing in one hour through a conductor passing one ampere. It is equal to 3600 coulombs.

**Ampère's law** A law of the form

$$dB = (\mu_o\, I \sin \theta\, dl)/4\pi r^2,$$

where d$B$ is the infinitesimal element of the magnitude of the magnetic flux density at a distance $r$ at a point P from the element length d$l$ of a conductor, $\mu_o$ is the magnetic *permeability of free space, $I$ is the current flowing through the conductor, and $\theta$ is the angle between the direction of the current and the line joining the element of the conductor and P. It is also called the **Ampère–Laplace law** after the French mathematician Pierre-Simon de Laplace (1749–1827).

**Ampère's rule** A rule that relates the direction of the electric current passing through a conductor and the magnetic field associated with it. The rule states that if the electric current is moving away from an observer, the direction of the lines of force of the magnetic field surrounding the conductor is clockwise and that if the electric current is moving towards an observer, the direction of the lines of force is counterclockwise. An equivalent statement to Ampère's rule is known as the **corkscrew rule**. A corkscrew, or screwdriver, is said to be right-handed if turning the corkscrew in a clockwise direction drives the screw into the object (such as the cork of a bottle). The corkscrew rule states that a right-handed corkscrew is analogous to an electric current and its magnetic field with the direction of the screw being analogous to electric current; the direction in which the corkscrew is being turned is analogous to the direction of lines of force of the field.

**ampere-turn** The SI unit of *magnetomotive force equal to the magnetomotive force produced when a current of one ampere flows through one turn of a magnetizing coil.

**amphetamine** A drug, 1-phenyl-2-aminopropane (or a derivative of this compound), that stimulates the central nervous system by causing the release of the transmitters noradrenaline and dopamine from nerve endings. It inhibits sleep, suppresses the appetite, and has variable effects on mood; prolonged use can lead to addiction.

**Amphibia** The class of vertebrate chordates (*see* CHORDATA) that contains the frogs, toads, newts, and salamanders. The amphibians evolved in the Devonian period (about 370 million years ago) as the first vertebrates to occupy the land, and many of their characteristics are adaptations to terrestrial life. All adult amphibians have a passage linking the roof of the mouth with the nostrils so they may breathe air and keep the mouth closed. The moist scaleless skin is used to supplement the lungs in gas exchange. They have no diaphragm, and therefore the muscles of the mouth and pharynx provide the pumping action for breathing. Fertilization is usually external and the eggs are soft and prone to desiccation, therefore reproduction commonly occurs in water. Amphibian larvae are aquatic, having gills for respiration; they undergo metamorphosis to the adult form.

**((()) SEE WEB LINKS**

• Hosted by the University of California, Berkeley, gives free access to information about amphibian biology and conservation

**amphiboles** A large group of rock-forming metasilicate minerals. They have a structure of silicate tetrahedra linked to form double endless chains, in contrast to the single chains of the *pyroxenes, to which they are closely related. They are present in many igneous and metamorphic rocks. The amphiboles show a wide range of compositional variation but all of them conform to the general formula: $X_{2-3}Y_5Z_8O_{22}(OH)_2$, where $X$ = Ca, Na, K, Mg, or $Fe^{2+}$; $Y$ = Mg, $Fe^{2+}$, $Fe^{3+}$, Al, Ti, or Mn; and $Z$ = Si or Al. The hydroxyl ions may be replaced by F, Cl, or O. Most amphiboles are monoclinic, including: cummingtonite,

$$(Mg,Fe^{2+})_7(Si_8O_{22})(OH)_2;$$

tremolite,

$$Ca_2Mg_5(Si_8O_{22})(OH,F)_2;$$

actinolite,

$$Ca_2(Mg,Fe^{2+})_5(Si_8O_{22})(OH,F)_2;$$

*hornblende,

$$NaCa_2(Mg,Fe^{2+},Fe^{3+},Al)_5((Si,Al)_8O_{22})(OH,F)_2;$$

edenite,

$$NaCa_2(Mg,Fe^{2+})_5(Si_7AlO_{22})(OH,F)_2;$$

riebeckite,

$$Na_2,Fe_3{}^{2+}(Si_8O_{22})(OH,F)_2.$$

The minerals anthophyllite,

$$(Mg,Fe^{2+})_7(Si_8O_{22})(OH,F)_2,$$

and gedrite,

$$(Mg,Fe^{2+})_6Al(Si,Al)_8O_{22})(OH,F)_2,$$

are orthorhombic amphiboles.

**amphimixis** True sexual reproduction, involving the fusion of male and female gametes and the formation of a zygote. *Compare* APOMIXIS.

**amphioxus** Another name for the lancelet: *see* CHORDATA.

**amphiprotic** *See* AMPHOTERIC; SOLVENT.

**ampholyte** A substance that can act as either an acid, in the presence of a strong base, or a base, when in the presence of a strong acid.

**ampholyte ion** *See* ZWITTERION.

**amphoteric** Describing a compound that can act as both an acid and a base (in the traditional sense of the term). For instance, aluminium hydroxide is amphoteric: as a base $Al(OH)_3$ it reacts with acids to form aluminium salts; as an acid $H_3AlO_3$ it reacts with alkalis to give *aluminates. Oxides of metals are typically basic and oxides of nonmetals tend to be acidic. The existence of amphoteric oxides is sometimes regarded as evidence that an element is a *metalloid. Compounds such as the amino acids, which contain both acidic and basic groups in their molecules, can also be described as amphoteric. Solvents, such as water, that can both donate and accept protons are usually described as **amphiprotic** (*see* SOLVENT).

**amplifier** A device that increases the strength of an electrical signal by drawing energy from a separate source to that of the signal. The original device used in electronic amplifiers was the *triode valve, in which the cathode–anode current is varied in accordance with the low-voltage signal applied to the valve's control grid. In the more recent *transistor, the emitter–collector current is controlled in much the same way by the signal applied to the transistor's base region. In the most modern devices the complete amplifier circuit is manufactured as a single *integrated circuit. The ratio of the output amplitude (of p.d. or current) of an amplifier (or stage of an amplifier) to the corresponding input amplitude is called the **gain** of the amplifier.

**amplitude** *See* WAVE.

**amplitude modulation (AM)** *See* MODU-
LATION; RADIO.

**ampulla** **1.** An enlargement at one end of
each of the *semicircular canals of the inner
ear. Each ampulla contains a group of receptors – sensory hair cells – embedded in a
gelatinous cap (**cupula**), which detects
movement in one particular dimension, corresponding to the plane of the canal. Movement of the head causes the cupula (and the
hairs within it) to bend in a direction opposite to that of the head movement; this stimulates nerve impulses in the receptors,
which are interpreted by the brain as movement in a particular dimension. **2.** Any small
vesicle or saclike process.

**AMT** *See* ALPHAMETHYLTRYPTAMINE.

**a.m.u.** *See* ATOMIC MASS UNIT.

**amylase** Any of a group of closely related
enzymes that degrade starch, glycogen, and
other polysaccharides. Plants contain both
α- and β-amylases; the name **diastase** is
given to the component of malt containing
β-amylase, important in the brewing industry. Animals possess only α-amylases, found
in pancreatic juice (as **pancreatic amylase**)
and also (in humans and some other
species) in saliva (as **salivary amylase** or
**ptyalin**). Amylases cleave the long polysaccharide chains, producing a mixture of glucose and maltose.

**amyl group** Formerly, any of several isomeric groups with the formula $C_5H_{11}-$.

**amylopectin** A *polysaccharide comprising highly branched chains of glucose molecules. It is one of the constituents (the other
being amylose) of *starch.

**amylose** A *polysaccharide consisting of
linear chains of between 100 and 1000 linked
glucose molecules. Amylose is a constituent
of *starch. In water, amylose reacts with iodine to give a characteristic blue colour.

**anabolic steroid** Any steroid compound
that promotes tissue growth, especially of
muscles. Naturally occurring anabolic
steroids include the male sex hormones
(*androgens). Synthetic forms of these are
used medically to help weight gain after de-
bilitating diseases; their use by athletes to
build up body muscles can cause liver damage and is banned by most athletic authorities.

**anabolism** The metabolic synthesis of

proteins, fats, and other constituents of living organisms from molecules or simple precursors. This process requires energy in the
form of ATP. Drugs that promote such metabolic activity are described as **anabolic** (*see*
ANABOLIC STEROID). *See* METABOLISM. *Compare* CATABOLISM.

**anaerobe** *See* ANAEROBIC RESPIRATION.

**anaerobic respiration** A type of *respiration in which foodstuffs (usually carbohydrates) are partially oxidized, with the
release of chemical energy, in a process not
involving atmospheric oxygen. Since the
substrate is never completely oxidized the
energy yield of this type of respiration is
lower than that of *aerobic respiration. It occurs in some yeasts and bacteria and in muscle tissue when oxygen is absent (*see* OXYGEN
DEBT). **Obligate anaerobes** are organisms
that cannot use free oxygen for respiration;
**facultative anaerobes** are normally aerobic
but can respire anaerobically during periods
of oxygen shortage. Alcoholic *fermentation
is a type of anaerobic respiration in which
one of the end products is ethanol.

**Analar reagent** A chemical reagent of
high purity with known contaminants for
use in chemical analyses.

**analgesic** A substance that reduces pain
without causing unconsciousness, either by
reducing the pain threshold or by increasing
pain tolerance. There are several categories
of analgesic drugs, including morphine and
its derivatives (*see* OPIATE), which produce
analgesia by acting on the central nervous
system; nonsteroidal anti-inflammatory
drugs (e.g. *aspirin); and local anaesthetics.

**analogous** Describing features of unrelated organisms that are superficially similar
but have evolved in very different ways. The
wings of butterflies and birds are analogous
organs.

**analyser** A device, used in the *polarization of light, that is placed in the eyepiece of
a *polarimeter to observe plane-polarized
light. The analyser, which may be a *Nicol
prism or *Polaroid, can be oriented in different directions to investigate in which plane
an incoming wave is polarized or if the light
is plane polarized. If there is one direction
from which light does not emerge from the
analyser when it is rotated, the incoming
wave is plane polarized. If the analyser is
horizontal when extinction of light takes
place, the polarization of light must have

been in the vertical plane. The intensity of a beam of light transmitted through an analyser is proportional to $\cos^2\theta$, where $\theta$ is the angle between the plane of polarization and the plane of the analyser. Extinction is said to be produced by 'crossing' the *polarizer and analyser.

**analysis** The determination of the components in a chemical sample. **Qualitative analysis** involves determining the nature of a pure unknown compound or the compounds present in a mixture. Various chemical tests exist for different elements or types of compound, and systematic analytical procedures can be used for mixtures. **Quantitative analysis** involves measuring the proportions of known components in a mixture. Chemical techniques for this fall into two main classes: *volumetric analysis and *gravimetric analysis. In addition, there are numerous physical methods of qualitative and quantitative analysis, including spectroscopic techniques, mass spectrometry, polarography, chromatography, activation analysis, etc.

**analytical geometry (coordinate geometry)** A form of geometry in which points are located in a two-dimensional, three-dimensional, or higher-dimensional space by means of a system of coordinates. Curves are represented by an equation for a set of such points. The geometry of figures can thus be analysed by algebraic methods. See CARTESIAN COORDINATES; POLAR COORDINATES.

**anamniote** A vertebrate that lacks an *amnion and whose embryos and larvae must therefore develop in water. Anamniotes comprise the agnathans, fishes, and amphibians. Compare AMNIOTE.

**anaphase** One of several stages of cell division. In *mitosis the chromatids of each chromosome move apart to opposite ends of the spindle. In the first anaphase of *meiosis, the paired homologous chromosomes separate and move to opposite ends; in the second anaphase the chromatids move apart, as in mitosis.

**anaphylaxis** An abnormal *immune response that occurs when an individual previously exposed to a particular *antigen is re-exposed to the same antigen. Anaphylaxis may follow an insect bite or the injection of a drug (such as penicillin). It is caused by the release of *histamine and similar substances and may produce a localized reaction or a

more generalized and severe one, with difficulty in breathing, pallor, or drop in blood pressure, unconsciousness, and possibly heart failure and death. See also ALLERGY.

**anastigmatic lens 1.** An objective lens for an optical instrument in which all *aberrations, including *astigmatism, are reduced greatly. **2.** A spectacle lens designed to correct astigmatism. It has different radii of curvature in the vertical and horizontal planes.

**anatomy** The study of the structure of living organisms, especially of their internal parts by means of dissection and microscopical examination. Compare MORPHOLOGY.

**anchor ring** See TORUS.

**and circuit** See LOGIC CIRCUITS.

**Anderson, Carl David** (1905–91) US physicist who became a professor at the California Institute of Technology, where he worked mainly in particle physics. In 1937 he discovered the *positron in *cosmic radiation, and four years later was awarded the Nobel Prize. In 1939 he discovered the mumeson (muon).

**Andreev reflection** A process occurring at the interface between a normal conductor of electricity and a *superconductor, in which charge is transferred from the normal conductor to the superconductor. This results in the formation of a Cooper pair in the superconductor and a reflected hole in the normal conductor. The process was first described by the Russian physicist Alexander Andreev in 1964. It has possible applications in spintronics and quantum computing.

**androecium** The male sex organs (*stamens) of a flower. Compare GYNOECIUM.

**androgen** One of a group of male sex hormones that stimulate development of the testes and of male *secondary sexual characteristics (such as growth of facial and pubic hair in men). **Testosterone** is the most important. Androgens are produced principally by the testes when stimulated with *luteinizing hormone but they are also secreted in smaller amounts by the adrenal glands and the ovaries. Injections of natural or synthetic androgens are used to treat hormonal disorders of the testes and breast cancer and to build up body tissue (see ANABOLIC STEROID).

**anechoic** Having a low degree of reverberation with little or no reflection of sound. An

**anechoic** chamber is one designed for experiments in acoustics. The walls are covered with small pyramids to avoid the formation of stationary waves between facing surfaces and the whole of the interior surface is covered with an absorbent material to avoid reflections.

**anemometer** An instrument for measuring the speed of the wind or any other flowing fluid. The simple **vane anemometer** consists of a number of cups or blades attached to a central spindle so that the air, or other fluid, causes the spindle to rotate. The instrument is calibrated to give a wind speed directly from a dial. It can be mounted to rotate about a vertical axis and in this form it also gives an indication of the direction of the wind. A **hot-wire anemometer** consists of an electrically heated wire that is cooled by the flow of fluid passing round it. The faster the flow the lower the temperature of the wire and the lower its resistance. Thus the rate of flow can be calculated by measuring the resistance of the wire.

**anemophily** Pollination of a flower in which the pollen is carried by the wind. Examples of anemophilous flowers are those of grasses and conifers. *Compare* ENTOMOPHILY; HYDROPHILY.

**aneroid barometer** *See* BAROMETER.

**ANFO** Ammonium nitrate–fuel oil. A mixture used extensively as a blasting agent in mining and quarrying. The proportions are approximately 94% ammonium nitrate and 6% fuel oil. ANFO has been used in terrorist attacks (e.g. an attack on the Murrah Federal Building, Oklahoma City, in 1995).

**angiosperms** *See* ANTHOPHYTA.

**angiotensin** Any of three related peptide hormones. Angiotensin I is derived, by the action of the enzyme *renin, from a protein (α-globulin) secreted by the liver into the bloodstream. As blood passes through the lungs, another enzyme (**angiotensin-converting enzyme; ACE**) splits angiotensin I, forming angiotensin II. This causes constriction of blood vessels and stimulates the release of *antidiuretic hormone and *aldosterone, which increase blood pressure. Angiotensin III, formed by removal of a single amino acid from angiotensin II, also stimulates aldosterone release by the adrenal gland.

**angle modulation** *See* MODULATION.

**angle of incidence 1.** The angle between a ray falling on a surface and the perpendicular (normal) to the surface at the point at which the ray strikes the surface. **2.** The angle between a wavefront and a surface that it strikes.

**angle of reflection 1.** The angle between a ray leaving a reflecting surface and the perpendicular (normal) to the surface at the point at which the ray leaves the surface. **2.** The angle between a wavefront and a surface that it leaves.

**angle of refraction 1.** The angle between a ray that is refracted at a surface between two different media and the perpendicular (normal) to the surface at the point of refraction. **2.** The angle between a wavefront and a surface at which it has been refracted.

**angle-resolved photoelectron spectroscopy (ARPES)** A technique for studying the composition and structure of surfaces by measuring both the kinetic energy and angular distribution of photoelectrons ejected from a surface by electromagnetic radiation. *See also* PHOTOELECTRON SPECTROSCOPY.

**anglesite** A mineral form of *lead(II) sulphate, $PbSO_4$.

**angstrom** Symbol Å. A unit of length equal to $10^{-10}$ metre. It was formerly used to measure wavelengths and intermolecular distances but has now been replaced by the nanometre. 1 Å = 0.1 nanometre. The unit is named after Anders Ångström.

**Ångström, Anders Jonas** (1814–74) Swedish astronomer and physicist who became professor of physics at the University of Uppsala from 1858 until his death. He worked mainly with emission *spectra, demonstrating the presence of hydrogen in the sun. He also worked out the wavelengths of *Fraunhofer lines. Since 1905 spectral wavelengths have been expressed in *angstroms.

**angular displacement, velocity, and acceleration** *See* ROTATIONAL MOTION.

**angular distance** A way of expressing sizes or distances in astronomy. Angular diameter is the size of a celestial object; angular distance is the distance between two objects. Both are measured as angles on the *celestial sphere (expressed in degrees, arc-minutes, or arc-seconds).

**angular frequency (pulsatance)** A quantity proportional to the *frequency of a periodic phenomenon but having the dimensions of angular velocity. The angular frequency in radians per second = frequency in hertz $\times 2\pi$ radians per cycle.

**angular magnification (magnifying power)** *See* MAGNIFICATION.

**angular momentum** Symbol $L$. The product of the angular velocity of a body and its *moment of inertia about the axis of rotation, i.e. $L = I\omega$.

**anharmonic oscillator** An oscillating system (in either *classical physics or *quantum mechanics) that is not oscillating in *simple harmonic motion. In general, the problem of an anharmonic oscillator is not exactly soluble, although many systems approximate to harmonic oscillators and for such systems the **anharmonicity** (the deviation of the system from being a *harmonic oscillator) can be calculated using *perturbation theory. If the anharmonicity is large other approximate or numerical techniques have to be used to solve the problem.

**anhydride** A compound that produces a given compound on reaction with water. For instance, sulphur trioxide is the (acid) anhydride of sulphuric acid

$$SO_3 + H_2O \rightarrow H_2SO_4$$

*See also* ACID ANHYDRIDES.

**anhydrite** An important rock-forming anhydrous mineral form of calcium sulphate, $CaSO_4$. It is chemically similar to *gypsum but is harder and heavier and crystallizes in the rhombic form (gypsum is monoclinic). Under natural conditions anhydrite slowly hydrates to form gypsum. It occurs chiefly in white and greyish granular masses and is often found in the caprock of certain salt domes. It is used as a raw material in the chemical industry and in the manufacture of cement and fertilizers.

**anhydrous** Denoting a chemical compound lacking water: applied particularly to salts lacking their water of crystallization.

**aniline** *See* PHENYLAMINE.

**anilinium ion** The ion $C_6H_5NH_3^+$, derived from *phenylamine.

**animal** Any member of the kingdom Animalia, which comprises multicellular organisms that develop from embryos formed by the fusion of haploid eggs and sperm. Unable to manufacture their own food, they feed on other organisms or organic matter (holozoic nutrition; *see* HETEROTROPHIC NUTRITION). Animals are therefore typically mobile (to search for food) and have evolved specialized sense organs for detecting changes in the environment; a *nervous system coordinates information received by the sense organs and enables rapid responses to environmental stimuli. Animal *cells lack the cellulose cells walls of *plant cells. Molecular systematics now includes animals in the opisthokonts, an assemblage that also contains the fungi. For a classification of the animal kingdom, see Appendix.

**animal behaviour** The activities that constitute an animal's response to its external environment. Certain categories of behaviour are seen in all animals (e.g. feeding, reproduction) but these activities involve different movements in different species and develop in different ways. Some movements are highly characteristic of a species (*see* INSTINCT) whereas others are more variable and depend on the interaction between innate tendencies and *learning during the individual's lifetime. Physiologists study how changes in the body (e.g. hormone levels) affect behaviour, psychologists study the mechanisms of learning, and ethologists study the behaviour of the whole animal: how this develops during the individual's lifetime and how it evolved through natural selection (*see* ETHOLOGY).

**animal charcoal** *See* CHARCOAL.

**animal starch** *See* GLYCOGEN.

**anion** A negatively charged *ion, i.e. an ion that is attracted to the *anode in *electrolysis. *Compare* CATION.

**anionic detergent** *See* DETERGENT.

**anionic resin** *See* ION EXCHANGE.

**anisogamy** Sexual reproduction involving the fusion of gametes that differ in size and sometimes also in form. *See also* OOGAMY. *Compare* ISOGAMY.

**anisotropic** Denoting a medium in which certain physical properties are different in different directions. Wood, for instance, is an anisotropic material: its strength along the grain differs from that perpendicular to the grain. Single crystals that are not cubic are anisotropic with respect to some physical properties, such as the transmission of electromagnetic radiation. *Compare* ISOTROPIC.

**annealing** A form of heat treatment applied to a metal to soften it, relieve internal stresses and instabilities, and make it easier to work or machine. It consists of heating the metal to a specified temperature for a specified time, both of which depend on the metal involved, and then allowing it to cool slowly. It is applied to both ferrous and nonferrous metals and a similar process can be applied to other materials, such as glass.

**Annelida** A phylum of invertebrates comprising the segmented worms (e.g. the earthworm). Annelids have cylindrical soft bodies showing *metameric segmentation, obvious externally as a series of rings separating the segments. Each segment is internally separated from the next by a membrane and bears stiff bristles (see CHAETA). Between the gut and other body organs there is a fluid-filled cavity called the *coelom, which acts as a hydrostatic skeleton. Movement is by alternate contraction of circular and longitudinal muscles in the body wall. The phylum contains three classes: *Polychaeta, *Oligochaeta, and *Hirudinea. Molecular systematics now places the annelids in the superphylum Lophotrochozoa, along with other wormlike phyla and the molluscs.

**annihilation** The destruction of a particle and its *antiparticle as a result of a collision between them. The **annihilation radiation** produced is carried away by *photons or *mesons. For example, in a collision between an electron and a positron the energy produced is carried away by two photons, each having an energy of 0.511 MeV, which is equivalent to the rest-mass energies of the annihilated particles plus their kinetic energies. When nucleons annihilate each other the energy is carried away by mesons.

**annual** A plant that completes its life cycle in one year, during which time it germinates, flowers, produces seeds, and dies. Examples are the sunflower and marigold. *Compare* BIENNIAL; EPHEMERAL; PERENNIAL.

**annual rhythm** The occurrence of a process or a function in a living organism on a yearly basis. Events that display an annual rhythm can include life cycles, such as those of *annual plants; mating behaviour; some kinds of movement, such as *migration; or growth patterns, such as the *growth rings of woody plant stems. *See also* BIORHYTHM.

**annual ring** *See* GROWTH RING.

**annular parallax** Symbol π. The *parallax of a celestial object resulting from the movement of the earth in its orbit during a year. It is equal to the semimajor axis of the parallatic ellipse described by the apparent movement of the object against the background of distant stars. Annular parallax (in arcseconds) is also approximately equal to the reciprocal of the distance to the object (in parsecs).

**annulenes** Organic hydrocarbons that have molecules containing simple single rings of carbon atoms linked by alternating single and double bonds. Such compounds have even numbers of carbon atoms. Cyclooctatetraene, $C_8H_8$, is the next in the series following benzene. Higher annulenes are usually referred to by the number of carbon atoms in the ring, as in [10]-annulene, $C_{10}H_{10}$, [12]-annulene, $C_{12}H_{12}$, etc. The lower members are not stable as a result of the interactions between hydrogen atoms inside the ring. This is true even for molecules that have the necessary number of pi electrons to be *aromatic compounds. Thus, [10]-annulene has $4n + 2$ pi electrons with $n = 2$, but is not aromatic because it is not planar. [14]-annulene also has a suitable number of pi electrons to be aromatic ($n = 3$) but is not planar because of interaction between the inner hydrogens.

The compound [18]-annulene is large enough to be planar and obeys the Hückel rule ($4n + 2 = 18$, with $n = 4$). It is a brownish red fairly stable reactive solid. NMR evidence shows that it has aromatic character. The annulene with $n = 7$, [30]-annulene, can also exist in a planar form but is highly unstable. *See also* PSEUDOAROMATIC.

**annulus** (*pl.* **annuli**) **1.** (in mathematics) The plane figure formed between two concentric circles of different radii, $R$ and $r$. Its area is $\pi(R^2 - r^2)$. **2.** (in botany) **a.** A ragged ring of tissue that remains on the stalk of a mushroom or toadstool. Also called a **velum**, it is formed from the ruptured membrane that originally covered the lower surface of the cap. **b.** The region of the wall of a fern sporangium that is specialized for spore dispersal. It consists of cells that are thickened except on their outer walls. On drying out, the cells contract and the sporangium ruptures, releasing the spores. The annulus springs back into position when the residual water in the cells vaporizes and any remaining spores are dispersed. **3.** (in zoology) Any of various ring-shaped structures in animals,

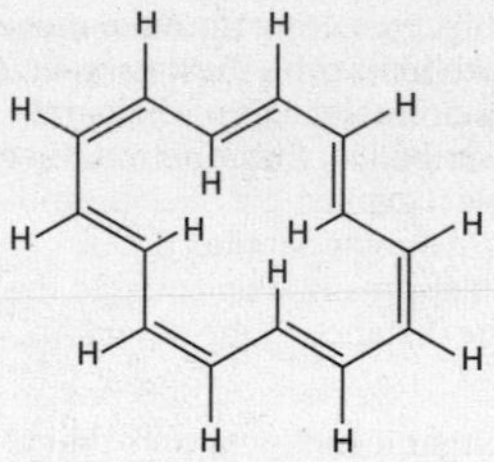

[14]-Annulene

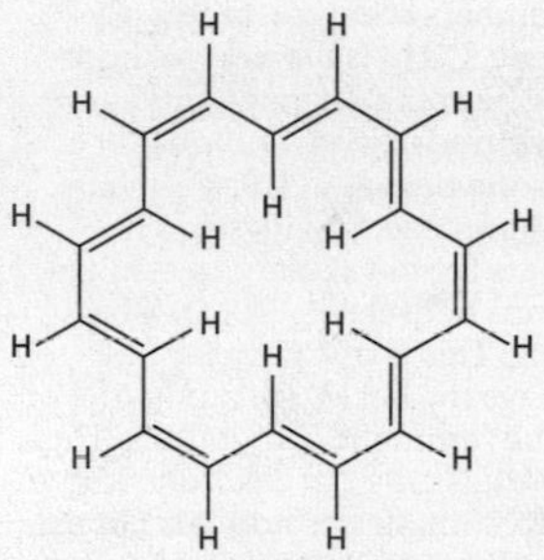

[30]-Annulene

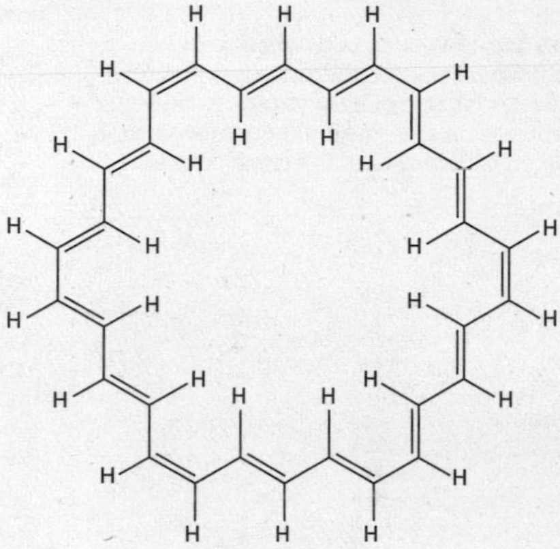

[18]-Annulene

**Annulenes.**

such as any of the segments of an earthworm or other annelid.

**anode** A positive electrode. In *electrolysis anions are attracted to the anode. In an electronic vacuum tube it attracts electrons from the *cathode and it is therefore from the anode that electrons flow out of the device. In these instances the anode is made positive by external means; however in a *voltaic cell the anode is the electrode that spontaneously becomes positive and therefore attracts electrons to it from the external circuit.

**anode sludge** *See* ELECTROLYTIC REFINING.

**anodizing** A method of coating objects made of aluminium with a protective oxide film, by making them the anode in an electrolytic bath containing an oxidizing electrolyte. Anodizing can also be used to produce a decorative finish by formation of an oxide layer that can absorb a coloured dye.

**anomaly** Any of three angles used to fix at a specific moment in time the position of a body in an elliptical orbit around another body located at one of the foci of the ellipse. For a planet in the solar system, the **true anomaly** is the angle formed by a line joining the planet to the sun and a line joining the sun to the planet's *perihelion; the angle is measured in the direction of the planet's motion. If a circle is superimposed upon the ellipse with its centre at the midpoint of the major axis of the ellipse, then the **eccentric anomaly** is the angle formed by a line connecting the centre of the circle to the perihelion and another joining the centre to a point on the circumference of the circle vertically above the planet. The **mean anomaly** is the angle formed by lines connecting the perihelion and the sun, and an imaginary planet having the same period as the real planet but assumed to be moving at constant speed.

**Anoplura** *See* SIPHUNCULATA.

**ANS** *See* AUTONOMIC NERVOUS SYSTEM.

**ANSI** American National Standards Institute: a US body that accredits organizations to write industrial standards – publicly available definitions, requirements, criteria, etc. – following the rules established by ANSI.

 SEE WEB LINKS
• The ANSI home page

**antagonism** 1. The interaction of two substances (e.g. drugs, hormones, or enzymes) having opposing effects in a system in such a way that the action of one partially or completely inhibits the effects of the other. For example, one group of anti-cancer drugs acts by antagonizing the effects of certain enzymes controlling the activities of the cancer cells. *See also* ANTAGONIST. 2. An interaction between two muscles, known as **antagonistic muscles**, in which contraction of one prevents that of the other. For example, the *biceps and triceps are an antagonistic pair. *See* VOLUNTARY MUSCLE. 3. An interaction between two organisms (e.g. moulds or bacteria) in which the growth of one is inhibited by the other. *Compare* SYNERGISM.

**antagonist** A drug that inhibits the effect of an *agonist in such a way that the combined biological effect of the two substances becomes smaller than the sum of their individual effects. **Competitive antagonists** act by binding to agonist receptors, while **noncompetitive antagonists** do not bind to the same receptor sites as the agonist. A **functional antagonist** binds to other receptors that elicit an effect opposite to that of the agonist.

**antenna** 1. (*pl.* **antennae**) (in zoology) A long whiplike jointed mobile paired appendage on the head of many arthropods, usually concerned with the senses of smell, touch, etc. In insects, millipedes, and centipedes they are the first pair of head appendages and are specialized and modified in many insects. In crustaceans they are the second pair of head appendages, the first pair (the **antennules**) having the sensory function, while the antennae are modified for swimming and for attachment. 2. (*pl.* **antennas**) (in radio) *See* AERIAL.

**antennule** *See* ANTENNA.

**anterior** 1. Designating the part of an animal that faces to the front, i.e. that leads when the animal is moving. In humans and other bipedal animals the anterior surface corresponds to the *ventral surface. 2. Designating the side of a flower or axillary bud that faces away from the flower stalk or main stem, respectively. *Compare* POSTERIOR.

**anther** The upper two-lobed part of a plant *stamen, usually yellow in colour. Each lobe contains two pollen sacs within which are numerous pollen grains, which are released when the anther ruptures.

**antheridium** The male sex organ of algae, fungi, bryophytes, and pteridophytes. It produces the male gametes (**antherozoids**). It may consist of a single cell or it may have a wall that is made up of one or several layers forming a sterile jacket around the developing gametes. *Compare* ARCHEGONIUM.

**antherozoid (spermatozoid)** The motile male gamete of algae, fungi, bryophytes, clubmosses, horsetails, ferns, and certain gymnosperms. Antherozoids usually develop in an *antheridium but in certain gymnosperms, such as *Ginkgo* and *Cycas*, they develop from a cell in the pollen tube.

**anthocyanin** One of a group of *flavonoid pigments. Anthocyanins occur in the cell vacuoles of various plant organs and are responsible for many of the blue, red, and purple colours in plants (particularly in flowers).

**Anthophyta (Angiospermophyta; Magnoliophyta)** A phylum comprising the flowering plants (angiosperms). Their gametes are produced within *flowers and the ovules (and the seeds into which they develop) are enclosed in a carpel (*compare* CONIFEROPHYTA). The angiosperms are the dominant plant forms of the present day. They show the most advanced structural organization in the plant kingdom, enabling them to inhabit a very diverse range of habitats. There are two classes within this group: the *Monocotyledoneae with one seed leaf (cotyledon) in the seed, and the *Dicotyledoneae with two seed leaves.

 SEE WEB LINKS
• A good overview of flowering plant diversity and systematics

**anthracene** A white crystalline solid, $C_{14}H_{10}$; r.d. 1.28; m.p. 215.8°C; b.p. 341.4°C. It is an aromatic hydrocarbon with three fused rings (see formula), and is obtained by the distillation of crude oils. The main use is in the manufacture of dyes.

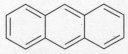

**Anthracene.**

**anthracite**  See COAL.

**anthraquinone**  A colourless stable crystalline solid, $C_{14}H_8O_2$, m.p. 285°C. It is made from benzene or naphthalene and used in the manufacture of a large range of dyes.

**anthropic principle**  The principle that the observable universe has to be as it is, rather than any other way, otherwise we would not be able to observe it. There are many versions of the anthropic principle. The **weak anthropic principle** is specifically concerned with the conditions necessary for conscious life on earth and asserts that numerical relations found for fundamental constants, such as the *gravitational constant, have to hold at the present epoch because at any other epoch there would be no intelligent lifeform to measure the constants. The **strong anthropic principle** is concerned with all possible universes and whether intelligent life could exist in any other universe, including the possibility of different fundamental constants and laws of physics. The anthropic principle is viewed with considerable scepticism by many physicists.

**antiaromatic**  See PSEUDOAROMATIC.

**antiatom**  An atom in which all the particles of an ordinary atom are replaced by their *antiparticles, i.e. electrons by positrons, protons by antiprotons, and neutrons by antineutrons. An antiatom cannot co-exist with an ordinary atom since the atom and the antiatom would annihilate each other with the production of energy in the form of high-energy *photons.

**antibiotics**  Substances that destroy or inhibit the growth of microorganisms, particularly disease-producing bacteria. Antibiotics are obtained from microorganisms (especially moulds) or synthesized. Common antibiotics include the *penicillins, **streptomycin**, and the **tetracyclines**. They are used to treat various infections but tend to weaken the body's natural defence mechanisms and can cause allergies. Overuse of antibiotics can lead to the development of resistant strains of microorganisms.

**antibody**  A protein (see IMMUNOGLOBULIN) produced by certain white blood cells (*plasma cells) in response to entry into the body of a foreign substance (*antigen) in order to render it harmless. An antibody–antigen reaction is highly specific. Antibody production is one aspect of the *immune response and is stimulated by such antigens as invading bacteria, foreign red blood cells (see ABO SYSTEM), inhaled pollen grains or dust, and foreign tissue *grafts. See also B CELL; IMMUNITY; MONOCLONAL ANTIBODY.

**antibonding orbital**  See ORBITAL.

**antibunching**  See BUNCHING.

**anticholinesterase**  Any substance that inhibits the enzyme *cholinesterase, which is responsible for the breakdown of the neurotransmitter acetylcholine at nerve synapses. Anticholinesterases, which include certain drugs, nerve gases, and insecticides, cause a build-up of acetylcholine within the synapses, leading to disruption of nerve and muscle function. In vertebrates, these agents often cause death by paralysing the respiratory muscles. See PESTICIDE.

**anticline**  See FOLD.

**anticoagulant**  A substance that prevents the formation of blood clots. *Heparin is a natural anticoagulant, which is extracted to treat such conditions as thrombosis and embolism. Synthetic anticoagulants include *warfarin.

**anticodon**  A sequence of three nucleotides (trinucleotide) on a strand of transfer *RNA that can form base pairs (see BASE PAIRING) with a specific trinucleotide sequence (see CODON) on a strand of messenger RNA during *translation. See also PROTEIN SYNTHESIS.

**anticyclone (high)**  An area of the atmosphere that is at a higher pressure than the surrounding air. It appears on synoptic charts with a roughly circular closed *isobar at its centre. Winds, which are generally light, circulate in a clockwise direction around the high-pressure centre in the northern hemisphere and in an anticlockwise direction in the southern hemisphere. An anticyclone is characteristically slow moving and is associated with settled weather.

**antidiuretic hormone (ADH; vasopressin)**  A hormone, secreted by the posterior *pituitary gland, that stimulates reabsorption of water by the kidneys and thus controls the concentration of body fluids. ADH is produced by specialized nerve cells in the hypothalamus of the brain and is transported to the posterior pituitary in the bloodstream. Deficiency of ADH results in a disorder known as **diabetes insipidus**, in which large volumes of urine are excreted; it is treated by administration of natural or synthetic hormone.

**antiferromagnetism** *See* MAGNETISM.

**antifreeze** A substance added to the liquid (usually water) in the cooling systems of internal-combustion engines to lower its freezing point so that it does not solidify at sub-zero temperatures. The commonest antifreeze is *ethane-1,2-diol (ethylene glycol).

**antifreeze molecule** Any substance produced by an organism in order to prevent freezing of its tissues or body fluids when subject to subzero environmental temperatures. Many animals living in cold climates produce relatively inert molecules, notably glycerol and other polyhydric alcohols (polyols), such as sorbitol and ribitol, which accumulate in their blood, thereby raising the osmotic concentration and so depressing the *supercooling point. Some families of teleost fish inhabiting polar regions manufacture **antifreeze peptides** or **antifreeze glycopeptides**, which bind to the edges of ice crystal lattices and prevent the addition of further water molecules.

**antigen** Any substance that the body regards as foreign and that therefore elicits an *immune response, particularly the formation of specific antibodies capable of binding to it. Antigens may be formed in, or introduced into, the body. They are usually proteins. **Histocompatibility antigens** are associated with the tissues and are involved in the rejection of tissue or organ *grafts (*see* HISTOCOMPATIBILITY); an example is the group of antigens encoded by the *HLA system. A graft will be rejected if the recipient's body regards such antigens on the donor's tissues as foreign. *See also* ANTIBODY.

**antigorite** *See* SERPENTINE.

**antihistamine** Any drug that inhibits the effects of *histamine in the body and is therefore used to relieve and prevent the symptoms associated with allergic reactions, such as hay fever. Since one of the side-effects produced by antihistamines is sleepiness, some are used to prevent motion sickness and induce sleep.

*anti*-**isomer** *See* ISOMERISM.

**antiknock agent** *See* KNOCKING.

**antilogarithm** *See* LOGARITHM.

**antimatter** *See* ANTIPARTICLE.

**antimonic compounds** Compounds of antimony in its +5 oxidation state; e.g. antimonic chloride is antimony(V) chloride $(SbCl_5)$.

**antimonous compounds** Compounds of antimony in its +3 oxidation state; e.g. antimonous chloride is antimony(III) chloride $(SbCl_3)$.

**antimony** Symbol Sb. An element belonging to *group 15 (formerly VB) of the periodic table; a.n. 51; r.a.m. 121.75; r.d. 6.68; m.p. 630.5°C; b.p. 1750°C. Antimony has several allotropes. The stable form is a bluish-white metal. Yellow antimony and black antimony are unstable nonmetallic allotropes made at low temperatures. The main source is stibnite $(Sb_2S_3)$, from which antimony is extracted by reduction with iron metal or by roasting (to give the oxide) followed by reduction with carbon and sodium carbonate. The main use of the metal is as an alloying agent in lead-accumulator plates, type metals, bearing alloys, solders, Britannia metal, and pewter. It is also an agent for producing pearlitic cast iron. Its compounds are used in flame-proofing, paints, ceramics, enamels, glass dyestuffs, and rubber technology. The element will burn in air but is unaffected by water or dilute acids. It is attacked by oxidizing acids and by halogens. It was first reported by Tholden in 1450.

(((⊕))) **SEE WEB LINKS**
• Information from the WebElements site

**antinode** *See* STATIONARY WAVE.

**antioxidants** Substances that slow the rate of oxidation reactions. Various antioxidants are used to preserve foodstuffs and to prevent the deterioration of rubber, synthetic plastics, and many other materials. Some antioxidants act as chelating agents to sequester the metal ions that catalyse oxidation reactions. Others inhibit the oxidation reaction by removing oxygen free radicals. Naturally occurring antioxidants include *vitamin E, β-*carotene, and *glutathione; they limit the cell and tissue damage caused by foreign substances, such as toxins and pollutants, in the body.

**antiparallel spins** Neighbouring spinning electrons in which the *spins, and hence the magnetic moments, of the electrons are aligned in the opposite direction. The interaction between the magnetic moments of electrons in atoms is dominated by exchange interactions. Under some circumstances the exchange interactions between magnetic moments favour *parallel spins, while under other conditions they favour antiparallel spins. The case of antiferromagnetism (see MAGNETISM) is an example of a system with antiparallel spins.

**antiparallel vectors** Vectors directed along the same line but in opposite directions.

**antiparticle** A subatomic particle that has the same mass as another particle and equal but opposite values of some other property or properties. For example, the antiparticle of the electron is the positron, which has a positive charge equal in magnitude to the electron's negative charge. The antiproton has a negative charge equal to the proton's positive charge. The neutron and the antineutron have *magnetic moments opposite in sign relative to their *spins. The existence of antiparticles is predicted by relativistic *quantum mechanics. When a particle and its corresponding antiparticle collide *annihilation takes place. **Antimatter** is postulated to consist of matter made up of antiparticles. For example, antihydrogen would consist of an antiproton with an orbiting positron. It appears that the universe consists overwhelmingly of (normal) matter, and explanations of the absence of large amounts of antimatter have been incorporated into cosmological models that involve the use of *grand unified theories of elementary particles.

**antipyretic** A drug that reduces fever by lowering body temperature. Certain *analgesic drugs, notably paracetamol, *aspirin, and phenylbutazone, also have antipyretic properties.

**antisense DNA** A single-stranded DNA molecule that can bind to a complementary base sequence in a particular messenger RNA (mRNA) molecule and so prevent synthesis of the protein encoded by the mRNA. Although antisense DNA has the potential to block the expression of a particular gene, this is achieved more effectively by short double-stranded RNA molecules (see RNA INTERFERENCE), which are not susceptible to degradation by DNase enzymes. See also ANTISENSE RNA.

**antisense RNA** A single-stranded RNA molecule whose base sequence is complementary to that of the RNA transcript of a gene, i.e. the 'sense' RNA, such as a messenger RNA (mRNA). Hence, an antisense RNA can undergo base pairing with its complementary mRNA sequence. This blocks gene expression, either by preventing access for ribosomes to translate the mRNA or by triggering degradation of the double-stranded RNA by ribonuclease enzymes. Like *antisense DNA, antisense RNA has therapeutic potential for modifying the activity of disease-causing genes. Also, genes encoding antisense RNAs can be used in genetic engineering to alter the makeup of organisms. For example, the FlavrSavr tomato was engineered with an artificial gene for antisense RNA that prevented expression of a gene for an enzyme involved in ripening, in order to retard spoilage.

**antiseptic** Any substance that kills or inhibits the growth of disease-causing microorganisms but is essentially nontoxic to cells of the body. Common antiseptics include hydrogen peroxide, the detergent cetrimide, and ethanol. They are used to treat minor wounds. *Compare* DISINFECTANT.

**antiserum** Serum containing antibodies, either raised against particular antigens, and hence of known specificity, or a broad mixture of antibodies. It is used to provide short-term passive immunity, e.g. against hepatitis A virus, and to treat an infection to which the patient has no immunity. Antisera may be obtained from large animals, such as horses, that have been inoculated with particular antigens, or from pooled donated human serum.

**antitail** A small tail on a *comet that points towards the sun, unlike the comet's main tail. It is thought to consist of large particles, which scatter sunlight, and is prominent only when observed from the plane of the comet's orbit.

**antitoxin** An antibody produced in response to a bacterial *toxin.

**antiviral** Describing a drug or other agent that kills or inhibits viruses and is used to treat viral infections. Several types of antiviral drug are now in use, such as *aciclovir, effective against herpesviruses, and **antiretroviral drugs** (e.g. *reverse transcriptase inhibitors and protease inhibitors); a combination of the latter, called **highly active antiretroviral therapy** (**HAART**), is used to treat HIV infection. The body's own natural antiviral agents, *interferons, can now be produced by genetic engineering and are sometimes used therapeutically. However, many antiviral agents are extremely toxic, and viruses evolve rapidly so that a drug's effectiveness can soon be lost.

**anti-Zeno effect** See QUANTUM ZENO EFFECT.

**anus** The terminal opening of the *alimentary canal in most animals, through which indigestible material (*faeces) is expelled.

**anyon** See QUANTUM STATISTICS.

**aorta** The major blood vessel in higher vertebrates through which oxygenated blood leaves the *heart from the left ventricle. The aorta branches to form many smaller arteries, which in turn branch many times to supply oxygen and essential nutrients to all living cells in the body.

**apastron** The point at which the distance between the two components of a *binary star is greatest. The point of closest approach is called periastron.

**apatite** A highly complex mineral form of *calcium phosphate, $Ca_5(PO_4)_3(OH,F,Cl)$; the commonest of the phosphate minerals. It has a hexagonal structure and occurs widely as an accessory mineral in igneous rocks (e.g. pegmatite) and often in regional and contact metamorphic rocks, especially limestone. Large deposits occur in the Kola Peninsula, Russia. It is used in the production of fertilizers and is a major source of phosphorus. The enamel of teeth is composed chiefly of apatite.

**aperture** The effective diameter of a lens or mirror. The ratio of the effective diameter to the focal length is called the **relative aperture**, which is commonly known as the aperture, especially in photographic usage. The reciprocal of the relative aperture is called the **focal ratio**. The numerical value of the focal ratio is known as the **f-number** of a lens. For example, a camera lens with a 40 mm focal length and a 10 mm aperture has a relative aperture of 0.25 and a focal ratio of 4. Its f-number would be f/4, often written f4.

The light-gathering power of a telescope depends on the area of the lens, i.e. it is related to the square of the aperture. However, the larger the relative aperture the greater the *aberrations. In microscopy large-aperture objectives (corrected for aberrations) are preferred, since they reduce the

blurring caused by *diffraction of light waves.

**aperture synthesis** See RADIO TELESCOPE.

**aphelion** The point in the solar orbit of a planet, comet, or other solar system object, natural or artificial, at which it is farthest from the sun. At the beginning of the 21st century, the earth is at aphelion on or about 4 July. Its distance from the sun at that point is 1.0167 astronomical units. Compare PERIHELION.

**aphotic zone (bathypelagic zone)** The region of a lake or sea where no light penetrates; it is situated beneath the *euphotic zone. The aphotic zone contains no algae or phytoplankton, and its inhabitants are exclusively carnivorous animals or organisms that feed on sediment or detritus, all reliant on energy inputs from the euphotic zone. It extends downwards from a depth of about 1000 m, or less in turbid waters, and includes the *abyssal zone.

**apical dominance** Inhibition of the growth of lateral buds in a plant by the presence of a growing apical bud. It is brought about by the action of auxins (produced by the apical bud) and abscisic acid.

**apical meristem** A region at the tip of each shoot and root of a plant in which cell divisions are continually occurring to produce new stem and root tissue, respectively. The new tissues produced are known collectively as the **primary tissues** of the plant. See also MERISTEM. Compare CAMBIUM.

**aplanatic lens** A lens that reduces both spherical *aberration and *coma.

**apocarpy** The condition in which the female reproductive organs (*carpels) of a flower are not joined to each other. It occurs, for example, in the buttercup. Compare SYNCARPY.

**apochromatic lens** See ACHROMATIC LENS.

**apocrine secretion** See SECRETION.

**apocynthion** The point in the orbit around the moon of a satellite launched from the earth that is furthest from the moon. For a satellite launched from the moon the equivalent point is the **apolune**. Compare PERICYNTHION.

**apoenzyme** An inactive enzyme that must associate with a specific *cofactor molecule or ion in order to function. *Compare* HOLOENZYME.

**apogee** The point in the orbit of the moon, or an artificial earth satellite, at which it is furthest from the earth. At apogee the moon is 406 700 km from the earth, some 42 000 km further away than at *perigee.

**apolune** *See* APOCYNTHION.

**apomixis (agamospermy)** A reproductive process in plants that superficially resembles normal sexual reproduction but in which there is no fusion of gametes. In apomictic flowering plants there is no fertilization by pollen, and the embryos develop simply by division of a *diploid cell of the ovule. *See also* PARTHENOCARPY; PARTHENOGENESIS.

**apoplast** An interconnected system in plants that consists of all the cell walls and the water that exists in them (the cell wall is composed of cellulose fibres, between which are spaces filled with water). The movement of water (and dissolved ions and solutes) through the cell walls is known as the **apoplastic pathway**. This is the main route by which water taken up by a plant travels across the root cortex to the *endodermis. *Compare* SYMPLAST.

**apoptosis (programmed cell death)** The process of cell death that occurs naturally as part of the normal development, maintenance, and renewal of tissues within an organism. During embryonic development it plays a vital role in determining the final size and form of tissues and organs. For example, the fingers are 'sculpted' on the spadelike embryonic hand by apoptosis of the cells between them; the tubules of the embryonic kidney are hollowed out by a similar process. Apoptosis involves the action of enzymes called **caspases**, cellular proteins that – when activated (e.g. by cell damage) – initiate a series of events that results in digestion of the cytoskeleton, DNA, and other cell components. Apoptosis is normally suppressed as long as cells continue to receive extracellular survival signals in the form of trophic factors (e.g. nerve growth factor: *see* NEUROTROPHIN). In the absence of such signals, the cell embarks on a 'suicide' programme. Sometimes, other cells, for example immune cells, release specific 'murder' signals, which activate apoptosis in target cells. Cancer is associated with the suppression of apoptosis, which also occurs when viruses infect

cells – in order to inhibit the activity of *natural killer cells. Apoptosis differs from cell necrosis, in which cell death may be stimulated by a toxic substance.

**(⊕) SEE WEB LINKS**
• Animations show apoptosis contrasted with necrosis

**aposematic coloration** *See* WARNING COLORATION.

**apparent expansivity** *See* EXPANSIVITY.

**appeasement** Behaviour that inhibits aggression from another animal of the same species, frequently taking the form of a special posture or *display emphasizing the weakness of the performer. Threatening structures (e.g. antlers) and markings are covered or turned away, and vulnerable parts of the body may be exposed. Appeasement is seen in *courtship, in greeting ceremonies, and often (from the loser) after a fight.

**appendicular skeleton** The components, collectively, of the vertebrate skeleton that are attached to the main supporting, or *axial, skeleton. The appendicular skeleton is made up of paired appendages (e.g. legs, wings, arms) together with the *pelvic girdle and *pectoral girdle.

**appendix (vermiform appendix)** An outgrowth of the *caecum in the alimentary canal. In humans it is a *vestigial organ containing lymphatic tissue and serves no function in normal digestive processes. Appendicitis is caused by inflammation of the appendix.

**Appleton layer** *See* EARTH'S ATMOSPHERE.

**applications software** Computer programs, or collections of programs, designed to meet the needs of the users of computer systems by directly contributing to the performance of specific roles. Examples include a word-processing or spreadsheet program, or a company's payroll package. In contrast, **systems software**, such as an operating system, are the group of programs required for effective use of a computer system.

**approximation technique** A method used to solve a problem in mathematics, or its physical applications, that does not give an exact solution but that enables an approximate solution to be found.

**aprotic** *See* SOLVENT.

**apsides** The two points in an astronomical orbit that lie closest to (**periapsis**) and farthest from (**apoapsis**) the centre of gravitational attraction. The **line of apsides** is the straight line that joins the two apsides. If the orbit is elliptical the line of apsides is the major axis of the ellipse.

**aqua acid** A type of acid in which the acidic hydrogen is on a water molecule coordinated to a metal ion. For example,

$$Al(OH_2)_6{}^{3+} + H_2O \rightarrow Al(OH_2)_5(OH)^{2+} + H_3O^+$$

**aquamarine** A variety of the mineral *beryl that has a transparent blue-green colour, valued as a gemstone. It is found as an accessory mineral in pegmatite, sometimes as very large crystals.

**aqua regia** A mixture of concentrated nitric acid and concentrated hydrochloric acid in the ratio 1:3 respectively. It is a very powerful oxidizing mixture and will dissolve all metals (except silver, which forms an insoluble chloride) including such noble metals as gold and platinum, hence its name ('royal water'). Nitrosyl chloride (NOCl) is believed to be one of the active constituents.

**aquation** The process in which water molecules solvate or form coordination complexes with ions.

**aqueous** Describing a solution in water.

**aqueous humour** The fluid that fills the space between the cornea and the lens of the vertebrate eye. In addition to supplying the cornea and lens with nutrients, the aqueous humour helps to maintain the shape of the eye. It is produced and renewed every four hours by the *ciliary body.

**aquifer** A deposit of rock that yields economic supplies of water to wells or springs as a result of its porosity or permeability. It may be, for example, a zone of sandstone, unconsolidated gravels, or jointed limestone.

**aquo ion** A hydrated positive ion present in a crystal or in solution.

*Arabidopsis* A genus of flowering plants of the family Cruciferae (Brassicaceae). The species *A. thaliana* (thale cress) is widely used as a research tool in molecular genetics and developmental biology because it has a small and simple genome (five pairs of chromosomes), over half of which codes for protein, and it can be easily cultured, having a

life cycle of only 6–8 weeks. Its full genome sequence was published in 2000.

**arachidonic acid** An unsaturated fatty acid, $CH_3(CH_2)_3(CH_2CH=CH)_4(CH_2)_3COOH$, that is essential for growth in mammals. It can be synthesized from *linoleic acid. Arachidonic acid acts as a precursor to several biologically active compounds, including *prostaglandins, and plays an important role in membrane production and fat metabolism. The release of arachidonic acid from membrane *phospholipids is triggered by certain hormones.

**Arachnida** A class of terrestrial *arthropods of the phylum Chelicerata, comprising about 65 000 species and including spiders, scorpions, harvestmen, ticks, and mites. An arachnid's body is divided into an anterior *cephalothorax (**prosoma**) and a posterior abdomen (**opisthosoma**). The prosoma bears a pair of grasping or piercing appendages (the **chelicerae**), a pair of **pedipalps** used for manipulation or as sensory structures, and four pairs of walking legs. The opisthosoma may bear various sensory or silk-spinning appendages (*see* SPINNERET). Arachnids are generally carnivorous, feeding on the body fluids of their prey or secreting enzymes to digest prey externally. Spiders immobilize their prey with poison injected by the fanglike chelicerae, while scorpions grasp their prey in large clawed pedipalps and may poison it using the posterior stinging organ. Ticks and some mites are parasitic but most arachnids are free-living. They breathe either via *tracheae (like insects) or by means of thin highly folded regions of the body wall called **lung books**.

(((🌐))) SEE WEB LINKS

• Descriptions and pictures for all orders of arachnids, plus societies, special spider pages, and even arachnids in literature

**arachnoid membrane** One of the three membranes (*meninges) that surround the brain and spinal cord of vertebrates. It lies between the *pia mater and the *dura mater. The arachnoid membrane is very delicate and carries *cerebrospinal fluid, which sustains and cushions the nervous tissue.

**arachno-structure** *See* BORANE.

**aragonite** A rock-forming anhydrous mineral form of calcium carbonate, $CaCO_3$. It is much less stable than *calcite, the commoner form of calcium carbonate, from which it may be distinguished by its greater hardness and specific gravity. Over time aragonite undergoes recrystallization to calcite. Aragonite occurs in cavities in limestone, as a deposit in limestone caverns, as a precipitate around hot springs and geysers, and in high-pressure low-temperature metamorphic rocks; it is also found in the shells of a number of molluscs and corals and is the main constituent of pearls. It is white or colourless when pure but the presence of impurities may tint it grey, blue, green, or pink.

**arbovirus** Obsolete name for any RNA-containing virus that is transmitted from animals to humans through the bite of mosquitoes and ticks (i.e. arthropods, hence *ar*thropod-*bor*ne viruses). They cause various forms of encephalitis (inflammation of the brain) and serious fevers, such as dengue and yellow fever.

**arccos, arcsin, arctan** *See* INVERSE FUNCTIONS.

**Archaea** A *domain of prokaryotic organisms containing the archaebacteria, including the *methanogens, which produce methane; the thermoacidophilic bacteria, which live in extremely hot and acidic environments (such as hot springs); and the halophilic bacteria, which can only function at high salt concentrations and are abundant in the world's oceans. The archaebacteria are grouped together principally on the basis of similarities in the base sequence of their ribosomal RNA. This molecular evidence shows them to be phylogenetically distinct and closer to the eukaryotes than the eubacteria. They are possibly descendants of the earliest forms of life, predating the earliest known microbial fossils.

**Archaean** The earliest eon of geological time, in which there is the first evidence of life on earth. It follows the *Hadean eon of pregeological time and extends from the time of the earliest known rocks, roughly 3900 million years ago, to the beginning of the *Proterozoic eon, about 2500 million years ago. Rock formations called *stromatolites, dated at 3500 million years old or older, are among the oldest of all fossil remains. They are thought to have been produced by the activities of microbial mats of filamentous purple and green bacteria. These prokaryotes performed photosynthesis anaerobically, perhaps using hydrogen sulphide as an electron donor instead of water. Some of their descendants evolved the ability to use water as an electron donor, producing oxygen as a by-product, and eventually brought about the change in atmospheric conditions necessary for aerobic life.

**archaebacteria** *See* ARCHAEA.

**Archaeplastida** *See* PLANT.

**archegonium** The multicellular flask-shaped female sex organ of bryophytes, clubmosses, horsetails, ferns, and many gymnosperms. Such plants are described as **archegoniate** to distinguish them from algae, which do not possess archegonia. The dilated base, the **venter**, contains the oosphere (female gamete). The cells of the narrow neck liquefy to allow the male gametes to swim towards the oosphere. The archegonium is thus an adaptation to the terrestrial environment as it provides a means for the male gametes to reach the female gamete. *Compare* ANTHERIDIUM.

**archenteron (gastrocoel)** A cavity within an animal embryo at the *gastrula stage of development. All or part of the archenteron eventually forms the cavity of the gut. It is connected to the outside by an opening (the **blastopore**), which becomes either the mouth, the mouth and anus, or the anal opening of the animal.

**Archimedes of Syracuse** (287–212 BC) Greek mathematician, who spent most of his life at his birthplace working on levers and other aspects of mechanics. In hydrostatics he devised a pump (the **Archimedian screw**) and formulated *Archimedes' principle. His method of successive approximation allowed him to determine the value of $\pi$ to a good approximation and some of his work came close to anticipating integral calculus. He was killed by a soldier in the Roman siege of Syracuse.

**Archimedes' principle** The weight of the liquid displaced by a floating body is equal to the weight of the body. The principle was not in fact stated by Archimedes, though it has some connection with his discoveries. The principle is often stated in the form: when a body is (partially or totally) immersed in a fluid, the upthrust on the body is equal to the weight of fluid displaced.

**arc lamp** *See* ELECTRIC LIGHTING.

**arcosh, arsinh, artanh** *See* INVERSE FUNCTIONS.

**area** A measure of the size of a two-dimensional shape. The areas of common geometrical figures are given below, where $l$ = length, $h$ = altitude or height, $r$ = radius, and $s$ = slant height:

| | |
|---|---|
| square | $l^2$ |
| rectangle | $lh$ |
| parallelogram | $lh$ |
| triangle | $\frac{1}{2}lh$ |
| circle | $\pi r^2$ |
| sphere | $4\pi r^2$ |
| cone | $\pi rs$ (curved surface) |
| | $\pi rs + \pi r^2$ (total surface) |
| cylinder | $2\pi rh$ (curved surface) |
| | $2\pi rh + 2\pi r^2$ (total surface) |

**arene complex** A complex in which an aromatic ring is bound to a metal atom by its pi-electrons. Examples of arene complexes are the *sandwich compounds $(C_6H_6)_2Cr$ and $(C_5H_5)_2Fe$.

**arenes** Aromatic hydrocarbons, such as benzene, toluene, and naphthalene.

**Argand diagram** *See* COMPLEX NUMBER.

**argentic compounds** Compounds of silver in its higher (+2) oxidation state; e.g. argentic oxide is silver(II) oxide (AgO).

**argentite** A sulphide ore of silver, $Ag_2S$. It crystallizes in the cubic system but most commonly occurs in massive form. It is dull grey-black in colour but bright when first cut and occurs in veins associated with other silver minerals. Important deposits occur in Mexico, Peru, Chile, Bolivia, and Norway.

**argentous compounds** Compounds of silver in its lower (+1) oxidation state; e.g. argentous chloride is silver(I) chloride.

**arginine** *See* AMINO ACID.

**argon** Symbol Ar. A monatomic noble gas present in air (0.93%); a.n. 18; r.a.m. 39.948; d. 0.00178 g $cm^{-3}$; m.p. $-189°C$; b.p. $-185°C$. Argon is separated from liquid air by fractional distillation. It is slightly soluble in water, colourless, and has no smell. Its uses include inert atmospheres in welding and special-metal manufacture (Ti and Zr), and (when mixed with 20% nitrogen) in gas-filled electric-light bulbs. The element is inert and has no true compounds. Lord Rayleigh and Sir William Ramsey identified argon in 1894.

(((•))) SEE WEB LINKS
• Information from the WebElements site

**argument** 1. A sequence of logical propositions based on a set of premises and leading to a conclusion. **2.** *See* COMPLEX NUMBER.

**aril** An outgrowth that grows around and may completely enclose the testa (seed coat) of a seed. It develops from the placenta, funicle, or micropyle of an ovule. The aril surrounding the nutmeg seed forms the spice mace. *See also* CARUNCLE.

**arithmetic/logic unit** *See* ALU.

**arithmetic mean** *See* MEAN.

**arithmetic series (arithmetic progression)** A series or progression of numbers in which there is a common difference between terms, e.g. 3, 9, 15, 21,... is an arithmetic series with a common difference of 6. The general formula for the $n$th term is

$$[a + (n-1)d]$$

and the sum of $n$ terms is

$$n[2a + (n-1)d]/2.$$

*Compare* GEOMETRIC SERIES.

**armature** Any moving part in an electrical machine in which a voltage is induced by a magnetic field, especially the rotating coils in an electric motor or generator and the ferromagnetic bar attracted by an electromagnet in a *relay.

**Arndt–Eisert synthesis** A method of converting a carboxylic acid into the next higher homologue acid (or one of its derivatives). Diazomethane is used to insert a $CH_2$ group.

**aromatic compound** An organic compound that contains a benzene ring in its molecules or that has chemical properties similar to benzene. Aromatic compounds are unsaturated compounds, yet they do not easily partake in addition reactions. Instead they undergo electrophilic substitution.

Benzene, the archetypal aromatic compound, has a hexagonal ring of carbon atoms and the classical formula (the Kekulé structure) would have alternating double and single bonds. In fact all the bonds in benzene are the same length intermediate between double and single C–C bonds. The properties arise because the electrons in the $\pi$-orbitals are delocalized over the ring, giving an extra stabilization energy of 150 kJ $mol^{-1}$ over the energy of a Kekulé structure. The condition for such delocalization is that a compound should have a planar ring with $(4n + 2)$ pi electrons – this is known as the **Hückel rule**. Aromatic behaviour is also found in heterocyclic compounds such as pyridine. Aromatic character can be detected by the presence of a ring current using NMR. *See also* ANNULENES; NONBENZENOID AROMATICS; PSEUDOAROMATIC.

**(⊕) SEE WEB LINKS**
• Information about IUPAC nomenclature

**aromaticity** The property characteristic of *aromatic compounds.

**arousal** A level of physiological and behavioural responsiveness in an animal, which tends to vary between sleep and full alertness. It is controlled by a particular part of the brain (the **reticular activating system**) and can be detected by changes in brain electrical activity, heart rate and muscle tone, responsiveness to new stimuli, and general activity.

**ARPES** *See* ANGLE-RESOLVED PHOTOELECTRON SPECTROSCOPY.

**Arrhenius equation** An equation of the form

$$k = A\exp(-E_a/RT)$$

where $k$ is the rate constant of a given reaction and $E_a$ the *activation energy. $A$ is a constant for a given reaction, called the **pre-exponential factor**. Often the equation is written in logarithmic form

$$\ln k = \ln A - E_a/RT$$

A graph of $\ln k$ against $1/T$ is a straight line with a gradient $-E_a/R$ and an intercept on the $\ln k$ axis of $\ln A$. It is named after Svante Arrhenius (1859–1927).

**Arrhenius theory** *See* ACID.

**arsenate(III)** *See* ARSENIC(III) OXIDE.

**arsenate(V)** *See* ARSENIC(V) OXIDE.

**arsenic** Symbol As. A metalloid element of *group 15 (formerly VB) of the periodic table; a.n. 33; r.a.m. 74.92; r.d. 5.7; sublimes at 613°C. It has three allotropes – yellow, black, and grey. The grey metallic form is the stable and most common one. Over 150 minerals contain arsenic but the main sources are as impurities in sulphide ores and in the minerals orpiment ($As_2S_3$) and realgar ($As_4S_4$). Ores are roasted in air to form arsenic oxide and then reduced by hydrogen or carbon to metallic arsenic. Arsenic compounds are used in insecticides and as doping agents in semiconductors. The element is included in some lead-based alloys to promote hardening. Confusion can arise because $As_4O_6$ is

often sold as white arsenic. Arsenic compounds are accumulative poisons. The element will react with halogens, concentrated oxidizing acids, and hot alkalis. Albertus Magnus is believed to have been the first to isolate the element in 1250.

**(⊕) SEE WEB LINKS**
• Information from the WebElements site

**arsenic acid** *See* ARSENIC(V) OXIDE.

**arsenic(III) acid** *See* ARSENIC(III) OXIDE.

**arsenic hydride** *See* ARSINE.

**arsenic(III) oxide (arsenic trioxide; arsenious oxide; white arsenic)** A white or colourless compound, $As_4O_6$, existing in three solid forms. The commonest has cubic or octahedral crystals (r.d. 3.87; sublimes at 193°C) and is soluble in water, ethanol, and alkali solutions. It occurs naturally as **arsenolite**. A vitreous form can be prepared by slow condensation of the vapour (r.d. 3.74); its solubility in cold water is more than double that of the cubic form. The third modification, which occurs naturally as **claudetite**, has monoclinic crystals (r.d. 4.15). Arsenic(III) oxide is obtained commercially as a byproduct from the smelting of nonferrous sulphide ores; it may be produced in the laboratory by burning elemental arsenic in air. The structure of the molecule is similar to that of $P_4O_6$, with a tetrahedral arrangement of As atoms edge linked by oxygen bridges. Arsenic(III) oxide is acidic; its solutions were formerly called **arsenious acid** (technically, **arsenic(III) acid**) It forms **arsenate(III)** salts (formerly called **arsenites**). Arsenic(III) oxide is extremely toxic and is used as a poison for vermin; trace doses are used for a variety of medicinal purposes. It is also used for producing opalescent glasses and enamels.

**arsenic(V) oxide (arsenic oxide)** A white amorphous deliquescent solid, $As_2O_5$; r.d. 4.32; decomposes at 315°C. It is soluble in water and ethanol. Arsenic(V) oxide cannot be obtained by direct combination of arsenic and oxygen; it is usually prepared by the reaction of arsenic with nitric acid followed by dehydration of the arsenic acid thus formed. It readily loses oxygen on heating to give arsenic(III) oxide. Arsenic(V) oxide is acidic, dissolving in water to give arsenic(V) acid (formerly called **arsenic acid**), $H_3AsO_4$; the acid is tribasic and slightly weaker than phosphoric acid and should be visualized as

(HO)$_3$AsO. It gives **arsenate(V)** salts (formerly called **arsenates**).

**arsenic trioxide** *See* ARSENIC(III) OXIDE.

**arsenious acid** *See* ARSENIC(III) OXIDE.

**arsenious oxide** *See* ARSENIC(III) OXIDE.

**arsenite** *See* ARSENIC(III) OXIDE.

**arsenolite** A mineral form of *arsenic(III) oxide, As$_4$O$_6$.

**arsine (arsenic hydride)** A colourless gas, AsH$_3$; m.p. −116.3°C; b.p. −55°C. It is soluble in water, chloroform, and benzene. Liquid arsine has a relative density of 1.69. Arsine is produced by the reaction of mineral acids with arsenides of electropositive metals or by the reduction of many arsenic compounds using nascent hydrogen. It is extremely poisonous and, like the hydrides of the heavier members of group 15 (formerly VB), is readily decomposed at elevated temperatures (around 260–300°C). Like ammonia and phosphine, arsine has a pyramidal structure.

Arsine gas has a very important commercial application in the production of modern microelectronic components. It is used in a dilute gas mixture with an inert gas and its ready thermal decomposition is exploited to enable other growing crystals to be doped with minute traces of arsenic to give *n*-type semiconductors.

**arteriole** A small muscular blood vessel that receives blood from the arteries and carries it to the capillaries.

**artery** A blood vessel that carries blood away from the heart towards the other body tissues. Most arteries carry oxygenated blood (the *pulmonary artery is an exception). The large arteries branch to form smaller ones, which in turn branch into *arterioles. All arteries have muscular walls, whose contraction aids in pumping blood around the body. The accumulation of fatty deposits in the walls of the arteries leads to **atherosclerosis**, which limits and may eventually block the flow of blood. *Compare* VEIN.

**artesian basin** A structural basin in the earth's crust in which a zone of permeable water-bearing rock (an *aquifer) is confined between impermeable beds. Water enters the aquifer where it reaches the surface and becomes trapped. As the point of intake of the water is above the level of the ground surface a well sunk into the aquifer through

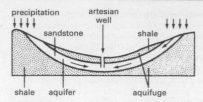

**Artesian basin.**

the overlying impermeable rock will result in water being forced up to the surface by hydrostatic pressure. However, the number of wells sunk into artesian basins often considerably lowers the level of water so that pumping is frequently required. Artesian wells were named after the region of Artois in France, where they were first observed. Other examples include the London Basin where chalk and Lower Eocene sandstones lie confined below the London clay and above the Gault clay.

**Arthrophyta** *See* SPHENOPHYTA.

**arthropod** Any invertebrate animal that characteristically possesses an outer body layer – the *cuticle – that functions as a rigid protective exoskeleton; growth is thus possible only by periodic moults (*see* ECDYSIS). There are over one million species of arthropods, inhabiting marine, freshwater, and terrestrial habitats worldwide. The arthropod body is composed of segments (*see* METAMERIC SEGMENTATION) usually forming distinct specialized body regions, e.g. head, thorax, and abdomen. These segments may possess hardened jointed appendages, modified variously as *mouthparts, limbs, wings, reproductive organs, or sense organs. The main body cavity, containing the internal organs, is a blood-filled *haemocoel, within which lies the heart. Although arthropods are generally placed in a single phylum, Arthropoda, the origins and relationships of the various groups of arthropods remain uncertain, and they are now usually assigned to several subphyla within the clade Ecdysozoa, notably *Crustacea (shrimps, barnacles, crabs, etc.); *Hexapoda (insects); Myriapoda (centipedes and millipedes); and Chelicerata, including the *Arachnida (spiders, scorpions, mites, and ticks).

**articulation** The attachment of two bones, usually by means of a *joint. The thigh bone (femur), for instance, articulates with the pelvic girdle.

**artificial chromosome** A type of cloning *vector that has some features of true chromosomes and is used to clone relatively large fragments of DNA. **Bacterial artificial chromosomes** (BACs) are based on the F (fertility) plasmid found naturally in *E. coli* bacteria with the addition of several bacterial genes necessary for replication of the plasmid by the host cell and a gene (usually for resistance to an antibiotic) that allows selection of BAC-containing cells. They can accommodate inserts of foreign DNA up to about 200 kilobase (kb) in length. Larger DNA fragments are cloned using **yeast artificial chromosomes** (YACs). These are linear vectors derived from a circular plasmid found naturally in baker's yeast (*Saccharomyces cerevisiae*) and capable of accommodating DNA inserts of up to 1000 kb. YACs have a *centromere, enabling them to undergo normal segregation in the yeast host during cell division, and *telomeres; thus they behave like mini-chromosomes. YACs are used for cloning eukaryotic genes or gene segments, for making *DNA libraries of organisms with large genomes (e.g. mammals), and for studying gene function.

**artificial insemination (AI)** The deposition of semen, using a syringe, at the mouth of the uterus to make conception possible. It is used in the selective *breeding of domestic animals and also in humans in some cases of impotence and infertility. It is timed to coincide with ovulation in the female.

**artificial intelligence (AI)** A field of computing concerned with the production of programs that perform tasks requiring intelligence when done by people. These tasks include playing games, such as chess or draughts, forming plans, understanding speech and natural languages, interpreting images, reasoning, and learning.

**artificial selection** The modification of species, traditionally by selective *breeding. Animals or plants with desirable characteristics are interbred with the aim of altering the *genotype and producing a new strain of the organism for a specific purpose. For example, sheep are bred by means of artificial selection in order to improve wool quality. Traditional breeding techniques have been supplemented, and in many cases supplanted, by genetic engineering, genetic testing, and embryo manipulation. Sequencing the genomes of commercially important animal and plant species has enabled the inheritance of desired genes to be monitored directly by molecular methods, instead of by phenotypic analysis. These methods have simultaneously opened up new approaches to selection and enabled it to become more refined and focused.

**artinite** A mineral form of basic *magnesium carbonate,

$$MgCO_3.Mg(OH)_2.3H_2O.$$

**Artiodactyla** An order of hooved mammals comprising the even-toed ungulates, in which the third and fourth digits are equally developed and bear the weight of the body. The order traditionally includes cattle and other ruminants (*see* RUMINANTIA), camels, hippopotamuses, and pigs. All except the latter are herbivorous, having an elongated gut and teeth with enamel ridges for grinding tough grasses. Evidence from molecular studies indicates that whales are closely related to hippos and represent a lineage that split off from ruminants about 55 million years ago. Hence whales and artiodactyls are now placed in a new superorder, **Cetartiodactyla**. *Compare* PERISSODACTYLA.

**aryl group** A group obtained by removing a hydrogen atom from an aromatic compound, e.g. phenyl group, $C_6H_5-$, derived from benzene.

**aryne** A compound that can be regarded as formed from an arene by removing two adjacent hydrogen atoms to convert a double bond into a triple bond. Arynes are transient intermediates in a number of reactions. The simplest example is *benzyne.

**asbestos** Any one of a group of fibrous amphibole minerals (amosite, crocidolite (blue asbestos), tremolite, anthophyllite, and actinolite) or the fibrous serpentine mineral chrysotile. Asbestos has widespread commercial uses because of its resistance to heat, chemical inertness, and high electrical resistance. The fibres may be spun and woven into fireproof cloth for use in protective clothing, curtains, brake linings, etc., or moulded into blocks. Since the 1970s short asbestos fibres have been recognized as a cause of asbestosis, a serious lung disorder, and mesothelioma, a fatal form of lung cancer. These concerns have limited its use and imposed many safety procedures when it is used. Canada is the largest producer of asbestos; others include Russia, South Africa, Zimbabwe, and China.

**a**

**ASCII** (pronounced askey) American standard code for information interchange: a standard scheme for encoding the letters A–Z, a–z, digits 0–9, punctuation marks, and other special and control characters in binary form. Originally developed in the US, it is widely used in many computers and for interchanging information between computers. Characters are encoded as strings of seven *bits, providing $2^7$ or 128 different bit patterns. International 8-bit codes that are extensions of ASCII have been published by the International Standards Organization; these allow the accented Roman letters used in European languages, as well as Cyrillic, Arabic, Greek, and Hebrew characters, to be encoded. *See also* UNICODE.

**Ascomycota** A phylum of fungi, formerly classified as a class (Ascomycetes) or a subdivision (Ascomycotina). It includes the *yeasts, some species of edible fungi, and *Claviceps purpurea*, which causes ergot in rye. Many are fungal partners of *lichens. Sexual reproduction is by means of **ascospores**, eight of which are characteristically produced within a spherical or cylindrical cell, the **ascus**. The asci are usually grouped together in an **ascocarp**.

**ascorbic acid** *See* VITAMIN C.

**asexual reproduction** Reproduction in which new individuals are produced from a single parent without the formation of gametes. It occurs chiefly in lower animals, microorganisms, and plants. In microorganisms and lower animals the chief methods are *fission (e.g. in protists), *fragmentation (e.g. in some aquatic annelid worms), and *budding (e.g. in cnidarians and yeasts). The principal methods of asexual reproduction in plants are by *vegetative propagation (e.g. bulbs, corms, tubers) and by the formation of *spores. Spore formation occurs in mosses, ferns, and other plants showing alternation of generations, as a dormant stage between sporophyte and gametophyte, and in some algae and fungi, to produce replicas of the organism. *Compare* SEXUAL REPRODUCTION.

**asparagine** *See* AMINO ACID.

**aspartic acid** *See* AMINO ACID.

**Aspect experiment** An experiment conducted by the French physicist Alain Aspect (1947– ) and his colleagues in the early 1980s to test Bell's inequality (*see* BELL'S THEOREM). The experiment involves producing pairs of photons from a source of excited calcium ions. The photons have different wavelengths and filters are used to ensure that the photons in a pair travel to different detectors in different directions. The photons are circularly polarized and the net angular momentum of the pair is zero. Two polarizing filters are used, each placed at an angle in the path of the photons. These filters either reflect or transmit photons of different linear polarization to one of four detectors (two for each polarizing filter). Coincidence measurements are made using these detectors and the experiment is organized so that the measurements apply only to photons separated to a point at which they cannot communicate by sending a signal at the speed of light. The results are generally believed to show that there are no local *hidden variables in quantum mechanics.

(((⊕))) SEE WEB LINKS

• The original 1982 paper in *Physical Review Letters*

**aspirin (acetylsalicylic acid)** an acetylated form of *salicylic acid (1-hydroxybenzoic acid), used extensively as a medicinal drug to alleviate pain, combat fever, and reduce inflammation; r.d. 1.4; m.p. 138–140°C; b.p. 140°C (with decomposition). It was first marketed in 1899 as an analgesic. The acid can be obtained from willow bark and it has long been known that the bark could be used for pain relief and for the reduction of fever. The name salicylic acid comes from the botanical name of the willow (*Salix alba*). Aspirin acts by suppressing the production of *prostaglandins, which are major factors in the inflammation process, by inhibiting the enzyme cyclooxygenase (COX). Consequently, it is known as a 'COX inhibitor'. It also reduces the aggregation of blood platelets and small doses are taken regularly to reduce the risk of heart attack. A common side effect of high doses is stomach bleeding and stomach ulcers. Aspirin is made industrially from phenol, which with concentrated sodium hydroxide and carbon dioxide gives sodium phenoxide:

$$C_6H_5OH + NaOH \rightarrow C_6H_5O^- Na^+ + H_2O.$$

The phenoxide ion undergoes electrophilic substitution to give sodium salicylate:

$$C_6H_5O^- + CO_2 + Na^+ \rightarrow C_6H_4 (OH)COO^- Na^+$$

With acid, this forms salicylic acid, which can be acetylated in the ortho position with ethanoic anhydride.

**assimilation** The utilization by a living organism of absorbed food materials in the processes of growth, reproduction, or repair.

**association 1.** (in ecology) An ecological unit in which two or more species occur in closer proximity to one another than would be expected on the basis of chance. Early plant ecologists recognized associations of fixed composition on the basis of the *dominant species present (e.g. a coniferous forest association). Associations now tend to be detected by using more objective statistical sampling methods. *See also* CONSOCIATION. **2.** (in chemistry) The combination of molecules of one substance with those of another to form chemical species that are held together by forces weaker than normal chemical bonds. For example, ethanol and water form a mixture (an **associated liquid**) in which hydrogen bonding holds the different molecules together.

**association centre** The part of the brain that links a primary sensory area (the part of the cerebral cortex that receives primary sensory impulses) with other parts of the brain, such as memory and motor areas, and deals with the interpretation and meaning of the primary sensory input. For example, the auditory association area interprets a 'moo' sound as that coming from a cow.

**associative law** The mathematical law stating that the value of an expression is independent of the grouping of the numbers, symbols, or terms in the expression. The **associative law for addition** states that numbers may be added in any order, e.g. $(x + y) + z = (x + y) + z$. The **associative law for multiplication** states that numbers can be multiplied in any order, e.g. $x(yz) = (xy)z$. Subtraction and division are not associative. *Compare* COMMUTATIVE LAW; DISTRIBUTIVE LAW.

**astatic galvanometer** A sensitive form of moving-magnet *galvanometer in which any effects of the earth's magnetic field are cancelled out. Two small oppositely directed magnets are suspended at the centres of two oppositely wound coils. As its resultant moment on the magnets is zero, the earth's field has no effect and the only restoring torque on the magnets is that provided by the suspending fibre. This makes a sensitive but delicate instrument.

**astatine** Symbol At. A radioactive *halogen element; a.n. 85; r.a.m. 211; m.p. 302°C; b.p. 337°C. It occurs naturally by radioactive decay from uranium and thorium isotopes. Astatine forms at least 20 isotopes, the most stable astatine–210 has a half-life of 8.3 hours. It can also be produced by alpha bombardment of bismuth–200. Astatine is stated to be more metallic than iodine; at least 5 oxidation states are known in aqueous solutions. It will form interhalogen compounds, such as AtI and AtCl. The existence of $At_2$ has not yet been established. The element was synthesized by nuclear bombardment in 1940 by D. R. Corson, K. R. MacKenzie, and E. Segrè at the University of California.

(((⊕))) SEE WEB LINKS
• Information from the WebElements site

**aster** A starlike arrangement of microtubules radiating from a *centrosome. Asters become conspicuous in animal cells at the ends of the *spindle when cell division starts. They are believed to help locate the spindle in relation to the cell's boundaries and to trigger cleavage of the cytoplasm when nuclear division is completed.

**asterism** A pattern of stars that forms a separate entity within a larger constellation. For example, the Plough forms an asterism within the constellation Ursa Major (Great Bear).

**asteroids (planetoids)** A number of *small solar system bodies that mostly revolve around the sun between the orbits of Mars and Jupiter in a zone extending from 1.7 to 4.0 astronomical units from the sun (the **asteroid belt** or **main belt**). The term 'asteroid' has also been used to describe many similar small objects whose orbits bring them close to the earth (**near-earth asteroids**) or cross the orbits of Jupiter and Saturn (the *centaurs). The **Trojans** are two groups of asteroids trapped in Jupiter's orbit at its *Lagrangian points. The size of asteroids varies from the largest, (1) Ceres (a spherical body now classed as a *dwarf planet, with a diameter of 933 km), to irregularly shaped objects less than 1 km in diameter. It is estimated that there are about 10 asteroids with diameters in excess of 250 km and some 120 bodies with diameters over 130 km.

**asthenosphere** A layer of the earth's mantle (*see* EARTH) that underlies the lithosphere at a depth of about 70 km. The velocity of *seismic waves is considerably reduced in the asthenosphere and it is thought to be a zone of partial melting. It extends to a depth of about 250 km where rocks are solid.

**astigmatism** A lens defect in which rays in one plane are in focus when those in another plane are not. In lenses and mirrors it occurs with objects not on the axis and is best controlled by reducing the *aperture to restrict the use of the lens or mirror to its central portion. The eye can also suffer from astigmatism, usually when the cornea is not spherical. It is corrected by using an *anastigmatic lens.

**Aston, Francis William** (1877–1945) British chemist and physicist, who until 1910 worked at Mason College (later Birmingham University) and then with J. J. *Thomson at Cambridge University. In 1919 Aston designed the mass spectrograph (*see* MASS SPECTROMETRY), for which he was awarded the Nobel Prize for chemistry in 1922. With it he discovered the *isotopes of neon, and was thus able to explain nonintegral atomic weights.

**astrobleme** A fairly ancient circular crater in the earth's crust formed by the impact of a meteorite. It is caused by the explosion of the meteorite on impact and characteristically gives rise to shatter cones in the adjacent rocks. The largest astroblemes range in size from 6 to 40 kilometres across.

**astrochemistry** The study of molecules in interstellar space. **Interstellar molecules** are usually detected by their spectra in the radio, microwave, or infrared regions of the electromagnetic spectrum. To date, over 140 different molecules have been detected. Of special interest in astrochemistry is the way in which these molecules are formed and the way in which they interact with clouds of interstellar dust.

**(⊕) SEE WEB LINKS**

• The website of the astrochemistry work group of the International Astronomical Union

**astrometry** The branch of astronomy concerned with the measurement of the positions of the celestial bodies on the *celestial sphere.

**astronomical observatory** An earth-based building or complex or a spacecraft or artificial satellite housing the equipment required for observing celestial objects and phenomena, including optical, infrared, and/or radio telescopes, as well as instruments for making spectrographic, photometric, or other similar measurements. On earth, most optical observatories are located away from city lights on high mountain tops, where the atmosphere is thin, or in dry high-altitude deserts, where observing conditions are stable. Radio telescopes must be isolated from earth-based radio and electrical interference. *See* TELESCOPE.

**astronomical telescope** *See* TELESCOPE.

**astronomical unit (AU)** The mean distance between the sun and the earth. It is equal to 149 597 870 km (499 light seconds).

**astronomy** The study of the universe beyond the earth's atmosphere. The main branches are *astrometry, *celestial mechanics, and *astrophysics.

**astrophysics** The study of the physical and chemical processes involving astronomical phenomena. Astrophysics deals with stellar structure and evolution (including the generation and transport of energy within stars), the properties of the interstellar medium and its interactions with stellar systems, and the structure and dynamics of systems of stars (such as clusters and galaxies), and of systems of galaxies. *See also* COSMOLOGY.

**asymmetric atom** *See* OPTICAL ACTIVITY.

**asymptote** A line that a curve approaches but only touches at infinity.

**asymptotic freedom** The consequence of certain *gauge theories, particularly *quantum chromodynamics, that the forces between such particles as quarks become weaker at shorter distances (i.e. higher energies) and vanish as the distance between particles tends to zero. Only non-Abelian gauge theories with unbroken gauge symmetries can have asymptotic freedom (*see* GROUP). In contrast, *quantum electrodynamics implies that the interaction between particles decreases as a result of dielectric screening; asymptotic freedom for quarks implies that antiscreening occurs. Physically, asymptotic freedom postulates that the *vacuum state for gluons is a medium that has colour paramagnetism, i.e. the vacuum antiscreen colour charges.

Asymptotic freedom explains the successes of the *parton model of pointlike objects inside hadrons and enables systematic corrections to the parton model to be calculated using perturbation theory. That the interaction between quarks increases as the distance between them increases has given rise to the hypothesis of *quark confinement. It appears that if a theory requires the presence of Higgs bosons, asymptotic freedom is destroyed. Thus, *electroweak theory does not have asymptotic freedom.

**asymptotic series** A series formed by the expansion of a function in the form $a_0 + a_1/x + a_2/x^2 + ... + a_n/x^n + ...$, such that the error resulting from terminating the series at the term $a_n/x^n$ tends to zero more rapidly than $1/x^n$ as $x$ tends to infinity. An asymptotic series expansion is not necessarily a *convergent series.

**atactic polymer** See POLYMER.

**atlas** The first *cervical vertebra, a ringlike bone that joins the skull to the vertebral column in terrestrial vertebrates. In advanced vertebrates articulation between the skull and atlas permits nodding movements of the head. See also AXIS.

**atm.** See ATMOSPHERE.

**atmolysis** The separation of a mixture of gases by means of their different rates of diffusion. Usually, separation is effected by allowing the gases to diffuse through the walls of a porous partition or membrane.

**atmosphere** **1.** Symbol atm. A unit of pressure equal to 101 325 pascals. This is equal to 760.0 mmHg. The actual *atmospheric pressure fluctuates around this value. The unit is usually used for expressing pressures well in excess of standard atmospheric pressure, e.g. in high-pressure chemical or physical processes. **2.** See EARTH'S ATMOSPHERE.

**atmospheric pressure** The pressure exerted by the weight of the air above it at any point on the earth's surface. At sea level the atmosphere will support a column of mercury about 760 mm high. This decreases with increasing altitude. The standard value for the atmospheric pressure at sea level in SI units is 101 325 pascals.

**atoll** A circular or elliptical coral reef that encloses a shallow central lagoon. It may be continuous or, more often, broken into closely spaced islets. The water outside the reef is deep. Atolls range in size from a few kilometres to more than 100 km across and are most often found in the Pacific Ocean. They represent the craters of volcanic islands that have sunk as coral grew on or around the rim.

**atom** The smallest part of an element that can exist. Atoms consist of a small dense nucleus of protons and neutrons surrounded by moving electrons. The number of electrons equals the number of protons so the overall charge is zero. The electrons may be thought of as moving in circular or elliptical orbits (see BOHR THEORY) or, more accurately, in regions of space around the nucleus (see ORBITAL).

The **electronic structure** of an atom refers to the way in which the electrons are arranged about the nucleus, and in particular the *energy levels that they occupy. Each electron can be characterized by a set of four quantum numbers, as follows:
(1) The **principal quantum number** $n$ gives the main energy level and has values 1, 2, 3, etc. (the higher the number, the further the electron from the nucleus). Traditionally, these levels, or the orbits corresponding to them, are referred to as **shells** and given letters K, L, M, etc. The K-shell is the one nearest the nucleus. The maximum number of electrons in a given shell is $2n^2$.
(2) The **orbital quantum number** $l$, which governs the angular momentum of the electron. The possible values of $l$ are $(n-1)$, $(n-2)$, ..., 1, 0. Thus, in the first shell ($n = 1$) the electrons can only have angular momentum zero ($l = 0$). In the second shell ($n = 2$) the values of $l$ can be 1 or 0, giving rise to two **subshells** of slightly different energy. In the third shell ($n = 3$) there are three subshells, with $l = 2$, 1, or 0. The subshells are denoted by letters $s$ ($l = 0$), $p$ ($l = 1$), $d$ ($l = 2$), $f$ ($l = 3$). The number of electrons in each subshell is written as a superscript numeral to the subshell symbol, and the maximum number of electrons in each subshell is $s^2$, $p^6$, $d^{10}$, and $f^{14}$. The orbital quantum number is sometimes called the **azimuthal quantum number**.
(3) The **magnetic quantum number** $m$, which governs the energies of electrons in an external magnetic field. This can take values of $+l$, $+(l-1)$, ..., 1, 0, −1, ..., $-(l-1)$, $-l$. In an $s$-subshell (i.e. $l = 0$) the value of $m = 0$. In a $p$-subshell ($l = 1$), $m$ can have values +1, 0, and −1; i.e. there are three $p$-orbitals in the $p$-subshell, usually designated $p_x$, $p_y$, and $p_z$. Under normal circumstances, these all have the same energy level.
(4) The **spin quantum number** $m_s$, which

a

gives the spin of the individual electrons and can have the values $+\frac{1}{2}$ or $-\frac{1}{2}$.

According to the *Pauli exclusion principle, no two electrons in the atom can have the same set of quantum numbers. The numbers define the **quantum state** of the electron, and explain how the electronic structures of atoms occur. See Chronology: Atomic Theory.

**atomic absorption spectroscopy (AAS)** An analytical technique in which a sample is vaporized and the nonexcited atoms absorb electromagnetic radiation at characteristic wavelengths.

**atomic bomb** See NUCLEAR WEAPONS.

**atomic clock** An apparatus for measuring or standardizing time that is based on periodic phenomena within atoms or molecules. See AMMONIA CLOCK; CAESIUM CLOCK.

**atomic emission spectroscopy** An analytical technique in which a sample is vaporized and the atoms present are detected by their emission of electromagnetic radiation at characteristic wavelengths.

**atomic energy** See NUCLEAR ENERGY.

**atomic force microscopy (AFM)** A variation of *scanning probe microscopy that measures the force of interaction between a fine-tipped probe and the surface of a sample. Capable of nanometre-scale resolution, it is suitable for imaging the topography of biomolecules such as DNA and proteins, cell surfaces, and cell organelles. Essentially the apparatus consists of a silicon-tipped probe, mounted on a flexible cantilever, which is moved across the sample surface. Deflec-

tions of the probe are detected by a laser beam focused onto the back of the cantilever and reflected to a photosensor position detector.

**⊕ SEE WEB LINKS**

• An account of the technique from Nanoscience Instruments Inc

**atomicity** The number of atoms in a given molecule. For example, oxygen ($O_2$) has an atomicity of 2, ozone ($O_3$) an atomicity of 3, benzene ($C_6H_6$) an atomicity of 12, etc.

**atomic mass unit (a.m.u.)** A unit of mass used to express *relative atomic masses. It is $1/12$ of the mass of an atom of the isotope carbon–12 and is equal to $1.660\,33 \times 10^{-27}$ kg. This unit superseded both the physical and chemical mass units based on oxygen–16 and is sometimes called the **unified mass unit** or the **dalton**.

**atomic number (proton number)** Symbol $Z$. The number of protons in the nucleus of an atom. The atomic number is equal to the number of electrons orbiting the nucleus in a neutral atom.

**atomic orbital** See ORBITAL.

**atomic pile** An early form of *nuclear reactor using graphite as a *moderator.

**atomic volume** The relative atomic mass of an element divided by its density.

**atomic weight** See RELATIVE ATOMIC MASS.

**ATP (adenosine triphosphate)** A nucleotide that is of fundamental importance as a carrier of chemical energy in all living organisms. It consists of adenine linked to D-ribose (i.e. adenosine); the D-ribose com-

ATP.

## ATOMIC THEORY

| | |
|---|---|
| c.430 BC | Greek natural philosopher Empedocles (d. c. 430 BC) proposes that all matter consists of four elements: earth, air, fire, and water. |
| c.400 BC | Greek natural philosopher Democritus of Abdera (c. 460–370 BC) proposes that all matter consists of atoms. |
| c.306 BC | Greek philosopher Epicurus (c. 342–270 BC) champions Democritus' atomic theory. |
| 1649 | French philosopher Pierre Gassendi (1592–1655) proposes an atomic theory (having read Epicurus). |
| 1803 | John Dalton proposes Dalton's atomic theory. |
| 1897 | J. J. Thomson discovers the electron. |
| 1904 | J. J. Thomson proposes his 'plum pudding' model of the atom, with electrons embedded in a nucleus of positive charges. Japanese physicist Hantaro Nagaoka (1865–1950) proposes a 'Saturn' model of the atom with a central nucleus having a ring of many electrons. |
| 1911 | Ernest Rutherford discovers the atomic nucleus. |
| 1913 | Niels Bohr proposes model of the atom with a central nucleus surrounded by orbiting electrons and the orbits characterized by quantum numbers. British physicist Henry Moseley (1887–1915) equates the positive charge on the nucleus with its atomic number. Frederick Soddy discovers isotopes. |
| 1916 | German physicist Arnold Sommerfield (1868–1951) modifies Bohr's model of the atom specifying elliptical orbits for the electrons; introduces azimuthal and magnetic quantum numbers in addition to the principal quantum number. |
| 1919 | Ernest Rutherford discovers the proton. |
| 1920 | Ernest Rutherford postulates the existence of the neutron. |
| 1925 | Wolfgang Pauli proposes a fourth quantum number, subsequently identified as spin, and proposes his exclusion principle. |
| 1926 | Erwin Schrödinger proposes a wave-mechanical model of the atom (with electrons represented as wave trains). |
| 1932 | James Chadwick discovers the neutron. Werner Heisenberg proposes a model of the atomic nucleus in which protons and neutrons exchange electrons to achieve stability. |
| 1936 | Niels Bohr proposes a 'liquid drop' model of the atomic nucleus. |
| 1948 | German-born US physicist Maria Goeppert-Mayer (1906–72) and German physicist Hans Jensen (1907–73) independently propose the 'shell' structure of the nucleus. |
| 1950 | Danish physicist Aage Bohr (1922–2009) and US physicists Benjamin Mottelson (1926– ) and Leo Rainwater (1917–86) combine the 'liquid-drop' and 'shell' models of the nucleus into a single theory. |

ponent bears three phosphate groups, linearly linked together by covalent bonds (see formula). These bonds can undergo hydrolysis to yield either a molecule of ADP (**adenosine diphosphate**) and inorganic phosphate or a molecule of AMP (**adenosine monophosphate**) and pyrophosphate (*see* ATP-ASE). Both these reactions yield a large amount of energy (about 30.6 kJ mol$^{-1}$) that is used to bring about such biological processes as muscle *contraction, the *active transport of ions and molecules across cell membranes, and the synthesis of biomolecules. The reactions bringing about these processes often involve the enzyme-catalysed transfer of the phosphate group to intermediate substrates. Most ATP-mediated reactions require Mg$^{2+}$ ions as *cofactors.

ATP is regenerated by the rephosphorylation of AMP and ADP using the chemical energy obtained from the oxidation of food. This takes place during *glycolysis and the *Krebs cycle but, most significantly, is also a result of the reduction–oxidation reactions of the *electron transport chain, which ultimately reduces molecular oxygen to water (*oxidative phosphorylation). ATP is also formed by the light-dependent reactions of *photosynthesis.

**ATPase** Any enzyme that brings about the hydrolysis of ATP. This results in the cleavage of either one phosphate group, with the formation of ADP and inorganic phosphate (P$_i$), or of two phosphate groups, with the formation of AMP and pyrophosphate (PP$_i$); the second reaction yields twice as much energy as the first. ATPase activity is associated with many energy-consuming processes; for example, in muscle contraction it is associated with *myosin when activated by actin.

**atrioventricular node (AVN)** A specialized group of *cardiac muscle fibres situated in the fibrous ring between the right atrium and ventricle of the heart. The AVN is the only pathway between the atria and the ventricles through which electrical impulses can pass. Thus, following the contraction of the atria, the AVN initiates a wave of contraction in the ventricles via the *bundle of His.

**atrium 1. (auricle)** A chamber of the *heart that receives blood from the veins and forces it by powerful muscular contraction into the *ventricle(s). Fish have a single atrium but all other vertebrates have two. **2.** Any of various cavities or chambers in animals, such as the chamber surrounding the

gill slits of the lancelet and other invertebrate chordates.

**atrophy** The degeneration or withering of an organ or part of the body.

**atropine** A poisonous crystalline alkaloid, C$_{17}$H$_{23}$NO$_3$; m.p. 118–119°C. It can be extracted from deadly nightshade and other solanaceous plants and is used in medicine to treat colic, to reduce secretions, and to dilate the pupil of the eye.

**ATRS** *See* ATTENUATED TOTAL REFLECTANCE SPECTROSCOPY.

**attenuated total reflectance spectroscopy (ATRS)** A variation of infrared spectroscopy in which the IR source is reflected from the sample and absorption occurs only in the surface layer. ATRS is used in forensic science for analysis of thin layers (e.g. paint).

**attenuation 1.** (in physics) **a.** A loss of intensity suffered by sound, radiation, etc., as it passes through a medium. It may be caused by absorption or scattering. **b.** The drop in voltage or current experienced by a signal as it passes through a circuit. **2.** (in medicine) A process of reducing the disease-producing ability of a microorganism. It can be achieved by chemical treatment, heating, drying, irradiation, by growing the organism under adverse conditions, or by serial passage through another organism. Attenuated bacteria or viruses are used for some *vaccines. **3.** (in mycology) The conversion by yeasts of carbohydrates to alcohol, as in brewing and wine and spirit production.

**atto-** Symbol a. A prefix used in the metric system to denote 10$^{-18}$. For example, 10$^{-18}$ second = 1 attosecond (as).

**attractor** The set of points in *phase space to which the representative point of a dissipative system (i.e. one with internal friction) tends as the system evolves. The attractor can be: a single point; a closed curve (a **limit cycle**), which describes a system with periodic behaviour; or a *fractal (or **strange attractor**), in which case the system exhibits *chaos. *See also* BRUSSELATOR.

**AU** *See* ASTRONOMICAL UNIT.

**audibility** The state of being perceptible to hearing. The limits of audibility of the human ear are between about 20 hertz (a low rumble) and 20 000 hertz (a shrill whistle). With increased age the upper limit falls quite considerably.

**audiofrequency** A frequency that is audible to the human ear. *See* AUDIBILITY.

**audiometer** An instrument that generates a sound of known frequency and intensity in order to measure an individual's hearing ability.

**auditory** Of or relating to the *ear. For example, the **auditory meatus** is the canal leading from the pinna to the tympanum (eardrum).

**auditory nerve** The nerve that transmits sensory information from the ear to the brain.

**Aufbau principle** A principle that gives the order in which orbitals are filled in successive elements in the periodic table. The order of filling is 1$s$, 2$s$, 2$p$, 3$s$, 3$p$, 4$s$, 3$d$, 4$p$, 5$s$, 4$d$, 5$p$, 6$s$, 4$f$, 5$d$, 6$p$, 7$s$, 5$f$, 6$d$. *See* ATOM.

**Auger effect** The ejection of an electron from an atom without the emission of an X- or gamma-ray photon, as a result of the de-excitation of an excited electron within the atom. This type of transition occurs in the X-ray region of the emission spectrum. The kinetic energy of the ejected electron, called an **Auger electron**, is equal to the energy of the corresponding X-ray photon minus the binding energy of the Auger electron. The effect was discovered by Pierre Auger (1899–1994) in 1925.

**auric compounds** Compounds of gold in its higher (+3) oxidation state; e.g. auric chloride is gold(III) chloride ($AuCl_3$).

**auricle** 1. *See* ATRIUM. 2. *See* PINNA.

**aurora** The luminous phenomena seen in the night sky in high latitudes, occurring most frequently near the earth's geomagnetic poles. The displays of aurora appear as coloured arcs, rays, bands, streamers, and curtains, usually green or red. The aurora is caused by the interaction of the atoms (mainly atomic oxygen) and molecules in the upper atmosphere (above about 100 km) with charged particles streaming from the sun, attracted to the auroral regions by the earth's magnetic field. The aurora is known as the **aurora borealis** (or northern lights) in the northern hemisphere and as the **aurora australis** (or southern lights) in the southern hemisphere.

**aurous compounds** Compounds of gold

in its lower (+1) oxidation state; e.g. aurous chloride is gold(I) chloride (AuCl).

**austenite** *See* STEEL.

***Australopithecus*** A genus of fossil primates that lived 4–2 million years ago, coexisting for some of this time with early forms of humans (*see* HOMO). They walked erect and had teeth resembling those of modern humans, but the brain capacity was less than one-third that of a modern human. Various finds have been made, chiefly in East and South Africa (hence the name, which means 'southern ape'). The earliest belong to the species *A. afarensis*, which includes the specimen of a female, dubbed 'Lucy', found at Laetoli in Tanzania. *Australopithecus* and related genera are known as **australopithecines**.

**autecology** The study of ecology at the level of the species. An autecological study aims to investigate the ecology of *populations or individuals of a particular species, including habitat, distribution, life cycle, etc. This should enable a full description of the *ecological niche of the organism to be made. *Compare* SYNECOLOGY.

**autocatalysis** *Catalysis in which one of the products of the reaction is a catalyst for the reaction. Reactions in which autocatalysis occurs have a characteristic S-shaped curve for reaction rate against time – the reaction starts slowly and increases as the amount of catalyst builds up, falling off again as the products are used up.

**autoclave** A strong steel vessel used for carrying out chemical reactions, sterilizations, etc., at high temperature and pressure.

**autogamy** 1. A type of reproduction that occurs in single isolated individuals of ciliate protozoans of the genus *Paramecium*. The nucleus divides into two genetically identical haploid nuclei, which then fuse to form a diploid zygote. The onset of autogamy is associated with changing environmental conditions and may be necessary to maintain cell vitality. 2. Self-fertilization in plants. *See* FERTILIZATION.

**autograft** *See* GRAFT.

**autoimmunity** A disorder of the body's defence mechanisms in which an *immune response is elicited against its own tissues, which are thereby damaged or destroyed. Rheumatoid arthritis, systemic lupus erythematosus, myasthenia gravis, and several forms of thyroid dysfunction are examples of autoimmune diseases.

**autolysis** The process of self-destruction of a cell, cell organelle, or tissue. It occurs by the action of enzymes within or released by *lysosomes. *See also* LYSIS.

**autonomic nervous system (ANS)** The part of the vertebrate *peripheral nervous system that supplies stimulation via motor nerves to the smooth and cardiac muscles (the involuntary muscles) and to the glands of the body. It is divided into the *parasympathetic and the *sympathetic nervous systems, which tend to work antagonistically on the same organs. The activity of the ANS is controlled principally by the *medulla oblongata and *hypothalamus of the brain.

**autopolyploid** A *polyploid organism in which the multiple sets of chromosomes are all derived from the same species. For example, doubling of the chromosome number during mitotic cell division, possibly induced by *colchicine, gives rise to a tetraploid known as an **autotetraploid**. *Compare* ALLOPOLYPLOID.

**autoradiography** An experimental technique in which a radioactive specimen is placed in contact with (or close to) a photographic plate, so as to produce a record of the distribution of radioactivity in the specimen. The film is darkened by the ionizing radiation from radioactive parts of the sample. Autoradiography has a number of applications, being used particularly to study the distribution of particular substances in living tissues, cells, and cultures. A radioactive isotope of the substance is introduced into the organism or tissue, which is killed, sectioned, and examined after enough time has elapsed for the isotope to be incorporated into the substance. Autoradiography is also used to locate radioactively labelled DNA probes or antibodies in such techniques as *Southern blotting and *Western blotting.

**autosome** Any of the chromosomes in a cell other than the *sex chromosomes.

**autotomy** The shedding by an animal of part of its body followed by the regeneration of the lost part. Autotomy is achieved by the contraction of muscles at specialized regions in the body. It serves as a protective mechanism if the animal is damaged or attacked (e.g. tail loss in certain reptiles) and is common as a method of asexual reproduction in polychaete worms, in which both new head and tail regions may be regenerated.

**autotrophic nutrition** A type of nutrition in which organisms synthesize the organic materials they require from inorganic sources. Chief sources of carbon and nitrogen are carbon dioxide and nitrates, respectively. All green plants are autotrophic and use light as a source of energy for the synthesis, i.e. they are **photoautotrophic** (*see* PHOTOSYNTHESIS). Some bacteria are also photoautotrophic; others are **chemoautotrophic**, using energy derived from chemical processes (*see* CHEMOSYNTHESIS). *Compare* HETEROTROPHIC NUTRITION.

**auxanometer** Any mechanical instrument or measuring device used to study the growth or movement of plant organs. One type of auxanometer consists of a recording device that translates any increase in stem height into movement of a needle across a scale.

**auxin** Any of a group of plant hormones responsible for such processes as the promotion of growth by cell enlargement, the maintenance of *apical dominance, and the initiation of root formation in cuttings. Auxins are also involved in suppressing the *abscission of leaves, fruits, or other plant organs and in the development of flowers and fruits. Naturally occurring auxins, principally **indoleacetic acid** (**IAA**), are synthesized in actively growing regions of the plant, from where they are transported to other parts. Synthetic auxins include **2,4-D**, which is used as a weedkiller, and **indolebutyric acid** and **naphthaleneacetic acid**, which are sold in preparations of 'rooting hormones'.

**auxochrome** A group in a dye molecule that influences the colour due to the *chromophore. Auxochromes are groups, such as $-OH$ and $-NH_2$, containing lone pairs of electrons that can be delocalized along with the delocalized electrons of the chromophore. The auxochrome intensifies the colour of the dye. Formerly, the term was also used of such groups as $-SO_2O^-$, which make the molecule soluble and affect its application.

**avalanche** A shower of ionized particles created by a single *ionization as a result of secondary ionizations caused by the original electron and ion being accelerated in an electric field. Each ionization leads to the formation of more electrons and ions, which themselves cause further ionizations. Such avalanches occur in a *Geiger counter.

**average** *See* MEAN. *See also* ROOT-MEAN-SQUARE VALUE.

**Aves** The birds: a class of bipedal vertebrate chordates (*see* CHORDATA) with *feathers, wings, and a beak. They evolved from reptilian ancestors, probably in the Jurassic period (190–136 million years ago), and modern birds still have scaly legs, like reptiles. Birds are warm-blooded (*see* HOMOIOTHERMY). The skin is dry and loose and has no sweat glands, so cooling is effected by panting. Their efficient lungs and four-chambered heart (which completely separates oxygenated and deoxygenated blood) ensure a good supply of oxygen to the tissues. Birds can therefore sustain a high body temperature and level of activity necessary for *flight. The breastbone bears a keel for the attachment of flight muscles. The skeleton is very light; many of the bones are tubular, having internal struts to provide strength and air sacs to reduce weight and provide extra oxygen in flight. Their feathers are vital for flight, streamlining the body, and insulation against heat loss.

Many birds show a high degree of social behaviour in forming large flocks and pair bonding for nesting, egg incubation, and rearing young. Fertilization is internal and the female lays hard-shelled eggs. *See also* RATITAE.

**(((:)) SEE WEB LINKS**

• Global partnership of organizations involved in bird conservation

**AVN** *See* ATRIOVENTRICULAR NODE.

**Avogadro, Amedeo** (1776–1856) Italian chemist and physicist. In 1811 he published his hypothesis (*see* AVOGADRO'S LAW), which provided a method of calculating molecular weights from vapour densities. The importance of the work remained unrecognized, however, until championed by Stanislao Cannizzaro (1826–1910) in 1860.

**Avogadro constant** Symbol $N_A$ or $L$. The number of atoms or molecules in one *mole of substance. It has the value 6.022 1367(36) $\times 10^{23}$. Formerly it was called **Avogadro's number**.

**Avogadro's law** Equal volumes of all gases contain equal numbers of molecules at the same pressure and temperature. The law, often called **Avogadro's hypothesis**, is

true only for ideal gases. It was first proposed in 1811 by Amedeo Avogadro.

**axenic culture** A *culture medium in which only one type of microorganism is growing. Such cultures are widely used in microbiology to determine the basic growth requirements or degree of inhibition by antibiotics or other chemicals of a particular species.

**axial period** The time taken for a celestial object to rotate once on its axis. For a planet, measured with respect to the direction of a fixed star, it is equal to a sidereal *day.

**axial skeleton** The main longitudinal section of the vertebrate *skeleton, including the *skull, the *vertebral column, and the rib cage. *Compare* APPENDICULAR SKELETON.

**axial tilt (obliquity)** The inclination angle of the axis of rotation of a celestial body, such as a planet or satellite, relative to the plane that is perpendicular to the plane of its orbit. In the case of the earth, this angle is currently (epoch 2000) 23° 26′ 21.448″. The earth's axial tilt, which results from the gravitational influences of the other planets, is responsible for the seasons, as first one hemisphere and then the other is turned to face the sun, causing annual variations in the amount of solar radiation falling on the earth and the angle at which it strikes the planet. The earth's axial tilt oscillates between 22.1° and 24.5° over a period of about 41 000 years and is currently decreasing.

**axial vector (pseudo-vector)** A *vector that does not reverse its sign when the coordinate system is changed to a new system by a reflection in the origin (i.e. $x'_i = -x_i$). An example of an axial vector is the *vector product of two *polar vectors, such as $L = r \times p$, where $L$ is the *angular momentum of a particle, $r$ is its position vector, and $p$ is its momentum vector. *Compare* PSEUDO-SCALAR.

**axil** The angle between a branch or leaf and the stem it grows from. **Axillary** (or **lateral**) **buds** develop in the axil of a leaf. The presence of axillary buds distinguishes a leaf from a leaflet.

**axillary bud** *See* AXIL.

**axion** A hypothetical elementary particle postulated to explain why there is no observed CP violation (*see* CP INVARIANCE) in the strong interaction (*see* FUNDAMENTAL INTERACTIONS). Axions have not been detected experimentally, although it has been possible to put limits on their mass and other properties from the effects that they would have on some astrophysical phenomena (e.g. the cooling of stars). It has also been suggested that they may account for some or all of the *missing mass in the universe.

**axis** 1. (in mathematics) **a.** One of a set of reference lines used to locate points on a graph or in a coordinate system. *See* CARTESIAN COORDINATES; POLAR COORDINATES. **b.** A line about which a figure, curve, or body is symmetrical (**axis of symmetry**) or about which it rotates (**axis of rotation**). 2. (in anatomy) The second *cervical vertebra, which articulates with the *atlas (the first cervical vertebra, which articulates with the skull). The articulation between the axis and atlas in reptiles, birds, and mammals permits side-to-side movement of the head. The body of the axis is elongated to form a peg (the **odontoid process**), which extends into the ring of the atlas and acts as a pivot on which the atlas (and skull) can turn.

**axon** The long threadlike part of a nerve cell (*neuron). It carries the nerve impulse (in the form of an *action potential) away from the *cell body of a neuron towards either an effector organ or the brain. *See also* NERVE FIBRE.

**azeotrope (azeotropic mixture; constant-boiling mixture)** A mixture of two liquids that boils at constant composition; i.e. the composition of the vapour is the same as that of the liquid. Azeotropes occur because of deviations in Raoult's law leading to a maximum or minimum in the *boiling-point–composition diagram. When the mixture is boiled, the vapour initially has a higher proportion of one component than is present in the liquid, so the proportion of this in the liquid falls with time. Eventually, the maximum and minimum point is reached, at which the two liquids distil together without change in composition. The composition of an azeotrope depends on the pressure.

**azeotropic distillation** A technique for separating components of an azeotrope by adding a third liquid to form a new azeotrope with one of the original components. It is most commonly used to separate ethanol from water, adding benzene to associate with the ethanol.

**azides** Compounds containing the ion $N_3^-$ or the group $-N_3$.

**azimuth** 1. In horizontal coordinate systems used in astronomy, the angular distance measured eastwards along the observer's horizon from the south point or sometimes the north point to the intersection of the vertical circle through a celestial body. 2. *See* POLAR COORDINATES.

**azimuthal quantum number** *See* ATOM.

**azine** An organic heterocyclic compound containing a six-membered ring formed from carbon and nitrogen atoms. Pyridine is an example containing one nitrogen atom ($C_5H_5N$). **Diazines** have two nitrogen atoms in the ring (e.g. $C_4H_4N_2$), and isomers exist depending on the relative positions of the nitrogen atoms. **Triazines** contain three nitrogen atoms.

**(())) SEE WEB LINKS**
• Information about IUPAC nomenclature

**azo compounds** Organic compounds containing the group $-N=N-$ linking two other groups. They can be formed by reaction of a diazonium ion with a benzene ring.

**(())) SEE WEB LINKS**
• Information about IUPAC nomenclature

**azo dye** *See* DYES.

**azoimide** *See* HYDROGEN AZIDE.

**azurite** A secondary mineral consisting of hydrated basic copper carbonate, $Cu_3(OH)_2(CO_3)_2$, in monoclinic crystalline form. It is generally formed in the upper zone of copper ore deposits and often occurs with *malachite. Its intense azure-blue colour made it formerly important as a pigment. It is a minor ore of copper and is used as a gemstone.

**Babbage, Charles** (1792–1871) British mathematician who is best known for his early work on mechanical calculating machines. He began work on his first 'difference engine' in 1823, but abandoned it ten years later due to lack of funds. His second machine of 1834 was a programmable punched-card 'analytical engine', but this too was never completed.

**Babbit metal** Any of a group of related alloys used for making bearings. They consist of tin containing antimony (about 10%) and copper (1–2%), and often lead. The original alloy was invented in 1839 by the US inventor Isaac Babbit (1799–1862).

**Babo's law** The vapour pressure of a liquid is decreased when a solute is added, the amount of the decrease being proportional to the amount of solute dissolved. The law was discovered in 1847 by the German chemist Lambert Babo (1818–99). *See also* RAOULT'S LAW.

**BAC** *See* BACTERIAL ARTIFICIAL CHROMOSOME.

**Bacillariophyta** A phylum of *algae comprising the diatoms. These marine or freshwater unicellular organisms have cell walls composed of pectin impregnated with silica and consisting of two halves, one overlapping the other. Diatoms are found in huge numbers in plankton and are important in the food chains of seas and rivers. Past deposition has resulted in diatomaceous earths (kieselguhr) and the oil reserves of these species have contributed to oil deposits.

**bacillus** Any rod-shaped bacterium. Generally, bacilli are large, Gram-positive, spore-bearing, and have a tendency to form chains and produce a *capsule. Some are motile, bearing flagella. They are ubiquitous in soil and air and many are responsible for food spoilage. The group also includes *Bacillus anthracis*, which causes anthrax.

**backbiting** A rearrangement that can occur in some polymerization reactions involving free radicals. A radical that has an unpaired electron at the end of the chain changes into a radical with the unpaired

electron elsewhere along the chain, the new radical being more stable than the one from which it originates. For example, the radical

$$RCH_2CH_2CH_2CH_2CH_2CH_2\cdot$$

may change into

$$RCH_2CH\cdot CH_2CH_2CH_2CH_3.$$

The rearrangement is equivalent to a hydrogen atom being transferred within the molecule. The new unpaired electron initiates further polymerization, with the production of polymers with butyl ($CH_3CH_2CH_2CH_2–$) side chains.

**backbone** *See* VERTEBRAL COLUMN.

**back cross** A mating between individuals of the parental generation (P) and the first generation of offspring ($F_1$) in order to identify hidden *recessive alleles. If an organism displays a *dominant characteristic, it may possess two dominant alleles (i.e. it is homozygous) or a dominant and a recessive allele for that characteristic (i.e. it is heterozygous). To find out which is the case, the organism is crossed with one displaying the recessive characteristic. If all the offspring show the dominant characteristic then the organism is homozygous, but if half show the recessive characteristic, then the organism is heterozygous. *See also* TEST CROSS.

**back donation** A form of chemical bonding in which a *ligand forms a sigma bond to an atom or ion by donating a pair of electrons, and the central atom donates electrons back by overlap of its *d*-orbitals with empty *p*- or *d*-orbitals on the ligand.

**back e.m.f.** An electromotive force that opposes the main current flow in a circuit. For example, when the coils of the armature in an electric motor are rotated a back e.m.f. is generated in these coils by their interaction with the field magnet (*see* INDUCTANCE). Also, in an electric cell, *polarization causes a back e.m.f. to be set up, in this case by chemical means.

**background radiation** Low intensity *ionizing radiation present on the surface of the earth and in the atmosphere as a result of *cosmic radiation and the presence of radioisotopes in the earth's rocks, soil, and atmosphere. The radioisotopes are either natural or the result of nuclear fallout or waste gas from power stations. Background counts must be taken into account when measuring the radiation produced by a specified source. *See also* MICROWAVE BACKGROUND RADIATION.

**backing store** Supplementary computer memory, usually in the form of magnetic disks, in which data and programs are held permanently for reference; small sections of this information can then be copied into the main memory (*RAM) of a computer when required for processing. Backing store is less costly and can hold more information than semiconductor RAM, but the speed of access of information in RAM is considerably faster. Use of a hierarchy of different memory devices, including backing store and RAM, greatly improves performance, efficiency, and economy in a computer.

**back titration** A technique in *volumetric analysis in which a known excess amount of a reagent is added to the solution to be estimated. The unreacted amount of the added reagent is then determined by titration, allowing the amount of substance in the original test solution to be calculated.

**backup** A resource that can be used as a substitute in the event of, say, a fault in a component or system or loss of data from a computer file. A backup file is a copy of a file taken in case the original is destroyed or unintentionally altered and the data lost.

**bacteria** A diverse group of ubiquitous microorganisms all of which consist of only a single *cell that lacks a distinct nuclear membrane and has a *cell wall of a unique composition (see illustration). Bacteria constitute the prokaryotic organisms of the living world. It is now recognized, on the basis of differences in ribosomal RNA structure and nucleotide sequences (*see* MOLECULAR SYSTEMATICS), that prokaryotes form two groups so evolutionarily distinct that they should each be raised to the status of *domain: *Archaea (the archaebacteria) and **Eubacteria**. Generally speaking, the term 'bacteria' includes both archaebacteria and eubacteria.

Bacteria can be characterized in a number of ways, for example by their reaction with *Gram's stain or on the basis of their metabolic requirements (e.g. whether or not they

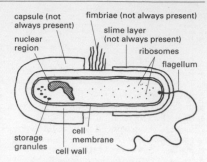

**Bacteria.** A generalized bacterial cell.

require oxygen: *see* AEROBIC RESPIRATION; ANAEROBIC RESPIRATION) and shape. A bacterial cell may be spherical (*see* COCCUS), rod-like (*see* BACILLUS), spiral (*see* SPIRILLUM), comma-shaped (*see* VIBRIO), corkscrew-shaped (*see* SPIROCHAETE), or filamentous, resembling a fungal cell. The majority of bacteria range in size from 0.5 to 5 μm. Many are motile, bearing *flagella, possess an outer slimy *capsule, and produce resistant spores (*see* ENDOSPORE). In general bacteria reproduce only asexually, by simple division of cells, but a few groups undergo a form of sexual reproduction (*see* CONJUGATION). Bacteria are largely responsible for decay and decomposition of organic matter, producing a cycling of such chemicals as carbon (*see* CARBON CYCLE), oxygen, nitrogen (*see* NITROGEN CYCLE), and sulphur (*see* SULPHUR CYCLE). A few bacteria obtain their food by means of *photosynthesis, including the *Cyanobacteria; some are saprotrophs; and others are parasites, causing disease. The symptoms of bacterial infections are produced by *toxins.

**(⊕) SEE WEB LINKS**

• Online textbook of bacteriology devised by Kenneth Todar, University of Wisconsin-Madison

**bacterial artificial chromosome** *See* ARTIFICIAL CHROMOSOME.

**bacterial growth curve** A curve on a graph that shows the changes in size of a bacterial population over time in a culture. The bacteria are cultured in sterile nutrient medium and incubated at the optimum temperature for growth. Samples are removed at intervals and the number of viable bacteria is counted. A logarithmic growth curve is plotted, which shows various phases (see graph).

In the **lag** (or **latent**) **phase** there is only a small increase in numbers as the bacteria

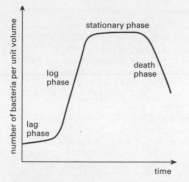

**Bacterial growth curve.**

imbibe water, and synthesize ribosomal RNA and subsequently enzymes, in adjusting to the new conditions. The length of this phase depends on which medium was used to culture the bacteria before the investigation and which phase the cells are already in. As the life span (generation time) of the cells decreases, they enter the **log** (or **exponential**) **phase**, in which the cells reach a maximum rate of reproduction and the number of bacteria increases directly with time, giving a straight slope on a logarithmic scale (*see* EX-PONENTIAL GROWTH). For example, the fastest generation time for *E. coli* is 21 minutes. Growth rate can be estimated in this phase. With time, as the population grows, it enters the **stationary phase**, when the nutrients and electron acceptors are depleted and the pH drops as carbon dioxide and other waste poisons accumulate. As the cell's energy stores are depleted the rate of cell division decreases. The **death** (or **final**) **phase** occurs when the rate at which the bacteria die exceeds the rate at which they are produced; the population declines as the levels of nutrients fall and toxin levels increase. *See also* POPULATION GROWTH.

**bactericidal** Capable of killing bacteria. Common bactericides are some *antibiotics, *antiseptics, and *disinfectants. *Compare* BACTERIOSTATIC.

**bacteriochlorophyll** A form of chlorophyll found in photosynthetic bacteria, notably the purple and green bacteria. There are several types, designated *a* to *g*. For example, bacteriochlorophyll *a* and bacteriochlorophyll *b* are structurally similar to the chlorophyll *a* and chlorophyll *b* found in

plants. Bacteriochlorophyll is located in specialized membrane systems (**chromatophores**).

**bacteriology** The study of bacteria, including their identification, form, function, reproduction, and classification. Much attention is focused on the role of bacteria as agents of disease in animals, including humans, and in plants, and on methods of controlling pathogenic bacteria in the food chain and elsewhere in the environment. However, bacteriologists also investigate the many benefits of bacteria, e.g. in the production of antibiotics, enzymes, and amino acids, and in sewage treatment.

**bacteriophage (phage)** A virus that is parasitic within a bacterium. Each phage is specific for only one type of bacterium. Most phages (**virulent phages**) infect, quickly multiply within, and destroy (lyse) their host cells. However, some (**temperate phages**) remain dormant in their hosts after initial infection: their nucleic acid becomes integrated into that of the host and multiplies with it, producing infected daughter cells. Lysis may eventually be triggered by environmental factors. Phages are used experimentally to identify bacteria, to control manufacturing processes (such as cheese production) that depend on bacteria, and, because they can alter the genetic make-up of bacterial cells, they are important tools in genetic engineering as cloning *vectors.

**bacteriostatic** Capable of inhibiting or slowing down the growth and reproduction of bacteria. Some *antibiotics are bacteriostatic. *Compare* BACTERICIDAL.

**Baer, Karl Ernst von** (1792–1876) Estonian-born German biologist. He studied medicine and comparative anatomy before becoming professor of zoology at Königsberg University in 1817. Ten years later he discovered the mammalian ovum, and traced its development from the Graafian follicle to the embryo. He also noted the similarities between the young embryos of widely different species (the biogenetic law).

**Baeyer test** A test for unsaturated compounds in which potassium permanganate is used. Alkenes, for example, are oxidized to glycols, and the permanganate loses its colour:

$$3R_2C{=}CR_2 + 2KMnO_4 + 4H_2O \rightarrow 2MnO_2 + 2KOH + 3R_2COHR_2COH$$

**Bakelite** A trade name for certain *phenol–formaldehyde resins, first introduced in 1909 by the Belgian–US chemist Leo Hendrik Baekeland (1863–1944).

**baker's yeast** Strains of the yeast *Saccharomyces cerevisiae* that are used in bread-making to enable the dough to rise. Water is added to flour, which activates the *amylase enzymes that hydrolyse the starch in flour to glucose. Baker's yeast is then added, which uses the glucose as a substrate for *aerobic respiration. The carbon dioxide produced from yeast respiration causes bubbles to form in the dough; these become larger during heating in an oven, giving bread its typical texture.

**baking soda** See SODIUM HYDROGENCARBONATE.

**balance** 1. An accurate weighing device. The simple **beam balance** consists of two pans suspended from a centrally pivoted beam. Known masses are placed on one pan and the substance or body to be weighed is placed in the other. When the beam is exactly horizontal the two masses are equal. An accurate laboratory balance weighs to the nearest hundredth of a milligram. Specially designed balances can be accurate to a millionth of a milligram. More modern **substitution balances** use the substitution principle. In this calibrated weights are removed from the single lever arm to bring the single pan suspended from it into equilibrium with a fixed counter weight. The substitution balance is more accurate than the two-pan device and enables weighing to be carried out more rapidly. In automatic electronic balances, mass is determined not by mechanical deflection but by electronically controlled compensation of an electric force. A scanner monitors the displacement of the pan support generating a current proportional to the displacement. This current flows through a coil forcing the pan support to return to its original position by means of a magnetic force. The signal generated enables the mass to be read from a digital display. The mass of the empty container can be stored in the balance's computer memory and automatically deducted from the mass of the container plus its contents. See also SPRING BALANCE. 2. (in animal physiology) Equilibrium in the posture of the body. In vertebrates balance is sensed and maintained by the *vestibular apparatus of the inner ear. 3. (in nutrition) See DIET.

**baleen** See WHALEBONE.

**ballistic galvanometer** A moving-coil *galvanometer designed for measuring charge by detecting a surge of current. It has a heavy coil with minimal damping. When a surge of current is passed through the coil, the initial maximum deflection (the 'throw') is proportional to the total charge that has passed.

**ballistic pendulum** A device used to measure the velocity of a projectile, such as a bullet. A large mass of relatively soft material is suspended from a horizontal bar and the angle through which this mass is displaced when it is struck by the projectile in flight enables the momentum and hence the velocity of the projectile to be calculated by successive application of the laws of conservation of linear momentum and of energy.

**ballistics** The study of the flight of projectiles, especially those that have a parabolic flight path from one point on the earth's surface to another.

**ball lightning** A luminous sphere that sometimes appears at ground level in a thunderstorm. Ball lightning is slow moving and usually disappears without detonation. The spheres vary in diameter up to a few metres. The phenomenon is still not fully understood. See also LIGHTNING.

**Balmer series** See HYDROGEN SPECTRUM.

**banana bond** Informal name for the type of electron-deficient three-centre bond holding the B–H–B bridges in *boranes and similar compounds.

**band spectrum** See SPECTRUM.

**band theory** See ENERGY BANDS.

**bandwidth** The frequency range over which a radio signal of specified frequency spreads. For example, in a *modulation system it is the range of frequencies occupied by the modulating signal on either side of the carrier wave. In an amplifier, it is the range of frequencies over which the power amplification falls within a specified fraction of the maximum value. In an aerial it is the range of frequencies that an aerial system can handle without mismatch.

**bar** A c.g.s. unit of pressure equal to $10^6$ dynes per square centimetre or $10^5$ pascals (approximately 750 mmHg or 0.987 atmosphere). The **millibar** (100 Pa) is commonly used in meteorology.

**barb** 1. (in zoology) Any one of the stiff filaments forming a row on each side of the longitudinal shaft of a feather (see illustration). Together the barbs form the expanded part (**vane**) of the feather. *See* BARBULE. 2. (in botany) A hooked hair.

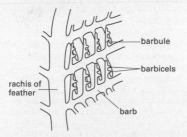

**Barb.** Interlocking barbs of a contour feather.

**Barbier–Wieland degradation** The stepwise degradation of a carboxylic acid to the next lower homologue. First the ester is converted into a tertiary alcohol using a Grignard reagent (PhMgX) and acid (HX):

$$RCH_2COOCH_3 \rightarrow RCH_2C(OH)Ph_2.$$

The secondary alcohol is then dehydrated using ethanoic anhydride ($CH_3COOCOCH_3$) to give an alkene:

$$RCH_2C(OH)Ph_2 \rightarrow RCH=CPh_2.$$

The alkene is oxidized with chromic acid:

$$RCH=CPh_2 \rightarrow RCOOH + Ph_2CO.$$

The result is conversion of an acid $RCH_2COOH$ to the lower acid $RCOOH$.

**barbiturate** Any one of a group of drugs derived from barbituric acid, which have a depressant effect on the central nervous system. Barbiturates were originally used as sedatives and sleeping pills but their clinical use is now limited due to their toxic side-effects; prolonged use can lead to addiction. Specific barbiturates still in clinical use include butobarbital, used to treat insomnia, and thiopental, used to induce general anaesthesia.

**barbule** Any of the minute filaments forming a row on each side of the *barb of a feather. In a *contour feather adjacent barbules interlock by means of hooks (**barbicels**) and grooves, forming a firm vane. Down feathers have no barbicels.

**barchan** *See* DUNE.

**Bardeen, John** (1908–91) US physicist, who worked at Harvard, Minnesota University, and Bell Telephone Labs before becoming a professor at the University of Illinois in 1951. At Bell, with Walter Brattain (1902–87) and William Shockley (1910–89), he developed the point-contact transistor. The three scientists shared the 1956 Nobel Prize for physics for this work. In 1956, with Leon Cooper (1930– ) and John Schrieffer (1931– ), he formulated the BCS theory of *superconductivity, for which they shared the 1972 Nobel Prize.

**Barfoed's test** A biochemical test to detect monosaccharide (reducing) sugars in solution, devised by the Swedish physician Christen T. Barfoed (1815–99). **Barfoed's reagent**, a mixture of ethanoic (acetic) acid and copper(II) acetate, is added to the test solution and boiled. If any reducing sugars are present a red precipitate of copper(II) oxide is formed. The reaction will be negative in the presence of disaccharide sugars as they are weaker reducing agents.

**barite** *See* BARYTES.

**barium** Symbol Ba. A silvery-white reactive element belonging to *group 2 (formerly IIA) of the periodic table; a.n. 56; r.a.m. 137.34; r.d. 3.51; m.p. 725°C; b.p. 1640°C. It occurs as the minerals barytes ($BaSO_4$) and witherite ($BaCO_3$). Extraction is by high-temperature reduction of barium oxide with aluminium or silicon in a vacuum, or by electrolysis of fused barium chloride. The metal is used as a getter in vacuum systems. It oxidizes readily in air and reacts with ethanol and water. Soluble barium compounds are extremely poisonous. It was first identified in 1774 by Karl Scheele, and was extracted by Humphry Davy in 1808.

**(((⊕))) SEE WEB LINKS**
• Information from the WebElements site

**barium bicarbonate** *See* BARIUM HYDROGENCARBONATE.

**barium carbonate** A white insoluble compound, $BaCO_3$; r.d. 4.43. It decomposes on heating to give barium oxide and carbon dioxide:

$$BaCO_3(s) \rightarrow BaO(s) + CO_2(g)$$

The compound occurs naturally as the mineral **witherite** and can be prepared by adding an alkaline solution of a carbonate to a solution of a barium salt. It is used as a raw material for making other barium salts, as a flux for ceramics, and as a raw material in the manufacture of certain types of optical glass.

**barium chloride** A white compound, $BaCl_2$. The anhydrous compound has two crystalline forms: an $\alpha$ form (monoclinic; r.d. 3.856), which transforms at 962°C to a $\beta$ form (cubic; r.d. 3.917; m.p. 963°C; b.p. 1560°C). There is also a dihydrate, $BaCl_2.2H_2O$ (cubic; r.d. 3.1), which loses water at 113°C. It is prepared by dissolving barium carbonate (witherite) in hydrochloric acid and crystallizing out the dihydrate. The compound is used in the extraction of barium by electrolysis.

**barium hydrogencarbonate (barium bicarbonate)** A compound, $Ba(HCO_3)_2$, which is only stable in solution. It can be formed by the action of carbon dioxide on a suspension of barium carbonate in cold water:

$$BaCO_3(s) + CO_2(g) + H_2O(l) \rightarrow$$
$$Ba(HCO_3)_2(aq)$$

On heating, this reaction is reversed.

**barium hydroxide (baryta)** A white solid, $Ba(OH)_2$, sparingly soluble in water. The common form is the octahydrate, $Ba(OH)_2.8H_2O$; monoclinic; r.d. 2.18; m.p. 78°C. It can be produced by adding water to barium monoxide or by the action of sodium hydroxide on soluble barium compounds and is used as a weak alkali in volumetric analysis.

**barium oxide** A white or yellowish solid, BaO, obtained by heating barium in oxygen or by the thermal decomposition of barium carbonate or nitrate; cubic; r.d. 5.72; m.p. 1923°C; b.p. 2000°C. When barium oxide is heated in oxygen the peroxide, $BaO_2$, is formed in a reversible reaction that was once used as a method for obtaining oxygen (the **Brin process**). Barium oxide is now used in the manufacture of lubricating-oil additives.

**barium peroxide** A dense off-white solid, $BaO_2$, prepared by carefully heating *barium oxide in oxygen; r.d. 4.96; m.p. 450°C. It is used as a bleaching agent. With acids, hydrogen peroxide is formed and the reaction is used in the laboratory preparation of hydrogen peroxide.

**barium sulphate** An insoluble white solid, $BaSO_4$, that occurs naturally as the mineral *barytes (or **heavy spar**) and can be prepared as a precipitate by adding sulphuric acid to barium chloride solution; r.d.

4.50; m.p. 1580°C. The rhombic form changes to a monoclinic form at 1149°C. It is used as a raw material for making other barium salts, as a pigment extender in surface coating materials (called **blanc fixe**), and in the glass and rubber industries. Barium compounds are opaque to X-rays, and a suspension of the sulphate in water is used in medicine to provide a contrast medium for X-rays of the stomach and intestine. Although barium compounds are extremely poisonous, the sulphate is safe to use because it is very insoluble.

**bark** The protective layer of mostly dead cells that covers the outside of woody stems and roots. It includes the living and dead tissues external to the xylem, including the phloem and periderm. The term can be used more specifically to describe the periderm together with other tissues isolated by the activity of the *cork cambium. In some species, such as birch, there is one persistent cork cambium but in the older stems of certain other species a second cork cambium becomes active beneath the periderm and further periderm layers are formed every few years. The result is a composite tissue called **rhytidome**, composed of cork, dead cortex, and dead phloem cells.

**Barkhausen effect** The magnetization of a ferromagnetic substance by an increasing magnetic field takes place in discontinuous steps rather than continuously. The effect results from the orientation of magnetic domains (*see* MAGNETISM). It was first observed by H. Barkhausen (1881–1956) in 1919.

**Barlow wheel** *See* HOMOPOLAR GENERATOR.

**barn** A unit of area sometimes used to measure *cross sections in nuclear interactions involving incident particles. It is equal to $10^{-28}$ square metre. The name comes from the phrase 'side of a barn' (something easy to hit).

**barograph** A meteorological instrument that records on paper variations in atmospheric pressure over a period. It often consists of an aneroid barometer operating a pen that rests lightly on a rotating drum to which the recording paper is attached.

**barometer** A device for measuring *atmospheric pressure. The **mercury barometer** in its simplest form consists of a glass tube about 80 cm long sealed at one end and filled with mercury. The tube is then inverted and the open end is submerged in a reservoir of mercury; the mercury column is held up by the pressure of the atmosphere acting on the surface of mercury in the reservoir. This type of device was invented by the Italian scientist Evangelista Torricelli (1608–47), who first noticed the variation in height from day to day, and constructed a barometer in 1644.

In such a device, the force exerted by the atmosphere balanced the weight of the mercury column. If the height of the column is $h$ and the cross-sectional area of the tube is $A$, then the volume of the mercury in the column is $hA$ and its weight is $hA\rho$, where $\rho$ is the density of mercury. The force is thus $hA\rho g$, where $g$ is the acceleration of free fall and the pressure exerted is this force divided by the area of the tube; i.e. $h\rho g$. Note that the height of the mercury is independent of the diameter of the tube. At standard atmospheric pressure the column is 760 mm high. The pressure is then expressed as 760 mmHg (101 325 pascals).

Mercury barometers of this type, with a reservoir of mercury, are known as **cistern barometers**. A common type is the **Fortin barometer**, in which the mercury is held in a leather bag so that the level in the reservoir can be adjusted. The height is read from a scale along the side of the tube in conjunction with a vernier scale that can be moved up and down. Corrections are made for temperature.

The second main type of barometer is the **aneroid barometer**, in which the cumbersome mercury column is replaced by a metal box with a thin corrugated lid. The air is removed from the box and the lid is supported by a spring. Variations in atmospheric pressure cause the lid to move against the spring. This movement is magnified by a system of delicate levers and made to move a needle around a scale. The aneroid barometer is less accurate than the mercury type but much more robust and convenient, hence its use in *altimeters.

**baroreceptor** A *receptor that responds to changes in pressure. The *carotid sinus in the carotid artery contains baroreceptors that respond to changes in arterial pressure and are therefore involved in the regulation of blood pressure and heart beat.

**Barr body** A structure consisting of a condensed X chromosome (*see* SEX CHROMOSOME) that is found in nondividing nuclei of female mammals. The presence of a Barr body is used to confirm the sex of athletes in sex determination tests. It is named after the Canadian anatomist M. L. Barr (1908–95), who identified it in 1949.

**barrel** A measurement of volume, widely used in the chemical industry, equal to 35 UK gallons (approximately 159 litres).

**barycentre** The *centre of mass of a system.

**barye** A c.g.s. unit of pressure equal to one dyne per square centimetre (0.1 pascal).

**baryon** A *hadron with half-integral spin. Nucleons comprise a subclass of baryons. According to currently accepted theory, baryons are made up of three quarks (**antibaryons** are made up of three antiquarks) held together by gluons (*see* ELEMENTARY PARTICLES). Baryons possess a quantum number, called the **baryon number**, which is +1 for baryons, –1 for antibaryons, 1/3 for quarks, –1/3 for antiquarks, and 0 for all other particles such as electrons, neutrinos, and photons. Baryon number has always appeared to have been conserved experimentally, but *grand unified theories postulate interactions at very high energies that allow it not to be conserved. It is thought that nonconservation of baryon number at the high energies characteristic of the early universe may provide an explanation for the asymmetry between matter and antimatter in the universe. *See* PROTON DECAY.

**baryta** *See* BARIUM HYDROXIDE.

**barytes (barite)** An orthorhombic mineral form of *barium sulphate, $BaSO_4$; the chief ore of barium. It is usually white but may also be yellow, grey, or brown. Large deposits occur in Andalusia, Spain, and in the USA.

**basal body** *See* UNDULIPODIUM.

**basal ganglia** Small masses of nervous tissue within the brain that connect the *cerebrum with other parts of the nervous system. They are involved with the subconscious regulation of voluntary movements.

**basal metabolic rate (BMR)** The rate of energy metabolism required to maintain an animal at rest. BMR is measured in terms of heat production per unit time and is usually expressed in kilojoules of heat released per square metre of body surface per hour ($kJ\ m^{-2}\ h^{-1}$). It indicates the energy consumed in order to sustain such vital functions as heartbeat, breathing, nervous activity, active transport, and secretion. Different tissues have different metabolic rates (e.g. the BMR of brain tissue is much greater than that of bone tissue) and therefore the tissue composition of an animal determines its overall BMR. For organisms generally, BMR is proportional to body weight; small animals tend to have a higher metabolic rate per unit weight than large ones.

**basalt** A fine-grained basic igneous rock. It is composed chiefly of calcium-rich plagioclase feldspar and pyroxene; other minerals present may be olivine, magnetite, and apatite. Basalt is the commonest type of lava.

**base** 1. (in chemistry) A compound that reacts with a protonic acid to give water (and a salt). The definition comes from the Arrhenius theory of acids and bases. Typically, bases are metal oxides, hydroxides, or compounds (such as ammonia) that give hydroxide ions in aqueous solution. Thus, a base may be either: (1) An insoluble oxide or hydroxide that reacts with an acid, e.g.

$$CuO(s) + 2HCl(aq) \rightarrow CuCl_2(aq) + H_2O(l)$$

Here the reaction involves hydrogen ions from the acid

$$CuO(s) + 2H^+(aq) \rightarrow H_2O(l) + Cu^{2+}(aq)$$

(2) A soluble hydroxide, in which case the solution contains hydroxide ions. The reaction with acids is a reaction between hydrogen ions and hydroxide ions:

$$H^+ + OH^- \rightarrow H_2O$$

(3) A compound that dissolves in water to produce hydroxide ions. For example, ammonia reacts as follows:

$$NH_3(g) + H_2O(l) \rightleftharpoons NH_4^+(aq) + {}^-OH$$

Similar reactions occur with organic *amines (*see also* NITROGENOUS BASE; AMINE SALTS). A base that dissolves in water to give hydroxide ions is called an **alkali**. Ammonia and sodium hydroxide are common examples.

The original Arrhenius definition of a base has been extended by the Lowry–Brønsted theory and by the Lewis theory. *See* ACID.
2. (in mathematics) **a.** The number of different symbols in a number system. In the decimal system the base is 10; in *binary notation it is 2. **b.** The number that when raised to a certain power has a *logarithm equal to that power. For example if 10 is raised to the power of 3 it is equal to 1000; 3 is then the (common) logarithm of 1000 to the base 10. In natural or Napierian logarithms the base is e. To change the base from common to natural logarithms the formula used is: $\log_{10}y = \log_e y \times \log_{10}e = 0.43429\log_e y$.
3. (in electronics) *See* TRANSISTOR.

**base dissociation constant** *See* DISSOCIATION.

**basement membrane** A thin sheet of fibrous proteins that supports the cells of an overlying epithelium or endothelium, separating this from underlying connective tissue. Such membranes also surround muscle cells, Schwann cells, and fat cells, and a thick basement membrane is found in the kidney glomerulus, where it acts as a filter (*see* ULTRAFILTRATION). Basement membranes are components of the *extracellular matrix and help to regulate passage of materials between epithelial cells and adjacent blood vessels. A basement membrane consists of two layers: a **reticular lamina**, containing a network of *collagen fibrils; and **basal lamina**, consisting largely of laminin proteins, which bind the collagen to neighbouring cells and the underlying connective tissue.

**base metal** A common relatively inexpensive metal, such as iron or lead, that corrodes, oxidizes, or tarnishes on exposure to air, moisture, or heat, as distinguished from precious metals, such as gold and silver.

**base pair** Symbol bp. A unit used at the molecular level for measuring distances along a duplex polynucleotide and corresponding to the number of paired bases in a particular segment of DNA (or duplex RNA). *See also* BASE PAIRING; KILOBASE.

**base pairing** The chemical linking of two complementary nitrogenous bases in *DNA and in certain types of *RNA molecules. Of the four such bases in DNA, adenine pairs with thymine and cytosine with guanine. In RNA, thymine is replaced by uracil. Base pairing is responsible for holding together the two strands of a DNA molecule to form a double helix and for faithful reproduction and reading of the *genetic code. The links between bases take the form of *hydrogen bonds.

**base unit** A unit that is defined arbitrarily rather than being defined by simple combinations of other units. For example, the ampere is a base unit in the SI system defined in terms of the force produced between two current-carrying conductors, whereas the coulomb is a **derived unit**, defined as the

**battery**

quantity of charge transferred by one ampere in one second.

**basic** **1.** Describing a compound that is a base. **2.** Describing a solution containing an excess of hydroxide ions; alkaline.

**BASIC** A high-level computer programming language. The name is short for *beginner's all-purpose symbolic instruction code*. It dates from the 1960s, is easy to learn and use, and is mainly employed in teaching programming. There are several variants, including **Visual Basic**, which is used to program graphic user interfaces.

**basic dye** *See* DYES.

**basicity constant** *See* DISSOCIATION.

**basic-oxygen process (BOP process)** A high-speed method of making high-grade steel. It originated in the **Linz–Donawitz (L–D) process**. Molten pig iron and scrap are charged into a tilting furnace, similar to the Bessemer furnace except that it has no tuyeres. The charge is converted to steel by blowing high-pressure oxygen onto the surface of the metal through a water-cooled lance. The excess heat produced enables up to 30% of scrap to be incorporated into the charge. The process has largely replaced the Bessemer and open-hearth processes.

**basic rock** An igneous rock containing a comparatively low amount of silica (up to 45–52%) but rich in calcium or magnesium and iron. Examples include basalt, dolerite, and gabbro.

**basic salt** A compound that can be regarded as being formed by replacing some of the oxide or hydroxide ions in a base by other negative ions. Basic salts are thus mixed salt–oxides (e.g. bismuth(III) chloride oxide, BiOCl) or salt–hydroxides (e.g. lead(II) chloride hydroxide, Pb(OH)Cl).

**basic slag** *Slag formed from a basic flux (e.g. calcium oxide) in a blast furnace. The basic flux is used to remove acid impurities in the ore and contains calcium silicate, phosphate, and sulphide. If the phosphorus content is high the slag can be used as a fertilizer.

**basic stains** *See* STAINING.

**Basidiomycota** A phylum of fungi, formerly classified as a class (Basidiomycetes) or a subdivision (Basidiomycotina). Sexual reproduction is by means of **basidiospores** (spores produced externally on a club-shaped or cylindrical cell, the **basidium**). Basidia are often grouped together forming fruiting structures, such as mushrooms, puffballs, and bracket fungi. Exceptions are the *rusts and *smuts, which do not produce obvious fruiting bodies.

**basophil** A type of white blood cell (*leucocyte) that has a lobed nucleus surrounded by granular cytoplasm (*see* GRANULOCYTE). Basophils are produced continually by stem cells in the red bone marrow; they are *phagocytes and – like *mast cells – produce histamine and heparin as part of the body's defences at the site of an infection or injury (*see* INFLAMMATION).

**Basov, Nikolai Gennediyevitch** (1922–2001) Russian physicist best known for the development of the *maser, the precursor of the laser. In 1955, while working as a research student with Aleksandr Prokhorov (1916–2000) at the Soviet Academy of Sciences, he devised a microwave amplifier based on ammonia molecules. The two scientists shared the 1964 Nobel Prize with American Charles Townes (1915–   ), who independently developed a maser.

**bast** An old name for *phloem.

**Bateson, William** (1861–1926) British geneticist, who worked at Cambridge University. In 1900 he translated and championed the rediscovered work of *Mendel and went on to study inheritance in chickens. He found that some traits are controlled by more than one gene. He also coined the term 'genetics'.

**batholith** A large *intrusion of igneous rock, often surrounded by metamorphic rock, with a surface area in excess of 100 sq km. It is an irregular mass, which may penetrate to a great depth (up to 30 km), and is usually associated with a mountain belt (*see* OROGENESIS). Many batholiths are composed of granite and there are often veins of useful ores in the surrounding country rock.

**bats** *See* CHIROPTERA.

**battery** A number of electric cells joined together. The common car battery, or *accumulator, usually consists of six secondary cells connected in series to give a total e.m.f. of 12 volts. A torch battery is usually a *dry cell, two of which are often connected in series. Batteries may also have cells connected in parallel, in which case they have the same e.m.f. as a single cell, but their capacity is increased, i.e. they will provide more total charge. The capacity of a battery is usually specified in ampere-hours, the ability to supply 1 A for 1 hr, or the equivalent.

**baud** A unit for measuring signal speed in a computer or communications system. When the signal is a sequence of *bits, the baud rate is given in bits per second (bps). The unit is named after Jean-Maurice-Emile Baudot (1845–1903).

**bauxite** The chief ore of aluminium, consisting of hydrous aluminium oxides and aluminous laterite. It is a claylike amorphous material formed by the weathering of silicate rocks under tropical conditions. The chief producers are Australia, Guinea, Jamaica, Russia, Brazil, and Surinam.

**Bayesian inference** A technique of statistical inference that estimates the probability of an event occurring in terms of the frequency at which the event occurred previously. It depends on *Bayes' theorem.

**Bayesian statistics and probability** A method of interpreting *probability in terms of a number indicating how much confidence is placed in a proposition, i.e. how true it is. The *entropy of a probability distribution is maximized to assign a probability. A hypothesis can be confirmed by an observation that is likely if the hypothesis is true and unlikely if it is false. *See also* BAYES' THEOREM.

**Bayes' theorem** A theorem that deals with conditional *probabilities of propositions or events. Given an event A the conditional probability of an event E is formulated as $P(A|E) = P(E)/P(A)$. Bayes' theorem enables prior estimates of probability to be updated when further information becomes available. It was named after the British mathematician Thomas Bayes (1702–61).

**b.c.c.** Body-centred cubic. *See* CUBIC CRYSTAL.

**B cell (B lymphocyte)** A *lymphocyte that is derived from stem cells in the bone marrow but does not mature in the thymus (*compare* T CELL); in birds it matures in the bursa of the cloaca (hence *B* cell). Each B cell has a unique set of receptor molecules on its surface, designed to recognize and bind to a specific antigen, which is taken into the cell. Here, the antigen is bound with MHC class II proteins (*see* HISTOCOMPATIBILITY); the antigen–MHC class II complex migrates to the cell surface and is recognized by helper T cells, which adhere to the B cell. This triggers the T cells to release lymphokines (*see* CYTOKINE), which cause the B cell to undergo repeated division to form a clone of cells. These mature into *plasma cells, capable of producing large amounts of specific antibody (*see* IMMUNOGLOBULIN), which circulates in the blood and lymph and binds to the corresponding antigen. After a few days of antibody production the plasma cells die. However, some cells from the clone remain in the form of **memory cells**, which initiate a more rapid immune response on subsequent exposure to the same antigen. *See also* CLONAL SELECTION THEORY.

**BCS theory** *See* SUPERCONDUCTIVITY.

**Beadle, George Wells** (1903–89) US geneticist who, after holding several professorships, went to Stanford University, where he worked with Edward Tatum (1909–75). Using moulds, they deduced that the function of genes is to control the production of enzymes, which in turn control metabolic processes. They found that mutant genes result in abnormal (and non-operative) enzymes. For this 'one gene–one enzyme' theory (*see* ONE GENE–ONE POLYPEPTIDE HYPOTHESIS), they were awarded the 1958 Nobel Prize in physiology or medicine.

**beam** A group of rays moving in an organized manner. It may consist of particles (e.g. an electron beam) or of electromagnetic radiation (e.g. a radar beam).

**beam balance** *See* BALANCE.

**beam hole** A hole through the shielding of a *nuclear reactor to enable a beam of neutrons or other particles to escape for experimental purposes.

**beats** A periodic increase and decrease in loudness heard when two notes of slightly different frequency are sounded at the same time. If a note of frequency $n$ is heard at the same time as a note of frequency $m$, the resulting note will have a frequency of about $(n + m)/2$. However the amplitude of this note will vary from the difference to the sum of the amplitudes of the $m$ and $n$ notes and the frequency (called the **beat frequency**) of this variation will be $(m - n)$. The beating sound produced occurs as the waves successively reinforce and oppose each other as they move in and out of phase. Beating also occurs with radio-frequency waves and is

| Beaufort number | description of wind | wind speed | |
|---|---|---|---|
| | | (knots) | (metres per second) |
| 0 | calm | <1 | 0.0– 0.2 |
| 1 | light air | 1– 3 | 0.3– 1.5 |
| 2 | light breeze | 4– 6 | 1.6– 3.3 |
| 3 | gentle breeze | 7–10 | 3.4– 5.4 |
| 4 | moderate breeze | 11–16 | 5.5– 7.9 |
| 5 | fresh breeze | 17–21 | 8.0–10.7 |
| 6 | strong breeze | 22–27 | 10.8–13.8 |
| 7 | near gale | 28–33 | 13.9–17.1 |
| 8 | gale | 34–40 | 17.2–20.7 |
| 9 | strong gale | 41–47 | 20.8–24.4 |
| 10 | storm | 48–55 | 24.5–28.4 |
| 11 | violent storm | 56–63 | 28.5–32.6 |
| 12 | hurricane | ≥64 | ≥32.7 |

**Beaufort wind scale.**

made use of in *heterodyne devices. *See also* INTERFERENCE.

**Beaufort wind scale** A scale of wind speed that was devised in the early 19th century by Rear Admiral Sir Francis Beaufort (1774–1857). Originally based on observations of the effect of various wind speeds on the sails of a full-rigged frigate, it has since been modified and is now based on observations of the sea surface or, on land, such easily observable indicators as smoke and tree movement. The scale ranges from 0 (calm) to 12 (hurricane).

**Becklin–Neugebauer object (BN object)** A point in the star-forming part of the Orion nebula that emits infrared radiation (but no visible radiation, probably because it is scattered by the dense dust of the nebula). It is thought to be a young near main-sequence star of type B spectral classification, and one of the youngest stars so far observed.

**Beckmann rearrangement** The chemical conversion of a ketone *oxime into an *amide, usually using sulphuric acid as a catalyst. The reaction, used in the manufacture of nylon and other polyamides, is named after the German chemist Ernst Beckmann (1853–1923).

**Beckmann thermometer** A thermometer for measuring small changes of temperature (see illustration). It consists of a mercury-in-glass thermometer with a scale covering only 5 or 6°C calibrated in hundredths of a degree. It has two mercury bulbs, the range of temperature to be measured is varied by running mercury from the upper bulb into the larger lower bulb. It is used particularly for measuring *depression of freezing point or *elevation of boiling point of liquids when solute is added, in order to find relative molecular masses.

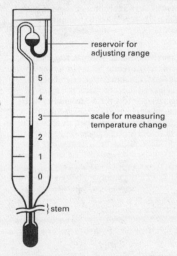

reservoir for adjusting range

scale for measuring temperature change

} stem

**Beckmann thermometer.**

**becquerel** Symbol Bq. The SI unit of activity (*see* RADIATION UNITS). The unit is named after A. H. Becquerel.

**Becquerel, Antoine Henri** (1852–1908) French physicist. His early researches were in optics; in 1896 he accidentally discovered *radioactivity in fluorescent salts of uranium. Three years later he showed that it consists of charged particles that are deflected by a magnetic field. For this work he was awarded the 1903 Nobel Prize for physics, which he shared with Pierre and Marie *Curie.

**bees** *See* HYMENOPTERA.

**beetles** *See* COLEOPTERA.

**beet sugar** *See* SUCROSE.

**behaviour** The sum of the responses of an organism to internal or external stimuli. The behaviour of an animal can be either instinctive (*see* INSTINCT) or learned. *See* ANIMAL BEHAVIOUR; LEARNING.

**behavioural genetics** The branch of genetics concerned with determining the relative importance of the genetic constitution of animals as compared to environmental factors in influencing animal behaviour.

**bel** Ten *decibels.

**bell metal** A type of *bronze used in casting bells. It consists of 60–85% copper alloyed with tin, often with some zinc and lead included.

**Bell's theorem** A theorem stating that no local *hidden-variables theory can make predictions in agreement with those of *quantum mechanics. Local hidden variables theories give rise to a result, called **Bell's inequality**, which is one of many similar results concerning the probabilities of two events both occurring in well-separated parts of a system. The British physicist John S. Bell (1928–90) showed in 1964 that quantum mechanics predicts a violation of the inequalities, which are consequences of local hidden-variables theories. Experiments are in agreement with quantum mechanics rather than local hidden variables theories by violating Bell's inequality, in accordance with Bell's theorem. *See also* LEGGETT'S THEOREM; QUANTUM ENTANGLEMENT.

**(((●))) SEE WEB LINKS**
• The original 1964 paper in *Physics*

**Belousov–Zhabotinskii reaction** *See* B–Z REACTION.

**Bénard cell** A structure associated with a layer of liquid that is confined by two horizontal parallel plates, in which the lateral dimensions are much larger than the width of the layer. Before heating the liquid is homogeneous. However, if after heating from below the temperatures of the plates are $T_1$ and $T_2$, at a critical value of the temperature gradient $\Delta T = T_1 - T_2$ the liquid abruptly starts to convect. The liquid spontaneously

organizes itself into a set of convection rolls, i.e. the liquid goes round in a series of 'cells'. The existence of such cells was discovered by a French scientist, Henri Bénard, around 1900. *See also* COMPLEXITY.

**bending moment** (about any point or section of a horizontal beam under load) The algebraic sum of the moments of all the vertical forces to either side of that point or section (*see* MOMENT OF A FORCE).

**Benedict's test** A biochemical test to detect *reducing sugars in solution, devised by the US chemist S. R. Benedict (1884–1936). **Benedict's reagent** – a mixture of copper(II) sulphate and a filtered mixture of hydrated sodium citrate and hydrated sodium carbonate – is added to the test solution and boiled. A high concentration of reducing sugars induces the formation of a red precipitate; a lower concentration produces a yellow precipitate. Benedict's test is a more sensitive alternative to *Fehling's test.

**beneficiation (ore dressing)** The separation of an ore into the valuable components and the waste material (gangue). This may be achieved by a number of processes, including crushing, grinding, magnetic separation, froth flotation, etc. The dressed ore, consisting of a high proportion of valuable components, is then ready for smelting or some other refining process.

**benthos** Flora and fauna occurring on the bottom of a sea or lake. Benthic organisms may crawl, burrow, or remain attached to a substrate. *Compare* PELAGIC.

**bent sandwich** *See* SANDWICH COMPOUND.

**benzaldehyde** *See* BENZENECARBALDEHYDE.

**benzene** A colourless liquid hydrocarbon, $C_6H_6$; r.d. 0.88; m.p. 5.5°C; b.p. 80.1°C. It is now made from gasoline from petroleum by catalytic reforming (formerly obtained from coal tar). Benzene is the archetypal *aromatic compound. It has an unsaturated molecule, yet will not readily undergo addition reactions. On the other hand, it does undergo substitution reactions in which hydrogen atoms are replaced by other atoms or groups. This behaviour occurs because of delocalization of *p*-electrons over the benzene ring, and all the C–C bonds in benzene are equivalent and intermediate in length between single and double bonds. It can be regarded as a resonance hybrid of Kekulé

Kekulé structures      Dewar structures

**Benzene.**

and Dewar structures (see formulae). In formulae it can be represented by a hexagon with a ring inside it.

**benzenecarbaldehyde (benzaldehyde)** A yellowish volatile oily liquid, $C_6H_5CHO$; r.d. 1.04; m.p. –26°C; b.p. 178.1°C. The compound occurs in almond kernels and has an almond-like smell. It is made from methylbenzene (by conversion to dichloromethyl benzene, $C_6H_5CHCl_2$, followed by hydrolysis). Benzenecarbaldehyde is used in flavourings, perfumery, and the dyestuffs industry.

**benzenecarbonyl chloride (benzoyl chloride)** A colourless liquid, $C_6H_5COCl$; r.d. 1.21; m.p. 0°C; b.p. 197.2°C. It is an *acyl halide, used to introduce benzenecarbonyl groups into molecules. *See* ACYLATION.

**benzenecarbonyl group (benzoyl group)** The organic group $C_6H_5CO$–.

**benzenecarboxylate (benzoate)** A salt or ester of benzenecarboxylic acid.

**benzenecarboxylic acid (benzoic acid)** A white crystalline compound, $C_6H_5COOH$; r.d. 1.27; m.p. 122.4°C; b.p 249°C. It occurs naturally in some plants and is used as a food preservative. Benzenecarboxylic acid has a carboxyl group bound directly to a benzene ring. It is a weak carboxylic acid ($K_a = 6.4 \times 10^{-5}$ at 25°C), which is slightly soluble in water. It also undergoes substitution reactions on the benzene ring.

**benzene-1,4-diol (hydroquinone; quinol)** A white crystalline solid, $C_6H_4(OH)_2$; r.d. 1.33; m.p. 173–174°C; b.p. 285°C. It is used in making dyes. *See also* QUINHYDRONE ELECTRODE.

**benzene hexachloride (BHC)** A crystalline substance, $C_6H_6Cl_6$, made by adding chlorine to benzene. It is used as a pesticide and, like *DDT, concern has been expressed at its environmental effects.

**benzenesulphonic acid** A colourless deliquescent solid, $C_6H_5SO_2OH$, m.p. 43–44°C, usually found as an oily liquid. It is made by treating benzene with concentrated sulphuric acid. Its alkyl derivatives are used as *detergents.

**benzil** 1,2-diphenylethan-1,2-dione. *See* BENZILIC ACID REARRANGEMENT.

**benzilic acid rearrangement** An organic rearrangement reaction in which **benzil** (1,2-diphenylethan-1,2-dione) is treated with hydroxide and then acid to give **benzilic acid** (2-hydroxy-2,2-diphenylethanoic acid):

$$C_6H_5.CO.CO.C_6H_5 \rightarrow (C_6H_5)_2C(OH).COOH$$

In the reaction a phenyl group ($C_6H_5$–) migrates from one carbon atom to another. The reaction was discovered in 1828 by Justus von Liebig; it was the first rearrangement reaction to be described.

**benzoate** *See* BENZENECARBOXYLATE.

**benzodiazepines** A group of related psychoactive drugs. They are used medically in the treatment of anxiety, insomnia, convulsions, and alcohol withdrawal. All are addictive and available only as prescription drugs in the UK. Common examples are *diazepam (Valium) and *flunitrazepam (Rohypnol).

**benzoic acid** *See* BENZENECARBOXYLIC ACID.

**benzopyrene** A crystalline aromatic hydrocarbon, $C_{20}H_{12}$; m.p. 179°C. It is found in coal tar and is highly carcinogenic.

**benzoquinone** *See* CYCLOHEXADIENE-1,4-DIONE.

**benzoylation** A chemical reaction in which a benzoyl group (benzenecarbonyl group, $C_6H_5CO$) is introduced into a molecule. *See* ACYLATION.

**benzoyl chloride** *See* BENZENECARBONYL CHLORIDE.

**benzoylecgonine (BZ)** A primary metabolite of cocaine, used in drug testing. It can be detected in the urine up to 48 hours after taking cocaine. BZ is tested for by immunoassay or by gas chromatography/mass spectrometry.

**benzoyl group** *See* BENZENECARBONYL GROUP.

**benzpyrene** A pale yellow solid, $C_{20}H_{12}$, m.p. 179°C, whose molecules consist of five fused benzene rings. It occurs in tars from coal and tobacco smoke and is a *carcinogen.

**benzvalene** A valence isomer of benzene, $C_6H_6$, with a bridged structure.

**benzyl alcohol** *See* PHENYLMETHANOL.

**benzylamine** (α-aminotoluene; phenylmethylamine) A colourless liquid, $C_6H_5CH_2NH_2$; r.d. 0.981; b.p. 185°C. It behaves in the same way as primary aliphatic amines.

**benzyne** A highly reactive short-lived compound, $C_6H_4$, having a hexagonal ring of carbon atoms containing two double bonds and one triple bond. Benzyne, which is the simplest example of an *aryne, is thought to be an intermediate in a number of reactions.

**Bergius, Friedrich Karl Rudolf** (1884–1949) German organic chemist. While working with Fritz *Haber in Karlsruhe, he become interested in reactions at high pressures. In 1912 he devised an industrial process for making light hydrocarbons by the high-pressure hydrogenation of coal or heavy oil. The work earned him a share of the 1931 Nobel Prize for chemistry with Carl Bosch (1874–1940). The Bergius process proved important for supplying petrol for the German war effort in World War II.

**Bergius process** A process for making hydrocarbon mixtures (for fuels) from coal by heating powdered coal mixed with tar and iron(III) oxide catalyst at 450°C under hydrogen at a pressure of about 200 atmospheres. In later developments of the process, the coal was suspended in liquid hydrocarbons and other catalysts were used. The process was developed by Friedrich Bergius during World War I as a source of motor fuel.

**beriberi** A disease caused by a low intake of vitamin $B_1$ (thiamine; *see* VITAMIN B COMPLEX), resulting in damage to peripheral nerves and heart failure. Beriberi is most common in regions of the Far East where the diet is based on polished white rice, which lacks the thiamine-rich seed coat.

**berkelium** Symbol Bk. A radioactive metallic transuranic element belonging to the *actinoids; a.n. 97; mass number of the most stable isotope 247 (half-life $1.4 \times 10^3$ years); r.d. (calculated) 14. There are eight known isotopes. It was first produced by G. T. Seborg and associates in 1949 by bombarding americium–241 with alpha particles.

((())) SEE WEB LINKS

• Information from the WebElements site

**Bernoulli, Daniel** (1700–82) Swiss mathematician. In 1724 he published a work on differential equations, which earned him a professorship at St Petersburg. He returned to Basel, Switzerland, in 1733 and began researches on hydrodynamics (*see* BERNOULLI THEOREM), the work for which he is best known. He also initiated the kinetic theory of matter.

**Bernoulli theorem** At any point in a pipe through which a fluid is flowing the sum of the pressure energy, the kinetic energy, and the potential energy of a given mass of the fluid is constant. This is equivalent to a statement of the law of the conservation of energy. The law was published in 1738 by Daniel Bernoulli.

**Bernoulli trial** An experiment in which there are two possible independent outcomes, for example, tossing a coin. It is named after the Swiss mathematician Jakob (or Jacques) Bernoulli (1654–1705), who made important contributions to the field of probability theory.

**berry** A fleshy fruit formed from either one carpel or from several fused together and containing many seeds. The fruit wall may have two or three layers but the inner layer is never hard and stony (as in some drupes). Examples of berries are grapes and tomatoes. A berry, such as a cucumber, that develops a hard outer rind is called a **pepo**. One that is segmented and has a leathery rind, such as a citrus fruit, is called a **hesperidium**. The rind contains oil glands and is lined by the white mesocarp, commonly called **pith**.

**Berthollide compound** *See* NONSTOICHIOMETRIC COMPOUND.

**beryl** A hexagonal mineral form of beryllium aluminium silicate, $Be_3Al_2Si_6O_{18}$; the chief ore of beryllium. It may be green, blue, yellow, or white and has long been used as a gemstone. Beryl occurs throughout the world in granite and pegmatites. *Emerald, the green gem variety, occurs more rarely and is of great value. Important sources of beryl are found in Brazil, Madagascar, and the USA.

**beryllate** A compound formed in solution when beryllium metal, or the oxide or hydroxide, dissolves in strong alkali. The reaction (for the metal) is often written

$$Be + 2OH^-(aq) \rightarrow BeO_2^{2-}(aq) + H_2(g)$$

The ion $BeO_2^{2-}$ is the beryllate ion. In fact, as with the *aluminates, the ions present are probably hydroxy ions of the type $Be(OH)_4^{2-}$ (the **tetrahydroxoberyllate(II) ion**) together with polymeric ions.

**beryllia** *See* BERYLLIUM OXIDE.

**beryllium** Symbol Be. A grey metallic element of *group 2 (formerly IIA) of the periodic table; a.n. 4; r.a.m. 9.012; r.d. 1.85; m.p. 1278°C; b.p. 2970°C. Beryllium occurs as beryl ($3BeO.Al_2O_3.6SiO_2$) and chrysoberyl ($BeO.Al_2O_3$). The metal is extracted from a fused mixture of $BeF_2/NaF$ by electrolysis or by magnesium reduction of $BeF_2$. It is used to manufacture Be–Cu alloys, which are used in nuclear reactors as reflectors and moderators because of their low absorption *cross section. Beryllium oxide is used in ceramics and in nuclear reactors. Beryllium and its compounds are toxic and can cause serious lung diseases and dermatitis. The metal is resistant to oxidation by air because of the formation of an oxide layer, but will react with dilute hydrochloric and sulphuric acids. Beryllium compounds show high covalent character. The element was isolated independently by F. Wohler and A. A. Bussy in 1828.

**(⊕) SEE WEB LINKS**
• Information from the WebElements site

**beryllium bronze** A hard, strong type of *bronze containing about 2% beryllium, in addition to copper and tin.

**beryllium hydroxide** A white crystalline compound, $Be(OH)_2$, precipitated from solutions of beryllium salts by adding alkali. Like the oxide, it is amphoteric and dissolves in excess alkali to give *beryllates.

**beryllium oxide (beryllia)** An insoluble solid compound, BeO; hexagonal; r.d. 3.01; m.p. 2530°C; b.p. 3900°C. It occurs naturally as **bromellite**, and can be made by burning beryllium in oxygen or by the decomposition of beryllium carbonate or hydroxide. It is an important amphoteric oxide, reacting with acids to form salts and with alkalis to form compounds known as *beryllates. Beryllium

oxide is used in the production of beryllium and beryllium–copper refractories, transistors, and integrated circuits.

**Berzelius, Jöns Jacob** (1779–1848) Swedish chemist. After moving to Stockholm he worked with mining chemists and, with them, discovered several elements, including *cerium (1803), *selenium (1817), *lithium (1818), *thorium (1828), and *vanadium (1830). He also worked on atomic weights and electrochemistry and devised the current notation for chemical elements.

**Bessel, Friedrich Wilhelm** (1784–1846) German astronomer. As an unqualified amateur he catalogued the positions of 50 000 stars and calculated the distance to 61 Cygni from its *parallax. In 1844 he discovered that Sirius is a binary star with a dark companion. Solving the complex mathematics in these studies led him to develop *Bessel functions.

**Bessel function** A type of function that occurs as a solution to problems involving waves in systems with cylindrical symmetry. Bessel functions have been extensively studied and tabulated and are used in many branches of mathematical physics. They are named after the German astronomer Friedrich Wilhelm Bessel (1784–1846).

**Bessemer process** A process for converting *pig iron from a *blast furnace into *steel. The molten pig iron is loaded into a refractory-lined tilting furnace (**Bessemer converter**) at about 1250°C. Air is blown into the furnace from the base and *spiegel is added to introduce the correct amount of carbon. Impurities (especially silicon, phosphorus, and manganese) are removed by the converter lining to form a slag. Finally the furnace is tilted so that the molten steel can be poured off. In the modern VLN (very low nitrogen) version of this process, oxygen and steam are blown into the furnace in place of air to minimize the absorption of nitrogen from the air by the steel. The process is named after the British engineer Sir Henry Bessemer (1813–98), who announced it in 1856. *See also* BASIC-OXYGEN PROCESS.

**beta adrenoceptor (beta adrenergic receptor)** *See* ADRENOCEPTOR.

**beta blocker (beta-adrenoceptor antagonist)** Any of a group of drugs that bind preferentially to beta *adrenoceptors and hence block their stimulation by the body's own neurotransmitters, adrenaline and noradrenaline. Beta blockers, such as propranolol, oxprenolol, and atenolol, are used to treat disorders of the cardiovascular system, including high blood pressure (hypertension), angina pectoris, and irregularities of heartbeat (arrhythmias). They are also effective in treating anxiety and glaucoma (as eye drops) and in preventing migraine. They tend to dampen the effects of exercise or stress on heart rate, heart output, and blood pressure, as well as improving the oxygenation of the heart muscles. The release of the enzyme *renin from the kidneys is also reduced, leading to an overall fall in arterial blood pressure.

**beta decay** A type of weak interaction (*see* FUNDAMENTAL INTERACTIONS) in which an unstable atomic nucleus changes into a nucleus of the same nucleon number ($A$) but different proton number ($Z$). There are three types of beta decay: negative beta decay, positive beta decay, and electron capture.
Negative beta decay:

$$^A_Z X \rightarrow {}^{\ A}_{Z+1} Y + {}^{\ 0}_{-1}e + {}^{0}_{0}\bar{\nu}$$

A neutron in the nucleus X has decayed into a proton forming a new nucleus Y with the emission of an electron and antineutrino. This process involves a decrease in mass and is energetically favourable; it can also occur outside the nucleus – free neutrons decay with a mean lifetime of about 15 minutes.
Positive beta decay:

$$^A_Z X \rightarrow {}^{\ A}_{Z-1} Y + {}^{0}_{1}e + {}^{0}_{0}\nu$$

A proton in the nucleus X transforms into a neutron and a new nucleus Y is formed with the emission of an antimatter electron (positron) and neutrino. This process involves an effective increase in mass for the proton and is not energetically favourable. It cannot occur outside the nucleus – free protons do not undergo this kind of interaction. The process is allowed within the environment of the nucleus because when the nucleus as a whole is taken into account the interaction represents an overall decrease in mass.
Electron capture:

$$^A_Z X + {}^{0}_{-1}e \rightarrow {}^{\ A}_{Z-1} Y + {}^{0}_{0}\nu$$

A proton in the nucleus X captures an electron from the atomic environment and becomes a neutron, emitting in the process a neutrino. This process also involves an effective increase in mass for the proton and is not energetically favourable; again, it cannot also occur outside the nucleus – free protons do not undergo this kind of interaction. The

process is allowed within the environment of the nucleus because, taking into account the whole nucleus, the interaction represents an overall decrease in mass.

**beta-iron** A nonmagnetic allotrope of iron that exists between 768°C and 900°C.

**beta particle** An electron or positron emitted during *beta decay. A stream of beta particles is known as **beta radiation**.

**betatron** A particle *accelerator for producing high-energy electrons (up to 340 MeV) for research purposes, including the production of high-energy X-rays. The electrons are accelerated by electromagnetic induction in a doughnut-shaped (toroidal) ring from which the air has been removed. This type of accelerator was first developed by D. W. Kerst (1911–93) in 1939; the largest such machine, at the University of Illinois, was completed in 1950.

**BET isotherm** An isotherm that takes account of the possibility that the monolayer in the *Langmuir adsorption isotherm can act as a substrate for further adsorption. The BET isotherm (named after S. Brunauer, P. Emmett, and E. Teller) has the form:

$$V/V_{mon} = cz/\{(1-z)[1-(1-c)z]\},$$

where $z = p/p^*$ ($p^*$ is the vapour pressure above a macroscopically thick layer of liquid on the surface), $V_{mon}$ is the volume that corresponds to the surface being covered by a monolayer, $V$ and $p$ are the volume and pressure of the gas respectively, and $c$ is a constant. In the BET isotherm, the isotherm rises indefinitely at high pressures (in contrast to the Langmuir isotherm). It provides a useful approximation over some ranges of pressure but underestimates adsorption for low pressures and overestimates adsorption for high pressures.

**Bevatron** A colloquial name for the proton *synchrotron at the Berkeley campus of the University of California. It produces energies up to 6 GeV.

**BHC** *See* BENZENE HEXACHLORIDE.

**biaxial crystal** *See* DOUBLE REFRACTION.

**bicarbonate** *See* HYDROGENCARBONATE.

**bicarbonate of soda** *See* SODIUM HYDROGENCARBONATE.

**biceps** A muscle that runs along the large bone of the upper arm (*humerus) and is connected to the *radius at one end and the shoulder bone (*scapula) at the other. Contraction of the biceps causes the arm to flex at the elbow joint (*see* FLEXOR). It works antagonistically with the triceps, which contracts to extend the arm (*see* ANTAGONISM). *See also* VOLUNTARY MUSCLE.

**biconcave** *See* CONCAVE.

**bicuspid valve (mitral valve)** A valve, consisting of two flaps, situated between the left atrium and the left ventricle of the heart of birds and mammals. When the left ventricle contracts, forcing blood into the aorta, the bicuspid valve closes the aperture to the left atrium, thereby preventing any backflow of blood. The valve reopens to allow blood to flow from the atrium into the ventricle. *Compare* TRICUSPID VALVE.

**biennial** A plant that requires two growing seasons to complete its life cycle. During the first year it builds up food reserves, which are used during the second year in the production of flowers and seeds. Examples are carrot and parsnip.

**big-bang theory** The cosmological theory that all the matter and energy in the universe originated from a state of enormous density and temperature that exploded at a finite moment in the past when space and time came into being. See Feature (pp 86–87).

**bilateral symmetry** A type of arrangement of the parts and organs of an animal in which the body can be divided into two halves that are mirror images of each other along one plane only (usually passing through the midline at right angles to the dorsal and ventral surfaces). Bilaterally symmetrical animals are characterized by a type of movement in which one end of the body always leads. In botany this type of symmetry is usually called **zygomorphy** when applied to flowers (e.g. foxglove and antirrhinum flowers are zygomorphic). *Compare* RADIAL SYMMETRY.

**bile (gall)** A bitter-tasting greenish-yellow alkaline fluid produced by the *liver, stored in the *gall bladder, and secreted into the *duodenum of vertebrates. It assists the digestion and absorption of fats by the action of **bile salts**, which chemically reduce fatty substances and decrease the surface tension of fat droplets so that they are broken down and emulsified. Bile may also stimulate gut muscle contraction (*peristalsis). Bile also contains the **bile pigments**, **bilirubin** and

**biliverdin**, which are produced by the breakdown of the blood pigment *haemoglobin.

**bile duct** The tube through which bile passes from the *liver or (when present) the *gall bladder to the duodenum.

**bilirubin** *See* BILE.

**biliverdin** *See* BILE.

**billion** **1.** (in the UK and Germany) One million million, $10^{12}$. **2.** (in the USA and France) One thousand million, $10^9$.

**bimetallic strip** A strip consisting of two metals of different *expansivity riveted or welded together so that the strip bends on heating. If one end is fixed the other end can be made to open and close an electric circuit, as in a *thermostat.

**bimolecular reaction** A step in a chemical reaction that involves two molecules. *See* MOLECULARITY.

**bimorph cell** A device consisting of two plates of piezoelectric material, such as Rochelle salt, joined together so that one expands on the application of a potential difference and the other contracts. The cell thus bends as a result of the applied p.d. The opposite effect is also used, in which the mechanical bending of the cell is used to produce a p.d., as in the crystal microphone and some types of record-player pickups.

**binary** Describing a compound or alloy formed from two elements.

**binary acid** An *acid in which the acidic hydrogen atom(s) are bound directly to an atom other than oxygen. Examples are hydrogen chloride (HCl) and hydrogen sulphide ($H_2S$). Such compounds are sometimes called **hydracids**. *Compare* OXOACID.

**binary fission** *See* FISSION.

**binary notation** A number system using only two different digits, 0 and 1. Instead of units, tens, hundreds, etc., as used in the decimal system, digits in the binary notation represent units, twos, fours, eights, etc. Thus one in decimal notation is represented by 0001, two by 0010, four by 0100, and eight by 1000. Because 0 and 1 can be made to correspond to off and on conditions in an electric circuit, the binary notation is widely used in computers.

**binary prefixes** A set of prefixes for binary powers designed to be used in data pro-

## THE BIG-BANG THEORY

Newton's work gave a mathematical basis for the universe on a large scale. However, the data available at the time suggested a static unchanging universe. This could not easily be explained in the context of the law of gravitation, since all bodies in the universe attract all other bodies with the force of gravity. Newton realized that there was only one solution to this problem: in a static universe, matter had to be uniformly spread throughout an infinitely large space. In 1826, Heinrich Olbers published a paper containing what is known as *Olbers' paradox; such a universe would lead to a perpetually bright sky on earth.

### Space–time

Cosmologists now believe that Newton's model was based on incorrect assumptions about the structure of space, time, and matter. Einstein in his general theory of *relativity (1915) proposed that the universe exists in four-dimensional space–time. This space–time is curved by the presence of matter, and the matter moves by following the resulting curves.

### The expanding universe

The discovery by Hubble in 1929 that the universe is expanding provided a starting point for the ideas on which our present understanding of the universe is based. Hubble made his discovery by analysing the *spectra of light from distant galaxies and noting a persistent *red shift, which he explained in terms of the *Doppler effect; an increase in observed wavelengths of light occurs because the light source is receding from the observer. The larger the speed, of recession, the larger the red shifts. Hubble discovered a pattern in his data: the further away the galaxy, the greater the speed of recession. Known as *Hubble's law, this provided the evidence that the universe is expanding and a resolution to Olbers' paradox. If the galaxies and the earth are moving apart, the radiation falling on the earth from the galaxies is reduced. The further galaxies are away from the earth, the smaller their contribution to the radiation falling on the earth.

This model might seem to place the earth at the centre of the universe again. However, it is space itself that is expanding and the galaxies are imbedded in this space. The ring (space) in the diagram has dots (the galaxies). The expansion of the ring means that the view from any one dot is that the other dots are receding at a speed proportional to their distance away. No single dot is at the centre of the system but all dots see the same thing.

### Age of the universe

Hubble's law may be stated in the form: $H_0 = v/d$, where $v$ is the speed of recession of the galaxy, $d$ is the earth–galaxy distance, and $H_0$ is called the **Hubble**

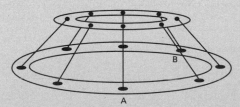

The expansion leads to the recession of B from A along the ring. The speed of recession will be directly proportional to the distance of B from A along the ring.

**constant**. Assuming that the galaxies have always been moving apart, the age of the universe (T) can be estimated, i.e. $T = 1/H_0$. On this basis the age of the universe would be between 15–18 billion years.

### The origin of the universe

This view of the origin of the universe is called the **big-bang theory**, first put forward by Georges Lamaître in the 1930s. The theory suggests that the universe originated as a minute but very hot body and that the temperature has been falling as the expansion has continued. In several papers in the 1940s George Gamow along with Ralph Alpher and Robert Herman predicted that there should be a *microwave background corresponding to a black-body temperature a few degrees above absolute zero. This microwave background was discovered 20 years later. The big-bang theory also explains the amount of helium in the universe.

In 1992, the COBE satellite discovered that there were very small variations in the microwave background. This discovery helped to explain why the universe formed into galaxies and stars. The non-uniformities that began the nucleation of galactic matter in the early universe now appear as the small variations in the microwave background. In 1998 it was discovered that the expansion of the universe is accelerating.

### Fundamental forces

It is thought that the four *fundamental interactions in the universe are all manifestations of the same force. This force existed when the big bang, occurred at a temperature above $10^{15}$K. As the universe cooled the forces separated as the original symmetries were broken. Gravity was the first to separate, followed by the strong nuclear force, and the weak and electromagnetic forces (see table).

| Time from big bang | Temperature (K) | State of the universe/forces |
| --- | --- | --- |
| 0 second | infinite | The universe is infinitesimally small and infinitely dense (i.e. a mathematical singularity). |
| $10^{-12}$ second | $10^{15}$ | Weak and electromagnetic forces begin to separate. |
| $10^{-6}$ second | $10^{14}$ | Quarks and leptons begin to form. |
| $10^{-3}$ second | $10^{12}$ | Quarks form the hadrons; quark confinement begins. |
| $10^2$ second | $10^7$ | Helium nuclei formed by fusion. |
| $10^5$ years | $10^4$ | Atomic era; atoms form as protons combine with electrons. |
| $10^6$ years | $10^3$ | Matter undergoes gravitational collapse. |
| $.5–1.8 \times 10^{10}$ years | 2.7 | Present day: cosmic background corresponds to about 2.7K. |

### The future

Research into the future of the universe is clearly speculative. Whether the universe will continue to expand indefinitely depends on its mean density. Below a critical level (the critical density), gravitational attraction will not be enough to stop the expansion. However, if the mean density is above the critical density the universe is **bound** and an eventual contraction will occur resulting in a **big crunch**. This may precede another big bang initiating the whole cycle again.

cessing and data transmission contexts. They were suggested in 1998 by the International Electrotechnical Commission (IEC) as a way of resolving the ambiguity in use of kilo-, mega-, giga-, etc., in computing. In scientific usage, these prefixes indicate $10^3$, $10^6$, $10^9$, etc. (*see* SI UNITS). In computing, it became common to use the prefix 'kilo-' to mean $2^{10}$, so one kilobit was 1024 bits (not 1000 bits). This was extended to larger prefixes, so 'mega-' in computing is taken to be $2^{20}$ (1 048 576) rather than $10^6$ (1 000 000). However, there is a variation in usage depending on the context. In discussing memory capacities megabyte generally means $2^{20}$ bytes, but in disk storage (and data transmission) megabyte is often taken to mean $10^6$ bytes. (In some contexts, as in the capacity of a floppy disk, it has even been quoted as 1 024 000 bytes, i.e. 1000 times a (binary) kilobyte.) The IEC attempted to resolve this confusion by introducing binary prefixes, modelled on the normal decimal prefixes, as follows:

kibi- $2^{10}$
mebi- $2^{20}$
gebi- $2^{30}$
tebi- $2^{40}$
pebi- $2^{50}$
exbi- $2^{60}$

These names are contractions of 'kilobinary', 'megabinary', etc., but are pronounced so that the second syllable rhymes with 'bee'. Using these prefixes, one gebibyte would be 1 073 741 824 bytes and one gigabyte would (unambiguously) be 1 000 000 000 bytes.

**binary stars** A pair of stars revolving about a common centre of mass. In a **visual binary** the stars are far enough apart to be seen separately by an optical telescope. In an **astrometric binary** one component is too faint to be seen and its presence is inferred from the perturbations in the motion of the other. In a **spectroscopic binary** the stars cannot usually be resolved by a telescope, but the motions can be detected by different Doppler shifts in the spectrum at each side of the binary, according to whether the components are approaching or receding from the observer. *See also* VARIABLE STAR.

**binding energy** The energy equivalent to the *mass defect when nucleons bind together to form an atomic nucleus. When a nucleus is formed some energy is released by the nucleons, since they are entering a more stable lower-energy state. Therefore the energy of a nucleus consists of the energy

equivalent of the mass of its individual nucleons minus the binding energy. The binding energy per nucleon plotted against the mass number provides a useful graph showing that up to a mass number of 50–60, the binding energy per nucleon increases rapidly, thereafter it falls slowly. Energy is released both by fission of heavy elements and by fusion of light elements because both processes entail a rearrangement of nuclei in the lower part of the graph to form nuclei in the higher part of the graph.

**binoculars** Any optical instrument designed to serve both the observer's eyes at once. **Binocular field glasses** consist of two refracting astronomical *telescopes inside each of which is a pair of prisms to increase the effective length and produce an upright image. Simpler binoculars, such as **opera glasses**, consist of two Galilean telescopes which produce upright images without prisms. Commonly, binoculars are specified by a pair of numbers, such as $10 \times 50$. The first number indicates the angular *magnification produced. The second is the diameter of the objective lens in millimetres, and indicates the amount of light gathered by the instrument. **Binocular microscopes** are used in biology and surgery to enable the observer to obtain a stereoscopic view of small objects or parts.

**binocular vision** The ability, found only in animals with forward-facing eyes, to produce a focused image of the same object simultaneously on the retinas of both eyes. This permits three-dimensional vision and contributes to distance judgment.

**binomial distribution** The distribution of the number of successes in an experiment in which there are two possible outcomes, success and failure. The probability of $k$ successes is:

$$b(k,n,p) = n!/k!(n-k)! \times p^n \times q^{n-k}$$

where $p$ is the probability of success and $q (= 1 - p)$ the probability of failure on each trial. These probabilities are given by the terms in the binomial expansion of $(p + q)^n$ (*see* BINOMIAL THEOREM). The distribution has a mean $np$ and variance $npq$. If $n$ is large and $p$ small it can be approximated by a *Poisson distribution with mean $np$. If $n$ is large and $p$ is not near 0 or 1, it can be approximated by a *normal distribution with mean $np$ and variance $npq$. *See also* PASCAL'S DISTRIBUTION.

**binomial nomenclature** The system of naming organisms using a two-part Latinized (or scientific) name that was devised by the Swedish botanist Linnaeus (Carl Linné); it is also known as the **Linnaean system**. The first part is the generic name (*see* GENUS), the second is the specific epithet or name (*see* SPECIES). The Latin name is usually printed in italics, starting with a capital letter. For example, in the scientific name of the common frog, *Rana temporaria*, *Rana* is the generic name and *temporaria* the specific name. The name of the species may be followed by an abbreviated form of the name of its discoverer; for example, the common daisy is *Bellis perennis* L. (for Linnaeus). There are several International Codes of Taxonomic Nomenclature that lay down the rules for naming organisms. *See also* CLASSIFICATION; TAXONOMY.

**binomial theorem** (binomial expansion) A rule for the expansion of a binomial expression (expression consisting of the sum of two variables raised to a given power). The general binomial expression $(x + y)^n$ expands to:

$$x^n + nx^{n-1}y + [n(n-1)/2!]x^{n-2}y^2 + \ldots y^n$$

**bioaccumulation** An increase in the concentration of chemicals, such as *pesticides, in organisms that live in environments contaminated by a wide variety of organic compounds. These compounds are not usually decomposed in the environment (i.e. they are not biodegradable) or metabolized by the organisms, so that their rate of absorption and storage is greater than their rate of excretion. The chemicals are normally stored in fatty tissues. *DDT is known as a **persistent pesticide**, as it is not easily broken down and bioaccumulates along *food chains, so that increasing concentrations occur in individual organisms at each trophic level – a process called **biomagnification**.

**bioassay** (biological assay) A controlled experiment for the quantitative estimation of a substance by measuring its effect in a living organism. For example, the amount of the plant hormone auxin can be estimated by observing its effect on the curvature of oat coleoptiles – the concentration of the hormone is proportional to the curvature of the coleoptile.

**biochemical evolution** (molecular evolution) The changes that occur at the molecular level in organisms over a period of time. Brought about by *mutations in genes or chromosomes, it results in functional changes to the proteins encoded by the genes, or even the evolution of novel genes and proteins.

**biochemical fuel cell** A system that exploits biological reactions for the conversion of biomass (chemical energy) to electricity (electrical energy). One potential application is the generation of electricity from industrial waste and *sewage. **Methyltrophic** organisms (i.e. organisms that use methane or methanol as their sole carbon sources) are being investigated for their potential use in biochemical fuel cells.

**biochemical oxygen demand** (BOD) The amount of oxygen taken up by microorganisms that decompose organic waste matter in water. It is therefore used as a measure of the amount of certain types of organic pollutant in water. BOD is calculated by keeping a sample of water containing a known amount of oxygen for five days at 20°C. The oxygen content is measured again after this time. A high BOD indicates the presence of a large number of microorganisms, which suggests a high level of pollution.

**biochemical taxonomy** *See* MOLECULAR SYSTEMATICS.

**biochemistry** The study of the chemistry of living organisms, especially the structure and function of their chemical components (principally proteins, carbohydrates, lipids, and nucleic acids). Biochemistry has advanced rapidly with the development, from the mid-20th century, of such techniques as chromatography, spectroscopy, X-ray diffraction, radioisotopic labelling, and electron microscopy. Using these techniques to separate and analyse biologically important molecules, the steps of the metabolic pathways in which they are involved (e.g. *glycolysis and the *Krebs cycle) have been determined. This has provided some knowledge of how organisms obtain and store energy, how they manufacture and degrade their biomolecules, how they sense and respond to their environment, and how all this information is carried and expressed by their genetic material. Biochemistry forms an important part of many other disciplines, especially physiology, nutrition, molecular biology, and genetics, and its discoveries have made a profound impact in medicine,

## BIOCHEMISTRY

| | |
|---|---|
| 1833 | French chemist Anselme Payen (1795–1871) discovers diastase (the first enzyme to be discovered). |
| 1836 | Theodor Schwann discovers the digestive enzyme pepsin. |
| c.1860 | Louis Pasteur demonstrates fermentation is caused by 'ferments' in yeasts and bacteria. |
| 1869 | German biochemist Johann Friedrich Miescher (1844–95) discovers nucleic acid. |
| 1877 | Pasteur's 'ferments' are designated as enzymes. |
| 1890 | German chemist Emil Fischer (1852–1919) proposes the 'lock-and-key' mechanism to explain enzyme action. |
| 1901 | Japanese chemist Jokichi Takamine (1854–1922) isolates adrenaline (the first hormone to be isolated). |
| 1903 | German biologist Eduard Buchner (1860–1917) discovers the enzyme zymase (causing fermentation). |
| 1904 | British biologist Arthur Harden (1865–1940) discovers coenzymes. |
| 1909 | Russian-born US biochemist Phoebus Levene (1869–1940) identifies ribose in RNA. |
| 1921 | Canadian physiologist Frederick Banting (1891–1941) and US physiologist Charles Best (1899–1978) isolate insulin. |
| 1922 | Alexander Fleming discovers the enzyme lysozyme. |
| 1925 | Russian-born British biologist David Keilin (1887–1963) discovers cytochrome. |
| 1926 | US biochemist James Sumner (1877–1955) crystallizes urease (the first enzyme to be isolated). |
| 1929 | German chemist Hans Fischer (1881–1945) determines the structure of haem (in haemoglobin).<br>K. Lohman isolates ATP from muscle. |
| 1930 | US biochemist John Northrop (1891–1987) isolates the stomach enzyme pepsin. |
| 1932 | Swedish biochemist Hugo Theorell (1903–82) isolates the muscle protein myoglobin. |
| 1937 | Hans Krebs discovers the Krebs cycle. |
| 1940 | German-born US biochemist Fritz Lipmann (1899–1986) proposes that ATP is the carrier of chemical energy in many cells. |
| 1943 | US biochemist Britton Chance (1913–  ) discovers how enzymes work (by forming an enzyme–substrate complex). |
| 1952 | US biologist Alfred Hershey (1908–97) proves that DNA carries genetic information. |
| 1953 | Francis Crick and James Watson discover the structure of DNA. |
| 1955 | Frederick Sanger discovers the amino acid sequence of insulin. |

b

1956    US biochemist Arthur Kornberg (1918–2007) discovers the enzyme DNA polymerase.

US molecular biologist Paul Berg (1926–   ) identifies the nucleic acid later known as transfer RNA.

1957    British biologist Alick Isaacs (1921–67) discovers interferon.

1959    Austrian-born British biochemist Max Perutz (1914–2002) determines the structure of haemoglobin.

1960    South African-born British molecular biologist Sydney Brenner (1927–   ) and French biochemist François Jacob (1920–   ) discover messenger RNA.

1961    British biochemist Peter Mitchell (1920–92) proposes the chemiosmotic theory.

Brenner and Crick discover that the genetic code consists of a series of base triplets.

1969    US biochemist Gerald Edelman (1929–   ) discovers the amino acid sequence of immunoglobulin G.

1970    US virologists Howard Temin (1934–94) and David Baltimore (1938–   ) discover the enzyme reverse transcriptase.

US molecular biologist Hamilton Smith (1931–   ) discovers restriction enzymes.

1973    US biochemists Stanley Cohen (1935–   ) and Herbert Boyer (1936–   ) use restriction enzymes to produce recombinant DNA.

1977    Sanger determines the complete base sequence of DNA in bacteriophage φX174.

1984    British biochemist Alec Jeffreys (1950–   ) devises DNA profiling.

1985    US biochemist Kary Mullis (1944–   ) invents the polymerase chain reaction for amplifying DNA.

1986    US pharmacologists Robert Furchgott (1916–2009) and Louis Ignarro (1941–   ) demonstrate the importance of nitric oxide as a signal molecule in the blood vascular system.

1988    US biochemist Peter Agre (1949–   ) identifies a water-channel protein (aquaporin) in the plasma membrane of cells.

1994    Beginnings of DNA chip technology.

1998    US biochemist Roderick MacKinnon (1956–   ) reveals detailed three-dimensional structure of potassium-ion channel in brain cells.

2001    US molecular biologist Harry Noller and colleagues produce first detailed X-ray crystallographic image of a complete ribosome.

2002    First synthetic virus created by Eckard Wimmer and associates, based on human poliovirus.

2004    A team led by David L. Spector produces the first real-time imaging of gene transcription in a living cell, using different fluorescent markers to tag nucleic acids and proteins.

agriculture, industry, and many other areas of human activity. See Chronology.

SEE WEB LINKS

- A virtual library of biochemistry, cell biology, and molecular biology

**biodegradable** *See* POLLUTION.

**biodiversity (biological diversity)** The existence of a wide variety of species (**species diversity**) or other taxa of plants, animals, and microorganisms in a natural community or habitat, or of communities within a particular environment (**ecological diversity**), or of genetic variation within a species (**genetic diversity**; *see* VARIATION). The maintenance of a high level of biodiversity is important for the stability of ecosystems. Certain habitats, especially rainforests, have a rich species diversity, which is threatened by the continued destruction of habitats (*see* DEFORESTATION; DESERTIFICATION; GREENHOUSE EFFECT). Such ecosystems typically support large numbers of rare species, and population sizes of individual species tend to be small; they are therefore especially vulnerable to habitat destruction. Biodiversity in natural habitats also represents an important pool of species and genetic material of potential use to human societies. For example, wild plants continue to be used as a source of new drugs and other products, and the development of new strains and varieties of crop plants with increased disease resistance usually depends on incorporating genetic material from wild plants.

**bioelement** Any chemical element that is found in the molecules and compounds that make up a living organism. In the human body the most common bioelements (in decreasing order of occurrence) are oxygen, carbon, hydrogen, nitrogen, calcium, and phosphorus. Other bioelements include sodium, potassium, magnesium, and copper. *See* ESSENTIAL ELEMENT.

**bioenergetics** The study of the flow and the transformations of energy that occur in living organisms. Typically, the amount of energy that an organism takes in (from food or sunlight) is measured and divided into the amount used for growth of new tissues; that lost through death, wastes, and (in plants) transpiration; and that lost to the environment as heat (through respiration).

**bioengineering** 1. The use of artificial tissues, organs, and organ components to replace parts of the body that are damaged, lost, or malfunctioning, e.g. artificial limbs, heart valves, and heart pacemakers. *See also* TISSUE ENGINEERING. 2. The application of engineering knowledge to medicine and zoology.

**biofeedback** The technique whereby a subject can learn to control certain body functions, such as heart rate or blood pressure, that are usually unconsciously regulated by the autonomic nervous system. It is facilitated by the use of monitoring devices, such as pulse monitors, electroencephalographs, and electromyographs, and can be useful in treating high blood pressure, migraine, epilepsy, and other disorders.

**biofilm** A colony of bacteria and other microorganisms that adheres to a substrate and is enclosed and protected by secreted slime. Biofilms readily form on virtually any surface, whether nonliving or living, where there is moisture and a supply of nutrients. They are important components of aquatic and terrestrial ecosystems, typically providing nutrients for small organisms at the base of food chains. Moreover, they form in the microenvironment (*rhizosphere) surrounding plant roots, where they assist the plant in absorbing nutrients from the soil. Biofilms can also occur in the body and in industrial installations, for example, on the surface of teeth (as dental plaque) or inside pipelines.

**biofuel** A gaseous, liquid, or solid fuel that contains an energy content derived from a biological source. The organic matter that makes up living organisms provides a potential source of trapped energy that is increasingly being exploited to supply worldwide energy demand. Biofuels are claimed to have a lower carbon footprint than fossil fuels and thus contribute less to the greenhouse effect. An example of a biofuel is rapeseed oil, which can be used in place of diesel fuel in modified engines. The methyl ester of this oil, **rapeseed methyl ester (RME)**, can be used in unmodified diesel engines and is sometimes known as **biodiesel**. Other biofuels include *biogas and *gasohol.

**biogas** A mixture of methane and carbon dioxide resulting from the anaerobic decomposition of such waste materials as domestic, industrial, and agricultural sewage. The decomposition is carried out by methanogenic bacteria (*see* METHANOGEN); these obligate anaerobes produce methane, the main component of biogas, which can be collected and used as an energy source for domestic processes, such as heating, cooking, and lighting. The production of biogas is carried out in special **digesters**, which are widely used in China and India. As well as providing a source of fuel, these systems also enable *sewage, which contains pathogenic bacteria, to be digested, thereby removing the danger to humans that could otherwise result from untreated domestic and agricultural waste.

**biogenesis** The principle that a living organism can only arise from other living organisms similar to itself (i.e. that like gives rise to like) and can never originate from nonliving material. *Compare* SPONTANEOUS GENERATION.

**biogeochemical cycle (nutrient cycle)** The cyclical movement of elements between living organisms (the biotic phase) and their nonliving (abiotic) surroundings (e.g. rocks, water, air). Examples of biogeochemical cycles are the *carbon cycle, *nitrogen cycle, *oxygen cycle, *phosphorus cycle, and *sulphur cycle.

**biogeography** The branch of biology that deals with the geographical distribution of plants and animals. *See* PLANT GEOGRAPHY; ZOOGEOGRAPHY.

**bioinformatics** The collection, storage, and analysis of DNA- and protein-sequence data using computerized systems. The data generated by genome projects and protein studies are held in various databanks and made available to researchers throughout the world via the Internet. Many computer programs have been developed to analyse sequence data, enabling the user to identify similarities between newly sequenced material and existing sequences. This allows, for example, predictions about the structure and function of a protein from its amino-acid sequence data or from the nucleotide sequence of its gene. Also, genome-wide sequence analysis allows comparisons to be made between genomes of different species, which provides information about their possible evolutionary relationships. *See also* GENOMICS.

**bioinorganic chemistry** Biochemistry involving compounds that contain metal atoms or ions. Two common examples of bioinorganic compounds are haemoglobin (which contains iron) and chlorophyll (which contains magnesium). Many enzymes contain metal atoms and bioinorganics are important in a number of biochemical processes, including oxygen transport, electron transfer, and protein folding.

**biological clock** The mechanism, presumed to exist within many animals and plants, that produces regular periodic changes in behaviour or physiology. Biological clocks underlie many of the *biorhythms seen in organisms (e.g. the sleep–wake cycle, hibernation in animals). They continue to run even when conditions are kept artificially constant, but eventually drift out of step with the natural environment without the specific signals that normally keep them synchronized. Studies in the fruit fly *Drosophila* have revealed the molecular basis of the biological clock, and similar mechanisms are thought to occur in other animals, including mammals, and in plants, fungi, and cyanobacteria.

**biological control** The control of *pests by biological (rather than chemical) means. This may be achieved, for example, by breeding disease-resistant crops or by introducing a natural enemy of the pest, such as a predator or a parasite. This technique, which may offer substantial advantages over the use of pesticides or herbicides, has been employed successfully on a number of occasions. Examples include the control of the prickly pear cactus (*Opuntia*) in Australia by introducing the cactus moth (*Cactoblastis cactorum*), whose caterpillars feed on the plant's growing shoots, and the use of the ladybird to prey upon the scale insect (*Icerya*), which kills citrus fruit trees. Insect pests have also been subjected to genetic control, by releasing large numbers of males of the pest species that have been sterilized by radiation: infertile matings subsequently cause a decline in the pest population. This method has been used to control the screw worm fly (*Cochliomyia hominivora*), which lays its eggs in the open wounds of domestic cattle. Biological control is considered to reduce a number of the problems associated with chemical control using *pesticides, but care should be taken to avoid upsetting the natural ecological balance; for example, a particular predator may also destroy harmless or beneficial species.

**biological rhythm** *See* BIORHYTHM.

**biological warfare** The military use of microorganisms, such as bacteria, viruses, fungi, and other microorganisms, including the agents for anthrax and botulism, to induce disease or death among humans, livestock, and crop plants. Though officially banned in most countries, research continues with the aim of developing virulent strains of existing microorganisms, using genetic engineering and other techniques.

**biology** The study of living organisms, which includes their structure (gross and microscopical), functioning, origin and evolution, classification, interrelationships, and distribution.

 **SEE WEB LINKS**

• This University of Arizona website contains useful links to various fields of biology, including biochemistry, cell biology, and human biology

**bioluminescence** The emission of light without heat (*see* LUMINESCENCE) by living organisms. The phenomenon occurs in glow-worms and fireflies, bacteria and fungi, and in many deep-sea fish (among others); in animals it may serve as a means of protection (e.g. by disguising the shape of a fish) or species recognition or it may provide mating signals. The light is produced during the oxidation of a compound called **luciferin** (the composition of which varies according to the species), the reaction being catalysed by an enzyme, **luciferase**. Fience may be continuous (e.g. in bacteria) or intermittent (e.g. in fireflies).

**biomass** The total mass of all the organisms of a given type and/or in a given area; for example, the world biomass of trees, or the biomass of elephants in the Serengeti National Park. It is normally measured in terms of grams of *dry mass per square metre. *See also* PYRAMID OF BIOMASS.

**biome** A major ecological community or complex of communities that extends over a large geographical area characterized by a dominant type of vegetation. The organisms of a biome are adapted to the climate conditions associated with the region. There are no distinct boundaries between adjacent biomes, which merge gradually with each other. Examples of biomes are *tundra, tropical *rainforest, *taiga, *grassland (temperate and tropical), and *desert.

**biomechanics** The application of the principles of *mechanics to living systems, particularly those living systems that have coordinated movements. Biomechanics also deals with the properties of biological materials, such as blood and bone. For example, biomechanics would be used to analyse the stresses on bones in animals, both when the animals are static and when they are moving. Other types of problems in biomechanics include the *fluid mechanics associated with swimming in fish and the *aerodynamics of birds flying. It is sometimes difficult to perform realistic calculations in biomechanics because of complexity in the shape of animals or the large number of *degrees of freedom that need to be considered (for example, the large number of muscles involved in the movement of a human leg).

**biomolecule** Any molecule that is involved in the maintenance and metabolic processes of living organisms (*see* METABOLISM). Biomolecules include carbohydrate, lipid, protein, nucleic acid, and water molecules; some biomolecules are *macromolecules.

**biophysics** The study of the physical aspects of biology, including the application of physical laws and the techniques of physics to study biological phenomena. *See also* PHYSICS.

 **SEE WEB LINKS**

• The website of the Biophysical Society

**biopoiesis** The development of living matter from complex organic molecules that are themselves nonliving but self-replicating. It is the process by which life is assumed to have begun. *See* ORIGIN OF LIFE.

**biopolymer** A polymer that occurs naturally, such as a *polysaccharide, *protein, or *nucleic acid.

**bioreactor (industrial fermenter)** A large stainless steel tank used to grow producer microorganisms in the industrial production of enzymes and other chemicals. After the tank is steam-sterilized, an inoculum of the producer cells is introduced into a medium that is maintained by probes at optimum conditions of temperature, pressure, pH, and oxygen levels for enzyme production. An **agitator** (stirrer) mixes the medium, which is constantly aerated. It is essential that the culture medium is sterile and contains the appropriate nutritional requirements for the microorganism. When the nutrients have been utilized the product is separated; if the product is an extracellular compound the medium can be removed during the growth phase of the microorganisms, but an intracellular product must be harvested when the batch culture growth stops. Some bioreactors are designed for *continuous culture.

**biorhythm (biological rhythm)** A roughly periodic change in the behaviour or physiology of an organism that is generated and maintained by a *biological clock. Well-known examples are the *annual and *circadian rhythms occurring in many animals and plants. **Infradian rhythms**, occurring in many cellular processes, have a periodicity of less than 24 hours. An example of an **ultradian rhythm**, with a periodicity greater than a day, is the reproductive cycle of many animals that corresponds with the 29.5-day lunar cycle.

**biosensor** A device that uses an immobilized agent to detect or measure a chemical compound. The agents include enzymes, antibiotics, organelles, or whole cells. A reaction between the immobilized agent and the molecule being analysed is transduced into an electronic signal. This signal may be produced in response to the presence of a reaction product, the movement of electrons, or the appearance of some other factor (e.g. light). Biosensors are used in diagnostic tests: these allow quick, sensitive, and specific analysis of a wide range of biological products, including antibiotics, vitamins, and other important biomolecules (such as glucose), as well as the determination of certain *xenobiotics, such as synthetic organic compounds.

**biosphere** The whole of the region of the earth's surface, the sea, and the air that is inhabited by living organisms.

**biostratigraphy** The characterization of rock strata on the basis of the fossils they contain. *See also* STRATIGRAPHY.

**biosynthesis** The production of molecules by a living cell, which is the essential feature of *anabolism.

**biosystematics** *See* SYSTEMATICS.

**biotechnology** The development of techniques for the application of biological processes to the production of materials of use in medicine and industry. For example, the production of antibiotics, cheese, and wine rely on the activity of various fungi and bacteria. *Genetic engineering can modify bacterial cells to synthesize completely new substances, e.g. hormones, vaccines, *monoclonal antibodies, etc., or introduce novel traits into plants or animals. *See also* GENETICALLY MODIFIED ORGANISMS (Feature).

**biotic factor** Any of the factors of an organism's environment that consist of other living organisms and together make up the **biotic environment**. These factors may affect an organism in many ways; for example, as competitors, predators, parasites, prey, or symbionts. In time, the distribution and abundance of the organism will be affected by its interrelationships with the biotic environment. *Compare* ABIOTIC FACTOR.

**biotin** A vitamin in the *vitamin B complex. It is the *coenzyme for various enzymes that catalyse the incorporation of carbon dioxide into various compounds. Adequate amounts are normally produced by the intestinal bacteria in animals although deficiency can be induced by consuming large amounts of raw egg white. This contains a protein, avidin, that specifically binds biotin, preventing its absorption from the gut. Other sources of biotin include cereals, vegetables, milk, and liver.

**biotite** An important rock-forming silicate mineral, a member of the *mica group of minerals, in common with which it has a sheetlike crystal structure. It is usually black, dark brown, or green in colour.

**bipolar outflow** A type of stellar wind (*see* SOLAR WIND) that leaves a star mainly from its poles. Such outflows are commonest in protostars, pre-main-sequence stars and red giants, and account for a great loss of stellar mass. They may take the form of radio-emitting molecules or long visible jets of gaseous material. It is thought that a disc of dense gas around the star's equatorial regions prevents the outflow of material elsewhere.

**biprism** A glass prism with an obtuse angle that functions as two acute-angle prisms placed base-to-base. A double image of a single object is thus formed; the device was used by Fresnel to produce two coherent beams for interference experiments.

**bipyramid** *See* COMPLEX.

**birds** *See* AVES.

**birefringence** *See* DOUBLE REFRACTION.

**Birkeland–Eyde process** A process for the fixation of nitrogen by passing air through an electric arc to produce nitrogen oxides. It was introduced in 1903 by the Norwegian chemists Kristian Birkeland (1867–1913) and Samuel Eyde (1866–1940). The process is economic only if cheap hydroelectricity is available.

**birth** *See* PARTURITION.

**birth control (contraception)** The intentional avoidance of pregnancy by methods that do not normally hinder sexual activity. The methods used can be 'natural' or 'artificial'. Natural methods, often used because of religious or moral objections to artificial methods, include the **rhythm method**, in which sexual intercourse is avoided during times when ovulation occurs; and **coitus interruptus**, an unreliable method in which the penis is withdrawn from the vagina before ejaculation. The rhythm method requires a monitoring of the woman's menstrual cycle and may be unsuitable in those women with irregular cycles. Artificial methods use devices or other agents (**contraceptives**) to prevent pregnancy. They include the **condom**, a rubber sheath placed over the penis to trap the sperm; and the **diaphragm**, a rubber cap placed over the cervix. Contraceptives that prevent *implantation include the **intrauterine device** (**IUD**), a metal or plastic coil placed in the uterus by a doctor (which may cause unacceptable bleeding in some women); and the 'morning-after pill', taken within three days after sexual intercourse. Other *oral contraceptives prevent ovulation. *Sterilization is usually considered to be irreversible but attempts at reversing the process are possible. For casual relationships, or relationships involving a partner whose sexual history is not known, health workers advise the use of condoms with all other forms of contraception as this provides the safest means of reducing the risk of infection by *sexually transmitted diseases.

**birth rate (natality)** The rate at which a particular species or population produces offspring. The birth rate of a species is used to measure its fecundity (reproductive capability). It is also an important factor in controlling the size of a population. *Compare* DEATH RATE.

**bisexual** (in biology) *See* HERMAPHRODITE.

**bismuth** Symbol Bi. A white crystalline metal with a pinkish tinge belonging to *group 15 (formerly VB) of the periodic table; a.n. 83; r.a.m. 208.98; r.d. 9.78; m.p. 271.3°C; b.p. 1560°C. The most important ores are bismuthinite ($Bi_2S_3$) and bismite ($Bi_2O_3$). Peru, Japan, Mexico, Bolivia, and Canada are major producers. The metal is extracted by carbon reduction of its oxide. Bismuth is the most diamagnetic of all metals and its ther-

mal conductivity is lower than any metal except mercury. The metal has a high electrical resistance and a high Hall effect when placed in magnetic fields. It is used to make low-melting-point casting alloys with tin and cadmium. These alloys expand on solidification to give clear replication of intricate features. It is also used to make thermally activated safety devices for fire-detection and sprinkler systems. More recent applications include its use as a catalyst for making acrylic fibres, as a constituent of malleable iron, as a carrier of uranium–235 fuel in nuclear reactors, and as a specialized thermocouple material. Bismuth compounds (when lead-free) are used for cosmetics and medical preparations. It is attacked by oxidizing acids, steam (at high temperatures), and by moist halogens. It burns in air with a blue flame to produce yellow oxide fumes. C. G. Junine first demonstrated that it was different from lead in 1753.

**((()) SEE WEB LINKS**
- Information from the WebElements site

**bisphosphonates (diphosphonates)** A class of medical drugs used in the treatment of osteoporosis and other conditions that involve fragile bones. The bisphosphonates attack osteoclasts (i.e. the bone cells that break down bone tissue). The general formula is $O_3P–C(R^1R^3)–PO_3$. The side chain $R^1$ is a simple group (–H, –OH, –Cl). $R^2$ is usually a longer chain (e.g. $–S–C_6H_4–Cl$ or $–(CH_2)_5–NH_2$).

**bistability** *See* OSCILLATING REACTION.

**bistable circuit** *See* FLIP-FLOP.

**bisulphate** *See* HYDROGENSULPHATE.

**bisulphite** *See* HYDROGENSULPHITE; ALDEHYDES.

**bit (binary digit)** Either of the digits 0 or 1 as used in the *binary notation. Bits are therefore the basic unit of information in a computer system.

**bite angle** *See* CHELATE.

**Bitnet** A computer network originally linking IBM mainframe systems located in North America and with backing from IBM. The network has been substantially extended to other parts of the world, usually on a region-by-region basis, and has been implemented on other computer systems. Complete messages of any length are transmitted from one computer system to the next, until the destination is reached.

**bittern** The solution of salts remaining when sodium chloride is crystallized from sea water.

**Bitter pattern** A microscopic pattern that forms on the surface of a ferromagnetic material that has been coated with a colloidal suspension of small iron particles. The patterns outline the boundaries of the magnetic domains (see MAGNETISM). They were first observed by F. Bitter in 1931.

**bitumen** See PETROLEUM.

**bituminous coal** See COAL.

**bituminous sand** See OIL SAND.

**biuret test** A biochemical test to detect proteins in solution, named after the substance **biuret** ($H_2NCONHCONH_2$), which is formed when urea is heated. Sodium hydroxide is mixed with the test solution and drops of 1% copper(II) sulphate solution are then added slowly. A positive result is indicated by a violet ring, caused by the reaction of *peptide bonds in the proteins or peptides. Such a result will not occur in the presence of free amino acids.

**bivalent** **1.** (in chemistry) **(divalent)** Having a valency of two. **2.** (in genetics) See PAIRING.

**Bivalvia (Pelecypoda; Lamellibranchia)** A class of aquatic molluscs (the bivalves) that include the oysters, mussels, and clams. They are characterized by a laterally flattened body and a shell consisting of two hinged valves (i.e. a bivalved shell). The enlarged gills are covered with cilia and have the additional function of filtering microscopic food particles from the water flowing over them. Bivalves live on the sea bed or lake bottom and are sedentary, so the head and foot are reduced.

**Black, Joseph** (1728–99) British chemist and physician, born in France. He studied at Glasgow and Edinburgh, where his thesis (1754) contained the first accurate description of the chemistry of carbon dioxide. In 1757 he discovered latent heat, and was the first to distinguish between heat and temperature.

**black body** A hypothetical body that absorbs all the radiation falling on it. It thus has an *absorptance and an *emissivity of 1. While a true black body is an imaginary concept, a small hole in the wall of an enclosure at uniform temperature is the nearest approach that can be made to it in practice.

**Black-body radiation** is the electromagnetic radiation emitted by a black body. It extends over the whole range of wavelengths and the distribution of energy over this range has a characteristic form with a maximum at a certain wavelength. The position of the maximum depends on temperature, moving to shorter wavelengths with increasing temperature. See STEFAN'S LAW; WIEN'S DISPLACEMENT LAW.

**blackdamp (choke damp)** Air left depleted in oxygen following the explosion of firedamp in a mine.

**black dwarf** A cold celestial object thought to be the remains of a dead star of low mass, that is formed after a *white dwarf star has radiated away all of its heat energy. Black dwarfs are extremely difficult to detect, and because white dwarfs take so long to cool down, it is possible that the universe may not yet be old enough for any black dwarfs to have formed.

**black earth** See CHERNOZEM.

**black hole** An object in space that has collapsed under its own gravitational forces to such an extent that its *escape velocity is equal to the speed of light. Black holes are believed to be formed in the gravitational collapse of massive stars at the ends of their lives (see STELLAR EVOLUTION; SUPERNOVA). If the mass of an evolved stellar core is greater than the analogue of the Chandrasekhar limit for neutron stars then neutron degeneracy pressure is unable to prevent contraction until the gravitational field is sufficiently strong to prevent the escape of electromagnetic radiation. The boundary of the black hole, which is known as the **event horizon**, is the surface in space at which the gravitational field reaches this critical value. Events occurring within this horizon (i.e. in the interior of the black hole) cannot be observed from outside.

The theoretical study of black holes involves the use of general *relativity. It has been shown that a black hole can be characterized uniquely by just three properties: its mass, angular momentum, and electrical charge (this is known as the no-hair theorem). Mathematical expressions have been derived for describing black holes; these are the **Schwarzschild solution** (uncharged nonrotating hole), the **Reissner–Nordstrøm**

**b**

solution (charged nonrotating hole), the **Kerr solution** (uncharged rotating hole), and the **Kerr–Newman solution** (charged rotating hole).

The ultimate fate of matter inside the black hole's event horizon is as yet unknown. General relativity predicts that at the centre of the hole there is a **singularity**, a point at which the density becomes infinite and the presently understood laws of physics break down. It is possible that a successful quantum theory of gravity could resolve this problem. However, since any singularity is hidden within the event horizon, it cannot influence the outside universe, so the normal laws of physics, including general relativity, can be used to describe processes outside the black hole.

Observational evidence of objects thought to be black holes comes from their effect on surrounding matter. Thus, if a black hole is part of a binary system with another star it will attract and capture matter from this star. The material leaving the star first forms a rotating **accretion disc** around the black hole, in which the matter becomes compressed and heated to such an extent that it emits X-rays. In the constellation Cygnus there is an X-ray source, Cygnus X-1, which consists of a supergiant star revolving around a small invisible companion with a mass of about ten times that of the sun, and therefore well above the Chandrasekhar limit. The companion is thought to be a black hole. Black holes have also been postulated as the power sources of *quasars and as possible generators of *gravitational waves. It appears that there may be very large black holes at the centres of all galaxies. It has been suggested that black-hole formation could be the cause of gamma-ray bursts, either by one dead star collapsing or by two neutron stars spiralling into each other.

Theoreticians have also postulated the existence of 'mini' black holes (with masses of about $10^{12}$ kilograms and radii about $10^{-15}$ metre). Such entities might have been formed shortly after the big bang when the universe was created. Quantum-mechanical effects are important for mini black holes, which emit Hawking radiation (*see* HAWKING PROCESS). *See also* SCHWARZSCHILD RADIUS.

**⊕ SEE WEB LINKS**

• A NASA website with information about black holes

**black lead** *See* CARBON.

**bladder** **1.** (in anatomy) **a.** A hollow muscular organ in most vertebrates, also known as the **urinary bladder**, in which urine is stored before being discharged. In mammals urine is conveyed from the *kidneys to the bladder by the *ureters and is discharged to the outside through the *urethra. **b.** Any of various other saclike organs in animals for the storage of liquid or gas. *See* GALL BLADDER; SWIM BLADDER. **2.** (in botany) **a.** A modified submerged leaf of certain aquatic insectivorous plants, such as the bladderwort (*Utricularia*). It forms a hollow with a single opening that is sealed by a valve to trap small aquatic invertebrates after they have been sucked in. **b.** An air-filled cavity in the thallus of certain seaweeds, such as the bladderwrack (*Fucus vesiculosus*).

**blanc fixe** *See* BARIUM SULPHATE.

**Blandford–Znajek process** An astrophysical process in which an external magnetic field is able to 'tap' the rotational energy of a rotating black hole, thereby making the black hole a powerful source of energy. There is some evidence that the process occurs around certain types of black hole. It was proposed by Roger Blandford and Roman Znajek in 1977. *See also* PENROSE PROCESS.

**blast furnace** A furnace for smelting iron ores, such as haematite ($Fe_2O_3$) or magnetite ($Fe_3O_4$), to make *pig iron. The furnace is a tall refractory-lined cylindrical structure that is charged at the top with the dressed ore (*see* BENEFICIATION), coke, and a flux, usually limestone. The conversion of the iron oxides to metallic iron is a reduction process in which carbon monoxide and hydrogen are the reducing agents. The overall reaction can be summarized thus:

$$Fe_3O_4 + 2CO + 2H_2 \rightarrow 3Fe + 2CO_2 + 2H_2O$$

The CO is obtained within the furnace by blasting the coke with hot air from a ring of tuyeres about two-thirds of the way down the furnace. The reaction producing the CO is:

$$2C + O_2 \rightarrow 2CO$$

In most blast furnaces hydrocarbons (oil, gas, tar, etc.) are added to the blast to provide a source of hydrogen. In the modern **direct-reduction process** the CO and $H_2$ may be produced separately so that the reduction process can proceed at a lower temperature. The pig iron produced by a blast furnace contains about 4% carbon and further refining is usually required to produce steel or cast iron.

**blasting gelatin** A high explosive made from nitroglycerine and gun cotton (cellulose nitrate).

**blastocoel** *See* BLASTULA.

**blastocyst** *See* BLASTULA; IMPLANTATION.

**blastula** The stage of *development of an animal embryo that results from *cleavage of a fertilized egg. This stage generally resembles a hollow ball with the dividing cells (**blastomeres**) of the embryo forming a layer (**blastoderm**) around a central cavity (**blastocoel**). Insect eggs have no blastula. In vertebrates the blastula forms a disc (**blastodisc**) on the surface of the yolk. In mammals the blastula stage is known as a **blastocyst**. *See also* GASTRULA.

**blazar** A very active type of galaxy, named from a combination of *BL* Lacertae object and qu*asar*. Its light output varies greatly, perhaps caused by jets of gas expelled from the nucleus of the galaxy at speeds approaching the speed of light. This effect is known as relativistic beaming and if it is directed towards the earth, it is seen as violent fluctuations in radiation.

**bleaching powder** A white solid regarded as a mixture of calcium chlorate(I), calcium chloride, and calcium hydroxide. It is prepared on a large scale by passing chlorine gas through a solution of calcium hydroxide. Bleaching powder is sold on the basis of available chlorine, which is liberated when it is treated with a dilute acid. It is used for bleaching paper pulps and fabrics and for sterilizing water.

**blende** A naturally occurring metal sulphide, e.g. zinc blende ZnS.

**blending inheritance** The early theory that assumed that hereditary substances from parents merge together in their offspring. Mendel showed that this does not occur (*see* MENDEL'S LAWS). In breeding experiments an appearance of blending may result from codominant alleles (*see* CODOMINANCE) and *quantitative inheritance but close study shows that the alleles retain their identity through successive generations. *Compare* PARTICULATE INHERITANCE.

**blind spot** The portion of the retina at which blood vessels and nerve fibres enter the optic nerve. There are no rods or cones

in this area, so no visual image can be transmitted from it.

**block** *See* PERIODIC TABLE.

**block copolymer** *See* POLYMER.

**blog (Web log)** A publicly accessible journal maintained on the Web by an individual or group. The topics covered in a blog are dictated by the author and so vary widely: some reflect personal interests and concerns, whereas others comment on aspects of current affairs or discuss work of public interest that their author is engaged in. Some have become respected sources of information or opinion, while others are vehicles for corporations or other bodies to disseminate information and obtain feedback. Many blogs include facilities for readers to post comments and to engage in debate; in this function they have tended to replace earlier forms of online discussion, such as Usenet and forums run by online service providers. It is very easy to create and maintain a blog, using freely available software and services, that can be read using a standard Web Browser; and blogs can contain links to other blogs, forming an extended online community. The term 'blog' was only coined in 1997, yet by the mid-2000s blogging had become a mainstream activity not confined only to Internet enthusiasts.

**blood** A fluid body tissue that acts as a transport medium within an animal. It is contained within a blood *vascular system and in vertebrates is circulated by means of contractions of the *heart. Oxygen and food are carried to tissues, and carbon dioxide and chemical (nitrogenous) waste are transported from tissues to excretory organs for disposal (*excretion). In addition blood carries *hormones and also acts as a defence system. Blood consists of a liquid (*see* BLOOD PLASMA) containing blood cells (*see* ERYTHROCYTE; LEUCOCYTE) and *platelets (see illustration overleaf).

**blood–brain barrier** The mechanism that controls the passage of substances from the blood to the cerebrospinal fluid bathing the brain and spinal cord. It takes the form of a semipermeable lipid membrane permitting the passage of solutions but excluding particles and large molecules. This barrier provides the central nervous system with a constant environment, while not interfering with the transport of essential metabolites.

**blood capillary** *See* CAPILLARY.

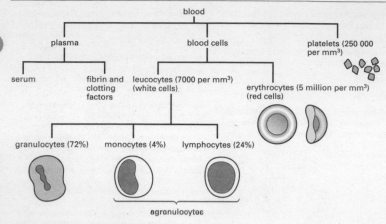

**Blood.** Composition of mammalian blood.

**blood cell (blood corpuscle)** Any of the cells that are normally found in the blood plasma. These include red cells (*see* ERYTHROCYTE) and white cells (*see* LEUCOCYTE).

**blood clotting (blood coagulation)** The production of a mass of semisolid material at the site of an injury that closes the wound, helping to prevent further blood loss and bacterial invasion. The clot is formed by the action of *clotting factors and *platelets. Damage to tissue triggers a series of reactions involving **thromboplastin** (a glycoprotein), calcium ions, phospholipids (from platelets), and clotting factors, which results in the conversion of *prothrombin in the blood to its enzymically active form **thrombin**. Thrombin catalyses the formation of the insoluble protein **fibrin** from soluble **fibrinogen**; the fibrin forms a fibrous network in which blood cells become enmeshed, producing a clot.

**blood groups** The many types into which an individual's blood may be classified, based on the presence or absence of certain antigenic proteins (*agglutinogens) on the surface of the red blood cells. Blood of one group contains *antibodies in the serum that react against the agglutinogens on the cells of other groups. **Incompatibility** between groups results in clumping of cells (*agglutination), so knowledge of blood groups is important for blood transfusions. In man, the two most important blood group systems are the *ABO system and the system involving the *rhesus factor.

**blood plasma** The liquid part of the *blood (i.e. excluding blood cells). It consists of water containing a large number of dissolved substances, including proteins, salts (especially sodium and potassium chlorides and bicarbonates), food materials (glucose, amino acids, fats), hormones, vitamins, and excretory materials. *See also* BLOOD SERUM; LYMPH.

**blood platelet** *See* PLATELET.

**blood pressure** The pressure exerted by the flow of blood through the major arteries of the body. This pressure is greatest during the contraction of the ventricles of the heart (**systolic pressure**; *see* SYSTOLE), which forces blood into the arterial system. Pressure falls to its lowest level when the heart is filling with blood (**diastolic pressure**; *see* DIASTOLE). Blood pressure is measured in millimetres of mercury using an instrument called a **sphygmomanometer**. Normal blood pressure for a young average adult human is in the region of 120/80 mmHg (the higher number is the systolic blood pressure; the lower number the diastolic blood pressure), but individual variations are common. Abnormally high blood pressure (**hypertension**) may be associated with disease or it may occur without an apparent cause.

**blood serum** Blood plasma from which the fibrin and clotting factors have been removed by centrifugation or vigorous stirring, so that it cannot clot. Serum containing a specific antibody or antitoxin may be used in the treatment or prevention of certain infections. Such serum is generally derived from a nonhuman mammal (e.g. a horse).

**blood vascular system** The tissues and organs of an animal that transport blood through the body. In vertebrates it consists of the heart and blood vessels. *See* VASCULAR SYSTEM.

**blood vessel** A tubular structure through which the blood of an animal flows. *See* ARTERY; ARTERIOLE; CAPILLARY; VENULE; VEIN.

**blooming** The process of depositing a transparent film of a substance, such as magnesium fluoride, on a lens to reduce (or eliminate) the reflection of light at the surface. The film is about one quarter of a wavelength thick and has a lower *refractive index than the lens. The anti-reflection effect is achieved by destructive interference.

**blue-green bacteria** *See* CYANOBACTERIA.

**blueshift** A general displacement of spectral lines toward the blue (shorter-wavelength) end of the spectrum. It is a manifestation of the *Doppler effect and is observed in the spectra of celestial objects that are approaching the earth. *Compare* REDSHIFT.

**Bluetooth** A wireless technology designed to replace cables between cell phones, laptops, and other devices. Bluetooth wireless technology works within a 1-, 10-, or 100-meter range and uses the 2.4 GHz band, which is unlicensed and can be used by many other types of devices, such as cordless phones and baby monitors.

**blue vitriol** *See* COPPER(II) SULPHATE.

**Blu-ray** An optical disk format. Developed by a consortium of over 180 companies led by Sony, Blu-ray disks are intended to supersede the *DVD both for the distribution of video and for data storage – they can hold up to 25 Gb on each of up to two layers. Blu-ray achieves this greater capacity by using a 'blue' (actually, a violet) laser rather than the red laser used for DVDs; the shorter wavelength allows the beam to be focused more precisely and so more data to be packed into a given space. Currently (2009) some high-definition DVDs have been released in Blu-ray format, and computer Blu-ray drives are becoming available.

**B lymphocyte** *See* B CELL.

**B-meson** Symbol $B^0$. A meson that consists of a down quark and an anti-bottom quark. It is electrically neutral, has spin zero, and a mass of 5.279 GeV. The antiparticle consists of a bottom quark and an anti-down quark. It is hoped that study of the decays of B-mesons will shed light on the problem of CP violation (*see* CP INVARIANCE).

**boat conformation** *See* CONFORMATION.

**BOD** *See* BIOCHEMICAL OXYGEN DEMAND.

**Bode, Johann Elert** (1747–1826) German astronomer, who became director of the Berlin Observatory. In 1766 his compatriot Johann Titius had discovered an apparently coincidental mathematical relationship involving the distances of the planets from the sun. If 4 is added to each number in the series 0, 3, 6, 12, 24,... and the answers divided by 10, the resulting sequence gives the distances of the planets in astronomical units (earth = 1). Known now as **Bode's law**, or the Titius-Bode law, the formula breaks down after Saturn.

**body cavity** The internal cavity of the body of an animal, which is present in most invertebrates and all vertebrates and contains the major organs. The body cavity of vertebrates and many invertebrates is the *coelom. In vertebrates the body cavity is divided by a transverse septum just posterior to the heart into the abdominal and thoracic cavities (*see* ABDOMEN; THORAX). In mammals the septum is the *diaphragm.

**body-centred cubic (b.c.c.)** *See* CUBIC CRYSTAL.

**body fluid** Any of the fluids found within animals, including blood, lymph, tissue fluid, urine, bile, sweat, and synovial fluid. Body fluids are generally involved with the processes of transport, excretion, or lubrication. They allow the distribution of oxygen and nutrients to the tissues and organs and the transport of waste products from the tissues, enabling their elimination from the body.

**boehmite** A mineral form of a mixed aluminium oxide and hydroxide, AlO.OH. It is named after the German scientist J. Böhm. *See* ALUMINIUM HYDROXIDE.

**Bohr, Niels Henrik David** (1885–1962) Danish physicist. In 1913 he published his explanation of how atoms, with electrons orbiting a central nucleus, achieve stability by assuming that their angular momentum is quantized. Movement of electrons from one orbit to another is accompanied by the absorption or emission of energy in the form of light, thus accounting for the series of lines in the emission *spectrum of hydrogen. For this work Bohr was awarded the 1922 Nobel Prize for physics. *See* BOHR THEORY.

**Bohr effect** The effect of pH on the dissociation of oxygen from haemoglobin, first discovered by the Danish physiologist Christian Bohr (1855–1911). An increase in carbon dioxide concentration makes the blood more acidic and decreases the efficiency of the uptake of oxygen by haemoglobin molecules. This shifts the *oxygen dissociation curve to the right and increases the tendency of haemoglobin to release oxygen. Thus in actively respiring tissues, where the concentration of carbon dioxide in the blood is high, haemoglobin readily releases its oxygen, while in the lungs, where blood carbon dioxide is low (due to its continual diffusion into the alveoli), haemoglobin readily binds oxygen.

**bohrium** Symbol Bh. A radioactive *transactinide element; a.n. 107. It was first made in 1981 by Peter Armbruster and a team in Darmstadt, Germany, by bombarding bismuth–209 nuclei with chromium–54 nuclei. Only a few atoms of bohrium have ever been detected.

**(((⊕))) SEE WEB LINKS**
• Information from the WebElements site

**Bohr theory** The theory published in 1913 by Niels Bohr to explain the line spectrum of hydrogen. He assumed that a single electron of mass $m$ travelled in a circular orbit of radius $r$, at a velocity $v$, around a positively charged nucleus. The *angular momentum of the electron would then be $mvr$. Bohr proposed that electrons could only occupy orbits in which this angular momentum had certain fixed values, $h/2\pi$, $2h/2\pi$, $3h/2\pi$,... $nh/2\pi$, where $h$ is the Planck constant. This means that the angular momentum is quantized, i.e. can only have certain values, each of which is a multiple of $n$. Each permitted value of $n$ is associated with an orbit of different radius and Bohr assumed that when the atom emitted or absorbed radiation of frequency $v$, the electron jumped from one orbit to another; the energy emitted or absorbed by each jump is equal to $hv$. This theory gave good results in predicting the lines observed in the spectrum of hydrogen and simple ions such as $He^+$, $Li^{2+}$, etc. The

idea of quantized values of angular momentum was later explained by the wave nature of the electron. Each orbit has to have a whole number of wavelengths around it; i.e. $n\lambda = 2\pi r$, where $\lambda$ is the wavelength and $n$ a whole number. The wavelength of a particle is given by $h/mv$, so $nh/mv = 2\pi r$, which leads to $mvr = nh/2\pi$. Modern atomic theory does not allow subatomic particles to be treated in the same way as large objects, and Bohr's reasoning is somewhat discredited. However, the idea of quantized angular momentum has been retained.

**boiling point (b.p.)** The temperature at which the saturated vapour pressure of a liquid equals the external atmospheric pressure. As a consequence, bubbles form in the liquid and the temperature remains constant until all the liquid has evaporated. As the boiling point of a liquid depends on the external atmospheric pressure, boiling points are usually quoted for standard atmospheric pressure (760 mmHg = 101 325 Pa).

**boiling-point–composition diagram** A graph showing how the boiling point and vapour composition of a mixture of two liquids depends on the composition of the mixture. The abscissa shows the range of compositions from 100% A at one end to 100% B at the other. The diagram has two curves: the lower one gives the boiling points (at a fixed pressure) for the different compositions. The upper one is plotted by taking the composition of vapour at each temperature on the boiling-point curve. The two curves would coincide for an ideal mixture, but generally they are different because of deviations from *Raoult's law. In some cases, they may show a maximum or minimum and coincide at some intermediate composition, explaining the formation of *azeotropes.

**boiling-point elevation** *See* ELEVATION OF BOILING POINT.

**boiling-water reactor** *See* NUCLEAR REACTOR.

**Bok globule** Any of a class of small near-spherical dark nebulae thought to be dense clouds of gas and dust in a late stage of gravitational collapse. The globules are visible against emission nebulae or a background of stars. They are thought to be the precursors of some low-mass protostars (*see* STELLAR EVOLUTION), sometimes accompanied by *bipolar outflows. They take their name from the US astronomer Bart Jason Bok (1906–83).

**bolide** An exceptionally bright meteor that appears to explode on its way through the earth's atmosphere: a detonating fireball.

**bolometer** A sensitive instrument used to measure radiant heat. The original form consists of two elements, each comprising blackened platinum strips (about $10^{-3}$ mm thick) arranged in series on an insulated frame to form a zigzag. The two elements are connected into the adjacent arms of a *Wheatstone bridge; one element is exposed to the radiation, the other is shielded from it. The change in the resistance of the exposed element, as detected by the bridge galvanometer, enables the heat reaching it to be calculated.

Modern semiconductor bolometers are now common, in which the platinum is replaced by a strip of semiconductor: this has a much greater (though usually negative) *temperature coefficient of resistance, and makes the system more sensitive.

**Boltzmann, Ludwig Eduard** (1844–1906) Austrian physicist. He held professorships in Graz, Vienna, Munich, and Leipzig, where he worked on the kinetic theory of gases (see MAXWELL–BOLTZMANN DISTRIBUTION) and on thermodynamics (see BOLTZMANN EQUATION). He suffered from depression and committed suicide.

**Boltzmann constant** Symbol $k$ or $k_B$. The ratio of the universal gas constant ($R$) to the Avogadro constant ($N_A$). It may be thought of therefore as the gas constant per molecule:

$$k = R/N_A = 1.380\,658(12) \times 10^{-23} \text{ J K}^{-1}$$

It is named after Ludwig Boltzmann.

**Boltzmann equation** An equation used in the study of a collection of particles in *non-equilibrium statistical mechanics, particularly their transport properties. The Boltzmann equation describes a quantity called the **distribution function**, $f$, which gives a mathematical description of the state and how it is changing. The distribution function depends on a position vector $r$, a velocity vector $v$, and the time $t$; it thus provides a statistical statement about the positions and velocities of the particles at any time. In the case of one species of particle being present, Boltzmann's equation can be written

$$\partial f/\partial t + a.(\partial f/\partial v) + v.(\partial f/\partial r) = (\partial f/\partial t)_{\text{coll}},$$

where $a$ is the acceleration of bodies between collisions and $(\partial f/\partial t)_{\text{coll}}$ is the rate of change of $f(r,v,t)$ due to collisions. The Boltzmann equation can be used to calculate *transport coefficients, such as *conductivity. The Boltzmann equation was proposed in 1872.

**Boltzmann formula** An equation concerning the entropy $S$ of a system derived from statistical mechanics. The formula is $S = k \ln W$, where $k$ is the Boltzmann constant and $W$ is the number of distinguishable ways of describing the system. It expresses in quantitative terms the concept that entropy is a measure of the disorder of a system.

**bolus** The ball of chewed food bound together with saliva that is formed in the mouth by the action of the tongue. The bolus is shaped to a size that allows it to pass into the oesophagus after being swallowed (see DEGLUTITION).

**bomb calorimeter** An apparatus used for measuring heats of combustion (e.g. calorific values of fuels and foods). It consists of a strong container in which the sample is sealed with excess oxygen and ignited electrically. The heat of combustion at constant volume can be calculated from the resulting rise in temperature.

**bond** See CHEMICAL BOND.

**bond energy** An amount of energy associated with a bond in a chemical compound. It is obtained from the heat of atomization. For instance, in methane the bond energy of the C–H bond is one quarter of the enthalpy of the process

$$CH_4(g) \rightarrow C(g) + 4H(g)$$

Bond energies (or **bond enthalpies**) can be calculated from the standard enthalpy of formation of the compound and from the enthalpies of atomization of the elements. Energies calculated in this way are called **average bond energies** or **bond–energy terms**. They depend to some extent on the molecule chosen; the C–H bond energy in methane will differ slightly from that in ethane. The **bond dissociation energy** is a different measurement, being the energy required to break a particular bond; e.g. the energy for the process:

$$CH_4(g) \rightarrow CH_3\cdot(g) + H\cdot(g)$$

**Bondi, Hermann** See HOYLE, SIR FRED.

**bonding orbital** See ORBITAL.

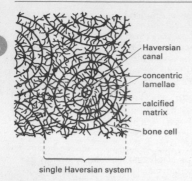

**Bone.** Structure of compact bone.

**bone** The hard connective tissue of which
the *skeleton of most vertebrates is formed.
It comprises a matrix of *collagen fibres
(30%) impregnated with bone salts (70%),
mostly calcium phosphate, in which are embedded bone cells (*see* OSTEOCYTE; OS-
TEOBLAST; OSTEOCLAST). Bone generally
replaces embryonic *cartilage and is of two
sorts – compact bone and spongy bone.
The outer **compact bone** is formed as concentric layers (**lamellae**) that surround small
holes (*Haversian canals): see illustration.
The inner **spongy bone** is chemically similar
but forms a network of bony bars. The
spaces between the bars may contain bone
marrow or (in birds) air for lightness. *See also*
CARTILAGE BONE; MEMBRANE BONE; PERIOS-
TEUM.

**bone black** *See* CHARCOAL.

**bone marrow** A soft tissue contained
within the central cavity and internal spaces
of a bone. At birth and in young animals
the marrow of all bones is concerned with
the formation of blood cells: it contains
*haemopoietic tissue and is known as **red
marrow**. In mature animals the marrow of
the long bones ceases producing blood cells
and is replaced by fat, being known as **yellow marrow**.

**bony fishes** *See* OSTEICHTHYES.

**bony labyrinth** *See* LABYRINTH.

**Boolean algebra** A form of symbolic
logic, devised by George Boole (1815–64) in
the middle of the 19th century, which provides a mathematical procedure for manipulating logical relationships in symbolic form.
For example in Boolean algebra $a + b$ means

$a$ or $b$, while $ab$ means $a$ and $b$. It makes use
of *set theory and is extensively used by the
designers of computers to enable the bits 0
and 1, as used in the binary notation, to relate to the logical functions the computer
needs in carrying out its calculations.

**borane (boron hydride)** Any of a group of
compounds of boron and hydrogen, many of
which can be prepared by the action of acid
on magnesium boride ($MgB_2$). Others are
made by pyrolysis of the products of this reaction in the presence of hydrogen and other
reagents. They are all volatile, reactive, and
oxidize readily in air, some explosively so.
The boranes are a remarkable group of compounds in that their structures cannot be described using the conventional two-electron
covalent bond model (*see* ELECTRON-DEFI-
CIENT COMPOUND). The simplest example is
**diborane** ($B_2H_6$): see formula. Other boranes
include $B_4H_{10}$, $B_5H_9$, $B_5H_{11}$, $B_6H_{10}$, and
$B_{10}H_4$. The larger borane molecules have
open or closed polyhedra of boron atoms.
In addition, there is a wide range of borane
derivatives containing atoms of other elements, such as carbon and phosphorus.
**Borohydride ions** of the type $B_6H_6{}^{2-}$ also
exist. Boranes and borohydride ions are
classified according to their structure. Those
with a complete polyhedron are said to have
a **closo-structure**. Those in which the polyhedron is incomplete by loss of one vertex
have a **nido-structure** (from the Greek for
'nest'). Those with open structures by removal of two or more vertices have an
**arachno- structure** (from the Greek for 'spider'). *See also* WADE'S RULES.

**Borane.** Diborane, the simplest of the boranes.

**borate** Any of a wide range of ionic compounds that have negative ions containing
boron and oxygen (see formulae). Lithium
borate, for example, contains the simple
anion $B(OH)_4{}^-$. Most borates, however, are
inorganic polymers with rings, chains, or
other networks based on the planar $BO_3$
group or the tetrahedral $BO_3(OH)$ group.
'Hydrated' borates are ones containing –OH
groups; many examples occur naturally. Anhydrous borates, which contain $BO_3$ groups,

$B_3O_6^{3-}$ as in $Na_3B_3O_6$

$(BO_2)_n^{n-}$ as in $CaB_2O_4$

$[B_4O_5(OH)_4]^{2-}$
as in borax $Na_2B_4O_7.10H_2O$

**Borate.** Structure of some typical borate ions.

can be made by melting together boric acid and metal oxides.

**borax (disodium tetraborate-10-water)** A colourless monoclinic solid, $Na_2B_4O_7.10H_2O$, soluble in water and very slightly soluble in ethanol; monoclinic; r.d. 1.73; loses $8H_2O$ at 75°C; loses $10H_2O$ at 320°C. The formula gives a misleading impression of the structure. The compound contains the ion $[B_4O_5(OH)_4]^{2-}$ (*see* BORATE). Attempts to recrystallize this compound above 60.8°C yield the pentahydrate. The main sources of borax are the borate minerals **kernite** ($Na_2B_4O_7.4H_2O$) and **tincal** ($Na_2B_4O_7.10H_2O$). The ores are purified by carefully controlled dissolution and recrystallization. On treatment with mineral acids borax gives boric acid.

Borax is a very important substance in the glass and ceramics industries as a raw material for making borosilicates. It is also important as a metallurgical flux because of the ability of molten borates to dissolve metal oxides. In solution it partially hydrolyses to boric acid and can thus act as a buffer. For this reason it is used as a laundry pre-soak. It is used medicinally as a mild alkaline antiseptic and astringent for the skin and mucous membranes.

Disodium tetraborate is the source of many industrially important boron compounds, such as barium borate (fungicidal paints), zinc borate (fire-retardant additive in plastics), and boron phosphate (heterogeneous acid catalyst in the petrochemicals industry).

**borax-bead test** A simple laboratory test for certain metal ions in salts. A small amount of the salt is mixed with borax and a molten bead formed on the end of a piece of platinum wire. Certain metals can be identified by the colour of the bead produced in the oxidizing and reducing parts of a Bunsen flame. For example, iron gives a bead that is red when hot and yellow when cold in the oxidizing flame and a green bead in the reducing flame.

**borax carmine** A red dye, used in optical microscopy, that stains nuclei and cytoplasm pink. It is frequently used to stain large pieces of animal tissue.

**borazon** *See* BORON NITRIDE.

**Bordeaux mixture** A mixture of copper(II) sulphate and calcium hydroxide in water, used as a fungicide.

**Borel sum** An integral that is defined so as to represent the sum of a divergent series. The concept enables many of the divergent series that occur in physics, such as perturbation series in quantum field theory, to be mathematically well-defined. It is named after the French mathematician Émile Borel (1871–1956).

**boric acid** Any of a number of acids containing boron and oxygen. Used without qualification the term applies to the compound $H_3BO_3$ (which is also called **orthoboric acid** or, technically, **trioxoboric(III) acid**). This is a white or colourless solid that is soluble in water and ethanol; triclinic; r.d. 1.435; m.p. 169°C. It occurs naturally in the condensate from volcanic steam vents (suffioni). Commercially, it is made by treating borate minerals (e.g. kernite, $Na_2B_4O_7$. $4H_2O$) with sulphuric acid followed by recrystallization.

In the solid there is considerable hydrogen bonding between $H_3BO_3$ molecules resulting in a layer structure, which accounts for the easy cleavage of the crystals. $H_3BO_3$ molecules also exist in dilute solutions but in more concentrated solutions polymeric acids and ions are formed (e.g. $H_4B_2O_7$; pyroboric acid or **tetrahydroxomonoxodiboric(III) acid**). The compound is a very weak acid but also acts as a Lewis *acid in accepting hydroxide ions:

$$B(OH)_3 + H_2O \rightleftharpoons B(OH)_4^- + H^+$$

If solid boric acid is heated it loses water and transforms to another acid at 300°C. This is given the formula $HBO_2$ but is in fact a polymer $(HBO_2)_n$. It is called **metaboric acid** or, technically, **polydioxoboric(III) acid**.

Boric acid is used in the manufacture of glass (borosilicate glass), glazes and enamels, leather, paper, adhesives, and explosives. It is widely used (particularly in the USA) in detergents, and because of the ability of fused boric acid to dissolve other metal oxides it is used as a flux in brazing and welding. Because of its mild antiseptic properties it is used in the pharmaceutical industry and as a food preservative.

**boride** A compound of boron with a metal. Most metals form at least one boride of the type MB, $MB_2$, $MB_4$, $MB_6$, or $MB_{12}$. The compounds have a variety of structures; in particular, the hexaborides contain clusters of $B_6$ atoms. The borides are all hard high-melting materials with metal-like conductivity. They can be made by direct combination of the elements at high temperatures (over 2000°C) or, more usually, by high-temperature reduction of a mixture of the metal oxide and boron oxide using carbon or aluminium. Chemically, they are stable to nonoxidizing acids but are attacked by strong oxidizing agents and by strong alkalis. Magnesium boride ($MgB_2$) is unusual in that it can be hydrolysed to boranes. Industrially, metal borides are used as refractory materials. The most important are CrB, $CrB_2$, $TiB_2$, and $ZnB_2$. Generally, they are fabricated using high-temperature powder metallurgy, in which the article is produced in a graphite die at over 2000°C and at very high pressure. Items are pressed as near to final shape as possible as machining requires diamond cutters and is extremely expensive.

**Born–Haber cycle** A cycle of reactions used for calculating the lattice energies of ionic crystalline solids. For a compound MX, the lattice energy is the enthalpy of the reaction

$$M^+(g) + X^-(g) \rightarrow M^+X^-(s) \; \Delta H_L$$

The standard enthalpy of formation of the ionic solid is the enthalpy of the reaction

$$M(s) + \tfrac{1}{2}X_2(g) \rightarrow M^+X^-(s) \; \Delta H_f$$

The cycle involves equating this enthalpy (which can be measured) to the sum of the enthalpies of a number of steps proceeding from the elements to the ionic solid. The steps are:

(1) Atomization of the metal:

$$M(s) \rightarrow M(g) \; \Delta H_1$$

(2) Atomization of the nonmetal:

$$\tfrac{1}{2}X_2(g) \rightarrow X(g) \; \Delta H_2$$

(3) Ionization of the metal:

$$M(g) \rightarrow M^+(g) + e \; \Delta H_3$$

This is obtained from the ionization potential.

(4) Ionization of the nonmetal:

$$X(g) + e \rightarrow X^-(g) \; \Delta H_4$$

This is the electron affinity.

(5) Formation of the ionic solids:

$$M^+(g) + X^-(g) \rightarrow M^+X^-(s) \; \Delta H_L$$

Equating the enthalpies gives:

$$\Delta H_f = \Delta H_1 + \Delta H_2 + \Delta H_3 + \Delta H_4 + \Delta H_L$$

from which $\Delta H_L$ can be found. It is named after the German physicist Max Born (1882–1970) and the chemist Fritz Haber.

**bornite** An important ore of copper composed of a mixed copper–iron sulphide, $Cu_5FeS_4$. Freshly exposed surfaces of the mineral are a metallic reddish-brown but a purplish iridescent tarnish soon develops – hence it is popularly known as **peacock ore**. Bornite is mined in Chile, Peru, Bolivia, Mexico, and the USA.

**Born–Oppenheimer approximation** An *adiabatic approximation used in molecular and solid-state physics in which the motion of atomic nuclei is taken to be so much slower than the motion of electrons that, when calculating the motions of electrons, the nuclei can be taken to be in fixed positions. This approximation was justified using *perturbation theory by Max Born and the US physicist Julius Robert Oppenheimer (1904–67) in 1927.

**borohydride ion** *See* BORANE.

**boron** Symbol B. An element of *group 13 (formerly IIIB) of the periodic table; a.n. 5; r.a.m. 10.81; r.d. 2.34–2.37 (amorphous); m.p. 2300°C; b.p. 2550°C. It forms two allotropes; amorphous boron is a brown powder but metallic boron is black. The metallic form is very hard (9.3 on Mohs' scale) and is a poor electrical conductor at room temperature. At least three crystalline forms are possible; two are rhombohedral and the other tetragonal. The element is never found free in nature. It occurs as orthoboric acid in volcanic springs in Tuscany, as borates in kernite ($Na_2B_4O_7.4H_2O$), and as colemanite ($Ca_2B_6O_{11}.5H_2O$) in California. Samples usually contain isotopes in the ratio of 19.78% boron–10 to 80.22% boron–11. Extraction is achieved by vapour-phase reduction of boron trichloride with hydrogen on electrically heated filaments. Amorphous boron can be obtained by reducing the trioxide with magnesium powder. Boron when heated reacts with oxygen, halogens, oxidizing acids, and hot alkalis. It is used in semiconductors and in filaments for specialized aerospace applications. Amorphous boron is used in flares, giving a green coloration. The isotope boron–10 is used in nuclear reactor control rods and shields. The element was discovered in 1808 by Sir Humphry Davy and by J. L. Gay-Lussac and L. J. Thenard.

**SEE WEB LINKS**
• Information from the WebElements site

**boron carbide** A black solid, $B_4C$, soluble only in fused alkali; it is extremely hard, over 9½ on Mohs' scale; rhombohedral; r.d. 2.52; m.p. 2350°C; b.p. >3500°C. Boron carbide is manufactured by the reduction of boric oxide with petroleum coke in an electric furnace. It is used largely as an abrasive, but objects can also be fabricated using high-temperature powder metallurgy. Boron nitride is also used as a neutron absorber because of its high proportion of boron–10.

**boron counter** A *counter tube containing a **boron chamber**, used for counting slow neutrons. The boron chamber is lined with boron or a boron compound or is filled with the gas boron trifluoride ($BF_3$). As natural boron contains about 18% of the isotope boron–10, and as this isotope absorbs neutrons with the emission of an alpha particle, the chamber can be coupled with a scaler to count the alpha particles emitted when neutrons enter the chamber.

**boron hydride** See BORANE.

**boron nitride** A solid, BN, insoluble in cold water and slowly decomposed by hot water; r.d. 2.25 (hexagonal); sublimes above 3000°C. Boron nitride is manufactured by

heating boron oxide to 800°C on an acid-soluble carrier, such as calcium phosphate, in the presence of nitrogen or ammonia. It is isoelectronic with carbon and, like carbon, it has a very hard cubic form (**borazon**) and a softer hexagonal form; unlike graphite this is a nonconductor. It is used in the electrical industries where its high thermal conductivity and high resistance are of especial value.

**boron trichloride** A colourless fuming liquid, $BCl_3$, which reacts with water to give hydrogen chloride and boric acid; r.d. 1.349; m.p. –107°C; b.p. 12.5°C. Boron trichloride is prepared industrially by the exothermic chlorination of boron carbide at above 700°C, followed by fractional distillation. An alternative, but more expensive, laboratory method is the reaction of dry chlorine with boron at high temperature. Boron trichloride is a Lewis *acid, forming stable addition compounds with such donors as ammonia and the amines and is used in the laboratory to promote reactions that liberate these donors. The compound is important industrially as a source of pure boron (reduction with hydrogen) for the electronics industry. It is also used for the preparation of boranes by reaction with metal hydrides.

**borosilicate** Any of a large number of substances in which $BO_3$ and $SiO_4$ units are linked to form networks with a wide range of structures. Borosilicate glasses are particularly important; the addition of boron to the silicate network enables the glass to be fused at lower temperatures than pure silica and also extends the plastic range of the glass. Thus such glasses as Pyrex have a wider range of applications than soda glasses (narrow plastic range, higher thermal expansion) or silica (much higher melting point). Borosilicates are also used in glazes and enamels and in the production of glass wools.

**Bosch, Carl** See BERGIUS, FRIEDRICH KARL RUDOLF; HABER PROCESS.

**Bosch process** See HABER PROCESS.

**Bose–Einstein condensation** A phenomenon occurring in a macroscopic system consisting of a large number of *bosons at a sufficiently low temperature, in which a significant fraction of the particles occupy a single quantum state of lowest energy (the ground state). Bose–Einstein condensation can only take place for bosons whose total number is conserved in collisions. Because of the Pauli exclusion principle, it is impossible for two or more fermions to occupy the same quantum state, and so there is no analogous condensation phenomenon for such particles. Bose–Einstein condensation is of fundamental importance in explaining the phenomenon of *superfluidity. At very low temperatures (around $2 \times 10^{-7}$ K) a Bose–Einstein condensate can form, in which several thousand atoms become a single entity (a **superatom**). This effect has been observed with atoms of rubidium and lithium and certain other atomic systems at very low temperature. The effect is named after the Indian physicist Satyendra Nath Bose (1894–1974) and Albert Einstein.

**Bose–Einstein statistics** *See* QUANTUM STATISTICS.

**boson** An *elementary particle (or bound state of an elementary particle, e.g. an atomic nucleus or an atom) with integral spin; i.e. a particle that conforms to Bose–Einstein statistics (*see* QUANTUM STATISTICS), from which it derives its name. *Compare* FERMION.

**botany** The scientific study of plants, including their anatomy, morphology, physiology, biochemistry, taxonomy, cytology, genetics, ecology, evolution, and geographical distribution.

**bottled gas** Gas supplied under pressure in metal cylinders. The term includes pressurized gas (e.g. oxygen and nitrogen cylinders) and gases liquefied under pressure (e.g. liquid butane for use as a fuel). Colour conventions are used to identify the type of gas or, in some cases, the specific gas. The colour indicating the contents is that of the shoulder of the cylinder at the top. The convention is not international, and practice differs in different countries. In the UK, the convention is:

Yellow for toxic or corrosive gases
Red for flammable gases
Light blue for oxidizing gases
Maroon for acetylene
Dark green for argon
Grey for carbon dioxide
Brown for helium
Blue for nitrous oxide
Black for nitrogen
White for oxygen

(🌐) SEE WEB LINKS
• Information on colour coding of gas containers

**botulinum toxin** A nerve toxin produced

by the bacterium *Clostridium botulinum*, which can cause fatal *food poisoning. It is the most toxic substance known. In minute doses it is used to treat certain conditions involving muscle dysfunction.

**boulder clay (till)** A mixture of rock and powdered rock formed beneath a moving glacier as it drags rocks beneath it. When the glacier subsequently melts, the boulder clay is left as a surface bed. Its components may be almost any type of rock in a wide range of sizes, from large angular boulders to tiny particles in clay. *See also* MORAINE.

**boundary conditions** In the general solution of a *differential equation, conditions that are imposed to allow the arbitrary constants to be determined. They thus permit a particular solution to be obtained.

**boundary layer** The thin layer of fluid formed around a solid body or surface relative to which the fluid is flowing. Adhesion between the molecules of the fluid and that of the body or surface causes the molecules of the fluid closest to the solid to be stationary relative to it. The transfer of heat or mass between a solid and a fluid flowing over it is largely controlled by the nature of the boundary layer.

**boundary slip** A boundary condition used in fluid mechanics. When a liquid flows over the surface of a solid, and layers of liquid close to the solid are assumed to be stationary relative to the solid; this is called a **no-slip boundary condition**. Such an assumption is a good approximation for macroscopic flow but is not accurate at the molecular scale. When this assumption is not made a **slip boundary condition** occurs. Boundary slip causes the viscosity at the interface to be different from the viscosity in the bulk of the liquid.

**bound state** A system in which two (or more) parts are bound together in such a way that energy is required to split them. An example of a bound state is a *molecule formed from two (or more) *atoms.

**Bourdon gauge** A pressure gauge consisting essentially of a C-shaped or spiral tube with an oval cross section. One end of the tube is connected to the fluid whose pressure is to be measured and the other end is sealed. As the pressure inside the tube is increased, the oval tube tends to become circular and this causes the tube to straighten. The movement of the end of the tube is transferred by a simple mechanism to a needle moving round a dial or to a digital display. With suitable design, Bourdon gauges can be used for high-pressure measurement and also for low pressures. It was invented by Eugène Bourdon (1804–88).

**bovine spongiform encephalopathy (BSE)** A degenerative disease of the brain that affects cattle and is caused by an abnormal form of a cellular protein (*see* PRION). Known colloquially as 'mad cow disease', it results in a build-up of fibrous tissue in the brain. The infective agent can be transmitted to other cattle via feed containing offal derived from infected animals. It can also, under certain circumstances, be transmitted to other species. *See also* CREUTZFELDT–JAKOB DISEASE.

**Bowman's capsule (renal capsule)** The cup-shaped end of a kidney *nephron. Its epithelium contains **podocytes**, cells that facilitate the passage of glomerular filtrate from the blood into the nephron. It is named after its discoverer, the British physician Sir William Bowman (1816–92).

**Boyle, Robert** (1627–91) English chemist and physicist, born in Ireland. After moving to Oxford in 1654 he worked on gases, using an air pump made by Robert *Hooke. With it he proved that sound does not travel in a vacuum. In 1662 he discovered *Boyle's law. In chemistry he worked on *flame tests and acid-base *indicators.

**SEE WEB LINKS**
• The Robert Boyle project at Birkbeck College, London

**Boyle's law** The volume (*V*) of a given mass of gas at a constant temperature is inversely proportional to its pressure (*p*), i.e. *pV* = constant. This is true only for an *ideal gas. This law was discovered in 1662 by Robert Boyle. On the continent of Europe it is known as **Mariotte's law** after E. Mariotte (1620–84), who discovered it independently in 1676. *See also* GAS LAWS.

**bp** *See* BASE PAIR.

**Brackett series** *See* HYDROGEN SPECTRUM.

**bract** A modified leaf with a flower or inflorescence in its axil. Bracts are often brightly coloured and may be mistaken for the petals of a flower. For example the showy 'flowers' of poinsettia and *Bougainvillea* are composed of bracts; the true flowers are comparatively inconspicuous. *See also* INVOLUCRE.

**bracteole** A reduced leaf that arises from the stalk of an individual flower.

**bradykinin** *See* KININ.

**Bragg, Sir William Henry** (1862–1942) British physicist, who with his son **Sir (William) Lawrence Bragg** (1890–1971) was awarded the 1915 Nobel Prize for physics for their pioneering work on *X-ray crystallography. He also constructed an X-ray spectrometer for measuring the wavelengths of X-rays. In the 1920s, while director of the Royal Institution in London, he initiated X-ray diffraction studies of organic molecules.

**Bragg's law** When a beam of X-rays (wavelength λ) strikes a crystal surface in which the layers of atoms or ions are separated by a distance *d*, the maximum intensity of the reflected ray occurs when $\sin\theta = n\lambda/2d$, where θ (known as the **Bragg angle**) is the complement of the angle of incidence and *n* is an integer. The law enables the structure of many crystals to be determined. It was discovered in 1912 by Sir Lawrence Bragg.

**Brahe, Tycho** (1546–1601) Danish astronomer, generally regarded as the most precise, systematic, and meticulous observer of the heavens in the era before the telescope. He observed the supernova of 1572 ('Tycho's star') in the constellation of Cassiopeia and showed that it was a fixed star, because its position relative to the other fixed stars remained unchanged and it showed no daily parallax. In 1577 Tycho moved to his own observatories on Hven Island (financed by King Frederick II), where, without the benefit of a telescope, he charted the positions of 777 stars. In 1599 he moved to Prague to join the court of the Holy Roman Emperor Rudolf II as imperial mathematician, with *Kepler as his assistant from 1600.

**brain** 1. The enlarged anterior part of the vertebrate central nervous system, which is encased within the cranium of the skull. Continuous with the spinal cord, the brain is surrounded by three membranes (*see* MENINGES) and bathed in cerebrospinal fluid, which fills internal cavities (*ventricles). It functions as the main coordinating centre for nervous activity, receiving information (in the form of nerve impulses) from sense organs, interpreting it, and transmitting 'in-

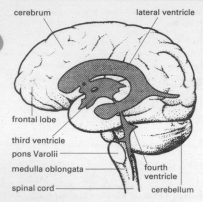

cerebrum — lateral ventricle

frontal lobe

third ventricle

pons Varolii

medulla oblongata

spinal cord

fourth ventricle

cerebellum

**The human brain.**

structions' to muscles and other *effectors. It is also the seat of intelligence and memory. The embryonic vertebrate brain is in three sections (see FOREBRAIN; HINDBRAIN; MID-BRAIN), which become further differentiated during development into specialized regions. The main parts of the adult human brain are a highly developed *cerebrum in the form of two cerebral hemispheres, a *cerebellum, *medulla oblongata, and *hypothalamus (see illustration). **2.** A concentration of nerve *ganglia at the anterior end of an invertebrate animal.

**(()) SEE WEB LINKS**

- Multilevel exploration of brain structure and function, sponsored by the Canadian Institutes of Health Research

**brain death** The permanent absence of vital functions of the brain, which is marked by cessation of breathing and other reflexes controlled by the *brainstem and by a zero reading on an *electroencephalogram. Organs may be removed for transplantation when brain death is established, which may not necessarily be associated with permanent absence of heart beat.

**brainstem** The part of the brain comprising the *medulla oblongata, the *midbrain, and the *pons. It resembles and is continuous with the spinal cord. The midbrain controls and integrates reflex activities (such as respiration) that originate in higher centres of the brain via a network of nerve pathways (the **reticular formation**).

**branched chain** See CHAIN.

**brane world** A theory in which the four

space–time dimensions of the universe that are apparent make up a surface, called the **brane**, in a higher-dimensional space–time, called the **bulk**. One of the appealing features of the brane world is that it can explain why gravity is much weaker than the other forces, with the nongravitational forces being localized to the brane and the gravitational force not limited to the brane. See also EKPYROTIC UNIVERSE; RANDALL–SUNDRUM SCENARIO.

**Brans–Dicke theory** A modification of the general theory of relativity that combined tensor and scalar qualities and allowed the *gravitational constant $G$ to vary with time in a way that is in accord with *Mach's principle. The Brans–Dicke theory was proposed by the American physicists Carl Brans (1935– ) and Robert Dicke (1916–97) in 1961. This type of theory has largely been abandonded since its predictions have not been supported by accurate experimental tests of the general theory of relativity.

**brass** A group of alloys consisting of copper and zinc. A typical yellow brass might contain about 67% copper and 33% zinc. See also DELTA-BRASS.

**Brattain, Walter** See BARDEEN, JOHN.

**Braun, Karl Ferdinand** (1850–1918) German physicist, who became professor of physics at Strasbourg in 1895. In the early 1900s he used crystals as diodes (later employed in crystal-set radios) and developed the *cathode-ray tube for use as an oscilloscope. He also worked on radio and in 1909 shared the Nobel Prize for physics with *Marconi.

**Bravais lattice** A lattice defined by the combination of one of the seven possible *crystal systems and one of the possible lattice centrings, i.e. (1) **primitive**, in which only the cell corners are occupled, (2) **body centred**, in which there is a point at the centre, (3) **face centred**, in which there are points at the centres of all the faces, (4) **centred at a single face**, in which there is a point at the centre of one of the faces. The Bravais lattice is named after the French physicist Auguste Bravais (1811–63), who demonstrated that in three spatial dimensions there are 14 possible such lattices.

**breakdown** The sudden passage of a current through an insulator. The voltage at which this occurs is the **breakdown voltage**.

**breaking stress**  *See* ELASTICITY.

**breathing**  *See* EXPIRATION; INSPIRATION; RESPIRATORY MOVEMENT.

**breed**  A domesticated *variety of an animal or, rarely, a cultivated variety of plant (cultivated plants are usually called varieties or, more correctly, *cultivars). Examples of animal breeds are Friesian cattle and Shetland sheepdogs.

**breeder reactor**  *See* NUCLEAR REACTOR.

**breeding**  The process of sexual *reproduction and bearing offspring. **Selective breeding** of both plants and animals is used in *agriculture to produce offspring that possess the beneficial characters of both parents (*see also* ARTIFICIAL INSEMINATION). *Inbreeding is the production of *homozygous phenotypically uniform offspring by mating between close relatives. Plants that self-fertilize, such as wheat and tomatoes, are inbreeders. *Outbreeding is the production of *heterozygous phenotypically variable offspring by mating between unrelated organisms.

**breeding season**  A specific season of the year in which many animals, including mammals and birds, mate, which ensures that offspring are produced only at a certain time of the year. This timing is important as it enables animals to give birth at a time of the year when environmental conditions and food supply are at their optimum. The breeding season of most animals is in the spring or summer. The stimulus to mate is the result of a photoperiodic response (*see* PHOTOPERIODISM), which is thought to be controlled by day length affecting levels of the hormone *melatonin.

**Bremsstrahlung**  (German: braking radiation) The X-rays emitted when a charged particle, especially a fast electron, is rapidly slowed down, as when it passes through the electric field around an atomic nucleus. The X-rays cover a whole continuous range of wavelengths down to a minimum value, which depends on the energy of the incident particles. Bremsstrahlung are produced by a metal target when it is bombarded by electrons.

**brewing**  The process by which beer is made. Fermentation of sugars from barley grain by the yeasts *Saccharomyces cerevisiae* and *S. uvarum* (or *S. carlsbergensis*) produces alcohol. In the first stage the barley grain is soaked in water, a process known as **malting**. The grain is then allowed to germinate and the natural enzymes of the grain (the amylases and the maltases) convert the starch to maltose and then to glucose. The next stage is **kilning** or **roasting**, in which the grains are dried and crushed. The colour of a beer depends on the temperature used for this process: the higher the temperature, the darker the beer. In the next stage, **mashing**, the crushed grain is added to water at a specific temperature and any remaining starch is converted to sugar; the resultant liquid is the raw material of brewing, called **wort**. The yeast is then added to the wort to convert the sugar to alcohol, followed by hops, which give beer its characteristic flavour. Hops are the female flowers of the vine *Humulus lupulus*; they contain resins (humulones, cohumulones, and adhumulones) that give beer its distinctive bitter taste.

**Brewster's law**  The extent of the polarization of light reflected from a transparent surface is a maximum when the reflected ray is at right angles to the refracted ray. The angle of incidence (and reflection) at which this maximum polarization occurs is called the **Brewster angle** or **polarizing angle**. For this angle $i_B$, the condition is that $\tan i_B = n$, where $n$ is the refractive index of the transparent medium. The law was discovered in 1811 by the British physicist David Brewster (1781–1868).

**bridge**  **1.** (in chemistry) An atom or group joining two other atoms in a molecule. *See* ALUMINIUM CHLORIDE; BORANE. **2.** (in physics) a type of electrical circuit in which four components are linked in a square, with inputs and outputs at pairs of opposite corners. *See* WHEATSTONE BRIDGE.

**bridge rectifier**  *See* RECTIFIER.

**brighteners**  Substances added to detergents or used to treat textiles or paper in order to brighten the colours or, particularly, to enhance whiteness. Blueing agents are used in laundries to give a slight blue cast to white material in order to counteract yellowing. Fluorescent brighteners are compounds that absorb visible or ultraviolet radiation and fluoresce in the blue region of the optical spectrum.

**Brinell hardness**  A scale for measuring the hardness of metals introduced around 1900 by the Swedish metallurgist Johann

Brinell (1849–1925). A small chromium-steel ball is pressed into the surface of the metal by a load of known weight. The ratio of the mass of the load in kilograms to the area of the depression formed in square millimetres is the **Brinell number**.

**Brin process** A process formerly used for making oxygen by heating barium oxide in air to form the peroxide and then heating the peroxide at higher temperature (>800°C) to produce oxygen

$$2BaO_2 \rightarrow 2BaO + O_2$$

**Britannia metal** A silvery alloy consisting of 80–90% tin, 5–15% antimony, and sometimes small percentages of copper, lead, and zinc. It is used in bearings and some domestic articles.

**British thermal unit (Btu)** The Imperial unit of heat, being originally the heat required to raise the temperature of 1lb of water by 1°F. 1 Btu is now defined as 1055.06 joules.

**broadband** Communication by a system that supports a wide range of frequencies, so that identical messages can be carried simultaneously. *See also* ADSL.

**broken symmetry** A situation in which the lowest-energy state of a many-body system or *vacuum state of a relativistic *quantum field theory has a lower symmetry than the equations defining the system. Examples in solid-state physics include ferromagnetism, antiferromagnetism, and superconductivity. In particle physics, the Weinberg–Salam model (*see* ELECTROWEAK THEORY) is an important example of a relativistic quantum field theory with broken symmetry.

A result associated with broken symmetry is **Goldstone's theorem**. This states that a relativistic quantum field theory having continuous symmetry that is broken must include the existence of massless particles called **Goldstone bosons**. In many-body theory Goldstone bosons are *collective excitations. An exception to Goldstone's theorem is provided in the case of broken *gauge theories, such as the Weinberg–Salam model, in which the Goldstone bosons become massive bosons known as *Higgs bosons. In many-body theory, long-range forces provide the analogous exception to Goldstone's theorem, with the Higgs bosons being excitations with a nonzero gap. Such Higgs bosons are found in superconductors.

**bromate** A salt or ester of a bromic acid.

**bromic(I) acid (hypobromous acid)** A yellow liquid, HBrO. It is a weak acid and a strong oxidizing agent.

**bromic(V) acid** A colourless liquid, $HBrO_3$, made by adding sulphuric acid to barium bromate. It is a strong acid.

**bromide** *See* HALIDE.

**bromination** A chemical reaction in which a bromine atom is introduced into a molecule. *See also* HALOGENATION.

**bromine** Symbol Br. A *halogen element; a.n. 35; r.a.m. 79.909; r.d. 3.13; m.p. –7.2°C; b.p. 58.78°C. It is a red volatile liquid at room temperature, having a red-brown vapour. Bromine is obtained from brines in the USA (displacement with chlorine); a small amount is obtained from sea water in Anglesey. Large quantities are used to make 1,2-dibromoethane as a petrol additive. It is also used in the manufacture of many other compounds. Chemically, it is intermediate in reactivity between chlorine and iodine. It forms compounds in which it has oxidation states of 1, 3, 5, or 7. The liquid is harmful to human tissue and the vapour irritates the eyes and throat. The element was discovered in 1826 by Antoine Balard.

**(⊕) SEE WEB LINKS**

• Information from the WebElements site

**bromoethane (ethyl bromide)** A colourless flammable liquid, $C_2H_5Br$; r.d. 1.46; m.p. –119°C; b.p. 38.4°C. It is a typical *haloalkane, which can be prepared from ethene and hydrogen bromide. Bromoethane is used as a refrigerant.

**bromoform** *See* TRIBROMOMETHANE; HALOFORMS.

**bromomethane (methyl bromide)** A colourless volatile nonflammable liquid, $CH_3Br$; r.d. 1.68; m.p. –93°C; b.p. 3.56°C. It is a typical *haloalkane.

***N*-bromosuccinimide (NBS)** A crystalline solid, $C_4O_2NBr$, used extensively as a reagent for electrophilic addition of bromine. It acts by producing a small constant supply of bromine in solution

$$C_4O_2NBr + H^+ + Br^- \rightarrow C_4O_2NH + Br_2.$$

**bromothymol blue** An acid–base *indicator that is yellow in acid solutions and blue in alkaline solutions. It changes colour over the pH range 6–8.

**bronchiole** A fine respiratory tube in the

lungs of reptiles, birds, and mammals. It is formed by the subdivision of a *bronchus and in reptiles and mammals it terminates in a number of *alveoli.

**bronchus (bronchial tube)** One of the major air tubes in the *lung. The *trachea divides into two main bronchi, one for each lung, which split into smaller bronchi and then into *bronchioles. The walls of the bronchi are stiffened by rings of cartilage.

**bronze** Any of a group of alloys of copper and tin, sometimes with lead and zinc present. The amount of tin varies from 1% to 30%. The alloy is hard and easily cast and extensively used in bearings, valves, and other machine parts. Various improved bronzes are produced by adding other elements; for instance, **phosphor bronzes** contain up to 1% phosphorus. In addition certain alloys of copper and metals other than tin are called bronzes – **aluminium bronze** is a mixture of copper and aluminium. Other special bronzes include *bell metal, *gun metal, and *beryllium bronze.

**Brown, Robert** (1773–1858) British botanist, born in Scotland. After serving as an army medical officer he met botanist Joseph Banks (1743–1820) in 1798. Three years later Banks recommended him as naturalist on a survey of the Australian coast, during which he collected 4000 plant specimens; it took him five years to classify them. During this work he was the first to distinguish between gymnosperms and angiosperms. Then in 1827, while observing pollen grains in water, he discovered *Brownian movement.

**brown algae** *See* PHAEOPHYTA.

**brown dwarf** An astronomical object with a mass intermediate between the mass of a giant planet and that of a small star. The mass of a brown dwarf is large enough to generate energy by gravitational pressure, but not large enough to sustain nuclear fusion. The energy is radiated as electromagnetic radiation. Brown dwarfs are faint objects, which are expected to shine for about 100 million years before cooling. Their masses lie between a few times the mass of Jupiter and 80 times the mass of Jupiter. It has been suggested that brown dwarfs may contribute to the *missing mass of the universe.

**brown earth (brown forest soil)** A type of soil that is characteristic of those mid-

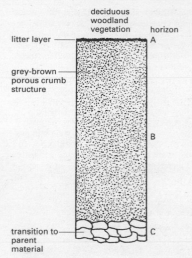

**Brown earth profile.**

latitude parts of the world that were originally covered with deciduous woodland. It is rich in organic matter derived from the annual leaf fall of deciduous trees and from associated shrubs, herbs, and grasses. Brown earths occur in the NE USA, N China, central Japan, and NW and central Europe. They are important agriculturally, possessing a good crumb structure, mild acidity, and free drainage, and consequently most of the original forest has long since been cleared for agricultural use.

**brown fat** A darker coloured region of *adipose tissue found in newborn and hibernating animals (in which it may also be called the **hibernating gland**). Compared to normal white *fat, deposits of brown fat are more richly supplied with blood vessels and have numerous mitochondria (hence the brown colour, due to the high concentrations of cytochrome oxidase). They can also be more rapidly converted to heat energy – a process that takes place in the fat cells themselves – especially during arousal from hibernation and during cold stress in young animals. Since the deposits are strategically placed near major blood vessels, the heat they generate warms the blood returning to the heart. Some types of obesity in humans may be linked to a lack of brown fat in affected individuals. *See* THERMOGENESIS.

**Brownian movement** The continuous random movement of microscopic solid particles (of about 1 micrometre in diameter) when suspended in a fluid medium. First observed by Robert Brown in 1827 when studying pollen grains in water, it was originally thought to be the manifestation of some vital force. It was later recognized to be a consequence of bombardment of the pollen by the continually moving molecules of the liquid. The smaller the particles the more extensive is the motion. The effect is also visible in particles of smoke suspended in a still gas and in the material of dead cells.

**(((∰))) SEE WEB LINKS**

• A description of the concept, using a series of simulations, from the University of Liverpool's Matter Initiative for Schools
• Jean Perrin's paper

**brown-ring test** A test for ionic nitrates. The sample is dissolved and iron(II) sulphate solution added in a test tube. Concentrated sulphuric acid is then added slowly so that it forms a separate layer. A brown ring (of $Fe(NO)SO_4$) at the junction of the liquids indicates a positive result.

**brucite** A mineral form of *magnesium hydroxide, $Mg(OH)_2$.

**brush** An electrical contact to a moving commutator on a motor or generator. It is made of a specially prepared form of carbon and is kept in contact with the moving part by means of a spring.

**brush border** A region of surface epithelium that possesses densely packed microvilli (see MICROVILLUS), rather like the bristles of a brush. This greatly increases the surface area of the epithelium and facilitates the absorption of materials. Brush borders are found in the convoluted tubules of the kidney and in the lining of the small intestine.

**brush discharge** A luminous discharge from a conductor that takes the form of luminous branching threads that penetrate into the surrounding gas. It is a form of *corona and it occurs when the electric field near the surface of the conductor exceeds a certain value but is not sufficiently high for a spark to appear.

**brusselator** A type of chemical reaction mechanism that leads to an *oscillating reaction. It involves the conversion of reactants A and B into products C and B by a series of four steps:

$$A \rightarrow X$$
$$2X + Y \rightarrow 3Y$$
$$B + X \rightarrow Y + C$$
$$X \rightarrow D$$

Autocatalysis occurs as in the *Lotka–Volterra mechanism and the *oregonator. If the concentrations of A and B are maintained constant, the concentrations of X and Y oscillate with time. A graph of the concentration of X against that of Y is a closed loop (the **limit cycle** of the reaction). The reaction settles down to this limit cycle whatever the initial concentrations of X and Y, i.e. the limit cycle is an **attractor** for the system. The reaction mechanism is named after the city of Brussels, where the research group that discovered it is based.

**Bryophyta** A phylum of simple plants – the mosses – possessing no vascular tissue and rudimentary rootlike organs (rhizoids). They grow in a variety of damp habitats, from fresh water to rock surfaces. Some use other plants for support. Mosses show a marked *alternation of generations between gamete-bearing forms (gametophytes) and spore-bearing forms (sporophytes): they possess erect or prostrate leafy stems (the gametophyte generation, which is *haploid); these give rise to leafless stalks bearing capsules (the sporophyte generation, which is *diploid), the latter being dependent on the former for water and nutrients. Spores formed in the capsules are released and grow to produce new plants.

Formerly, this phylum also included the liverworts (see HEPATOPHYTA) and the mosses were classified as a class (Musci) of the Bryophyta. The term 'bryophytes' is still used informally to refer to both the mosses and the liverworts.

**(((∰))) SEE WEB LINKS**

• A resource devoted to mosses, liverworts, and hornworts from Southern Illinois University, Carbondale

**Bryozoa (Ectoprocta)** A phylum of aquatic, mainly marine, invertebrates – the moss animals and sea mats. They live in colonies, 50 cm or more across, which are attached to rocks, seaweeds, or shells. The individuals making up the colonies are about 1 mm long and superficially resemble cnidarian *polyps, with a mouth surrounded by ciliated tentacles that trap minute particles of organic matter in the water. Some have a horny or calcareous outer skeleton into which the body can be withdrawn. Bryozoans are placed in the clade Lophotrochozoa.

**BSE** *See* BOVINE SPONGIFORM ENCEPHALOPATHY.

**bubble chamber** A device for detecting ionizing radiation. It consists of a chamber containing a liquid, often hydrogen, kept at slightly above its boiling point at a preliminary pressure that is high enough to prevent boiling. Immediately before the passage of the ionizing particles the pressure is reduced, and the particles then act as centres for the formation of bubbles, which can be photographed to obtain a record of the particles' tracks. The device was invented in 1952 by Donald Glaser. *Compare* CLOUD CHAMBER.

**buccal cavity (oral cavity)** The mouth cavity: the beginning of the *alimentary canal, which leads to the pharynx and (in vertebrates) to the oesophagus. In vertebrates it is separated from the nasal cavity by the *palate. In mammals it contains the tongue and teeth, which assist in the mechanical breakdown of food, and the openings of the *salivary gland ducts.

**Buchner funnel** A type of funnel with an internal perforated tray on which a flat circular filter paper can be placed, used for filtering by suction. It is named after the German chemist Eduard Buchner (1860–1917).

**buckminsterfullerene** A form of carbon composed of clusters of 60 carbon atoms bonded together in a polyhedral structure composed of pentagons and hexagons (see illustration). Originally it was identified in 1985 in products obtained by firing a high-power laser at a graphite target. It can be made by an electric arc struck between graphite electrodes in an inert atmosphere. The molecule, $C_{60}$, was named after the US architect Richard Buckminster Fuller (1895–1983) because of the resemblance of the structure to the geodesic dome, which Fuller invented. The molecules are informally called **buckyballs**; more formally, the substance itself is also called **fullerene**. The substance is a yellow crystalline solid (**fullerite**), soluble in benzene.

Various fullerene derivatives are known in which organic groups are attached to carbon atoms on the sphere. In addition, it is possible to produce novel enclosure compounds by trapping metal ions within the $C_{60}$ cage. Some of these have semiconducting properties. The electric-arc method of producing $C_{60}$ also leads to a smaller number of fullerenes such as $C_{70}$, which have less symmetrical molecular structures. It is also possible to produce forms of carbon in which the atoms are linked in a cylindrical, rather than spherical, framework with a diameter of a few nanometres. They are known as **buckytubes** (or **nanotubes**).

**(((⊕))) SEE WEB LINKS**

• Information about IUPAC nomenclature and representation of fullerenes and related compounds

**buckyball** *See* BUCKMINSTERFULLERENE.

**buckytube** *See* BUCKMINSTERFULLERENE.

**bud 1.** (in botany) A condensed immature shoot with a short stem bearing small folded or rolled leaves. The outer leaves of a bud are often scalelike and protect the delicate inner leaves. A **terminal** (or **apical**) **bud** exists at the tip of a stem or branch while **axillary** (or **lateral**) **buds** develop in the *axils of leaves. However, in certain circumstances buds can be produced anywhere on the surface of a plant. Some buds remain dormant, but may become active if the terminal bud is removed. It is common gardening practice to remove the terminal buds of some shoots to induce the development of lateral shoots from axillary buds. *See also* APICAL DOMINANCE. **2.** (in biology) An outgrowth from a parent organism that breaks away and develops into a new individual in the process of *budding.

**budding 1.** (in biology) A method of asexual reproduction in which a new individual is derived from an outgrowth (**bud**) that be-

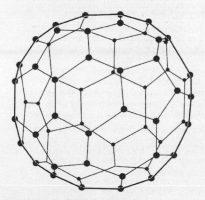

**Buckminsterfullerene structure.**

comes detached from the body of the parent. In animals the process is also called **gemmation**; it is common in cnidarians (e.g. *Hydra*) and also occurs in some sponges and other invertebrates. Among fungi, budding is characteristic of the yeasts. **2.** (in horticulture) A method of grafting in which a bud of the scion is inserted onto the stock, usually beneath the bark.

**buffer** A solution that resists change in pH when small amounts of an acid or alkali are added over a certain range or when the solution is diluted. Acidic buffers consist of a weak acid with a salt of the acid. The salt provides the negative ion A⁻, which is the conjugate base of the acid HA. An example is carbonic acid and sodium hydrogencarbonate. Basic buffers have a weak base and a salt of the base (to provide the conjugate acid). An example is ammonia solution with ammonium chloride.

In an acidic buffer, for example, molecules HA and ions A⁻ are present. When acid is added most of the extra protons are removed by the base:

$$A^- + H^+ \rightarrow HA$$

When base is added, most of the extra hydroxide ions are removed by reaction with undissociated acid:

$$OH^- + HA \rightarrow A^- + H_2O$$

Thus, the addition of acid or base changes the pH very little. The hydrogen-ion concentration in a buffer is given by the expression

$$K_a = [H^+] = [A^-]/[HA]$$

i.e. it depends on the ratio of conjugate base to acid. As this is not altered by dilution, the hydrogen-ion concentration for a buffer does not change much during dilution.

In the laboratory, buffers are used to prepare solutions of known stable pH. Natural buffers occur in living organisms, where the biochemical reactions are very sensitive to change in pH (*see* ACID–BASE BALANCE). The main natural buffers are $H_2CO_3/HCO_3^-$ and $H_2PO_4^-/HPO_4^{2-}$. Buffer solutions are also used in medicine (e.g. in intravenous injections), in agriculture, and in many industrial processes (e.g. dyeing, fermentation processes, and the food industry).

**bugs** *See* HEMIPTERA.

**bulb** An underground plant organ that enables a plant to survive from one growing season to the next. It is a modified shoot with a short flattened stem. A terminal bud develops at the centre of its upper surface, sur-

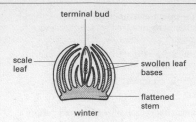

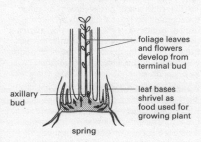

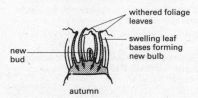

**Development of a bulb.**

rounded by swollen leaf bases that contain food stored from the previous growing season. Papery brown scale leaves cover the outside of the bulb. The stored food is used in the growing season when the terminal bud produces foliage leaves and flowers. The new leaves photosynthesize and some of the manufactured food passes into the leaf bases forming a new bulb (see illustration). If more than one bud develops, then additional bulbs form, resulting in vegetative propagation. Examples of bulb-forming plants are daffodil, onion, and tulip. *Compare* CORM.

**bulbil** A small bulblike organ that may develop in place of a flower, from an axillary bud, or at the base of a stem in certain plants. If it becomes detached it develops into a new plant.

**bulk modulus** *See* ELASTIC MODULUS.

**bulla** The rounded hollow projection of

bone from the skull that encloses the *middle ear in mammals.

**bumping** Violent boiling of a liquid caused by superheating so that bubbles form at a pressure above atmospheric pressure. It can be prevented by putting pieces of porous pot in the liquid to enable bubbles of vapour to form at the normal boiling point.

**buna rubber** A type of synthetic rubber based on polymerization of butadiene (buta-1,3-diene). The name comes from Bu (for butadiene) and Na (for sodium, which was used as a catalyst in the original polymerization reaction). An improved form, known as **Buna-S**, was developed by copolymerizing butadiene with styrene. In 1934, **Buna-N** was invented, in which the styrene was replaced by acrylonitrile, giving a product with better oil resistance (*see* NITRILE RUBBER).

**bunching** In quantum optics, the arrival of photons at a detector grouped more closely together than they would be if grouped randomly. Bunching can also be described without quantum effects in terms of fluctuations of the classical electromagnetic field. The arrival of photons at a detector less closely together than randomly is called **antibunching** and is a specifically quantum-mechanical phenomenon.

**bundle of His** The specialized cardiac muscle fibres in the mammalian heart that receive electrical stimuli from the *atrioventricular node and transmit them throughout the network of *Purkyne fibres. This allows the excitation to reach all parts of the ventricles rapidly and initiates a wave of contraction to expel blood into the aorta and pulmonary artery. It is named after Wilhelm His (1831–1904).

**Bunsen, Robert Wilhelm** (1811–99) German chemist, who held professorships at Kassel, Marburg, and Heidelberg. His early researches on arsenic-containing compounds cost him an eye in an explosion. He then turned to gas analysis and spectroscopy, enabling him and *Kirchhoff to discover the elements *caesium (1860) and *rubidium (1861). He also popularized the use of the *Bunsen burner.

**Bunsen burner** A laboratory gas burner having a vertical metal tube into which the gas is led, with a hole in the side of the base of the tube to admit air. The amount of air can be regulated by a sleeve on the tube. When no air is admitted the flame is luminous and smoky. With air, it has a faintly visible hot outer part (the oxidizing part) and an inner blue cone where combustion is incomplete (the cooler reducing part of the flame). The device is named after Robert Bunsen, who used a similar device (without a regulating sleeve) in 1855.

**Bunsen cell** A *primary cell consisting of a zinc cathode immersed in dilute sulphuric acid and a carbon anode immersed in concentrated nitric acid. The electrolytes are separated by a porous pot. The cell gives an e.m.f. of about 1.9 volts.

**buoyancy** The upward thrust on a body immersed in a fluid. This force is equal to the weight of the fluid displaced (*see* ARCHIMEDES' PRINCIPLE).

**burette** A graduated glass tube with a tap at one end leading to a fine outlet tube, used for delivering known volumes of a liquid (e.g. in titration).

**Burgess shale** A fossil-rich deposit of shale and slate dating from the mid-Cambrian period (about 520 million years ago) and located in the Burgess Pass in the Rocky Mountains of British Columbia, Canada. First excavated in 1909, it has since yielded one of the world's oldest assemblages of fossilized marine invertebrates and early vertebrate animals, as well as algae and sponges. The species described include brachiopods, crustaceans, trilobites, and other arthropods, plus a remarkably well-preserved array of fossil worms and other soft-bodied animals. Many forms are of uncertain classification and have been assigned to new phyla. Such great diversity of fossils in deposits from this period is seen by some biologists as evidence for a burst of rapid evolution (*see* CAMBRIAN EXPLOSION).

**Burnet, Sir Frank Macfarlane** (1899–1985) Australian virologist, who spent his working life at the Walter and Eliza Hall Institute in Melbourne. In the early 1930s he developed a method of growing influenza virus in chick embryos. He later discovered that immunological tolerance (failure of the immune response) required repeated exposure to the antigen. For this work he shared the 1960 Nobel Prize for physiology or medicine with Sir Peter *Medawar. He also proposed the *clonal selection theory.

**bus** A set of conducting paths – wires or optical fibres – connecting several components of a computer system and allowing the components to send signals to each other. The components take it in turns to transmit.

**buta-1,3-diene (butadiene)** A colourless gaseous hydrocarbon, $CH_2:CHCH:CH_2$; m.p. $-109°C$; b.p. $-4.5°C$. It is made by catalytic dehydrogenation of butane (from petroleum or natural gas) and polymerized in the production of synthetic rubbers. The compound is a conjugated *diene in which the electrons in the pi orbitals are partially delocalized over the whole molecule. It can have trans and cis forms, the latter taking part in *Diels–Alder reactions.

**butanal (butyraldehyde)** A colourless flammable liquid aldehyde, $C_3H_7CHO$; r.d. 0.8; m.p. $-99°C$; b.p. $75.7°C$.

**butane** A gaseous hydrocarbon, $C_4H_{10}$; d. $0.58$ g cm$^{-3}$; m.p. $-138°C$; b.p. $0°C$. Butane is obtained from petroleum (from refinery gas or by cracking higher hydrocarbons). The fourth member of the *alkane series, it has a straight chain of carbon atoms and is isomeric with 2-methylpropane ($CH_3CH(CH_3)CH_3$, formerly called **isobutane**). It can easily be liquefied under pressure and is supplied in cylinders for use as a fuel gas. It is also a raw material for making buta-1,3-diene (for synthetic rubber).

**butanedioic acid (succinic acid)** A colourless crystalline fatty acid, $(CH_2)_2(COOH)_2$; r.d. 1.6; m.p. 185°C; b.p. 235°C. A weak carboxylic acid, it is produced by fermentation of sugar or ammonium tartrate and used as a sequestrant and in making dyes. It occurs in living organisms as an intermediate in metabolism, especially in the *Krebs cycle.

**butanoic acid (butyric acid)** A colourless liquid water-soluble acid, $C_3H_7COOH$; r.d. 0.96; b.p. 163°C. It is a weak acid ($K_a = 1.5 \times 10^{-5}$ mol dm$^{-3}$ at 25°C) with a rancid odour. Its esters are present in butter and in human perspiration. The acid is used to make esters for flavourings and perfumery.

**butanol** Either of two aliphatic alcohols with the formula $C_4H_9OH$. **Butan-1-ol**, $CH_3(CH_2)_3OH$, is a primary alcohol; r.d. 0.81; m.p. $-89.5°C$; b.p. 117.3°C. **Butan-2-ol**, $CH_3CH(OH)C_2H_5$, is a secondary alcohol; r.d. 0.81; m.p. $-114.7°C$; b.p. 100°C. Both are colourless volatile liquids obtained from butane and are used as solvents.

**butanone (methyl ethyl ketone)** A colourless flammable water-soluble liquid,

$CH_3COC_2H_5$; r.d. 0.8; m.p. $-86.4°C$; b.p. 79.6°C. It can be made by the catalytic oxidation of butane and is used as a solvent.

**butenedioic acid** Either of two isomers with the formula $HCOOHC:CHCOOH$. Both compounds can be regarded as derivatives of ethene in which a hydrogen atom on each carbon has been replaced by a $-COOH$ group. The compounds show cis–trans isomerism. The trans form is **fumaric acid** (r.d. 1.64; sublimes at 165°C) and the cis form is **maleic acid** (r.d. 1.59; m.p. 139–140°C). Both are colourless crystalline compounds used in making synthetic resins. The cis form is rather less stable than the trans form and converts to the trans form at 120°C. Unlike the trans form it can combine water on heating to form a cyclic anhydride containing a $-CO.O.CO-$ group (**maleic anhydride**). Fumaric acid is an intermediate in the *Krebs cycle.

**butterflies** *See* LEPIDOPTERA.

**butterfly effect** *See* CHAOS.

**buttress root** *See* PROP ROOT.

**butyl group** The organic group $CH_3(CH_2)_3-$.

**butyl rubber** A type of synthetic rubber obtained by copolymerizing 2-methylpropene ($CH_2:C(CH_3)CH_3$; isobutylene) and methylbuta-1,3-diene ($CH_2:C(CH_3)CH:CH_2$, isoprene). Only small amounts of isoprene (about 2 mole %) are used. The rubber can be vulcanized. Large amounts were once used for tyre inner tubes.

**butyraldehyde** *See* BUTANAL.

**butyric acid** *See* BUTANOIC ACID.

**Buys Ballot's law** A law relating to winds stating that observers with their backs to the wind will experience a lower pressure on the left than on the right in the northern hemisphere, and lower on the right than on the left in the southern hemisphere. The law was propounded by the Dutch meteorologist Christoph Buys Ballot (1817–90) in 1857.

**by-product** A compound formed during a chemical reaction at the same time as the main product. Commercially useful by-products are obtained from a number of industrial processes. For example, calcium chloride is a by-product of the *Solvay process for making sodium carbonate. Propanone is a by-product in the manufacture of *phenol.

**byte** A subdivision of a *word in a computer, it usually consists of eight *bits. A kilobyte is 1024 bytes (not 1000 bytes).

**BZ** *See* BENZOYLECGONINE.

**B–Z reaction (Belousov–Zhabotinskii reaction)** A chemical reaction that shows a periodic colour change between magenta and blue with a period of about one minute. It occurs with a mixture of sulphuric acid, potassium bromate(V), cerium sulphate, and propanedioic acid. The colour change is caused by alternating oxidation–reductions in which cerium changes its oxidation state ($Ce^{3+}$ gives a magenta solution while $Ce^{4+}$ gives a blue solution). The B–Z reaction is an example of a chemical *oscillating reaction – a reaction in which there is a regular periodic change in the concentration of one or more reactants. The mechanism is highly complicated, involving a large number of steps. *See* BRUSSELATOR.

**C** A high-level general-purpose computer language developed in 1972. It is fast and can be used as an alternative to assembly language, with which it shares some features. It was superseded in the mid-1980s by an object-oriented version known as C++, which is better suited to the design of modular programs.

 **SEE WEB LINKS**
• The current C standard

**cadmium** Symbol Cd. A soft bluish metal belonging to *group 12 (formerly IIB) of the periodic table; a.n. 48; r.a.m. 112.41; r.d. 8.65; m.p. 320.9°C; b.p. 765°C. The element's name is derived from the ancient name for calamine, zinc carbonate $ZnCO_3$, and it is usually found associated with zinc ores, such as sphalerite (ZnS), but does occur as the mineral greenockite (CdS). Cadmium is usually produced as an associate product when zinc, copper, and lead ores are reduced. Cadmium is used in low-melting-point alloys to make solders, in Ni–Cd batteries, in bearing alloys, and in electroplating (over 50%). Cadmium compounds are used as phosphorescent coatings in TV tubes. Cadmium and its compounds are extremely toxic at low concentrations; great care is essential where solders are used or where fumes are emitted. It has similar chemical properties to zinc but shows a greater tendency towards complex formation. The element was discovered in 1817 by F. Stromeyer.

**SEE WEB LINKS**
• Information from the WebElements site

**cadmium cell** *See* WESTON CELL.

**cadmium sulphide** A water-insoluble compound, CdS; r.d. 4.82. It occurs naturally as the mineral **greenockite** and is used as a pigment and in semiconductors and fluorescent materials.

**caecum** A pouch in the alimentary canal of vertebrates between the *small intestine and *colon. The caecum (and its *appendix) is large and highly developed in herbivorous animals (e.g. rabbits and cows), in which it contains a large population of bacteria essential for the breakdown of cellulose. In humans the caecum is a *vestigial organ and is poorly developed.

**caesium** Symbol Cs. A soft silvery-white metallic element belonging to *group 1 (formerly IA) of the periodic table; a.n. 55; r.a.m. 132.905; r.d. 1.88; m.p. 28.4°C; b.p. 678°C. It occurs in small amounts in a number of minerals, the main source being carnallite ($KCl.MgCl_2.6H_2O$). It is obtained by electrolysis of molten caesium cyanide. The natural isotope is caesium–133. There are 15 other radioactive isotopes. Caesium–137 (half-life 33 years) is used as a gamma source. As the heaviest alkali metal, caesium has the lowest ionization potential of all elements, hence its use in photoelectric cells, etc.

**SEE WEB LINKS**
• Information from the WebElements site

**caesium chloride structure** A type of ionic crystal structure in which the anions are at the eight corners of a cubic unit cell with one cation at the centre of the cell. It can equivalently by described as cations at the corners of the cell with an anion at the centre. Each type of ion has a coordination number of 8. Examples of compounds with this structure are CsCl, CsBr, CsI, CsCN, CuZn, and $NH_4Cl$.

**caesium clock** An *atomic clock that depends on the energy difference between two states of the caesium–133 nucleus when it is in a magnetic field. In one type, atoms of caesium–133 are irradiated with *radio-frequency radiation, whose frequency is chosen to correspond to the energy difference between the two states. Some caesium nuclei absorb this radiation and are excited to the higher state. These atoms are deflected by a further magnetic field, which causes them to hit a detector. A signal from this detector is fed back to the radio-frequency oscillator to prevent it drifting from the resonant frequency of 9 192 631 770 hertz. In this way the device is locked to this frequency with an accuracy better than 1 part in $10^{13}$.

The caesium clock is used in the *SI unit definition of the second.

**caffeine (1,3,7-trimethylxanthine)** An alkaloid, $C_8H_{10}N_4O_2$; m.p. 235°C; sublimes at 176°C. It is a stimulant and diuretic and is present in coffee, tea, and some soft drinks. *See* METHYLXANTHINES.

**cage compound** *See* CLATHRATE.

**Cainozoic** *See* CENOZOIC.

**calciferol** *See* VITAMIN D.

**calcination** The formation of a calcium carbonate deposit from hard water. *See* HARDNESS OF WATER.

**calcinite** A mineral form of *potassium hydrogencarbonate, $KHCO_3$.

**calcite** One of the most common and widespread minerals, consisting of crystalline calcium carbonate, $CaCO_3$. Calcite crystallizes in the rhombohedral system; it is usually colourless or white and has a hardness of 3 on the Mohs' scale. It has the property of double refraction, which is apparent in Iceland spar – the transparent variety of calcite. It is an important rock-forming mineral and is a major constituent in limestones, marbles, and carbonatites.

**calcitonin (thyrocalcitonin)** A peptide hormone in vertebrates that lowers the concentration of calcium (and phosphate) in the blood. It operates in opposition to *parathyroid hormone. Calcitonin is produced by the *C cells, which in mammals are located in the *thyroid gland.

**calcitriol** *See* VITAMIN D.

**calcium** Symbol Ca. A soft grey metallic element belonging to *group 2 (formerly IIA) of the periodic table; a.n. 20; r.a.m. 40.08; r.d. 1.54; m.p. 839°C; b.p. 1484°C. Calcium compounds are common in the earth's crust; e.g. limestone and marble ($CaCO_3$), gypsum ($CaSO_4.2H_2O$), and fluorite ($CaF_2$). The element is extracted by electrolysis of fused calcium chloride and is used as a getter in vacuum systems and a deoxidizer in producing nonferrous alloys. It is also used as a reducing agent in the extraction of such metals as thorium, zirconium, and uranium.

Calcium is an *essential element for living organisms, being required for normal growth and development. In animals it is an important constituent of bones and teeth and is present in the blood, being required for muscle contraction and other metabolic processes. In plants it is a constituent (in the form of calcium pectate) of the *middle lamella.

**⊕ SEE WEB LINKS**

• Information from the WebElements site

**calcium acetylide** *See* CALCIUM DICARBIDE.

**calcium bicarbonate** *See* CALCIUM HYDROGENCARBONATE.

**calcium carbide** *See* CALCIUM DICARBIDE.

**calcium carbonate** A white solid, $CaCO_3$, which is only sparingly soluble in water. Calcium carbonate decomposes on heating to give *calcium oxide (quicklime) and carbon dioxide. It occurs naturally as the minerals *calcite (rhombohedral; r.d. 2.71) and *aragonite (rhombic; r.d. 2.93). Rocks containing calcium carbonate dissolve slowly in acidified rainwater (containing dissolved $CO_2$) to cause temporary hardness. In the laboratory, calcium carbonate is precipitated from *limewater by carbon dioxide. Calcium carbonate is used in making lime (calcium oxide) and is the main raw material for the *Solvay process.

**calcium chloride** A white deliquescent compound, $CaCl_2$, which is soluble in water; r.d. 2.15; m.p. 782°C; b.p. >1600°C. There are a number of hydrated forms, including the monohydrate, $CaCl_2.H_2O$, the dihydrate, $CaCl_2.2H_2O$ (r.d. 0.84), and the hexahydrate, $CaCl_2.6H_2O$ (trigonal; r.d. 1.71; the hexahydrate loses $4H_2O$ at 30°C and the remaining $2H_2O$ at 200°C). Large quantities of it are formed as a byproduct of the *Solvay process and it can be prepared by dissolving calcium carbonate or calcium oxide in hydrochloric acid. Crystals of the anhydrous salt can only be obtained if the hydrated salt is heated in a stream of hydrogen chloride. Solid calcium chloride is used in mines and on roads to reduce dust problems, whilst the molten salt is the electrolyte in the extraction of calcium. An aqueous solution of calcium chloride is used in refrigeration plants.

**calcium cyanamide** A colourless solid, $CaCN_2$, which sublimes at 1300°C. It is prepared by heating calcium dicarbide at 800°C in a stream of nitrogen:

$$CaC_2(s) + N_2(g) \rightarrow CaCN_2(s) + C(s)$$

The reaction has been used as a method of fixing nitrogen in countries in which cheap electricity is available to make the calcium dicarbide (the **cyanamide process**). Calcium cyanamide can be used as a fertilizer because it reacts with water to give ammonia and calcium carbonate:

$$CaCN_2(s) + 3H_2O(l) \rightarrow CaCO_3(s) + 2NH_3(g)$$

It is also used in the production of melamine, urea, and certain cyanide salts.

**calcium dicarbide (calcium acetylide; calcium carbide; carbide)** A colourless solid compound, $CaC_2$; tetragonal; r.d. 2.22; m.p. 450°C; b.p. 2300°C. In countries in which electricity is cheap it is manufactured by heating calcium oxide with either coke or ethyne at temperatures above 2000°C in an electric arc furnace. The crystals consist of $Ca^{2+}$ and $C_2^-$ ions arranged in a similar way to the ions in sodium chloride. When water is added to calcium dicarbide, the important organic raw material ethyne (acetylene) is produced:

$$CaC_2(s) + 2H_2O(l) \rightarrow Ca(OH)_2(s) + C_2H_2(g)$$

**calcium fluoride** A white crystalline solid, $CaF_2$; r.d. 3.2; m.p. 1423°C; b.p. 2500°C. It occurs naturally as the mineral *fluorite (or fluorspar) and is the main source of fluorine. The **calcium fluoride structure (fluorite structure)** is a crystal structure in which the calcium ions are each surrounded by eight fluoride ions arranged at the corners of a cube. Each fluoride ion is surrounded by four calcium ions at the corners of a tetrahedron.

**calcium hydrogencarbonate (calcium bicarbonate)** A compound, $Ca(HCO_3)_2$, that is stable only in solution and is formed when water containing carbon dioxide dissolves calcium carbonate:

$$CaCO_3(s) + H_2O(l) + CO_2(g) \rightarrow$$
$$Ca(HCO_3)_2(aq)$$

It is the cause of temporary *hardness of water, because the calcium ions react with soap to give scum. Calcium hydrogencarbonate is unstable when heated and decomposes to give solid calcium carbonate. This explains why temporary hardness is removed by boiling and the formation of 'scale' in kettles and boilers.

**calcium hydroxide (slaked lime)** A white solid, $Ca(OH)_2$, which dissolves sparingly in water (*see* LIMEWATER); hexagonal; r.d. 2.24. It is manufactured by adding water to calcium oxide, a process that evolves much heat and is known as slaking. It is used as a cheap alkali to neutralize the acidity in certain soils

and in the manufacture of mortar, whitewash, bleaching powder, and glass.

**calcium nitrate** A white deliquescent compound, $Ca(NO_3)_2$, that is very soluble in water; cubic; r.d. 2.50; m.p. 561°C. It can be prepared by neutralizing nitric acid with calcium carbonate and crystallizing it from solution as the tetrahydrate $Ca(NO_3)_2.4H_2O$, which exists in two monoclinic crystalline forms (α, r.d. 1.9; β, r.d. 1.82). There is also a trihydrate, $Ca(NO_3)_2.3H_2O$. The anhydrous salt can be obtained from the hydrate by heating but it decomposes on strong heating to give the oxide, nitrogen dioxide, and oxygen. Calcium nitrate is sometimes used as a nitrogenous fertilizer.

**calcium octadecanoate (calcium stearate)** An insoluble white salt, $Ca(CH_3(CH_2)_{16}COO)_2$, which is formed when soap is mixed with water containing calcium ions and is the scum produced in hard-water regions.

**calcium oxide (quicklime)** A white solid compound, CaO, formed by heating calcium in oxygen or by the thermal decomposition of calcium carbonate; cubic; r.d. 3.35; m.p. 2580°C; b.p. 2850°C. On a large scale, calcium carbonate in the form of limestone is heated in a tall tower (lime kiln) to a temperature above 550°C:

$$CaCO_3(s) \rightleftharpoons CaO(s) + CO_2(g)$$

Although the reaction is reversible, the carbon dioxide is carried away by the upward current through the kiln and all the limestone decomposes. Calcium oxide is used to make calcium hydroxide, as a cheap alkali for treating acid soil, and in extractive metallurgy to produce a slag with the impurities (especially sand) present in metal ores.

**calcium phosphate(V)** A white insoluble powder, $Ca_3(PO_4)_2$; r.d. 3.14. It is found naturally in the mineral *apatite, $Ca_5(PO_4)_3(OH,F,Cl)$, and as rock phosphate. It is also the main constituent of animal bones. Calcium phosphate can be prepared by mixing solutions containing calcium ions and hydrogenphosphate ions in the presence of an alkali:

$$HPO_4{}^{2-} + OH^- \rightarrow PO_4{}^{3-} + H_2O$$
$$3Ca^{2+} + 2PO_4{}^{3-} \rightarrow Ca_3(PO_4)_2$$

It is used extensively as a fertilizer. The compound was formerly called **calcium orthophosphate** (*see* PHOSPHATES).

**calcium stearate** *See* CALCIUM OCTADE-CANOATE.

**calcium sulphate** A white solid compound, $CaSO_4$; r.d. 2.96; 1450°C. It occurs naturally as the mineral *anhydrite, which has a rhombic structure, transforming to a monoclinic form at 200°C. More commonly, it is found as the dihydrate, *gypsum, $CaSO_4.2H_2O$ (monoclinic; r.d. 2.32). When heated, gypsum loses water at 128°C to give the hemihydrate, $2CaSO_4.H_2O$, better known as *plaster of Paris. Calcium sulphate is sparingly soluble in water and is a cause of permanent *hardness of water. It is used in the manufacture of certain paints, ceramics, and paper. The naturally occurring forms are used in the manufacture of sulphuric acid.

**calculus** A series of mathematical techniques developed independently by Isaac Newton and Gottfried Leibniz (1646–1716). **Differential calculus** treats a continuously varying quantity as if it consisted of an infinitely large number of infinitely small changes. For example, the velocity $v$ of a body at a particular instant can be regarded as the infinitesimal distance, written $ds$, that it travels in the vanishingly small time interval, $dt$; the instantaneous velocity $v$ is then $ds/dt$, which is called the **derivative** of $s$ with respect to $t$. If $s$ is a known function of $t$, $v$ at any instant can be calculated by the process of *differentiation. The differential calculus is a powerful technique for solving many problems concerned with rate processes, maxima and minima, and similar problems.

**Integral calculus** is the opposite technique. For example, if the velocity of a body is a known function of time, the infinitesimal distance $ds$ travelled in the brief instant $dt$ is given by $ds = vdt$. The measurable distance $s$ travelled between two instants $t_1$ and $t_2$ can then be found by a process of summation, called *integration, i.e.

$$s = \int_{t_2}^{t_1} vdt$$

The technique is used for finding areas and volumes and other problems involving the summation of infinitesimals.

**caldera** A crater or large depression at the top of an inactive shield *volcano, which may be 1–20 km in diameter. It forms when magma subsides from the summit, sometimes aided by the explosive ejection of material. Often a caldera fills with water, forming a crater lake. Large calderas appear to be characteristic features of the landscapes of Mars and Venus.

**Calgon** Trade name for a water-softening agent. *See* HARDNESS OF WATER.

**caliche** A mixture of salts found in deposits between gravel beds in the Atacama and Tarapaca regions of Chile. They vary from 4 m to 15 cm thick and were formed by periodic leaching of soluble salts during wet geological epochs, followed by drying out of inland seas in dry periods. They are economically important as a source of nitrates. A typical composition is $NaNO_3$ 17.6%, NaCl 16.1%, $Na_2SO_4$ 6.5%, $CaSO_4$ 5.5%, $MgSO_4$ 3.0%, $KNO_3$ 1.3%, $Na_2B_4O_7$ 0.94%, $KClO_3$ 0.23%, $NaIO_3$ 0.11%, sand and gravel to 100%.

**californium** Symbol Cf. A radioactive metallic transuranic element belonging to the *actinoids; a.n. 98; mass number of the most stable isotope 251 (half-life about 700 years). Nine isotopes are known; californium–252 is an intense neutron source, which makes it useful in neutron *activation analysis and potentially useful as a radiation source in medicine. The element was first produced by Glenn Seaborg (1912–99) and associates in 1950.

(((●))) SEE WEB LINKS
• Information from the WebElements site

**calixarenes** Compounds that have molecules with a cuplike structure (the name comes from the Greek *calix*, cup). The simplest, has four phenol molecules joined by four –$CH_2$– groups into a ring (forming the base of the 'cup'). The four phenol hexagons point in the same direction to form a cavity that can bind substrate molecules. Interest has been shown in the potential ability of calixarene molecules to mimic enzyme action.

**callus** 1. (in botany) A protective tissue, consisting of parenchyma cells, that develops over a cut or damaged plant surface. Callus tissue can also be induced to form in cell cultures by hormone treatment. 2. (in pathology) A thick hard area of skin that commonly forms on the palms of the hands and soles of the feet as a result of continuous pressure or friction. 3. (in physiology) Hard tissue formed round bone ends following a fracture, which is gradually converted to new bone.

**calomel** *See* MERCURY(I) CHLORIDE.

**calomel half cell (calomel electrode)** A type of half cell in which the electrode is mercury coated with calomel (HgCl) and the electrolyte is a solution of potassium chloride and saturated calomel. The standard electrode potential is –0.2415 volt (25°C). In the calomel half cell the reactions are

$$HgCl(s) \rightleftharpoons Hg^+(aq) + Cl^-(aq)$$

$$Hg^+(aq) + e \rightleftharpoons Hg(s)$$

The overall reaction is

$$HgCl(s) + e \rightleftharpoons Hg(s) + Cl^-(aq)$$

This is equivalent to a $Cl_2(g)|Cl^-(aq)$ half cell.

**caloric theory** A former theory concerning the nature of heat, which was regarded as a weightless fluid (called **caloric**). It was unable to account for the fact that friction could produce an unlimited quantity of heat and it was abandoned when Joule showed that heat is a form of energy.

**calorie** The quantity of heat required to raise the temperature of 1 gram of water by 1°C (1 K). The calorie, a c.g.s. unit, is now largely replaced by the *joule, an *SI unit. 1 calorie = 4.186 8 joules.

**Calorie (kilogram calorie; kilocalorie)** 1000 calories. This unit is still in limited use in estimating the energy value of foods, but is obsolescent.

**calorific value** The heat per unit mass produced by complete combustion of a given substance. Calorific values are used to express the energy values of fuels; usually these are expressed in megajoules per kilogram ($MJ\ kg^{-1}$). They are also used to measure the energy content of foodstuffs; i.e. the energy produced when the food is oxidized in the body. The units here are kilojoules per gram ($kJ\ g^{-1}$), although Calories (kilocalories) are often still used in nontechnical contexts. Calorific values are measured using a *bomb calorimeter.

**calorimeter** Any of various devices used to measure thermal properties, such as *calorific value, specific *heat capacity, specific *latent heat, etc. *See* BOMB CALORIMETER.

**Calvin, Melvin** (1911–97) US biochemist. After World War II, at the Lawrence Radiation Laboratory, Berkeley, he investigated the light-independent reactions of *photosynthesis. Using radioactive carbon-14 to label carbon dioxide, he discovered the

*Calvin cycle, for which he was awarded the 1961 Nobel Prize for chemistry.

**Calvin cycle** The metabolic pathway of the light-independent stages of *photosynthesis, which occurs in the stroma of the chloroplasts. The pathway was elucidated by Melvin Calvin and his co-workers and involves the fixation of carbon dioxide and its subsequent reduction to carbohydrate. During the cycle, carbon dioxide combines with *ribulose bisphosphate, through the mediation of the enzyme ribulose bisphosphate carboxylase/oxygenase (rubisco), to form an unstable six-carbon compound that breaks down to form two molecules of the three-carbon compound glycerate 3-phosphate. This is converted to glyceraldehyde 3-phosphate, which is used to regenerate ribulose bisphosphate and to produce glucose and fructose.

**calx** A metal oxide formed by heating an ore in air.

**calyptra** **1.** A layer of cells that covers the developing sporophyte of mosses, liverworts, clubmosses, horsetails, and ferns. In mosses it forms a hood over the *capsule and in liverworts it forms a sheath at the base of the capsule stalk. **2.** *See* ROOT CAP.

**calyptrogen** The region within the root *apical meristem that divides to produce the *root cap (calyptra).

**calyx** The *sepals of a flower, collectively, forming the outer whorl of the *perianth. It encloses the petals, stamens, and carpels and protects the flower in bud. *See also* PAPPUS.

**cambium (lateral meristem)** A plant tissue consisting of actively dividing cells (*see* MERISTEM) that is responsible for increasing the girth of the plant, i.e. it causes secondary growth. The two most important cambia are the **vascular** (or **fascicular**) **cambium** and the *cork cambium. The vascular cambium occurs in the stem and root; it divides to produce secondary *xylem and secondary *phloem (new food- and water-conducting tissues). In mature stems the vascular cambium is extended laterally to form a complete ring: the sections of this ring between the vascular bundles comprises the **interfascicular cambium**. *Compare* APICAL MERISTEM.

**Cambrian** The earliest geological period of the Palaeozoic era. It is estimated to have begun about 542 million years ago and lasted for some 54 million years. During this period marine animals with mineralized shells made their first appearance and Cambrian rocks are the first to contain an abundance of fossils. Cambrian fossils are chiefly of marine animals; they include *trilobites, which dominated the Cambrian seas, echinoderms, brachiopods, molluscs, and primitive *graptolites (from the mid Cambrian). Trace *fossils also provide evidence for a variety of worms.

**Cambrian explosion** A relatively short interval of rapid intense evolution that supposedly occurred in the early to mid-Cambrian period, some 540 to 520 million years ago. The supposition is based on the sudden appearance in the fossil record from this time of many diverse and novel forms, particularly marine animals, among which can be found representatives of all major modern groups. *See* BURGESS SHALE.

**camcorder** *See* VIDEO CAMERA.

**camera** **1.** An optical device for obtaining still photographs or for exposing cinematic film. It consists of a light-proof box with a lens at one end and a plate or film at the other. To make an exposure the shutter is opened and an image of the object to be photographed is formed on the light-sensitive film. The length of the exposure is determined by the intensity of light available, the film speed, and the *aperture of the lens. In the simpler cameras the shutter speed and aperture are controlled manually, but in automatic cameras the iris over the lens or the shutter is adjusted on the basis of information provided by a built-in *exposure meter. In ciné cameras the shutter automatically opens as the film comes to rest behind the lens for each frame; the film passes through the camera so that a set number (commonly 16, 18, or 24) of frames are exposed every second. **2.** A similar device (a **digital camera**) in which the film is replaced by a semiconductor array, which records the picture and stores it within the camera in a (usually) replaceable memory module. Moving pictures can be similarly recorded using a *video camera. **3.** The part of a television system that converts optical images into electronic signals. It consists of a lens system, which focuses the image to be televised on the photosensitive mosaic of the camera tube, causing localized discharge of those of its elements that are illuminated. This mo-

saic is scanned from behind by an electron beam so that the beam current is varied as it passes over areas of light and shade. The signal so picked up by the scanning beam is preamplified in the camera and passed to the transmitter with sound and synchronization signals. In *colour television three separate camera tubes are used, one for each *primary colour.

**camouflage** A high degree of similarity between an animal and its visual environment, which enables it to be disguised or concealed. By blending into the background the animal can elude predators or remain invisible to potential prey. *See also* CRYPTIC COLORATION; MIMICRY. *Compare* WARNING COLORATION.

**cAMP** *See* CYCLIC AMP.

**camphor** A white crystalline cyclic ketone, $C_{10}H_{16}O$; r.d. 0.99; m.p. 179°C; b.p. 204°C. It was formerly obtained from the wood of the Formosan camphor tree, but can now be synthesized. The compound has a characteristic odour associated with its use in mothballs. It is a plasticizer in celluloid.

**Canada balsam** A yellow-tinted resin used for mounting specimens in optical microscopy. It has similar optical properties to glass.

**canaliculus** A very small channel that occurs between the cells of the liver and bone. In the liver the bile canaliculi carry bile to the bile ducts; in bone, canaliculi connect lacunae, the cavities containing bone cells.

**canal rays** Streams of positive ions produced in a *discharge tube by boring holes (canals) in the cathode. The positive ions attracted to the cathode pass through the holes and emerge on the other side as positive rays.

**cancer** Any disorder of cell growth that results in invasion and destruction of surrounding healthy tissue by abnormal cells. Cancer cells arise from normal cells whose nature is permanently changed. They multiply more rapidly than healthy body cells and do not seem subject to normal control by nerves and hormones. They may spread via the bloodstream or lymphatic system to other parts of the body, where they produce further tissue damage (**metastases**). **Malignant tumour** is another name for cancer. A cancer that arises in epithelium is called a **carcinoma**; one that arises in connective tissue is called a **sarcoma**. **Leukaemia** is cancer of white blood cells; **lymphoma** is cancer of *lymphoid tissue; and **myeloma** is cancer of *plasma cells of the bone marrow. Causative agents (**carcinogens**) include various chemicals (including those in tobacco smoke), ionizing radiation, silica and asbestos particles, and *oncogenic viruses (*see also* ONCOGENE). Hereditary factors and stress may also play a role. Whatever the initiating factor, the mechanism by which cancer arises is mutation of genes in somatic cells.

(((●))) SEE WEB LINKS

- Extensive multimedia coverage of cancer biology and different types of cancer, produced by Emory University.

**candela** Symbol Cd. The *SI unit of luminous intensity equal to the luminous intensity in a given direction of a source that emits monochromatic radiation of frequency $540 \times 10^{12}$ Hz and has a radiant intensity in that direction of 1/683 watt per steradian.

**candle power** Luminous intensity as formerly expressed in terms of the international candle but now expressed in candela.

**cane sugar** *See* SUCROSE.

**canine tooth** A sharp conical *tooth in mammals that is large and highly developed in carnivores (e.g. dogs) for tearing meat. There are two canines in each jaw, each situated between the second *incisor and the first *premolar. In some animals (e.g. herbivores, such as giraffes and rabbits) canine teeth are absent.

**cannabinoids** A large group of structurally related phenolic compounds found in the plant *Cannabis sativa*. The main one is tetrahydrocannabinol (THC), which is the compound responsible for the effects of cannabis. It effects cannabinoid receptors found in the brain (and also in the spleen). The name 'cannabinoid' is also applied to structurally unrelated compounds found naturally in animal tissue and having an effect on the cannabinoid receptors. These **endocannabinoids** (endogenous cannabinoids) are believed to act as 'messengers' between cells.

**cannabis** An illegal drug produced from the plant *Cannabis sativa*. The dried inflorescences of the plant are known as **marijuana** and the thick resin produced from the plant is known as **hashish**. The normal method of using cannabis is to smoke it, although it can also be taken in certain foods. Cannabis is a class B controlled substance in the UK. It contains a large number of related compounds known as herbal *cannabinoids. The main active component is tetrahydrocannabinol (THC).

**Cannizzaro reaction** A reaction of aldehydes to give carboxylic acids and alcohols. It occurs in the presence of strong bases with aldehydes that do not have alpha hydrogen atoms. For example, benzenecarbaldehyde gives benzenecarboxylic acid and benzyl alcohol:

$$2C_6H_5CHO \rightarrow C_6H_5COOH + C_6H_5CH_2OH$$

Aldehydes that have alpha hydrogen atoms undergo the *aldol reaction instead. The Cannizzaro reaction is an example of a *disproportionation. It was discovered in 1853 by the Italian chemist Stanislao Cannizzaro (1826–1910).

**canonical form** One of the possible structures of a molecule that together form a *resonance hybrid.

**capacitance** The property of a conductor or system of conductors that describes its ability to store electric charge. The capacitance ($C$) is given by $Q/V$, where $Q$ is stored charge on one conductor and $V$ the potential difference between the two conductors (or between a single conductor and earth); it is measured in farads (or, in practice, microfarads).

An isolated sphere has a capacitance of $4\pi\varepsilon r$, where $r$ is the radius and $\varepsilon$ the *permittivity of the medium surrounding it. Capacitance is more commonly applied to systems of conductors (or semiconductors) separated by insulators (*see* CAPACITOR).

**capacitation** The final stage in the maturation process of a spermatozoon. This takes place inside the genital tract as the sperm penetrates the ovum.

**capacitor** An arrangement of conductors separated by an insulator (dielectric) used to store charge or introduce *reactance into an alternating-current circuit. The earliest form was the *Leyden jar. Capacitors used as circuit elements have two conducting plates separated by the dielectric. The dielectric may be air, paper impregnated with oil or wax, plastic film, or ceramic. The simplest form has two parallel rectangular conducting plates (area $A$) separated by a dielectric (thickness $d$, permittivity $\varepsilon$). The capacitance of such a capacitor is $A\varepsilon/d$. **Electrolytic capacitors** are devices in which a thin layer of an oxide is deposited on one of the electrodes to function as the dielectric.

**capacitor microphone** A microphone consisting of a *capacitor with a steady voltage applied across its parallel plates. One plate is fixed, the other is a thin diaphragm that is moved by the pressure of the sound waves. The movements of the diaphragm cause a variation in the spacing and therefore in the *capacitance of the device. This variation in capacitance is, in turn, reflected in a similar variation in the charge carried by each plate. The consequent current to and from one plate is carried by a resistor, the varying potential difference across which constitutes the device's output signal. It was formerly known as a **condenser microphone**.

**capillarity** See SURFACE TENSION.

**capillary** 1. A tube of small diameter. 2. **(blood capillary)** The narrowest type of blood vessel in the vertebrate circulatory system. Capillaries conduct blood from *arterioles to all living cells: their walls are only one cell layer thick, so that oxygen and nutrients can pass through them into the surrounding tissues. Capillaries also transport waste material (e.g. urea and carbon dioxide) to venules for ultimate excretion. Capillaries can be constricted or dilated, according to local tissue requirements.

**capillary electrophoresis (CE)** A technique for investigating mixtures of charged species. In **capilliary zone electrophoresis (CZE)**, the sample is introduced into a fine capillary tube, with each end of the capillary placed in a reservoir containing an electrolyte (e.g. a buffer solution). The source reservoir contains a positive electrode and the destination reservoir contains a negative electrode. A high potential difference is maintained between the electrodes. Components of the sample flow through the buffer solution in the capillary under the influence of the electric field. Their mobility depends on their charge, size, and shape, and the components separate as they move through the tube. They are detected close to the end of the capillary, usually by ultraviolet absorption. **Capilliary gel electrophoresis (CGE)** is used for separating large charged species, such as DNA fragments. The capilliary is filled with a gel and separation depends on the size of the species. In CE, a

graph of detector output against time is known as an **electropherogram**. Individual components can be identified by their retention times in the capillary. The technique can be highly sensitive and is widely used in forensic laboratories.

**capillary gel electrophoresis (CGE)** See CAPILLARY ELECTROPHORESIS.

**capillary zone electrophoresis (CZE)** See CAPILLARY ELECTROPHORESIS.

**capitulum** A type of flowering shoot (see RACEMOSE INFLORESCENCE) characteristic of plants of the family Compositae (Asteraceae), e.g. daisy and dandelion. The tip of the shoot is flattened and bears many small stalkless flowers (**florets**) surrounded by an involucre (ring) of bracts. This arrangement gives the appearance of a single flower.

**capric acid** See DECANOIC ACID.

**caproic acid** See HEXANOIC ACID.

**caprolactam (6-hexanelactam)** A white crystalline substance, $C_6H_{11}NO$; r.d. 1.02; m.p. 69–71°C; b.p. 139°C. It is a *lactam containing the –NH.CO– group with five $CH_2$ groups making up the rest of the seven-membered ring. Caprolactam is used in making *nylon.

**caprylic acid** See OCTANOIC ACID.

**capsid** The protein coat of a *virus. The chemical nature of the capsid is important in stimulating the body's *immune response against the invading virus.

**capsule** 1. (in botany) a. A dry fruit that releases its seeds when ripe; it is formed from several fused carpels and contains many seeds. The seeds may be dispersed through pores (as in the poppy), through a lid (as in plantain), or by the splitting and separation of the individual carpels (as in the crocus). Various other forms of capsules include the *silicula and *siliqua. b. The part of the sporophyte of mosses and liverworts in which the haploid spores are produced. It is borne on a long stalk (**seta**) and sheds its spores when mature (see PERISTOME). 2. (in microbiology) A thick gelatinous layer completely surrounding the cell wall of certain bacteria. It appears to have a protective function, making ingestion of the bacterial cell by *phagocytes more difficult and preventing desiccation. 3. (in animal anatomy) a. The membranous or fibrous envelope that surrounds certain organs, e.g. the kidneys, spleen, and lymph nodes. **b.** The ligamentous sheath of connective tissue that surrounds various skeletal joints.

**capture** Any of various processes in which a system of particles absorbs an extra particle. There are several examples in atomic and nuclear physics. For instance, a positive ion may capture an electron to give a neutral atom or molecule. Similarly, a neutral atom or molecule capturing an electron becomes a negative ion. An atomic nucleus may capture a neutron to produce a different (often unstable) nucleus. Another type of nuclear capture is the process in which the nucleus of an atom absorbs an electron from the innermost orbit (the K shell) to transform into a different nucleus. In this process (called **K capture**) the atom is left in an excited state and generally decays by emission of an X-ray photon.

**Radiative capture** is any such process in which the capture results in an excited state that decays by emission of photons. A common example is neutron capture to yield an excited nucleus, which decays by emission of a gamma ray.

**carapace** **1.** The dorsal part of the *exoskeleton of some crustaceans (e.g. crabs), which spreads like a shield over several segments of the head and thorax. **2.** The domed dorsal part of the shell of tortoises and turtles, formed of bony plates fused with the ribs and vertebrae and covered by a horny epidermal layer. The ventral part of the shell (**plastron**) is similar but flatter.

**carat** **1.** A measure of fineness (purity) of gold. Pure gold is described as 24-carat gold. 14-carat gold contains 14 parts in 24 of gold, the remainder usually being copper. **2.** A unit of mass equal to 0.200 gram, used to measure the masses of diamonds and other gemstones.

**carbamide** *See* UREA.

**carbanion** An organic ion with a negative charge on a carbon atom; i.e. an ion of the type $R_3C^-$. Carbanions are intermediates in certain types of organic reaction (e.g. the *aldol reaction).

**carbene** A species of the type $R_2C:$, in which the carbon atom has two electrons that do not form bonds. **Methylene**, $:CH_2$, is the simplest example. Carbenes are highly reactive and exist only as transient intermediates in certain organic reactions. They attack double bonds to give cyclopropane

derivatives. They also cause insertion reactions, in which the carbene group is inserted between the carbon and hydrogen atoms of a C–H bond:

$$C–H + :CR_2 \rightarrow C–CR_2–H$$

**carbenium ion** *See* CARBOCATION.

**carbide** Any of various compounds of carbon with metals or other more electropositive elements. True carbides contain the ion $C^{4-}$ as in $Al_4C_3$. These are saltlike compounds giving methane on hydrolysis, and were formerly called **methanides**. Compounds containing the ion $C_2^{2-}$ are also saltlike and are known as **dicarbides**. They yield ethyne (acetylene) on hydrolysis and were formerly called **acetylides**. The above types of compound are ionic but have partially covalent bond character, but boron and silicon form true covalent carbides, with giant molecular structures. In addition, the transition metals form a range of interstitial carbides in which the carbon atoms occupy interstitial positions in the metal lattice. These substances are generally hard materials with metallic conductivity. Some transition metals (e.g. Cr, Mn, Fe, Co, and Ni) have atomic radii that are too small to allow individual carbon atoms in the interstitial holes. These form carbides in which the metal lattice is distorted and chains of carbon atoms exist (e.g. $Cr_3C_2$, $Fe_3C$). Such compounds are intermediate in character between interstitial carbides and ionic carbides. They give mixtures of hydrocarbons on hydrolysis with water or acids.

**carbocation** An ion with a positive charge that is mostly localized on a carbon atom. There are two types:

**Carbonium ions** have five bonds to the carbon atom and a complete outer shell of eight electrons. A simple example is the ion $CH_5^+$, which has a trigonal bipyramidal shape. Ions of this type are transient species. They can be produced by electron impact and detected by mass spectroscopy.

**Carbenium ions** have three bonds to the carbon atom and are planar, with six outer electrons and a vacant *p*-orbital. Ions of this type are intermediates in a number of organic reactions (for example, in the $S_N1$ mechanism of *nucleophilic substitution). Certain carbenium ions are stabilized by delocalization of the charge. An example is the orange-red salt $(C_6H_5)_3C^+Cl^-$. Carbenium ions can be produced by superacids.

**carbocyclic** *See* CYCLIC.

**carbohydrate** One of a group of organic compounds based on the general formula $C_x(H_2O)_y$. The simplest carbohydrates are the *sugars (saccharides), including glucose and sucrose. *Polysaccharides are carbohydrates of much greater molecular weight and complexity; examples are starch, glycogen, and cellulose. Carbohydrates perform many vital roles in living organisms. Sugars, notably glucose, and their derivatives are essential intermediates in the conversion of food to energy. Starch and other polysaccharides serve as energy stores in plants, particularly in seeds, tubers, etc., which provide a major energy source for animals, including man. Cellulose, lignin, and others form the supporting cell walls and woody tissue of plants. Chitin is a structural polysaccharide found in the body shells of many invertebrate animals. Carbohydrates also occur in the surface coat of animal cells and in bacterial cell walls.

(((●))) **SEE WEB LINKS**
• Information about IUPAC nomenclature

**carbolic acid** *See* PHENOL.

**carbon** Symbol C. A nonmetallic element belonging to *group 14 (formerly IVB) of the periodic table; a.n. 6; r.a.m. 12.011; m.p. ~3550°C; b.p. ~4827°C. Carbon has three main allotropic forms (*see* ALLOTROPY).

*Diamond (r.d. 3.52) occurs naturally and can be produced synthetically. It is extremely hard and has highly refractive crystals. The hardness of diamond results from the covalent crystal structure, in which each carbon atom is linked by covalent bonds to four others situated at the corners of a tetrahedron. The C–C bond length is 0.154 nm and the bond angle is 109.5°.

Graphite (r.d. 2.25) is a soft black slippery substance (sometimes called **black lead** or **plumbago**). It occurs naturally and can also be made by the *Acheson process. In graphite the carbon atoms are arranged in layers, in which each carbon atom is surrounded by three others to which it is bound by single or double bonds. The layers are held together by much weaker van der Waals' forces. The carbon–carbon bond length in the layers is 0.142 nm and the layers are 0.34 nm apart. Graphite is a good conductor of heat and electricity. It has a variety of uses including electrical contacts, high-temperature equipment, and as a solid lubricant. Graphite mixed with clay is the

'lead' in pencils (hence its alternative name). The third crystalline allotrope is fullerite (*see* BUCKMINSTERFULLERENE). There are also several amorphous forms of carbon, such as *carbon black and *charcoal.

There are two stable isotopes of carbon (proton numbers 12 and 13) and four radioactive ones (10, 11, 14, 15). Carbon–14 is used in *carbon dating.

Carbon occurs in all organic compounds and is therefore fundamental to the structure of all living organisms. It is an *essential element for plants and animals, being ultimately derived from atmospheric carbon dioxide assimilated by plants during photosynthesis (*see* CARBON CYCLE). The ubiquitous nature of carbon in living organisms is due to its unique ability to form stable covalent bonds with other carbon atoms and also with hydrogen, oxygen, nitrogen, and sulphur atoms, resulting in the formation of a variety of compounds containing chains and rings of carbon atoms.

(((●))) **SEE WEB LINKS**
• Information from the WebElements site

**carbon assimilation** The incorporation of carbon from atmospheric carbon dioxide into organic molecules. This process occurs during *photosynthesis. *See* CARBON CYCLE.

**carbonate** A salt of carbonic acid containing the carbonate ion, $CO_3^{2-}$. The free ion has a plane triangular structure. Metal carbonates may be ionic or may contain covalent metal–carbonate bonds (complex carbonates) via one or two oxygen atoms. The carbonates of the alkali metals are all soluble but other carbonates are insoluble; they all react with mineral acids to release carbon dioxide.

**carbonate minerals** A group of common rock-forming minerals containing the anion $CO_3^{2-}$ as the fundamental unit in their structure. The most important carbonate minerals are *calcite, *dolomite, and *magnesite. *See also* ARAGONITE.

**carbonation** The solution of carbon dioxide in a liquid under pressure.

**carbon bisulphide** *See* CARBON DISULPHIDE.

**carbon black** A fine carbon powder made by burning hydrocarbons in insufficient air. It is used as a pigment and a filler (e.g. for rubber).

**carbon capture and storage** (CCS) The removal of carbon dioxide from a point source of pollution and its subsequent transfer to a repository so that it does not enter the atmosphere. Methods of CCS are being developed as ways of reducing emissions of greenhouse gases (*see* GREENHOUSE EFFECT) and mitigating the environmental impact of burning coal, natural gas, or oil. For example, it is feasible to remove carbon dioxide from power station flue gases by scrubbing, using activated carbon filters. However, this significantly increases the operating cost of the power station. Once removed from a point source, the carbon dioxide is piped to a suitable storage site, often located in porous underground rock formations, notably depleted oil and gas fields. Other possibilities are storage in deep ocean sites and chemical reaction with metal oxides to create stable carbonates.

**carbon cycle** **1.** (in biology) One of the major cycles of chemical elements in the environment (*see* BIOGEOCHEMICAL CYCLE). Carbon (as carbon dioxide) is taken up from the atmosphere and incorporated into the tissues of plants in *photosynthesis. It may then pass into the bodies of animals as the plants are eaten (*see* FOOD CHAIN). During the respiration of plants, animals, and organisms that bring about decomposition, carbon dioxide is returned to the atmosphere.

The combustion of fossil fuels (e.g. coal and peat) also releases carbon dioxide into the atmosphere. See illustration. *See also* GREENHOUSE EFFECT.

**2.** (in physics) A series of nuclear reactions in which four hydrogen nuclei combine to form a helium nucleus with the liberation of energy, two positrons, and two neutrinos. The process is believed to be the source of energy in many stars and to take place in six stages. In this series carbon–12 acts as if it were a catalyst, being reformed at the end of the series:

$$^{12}_{6}C + ^{1}_{1}H \rightarrow ^{13}_{7}N + \gamma$$

$$^{13}_{7}N \rightarrow ^{13}_{6}C + e^+ + \nu_e$$

$$^{13}_{6}C + ^{1}_{1}H \rightarrow ^{14}_{7}N + \gamma$$

$$^{14}_{7}N + ^{1}_{1}H \rightarrow ^{15}_{8}O + \gamma$$

$$^{15}_{8}O \rightarrow ^{15}_{7}N + e^+ + \nu_e$$

$$^{15}_{7}N + ^{1}_{1}H \rightarrow ^{12}_{6}C + ^{4}_{2}He.$$

*See* STELLAR EVOLUTION.

**carbon dating (radiocarbon dating)** A method of estimating the ages of archaeological specimens of biological origin. As a result of *cosmic radiation a small number of atmospheric nitrogen nuclei are continuously being transformed by neutron

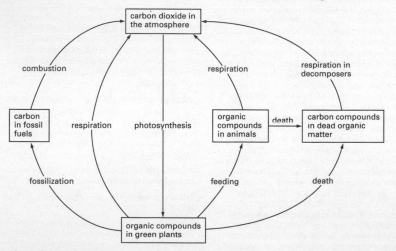

**The carbon cycle in nature.**

bombardment into radioactive nuclei of carbon–14:

$$^{14}_{7}N + n \rightarrow ^{14}_{6}C + p$$

Some of these radiocarbon atoms find their way into living trees and other plants in the form of carbon dioxide, as a result of *photosynthesis. When the tree is cut down photosynthesis stops and the ratio of radiocarbon atoms to stable carbon atoms begins to fall as the radiocarbon decays. The ratio $^{14}C/^{12}C$ in the specimen can be measured and enables the time that has elapsed since the tree was cut down to be calculated. The method has been shown to give consistent results for specimens up to some 40 000 years old, though its accuracy depends upon assumptions concerning the past intensity of the cosmic radiation. The technique was developed by Willard F. Libby (1908–80) and his coworkers in 1946–47.

**carbon dioxide** A colourless odourless gas, $CO_2$, soluble in water, ethanol, and acetone; d. 1.977 g $dm^{-3}$ (0°C); m.p. –56.6°C; b.p. –78.5°C. It occurs in the atmosphere (0.04% by volume) but has a short residence time in this phase as it is both consumed by plants during *photosynthesis and produced by *respiration and by combustion. It is readily prepared in the laboratory by the action of dilute acids on metal carbonates or of heat on heavy-metal carbonates. Carbon dioxide is a by-product from the manufacture of lime and from fermentation processes.

Carbon dioxide has a small liquid range and liquid carbon dioxide is produced only at high pressures. The molecule $CO_2$ is linear with each oxygen making a double bond to the carbon. Chemically, it is unreactive and will not support combustion. It dissolves in water to give *carbonic acid.

Large quantities of solid carbon dioxide (**dry ice**) are used in processes requiring large-scale refrigeration. It is also used in fire extinguishers as a desirable alternative to water for most fires, and as a constituent of medical gases as it promotes exhalation. It is also used in carbonated drinks.

The level of carbon dioxide in the atmosphere has increased by some 30% since the Industrial Revolution, mainly because of extensive burning of fossil fuels and the destruction of large areas of rainforest. This has been postulated as the main cause of the average increase of 0.74°C in global temperatures over the last 100 years, through the *greenhouse effect. Atmospheric $CO_2$ concentration continues to rise, in spite of tentative steps to control emissions, giving the prospect of accelerated *global warming in the foreseeable future.

**carbon disulphide (carbon bisulphide)** A colourless highly refractive liquid, $CS_2$, slightly soluble in water and soluble in ethanol and ether; r.d. 1.261; m.p. –110°C; b.p. 46.3°C. Pure carbon disulphide has an ethereal odour but the commercial product is contaminated with a variety of other sulphur compounds and has a most unpleasant smell. It was previously manufactured by heating a mixture of wood, sulphur, and charcoal; modern processes use natural gas and sulphur. Carbon disulphide is an excellent solvent for oils, waxes, rubber, sulphur, and phosphorus, but its use is decreasing because of its high toxicity and its flammability. It is used for the preparation of xanthates in the manufacture of viscose yarns.

**carbon fibres** Fibres of carbon in which the carbon has an oriented crystal structure. Carbon fibres are made by heating textile fibres and are used in strong composite materials for use at high temperatures.

**carbon footprint** The total amount of greenhouse gases (GHGs) emitted to the atmosphere as a result of the activities of an individual, household, business, or other entity. It is usually measured as the mass of carbon dioxide, or carbon dioxide equivalent (i.e. including methane, nitrogen oxides, and other GHGs), emitted per year, and indicates the environmental impact of those activities through their contribution to the *greenhouse effect and hence global warming. Calculation of a full carbon footprint involves both direct and indirect emissions. Direct emissions include those arising from activities that are controlled directly by an individual or organization, such as the combustion of fuel for heating or transport. Indirect emissions are those arising from all goods and services used by the individual or organization, but not directly controlled by them, such as the energy used in the extraction and processing of raw materials and manufacture of goods.

**carbonic acid** A dibasic acid, $H_2CO_3$, formed in solution when carbon dioxide is dissolved in water:

$$CO_2(aq) + H_2O(l) \rightleftharpoons H_2CO_3(aq)$$

The acid is in equilibrium with dissolved carbon dioxide, and also dissociates as follows:

$$H_2CO_3 \rightleftharpoons H^+ + HCO_3^-$$

$K_a = 4.5 \times 10^{-7} \text{ mol dm}^{-3}$

$HCO_3^- \rightleftharpoons CO_3^{2-} + H^+$

$K_a = 4.8 \times 10^{-11} \text{ mol dm}^{-3}$

The pure acid cannot be isolated, although it can be produced in ether solution at $-30°C$. Carbonic acid gives rise to two series of salts: the *carbonates and the *hydrogen-carbonates.

**carbonic anhydrase** An enzyme, present in red blood cells and kidney cells, that catalyses the reaction between carbon dioxide with water:

$CO_2 + H_2O \rightleftharpoons H_2CO_3$

$H_2CO_3 \rightleftharpoons H^+ + HCO_3^-$.

This reaction is one of the fastest known and controls the elimination of carbon dioxide from the body and the pH of urine. It also facilitates the transfer of carbon dioxide from the tissues to the blood and from the blood to the alveoli (air sacs) of the lungs. *See also* CHLORIDE SHIFT.

**Carboniferous** A geological period in the Palaeozoic era. It began about 360 million years ago, following the Devonian period, and extended until the beginning of the Permian period, about 299 million years ago. In Europe the period is divided into the Lower and Upper Carboniferous, which roughly correspond to the Mississippian and Pennsylvanian periods, respectively, of North America. During the Lower Carboniferous a marine transgression occurred and the characteristic rock of this division – the Carboniferous limestone – was laid down in the shallow seas. Fauna included foraminiferans, corals, bryozoans, brachiopods, blastoids, and other invertebrates. The Upper Carboniferous saw the deposition of the millstone grit, a mixture of shale and sandstone formed in deltaic conditions, followed by the coal measures, alternating beds of coal, sandstone, shale, and clay. The coal was formed from the vast swamp forests composed of seed ferns, lycopsids, and other plants. During the period fishes continued to diversify and amphibians became more common.

**carbonium ion** *See* CARBOCATION.

**carbonize (carburize)** To change an organic compound into carbon by heating, or to coat something with carbon in this way.

**carbon monoxide** A colourless odourless gas, CO, sparingly soluble in water and solu-

$$
\begin{array}{ll}
C \equiv O & \text{carbon monoxide} \\[4pt]
O = C = O & \text{carbon dioxide} \\[4pt]
O = C = C = C = O & \text{tricarbon dioxide} \\
& \text{(carbon suboxide)}
\end{array}
$$

**Carbon monoxide.** Oxides of carbon.

ble in ethanol and benzene; d. $1.25 \text{ g dm}^{-3}$ ($0°C$); m.p. $-199°C$; b.p. $-191.5°C$. It is flammable and highly toxic. In the laboratory it can be made by the dehydration of methanoic acid (formic acid) using concentrated sulphuric acid. Industrially it is produced by the oxidation of natural gas (methane) or (formerly) by the water-gas reaction. It is formed by the incomplete combustion of carbon and is present in car-exhaust gases.

It is a neutral oxide, which burns in air to give carbon dioxide, and is a good reducing agent, used in a number of metallurgical processes. It has the interesting chemical property of forming a range of transition metal carbonyls, e.g. $Ni(CO)_4$. Carbon monoxide is able to use vacant $p$-orbitals in bonding with metals; the stabilization of low oxidation states, including the zero state, is a consequence of this. This also accounts for its toxicity, which is due to the binding of the CO to the iron in haemoglobin, thereby blocking the uptake of oxygen.

**carbon suboxide** *See* TRICARBON DIOXIDE.

**carbon tetrachloride** *See* TETRA-CHLOROMETHANE.

**carbonyl chloride (phosgene)** A colourless gas, $COCl_2$, with an odour of freshly cut hay. It is used in organic chemistry as a chlorinating agent, and was formerly used as a war gas.

**carbonyl compound** A compound containing the carbonyl group >C=O. Aldehydes, ketones, and carboxylic acids are examples of organic carbonyl compounds. Inorganic carbonyls are complexes in which carbon monoxide has coordinated to a metal atom or ion, as in *nickel carbonyl, $Ni(CO)_4$. *See also* LIGAND.

**carbonyl group** The group >C=O, found in aldehydes, ketones, carboxylic acids, amides, etc., and in inorganic carbonyl complexes (*see* CARBONYL COMPOUND).

**carboranes** Compounds similar to the *boranes, but with one or more boron atoms replaced by carbon atoms.

**carborundum** *See* SILICON CARBIDE.

**carboxyhaemoglobin** The highly stable product formed when *haemoglobin combines with carbon monoxide. Carbon monoxide competes with oxygen for haemoglobin, with which it binds strongly: the affinity of haemoglobin for carbon monoxide is 250 times greater than that for oxygen. This reduces the availability of haemoglobin for combination with (and transport of) oxygen and accounts for the toxic effects of carbon monoxide on the respiratory system.

**carboxylate** An anion formed from a *carboxylic acid. For example, ethanoic acid gives rise to the ethanoate ion, $CH_3COO^-$.

**carboxyl group** The organic group –COOH, present in *carboxylic acids.

**carboxylic acids** Organic compounds containing the group –COOH (the **carboxyl group**; i.e. a carbonyl group attached to a hydroxyl group). In systematic chemical nomenclature carboxylic-acid names end in the suffix *-oic*, e.g. ethanoic acid, $CH_3COOH$. They are generally weak acids. Many long-chain carboxylic acids occur naturally as esters in fats and oils and are therefore also known as *fatty acids. *See also* GLYCERIDE.

**⊕ SEE WEB LINKS**

• Information about IUPAC nomenclature

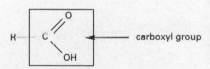

**Carboxylic acid structure.**

**carboxypeptidase** An *exopeptidase enzyme in pancreatic juice that is secreted into the duodenum. The enzyme is secreted as an inactive precursor, **procarboxypeptidase**, which is activated by another pancreatic protease, *trypsin. *See also* CHYMOTRYPSIN.

**carburize** *See* CARBONIZE.

**carbylamine reaction** *See* ISOCYANIDE TEST.

**carcerulus** A dry fruit that is a type of *schizocarp. It consists of a number of one-seeded fragments (**mericarps**) that adhere to a central axis. It is characteristic of mallow.

**carcinogen** Any agent that produces *cancer, e.g. tobacco smoke, certain industrial chemicals, and *ionizing radiation (such as X-rays and ultraviolet rays).

**carcinoma** *See* CANCER.

**cardiac** **1.** Relating to the heart. **2.** Relating to the part of the stomach nearest to the oesophagus.

**cardiac cycle** The sequence of events that occurs in the heart during one full heartbeat. These events comprise contraction (*see* SYSTOLE) and relaxation (*see* DIASTOLE) of the chambers of the heart, associated with opening and closing of the heart valves. When both the atria and the ventricles are relaxed, pressure in the heart is low and blood flows from the vena cava and pulmonary vein into the atria and through to the ventricles. The aortic and pulmonary valves, at the junction between the left ventricle and aorta and the right ventricle and pulmonary artery, respectively, are closed; therefore, blood can enter but not leave the heart, which increases the pressure in the chambers. As the pressure in the heart increases, the atria begin to contract, forcing the blood into the ventricles and closing the *tricuspid valve and the *bicuspid valve. A wave of ventricular contraction follows, expelling the blood into the aorta and pulmonary artery to complete the cardiac cycle. At a resting heart rate, the human cardiac cycle lasts approximately 0.85 second.

**cardiac muscle** A specialized form of *muscle that is peculiar to the vertebrate heart. There are two types of cardiac muscle fibres: contractile fibres, which are striated and contain numerous myofibrils; and conducting fibres, or *Purkyne fibres, which branch extensively and conduct electrical signals throughout the muscle. The muscle itself shows spontaneous contraction and does not need nervous stimulation (*see* PACEMAKER). The vagus nerve to the heart can, however, affect the rate of contraction.

**cardiac output** The volume of blood pumped per minute by each ventricle, which is also the total blood flow through the pulmonary circuit. At rest, normal human cardiac output is approximately 5 litres per minute, rising to 22 litres per minute during maximum physical exertion. The cardiac output can be calculated from heart rate (number of beats per minute) and stroke volume (volume of blood expelled from the heart per beat).

**cardiovascular centre** One of the areas in the brain that are responsible for the modification of the cardiovascular system based upon the integration of sensory information from the autonomic nervous system. These centres influence the heart rate via the sympathetic and parasympathetic nerves and by the action of certain hormones.

**Carius method** A method of determining the amount of sulphur and halogens in an organic compound, by heating the compound in a sealed tube with silver nitrate in concentrated nitric acid. The compound is decomposed and silver sulphide and halides are precipitated, separated, and weighed.

**carnallite** A mineral consisting of a hydrated mixed chloride of potassium and magnesium, $KCl.MgCl_2.6H_2O$.

**carnassial teeth** Molar and premolar teeth modified for shearing flesh by having cusps with sharp cutting edges. They are typical of animals of the order *Carnivora (e.g. tigers, wolves), in which they are the first molars in the lower jaw and the last premolars in the upper.

**Carnivora** An order of mainly flesh-eating mammals that includes the dogs, wolves, bears, badgers, weasels, and cats. Carnivores typically have very keen sight, smell, and hearing. The hinge joint between the lower jaw and skull is very tight, allowing no lateral movement of the lower jaw. This – together with the arrangement of jaw muscles – enables a very powerful bite. The teeth are specialized for stabbing and tearing flesh: canines are large and pointed and some of the cheek teeth are modified for shearing (*see* CARNASSIAL TEETH).

**carnivore** An animal that eats meat, especially a member of the order *Carnivora (e.g. tigers, wolves). Carnivores are specialized by having strong powerful jaws and well-developed canine teeth. They may be *predators or carrion eaters. *See also* CONSUMER. *Compare* HERBIVORE; OMNIVORE.

**carnivorous plant (insectivorous plant)** Any plant that supplements its supply of nitrates in conditions of nitrate deficiency by digesting small animals, especially insects. Such plants are adapted in various ways to attract and trap the insects and produce proteolytic enzymes to digest them. Venus' fly trap (*Dionaea*), for example, has spiny-margined hinged leaves that snap shut on an alighting insect. Sundews (*Drosera*) trap and digest insects by means of glandular leaves that secrete a sticky substance, and pitcher plants (families Nepenthaceae and Sarraceniaceae) have leaves modified as pitchers into which insects fall, drowning in the water and digestive enzymes at the bottom.

**Carnot, Nicolas Léonard Sadi**
(1796–1832) French physicist, who first worked as a military engineer. He then turned to scientific research and in 1824 published his analysis of the efficiency of heat engines. The key to this analysis is the thermodynamic *Carnot cycle. He died at an early age of cholera.

**Carnot cycle** The most efficient cycle of operations for a reversible *heat engine. Published in 1824 by Nicolas Carnot, it consists of four operations on the working substance in the engine (see illustration):
 **a.** Isothermal expansion at thermodynamic temperature $T_1$ with heat $Q_1$ taken in.
 **b.** Adiabatic expansion with a fall of temperature to $T_2$.
 **c.** Isothermal compression at temperature $T_2$ with heat $Q_2$ given out.
 **d.** Adiabatic compression with a rise of temperature back to $T_1$.

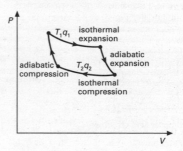

**Carnot cycle.**

According to the **Carnot principle**, the efficiency of any reversible heat engine depends only on the temperature range through which it works, rather than the properties of the working substances. In any reversible engine, the efficiency ($\eta$) is the ratio of the work done ($W$) to the heat input ($Q_1$), i.e. $\eta = W/Q_1$. As, according to the first law of *thermodynamics, $W = Q_1 - Q_2$, it follows that $\eta = (Q_1 - Q_2)/Q_1$. For the Kelvin temperature scale, $Q_1/Q_2 = T_1/T_2$ and $\eta = (T_1 - T_2)/T_1$. For maximum efficiency $T_1$ should be as high as possible and $T_2$ as low as possible.

**carnotite** A radioactive mineral consisting of hydrated uranium potassium vanadate, $K_2(UO_2)_2(VO_4)_2 \cdot nH_2O$. It varies in colour from bright yellow to lemon- or greenish-yellow. It is a source of uranium, radium, and vanadium. The chief occurrences are in the Colorado Plateau, USA; Radium Hill, Australia; and Katanga, Democratic Republic of Congo.

**Caro's acid** See PEROXOSULPHURIC(VI) ACID.

**carotene** A member of a class of *carotenoid pigments. Examples are β-carotene and lycopene, which colour carrot roots and ripe tomato fruits respectively. α- and β-carotene yield vitamin A when they are broken down during animal digestion.

**carotenoid** Any of a group of yellow, orange, red, or brown plant pigments chemically related to terpenes. Carotenoids are responsible for the characteristic colour of many plant organs, such as ripe tomatoes, carrots, and autumn leaves. They also function as accessory *photosynthetic pigments in the light-dependent reactions of *photosynthesis. See CAROTENE; XANTHOPHYLL.

**Carothers, Wallace Hume** (1896–1937) US industrial chemist, who joined the Du Pont company where he worked on polymers. In 1931 he produced *neoprene, a synthetic rubber. His greatest success came in 1935 with the discovery of the polyamide that came to be known as *nylon. Carothers, who suffered from depression, committed suicide.

**carotid artery** The major artery that supplies blood to the head. A pair of **common carotid arteries** arise from the aorta (on the left) and the innominate artery (on the right) and run up the neck; each branches into an **external** and an **internal carotid artery**, which supply the head.

**carotid body** One of a pair of tissue masses adjacent to the *carotid sinus. Each contains receptors that are sensitive to oxygen and pH levels (acidity) in the blood. High levels of carbon dioxide in the blood lower the pH (i.e. increase the acidity). By responding to fluctuations in pH, the carotid body coordinates reflex changes in respiration rate.

**carotid sinus** An enlarged region of the *carotid artery at its major branching point in the neck. Its walls contain many receptors

that are sensitive to changes in pressure and it regulates blood pressure by initiating reflex changes in heart rate and dilation of blood vessels.

**carpal (carpal bone)** One of the bones that form the wrist (*see* CARPUS) in terrestrial vertebrates.

**carpel** The female reproductive organ of a flower. Typically it consists of a *stigma, *style, and *ovary. It is thought to have evolved by the fusion of the two edges of a flattened megasporophyll (*see* SPOROPHYLL). Each flower may have one carpel (**monocarpellary**) or many (**polycarpellary**), either free (**apocarpous**) or fused together (**syncarpous**). *See also* PISTIL.

**carpus** The wrist (or corresponding part of the forelimb) in terrestrial vertebrates, consisting of a number of small bones (**carpals**). The number of carpal bones varies with the species. The rabbit, for example, has two rows of carpals, the first (proximal) row containing three bones and the second (distal) row five. In humans there are also eight carpals. This large number of bones enables flexibility at the wrist joint, between the hand and forelimb. *See also* PENTADACTYL LIMB.

**carrier 1.** (in radio) See CARRIER WAVE. **2.** (in physics) See CHARGE CARRIER. **3.** (in medicine) An individual who harbours a particular disease-causing microorganism without ill-effects and who can transmit the microorganism to others. *Compare* VECTOR. **4.** (in genetics) An individual with an *allele for some defective condition that is masked by a normal *dominant allele. Such individuals therefore do not suffer from the condition themselves but they may pass on the defective allele to their offspring. In humans, women may be carriers of such conditions as red–green colour blindness and haemophilia, the alleles for which are carried on the X chromosomes (*see* SEX LINKAGE). **5.** (in biochemistry) See CARRIER MOLECULE; HYDROGEN CARRIER.

**carrier gas** The gas that carries the sample in *gas chromatography.

**carrier molecule 1.** A molecule that plays a role in transporting electrons through the *electron transport chain. Carrier molecules are usually proteins bound to a nonprotein group; they can undergo oxidation and reduction relatively easily, thus allowing electrons to flow through the system. There are four types of carrier: flavoproteins (e.g. *FAD), *cytochromes, iron–sulphur proteins (e.g. ferredoxin), and *ubiquinone. **2.** A lipid-soluble molecule that can bind to lipid-insoluble molecules and transport them across membranes. Carrier molecules have specific sites that interact with the molecules they transport. Several different molecules may compete for transport by the same carrier. *See* TRANSPORT PROTEIN.

**carrier wave** An electromagnetic wave of specified frequency and amplitude that is emitted by a radio transmitter in order to carry information. The information is superimposed onto the carrier by means of *modulation.

**carrying capacity** Symbol $K$. The maximum population of a particular species that can be supported indefinitely by a given habitat or area without damage to the environment. It can be manipulated by human intervention. For example, the carrying capacity for grazing mammals could be increased by boosting the yield of their grassland habitat by the application of fertilizer. *See also* K SELECTION.

**Cartesian coordinates** A system used in analytical geometry to locate a point $P$, with reference to two or three **axes** (see graphs). In a two-dimensional system the vertical axis is the $y$-axis and the horizontal axis is the $x$-axis. The point at which the axes intersect each other is called the **origin**, $O$. Values of $y$ <0 fall on the $y$-axis below the origin, values of $x$ <0 fall on the $x$-axis to the left of the origin. Any point $P$ is located by its perpendicular distances from the two axes. The distance from the $x$-axis is called the **ordinate**; the distance from the $y$-axis is the **abscissa**. The position is indicated numerically by enclosing the values of the abscissa and the ordinate in parentheses and separating the two by means of a comma, e.g. $(x,y)$. In three dimensions the system can be used to locate a point with reference to a third, $z$-axis. It is named after René Descartes (1596–1650).

**cartilage (gristle)** A firm flexible connective tissue that forms the adult skeleton of cartilaginous fish (e.g. sharks). In other vertebrates cartilage forms the skeleton of the embryo, being largely replaced by *bone in mature animals (although it persists in certain areas). Cartilage comprises a matrix consisting chiefly of a glycosaminoglycan (mucopolysaccharide) called **chondroitin sulphate** secreted by cells (**chondroblasts**)

that become embedded in the matrix as **chondrocytes**. It also contains collagenous and elastic fibres. **Hyaline cartilage** consists largely of glycosaminoglycan, giving it a shiny glasslike appearance; this type of cartilage gives flexibility and support at the joints. **Fibrocartilage**, in which bundles of collagen fibres predominate, is stronger and less elastic than hyaline cartilage; it is found in such areas as the intervertebral discs. **Elastic cartilage** has a yellow appearance due to the presence of numerous elastic fibres (*see* ELASTIN). This cartilage maintains the shape of certain organs, such as the pinna of the ear.

**cartilage bone (replacing bone)** *Bone that is formed by replacing the cartilage of an embryo skeleton. The process, called **ossification**, is brought about by the cells (osteoblasts) that secrete bone. *Compare* MEMBRANE BONE.

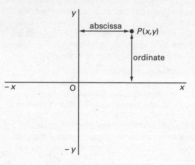

two-dimensional system

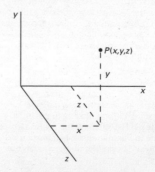

three-dimensional system

**Cartesian coordinates.**

**cartilaginous fishes** *See* CHON-
DRICHTHYES.

**cartography** The science of the production of maps and charts. Maps may be based
on original surveys, aerial photographs (pho-
togrammetry), or compiled from existing
maps and records. Computer-based information systems are increasingly used in the
production of maps (**digital cartography**) in
place of more traditional methods. *See also*
MAP PROJECTIONS.

**caruncle** A small outgrowth from the testa
of a seed that develops from the placenta, funicle, or micropyle. Examples include the
warty outgrowth from the castor-oil seed
and the tuft of hairs on the testa of the seed
of willowherb. *See also* ARIL.

**caryopsis** A dry single-seeded indehiscent
fruit that differs from an *achene in that the
fruit wall is fused to the testa of the seed. It is
the grain of cereals and grasses.

**cascade liquefier** An apparatus for liquefying a gas of low *critical temperature.
Another gas, already below its critical temperature, is liquified and evaporated at a reduced pressure in order to cool the first gas
to below its critical temperature. In practice
a series of steps is often used, each step enabling the critical temperature of the next
gas to be reached.

**cascade process** Any process that takes
place in a number of steps, usually because
the single step is too inefficient to produce
the desired result. For example, in some ura-
nium-enrichment processes the separation
of the desired isotope is only poorly achieved
in a single stage; to achieve better separation
the process has to be repeated a number of
times, in a series, with the enriched fraction
of one stage being fed to the succeeding
stage for further enrichment. Another example of cascade process is that operating in a
*cascade liquefier.

**case hardening** The hardening of the surface layer of steel, used for tools and certain
mechanical components. The commonest
method is to carburize the surface layer by
heating the metal in a hydrocarbon or by
dipping the red hot metal into molten
sodium cyanide. Diffusion of nitrogen into
the surface layer to form nitrides is also used.

**casein** One of a group of phosphate-
containing proteins (phosphoproteins)
found in milk. Caseins are easily digested

by the enzymes of young mammals and represent a major source of phosphorus. *See*
RENNIN.

**CAS registry** A database of chemical compounds, certain mixtures, and biological sequences maintained by the Chemical
Abstracts Service of the American Chemical
Society. In the registry every entry has a
unique CAS registry number (**CASRN**). This
has three parts: up to six digits, followed by
two digits, followed by one digit. For example, the CAS number of water is 7732-18-5.
The final digit is a check number. CAS registry numbers are used for searching chemical databases. The size of the registry is
immense, with over 32 million substances
and 59 million sequences. It identifies every
chemical that has been described in the literature since 1957 and around 50 000 new
numbers are added each week.

**( ) SEE WEB LINKS**
- Further information from the CAS site
- A free service for finding CAS numbers from
  CambridgeSoft

**Cassegrainian telescope** *See* TELESCOPE.

**Cassini, Giovanni Domenico** (1625–
1712) Italian-born French astronomer, who
was professor of astronomy at Bologna. In
1669 he moved to Paris to run the new observatory there, becoming a French citizen
in 1673. He is best known for his discovery
(1675) of the gap that divides Saturn's ring
system into two parts, now called the **Cassini
division**. He also discovered four new satellites of Saturn.

**cassiterite** A yellow, brown, or black form
of tin(IV) oxide, $SnO_2$, that forms tetragonal,
often twinned, crystals; the principal ore of
tin. It occurs in hydrothermal veins and
metasomatic deposits associated with acid
igneous rocks and in alluvial (placer) deposits. The chief producers are Malaysia, Indonesia, Democratic Republic of Congo, and
Nigeria.

**caste** A division found in social insects,
such as the *Hymenoptera (ants, bees,
wasps) and the Isoptera (termites), in which
the individuals are structurally and physiologically specialized to perform a particular
function. For example, in honeybees there
are queens (fertile females), workers (sterile
females), and drones (males). There are several different castes of workers (all sterile females) among ants.

**cast iron** A group of iron alloys containing 1.8 to 4.5% of carbon. It is usually cast into specific shapes ready for machining, heat treatment, or assembly. It is sometimes produced direct from the *blast furnace or it may be made from remelted *pig iron.

**castor oil** A pale-coloured oil extracted from the castor-oil plant. It contains a mixture of glycerides of fatty acids, the predominant acid being ricinoleic acid, $C_{17}H_{32}(OH)COOH$. It is used as a *drying oil in paints and varnishes and medically as a laxative.

**catabolism** The metabolic breakdown of large molecules in living organisms to smaller ones, with the release of energy. Respiration is an example of a catabolic series of reactions. *See* METABOLISM. *Compare* ANABOLISM.

**catalysis** The process of changing the rate of a chemical reaction by use of a *catalyst.

**catalyst** A substance that increases the rate of a chemical reaction without itself undergoing any permanent chemical change. Catalysts that have the same phase as the reactants are **homogeneous catalysts** (e.g. *enzymes in biochemical reactions or transition-metal complexes used in the liquid phase for catalysing organic reactions). Those that have a different phase are **heterogeneous catalysts** (e.g. metals or oxides used in many industrial gas reactions). The catalyst provides an alternative pathway by which the reaction can proceed, in which the activation energy is lower. It thus increases the rate at which the reaction comes to equilibrium, although it does not alter the position of the equilibrium. The catalyst itself takes part in the reaction and consequently may undergo physical change (e.g. conversion into powder). In certain circumstances, very small quantities of catalyst can speed up reactions. Most catalysts are also highly specific in the type of reaction they catalyse, particularly enzymes in biochemical reactions. Generally, the term is used for a substance that increases reaction rate (a **positive catalyst**). Some reactions can be slowed down by **negative catalysts** (*see* INHIBITION).

**catalytic activity** The increase in the rate of a specified chemical reaction caused by an enzyme or other catalyst under specified assay conditions. It is measured in *katals or in moles per second.

**catalytic converter** A device used in the exhaust systems of motor vehicles to reduce atmospheric pollution. The three main pollutants produced by petrol engines are: unburnt hydrocarbons, carbon monoxide produced by incomplete combustion of hydrocarbons, and nitrogen oxides produced by nitrogen in the air reacting with oxygen at high engine temperatures. Hydrocarbons and carbon monoxide can be controlled by a higher combustion temperature and a weaker mixture. However, the higher temperature and greater availability of oxygen arising from these measures encourage formation of nitrogen oxides. The use of three-way catalytic converters solves this problem by using platinum and palladium catalysts to oxidize the hydrocarbons and the CO and rhodium catalysts to reduce the nitrogen oxides back to nitrogen. These three-way catalysts require that the air–fuel ratio is strictly stochiometric. Some catalytic converters promote oxidation reactions only, leaving the nitrogen oxides unchanged. Three-way converters can reduce hydrocarbons and CO emissions by some 85%, at the same time reducing nitrogen oxides by 62%.

**catalytic cracking** *See* CRACKING.

**catalytic rich gas process** *See* CRG PROCESS.

**catalytic RNA** *See* RIBOZYME.

**cataphoresis** *See* ELECTROPHORESIS.

**catastrophe theory** A branch of mathematics dealing with the sudden emergence of discontinuities, in contrast to *calculus, which is concerned with continuous quantities. Catastrophe theory originated in *topology in work by the French mathematician René Thom (1923–2002) and was developed by Thom and the Russian mathematician Vladimir Igorevich Arnold (1937–2010). There are physical applications of catastrophe theory in *optics and in systems involving *complexity, including biological systems.

**catechol** *See* 1,2-DIHYDROXYBENZENE.

**catecholamine** Any of a class of amines that possess a catechol ($C_6H_4(OH)_2$) ring. Including *dopamine, *adrenaline, and *noradrenaline, they function as *neurotransmitters and/or hormones.

**category** (in taxonomy) *See* RANK.

**catenane** A type of compound consisting of two or more large rings that are interlocked like the links of a chain. In a catenane, there is no chemical bonding between the rings; the rings are held together by *mechanical bonding.

**catenary** A curve formed when a chain or rope of uniform density hangs from two fixed points. If the lowest point on the curve passes through the origin, the equation is $y = c(\cosh x/c)$, where $c$ is the distance between the $x$-axis and the directrix.

**catenation** 1. The formation of chains of atoms in chemical compounds. 2. The formation of a *catenane compound by mechanical bonding.

**cathetometer** A telescope or microscope fitted with crosswires in the eyepiece and mounted so that it can slide along a graduated scale. Cathetometers are used for accurate measurement of lengths without mechanical contact. The microscope type is often called a **travelling microscope**.

**cathine ($\beta$-hydroxyamphetamine)** An alkaloid, $C_9H_{13}NO$. It may contribute to the stimulant activity of *khat.

**cathinone ($\beta$-ketoamphetamine)** An alkaloid, $C_9H_{11}NO$. It is the main active ingredient in fresh *khat.

**cathode** A negative electrode. In *electrolysis cations are attracted to the cathode. In vacuum electronic devices electrons are emitted by the cathode and flow to the *anode. It is therefore from the cathode that electrons flow into these devices. However, in a primary or secondary cell the cathode is the electrode that spontaneously becomes negative during discharge, and from which therefore electrons emerge.

**cathode-ray oscilloscope (CRO)** An instrument based on the *cathode-ray tube that provides a visual image of electrical signals. The horizontal deflection is usually provided by an internal *timebase, which causes the beam to sweep across the screen at a specified rate. The signal to be investigated is fed to the vertical deflection plates after amplification. Thus the beam traces a graph of the signal amplitude against time.

**cathode rays** Streams of electrons emitted at the cathode in an evacuated tube containing a cathode and an anode. They were first observed in gas *discharge tubes operated at low pressure. Under suitable conditions electrons produced by secondary emission at the cathode are accelerated down the tube to the anode. In such devices

as the *cathode-ray tube the electrons are produced by *thermionic emission from a hot cathode in a vacuum.

**cathode-ray tube (CRT)** The device that provides the viewing screen in the television tube, the radar viewer, and the *cathode-ray oscilloscope. The cathode-ray tube consists of an evacuated tube containing a heated cathode and two or more ring-shaped anodes through which the cathode rays can pass so that they strike the enlarged end of the tube (see illustration). This end of the tube is coated with fluorescent material so that it provides a screen. Any point on the screen that is struck by the cathode ray becomes luminous. A *control grid between the cathode and the anode enables the intensity of the beam to be varied, thus controlling the brightness of the illumination on the screen. The assembly of cathode, control grid, and anode is called the *electron gun. The beam emerging from the electron gun is focused and deflected by means of plates providing an electric field or coils providing a magnetic field. This enables the beam to be focused to a small point of light and deflected to produce the illusion of an illuminated line as this point sweeps across the tube.

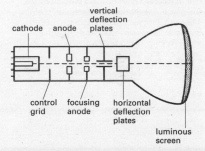

**Cathode-ray tube.**

The television tube is a form of cathode-ray tube in which the beam is made to scan the screen 625 times to form a frame, with 25 new frames being produced every second. (These are the figures for standard television tubes in the UK). Each frame creates a picture by variations in the intensity of the beam as it forms each line.

**cathodic protection** *See* SACRIFICIAL PROTECTION.

**cation** A positively charged ion, i.e. an ion

that is attracted to the cathode in *electrolysis. *Compare* ANION.

**cationic detergent** *See* DETERGENT.

**cationic dye** *See* DYES.

**cationic resin** *See* ION EXCHANGE.

**catkin** A type of flowering shoot (*see* RACEMOSE INFLORESCENCE) in which the axis, which is often long, bears many small stalkless unisexual flowers. Usually the male catkins hang down from the stem; the female catkins are shorter and often erect. Examples include birch and hazel. Most plants with catkins are adapted for wind pollination, the male flowers producing large quantities of pollen; willows are an exception, having nectar-secreting flowers and being pollinated by insects.

**caudal vertebrae** The bones (*see* VERTEBRA) of the tail, which articulate with the *sacral vertebrae. The number of caudal vertebrae varies with the species. Rabbits, for example, have 15 caudal vertebrae, while in humans these vertebrae are fused to form a single bone, the *coccyx.

**causality** The principle that effect cannot precede cause. The principle is particularly useful when combined with the principle that the highest attainable speed in the universe is the *speed of light in a vacuum. Causality is used to analyse the results of scattering experiments and in optics.

**caustic** 1. (in chemistry) Describing a substance that is strongly alkaline (e.g. caustic soda). 2. (in optics) The curve or surface formed by the reflection of parallel rays of light in a large-aperture concave mirror. The apex of the caustic lies at the principal focus of the mirror. Such a curve can sometimes be seen on the surface of the liquid in a cup as a result of reflection by the curved walls of the cup. A similar curve is formed by a convex lens with spherical surfaces refracting parallel rays of light.

**caustic potash** *See* POTASSIUM HYDROXIDE.

**caustic soda** *See* SODIUM HYDROXIDE.

**Cavendish, Henry** (1731–1810) British chemist and physicist, born in France. Although untrained, his inheritance from his grandfather, the Duke of Devonshire, enabled him to live as a recluse and study science. In his experiments with gases (1766), he correctly distinguished between hydrogen and carbon dioxide, and in 1781 synthe-

sized water by exploding hydrogen in oxygen. He also constructed a torsion balance in 1798, with which he measured the mean density (and hence mass) of the earth.

**cavitation** The formation of gas- or vapour-filled cavities in liquids in motion when the pressure is reduced to a critical value while the ambient temperature remains constant. If the velocity of the flowing liquid exceeds a certain value, the pressure can be reduced to such an extent that the *Bernoulli theorem breaks down. It is at this point that cavitation occurs, causing a restriction on the speed at which hydraulic machinery can be run without noise, vibration, erosion of metal parts, or loss of efficiency.

**cavity resonator** *See* RESONANT CAVITY.

**C cell (parafollicular cell)** Any one of a group of calcium-secreting cells in vertebrates that are derived from the terminal pair of gill pouches. In mammals these cells are incorporated into the *thyroid gland and the *parathyroid gland.

**c.c.p.** Cubic close packing. *See* CLOSE PACKING.

**CD** 1. (cluster of differentiation) Any group of antigens that is associated with a specific subpopulation of human *T cells. The differentiation antigens expressed by a T cell vary with its stage of development and thus with its role in the immune response. Hence, for example, CD4 antigens are expressed by helper T cells, whereas CD8 antigens are expressed by cytotoxic T cells. The antigens are glycoproteins and are characterized using *monoclonal antibodies. 2. *See* COMPACT DISK. 3. *See* CIRCULAR DICHROISM.

**CD-I** CD interactive. A variant of *CD-ROM in which data, sound, and images can be interleaved on the same disk, i.e. it is a *multimedia disk. It was designed as a 'buy and play' system for the home.

**cDNA** *See* COMPLEMENTARY DNA.

**CD-ROM** CD read-only memory. A device that is based on the audio *compact disk and provides read-only access to a large amount of data (up to 640 megabytes) for use on computer systems. The term also refers to the medium in general. A **CD-ROM drive** must be used with the computer system to read the data from disk; the data cannot normally be rewritten. Most drives can also play CD audio disks, but audio disk players cannot handle CD-ROMs. The data may be in any form – text, sound, images, or binary data, or a mixture – and various CD-ROM format standards exist to handle these. CD-ROM is widely used for the distribution of data, images, and software and for archiving data.

**CD-RW** CD-rewritable, a CD format launched around 1997 that enabled recording and re-use of CDs. CD-RW uses a phase change to record data. The recording layer is a special alloy (typically silver/indium/antimony/tellurium). The laser in the CD drive has three power levels. The highest level melts small regions of the recording layer and these cool quickly to an amorphous form, thereby creating small pits in the recording surface. This level is used for writing data to the disk. The intermediate power level heats the surface to a temperature below the melting point, but high enough to cause recrystallization of the amorphous pits. This is used for erasing data. The lowest power level is used for reading data from the disk in the same way that data is read from a CD-ROM.

**CD spectrum (circular dichroism spectrum)** The spectrum obtained by plotting the variable $I_R - I_L$ against frequency of the incident electromagnetic radiation, where $I_R$ and $I_L$ are the absorption intensities for right- and left-circularly polarized light, respectively. One application of CD spectroscopy is to determine the configurations of complexes of transition metals in inorganic chemistry.

**CE** *See* CAPILLARY ELECTROPHORESIS.

**celestial equator** *See* EQUATOR.

**celestial mechanics** The study of the motions of and forces between the celestial bodies. It is based on *Newton's laws of motion and *Newton's law of gravitation. Refinements based on the general theory of *relativity are also included in the study, although the differences between the two theories are only important in a few cases.

**celestial sphere** The imaginary sphere on the inside of which all celestial bodies appear to be projected. The earth, and the observer, are visualized as being at the centre of the sphere and the sphere as rotating from east to west once every sidereal *day (see illustration). The sphere is used to describe

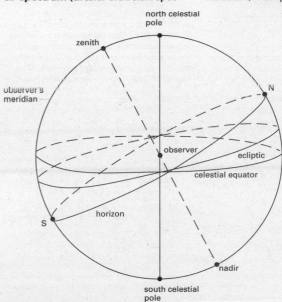

Celestial sphere.

the position of celestial bodies with respect to the earth.

**celestine** A mineral form of strontium sulphate, $SrSO_4$.

**cell** (in physical chemistry) **1.** A system in which two electrodes are in contact with an electrolyte. The electrodes are metal or carbon plates or rods or, in some cases, liquid metals (e.g. mercury). In an *electrolytic cell a current from an outside source is passed through the electrolyte to produce chemical change (see ELECTROLYSIS). In a *voltaic cell, spontaneous reactions between the electrodes and electrolyte(s) produce a potential difference between the two electrodes.

Voltaic cells can be regarded as made up of two *half cells, each composed of an electrode in contact with an electrolyte. For instance, a zinc rod dipped in zinc sulphate solution is a $Zn|Zn^{2+}$ half cell. In such a system zinc atoms dissolve as zinc ions, leaving a negative charge on the electrode

$$Zn(s) \rightarrow Zn^{2+}(aq) + 2e$$

The solution of zinc continues until the charge build-up is sufficient to prevent further ionization. There is then a potential difference between the zinc rod and its solution. This cannot be measured directly, since measurement would involve making contact with the electrolyte, thereby introducing another half cell (see ELECTRODE POTENTIAL). A rod of copper in copper sulphate solution comprises another half cell. In this case the spontaneous reaction is one in which copper ions in solution take electrons from the electrode and are deposited on the electrode as copper atoms. In this case, the copper acquires a positive charge.

The two half cells can be connected by using a porous pot for the liquid junction (as in the *Daniell cell) or by using a salt bridge. The resulting cell can then supply current if the electrodes are connected through an external circuit. The cell is written

$$Zn(s)|Zn^{2+}(aq)|Cu^{2+}(aq)|Cu$$

$$E = 1.10 \text{ V}$$

Here, $E$ is the e.m.f. of the cell equal to the potential of the right-hand electrode minus that of the left-hand electrode for zero current. Note that 'right' and 'left' refer to the cell as written. Thus, the cell could be written

$$Cu(s)|Cu^{2+}(aq)|Zn^{2+}(aq)|Zn(s)$$

$$E = -1.10 \text{ V}$$

The overall reaction for the cell is

$$Zn(s) + Cu^{2+}(aq) \rightarrow Cu(s) + Zn^{2+}(aq)$$

This is the direction in which the cell reaction occurs for a positive e.m.f.

The cell above is a simple example of a **chemical cell**; i.e. one in which the e.m.f. is produced by a chemical difference. **Concentration cells** are cells in which the e.m.f. is caused by a difference of concentration. This may be a difference in concentration of the electrolyte in the two half cells. Alternatively,

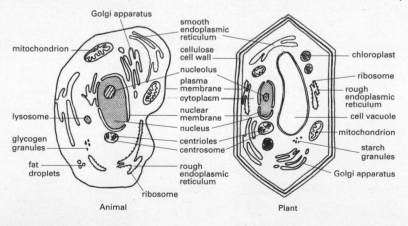

**Cell.** Generalized eukaryotic cells.

it may be an electrode concentration difference (e.g. different concentrations of metal in an amalgam, or different pressures of gas in two gas electrodes). Cells are also classified into cells **without transport** (having a single electrolyte) and **with transport** (having a liquid junction across which ions are transferred). Various types of voltaic cell exist, used as sources of current, standards of potential, and experimental set-ups for studying electrochemical reactions. *See also* DRY CELL; PRIMARY CELL; SECONDARY CELL; LITHIUM BATTERY.
**2.** *See* PHOTOELECTRIC CELL. **3.** *See* SOLAR CELL. **4.** *See* KERR EFFECT (for Kerr cell).

**cell** (in biology) The structural and functional unit of most living organisms (*compare* COENOCYTE; SYNCYTIUM). Cell size varies, but most cells are microscopic (average diameter 0.01–0.1 mm). Cells may exist as independent units of life, as in bacteria and certain protoctists, or they may form colonies or tissues, as in all plants and animals. Each cell consists of a mass of protein material that is differentiated into *cytoplasm and a *nucleus, which contains DNA. The cell is bounded by a *plasma membrane, which in the cells of plants, fungi, algae, and bacteria is surrounded by a *cell wall. There are two main types of cell. **Prokaryotic cells** (bacteria) are the more primitive. The nuclear material is not bounded by a membrane and chemicals involved in cell metabolism are associated with the plasma membrane. The shape and internal organization of cells depends on the *cytoskeleton. Reproduction is generally asexual and involves simple cell cleavage. In **eukaryotic cells** the nucleus is bounded by a nuclear envelope and the cytoplasm is divided by membranes into a system of interconnected cavities and separate compartments (**organelles**), e.g. *mitochondria, *endoplasmic reticulum, *Golgi apparatus, *lysosomes, and *ribosomes (see illustration). The shape and internal organization of cells depends on the *cytoskeleton. Reproduction can be either asexual (*see* MITOSIS) or sexual (*see* MEIOSIS). Plants and animals consist of eukaryotic cells but plant cells possess *chloroplasts and other *plastids and bear a rigid cellulose cell wall. See Chronology: Cell Biology.

(((●))) **SEE WEB LINKS**

• Overview of cell biology, with animations and illustrations, from Cells Alive!

**cell body (perikaryon)** The part of a *neuron that contains the nucleus. The cell processes that are involved in the transmission and reception of nervous impulses (the axon and the dendrites respectively) develop as extensions from the cell body.

**cell cycle** The sequence of stages that a cell passes through between one cell division and the next. The cell cycle can be divided into four main stages: (1) the **M phase**, which consists of *mitosis (nuclear division) and **cytokinesis** (cytoplasmic division); (2) the $G_1$ **phase**, in which there is a high rate of biosynthesis and growth; (3) the **S phase**, in which the DNA content of the cell doubles and the chromosomes replicate; (4) the $G_2$ **phase**, during which the final preparations for cell division are made. *Interphase consists of the $G_1$, S, and $G_2$ phases, which comprise about 90% (16–24 hours) of the total time of the cell cycle in rapidly dividing cells. The M phase lasts about 1–2 hours. A point is reached in the $G_1$ phase, known as the **restriction point**, after which the cell becomes committed to passing through the remainder of the cell cycle regardless of the external conditions.

**cell division** The formation of two or more daughter cells from a single mother cell. The nucleus divides first and this is followed by the formation of a plasma membrane between the daughter nuclei. *Mitosis produces two daughter nuclei that are identical to the original nucleus; *meiosis results in four daughter nuclei each with half the number of chromosomes in the mother cell nucleus. *See also* CELL CYCLE.

(((●))) **SEE WEB LINKS**

• Animation comparing mitosis with meiosis

**cell fusion (somatic cell hybridization)** The technique of combining two cells from different tissues or species in a cell culture. The cells fuse and coalesce but their nuclei generally remain separate. However, during cell division a single spindle is formed so that each daughter cell has a single nucleus containing sets of chromosomes from each parental line. Subsequent division of the hybrid cells often results in the loss of chromosomes (and therefore genes), so that absence of a gene product in the culture can be related to the loss of a particular chromosome. Thus the technique is used to determine the control of characteristics exerted by specific chromosomes. Hybrid cells resulting from

## CELL BIOLOGY

| | |
|---|---|
| 1665 | English physicist Robert Hooke (1635–1703) coins the word 'cell'. |
| 1831 | Robert Brown discovers the nucleus in plant cells. |
| 1838 | German botanist Matthias Schleiden (1804–81) proposes that plants are composed of cells. |
| 1839 | Theodor Schwann states that animals are composed of cells and concludes that all living things are made up of cells. |
| 1846 | German botanist Hugo von Mohl (1805–72) coins the word 'protoplasm' for the living material of cells. |
| 1858 | German pathologist Rudolf Virchow (1821–1902) postulates that all cells arise from other cells. |
| 1865 | German botanist Julius von Sachs (1832–97) discovers the chlorophyll-containing bodies in plant cells later named chloroplasts. |
| 1876–80 | German cytologist Eduard Strasburger (1844–1912) describes cell division in plants and states that new nuclei arise from division of existing nuclei. |
| 1882 | German cytologist Walther Flemming (1843–1905) describes the process of cell division in animal cells, for which he coins the term 'mitosis'. |
| | Strasburger coins the words 'cytoplasm' and 'nucleoplasm'. |
| 1886 | German biologist August Weismann (1834–1914) proposes his theory of the continuity of the germ plasm. |
| 1887 | Belgian cytologist Edouard van Beneden (1846–1910) discovers that the number of chromatin-containing threadlike bodies (subsequently named chromosomes) in the cells of a given species is always the same and that the sex cells contain half this number. |
| 1888 | German anatomist Heinrich von Waldeyer (1836–1921) coins the word 'chromosome'. |
| 1898 | Camillo Golgi discovers the Golgi apparatus. |
| 1901 | US biologist Clarence McClung (1870–1946) discovers the sex chromosomes. |
| 1911 | Thomas Hunt Morgan produces the first chromosome map. |
| 1949 | Canadian geneticist Murray Barr (1908–95) discovers Barr bodies. |
| 1955 | Belgian biochemist Christian de Duve (1917–   ) discovers lysosomes and peroxisomes. |
| 1956 | Romanian-born US physiologist George Palade (1912–2008) discovers the role of microsomes (later renamed ribosomes). |
| 1956 | US biochemist Arthur Kornberg (1918–2007) discovers DNA polymerase. |
| 1957 | US biochemist Melvin Calvin (1911–97) publishes details of the photosynthetic carbon-fixation cycle (Calvin cycle). |

cell fusion are used to produce *monoclonal antibodies.

**cell junction** Any of various kinds of connection between cells. **Tight junctions** form a seal between adjacent cells, particularly in epithelia, to prevent the passage of materials between cells. A primarily structural bond between cells is provided by **adherens junctions** and **desmosomes**, whereas communication between adjacent cells is facilitated by **gap junctions** in animal cells and *plasmodesmata in plant cells.

| | |
|---|---|
| 1961 | South African-born British biochemist Sydney Brenner (1927– ) discovers messenger RNA, in conjunction with François Jacob (1920– ) and Mathew S. Meselson (1930– ). |
| 1964 | US microbiologists Keith Porter and Thomas F. Roth discover the first cell receptors. |
| 1970 | US biologist Lynn Margulis (1938– ) proposes the endosymbiont theory for the origin of eukaryote cellular organelles. |
| 1971 | German-born US cell biologist Günter Blobel (1936– ) proposes the signal hypothesis to explain how proteins are delivered to their correct destinations within cells. |
| 1975 | British biologists J. A. Lucy and E. C. Cocking achieve successful fusion of plant and animal cells. |
| 1979 | The first 'test-tube baby', Louise Brown, is born in the UK using *in vitro* fertilization. |
| 1982 | British cell biologist Timothy Hunt (1943– ) discovers cyclins, proteins that control the cell cycle. US neurologist Stanley Pruisner (1942– ) discovers prions. |
| 1983 | A mouse embryo is engineered to include the gene for human growth hormone, creating a 'supermouse'. |
| 1984 | Sheep embryos are cloned for the first time. |
| 1986 | US cell biologist Robert Horvitz (1947– ) identifies genes involved in programmed cell death in the nematode *Caenorhabditis elegans*. First licence granted in USA for marketing a genetically engineered organism. |
| 1993 | First successful cloning of human embryos. |
| 1997 | Birth of Dolly the sheep, the first mammal to be cloned from adult body cells. |
| 1998 | Approval given in USA for therapeutic use of a synthetic skin containing live cultured human tissue cells. |
| 2000 | The embryo of a gaur, an endangered mammal, is cloned from skin cells of an adult and develops inside the womb of a cow. |
| 2002 | A pluripotent stem cell is isolated from adult human bone marrow. Discovery of new mechanism for regulating gene expression, called a riboswitch. |
| 2004 | World's first bank for stem cells opens in north London. |
| 2005 | First cloned dog (an Afghan hound called Snuppy) is created, using somatic cell transfer, by Korean researchers led by Woo Suk Huang. |

**cell membrane** Any membrane that is found in a living cell, especially the *plasma membrane, which forms the cell boundary. Other cell membranes include the nuclear envelope (*see* NUCLEUS); the *tonoplast, which encloses the vacuole of plant cells; and the membranes of the various cell organelles, such as the endoplasmic reticulum, Golgi apparatus, mitochondria, chloroplasts, and lysosomes.

**cell sap** The solution that fills the vacuoles of plant cells. It contains sugars, amino acids, waste substances (such as tannins), and mineral salts.

**cell theory** The theory that was born of the findings of Matthias Schleiden (1804–81) in 1838 and Theodor *Schwann in 1839, who postulated, respectively, that plants and animals were made up of cells and that these units were basic to the structure and function of all organisms. Previously, in 1665, Robert Hooke, while examining cork under the microscope, had observed that its structure consisted of hollow boxlike units, which he called 'cells'. At the time, however, he did not realize the significance of these units, which were in fact dead cells.

**cellular plastic** See EXPANDED PLASTIC.

**cellulase** A carbohydrate-digesting enzyme (a **carbohydrase**) that hydrolyses cellulose to sugars, including **cellobiose** (a disaccharide consisting of two β-(1,4) linked molecules of glucose) and glucose. Cellulase breaks the β-glycosidic links that join the constituent sugar units of cellulose. See also RUMINANTIA.

**celluloid** A transparent highly flammable substance made from cellulose nitrate with a camphor plasticizer. It was formerly widely used as a thermoplastic material, especially for film (a use now discontinued owing to the flammability of celluloid).

**cellulose** A polysaccharide that consists of a long unbranched chain of glucose units. It is the main constituent of the cell walls of all plants, many algae, and some fungi and is responsible for providing the rigidity of the cell wall. It is an important constituent of dietary *fibre. The fibrous nature of extracted cellulose has led to its use in the textile industry for the production of cotton, artificial silk, etc.

**cellulose acetate** See CELLULOSE ETHANOATE.

**cellulose ethanoate (cellulose acetate)** A compound prepared by treating cellulose (cotton linters or wood pulp) with a mixture of ethanoic anhydride, ethanoic acid, and concentrated sulphuric acid. Cellulose in the cotton is ethanoylated and when the resulting solution is treated with water, cellulose ethanoate forms as a flocculent white mass. It is used in lacquers, nonshatterable glass, varnishes, and as a fibre (see also RAYON).

**cellulose nitrate** A highly flammable material made by treating cellulose (wood pulp) with concentrated nitric acid. Despite the alternative name **nitrocellulose**, the compound is in fact an ester (containing $CONO_2$ groups), not a nitro compound (which would contain $C–NO_2$). It is used in explosives (as **guncotton**) and celluloid.

**cell wall** A rigid outer layer that surrounds the plasma membrane of plant, fungal, algal and bacterial (but not animal) cells. It protects and/or gives shape to a cell, and in herbaceous plants provides mechanical support for the plant body. Most plant cell walls are composed of the polysaccharide *cellulose and may be strengthened by the addition of *lignin. The cell walls of fungi consist mainly of *chitin. Bacterial cell walls consist of complex polymers of polysaccharides and amino acids.

**Celsius scale** A *temperature scale in which the fixed points are the temperatures at standard pressure of ice in equilibrium with water (0°C) and water in equilibrium with steam (100°C). The scale, between these two temperatures, is divided in 100 degrees. The degree Celsius (°C) is equal in magnitude to the *kelvin. This scale was formerly known as the **centigrade scale**; the name was officially changed in 1948 to avoid confusion with a hundredth part of a grade. It is named after the Swedish astronomer Anders Celsius (1701–44), who devised the inverted form of this scale (ice point 100°, steam point 0°) in 1742.

**cement** 1. Any of various substances used for bonding or setting to a hard material. Portland cement is a mixture of calcium silicates and aluminates made by heating limestone ($CaCO_3$) with clay (containing aluminosilicates) in a kiln. The product is ground to a fine powder. When mixed with water it sets in a few hours and then hardens over a longer period of time due to the formation of hydrated aluminates and silicates. 2. (**cementum**) A thin layer of bony material that fixes teeth to the jaw. It covers the dentine of the root of a *tooth, below the level of the gum, and is attached to the *periodontal membrane lining the tooth socket in the jawbone.

**cementation** Any metallurgical process in which the surface of a metal is impregnated by some other substance, especially an obsolete process for making steel by heating bars of wrought iron to red heat for several days in a bed of charcoal. See also CASE HARDENING.

**cementite** See STEEL.

**Cenozoic (Cainozoic; Kainozoic)** The geological era that began about 65 million years ago and extends to the present. It followed the *Mesozoic era and is subdivided into the *Palaeogene and *Neogene periods. The Cenozoic is often known as the **Age of Mammals** as these animals evolved to become an abundant, diverse, and dominant group. Birds and flowering plants also flourished. The era saw the formation of the major mountain ranges of the Himalayas and the Alps.

**centaur** Any of a class of *minor planets that appear to be half-way between an asteroid and a comet and follow an unstable eccentric solar orbit between Jupiter and Neptune. Centaurs are probably *scattered disc objects that have been perturbed into their present orbits.

**centi-** Symbol c. A prefix used in the metric system to denote one hundredth. For example, 0.01 metre = 1 centimetre (cm).

**centigrade scale** See CELSIUS SCALE.

**centipedes** See CHILOPODA.

**Central Dogma** The basic belief originally held by molecular geneticists, that flow of genetic information can only occur from *DNA to *RNA to proteins. It is now known, however, that information contained within RNA molecules can also flow back to DNA, for example during the replication of *retroviruses. See also GENETIC CODE.

**central nervous system (CNS)** The part of the nervous system that coordinates all neural functions. In invertebrates it may comprise simply a few *nerve cords and their associated *ganglia. In vertebrates it consists of the *brain and the *spinal cord. The vertebrate CNS contains *reflex arcs, which produce automatic and rapid responses to particular stimuli.

**central processing unit** See CPU; COMPUTER.

**centre** (in neurology) A part of the nervous system, consisting of a group of nerve cells, that coordinates a particular process. An example is the respiratory centre in the vertebrate brainstem, which controls breathing movements. The stimulation of a centre will initiate the process, while destruction of the centre will prevent or impair it.

**centre of curvature** The centre of the sphere of which a *lens surface or curved *mirror forms a part. The **radius of curvature** is the radius of this sphere.

**centre of gravity** See CENTRE OF MASS.

**centre of mass** The point at which the whole mass of a body may be considered to be concentrated. This is the same as the **centre of gravity**, the point at which the whole weight of a body may be considered to act, if the body is situated in a uniform gravitational field.

**centrifugal force** See CENTRIPETAL FORCE.

**centrifugal pump** See PUMP.

**centrifuge** A device in which solid or liquid particles of different densities are separated by rotating them in a tube in a horizontal circle. The denser particles tend to move along the length of the tube to a greater radius of rotation, displacing the lighter particles to the other end.

**centriole** A cylindrical structure associated with the *centrosome in animal cells but not normally found in plant cells. Centrioles occur in pairs, orientated at right angles to each other, and are composed of *microtubules. During cell division the pair separates: a centriole migrates with each centrosome to opposite poles of the cell. Centrioles seem to act as an orientational device in the assembly of centrosomes; they are also essential for the assembly of *undulipodia.

**centripetal force** A force acting on a body causing it to move in a circular path. If the mass of the body is $m$, its constant speed $v$, and the radius of the circle $r$, the magnitude of the force is $mv^2/r$ and it is directed towards the centre of the circle. Even though the body is moving with a constant speed $v$, its velocity is changing, because its direction is constantly changing. There is therefore an acceleration $v^2/r$ towards the centre of the circle. For example, when an object is tied to a string and swung in a horizontal circle there is a tension in the string equal to $mv^2/r$. If the string breaks, this restraining force disappears and the object will move off in a straight line along the tangent to the circle in which it was previously moving.

In the case of a satellite (mass $m$) orbiting the earth (mass $M$), the centripetal force holding the satellite in orbit is the gravitational force, $GmM/d^2$, where $G$ is the gravitational constant and $d$ is the height of the satellite above the centre of the earth. Therefore $GmM/d^2 = mv^2/d$. This equation enables the height of the orbit to be calculated for a given orbital velocity.

Another way of looking at this situation, which was once popular, is to assume that the centripetal force is balanced by an equal and opposite force, acting away from the centre of the circle, called the **centrifugal force**. One could then say that the satellite stays in orbit when the centrifugal force balances the gravitational force. This is, however, a confusing and misleading argument because the centrifugal force is fictitious – it does not exist. The gravitational force is not balanced by the centrifugal force: it *is* the centripetal force.

Another example is that of a car rounding a bend. To an observer in the car, a tennis ball lying on the back shelf will roll across the shelf as if it was acted on by an outward centrifugal force. However, to an observer outside the car it can be seen that the ball, because of its almost frictionless contact with the car, is continuing in its straight line motion, uninfluenced by the centripetal force. Occasionally the concept of a centrifugal force can be useful, as long as it is recognized as a fictitious force. A true centrifugal force is exerted, as by a *reaction, by the rotating object on whatever is providing its centripetal force.

**centroid** The point within an area or volume at which the centre of mass would be if the surface or body had a uniform density. For a symmetrical area or volume it coincides with the centre of mass. For a nonsymmetrical area or volume it has to be found by integration.

**centromere (kinomere; spindle attachment)** The part of a *chromosome that attaches to the *spindle during cell division (*see* MEIOSIS; MITOSIS), attachment being via a platelike structure called the *kinetochore. The centromere usually appears as a constriction when chromosomes contract during cell division. The position of the centromere is a distinguishing feature of individual chromosomes.

**centrosome (cell centre; centrosphere)** A specialized region of all eukaryote cells except fungi, situated next to the nucleus, that organizes the microtubules of the *spindle during cell division. In animal cells it is also the main region of the cell from which the microtubules of the *cytoskeleton radiate.

The centrosomes of most animal cells contain a pair of *centrioles. During *metaphase of mitosis and meiosis, the centrosome separates into two regions, each containing one of the centrioles (where present). The two regions move to opposite ends of the cell and a spindle forms between them. *See also* ASTER.

**centrum** *See* VERTEBRA.

**cephalization** The tendency among animal groups for the major sense organs, mouth, and brain to be grouped together at the front (anterior) end of the body. These are usually contained in a specialized cephalic region – the head.

**Cephalopoda** The most advanced class of molluscs, containing the squids, cuttlefishes, octopuses, and the extinct *ammonites. Cephalopods have a highly concentrated central nervous system within a protective cartilaginous case. The eye has a well-developed retina and is comparable to that of vertebrates. All cephalopods are predacious carnivores capable of swimming by jet propulsion; they have highly mobile tentacles for catching and holding prey.

**cephalothorax** The fused head and thorax of crustaceans and arachnids (spiders, scorpions), which is connected to the abdomen.

**Cepheid variable** A type of pulsating star that has departed from the main sequence (*see* HERTZSPRUNG–RUSSELL DIAGRAM) and is either a young star belonging to population I, (the **classical cepheids**) or an old star belonging to population II (the **W Virginis stars**). Cepheids contract and expand with great regularity over periods that range from 1 to 50 days. There is a direct correlation between a cepheid's period and its luminosity: those with the longest periods are intrinsically the brightest. This fact has allowed astronomers to use these stars to determine stellar and galactic distances with great accuracy. Most Cepheids are yellow supergiants destined to become *red giants. There are more than 700 known Cepheids within the Milky Way Galaxy and thousands more in the galaxies of the Local Group. Cepheids take their name from their prototype, Delta Cephei in the northern constellation Cepheus. *See also* POPULATION TYPE.

**ceramics** Inorganic materials, such as pottery, enamels, and refractories. Ceramics are metal silicates, oxides, nitrides, etc.

**cerebellum** The part of the vertebrate *brain concerned with the coordination and regulation of muscle activity and the maintenance of muscle tone and balance. In mammals it consists of two connected hemispheres, composed of a core of white matter and a much-folded outer layer of grey matter, and is situated above the medulla oblongata and partly beneath the cerebrum.

**cerebral cortex (pallium)** The layer of *grey matter that forms the outer layer of the hemispheres of the *cerebrum in many vertebrates. It is most highly developed in mammals. The cortex is responsible for the control and integration of voluntary movement and the senses of vision, hearing, touch, etc.; it also contains centres concerned with memory, language, thought, and intellect.

**cerebral hemisphere** Either of the two halves of the vertebrate *cerebrum.

**cerebrospinal fluid (CSF)** The fluid, similar in composition to *lymph, that bathes the central nervous system of vertebrates. It is secreted by the *choroid plexus into the *ventricles of the brain, filling these and other cavities in the brain and spinal cord, and is reabsorbed by veins on the brain surface. Its function is to protect the central nervous system from mechanical injury.

**cerebrum** The largest part of the vertebrate *brain. It consists of two **cerebral hemispheres**, which develop from the embryonic *forebrain. The hemispheres have an outer convoluted layer of grey matter – the *cerebral cortex – which contains an estimated ten billion nerve cells. Underneath this is *white matter. The two halves of the cerebrum are linked by the *corpus callosum. The function of the cerebrum is to integrate complex sensory and neural functions. The cerebrum plays a critical role in the process of learning, which involves both short-term and long-term memory.

**Cerenkov, Pavel Alekseyevich** (1904–90) Soviet physicist, who became a professor at the Lebedev Institute of Physics in Moscow. In 1934, while observing radioactive radiation underwater, he discovered *Cerenkov radiation. The explanation of the phenomenon was provided by Igor Tamm (1895–1971) and Ilya Frank (1908–90), and in 1958 the three scientists shared the Nobel Prize for physics.

**Cerenkov counter (Cerenkov detector)** A type of *counter for detecting and counting high-energy charged particles. The particles pass through a liquid and the light emitted as *Cerenkov radiation is registered by a *photomultiplier tube.

**Cerenkov radiation** Electromagnetic radiation, usually bluish light, emitted by a beam of high-energy charged particles passing through a transparent medium at a speed greater than the speed of light in that medium. It was discovered in 1934 by Pavel Cerenkov. The effect is similar to that of a *sonic boom when an object moves faster than the speed of sound; in this case the radiation is a shock wave set up in the electromagnetic field. Cerenkov radiation is used in the *Cerenkov counter.

**cerium** Symbol Ce. A silvery metallic element belonging to the *lanthanoids; a.n. 58; r.a.m. 140.12; r.d. 6.77 (20°C); m.p. 799°C; b.p. 3426°C. It occurs in allanite, bastnasite, cerite, and monazite. Four isotopes occur naturally: cerium–136, –138, –140, and –142; fifteen radioisotopes have been identified. Cerium is used in mischmetal, a rare-earth metal containing 25% cerium, for use in lighter flints. The oxide is used in the glass industry. It was discovered by Martin Klaproth (1743–1817) in 1803.

( ) **SEE WEB LINKS**
• Information from the WebElements site

**cermet** A composite material consisting of a ceramic in combination with a sintered metal, used when a high resistance to temperature, corrosion, and abrasion is needed.

**CERN (Conseil Européen pour la Recherche Nucléaire)** The European Laboratory for Particle Physics, formerly known as the European Organization for Nuclear Research, which is situated close to Geneva in Switzerland and is supported by a number of European nations. It runs the **Super Proton Synchrotron** (**SPS**), which has a 7-kilometre underground tunnel enabling protons to be accelerated to 400 GeV, and the **Large Electron-Positron Collider** (**LEP**), in which 50 GeV electron and positron beams are collided. The *Large Hadron Collider began operation in September 2008.

( ) **SEE WEB LINKS**
• The CERN public website

**certificate (public key certificate; digital certificate)** In computing, a means of authenticating public keys. *Public key encryption is a very powerful system but has an important security hole: there is no intrinsic guarantee that the people or organizations distributing public keys are who they claim to be. A certificate is a file issued by a trusted third party – a certificate authority – that contains both a public key and details of the person or organization to whom it belongs, which the third party declares to be correct. Crucially, the certificate is digitally signed by the third party. A recipient can verify that the certificate itself is genuine by using the third party's public key, and can then be confident in using the public key it contains. Certificate files comply with the X509 standard and their use on the Internet is governed by RFC 3280.

**(((⊕))) SEE WEB LINKS**
• The Internet X509 specification

**cerussite** An ore of lead consisting of lead carbonate, $PbCO_3$. It is usually of secondary origin, formed by the weathering of *galena. Pure cerussite is white but the mineral may be grey due to the presence of impurities. It forms well-shaped orthorhombic crystals. It occurs in the USA, Spain, and SW Africa.

**cervical vertebrae** The *vertebrae of the neck. The number of cervical vertebrae varies with the vertebrate group: most mammals (including humans) have seven. Their main functions are to support the head and to provide articulating surfaces against which it can move relative to the backbone. *See* ATLAS; AXIS.

**cervix** A narrow or necklike part of an organ. The cervix of the uterus (**cervix uteri**) leads to the vagina. Glands in its walls produce mucus, whose viscosity changes according to the oestrous cycle. During labour, the cervix enlarges greatly to allow passage of the fetus.

**Cestoda** A class of flatworms (*see* PLATY-HELMINTHES) comprising the tapeworms – ribbon-like parasites within the gut of vertebrates. Tapeworms are surrounded by partially digested food in the host gut so they are able to absorb nutrients through their whole body surface. The body consists of a **scolex** (head), bearing suckers and hooks for attachment, and a series of **proglottids**, which contain male and female reproductive systems. The life cycle of a tapeworm requires two hosts, the primary host usually being a predator of the secondary host. *Taenia solium* has humans for its primary host

and the pig as its secondary host. Mature proglottids, containing thousands of fertilized eggs, leave the primary host with its faeces and develop into embryos and then larvae that continue the life cycle in the gut of a secondary host.

**Cetacea** An order of marine mammals comprising the whales, which includes what is probably the largest known animal – the blue whale (*Balaenoptera musculus*), over 30 m long and over 150 tonnes in weight. The forelimbs of whales are modified as short stabilizing flippers and the skin is very thin and almost hairless. A thick layer of blubber insulates the body against heat loss and is an important food store. Whales breathe through a dorsal blowhole, which is closed when the animal is submerged. The toothed whales (suborder Odontoceti), such as the dolphins and killer whale, are carnivorous; whalebone whales (suborder Mysticeti), such as the blue whale, feed on plankton filtered by *whalebone plates. Molecular systematics now indicates that whales are closely related to hippos and should be classified with them and other artiodactyls in the superorder Cetartiodactyla.

**cetane** *See* HEXADECANE.

**cetane number** A number that provides a measure of the ignition characteristics of a Diesel fuel when it is burnt in a standard Diesel engine. It is the percentage of cetane (hexadecane) in a mixture of cetane and 1-methylnaphthalene that has the same ignition characteristics as the fuel being tested. *Compare* OCTANE NUMBER.

**Cetartiodactyla** *See* ARTIODACTYLA.

**CFC** *See* CHLOROFLUOROCARBON.

**CGE** Capillary gel electrophoresis. *See* CAPILLARY ELECTROPHORESIS.

**c.g.s. units** A system of *units based on the centimetre, gram, and second. Derived from the metric system, it was not well suited for use with thermal quantities (based on the inconsistently defined *calorie) and with electrical quantities (in which two systems, based respectively on unit permittivity and unit permeability of free space, were used). For many scientific purposes c.g.s. units have now been replaced by *SI units.

**Chadwick, Sir James** (1891–1974) British physicist. After working at Manchester University under *Rutherford, he went to work with Hans *Geiger in Leipzig in 1913. Interned for the duration of World War I, he joined Rutherford in Cambridge after the war. In 1932 he discovered the *neutron, as predicted by Rutherford. In 1935 he was awarded the Nobel Prize, the same year in which he built Britain's first *cyclotron at Liverpool University.

**chaeta** A bristle, made of *chitin, occurring in annelid worms. In the earthworm they occur in small groups projecting from the skin in each segment and function in locomotion. The chaetae of polychaete worms (e.g. ragworm) are borne in larger groups on paddle-like appendages (**parapodia**).

**chain** A line of atoms of the same type in a molecule. In a **straight chain** the atoms are attached only to single atoms, not to groups. Propane, for instance, is a straight-chain alkane, $CH_3CH_2CH_3$, with a chain of three carbon atoms. A **branched chain** is one in which there are side groups attached to the chain. Thus, 3-ethyloctane, $CH_3CH_2CH(C_2H_5)C_5H_{11}$, is a branched-chain alkane in which there is a **side chain** ($C_2H_5$) attached to the third carbon atom. A **closed chain** is a *ring of atoms in a molecule; otherwise the molecule has an **open chain**.

**Chain, Sir Ernst Boris** (1906–79) German-born British biochemist, who began his research career at Cambridge University in 1933. Two years later he joined *Florey at Oxford, where they isolated and purified *penicillin. They also developed a method of producing the drug in large quantities and carried out its first clinical trials. The two men shared the 1945 Nobel Prize for physiology or medicine with penicillin's discoverer, Alexander *Fleming.

**chain reaction** A reaction that is self-sustaining as a result of the products of one step initiating a subsequent step.

In nuclear chain reactions the succession depends on production and capture of neutrons. Thus, one nucleus of the isotope uranium–235 can disintegrate with the production of two or three neutrons, which cause similar fission of adjacent nuclei. These in turn produce more neutrons. If the total amount of material exceeds a *critical mass, the chain reaction may cause an explosion.

Chemical chain reactions usually involve free radicals as intermediates. An example is the reaction of chlorine with hydrogen initi-

ated by ultraviolet radiation. A chlorine molecule is first split into atoms:

$$Cl_2 \rightarrow Cl\cdot + Cl\cdot$$

These react with hydrogen as follows

$$Cl\cdot + H_2 \rightarrow HCl + H\cdot$$

$$H\cdot + Cl_2 \rightarrow HCl + Cl\cdot, \text{ etc.}$$

Combustion and explosion reactions involve similar free-radical chain reactions.

**chair conformation** *See* CONFORMATION.

**chalaza** 1. A twisted strand of fibrous albumen in a bird's egg that is attached to the membrane at either end of the yolk and thus holds the yolk in position in the albumen. 2. The part of a plant *ovule where the nucellus and integuments merge.

**chalcedony** A mineral consisting of a microcrystalline variety of *quartz. It occurs in several forms, including a large number of semiprecious gemstones; for example, sard, carnelian, jasper, onyx, chrysoprase, agate, and tiger's-eye.

**chalcogens** *See* GROUP 16 ELEMENTS.

**chalconides** Binary compounds formed between metals and group 16 elements; i.e. oxides, sulphides, selenides, and tellurides.

**chalcopyrite (copper pyrites)** A brassy yellow mineral consisting of a mixed copper–iron sulphide, $CuFeS_2$, crystallizing in the tetragonal system; the principal ore of copper. It is similar in appearance to pyrite and gold. It crystallizes in igneous rocks and hydrothermal veins associated with the upper parts of acid igneous intrusions. Chalcopyrite is the most widespread of the copper ores, occurring, for example, in Cornwall (UK), Sudbury (Canada), Chile, Tasmania (Australia), and Rio Tinto (Spain).

**chalk** A very fine-grained white rock composed of the fossilized skeletal remains of marine plankton known as **coccoliths** and consisting largely of *calcium carbonate ($CaCO_3$). It is used in toothpaste and cosmetics and is the characteristic rock of the *Cretaceous period. It should not be confused with blackboard 'chalk', which is made from calcium sulphate.

**Chandrasekhar limit** The maximum possible mass of a star that is prevented from collapsing under its own gravity by the *degeneracy pressure of electrons. For white dwarfs the **Chandrasekhar mass** is about 1.4 times the mass of the sun. There is an analogue of the Chandrasekhar limit for neutron stars. For neutron stars its value is less precisely known because of uncertainties regarding the equation of state of neutron matter, but it is generally taken to be in the range of 1.5 to 3 (and almost certainly no more than 5) times the mass of the sun. It is named after Subrahmanyan Chandrasekhar (1910–95).

**change of phase** (change of state) A change of matter in one physical *phase (solid, liquid, or gas) into another. The change is invariably accompanied by the evolution or absorption of energy, even if it takes place at constant temperature (*see* LATENT HEAT).

**channel** **1.** The region between the source and the drain in a field-effect *transistor. The conductivity of the channel is controlled by the voltage applied to the gate. **2.** A path, or a specified frequency band, along which signals, information, or data flow. **3.** A pore formed by a protein molecule in a plasma membrane that aids the diffusion of certain substances into and out of the cell. These substances are usually charged ions (*see* ION CHANNEL) or lipid-insoluble molecules. *See also* TRANSPORT PROTEIN.

**chaos** Unpredictable and seemingly random behaviour occurring in a system that should be governed by deterministic laws. In such systems, the equations that describe the way the system changes with time are nonlinear and involve several variables. Consequently, they are very sensitive to the initial conditions, and a very small initial difference may make an enormous change to the future state of the system. Originally, the theory was introduced to describe unpredictability in meteorology, as exemplified by the **butterfly effect**. It has been suggested that the dynamical equations governing the weather are so sensitive to the initial data that whether or not a butterfly flaps its wings in one part of the world may make the difference between a tornado occurring or not occurring in some other part of the world. Chaos theory has subsequently been extended to other branches of science; for example to turbulent flow, planetary dynamics, and electrical oscillations in physics, and to combustion processes and *oscillating reactions in chemistry. *See also* ATTRACTOR; FRACTAL.

**chaotic reaction** A type of chemical reaction in which the concentrations of reactants show chaotic behaviour. This may occur when the reaction involves a large number of complex interlinked steps. Under such conditions, it is possible for the reaction to display unpredictable changes with time. *See also* OSCILLATING REACTION.

**character** (trait) A distinctive inherited feature of an organism. Organisms in a population may display different aspects of a particular character, e.g. the A, B, and O human blood groups (*see* ABO SYSTEM) are different aspects of the blood group character.

**characteristic** *See* LOGARITHM.

**charcoal** A porous form of carbon produced by the destructive distillation of organic material. Charcoal from wood is used as a fuel. All forms of charcoal are porous and are used for adsorbing gases and purifying and clarifying liquids. There are several types depending on the source. Charcoal from coconut shells is a particularly good gas adsorbent. **Animal charcoal** (or **bone black**) is made by heating bones and dissolving out the calcium phosphates and other mineral salts with acid. It is used in sugar refining. **Activated charcoal** is charcoal that has been activated for adsorption by steaming or by heating in a vacuum.

**charge** A property of some *elementary particles that gives rise to an interaction between them and consequently to the host of material phenomena described as electrical. Charge occurs in nature in two forms, conventionally described as **positive** and **negative** in order to distinguish between the two kinds of interaction between particles. Two particles that have similar charges (both negative or both positive) interact by repelling each other; two particles that have dissimilar charges (one positive, one negative) interact by attracting each other. The size of the interaction is determined by *Coulomb's law.

The natural unit of negative charge is the charge on an *electron, which is equal but opposite in effect to the positive charge on the proton. Large-scale matter that consists of equal numbers of electrons and protons is electrically neutral. If there is an excess of electrons the body is negatively charged; an excess of protons results in a positive charge. A flow of charged particles, especially a flow of electrons, constitutes an electric current. Charge is measured in coulombs, the charge on an electron being $1.602 \times 10^{-19}$ coulombs.

**charge carrier** The entity that transports electric charge in an electric current. The nature of the carrier depends on the type of conductor: in metals, the charge carriers are electrons; in *semiconductors the carriers are electrons ($n$-type) or positive *holes ($p$-type); in gases the carriers are positive ions and electrons; in electrolytes they are positive and negative ions.

**charge conjugation** Symbol $C$. A property of elementary particles that determines the difference between a particle and its *antiparticle. The property is not restricted to electrically charged particles (i.e. it applies to neutral particles such as the neutron). *See* CP INVARIANCE.

**charge density** 1. The electric charge per unit volume of a medium or body (**volume charge density**). 2. The electric charge per unit surface area of a body (**surface charge density**).

**charge-transfer complex** A chemical compound in which there is weak coordination involving the transfer of charge between two molecules. An example is phenoquinone, in which the phenol and quinone molecules are not held together by formal chemical bonds but are associated by transfer of charge between the compounds' aromatic ring systems.

**Charles, Jacques Alexandre César** (1746–1823) French chemist and physicist, who became professor of physics at the Paris Conservatoire des Arts et Métiers. He is best remembered for discovering *Charles' law (1787), relating to the volume and temperature of a gas. In 1783 he became the first person to make an ascent in a hydrogen balloon.

**Charles' law** The volume of a fixed mass of gas at constant pressure expands by a constant fraction of its volume at 0°C for each Celsius degree or kelvin its temperature is raised. For any *ideal gas the fraction is approximately 1/273. This can be expressed by the equation

$$V = V_0(1 + t/273),$$

where $V_0$ is the volume at 0°C and $V$ is its volume at $t$°C. This is equivalent to the statement that the volume of a fixed mass of gas at constant pressure is proportional to its thermodynamic temperature, $V = kT$, where $k$ is a constant. The law resulted from experiments begun around 1787 by Jacques Charles but was properly established only by the more accurate results published in 1802 by the French scientist Joseph Gay-Lussac (1778–1850). Thus the law is also known as **Gay-Lussac's law**. An equation similar to that given above applies to pressures for ideal gases:

$$p = p_0(1 + t/273),$$

a relationship known as **Charles' law of pressures**. *See also* GAS LAWS.

**charm** A property of certain *elementary particles that is expressed as a quantum number and is used in the quark model. It was originally suggested to account for the unusually long lifetime of the *psi particle. In this theory the three original quark–antiquark pairs were supplemented by a fourth pair – the charmed quark and its antiquark. The psi particle itself is a meson having zero charm as it consists of the charmed pair. However, charmed *hadrons do exist; they are said to possess **naked charm**. Charm is thought to be conserved in strong and electromagnetic interactions.

**cheddite** Any of a group of high explosives made from nitro compounds mixed with sodium or potassium chlorate.

**chelate** An inorganic complex in which a *ligand is coordinated to a metal ion at two (or more) points, so that there is a ring of atoms including the metal (see formula). The process is known as **chelation**. A ligand such as diaminoethane, which coordinates at two points, is said to be **bidentate** ('having two teeth'). Other ligands are **tridentate**, **tetradentate**, etc. The angle made by two bonds coordinating to the metal atom is the **bite angle** of the ligand. *See also* SEQUESTRATION.

**Chelate.** Chelate formed by coordination of two molecules of H$_2$N(CH$_2$)$_2$NH$_2$.

**chelate effect** The effect in which a chelate complex is generally more stable than the analogous complex formed with monodentate ligands. For example, the complex ion [Cu(en) (OH$_2$)$_4$]$^{2+}$ is more stable than the complex ion $[Cu(NH_3)_2 (OH_2)_4]^{2+}$. Here, en denotes the bidentate ethylene diamine (1,2-diaminoethane) ligand. The main cause of the chelate effect is the effect of reaction entropy when the complex is formed. Thus, the reaction

$$[Cu(OH_2)_6]^{2+} + en \rightarrow [Cu(en)(OH_2)_4 + 2H_2O$$

results in a net increase in the number of molecules (from 2 to 3). The reaction

$$[Cu(OH_2)_6]^{2+} + 2NH_3 \rightarrow Cu(NH_3)2(OH_2)_4 + 2H_2O$$

involves no net increase in the number of molecules. As a result, the chelate reaction has a larger reaction entropy and is more favourable.

**chelicerae** The first pair of appendages on the head of arachnids and other *arthropods of the phylum Chelicerata. These appendages take the form of pincers or claws and are used for grasping or tearing food.

**ChemDraw** A widely used chemical drawing and modelling program produced by CambridgeSoft.

(((🌐))) SEE WEB LINKS

• The CambridgeSoft website, giving details about ChemDraw and associated software

**chemical bond** A strong force of attraction holding atoms together in a molecule or crystal. Typically chemical bonds have energies of about 1000 kJ mol⁻¹ and are distinguished from the much weaker forces between molecules (*see* VAN DER WAALS' FORCE). There are various types.

**Ionic** (or **electrovalent**) bonds can be formed by transfer of electrons. For instance, the calcium atom has an electron configuration of $[Ar]4s^2$, i.e. it has two electrons in its outer shell. The chlorine atom is $[Ne]3s^23p^5$, with seven outer electrons. If the calcium atom transfers two electrons, one to each chlorine atom, it becomes a $Ca^{2+}$ ion with the stable configuration of an inert gas $[Ar]$. At the same time each chlorine, having gained one electron, becomes a $Cl^-$ ion, also with an inert-gas configuration $[Ar]$. The bonding in calcium chloride is the electrostatic attraction between the ions.

**Covalent** bonds are formed by sharing of valence electrons rather than by transfer. For instance, hydrogen atoms have one outer electron ($1s^1$). In the hydrogen molecule, $H_2$, each atom contributes 1 electron to the bond. Consequently, each hydrogen atom has control of 2 electrons – one of its own

and the second from the other atom – giving it the electron configuration of an inert gas [He]. In the water molecule, $H_2O$, the oxygen atom, with six outer electrons, gains control of an extra two electrons supplied by the two hydrogen atoms. This gives it the configuration [Ne]. Similarly, each hydrogen atom gains control of an extra electron from the oxygen, and has the [He] electron configuration.

A particular type of covalent bond is one in which one of the atoms supplies both the electrons. These are known as **coordinate** (**semipolar** or **dative**) bonds, and written A→B, where the direction of the arrow denotes the direction in which electrons are donated.

Covalent or coordinate bonds in which one pair of electrons is shared are **electron-pair bonds** and are known as **single bonds**. Atoms can also share two pairs of electrons to form **double bonds** or three pairs in **triple bonds**. *See* ORBITAL.

In a compound such as sodium chloride, $Na^+Cl^-$, there is probably complete transfer of electrons in forming the ionic bond (the bond is said to be **heteropolar**). Alternatively, in the hydrogen molecule H–H, the pair of electrons is equally shared between the two atoms (the bond is **homopolar**). Between these two extremes, there is a whole range of **intermediate bonds**, which have both ionic and covalent contributions. Thus, in hydrogen chloride, H–Cl, the bonding is predominantly covalent with one pair of electrons shared between the two atoms. However, the chlorine atom is more electronegative than the hydrogen and has more control over the electron pair; i.e. the molecule is polarized with a positive charge on the hydrogen and a negative charge on the chlorine, forming a *dipole. *See also* BANANA BOND; HYDROGEN BOND; METALLIC BOND; MULTICENTRE BOND; MULTIPLE BOND.

**chemical cell** *See* CELL.

**chemical combination** The combination of elements to give compounds. There are three laws of chemical combination.
(1) The **law of constant composition** states that the proportions of the elements in a compound are always the same, no matter how the compound is made. It is also called the **law of constant proportions** or **definite proportions**.
(2) The **law of multiple proportions** states that when two elements A and B combine to form more than one compound, then the

masses of B that combine with a fixed mass of A are in simple ratio to one another. For example, carbon forms two oxides. In one, 12 grams of carbon is combined with 16 grams of oxygen (CO); in the other 12 grams of carbon is combined with 32 grams of oxygen ($CO_2$). The oxygen masses combining with a fixed mass of carbon are in the ratio 16:32, i.e. 1:2.

(3) The **law of equivalent proportions** states that if two elements A and B each form a compound with a third element C, then a compound of A and B will contain A and B in the relative proportions in which they react with C. For example, sulphur and carbon both form compounds with hydrogen. In methane 12 g of carbon react with 4 g of hydrogen. In hydrogen sulphide, 32 g of sulphur react with 2 g of hydrogen (i.e. 64 g of S for 4 g of hydrogen). Sulphur and carbon form a compound in which the C:S ratio is 12:64 (i.e. $CS_2$). The law is sometimes called the law of **reciprocal proportions**.

**chemical control** The use of chemicals to kill pests (*see* PESTICIDE). *Compare* BIOLOGICAL CONTROL.

**chemical dating** An absolute *dating technique that depends on measuring the chemical composition of a specimen. Chemical dating can be used when the specimen is known to undergo slow chemical change at a known rate. For instance, phosphate in buried bones is slowly replaced by fluoride ions from the ground water. Measurement of the proportion of fluorine present gives a rough estimate of the time that the bones have been in the ground. Another, more accurate, method depends on the fact that amino acids in living organisms are L-optical isomers. After death, these racemize and the age of bones can be estimated by measuring the relative amounts of D- and L-amino acids present.

**chemical engineering** The study of the design, manufacture, and operation of plant and machinery in industrial chemical processes.

**chemical equation** A way of denoting a chemical reaction using the symbols for the participating particles (atoms, molecules, ions, etc.); for example,

$$xA + yB \rightarrow zC + wD$$

The single arrow is used for an irreversible reaction; double arrows ($\rightleftharpoons$) are used for reversible reactions. When reactions involve different phases it is usual to put the phase in brackets after the symbol (s = solid; l = liquid; g = gas; aq = aqueous). The numbers $x$, $y$, $z$, and $w$, showing the relative numbers of molecules reacting, are called the **stoichiometric coefficients**. The sum of the coefficients of the reactants minus the sum of the coefficients of the products ($x + y - z - w$ in the example) is the **stoichiometric sum**. If this is zero the equation is balanced. Sometimes a generalized chemical equation is considered

$$\nu_1 A_1 + \nu_2 A_2 + \ldots \rightarrow \ldots \nu_n A_n + \nu_{n+1} A_{n+1} \ldots$$

In this case the reaction can be written $\Sigma \nu_i A_i = 0$, where the convention is that stoichiometric coefficients are positive for reactants and negative for products. The stoichiometric sum is $\Sigma \nu_i$.

**chemical equilibrium** A reversible chemical reaction in which the concentrations of reactants and products are not changing with time because the system is in thermodynamic equilibrium. For example, the reversible reaction

$$3H_2 + N_2 \rightleftharpoons 2NH_3$$

is in chemical equilibrium when the rate of the **forward reaction**

$$3H_2 + N_2 \rightarrow 2NH_3$$

is equal to the rate of the **back reaction**

$$2NH_3 \rightarrow 3H_2 + N_2$$

*See also* EQUILIBRIUM CONSTANT.

**chemical equivalent** *See* EQUIVALENT WEIGHT.

**chemical fossil** Any of various organic compounds found in ancient geological strata that appear to be biological in origin and are assumed to indicate that life existed when the rocks were formed. The presence of chemical fossils in Archaean strata indicates that life existed over 3500 million years ago.

**chemical potential** Symbol: $\mu$. For a given component in a mixture, the coefficient $\partial G/\partial n$, where $G$ is the Gibbs free energy and $n$ the amount of substance of the component. The chemical potential is the change in Gibbs free energy with respect to change in amount of the component, with pressure, temperature, and amounts of other components being constant. Components are in equilibrium if their chemical potentials are equal.

**chemical reaction** A change in which one

or more chemical elements or compounds (the **reactants**) form new compounds (the **products**). All reactions are to some extent **reversible**; i.e. the products can also react to give the original reactants. However, in many cases the extent of this back reaction is negligibly small, and the reaction is regarded as **irreversible**.

**chemical shift** A change in the normal wavelength of absorption or emission of electromagnetic wavelength in a process in which there is a nuclear energy change (as in the *Mössbauer effect and *nuclear magnetic resonance) or a change in electron energy levels in the inner shells of an atom (as in X-ray *photoelectron spectroscopy).

**chemical warfare** The use of toxic chemical substances in warfare or military operations. A large number of chemicals have been designed or used for warfare, including pulmonary agents (chlorine, *carbonyl chloride, *diphosgene), blister agents (lewisite, *sulphur mustard, *nitrogen mustard), and the *nerve agents. Chemical warfare agents are classified as weapons of mass destruction by the United Nations.

**chemiluminescence** *See* LUMINESCENCE.

**chemiosmotic theory** A theory postulated by the British biochemist Peter Mitchell (1920–92) to explain the formation of ATP in the mitochondrial *electron transport chain. As electrons are transferred along the electron carrier system in the inner mitochondrial membrane, hydrogen ions (protons) are actively transported (via *hydrogen carriers) into the space between the inner and outer mitochondrial membranes, which thus contains a higher concentration of protons than the matrix. This creates an electrochemical gradient across the inner membrane, down which protons move back into the matrix. This movement occurs through special channels associated with ATP synthetase, the enzyme that catalyses the conversion of ADP to ATP, and is coupled with the phosphorylation of ADP. A similar gradient is created across the thylakoid membranes of chloroplasts during the light-dependent reactions of *photosynthesis (*see* PHOTOPHOSPHORYLATION).

**chemisorption** *See* ADSORPTION.

**chemistry** The study of the elements and the compounds they form. Chemistry is mainly concerned with effects that depend on the outer electrons in atoms. *See* BIO-

CHEMISTRY; GEOCHEMISTRY; INORGANIC CHEMISTRY; ORGANIC CHEMISTRY; PHYSICAL CHEMISTRY.

**chemoautotroph** *See* AUTOTROPHIC NUTRITION; CHEMOSYNTHESIS.

**chemoinformatics** The branch of chemistry concerned with methods of representing molecules and reactions and with the design and use of databases for storing chemical information.

**chemoorganotroph** An organism, especially a microorganism, that obtains its energy by the oxidation of organic compounds.

**chemoreceptor** A *receptor that detects the presence of particular chemicals and (in multicellular organisms) transmits this information to sensory nerves. Examples include the *taste buds and the receptors in the *carotid body.

**chemosynthesis** A type of *autotrophic nutrition in which organisms (called **chemoautotrophs**) synthesize organic materials using energy derived from the oxidation of inorganic chemicals, rather than from sunlight. Most chemoautotrophs are bacteria, including *Nitrosomonas*, which oxidizes ammonium to nitrite, and *Thiobacillus*, which oxidizes sulphur to sulphate.

**chemosystematics** *See* SYSTEMATICS.

**chemotaxis** *See* TAXIS.

**chemotaxonomy** The *classification of plants and microorganisms based on similarities and differences in their natural products and the biochemical pathways involved in their manufacture. *See also* TAXONOMY.

**chemotherapy** The use of chemicals, especially drugs, in the treatment of disease. The term is often used specifically to denote drug therapy for cancer, as distinct from treatments with radiation (radiotherapy).

**chemotropism** The growth or movement of a plant or plant part in response to a chemical stimulus. An example is the growth of a pollen tube down the style during fertilization in response to the presence of sugars in the style.

**ChemSketch** A commonly used chemical drawing program for 2D and 3D structures, copyright of Advanced Chemistry Development, Inc. The program has certain additional features including calculation of molecular weight, calculation of percentages of elements present, IUPAC name generation, and viewing in RasMol.

- Details and download from Advanced Chemistry Development website

**chernozem (black earth)** A type of soil that is characteristic of the continental interiors of the mid-latitudes, in which grassland formed the natural vegetation. Chernozems occur across the Russian steppes and parts of Romania and Hungary; these soils also occur in North America. The deep surface layer (A horizon) of a chernozem is black and rich in alkaline humus derived from the decomposition of the natural grassland vegetation. The underlying horizon contains calcium carbonate concretions. Chernozems are important agriculturally and most have been ploughed up for cereal production.

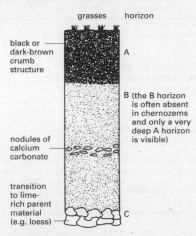

grasses     horizon

black or dark-brown crumb structure    A

B (the B horizon is often absent in chernozems and only a very deep A horizon is visible)

nodules of calcium carbonate

transition to lime-rich parent material (e.g. loess)    C

**Chernozem soil profile.**

**chert** *See* FLINT.

**chiasma** (*pl.* **chiasmata**) The point at which paired *homologous chromosomes remain in contact as they begin to separate during the first prophase of *meiosis, forming a cross shape. A number of chiasmata can usually be identified and at these points *crossing over occurs.

**Chile saltpetre** A commercial mineral largely composed of *sodium nitrate from the caliche deposits in Chile. Before the ammonia-oxidation process for nitrates

most imported Chilean saltpetre was used by the chemical industry; its principal use today is as an agricultural source of nitrogen.

**Chilopoda** A class of terrestrial *arthropods belonging to the subphylum *Myriapoda and comprising the centipedes, characterized by a distinct head bearing one pair of poison jaws and 15–177 body segments, each bearing one pair of similar legs. Centipedes are fast-moving predators.

**chimaera** An organism composed of tissues that are genetically different. Chimaeras can develop if a *mutation occurs in a cell of a developing embryo. All the cells arising from it have the mutation and therefore produce tissue that is genetically different from adjacent tissue, e.g. brown patches in otherwise blue eyes in humans. *Graft hybrids are examples of plant chimaeras.

**Chime** A plug-in that allows chemical structures to be displayed within a web page. The structures, which are embedded as mol files, can be viewed interactively; i.e. they can be rotated, expanded, exported, and rendered in *RasMol. The program is the copyright of MDL Information Systems, Inc., and is available free (subject to a license agreement). Chime is being phased out and will not be available at all from the end of 2012. It has effectively been superseded by JMOL.

**china clay** *See* KAOLIN.

**Chinese white** *See* ZINC OXIDE.

**chip** *See* SILICON CHIP.

**chirality** The property of existing in left- and right-handed structural forms. *See* OPTICAL ACTIVITY.

**chirality element** The part of a molecule that makes it exist in left- and right-handed forms. In most cases this is a **chirality centre** (i.e. an asymmetric atom). In certain cases the element is a **chirality axis**. For example, in allenenes of the type $R_1R_2C=C=CR_3R_4$ the $C=C=C$ chain is a chirality axis. Certain ring compounds may display chirality as a result of a **chirality plane** in the molecule.

**Chiron** A minor planet discovered in 1977. It has an orbit of 50.68 years that, unlike other known minor planets, lies almost entirely outside that of Saturn. Its diameter is uncertain, but appears to be of the order of 300 km.

**chirooptical spectroscopy** Spectroscopy making use of the properties of chiral substances when they interact with polarized light of various wavelengths. **Optical rotatory dispersion** (the change of optical rotation with wavelength) and circular dichroism are old examples of chirooptical spectroscopy. More recent types of chirooptical spectroscopy involve infrared radiation and Raman spectroscopy. Chirooptical spectroscopy is used in the analysis of the structure of molecules.

**Chiroptera** An order of flying mammals comprising the bats. Their membranous wings are supported by very elongated forelimbs and digits and stretch along the sides of the body to the hindlimbs and tail. Whenever bats rest they allow their body temperature to fall, hibernating in winter when food is scarce. Most bats are nocturnal; their ears are enlarged and specialized for *echolocation, which they use to hunt prey and avoid obstacles. Bats feed variously on insects, fruit, nectar, or blood.

**chi-square distribution** The distribution of the sum of the squares of random variables with standard normal distributions. For example, if $x_1, x_2, ..., x_i, ...$ are independent variables with standard normal distribution, then

$$\chi^2 = \Sigma x_i^2$$

has a chi-square distribution with $n$ degrees of freedom, written $\chi_n^r$. The mean and variance are $n$ and $2n$, respectively. The values $\chi_n^2(\alpha)$ for which

$$P(\chi^2 \leq \chi_n^2(\alpha)) = \alpha$$

are tabulated for various values of $n$.

**chi-squared test** A procedure in *statistics to test how well a frequency distribution matches one predicted theoretically. Statistical tables are used to assess the significance of the result obtained by calculating $\chi^2 = \Sigma(O - E)2/E$, where $O$ are the observed frequencies and $E$ are the predicted frequencies.

**chitin** A *polysaccharide comprising chains of $N$-acetyl-D-glucosamine, a derivative of glucose. Chitin is structurally very similar to cellulose and serves to strengthen the supporting structures of various invertebrates. It also occurs in fungi.

**chloral** See TRICHLOROETHANAL.

**chloral hydrate** See 2,2,2-TRICHLOROETHANEDIOL.

**chlorates** Salts of the chloric acids; i.e. salts containing the ions $ClO^-$ (chlorate(I) or **hypochlorite**), $ClO_2^-$ (chlorate(III) or **chlorite**), $ClO_3^-$ (chlorate(V)), or $ClO_4^-$ (chlorate(VII) or **perchlorate**). When used without specification of an oxidation state the term 'chlorate' refers to a chlorate(V) salt.

**chlorenchyma** *Parenchyma tissue that contains chloroplasts and is photosynthetic. Chlorenchyma makes up the mesophyll tissue of plant leaves and is also found in the stems of certain plant species. *Compare* COLLENCHYMA (*see* GROUND TISSUES); SCLERENCHYMA.

**chloric acid** Any of the oxoacids of chlorine: *chloric(I) acid, *chloric(III) acid, *chloric(V) acid, and *chloric(VII) acid. The term is commonly used without specification of the oxidation state of chlorine to mean chloric(V) acid, $HClO_3$.

**chloric(I) acid (hypochlorous acid)** A liquid acid that is stable only in solution, HOCl. It may be prepared by the reaction of chlorine with an agitated suspension of mercury(II) oxide. Because the disproportionation of the ion $ClO^-$ is slow at low temperatures chloric(I) acid may be produced, along with chloride ions by the reaction of chlorine with water at 0°C. At higher temperatures disproportionation to the chlorate(V) ion, $ClO_3^-$, takes place. Chloric(I) acid is a very weak acid but is a mild oxidizing agent and is widely used as a bleaching agent.

**chloric(III) acid (chlorous acid)** A paleyellow acid known only in solution, $HClO_2$. It is formed by the reaction of chlorine dioxide and water and is a weak acid and an oxidizing agent.

**chloric(V) acid (chloric acid)** A colourless unstable liquid, $HClO_3$; r.d. 1.2; m.p. <−20°C; decomposes at 40°C. It is best prepared by the reaction of barium chlorate with sulphuric acid although chloric(V) acid is also formed by the disproportionation of chloric(I) acid in hot solutions. It is both a strong acid and a powerful oxidizing agent; hot solutions of the acid or its salts have been known to detonate in contact with readily oxidized organic material.

**chloric(VII) acid (perchloric acid)** An unstable liquid acid, $HClO_4$; r.d. 1.76; m.p. −112°C; b.p. 39°C (50 mmHg); explodes at about 90°C at atmospheric pressure. There is also a monohydrate (r.d. 1.88 (solid), 1.77 (liquid); m.p. 48°C; explodes at about 110°C) and a dihydrate (r.d. 1.65; m.p. –17.8°C; b.p. 200°C). Commercial chloric(VII) acid is a water azeotrope, which is 72.5% $HClO_4$, boiling at 203°C. The anhydrous acid may be prepared by vacuum distillation of the concentrated acid in the presence of magnesium perchlorate as a dehydrating agent. Chloric(VII) acid is both a strong acid and a strong oxidizing agent. It is widely used to decompose organic materials prior to analysis, e.g. samples of animal or vegetable matter requiring heavy-metal analysis.

**chloride** *See* HALIDE.

**chloride shift** The movement of chloride ions (Cl⁻) into red blood cells. Carbon dioxide reacts with water to form carbonic acid in the red blood cells (*see* CARBONIC ANHYDRASE). The carbonic acid then dissociates into hydrogencarbonate ions ($HCO_3^-$) and hydrogen ions (H⁺). The plasma membrane is relatively permeable to negative ions. Therefore the hydrogencarbonate ions diffuse out of the cell into the blood plasma, leaving the hydrogen ions, which create a net positive charge; this is neutralized by the diffusion of chloride ions from the plasma into the cell.

**chlorinating agent** A chemical reagent that introduces chlorine atoms into a compound or substitutes chlorine for some other group. Examples include phosphorus pentachloride, $PCl_5$, and sulphur dichloride oxide (thionyl chloride, $SOCl_2$).

**chlorination** 1. A chemical reaction in which a chlorine atom is introduced into a compound. *See* HALOGENATION. 2. The treatment of water with chlorine to disinfect it.

**chlorine** Symbol Cl. A *halogen element; a.n. 17; r.a.m. 35.453; d. 3.214 g dm⁻³; m.p. –100.98°C; b.p. –34.6°C. It is a poisonous greenish-yellow gas and occurs widely in nature as sodium chloride in seawater and as halite (NaCl), carnallite ($KCl.MgCl_2.6H_2O$), and sylvite (KCl). It is manufactured by the electrolysis of brine and also obtained in the *Downs process for making sodium. It has many applications, including the chlorination of drinking water, bleaching, and the manufacture of a large number of organic chemicals.

It reacts directly with many elements and compounds and is a strong oxidizing agent. Chlorine compounds contain the element in the 1, 3, 5, and 7 oxidation states. It was discovered by Karl Scheele in 1774 and Humphry Davy confirmed it as an element in 1810.

(((●))) **SEE WEB LINKS**
• Information from the WebElements site

**chlorine dioxide** A yellowish-red explosive gas, $ClO_2$; d. 3.09 g dm⁻³; m.p. –59.5°C; b.p. 9.9°C. It is soluble in cold water but decomposed by hot water to give chloric(VII) acid, chlorine, and oxygen. Because of its high reactivity, chlorine dioxide is best prepared by the reaction of sodium chlorate and moist oxalic acid at 90°–100°C, as the product is then diluted by liberated carbon dioxide. Commercially the gas is produced by the reaction of sulphuric acid containing chloride ions with sulphur dioxide. Chlorine dioxide is widely used as a bleach in flour milling and in wood pulping and also finds application in water purification.

**chlorine monoxide** *See* DICHLORINE OXIDE.

**chlorite** 1. *See* CHLORATES. 2. A group of layered silicate minerals, usually green or white in colour, that are similar to the micas in structure and crystallize in the monoclinic system. Chlorites are composed of complex silicates of aluminium, magnesium, and iron in combination with water, with the formula $(Mg,Al,Fe)_{12}(Si,Al)_8O_{20}(OH)_{16}$. They are most common in low-grade metamorphic rocks and also occur as secondary minerals in igneous rocks as alteration products of pyroxenes, amphiboles, and micas. The term is derived from *chloros*, the Greek word for green.

**chloroacetic acids** *See* CHLOROETHANOIC ACIDS.

**chlorobenzene** A colourless highly flammable liquid, $C_6H_5Cl$; r.d. 1.106; m.p. –45.43°C; b.p. 131.85°C. It is prepared by the direct chlorination of benzene using a halogen carrier (*see* FRIEDEL–CRAFTS REACTION), or manufactured by the *Raschig process. It is used mainly as an industrial solvent.

**2-chlorobuta-1,3-diene (chloroprene)** A colourless liquid chlorinated diene, $CH_2{:}CClCH{:}CH_2$; r.d. 0.96; b.p. 59°C. It is polymerized to make synthetic rubbers (e.g. neoprene).

**chloroethane (ethyl chloride)** A colourless flammable gas, $C_2H_5Cl$; m.p. –136.4°C; b.p. 12.3°C. It is made by reaction of ethene and hydrogen chloride and used in making lead tetraethyl for petrol.

**chloroethanoic acids (chloroacetic acids)** Three acids in which hydrogen atoms in the methyl group of ethanoic acid have been replaced by chlorine atoms. They are: **monochloroethanoic acid** ($CH_2ClCOOH$); **dichloroethanoic acid** ($CHCl_2COOH$); **trichloroethanoic acid** ($CCl_3COOH$). The presence of chlorine atoms in the methyl group causes electron withdrawal from the COOH group and makes the chloroethanoic acids stronger acids than ethanoic acid itself. The $K_a$ values (in moles $dm^{-3}$ at 25°C) are

$CH_3COOH$ $1.7 \times 10^{-5}$

$CH_2ClCOOH$ $1.3 \times 10^{-3}$

$CHCl_2COOH$ $5.0 \times 10^{-2}$

$CCl_3COOH$ $2.3 \times 10^{-1}$

**chloroethene (vinyl chloride)** A gaseous compound, $CH_2{:}CHCl$; r.d. 0.911; m.p. −153.8°C; b.p. −13.37°C. It is made by chlorinating ethene to give dichloroethane, then removing HCl:

$C_2H_4 + Cl_2 \rightarrow CH_2ClCH_2Cl \rightarrow CH_2CHCl$

The compound is used in making PVC.

**chlorofluorocarbon (CFC)** A type of compound in which some or all of the hydrogen atoms of a hydrocarbon (usually an alkane) have been replaced by chlorine and fluorine atoms (*see* HALOCARBONS). CFCs are used in oils, polymers, and solvents, and in the manufacture of rigid packaging foam. A commonly encountered commercial name for these compounds is **freon**. Most chlorofluorocarbons are chemically unreactive and are stable at high temperatures. Because of this, they can diffuse unchanged into the upper atmosphere. Their former widespread use in aerosols and refrigerator coolants led to increased concentrations in the upper atmosphere, where photochemical reactions cause them to break down and react with ozone (*see* OZONE LAYER). They also contribute to the greenhouse effect. For these reasons, CFCs have now largely been replaced by less damaging alternatives, such as hydrofluorcarbons. *See also* POLLUTION.

**chloroform** *See* TRICHLOROMETHANE.

**chloromethane (methyl chloride)** A colourless flammable gas, $CH_3Cl$; r.d. 0.916; m.p. −97.1°C; b.p. −24.2°C. It is a *halo-alkane, made by direct chlorination of methane and used as a local anaesthetic and refrigerant.

**chlorophyll** One of two pigments (**chlorophyll a** and **chlorophyll b**) responsible for the green colour of most plants. Chlorophyll molecules are the principal sites of light absorption in the light-dependent reactions of *photosynthesis. They are magnesium-containing *porphyrins, chemically related to *cytochrome and *haemoglobin. *See also* BACTERIOCHLOROPHYLL.

**Chlorophyta (green algae)** A large phylum of *algae, the members of which possess chlorophylls *a* and *b*, store food reserves as starch, and have cellulose cell walls. In these respects, and on the basis of molecular studies, they resemble plants more closely than do any of the other algal phyla and are placed with them in the supergroup Archaeplastida (*see* PLANT). The Chlorophyta are widely distributed and diverse in form. Unicellular forms may occur singly (sometimes with *flagella for motility) or in colonies, while multicellular forms may be filamentous (e.g. *Spirogyra*) or platelike (e.g. *Ulva*).

**chloroplast** Any of the chlorophyll-containing organelles (*see* PLASTID) that are found in large numbers in those plant cells undergoing *photosynthesis. Chloroplasts are typically lens-shaped and bounded by a double membrane. They contain membranous structures called **thylakoids**, which are piled up into stacks (*see* GRANUM), surrounded by a gel-like matrix (**stroma**). The light-dependent reactions of photosynthesis occur on the thylakoid membranes, while the light-independent reactions take place in the stroma.

**chloroplatinic acid** A reddish crystalline compound, $H_2PtCl_6$, made by dissolving platinum in aqua regia.

**chloroprene** *See* 2-CHLOROBUTA-1,3-DIENE.

**chlorosis** The abnormal condition in plant stems and leaves in which synthesis of the green pigment chlorophyll is inhibited, resulting in a pale yellow coloration. This may be caused by lack of light, mineral deficiency, infection (particularly by viruses), or genetic factors.

**chlorosulphanes** *See* DISULPHUR DICHLORIDE.

**chlorous acid** *See* CHLORIC(III) ACID.

**chloroxybacteria (grass-green bacteria)**
*See* Cyanobacteria.

**choanae (internal nares)** *See* nares.

**choke** A coil of wire with high inductance and low resistance. It is used in radio circuits to impede the passage of audio-frequency or radio-frequency currents or to smooth the output of a rectifying circuit.

**choke damp** *See* blackdamp.

**cholecalciferol** *See* vitamin D.

**cholecystokinin (CCK; pancreozymin)** A hormone, produced by the duodenal region of the small intestine, that induces the gall bladder to contract and eject bile into the intestine and stimulates the pancreas to secrete its digestive enzymes. Cholecystokinin output is stimulated by contact with the contents of the stomach.

**cholesteric crystal** *See* liquid crystal.

**cholesterol** A *sterol (see also* steroid) occurring widely in animal tissues and also in some plants and algae. It can exist as a free sterol or esterified with a long-chain fatty acid. Cholesterol is absorbed through the intestine or manufactured in the liver. It serves principally as a constituent of blood plasma *lipoproteins and of the lipid–protein complexes that form plasma membranes. It is also important as a precursor of various steroids, especially the bile acids, sex hormones, and adrenocorticoid hormones. The derivative 7-dehydrocholesterol is converted to vitamin $D_3$ by the action of sunlight on skin. Increased levels of dietary and blood cholesterol have been associated with **atherosclerosis**, a condition in which lipids accumulate on the inner walls of arteries and eventually obstruct blood flow. It is now thought that this damage to blood vessels is caused by high concentrations of low-density lipoproteins in the blood.

**choline** An amino alcohol, $CH_2OHCHCH_2N(CH_3)_3OH$. It occurs widely in living organisms as a constituent of certain types of phospholipids – the *lecithins and sphingomyelins – and in the neurotransmitter *acetylcholine. It is sometimes classified as a member of the *vitamin B complex.

**cholinergic** Describing a nerve fibre that either releases *acetylcholine when stimulated or is itself stimulated by acetylcholine. *Compare* adrenergic.

**cholinesterase (acetylcholinesterase)**

An enzyme that hydrolyses the neurotransmitter *acetylcholine to choline and acetate. Cholinesterase is secreted by nerve cells at *synapses and by muscle cells at *neuromuscular junctions. Organophosphorus insecticides (*see* pesticide) act as *anticholinesterases by inhibiting the action of cholinesterase.

**Chondrichthyes** A class of vertebrates comprising the fishes with cartilaginous skeletons. The majority belong to the subclass Elasmobranchii (skates, rays, and sharks – *see* Selachii). Most cartilaginous fishes are marine carnivores with powerful jaws. Unlike bony fishes, they have no swim bladder, and therefore avoid sinking only by constant swimming with the aid of an asymmetrical (**heterocercal**) tail. There is no operculum covering the gill slits, the first of which is modified as a *spiracle. Fertilization is internal so the few eggs produced are consequently yolky, large, and well-protected. Some cartilaginous fishes show viviparous development of the young (*see* viviparity).

**chondrin** The matrix of *cartilage, which is made up of chondrocytes embedded in chondroitin sulphate.

**chondrocyte** Any of the cells that make up the matrix of *cartilage.

**Chordata** A phylum of animals characterized by a hollow dorsal nerve cord and, at some stage in their development, a flexible skeletal rod (the *notochord) and *gill slits opening from the pharynx. There are three subphyla: the Urochordata (sea squirts), Cephalochordata (lancelets), and *Craniata.

**chorion** 1. A membrane enclosing the embryo, yolk sac, and allantois of reptiles, birds, and mammals. In mammals a section of the chorion becomes the embryonic part of the *placenta. *See* extraembryonic membranes. 2. The protective shell of an insect egg, produced by the ovary. It is pierced by a small pore (**micropyle**) that allows the entry of spermatozoa for fertilization. *See also* egg membrane.

**chorionic gonadotrophin** *See* gonadotrophin.

**choroid** A pigmented layer, rich in blood vessels, that lies between the retina and the sclerotic of the vertebrate eye. At the front of the eye the choroid is modified to form the *ciliary body and the *iris.

**choroid plexus** A membrane rich in

blood vessels that lines the *ventricles of the brain. It is an extension of the *pia mater and secretes *cerebrospinal fluid into the ventricles; it also controls exchange of materials between the blood and cerebrospinal fluid.

**chromate** A salt containing the ion $CrO_4^{2-}$.

**chromatic aberration** *See* ABERRATION.

**chromaticity** An objective description of the colour quality of a visual stimulus that does not depend on its luminance but which, together with its luminance, completely specifies the colour. The colour quality is defined in terms of **chromaticity coordinates**, $x, y$, and $z$, where

$$x = X/(X + Y + Z)$$
$$y = Y/(X + Y + Z)$$
$$\text{and } z = Z/(X + Y + Z)$$

$X$, $Y$, and $Z$ are the **tristimulus values** of a light, i.e. they are the amounts of three reference stimuli needed to match exactly the light under consideration in a trichromatic system.

**chromatid** A threadlike strand formed from a *chromosome during the early stages of cell division. Each chromosome divides along its length into two chromatids, which are at first held together at the centromere. They separate completely at a later stage. The DNA of the chromosome reproduces itself exactly so that each chromatid has the complete amount of DNA and becomes a daughter chromosome with exactly the same genes as the original chromosome from which it was formed.

**chromatin** The substance of which eukaryotic *chromosomes are composed. It consists of proteins (principally histones), DNA, and small amounts of RNA and can be observed microscopically in two forms. In the *interphase of the cell cycle, chromatin is mainly in a condensed form, **heterochromatin**, which stains densely with basic stains and cannot undergo transcription. During the *metaphase of cell division most of the chromatin is in an expanded, lighter staining form, **euchromatin**, in which genes are available for transcription.

**chromatogram** A record obtained by chromatography. The term is applied to the developed records of *paper chromatography and *thin-layer chromatography and also to the graphical record produced in *gas chromatography.

**chromatography** A technique for analysing or separating mixtures of gases, liquids, or dissolved substances. The original technique (invented by the Russian botanist Mikhail Tsvet (1872–1919) in 1906) is a good example of **column chromatography**. A vertical glass tube is packed with an adsorbing material, such as alumina. The sample is poured into the column and continuously washed through with a solvent (a process known as **elution**). Different components of the sample are adsorbed to different extents and move down the column at different rates. In Tsvet's original application, plant pigments were used and these separated into coloured bands in passing down the column (hence the name chromatography). The usual method is to collect the liquid (the **eluate**) as it passes out from the column in fractions.

In general, all types of chromatography involve two distinct phases – the **stationary phase** (the adsorbent material in the column in the example above) and the **moving phase** (the solution in the example). The separation depends on competition for molecules of sample between the moving phase and the stationary phase. The form of column chromatography above is an example of **adsorption chromatography**, in which the sample molecules are adsorbed on the alumina. In **partition chromatography**, a liquid (e.g. water) is first absorbed by the stationary phase and the moving phase is an immiscible liquid. The separation is then by *partition between the two liquids. In ion-exchange chromatography (*see* ION EXCHANGE), the process involves competition between different ions for ionic sites on the stationary phase. *Gel filtration is another chromatographic technique in which the size of the sample molecules is important.

*See also* GAS CHROMATOGRAPHY; HIGH-PERFORMANCE LIQUID CHROMATOGRAPHY; PAPER CHROMATOGRAPHY; $R_F$ VALUE; THIN-LAYER CHROMATOGRAPHY.

**chromatophore** 1. A pigment-containing cell found in the skin of many lower vertebrates (e.g. chameleon) and in the integument of crustaceans. Concentration or dispersion of the pigment granules in the cytoplasm of the cell causes the colour of the animal to alter to match its surroundings. A common type of chromatophore is the **melanophore**, which contains the pigment *melanin. 2. A membrane-bound structure in photosynthetic bacteria that contains photosynthetic pigments.

**chrome alum** *See* POTASSIUM CHROMIUM SULPHATE.

**chrome iron ore** A mixed iron–chromium oxide, $FeO.Cr_2O_3$, used to make ferrochromium for chromium steels.

**chrome red** A basic lead chromate, $PbO.PbCrO_4$, used as a red pigment.

**chrome yellow** Lead chromate, $PbCrO_4$, used as a pigment.

**chromic acid** A hypothetical acid, $H_2CrO_4$, known only in chromate salts.

**chromic anhydride** *See* CHROMIUM(VI) OXIDE.

**chromic compounds** Compounds containing chromium in a higher (+3 or +6) oxidation state; e.g. chromic oxide is chromium(VI) oxide ($CrO_3$).

**chromite** A spinel mineral, $FeCr_2O_4$; the principal ore of chromium. It is black with a metallic lustre and usually occurs in massive form. It is a common constituent of peridotites and serpentines. The chief producing countries are Turkey, South Africa, Russia, the Philippines, and Zimbabwe.

**chromium** Symbol Cr. A hard silvery *transition element; a.n. 24; r.a.m. 52.00; r.d. 7.19; m.p. 1857°C; b.p. 2672°C. The main ore is chromite ($FeCr_2O_4$). The metal has a body-centred-cubic structure. It is extracted by heating chromite with sodium chromate, from which chromium can be obtained by electrolysis. Alternatively, chromite can be heated with carbon in an electric furnace to give ferrochrome, which is used in making alloy steels. The metal is also used as a shiny decorative electroplated coating and in the manufacture of certain chromium compounds.

At normal temperatures the metal is corrosion-resistant. It reacts with dilute hydrochloric and sulphuric acids to give chromium(II) salts. These readily oxidize to the more stable chromium(III) salts. Chromium also forms compounds with the +6 oxidation state, as in chromates, which contain the $CrO_4^{2-}$ ion. The element was discovered in 1797 by Vauquelin.

(⊕) SEE WEB LINKS
• Information from the WebElements site

**chromium(II) oxide** A black insoluble

powder, CrO. Chromium(II) oxide is prepared by oxidizing chromium amalgam with air. At high temperatures hydrogen reduces it to the metal.

**chromium(III) oxide** (**chromium sesquioxide**) A green crystalline water-insoluble salt, $Cr_2O_3$; r.d. 5.21; m.p. 2435°C; b.p. 4000°C. It is obtained by heating chromium in a stream of oxygen or by heating ammonium dichromate. The industrial preparation is by reduction of sodium dichromate with carbon. Chromium(III) oxide is amphoteric, dissolving in acids to give chromium(III) ions and in concentrated solutions of alkalis to give **chromites**. It is used as a green pigment in glass, porcelain, and oil paint.

**chromium(IV) oxide** (**chromium dioxide**) A black insoluble powder, $CrO_2$; m.p. 300°C. It is prepared by the action of oxygen on chromium(VI) oxide or chromium(III) oxide at 420–450°C and 200–300 atmospheres. The compound is unstable.

**chromium(VI) oxide** (**chromium trioxide; chromic anhydride**) A red compound, $CrO_3$; rhombic; r.d. 2.70; m.p. 196°C. It can be made by careful addition of concentrated sulphuric acid to an ice-cooled concentrated aqueous solution of sodium dichromate with stirring. The mixture is then filtered through sintered glass, washed with nitric acid, then dried at 120°C in a desiccator.

Chromium(VI) oxide is an extremely powerful oxidizing agent, especially to organic matter; it immediately inflames ethanol. It is an acidic oxide and dissolves in water to form 'chromic acid', a powerful oxidizing agent and cleansing fluid for glassware. At 400°C, chromium(VI) oxide loses oxygen to give chromium(III) oxide.

**chromium potassium sulphate** A red crystalline solid, $K_2SO_4.Cr_2(SO_4)_3.24H_2O$; r.d. 1.91. It is used as a mordant. *See also* ALUMS.

**chromium sesquioxide** *See* CHROMIUM(III) OXIDE.

**chromium steel** Any of a group of *stainless steels containing 8–25% of chromium. A typical chromium steel might contain 18% of chromium, 8% of nickel, and 0.15% of carbon. Chromium steels are highly resistant to corrosion and are used for cutlery, chemical plant, ball bearings, etc.

**chromophore** A group causing coloration

in a *dye. Chromophores are generally groups of atoms having delocalized electrons.

**chromoplast** Any of various pigment-containing *plastids in plant cells. Red, orange, and yellow chromoplasts contain carotenoid pigments and are responsible for the coloration of fruits and flowers. *See also* PLASTOGLOBULUS. *Compare* CHLOROPLAST; LEUCOPLAST.

**chromosome** A threadlike structure several to many of which are found in the nucleus of plant and animal (eukaryotic) cells. Chromosomes are composed of *chromatin and carry the *genes in a linear sequence; these determine the individual characteristics of an organism. When the nucleus is not dividing, individual chromosomes cannot be identified with a light microscope. During the first stage of nuclear division, however, the chromosomes contract and, when stained, can be clearly seen under a microscope. Each consists of two *chromatids held together at the *centromere (*see also* MEIOSIS; MITOSIS). The number of chromosomes in each cell is constant for and characteristic of the species concerned. In the normal body cells of *diploid organisms the chromosomes occur in pairs (*see* HOMOLOGOUS CHROMO-SOMES); in the gamete-forming germ cells, however, the diploid number is halved and each cell contains only one member of each chromosome pair. Thus in man each body cell contains 46 chromosomes (22 matched pairs and one pair of *sex chromosomes) and each germ cell 23. Abnormalities in the number or structure of chromosomes may give rise to abnormalities in the individual; *Down's syndrome is the result of one such abnormality. *See* CHROMOSOME MUTATION.

Bacterial cells contain only a single circular chromosome, aggregated into a *nucleoid. Viral chromosomes can consist of one or several single- or double-stranded nucleic acid molecules. *See also* ARTIFICIAL CHROMOSOME.

**chromosome map** Any plan that shows the positions of genes, genetic markers, or other landmarks along the length of a chromosome. There are essentially two complementary types of map: *linkage maps, which give the relative positions of genetic sites (loci) determined by studies of how frequently recombination occurs between the loci; and *physical maps, which show the arrangement of the chromosomal material.

Accumulated data for the chromosomes of many species of organism are now held in databases and available freely via the Internet for geneticists and others worldwide.

**chromosome mutation** A change in the gross structure of a chromosome, which usually causes severely deleterious effects in the organism. Chromosome mutations often occur due to an error in pairing during the *crossing over stage of meiosis. The main types of chromosome mutation include *translocation, *duplication, *deletion, and *inversion. *Compare* POINT MUTATION. *See also* MUTATION.

**chromosome painting** A technique based on *fluorescence in situ hybridization (FISH) that uses a palette of fluorescently labelled probes to identify specific chromosomes or chromosomal regions by 'painting' them in different colours. It is used diagnostically in clinical cytogenetics to screen for translocations or other structural aberrations, for example in hereditary diseases and cancer, and in comparative cytogenetics to determine the structural changes in genomes occurring during evolution.

**chromosphere** The layer of the *sun's atmosphere immediately above the *photosphere. The chromosphere is normally only visible when the photosphere is totally eclipsed by the moon. The chromosphere is about 10 000 kilometres thick and the temperature in it rises from 4000 K, where it merges with the photosphere, to about 50 000 K, where it reaches the transition region below the *corona.

**chromous compounds** Compounds containing chromium in its lower (+2) oxidation state; e.g. chromous chloride is chromium(II) chloride ($CrCl_2$).

**chromyl chloride (chromium oxychloride)** A dark red liquid, $CrO_2Cl_2$; r.d. 1.911; m.p. $-96.5°C$; b.p. $117°C$. It is evolved as a dark-red vapour on addition of concentrated sulphuric acid to a mixture of solid potassium dichromate and sodium chloride; it condenses to a dark-red covalent liquid, which is immediately hydrolysed by solutions of alkalis to give the yellow chromate. Since bromides and iodides do not give analogous compounds this is a specific test for chloride ions. The compound is a powerful oxidizing agent, exploding on contact with phosphorus and inflaming sulphur and many organic compounds.

**chronology protection conjecture** A conjecture put forward by Stephen *Hawking in the early 1990s that asserts that the fundamental laws of physics should forbid time travel. There is some theoretical evidence in favour of this idea.

**chrysalis** *See* PUPA.

**chrysotile** *See* SERPENTINE.

**chyle** A milky fluid consisting of *lymph that contains absorbed food materials (especially emulsified fats). Most chyle occurs in the lymphatic ducts (*lacteals) in the *villi of the small intestine during the absorption of fat.

**chyme** The semisolid and partly digested food that is discharged from the stomach into the duodenum.

**chymosin** *See* RENNIN.

**chymotrypsin** An *endopeptidase enzyme in pancreatic juice that is secreted into the duodenum. The enzyme is secreted as an inactive precursor, **chymotrypsinogen**, which is activated by another pancreatic protease, *trypsin.

**chymotrypsinogen** *See* CHYMOTRYPSIN.

**ciliary body** The circular band of tissue surrounding and supporting the *lens of the vertebrate eye. It contains the **ciliary muscles**, which bring about changes in the shape of the lens (*see also* ACCOMMODATION). The ciliary body produces the *aqueous humour.

**ciliary feeding** A method of feeding used by lancelets and many other aquatic invertebrates. The movement of cilia causes a current of water to be drawn towards and through the animal, and microorganisms in the water are filtered out by the cilia.

**ciliary muscle** *See* CILIARY BODY.

**cilium** A short minute hairlike structure (up to 10 μm long) present on the surface of many cells, notably in certain protozoans and some types of vertebrate *epithelium. Cilia usually occur in large groups. Beating of cilia can produce cell movement or create a current in fluid surrounding a cell. Cilia and eukaryotic flagella have the same structure and are collectively called undulipodia (*see* UNDULIPODIUM).

**cinnabar** A bright red mineral form of mercury(II) sulphide, HgS, crystallizing in the hexagonal system; the principal ore of mercury. It is deposited in veins and impreg-

nations near recent volcanic rocks and hot springs. The chief sources include Spain, Italy, and the former Yugoslavia.

**cinnamic acid (3-phenylpropenoic acid)** A white crystalline aromatic *carboxylic acid, $C_6H_5CH{:}CHCOOH$; r.d. 1.248 (trans isomer); m.p. 135–136°C; b.p. 300°C. Esters of cinnamic acid occur in some essential oils.

**CIP system (Cahn–Ingold–Prelog system)** A system for the unambiguous description of stereoisomers used in the $R$–$S$ convention (*see* ABSOLUTE CONFIGURATION) and in the *$E$–$Z$ convention. The system involves a **sequence rule** for determining a conventional order of ligands. The rule is that the atom bonded directly to the chiral centre or double bond is considered and the ligand in which this atom has the highest proton number takes precedence. So, for example, I takes precedence over Cl. If two ligands have bonding atoms with the same proton number, then substituents are taken into account (with the substituent of highest proton number taking precedence). Thus, $-C_2H_5$ has a higher precedence than $-CH_3$. If a double (or triple) bond occurs to a substituent, then the substituent is counted twice (or three times). An isotope of high nucleon number takes precedence over one of lower nucleon number. Hydrogen always has lowest priority in this system. For example, the sequence for some common ligands is I, Br, Cl, $SO_3H$, $OCOCH_3$, $OCH_3$, OH, $NO_2$, $NH_2$, $COOCH_3$, $CONH_2$, $COCH_3$, CHO, $CH_2OH$, $C_6H_5$, $C_2H_5$, $CH_3$, H. The system was jointly developed by the British chemists Robert Sidney Cahn (1899–1981) and Sir Christopher Kelk Ingold (1893–1970) and the Bosnian–Swiss chemist Vladimir Prelog (1906–98).

**circadian rhythm (diurnal rhythm)** Any 24-hour periodicity in the behaviour or physiology of animals or plants. Examples are the sleep/activity cycle in many animals and the growth movements of plants. Circadian rhythms are generally controlled by *biological clocks.

**circle** A closed curve every point on which is a fixed distance (the **radius**) from a point (the **centre**) within the curve (see illustration). The **diameter** is a line that joins two points on the **circumference** and passes through the centre: the diameter is twice the radius ($r$). The circumference of a circle is equal to $2\pi r$; the area is $\pi r^2$, where $\pi$ is a constant with the value 3.141 592. In analytical geometry the equation of a circle, centred at the origin, is $x^2 + y^2 = r^2$.

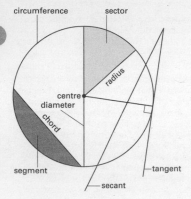

**A circle.**

**circular birefringence** A phenomenon in which there is a difference between the refractive indices of the molecules of a substance for right- and left-circularly polarized light. Circular birefringence depends on the way in which the electromagnetic field interacts with the molecule, and is affected by the handedness of the molecule, and hence its polarizability. If a molecule has the shape of a helix, the polarizability is dependent on whether or not the electric field of the electromagnetic field rotates in the same sense as the helix, thus giving rise to circular birefringence.

**circular dichroism (CD)** The production of an elliptically polarized light wave when a linearly polarized light wave passes through a substance that has differences in the extinction coefficients for left- and right-handed polarized light. The size of this effect is given by $\phi = \pi/\lambda(\eta_l - \eta_r)$, where $\phi$ is the ellipticity of the beam that emerges (in radians), $\lambda$ is the wavelength of the light, and $\eta_l$ and $\eta_r$ are the absorption indices of the left- and right-handed circularly polarized light, respectively. Circular dichroism is a property of optically active molecules and is used to obtain information about proteins. *See also* CD SPECTRUM.

**circular measure** A method of measuring angles by treating them as the angle formed by a sector of a circle at the circle's centre. The unit of measure is the **radian**, the angle subtended at the centre of a circle by an arc

of equal length to the radius. Since an arc of length $r$ subtends an angle of 1 radian, the whole circumference, length $2\pi r$, will subtend an angle of $2\pi r/r = 2\pi$ radians. Thus, $360° = 2\pi$ radians; 1 radian = 57.296°.

**circular polarization** *See* POLARIZATION OF LIGHT.

**circulation** The mass flow of fluid (e.g. blood or lymph) through the tissues and organs of an animal, allowing the transport and exchange of such materials as oxygen, nutrients, and waste products (*see also* VASCULAR SYSTEM; LYMPHATIC SYSTEM). Smaller animals (e.g. arthropods and most molluscs) have an **open circulation**, i.e. the blood is pumped into the body cavity, in which the internal organs are suspended. In open circulatory systems the tissues are in direct contact with the blood and materials are exchanged directly by diffusion. In a **closed circulation**, found in larger animals, the blood flows in vessels, which usually contain a series of one-way valves to maintain the flow in one direction. *See also* DOUBLE CIRCULATION; SINGLE CIRCULATION.

**circulatory system** The heart, blood vessels, blood, lymphatic vessels, and lymph, which together serve to transport materials throughout the body. *See also* DOUBLE CIRCULATION; SINGLE CIRCULATION; VASCULAR SYSTEM.

**circumnutation** *See* NUTATION.

**cirque** A steep semicircular hollow formed high on a mountain slope by the erosive action of a glacier. Many cirques fill with water to form lakes (called tarns). In Britain cirques are also called corries or cwms.

**cis-isomer** *See* ISOMERISM.

**cisplatin** A platinum complex, *cis*-[PtCl$_2$(NH$_3$)$_2$], used in cancer treatment to inhibit the growth of tumours. It acts by binding between strands of DNA.

**cis–trans isomerism** *See* ISOMERISM.

**citrate** A salt or ester of citric acid.

**citric acid** A white crystalline hydroxy-carboxylic acid, HOOCCH$_2$C(OH)(COOH)-CH$_2$COOH; r.d. 1.54; m.p. 153°C. It is present in citrus fruits and is an intermediate in the *Krebs cycle in plant and animal cells.

**citric-acid cycle** *See* KREBS CYCLE.

**CJD** *See* CREUTZFELDT–JAKOB DISEASE.

**CL20** *See* HNIW.

**cladding 1.** A thin coating of an expensive metal rolled on to a cheaper one. **2.** A thin covering of a metal around a fuel element in a nuclear reactor to prevent corrosion of the fuel elements by the coolant.

**clade** A group of organisms that share a common ancestor. *See* CLADISTICS.

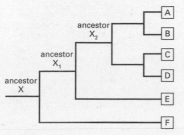

**Cladistics.** A cladogram showing the relationships of six species (A–F).

**cladistics** A method of classification in which animals and plants are placed into taxonomic groups called **clades** strictly according to their evolutionary relationships. These relationships are deduced on the basis of certain shared *homologous characters that are thought to indicate common ancestry (*see* MONOPHYLETIC). Implicit in this is the assumption that two new species are formed suddenly, by splitting from a common ancestor, and not by gradual evolutionary change. A diagram indicating these relationships (called a **cladogram**) therefore consists of a system of dichotomous branches: each point of branching represents divergence from a common ancestor, as shown in the diagram. Thus the species A to F form a clade as they share the common ancestor X, and species A to D form a clade of a different taxonomic rank, sharing the ancestor $X_2$. Species C to F do not form a clade, since the latter must include *all* the descendants of a common ancestor.

**cladode** A flattened stem or internode that resembles and functions as a leaf. It is an adaptation to reduce water loss, since it contains fewer *stomata than a leaf. An example of a plant with cladodes is asparagus.

**cladogram** *See* CLADISTICS.

**Claisen condensation** A reaction of es-

ters in which two molecules of the ester react to give a keto ester, e.g.

$$2CH_3COOR \rightarrow CH_3COCH_2COOR + ROH$$

The reaction is catalysed by sodium ethoxide, the mechanism being similar to that of the *aldol reaction. It is named after Ludwig Claisen (1851–1930).

**Clapeyron–Clausius equation** A differential equation that describes the relationship between variables when there is a change in the state of a system. In a system that has two *phases of the same substance, for example solid and liquid, heat is added or taken away very slowly so that one phase can change reversibly into the other while the system remains at equilibrium. If the two phases are denoted A and B, the Clapeyron–Clausius equation is:

$$dp/dT = L/T(V_B - V_A),$$

where $p$ is the pressure, $T$ is the thermodynamic *temperature, $L$ is the heat absorbed per mole in the change from A to B, and $V_B$ and $V_A$ are the volumes of B and A respectively. In the case of a transition from liquid to vapour, the volume of the liquid can be ignored. Taking the vapour to be an *ideal gas, the Clapeyron–Clausius equation can be written:

$$d\log_e p / dT = L/RT^2.$$

The Clapeyron–Clausius equation is named after the French engineer Benoit-Pierre-Émile Clapeyron (1799–1864) and Rudolf *Clausius.

**Clark cell** A type of *voltaic cell consisting of an anode made of zinc amalgam and a cathode of mercury both immersed in a saturated solution of zinc sulphate. The Clark cell was formerly used as a standard of e.m.f.; the e.m.f. at 15°C is 1.4345 volts. It is named after the British scientist Hosiah Clark (d. 1898).

**Clark process** *See* HARDNESS OF WATER.

**class** A category used in the *classification of organisms that consists of similar or closely related orders. Similar classes are grouped into a phylum. Examples include Mammalia (mammals), Aves (birds), and Dicotyledoneae (dicots).

**classical field theory** A theory that describes a *field in terms of *classical physics rather than *quantum mechanics. Examples of classical field theories include classical *electrodynamics, described by *Maxwell's equations, and the general theory of *relativity, describing classical gravitation. A classical field theory emerges as a limit of the corresponding *quantum field theory. In order for a classical field theory to apply on a macroscopic scale it is necessary for the interactions to be long range, as they are in electrodynamics and gravitation, rather than short range, as in nuclear forces. Classical field theory is also used for mathematical convenience to describe the physics of continuous media, such as fluids.

**classical physics** Theoretical physics up to approximately the end of the 19th century, before the concepts of *quantum theory (1900) and special *relativity (1905). Classical physics relied largely on Newton's mechanics and James Clerk Maxwell's theory of electromagnetism. It may still be applied with high precision to large-scale phenomena involving no very rapid relative motion.

**classification** The arrangement of organisms into a series of groups based on physiological, biochemical, anatomical, or other relationships. An **artificial classification** is based on one or a few characters simply for ease of identification or for a specific purpose; for example, birds are often arranged according to habit and habitat (seabirds, songbirds, birds of prey, etc.) while fungi may be classified as edible or poisonous. Such systems do not reflect evolutionary relationships. A **natural classification** is based on resemblances and is a hierarchical arrangement. The smallest group commonly used is the *species. Species are grouped into genera (*see* GENUS), the hierarchy continuing up through *tribes, *families, *orders, *classes, and phyla (*see* PHYLUM) to *kingdoms and – in some systems – *domains. (In traditional systems of plant classification the phylum was replaced by the *division.) Higher up in the hierarchy the similarities between members of a group become fewer. Present-day natural classifications try to take into account as many features as possible and in so doing aim to reflect evolutionary relationships (*see* CLADISTICS). Natural classifications are also predictive. Thus if an organism is placed in a particular genus because it shows certain features characteristic of the genus, then it can be assumed it is very likely to possess most (if not all) of the other features of that genus. *See also* BINOMIAL NOMENCLATURE; MOLECULAR SYSTEMATICS; TAXONOMY.

**clastic rock** A rock composed of fragments (clasts) of other older rocks or their minerals. The fragments, generally the products of erosion, may vary in size from large boulders to the tiny particles in silt. They have often been transported from their previous location, and are commonly found as sedimentary rocks along coastlines. Typical consolidated clastic rocks include sandstone and shale.

**clathrate** A solid mixture in which small molecules of one compound or element are trapped in holes in the crystal lattice of another substance. Clathrates are sometimes called **enclosure compounds** or **cage compounds**, but they are not true compounds (the molecules are not held by chemical bonds). Quinol and ice both form clathrates with such substances as sulphur dioxide and xenon.

**Claude process** A process for liquefying air on a commercial basis. Air under pressure is used as the working substance in a piston engine, where it does external work and cools adiabatically. This cool air is fed to a counter-current *heat exchanger, where it reduces the temperature of the next intake of high-pressure air. The same air is recompressed and used again, and after several cycles eventually liquefies. The process was perfected in 1902 by the French scientist Georges Claude (1870–1960).

**claudetite** A mineral form of *arsenic(III) oxide, $As_4O_6$.

**Clausius, Rudolf Julius Emmanuel** (1822–88) German physicist, who held teaching posts in Berlin and Zurich, before going to Würzburg in 1869. He is best known for formulating the second law of *thermodynamics in 1850, independently of William Thomson (Lord *Kelvin). In 1865 he introduced the concept of *entropy, and later contributed to electrochemistry and electrodynamics.

**Clausius–Mossotti equation** A relation between the *polarizability $\alpha$ of a molecule and the dielectric constant $\varepsilon$ of a dielectric substance made up of molecules with this polarizability. The Clausius–Mossotti equation can be written in the form $\alpha = (3/4\pi N)/[(\varepsilon - 1)/(\varepsilon - 2)]$, where $N$ is the number of molecules per unit volume. The equation provides a link between a microscopic quantity (the polarizability) and a macroscopic quantity (the dielectric constant); it was derived using macroscopic electrostatics by the Italian physicist Ottaviano Fabrizio Mossotti (1791–1863) in 1850 and independently by Rudolf Clausius in 1879. It works best for gases and is only approximately true for liquids or solids, particularly if the dielectric constant is large. *Compare* LORENTZ–LORENZ EQUATION.

**Claus process** A process for obtaining sulphur from hydrogen sulphide (from natural gas or crude oil). It involves two stages. First, part of the hydrogen sulphide is oxidized to sulphur dioxide:

$$2H_2S + 3O_2 \rightarrow 2SO_2 + 2H_2O.$$

Subsequently, the sulphur dioxide reacts with hydrogen sulphide to produce sulphur:

$$SO_2 + 2H_2S \rightarrow 3S + 2H_2O.$$

The second stage occurs at 300°C and needs an iron or aluminium oxide catalyst.

**clavicle** A bone that forms part of the *pectoral (shoulder) girdle, linking the *scapula (shoulder blade) to the sternum (breastbone). In man it forms the collar bone and serves as a brace for the shoulders.

**clay** A fine-grained deposit consisting chiefly of *clay minerals. It is characteristically plastic and virtually impermeable when wet and cracks when it dries out. In geology the size of the constituent particles is usually taken to be less than 1/256 mm. In soil science clay is regarded as a soil with particles less than 0.002 mm in size.

**clay minerals** Very small particles, chiefly hydrous silicates of aluminium, sometimes with magnesium and/or iron substituting for all or part of the aluminium, that are the major constituents of clay materials. The particles are essentially crystalline (either platy or fibrous) with a layered structure, but may be amorphous or metalloidal. The clay minerals are responsible for the plastic properties of clay; the particles have the property of being able to hold water. The chief groups of clay minerals are:
**kaolinite**, $Al_4Si_4O_{10}(OH)_8$, the chief constituent of *kaolin;
**halloysite**, $Al_4Si_4(OH)_8O_{10}.4H_2O$;
**illite**, $KAl_4(Si,Al)_8O_{18}.2H_2O$;
**montmorillonite**,

$$(Na,Ca)_{0.33}(Al,Mg)_2Si_4O_{10}(OH)_2.nH_2O,$$

formed chiefly through alteration of volcanic ash;
**vermiculite**,

$$(Mg,Fe,Al)_3(Al,Si)_4O_{10}(OH)_2.4H_2O,$$

used as an insulating material and potting soil.

**cleavage 1.** (in embryology) The series of cell divisions by which a single fertilized egg cell is transformed into a multicellular body, the *blastula. Characteristically no growth occurs during cleavage, the shape of the embryo is unchanged except for the formation of a central cavity (the blastocoel), and the ratio of nuclear material (DNA) to cytoplasm increases. **2.** (in crystallography) The splitting of a crystal along planes of atoms in the lattice.

**Clemmensen reduction** A method of reducing a *carbonyl group (C=O) to $CH_2$, using zinc amalgam and concentrated hydrochloric acid. It is used as a method of ascending a homologous series. The reaction is named after Erik Clemmensen (1876–1941).

**climate** The characteristic pattern of weather elements in an area over a period. The main weather elements include temperature, precipitation, humidity, solar insolation, and wind. The climate of a large area is determined by several climatic controls: (1) the latitude of the area, which accounts for the amount of solar radiation it receives; (2) the distribution of land and sea masses; (3) the altitude and topography of the area; and (4) the location of the area in relation to the ocean currents. Climates may be classified into groups; the classification devised by the meteorologist Vladimir Köppen (1846–1940) in 1910, with subsequent modifications, is the most frequently used today. The scientific study of climate is **climatology.**

**climate change** A long-term change in the elements of climate, such as temperature, precipitation, wind, and pressure, measured over a period of time of at least several decades. Throughout the geological history of the earth there have been periodic fluctuations between warmer and cooler periods on a wide range of time scales. The causes of climate change are complex. Factors include the external processes of variations of solar emissions, variations of the earth's orbit, volcanic eruptions, mountain building, and tectonic movements; anthropogenic (human-induced) processes include the increase of greenhouse gases as a result of activities that include the burning of fossil fuels, changes in land use, and the emission of aerosols. Internal processes, natural interactions within the climate system, are also factors. The term climate change is also used synonymously with *global warming.

(((⊕))) **SEE WEB LINKS**

• Overview of climate science and the mechanisms of climate change, from the UK Met Office.

**climax community** A relatively stable ecological *community that is achieved at the end of a *succession.

**cline** A gradual variation in the characteristics of a species or population over its geographical range. It occurs in response to varying environmental factors, such as soil type or climate.

**clinostat** A mechanical device that rotates whole plants (usually seedlings), so removing the effect of any stimulus that normally acts in one particular direction. It is most often used to study the growth of plant organs when the influence of gravity has been removed.

**clitoris** An erectile rod of tissue in female mammals (and also some reptiles and birds) that is the equivalent of the male penis. It lies in front of the *urethra and *vagina.

**cloaca** The cavity in the pelvic region into which the terminal parts of the alimentary canal and the urinogenital ducts open in most vertebrates. Placental mammals, however, have a separate anus and urinogenital opening.

**clock reaction** See OSCILLATING REACTION.

**clonal selection theory** A theory explaining how the cells of the immune system produce large quantities of the right antibody at the right time, i.e. when the appropriate antigen is encountered. It proposes that there is a pre-existing pool of lymphocytes (*B cells) consisting of numerous small subsets. Each subset carries a unique set of surface antibody molecules with its own particular binding characteristics. If a cell encounters and binds the corresponding antigen it is 'selected' – stimulated to divide repeatedly and produce a large clone of identical cells, all secreting the antibody. The involvement of helper T cells (see T CELL) is essential for activation of the B cell.

**clone** 1. A group of cells, an organism, or a population of organisms arising from a single ancestral cell. All members of a particular clone are genetically identical. In nature clones are produced by asexual reproduction, for example by the formation of bulbs

and tubers in plants or by *parthenogenesis in certain animals. New techniques of cell manipulation and tissue culture have enabled the cloning of many plants and some animals. A wide range of commercially important plant species, including potatoes, tulips, and certain forest trees, are now cloned by *micropropagation, resulting in more uniform crops. Cloning in animals is more complex, but has been accomplished successfully in several species. The first mammal to be cloned experimentally from the body cell of an adult was a sheep ('Dolly') born in 1997. The nucleus containing DNA was extracted from an udder cell (which had been deprived of nutrients) and inserted into an 'empty' egg cell (from which the nucleus had been removed) using the technique of *nuclear transfer. This reconstituted egg cell was then stimulated to divide by an electric shock and implanted into the uterus of a surrogate mother ewe, which subsequently gave birth to a clone of the original sheep. This breakthrough has led to the production of exact replicas of animals with certain genetically engineered traits, for example to manufacture drugs in their milk or provide organs for human transplantation. **2. (gene clone)** An exact replica of a gene. See GENE CLONING.

**cloning vector** See VECTOR.

**closed chain** See CHAIN; RING.

**close packing** The packing of spheres so as to occupy the minimum amount of space. In a single plane, each sphere is surrounded by six close neighbours in a hexagonal arrangement. The spheres in the second plane fit into depressions in the first layer, and so on. Each sphere has 12 other touching spheres. There are two types of close packing. In **hexagonal close packing** the spheres in the third layer are directly over those in the first, etc., and the arrangement of planes is ABAB.... In **cubic close packing** the spheres in the third layer occupy a different set of depressions than those in the first. The arrangement is ABCABC.... See also CUBIC CRYSTAL.

(((⊕))) **SEE WEB LINKS**

• An interactive version of cubic close packing
• An interactive version of hexagonal close packing

**closo-structure** See BORANE.

**clotting factors (coagulation factors)** A group of substances present in blood plasma that, under certain circumstances, undergo a series of chemical reactions leading to the conversion of blood from a liquid to a solid state (*see* BLOOD CLOTTING). Although they have specific names, most coagulation factors are referred to by an agreed set of Roman numerals (e.g. *Factor VIII, Factor IX). Lack of any of these factors in the blood results in the inability of the blood to clot. *See also* HAEMOPHILIA.

**cloud** A mass of minute water droplets or ice crystals held in suspension in the atmosphere that appears as an opaque drifting body. The droplets or crystals are formed by the condensation of water vapour in the atmosphere in the presence of condensation nuclei – minute particles, such as smoke or salt. A number of cloud classifications have been devised; that most commonly used is based on cloud appearance and height. The high clouds (above 5000 metres) comprise cirrocumulus, cirrostratus, and cirrus (mares' tails) clouds; the medium clouds (approximately 2000–5000 metres) comprise altocumulus and altostratus; and the low clouds (below 2000 metres) are nimbostratus, stratocumulus, and stratus. Some clouds with great vertical development cannot be confined to these height categories; these are cumulus and cumulonimbus clouds.

**cloud chamber** A device for making visible the paths of particles of *ionizing radiation. The **Wilson (expansion) cloud chamber** consists of a container containing air and ethanol vapour, which is cooled suddenly by adiabatic expansion, causing the vapour to become supersaturated. The excess moisture in the vapour is then deposited in drops on the tracks of ions created by the passage of the ionizing radiation. The resulting row of droplets can be photographed. If the original moving particle was being deflected by electric or magnetic fields, the extent of the deflection provides information on its mass and charge. This device was invented in 1911 by C. T. R. Wilson.

A simpler version of this apparatus is the **diffusion cloud chamber**, developed by Cowan, Needels, and Nielsen in 1950, in which supersaturation is achieved by placing a row of felt strips soaked in a suitable alcohol at the top of the chamber. The lower part of the chamber is cooled by solid carbon dioxide. The vapour continuously diffuses downwards, and that in the centre (where it becomes supersaturated) is almost continuously sensitive to the presence of ions created by the radiation.

**club drug** A drug typically used by young people at clubs, parties, etc. Examples are ecstasy, gammahydroxybutyric acid (GHB), and methamphetamines.

**clubmoss** *See* LYCOPHYTA.

**Clusius column** A device for separating isotopes by *thermal diffusion. One form consists of a vertical column some 30 metres high with a heated electric wire running along its axis. The lighter isotopes in a gaseous mixture of isotopes diffuse faster than the heavier isotopes. Heated by the axial wire, and assisted by natural convection, the lighter atoms are carried to the top of the column, where a fraction rich in lighter isotopes can be removed for further enrichment.

**cluster** *See* GALAXY CLUSTER; STAR CLUSTER.

**cluster compound** A compound in which groups of metal atoms are joined together by metal–metal bonds. The formation of such compounds is a feature of the chemistry of certain transition elements, particularly molybdenum and tungsten, but also vanadium, tantalum, niobium, and uranium. **Isopoly compounds** are ones in which the cluster contains atoms of the same element; **heteropoly compounds** contain a mixture of different elements.

**cluster model** A model of atomic nuclei in which nuclei are regarded as being made up of clusters of nucleons, particularly alpha particles, with the clusters having a brief existence before splitting. Cluster models have had considerable success in describing nuclear structure and reactions.

**cluster of differentiation** *See* CD.

**CMR** *See* COLOSSAL MAGNETORESISTANCE.

**CNB** *See* COSMIC NEUTRINO BACKGROUND.

**Cnidaria** A phylum of aquatic invertebrates (sometimes known as coelenterates) that includes *Hydra*, jellyfish, sea anemones, and *corals. A cnidarian's body is *diploblastic, with two cell layers of the body wall separated by *mesoglea, and shows *radial symmetry. The body cavity (**gastrovascular cavity**) is sac-shaped, with one opening acting as both mouth and anus. This opening is surrounded by tentacles bearing *thread cells. Cnidarians exist both as free-swimming *medusae (e.g. jellyfish) and as sedentary *polyps. The latter may be colonial (e.g. corals) or solitary (e.g. sea anemones and

*Hydra*). In many cnidarians the life cycle alternates between these two forms (*see* ALTERNATION OF GENERATIONS).

**(⊕) SEE WEB LINKS**

• Part of the comprehensive website of the University of California Museum of Paleontology, providing a useful summary of cnidarian biology and systematics

**cnidoblast** *See* THREAD CELL.

**CNS** *See* CENTRAL NERVOUS SYSTEM.

**CoA** *See* COENZYME A.

**coacervate** An aggregate of macromolecules, such as proteins, lipids, and nucleic acids, that form a stable *colloid unit with properties that resemble living matter. Many are coated with a lipid membrane and contain enzymes that are capable of converting such substances as glucose into more complex molecules, such as starch. Coacervate droplets arise spontaneously under appropriate conditions and have been suggested as possible prebiological systems from which living organisms originated.

**coadaptation** The mutual adaptation of two species that occurs during *coevolution.

**coagulation** The process in which colloidal particles come together irreversibly to form larger masses. Coagulation can be brought about by adding ions to change the ionic strength of the solution and thus destabilize the colloid (*see* FLOCCULATION). Ions with a high charge are particularly effective (e.g. alum, containing $Al^{3+}$, is used in styptics to coagulate blood). Another example of ionic coagulation is in the formation of river deltas, which occurs when colloidal silt particles in rivers are coagulated by ions in sea water. Alum and iron (III) sulphate are also used for coagulation in sewage treatment. Heating is another way of coagulating certain colloids (e.g. boiling an egg coagulates the albumin). *See also* BLOOD CLOTTING.

**coal** A brown or black carbonaceous deposit derived from the accumulation and alteration of ancient vegetation, which originated largely in swamps or other moist environments. As the vegetation decomposed it formed layers of peat, which were subsequently buried (for example, by marine sediments following a rise in sea level or subsidence of the land). Under the increased pressure and resulting higher temperatures the peat was transformed into coal. Two types of coal are recognized: **humic** (or

**woody**) **coals**, derived from plant remains; and **sapropelic coals**, which are derived from algae, spores, and finely divided plant material.

As the processes of coalification (i.e. the transformation resulting from the high temperatures and pressures) continue, there is a progressive transformation of the deposit: the proportion of carbon relative to oxygen rises and volatile substances and water are driven out. The various stages in this process are referred to as the **ranks** of the coal. In ascending order, the main ranks of coal are: **lignite** (or **brown coal**), which is soft, brown, and has a high moisture content; **subbituminous coal**, which is used chiefly by generating stations; **bituminous coal**, which is the most abundant rank of coal; **semibituminous coal**; **semianthracite coal**, which has a fixed carbon content of between 86% and 92%; and **anthracite coal**, which is hard and black with a fixed carbon content of between 92% and 98%.

Most deposits of coal were formed during the Carboniferous and Permian periods. More recent periods of coal formation occurred during the early Jurassic and Palaeogene periods. Coal deposits occur in all the major continents, and coal is used as a fuel and in the chemical industry; by-products include coke and coal tar. Combustion of coal is a major source of greenhouse gases worldwide, and efforts are underway to develop 'clean coal' technology.

**coal gas** A fuel gas produced by the destructive distillation of coal. In the late 19th and early 20th centuries coal gas was a major source of energy and was made by heating coal in the absence of air in local gas works. Typically, it contained hydrogen (50%), methane (35%), and carbon monoxide (8%). By-products of the process were *coal tar and coke. The use of this type of gas declined with the increasing availability of natural gas, although since the early 1970s interest has developed in using coal in making *SNG.

**coal tar** A tar obtained from the destructive distillation of coal. Formerly, coal tar was obtained as a by-product in manufacturing *coal gas. Now it is produced in making coke for steel making. The crude tar contains a large number of organic compounds, such as benzene, naphthalene, methylbenzene, phenols, etc., which can be obtained by distillation. The residue is **pitch**. Coal tar was once the major source of organic chemicals, most of which are now derived from petroleum and natural gas.

**coaxial cable** A cable consisting of a central conductor surrounded by an insulator, which is in turn contained in an earthed sheath of another conductor. The central conductor and the outer conductor are coaxial (i.e. have the same axis). They are used to transmit high-frequency signals as they produce no external fields and are not influenced by them.

**cobalamin (vitamin B$_{12}$)** *See* VITAMIN B COMPLEX.

**cobalt** Symbol Co. A light-grey *transition element; a.n. 27; r.a.m. 58.933; r.d. 8.9; m.p. 1495°C; b.p. 2870°C. Cobalt is ferromagnetic below its Curie point of 1150°C. Small amounts of metallic cobalt are present in meteorites but it is usually extracted from ore deposits worked in Canada, Morocco, and the Democratic Republic of Congo. It is present in the minerals cobaltite, smaltite, and erythrite but also associated with copper and nickel as sulphides and arsenides. Cobalt ores are usually roasted to the oxide and then reduced with carbon or water gas. Cobalt is usually alloyed for use. Alnico is a well-known magnetic alloy and cobalt is also used to make stainless steels and in high-strength alloys that are resistant to oxidation at high temperatures (for turbine blades and cutting tools).

The metal is oxidized by hot air and also reacts with carbon, phosphorus, sulphur, and dilute mineral acids. Cobalt salts, usual oxidation states II and III, are used to give a brilliant blue colour in glass, tiles, and pottery. Anhydrous cobalt(II) chloride paper is used as a qualitative test for water and as a heat-sensitive ink. Small amounts of cobalt salts are essential in a balanced diet for mammals (*see* ESSENTIAL ELEMENT). Artificially produced cobalt–60 is an important radioactive tracer and cancer-treatment agent. The element was discovered by Georg Brandt (1694–1768) in 1737.

(((⊕))) **SEE WEB LINKS**
• Information from the WebElements site

**cobalt(II) oxide** A pink solid, CoO; cubic; r.d. 6.45; m.p. 1935°C. The addition of potassium hydroxide to a solution of cobalt(II) nitrate gives a bluish-violet precipitate, which on boiling is converted to pink impure cobalt(II) hydroxide. On heating this in the absence of air, cobalt(II) oxide is formed.

The compound is readily oxidized in air to form tricobalt tetroxide, $Co_3O_4$, and is readily reduced by hydrogen to the metal.

**cobalt(III) oxide (cobalt sesquioxide)** A black grey insoluble solid, $Co_2O_3$; hexagonal or rhombic; r.d. 5.18; decomposes at 895°C. It is produced by the ignition of cobalt nitrate; the product however never has the composition corresponding exactly to cobalt(III) oxide. On heating it readily forms $Co_3O_4$, which contains both Co(II) and Co(III), and is easily reduced to the metal by hydrogen. Cobalt(III) oxide dissolves in strong acid to give unstable brown solutions of trivalent cobalt salts. With dilute acids cobalt(II) salts are formed.

**cobalt steel** Any of a group of alloy *steels containing 5–12% of cobalt, 14–20% of tungsten, usually with 4% of chromium and 1–2% of vanadium. They are very hard but somewhat brittle. Their main use is in high-speed tools.

**cobalt thiocyanate test** *See* SCOTT'S TEST.

**COBE** Cosmic Background Explorer; an orbiting satellite launched in November 1989 for cosmological research. In 1992, statistical studies of measurements on the *microwave background radiation indicated the presence of weak temperature fluctuations thought to be imprints of quantum fluctuations in the *early universe. *See also* WMAP.

(((⊕))) **SEE WEB LINKS**
• The NASA website for the COBE project

**COBOL** A high-level computer language developed in the early 1960s. The name is short for *common business-oriented language*. COBOL is employed widely for data processing in commerce and government. Although programs in COBOL tend to be long and wordy, they are easy to read and easy to modify (even by someone who is not the original author).

**cocaine** A powerful drug present in the leaves of the coca plant (*Erythroxylon coca*). It stimulates the central nervous system and has effects similar to the amphetamines. It was originally used as a local anaesthetic. The illegal drug is usually the soluble hydrochloride. This can be converted into the free-base form (known as **crack cocaine**) by dissolving in water and heating with sodium bicarbonate. Cocaine is a class A drug in the UK. It can be detected by *Scott's test.

**coccus** Any spherical bacterium. Cocci may occur singly, in pairs, in groups of four or more, in cubical packets, in grapelike clusters (*Staphylococcus*), or in chains (*Streptococcus*). Staphylococci and streptococci include pathogenic species. They are generally nonmotile and do not form spores.

**coccyx** The last bone in the *vertebral column in apes and man (i.e. tailless primates). It is formed by the fusion of 3–5 *caudal vertebrae.

**cochlea** Part of the *inner ear of mammals, birds, and some reptiles that transforms sound waves into nerve impulses. In mammals it is coiled, resembling a snail shell. The cochlea is lined with sensitive cells that bear tiny hairs; it is filled with fluid (**endolymph**) and surrounded by fluid (**perilymph**). Sound-induced vibrations of the *oval window are transmitted through the perilymph and endolymph and stimulate the hair cells that line the cochlea. These in turn stimulate nerve cells that transmit information, via the *auditory nerve, to the brain for interpretation of the sounds.

**Cockcroft, Sir John Douglas** (1897–1967) British physicist, who joined *Rutherford at the Cavendish Laboratory in Cambridge, where with Ernest Walton (1903–95) he built a linear accelerator (*see* COCKCROFT–WALTON GENERATOR). In 1932, using the apparatus to bombard lithium nuclei with protons, they produced the first artificial nuclear transformation. For this work they were awarded the 1951 Nobel Prize.

**Cockcroft–Walton generator** The first proton accelerator; a simple *linear accelerator producing a potential difference of some 800 kV (d.c.) from a circuit of rectifiers and capacitors fed by a lower (a.c.) voltage. The experimenters, Sir John Cockcroft and E. T. S. Walton (1903–95), used this device in 1932 to achieve the first artificially induced nuclear reaction by bombarding lithium with protons to produce helium:

$$^{1}_{1}H + ^{7}_{3}Li \rightarrow ^{4}_{2}He + ^{4}_{2}He$$

**cockroaches** *See* DICTYOPTERA.

**cocoon** A protective covering for eggs and/or larvae produced by many invertebrates. For example, the larvae of many insects spin a cocoon in which the pupae develop (that of the silkworm moth produces silk), and earthworms secrete a cocoon for the developing eggs.

**cocurrent flow** Flow of two fluids in the same direction with transfer of matter or heat between them. *Compare* COUNTERCURRENT FLOW.

**codominance** The condition that arises when both alleles in a *heterozygous organism are dominant and are fully expressed in the *phenotype. For example, the human blood group AB is the result of two alleles, A and B, both being expressed. A is not dominant to B, nor vice versa. *Compare* INCOMPLETE DOMINANCE.

**codon** A triplet of nucleotides within a molecule of messenger *RNA that functions as a unit of genetic coding (the **triplet code**), usually by specifying a particular amino acid during the synthesis of proteins in a cell (*see* GENETIC CODE). A few codons specify instructions during this process (*see* START CODON; STOP CODON). The term codon may also refer to any of the corresponding nucleotide triplets of DNA that are transcribed into codons. *See also* READING FRAME. *Compare* ANTICODON.

**coefficient** **1.** (in mathematics) A number or other known factor by which a variable quantity is multiplied, e.g. in $ax^2 + bx + c = 0$, $a$ is the coefficient of $x^2$ and $b$ is the coefficient of $x$. **2.** (in physics) A measure of a specified property of a particular substance under specified conditions, e.g. the coefficient of *friction of a substance.

**coefficient of expansion** *See* EXPANSIVITY.

**coefficient of friction** *See* FRICTION.

**coelacanth** A bony fish of the genus *Latimeria*, thought to be extinct until the first modern specimen of *L. chalumnae* was discovered in 1938 in the Indian Ocean around the Comoros Islands, off the SE coast of Africa. A second species, *L. menadoensis*, was discovered in 1999 in the Celebes Sea, SE Asia. The coelacanth belongs to the same order (Crossopterygii – lobe-finned fishes) as the ancestors of the amphibians. It is a large fish, 1–2 m long and weighing 80 kg or more, with a three-lobed tail fin. The body is covered in rough heavy scales and the pectoral fins can be used like crutches to help movement across the sea bed. The young are born alive. Fossil coelacanths are most abundant in deposits about 400 million years old and no fossils less than 70 million years old have been found.

**coelenterates**  *See* CNIDARIA.

**coelom**  A fluid-filled cavity that forms the main *body cavity of vertebrate and most invertebrate animals. It is formed by the splitting of the *mesoderm. Ciliated ducts (**coelomoducts**) connect the coelom to the exterior allowing the exit of waste products and gametes; in higher animals these are specialized as oviducts, etc. The coelom is large and often subdivided in annelid worms (in which it functions as a hydrostatic skeleton) and vertebrates. In arthropods it is restricted to the cavities of the gonads and excretory organs, the body cavity being a blood-filled *haemocoel.

**coelomoduct**  *See* COELOM.

**coelostat**  A device that enables light from the same area of the sky (i.e., from a selected celestial object, such as the sun) to be continuously reflected into the field of view of an astronomical telescope or other instrument. It consists of a plane mirror driven by a clockwork or electrical mechanism so that it rotates from east to west to compensate for the west-to-east rotation of the earth.

**coenocyte**  A mass of cytoplasm surrounding many nuclei and enclosed by a cell wall. It is found in certain algae and fungi. *Compare* CELL; SYNCYTIUM.

**coenzyme**  An organic nonprotein molecule that associates with an enzyme molecule in catalysing biochemical reactions. Coenzymes usually participate in the substrate–enzyme interaction by donating or accepting certain chemical groups. Many vitamins are precursors of coenzymes. *See also* COFACTOR.

**coenzyme A (CoA)**  A complex organic compound that acts in conjunction with enzymes involved in various biochemical reactions, notably the oxidation of pyruvate via the *Krebs cycle and fatty-acid oxidation and synthesis. It comprises principally the B vitamin *pantothenic acid, the nucleotide *adenine, and a ribose–phosphate group.

**coenzyme Q**  *See* UBIQUINONE.

**coercive force (coercivity)**  The magnetizing force necessary to reduce the flux density in a magnetic material to zero. *See* HYSTERESIS.

**coevolution**  The evolution of complementary adaptations in two species caused by the *selection pressures that each exerts on the other. It is common in symbiotic associations (*see* SYMBIOSIS). For example, many insect-pollinated plants have evolved flowers whose shapes, colours, etc., make them attractive to particular insects; at the same time the pollinating insects have evolved sense organs and mouthparts specialized for quickly locating, and extracting nectar from, particular species of plants.

**cofactor**  A nonprotein component essential for the normal catalytic activity of an enzyme. Cofactors may be organic molecules (*coenzymes) or inorganic ions. They may activate the enzyme by altering its shape or they may actually participate in the chemical reaction.

**coherent radiation**  Electromagnetic radiation in which two or more sets of waves have a constant phase relationship, i.e. with peaks and troughs always similarly spaced.

**coherent scattering**  Scattering for which there is a well-defined relationship between the phase of the incoming wave and the phase of the outgoing wave. Scattering for which there is no well defined such relationship is called **incoherent scattering**.

**coherent state**  The quantum-mechanical state of a system that most closely resembles the corresponding classical state of the system. An example of a coherent state is the quantum state of a harmonic oscillator or of an electromagnetic field. The concept of a coherent state is very important in quantum optics.

**coherent units**  A system of *units of measurement in which derived units are obtained by multiplying or dividing base units without the use of numerical factors. *SI units form a coherent system; for example the unit of force is the newton, which is equal to 1 kilogram metre per second squared ($kg\ m\ s^{-2}$), the kilogram, metre, and second all being base units of the system.

**cohesion**  1. The force of attraction between like molecules. Cohesion provides the force that holds up a column of water in the xylem tissue of plants without it breaking. The **cohesion–tension theory** is the most widely accepted explanation for the continual flow of water upwards through the xylem of a plant. Water is removed from the plant by the process of *transpiration, which creates a tension that pulls the water in the xylem upwards as a single column held together by cohesive forces. **2.** (in botany) The union of like parts, such as the fusion of petals that occurs in some flowers.

**coinage metals** A group of three malleable ductile transition metals forming group 11 (formerly IB) of the *periodic table: copper (Cu), silver (Ag), and gold (Au). Their outer electronic configurations have the form $nd^{10}(n+1)s^1$. Although this is similar to that of alkali metals, the coinage metals all have much higher ionization energies and higher (and positive) standard electrode potentials. Thus, they are much more difficult to oxidize and are more resistant to corrosion. In addition, the fact that they have $d$-electrons makes them show variable valency ($Cu^I$, $Cu^{II}$, and $Cu^{III}$; $Ag^I$ and $Ag^{II}$; $Au^I$ and $Au^{III}$) and form a wide range of coordination compounds. They are generally classified with the *transition elements.

**coincidence circuit** An electronic logic device that gives an output only if two input signals are fed to it simultaneously or within a specified time of each other. A **coincidence counter** is an electronic counter incorporating such a device.

**coitus** *See* SEXUAL INTERCOURSE.

**coke** A form of carbon made by the destructive distillation of coal. Coke is used for blast-furnaces and other metallurgical and chemical processes requiring a source of carbon. Lower-grade cokes, made by heating the coal to a lower temperature, are used as smokeless fuels for domestic heating.

**colchicine** An *alkaloid derived from the autumn crocus, *Colchicum autumnale*. It inhibits *spindle formation in cells during mitosis so that chromosomes cannot separate during anaphase, thus inducing multiple sets of chromosomes (*see* POLYPLOID). Colchicine is used in genetics, cytology, and plant breeding research and also in cancer therapy to inhibit cell division.

**cold-blooded animal** *See* ECTOTHERM.

**cold emission** The emission of electrons by a solid without the use of high temperature, either as a result of field emission (*see* FIELD-EMISSION MICROSCOPE) or *secondary emission.

**cold front** *See* FRONT.

**cold fusion** *See* NUCLEAR FUSION.

**Coleoptera** An order of insects comprising the beetles and weevils and containing about 330 000 known species – the largest order in the animal kingdom. The forewings are hardened and thickened to form **elytra**, which meet at a precise mid-dorsal line and protect the underlying pair of hindwings and abdomen. The mouthparts are generally modified for biting and in some species assume antler-like proportions. Beetles occur in a wide variety of terrestrial and aquatic habitats; many feed on decaying organic matter, some eat living vegetation, while others prey on other arthropods. A number of beetles and weevils are economically important pests of stored grain, timber, and crops. The young emerge as larvae and generally undergo metamorphosis via a pupal stage to form the adult beetle.

**coleoptile** A protective sheath that covers the young shoot of the embryo in plants of the grass family. It bursts open when the first leaves develop. Experiments investigating growth movements of the oat coleoptile led to the discovery of the plant growth substance indoleacetic acid (IAA).

**coleorhiza** A protective sheath that covers the young root of the embryo in plants of the grass family.

**collagen** An insoluble fibrous protein found extensively in the connective tissue of skin, tendons, and bone. The polypeptide chains of collagen (containing the amino acids glycine and proline predominantly) form triple-stranded helical coils that are bound together to form fibrils, which have great strength and limited elasticity. Collagen accounts for over 30% of the total body protein of mammals.

**collapsar** A *neutron star, *white dwarf, or any other star composed of degenerate matter: *black holes are also regarded as collapsars. They share the property that they have all collapsed under the effect of a strong gravitational field and ceased to produce energy within their cores.

**collecting duct** Any of the ducts in the mammalian *kidney that drains into the renal pelvis, which leads to the ureter. They are the main sites of water reabsorption from the glomerular filtrate, which drains into the ducts from the *distal convoluted tubules of the *nephrons. The cells of the collecting ducts are relatively impermeable to water. However, the influence of *antidiuretic hormone increases the permeability of the collecting ducts, allowing the reabsorption of water and controlling the final urine concentration according to the body's state of hydration.

**collective excitation** A quantized mode in a many-body system, occurring because of cooperative motion of the whole system as a result of interactions between particles. *Plasmons and *phonons in solids are examples of collective excitations. Collective excitations obey Bose–Einstein statistics (*see* QUANTUM STATISTICS).

**collector** *See* TRANSISTOR.

**collenchyma** *See* GROUND TISSUES.

**colligative properties** Properties that depend on the concentration of particles (molecules, ions, etc.) present in a solution, and not on the nature of the particles. Examples of colligative properties are osmotic pressure (*see* OSMOSIS), *lowering of vapour pressure, *depression of freezing point, and *elevation of boiling point.

**collimator** 1. Any device for producing a parallel beam of particle or wave radiation. A common arrangement used for light consists of a convex achromatic lens fitted to one end of a tube with an adjustable slit at the other end, the slit being at the principal focus of the lens. Light rays entering the slit leave the lens as a parallel beam. Collimators for particle beams and other types of electromagnetic radiation utilize a system of slits or apertures. 2. A small fixed telescope attached to a large astronomical telescope to assist in lining up the large one onto the desired celestial body.

**collision density** The number of collisions that occur in unit volume in unit time when a given neutron flux passes through matter.

**collision quenching** *See* EXTERNAL CONVERSION.

**collodion** A thin film of cellulose nitrate made by dissolving the cellulose nitrate in ethanol or ethoxyethane, coating the surface, and evaporating the solvent.

**colloids** As originally defined by Thomas Graham in 1861, substances, such as starch or gelatin, that will not diffuse through a membrane. Graham distinguished colloids from **crystalloids** (e.g. inorganic salts), which would pass through membranes. Later it was recognized that colloids were

distinguished from true solutions by the presence of particles that were too small to be observed with a normal microscope yet were much larger than normal molecules. Colloids are now regarded as systems in which there are two or more phases, with one (the **dispersed phase**) distributed in the other (the **continuous phase**). Moreover, at least one of the phases has small dimensions (in the range $10^{-9}$–$10^{-6}$ m). Colloids are classified in various ways.

**Sols** are dispersions of small solid particles in a liquid. The particles may be macromolecules or may be clusters of small molecules. **Lyophobic sols** are those in which there is no affinity between the dispersed phase and the liquid. An example is silver chloride dispersed in water. In such colloids the solid particles have a surface charge, which tends to stop them coming together. Lyophobic sols are inherently unstable and in time the particles aggregate and form a precipitate. **Lyophilic sols**, on the other hand, are more like true solutions in which the solute molecules are large and have an affinity for the solvent. Starch in water is an example of such a system. **Association colloids** are systems in which the dispersed phase consists of clusters of molecules that have lyophobic and lyophilic parts. Soap in water is an association colloid (*see* MICELLE).

**Emulsions** are colloidal systems in which the dispersed and continuous phases are both liquids, e.g. oil-in-water or water-in-oil. Such systems require an emulsifying agent to stabilize the dispersed particles.

**Gels** are colloids in which both dispersed and continuous phases have a three-dimensional network throughout the material, so that it forms a jelly-like mass. Gelatin is a common example. One component may sometimes be removed (e.g. by heating) to leave a rigid gel (e.g. silica gel).

Other types of colloid include *aerosols (dispersions of liquid or solid particles in a gas, as in a mist or smoke) and foams (dispersions of gases in liquids or solids). Colloids are analysed theoretically in terms of intermolecular forces.

**cologarithm** The logarithm of the reciprocal of a number.

**colon** The section of the vertebrate *large intestine that lies between the *caecum and the *rectum. Its prime function is to absorb water and minerals from indigestible food residues passing from the small intestine, which results in the formation of *faeces.

**colony** **1.** (in zoology) A group of animals of the same species living together and dependent upon each other. Some, such as the corals and sponges, are physically connected and function as a single unit. Others, such as insect colonies, are not physically joined but show a high level of social organization with members specialized for different functions (*see* CASTE). **2.** (in microbiology) A group of microorganisms, usually bacteria or yeasts, that are considered to have developed from a single parent cell. Colonies that grow on *agar plates differ in shape, colour, surface texture, and translucency and can therefore be used as a means of identification.

**colony-stimulating factor (CSF)** Any of several *cytokines that stimulate development of certain types of blood cells from haemopoietic *stem cells. They include **GM-CSF**, a glycoprotein that causes the stem cells to develop into mixed colonies of granulocytes and monocytes/macrophages (hence the name); **G-CSF**, which stimulates production of granulocytes only; and **M-CSF**, which promotes only monocyte/macrophage cell production. *Interleukin-3 (IL-3) is sometimes called the 'multi-CSF' because it stimulates the production of all types of lymphocytes and also erythrocytes.

**colorimeter** Any instrument for comparing or reproducing colours. Monochromatic colorimeters match a *colour with a mixture of monochromatic and white lights. Trichromatic colorimeters use a mixture of three *primary colours.

**colorimetric analysis** Quantitative analysis of solutions by estimating their colour, e.g. by comparing it with the colours of standard solutions.

**colossal magnetoresistance (CMR)** A property of certain materials in which electrical resistance can change by several orders of magnitude in a magnetic field. It is observed particularly in certain manganese oxides and is likely to have applications in magnetic storage and spintronics. At present, there is no complete quantitative theory of the phenomenon. *See also* GIANT MAGNETORESISTANCE; MAGNETORESISTANCE.

**colostrum** A liquid with a high content of nitrogen, antibodies, and vitamins that is secreted from the mammary glands before and just after giving birth. The change of secretion from colostrum to proper milk takes place gradually during the days after birth.

**colour** The sensation produced when light of different wavelengths falls on the human eye. Although the visible spectrum covers a continuously varying range of colours from red to violet it is usually split into seven colours (the **visible spectrum**) with the following approximate wavelength ranges:

red 740–620 nm
orange 620–585 nm
yellow 585–575 nm
green 575–500 nm
blue 500–445 nm
indigo 445–425 nm
violet 425–390 nm

A mixture of all these colours in the proportions encountered in daylight gives white light; other colours are produced by varying the proportions or omitting components.

A coloured light has three attributes: its **hue**, depending on its wavelength; its **saturation**, depending on the degree to which it departs from white light; and its *luminosity. Coloured objects that owe their colour to pigments or dyes absorb some components of white light and reflect the rest. For example, a red book seen in white light absorbs all the components except the red, which it reflects. This is called a **subtractive process** as the final colour is that remaining after absorption of the rest. This is the basis of the process used in *colour photography. Combining coloured lights, on the other hand, is an **additive process** and this is the method used in *colour television. *See also* PRIMARY COLOUR.

**colour blindness** Any disorder of vision in which colours are confused. The most common type is red–green colour blindness. This is due to a recessive gene carried on the X chromosome (*see* SEX LINKAGE), and therefore men are more likely to show the defect although women may be *carriers. It results in absence or malfunctioning of one or more of the three types of cone cell responsible for *colour vision.

**colour centre** A defect in a crystal that changes the way in which it absorbs light or other electromagnetic radiation. Impurities in the crystal affect the bind structure and allow transitions in different regions of the spectrums. Impurity colour centres are responsible for the characteristic colours of many gemstones. A particular type of colour centre is an **F-centre**. This is a missing negative ion in an ionic crystal, where the overall charge neutrality occurs by trapping an electron in the vacancy. The electron has energy levels similar to those of a particle in a box. F-centres can be produced by chemical activity or by irradiation.

**colour charge** *See* ELEMENTARY PARTICLES.

**colour photography** Any of various methods of forming coloured images on film or paper by photographic means. One common process is a subtractive reversal system that utilizes a film with three layers of light-sensitive emulsion, one responding to each of the three *primary colours. On development a black image is formed where the scene is blue. The white areas are dyed yellow, the *complementary colour of blue, and the blackened areas are bleached clean. A yellow filter between this emulsion layer and the next keeps blue light from the second emulsion, which is green-sensitive. This is dyed magenta where no green light has fallen. The final emulsion is red-sensitive and is given a cyan (blue-green) image on the negative after dying. When white light shines through the three dye layers the cyan dye subtracts red where it does not occur in the scene, the magenta subtracts green, and the yellow subtracts blue. The light projected by the negative therefore reconstructs the original scene either as a transparency or for use with printing paper.

**colour television** A television system in which the camera filters the light from the scene into the three component *primary colours, red, blue, and green, which are detected by separate camera tubes. The separate information so obtained relating to the colour of the image is combined with the sound and synchronization signals and transmitted using one of three systems, the American, British, or French. At the receiver, the signal is split again into red, blue, and green components, each being fed to a separate *electron gun in the cathode-ray tube of the receiver. By an additive process (*see* COLOUR) the picture is reconstituted by the beam from each gun activating a set of phosphor dots of that colour on the screen.

**colour temperature** The temperature of a non-black body as indicated by the temperature of a black body having approximately the same spectral distribution.

**colour vision** The ability of the eye to detect different wavelengths of light and to distinguish between these different wavelengths and their corresponding colours. In the mammalian eye this is achieved by the *cone cells, which are located in and around the *fovea near to the centre of the retina. The cone cells contain the light-sensitive pigment **iodopsin**, which – according to the **trichromatic theory** – exists in three forms, each form occurring in a different cone cell. Each form of iodopsin is sensitive to a different range of wavelengths of light. The relative stimulation of each type of cone will determine the colour that is interpreted by the brain. *See also* COLOUR BLINDNESS.

The *compound eye of certain insects is also capable of colour vision.

**columbium** A former name for the element *niobium.

**column chromatography** *See* CHROMATOGRAPHY.

**coma** **1.** A nebulous cloud of gas and dust that surrounds the nucleus of a *comet. **2.** An *aberration of a lens or mirror in which the image of a point lying off the axis has a comet-shaped appearance.

**combinations** *See* PERMUTATIONS AND COMBINATIONS.

**combined cycle** *See* FLUIDIZATION.

**combustion** A chemical reaction in which a substance reacts rapidly with oxygen with the production of heat and light. Such reactions are often free-radical chain reactions, which can usually be summarized as the oxidation of carbon to form its oxides and the oxidation of hydrogen to form water. *See also* FLAME.

**comet** A type of *small solar system body that travels around the sun in an eccentric orbit and typically has a **nucleus** consisting of ice and dust, surrounded by a *coma of dust and frozen gases that become volatile near the sun and are blown off by the *solar wind to form a **comet tail**. **Short-period comets** have orbital periods of less than 150 years. The others have very long periods, some exceeding 100 000 years, or visit the inner solar system only once before being ejected into interstellar space. The nuclei of most comets are thought to be 'dirty snowballs' about one kilometre in diameter, although the solar system has a few comets with nuclei exceeding 10 km in diameter. The coma may be $10^4$–$10^5$ km in diameter, and the tail can be $10^7$ km in length. Debris from comets produces annual meteor showers as the earth passes through it on its annual solar orbit. *See also* HALLEY'S COMET.

**commensalism** An interaction between two animal or plant species that habitually live together in which one species (the **commensal**) benefits from the association while the other is not significantly affected. For example, the burrows of many marine worms contain commensals that take advantage of the shelter provided but do not affect the worm.

**common-collector connection** A technique used in the operation of some *transistors, in which the collector is common to both the input and output circuits, the input terminal is the *base, and the output terminal is the collector.

**common logarithm** *See* LOGARITHM.

**common salt** *See* SODIUM CHLORIDE.

**communication** An interaction between two organisms in which information is conveyed from one to the other. Communication can occur between individuals of the same species (**intraspecific communication**) or between members of different species (**interspecific communication**). It generally involves the transmission of a signal from one organism to another; signals can be visual, chemical, or tactile or they can take the form of sounds. Visual signals between members of the same species are widely used by animals in such activities as defining and protecting *territories and finding suitable mates (*see* COURTSHIP; DISPLAY BEHAVIOUR; BIOLUMINESCENCE). Chemical and tactile signals also play an important role in these activities (*see* PHEROMONE). Social species rely heavily on all three types of signalling, the classic example being provided by the *dance of the bees, in which information about the distance and direction of a food source is conveyed to other members of the colony. Visual signals, in the form of body coloration, are the principal means of communication between animals of different species (*see* MIMICRY; WARNING COLORATION). Sounds are more effective than visual signals for intraspecific communication over long distances and at night. Certain insects produce sounds by *stridulation, while birdsong and language are sophisticated examples of sound signals in birds and humans, respectively. Among plants, visual and chemical signals are important in communication. Flowering plants whose flowers are pollinated by insects or other animals depend on the colour, shape, and scent of their flowers to attract suitable pollinating agents. Some plants produce chemical signals to deter competitors and predators (*see* ALLELOPATHY).

**communication satellite** An unmanned artificial satellite sent by rocket into a geostationary orbit (*see* SYNCHRONOUS ORBIT) around the earth to enable television broadcasts and telephone communications to be made between points on the earth's surface that could not otherwise communicate by radio owing to the earth's curvature. Modulated *microwaves are transmitted to the satellite, which amplifies them and retransmits them at a different frequency to the receiving station. The satellites are powered by *solar cells. Three or more satellites in equatorial orbits can provide a world-wide communications linkage. The satellites are placed well above the ionosphere and therefore the carrier waves used have to be in the microwave region of the spectrum in order to pass through the ionosphere.

**community** A naturally occurring assemblage of plant and animal species living within a defined area or habitat. Communities are named after one of their *dominant species (e.g. a pine community) or the major physical characteristics of the area (e.g. a freshwater pond community). Members of a community interact in various ways (e.g. through *food chains and *competition). Large communities may be divided into smaller component communities. *See* ASSOCIATION.

**commutative law** The mathematical law stating that the value of an expression is independent of the order of combination of the numbers, symbols, or terms in the expression. The **commutative law for addition** applies if $x + y = y + x$. The **commutative law of multiplication** applies if $x \times y = y \times x$. Subtraction and division are not commutative. *Compare* ASSOCIATIVE LAW; DISTRIBUTIVE LAW.

**commutator** The part of the armature of an electrical motor or generator through which connections are made to external circuits. It consists of a cylindrical assembly of insulated copper conductors, each of which is connected to one point in the armature winding. Spring-loaded carbon brushes are positioned around the commutator to carry the current to or from it.

**compact disk (CD)** A 120 mm metal disk on which there is a *digital recording of audio information, providing high-quality recording and reproduction of music, speech, etc. The recording is protected by a layer of clear plastic. The information is permanently encoded in the form of a spiral track of minute pits impressed on one surface of the disk during manufacture; these impressions correspond to a changing sequence of *bits. The CD is rotated at constant linear velocity (CLV) in a CD player; the rotation rate is varied according to the radius of the track accessed. Information is retrieved from the rotating disk by means of a low-power laser focused on the track and modulated by the binary code impressed on the track.

**companion cell** A type of cell found within the *phloem of flowering plants. Each companion cell is usually closely associated with a *sieve element. It regulates the activity of the adjacent sieve element and takes part in loading and unloading sugar into the sieve element. In gymnosperms a similar function is attributed to **albuminous cells**, which are found closely associated with gymnosperm sieve elements.

**compass** A small magnet pivoted at its central point to revolve in a horizontal plane. In the earth's magnetic field the magnet (called the compass needle) aligns itself so that its north-seeking end points to the earth's magnetic north pole. A scale (called a compass card) is placed below the needle for use in navigation. In some navigation compasses the entire card is pivoted, indicating direction by a fixed mark on the casing. Such compasses are often filled with alcohol to provide damping. Magnetic compasses suffer from being affected by magnetic metals in their vicinity and to a large extent they have been replaced by *gyrocompasses.

**compass plant** A plant that has its leaves permanently orientated in a north–south direction. Such an arrangement enables the plant to take full advantage of morning and evening sun, while avoiding the stronger midday sunlight. An example is the compass plant of the prairies (*Silphium laciniatum*).

**competent** Describing embryonic tissue that is capable of developing into a specialized tissue when suitably stimulated. *See* INDUCTION; EVOCATION.

**competition** The interaction that occurs between two or more organisms, populations, or species that share some environmental resource when this is in short supply. Competition is an important force in evolution: plants, for example, become tall to compete for light, and animals evolve various foraging methods to compete for food. There may be a direct confrontation between competitors, as occurs between barnacles competing for space on a rock, or the numbers or fecundity of the competitors are indirectly reduced through joint dependence on limited resources. Competition occurs both between members of a species (**intraspecific competition**) and between different species (**interspecific competition**). Interspecific competition often results in the dominance of one species over another (*see* DOMINANT). Since competition ultimately results in the displacement by one competitor of the others, it is to the advantage of the competitors to avoid one another wherever possible. Thus in time the competitors become separated from each other geographically or ecologically, which promotes evolutionary change. Competition for mates may lead to *sexual selection.

**competitive inhibition** *See* INHIBITION.

**complement** A group of proteins present in blood plasma and tissue fluid that aids the body's defences following an *immune response; the genes encoding it form part of the *major histocompatibility complex. Following an antibody–antigen reaction, complement is activated chemically and becomes bound to the antibody–antigen complex (**complement fixation**); it can cause *lysis of certain types of bacteria, or it can render the target cell more susceptible to *phagocytosis.

**complemental males** The small males of certain animals that live in or on the females and are usually more or less degenerate apart from the reproductive organs. They occur in certain crustaceans (e.g. some barnacles), in which the normal individuals are hermaphrodite but the complemental males have suppressed ovaries, lose their alimentary canal, and lead a semiparasitic existence in the mantle cavity of the larger partner. This may ensure that cross fertilization occurs.

**complementarity** The concept that a single model may not be adequate to explain all the observations made of atomic or subatomic systems in different experiments. For example, *electron diffraction is best explained by assuming that the electron is a wave (*see* DE BROGLIE WAVELENGTH), whereas the *photoelectric effect is described by assuming that it is a particle. The idea of two different but complementary concepts to treat quantum phenomena was first put forward by the Danish physicist Niels Bohr (1855–1962) in 1927. *See also* LIGHT.

**complementary colours** A pair of coloured lights of specific hue (*see* COLOUR) that produce the sensation of white when mixed in appropriate intensities. There is an infinite number of such pairs, an example (with wavelengths) is orange (608 nm) and blue (490 nm).

**complementary DNA (cDNA)** A form of DNA prepared in the laboratory using messenger *RNA (mRNA) as template, i.e. the reverse of the usual process of *transcription in cells; the synthesis is catalysed by *reverse transcriptase. cDNA thus has a base sequence that is complementary to that of the mRNA template; unlike genomic DNA, it contains no noncoding sequences (*introns). cDNA is used in *gene cloning for the expression of eukaryote genes in prokaryote host cells or as a *gene probe.

**complementary genes** Two (or more) genes that are interdependent, such that the dominant *allele from either gene can only produce an effect on the *phenotype of an organism if the dominant allele from the other gene is also present.

**complex** A compound in which molecules or ions form coordinate bonds to a metal atom or ion (see illustration overleaf). Often complexes occur as positive or negative **complex ions**, such as $[Cu(H_2O)_6]^{2+}$ and $Fe[(CN)_6]^{3-}$. A complex may also be a neutral molecule (e.g. $PtCl_2(NH_3)_2$). The formation of such coordination complexes is typical behaviour of transition metals. The complexes formed are often coloured and have unpaired electrons (i.e. are paramagnetic). *See also* LIGAND; CHELATE.

**complexity** The levels of *self-organization of a system. In physical systems, complexity is associated with *broken symmetry and the ability of a system to have different states between which it can make *phase transitions. It is also associated with having coherence in space over a long range. Examples of complexity include *superconductivity, *superfluidity, *lasers, and ordered

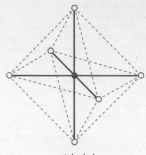

octahedral

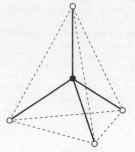

tetrahedral

square-planar

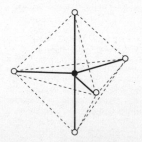

trigonal-bipyramid

**Complex.** Some common shapes of coordination complexes.

phases that arise when a system is driven far from thermal equilibrium (*see* BÉNARD CELL). It is not necessary for a system to have a large number of degrees of freedom in order for complexity to occur. The study of complexity is greatly aided by computers in systems that cannot be described analytically. Complexity is also very important in a number of other fields, including theoretical biology.

**complex number** A number that has a real part, $x$, and an imaginary part, i$y$, where i = $\sqrt{-1}$ and $x$ and $y$ are real ($x$ can also equal 0). The complex number therefore has the form $x + iy$, which can also be written in the polar form $r\cos\theta + ir\sin\theta$, where $r$ is the **modulus** and $\theta$ is the **argument** (or **amplitude**). A complex number can be represented on an **Argand diagram**, devised by Jean-Robert Argand (1768–1822), in which the horizontal axis represents the real part of the number and the vertical axis the imaginary part (see illustration). In the polar form the modulus is the line joining the origin to the point representing the complex number and the argument is the angle between the modulus and the $x$-axis.

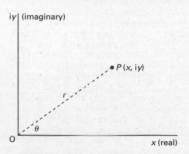

**Complex number.** Argand diagram.

**complexometric analysis** A type of volumetric analysis in which the reaction involves the formation of an inorganic *complex.

**component** A distinct chemical species in a mixture. If there are no reactions taking place, the number of components is the number of separate chemical species. A mixture of water and ethanol, for instance, has two components (but is a single phase). A mixture of ice and water has two phases but one component ($H_2O$). If an equilibrium reaction occurs, the number of components is

taken to be the number of chemical species minus the number of reactions. Thus, in
$$H_2 + I_2 \rightleftharpoons 2HI$$
there are two components. *See also* PHASE RULE.

**component vectors** Two or more vectors that produce the same effect as a given vector; the vectors that combine to produce the effect of a resultant vector. A component vector in a given direction is the projection of the given vector (*V*) along that direction, i.e. $V\cos\theta$, where $\theta$ is the angle between the given vector and the direction.

**composite fruit** A type of fruit that develops from an inflorescence rather than from a single flower. *See* PSEUDOCARP; SOROSIS; STROBILUS; SYCONUS.

**compost** A mixture of decaying organic matter, such as vegetation and manure, that is used as a *fertilizer. The organic material is decomposed by aerobic saprotrophic organisms, mostly fungi and bacteria. Some decomposition is also carried out by *detritivores. Compost is used mainly on a domestic scale.

**compound** A substance formed by the combination of elements in fixed proportions. The formation of a compound involves a chemical reaction; i.e. there is a change in the configuration of the valence electrons of the atoms. Compounds, unlike mixtures, cannot be separated by physical means. *See also* MOLECULE.

**compound eye** The eye of insects and crustaceans, which consists of numerous visual units, the **ommatidia**. Each ommatidium consists of an outer cuticle covering a lens, beneath which are 6–8 retinal cells surrounding a light-sensitive **rhabdom**. Adjacent ommatidia are separated by pigment cells. The eye is convex, with nerve fibres from the retinal cells converging onto the optic nerve. There are two types of compound eye. In **apposition eyes**, typical of diurnal insects, each ommatidium focuses rays parallel to its long axis so that each gives an image of a minute part of the visual field, producing a detailed mosaic image. In **superposition eyes**, typical of nocturnal insects, the pigment separating ommatidia migrates to the ends of the cells, so that each ommatidium receives light from a larger part of the visual field and the image may overlap with those received by many neighbouring ommatidia. This produces an image that is bright but lacks sharpness of detail.

**compound microscope** *See* MICROSCOPE.

**compressibility** The reciprocal of bulk modulus (*see* ELASTIC MODULUS). The compressibility ($k$) is given by $-V^{-1}dV/dp$, where $dV/dp$ is the rate of change of volume ($V$) with pressure.

**compression ratio** The ratio of the total volume enclosed in the cylinder of an *internal-combustion engine at the beginning of the compression stroke to the volume enclosed at the end of the compression stroke. For petrol engines the compression ratio is 8.5–9:1, with a recent tendency to the lower end of the range in order to make use of unleaded petrols. For Diesel engines the compression ratio is in the range 12–25:1.

**comproportionation** A reaction in which an element in a higher oxidation state reacts with the same element in a lower oxidation state to give the element in an intermediate oxidation state. An example is

$$Ag^{2+}(aq) + Ag(s) \rightarrow 2Ag^+(aq).$$

It is the reverse of *disproportionation.

**Compton, Arthur Holly** (1892–1962) US physicist, who became professor of physics at the University of Chicago in 1923. He is best known for his discovery (1923) of the *Compton effect, for which he shared the 1927 Nobel Prize for physics with C. T. R. *Wilson. In 1938 he demonstrated that *cosmic radiation consists of charged particles.

**Compton effect** The reduction in the energy of high-energy (X-ray or gamma-ray) photons when they are scattered by free electrons, which thereby gain energy. The phenomenon, first observed in 1923 by Arthur Compton, occurs when the photon collides with an electron; some of the photon's energy is transferred to the electron and consequently the photon loses energy $h(v_1 - v_2)$, where $h$ is the *Planck constant and $v_1$ and $v_2$ are the frequencies before and after collision. As $v_1 > v_2$, the wavelength of the radiation increases after the collision. This type of inelastic scattering is known as **Compton scattering** and is similar to the *Raman effect. *See also* INVERSE COMPTON EFFECT.

(🌐) **SEE WEB LINKS**

• Compton's original 1923 paper in *The Physical Review*

**Compton wavelength** The length below which a particle's quantum-mechanical properties become relevant in relativistic *quantum mechanics. For a particle of rest mass $m$ the Compton wavelength is $\hbar/mc$, where $\hbar$ is the rationalized Planck constant and $c$ is the speed of light. The Compton wavelength is so named because of its occurrence in the theory of the *Compton effect, where its value for the electron is $3.8616 \times 10^{-13}$ m. The Compton wavelength is sometimes defined as $h/mc$, with $h$ being the Planck constant, in which case the electron value is $2.4263 \times 10^{-12}$ m.

**computable structure** A mathematical structure for which a systematic algorithm exists for solving all problems with that structure. It has been postulated that fundamental physics should involve computable structures, although some approaches to *quantum gravity involve mathematical structures that are not computable.

**computer** An electronic device that processes information according to a set of instructions, called the **program**. The most versatile type of computer is the **digital computer**, in which the input is in the form of characters, represented within the machine in *binary notation. Central to the operation of a computer is the **central processing unit** (**CPU**), which contains circuits for manipulating the information (*see* LOGIC CIRCUITS). The CPU contains the arithmetic/logic unit (ALU), which performs operations, and a control unit. It is supported by a short-term **memory**, in which data is stored in electronic circuits (*see* RAM). Associated storage usually involves *magnetic disks or *CD-ROM. There are also various peripheral input and output devices, such as a keyboard, visual-display unit (VDU), magnetic tape unit, and *printer. Computers range in size from the **microprocessor** with a few thousand logic elements, to the large **mainframe computer** with millions of logic circuits.

The **analog computer** is used in scientific experiments, industrial control, etc. In this type of device the input and output are continuously varying quantities, such as a voltage, rather than the discrete digits of the more commercially useful digital device. **Hybrid computers** combine the properties of both digital and analog devices. Input is usually in analog form, but processing is carried out digitally in a CPU.

Computer **hardware** consists of the actual

electronic or mechanical devices used in the system; the **software** consists of the programs and data. *See also* ROM.

**concave** Curving inwards. A **concave mirror** is one in which the reflecting surface is formed from the interior surface of a sphere or paraboloid. A **concave lens** has at least one face formed from the interior surface of a sphere. A **biconcave lens** has both faces concave and is therefore thinnest at its centre. The **plano-concave lens** has one plane face and one concave face. The **concavo-convex lens** (also called a **meniscus**) has one concave face and one *convex face. *See* LENS.

**concavo-convex** *See* CONCAVE.

**concentrated** Describing a solution that has a relatively high concentration of solute.

**concentration** The quantity of dissolved substance per unit quantity of solvent in a solution. Concentration is measured in various ways. The amount of substance dissolved per unit volume (symbol $c$) has units of $mol\ dm^{-3}$ or $mol\ l^{-1}$. It is now called **amount concentration** (formerly **molarity**). The **mass concentration** (symbol $\rho$) is the mass of solute per unit volume of solvent. It has units of $kg\ dm^{-3}$, $g\ cm^{-3}$, etc. The **molal concentration** (or **molality**; symbol $m$) is the amount of substance per unit mass of solvent, commonly given in units of $mol\ kg^{-1}$. *See also* MOLE FRACTION.

**concentration cell** *See* CELL.

**concentration gradient (diffusion gradient)** The difference in concentration between a region of a solution or gas that has a high density of particles and a region that has a relatively lower density of particles. By random motion, particles will move from the area of high concentration towards the area of low concentration, by the process of *diffusion, until the particles are evenly distributed in the solution or gas.

**conceptacle** A flask-shaped cavity with a small opening (the **ostiole**) that is found in the swollen tip of certain brown algae, such as *Fucus*. It contains the sex organs.

**conception** The fertilization of a mammalian egg cell by a sperm cell, which occurs in the fallopian tube. Conception is followed by *implantation.

**concerted reaction** A type of reaction in which there is only one stage rather than a

series of steps. The $S_N2$ mechanism in *nucleophilic substitutions is an example.

**conchoidal fracture** Fracture of a solid in which the surface of the material is curved and marked by concentric rings. It occurs particularly in amorphous materials.

**condensation** The change of a vapour or gas into a liquid. The change of phase is accompanied by the evolution of heat (*see* LATENT HEAT).

**condensation polymerization** *See* POLYMER.

**condensation pump** *See* DIFFUSION PUMP.

**condensation reaction** A chemical reaction in which two molecules combine to form a larger molecule with elimination of a small molecule (e.g. $H_2O$). *See* ALDEHYDES; KETONES.

**condensed-matter physics** *See* SOLID-STATE PHYSICS.

**condenser** 1. A mirror or set of lenses used in optical instruments, such as a microscope or film projector, to concentrate the light diverging from a compact source. A common form consists of two plano-convex lenses with the plane faces pointing outwards. 2. A device used to cool a vapour to cause it to condense to a liquid. In a steam engine the condenser acts as a reservoir that collects the part of the steam's internal energy that has not been used in doing work on the piston. The cooling water passed through the condenser is warmed and is used as fresh feedwater for the boiler. *See also* LIEBIG CONDENSER. 3. *See* CAPACITOR.

**condenser microphone** *See* CAPACITOR MICROPHONE.

**conditional response (conditioned reflex)** A learned response that develops to an initially ineffective stimulus in classical *conditioning.

**conditioning** A process by which animals learn about a relation between two events. In **classical** (or **Pavlovian**) **conditioning**, repeated presentations of a neutral stimulus (e.g. the sound of a bell or buzzer) are followed each time by a biologically important stimulus (such as food or electric shock), which elicits a response (e.g. salivation). Eventually the neutral stimulus presented by itself produces a response (the **conditional response**, or **conditioned reflex**) similar to that originally evoked by the biologically important stimulus. In **instrumental** (or **operant**) **conditioning** the animal is rewarded (or punished) each time it makes a particular response; this eventually causes the frequency of the response to increase (or decrease). *See* LEARNING (Feature); REINFORCEMENT.

**conductance** The reciprocal of electrical resistance in a direct-current circuit. The ratio of the resistance to the square of the *impedance in an alternating-current circuit. The SI unit is the siemens, formerly called the mho or reciprocal ohm.

**conducting polymer** An organic polymer that conducts electricity. Conducting polymers have a crystalline structure in which chains of conjugated unsaturated carbon–carbon bonds are aligned. Examples are polyacetylene and polypyrrole. There has been considerable interest in the development of such materials because they would be cheaper and lighter than metallic conductors. They do, however, tend to be chemically unstable and, so far, no commercial conducting polymers have been developed.

**conductiometric titration** A type of titration in which the electrical conductivity of the reaction mixture is continuously monitored as one reactant is added. The equivalence point is the point at which this undergoes a sudden change. The method is used for titrating coloured solutions, which cannot be used with normal indicators.

**conduction** 1. (thermal conduction) The transmission of heat through a substance from a region of high temperature to a region of lower temperature. In gases and most liquids, the energy is transmitted mainly by collisions between atoms and molecules with those possessing lower kinetic energy. In solid and liquid metals, heat conduction is predominantly by migration of fast-moving electrons, followed by collisions between these electrons and ions. In solid insulators the absence of free *electrons restricts heat transfer to the vibrations of atoms and molecules within crystal lattices. *See* CONDUCTIVITY. 2. (electrical conduction) The passage of electric charge through a substance under the influence of an electric field. *See also* CHARGE CARRIER; ENERGY BANDS.

**conduction band** *See* ENERGY BANDS.

**conductivity** 1. (thermal conductivity) A measure of the ability of a substance to

conduct heat. For a block of material of cross section $A$, the energy transferred per unit time $E/t$, between faces a distance, $l$, apart is given by $E/t = \lambda A(T_2 - T_1)/l$, where $\lambda$ is the conductivity and $T_2$ and $T_1$ are the temperatures of the faces. This equation assumes that the opposite faces are parallel and that there is no heat loss through the sides of the block. The SI unit is therefore $J\ s^{-1}\ m^{-1}\ K^{-1}$. 2. (electrical conductivity) The reciprocal of the *resistivity of a material. It is measured in siemens per metre in SI units. When a fluid is involved the electrolytic conductivity is given by the ratio of the current density to the electric field strength.

**(⊕) SEE WEB LINKS**

- Values of thermal conductivity for a range of materials at the NPL

**conductivity water** *See* DISTILLED WATER.

**conductometric titration** A type of *titration in which the end point is determined by detecting a sudden change in the conductivity of the solution. It is particularly useful in titrating weak acids against weak bases or for coloured solutions, for which indicators cannot be used.

**conductor** 1. A substance that has a high thermal *conductivity. Metals are good conductors on account of the high concentration of free *electrons they contain. Energy is transmitted through a metal predominantly by means of collisions between electrons and ions. Most nonmetals are poor conductors (good **thermal insulators**) because there are relatively few free electrons. 2. A substance that has a high electrical conductivity. Again conduction results from the movement of free electrons. *See* ENERGY BANDS.

**condyle** A smooth round knob of bone that fits into a socket on an adjoining bone, forming a *joint. Such a joint permits up-and-down or side-to-side movement but does not allow rotation. There are condyles where the lower jawbone (mandible) is attached to the skull, which permits chewing movements. *See also* OCCIPITAL CONDYLE.

**Condy's fluid** A mixture of calcium and potassium permanganates (manganate(VII)) used as an antiseptic.

**cone** 1. (in botany) A reproductive structure occurring in gymnosperms, known technically as a **strobilus**. It consists of *sporophylls bearing the spore-producing sporangia. Gymnosperms produce different male and female cones. The large woody female cones of pines, firs, and other conifers are made up of structures called **ovuliferous scales**, which bear the ovules. Cones are also produced by clubmosses and horsetails. **2.** (in animal anatomy) A type of light-sensitive receptor cell, found in the *retinas of all diurnal vertebrates. Cones are specialized to transmit information about colour (*see* COLOUR VISION) and are responsible for the *visual acuity of the eye. They function best in bright light. They are not evenly distributed on the retina, being concentrated in the *fovea and absent on the margin of the retina. *Compare* ROD. **3.** (in mathematics) A solid figure generated by a line (the **generator**) joining a point on the perimeter of a closed plane curve (the **directrix**) to a point (the **vertex**) outside this plane, as the line moves round the directrix. If the directrix is a circle, the figure is a **circular cone** standing on a circular **base**. If the line joining the vertex to the centre of the base (the **axis**) is perpendicular to the base the figure is a **right circular cone**, which has a volume $\pi r^2 h/3$, where $r$ is the radius of the base and $h$ is the height of the vertex above the base. If the axis of the cone is not perpendicular to the base, the figure is an **oblique cone**. In general, the volume of any cone is one third of its base area multiplied by the perpendicular distance of the vertex from the base.

**configuration** **1.** The arrangement of atoms or groups in a molecule. **2.** The arrangement of electrons in atomic *orbitals in an atom.

**configuration space** The $n$-dimensional space with coordinates $(q_1, q_2, \ldots, q_n)$ associated with a system that has $n$ *degrees of freedom, where the values $q$ describe the degrees of freedom. For example, in a gas of $N$ atoms each atom has three positional coordinates, so the configuration space is $3N$-dimensional. If the particles also have internal degrees of freedom, such as those caused by vibration and rotation in a molecule, then these must be included in the configuration space, which is consequently of a higher dimension. *See also* PHASE SPACE.

**confinement** *See* QUANTUM CHROMODYNAMICS; QUARK CONFINEMENT.

**confocal fluorescence microscopy** A light microscopic technique that produces high-resolution images of fluorescently stained specimens without requiring elaborate preparation of the sample. The fluorescent markers, generally fluorescently labelled antibodies, are excited by light from a laser focused by the objective lens of the microscope so that it scans a single plane in the specimen, creating an optical section, under computer control. The emitted fluorescent light is captured by a photomultiplier and assembled into digital images by a computer. Serial scanning of, say, an entire cell can thus visualize successive sections through the cell or create three-dimensional, or even time-lapse, images. Moreover, numerous fluorescent probes are available for labelling different components of cells or other material.

**conformal field theory** A field theory that has *conformal invariance. There are important applications of conformal field theory in *string theory and *statistical mechanics.

**conformal invariance** Invariance under *conformal transformations and under transformations of scale. Conformal invariance is important in theories with massless particles and in the theory of phase transitions.

**conformal transformation** A transformation that preserves the angles between curves. A well-known example is Mercator's projection in cartography, in which any angle between a line on the spherical surface and a line of latitude or longitude will be the same on the map. Conformal transformations are used in a number of areas in physics, particularly in dealing with electromagnetic and gravitational fields and in fluid mechanics. Problems can often be simplified by applying conformal transformations to change a complicated geometrical arrangement to a simpler one.

**conformation** Any of the large number of possible shapes of a molecule resulting from rotation of one part of the molecule about a single bond. See illustration overleaf.

**congeners** Elements that belong to the same group in the periodic table.

**congenital** Present at birth. Congenital disorders of the body may be due to genetic factors, e.g. *Down's syndrome, or caused by injury or environmental factors, e.g. drugs (such as thalidomide), chemicals (such as dioxin), and infections (such as those caused by *_Listeria_ and *cytomegalovirus).

| eclipsed conformation | anti conformation | gauche conformation |

● = methyl group

Conformations of butane (sawhorse projection)

bisecting conformation     eclipsed conformation

Conformations of $R_3CHO$ (Newman projection)

**Conformation.**

**conic** A figure formed by the intersection of a plane and a *cone. If the intersecting plane is perpendicular to the axis of a right circular cone, the figure formed is a *circle. If the intersecting plane is inclined to the axis at an angle in excess of half the apex angle of the cone it is an *ellipse. If the plane is parallel to the sloping side of the cone, the figure is a *parabola. If the plane cuts both halves of the cone a *hyperbola is formed.

A conic can be defined as a plane curve in which for all points on the curve the ratio of the distance from a fixed point (the **focus**) to the perpendicular distance from a straight line (the **directrix**) is a constant called the **eccentricity** $e$. For a parabola $e = 1$, for an ellipse $e < 1$, and for a hyperbola $e > 1$.

**conidiospore** See CONIDIUM.

**conidium (conidiospore)** A spore of certain fungi, such as moulds, that is produced by the constriction of the tip of a specialized hypha, the **conidiophore**. Chains of conidia may be formed in this way; they are cut off, one at a time, from the tip of the hypha.

**Coniferophyta** A phylum of seed-bearing plants comprising the conifers, including the pines, firs, and spruces. Conifers have an extensive fossil record going back to the late Devonian. The gametes are carried in male

and female *cones, fertilization usually being achieved by wind-borne pollen. The ovules and the seeds into which they develop are borne unprotected (rather than enclosed in a carpel, as are those of the *Anthophyta). Internal tissue and cell structure of these species is not as advanced as in the angiosperms. Conifers are typically evergreen trees inhabiting cool temperate regions and have leaves reduced to needles or scales. The wood of conifers, which is called **softwood** in contrast to the **hardwood** of angiosperm trees, is widely used for timber and pulp. See also GYMNOSPERM.

**⊕ SEE WEB LINKS**

• Well-illustrated review of conifer genera, hosted by the Tree of Life project

**conjugate acid (conjugate base)** See ACID.

**conjugated** Describing double or triple bonds in a molecule that are separated by one single bond. For example, the organic compound buta-1,3-diene, $H_2C=CH-CH=CH_2$, has conjugated double bonds. In such molecules, there is some delocalization of electrons in the pi orbitals between the carbon atoms linked by the single bond.

**conjugate points** Two points in the

vicinity of a *lens or *mirror such that a bright object placed at one will form an image at the other.

**conjugation 1.** The fusion of two reproductive cells, particularly when these are both the same size (*see* ISOGAMY). **2.** A form of sexual reproduction seen in some algae (e.g. *Spirogyra*), some bacteria (e.g. *Escherichia coli*), and ciliate protozoans. Two individuals are united by a tube formed by outgrowths from one or both of the cells. Genetic material from one cell (designated the male) then passes through the tube to unite with that in the other (female) cell.

**conjunction** The alignment of two celestial bodies within the solar system so that they have the same celestial longitude as seen from the earth. A planet that orbits between the sun and the earth (Venus and Mercury) is in **superior conjunction** when it is in line with the sun and the earth but on the opposite side of the sun to the earth. It is in **inferior conjunction** when it lies between the earth and the sun. Conjunction may also occur between two planets or a moon and a planet.

**conjunctiva** The delicate membrane that covers the cornea and lines the inside of the eyelid of a vertebrate eye. It is kept clean by secretions of the *lacrimal (tear) gland and the reflex blink mechanism.

**connection table** A way of representing a molecule as a table showing the atoms, their coordinates, and the links between them. The widely used MDL molfile format uses a connection table in its representation of structure. Connection tables are a useful way of storing molecular data, both from the point of view of graphics programs and also for database searches, in which it is possible to use the table to look for substructures.

**connective tissue** An animal tissue consisting of a small number of cells (e.g. *fibroblasts and *mast cells) and fibres and a large amount of *extracellular matrix (ground substance). It is widely distributed and has many functions, including support, packing, defence, and repair. The individual constituents vary, depending on the function of the tissue. Different types of connective tissue include **mesenchyme** in the embryo, *adipose tissue, loose **areolar connective tissue** for packing and support, *blood, lymph, cartilage, and bone.

**consensus sequence** A sequence of nucleotides found in comparable regions of DNA or RNA, e.g. in the promoter regions (*see* OPERON) of different genes, in which certain bases occur with a frequency significantly greater than that expected by chance. Although such sequences may vary from case to case, it is possible to derive the most likely sequence overall. An example is the *Pribnow box of prokaryote promoters. The term is also applied to sequences of amino acids in polypeptides.

**conservation** The sensible use of the earth's natural resources in order to avoid excessive degradation and impoverishment of the environment (*see* DESERTIFICATION). It should include the search for alternative food and fuel supplies when these are endangered (as by *deforestation and overfishing); an awareness of the dangers of *pollution; and the maintenance and preservation of natural habitats and the creation of new ones (e.g. nature reserves, national parks, and *SSSIs).

**conservation law** A law stating that the total magnitude of a certain physical property of a system, such as its mass, energy, or charge, remains unchanged even though there may be exchanges of that property between components of the system. For example, imagine a table with a bottle of salt solution (NaCl), a bottle of silver nitrate solution (AgNO$_3$), and a beaker standing on it. The mass of this table and its contents will not change even when some of the contents of the bottles are poured into the beaker. As a result of the reaction between the chemicals two new substances (silver chloride and sodium nitrate) will appear in the beaker:

$$NaCl + AgNO_3 \rightarrow AgCl + NaNO_3,$$

but the total mass of the table and its contents will not change. This **conservation of mass** is a law of wide and general applicability, which is true for the universe as a whole, provided that the universe can be considered a closed system (nothing escaping from it, nothing being added to it). According to Einstein's mass–energy relationship, every quantity of energy ($E$) has a mass ($m$), which is given by $E/c^2$, where $c$ is the speed of light. Therefore if mass is conserved, the law of **conservation of energy** must be of equally wide application. The laws of **conservation of linear momentum** and **angular momentum** also are believed to be universally true.

Because no way is known of either creating or destroying electric charge, the law of

**conservation of charge** is also a law of universal application. Other quantities are also conserved in reactions between elementary particles.

**conservative field** A field of force in which the work done in moving a body from one point to another is independent of the path taken. The force required to move the body between these points in a conservative field is called a **conservative force**.

**conserved sequence** Any sequence of bases (or amino acids) in comparable segments of different nucleotides (or proteins) that tends to show similarity greater than that due to chance alone. The degree to which sequences are conserved can indicate the extent of structural and functional similarities between different genes or between different proteins and provides clues to their possible evolutionary relations.

**consistent histories** An interpretation of quantum mechanics that makes use of the concept of *decoherence to explain how the classical world emerges from quantum mechanics. The consistent-histories interpretation avoids the problem of observers and has greatly clarified our understanding of the problem of measurement in quantum mechanics.

**consociation** A climax plant *community that is dominated by one particular species, e.g. a pine forest. See DOMINANT. Compare ASSOCIATION.

**consolute temperature** The temperature at which two partially miscible liquids become fully miscible as the temperature is increased.

**constant** 1. A component of a relationship between variables that does not change its value, e.g. in $y = ax + b$, $b$ is a constant. 2. A fixed value that has to be added to an indefinite integral. Known as the **constant of integration**, it depends on the limits between which the integration has been performed. 3. See FUNDAMENTAL CONSTANTS.

**constantan** An alloy having an electrical resistance that varies only very slightly with temperature (over a limited range around normal room temperatures). It consists of copper (50–60%) and nickel (40–50%) and is used in resistance wire, thermocouples, etc.

**constant-boiling mixture** See AZEOTROPE.

**constant proportions** See CHEMICAL COMBINATION.

**constellation** A collection of stars arbitrarily grouped into a recognizable pattern, more than half of which were named by the ancients after animals and mythological characters; the rest, discovered when mariners explored the southern hemisphere, received names derived from navigational and scientific instruments. There are 88 constellations, which divide up the celestial sphere into regions named after them. The smallest is Crux (the [Southern] Cross) and the largest Hydra (the Sea Monster). Stars within a constellation are named according to a number of conventional systems. The brightest stars have individual names from Greek or Latin (Sirius, Arcturus) or Arabic (Aldebaran, Rigel), But for scientific and cataloguing purposes, most stars are named according to a system invented by a German lawyer and astronomer Johann Bayer (1572–1625), published in 1603. It uses a Greek letter and the genitive form of the constellation's Latin name (e.g. Alpha Crucis, Delta Cephei). Generally, α (alpha) signifies the brightest star, β (beta) the second brightest, etc. When the 24 letters of the Greek alphabet are exhausted, the Bayer system uses lower- and then upper-case Roman letters. Variable stars have their own designations, as in RR Lyrae or T Tauri.

**(((ⵙ))) SEE WEB LINKS**

- Information about constellations, including clear star charts, from the International Astronomical Union
- A discussion of how the present constellation boundaries were established by the IAU in 1930

**constitutive equations** The equations $D = \varepsilon E$ and $B = \mu H$, where $D$ is the electric displacement, ε is the *permittivity of the medium, $E$ is the electric field intensity, $B$ is the magnetic flux density, μ is the *permeability of the medium, and $H$ is the *magnetic field strength.

**consumer** An organism that feeds upon those below it in a *food chain (i.e. at the preceding *trophic level). Herbivores, which feed upon green plants, are **primary consumers**; a carnivore that feeds only upon herbivores is a **secondary consumer**; a **tertiary consumer** is a carnivore that feeds on other carnivores. The consumer at the end of a food chain is known as the **top carnivore**. Compare PRODUCER.

**contact potential difference** The potential difference that occurs between two electrically connected metals or between the base regions of two semiconductors. If two metals with work functions $\phi_1$ and $\phi_2$ are brought into contact, their Fermi levels will coincide. If $\phi_1 > \phi_2$ the first metal will acquire a positive surface charge with respect to the other at the area of contact. As a result, a contact potential difference occurs between the two metals or semiconductors.

**contact process** A process for making sulphuric acid from sulphur dioxide ($SO_2$), which is made by burning sulphur or by roasting sulphide ores. A mixture of sulphur dioxide and air is passed over a hot catalyst

$$2SO_2 + O_2 \rightarrow 2SO_3$$

The reaction is exothermic and the conditions are controlled to keep the temperature at an optimum 450°C. Formerly, platinum catalysts were used but vanadium–vanadium oxide catalysts are now mainly employed (although less efficient, they are less susceptible to poisoning). The sulphur trioxide is dissolved in sulphuric acid

$$H_2SO_4 + SO_3 \rightarrow H_2S_2O_7$$

and the oleum is then diluted.

**containment 1.** The prevention of the escape of radioactive materials from a *nuclear reactor. **2.** The process of preventing the plasma in a *thermonuclear reactor from touching the walls of the vessel by means of magnetic fields.

**contig map** *See* PHYSICAL MAP.

**continent** A large landmass that rises above the deep ocean floor. Geologically, the boundary of a continent lies offshore at the edge of the gentle slope of the continental shelf. The British Isles and other offshore islands consequently are parts of the nearby continents. It is generally accepted that there are seven continents – Asia, Africa, North America, South America, Europe, Australia, and Antarctica – occupying about 29% of the earth's surface.

**continental drift** The theory that the earth's continents once formed a single mass and have since moved relative to each other. It was first postulated by A. Snider in 1858 and greatly developed by Alfred Wegener in 1915. He used evidence, such as the fit of South America into Africa and the distribution of rock types, flora, fauna, and geological structures, to suggest that the present

(a) 200 million years ago

(b) 135 million years ago

(c) 65 million years ago

**Continental drift.**

distribution of the continents results from the breaking up of one or two greater land masses. The original land mass was named Pangaea and it was suggested that this broke up into the northerly **Laurasia** and the southerly **Gondwanaland** (see illustration). The theory was not accepted for about 50 years by the majority of geologists but during the early 1960s, the seafloor-spreading hypothesis of Harry Hess (1906–69) and the subsequent development of *plate tectonics produced a mechanism to explain the drift of the continents.

**((⊕)) SEE WEB LINKS**
- Animation of continental drift during the earth's history

**continuous culture** A technique used to grow microorganisms or cells continually in a particular phase of growth. For example, if a constant supply of cells is required, a cell culture maintained in the log phase is best; the conditions must therefore be continually monitored and adjusted accordingly so that the cells do not enter the stationary phase (*see* BACTERIAL GROWTH CURVE). Growth may also have to be maintained in a particular growth phase if an enzyme or chemical product is produced only during that phase.

**continuous function** A function f(*x*) is continuous at *x* = *a* if the limit of f(*x*) as *x* approaches *a* is f(*a*). A function that does not satisfy this condition is said to be a **discontinuous function**.

**continuous phase** *See* COLLOIDS.

**continuous spectrum** *See* SPECTRUM.

**continuous variation (quantitative variation)** The range of differences that can be observed in many characteristics in a population. Characteristics resulting from polygenic inheritance (*see* QUANTITATIVE INHERITANCE) show continuous variation, e.g. the wide range of foot sizes in an adult human population. *Compare* DISCONTINUOUS VARIATION.

**continuous wave** A wave that is transmitted continuously rather than in pulses.

**continuum** A system of axes that form a *frame of reference. The three dimensions of space and the dimension of time together can be taken to form a four-dimensional continuum; this was suggested by Minkowski in connection with special *relativity.

**contour** A line drawn on a map or chart that joins points with equal elevation above (or below) a level (usually mean sea level). Contours thus show the relief of the land surface or sea bed (below sea level the line is called a **submarine contour**). The difference in height between two consecutive contours is the **contour interval**.

**contour feathers** *Feathers that are arranged in regular rows on a bird's body, giving the body its streamlined shape. Each has a central horny shaft (the **rachis**) with a flattened **vane** on each side. Each vane is composed of two rows of filament-like *barbs, which are connected to each other by means of hooked *barbules to form a smooth surface. There is often a small second vane, the **aftershaft**, near the base of the feather.

**contraception** *See* BIRTH CONTROL.

**contractile root** Any of the modified adventitious roots that develop from the base of the stem of a bulb or corm. The new bulb or corm develops at a higher level in the soil

than the old one. The contractile roots shorten and pull it down to a suitable level.

**contractile vacuole** A membrane-surrounded cavity in a cell that periodically expands, filling with water, and then suddenly contracts, expelling its contents to the cell's exterior. It is thus an organ of *osmoregulation and excretion. Contractile vacuoles are common in freshwater sponges and typical of freshwater protists, such as *Amoeba* (which has one spherical vacuole) and *Paramecium* (in which a number of accessory vacuoles are attached to a main vacuole).

**contraction** (in animal physiology) The shortening of muscle fibres in order to exert a force on a tissue or organ of the body. In striated muscle contraction is brought about by interaction of actin and myosin filaments (*see* SARCOMERE; VOLUNTARY MUSCLE): it provides a force for *locomotion and plays a role in maintaining the balance and posture of the animal. *See also* INVOLUNTARY MUSCLE.

**control** **1.** The part of an experiment that acts as a standard by which to compare experimental observations. **2.** The natural regulation of biological processes. *See* CONTROL MECHANISM. **3.** *See* BIOLOGICAL CONTROL; CHEMICAL CONTROL.

**control grid** A wire-mesh electrode placed between the cathode and anode in a *thermionic valve or a *cathode-ray tube to control the flow of electrons from one to the other. A fluctuating potential signal fed to the control grid produces a current signal at the anode with similar but amplified fluctuations. It thus forms the basis of the electronic valve amplifier. In a cathode-ray tube the grid controls the intensity of the electron beam and hence the brightness of the image on the screen.

**control mechanism** (in biology) Any mechanism that regulates a biological process, such as a metabolic pathway or enzyme-controlled reaction, or that helps to maintain the *internal environment (*see* HOMEOSTASIS). *See also* FEEDBACK.

**control rod** One of a number of rods of a material, such as boron or cadmium, that absorbs neutrons. Control rods can be moved into or out of the core of a *nuclear reactor to control the rate of the reaction taking place within it.

**control unit (CU)** The part of the central

processor of a *computer that supervises the execution of a computer program.

**convection** A process by which heat is transferred from one part of a fluid to another by movement of the fluid itself. In **natural convection** the movement occurs as a result of gravity; the hot part of the fluid expands, becomes less dense, and is displaced by the colder denser part of the fluid as this drops below it. This is the process that occurs in most domestic hot-water systems between the boiler and the hot-water cylinder. A natural convection current is set up transferring the hot water from the boiler up to the cylinder (always placed above the boiler) so that the cold water from the cylinder can move down into the boiler to be heated. In some modern systems, where small-bore pipes are used or it is inconvenient to place the cylinder above the boiler, the circulation between boiler and hot-water cylinder relies upon a pump. This is an example of **forced convection**, where hot fluid is transferred from one region to another by a pump or fan.

**conventional current** A 19th-century convention, still in use, that treats any electrical current as a flow of positive charge from a region of positive potential to one of negative potential. The real motion, however, in the case of electrons flowing through a metal conductor, is in the opposite direction, from negative to positive. In semiconductors *hole conduction is in the direction of the conventional current, electron conduction is in the opposite direction.

**convergent evolution** The development of superficially similar structures in unrelated organisms, usually because the organisms live in the same kind of environment. Examples are the wings of insects and birds and the streamlined bodies of whales and fish. *Compare* ADAPTIVE RADIATION.

**convergent series** A series $a_1 + a_2 + \ldots + a_i + \ldots$, for which a partial sum $S_n = a_1 + a_2 + \ldots + a_n$ tends to a finite (or zero) limit as $n$ tends to infinity. This limit is the **sum** of the series. For example, the series $1 + 1/2 + 1/4 + 1/8 + \ldots$ (with the general term $a_i$ equal to $(1/2)^{i-1}$) tends to the limit 2. A series that is not convergent is said to be a **divergent series**. In such a series the partial sum tends to plus or minus infinity or may oscillate. For example, the series $1 + 1/2 + 1/3 + 1/4 + \ldots$ (with $a_i$ equal to $1/i$) is divergent. As can be seen from this latter example, a series may be divergent even if the individual terms $a_i$ tend to zero as $i$ tends to infinity.

**converging lens or mirror** A lens or mirror that can refract or reflect a parallel beam of light so that it converges at a point (the principal focus). Such a mirror is concave; a converging lens is thicker at its centre than at its edges (i.e. it is biconvex, planoconvex, or convexo-concave). *Compare* DIVERGING LENS OR MIRROR.

**conversion electron** *See* INTERNAL CONVERSION.

**converter** **1.** An electrical machine for converting alternating current into direct current, or less frequently, vice versa. **2.** The reaction vessel in the *Bessemer process or some similar steel-making process. **3.** A computer device for converting information coded in one form into some other form.

**converter reactor** A *nuclear reactor that converts fertile material (e.g. thorium–232) into *fissile material (e.g. uranium–233). A converter reactor can also be used to produce electrical power.

**convex** Curving outwards. A **convex mirror** is one in which the reflecting surface is formed from the exterior surface of a sphere or paraboloid. A **convex lens** has at least one face formed from the exterior surface of a sphere. A **biconvex lens** has both faces convex and is therefore thickest at its centre. The **plano-convex lens** has one plane face and one convex face. The **convexo-concave lens** (also called a **meniscus**) has one convex face and one *concave face. *See* LENS.

**convoluted tubule** *See* DISTAL CONVOLUTED TUBULE; PROXIMAL CONVOLUTED TUBULE; NEPHRON.

**coolant** A fluid used to remove heat from a system by *convection (usually forced), either to control the temperature or to extract energy. In a water-cooled car engine the coolant is water (or water and antifreeze), which is pumped around the engine and cooled in the radiator. In a *nuclear reactor the coolant is used to transfer the heat of the reaction from the core to a heat exchanger or to the steam-raising plant. In gas-cooled reactors the coolant is usually carbon dioxide. Pressurized water or boiling water is used as both coolant and *moderator in several types of reactor. In fast reactors, liquid sodium is used as the coolant.

**Cooper, Leon** *See* BARDEEN, JOHN.

**cooperative phenomenon** A phenomenon in which the constituents of a system cannot be regarded as acting independently from each other. Cooperative phenomena result from interactions between the constituents. Phenomena that can be described by the *liquid-drop model of nuclei, such as nuclear fission, are examples of cooperative phenomena because they involve the *nucleus as a whole rather than individual nucleons. Other examples of cooperative phenomena occur when a substance undergoes a *phase transition, as in the phenomena of ferromagnetism (*see* MAGNETISM) or *superconductivity.

**Cooper pairs** *See* SUPERCONDUCTIVITY.

**coordinate** *See* CARTESIAN COORDINATES; POLAR COORDINATES.

**coordinate bond** *See* CHEMICAL BOND.

**coordinate geometry** *See* ANALYTICAL GEOMETRY.

**coordinate system** A system that uniquely specifies points in a plane or in three-dimensional space. The simplest coordinate system is the *Cartesian coordinate system. In a plane two coordinates are necessary to specify a point. In three-dimensional space three coordinates are required. Many coordinate systems can be used to specify a point; however, sometimes one particular coordinate system is more convenient than others; indeed, certain problems can be solved in one coordinate system but not in others. For example, the Schrödinger equation for the hydrogen atom can be solved using spherical *polar coordinates but not using Cartesian coordinates.

**coordination** (in animal physiology) The processes involved in the reception of sensory information, the integration of that information, and the subsequent response of the organism. Coordination is controlled by regions of the brain that deal with specific functions, such as locomotion and breathing, and is carried out by the nervous system.

**coordination compound** A compound in which coordinate bonds are formed (*see* CHEMICAL BOND). The term is used especially for inorganic *complexes.

**⊕ SEE WEB LINKS**
• Information about IUPAC nomenclature

**coordination number** The number of groups, molecules, atoms, or ions surrounding a given atom or ion in a complex or crystal. For instance, in a square-planar complex the central ion has a coordination number of four. In a close-packed crystal (*see* CLOSE PACKING) the coordination number is twelve.

**Copepoda** A class of crustaceans occurring in marine and freshwater habitats. Copepods are usually 0.5–2 mm long and lack both a carapace and compound eyes. Copepods are important members of plankton: some are free-living, feeding on microscopic organisms; others are parasitic. A familiar freshwater genus is *Cyclops*, so named because the members have a single median eye.

**Copernican astronomy** The system of astronomy that was proposed by the Polish astronomer Nicolaus Copernicus in his book *De revolutionibus orbium coelestium*, which was published in the month of his death (in 1543) and first seen by him on his deathbed. It used some elements of *Ptolemaic astronomy, but rejected the notion, then current, that the earth was a stationary body at the centre of the universe. Instead, Copernicus proposed the apparently unlikely concept that the sun was at the centre of the universe and that the earth travelled in a circular orbit about it. In addition Copernicus revived the idea that the movement of the sun and the fixed stars was due to the daily axial rotation of the earth. Galileo's attempts, some 70 years later, to convince the Catholic church that in spite of scriptural authority to the contrary, the Copernican system was correct, met with such stern resistance that *De revolutionibus* was placed on the church's list of forbidden books, where it remained until 1835.

**Copernicus, Nicolaus** (Mikolaj Kopernik; 1473–1543) Polish astronomer, who studied mathematics and optics. By 1514 he had formulated his proposal that the planets, including the earth, orbit the sun in circular paths, although it was not formally published until the year he died. This refutation of an earth-centred universe raised hostile opposition from the church as well as from other astronomers.

**copolymer** *See* POLYMER.

**copper** Symbol Cu. A red-brown *transition element; a.n. 29; r.a.m. 63.546; r.d. 8.92; m.p. 1083.4°C; b.p. 2567°C. Copper has been extracted for thousands of years; it was known to the Romans as cuprum, a name linked with the island of Cyprus. The metal is malleable and ductile and an excellent conductor of heat and electricity. Copper-containing minerals include cuprite ($Cu_2O$) as well as azurite ($2CuCO_3.Cu(OH)_2$), chalcopyrite ($CuFeS_2$), and malachite ($CuCO_3$. $Cu(OH)_2$). Native copper appears in isolated pockets in some parts of the world. The large mines in the USA, Chile, Canada, Zambia, Democratic Republic of Congo, and Peru extract ores containing sulphides, oxides, and carbonates. They are usually worked by smelting, leaching, and electrolysis. Copper metal is used to make electric cables and wires. Its alloys, brass (copper–zinc) and bronze (copper–tin), are used extensively.

Water does not attack copper but in moist atmospheres it slowly forms a characteristic green surface layer (patina). The metal will not react with dilute sulphuric or hydrochloric acids, but with nitric acid oxides of nitrogen are formed. Copper compounds contain the element in the +1 and +2 oxidation states. Copper(I) compounds are mostly white (the oxide is red). Copper(II) salts are blue in solution. The metal also forms a large number of coordination complexes.

**SEE WEB LINKS**
• Information from the WebElements site

**copperas** See IRON(II) SULPHATE.

**copper(I) chloride** A white solid compound, CuCl; cubic; r.d. 4.14; m.p. 430°C; b.p. 1490°C. It is obtained by boiling a solution containing copper(II) chloride, excess copper turnings, and hydrochloric acid. Copper(I) is present as the $[CuCl_2]^-$ complex ion. On pouring the solution into air-free distilled water copper(I) chloride precipitates. It must be kept free of air and moisture as it oxidizes to copper(II) chloride under those conditions.

Copper(I) chloride is essentially covalent and its structure is similar to that of diamond; i.e. each copper atom is surrounded tetrahedrally by four chlorine atoms and vice versa. In the vapour phase, dimeric and trimeric species are present. Copper(I) chloride is used in conjunction with ammonium chloride as a catalyst in the dimerization of ethyne to but-1-ene-3-yne (vinyl acetylene), which is used in the production of synthetic rubber. In the laboratory a mixture of copper(I) chloride and hydrochloric acid is used for converting benzene diazonium chloride to chlorobenzene – the Sandmeyer reaction.

**copper(II) chloride** A brown-yellow powder, $CuCl_2$; r.d. 3.386; m.p. 620°C. It exists as a blue-green dihydrate (rhombic; r.d. 2.54; loses $H_2O$ at 100°C). The anhydrous solid is obtained by passing chlorine over heated copper. It is predominantly covalent and adopts a layer structure in which each copper atom is surrounded by four chlorine atoms at a distance of 0.23 nm and two more at a distance of 0.295 nm. A concentrated aqueous solution is dark brown in colour due to the presence of complex ions such as $[CuCl_4]^{2-}$. On dilution the colour changes to green and then blue because of successive replacement of chloride ions by water molecules, the final colour being that of the $[Cu(H_2O)_6]^{2+}$ ion. The dihydrate can be obtained by crystallizing the solution.

**copper glance** A mineral form of copper(I) sulphide, $Cu_2S$.

**copper(II) nitrate** A blue deliquescent solid, $Cu(NO_3)_2.3H_2O$; r.d. 2.32; m.p. 114.5°C. It may be obtained by reacting either copper(II) oxide or copper(II) carbonate with dilute nitric acid and crystallizing the resulting solution. Other hydrates containing 6 or 9 molecules of water are known. On heating it readily decomposes to give copper(II) oxide, nitrogen dioxide, and oxygen. The anhydrous form can be obtained by reacting copper with a solution of nitrogen dioxide in ethyl ethanoate. It sublimes on heating suggesting that it is appreciably covalent.

**copper(I) oxide** A red insoluble solid, $Cu_2O$; r.d. 6.0; m.p. 1235°C. It is obtained by reduction of an alkaline solution of copper(II) sulphate. Since the addition of alkalis to a solution of copper(II) salt results in the precipitation of copper(II) hydroxide the copper(II) ions are complexed with tartrate ions; under such conditions the concentration of copper(II) ions is so low that the solubility product of copper(II) hydroxide is not exceeded.

When copper(I) oxide reacts with dilute sulphuric acid a solution of copper(II) sulphate and a deposit of copper results, i.e. disproportionation occurs.

$$Cu_2O + 2H^+ \rightarrow Cu^{2+} + Cu + H_2O$$

When dissolved in concentrated hydrochloric acid the $[CuCl_2]^-$ complex ion is formed. Copper(I) oxide is used in the manufacture of rectifiers and the production of red glass.

**copper(II) oxide** A black insoluble solid, CuO; monoclinic; r.d. 6.3; m.p. 1326°C. It is obtained by heating either copper(II) carbonate or copper(II) nitrate. It decomposes on heating above 800°C to copper(I) oxide and oxygen. Copper(II) oxide reacts readily with mineral acids with the formation of copper(II) salts; it is also readily reduced to copper on heating in a stream of hydrogen. Copper(II) oxide is soluble in dilute acids forming blue solutions of cupric salts.

**copper pyrites** See CHALCOPYRITE.

**copper(II) sulphate** A blue crystalline solid, $CuSO_4.5H_2O$; triclinic; r.d. 2.284. The pentahydrate loses $4H_2O$ at 110°C and the fifth $H_2O$ at 150°C to form the white anhydrous compound (rhombic; r.d. 3.6; decomposes above 200°C). The pentahydrate is prepared either by reacting copper(II) oxide or copper(II) carbonate with dilute sulphuric acid; the solution is heated to saturation and the blue pentahydrate crystallizes out on cooling (a few drops of dilute sulphuric acid are generally added to prevent hydrolysis). It is obtained on an industrial scale by forcing air through a hot mixture of copper and dilute sulphuric acid. In the pentahydrate each copper(II) ion is surrounded by four water molecules at the corner of a square, the fifth and sixth octahedral positions are occupied by oxygen atoms from the sulphate anions, and the fifth water molecule is held in place by hydrogen bonding. Copper(II) sulphate has many industrial uses, including the preparation of the Bordeaux mixture (a fungicide) and the preparation of other copper compounds. It is also used in electroplating and textile dying and as a timber preservative. The anhydrous form is used in the detection of traces of moisture.

Copper(II) sulphate pentahydrate is also known as **blue vitriol**.

**coprecipitation** The removal of a substance from solution by its association with a precipitate of some other substance. For example, if A and B are present in solution and a reagent is added such that A forms an insoluble precipitate, then B may be carried down with the precipitate of A, even though it is soluble under the conditions. This can occur by occlusion or absorption.

**copulation** See SEXUAL INTERCOURSE.

**coral** Any of a group of sedentary colonial marine invertebrates belonging to the class Anthozoa of the phylum *Cnidaria. A coral colony consists of individual *polyps within a protective skeleton that they secrete: this skeleton may be soft and jelly-like, horny, or stony. The horny skeleton secreted by corals of the genus *Corallium*, especially *C. rubrum*, constitutes the red, or precious, coral used as a gemstone. The skeleton of stony, or true, corals consists of almost pure calcium carbonate and forms the coral reefs common in tropical seas.

**cordite** An explosive mixture of cellulose nitrate and nitroglycerin, with added plasticizers and stabilizers, used as a propellant for guns.

**core** 1. A rod or frame of magnetic material that increases the inductance of a coil through which it passes. Cores are used in transformers, electromagnets, and the rotors and stators of electrical machines. It may consist of laminated metal, ferrite, or compressed ferromagnetic particles in a matrix of an insulating binder (**dust core**). 2. The inner part of a *nuclear reactor in which the nuclear reaction takes place. 3. The devices that make up the memory in certain types of computer. 4. The central region of a star or planet.

**Coriolis force** A fictitious force sometimes used to simplify calculations involving rotating systems, such as the movement of air, water, and projectiles over the surface of the rotating earth. The concept was first used in 1835 by Gaspard de Coriolis (1792–1843), a French physicist. The daily rotation of the earth means that in 24 hours a point on its equator moves a distance of some 40 000 kilometres, giving it a tangential velocity of about 1670 kilometres per hour. A point at the latitude of, say, Rome, travels a shorter distance in the same time and therefore has a lower tangential velocity – about 1340 km/hr. Air over the equator has the full tangential velocity of 1670 km/hr and as it travels north, say, it will retain this velocity; to an observer outside the earth this would be clear. However, to an observer in Rome it appears to be moving eastwards, because the earth at that point is moving eastwards more slowly than the air. The Coriolis force (which is quite fictitious) is the force that a naive observer thinks is needed to push the air eastwards.

**cork (phellem)** A protective waterproof plant tissue produced by the *cork cambium. It develops in plants undergoing *secondary growth and replaces the epidermis. Its cells, whose walls are impregnated with *suberin, are arranged in radial rows and fit closely together except where the cork is interrupted by *lenticels. Some cork cells become air-filled while others contain deposits of lignin, tannins, and fatty acids, which give the cork a particular colour. The cork oak (*Quercus suber*) produces cork that can be used commercially.

**cork cambium (phellogen)** A type of *cambium arising within the outer layers of the stems of woody plants, usually as a complete ring surrounding the inner tissues. The cells of the cork cambium divide to produce an outer corky tissue (*cork or **phellem**) and an inner secondary cortex (**phelloderm**). Cork, cork cambium, and phelloderm together make up the **periderm**, an impermeable outer layer that protects the inner stem tissues if the outer tissues split as the stem girth increases with age. It thus takes over the functions of the epidermis.

**corm** An underground organ formed by certain plants, e.g. crocus and gladiolus, that enables them to survive from one growing season to the next (see illustration). It consists of a short swollen food-storing stem surrounded by protective scale leaves. One or more buds in the axils of scale leaves produce new foliage leaves and flowers in the subsequent season, using up the food stored in the stem. *Compare* BULB.

**cornea** A transparent layer of tissue, continuous with the *sclerotic, that forms the front part of the vertebrate eye, over the iris and lens. The cornea refracts light entering the eye onto the lens, thus assisting in the focusing of images onto the *retina. *See also* ASTIGMATISM.

**cornification** *See* KERATINIZATION.

**CORN rule** *See* ABSOLUTE CONFIGURATION.

**corolla** The *petals of a flower, collectively, forming the inner whorl of the *perianth. It encircles the stamens and carpels. The form of the corolla is very variable. The petals may either be free (**polypetalous**) or united to form a tube (**gamopetalous** or **sympetalous**).

**corona 1.** The outer part of the sun's atmosphere. Its two main components are the K-corona (or inner corona), with a temperature of about $2 \times 10^6$ K at a height of some 75 000 km, and the F-corona (or outer

corona), which is considerably cooler and extends for several million kilometres into space. **2.** A glowing region of the air surrounding a conductor when the potential gradient near it exceeds a critical value. It is caused by ionization of the air and may be accompanied by hissing sounds. **Corona discharge** (or **point discharge**) occurs at sharp points where the surface charge density is high by the attraction, charging, and consequent repulsion of air molecules.

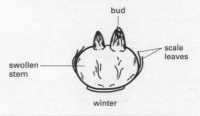
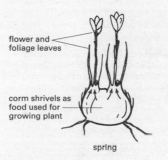
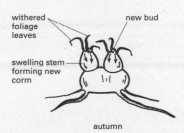

**Corm development.**

**coronal mass ejection (CME; coronal transient)** A huge bubble of energetic gas interlaced with magnetic field lines that is hurled out of the sun's corona into interplanetary space, reaching speeds of between

200 and 1000 km s$^{-1}$ as it expands over several hours. CMEs are often connected with *solar flares or solar *prominences but may occur alone. The frequency of these events reaches its peak at *sunspot maximum (*see* SOLAR CYCLE). On earth CMEs can produce magnetic storms and spectacular aurorae, disrupt radio communications and electricity supplies, and damage power transmission lines and satellites.

**coronary vessels** Two pairs of blood vessels (the coronary arteries and coronary veins) that supply the muscles of the heart itself. The coronary arteries arise from the aorta and divide into branches that encircle the heart. A blood clot in a coronary artery (**coronary thrombosis**) is one of the causes of a 'heart attack'.

**corpus callosum** The sweeping band (commissure) of *white matter that provides a connection between the two halves of the cerebrum in the brain. It enables the transfer of information and learning from one cerebral hemisphere to the other.

**corpuscular theory** *See* LIGHT.

**corpus luteum (yellow body)** The yellowish mass of tissue that forms in the cavity of a *Graafian follicle in the ovary of a mammal after the release of the egg cell. It secretes the hormone *progesterone. Some species of sharks, reptiles, and birds have similar structures in their ovaries but the function of these is less well understood.

**corrosion** Chemical or electrochemical attack on the surface of a metal. *See also* ELECTROLYTIC CORROSION; RUSTING.

**cortex** **1.** (in botany) The tissue between the epidermis and the vascular system in plant stems and roots. It is composed of *parenchyma cells and shows little or no structural differentiation. Cortex is produced by activity of the *apical meristem. *See also* ENDODERMIS. **2.** (in zoology) The outermost layer of tissue of various organs, including the adrenal glands (**adrenal cortex**), kidneys (**renal cortex**), and cerebral hemispheres (*see* CEREBRAL CORTEX).

**corticosteroid** Any of several hormones produced by the cortex of the *adrenal glands. **Glucocorticoids** regulate the use of carbohydrates, proteins, and fats in the body and include *cortisol and *cortisone. **Mineralocorticoids** regulate salt and water balance (*see* ALDOSTERONE).

**corticotrophin** *See* ACTH.

**cortisol (hydrocortisone)** A hormone (*see* CORTICOSTEROID), produced by the adrenal glands, that promotes the synthesis and storage of glucose and is therefore important in the normal response to stress, suppresses or prevents inflammation, and regulates deposition of fat in the body. It is used as treatment for various allergies and for rheumatic fever, certain skin conditions, and adrenal failure (Addison's disease).

**cortisone** A biologically inactive *corticosteroid produced in the adrenal glands from the active hormone *cortisol, which is structurally very similar to it. Cortisone is reconverted to cortisol in the liver and other organs. Cortisone may be administered therapeutically as an inactive precursor (prodrug) of cortisol.

**corundum** A mineral form of aluminium oxide, $Al_2O_3$. It crystallizes in the trigonal system and occurs as well-developed hexagonal crystals. It is colourless and transparent when pure but the presence of other elements gives rise to a variety of colours. *Ruby is a red variety containing chromium; *sapphire is a blue variety containing iron and titanium. Corundum occurs as a rock-forming mineral in both metamorphic and igneous rocks. It is chemically resistant to weathering processes and so also occurs in alluvial (placer) deposits. The second hardest mineral after diamond (it has a hardness of 9 on the Mohs' scale), it is used as an abrasive.

**corymb** A type of flowering shoot (*see* RACEMOSE.INFLORESCENCE) in which the lower flower stalks are longer than the higher ones, resulting in a flat-topped cluster of flowers. Examples are candytuft and wallflower.

**cos** *See* TRIGONOMETRIC FUNCTIONS.

**cosecant** *See* TRIGONOMETRIC FUNCTIONS.

**cosech** *See* HYPERBOLIC FUNCTIONS.

**cosh** *See* HYPERBOLIC FUNCTIONS.

**cosine** *See* TRIGONOMETRIC FUNCTIONS.

**cosine rule** In any triangle, with sides of length $a$, $b$, and $c$, $c^2 = a^2 + b^2 - 2ab\cos\theta$, where $\theta$ is the angle between sides $a$ and $b$.

**cosmic censorship** A hypothesis concerning singularities and *black holes in the general theory of *relativity. It was suggested in 1969 by the British physicist Roger Penrose (1931– ). The **cosmic censorship conjecture** asserts that all singularities in general relativity are hidden behind an event horizon. The conjecture has never been proved mathematically, although there is some evidence for it in many situations. Even if cosmic censorship is not correct, singularities would not be seen experimentally if the singularities are removed by *quantum gravity. It may be that in classical general relativity the cosmic censorship hypothesis is true for 'reasonable' physical situations but that it is possible to construct counter-examples to it for various special situations.

**cosmic neutrino background (CNB)** The neutrino analogue of the cosmic *microwave background radiation, i.e. a background radiation consisting of neutrinos. The CNB has not yet been detected; it is thought that it would have a temperature of about 1.9 kelvin.

**cosmic radiation** High-energy particles that fall on the earth from space. **Primary cosmic rays** consist of nuclei of the most abundant elements, with *protons (hydrogen nuclei) forming by far the highest proportion; electrons, positrons, neutrinos, and gamma-ray photons are also present. The particle energies range from $10^{-11}$ J to 10 J ($10^8$ to $10^{20}$ eV) and as they enter the earth's atmosphere they collide with oxygen and nitrogen nuclei producing **secondary cosmic rays**. The secondary rays consist of elementary particles and gamma-ray photons. A single high-energy primary particle can produce a large **shower** of secondary particles. The sources of the primary radiation are not all known, although the sun is believed to be the principal source of particles with energies up to about $10^{10}$ eV. It is believed that all particles with energies of less than $10^{18}$ eV originate within the Galaxy.

**cosmic string** *See* STRING.

**cosmid** A hybrid *vector, used in *gene cloning, that includes the *cos* gene (from the lambda bacteriophage). It also contains drug resistance *marker genes and other plasmid genes. Cosmids can incorporate larger DNA fragments than either phage or plasmid vectors alone and are especially suitable for cloning large mammalian genes or multigene fragments.

**cosmoid scale** *See* SCALES.

**cosmological constant** A term that can be added to Einstein's field equation for general *relativity theory. The cosmological constant is independent of space and time. It was put forward by Einstein in 1917 to allow for the possibility of a static universe. Although the discovery of the *expansion of the universe removed the original motivation for the cosmological constant, the discovery that the expansion of the universe is accelerating suggests that the constant has a non-zero value, albeit by a factor of $10^{120}$ smaller than expected theoretically. Explaining this small non-zero value is one of the main challenges for theoretical physics at the present time.

**cosmological principle** The claim that on extremely large scales, i.e. much greater scales than those associated with *large-scale structure, the universe is homogeneous and isotropic. There is some evidence that the cosmological principle is valid, notably from the cosmic microwave background radiation, but it cannot be said to have been demonstrated conclusively.

**cosmology** The study of the nature, origin, and evolution of the universe. Various theories concerning the origin and evolution of the universe exist. See Chronology. *See also* BIG-BANG THEORY; STEADY-STATE THEORY; EARLY UNIVERSE.

(((•))) **SEE WEB LINKS**

• The home page of the Caltech Observational Cosmology Group
• A cosmology website run by the University of Cambridge

**cotangent** *See* TRIGONOMETRIC FUNCTIONS.

**coth** *See* HYPERBOLIC FUNCTIONS.

**Cottrell precipitator** An electrostatic precipitator used to remove dust particles from industrial waste gases, by attracting them to charged grids or wires.

**cotyledon (seed leaf)** A part of the embryo in a seed plant. The number of cotyledons is an important feature in classifying plants. Among the flowering plants, the class known as *Monocotyledoneae have a single cotyledon and *Dicotyledoneae have two. Conifers have either two cotyledons, as in *Taxus* (yews), or five to ten, as in *Pinus* (pines). In seeds without an *endosperm, e.g. garden pea and broad bean, the cotyledons store food, which is used in germination. In seeds showing *epigeal germination,

## COSMOLOGY

260 BC — Greek astronomer Aristarchus of Samos (c. 320–230 BC) proposes a sun-centred universe.

c.150 AD — Greek-Egyptian astronomer Ptolemy (2nd century AD) proposes an earth-centred universe.

1543 — Copernicus publishes his sun-centred theory of the universe (solar system).

1576 — English mathematician Thomas Digges (c. 1546–95) proposes that the universe is infinite (because stars are at varying distances).

1584 — Italian philosopher Giordano Bruno (1548–1600) states that the universe is infinite.

1633 — Galileo champions Copernicus's sun-centred universe, but is forced by the Roman Catholic Inquisition to recant.

1854 — Helmholtz predicts the heat death of the universe, based on thermodynamics.

1917 — Einstein proposes a static universe theory.

1922 — Russian astronomer Alexander Friedmann proposes the expanding universe theory.

1927 — George Lemaître proposes the big-bang theory of the universe.

1929 — Edwin Hubble demonstrates the expansion of the universe.

1948 — US physicists George Gamow (1904–68), Ralph Alpher (1921–2007), and Hans Bethe (1906–2005) develop the big-bang theory, and the $\alpha$-$\beta$-$\gamma$ theory of the origin of the elements; Alpher also predicts that the big bang would have produced a microwave background.

British astronomers Hermann Bondi (1919–2005), Thomas Gold (1920–2004), and Fred Hoyle (1915–2001) propose a steady-state theory of the universe in which matter is continuously being formed.

1965 — US astrophysicists Arno Penzias (1933–  ) and Robert Wilson (1936–  ) discover the microwave background radiation.

1980 — US physicist Allan Guth (1947–  ) proposes the inflationary theory of the universe.

1992 — US COBE (Cosmic Background Explorer) astronomical satellite detects ripples in residual cosmic radiation (cited as evidence of the big bang).

2001 — US WMAP (Wilkinson Microwave Anisotropy Probe) launched.

The ekpyrotic universe model is proposed as an alternative to the inflationary model.

2003 — WMAP results give a detailed map of the cosmic microwave background and support the inflationary theory.

---

e.g. runner bean, they emerge above the soil surface and become the first photosynthetic leaves.

**coudé system** *See* TELESCOPE.

**coulomb** Symbol C. The *SI unit of electric charge. It is equal to the charge trans-ferred by a current of one ampere in one second. The unit is named after Charles de Coulomb.

**Coulomb, Charles Augustin de** (1736–1806) French physicist, who served as an army engineer in Martinique before returning to France. He is best known for his 1785 proposal of *inverse-square laws to describe the interaction between electrical charges and between magnets (*see* COULOMB'S LAW), which he proved experimentally using a *torsion balance.

**Coulomb explosion** The sudden disruption of a molecule from which the electrons have been stripped to leave only the nuclei, which repel each other because of their electric charge. The technique of **coulomb explosion imaging** uses this effect to investigate the shape of molecules. A beam of high-energy neutral molecules is produced by first adding electrons, accelerating the ions in an electric field, and then removing the electrons. The beam collides with a thin metal foil having a thickness of about 30 atoms. As the molecules pass through this foil their electrons are scattered and only the nuclei of the molecules emerge. The process occurs within a very short period of time, shorter than the time required for a complete molecular vibration, and consequently the nuclei retain the molecular shape until they are suddenly repulsed by the like charges. The nuclei then impinge on a detector that records their velocity and direction, thus enabling the spatial arrangement of the original molecule to be derived.

**Coulomb field** *See* COULOMB'S LAW.

**Coulomb force** *See* COULOMB'S LAW.

**Coulomb's law** The force (sometimes called the **Coulomb force**) between two charged particles, regarded as point charges $Q_1$ and $Q_2$ a distance $d$ apart, is proportional to the product of the charges and inversely proportional to the square of the distance between them. The law is now usually stated in the form $F = Q_1 Q_2 / 4\pi\varepsilon d^2$, where $\varepsilon$ is the absolute *permittivity of the intervening medium. $\varepsilon = \varepsilon_r \varepsilon_0$, where $\varepsilon_r$ is the relative permittivity (the dielectric constant) and $\varepsilon_0$ is the electric constant. The electric field surrounding a point charge is called the **Coulomb field** and the scattering of charged particles by the Coulomb field surrounding an atomic nucleus is called **Coulomb scattering**. The law was first published by Charles de Coulomb in 1785.

**counter** Any device for detecting and counting objects or events, often incident charged particles or photons. The latter devices usually work by allowing the particle to cause ionization, which creates a current or

voltage pulse. The pulses are then counted electronically. *See* CERENKOV COUNTER; CRYSTAL COUNTER; GEIGER COUNTER; PROPORTIONAL COUNTER; SCINTILLATION COUNTER; SPARK COUNTER. These names are often applied merely to the actual detectors; the ancillary counting mechanism is then called a *scaler.

**countercurrent flow** Flow of two fluids in opposite directions with transfer of heat or matter between them. *Compare* COCURRENT FLOW.

**countercurrent heat exchange** A *counterflow mechanism that enables fluids at different temperatures flowing in channels in opposite directions to exchange their heat content without mixing. An example of countercurrent heat exchange occurs in the feet of penguins, in which heat from blood in the arteries supplying the feet is transferred to blood returning to the body's core in veins that lie close to these arteries. This helps to maintain the core temperature in freezing conditions.

**counterflow** The flow of two fluids in apposed vessels in opposite directions. In biological systems such an arrangement enables the efficient transfer of heat, ions, molecules, etc., from fluids that are rich in these resources to fluids that are deficient in them.

**counter ion** An ion of opposite charge to a given ion. For example, in a crystal of sodium chloride, the chloride ions can be regarded as counter ions to the sodium ions. In certain colloids, the charge on the surface of colloidal particles is neutralized by oppositely charged counter ions in the surrounding solution.

**country rock (host rock)** Older rock that surrounds veins of minerals or an igneous magma *intrusion, such as a *batholith. The extreme heat of the intrusion may cause changes (contact metamorphism) in the composition of the adjacent country rock.

**couple** Two equal and opposite parallel forces applied to the same body that do not act in the same line. The forces create a torque, the *moment of which is equal to the product of the force and the perpendicular distance between them.

**coupling** 1. (in physics) An interaction between two different parts of a system or between two or more systems. Examples of coupling in the *spectra of atoms and nuclei are *Russell–Saunders coupling, *j-j coupling, and spin–orbit coupling. In the spectra of molecules there are five idealized ways (called the **Hund coupling cases**) in which the different types of angular momentum in a molecule (the electron orbital angular momentum $L$, the electron spin angular momentum $S$, and the angular momentum of nuclear rotation $N$) couple to form a resultant angular momentum $J$. (In practice, the coupling for many molecules is intermediate between Hund's cases due to interactions, which are ignored in the idealized cases.) In *solid-state physics an example of coupling is electron–phonon coupling, the analysis of which gives the theories of electrical *conductivity and *superconductivity. *See also* COUPLING CONSTANT. **2.** (in chemistry) A type of chemical reaction in which two molecules join together; for example, the formation of an *azo compound by coupling of a diazonium ion with a benzene ring.

**coupling constant** A physical constant that is a measure of the strength of interaction between two parts of a system or two or more systems. In the case of a *field theory, the coupling constant is a measure of the magnitude of the force exerted on a particle by a field. In the case of a *quantum field theory, a coupling constant is not constant but is a function of energy, the dependence on energy being described by the *renormalization group. *See also* COUPLING; ASYMPTOTIC FREEDOM.

**courtship** Behaviour in animals that plays a part in the initial attraction of a mate or as a prelude to copulation. Courtship often takes the form of *displays that have evolved through *ritualization; some are derived from other contexts (e.g. food begging in some birds). Chemical stimuli (*see* PHEROMONE) are also important in many mammals and insects.

As well as ensuring that the prospective mate is of the same species, the male's courtship performance allows females to choose between different males. The later stages of courtship may involve both partners in an alternating series of displays that inhibit *aggression and fear responses and ensure synchrony of sexual arousal.

**COV** *See* CROSSOVER VALUE.

**covalent bond** *See* CHEMICAL BOND.

**covalent crystal** A crystal in which the atoms are held together by covalent bonds.

Covalent crystals are sometimes called **macromolecular** or **giant-molecular crystals**. They are hard high-melting substances. Examples are diamond and boron nitride.

**covalent radius** An effective radius assigned to an atom in a covalent compound. In the case of a simple diatomic molecule, the covalent radius is half the distance between the nuclei. Thus, in $Cl_2$ the internuclear distance is 0.198 nm so the covalent radius is taken to be 0.099 nm. Covalent radii can also be calculated for multiple bonds; for instance, in the case of carbon the values are 0.077 nm for single bonds, 0.0665 nm for double bonds, and 0.0605 nm for triple bonds. The values of different covalent radii can sometimes be added to give internuclear distances. For example, the length of the bond in interhalogens (e.g. ClBr) is nearly equal to the sum of the covalent radii of the halogens involved. This, however, is not always true because of other effects (e.g. ionic contributions to the bonding).

**covariance** In *statistics, a measure of the association between a pair of random variables. It equals the expected value of the product of their deviations (from the mean value). For two sets of observations $(x_1,y_1),\ldots,(x_n,y_n)$, where $\bar{x}$ is the mean of $x_i$ and $\bar{y}$ is the mean of $y_i$, it is given by

$$(1/n) \sum_{i=1}^{n} (x_i - \bar{x})(y_i - \bar{y})$$

*See also* VARIANCE.

**Cowan, Clyde** *See* PAULI, WOLFGANG ERNST.

**Cowper's glands (bulbourethral glands)** A pair of pea-sized glands that lie beneath the prostate gland. Cowper's glands secrete an alkaline fluid that forms part of the *semen. This fluid neutralizes the acidic environment of the urethra, thereby protecting the sperm. The glands are named after William Cowper (1666–1709). *See also* SEMINAL VESICLE.

**coxa** The first segment, attached to the thorax, of an insect's leg. *See also* FEMUR; TROCHANTER.

**CP invariance** The symmetry generated by the combined operation of changing *charge conjugation ($C$) and *parity ($P$). **CP violation** occurs in weak interactions in kaon decay and in *B-mesons. *See also* CPT THEOREM; TIME REVERSAL.

**CPT theorem** The theorem that the combined operation of changing *charge conjugation *C*, *parity *P*, and *time reversal *T*, denoted **CPT**, is a fundamental *symmetry of relativistic *quantum field theory. No violation of the CPT theorem is known experimentally. When *C*, *P*, and *T* (or any two of them) are violated, the principles of relativistic quantum field theory are not affected; however, violation of **CPT invariance** would drastically alter the fundamentals of relativistic quantum field theory. It is not known whether *superstrings obey versions of the CPT theorem.

**CPU (central processing unit)** The main operating part of a *computer; it includes the **control unit** (CU) and the arithmetic/logic unit (*see* ALU). Its function is to fetch instructions from memory, decode them, and execute the program. It also provides timing signals. An *integrated circuit that has a complete CPU on a single silicon chip is called a microprocessor.

**crack cocaine** *See* COCAINE.

**cracking** The process of breaking down chemical compounds by heat. The term is applied particularly to the cracking of hydrocarbons in the kerosine fraction obtained from *petroleum refining to give smaller hydrocarbon molecules and alkenes. It is an important process, both as a source of branched-chain hydrocarbons suitable for gasoline (for motor fuel) and as a source of ethene and other alkenes. **Catalytic cracking** is a similar process in which a catalyst is used to lower the temperature required and to modify the products obtained.

**cranial nerves** Ten to twelve pairs of nerves in vertebrates that emerge directly from the brain. They supply the sense organs and muscles of the head, neck, and viscera. Examples of cranial nerves include the *optic nerve (II) and the *vagus nerve (X). With the *spinal nerves, the cranial nerves form an important part of the *peripheral nervous system.

**cranial reflex** *See* REFLEX.

**Craniata** A clade or subphylum of chordate animals characterized by a cartilaginous or bony skull protecting the brain and major sense organs. Craniates comprise two major clades: the Myxini (hagfishes) and the Vertebrata (lampreys and jawed vertebrates). Hagfishes have long eel-like bodies lacking fins, vertebrae, and a lower jaw; the notochord is retained into adulthood. Hagfishes

scavenge for food on the seabed, often feeding on dead fish and whales. In vertebrates the notochord is present only in the embryo or larva and becomes replaced by the *vertebral column (backbone) before birth or metamorphosis. This has permitted the vertebrates a greater degree of movement and subsequent improvement in the sense organs and enlargement of the brain. Lampreys (clade Hyperoartia; order Petromyzontiformes) are slender wormlike animals that lack jaws but possess rudimentary vertebrae in adult life. They chiefly feed as parasites on fish using their toothed circular mouth as a sucker to bore into their host and suck its blood. The jawed vertebrates comprise the *Gnathostomata.

**cranium (brain case)** The part of the vertebrate *skull that encloses and protects the brain. It is formed by the fusion of several flattened bones, which have immovable joints (sutures) between them.

**C-reactive protein (CRP)** A protein secreted into blood plasma by the liver in response to inflammation and infection. It binds to the cell walls of certain bacteria and fungi, thereby increasing susceptibility of the target cell to ingestion by phagocytes; it can also activate the *complement cascade, hence triggering destruction of pathogens by this means.

**cream of tartar** *See* POTASSIUM HYDROGENTARTRATE.

**creatine** A compound, synthesized from the amino acids arginine, glycine, and methionine, that occurs in muscle. In the form of **creatine phosphate** (or **phosphocreatine**), it is an important reserve of energy for muscle contraction, which is released when creatine phosphate loses its phosphate and is converted to **creatinine**, which is excreted in the urine (at a rate of 1.2–1.5 g/day in humans). *See also* PHOSPHAGEN.

**creatinine** *See* CREATINE.

**creationist** A proponent of the theory of *special creation.

**creep** The continuous deformation of a solid material, usually a metal, under a constant stress that is well below its yield point. It usually only occurs at high temperatures and the creep characteristics of any material destined to be used under conditions of high stress at high temperatures must be investigated.

**cremocarp** A dry fruit that is a type of
*schizocarp formed from two one-seeded
carpels. The carpels remain separate and
form indehiscent **mericarps** that are attached to a central supporting strand (**car-
pophore**) for some time before dispersal. It
is characteristic of the Umbelliferae (Apiaceae; carrot family).

**crenation** The shrinkage of cells that occurs when the surrounding solution is *hypertonic to the cellular cytoplasm. Water
leaves the cells by *osmosis, which causes
the plasma membrane to wrinkle and the
cellular contents to condense.

**creosote 1. (wood creosote)** An almost
colourless liquid mixture of phenols obtained by distilling tar obtained by the destructive distillation of wood. It is used
medically as an antiseptic and expectorant.
**2. (coal-tar creosote)** A dark liquid mixture
of phenols and cresols obtained by distilling
coal tar. It is used for preserving timber.

**cresols** *See* METHYLPHENOLS.

**Cretaceous** The final geological period of
the Mesozoic era. It extended from about
144 million years ago, following the Jurassic,
to about 65 million years ago, when it was
succeeded by the Palaeogene period. The
name of the period is derived from *creta*
(Latin: chalk) and the Cretaceous was characterized by the deposition of large amounts
of *chalk in western Europe. The Cretaceous
was the time of greatest flooding in the
Mesozoic. Angiosperm plants made their
first appearance on land and in the early
Cretaceous Mesozoic reptiles reached their
peak. At the end of the period there was a
*mass extinction of the dinosaurs, flying rep-
tiles, and ammonites, the cause of which
may be related to environmental changes resulting from collisions of the earth with large
meteorites (*see* ALVAREZ EVENT; IRIDIUM
ANOMALY).

**Creutzfeldt–Jakob disease (CJD)** A disease of humans characterized by dementia
and destruction of brain tissue, first described by the German psychiatrists H. G.
Creutzfeldt (1885–1964) and A. M. Jakob
(1884–1931). It is now known to be caused by
an abnormal *prion protein and is transmissible, although there is also an inherited fa-
milial form. This rare disease typically affects
middle-aged and elderly people and leads to
rapid mental deterioration and death. The
abnormal prion interferes with the structure

of normal prion protein in brain tissue, resulting in accumulations of the protein and
consequent tissue damage. In most cases the
source of infection is unknown. However, it
is well established that infection can result,
for example, via injections of growth hormone derived from infected human cadavers. During the 1990s a novel form of the
disease emerged, called variant CJD, which
typically affects young healthy individuals.
This is thought to be caused by consumption
of beef products derived from cattle infected
with *bovine spongiform encephalopathy.

**CRG process (catalytic rich gas process)**
An industrial process for producing fuel gas
from naphtha and other hydrocarbon
sources. It involves a nickel-based catalyst,
pressures of up to 70 bar, and temperatures
between 250°C and 650°C depending on the
feedstock. The reactions are:

$$C_nH_{2n+2} + nH_2O \rightarrow nCO + (2n+1)H_2$$
$$CO + 3H_2 \rightarrow CH_4 + H_2O$$
$$CO + H_2O \rightarrow CO_2 + H_2$$

The result is a mixture of methane, carbon
monoxide, carbon dioxide, and trace
amounts of ethane and other hydrocarbons.
With partial carbon dioxide removal it is
possible to produce town gas with medium
calorific value containing about 30% $CH_4$,
30% $H_2$, and 2% CO. The process can be used
to produce SNG. In this case there are multiple methanation stages and complete removal of $CO_2$ to give a product containing
about 98.5% $CH_4$, 0.9% $H_2$, and 0.1% CO.

**Crick, Francis Harry Compton**
(1916–2004) British molecular biologist, who
in 1951 teamed up with James *Watson at
Cambridge University to try to find the structure of *DNA. This they achieved in 1953,
using the X-ray diffraction data of Rosalind
Franklin (1920–58) and Maurice Wilkins
(1916–2004). Crick went on to investigate
*codons and the role of transfer *RNA. Crick,
Watson, and Wilkins shared a Nobel Prize in
1962.

**crista 1.** *See* SEMICIRCULAR CANALS. **2.** *See*
MITOCHONDRION.

**cristobalite** A mineral form of *silicon(IV)
oxide, $SiO_2$.

**critical angle** *See* TOTAL INTERNAL REFLEC-
TION.

**critical damping** *See* DAMPING.

**critical density** In astronomy, the mean

density of the universe below which value it is an expanding and continuously open system. The luminous material in the universe (galaxies, etc.) is estimated to account for about 10% of this figure. The remainder is thought to consist mainly of dark matter (*see* MISSING MASS).

**critical group** A large group of related organisms that, although variations exist between them, cannot be divided into smaller groups of equivalent taxonomic rank to the parent group. Critical groups are found among plants that reproduce by *apomixis; for example, the 400 or so species of *Rubus* (brambles, etc.) are regarded as a critical group.

**critical mass** The minimum mass of fissile material that will sustain a nuclear *chain reaction. For example, when a nucleus of uranium–235 disintegrates two or three neutrons are released in the process, each of which is capable of causing another nucleus to disintegrate, so creating a chain reaction. However, in a mass of U–235 less than the critical mass, too many neutrons escape from the surface of the material for the chain reaction to proceed. In the atom bomb, therefore, two or more subcritical masses have to be brought together to make a mass in excess of the critical mass before the bomb will explode.

**critical pressure** The pressure of a fluid in its *critical state; i.e. when it is at its critical temperature and critical volume.

**critical reaction** A nuclear *chain reaction in which, on average, one transformation causes exactly one other transformation so that the chain reaction is self-sustaining. If the average number of transformations caused by one transformation falls below one, the reaction is **subcritical** and the chain reaction ceases; if it exceeds one the reaction is **supercritical** and proceeds explosively.

**critical state** The state of a fluid in which the liquid and gas phases both have the same density. The fluid is then at its *critical temperature, *critical pressure, and *critical volume.

**critical temperature 1.** The temperature above which a gas cannot be liquefied by an increase of pressure. *See also* CRITICAL STATE. **2.** *See* TRANSITION POINT.

**critical volume** The volume of a fixed mass of a fluid in its *critical state; i.e. when

it is at its critical temperature and critical pressure. The **critical specific volume** is its volume per unit mass in this state: in the past this has often been called the critical volume.

**CRO** *See* CATHODE-RAY OSCILLOSCOPE.

**Cromagnon man** The earliest form of modern humans (*Homo sapiens*), which is believed to have appeared in Europe about 35 000 years ago and possibly at least 70 000 years ago in Africa and Asia. Fossils indicate that these hominids were taller and more delicate than *Neanderthal man, which they replaced. They used intricately worked tools of stone and bone and left the famous cave drawings at Lascaux in the Dordogne. The name is derived from the site at Cromagnon, France, where the first fossils were found in 1868.

**Crookes, Sir William** (1832–1919) British chemist and physicist, who in 1861 used *spectroscopy to discover *thallium and in 1875 invented the radiometer. He also developed an improved vacuum tube (**Crookes' tube**) for studying gas discharges.

**crop 1.** A plant that is cultivated for the purpose of harvesting its seeds, roots, leaves, or other parts that are useful to humans. *See* AGRICULTURE. **2.** An enlarged portion of the anterior section of the alimentary canal in some animals, in which food may be stored and/or undergo preliminary digestion. The term is most commonly applied to the thin-walled sac in birds between the oesophagus and the *proventriculus. In female pigeons the crop contains glands that secrete **crop milk**, used to feed nestlings.

**crop rotation** An agricultural practice in which different crops are cultivated in succession on the same area of land over a period of time so as to maintain soil fertility and reduce the adverse effects of pests. Legumes are important in the rotation as they are a source of nitrogen for the soil (*see* NITROGEN FIXATION; ROOT NODULE). In the UK, other crops that may be included in a typical four-stage rotation are wheat, barley, and root crops. However, the use of pesticides enables the monoculture of crops in modern farming systems (*see* AGRICULTURE).

**cross 1.** A mating between two selected individuals. Controlled crosses are made for many reasons, e.g. to investigate the inheritance of a particular characteristic or to improve a livestock or crop variety. *See also*

BACK CROSS; RECIPROCAL CROSS; TEST CROSS.
**2.** An organism resulting from such a mating.

**cross-fertilization** *See* FERTILIZATION.

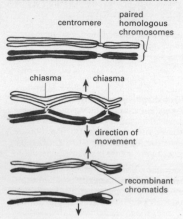

**Crossing over.** At two chiasmata in a pair of homologous chromosomes.

**crossing over** An exchange of portions of chromatids between *homologous chromosomes. As the chromosomes begin to move apart at the end of the first prophase of *meiosis, they remain in contact at a number of points (*see* CHIASMA). At these points the chromatids break and rejoin in such a way that sections are exchanged (see illustration). Crossing over thus alters the pattern of genes in the chromosomes. *See* RECOMBINATION.

**cross linkage** A short side chain of atoms linking two longer chains in a polymeric material.

**crossover value (COV)** The percentage of linked genes (*see* LINKAGE) that are exchanged during the process of *crossing over during the first prophase of *meiosis. The COV can be calculated by the percentage of offspring that show *recombination and is used to map the genes on a chromosome (*see* CHROMOSOME MAP). A small COV for a given pair of genes indicates that the genes are situated close together on the chromosome.

**cross-pollination** *See* POLLINATION.

**cross product** *See* VECTOR PRODUCT.

**cross section** **1.** A plane surface formed by cutting a solid, especially by cutting at right angles to its longest axis. **2.** The area of such a surface. **3.** A measure of the probability that a collision will occur between a beam of radiation and a particular particle, expressed as the effective area presented by the particle in that particular process. It is measured in square metres or *barns. The **elastic cross section** amounts for all elastic scattering in which the radiation loses no energy to the particle. The **inelastic cross section** accounts for all other collisions. It is further subdivided to account for specific interactions, such as the **absorption cross section**, **fission cross section**, **ionization cross section**, etc.

**crown ethers** Organic compounds with molecules containing large rings of carbon and oxygen atoms. The crown ethers are

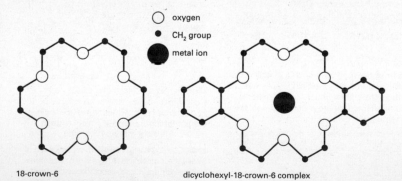

○ oxygen
● CH₂ group
⬤ metal ion

18-crown-6          dicyclohexyl-18-crown-6 complex

**Crown ethers.**

macrocyclic polyethers. The first to be synthesized was the compound 18-crown-6, which consists of a ring of six –$CH_2$–$CH_2$–O– units (i.e. $C_{12}H_{24}O_6$). The general method of naming crown ethers is to use the form $n$-crown-$m$, where $n$ is the number of atoms in the ring and $m$ is the number of oxygen atoms. Substituted crown ethers can also be made. The crown ethers are able to form strongly bound complexes with metal ions by coordination through the oxygen atoms. The stability of these complexes depends on the size of the ion relative to the cavity available in the ring of the particular crown ether. Crown ethers also form complexes with ammonium ions ($NH_4^+$) and alkyl ammonium ions ($RNH_3^+$). They can be used for increasing the solubility of ionic salts in nonpolar solvents. For example, dicyclohexyl-18-crown-6 complexes with the potassium ion of potassium permanganate and allows it to dissolve in benzene, giving a purple neutral solution that can oxidize many organic compounds. They also act as catalysts in certain reactions involving organic salts by complexing with the positive metal cation and thereby increasing its separation from the organic anion, which shows a consequent increase in activity. Some of the uses of crown ethers depend on their selectivity for specific sizes of anions. Thus they can be used to extract specific ions from mixtures and enrich isotope mixtures. Their selectivity also makes them useful analytical reagents. *See also* CRYPTANDS.

**crucible** A dish or other vessel in which substances can be heated to a high temperature.

**crude oil** *See* PETROLEUM.

**Crustacea** A subphylum of *arthropods containing over 35 000 species distributed worldwide, mainly in freshwater and marine habitats, where they constitute a major component of plankton. Crustaceans include shrimps, crabs, lobsters, etc. (*see* DECAPODA) and the terrestrial woodlice, all of which belong to the class Malacostraca; the barnacles (class Cirripedia); the water fleas (*see* DAPHNIA), fairy shrimps, and tadpole shrimps (class Branchiopoda); and the copepods (*see* COPEPODA). The segmented body usually has a distinct head (bearing *compound eyes, two pairs of *antennae, and various mouthparts), thorax, and abdomen, and is protected by a shell-like carapace. Each body segment may bear a pair of branched (**bira-**

**mous**) appendages used for locomotion, as gills, and for filtering food particles from the water. Appendages in the head region are modified to form jaws and in the abdominal region are often reduced or absent. Typically, the eggs hatch to produce a free-swimming **nauplius** larva. This develops either by a series of moults or undergoes metamorphosis to the adult form.

**((⊕)) SEE WEB LINKS**

- Overview of crustacean phylogeny at the Tree of Life web project

**cryobiology** The study of the effects of very low temperatures on organisms, tissues, and cells. The ability of some animal tissues to remain viable in a frozen state (**cryopreservation**) enables them to be preserved by freezing for future use as *grafts.

**cryogenic pump** A *vacuum pump in which pressure is reduced by condensing gases on surfaces maintained at about 20 K by means of liquid hydrogen or at 4 K by means of liquid helium. Pressures down to $10^{-8}$ mmHg ($10^{-6}$ Pa) can be maintained; if they are used in conjunction with a *diffusion pump, pressures as low as $10^{-15}$ mmHg ($10^{-13}$ Pa) can be reached.

**cryogenics** The study of very low temperatures and the techniques for producing them. Objects are most simply cooled by placing them in a bath containing liquefied gas maintained at a constant pressure. In general, a liquefied gas can provide a constant bath temperature from its triple point to its critical temperature and the bath temperature can be varied by changing the pressure above the liquid. The lowest practical temperature for a liquid bath is 0.3 K. Refrigerators (*see* REFRIGERATION) consist essentially of devices operating on a repeated cycle, in which a low-temperature reservoir is a continuously replenished liquid bath. Above 1 K they work by compressing and expanding suitable gases. Below this temperature liquids or solids are used and by *adiabatic demagnetization it is possible to reach $10^{-6}$ K.

**cryohydrate** A eutectic mixture of ice and some other substance (e.g. an ionic salt) obtained by freezing a solution.

**cryolite** A rare mineral form of sodium aluminofluoride, $Na_3AlF_6$, which crystallizes in the monoclinic system. It is usually white but may also be colourless. The only important occurrence of the mineral is in Green-

C

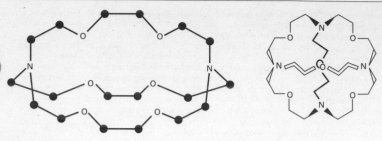

(2,2,2) cryptand                    spherical cryptand

**Cryptands.**

land. It is used chiefly to lower the melting point of alumina in the production of aluminium.

**cryometer** A thermometer designed to measure low temperatures. *Thermocouples can be used down to about 1 K and *resistance thermometers can be used at 0.01 K. Below this magnetic thermometers (0.001 K) and nuclear-resonance thermometers ($3 \times 10^{-7}$ K) are required.

**cryophyte** An organism that can live in ice and snow. Most cryophytes are algae, including the green alga *Chlamydomonas nivalis* and some diatoms, but they also include certain dinoflagellates, mosses, bacteria, and fungi.

**cryoscopic constant** *See* DEPRESSION OF FREEZING POINT.

**cryoscopy** The use of *depression of freezing point to determine relative molecular masses.

**cryostat** A vessel enabling a sample to be maintained at a very low temperature. The *Dewar flask is the most satisfactory vessel for controlling heat leaking in by radiation, conduction, or convection. Cryostats usually consist of two or more Dewar flasks nesting in each other. For example, a liquid nitrogen bath is often used to cool a Dewar flask containing a liquid helium bath.

**cryotron** A switch that relies on *superconductivity. It consists of a coil of wire of one superconducting material surrounding a straight wire of another superconducting material; both are immersed in a liquid-helium bath. A current passed through the coil creates a magnetic field, which alters the superconducting properties of the central

wire, switching its resistance from zero to a finite value. Cryotron switches can be made very small and take very little current.

**cryptands** Compounds with large three-dimensional molecular structures containing ether chains linked by three-coordinate nitrogen atoms. Thus cryptands are macropolycyclic polyaza-polyethers. For example, the compound (2,2,2)-cryptand has three chains of the form

$$-CH_2CH_2OCH_2CH_2OCH_2CH_2-.$$

These chains are linked at each end by a nitrogen atom. Cryptands, like the *crown ethers, can form coordination complexes with ions that can fit into the cavity formed by the open three-dimensional structure, i.e. they can 'cryptate' the ion. Various types of cryptand have been produced having both spherical and cylindrical cavities. The cryptands have the same kind of properties as the crown ethers and the same uses. In general, they form much more strongly bound complexes and can be used to stabilize unusual ionic species. For example, it is possible to produce the negative $Na^-$ ion in the compound [(2,2,2)-cryptand-Na]$^+$Na$^-$, which is a gold-coloured crystalline substance stable at room temperature. Cluster ions, such as $Pb_5^{2-}$, can be similarly stabilized.

**cryptic coloration** The type of colouring or marking of an animal that helps to camouflage it in its natural environment. It may enable the animal to blend with its background or, like the stripes of zebras and tigers, help to break up the outline of its body.

**crypts of Lieberkühn (intestinal glands)** Tubular glands that lie between the finger-like projections (*see* VILLUS) of the inner surface of the small intestine. The cells of these glands (called **Paneth cells**) secrete *intestinal juice as they gradually migrate along the side of the crypt and the villus; they are eventually shed into the lumen of the intestine.

**crystal** A solid with a regular polyhedral shape. All crystals of the same substance grow so that they have the same angles between their faces. However, they may not have the same external appearance because different faces can grow at different rates, depending on the conditions. The external form of the crystal is referred to as the **crystal habit**. The atoms, ions, or molecules forming the crystal have a regular arrangement and this is the **crystal structure**.

**crystal counter** A type of solid-state *counter in which a potential difference is applied across a crystal; when the crystal is struck by an elementary particle or photon, the electron–ion pairs created cause a transient increase in conductivity. The resulting current pulses are counted electronically.

**crystal defect** An imperfection in the regular lattice pattern of a crystal. See Feature (pp 210–211).

**crystal-field theory** A theory of the electronic structures of inorganic *complexes, in which the complex is assumed to consist of a central metal atom or ion surrounded by ligands that are ions. For example, the complex $[PtCl_4]^{2-}$ is thought of as a $Pt^{2+}$ ion surrounded by four $Cl^-$ ions at the corners of a square. The presence of these ions affects the energies of the $d$-orbitals, causing a splitting of energy levels. The theory can be used to explain the spectra of complexes and their magnetic properties. Ligand-field theory is a development of crystal-field theory in which the overlap of orbitals is taken into account. Crystal-field theory was initiated in 1929 by the German-born US physicist Hans Albrecht Bethe (1906–2005) and extensively developed in the 1930s.

**crystal habit** *See* CRYSTAL.

**crystal lattice** The regular pattern of atoms, ions, or molecules in a crystalline substance. A crystal lattice can be regarded as produced by repeated translations of a **unit cell** of the lattice. *See also* CRYSTAL SYSTEM.

(globe icon) SEE WEB LINKS
• An interactive version of the structures of a range of crystals from a US Navy website
• Crystal lattice structures

**crystalline** Having the regular internal arrangement of atoms, ions, or molecules characteristic of crystals. Crystalline materials need not necessarily exist as crystals; all metals, for example, are crystalline although they are not usually seen as regular geometric crystals.

**crystallite** A small crystal, e.g. one of the small crystals forming part of a microcrystalline substance.

**crystallization** The process of forming crystals from a liquid or gas.

**crystallography** The study of crystal form and structure. *See also* X-RAY CRYSTALLOGRAPHY.

**crystalloids** *See* COLLOIDS.

**crystal meth** *See* AMPHETAMINE.

**crystal microphone** A microphone in which the sound waves fall on a plate of Rochelle salt or similar material with piezoelectric properties, the variation in pressure being converted into a varying electric field by the *piezoelectric effect. Crystal microphones have a good high-frequency response and are nondirectional; they are now rarely used except when their cheapness is important.

**crystal oscillator (piezoelectric oscillator)** An oscillator in which a piezoelectric crystal is used to determine the frequency. An alternating electric field applied to two metallic films sputtered onto the parallel faces of a crystal, usually of quartz, causes it to vibrate at its natural frequency; this frequency can be in the kilohertz or megahertz range, depending on how the crystal is cut. The mechanical vibrations in turn create an alternating electric field across the crystal that does not suffer from frequency drift. The device can be used to replace the tuned circuit in an oscillator by providing the resonant frequency or it can be coupled to the oscillator circuit, which is tuned approximately to the crystal frequency. In this type, the crystal prevents frequency drift. The device is widely used in *quartz clocks and watches.

**crystal pick-up** A pick-up in a record player in which the mechanical vibrations

## CRYSTAL DEFECTS

A crystal *lattice is formed by a repeated arrangement of atoms, ions, or molecules. Within one cubic centimetre of material one can expect to find up to $10^{22}$ atoms and it is extremely unlikely that all of these will be arranged in perfect order. Some atoms will not be exactly in the right place with the result that the lattice will contain *defects. The presence of defects within the crystal structure has profound consequences for certain bulk properties of the solid, such as the electrical resistance and the mechanical strength.

### Point defects

Local crystal defects called **point defects**, appear as either impurity atoms or gaps in the lattice. Impurity atoms can occur in the lattice either at **interstitial sites** (between atoms in a non-lattice site) or at **substitutional sites** (replacing an atom in the host lattice). Lattice gaps are called **vacancies** and arise when an atom is missing from its site in the lattice. Vacancies are sometimes called **Schottky defects**. A vacancy in which the missing atom has moved to an interstitial position is known as a **Frenkel defect**.

### Colour centres

In ionic crystals, the ions and vacancies always arrange themselves so that there is no build-up of one type of charge in any small volume of the crystal. If ions or charges are introduced into or removed from the lattice, there will, in general, be an accompanying rearrangement of the ions and their outer valence electrons. This rearrangement is called **charge compensation** and is most dramatically observed in **colour centres**. If certain crystals are irradiated with X-rays, gamma rays, neutrons, or electrons a colour change is observed. For example, diamond may be coloured blue by electron bombardment and quartz may be coloured brown by irradiation with neutrons. The high-energy radiation produces defects in the lattice and, in an attempt to maintain charge neutrality, the crystal undergoes some measure of charge compensation. Just as electrons around an atom have a series of discrete permitted energy levels, so charges residing at point defects exhibit sets of discrete levels, which are separated from one another by energies corresponding to wavelengths in the visible region of the spectrum. Thus light of certain wavelengths can be absorbed at the defect sites, and the material appears to be coloured. Heating the irradiated crystal can, in many cases, repair the irradiation damage and the crystal loses its coloration.

### Dislocations

Non-local defects may involve entire planes of atoms. The most important of these is called a **dislocation.** Dislocations are essentially **line-defects**; that is,

Formation of a Schottky defect          Formation of a Frenkel defect

Point defects in a two-dimensional crystal

there is an incomplete plane of atoms in the crystal lattice. In 1934, Taylor, Orowan, and Polanyi independently proposed the concept of the dislocation to account for the mechanical strength of metal crystals. Their microscopic studies revealed that when a metal crystal is plastically deformed, the deformation does not occur by a separation of individual atoms but rather by a slip of one plane of atoms over another plane. Dislocations provide a mechanism for this slipping of planes that does not require the bulk movement of crystal material. The passage of a dislocation in a crystal is similar to the movement of a ruck in a carpet. A relatively large force is required to slide the carpet as a whole. However, moving a ruck over the carpet can inch it forward without needing such large forces. This movement of dislocations is called **plastic flow**.

## Strength of materials

In practice most metal samples are **polycrystalline;** that is they consist of many small crystals or grains at different angles to each other. The boundary between two such grains is called a **grain boundary**. The plastic flow of dislocations may be hindered by the presence of grain boundaries, impurity atoms, and other dislocations. Pure metals produced commercially are generally too weak to be of much mechanical use. The weakness of these samples can be attributed to the ease with which the dislocations are able to move within the sample. Slip, and therefore deformation, can then occur under relatively low stresses. Impurity atoms, other dislocations, and grain boundaries can all act as obstructions to the slip of atomic planes. Traditionally, methods of making metals stronger involved introducing defects that provide regions of disorder in the material. For example, in an alloy, such as steel, impurity atoms (e.g. carbon) are introduced into the lattice during the forging process. The perfection of the iron lattice structure is disturbed and the impurities oppose the dislocation motion. This makes for greater strength and stiffness.

The complete elimination of dislocations may seem an obvious way to strengthen materials. However, this has only proved possible for hair-like single crystal specimens called **whiskers**. These whiskers are only a few micrometers thick and are seldom more than a few millimetres long, nevertheless their strength approaches the theoretical value.

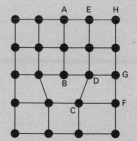 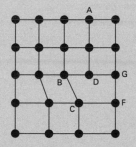

**Dislocation in a two-dimensional crystal.** The extra plane of atoms AB causes strain at bond CD. On breaking, the bond flips across to form CB. This incremental movement shifts the dislocation across so that the overall effect is to slide the two planes BDG and CF over each other.

produced by undulations in the record groove are transmitted to a piezoelectric crystal, which produces a varying electric field of the same frequency as the sound. This signal is amplified and fed to loud-speakers in order to recreate the sound.

**crystal structure** *See* CRYSTAL.

**crystal system** A method of classifying crystalline substances on the basis of their unit cell. There are seven crystal systems. If the cell is a parallelopiped with sides $a$, $b$, and $c$ and if $\alpha$ is the angle between $b$ and $c$, $\beta$ the angle between $a$ and $c$, and $\gamma$ the angle between $a$ and $b$, the systems are:
(1) **cubic** $a=b=c$ and $\alpha=\beta=\gamma=90°$
(2) **tetragonal** $a=b\neq c$ and $\alpha=\beta=\gamma=90°$
(3) **rhombic** (or **orthorhombic**) $a\neq b\neq c$ and $\alpha=\beta=\gamma=90°$
(4) **hexagonal** $a=b\neq c$ and $\alpha=\beta=\gamma=90°$
(5) **trigonal** $a=b\neq c$ and $\alpha=\beta=\gamma\neq90°$
(6) **monoclinic** $a\neq b\neq c$ and $\alpha=\gamma=90°\neq\beta$
(7) **triclinic** $a=b=c$ and $\alpha\neq\beta\neq\gamma$

**crystal test** A type of *presumptive test in which a substance is identified by the forma-tion of characteristic crystals when a certain reagent is added. Usually, such tests are con-ducted using a microscope (**microcrystal test**). An example is the *acetone–chlor–haemin test for blood.

**CSF** 1. *See* CEREBROSPINAL FLUID. 2. *See* COLONY-STIMULATING FACTOR.

**CS gas** The vapour from a white solid, $C_6H_4(Cl)CH:C(CN)_2$, causing tears and chok-ing, used in 'crowd control'.

**CT scanner (computerized tomography scanner)** *See* TOMOGRAPHY.

**CU** *See* CONTROL UNIT.

**cubane** A crystalline hydrocarbon, $C_8H_8$;

r.d. 1.29; m.p. 131°C. It has a novel structure with eight carbon atoms at the corners of a cube, each attached to a hydrogen. Cubane was first synthesized in 1964 by Philip Eaton. The C–C–C bond angle of 90° is highly strained and cubane and its derivatives have been investigated as high-energy fuels and explosives. In particular, **octanitrocubane**, in which the hydrogen atoms are replaced by $-NO_2$ groups, is possibly the most powerful chemical explosive known, although, so far, only small amounts have been synthesized. It decomposes to carbon dioxide and nitro-gen:

$$C_8(NO_2)_8 \rightarrow 8CO_2 + 4N_2$$

**cubewano** A *Kuiper belt object (KBO) that is in direct orbit around the sun but is not held in orbital resonance with Neptune or any other planet. Cubewanos take their name from QB$_1$ – now (15760) QB$_1$ – the first known KBO, which was discovered in 1992. Cubewanos orbit the sun at mean distances of between 42 and 48 astronomical units. The dwarf planet (136472) Makemake (at an estimated 1500 km in diameter) is the largest cubewano so far known.

**cubic close packing** *See* CLOSE PACKING.

**cubic crystal** A crystal in which the unit cell is a cube (*see* CRYSTAL SYSTEM). There are three possible packings for cubic crystals: **simple cubic**, **face-centred cubic**, and **body-centred cubic**. See illustration.

((())) SEE WEB LINKS
• An interactive version of a simple cubic crystal
• An interactive version of a body-centred cubic crystal
• An interactive version of a face-centred cubic crystal

**cubic equation** An equation in which the

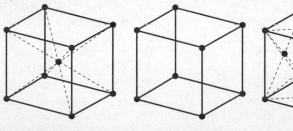

body-centred          simple cubic          face-centred

**Cubic crystal structures.**

highest power of the variable is three. It has the general form $ax^3 + bx^2 + cx + d = 0$ and, in general, is satisfied by three values of $x$.

**cubic expansivity** *See* EXPANSIVITY.

**cubic zircona (CZ)** A crystal form of zircon(IV) oxide (zircon dioxide, $ZrO_2$) made by fusing $ZrO_2$ and allowing it to cool under controlled conditions. It is used as an inexpensive diamond substitute in jewellery. Often it is erroneously called *zircon.

**cultivar** A plant that has been developed and maintained by cultivation as a result of agricultural or horticultural practices. The term is derived from *culti*vated *var*iety.

**cultivation** The planting and breeding of crop plants in *agriculture and horticulture. It involves the investigation of new means of increasing crop yield and quality.

**culture** A batch of cells, which can be microorganisms or of animal or plant origin, that are grown under specific conditions of nutrient levels, temperature, pH, oxygen levels, osmotic factors, light, pressure, and water content. Cultures of cells are prepared in the laboratory for a wide spectrum of scientific research. A *culture medium provides the appropriate conditions for growth. *See also* CONTINUOUS CULTURE; TISSUE CULTURE.

**culture medium** A nutrient material, either solid or liquid, used to support the growth and reproduction of microorganisms or to maintain tissue or organ cultures. *See also* AGAR.

**cumene process** An industrial process for making phenol from benzene. A mixture of benzene vapour and propene is passed over a phosphoric acid catalyst at 250°C and high pressure

$$C_6H_6 + CH_3CH:CH_2 \rightarrow C_6H_5CH(CH_3)_2$$

The product is called **cumene**, and it can be oxidized in air to a peroxide, $C_6H_5C(CH_3)_2O_2H$. This reacts with dilute acid to give phenol ($C_6H_5OH$) and propanone (acetone, $CH_3OCH_3$), which is a valuable by-product.

**cupellation** A method of separating noble metals (e.g. gold or silver) from base metals (e.g. lead) by melting the mixture with a blast of hot air in a shallow porous dish (the **cupel**). The base metals are oxidized, the oxide being carried away by the blast of air or absorbed by the porous container.

**cuprammonium ion** The tetraam-minecopper(II) ion $[Cu(NH_3)_4]^{2+}$. *See* AMMINE.

**cupric compounds** Compounds containing copper in its higher (+2) oxidation state; e.g. cupric chloride is copper(II) chloride ($CuCl_2$).

**cuprite** A red mineral cubic form of copper(I) oxide, $Cu_2O$; an important ore of copper. It occurs where deposits of copper have been subjected to oxidation. The mineral has been mined as a copper ore in Chile, Democratic Republic of Congo, Bolivia, Australia, Russia, and the USA.

**cupronickel** A type of corrosion-resistant alloy of copper and nickel containing up to 45% nickel.

**cuprous compounds** Compounds containing copper in its lower (+1) oxidation state; e.g. cuprous chloride is copper(I) chloride ($CuCl$).

**cupule** 1. A hard or membranous cup-shaped structure formed from bracts and enclosing various fruits, such as the hazelnut and acorn. 2. A structure in club mosses (*Lycopodium* species) that protects the gemma (resting bud) during its development. It is composed of six leaflike structures. 3. The bright red tissue around the seed of yew (*Taxus*), forming the yew 'berry'.

**curare** A resin obtained from the bark of South American trees of the genera *Strychnos* and *Chondrodendron* that causes paralysis of voluntary muscle. It acts by blocking the action of the neurotransmitter *acetylcholine at *neuromuscular junctions. Curare is used as an arrow poison by South American Indians and was formerly used as a muscle relaxant in surgery.

**curd** The solid component produced by the coagulation of milk during the manufacture of cheese. After being pasteurized, milk is cooled down and a culture of lactic acid bacteria is added to ferment the milk sugar, lactose, to lactic acid. The resulting decrease in pH causes casein, a milk protein, to coagulate, a process known as **curdling**. The solid curds are then separated from the liquid component, known as **whey**, and inoculated with different types of microbes to produce different cheeses.

**curie** The former unit of *activity (*see* RADIATION UNITS). It is named after Pierre Curie.

**Curie, Marie** (Marya Sklodowska;

1867–1934) Polish-born French chemist, who went to Paris in 1891. She married the physicist **Pierre Curie** (1859–1906) in 1895 and soon began work on seeking radioactive elements other than uranium in pitchblende (to account for its unexpectedly high radioactivity). By 1898 she had discovered *radium and *polonium, although it took her four years to purify them. In 1903 the Curies shared the Nobel Prize for physics with Henri *Becquerel, who had discovered radioactivity. In 1911 Marie Curie was awarded the Nobel Prize for chemistry.

(((⊕))) SEE WEB LINKS

• A comprehensive website about her life and work run by The American Institute of Physics Center for History of Physics

**Curie point (Curie temperature)** The temperature at which a ferromagnetic substance loses its ferromagnetism and becomes only paramagnetic. For iron the Curie point is 760°C and for nickel 356°C.

**Curie's law** The susceptibility ($\chi$) of a paramagnetic substance is proportional to the thermodynamic temperature ($T$), i.e. $\chi = C/T$, where $C$ is the Curie constant. A modification of this law, the **Curie–Weiss law**, is more generally applicable. It states that $\chi = C/(T - \theta)$, where $\theta$ is the Weiss constant, a characteristic of the material. The law was first proposed by Pierre Curie and modified by another French physicist, Pierre-Ernest Weiss (1865–1940).

**curium** Symbol Cm. A radioactive metallic transuranic element belonging to the *actinoids; a.n. 96; mass number of the most stable isotope 247 (half-life $1.64 \times 10^7$ years); r.d. (calculated) 13.51; m.p. 1340±40°C. There are nine known isotopes. The element was first identified by Glenn Seaborg (1912–99) and associates in 1944 and first produced by L. B. Werner and I. Perlman in 1947 by bombarding americium–241 with neutrons.

(((⊕))) SEE WEB LINKS

• Information from the WebElements site

**curl (rot)** The *vector product of the *gradient operator with a vector. For a vector $u$ that has components $u_1$, $u_2$, and $u_3$ in the $x$, $y$, and $z$ directions (with respective unit vectors $i$, $j$, and $k$), and is a function of $x$, $y$, and $z$, the curl is given by:

curl $u = \nabla \times u$

$= (\partial u_3/\partial y - \partial u_2/\partial z)\,i +$

$(\partial u_1/\partial z - \partial u_3/\partial x)\,j +$

$(\partial u_2/\partial x - \partial u_1/\partial y)\,k.$

*See also* DIVERGENCE.

**current** Symbol $I$. A flow of electric charge through a conductor. The current at a particular cross section is the rate of flow of charge. The charge may be carried by electrons, ions, or positive holes (*see* CHARGE CARRIER). The unit of current is the ampere. *See also* CONVENTIONAL CURRENT.

**current balance** An instrument used to measure a current absolutely, on the basis of the definition of the ampere. An accurate form consists of a beam balance with similar coils attached to the ends of the balance arms. Fixed coils are situated above and below these two coils. The six coils are then connected in series so that a current passing through them creates a torque on the beam, which is restored to the horizontal by means of a rider. From the position and weight of the rider, and the geometry of the system, the current can be calculated.

**current density** 1. The current flowing through a conductor per unit cross-sectional area, measured in amperes per square metre. 2. The current flowing through an electrolyte per unit area of electrode.

**cusp** 1. (in dentistry) A sharp raised protuberance on the surface of a *molar tooth. The cusps of opposing molars (i.e. on opposite jaws) are complementary to each other, which increases the efficiency of grinding food during chewing. 2. (in anatomy) A flap forming part of a *valve. 3. (in mathematics) A point at which two arcs of a curve intersect.

**cuticle** 1. (in botany) The continuous waxy layer that covers the aerial parts of a plant. Composed of *cutin, it is secreted by the *epidermis and its primary function is to prevent water loss. 2. (in zoology) A layer of horny noncellular material covering, and secreted by, the epidermis of many invertebrates. It is usually made of a collagen-like protein or of *chitin and its main function is protection. In arthropods it is also strong enough to act as a skeleton (*see* EXOSKELETON) and in insects it reduces water loss. Growth is allowed by moulting of the cuticle (*see* ECDYSIS).

**cuticularization** The secretion by the outer (epidermal) layer of cells of plants and many invertebrates of substances that then harden to form a *cuticle.

**cutin** A polymer of long-chain fatty acids that forms the main constituent of the *cuticle of mature epidermal plant cells. Cutin polymers are cross-linked to form a network, which is embedded in a matrix of waxes. The deposition of cutin (**cutinization**) reduces water loss by the plant and helps prevent the entry of pathogens. *See also* SUBERIN.

**cutinization** The deposition of *cutin in plant cell walls, principally in the outermost layers of leaves and young stems.

**cutis** *See* DERMIS.

**cutting** A part of a plant, such as a bud, leaf, or a portion of a root or shoot, that, when detached from the plant and inserted in soil, can take root and give rise to a new daughter plant. Taking or striking cuttings is a horticultural method for propagating plants. *See also* VEGETATIVE PROPAGATION.

**cutting agent (adulterant)** A substance used to dilute illegal drugs such as heroin and cocaine. Examples include flour, starch, sugar, and caffeine.

**Cuvier, George Léopold Chrétien Frédéric Dagobert** (1769–1832) French anatomist, who became professor at the Collège de France in 1799, moving in 1802 to the Jardin de Plantes. Cuvier extended the classification system of *Linnaeus, adding the category *phylum and concentrating on the taxonomy of fishes. He also initiated the classification of fossils and established the science of palaeontology.

**cyanamide** 1. An inorganic salt containing the ion $CN_2^{2-}$. *See* CALCIUM CYANAMIDE. 2. A colourless crystalline solid, $H_2NCN$, made by the action of carbon dioxide on hot sodamide. It is a weakly acidic compound (the parent acid of cyanamide salts) that is soluble in water and ethanol. It is hydrolysed to urea in acidic solutions.

**cyanamide process** *See* CALCIUM CYANAMIDE.

**cyanate** *See* CYANIC ACID.

**cyanic acid** An unstable explosive acid, HOCN. The compound has the structure H–O–C≡N, and is also called **fulminic acid**. Its salts and esters are **cyanates** (or **fulminates**). The compound is a volatile liquid, which readily polymerizes. In water it hydrolyses to ammonia and carbon dioxide. It

is isomeric with another acid, H–N=C=O, which is known as **isocyanic acid**. Its salts and esters are **isocyanates**.

**cyanide** 1. An inorganic salt containing the cyanide ion $CN^-$. Cyanides are extremely poisonous because of the ability of the $CN^-$ ion to coordinate with the iron in haemoglobin, thereby blocking the uptake of oxygen by the blood. 2. A metal coordination complex formed with cyanide ions.

**cyanide process** A method of extracting gold by dissolving it in potassium cyanide (to form the complex ion $[Au(CN)_2]^-$). The ion can be reduced back to gold with zinc.

**cyanine dyes** A class of dyes that contain a –CH= group linking two nitrogen-containing heterocyclic rings. They are used as sensitizers in photography.

**Cyanobacteria** A phylum consisting of two groups of photosynthetic eubacteria. The **blue-green bacteria** (formerly known as blue-green algae, or Cyanophyta), which comprise the vast majority of members, contain the photosynthetic pigment chlorophyll *a* plus accessory pigments: phycocyanin, responsible for their blue colour, and (in some) red pigments (phycoerythrins). Blue-green bacteria are unicellular but sometimes become joined in colonies or filaments by a sheath of mucilage. They occur in all aquatic habitats. A few species fix atmospheric nitrogen and thus contribute to soil fertility (*see* NITROGEN FIXATION). Others exhibit symbiosis (*see* LICHENS). The **chloroxybacteria** (grass-green bacteria or prochlorophytes) have been found in marine and freshwater habitats. They differ from the blue-green bacteria in containing chlorophyll *a* and chlorophyll *b* but no blue or red pigments – a combination like that found in plant chloroplasts, which they resemble.

**cyanocobalamin** *See* VITAMIN B COMPLEX.

**cyanogen** A colourless gas, $(CN)_2$, with a pungent odour; soluble in water, ethanol, and ether; d. 2.335 g dm$^{-3}$; m.p. –27.9°C; b.p. –20.7°C. The compound is very toxic. It may be prepared in the laboratory by heating mercury(II) cyanide; industrially it is made by gas-phase oxidation of hydrogen cyanide using air over a silver catalyst, chlorine over activated silicon(IV) oxide, or nitrogen dioxide over a copper(II) salt. Cyanogen is an important intermediate in the preparation of various fertilizers and is also used as a stabilizer in making nitrocellulose. It is an example of a *pseudohalogen.

**cyano group** The group –CN in a chemical compound. *See* NITRILES.

**cyanohydrins** Organic compounds formed by the addition of hydrogen cyanide to aldehydes or ketones (in the presence of a base). The first step is attack by a $CN^-$ ion on the carbonyl carbon atom. The final product is a compound in which a –CN and –OH group are attached to the same carbon atom. For example, ethanal reacts as follows

$$CH_3CHO + HCN \rightarrow CH_3CH(OH)(CN)$$

The product is 2-hydroxypropanonitrile. Cyanohydrins of this type can be oxidized to α-hydroxy carboxylic acids.

**cyanuric acid** A white crystalline water-soluble trimer of cyanic acid, $(HNCO)_3$. It is a cyclic compound having a six-membered ring made of alternating imide (NH) and carbonyl (CO) groups (i.e. three –NH–C(O)– units). It can also exist in a phenolic form (three –N=C(OH)– units).

**Cycadofilicales (Pteridospermales; seed ferns)** An extinct order of gymnosperms that flourished in the Carboniferous period. They possessed characteristics of both the ferns and the seed plants in reproducing by means of seeds and yet retaining fernlike leaves. Their internal anatomy combined both fern and seed-plant characteristics.

**Cycadophyta** A phylum of seed plants (*see* GYMNOSPERM) that contains many extinct species; the few modern representatives of the group include *Cycas* and *Zamia*. Cycads inhabit tropical and subtropical regions, sometimes growing to a height of 20 m. The stem bears a crown of fernlike leaves. These species are among the most primitive of living seed plants.

**cyclamates** Salts of the acid, $C_6H_{11}.NH.SO_3H$, where $C_6H_{11}$– is a cyclohexyl group. Sodium and calcium cyclamates were formerly used as sweetening agents in soft drinks, etc., until their use was banned when they were suspected of causing cancer.

**cycle** A regularly repeated set of changes to a system that brings back all its parameters to their original values once in every set of changes. The duration of one cycle is called its *period and the rate of repetition of cycle, called the *frequency, is measured in *hertz. *See* SIMPLE HARMONIC MOTION.

**cyclic** Describing a compound that has a ring of atoms in its molecules. In **homocyclic** compounds all the atoms in the ring are the same type, e.g. benzene ($C_6H_6$) and cyclohexane ($C_6H_{12}$). These two examples are also examples of **carbocyclic** compounds; i.e. the rings are of carbon atoms. If different atoms occur in the ring, as in pyridine ($C_5H_5N$), the compound is said to be **heterocyclic**.

**cyclic AMP (cAMP; cyclic adenosine monophosphate)** A derivative of *ATP that is widespread in cells as a *second messenger in many biochemical reactions induced by hormones. Binding of the hormone to its receptor on the cell surface activates *G proteins, which in turn activate *adenylate cyclase, the enzyme that catalyses cyclic AMP production. Cyclic AMP controls activity of protein kinase A, enabling it to activate intracellular proteins that mediate the ultimate effects of the hormone on the cell. Cyclic AMP is also involved in controlling gene expression and cell division, in immune responses, and in nervous transmission.

**cyclic model** (of the universe) *See* EKPYROTIC UNIVERSE.

**cyclic phosphorylation (cyclic photophosphorylation)** *See* PHOTOPHOSPHORYLATION.

**cyclization** The formation of a cyclic compound from an open-chain compound. *See* RING.

**cyclo-** Prefix designating a cyclic compound, e.g. a cycloalkane or a cyclosilicate.

**cycloalkanes** Cyclic saturated hydrocarbons containing a ring of carbon atoms joined by single bonds. They have the general formula $C_nH_{2n}$, for example cyclohexane, $C_6H_{12}$, etc. In general they behave like the *alkanes but are rather less reactive.

(((●))) SEE WEB LINKS

• Information about IUPAC nomenclature

**cyclohexadiene-1,4-dione (benzoquinone; quinone)** A yellow solid, $C_6H_4O_2$; r.d. 1.3; m.p. 116°C. It has a six-membered ring of carbon atoms with two opposite carbon atoms linked to oxygen atoms (C=O) and the other two pairs of carbon atoms linked by double bonds (HC=CH). The compound is used in making dyes. *See also* QUINHYDRONE ELECTRODE.

**cyclohexane** A colourless liquid *cycloalkane, $C_6H_{12}$; r.d. 0.78; m.p. 6.5°C; b.p.

81°C. It occurs in petroleum and is made by passing benzene and hydrogen under pressure over a heated Raney nickel catalyst at 150°C, or by the reduction of cyclohexanone. It is used as a solvent and paint remover and can be oxidized using hot concentrated nitric acid to hexanedioic acid (adipic acid). The cyclohexane ring is not planar and can adopt boat and chair *conformations; in formulae it is represented by a single hexagon.

**cycloid** The curve traced by a point on the circumference of a circle as it rolls without slipping along a straight line. The length of the arc formed by one revolution of the circle is $8r$, where $r$ is the radius of the circle. The horizontal distance between cusps is $2\pi r$.

**cyclone** An area of low pressure in the atmosphere. Winds rotate about the low-pressure centre in an anti-clockwise direction in the northern hemisphere and in a clockwise direction in the southern hemisphere. In the mid- and high-latitudes these low-pressure systems are now commonly referred to as *depressions, or lows, and the term cyclone is avoided. *See also* TROPICAL CYCLONE.

**cyclonite** A highly explosive nitro compound, $(CH_2N.NO_2)_3$. It has a cyclic structure with a six-membered ring of alternating $CH_2$ groups and nitrogen atoms, with each nitrogen being attached to a $NO_2$ group. It is made by nitrating hexamine, $C_6H_{12}N_4$, which is obtained from ammonia and methanal. Cyclonite is a very powerful explosive used mainly for military purposes. It is also called **RDX**. The abbreviation is for 'Research Department composition X', used at the Chemical Research and Development Department, Woolwich.

**cyclopentadiene** A colourless liquid cyclic *alkene, $C_5H_6$; r.d. 0.8021; m.p. −97.2°C; b.p. 40.0°C. It is prepared as a byproduct during the fractional distillation of crude benzene from coal tar. It undergoes condensation reactions with ketones to give highly coloured compounds (fulvenes) and readily undergoes polymerization at room temperature to give the dimer, dicyclopentadiene. The compound itself is not aromatic because it does not have the required number of pi electrons. However, removal of a hydrogen atom produces the stable **cyclopentadienyl ion**, $C_5H_5^-$, which does have aromatic properties. In particular, the ring can coordinate to positive ions in such compounds as *ferrocene.

**cyclopentadienyl ion** *See* CYCLOPENTADIENE.

**cyclophane** A compound consisting of one or more aromatic rings forming part of a larger ring system in which aliphatic chains of the $CH_2$ groups link the aromatic rings. Compounds of this type have the suffix **phane** in their names. Depending on the sizes of the $(CH_2)_n$ chains, the aromatic rings may not be planar.

**(((⊕))) SEE WEB LINKS**

• Information about IUPAC nomenclature

**cyclopropane** A colourless gas, $C_3H_6$, b.p. −34.5°C, whose molecules contain a triangular ring of carbon atoms. It is made by treating 1,3-dibromopropane with zinc metal, and is used as a general anaesthetic.

**cyclosarin** A highly toxic colourless liquid, $C_7H_{14}FO_2P$; r.d. 1.13; m.p. −30°C; b.p. 239°C. it is a fluorinated organophosphorus compound, (fluoromethylphosphoryl)oxycyclohexane. Cyclosarin was discovered in 1949 and belongs to the G-series of *nerve agents (GF).

**cyclosis** *See* CYTOPLASMIC STREAMING.

**cyclotetramethylenetetranitramine** *See* HMX.

**cyclotron** A cyclic particle *accelerator in which charged particles fed into the centre of the device are accelerated in an outward spiral path inside two hollow D-shaped conductors placed to form a split circle. A magnetic field is applied at right-angles to the plane of the dees and an alternating potential difference is applied between them. The frequency of the alternating p.d. is arranged so that the particles are accelerated each time they reach the evacuated gap between the dees. The magnetic field makes them follow curved paths. After several thousand revolutions inside the dees the particles reach the perimeter of the dees, where a deflecting field directs them onto the target. In this device protons can achieve an energy of $10^{-12}$ J (10 MeV). The first working cyclotron was produced in 1931 by the US physicist Ernest Lawrence (1901–58). *See also* SYNCHROCYCLOTRON.

**(((⊕))) SEE WEB LINKS**

• The US National Superconducting Cyclotron Society's You Tube channel

**cylinder gas** *See* BOTTLED GAS.

**cylindrical polar coordinates** *See* POLAR COORDINATES.

**cyme** *See* CYMOSE INFLORESCENCE.

buttercup    forget-me-not

monochasial cymes

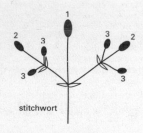

stitchwort

dichasial cyme

1 = oldest flower

**Cymose inflorescence.** Different types.

**cymose inflorescence (cyme; definite inflorescence)** A type of flowering shoot (*see* INFLORESCENCE) in which the first-formed flower develops from the growing region at the top of the flower stalk (see illustration). Thus no new flower buds can be produced at the tip and other flowers are produced from lateral buds beneath. In a **monochasial cyme** (or **monochasium**), the development of the flower at the tip is followed by a new flower axis growing from a single lateral bud. Subsequent new flowers may develop from the same side of the lateral shoots, as in the buttercup, or alternately on opposite sides, as in forget-me-not. In a **dichasial cyme** (or **dichasium**), the development of the flower at the apex is followed by two new flower axes developing from buds opposite one another, as in plants of the family Caryophyllaceae (such as stitchwort). *Compare* RACEMOSE INFLORESCENCE.

**cypsela** A dry single-seeded fruit that does not split open during seed dispersal and is formed from a double ovary in which only one ovule develops into a seed. It is similar to an *achene and characteristic of members of the family Compositae (Asteraceae), such as the dandelion. *See also* PAPPUS.

**cysteine** *See* AMINO ACID.

**cystine** A molecule resulting from the oxidation reaction between the sulphydryl (–SH) groups of two cysteine molecules (*see* AMINO ACID). This often occurs between adjacent cysteine residues in polypeptides. The resultant *disulphide bridges (–S–S–) are important in stabilizing the structure of protein molecules.

**cytidine** A nucleoside comprising one cytosine molecule linked to a D-ribose sugar molecule. The derived nucleotides, cytidine mono-, di-, and triphosphate (CMP, CDP, and CTP respectively), participate in various biochemical reactions, notably in phospholipid synthesis.

**cytochrome** Any of a group of proteins, each with an iron-containing *haem group, that form part of the *electron transport chain in mitochondria and chloroplasts. Electrons are transferred by reversible changes in the iron atom between the reduced Fe(II) and oxidized Fe(III) states. *See also* CYTOCHROME OXIDASE.

**cytochrome oxidase** An enzyme complex comprising the terminal two cytochromes of the respiratory chain in the mitochondria (*see* ELECTRON TRANSPORT CHAIN). It is responsible for the reduction of oxygen to form water.

**cytogenetics** The study of inheritance in relation to the structure and function of cells. For example, the results of breeding experiments can be explained in terms of the behaviour of chromosomes during the formation of the reproductive cells.

**cytokine** Any of numerous small proteins released from a variety of cell types that affect cell behaviour. Cytokines can influence the cells releasing them or nearby cells; in some cases they can enter the bloodstream to influence distant cells. Cytokines are crucial to many aspects of cell proliferation, differentiation, migration, and function, and play a central role in immune responses and inflammation. Cytokines produced by lymphocytes are termed **lymphokines**, and those affecting cell migration, particularly of immune cells, form a large group of **chemokines**. Cytokines fall into two main families: the haemopoietins, which include some of the interleukins, erythropoietin, and the *colony-stimulating factors; and the tumour necrosis factor (TNF) family.

**cytokinesis** *See* CELL CYCLE; MITOSIS.

**cytokinin (kinin)** Any of a group of plant hormones chemically related to the purine adenine. Among other roles, cytokinins stimulate cell division in the presence of *auxin and have also been found to delay senescence, overcome *apical dominance, and promote cell expansion. Zeatin is a naturally occurring cytokinin.

**cytology** The study of the structure and function of cells. The development of the light and electron microscopes has enabled the detailed structure of the nucleus (including the chromosomes) and other organelles to be elucidated. Microscopic examination of cells, either live or as stained sections on a slide, is also used in the detection and diagnosis of various diseases, especially *cancer.

**cytolysis** The breakdown of cells, usually as a result of destruction or dissolution of their outer membranes. Certain drugs (**cytotoxic drugs**) have this effect and are used in the treatment of some forms of cancer.

**cytomegalovirus** A virus belonging to the herpes group (*see* HERPESVIRUS). In humans it normally causes symptoms that are milder than the common cold, but it can produce more serious symptoms in those whose *immune response is disturbed (e.g. cancer patients and people who are HIV-positive). Infection in pregnant women may cause congenital handicap in their children.

**cytoplasm** The material surrounding the nucleus of a *cell. It can be differentiated into dense outer **ectoplasm**, which is concerned primarily with cell movement, and less dense **endoplasm**, which contains most of the cell's structures.

**cytoplasmic inheritance** The inheritance of genes contained in the cytoplasm of a cell, rather than the nucleus. Only a very small number of genes are inherited in this way. The phenomenon occurs because certain organelles, the *mitochondria and (in plants) the *chloroplasts, contain their own genes and can reproduce independently.

The female reproductive cell (the egg) has a large amount of cytoplasm containing many such organelles, which are consequently incorporated into the cytoplasm of all the cells of the embryo. The male reproductive cells (sperm or pollen), however, consist almost solely of a nucleus. Cytoplasmic organelles are thus not inherited from the male parent. In plants, male sterility can be inherited via the cytoplasm. The inheritance of any such factors does not follow Mendelian laws.

**cytoplasmic streaming (cyclosis)** The directional movement of cytoplasm in certain cells, which allows movement of substances through the cell, especially around the cell's periphery. It has been observed most clearly in large cells, such as plant sieve elements and unicellular algae, in which simple diffusion is ineffective as a means of local transport in the cell. The mechanism involves the interaction of *myosin proteins (attached to organelles) with *actin microfilaments parallel to the direction of flow and requires energy from ATP. A similar streaming of cytoplasm is responsible for *amoeboid movement.

**cytosine** A *pyrimidine derivative. It is one of the principal component bases of *nucleotides and the nucleic acids *DNA and *RNA.

**cytoskeleton** A network of fibres permeating the matrix of eukaryotic cells that provides a supporting framework for organelles, anchors the cell membrane, facilitates cellular movement, and provides a suitable surface for chemical reactions to take place. The fibres are composed of *microtubules and *actin microfilaments.

**cytosol** The semifluid soluble part of the cytoplasm of cells, which contains the components of the *cytoskeleton. The cell's organelles are suspended in the cytosol.

**cytotaxonomy** *See* TAXONOMY.

**cytotoxic** Destructive to living cells. The term is applied particularly to a class of drugs that inhibit cell division and are therefore used in chemotherapy to destroy cancer cells and to a group of *T cells that destroy virus-infected cells.

**CZ** *See* CUBIC ZIRCONA.

**CZE** Capillary zone electrophoresis. *See* CAPILLARY ELECTROPHORESIS.

**2,4-D** 2,4-dichlorophenoxyacetic acid (2,4-dichlorophenoxyethanoic acid): a synthetic *auxin used as a weedkiller of broad-leaved weeds. *See* PESTICIDE.

**D'Alembertian** Symbol □ (sometimes printed □²). An operator that is the analogue of the Laplace operator in four-dimensional Minkowski space–time, i.e.

$$\square = \partial^2/\partial x^2 + \partial^2/\partial y^2 + \partial^2/\partial z^2 - (1/c^2)\,\partial^2/\partial t^2,$$

where $c$ is the speed of light. The D'Alembertian is very useful in the special theory of relativity.

**D'Alembert's principle** A formulation of Newtonian mechanics that reduces dynamics to an equilibrium condition like statics by extending Newton's third law to the case of forces acting on bodies rather than just bodies in static equilibrium, i.e. it postulates that bodies are always in equilibrium even when they are acted on by a force because the force applied to the body minus the rate of change of the momentum with respect to time is always zero. It was proposed by the French mathematician Jean le Rond D'Alembert (1717–83) in 1743.

**dalton** *See* ATOMIC MASS UNIT.

**Dalton, John** (1766–1844) British chemist and physicist. In 1801 he formulated his law of partial pressures (*see* DALTON'S LAW), but he is best remembered for *Dalton's atomic theory, which he announced in 1803. Dalton also studied colour blindness (a condition, once called Daltonism, that he shared with his brother).

**Dalton's atomic theory** A theory of *chemical combination, first stated by John Dalton in 1803. It involves the following postulates:
(1) Elements consist of indivisible small particles (atoms).
(2) All atoms of the same element are identical; different elements have different types of atom.
(3) Atoms can neither be created nor destroyed.
(4) 'Compound elements' (i.e. compounds) are formed when atoms of different elements join in simple ratios to form 'compound atoms' (i.e. molecules).

Dalton also proposed symbols for atoms of different elements (later replaced by the present notation using letters).

**Dalton's law** The total pressure of a mixture of gases or vapours is equal to the sum of the partial pressures of its components, i.e. the sum of the pressures that each component would exert if it were present alone and occupied the same volume as the mixture of gases. Strictly speaking, the principle is true only for ideal gases.

**dam** A structure built across a river to impound or divert the flow of water. A dam may be constructed for one or a number of purposes. It may raise the water level to increase the depth for navigation purposes; divert water; provide a head of water for the generation of hydroelectric power; or store water for industrial or domestic use, irrigation, flood control, or power production. In addition, water stored by dams is often used for recreational purposes. Most dams are constructed of either concrete or of earth and rock. **Gravity dams** depend on the weight of their bulk for stability and usually have a flat vertical face upstream. **Arch dams** are curved concrete structures with a convex curve upstream. Pressure is transmitted to the sides of the dam.

**damping** A decrease in the amplitude of an oscillation as a result of energy being drained from the oscillating system to overcome frictional or other resistive forces. For example, a pendulum soon comes to rest unless it is supplied with energy from an outside source; in a pendulum clock, energy is supplied through an *escapement from a wound spring or a falling mass to compensate for the energy lost through friction. Damping is introduced intentionally in measuring instruments of various kinds to overcome the problem of taking a reading from an oscillating needle. A measuring instrument is said to be **critically damped** if the system just fails to oscillate and the system comes to rest in the shortest possible time. If it is **underdamped** it will oscillate repeatedly before coming to rest; if it is **over-damped** it will not oscillate but it will take longer to come to rest than it would if it was critically damped. An instrument, such as a galvanometer, that is critically damped is often called a **deadbeat** instrument.

**dance of the bees** A celebrated example of *communication in animals, first investigated by Karl von Frisch (1886–1982). Honeybee workers on returning to the hive after a successful foraging expedition perform a 'dance' on the comb that contains coded information about the distance and direction of the food source. For example the **waggle dance**, characterized by tail-wagging movements, indicates the direction of a food source at a distance of more than 100 metres. Other workers, sensing vibrations from the dance, follow the instructions to find the food source.

**Daniell cell** A type of primary *voltaic cell with a copper positive electrode and a negative electrode of a zinc amalgam. The zinc-amalgam electrode is placed in an electrolyte of dilute sulphuric acid or zinc sulphate solution in a porous pot, which stands in a solution of copper sulphate in which the copper electrode is immersed. While the reaction takes place ions move through the porous pot, but when it is not in use the cell should be dismantled to prevent the diffusion of one electrolyte into the other. The e.m.f. of the cell is 1.08 volts with sulphuric acid and 1.10 volts with zinc sulphate. It was invented in 1836 by the British chemist John Daniell (1790–1845).

***Daphnia*** A genus of crustaceans belonging to the class Branchiopoda and order Cladocera (water fleas). *Daphnia* species have a transparent carapace and a protruding head with a pair of highly branched antennae for swimming and a single median compound eye. The five pairs of thoracic appendages form an efficient filter-feeding mechanism. Reproduction can take place without mating, i.e. by *parthenogenesis.

**dark energy** Energy in the universe associated with the fact that the expansion of the universe is accelerating and the *cosmological constant could have a non-zero value. Analysis of data from *WMAP indicates that about 70% of the energy of the universe is in the form of dark energy. The nature of dark energy is not known.

**dark galaxy** A galaxy that is composed largely of dark matter. There is some evidence for the existence of such galaxies. It is thought that they should be very common, particularly since theories of large-scale structure work much better if the existence of plentiful dark galaxies is assumed.

**dark matter** *See* MISSING MASS.

(((•))) **SEE WEB LINKS**
• The home page of the Cryogenic Dark Matter Search Experiment

**dark period** (in botany) The period considered to be critical in the responses of plants to changes in day length (*see* PHOTOPERIODISM). It is believed that such responses, which include the onset of flowering, are determined by the length of the period of darkness that occurs between two periods of light.

**dark reaction** *See* PHOTOSYNTHESIS.

**darmstadtium** Symbol Ds. A radioactive transactinide element; a.n. 110. It has several isotopes; the most stable being $^{281}$Ds, with a half-life of about 1.6 minutes. It can be produced by bombarding a plutonium target with sulphur nuclei or by bombarding a lead target with nickel nuclei. Darmstadtium was named after the German city of Darmstadt, the location of the Institute for Heavy Ion Research where it was first produced.

(((•))) **SEE WEB LINKS**
• Information from the WebElements site

**Darwin, Charles** (1809–82) British naturalist, who studied medicine in Edinburgh followed by theology at Cambridge University, intending a career in the Church. However, his interest in natural history led him to accept an invitation in 1831 to join HMS *Beagle* as naturalist on a round-the-world voyage. After his return five years later he published works on the geology he had observed. He was also formulating his theory of *evolution by means of *natural selection, but it was to be 20 years before he published *On the Origin of Species* (1859), prompted by similar views expressed by Alfred Russel *Wallace. Among his later works was *The Descent of Man* (1871). *See also* DARWINISM.

(((•))) **SEE WEB LINKS**
• The complete works of Charles Darwin online

**Darwinism** The theory of *evolution proposed by Charles Darwin in *On the Origin of Species* (1859), which postulated that present-day species have evolved from simpler ancestral types by the process of *natural selection acting on the variability found within populations. *On the Origin of Species* caused a furore when it was first published because it suggested that species are not immutable nor were they specially created – a view directly opposed to the doctrine of *special creation. However the wealth of evidence presented by Darwin gradually convinced most people and the only major unresolved problem was to explain how the variations in populations arose and were maintained from one generation to the next. This became clear with the rediscovery of Mendel's work on classical genetics in the 1900s and led to the present theory known as *neo-Darwinism.

**Darwin's finches** (Galapagos finches) The 14 species of finch, unique to the Galapagos Islands, that Charles Darwin studied during his journey on HMS *Beagle*. Each is adapted to exploit a different food source. They are not found on the mainland because competition there for these food sources from other birds is fiercer. Darwin believed all the Galapagos finches – basically similar but differing in bill shape – to be descendants of a few that strayed from the mainland, and this provided important evidence for his theory of evolution. *See also* ADAPTIVE RADIATION.

**DAT (digital audio tape)** A type of magnetic tape originally designed for audio recording but now adapted for computer storage and backup use. The recording method allows a capacity of about 1 gigabyte.

**database** A large collection of information that has been coded and stored in a computer in such a way that it can be extracted under a number of different category headings.

**date-rape drugs** Drugs that are used to render the victim unable to resist a sexual assault. Particular drugs used for this purpose are flunitrazepam and gammahydroxybutyric acid (GHB).

**dating techniques** Methods of estimating the age of rocks, palaeontological specimens, archaeological sites, etc. **Relative dating techniques** date specimens in relation to one another; for example, *stratigraphy is used to establish the succession of fossils. **Absolute** (or **chronometric**) **techniques** give an absolute estimate of the age and fall into two main groups. The first de-

pends on the existence of something that develops at a seasonally varying rate, as in *dendrochronology and *varve dating. The other uses some measurable change that occurs at a known rate, as in *chemical dating, **radioactive** (or **radiometric**) **dating** (*see* CARBON DATING; FISSION-TRACK DATING; POTASSIUM–ARGON DATING; RUBIDIUM–STRONTIUM DATING; URANIUM–LEAD DATING), and *thermoluminescence.

**dative bond** *See* CHEMICAL BOND.

**daughter** **1.** A nuclide produced by radioactive *decay of some other nuclide (the **parent**). **2.** An ion or free radical produced by dissociation or reaction of some other (**parent**) ion or radical.

**Davy, Sir Humphry** (1778–1829) British chemist, who studied gases at the Pneumatic Institute in Bristol, where he discovered the anaesthetic properties of *dinitrogen oxide (nitrous oxide). He moved to the Royal Institution, London, in 1801 and five years later isolated potassium and sodium by electrolysis. He also prepared barium, boron, calcium, and strontium as well as proving that chlorine and iodine are elements. In 1816 he invented the *Davy lamp.

**Davy lamp** An oil-burning miner's safety lamp invented by Sir Humphry Davy in 1816 when investigating firedamp (methane) explosions in coal mines. The lamp has a metal gauze surrounding the flame, which cools the hot gases by conduction and prevents ignition of gas outside the gauze. If firedamp is present it burns within the gauze cage, and lamps of this type are still used for testing for gas.

**day** The time taken for the earth to complete one rotation on its axis. The **solar day** is the interval between two successive returns of the sun to the *meridian. The **mean solar day** of 24 hours is the average value of the solar day for one year. The **sidereal day** is measured with respect to the fixed stars and is 4.09 minutes shorter than the mean solar day as a result of the imposition of the earth's orbital motion on its rotational motion.

**day-neutral plant** A plant in which flowering can occur irrespective of the day length. Examples are cucumber and maize. *See* PHOTOPERIODISM. *Compare* LONG-DAY PLANT; SHORT-DAY PLANT.

***d*-block elements** The block of elements

in the *periodic table consisting of scandium, yttrium, and lanthanum together with the three periods of transition elements: titanium to zinc, zirconium to cadmium, and hafnium to mercury. These elements all have two outer $s$-electrons and have $d$-electrons in their penultimate shell; i.e. an outer electron configuration of the form $(n-1)d^{x}ns^{2}$, where $x$ is 1 to 10. *See also* TRANSITION ELEMENTS.

**d.c.** *See* DIRECT CURRENT.

**DDT** Dichlorodiphenyltrichloroethane; a colourless organic crystalline compound, $(ClC_6H_4)_2CH(CCl_3)$, made by the reaction of trichloromethanal with chlorobenzene. DDT is the best known of a number of chlorine-containing pesticides used extensively in agriculture in the 1940s and 1950s. The compound is stable, accumulates in the soil, and concentrates in fatty tissue, reaching dangerous levels in carnivores high in the food chain. Restrictions are now placed on the use of DDT and similar pesticides.

**Deacon process** A former process for making chlorine by oxidizing hydrogen chloride in air at 450°C using a copper chloride catalyst. It was patented in 1870 by Henry Deacon (1822–76).

**deactivation** A partial or complete reduction in the reactivity of a substance, as in the poisoning of a catalyst.

**deadbeat** *See* DAMPING.

**deamination** The removal of an amino group ($-NH_2$) from a compound. Enzymatic deamination occurs in the liver and is important in amino-acid metabolism, especially in their degradation and subsequent oxidation. The amino group is removed as ammonia and excreted, either unchanged or as urea or uric acid.

**death** The point at which the processes that maintain an organism alive no longer function. In humans it is diagnosed by permanent cessation of the heartbeat; however, the heart can continue beating after a large part of the brain ceases to function (*see* BRAIN DEATH). The death of a cell due to external damage or the action of toxic substances is known as **necrosis**. This must be distinguished from programmed cell death (*see* APOPTOSIS), which is a normal part of the developmental process.

**death phase** *See* BACTERIAL GROWTH CURVE.

**death rate (mortality)** The rate at which a particular species or population dies, whatever the cause. The death rate is an important factor in controlling the size of a population. *Compare* BIRTH RATE.

**de Broglie, Louis-Victor Pierre Raymond** (1892–1987) French physicist, who taught at the Sorbonne in Paris for 34 years. He is best known for his 1923 theory of *wave–particle duality (*see also* DE BROGLIE WAVELENGTH), which reconciled the corpuscular and wave theories of *light and proved important in *quantum theory. For this work he was awarded the 1929 Nobel Prize.

**de Broglie wavelength** The wavelength of the wave associated with a moving particle. The wavelength ($\lambda$) is given by $\lambda = h/mv$, where $h$ is the Planck constant, $m$ is the mass of the particle, and $v$ its velocity. The **de Broglie wave** was first suggested by Louis de Broglie in 1923 on the grounds that electromagnetic waves can be treated as particles (*photons) and one could therefore expect particles to behave in some circumstances like waves (*see* COMPLEMENTARITY). The subsequent observation of *electron diffraction substantiated this argument and the de Broglie wave became the basis of *wave mechanics.

**debye** A unit of electric *dipole moment in the electrostatic system, used to express dipole moments of molecules. It is the dipole moment produced by two charges of opposite sign, each of 1 statcoulomb and placed $10^{-18}$ cm apart, and has the value $3.335\ 64 \times 10^{-30}$ coulomb metre.

**Debye–Hückel theory** A theory to explain the nonideal behaviour of electrolytes, published in 1923 by Peter Debye (1884–1966) and Erich Hückel (1896– ). It assumes that electrolytes in solution are fully dissociated and that nonideal behaviour arises because of electrostatic interactions between the ions. The theory shows how to calculate the extra free energy per ion resulting from such interactions, and consequently the activity coefficient. It gives a good description of nonideal electrolyte behaviour for very dilute solutions, but cannot be used for more concentrated electrolytes.

**Debye–Scherrer method** A method of X-ray diffraction in which a beam of X-rays is diffracted by material in the form of powder. Since the powder consists of very small crystals of the material in all possible orientations, the diffraction pattern is a series of concentric circles. This type of pattern allows the unit cell to be found with great precision. This method was first used by Peter Debye and Paul Scherrer in 1916 and independently by Albert Hull in 1917.

**deca-** Symbol da. A prefix used in the metric system to denote ten times. For example, 10 coulombs = 1 decacoulomb (daC).

**decahydrate** A crystalline hydrate containing ten molecules of water per molecule of compound.

**decalescence** *See* RECALESCENCE.

**decanedioic acid (sebacic acid)** A white crystalline dicarboxylic acid, $HOOC(CH_2)_8COOH$; r.d. 1.12; m.p. 131–134.5°C; b.p. 294.4°C (100 mmHG). Obtained from castor oil, it is used in plasticizers, lubricants, and cosmetics and in the production of other organic chemicals.

**decanoic acid (capric acid)** A white crystalline straight-chain saturated *carboxylic acid, $CH_3(CH_2)_8COOH$; m.p. 31.5°C. Its esters are used in perfumes and flavourings.

**decantation** The process of separating a liquid from a settled solid suspension or from a heavier immiscible liquid by carefully pouring it into a different container.

**Decapoda** An order of crustaceans of the class Malacostraca that are distributed worldwide, mainly in marine habitats. Decapods comprise swimming forms (shrimps and prawns) and crawling forms (crabs, lobsters, and crayfish). All are characterized by five pairs of walking legs, the first pair of which are highly modified in crawling forms to form powerful grasping pincers. The carapace is fused with the thorax and head forming a *cephalothorax. The antennae are especially long in shrimps and prawns, which also possess several pairs of well-developed swimming appendages (**pleopods**) posterior to the walking legs. Following fertilization by the male, females usually carry the eggs until they hatch. The larvae undergo several transformations before attaining adult form.

**decarboxylation** The removal of carbon dioxide from a molecule. Decarboxylation is an important reaction in many biochemical processes, such as the *Krebs cycle and the synthesis of *fatty acids. *See also* OXIDATIVE DECARBOXYLATION.

**decay** 1. *See* DECOMPOSITION. 2. The spontaneous transformation of one radioactive nuclide into a daughter nuclide, which may be radioactive or may not, with the emission of one or more particles or photons. The decay of $N_0$ nuclides to give $N$ nuclides after time $t$ is given by $N = N_0\exp(-\gamma t)$, where $\gamma$ is called the **decay constant** or the **disintegration constant**. The reciprocal of the decay constant is the **mean life**. The time required for half the original nuclides to decay (i.e. $N = \frac{1}{2}N_0$) is called the **half-life** of the nuclide. The same terms are applied to elementary particles that spontaneously transform into other particles. For example, a free neutron decays into a proton and an electron (*see* BETA DECAY). *See also* ALPHA PARTICLE.

**(((⊕))) SEE WEB LINKS**
• Values of radionuclide half-lives at the NIST website

**deci-** Symbol d. A prefix used in the metric system to denote one tenth. For example, 0.1 coulomb = 1 decicoulomb (dC); 0.1 metre = 1 decimetre (dm).

**decibel** A unit used to compare two power levels, usually applied to sound or electrical signals. Although the decibel is one tenth of a **bel**, it is the decibel, not the bel, that is invariably used. Two power levels $P$ and $P_0$ differ by $n$ decibels when $n = 10\log_{10}P/P_0$. If $P$ is the level of sound intensity to be measured, $P_0$ is a reference level, usually the intensity of a note of the same frequency at the threshold of audibility.

The logarithmic scale is convenient as human audibility has a range of 1 (just audible) to $10^{12}$ (just causing pain) and one decibel, representing an increase of some 26%, is about the smallest change the ear can detect.

**deciduous** Describing plants in which all the leaves are shed at the end of each growing season, usually the autumn in temperate regions or at the beginning of a dry season in the tropics. This seasonal leaf fall helps the plant retain water that would otherwise be lost by transpiration from the leaves. Examples of deciduous plants are rose and horse chestnut. *Compare* EVERGREEN.

**deciduous teeth (milk teeth)** The first of two sets of teeth of a mammal. These teeth are smaller than those that replace them (the *permanent teeth) and fewer in number, since there are no deciduous *molars. *See also* DIPHYODONT.

**decimal system** A number system based

on the number 10; the number system in common use. All rational numbers can be written as a **finite decimal** (e.g. ¼ = 0.25) or a **repeating decimal** (e.g. 5/27 = 0.185 185 185…). An *irrational number can be written to any number of decimal places, but can never be given exactly (e.g. √3 = 1.732 050 8…).

**declination** 1. The angle between the magnetic meridian and the geographic meridian at a point on the surface of the earth. *See* GEOMAGNETISM. 2. The angular distance of a celestial body north (positive) or south (negative) of the celestial *equator.

**decoction** A solution made by boiling material (e.g. plant substances) in water, followed by filtration.

**decoherence** A process in which a quantum mechanical state of a system is altered by the interaction between the system and its environment. The process of decoherence has been detected experimentally. Decoherence was postulated in the 1980s and has been used to clarify discussions of the foundations of quantum mechanics.

**decomposer** An organism that obtains energy from the chemical breakdown of dead organisms or animal or plant wastes. Decomposers, most of which are bacteria and fungi, secrete enzymes onto dead matter and then absorb the breakdown products (*see* SAPROTROPH). Many decomposers (e.g. nitrifying bacteria) are specialized to break down organic materials that are difficult for other organisms to digest. Decomposers fulfil a vital role in the *ecosystem, returning the constituents of organic matter to the environment in inorganic form so that they can again be assimilated by plants. *Compare* DETRITIVORE. *See also* CARBON CYCLE; NITROGEN CYCLE.

**decomposition** 1. **(decay)** The chemical breakdown of organic matter into its constituents by the action of *decomposers. 2. A chemical reaction in which a compound breaks down into simpler compounds or into elements.

**deconfinement temperature** *See* QUARK CONFINEMENT.

**decrepitation** A crackling noise produced when certain crystals are heated, caused by changes in structure resulting from loss of water of crystallization.

**deep-sky object** Any object located out-

side our solar system that is of interest to an astronomical observer. The term usually applies to a *star cluster, *nebula, *galaxy, or *galaxy cluster, but not a star.

**defecation** The expulsion of faeces from the rectum due to contractions of muscles in the rectal wall. A sphincter muscle, which is under voluntary control, is situated at the end of the rectum (the anus); relaxation of this muscle allows defecation to occur. In babies control of the anal sphincter muscle has not been developed and defecation occurs automatically as a reflex response to the presence of faeces in the rectum.

**defect** 1. *See* CRYSTAL DEFECT. 2. *See* MASS DEFECT. 3. *See* TOPOLOGICAL DEFECT.

**deficiency disease** Any disease caused by an inadequate intake of an essential nutrient in the diet, primarily vitamins, minerals, and amino acids. Examples are scurvy (lack of vitamin C), rickets (lack of vitamin D), and iron-deficiency anaemia.

**definite inflorescence** *See* CYMOSE INFLORESCENCE.

**definite integral** *See* INTEGRATION.

**definite proportions** *See* CHEMICAL COMBINATION.

**deflagration** A type of explosion in which the shock wave arrives before the reaction is complete (because the reaction front moves more slowly than the speed of sound in the medium).

**deforestation** The extensive cutting down of forests for the purpose of extracting timber or fuel wood or to clear the land for mining or agriculture. Forests are often situated in upland areas and are important in trapping rainwater. Deforestation in these areas, particularly in India and Bangladesh, has resulted in the flooding of low-lying plains; it has also led to an increase in soil erosion and hence desert formation (*see* DESERTIFICATION), resulting in crop loss and economic problems for local communities. The felling and burning of trees releases large amounts of carbon dioxide, thereby increasing global carbon dioxide levels and contributing to the *greenhouse effect. Rainforests, particularly those of South America, are rich in both fauna and flora; their removal leads to an overall decrease in *biodiversity and the loss of plant species that have potentially beneficial pharmaceutical effects. Despite movements to reduce deforestation, economic pressures ensure that the process still continues.

**degassing** The removal of dissolved, absorbed, or adsorbed gases from a liquid or solid. Degassing is important in vacuum systems, where gas absorbed in the walls of the vacuum vessel starts to desorb as the pressure is lowered.

**degaussing** The process of neutralizing the magnetization in an object that has inadvertently become magnetized. For example, ferromagnetic components of TV sets may become magnetized and misdirect the electron beams. A degaussing coil is often provided and fed with a diminishing alternating current each time the set is switched on. Ships can be degaussed by surrounding them with current-carrying cables that set up an equal and opposite field. This prevents the ships from detonating magnetic mines. Degaussing is used to protect scientific and other electronic devices from strong magnetic fields; usually a system of coils is designed to neutralize such fields over the important region of the equipment or the equipment is surrounded by a shield of suitable alloy (e.g. Mumetal).

**degeneracy pressure** The pressure in a *degenerate gas of fermions caused by the Pauli exclusion principle and the Heisenberg uncertainty principle. Because of the exclusion principle, fermions at a high density, with small interparticle spacing, must have different momenta. Because of the uncertainty principle, the momentum difference must be inversely proportional to the spacing. Consequently, in a high-density gas (small spacing) the particles have high relative momenta, which leads to a degeneracy pressure much greater than the thermal pressure. *White dwarfs and *neutron stars are supported against collapse under their own gravitational fields by the degeneracy pressure of electrons and neutrons respectively.

**degenerate 1.** Having quantum states with the same energy. For example, the five $d$-orbitals in an isolated transition-metal atom have the same energy (although they have different spatial arrangements) and are thus **degenerate levels**. The application of a magnetic or electric field may cause the quantum states to have different energies (see CRYSTAL-FIELD THEORY). In this case, the degeneracy is said to be 'lifted'. **2.** Having a high particle concentration such that the *Maxwell–Boltzmann distribution does not apply and the behaviour of the particles is governed by *quantum statistics. Systems in which the particles are degenerate are called **degenerate gases**; examples are the conduction electrons in a metal (or *degenerate semiconductor), the electrons in a *white dwarf, and the neutrons in a *neutron star. See also DEGENERACY PRESSURE.

**degenerate semiconductor** A heavily doped *semiconductor in which the *Fermi level is located in either the valence band or the conduction band (see ENERGY BANDS) causing the material to behave as a metal.

**degenerate states** *Quantum states of a system that have the same energy.

**degeneration 1.** Changes in cells, tissues, or organs due to disease, etc., that result in an impairment or loss of function and possibly death and breakdown of the affected part. **2.** The reduction in size or complete loss of organs during evolution. The human appendix has undergone this process and performs no obvious function in man. Degeneration of external organs may cause animals to appear to be more primitive than they really are; for example, early zoologists believed whales were fish rather than mammals because of the degeneration of their limbs. See also VESTIGIAL ORGAN.

**deglutition (swallowing)** A reflex action initiated by the presence of food in the pharynx. During deglutition, the soft *palate is raised, which prevents food from entering the nasal cavity; the *epiglottis closes, which blocks the entrance to the windpipe; and the oesophagus starts to contract (see PERISTALSIS), which ensures that food is conveyed to the stomach.

**degradation** A type of organic chemical reaction in which a compound is converted into a simpler compound. An example is the *Barbier–Wieland degradation.

**degree 1.** A unit of plane angle equal to 1/360th of a complete revolution. **2.** A division on a *temperature scale. **3.** The power to which a variable is raised. If one expression contains several variables the overall degree of the expression is the sum of the powers. For example, the expression $p^2q^3r^4$ has a degree of 9 overall (it is a second-degree expression in $p$). The degree of a polynomial is the degree of the variable with the highest power, e.g. $ax^5 + bx^4 + c$ has a degree of 5. **4.** The highest power to which the derivative of the highest order is raised in a *differential equation. For example, $(d^2y/dx^2)^3 + dy/dx = c$ is a differential equation of the third degree (but second order).

**degrees of freedom** **1.** The number of independent parameters required to specify the configuration of a system. This concept is applied in the *kinetic theory to specify the number of independent ways in which an atom or molecule can take up energy. There are however various sets of parameters that may be chosen, and the details of the consequent theory vary with the choice. For example, in a monatomic gas each atom may be allotted three degrees of freedom, corresponding to the three coordinates in space required to specify its position. The mean energy per atom for each degree of freedom is the same, according to the principle of the *equipartition of energy, and is equal to $kT/2$ for each degree of freedom (where $k$ is the *Boltzmann constant and $T$ is the thermodynamic temperature). Thus for a monatomic gas the total molar energy is $3LkT/2$, where $L$ is the Avogadro constant (the number of atoms per mole). As $k = R/L$, where $R$ is the molar gas constant, the total molar energy is $3RT/2$.

In a diatomic gas the two atoms require six coordinates between them, giving six degrees of freedom. Commonly these are interpreted as six independent ways of storing energy: on this basis the molecule has three degrees of freedom for different directions of translational motion, and in addition there are two degrees of freedom for rotation of the molecular axis and one vibrational degree of freedom along the bond between the atoms. The rotational degrees of freedom each contribute $kT/2$, to the total energy; similarly the vibrational degree of freedom has an equal share of kinetic energy and must on average have as much potential energy (*see* SIMPLE HARMONIC MOTION). The total energy per molecule for a diatomic gas is therefore $3kT/2$ (for translational energy of the whole molecule) plus $2kT/2$ (for rotational energy of each atom) plus $2kT/2$ (for vibrational energy), i.e. a total of $7kT/2$.
**2.** The least number of independent variables required to define the state of a system in the *phase rule. In this sense a gas has two degrees of freedom (e.g. temperature and pressure).

**dehiscence** The spontaneous and often violent opening of a fruit, seed pod, or anther to release and disperse the seeds or pollen.

Examples are the splitting of laburnum pods and primrose capsules; such structures are described as **dehiscent** (*compare* INDEHISCENT).

**dehydration** **1.** Removal of water from a substance. **2.** A chemical reaction in which a compound loses hydrogen and oxygen in the ratio 2:1. For instance, ethanol passed over hot pumice undergoes dehydration to ethene:

$$C_2H_5OH - H_2O \rightarrow CH_2{:}CH_2$$

Substances such as concentrated sulphuric acid, which can remove $H_2O$ in this way, are known as **dehydrating agents**. For example, with sulphuric acid, methanoic acid gives carbon monoxide:

$$HCOOH - H_2O \rightarrow CO$$

**dehydrogenase** Any enzyme that catalyses the removal of hydrogen atoms (**dehydrogenation**) in biological reactions. Dehydrogenases occur in many biochemical pathways but are particularly important in driving the reactions of the *electron-transport chain in cell respiration. They work in conjunction with the hydrogen-accepting coenzymes *NAD and *FAD.

**dehydrogenation** A chemical reaction in which hydrogen is removed from a compound. Dehydrogenation of organic compounds converts single carbon–carbon bonds into double bonds. It is usually effected by means of a metal catalyst.

**dehydrohalogenation** A type of chemical reaction in which a hydrogen halide is removed from a molecule with formation of a double bond. A simple example is the formation of ethene from chloroethane using alcoholic potassium hydroxide:

$$CH_3CH_2Cl + KOH \rightarrow CH_2{=}CH_2 + KCl + H_2O.$$

**deionized water** Water from which dissolved ionic salts have been removed by *ion-exchange techniques. It is used for many purposes as an alternative to distilled water.

**dekatron** A neon-filled tube with a central anode surrounded by ten cathodes and associated transfer electrodes. As voltage pulses are received by the tube a glow discharge moves from one set of electrodes to the next, enabling the device to be used as a visual counting tube in the decimal system. The tube can also be used for switching.

**delayed neutrons** The small proportion of neutrons that are emitted with a measurable time delay in a nuclear fission process. *Compare* PROMPT NEUTRONS.

**delay line** A component in an electronic circuit that is introduced to provide a specified delay in transmitting the signal. Coaxial cable or inductor-capacitor networks can be used to provide a short delay but for longer delays an **acoustic delay line** is required. In this device the signal is converted by the *piezoelectric effect into an acoustic wave, which is passed through a liquid or solid medium, before reconversion to an electronic signal.

**deletion** (in genetics) **1.** A *point mutation involving the removal of one or more base pairs in the DNA sequence. **2.** A frequently lethal *chromosome mutation that arises from an inequality in *crossing over during meiosis such that one of the chromatids loses more genetic information than it receives.

**deliquescence** The absorption of water from the atmosphere by a hygroscopic solid to such an extent that a concentrated solution of the solid eventually forms.

**delocalization** In certain chemical compounds the valence electrons cannot be regarded as restricted to definite bonds between the atoms but are 'spread' over several atoms in the molecule. Such electrons are said to be **delocalized**. Delocalization occurs particularly when the compound contains alternating (conjugated) double or triple bonds, the delocalized electrons being those in the pi *orbitals. The molecule is then more stable than it would be if the electrons were localized, an effect accounting for the properties of benzene and other aromatic compounds. Another example is in the ions of carboxylic acids, containing the carboxylate group $-COO^-$. In terms of a simple model of chemical bonding, this group would have the carbon joined to one oxygen by a double bond (i.e. $C=O$) and the other joined to $O^-$ by a single bond ($C-O^-$). In fact, the two $C-O$ bonds are identical because the extra electron on the $O^-$ and the electrons in the pi bond of $C=O$ are delocalized over the three atoms.

**delta** A fan-shaped area of sediment deposited at the mouth of a river, where it enters a lake or the sea. Usually the river divides and subdivides into many smaller channels (distributaries), sometimes depositing bars and building up levees. Marshes may border the delta. The sediment may range in particle size, consisting of gravel, sand, silt, or clay. The larger material is deposited first as the load-carrying capacity of the river diminishes, with the clay being deposited last.

**delta-brass** A strong hard type of *brass that contains, in addition to copper and zinc, a small percentage of iron. It is mainly used for making cartridge cases.

**deltahedron** A polyhedron that has triangular faces. *See* WADE'S RULES.

**delta-iron** *See* IRON.

**demagnetization** The removal of the ferromagnetic properties of a body by disordering the domain structure (*see* MAGNETISM). One method of achieving this is to insert the body within a coil through which an alternating current is flowing; as the magnitude of the current is reduced to zero, the domains are left with no predominant direction of magnetization.

**deme** A group of organisms in the same *taxon. The term is used with various prefixes that denote how the group differs from other groups. For example, an **ecodeme** occurs in a particular ecological habitat, **cytodemes** differ from each other cytologically, and **genodemes** differ genetically.

**demodulation** The process of extracting the information from a modulated carrier wave (*see* MODULATION; RADIO). The device used is called a **demodulator** or a **detector**.

**de Moivre's theorem** For *complex numbers in polar form, the equation

$$(\cos\theta + i\sin\theta)^n = \cos n\theta + i\sin n\theta.$$

It is true for all integer values of $n$. There is at least one value of the left-hand side of the equation that makes it also true for non-integer values of $n$. It was named after the French mathematician Abraham de Moivre (1667–1754).

**denature** **1.** To add a poisonous or unpleasant substance to ethanol to make it unsuitable for human consumption (*see* METHYLATED SPIRITS). **2.** To produce a structural change in a protein or nucleic acid that results in the reduction or loss of its biological properties. Denaturation is caused by heat, chemicals, and extremes of pH. The differences between raw and boiled eggs are largely a result of denaturation. **3.** To add another isotope to a fissile material to make it unsuitable for use in a nuclear weapon.

**dendrimer (dendritic polymer)** A type of macromolecule in which a number of chains radiate out from a central atom or cluster of atoms. Dendritic polymers have a number of possible applications. *See also* SUPRAMOLEC-ULAR CHEMISTRY.

**dendrite 1.** (in chemistry) A crystal that has branched in growth into two parts. Crystals that grow in this way (**dendritic growth**) have a branching treelike appearance. **2.** (in neurology) Any of the slender branching processes that arise from the *dendrons of the cell body of a motor *neuron. It forms connections (*see* SYNAPSE) with the axons of other neurons and transmits nerve impulses from these to the cell body.

**dendrochronology** An absolute *dating technique using the *growth rings of trees. It depends on the fact that trees in the same locality show a characteristic pattern of growth rings resulting from climatic conditions. Thus it is possible to assign a definite date for each growth ring in living trees, and to use the ring patterns to date fossil trees or specimens of wood (e.g. used for buildings or objects on archaeological sites) with lifespans that overlap those of living trees. The bristlecone pine (*Pinus aristata*), which lives for up to 5000 years, has been used to date specimens over 8000 years old. Fossil specimens accurately dated by dendrochronology have been used to make corrections to the *carbon-dating technique. Dendrochronology is also helpful in studying past climatic conditions. Analysis of trace elements in sections of rings can also provide information on past atmospheric pollution.

**dendron** Any of the major cytoplasmic processes that arise from the cell body of a motor neuron. A dendron usually branches into *dendrites.

**denitrification** A chemical process in which nitrates in the soil are reduced to molecular nitrogen, which is released into the atmosphere. This process is effected by denitrifying bacteria (e.g. *Pseudomonas denitrificans*), which use nitrates as a source of energy for other chemical reactions in a manner similar to respiration in other organisms. *Compare* NITRIFICATION. *See* NITROGEN CYCLE.

**densitometer** An instrument used to measure the *photographic density of an image on a film or photographic print. Densitometers work by letting the specimen transmit or reflect a beam of light and monitoring the transmitted or reflected intensity. They originally consisted of visual *photometers but most instruments are now photoelectric. The simplest transmission densitometer consists of a light source, a photosensitive cell, and a microammeter: the density is measured in terms of the meter readings with and without the sample in place. They have a variety of uses, including detecting the sound track on a cinematic film, measuring intensities in spectrographic records, and checking photographic prints.

**density 1.** The mass of a substance per unit of volume. In *SI units it is measured in $kg\ m^{-3}$. *See also* RELATIVE DENSITY; VAPOUR DENSITY. **2.** *See* CHARGE DENSITY. **3.** *See* PHOTOGRAPHIC DENSITY.

(((•))) SEE WEB LINKS
• Values of densities of commonly used materials at the NPL website

**density functional theory** A theory used to describe many-fermion systems in which the energy is a *functional of the density of fermions. Density functional theory has been used extensively in the theory of electrons in atoms, molecules, and solids, and the theory of nucleons in nuclei.

**dental caries** Tooth decay, which involves the destruction of the enamel layer of the tooth by acids produced by the action of bacteria on sugar. Bacteria can bind to teeth on **dextran**, a sticky substance derived from sucrose. The bacterial cells and other waste attached to dextran gives rise to *plaque. If dental caries is not treated it can spread to the dentine and pulp of the tooth, which leads to infection and death of the tooth.

**dental formula** A representation of the dentition of an animal. A dental formula consists of eight numbers, four above and four below a horizontal line. The numbers represent (from left to right) the numbers of incisors, canines, premolars, and molars in either half of the upper and lower jaws. The total number of teeth in both jaws is therefore obtained by adding up all the numbers in the dental formula and multiplying by 2. Representative dental formulas are shown in the illustration overleaf. *See also* PERMANENT TEETH.

| 2 | 1 | 2 | 3 | human | 2 | 0 | 3 | 3 | rabbit | 3 | 1 | 4 | 2 | bear |
| 2 | 1 | 2 | 3 | (32 teeth) | 1 | 0 | 2 | 3 | (28 teeth) | 3 | 1 | 4 | 3 | (42 teeth) |

**Representative dental formulas.**

**denticle (placoid scale)** *See* SCALES.

**dentine** The bony material that forms the bulk of a *tooth. Dentine is similar in composition to bone but is perforated with many tiny canals for nerve fibres, blood capillaries, and processes of the dentine-forming cells (*odontoblasts). Ivory, the material that forms elephant tusks, is made of dentine.

**dentition** The type, number, and arrangement of teeth in a species. This can be represented concisely by a *dental formula. *See also* PERMANENT TEETH; DIPHYODONT; MONOPHYODONT; POLYPHYODONT; HETERODONT; HOMODONT.

**deoxyribonuclease** *See* DNASE.

**deoxyribonucleic acid** *See* DNA.

**deoxyribose (2-deoxyribose)** A pentose (five-carbon) sugar, a derivative of *ribose, that is a component of the nucleotides (deoxyribonucleotides) that form the building blocks of *DNA.

**depleted** Denoting a material that contains less of a particular isotope than it normally contains, especially a residue from a nuclear reactor or isotope-separation plant.

**depleted uranium** Uranium mostly consisting of uranium–238, obtained as a by-product of enriching natural uranium or obtained from reprocessing plants. It has a high density and is used in armour-piercing shells.

**depletion layer** A region in a *semiconductor that has a lower-than-usual number of mobile charge carriers. A depletion layer forms at the interface between two dissimilar regions of conductivity (e.g. a $p$–$n$ junction). *See* DIODE.

**depolarization** 1. (in physics) The prevention of *polarization in a *primary cell. For example, maganese(IV) oxide (the **depolarizer**) is placed around the positive electrode of a *Leclanché cell to oxidize the hydrogen released at this electrode. 2. (in neurophysiology) A reduction in the difference of electrical potential that exists across the plasma membrane of a nerve or muscle cell. Depolarization of a nerve-cell membrane occurs during the passage of an *action potential along the axon when the nerve is transmitting an impulse.

**deposition** The laying down of material on the earth's surface. This includes materials that have been eroded elsewhere and transported by natural agents, such as rivers, wind, ice (e.g. glaciers), and the tides and currents in the sea. It also includes the deposition of materials resulting from chemical precipitation, the formation of crusts on the earth's surface through evaporation, and the growth, accumulation, and decay of natural organisms.

**depression (low; disturbance)** An area of low atmospheric pressure in the mid- and high-latitudes that is shown on a synoptic chart (weather map) surrounded by several closed isobars. Depressions generally move towards the northeast in the northern hemisphere and towards the southeast in the southern hemisphere. Unsettled weather conditions are associated with depressions and they form the main source of precipitation in most lowland parts of the mid-latitudes.

**depression of freezing point** The reduction in the freezing point of a pure liquid when another substance is dissolved in it. It is a *colligative property – i.e. the lowering of the freezing point is proportional to the number of dissolved particles (molecules or ions), and does not depend on their nature. It is given by $\Delta t = K_f C_m$, where $C_m$ is the molar concentration of dissolved solute and $K_f$ is a constant (the **cryoscopic constant**) for the solvent used. Measurements of freezing-point depression (using a Beckmann thermometer) can be used for finding relative molecular masses of unknown substances.

(((🌐))) **SEE WEB LINKS**

• Raoult's original paper

**depsides** A class of compounds formed by condensation of a phenolic carboxylic acid (such as gallic acid) with a similar compound, the reaction being between the carboxylic acid group on one molecule and a phenolic –OH group on the other. Depsides are similar to esters, except that the –OH group is linked directly to an aromatic ring. In such compounds, the –O–CO– group is called a **depside linkage**. Depsides are found in tannins and other natural products.

**depth of field** The range of distance in front of and behind an object that is being focused by an optical instrument, such as a microscope or camera, within which other objects will be in focus. The **depth of focus** is the amount by which the distance between the camera and the film can be changed without upsetting the sharpness of the image.

**derivative** 1. (in chemistry) A compound that is derived from some other compound and usually maintains its general structure, e.g. trichloromethane (chloroform) is a derivative of methane. 2. (in mathematics) *See* DIFFERENTIATION; CALCULUS.

**derived unit** *See* BASE UNIT.

**dermal bone** *See* MEMBRANE BONE.

**Dermaptera** An order of insects comprising the earwigs. Earwigs typically have long thin cylindrical bodies with biting mouthparts and a stout pair of curved forceps at the tip of the abdomen, used for catching prey and in courtship. Some species have a single pair of wings, which at rest are folded back over the abdomen like a fan; others are wingless. Most earwigs are nocturnal and omnivorous.

**dermis (corium; cutis)** The thicker and innermost layer of the *skin of vertebrates, the other layer being the *epidermis. The dermis consists of fibrous connective tissue in which are embedded blood vessels, sensory nerve endings, and (in mammals) hair follicles, sebaceous glands, and sweat ducts. Beneath the dermis lies the *subcutaneous tissue.

**desalination** The removal of salt from sea water for irrigation of the land or to provide drinking water. The process is normally only economic if a cheap source of energy, such as the waste heat from a nuclear power station, can be used. Desalination using solar energy has the greatest economic potential since shortage of fresh water is most acute in hot regions. The methods employed include evaporation, often under reduced pressure (flash evaporation); freezing (pure ice forms from freezing brine); *reverse osmosis; *electrodialysis; and *ion exchange.

**desert** A major terrestrial *biome characterized by low rainfall. Hot deserts, such as the Sahara and Kalahari deserts of Africa,

have a rainfall of less than 25 cm a year and extremely high daytime temperatures (up to 36°C). Vegetation is sparse, and desert plants are adapted to conserve water and take advantage of the rain when it falls. The perennials include xerophytic trees and shrubs (*see* XEROPHYTE) and *succulents, such as cacti. Annual plants are *ephemerals, lying dormant as seeds for most of the year and completing their life cycle in the brief rainy periods. Desert animals are typically nocturnal or active at dawn and dusk, thus avoiding the extreme daytime temperatures.

**desertification** The gradual conversion of fertile land into desert, usually as a result of human activities. Loss of topsoil leads to further soil erosion until the land can no longer be used to grow crops or support livestock. A major factor contributing to desertification is bad management of farmland. Overgrazing of livestock removes the plant cover and exposes the soil, making it vulnerable to erosion. Overintensive cultivation of crop plants, especially monoculture (*see* AGRICULTURE), depletes the soil of nutrients and organic matter, resulting in loss of fertility and increasing its susceptibility to erosion. In many Third World countries it is difficult to control the process of desertification as the livelihood of the people often depends on practices that contribute to soil erosion. Another major cause of desertification is *deforestation. Loss of vegetation also results from reductions in rainfall resulting from climate change.

**desiccation** A method of preserving organic material by the removal of its water content. Cells and tissues can be preserved by desiccation after lowering the samples to freezing temperatures; thereafter they can be stored at room temperature.

**desiccator** A container for drying substances or for keeping them free from moisture. Simple laboratory desiccators are glass vessels containing a drying agent, such as silica gel. They can be evacuated through a tap in the lid.

**desmids** Unicellular mainly freshwater green algae that belong to the class Desmidioideae. Like *Spirogyra*, they have an elaborate chloroplast. The cells of desmids are characteristically split into two halves joined by a narrow neck, each half being a mirror image of the other. The outer wall of the cell is patterned with various protuberances and covered with a mucilaginous sheath, which may function in the cell's gliding movement.

**desmotubule** *See* PLASMODESMATA.

**desorption** The removal of adsorbed atoms, molecules, or ions from a surface.

**destructive distillation** The process of heating complex organic substances in the absence of air so that they break down into a mixture of volatile products, which are condensed and collected. At one time the destructive distillation of coal (to give coke, coal tar, and coal gas) was the principal source of industrial organic chemicals.

**detailed balance** The cancellation of the effect of one process by another process that operates at the same time with the opposite effect. An example of detailed balance is provided by a chemical reaction between two molecular species A and B, which results in the formation of the molecular species C and D. Detailed balance for this chemical reaction occurs if the rate at which the reaction A + B → C + D occurs is equal to the rate at which the reaction C + D → A + B occurs. The equilibrium state in thermodynamics is characterized by detailed balance. When there is detailed balance in a system, the *self-organization far from equilibrium associated with *non-equilibrium statistical mechanics cannot occur.

**detector** 1. *See* DEMODULATION. 2. *See* COUNTER.

**detergent** A substance added to water to improve its cleaning properties. Although water is a powerful solvent for many compounds, it will not dissolve grease and natural oils. Detergents are compounds that cause such nonpolar substances to go into solution in water. *Soap is the original example, owing its action to the presence of ions formed from long-chain fatty acids (e.g. the octadecanoate (stearate) ion, $CH_3(CH_2)_{16}COO^-$). These have two parts: a nonpolar part (the hydrocarbon chain), which attaches to the grease; and a polar part (the $-COO^-$ group), which is attracted to the water. A disadvantage of soap is that it forms a scum with hard water (*see* HARDNESS OF WATER) and is relatively expensive to make. Various synthetic ('soapless') detergents have been developed from petrochemicals. The commonest, used in washing powders, is sodium dodecylbenzene-sulphonate, which contains $CH_3(CH_2)_{11}C_6H_4SO_2O^-$ ions. This, like soap,

is an example of an **anionic detergent**, i.e. one in which the active part is a negative ion. **Cationic detergents** have a long hydrocarbon chain connected to a positive ion. Usually they are amine salts, as in $CH_3(CH_2)_{15}N(CH_3)_3^+Br^-$, in which the polar part is the $-N(CH_3)_3^+$ group. **Nonionic detergents** have nonionic polar groups of the type $-C_2H_4-O-C_2H_4-OH$, which form hydrogen bonds with the water. Synthetic detergents are also used as wetting agents, emulsifiers, and stabilizers for foam.

**determinant** A *scalar quantity written |A| or det A. It represents a particular defined sum of products of a square *matrix. The determinant |A| expands as

$$\left|\begin{smallmatrix} a & b \\ c & d \end{smallmatrix}\right| = ad - bc.$$

**determined** Describing embryonic tissue at a stage when it can develop only as a certain kind of tissue (rather than as any kind).

**detonating gas** A mixture of hydrogen and oxygen in the ratio two parts hydrogen to one part oxygen by volume, produced by electrolysis of water. When sparked or ignited it explodes to produce water.

**detoxification (detoxication)** The process by which harmful compounds, such as drugs and poisons, are converted to less toxic compounds in the body. Detoxification is an important function of the *liver. *See also* PHASE I METABOLISM; PHASE II METABOLISM.

**detritivore** An animal that feeds on *detritus. Examples of detritivores are earthworms, blowflies, maggots, and woodlice. Detritivores play an important role in the breakdown of organic matter from decomposing animals and plants (*see* DECOMPOSER).

**detritus** Particles of organic material derived from dead and decomposing organisms, resulting from the activities of the *decomposers. Detritus is the source of food for *detritivores, which can themselves be eaten by carnivores in a **detritus food chain**:

detritus → detritivore → carnivore.

**deuterated compound** A compound in which some or all of the hydrogen–1 atoms have been replaced by deuterium atoms.

**deuterium (heavy hydrogen)** Symbol D. The isotope of hydrogen that has a mass number 2 (r.a.m. 2.0144). Its nucleus contains one proton and one neutron. The abundance of deuterium in natural hydrogen is about 0.015%. It is present in water as the oxide HDO (*see also* HEAVY WATER), from which it is usually obtained by electrolysis or fractional distillation. Its chemical behaviour is almost identical to hydrogen although deuterium compounds tend to react rather more slowly than the corresponding hydrogen compounds. Its physical properties are slightly different from those of hydrogen, e.g. b.p. 23.6 K (hydrogen 20.4 K).

**deuterium oxide** See HEAVY WATER.

**Deuteromycota** A phylum formerly used in some classifications to include all fungi in which sexual reproduction is absent. Also known as Fungi Imperfecti ('imperfect fungi'). They are now regarded as ascomycetes or basidiomycetes that have lost the ability to produce asci or basidia, respectively. Examples of these fungi are the *Penicillium* moulds.

**deuteron** A nucleus of a deuterium atom, consisting of a proton and a neutron bound together.

**Devarda's alloy** An alloy of copper (50%), aluminium (45%) and zinc (5%), used in chemical tests for the nitrate ion (in alkaline solutions it reduces a nitrate to ammonia).

**development** (in biology) The complex process of growth and maturation that occurs in living organisms. Cell division and differentiation are important processes in development. In vertebrate animals there are three developmental stages: (1) *cleavage, in which the zygote divides to form a ball of cells, the *blastula; (2) gastrulation, in which the cells become arranged in three primary *germ layers (*see* GASTRULA); (3) **organogenesis** (or **organogeny**), in which further cell division and differentiation results in the formation of organs. The development of many invertebrates (e.g. insects) and amphibians involves the process of *metamorphosis. *See also* MORPHOGENESIS; PRIMARY GROWTH.

**deviation** 1. **(angle of deviation)** The angle formed between a ray of light falling on a surface or transparent body and the ray leaving it. 2. The difference between one of an observed set of values and the true value, usually represented by the mean of all the observed values. The **mean deviation** is the mean of all the individual deviations of the set. *See* STANDARD DEVIATION.

**devitrification** Loss of the amorphous nature of glass as a result of crystallization.

**Devonian** A geological period in the Palaeozoic era that extended from the end of the Silurian (about 416 million years ago) to the beginning of the Carboniferous (about 360 million years ago). It was named by Adam Sedgwick (1785–1873) and Roderick Murchison (1792–1871) in 1839. The Devonian is divided into seven stages based on invertebrate fossil remains, such as corals, brachiopods, ammonoids, and crinoids, found in marine deposits. There were also extensive continental deposits consisting of conglomerates, red silts, and sandstones, forming the Old Red Sandstone facies. Fossils in the Old Red Sandstone include fishes and the earliest land plants. Graptolites became extinct early in the Devonian and the trilobites declined.

**de Vries, Hugo** *See* MENDEL, JOHANN GREGOR.

**dew** *See* PRECIPITATION.

**Dewar, Sir James** (1842–1923) British chemist and physicist, born in Scotland. In 1875 he became a professor at Cambridge University, while carrying out much of his experimental work at the Royal Institution in London. He began studying gases at low temperatures and in 1872 invented the *Dewar flask. In 1891, together with Frederick Abel (1827–1902), he developed the smokeless propellant explosive *cordite, and in 1898 was the first to liquefy hydrogen.

**Dewar benzene** An isomer of benzene, $C_6H_6$. Dewar benzene has the systematic name bicyclo[2,2,0]hexa-2,5-diene, and is a nonplanar bicyclic molecule. It has a high bond strain and will spontaneously slowly convert to benzene. The unsubstituted compound was first synthesized in 1963.

**Dewar flask** A vessel for storing hot or cold liquids so that they maintain their temperature independently of the surroundings. Heat transfer to the surroundings is reduced to a minimum: the walls of the vessel consist of two thin layers of glass (or, in large vessels, steel) separated by a vacuum to reduce conduction and convection; the inner surface of a glass vessel is silvered to reduce radiation; and the vessel is stoppered to prevent evaporation. It was devised around 1872 by Sir James Dewar and is also known by its first trade name **Thermos flask**. *See also* CRYOSTAT.

**Dewar structure** A proposed structure of *benzene, having a hexagonal ring of six carbon atoms with two opposite atoms joined by a long single bond across the ring and with two double C–C bonds, one on each side of the hexagon. Dewar structures contribute to the resonance hybrid of benzene.

**dew point** The temperature at which the water vapour in the air is saturated. As the temperature falls the dew point is the point at which the vapour begins to condense as droplets of water.

**dew-point hygrometer** *See* HYGROMETER.

**dextrin** An intermediate polysaccharide compound resulting from the hydrolysis of starch to maltose by amylase enzymes.

**dextrorotatory** Denoting a chemical compound that rotates the plane of polarization of plane-polarized light to the right (clockwise as observed by someone facing the oncoming radiation). *See* OPTICAL ACTIVITY.

**dextrose** *See* GLUCOSE.

**diabetes** *See* ANTIDIURETIC HORMONE; INSULIN.

**diacylglycerol** *See* INOSITOL.

**diagenesis** The processes, both chemical and physical, that modify a sediment once it has become buried. A material, such as calcite or silica, may crystallize and form a cement that binds together the particles of a sedimentary rock. Changes in temperature and pressure may convert one type of clay into another. New minerals may be deposited or formed by recrystallization, such as the formation of calcite from aragonite.

**diagonal relationship** A relationship within the periodic table by which certain elements in the second period have a close chemical similarity to their diagonal neighbours in the next group of the third period. This is particularly noticeable with the following pairs.

Lithium and magnesium:
(1) both form chlorides and bromides that hydrolyse slowly and are soluble in ethanol;
(2) both form colourless or slightly coloured crystalline nitrides by direct reaction with nitrogen at high temperatures;
(3) both burn in air to give the normal oxide only;

(4) both form carbonates that decompose on heating.

Beryllium and aluminium:
(1) both form highly refractory oxides with polymorphs;
(2) both form crystalline nitrides that are hydrolysed in water;
(3) addition of hydroxide ion to solutions of the salts gives an amphoteric hydroxide, which is soluble in excess hydroxide giving beryllate or aluminate ions $[Be(OH)_4]^{2-}$ and $[Al(OH)_4]^-$;
(4) both form covalent halides and covalent alkyl compounds that display bridging structures;
(5) both metals dissolve in alkalis.

Boron and silicon:
(1) both display semiconductor properties;
(2) both form hydrides that are unstable in air and chlorides that hydrolyse in moist air;
(3) both form acidic oxides with covalent crystal structures, which are readily incorporated along with other oxides into a wide range of glassy materials.

The reason for this relationship is a combination of the trends to increase size down a group and to decrease size along a period, and a similar, but reversed, effect in electronegativity, i.e. decrease down a group and increase along a period.

**diakinesis** The period at the end of the first prophase of *meiosis when the separation of *homologous chromosomes is almost complete and *crossing over has occurred.

**dialysis** A method by which large molecules (such as starch or protein) and small molecules (such as glucose or amino acids) in solution may be separated by selective diffusion through a semipermeable membrane. For example, if a mixed solution of starch and glucose is placed in a closed container made of a semipermeable substance (such as Cellophane), which is then immersed in a beaker of water, the smaller glucose molecules will pass through the membrane into the water while the starch molecules remain behind. The plasma membranes of living organisms are semipermeable, and dialysis takes place naturally in the kidneys for the excretion of nitrogenous waste. An artificial kidney (**dialyser**) utilizes the principle of dialysis by taking over the functions of diseased kidneys.

**diamagnetism** *See* MAGNETISM.

**1,6-diaminohexane (hexamethylenediamine)** A solid colourless amine,

$H_2N(CH_2)_6NH_2$; m.p. 41°C; b.p. 204°C. It is made by oxidizing cyclohexane to hexane-dioic acid, reacting this with ammonia to give the ammonium salt, and dehydrating the salt to give hexanedionitrile ($NC(CH_2)_6CN$). This is reduced with hydrogen to the diamine. The compound is used, with hexanedioic acid, for producing *nylon 6,6.

**diamond** The hardest known mineral (with a hardness of 10 on Mohs' scale). It is an allotropic form of pure *carbon that has crystallized in the cubic system, usually as octahedra or cubes, under great pressure. Diamond crystals may be colourless and transparent or yellow, brown, or black. They are highly prized as gemstones but also have extensive uses in industry, mainly for cutting and grinding tools. Diamonds occur in ancient volcanic pipes of kimberlite; the most important deposits are in South Africa but others are found in Tanzania, the USA, Russia, and Australia. Diamonds also occur in river deposits that have been derived from weathered kimberlite, notably in Brazil, Democratic Republic of Congo, Sierra Leone, and India. Industrial diamonds are increasingly being produced synthetically.

**diapause** A period of suspended development or growth occurring in many insects and other invertebrates during which metabolism is greatly decreased. It is often triggered by seasonal changes and regulated by an inborn rhythm and enables the animal to survive unfavourable environmental conditions so that its offspring may be produced in more favourable ones. The egg is the most common diapausal stage.

**diaphragm** **1.** (in optics) An opaque disc with a circular aperture at its centre. Diaphragms of different sizes are used to control the total light flux passing through an optical system or to reduce aberration by restricting the light passing through a system to the central portion. An **iris diaphragm** consists of a number of overlapping crescent-shaped discs arranged so that the central aperture can be continuously varied in diameter. **2.** (in anatomy) The muscular membrane that divides the thorax (chest) from the abdomen in mammals. It plays an essential role in breathing (*see* RESPIRATORY MOVEMENT), being depressed during inhalation and raised during exhalation.

**diaphysis** The shaft of a mammalian limb bone, which in immature animals is sepa-

rated from the ends of the bone (*see* EPIPHYSIS) by cartilage.

**diaspore** A mineral form of a mixed aluminium oxide and hydroxide, AlO.OH. *See* ALUMINIUM HYDROXIDE.

**diastase** *See* AMYLASE.

**diastema** The gap that separates the biting teeth from the grinding teeth in herbivores. It creates a space in which food can be held in readiness for the grinding action of the teeth. This space is filled by large canine teeth in carnivores.

**diastereoisomers** Stereoisomers that are not identical and yet not mirror images. For instance, the *d*-isomer of tartaric acid and the meso-isomer constitute a pair of diastereoisomers. *See* OPTICAL ACTIVITY.

**diastole** The phase of a heart beat that occurs between two contractions of the heart, during which the heart muscles relax and the ventricles fill with blood. *Compare* SYSTOLE. *See* BLOOD PRESSURE.

**diatomic molecule** A molecule formed from two atoms (e.g. $H_2$ or HCl).

**diatoms** *See* BACILLARIOPHYTA.

**diazepam** A *benzodiazepine used medically to treat anxiety, convulsions, insomnia, and alcohol withdrawal. It is widely used, and often known under its tradename **Valium**.

**diazine** *See* AZINE.

**diazo compounds** Organic compounds containing two linked nitrogen compounds. The term includes *azo compounds, diazonium compounds, and also such compounds as diazomethane, $CH_2N_2$.

(((•))) SEE WEB LINKS
• Information about IUPAC nomenclature

**diazonium salts** Unstable salts containing the ion $C_6H_5N_2^+$ (the **diazonium ion**: see formula). They are formed by *diazotization reactions.

**Diazonium salts.** Structure of diazonium ion $C_6H_5N_2^+$.

**diazotization** The formation of a *diazonium salt by reaction of an aromatic amine with nitrous acid at low temperature (below 5°C). The nitrous acid is produced in the reaction mixture from sodium nitrite and hydrochloric acid:

$$ArNH_2 + NaNO_2 + HCl \rightarrow ArN^+N + Cl^- + Na^+ + OH^- + H_2O$$

**dibasic acid** An *acid that has two acidic hydrogen atoms in its molecules. Sulphuric ($H_2SO_4$) and carbonic ($H_2CO_3$) acids are common examples.

**1,2-dibromoethane** A colourless liquid *haloalkane, $BrCH_2CH_2Br$; r.d. 2.2; m.p. 9.79°C; b.p. 131.36°C. It is made by addition of bromine to ethene and used as an additive in petrol to remove lead during combustion as the volatile lead bromide.

**dicarbide** *See* CARBIDE.

**dicarboxylic acid** A *carboxylic acid having two carboxyl groups in its molecules. In systematic chemical nomenclature, dicarboxylic acids are denoted by the suffix *-dioic*; e.g. hexanedioic acid, $HOOC(CH_2)_4COOH$.

**dichasium** *See* CYMOSE INFLORESCENCE.

**dichlorine oxide (chlorine monoxide)** A strongly oxidizing orange gas, $Cl_2O$, made by oxidation of chlorine using mercury(II) oxide. It is the acid anhydride of chloric(I) acid.

**dichloroethanoic acid** *See* CHLORO-ETHANOIC ACIDS.

**dichloromethane (methylene chloride)** A colourless, slightly toxic liquid, $CH_2Cl_2$, b.p. 41°C. It has a characteristic odour similar to that of trichloromethane (chloroform), from which it is made by heating with zinc and hydrochloric acid. It is used as a refrigerant and solvent (for paint stripping and degreasing).

**2,4-dichlorophenoxyacetic acid** *See* 2,4-D.

**dichogamy** The condition in which the male and female reproductive organs of a flower mature at different times, thereby ensuring that self-fertilization does not occur. *Compare* HOMOGAMY. *See also* PROTANDRY; PROTOGYNY.

**dichotomous** Describing the type of branching in plants that results when the growing point (apical bud) divides into two equal growing points, which in turn divide in a similar manner after a period of growth,

and so on. Dichotomous branching is common is ferns and mosses.

**dichroism** The property of some crystals, such as tourmaline, of selectively absorbing light vibrations in one plane while allowing light vibrations at right angles to this plane to pass through. Polaroid is a synthetic dichroic material. *See* POLARIZATION.

**dichromate(VI)** A salt containing the ion $Cr_2O_7^-$. Solutions containing dichromate(VI) ions are strongly oxidizing.

**Dicotyledoneae** In traditional classifications, one of the two classes of flowering plants (*see* ANTHOPHYTA), distinguished by having two seed leaves (*cotyledons) within the seed. The dicotyledons usually have leaf veins in the form of a net, a ring of vascular bundles in the stem, and flower parts in fours or fives or multiples of these. They include many food plants (e.g. potatoes, peas, beans), ornamentals (e.g. roses, ivies, honeysuckles), and hardwood trees (e.g. oaks, limes, beeches). Dicots are no longer considered a taxonomically valid group. *See* EUDICOT. *Compare* MONOCOTYLEDONEAE.

**Dictyoptera** An order of insects (sometimes classified as *Orthoptera) comprising the cockroaches (suborder Blattaria) and the mantids (suborder Mantodea), occurring mainly in tropical regions. Cockroaches are oval and flattened in shape; some have a single well-developed pair of wings, folded back over the abdomen at rest, while in others the wings may be reduced or absent. They are usually found in forest litter, feeding on dead organic matter, but some species, e.g. the American cockroach (*Periplaneta americana*), are major household pests, scavenging on starchy foods, fruits, etc. In most species the females produce capsules (**oothecae**) containing 16–40 eggs. These are either deposited or carried by the female during incubation.

**dielectric** A nonconductor of electric charge in which an applied electric field causes a *displacement of charge but not a flow of charge. Electrons within the atoms of a dielectric are, on average, displaced by an applied field with respect to the nucleus, giving rise to a dipole that has an electric moment in the direction of the field. The resulting stress within the dielectric is known as the **electric polarization** ($P$) and is defined by $P = D - E\varepsilon_0$, where $D$ is the displacement, $E$ is the electric field strength, and $\varepsilon_0$ is the electric constant.

The **dielectric constant** is now called the relative *permittivity. The **dielectric strength** is the maximum potential gradient that can be applied to a material without causing it to break down. It is usually expressed in volts per millimetre. *See also* CAPACITOR.

**dielectric constant**  *See* PERMITTIVITY.

**dielectric heating**  The heating of a dielectric material, such as a plastic, by applying a radio-frequency electric field to it. The most common method is to treat the material as the dielectric between the plates of a capacitor. The heat produced is proportional to $V^2 fA\phi/t$, where $V$ is the applied potential difference, $f$ its frequency, $A$ is the area of the dielectric, $t$ its thickness, and $\phi$ is the loss factor of the material (related to its *permittivity).

**Diels–Alder reaction**  A type of chemical reaction in which a compound containing two double bonds separated by a single bond (i.e. a conjugated *diene) adds to a suitable compound containing one double bond (known as the **dienophile**) to give a ring compound. In the dienophile, the double bond must have a carbonyl group on each side. It is named after the German chemists Otto Diels (1876–1954) and Kurt Alder (1902–58), who discovered it in 1928.

**diene**  An *alkene that has two double bonds in its molecule. If the two bonds are separated by one single bond, as in buta-1,3-diene $CH_2{:}CHCH{:}CH_2$, the compound is a **conjugated diene**.

**dienophile**  *See* DIELS–ALDER REACTION.

**Diesel engine**  *See* INTERNAL-COMBUSTION ENGINE.

**diet**  The food requirements of an organism. The foods that constitute the human diet should contain vitamins, mineral salts (*see* ESSENTIAL ELEMENT), and dietary *fibre as well as water, carbohydrates and fats (which provide energy), and proteins (required for growth and maintenance). A balanced diet contains of the correct proportions of these *nutrients, which will vary depending on the age, sex, body size, and the level of activity of the individual. An inadequate supply of different food types in the diet can lead to *malnutrition.

**dietary fibre**  *See* FIBRE.

**diethyl ether**  *See* ETHOXYETHANE.

**differential calculus**  *See* CALCULUS.

**differential equation**  An equation in which a derivative of $y$ with respect to $x$ appears as well as the variables $x$ and $y$. The **order** of a differential equation is the order of its highest derivative. The **degree** of the equation is the highest power present of the highest-order derivative. There are many types of differential equation, each having its own method of solution. The simplest type has separable variables, enabling each side of the equation to be integrated separately.

**differential scanning calorimetry (DSC)**  *See* THERMAL ANALYSIS.

**differential thermal analysis (DTA)**  *See* THERMAL ANALYSIS.

**differentiation**  **1.** (in mathematics) The process of finding the **derivative** of a function in differential *calculus. If $y = f(x)$, the derivative of $y$, written $dy/dx$ or $f'(x)$, is equal to the limit as $\Delta x \to 0$ of $[f(x + \Delta x) - f(x)]/\Delta x$. In general, if $y = x^n$, then $dy/dx = nx^{n-1}$. On a graph of $y = f(x)$, the derivative $dy/dx$ is the gradient of the tangent to the curve at the point $x$. **2.** (in biology) The changes from simple to more complex forms undergone by developing tissues and organs so that they become specialized for particular functions. Differentiation occurs during embryonic development, *regeneration, and (in plants) meristematic activity (*see* MERISTEM). *See also* HOMEOTIC GENES.

**diffraction**  The spreading or bending of waves as they pass through an aperture or round the edge of a barrier. The diffracted waves subsequently interfere with each other (*see* INTERFERENCE) producing regions of reinforcement and weakening. First noticed as occurring with light by Francesco Grimaldi (1618–63), the phenomenon gave considerable support to the wave theory of light. Diffraction also occurs with streams of particles such as electrons because of the quantum-mechanical wave nature of such particles. *See also* FRESNEL DIFFRACTION; FRAUNHOFER DIFFRACTION; ELECTRON DIFFRACTION.

**diffraction grating**  A device for producing spectra by diffraction and interference. The usual grating consists of a glass or speculum-metal sheet with a very large number of equidistant parallel lines ruled on it (usually of the order of 1000 per mm). Diffracted light after transmission through the glass or reflection by the speculum produces maxima of illumination (spectral lines) according to the equation $m\lambda = d(\sin i + \sin\theta)$, where $d$ is the distance between grating lines, $\lambda$ is the wavelength of the light, $i$ is the angle of incidence, $\theta$ the direction of the diffracted maximum, and $m$ is the 'order' of the spectral line. Reflection gratings are also used to produce spectra in the ultraviolet region of the electromagnetic spectrum.

**diffusion** 1. The process by which different substances mix as a result of the random motions of their component atoms, molecules, and ions. In gases, all the components are perfectly miscible with each other and mixing ultimately becomes nearly uniform, though slightly affected by gravity. The diffusion of a solute through a solvent to produce a solution of uniform concentration is slower, but otherwise very similar to the process of gaseous diffusion. Diffusion of small molecules and ions across a *cell membrane is known as **passive transport**. In solids, diffusion occurs very slowly at normal temperatures. *See also* FICK'S LAW; GRAHAM'S LAW. 2. The scattering of a beam of light by reflection at a rough surface or by transmission through a translucent (rather than transparent) medium, such as frosted glass. 3. The passage of elementary particles through matter when there is a high probability of scattering and a low probability of capture.

**diffusion cloud chamber** *See* CLOUD CHAMBER.

**diffusion gradient** *See* CONCENTRATION GRADIENT.

**diffusion pump (condensation pump)** A *vacuum pump in which oil or mercury vapour is diffused through a jet, which entrains the gas molecules from the container in which the pressure is to be reduced. The diffused vapour and entrained gas molecules are condensed on the cooled walls of the pump. Pressures down to $10^{-7}$ Pa can be reached by sophisticated forms of the diffusion pump.

**digestion** The breakdown by a living organism of ingested food material into chemically simpler forms that can be readily absorbed and assimilated by the body. This process requires the action of digestive enzymes and may take place extracellularly (i.e. in the *alimentary canal), as is the case in

most animals; or intracellularly (e.g. by engulfing phagocytic cells), as occurs in protozoans and cnidarians.

**digestive system** The system of organs that are involved in the process of *digestion. The digestive system of mammals is divided into the **gastrointestinal tract** (*see* ALIMENTARY CANAL) and accessory structures, such as teeth, tongue, liver, pancreas, and gall bladder.

**digit** 1. (in mathematics) A symbol used to represent a single number. For example, the number 479 consists of three digits. 2. (in anatomy) A finger or toe. In the basic limb structure of terrestrial vertebrates there are five digits (*see* PENTADACTYL LIMB). This number is retained in man and other primates, but in some other species the number of digits is reduced. Frogs, for example, have four fingers and five toes, and in ungulate (hooved) mammals, the digits are reduced and their tips are enclosed in horn, forming hooves.

**digital audio tape** *See* DAT.

**digital camera** *see* CAMERA.

**digital certificate** *See* CERTIFICATE.

**digital computer** *See* COMPUTER.

**digital display** A method of indicating a reading of a measuring instrument, clock, etc., in which the appropriate numbers are generated on a fixed display unit by the varying parameter being measured rather than fixed numbers on a scale being indicated by a moving pointer or hand. *See* DIGITRON; LCD; LIGHT-EMITTING DIODE.

**digital recording** A method of recording or transmitting sound in which the sound itself is not transmitted or recorded. Instead the pressure in the sound wave is sampled at least 30 000 times per second and the successive values represented by numbers, which are then transmitted or recorded. Afterwards they are restored to analogue form in the receiver or player. This method is used for very high fidelity recordings as no distortion or interference occurs during transmission or in the recording process.

**digitigrade** Describing the gait of most fast-running animals, such as dogs and cats, in which only the toes are on the ground and the rest of the foot is raised off the ground. *Compare* PLANTIGRADE; UNGULIGRADE.

**digitron** An electronic gas-discharge tube

that provides a *digital display in calculators, counters, etc. It usually has 10 cold cathodes shaped into the form of the digits 0–9. The cathode selected receives a voltage pulse causing a glow discharge to illuminate the digit. It has now largely been superseded by *light-emitting diodes and liquid-crystal displays (*see* LCD).

**dihedral (dihedron)** An angle formed by the intersection of two planes (e.g. two faces of a polyhedron). The **dihedral angle** is the angle formed by taking a point on the line of intersection and drawing two lines from this point, one in each plane, perpendicular to the line of intersection.

**dihybrid cross** A genetic cross between parents that differ in two characteristics, controlled by genes at different loci. Mendel performed a dihybrid cross using pea plants and the characteristics of seed colour and texture: the parental plants had either smooth yellow seeds (SSYY) – the dominant characteristics – or wrinkled green seeds (ssyy) – the recessive characteristics. All the offspring had smooth yellow seeds, being heterozygous (SsYy) for the two alleles. Crossing between these offspring produced an $F_2$ generation of plants with smooth yellow, smooth green, wrinkled yellow, and wrinkled green seeds in the ratio 9:3:3:1 (see illustration). Mendel used these results as the basis for his Law of Independent Assortment (*see* MENDEL'S LAWS). *Compare* MONO-HYBRID CROSS.

**dihydrate** A crystalline hydrate containing two moles of water per mole of compound.

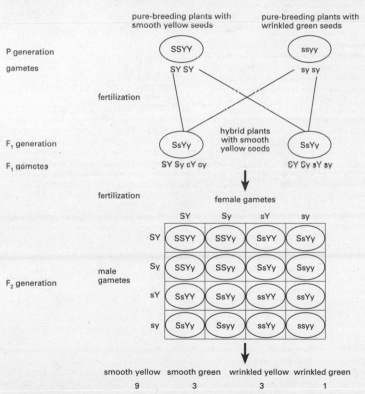

**Dihybrid cross.**

**dihydric alcohol** *See* DIOL.

**dihydrogen** The normal form of molecular hydrogen, $H_2$, used to distinguish it from hydrogen atoms.

**1,2-dihydroxybenzene (catechol)** A colourless crystalline phenol, $C_6H_4(OH)_2$; r.d. 1.15; m.p. 105°C; b.p. 245°C. It is used as a photographic developer.

**2,3-dihydroxybutanedioic acid** *See* TARTARIC ACID.

**dikaryon** A cell of a fungal hypha or mycelium containing two haploid nuclei of different strains. The nuclei associate in pairs but do not fuse, therefore the cell is not truly diploid. Dikaryosis occurs in the Basidiomycota and Ascomycota.

**dilatancy** *See* NEWTONIAN FLUID.

**dilation (dilatation)** 1. An increase in volume. *See also* VASODILATION. 2. *See* TIME DILATION.

**dilatometer** A device for measuring the cubic *expansivities of liquids. It consists of a bulb of known volume joined to a graduated capillary tube, which is closed at the top to prevent evaporation. A known mass of liquid is introduced into the device, which is submerged in a bath maintained at different temperatures $t_1$ and $t_2$. The two volumes corresponding to these temperatures, $V_1$ and $V_2$, are read off the calibrated stem. The value of the cubic expansivity ($\gamma$) is then given by

$$\gamma = (V_2 - V_1) / V_1(t_2 - t_1).$$

**dilead(II) lead(IV) oxide** A red amorphous powder, $Pb_3O_4$; r.d. 9.1; decomposes at 500°C to lead(II) oxide. It is prepared by heating lead(II) oxide to 400°C and has the unusual property of being black when hot and red-orange when cold. The compound is nonstoichiometric, generally containing less oxygen than implied by the formula. It is largely covalent and has $Pb(IV)O_6$ octahedral groups linked together by Pb(II) atoms, each joined to three oxygen atoms. It is used in glass making but its use in the paint industry has largely been discontinued because of the toxicity of lead. Dilead(II) lead(IV) oxide is commonly called **red lead** or, more accurately, **red lead oxide**.

**Dillie–Koppanyi test** A presumptive test for barbituates. The **Dillie–Koppanyi test reagent** has two solutions: a 1% solution of cobalt acetate in methanol, followed by a 5% solution of isopropylamine

($CH_3CH(CH_3)NH_2$) in methanol. Barbiturates give a reddish-violet colour.

**diluent** A substance added to dilute a solution or mixture (e.g. a *filler).

**dilute** Describing a solution that has a relatively low concentration of solute.

**dilution** The volume of solvent in which a given amount of solute is dissolved.

**dilution law** *See* OSTWALD'S DILUTION LAW.

**dimensional analysis** A method of checking an equation or a solution to a problem by analysing the dimensions in which it is expressed. It is also useful for establishing the form, but not the numerical coefficients, of an empirical relationship. If the two sides of an equation do not have the same dimensions, the equation is wrong. If they do have the same dimensions, the equation may still be wrong, but the error is likely to be in the arithmetic rather than the method of solution.

**dimensionless units** *See* SI UNITS.

**dimensions** The product or quotient of the basic physical quantities, raised to the appropriate powers, in a derived physical quantity. The basic physical quantities of a mechanical system are usually taken to be mass ($M$), length ($L$), and time ($T$). Using these dimensions, the derived physical quantity velocity will have the dimensions $L/T$ and acceleration will have the dimensions $L/T^2$. As force is the product of a mass and an acceleration (*see* NEWTON'S LAWS OF MOTION), force has the dimensions $MLT^{-2}$. In electrical work in *SI units, current, $I$, can be regarded as dimensionally independent and the dimensions of other electrical units can be found from standard relationships. Charge, for example, is measured as the product of current and time. It therefore has the dimension $IT$. Potential difference is given by the relationship $P = VI$, where $P$ is power. As power is force × distance ÷ time ($MLT^{-2} \times L \times T^{-1} = ML^2T^{-3}$), voltage $V$ is given by $V = ML^2T^{-3}I^{-1}$.

**dimer** An association of two identical molecules linked together. The molecules may react to form a larger molecule, as in the formation of dinitrogen tetroxide ($N_2O_4$) from nitrogen dioxide ($NO_2$), or the formation of an *aluminium chloride dimer ($Al_2Cl_6$) in the vapour. Alternatively, they may be held by hydrogen bonds. For example, carboxylic acids form dimers in organic solvents, in which hydrogen bonds exist between the O of the C=O group and the H of the –O–H group.

**dimethylbenzenes (xylenes)** Three compounds with the formula $(CH_3)_2C_6H_4$, each having two methyl groups substituted on the benzene ring. 1,2-dimethylbenzene is ortho-xylene, etc. A mixture of the isomers (b.p. 135–145°C) is obtained from petroleum and is used as a clearing agent in preparing specimens for optical microscopy.

**dimethylformamide (DMF)** A colourless liquid compound, $(CH_3)_2NCHO$; r.d. 0.944; m.p. –61°C; b.p. 153°C. The systematic name is N,N-dimethylmethanamide. It can be made from methanoic acid (formic acid) and dimethylamine, and is widely used as a solvent.

**dimethylglyoxime (DMG)** A colourless solid, $(CH_3CNOH)_2$, m.p. 234°C. It sublimes at 215°C and slowly polymerizes if left to stand. It is used in chemical tests for nickel, with which it forms a dark-red complex.

**dimethyl sulphoxide (DMSO)** A colourless solid, $(CH_3)_2SO$; m.p. 18°C; b.p. 189°C. It is used as a solvent and as a reagent in organic synthesis.

**1,3-dimethylxanthine** *See* THEOPHYLLINE.

**3,7-dimethylxanthine** *See* THEOBROMINE.

**dimictic lake** A lake that is stratified by a *thermocline that is not permanent but is turned over twice during one year. The thermocline is disrupted due to seasonal changes in the climate. A **meromictic lake** is one in which there is a permanent stratification.

**dimorphism** 1. (in biology) The existence of two distinctly different types of individual within a species. An obvious example is **sexual dimorphism** in certain animals, in which the two sexes differ in colouring, size, etc. Dimorphism also occurs in some lower plants, such as mosses and ferns, that show an *alternation of generations. 2. (in chemistry) *See* POLYMORPHISM.

**dinitrogen** The normal form of molecular nitrogen, $N_2$, used to distinguish it from nitrogen atoms.

**dinitrogen oxide (nitrous oxide)** A colourless gas, $N_2O$, d. 1.97 g dm$^{-3}$; m.p. –90.8°C; b.p. –88.5°C. It is soluble in water, in

ethanol, and sulphuric acid. It may be prepared by the controlled heating of ammonium nitrate (chloride free) to 250°C and passing the gas produced through solutions of iron(II) sulphate to remove impurities of nitrogen monoxide. It is relatively unreactive, being inert to halogens, alkali metals, and ozone at normal temperatures. It is decomposed on heating above 520°C to nitrogen and oxygen and will support the combustion of many compounds. Dinitrogen oxide is used as an anaesthetic gas ('laughing gas') and as an aerosol propellant.

**dinitrogen tetroxide** A colourless to pale yellow liquid or a brown gas, $N_2O_4$; r.d. 1.45 (liquid); m.p. –11.2°C; b.p. 21.2°C. It dissolves in water with reaction to give a mixture of nitric acid and nitrous acid. It may be readily prepared in the laboratory by the reaction of copper with concentrated nitric acid; mixed nitrogen oxides containing dinitrogen oxide may also be produced by heating metal nitrates. The solid compound is wholly $N_2O_4$ and the liquid is about 99% $N_2O_4$ at the boiling point; $N_2O_4$ is diamagnetic. In the gas phase it dissociates to give **nitrogen dioxide**

$$N_2O_4 \rightleftharpoons 2NO_2$$

Because of the unpaired electron this is paramagnetic and brown. Liquid $N_2O_4$ has been widely studied as a nonaqueous solvent (self-ionizes to $NO^+$ and $NO_3^-$). Dinitrogen tetroxide, along with other nitrogen oxides, is a product of combustion engines and is thought to be involved in the depletion of stratospheric ozone.

**Dinomastigota (Dinoflagellata)** A phylum of mostly single-celled protists. They are abundant in the marine plankton; many are *photoautotrophs, containing brown xanthophyll pigments in addition to chlorophyll. Dinoflagellates characteristically have two undulipodia (flagella) for locomotion and most have a rigid cell wall of cellulose encrusted with silica. Some species (e.g. *Noctiluca miliaris*) are bioluminescent.

**dinosaur** An extinct terrestrial reptile belonging to a group that constituted the dominant land animals of the Jurassic and Cretaceous periods, 190–65 million years ago. There were two orders. The Ornithischia were typically quadrupedal herbivores, many with heavily armoured bodies, and included *Stegosaurus*, *Triceratops*, and *Iguanodon*. They were all characterized by birdlike pelvic girdles. The Saurischia included many bipedal carnivorous forms, such as *Tyrannosaurus* (the largest known carnivore), and some quadrupedal herbivorous forms, such as *Apatosaurus* (*Brontosaurus*) and *Diplodocus*. They all had lizard-like pelvic girdles. Many of the herbivorous dinosaurs were amphibious or semiaquatic.

( SEE WEB LINKS )

- A Dino Directory produced by the Natural History Museum, London; contains a wealth of information and images

**dinucleotide** A compound consisting of two *nucleotides.

**diode** An electronic device with two electrodes. In the obsolescent **thermionic diode** a heated cathode emits electrons, which flow across the intervening vacuum to the anode when a positive potential is applied to it. The device permits flow of current in one direction only as a negative potential applied to the anode repels the electrons. This property of diodes was made use of in the first thermionic radios, in which the diode was used to demodulate the transmitted signal (*see* MODULATION). In the **semiconductor diode**, a *p*–*n* junction performs a similar function. The forward current increases with increasing potential difference whereas the reverse current is very small indeed. *See* SEMICONDUCTOR; TRANSISTOR.

**dioecious** Describing plant species that have male and female flowers on separate plants. Examples of dioecious plants are willows. *Compare* MONOECIOUS.

**diol (dihydric alcohol)** An *alcohol containing two hydroxyl groups per molecule.

**dioptre** A unit for expressing the power of a lens or mirror equal to the reciprocal of its focal length in metres. Thus a lens with a focal length of 0.5 metre has a power of $1/0.5 = 2$ dioptres. The power of a converging lens is usually taken to be positive and that of a diverging lens negative. Because the power of a lens is a measure of its ability to cause a beam to converge, the dioptre is now sometimes called the radian per metre.

**dioxan** A colourless toxic liquid, $C_4H_8O_2$; r.d. 1.03; m.p. 11°C; b.p. 101.5°C. The molecule has a six-membered ring containing four $CH_2$ groups and two oxygen atoms at opposite corners. It can be made from ethane-1,2-diol and is used as a solvent.

**dioxin (2,4,7,8-tetrachlorodibenzo-*p*-dioxin)** A toxic solid, formed in the manufacture of the herbicide *2,4,5-T; it was present as an impurity in Agent Orange, used as a defoliant during the Vietnam War. It is the most toxic member of a group of compounds (called **dioxins**) that occur widely as environmental pollutants, being produced during combustion processes and as byproducts in various industrial manufacturing processes. Dioxins decompose very slowly and may be concentrated in the food chain; in animals they are stored in fat. Exposure to high levels of dioxins can cause skin disfigurement (chloracne) and may result in fetal defects. Because of their toxicity, many countries have imposed strict controls to reduce industrial emissions of dioxins.

**dioxonitric(III) acid** *See* NITROUS ACID.

**dioxygen** The normal form of molecular oxygen, $O_2$, used to distinguish it from oxygen atoms or from ozone ($O_3$).

**dioxygenyl compounds** Compounds containing the positive ion $O_2^+$, as in dioxygenyl hexafluoroplatinate $O_2PtF_6$ – an orange solid that sublimes in vacuum at 100°C. Other ionic compounds of the type $O_2^+[MF_6]^-$ can be prepared, where M is P, As, or Sb.

**dip** *See* GEOMAGNETISM.

**dipeptide** A compound consisting of two amino acid units joined at the amino (–$NH_2$) end of one and the carboxyl (–COOH) end of the other. This peptide bond (*see* PEPTIDE) is formed by a condensation reaction that involves the removal of one molecule of water.

**diphenylmethanone (benzophenone)** A colourless solid, $C_6H_5COC_6H_5$, m.p. 49°C. It has a characteristic smell and is used in making perfumes. It is made from benzene and benzoyl chloride using the *Friedel–Crafts reaction with aluminium chloride as catalyst.

**diphosgene** A colourless liquid, $ClCO.O.CCl_3$, originally used in 1916 by Germany in World War I as a chemical warfare agent. It is now used as a reagent in organic synthesis. *See also* CARBONYL CHLORIDE.

**diphosphane (diphosphine)** A yellow liquid, $P_2H_4$, which is spontaneously flammable in air. It is obtained by hydrolysis of calcium phosphide. Many of the references to the spontaneous flammability of phosphine ($PH_3$) are in fact due to traces of $P_2H_4$ as impurities.

**diphosphine** *See* DIPHOSPHANE.

**diphosphonates** *See* BISPHOSPHONATES.

**diphyodont** Describing a type of dentition that is characterized by two successive sets of teeth: the *deciduous (milk) teeth, which are followed by the *permanent (adult) teeth. Mammals have a diphyodont dentition. *Compare* MONOPHYODONT; POLYPHYODONT.

**diploblastic** Describing an animal with a body wall composed only of two layers, *ectoderm and *endoderm, sometimes with a noncellular *mesoglea between them. Cnidarians are diploblastic. *Compare* TRIPLOBLASTIC.

**diploid** Describing a nucleus, cell, or organism with twice the *haploid number of chromosomes characteristic of the species. The diploid number is designated as 2$n$. Two sets of chromosomes are present, one set being derived from the female parent and the other from the male. In animals, all the cells except the reproductive cells are diploid.

**Diplopoda** A class of terrestrial *arthropods comprising the millipedes and belonging to the subphylum *Myriapoda. Diplopods are characterized by 20 to over 60 body segments each bearing two pairs of legs. They are slow moving and feed on decaying leaves.

**diplotene** The period in the first prophase of *meiosis when paired *homologous chromosomes begin to move apart. They remain attached at a number of points (*see* CHIASMA).

**Dipnoi** A subclass or order of bony fishes that contains the lungfishes, which have lungs and breathe air. They are found in Africa, Australia, and South America, where they live in freshwater lakes and marshes that tend to become stagnant or even dry up in summer. They survive in these conditions by burrowing into the mud, leaving a small hole for breathing air, and entering a state of *aestivation, in which they can remain for six months or more. The Dipnoi date from the Devonian era (416–360 million years ago) and share many features with the modern *Amphibia.

**dipole 1.** A pair of separated opposite electric charges. The **dipole moment** (symbol μ) is the product of the positive charge and the distance between the charges. Dipole moments are often stated in *debyes; the SI unit is the coulomb metre. In a diatomic molecule, such as HCl, the dipole moment is a measure of the polar nature of the bond (*see* POLAR MOLECULE); i.e. the extent to which the average electron charge is displaced towards one atom (in the case of HCl, the electrons are attracted towards the more electronegative chlorine atom). In a polyatomic molecule, the dipole moment is the vector sum of the dipole moments of the individual bonds. In a symmetrical molecule, such as tetrachloromethane ($CCl_4$), there is no overall dipole moment, although the individual C–Cl bonds are polar. **2.** An aerial commonly used for frequencies below 30 megahertz, although some are in use above this frequency. It consists of a rod, fed or tapped at its centre. It may be half a wavelength or a full wavelength long.

**dipole radiation** *See* FORBIDDEN TRANSITIONS.

**Diptera** An order of insects comprising the true, or two-winged, flies. Flies possess only one pair of wings – the forewings; the hindwings are modified to form small clublike **halteres** that function as balancing organs. Typically fluid feeders, flies have mouthparts adapted for piercing and sucking or for lapping; the diet includes nectar, sap, decaying organic matter, and blood. Some species prey on insects; others are parasitic. Dipteran larvae (**maggots**) are typically wormlike with an inconspicuous head. They undergo metamorphosis via a pupal stage to the adult form. Many flies or their larvae are serious pests, either by feeding on crops (e.g. fruit flies) or as vectors of disease organisms (e.g. the house fly (*Musca domestica*) and certain mosquitoes).

**Dirac, Paul Adrien Maurice** (1902–84) British physicist, who shared the 1933 Nobel Prize for physics with Erwin *Schrödinger for developing Schrödinger's non-relativistic wave equations to take account of relativity. This modified equation predicted the existence and properties of the *positron. Dirac also invented, independently of Enrico Fermi, the form of *quantum statistics known as Fermi–Dirac statistics.

**Dirac constant** *See* PLANCK CONSTANT.

**Dirac equation** A version of the nonrelativistic *Schrödinger equation taking special relativity theory into account. The Dirac equation is needed to discuss the quantum mechanics of electrons in heavy atoms and, more generally, to discuss fine-structure features of atomic spectra. The equation was put forward by Paul Dirac in 1928. It can be solved exactly in the case of the hydrogen atom but can only be solved using approximation techniques for more complicated atoms.

**direct current (d.c.)** An electric current in which the net flow of charge is in one direction only. *Compare* ALTERNATING CURRENT.

**direct-current motor** *See* ELECTRIC MOTOR.

**direct dye** *See* DYES.

**direct motion** 1. The apparent motion of a planet from west to east as seen from the earth against the background of the stars. 2. The anticlockwise rotation of a planet, as seen from its north pole. *Compare* RETROGRADE MOTION.

**directrix** 1. A plane curve defining the base of a *cone. 2. A straight line from which the distance to any point on a *conic is in a constant ratio to the distance from that point to the focus.

**disaccharide** A sugar consisting of two linked *monosaccharide molecules. For example, sucrose comprises one glucose molecule and one fructose molecule bonded together.

**discharge** 1. The conversion of the chemical energy stored in a *secondary cell into electrical energy. 2. The release of electric charge from a capacitor in an external circuit. 3. The passage of charge carriers through a gas at low pressure in a **discharge tube**. A potential difference applied between cathode and anode creates an electric field that accelerates any free electrons and ions to their appropriate electrodes. Collisions between electrons and gas molecules create more ions. Collisions also produce excited ions and molecules (*see* EXCITATION), which decay with emission of light in certain parts of the tube.

**discontinuous function** *See* CONTINUOUS FUNCTION.

**discontinuous variation (qualitative variation)** Clearly defined differences in a characteristic that can be observed in a population. Characteristics that are determined by different *alleles at a single locus show

discontinuous variation, e.g. garden peas are either wrinkled or smooth. *Compare* CONTINUOUS VARIATION.

**disease** A condition in which the normal function of some part of the body (cells, tissues, or organs) is disturbed. A variety of microorganisms and environmental agents are capable of causing disease. The functional disturbances are often accompanied by structural changes in tissue.

**disilane** *See* SILANE.

**disinfectant** Any substance that kills or inhibits the growth of disease-producing microorganisms and is in general toxic to human tissues. Disinfectants include cresol, bleaching powder, and phenol. They are used to cleanse surgical apparatus, sickrooms, and household drains and if sufficiently diluted can be used as *antiseptics.

**disintegration** Any process in which an atomic nucleus breaks up spontaneously into two or more fragments in a radioactive decay process or breaks up as a result of a collision with a high-energy particle or nuclear fragment.

**disintegration constant** *See* DECAY.

**dislocation** *See* CRYSTAL DEFECTS (Feature).

**disodium hydrogenphosphate(V) (disodium orthophosphate)** A colourless crystalline solid, $Na_2HPO_4$, soluble in water and insoluble in ethanol. It is known as the dihydrate (r.d. 2.066), heptahydrate (r.d. 1.68), and dodecahydrate (r.d. 1.52). It may be prepared by titrating phosphoric acid with sodium hydroxide to an alkaline end point (phenolphthalein) and is used in treating boiler feed water and in the textile industry.

**disodium orthophosphate** *See* DISODIUM HYDROGENPHOSPHATE(V).

**disodium tetraborate-10-water** *See* BORAX.

**d-isomer** *See* OPTICAL ACTIVITY.

**D-isomer** *See* ABSOLUTE CONFIGURATION.

**disordered solid** A material that neither has the structure of a perfect *crystal lattice nor of a crystal lattice with isolated *crystal defects. In a **random alloy**, one type of disordered solid, the order of the different types of atom occurs at random. Another type of disordered solid is formed by introducing a high concentration of defects, with the defects distributed randomly throughout the solid. In an *amorphous solid, such as glass, there is a random network of atoms with no lattice. The theory of disordered solids is more complicated than the theory of crystals, requiring such concepts as *localization and *spin glasses.

**dispersal** The dissemination of offspring of plants or sessile animals. Dispersal provides organisms that are not mobile with a better chance of survival by reducing *competition among offspring and parents. It also promotes the colonization of new habitats. Flowering plants produce fruits or seeds that are dispersed by such agents as wind, water, or animals. Specialized structures have evolved in many species to aid dispersal (*see* FRUIT).

**disperse dye** *See* DYES.

**disperse phase** *See* COLLOIDS.

**dispersion** The splitting up of a ray of light of mixed wavelengths by refraction into its components. Dispersion occurs because the *deviation for each wavelength is different on account of the different speeds at which waves of different wavelengths pass through the refracting medium. If a ray of white light strikes one face of a prism and passes out of another face, the white light will be split into its components and the full visible spectrum will be formed. The **dispersive power** of a prism (or other medium) for white light is defined by

$$(n_b - n_r)/(n_y - 1),$$

where $n_b$, $n_r$, and $n_y$ are the *refractive indexes for blue, red, and yellow light respectively. The term is sometimes applied to the separation of wavelengths produced by a *diffraction grating.

**dispersion forces** *See* VAN DER WAALS' FORCE.

**dispersive power** *See* DISPERSION.

**displacement** 1. Symbol $s$. A specified distance in a specified direction. It is the vector equivalent of the scalar distance. 2. *See* ELECTRIC DISPLACEMENT.

**displacement activity** An activity shown by an animal that appears to be irrelevant to its situation. Displacement activities are frequently observed when there is conflict between opposing tendencies. For example, birds in aggressive situations, in which there are simultaneous tendencies to attack and to

flee, may preen their feathers as a displacement activity.

**displacement current** A term of the form $\partial D/\partial t$, where $D$ is the *electric displacement, which is added to the electric current density $J$ to modify *Ampère's law in the fourth of *Maxwell's equations. The necessity for the displacement current term was postulated by Maxwell when he put forward his equations to introduce a degree of symmetry between electricity and magnetism.

**displacement reaction** *See* SUBSTITUTION REACTION.

**display behaviour** Stereotyped movement or posture that serves to influence the behaviour of another animal. Many displays in *courtship and *aggression are conspicuous and characteristic of the species; special markings or parts of the body may be prominently exhibited (for example, the male peacock spreads its tail in courtship). Other displays are cryptic and make it harder for a predator to recognize the displaying animal as potential prey. For example, geometer moth caterpillars, which look like twigs, hold themselves on plant stems with one end sticking into the air.

**disproportionation** A type of chemical reaction in which the same compound is simultaneously reduced and oxidized. For example, copper(I) chloride disproportionates thus:

$$2CuCl \rightarrow Cu + CuCl_2$$

The reaction involves oxidation of one molecule

$$Cu^I \rightarrow Cu^{II} + e$$

and reduction of the other

$$Cu^I + e \rightarrow Cu$$

The reaction of halogens with hydroxide ions is another example of a disproportionation reaction, for example

$$Cl_2(g) + 2OH^-(aq) \rightleftharpoons Cl^-(aq) + ClO^-(aq) + H_2O(l)$$

The reverse process is *comproportionation.

**dissipative structure** A state of matter that occurs when a system is driven away from thermal *equilibrium by external constraints that have exceeded certain critical values. A dissipative structure, which is associated with *broken symmetry, is an example of *complexity and *self-organization. An example of dissipative structure is a *Bénard cell. The order in a dissipative structure that is not in thermal equilibrium occurs as a response to such parameters as heat.

**dissipative system** A system that involves irreversible processes (*see* IRREVERSIBILTY). All real systems are dissipative (in contrast to such idealized systems as the frictionless pendulum, which is invariant under time reversal). In a dissipative system the system is moving towards a state of equilibrium, which can be regarded as moving toward a point *attractor in *phase space; this is equivalent to moving towards the minimum of the *free energy, $F$.

**dissociation** The breakdown of a molecule, ion, etc., into smaller molecules, ions, etc. An example of dissociation is the reversible reaction of hydrogen iodide at high temperatures

$$2HI(g) \rightleftharpoons H_2(g) + I_2(g)$$

The *equilibrium constant of a reversible dissociation is called the **dissociation constant**. The term 'dissociation' is also applied to ionization reactions of *acids and *bases in water; for example

$$HCN + H_2O \rightleftharpoons H_3O^+ + CN^-$$

which is often regarded as a straightforward dissociation into ions

$$HCN \rightleftharpoons H^+ + CN^-$$

The equilibrium constant of such a dissociation is called the **acid dissociation constant** or **acidity constant**, given by

$$K_a = [H^+][A^-]/[HA]$$

for an acid HA (the concentration of water $[H_2O]$ can be taken as constant). $K_a$ is a measure of the strength of the acid. Similarly, for a nitrogenous base B, the equilibrium

$$B + H_2O \rightleftharpoons BH^+ + OH^-$$

is also a dissociation; with the **base dissociation constant**, or **basicity constant**, given by

$$K_b = [BH^+][OH^-]/[B]$$

For a hydroxide MOH,

$$K_b = [M^+][OH^-]/[MOH]$$

**dissociation pressure** When a solid compound dissociates to give one or more gaseous products, the dissociation pressure is the pressure of gas in equilibrium with the solid at a given temperature. For example, when calcium carbonate is maintained at a constant high temperature in a closed container, the dissociation pressure at that temperature is the pressure of carbon dioxide from the equilibrium

$$CaCO_3(s) \rightleftharpoons CaO(s) + CO_2(g)$$

**distal** Describing the part of an organ that is farthest from the organ's point of attachment to the rest of the body. For example, hands and feet are at the distal ends of arms and legs, respectively. *Compare* PROXIMAL.

**distal convoluted tubule (second convoluted tubule)** The part of a *nephron that leads from the thick ascending limb of the *loop of Henle and drains into a *collecting duct. The main function of the distal tubule is to absorb sodium chloride and other inorganic salts while retaining water.

**distillation** The process of boiling a liquid and condensing and collecting the vapour. The liquid collected is the **distillate**. It is used to purify liquids and to separate liquid mixtures (*see* FRACTIONAL DISTILLATION; STEAM DISTILLATION). *See also* DESTRUCTIVE DISTILLATION; EXTRACTIVE DISTILLATION.

**distilled water** Water purified by distillation so as to free it from dissolved salts and other compounds. Distilled water in equilibrium with the carbon dioxide in the air has a conductivity of about $0.8 \times 10^{-6}$ siemens $cm^{-1}$. Repeated distillation in a vacuum can bring the conductivity down to $0.043 \times 10^{-6}$ siemens $cm^{-1}$ at 18°C (sometimes called **conductivity water**). The limiting conductivity is due to self-ionization: $H_2O \rightleftharpoons H^+ + OH^-$.

**distortion** The extent to which a system fails to reproduce the characteristics of its input in its output. It is most commonly applied to electronic amplifiers and to optical systems. *See* ABERRATION.

**distribution function** A function of some variable, such as velocity, that expresses the probability that a particle, such as a molecule of gas, will have a particular value of that variable.

**distributive law** The mathematical law stating that one operation is independent of being carried out before or after another operation. For example, multiplication is distributive with respect to addition and subtraction, i.e. $x(y + z) = xy + xz$. *Compare* ASSOCIATIVE LAW; COMMUTATIVE LAW.

**disulphide bridge (sulphur bridge)** A covalent bond (S–S) formed between the thiol groups (–SH) of two cysteine residues, usually in the polypeptide chains of proteins. Easily hydrolysed and prone to rearrangement, these bonds contribute to the tertiary structure of *proteins.

**disulphur dichloride (sulphur mono-chloride)** An orange–red liquid, $S_2Cl_2$, which is readily hydrolysed by water and is soluble in benzene and ether; r.d. 1.678; m.p. –80°C; b.p. 136°C. It may be prepared by passing chlorine over molten sulphur; in the presence of iodine or metal chlorides **sulphur dichloride**, $SCl_2$, is also formed. In the vapour phase $S_2Cl_2$ molecules have Cl–S–S–Cl chains. The compound is used as a solvent for sulphur and can form higher **chlorosulphanes** of the type Cl–(S)$_n$–Cl ($n <$ 100), which are of great value in *vulcanization processes.

**disulphuric(VI) acid (pyrosulphuric acid)** A colourless hygroscopic crystalline solid, $H_2S_2O_7$; r.d. 1.9; m.p. 35°C. It is commonly encountered mixed with sulphuric acid as it is formed by dissolving sulphur trioxide in concentrated sulphuric acid. The resulting fuming liquid, called **oleum** or **Nordhausen sulphuric acid**, is produced during the *contact process and is also widely used in the *sulphonation of organic compounds. *See also* SULPHURIC ACID.

**dithionate** A salt of dithionic acid, containing the ion $S_2O_6^{2-}$, usually formed by the oxidation of a sulphite using manganese(IV) oxide. The ion has neither pronounced oxidizing nor reducing properties.

**dithionic acid** An acid, $H_2S_2O_6$, known in the form of its salts (dithionates).

**dithionite** *See* SULPHINATE.

**dithionous acid** *See* SULPHINIC ACID.

**diuretic** A drug or other agent that increases the rate of urine formation and hence the rate at which water and certain salts are lost from the body. Many diuretic drugs work by decreasing the reabsorption of sodium and chloride ions from the filtrate in the kidney tubules, so that less water is reabsorbed. They are used to treat fluid retention (oedema) arising from disorders of the heart, kidneys, or other organs, and are used in helping to reduce high blood pressure (hypertension). There are several groups of diuretic drugs, with different modes of action. The most powerful are **loop diuretics**, such as frusemide, which act primarily by blocking $Na^+/K^+/Cl^-$ carriers in cells of the *loop of Henle. Another group consists of the thiazides, such as metolazone, which inhibit

$Na^+/Cl^-$ transport in the *distal convoluted tubule. Spironolactone exerts its diuretic effect by blocking the binding of the hormone *aldosterone to its receptors. The **osmotic diuretics**, such as mannitol, act by increasing the osmolarity of the filtrate, and hence increasing urine volume.

**diurnal** Daily; denoting an event that happens once every 24 hours.

**diurnal rhythm** *See* CIRCADIAN RHYTHM.

**divalent (bivalent)** Having a valency of two.

**divergence (div)** The *scalar product of the *gradient operator ∇ with a vector. For a vector $u$ that has components $u_1$, $u_2$, and $u_3$ in the $x$, $y$, and $z$ directions, and is a function of $x$, $y$, and $z$, the divergence is given by:

$$\mathrm{div}\,u = \nabla.u = \partial u_1/\partial x + \partial u_2/\partial y + \partial u_3/\partial z.$$

The divergence of a vector at a given point represents the flux of the vector per unit volume in the neighbourhood of that point. *See also* CURL; LAPLACE EQUATION.

**divergence theorem** A theorem that gives the relation between the total flux of a vector $F$ out of a surface $S$, which surrounds the volume $V$, to the vector inside the volume. The divergence theorem states that

$$\int_v \mathrm{div}\,F dV = \int_s F.dS.$$

The divergence theorem is also known as **Gauss' theorem** and **Ostrogradsky's theorem** (named after the Russian mathematician Michel Ostrogradsky (1801–61), who stated it in 1031). *Gauss' law for electric fields is a particular case of the divergence theorem.

**divergent evolution** *See* ADAPTIVE RADIATION.

**divergent series** *See* CONVERGENT SERIES.

**diverging lens or mirror** A lens or mirror that can refract or reflect a parallel beam of light into a diverging beam. A diverging lens is predominantly concave; a diverging mirror is convex. *Compare* CONVERGING LENS OR MIRROR.

**diverticulum** A saclike or tubular outgrowth from a tubular or hollow internal organ. Diverticula may occur as normal structures (e.g. the *caecum and *appendix in the alimentary canal) or abnormally, from a weakened area of the organ.

**division** A category used traditionally in the *classification of plants that consists of one or several similar classes. Division names end in -phyta; an example is the Spermatophyta (seed-bearing plants).

**dizygotic twins** *See* FRATERNAL TWINS.

**D-lines** Two close lines in the yellow region of the visible spectrum of sodium, having wavelengths 589.0 and 589.6 nm. As they are prominent and easily recognized they are used as a standard in spectroscopy.

***dl*-isomer** *See* OPTICAL ACTIVITY; RACEMIC MIXTURE.

**DMF** *See* DIMETHYLFORMAMIDE.

**DMG** *See* DIMETHYLGLYOXIME.

**DMSO** *See* DIMETHYL SULPHOXIDE.

**DNA (deoxyribonucleic acid)** The genetic material of most living organisms, which is a major constituent of the *chromosomes within the cell nucleus and plays a central role in the determination of hereditary characteristics by controlling *protein synthesis in cells (*see also* GENETIC CODE). It is also found in chloroplasts and mitochondria (*see* CYTOPLASMIC INHERITANCE; MITOCHONDRIAL DNA). DNA is a nucleic acid composed of two chains of *nucleotides in which the sugar is **deoxyribose** and the bases are *adenine, *cytosine, *guanine, and *thymine (*compare* RNA). The two chains are wound round each other and linked together by hydrogen bonds between specific complementary bases (*see* BASE PAIRING) to form a spiral ladder-shaped molecule (**double helix**). See illustration.

When the cell divides, its DNA also replicates in such a way that each of the two

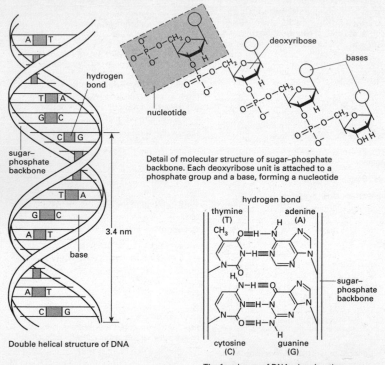

Double helical structure of DNA

Detail of molecular structure of sugar–phosphate backbone. Each deoxyribose unit is attached to a phosphate group and a base, forming a nucleotide

The four bases of DNA, showing the hydrogen bonding between base pairs

**DNA.** Its molecular structure.

daughter molecules is identical to the parent molecule (*see* DNA REPLICATION). *See also* COMPLEMENTARY DNA.

**( ) SEE WEB LINKS**

• An animated primer on basic aspects of DNA, genes, and heredity; produced for the Cold Spring Harbor Laboratory.

**DNAase** *See* DNASE.

**DNA blotting** *See* SOUTHERN BLOTTING.

**DNA chip** *See* DNA MICROARRAY.

**DNA cloning** *See* GENE CLONING.

**DNA-dependent RNA polymerase** *See* POLYMERASE.

**DNA fingerprinting** *See* DNA PROFILING.

**DNA hybridization** A method of determining the similarity of DNA from different sources. Single strands of DNA from two sources, e.g. different bacterial species, are put together and the extent to which double hybrid strands are formed is estimated. The greater the tendency to form these hybrid molecules, the greater the extent of complementary base sequences, i.e. gene similarity. The method is one way of determining the genetic relationships of species.

**DNA library (gene library; gene bank)** A collection of cloned DNA fragments representing the entire genetic material of an organism. This facilitates screening and isolation of any particular gene. DNA libraries are created by fractionating the genomic DNA into fragments using *restriction enzymes and/or physical methods. These fragments are cloned (*see* GENE CLONING) and the host cells containing the recombinant fragments are centrifuged and frozen; alternatively, the phage *vectors are maintained in culture. Individual genes in the library are identified using specific *gene probes with the *Southern blotting technique or, via their protein products, using *Western blotting. DNA libraries are thus repositories of raw material for use in genetic engineering. A large genome, such as that of humans, is most conveniently cloned using vectors that can accommodate large fragments of DNA, such as yeast *artificial chromosomes, maintained in cell culture.

**DNA ligase** An enzyme that is able to join together two portions of DNA and therefore plays an important role in *DNA repair. DNA ligase is also used in recombinant DNA technology (*see* GENETIC ENGINEERING) as it en-

sures that the foreign DNA (e.g. the complementary DNA used in *gene cloning) is bound to the plasmid into which it is incorporated.

**DNA methylation** The addition of methyl groups to constituent bases of DNA. In both prokaryotes and eukaryotes certain bases of the DNA generally occur in a methylated form. In bacteria this methylation protects the cell's DNA from attack by its own restriction enzymes, which cleave foreign unmethylated DNA and thereby help to eliminate viral DNA from the bacterial chromosome. Methylation is also important in helping *DNA repair enzymes to distinguish the parent strand from the progeny strand when repairing mismatched bases in newly replicated DNA, and it also plays a role in controlling the transcription of DNA.

**DNA microarray (DNA chip)** A *microarray containing numerous small DNA molecules. DNA microarrays consist of thousands of short synthetic single-stranded DNA molecules, each comprising 20–25 nucleotides and all with unique sequences designed to complement and bind to specific target nucleotide sequences. They can be used to quantify gene expression by determining the total output of messenger RNAs (mRNAs) (i.e. the *transcriptome) of a cell or tissue. This involves adding fluorescent labels to the mRNAs, then incubating these with the microarray so that they bind to complementary oligonucleotides on it. A computerized scanner measures the intensity of fluorescence and thereby the amount of bound mRNA. DNA microarrays can also detect mutations in particular genes, for example the *BRCA* genes involved in hereditary forms of breast cancer. An individual's DNA is denatured, and its binding to a microarray is compared with that of normal (control) DNA on the same microarray. Any disparities between the two binding patterns will pinpoint sequences from the individual with possible abnormalities, enabling closer examination.

**DNA polymerase** *See* POLYMERASE.

**DNA probe** *See* GENE PROBE.

**DNA profiling (genetic fingerprinting)** A technique in which an individual's DNA is analysed to reveal the pattern of repetition of certain short nucleotide sequences, called short tandem repeats: (STRs), throughout the genome. This pattern is claimed to be unique to the individual concerned, and the technique is therefore used for identification purposes in forensic science and paternity disputes, and in veterinary science. Sufficient DNA can be obtained from very small samples of body tissue, such as blood, semen, or hair.

**DNA repair** A variety of mechanisms that help to ensure that the genetic sequence, as expressed in the DNA, is maintained and that errors that occur during *DNA replication, by mutation, are not allowed to accumulate. An error in the genetic sequence could cause cell death by interfering with the replication process. The mechanisms work because DNA is made up of two strands, each of which contains a copy of the genetic sequence. A damaged section of a strand, or a mismatched base, can be recognized and removed and replaced by the correct form by DNA *polymerases. The phosphodiester backbone is then sealed by *DNA ligase.

**DNA replication** The process whereby DNA makes exact copies of itself, which is controlled by the enzyme DNA *polymerase. Replication occurs at rates of between 50 nucleotides per second (in mammals) and 500 nucleotides per second (in bacteria). The hydrogen bonds between the complementary bases on the two strands of the parent DNA molecule break and the strands unwind, each strand acting as a template for the synthesis of a new one complementary to itself. DNA polymerases move down the two single strands linking free nucleotides to their complementary bases (*see* BASE PAIRING) on the templates. The process continues until all the nucleotides on the templates have joined with appropriate free nucleotides and two identical molecules of DNA have been formed. This process is known as **semiconservative replication** as each new molecule contains half of the original parent DNA molecule. Sometimes mutations occur that may cause the exact sequence of the parent DNA not to be replicated. However, *DNA repair mechanisms reduce this possibility.

**DNase (DNAase; deoxyribonuclease)** An enzyme that catalyses the cleavage of DNA. DNase I is a digestive enzyme, secreted by the pancreas, that degrades DNA into shorter nucleotide fragments. Many other *endonucleases and *exonucleases cleave DNA, including the *restriction enzymes and

enzymes involved in DNA repair and replication.

**DNA sequencing (gene sequencing)** The process of elucidating the nucleotide sequence of a DNA fragment. The method now used almost universally is the Sanger method (named after Frederick *Sanger), also called the dideoxy method, which involves synthesizing a new DNA strand using as template single-stranded DNA from the gene being sequenced. Synthesis of the new strand can be stopped at any of the four bases by adding the corresponding dideoxy (dd) derivative of the deoxyribonucleoside phosphates; for example, by adding ddATP the synthesis terminates at an adenosine; by adding ddGTP it terminates at a guanosine, etc. The fragments, which comprise fluorescently labelled nucleotides, are subjected to electrophoresis and scanned by a fluorescence detector. A big advantage of the Sanger method is that it can easily be adapted to sequencing RNA, by making single-stranded DNA from the RNA template using the enzyme *reverse transcriptase. This enables, for example, sequencing of ribosomal RNA for use in *molecular systematics. After separation of the fragments, the products of all four reactions are detected by fluorescence spectroscopy and analysed by computer, which gives a printout of the base sequence. DNA sequencing using fully automated sequencers is now employed on a major scale, for example in determining the nucleotide sequence of entire genomes (*see* GENOME PROJECT).

**DNS** *See* DOMAIN NAME SYSTEM.

**Döbereiner's triads** A set of triads of chemically similar elements noted by Johann Döbereiner (1780–1849) in 1817. Even with the inaccurate atomic mass data of the day it was observed that when each triad was arranged in order of increasing atomic mass, then the mass of the central member was approximately the average of the values for the other two. The chemical and physical properties were similarly related. The triads are now recognized as consecutive members of the groups of the periodic table. Examples are: lithium, sodium, and potassium; calcium, strontium, and barium; and chlorine, bromine, and iodine.

 **SEE WEB LINKS**
• Döbereiner's original paper

**Document Object Model (DOM)** A standard interface for representing *XML and *HTML documents. The data is parsed into a tree of objects, which a programmer can navigate and manipulate. It is especially relevant to web browsers, and in particular to the implementation of dynamic HTML. The World Wide Web Consortium released the first version of the DOM (level 1) in 1998, although a previous intermediate version (level 0) had already been incorporated in HTML 4; levels 2 and 3 have subsequently been released. Adoption of the DOM by web browsers – which had previously implemented idiosyncratic and incompatible models – was initially slow and is still not uniform.

**SEE WEB LINKS**
• The WSC's DOM page

**dodecanoic acid (lauric acid)** A white crystalline *fatty acid, $CH_3(CH_2)_{10}COOH$; r.d. 0.868; m.p. 44°C; b.p. 131°C. Glycerides of the acid are present in natural fats and oils (e.g. coconut and palm-kernel oil).

**dodecene** A straight-chain alkene, $CH_3(CH_2)_9CH:CH_2$, obtained from petroleum and used in making *dodecylbenzene.

**dodecylbenzene** A hydrocarbon, $CH_3(CH_2)_{11}C_6H_5$, manufactured by a Friedel–Crafts reaction between dodecene $(CH_3(CH_2)_9CH:CH_2)$ and benzene. It can be sulphonated, and the sodium salt of the sulphonic acid is the basis of common *detergents.

**dolomite** A carbonate mineral consisting of a mixed calcium–magnesium carbonate, $CaCO_3.MgCO_3$, crystallizing in the rhombohedral system. It is usually white or colourless. The term is also used to denote a rock with a high ratio of magnesium to calcium carbonate. See LIMESTONE.

**DOM** See DOCUMENT OBJECT MODEL.

**Domagk, Gerhard** (1895–1964) German biochemist who went to work for IG Farbenindustrie to investigate new drugs. In 1935 he discovered the antibacterial properties of a dye, Prontosil, which became the first sulpha drug (see SULPHONAMIDES). He was offered the 1939 Nobel Prize for physiology or medicine but was forced by Hitler to refuse; he finally received the award in 1947.

**domain** 1. (in taxonomy) The highest taxonomic category, consisting of one or more *kingdoms. Living organisms are divided into three domains: *Archaea (archaebacte-

ria), Eubacteria (see BACTERIA), and Eukarya (see EUKARYOTE). 2. (in physics) See MAGNETISM.

**domain name system (DNS)** In computing, a system that provides mappings between the human-oriented names of users or services in a network, and the machine-oriented network addresses of the named entity. It is used primarily on *TCP/IP networks, primarily the Internet, to map such human-oriented names as www.oup.co.uk to the equivalent IP address; however, other networks have similar facilities. Names are usually hierarchical, and in general terms the boundaries of a *domain will coincide with some form of natural boundary within the network environment, such as a country, a community of users within a country, or the users on a site. The above example consists of the four domains 'uk' (the top-level domain), 'co', 'oup', and 'www'. While the arrangement of IP addresses is also usually hierarchical, there is no assumption in the mapping between the two that the hierarchies are in any way equivalent.

**SEE WEB LINKS**
• Introduction to the Internet domain name system
• Domain names – implementation and specification

**dominance hierarchy** See DOMINANT.

**dominant** 1. (in genetics) Describing the *allele that is expressed in the *phenotype when two different alleles of a gene are present in the cells of an organism. For example, the height of garden peas is controlled by two alleles, 'tall' (T) and 'dwarf' (t). When both are present (Tt), i.e. when the cells are *heterozygous, the plant is tall since T is dominant and t is *recessive. See also CODOMINANCE; INCOMPLETE DOMINANCE. 2. (in ecology) Describing the most conspicuously abundant and characteristic species in a *community. The term is usually used of a plant species in plant ecology; for example, pine trees in a pine forest. 3. (in animal behaviour) Describing an animal that is allowed priority in access to food, mates, etc., by others of its species because of its success in previous aggressive encounters. Less dominant animals frequently show *appeasement behaviour towards a more dominant individual, so overt *aggression is minimized. In a stable group there may be a linear **dominance hierarchy** or **peck order** (so called because it was first observed in domestic fowl), with each animal being subservient to those above it in the hierarchy and taking precedence over those below it.

**donor** 1. (in chemistry) An atom, ion, or molecule that provides a pair of electrons in forming a coordinate bond. 2. (in electronics) A substance added as an impurity to a *semiconductor because it can donate electrons to the conduction bands, causing *n*-type conduction by electrons. *Compare* ACCEPTOR. 3. (in medicine) An individual whose tissues or organs are transferred to another (the **recipient**). Donors may provide blood for transfusion or a kidney or heart for transplantation. 4. (in genetics) A cell that contributes genetic material for insertion into another cell, for example to produce a transgenic cell by genetic engineering.

**donor levels** Energy levels of a donor atom in a *semiconductor, such as arsenic in silicon. These energy levels are very near the bottom of the conduction band, thus causing *n*-type conduction. *See also* ENERGY BANDS.

**dopa (dihydroxyphenylalanine)** A derivative of the amino acid tyrosine. It is found in particularly high levels in the adrenal glands and is a precursor in the synthesis of *dopamine, *noradrenaline, and *adrenaline. The laevorotatory form, **L-dopa**, is administered in the treatment of Parkinson's disease, in which brain levels of dopamine are reduced.

**dopamine** A *catecholamine that is a precursor in the synthesis of *noradrenaline and *adrenaline. It also functions as a neurotransmitter, especially in the brain.

**doping** *See* SEMICONDUCTOR.

**Doppler broadening** Broadening of a spectral line caused by the *Doppler effect. Shifts in frequency occur according to whether the emitting atom or molecule is moving towards or away from the observer. The effect depends on temperature and is used for temperature measurement in astronomy.

**Doppler cooling** *See* LASER COOLING.

**Doppler effect** The apparent change in the observed frequency of a wave as a result of relative motion between the source and the observer. For example, the sound made by a low-flying aircraft as it approaches appears to fall in pitch as it passes and flies away. In fact, the frequency of the aircraft

engine remains constant but as it is approaching more sound waves per second impinge on the ear and as it recedes fewer sound waves per second impinge on the ear. The apparent frequency, $F$, is given by

$$F = f(c - u_o)/(c - u_s),$$

where $f$ is the true frequency, $c$ is the speed of sound, and $u_o$ and $u_s$ are the speeds of the observer and the source, respectively.

Although the example of sound is most commonly experienced, the effect was suggested by Christian Johann Doppler (1803–53), an Austrian physicist, as an attempt to explain the coloration of stars. In fact the Doppler effect cannot be observed visually in relation to the stars, although the effect does occur with electromagnetic radiation and the *redshift of light from receding stars can be observed spectroscopically. The Doppler effect is also used in radar to distinguish between stationary and moving targets and to provide information regarding the speed of moving targets by measuring the frequency shift between the emitted and reflected radiation.

For electromagnetic radiation, the speed of light, $c$, features in the calculation and as there is no fixed medium to provide a frame of reference, relativity has to be taken into account, so that

$$F = f\sqrt{[(1 - v/c)/(1 + v/c)]},$$

where $v$ is the speed at which source and observer are moving apart. If $v^2/c^2$ is small compared to 1, i.e. if the speed of separation is small compared to the speed of light, this equation simplifies to

$$F = f(1 - v/c).$$

**d-orbital** *See* ORBITAL.

**dormancy** An inactive period in the life of an animal or plant during which growth slows or completely ceases. Physiological changes associated with dormancy help the organism survive adverse environmental conditions. Annual plants survive the winter as dormant seeds while many perennial plants survive as dormant tubers, rhizomes, or bulbs. *Hibernation and *aestivation in animals help them survive extremes of cold and heat, respectively.

**dorsal** Describing the surface of a plant or animal that is farthest from the ground or other support, i.e. the upper surface. In vertebrates, the dorsal surface is that down which the backbone runs. Thus in upright (bipedal) mammals, such as man and kangaroos, it is the backward-directed (*posterior) surface. *Compare* VENTRAL.

**dorsal root** The part of a spinal nerve that enters the *spinal cord on the dorsal side and contains only sensory fibres. The cell bodies of these fibres form the **dorsal root ganglion** (*see* GANGLION), a swelling in the root that lies just outside the cord. *Compare* VENTRAL ROOT.

**dose** A measure of the extent to which matter has been exposed to *ionizing radiation. The **absorbed dose** is the energy per unit mass absorbed by matter as a result of such exposure. The SI unit is the gray, although it is often measured in rads (1 rad = 0.01 gray; *see* RADIATION UNITS). The **maximum permissible dose** is the recommended upper limit of absorbed dose that a person or organ should receive in a specified period according to the International Commission on Radiological Protection. *See also* LINEAR ENERGY TRANSFER.

**dosimeter** Any device used to measure absorbed *dose of ionizing radiation. Methods used include the *ionization chamber, photographic film, or the rate at which certain chemical reactions occur in the presence of ionizing radiation.

**dot product** *See* SCALAR PRODUCT.

**double bond** *See* CHEMICAL BOND.

**double circulation** The type of circulatory system that occurs in mammals, in which the blood passes through the heart twice before completing a full circuit of the body (see illustration). Blood is pumped

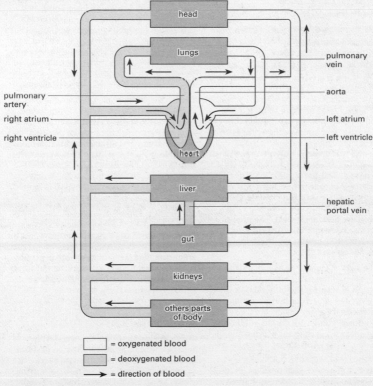

**Double circulation.**

from the heart to the lungs and returns to the heart before being distributed to the other organs and tissues of the body. The heart is divided into two separate compartments to prevent oxygenated blood returning from the lungs from mixing with deoxygenated blood from the other parts of the body. *See also* PULMONARY CIRCULATION; SYSTEMIC CIRCULATION. *Compare* SINGLE CIRCULATION.

**double decomposition**  *See* METATHESIS.

**double fertilization**  A process, unique to flowering plants, in which two male nuclei, which have travelled down the pollen tube, separately fuse with different female nuclei in the *embryo sac. The first male nucleus fuses with the egg cell to form the zygote; the second male nucleus fuses with the two polar nuclei to form a triploid nucleus that develops into the endosperm.

**double helix**  *See* DNA.

**double recessive**  An organism with two *recessive alleles for a particular characteristic.

**double refraction**  The property, possessed by certain crystals (notably calcite), of forming two refracted rays from a single incident ray. The **ordinary ray** obeys the normal laws of refraction. The other refracted ray, called the **extraordinary ray**, follows different laws. The light in the ordinary ray is polarized at right angles to the light in the extraordinary ray. Along an *optic axis the ordinary and extraordinary rays travel with the same speed. Some crystals, such as calcite, quartz, and tourmaline, have only one optic axis; they are **uniaxial crystals**. Others, such as mica and selenite, have two optic axes; they are **biaxial crystals**. The phenomenon is also known as **birefringence** and the double-refracting crystal as a **birefringent crystal**. *See also* POLARIZATION.

**double salt**  A crystalline salt in which there are two different anions and/or cations. An example is the mineral dolomite, $CaCO_3.MgCO_3$, which contains a regular arrangement of $Ca^{2+}$ and $Mg^{2+}$ ions in its crystal lattice. *Alums are double sulphates. Double salts only exist in the solid; when dissolved they act as a mixture of the two separate salts. **Double oxides** are similar.

**doublet**  **1.** A pair of optical lenses of different shapes and made of different materials used together so that the chromatic aberration produced by one is largely cancelled by the reverse aberration of the other. **2.** A pair of associated lines in certain spectra, e.g. the two lines that make up the sodium D-lines.

**down feathers (plumules)**  Small soft feathers that cover and insulate the whole body of a bird. In nestlings they are the only feathers; in adults they lie between and beneath the *contour feathers. Down feathers have a fluffy appearance as their *barbs are not joined together to form a smooth vane.

**Downs process**  A process for extracting sodium by the electrolysis of molten sodium chloride. The **Downs cell** has a central graphite anode surrounded by a cylindrical steel cathode. Chlorine released is led away through a hood over the anode. Molten sodium is formed at the cathode and collected through another hood around the top of the cathode cylinder (it is less dense than the sodium chloride). The two hoods and electrodes are separated by a coaxial cylindrical steel gauze. A small amount of calcium chloride is added to the sodium chloride to lower its melting point. The sodium chloride is melted electrically and kept molten by the current through the cell. More sodium chloride is added as the electrolysis proceeds.

**Down's syndrome**  A congenital form of mental retardation due to a chromosome defect in which there are three copies of chromosome no. 21 instead of the usual two (*see* TRISOMY). The affected individual has a short broad face and slanted eyes (as in the Mongolian races), short fingers, and weak muscles. Down's syndrome can be detected before birth by *amniocentesis. It is named after the British physician John Down (1828–96), who first studied the incidence of the disorder.

**Dragendorff test**  A *presumptive test for alkaloids. The **Dragendorff reagent** has two solutions. One is bismuth nitrate in acetic acid. This is followed by a sodium nitrate solution. Alkaloids are indicated by a reddish-brown deposit.

**dragonflies**  *See* ODONATA.

**drain**  *See* TRANSISTOR.

**dreikanter**  (German: three edges) A faceted stone or pebble formed in desert regions by erosion by wind-blown sand. The stones are too heavy to be blown along the desert floor and are merely battered back and forth. Most have three curved facets.

**drift-tube accelerator** *See* LINEAR ACCELERATOR.

**drone** A fertile male in a colony of social bees, especially the honeybee (*Apis mellifera*). The drones die after mating with the queen bee as the male reproductive organs explode within the female.

***Drosophila*** A genus of fruit flies often used in genetic research because the larvae possess **giant chromosomes** in their salivary glands. These chromosomes have resulted from repeated duplication without separation of the chromatids; they have conspicuous transverse bands, which can be studied microscopically to reveal gene activity. Fruit flies have a short life cycle and produce a large number of offspring, which also makes them a good model animal for genetic research.

**drug** Any chemical substance that alters the physiological state of a living organism. Drugs are widely used in medicine for the prevention, diagnosis, and treatment of diseases; they include *analgesics, *antibiotics, anaesthetics, *antihistamines, and *anticoagulants. Some drugs are taken solely for the pleasurable effects they induce; these include *narcotics; stimulants, such as cocaine and *amphetamine; *hallucinogens, such as *LSD; and some tranquillizers. Many of these drugs are habit-forming and their use is illegal.

**drupe (pyrenocarp)** A fleshy fruit that develops from either one or several fused carpels and contains one or many seeds. The seeds are enclosed by the hard protective endocarp (*see* PERICARP) of the fruit. Thus the stone of a peach is the endocarp containing the seed. Plums, cherries, coconuts, and almonds are other examples of one-seeded drupes; holly and elder fruits are examples of many-seeded drupes. *See also* ETAERIO.

**dry cell** A primary or secondary cell in which the electrolytes are restrained from flowing in some way. Many torch, radio, and calculator batteries are *Leclanché cells in which the electrolyte is an ammonium chloride paste and the container is the negative zinc electrode (with an outer plastic wrapping). Various modifications of the Leclanché cell are used in dry cells. In the **zinc chloride cell**, the electrolyte is a paste of zinc chloride rather than ammonium chloride. The electrical characteristics are similar to those of the Leclanché cell but the cell

works better at low temperatures and has more efficient depolarization characteristics. A number of alkaline secondary cells can be designed for use as dry cells. In these, the electrolyte is a liquid (sodium or potassium hydroxide) held in a porous material or in a gel. Alkaline dry cells typically have zinc–manganese dioxide, silver oxide–zinc, nickel–cadmium, or nickel–iron electrode systems (*see* NICKEL–IRON ACCUMULATOR). For specialized purposes, dry cells and batteries have been produced with solid electrolytes. These may contain a solid crystalline salt, such as silver iodide, an ion-exchange membrane, or an organic wax with a small amount of dissolved ionic material. Such cells deliver low currents. They are used in miniature cells for use in electronic equipment.

**dry ice** Solid carbon dioxide used as a refrigerant. It is convenient because it sublimes at −78°C (195 K) at standard pressure rather than melting.

**drying oil** A natural oil, such as linseed oil, that hardens on exposure to the air. Drying oils contain unsaturated fatty acids, such as linoleic and linolenic acids, which polymerize on oxidation. They are used in paints, varnishes, etc.

**dry mass** The mass of a biological sample after the water content has been removed, usually by placing the sample in an oven. The dry mass is used as a measure of the *biomass of a sample.

***Dryopithecus*** A genus of extinct apes, fossils of which have been found in Europe and Asia and dated to the mid-Miocene (about 9–13 million years ago). Fossils of *Dryopithecus* and of the similar genus *Proconsul* are often referred to as **dryopithecines**. Dryopithecines are believed to have split into several lines, some of which survived to give rise to the chimpanzees, gorillas, early hominids, and orang-utans.

**DSC** Differential scanning calorimetry. *See* THERMAL ANALYSIS.

**D-series** *See* ABSOLUTE CONFIGURATION.

**DTA** Differential thermal analysis. *See* THERMAL ANALYSIS.

**DTD** Document type definition. A file associated with an *SGML or *XML document, giving a formalized description of the tags used and the structure of the document, as well as special entities that may be present.

DTDs are optional for XML documents but compulsory for SGML documents. To **validate** a document is to check its structure against its DTD; a document that passes this test without errors is said to be 'valid' or to **conform to** its DTD.

**dubnium** Symbol Db. A radioactive *transactinide element; a.n. 105. It was first reported in 1967 by a group at Dubna near Moscow and was confirmed in 1970 at Dubna and at Berkeley, California. It can be made by bombarding californium-249 nuclei with nitrogen-15 nuclei. Only a few atoms have ever been made.

**(⊕) SEE WEB LINKS**
• Information from the WebElements site

**duct** A tube or passage in an organism that is involved in the secretion or excretion of substances (*see* GLAND).

**ductility** The ability of certain metals, such as copper, to retain their strength when their shape is changed, especially the ability of such metals to be drawn into a thin wire without cracking or breaking.

**ductless gland** See ENDOCRINE GLAND.

**ductus arteriosus** A channel that connects the pulmonary artery with the aorta in the mammalian fetus and therefore allows blood to bypass the inactive lungs of the fetus. It normally closes soon after birth.

**Dulong and Petit's law** For a solid element the product of the relative atomic mass and the specific heat capacity is a constant equal to about 25 J mol$^{-1}$ K$^{-1}$. Formulated in these terms in 1819 by the French scientists Pierre Dulong (1785–1838) and Alexis Petit (1791–1820), the law in modern terms states: the molar heat capacity of a solid element is approximately equal to 3$R$, where $R$ is the *gas constant. The law is only approximate but applies with fair accuracy at normal temperatures to elements with a simple crystal structure.

**(⊕) SEE WEB LINKS**
• A translation of the original 1819 paper in *Annals of Philosophy*

**Dumas, Jean Baptiste André** (1800–84) French chemist, who became an apothecary in Geneva, where in 1818 he investigated the use of iodine to treat goitre. He then took up chemistry and moved to Paris. In 1826 he devised a method of measuring *vapour density. He went on to discover various organic compounds, including anthracene (1832), urethane (1833), and methanol (1834), which led him in 1840 to propose the theory of types (functional groups).

**Dumas' method** **1.** A method of finding the amount of nitrogen in an organic compound. The sample is weighed, mixed with copper(II) oxide, and heated in a tube. Any nitrogen present in the compound is converted into oxides of nitrogen, which are led over hot copper to reduce them to nitrogen gas. This is collected and the volume measured, from which the mass of nitrogen in a known mass of sample can be found. **2.** A method of finding the relative molecular masses of volatile liquids by weighing. A thin-glass bulb with a long narrow neck is used. This is weighed full of air at known temperature, then a small amount of sample is introduced and the bulb heated (in a bath) so that the liquid is vaporized and the air is driven out. The tip of the neck is sealed and the bulb cooled and weighed at known (room) temperature. The volume of the bulb is found by filling it with water and weighing again. If the density of air is known, the mass of vapour in a known volume can be calculated.

The techniques are named after J. B. A. Dumas.

**dune** A mound or ridge of unconsolidated sand formed by the action of wind. Dunes are characteristic of desert regions and some coastlines. Coastal dunes are usually anchored by vegetation, whereas desert dunes generally move gradually in the direction of the prevailing wind. There are various types, often named after their shapes. Crescent-shaped dunes are called **barchans**, and longitudinal dunes are **seif dunes**; others include sinuous anklé dunes, star dunes, whaleback dunes and tail dunes, which form in the lee of an obstacle. A large seif dune may be up to 100 m high and 10 km long. Similar structures to dunes may also form underwater by the action of currents.

**duodenum** The first section of the *small intestine of vertebrates. It is the site where food from the stomach is subjected to the action of bile (from the bile duct) and pancreatic enzymes (from the pancreatic duct) as well as the enzymes secreted by digestive glands in the duodenum itself, which are required in the breakdown of proteins, carbohydrates, and fats. By neutralizing the acidic secretions of the stomach, the duodenum provides an alkaline environment necessary for the action of the intestinal enzymes. *See also* INTESTINAL JUICE.

**duplet** A pair of electrons in a covalent chemical bond.

**duplex** Describing a biological molecule comprising two cross-linked polymeric chains oriented lengthways side by side. The term is applied particularly to the double-stranded structure of *DNA.

**duplication** (in genetics) The doubling or repetition of part of a chromosome, which generally originates during the *crossing over phase of meiosis. Occasionally this type of *chromosome mutation may have beneficial effects on a population. For example, a beneficial duplication resulted in the evolution of four types of haemoglobin in man and apes from a single form. One of these types of haemoglobin (gamma or fetal haemoglobin) has a greater affinity for oxygen and maximizes fetal uptake of oxygen from the mother's blood.

**Duralumin** Trade name for a class of strong lightweight aluminium alloys containing copper, magnesium, manganese, and sometimes silicon. Duralumin alloys combine strength with lightness and are extensively used in aircraft, racing cars, etc.

**dura mater** The outermost and toughest of the three membranes (*meninges) that surround the central nervous system in vertebrates. It lies adjacent to the skull and its purpose is to protect the delicate inner meninges (the *arachnoid membrane and the *pia mater).

**duramen** *See* HEARTWOOD.

**dust core** *See* CORE.

**Dutch metal** An alloy of copper and zinc, which can be produced in very thin sheets and used as imitation gold leaf. It spontaneously inflames in chlorine.

**DVD** Digital versatile disk: a disk format similar to a compact disk (*see* CD-ROM) but containing much more data. It was introduced in 1996. DVD disks are the same 120 mm diameter as CDs with potential capacities of up to 4.7 gigabytes for a single-sided single-layer disk. The technology involved in DVD storage is similar to that in compact disks, but more precise. The extra capacity is achieved in a number of ways. The tracks on a DVD are closer and the pits

are smaller, allowing more pits per unit area. The key to this was the use of a shorter wavelength laser (typically 635 or 650 nm in the red region for DVDs as opposed to 780 nm in the infrared for CDs). Moreover, a DVD can have two layers on the same side of the disk. The top layer is translucent and the bottom layer opaque. Data can be read from either layer by refocusing the laser. In addition DVDs may be double-sided. DVD formats also have a more efficient error-correction system. The potential capacity of a double-sided double-layer DVD is up to 17 gigabytes. DVDs have been increasingly used in computing as a higher-capacity version of compact disks. As with compact disks, there are various types. **DVD-ROM** (DVD read-only memory) is similar to CD-ROM. **DVD-R** (DVD-recordable) is similar to CD-R. There are also different rewritable formats: **DVD-RAM**, **DVD+RW**, and **DVD-RW**.

(((●))) SEE WEB LINKS
• The DVD Forum home page

**dwarf galaxy** A galaxy made up of only a few billion stars rather than the 200 to 400 billion that form the Milky Way System. Dwarf galaxies are classified like ordinary galaxies. In the *Local Group, several dwarf galaxies are satellites of the group's three largest galaxies.

**dwarf planet** A *small solar system body that is in direct orbit around the sun, not around another planet, and has sufficient mass to have contracted to a spherical or nearly spherical shape but has not cleared its orbital zone of *planetesimals. Five objects have been classified as dwarf planets since 2006, including *Pluto (formerly the ninth planet of the solar system and now recognized as a *Kuiper belt object [KBO]), *Eris (the largest dwarf planet so far known), the asteroid (1) Ceres, and the KBOs (136472) Makemake and (136108) Haumea.

(((●))) SEE WEB LINKS
• International Astronomical Union 2006 General Assembly: Result of the IAU resolution votes. This press release carries the official definition of a planet & dwarf planet
• NASA's explanation of the IAU's definitions of 'planet' and 'dwarf planet'

**dwarf star** A star, such as the sun, that lies on the main sequence in a *Hertzsprung–Russell diagram and is of average or normal size in relation to its mass. *See also* BLACK DWARF; BROWN DWARF; RED DWARF; WHITE DWARF.

**dye laser** A type of laser in which the active material is a dye dissolved in a suitable solvent (e.g. Rhodanine G in methanol). The dye is excited by an external source. The solvent broadens the states into bands and consequently laser action can be obtained over a range of wavelengths. This allows one to select a specific wavelength (using a grating) and to change the wavelength of the laser. Such a device is called a **tuneable laser**. Dye lasers are also used in producing very short pulses of radiation. The technique is to use a dye that stops absorbing radiation when a high proportion of its molecules become excited. The cavity then becomes resonant and a pulse of radiation is produced. This technique can give pulses of about 10 nanoseconds duration and is used in *femtochemistry.

**dyes** Substances used to impart colour to textiles, leather, paper, etc. Compounds used for dyeing (**dyestuffs**) are generally organic compounds containing conjugated double bonds. The group producing the colour is the *chromophore; other noncoloured groups that influence or intensify the colour are called *auxochromes. Dyes can be classified according to the chemical structure of the dye molecule. For example, **azo dyes** contain the –N=N– group (*see* AZO COMPOUNDS). In practice, they are classified according to the way in which the dye is applied or is held on the substrate.

   **Acid dyes** are compounds in which the chromophore is part of a negative ion (usually an organic sulphonate $RSO_2O^-$). They can be used for protein fibres (e.g. wool and silk) and for polyamide and acrylic fibres. Originally, they were applied from an acidic bath. **Metallized dyes** are forms of acid dyes in which the negative ion contains a chelated metal atom. **Basic dyes** have chromophores that are part of a positive ion (usually an amine salt or ionized imino group). They are used for acrylic fibres and also for wool and silk, although they have only moderate fastness with these materials.

   **Direct dyes** are dyes that have a high affinity for cotton, rayon, and other cellulose fibres. They are applied directly from a neutral bath containing sodium chloride or sodium sulphate. Like acid dyes, they are usually sulphonic acid salts but are distinguished by their greater substantivity (affinity for the substrate), hence the alternative name **substantive dyes**.

   **Vat dyes** are insoluble substances used for cotton dyeing. They usually contain keto groups, C=O, which are reduced to C–OH groups, rendering the dye soluble (the **leuco form** of the dye). The dye is applied in this form, then oxidized by air or oxidizing agents to precipitate the pigment in the fibres. Indigo and anthroquinone dyes are examples of vat dyes. **Sulphur dyes** are dyes applied by this technique using sodium sulphide solution to reduce and dissolve the dye. Sulphur dyes are used for cellulose fibres.

   **Disperse dyes** are insoluble dyes applied in the form of a fine dispersion in water. They are used for cellulose acetate and other synthetic fibres.

   **Reactive dyes** are compounds that contain groups capable of reacting with the substrate to form covalent bonds. They have high substantivity and are used particularly for cellulose fibres.

**dyke** A sheetlike vertical *intrusion of igneous rock cutting across the strata of older rocks. Dykes vary in thickness from a few centimetres to several metres. Several dykes may be grouped, radially or in parallel, as a dyke swarm. Depending on composition and how long it took to cool, a dyke's grain structure may be coarse, medium, or fine.

**dynamical meteorology** The branch of meteorology concerned with motions in the atmosphere. It is based on hydrodynamics and thermodynamics.

**dynamic equilibrium** See EQUILIBRIUM.

**dynamics** The branch of mechanics concerned with the motion of bodies under the action of forces. Time intervals, distances, and masses are regarded as fundamental and bodies are assumed to possess *inertia. Bodies in motion have an attribute called *momentum (*see* NEWTON'S LAWS OF MOTION), which can only be changed by the application of a force. *Compare* KINETICS; STATICS.

**dynamite** Any of a class of high explosives based on nitroglycerin. The original form, invented in 1867 by Alfred Nobel, consisted of nitroglycerin absorbed in kieselguhr. Modern dynamites, which are used for blasting, contain sodium or ammonium nitrate sensitized with nitroglycerin and use other absorbers (e.g. wood pulp).

**dynamo** An electric *generator, especially one designed to provide *direct current. Alternating-current generators can be called dynamos but are more often called alternators.

**dynamo action** The generation of electrical current and magnetic field by the motion of an electrically conducting fluid. It is generally believed that the magnetic fields of the earth and the sun are produced by dynamo action in the molten iron–nickel core of the earth and in the plasma of the solar interior.

**dynamometer** **1.** An instrument used to measure a force, often a spring balance. **2.** A device used to measure the output power of an engine or motor. **3. (current dynamometer)** A variety of *current balance, for measuring electric current.

**dyne** The unit of force in the *c.g.s. system; the force required to give a mass of one gram an acceleration of 1 cm s$^{-2}$. 1 dyne = 10$^{-5}$ newton.

**dysprosium** Symbol Dy. A soft silvery metallic element belonging to the *lan-thanoids; a.n. 66; r.a.m. 162.50; r.d. 8.551 (20°C); m.p. 1412°C; b.p. 2562°C. It occurs in apatite, gadolinite, and xenotime, from which it is extracted by an ion-exchange process. There are seven natural isotopes and twelve artificial isotopes have been identified. It finds limited use in some alloys as a neutron absorber, particularly in nuclear technology. It was discovered by Paul Lecoq de Boisbaudran (1838–1912) in 1886.

((⊕)) SEE WEB LINKS
• Information from the WebElements site

**dystectic mixture** A mixture of substances that has a constant maximum melting point.

**dystrophic** Describing a body of water, such as a lake, that contains large amounts of undecomposed organic matter derived from terrestrial plants. Dystrophic lakes are poor in dissolved nutrients and therefore unproductive; they are common in peat areas and may develop into peat bogs.

**e** The irrational number defined as the limit as $n$ tends to infinity of $(1 + 1/n)^n$. It has the value 2.718 28.... It is used as the base of natural *logarithms and occurs in the *exponential function, $e^x$.

**EAC** Emergency action code. *See* HAZCHEM CODE.

**ear** The sense organ in vertebrates that is specialized for the detection of sound and the maintenance of balance. It can be divided into the *outer ear and *middle ear, which collect and transmit sound waves, and the *inner ear, which contains the organs of balance and (except in fish) hearing (see illustration).

**eardrum** *See* TYMPANUM.

**early universe** The study of *cosmology at the time very soon after the *big bang. Theories of the early universe have led to a mutually beneficial interaction between cosmology and the theory of *elementary particles, particularly *grand unified theories.

Because there were very high tempera-

tures in the early universe many of the *broken symmetries in *gauge theories were unbroken symmetries. As the universe cooled after the big bang there is thought to have been a sequence of transitions to broken symmetry states.

Combining cosmology with grand unified theories helps to explain why the observed universe appears to consist of matter with no antimatter. This means that one has a nonzero *baryon number for the universe. This solution relies on the fact that there were nonequilibrium conditions in the early universe due to its rapid expansion after the big bang.

An important idea in the theory of the early universe is that of **inflation** – the idea that the nature of the *vacuum state gave rise, after the big bang, to an exponential expansion of the universe. The hypothesis of the **inflationary universe** solves several long-standing problems in cosmology, such as the flatness and homogeneity of the universe. In particular, it is thought that quantum fluctuations in the early universe

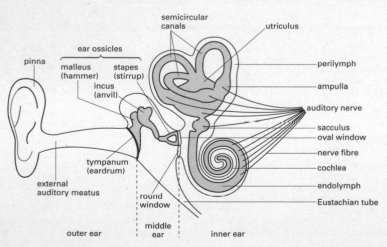

**Ear.** Structure of the mammalian ear.

were responsible for the emergence of large-scale structures in the universe, such as galaxies.

**ear ossicles** Three small bones – the **incus** (**anvil**), **malleus** (**hammer**), and **stapes** (**stirrup**) – that lie in the mammalian *middle ear, forming a bridge between the tympanum (eardrum) and the *oval window. The function of the ossicles is to transmit (and amplify) vibrations of the tympanum across the middle ear to the oval window, which transfers them to the *inner ear. Muscles of the middle ear constrict the movement of the ossicles. This serves to safeguard the ear from damage caused by excessively loud noise.

**earth** The planet that orbits the sun between the planets Venus and Mars at a mean distance from the sun of 149 600 000 km. It has a mass of about $5.974 \times 10^{24}$ kg and an equatorial diameter of 12 756.3 km. The earth consists of three layers: the gaseous atmosphere (*see* EARTH'S ATMOSPHERE), the liquid *hydrosphere, and the solid *lithosphere. The solid part of the earth also consists of three layers: the **crust** with a mean thickness of about 32 km under the land and 10 km under the seas; the **mantle**, which extends some 2900 km below the crust; and the **core**, part of which is believed to be liquid. The crust has a relative density of about 2.7 to 3.0 and consists largely of sedimentary rocks overlaying igneous rocks. The composition of the crust is: oxygen 47%, silicon 28%, aluminium 8%, iron 4.5%, calcium 3.5%, sodium and potassium 2.5% each, and magnesium 2.2%. Hydrogen, carbon, phosphorus, and sulphur are all present to an extent of less than 1%. The mantle reaches a relative density of 5.5 at its maximum depth and is believed to consist mainly of silicate rocks. The core is believed to have a maximum relative density of 13 and a maximum temperature of 6400 K. *See also* GEOMAGNETISM; PLATE TECTONICS.

**(( )) SEE WEB LINKS**
• NASA's earth observatory

**earthquake** A sudden movement or fracturing within the earth's lithosphere, causing a series of shocks. This may range from a mild tremor to a large-scale earth movement causing extensive damage over a wide area. The point at which the earthquake originates is known as the **seismic focus**; the point on the earth's surface directly above this is the **epicentre** (or **hypocentre**). *See* SEISMIC

WAVES. Earthquakes result from a build-up of stresses within the rocks until they are strained to the point beyond which they will fracture. They occur in narrow continuous belts of activity, which correspond with the junction of lithospheric plates, including the circum-Pacific belt, the Alpine–Himalayan belt, and mid-ocean ridges. The scale of the shock of an earthquake is known as the magnitude; the most commonly used scale for comparing the magnitude of earthquakes is the logarithmic *Richter scale (9.5 is the highest recorded magnitude on the scale).

**earth's atmosphere** The gas that surrounds the earth. The composition of dry air at sea level is: nitrogen 78.08%, oxygen 20.95%, argon 0.93%, carbon dioxide 0.03%, neon 0.0018%, helium 0.0005%, krypton 0.0001%, and xenon 0.00001%. In addition to water vapour, air in some localities contains sulphur compounds, hydrogen peroxide, hydrocarbons, and dust particles.

The lowest level of the atmosphere, in which most of the weather occurs, is called the **troposphere**. Its thickness varies from about 7 km at the poles to 28 km at the equator and in this layer temperature falls with increasing height. The next layer is the **stratosphere**, which goes up to about 50 km. Here the temperature remains approximately constant. Above this is the **ionosphere**, which extends to about 1000 km, with the temperature rising and the composition changing substantially. At about 100 km and above most of the oxygen has dissociated into atoms; at above 150 km the percentage of nitrogen has dropped to nil. In the ionosphere the gases are ionized by the absorption of solar radiation. This enables radio transmissions to be made round the curved surface of the earth as the ionized gas acts as a reflector for certain wavelengths. The ionosphere is divided into three layers. The D-layer (50–90 km) contains a low concentration of free electrons and reflects low-frequency radio waves. The E-layer (90–150 km) is also called the **Heaviside layer** or **Heaviside–Kennelly layer** as its existence was predicted independently by Oliver Heaviside (1850–1925) and Arthur E. Kennelly (1861–1939). This layer reflects medium-frequency waves. The F-layer (150–1000 km) is also called the **Appleton layer** after its discoverer Sir Edward Appleton (1892–1965). It has the highest concentration of free electrons and is the most useful for radio transmission. Wavelengths between 8 mm and 20 m are not reflected by the ionosphere but escape into space. Therefore television transmissions, which utilize this range, require artificial *satellites for reflection (or reception, amplification, and retransmission). From about 400 km, the outermost region of the atmosphere is also called the **exosphere**. See illustration.

**earth sciences** A group of sciences concerned with the study of the earth. The chief earth sciences are geology, physical geography, oceanography, meteorology, geophysics, and geochemistry.

**earthshine** Sunlight reflected from the surface of the earth. An observer in space may see nearby objects dimly illuminated by earthshine, as things on earth may be illuminated by moonlight. Under certain conditions near new moon the dark disc of the moon can be seen faintly illuminated by earthshine – a phenomenon called 'the old moon in the new moon's arms'.

**earthslide** The movement of a layer of dry soil down a slope. The soil layer is inherently unstable and shears, sometimes brought about by the action of water. But if the soil becomes very wet, and the slope is steep, an earthflow may occur. Contributory factors are sparse vegetation and sudden rainfall.

**earth's magnetic field** See GEOMAGNETISM.

**earwigs** See DERMAPTERA.

**ebonite** See VULCANITE.

**ebullioscopic constant** See ELEVATION OF BOILING POINT.

**ebullioscopy** The use of *elevation of boiling point to determine relative molecular masses.

**eccentricity** See CONIC.

**Eccles, Sir John Carew** (1903–97) Australian physiologist, who was educated in Melbourne and Oxford, and held appointments in Britain, Australia, New Zealand and, finally, the USA. While in Australia he carried out his best-known work, on the transmission of nerve impulses across

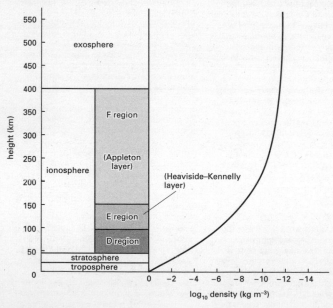

**Earth's atmosphere.**

synapses, which he attributed to the neurotransmitter acetylcholine. He shared the 1963 Nobel Prize for physiology or medicine with Sir Alan Hodgkin (1914–98) and Sir Andrew Huxley (1917–   ), who worked in the same area of biology.

**ecdysis (moulting)** **1.** The periodic loss of the outer cuticle of arthropods. It starts with the reabsorption of some materials in the inner part of the old cuticle and the formation of a new soft cuticle. The remains of the old cuticle then split; the animal emerges and absorbs water or swallows air and increases in size while the new cuticle is still soft. This cuticle is then hardened with chitin and lime salts. In insects and crustaceans ecdysis is controlled by the hormone *ecdysone. **2.** The periodic shedding of the outer layer of the epidermis of reptiles (except crocodiles) to allow growth to occur.

**ecdysone** A steroid hormone, produced by insects and crustaceans, that stimulates moulting (*see* ECDYSIS) and metamorphosis. It acts on specific gene loci, stimulating the synthesis of proteins involved in these bodily changes.

**e-cell** A computer simulation of a living cell based on information obtained from genomics, proteomics, metabolomics, and other data drawn from real cells.

**ECG** *See* ELECTROCARDIOGRAM.

**echelon** A form of *interferometer consisting of a stack of glass plates arranged stepwise with a constant offset. It gives a high resolution and is used in spectroscopy to study hyperfine line structure. In the **transmission echelon** the plates are made equal in optical thickness to introduce a constant delay between adjacent parts of the wavefront. The **reflecting echelon** has the exposed steps metallized and acts like an exaggerated *diffraction grating.

**Echinodermata** A phylum of marine invertebrates that includes the sea urchins, starfish, brittlestars, and sea cucumbers. Echinoderms have an exoskeleton (**test**) of calcareous plates embedded in the skin. In many species (e.g. sea urchins) spines protrude from the test. A system of water-filled canals (the **water vascular system**) provides hydraulic power for thousands of **tube feet**: saclike protrusions of the body wall used for locomotion, feeding, and respiration. Echinoderms have a long history: fossils of primitive echinoderms are known from rocks over 500 million years old.

**(((⊕))) SEE WEB LINKS**

- Overview of echinoderm characteristics and phylogeny from the Tree of Life Project

**echo** The reflection of a wave by a surface or object so that a weaker version of it is detected shortly after the original. The delay between the two is an indication of the distance of the reflecting surface. An **echo sounder** is an apparatus for determining the depth of water under a ship. The ship sends out a sound wave and measures the time taken for the echo to return after reflection by the sea bottom. **Sonar** (*so*und *na*vigation *ranging*) is a technique for locating underwater objects by a similar method. Echoes also occur with radio waves; reflection of waves causes an echo in radio transmission and ghosts in television pictures. *See also* RADAR.

**echolocation** **1.** *See* RADAR; ECHO. **2.** A method used by some animals (such as bats, dolphins, and certain birds) to detect objects in the dark. The animal emits a series of high-pitched sounds that echo back from the object and are detected by the ear or some other sensory receptor. From the direction of the echo and from the time between emission and reception of the sounds the object is located, often very accurately.

**ECL** *See* EMITTER-COUPLED LOGIC.

**eclipse** The obscuring of light from a celestial body as it passes behind or through the shadow of another body. In a **total eclipse**, the light is completely cut off; in a **partial eclipse**, the light is only partly hidden. A **lunar eclipse** occurs when the sun, earth, and moon are in a straight line and the shadow of the earth falls on the moon. A **solar eclipse** occurs when the sun, moon, and earth are aligned in such a way that the moon blocks out light from the sun's *photosphere and casts its shadow on the earth. An **annular eclipse** occurs when the moon is too far away for its shadow to reach the earth: during an eclipse, the rim of the photosphere is visible as a fiery ring (**annulus**) around the dark silhouette of the moon. See illustrations overleaf.

**(((⊕))) SEE WEB LINKS**

- NASA's eclipse Website, a fully comprehensive site maintained by Fred Espenak at the Goddard Space Flight Center, near Washington DC

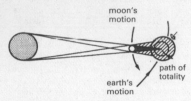

Solar and lunar eclipses

Moon's shadow in solar eclipse

**Eclipse.**

**eclipsed conformation** *See* CONFORMA-
TION.

**ecliptic** The *great circle on the *celestial
sphere that traces the sun's apparent passage through the sky against the background
of the *zodiac constellations over the course
of a year. In reality, it is the earth that is travels around the sun, and so the ecliptic delin-
eates the plane of the earth's orbit. *See also*
EQUINOX; SOLSTICE.

**ECM** **1.** *See* EXTRACELLULAR MATRIX. **2.** *See*
ELECTRON CYCLOTRON MASER.

**ECMAScript** *See* JAVASCRIPT.

**E. coli** *See* ESCHERICHIA COLI.

**ecological niche** The status or role of an
organism in its environment. An organism's
niche is defined by its food supply, predators, temperature tolerances, etc. Two
species cannot coexist stably if they occupy
identical niches.

**ecology** The study of the interrelationships between organisms and their natural
environment, both living and nonliving. For
this purpose, ecologists study organisms in
the context of the *populations and *communities in which they can be grouped and
the *ecosystems of which they form a part.
The study of ecological interactions provides
important information on the nature and
mechanisms of evolutionary change. Advances made in ecology over the last 30 years
have led to increased concern about the
effects of human activities on the environment (notably the effects of *pollution),
which has resulted in a greater awareness of
the importance of *conservation.

**ecosystem** A biological *community and
the physical environment associated with it.
Nutrients pass between the different organisms in an ecosystem in definite pathways;
for example, nutrients in the soil are taken
up by plants, which are then eaten by herbivores, which in turn may be eaten by carnivores (*see* FOOD CHAIN). Organisms are
classified on the basis of their position in an
ecosystem into various *trophic levels. Nutrients and energy move round ecosystems
in loops or cycles (in the case above, for example, nutrients are returned to the soil via
animal wastes and decomposition). *See* CAR-
BON CYCLE; NITROGEN CYCLE.

**(((⊕))) SEE WEB LINKS**
• An overview of ecosystems, biomes, and habitats from the Franklin Institute, Philadelphia

**ecosystem services** The benefits that
people receive from ecosystems. These
benefits can be classified into several categories. **Provisioning services** include the
products that humans derive from ecosystems, such as food, fibres, fuel, drinking
water, and medicinal products. **Regulating
services** encompass such processes as control of the climate, purification of air and
water, amelioration of erosion, reduction of
pests and diseases, and provision of pollination mechanisms. **Cultural services** comprise the nonmaterial benefits that accrue
from the landscape, such as recreational
pursuits, spiritual renewal, and aesthetic experiences. Underlying all these are the fundamental **supporting services** of all
ecosystems, such as primary production (i.e.
photosynthesis by green plants, algae, etc.),
nutrient cycling (e.g. carbon, nitrogen, sulphur cycles), water cycling, and soil formation. In recent decades the widespread
degradation of many of the world's ecosystems has seen the concept of ecosystem ser-

vices acquire greater prominence, chiefly as a means of assessing the economic value of particular ecological benefits and the cost of their destruction.

**ecotype** A subgroup of a species that has characteristic genetically determined adaptations to its local environment. In some cases individuals belonging to different ecotypes cannot interbreed, for example where accumulated genetic differences are too great.

**ecstasy (methylenedioxymethamphetamine; MDMA)** A designer drug based on methamphetamine (*see* AMPHETAMINE). Originally intended as an appetite suppressant, it produces a feeling of euphoria and is widely used as a club drug. It is a class A drug in the UK.

**ectoderm** The external layer of cells of the *gastrula, which will develop into the epidermis and the nervous system in the adult. *See also* GERM LAYERS.

**ectoparasite** A parasite that lives on the outside of its host's body. *See* PARASITISM.

**ectoplasm** *See* CYTOPLASM.

**Ectoprocta** *See* BRYOZOA.

**ectotherm (poikilotherm)** An animal that maintains its body temperature by absorbing heat from the surrounding environment. All animals except mammals and birds are ectotherms; they are often described as being **cold-blooded** and are unable to regulate their body temperature metabolically. *See* POIKILOTHERMY. *Compare* ENDOTHERM.

**edaphic factor** A factor relating to the physical or chemical composition of the soil found in a particular area. For example, very alkaline soil may be an edaphic factor limiting the variety of plants growing in a region.

**Eddington limit** A limit for the maximum value of the brightness of a star of a given mass. This limit exists because the radiation pressure caused by the nuclear fusion reactions powering the star has to counter, but not exceed, the gravitational force that would cause gravitational collapse of the star. The existence of this limit was first pointed out by the English astrophysicist Sir Arthur Stanley Eddington (1882–1944).

**eddy current** A current induced in a conductor situated in a changing magnetic field or moving in a fixed one. Any imagined circuit within the conductor will change its

magnetic flux linkage, and the consequent induced e.m.f. will drive current around the circuit. In a substantial block of metal the resistance will be small and the current therefore large. Eddy currents occur in the cores of transformers and other electrical machines and represent a loss of useful energy (the **eddy-current loss**). To reduce this loss to a minimum metal cores are made of insulated sheets of metal, the resistance between these laminations reducing the current. In high-frequency circuits *ferrite cores can be used. Eddy currents in a moving conductor interact with the magnetic field producing them to retard the motion of the conductor. This enables some electrical instruments (moving-coil type) to utilize eddy currents to create damping. Eddy currents are also used in *induction heating.

**Edison cell** *See* NICKEL–IRON ACCUMULATOR.

**EDTA** Ethylenediaminetetraacetic acid,

$$(HOOCCH_2)_2N(CH_2)_2N(CH_2COOH)_2$$

A compound that acts as a chelating agent, reversibly binding with iron, magnesium, and other metal ions. It is used in certain culture media bound with iron, which it slowly releases into the medium, and also in some forms of quantitative analysis.

**EEG** *See* ELECTROENCEPHALOGRAM.

**effective temperature** *See* LUMINOSITY.

**effective value** *See* ROOT-MEAN-SQUARE VALUE.

**effector** A cell or organ that produces a physiological response when stimulated by a nerve impulse. Examples include muscles and glands.

**effector neuron** A nerve cell, such as a motor neuron, that transmits impulses from the central nervous system to an *effector in order to bring about a physiological response to changes in the environment.

**efferent** Carrying (nerve impulses, blood, etc.) away from the centre of a body or organ towards peripheral regions. The term is usually applied to types of nerve fibres or blood vessels. *Compare* AFFERENT.

**effervescence** The formation of gas bubbles in a liquid by chemical reaction.

**efficiency** A measure of the performance of a machine, engine, etc., being the ratio of the energy or power it delivers to the energy or power fed to it. In general, the efficiency of a machine varies with the conditions under which it operates and there is usually a load at which it operates with the highest efficiency. The **thermal efficiency** of a heat engine is the ratio of the work done by the engine to the heat supplied by the fuel. For a reversible heat engine this efficiency equals $(T_1 - T_2)/T_1$, where $T_1$ is the thermodynamic temperature at which all the heat is taken up and $T_2$ is the thermodynamic temperature at which it is given out (*see* CARNOT CYCLE). For real engines it is always less than this.

**efflorescence** The process in which a crystalline hydrate loses water, forming a powdery deposit on the crystals.

**effusion** The flow of a gas through a small aperture. The relative rates at which gases effuse, under the same conditions, is approximately inversely proportional to the square roots of their densities.

**egestion** The expulsion from the body of waste food materials that have never left the gut, particularly the expulsion of undigested materials from the gut through the anus (*see* DEFECATION). Egestion should not be confused with *excretion, in which the waste materials are produced by metabolic activity in the body's tissues.

**egg** **1.** The fertilized ovum (*zygote) in egg-laying animals, e.g. birds and insects, after it emerges from the body. The egg is covered by *egg membranes that protect it from environmental damage, such as drying. **2. (egg cell)** The mature female reproductive cell in animals and plants. *See* OOSPHERE; OVUM.

**egg membrane** The layer of material that covers an animal egg cell. **Primary membranes** develop in the ovary and cover the egg surface in addition to the normal plasma membrane. The primary membrane is called the **vitelline membrane** in insects, molluscs, birds, and amphibians, the **chorion** in tunicates and fish, and the **zona pellucida** in mammals. Insects have a second thicker membrane, also called the chorion. **Secondary membranes** are secreted by the oviducts and parts of the genital system while the egg is passing to the outside. They include the jelly coat of frogs' eggs and the albumen and shell of birds' eggs.

**Ehrlich, Paul** (1854–1915) German bacteriologist, who graduated as a physician in 1878. After working in a Berlin hospital for nine years he taught at Berlin University (un-

paid because he was a Jew). In 1890 he went to work with Robert Koch (1843–1910) to study tuberculosis, cholera, and other diseases. In 1910 he discovered Salvarsan, an arsenical drug effective against syphilis. He was awarded the 1908 Nobel Prize for physiology or medicine for his earlier work on serum therapy.

**eigenfunction** An allowed *wave function of a system in quantum mechanics. The associated energies are **eigenvalues**.

**Einstein, Albert** (1879–1955) German-born US physicist, who took Swiss nationality in 1901. A year later he went to work in the Bern patent office. In 1905 he published five enormously influential papers, one on *Brownian movement, one on the *photoelectric effect, one on the special theory of *relativity, and one on energy and inertia (which included the famous expression $E = mc^2$). In 1915 he published the general theory of relativity, concerned mainly with gravitation. In 1921 he was awarded the Nobel Prize for physics. In 1933, as a Jew, Einstein decided to remain in the USA (where he was lecturing), as Hitler had come to power. For the remainder of his life he sought a unified field theory. In 1939 he informed President Roosevelt that an atom bomb was feasible and that Germany might be able to make one.

(((●))) **SEE WEB LINKS**

- A comprehensive website about his life and work run by The American Institute of Physics Center for History of Physics

**Einstein coefficients** Coefficients used in the *quantum theory of radiation, related to the probability of a transition occurring between the ground state and an excited state (or vice versa) in the processes of *induced emission and *spontaneous emission. For an atom exposed to *electromagnetic radiation, the rate of absorption $R_a$ is given by $R_a = B\rho$, where $\rho$ is the density of electromagnetic radiation and $B$ is the **Einstein B coefficient** associated with absorption. The rate of induced emission is also given by $B\rho$, with the coefficient B of induced emission being equal to the coefficient of absorption. The rate of spontaneous emission is given by $A$, where $A$ is the **Einstein A coefficient of spontaneous emission**. The $A$ and $B$ coefficients are related by $A = 8\pi h\nu^3 B/c^3$, where $h$ is the *Planck constant, $\nu$ is the frequency of electromagnetic radiation, and $c$ is the speed of light. The coefficients were put forward by

Albert Einstein in 1916–17 in his analysis of the quantum theory of radiation.

**Einstein equation** 1. The mass–energy relationship announced by Einstein in 1905 in the form $E = mc^2$, where $E$ is a quantity of energy, $m$ its mass, and $c$ is the speed of light. It presents the concept that energy has mass. *See also* RELATIVITY. **2.** The relationship

$$E_{max} - hf - W,$$

where $E_{max}$ is the maximum kinetic energy of the electrons emitted in the photoemissive effect, $h$ is the Planck constant, $f$ the frequency of the incident radiation, and $W$ the *work function of the emitter. This is also written $E_{max} = hf - \phi e$, where $e$ is the electronic charge and $\phi$ a potential difference, also called the work function. (Sometimes $W$ and $\phi$ are distinguished as **work function energy** and **work function potential**.) The equation can also be applied to photoemission from gases, when it has the form $E = hf - I$, where $I$ is the ionization potential of the gas.

**einsteinium** Symbol Es. A radioactive metallic transuranic element belonging to the *actinoids; a.n. 99; mass number of the most stable isotope 254 (half-life 270 days). Eleven isotopes are known. The element was first identified by Albert Ghiorso and associates in debris from the first hydrogen bomb explosion in 1952. Microgram quantities of the element did not become available until 1961.

**(((⊕))) SEE WEB LINKS**

• Information from the WebElements site

**Einstein shift** *See* REDSHIFT.

**ejaculation** The propulsion of semen out of the erect penis due to powerful rhythmic contractions of the urethra. An ejaculation coincides with the peak of sexual excitement (**orgasm**) and is accompanied by various physiological effects in the body, such as increased respiration rate and heart rate.

**ejecta** Rocks and other material thrown up when a crater is formed. Some craters and their associated ejecta are volanic in origin, but most (especially on the moon, Mercury, and other planets and their satellites) are caused by the impact of meteorites. Often these are surrounded by an ejecta blanket, which is thickest near the crater's rim.

**ekpyrotic universe** A model of the universe that postulates that the big bang oc-

curred because of the collision of two parallel branes of a type predicted to occur in string theory (*see* BRANE WORLD). The term 'ekpyrotic' means 'coming out of fire'. It has been developed into a **cyclic model** of the universe, which avoids the problems of entropy associated with the *oscillatory universe. There are some differences between the predictions of the ekpyrotic theory and those of the more established inflationary universe (*see* EARLY UNIVERSE).

**Elasmobranchii** *See* CHONDRICHTHYES.

**elastance** The reciprocal of *capacitance. It is measured in farad$^{-1}$ (sometimes called a 'daraf').

**elastic cartilage** *See* CARTILAGE.

**elastic collision** A collision in which the total kinetic energy of the colliding bodies after collision is equal to their total kinetic energy before collision. Elastic collisions occur only if there is no conversion of kinetic energy into other forms, as in the collision of atoms. In the case of macroscopic bodies this will not be the case as some of the energy will become heat. In a collision between polyatomic molecules, some kinetic energy may be converted into vibrational and rotational energy of the molecules, but otherwise molecular collisions appear to be elastic.

**elastic fibres** *See* ELASTIN.

**elasticity** The property of certain materials that enables them to return to their original dimensions after an applied *stress has been removed. In general, if a stress is applied to a wire, the *strain will increase in proportion (see $OA$ on the illustration) until a certain point called the **limit of proportionality** is reached. This is in accordance with *Hooke's law. Thereafter there is at first a slight increase in strain with increased load until a point $L$ is reached. This is the **elastic**

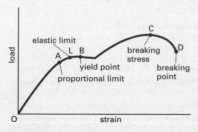

**Elasticity.**

limit; up to this point the deformation of the specimen is elastic, i.e. when the stress is removed the specimen returns to its original length. Beyond the point $L$ there is permanent deformation when the stress is removed, i.e. the material has ceased to be **elastic** and has become **plastic**. In the plastic stages individual materials vary somewhat; in general, however, at a point $B$ there is a sudden increase in strain with further increases of stress – this is the **yield point**. Beyond the point $C$, the **breaking stress**, the wire will snap (which occurs at point $D$).

**elastic modulus** The ratio of the *stress applied to a body to the *strain produced. The **Young modulus of elasticity**, named after Thomas Young, refers to longitudinal stress and strain. The **bulk modulus** is the ratio of the pressure on a body to its fractional decrease in volume. The **shear** (or **rigidity**) **modulus** is the tangential force per unit area divided by the angular deformation in radians.

**elastin** A fibrous protein that is the major constituent of the yellow **elastic fibres** of *connective tissue. It is rich in glycine, alanine, proline, and other nonpolar amino acids that are cross-linked, making the protein relatively insoluble. Elastic fibres can stretch to several times their length and then return to their original size. Elastin is particularly abundant in elastic *cartilage, blood-vessel walls, ligaments, and the heart.

**elastomer** A natural or synthetic rubber or rubberoid material, which has the ability to undergo deformation under the influence of a force and regain its original shape once the force has been removed.

**electret** A permanently electrified substance or body that has opposite charges at its extremities. Electrets resemble permanent magnets in many ways. An electret can be made by cooling certain waxes in a strong electric field.

**electrical energy** A form of energy related to the position of an electric charge in an electric field. For a body with charge $Q$ and an electric potential $V$, its electrical energy is $QV$. If $V$ is a potential difference, the same expression gives the energy transformed when the charge moves through the p.d.

**electric arc** A luminous discharge between two electrodes. The discharge raises the electrodes to incandescence, the resulting thermal ionization largely providing the carriers to maintain the high current between the electrodes.

**electric-arc furnace** A furnace used in melting metals to make alloys, especially in steel manufacture, in which the heat source is an electric arc. In the direct-arc furnace, such as the Héroult furnace, an arc is formed between the metal and an electrode. In the indirect-arc furnace, such as the Stassano furnace, the arc is formed between two electrodes and the heat is radiated onto the metal.

**electric bell** A device in which an electromagnetically operated hammer strikes a bell (see illustration). Pressing the bell-push closes a circuit, causing current to flow from a battery or mains step-down transformer through an electromagnet. The electromagnet attracts a piece of soft iron attached to the hammer, which strikes the bell and at the same time breaks the circuit. The hammer springs back into its original position again, closing the circuit and causing the magnet to attract the soft iron. This process continues until the bell-push is released.

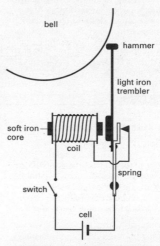

**Electric bell.**

**electric charge** *See* CHARGE.

**electric constant** *See* PERMITTIVITY.

**electric current** *See* CURRENT.

**electric displacement (electric flux den-**

**sity)** Symbol $D$. The charge per unit area that would be displaced across a layer of conductor placed across an *electric field. This describes also the charge density on an extended surface that could be causing the field.

**electric field** A region in which an electric charge experiences a force usually because of a distribution of other charges. The **electric field strength** or **electric intensity** ($E$) at any point in an electric field is defined as the force per unit charge experienced by a small charge placed at that point. This is equivalent to a potential gradient along the field and is measured in volts per metre. The strength of the field can alternatively be described by its *electric displacement $D$. The ratio $D/E$ for measurements in a vacuum is the electric constant $\varepsilon_0$. In a substance the observed potential gradient is reduced by electron movement so that $D/E$ appears to increase: the new ratio ($\varepsilon$) is called the *permittivity of the substance. An electric field can be created by an isolated electric charge, in which case the field strength at a distance $r$ from a point charge $Q$ is given by $E = Q/4\pi r^2\varepsilon$, where $\varepsilon$ is the permittivity of the intervening medium (*see* COULOMB'S LAW). An electric field can also be created by a changing magnetic field.

**electric flux** Symbol $\Psi$. In an *electric field, the product of the electric flux density and the relevant area. *See* ELECTRIC DISPLACEMENT.

**electric flux density** *See* ELECTRIC DISPLACEMENT.

**electricity** Any effect resulting from the existence of stationary or moving electric charges.

**electric lighting** Illumination provided by electric currents. The devices used are the **arc lamp**, the **light bulb** (incandescent filament lamp), and the **fluorescent tube**. In the arc lamp, which is no longer used as a general means of illumination, an electric current flows through a gap between two carbon electrodes, between which a high potential difference is maintained. The current is carried by electrons and ions in the vapour produced by the electrodes and a mechanism is required to bring the electrodes closer together as they are vaporized. The device produces a strong white light but has many practical disadvantages. However, arcs enclosed in an inert gas (usually xenon) are

increasingly used for such purposes as cinema projectors. The common light bulb is a glass bulb containing a tungsten filament and usually an inert gas. The passage of an electric current through the filament heats it to a white heat. Inert gas is used in the bulb to minimize blackening of the glass by evaporation of tungsten. In the fluorescent tube a glass tube containing mercury vapour (or some other gas) at a low pressure has its inner surface coated with a fluorescent substance. A discharge is created within the tube between two electrodes. Electrons emitted by the cathode collide with gas atoms or molecules and raise them to an excited state (*see* EXCITATION). When they fall back to the *ground state they emit photons of ultraviolet radiation, which is converted to visible light by the coating of phosphor on the inner walls of the tube. In some lamps, such as the *sodium-vapour and *mercury-vapour lamps used in street lighting, no fluorescent substance is used, the light being emitted directly by the excited atoms of sodium or mercury. Vapour lights are more efficient than filament lights as less of the energy is converted into heat.

**electric motor** A machine for converting electrical energy into mechanical energy. They are quiet, clean, and have a high efficiency (75–95%). They work on the principle that a current passing through a coil within a magnetic field will experience forces that can be used to rotate the coil. In the **induction motor**, alternating current is fed to a stationary coil (the **stator**), which both creates the magnetic field and induces a current in the rotating coil (**rotor**), which it surrounds. The advantage of this kind of motor is that current does not have to be fed through a commutator to a moving part. In the **synchronous motor**, alternating current fed to the stator produces a magnetic field that rotates and locks with the field of the rotor, in this case an independent magnet, causing the rotor to rotate at the same speed as the stator field rotates. The rotor is either a permanent magnet or an electromagnet fed by a direct current through slip rings. In the **universal motor**, current is fed to the stator and, through a commutator, to the rotor. In the series-wound motor the two are in series; in the shunt-wound motor they are in parallel. These motors can be used with either a.c. or d.c. but some small motors use a permanent magnet as the stator and require d.c. for the rotor (via the commutator). *See also* LINEAR MOTOR.

**electric organ** An organ occurring on the body or tail of certain fish, such as the electric ray (*Torpedo*) and electric eel (*Electrophorus electricus*). It gives an electric shock when touched and is used either to stun prey or predators or, in some species, to maintain a weak electric field in the surrounding water that is used in navigation. The organ is composed of modified muscle cells (**electroplate cells**), nervous stimulation of which greatly increases the potential difference across the cell. The electroplates are in series so a high overall voltage can be achieved.

**electric polarization** *See* DIELECTRIC.

**electric potential** Symbol $V$. The energy required to bring unit electric charge from infinity to the point in an electric field at which the potential is being specified. The unit of electric potential is the volt. The **potential difference (p.d.)** between two points in an electric field or circuit is the difference in the values of the electric potentials at the two points, i.e. it is the work done in moving unit charge from one point to the other.

**electric power** The rate of expending energy or doing work in an electrical system. For a direct-current circuit, it is given by the product of the current passing through a system and the potential difference across it. In alternating-current circuits, the power is given by $VI\cos\phi$, where $V$ and $I$ are the RMS values and $\phi$ is the *phase angle. Cos$\phi$ is called the **power factor** of the circuit.

**electric spark** The transient passage of an electric current through a gas between two points of high opposite potential, with the emission of light and sound. *Lightning consists of a spark between a cloud and earth or between two oppositely charged parts of the same cloud.

**electric susceptibility** *See* SUSCEPTIBILITY.

**electrocardiogram (ECG)** A tracing or graph of the electrical activity of the heart. Recordings are made from electrodes fastened over the heart and usually on both arms and a leg. Changes in the normal pattern of an ECG may indicate heart irregularities or disease.

**electrochemical cell** *See* CELL.

**electrochemical equivalent** Symbol $z$. The mass of a given element liberated from a solution of its ions in electrolysis by one coulomb of charge. *See* FARADAY'S LAWS (of electrolysis).

**electrochemical series** *See* ELECTROMOTIVE SERIES.

**electrochemistry** The study of chemical properties and reactions involving ions in solution, including electrolysis and electric cells.

**electrochromatography** *See* ELECTROPHORESIS.

**electrode** 1. A conductor that emits or collects electrons in a cell, thermionic valve, semiconductor device, etc. The **anode** is the positive electrode and the **cathode** is the negative electrode. 2. *See* HALF CELL.

**electrodeposition** The process of depositing one metal on another by electrolysis, as in *electroforming and *electroplating.

**electrode potential** The potential difference produced between the electrode and the solution in a *half cell. It is not possible to measure this directly since any measurement involves completing the circuit with the electrolyte, thereby introducing another half cell. **Standard electrode potentials** $E^{\ominus}$ are defined by measuring the potential relative to a standard *hydrogen half cell using 1.0 molar solution at 25°C. The convention is to designate the cell so that the oxidized form is written first. For example,

$$Pt(s)|H_2(g)H^+(aq)|Zn^{2+}(aq)|Zn(s)$$

The e.m.f. of this cell is –0.76 volt (i.e. the zinc electrode is negative). Thus the standard electrode potential of the $Zn^{2+}|Zn$ half cell is –0.76 V. Electrode potentials are also called **reduction potentials**. *See also* ELECTROMOTIVE SERIES.

**electrodialysis** A method of obtaining pure water from water containing a salt, as in *desalination. The water to be purified is fed into a cell containing two electrodes. Between the electrodes is placed an array of *semipermeable membranes alternately semipermeable to positive ions and negative ions. The ions tend to segregate between alternate pairs of membranes, leaving pure water in the other gaps between membranes. In this way, the feed water is separated into two streams: one of pure water and the other of more concentrated solution.

**electrodynamics** The study of electric charges in motion, the forces created by electric and magnetic fields, and the relationship between them. *Compare* ELECTROSTATICS.

**electroencephalogram (EEG)** A tracing or graph of the electrical activity of the brain. Electrodes taped to the scalp record electrical waves from different parts of the brain. The pattern of an EEG reflects an individual's level of consciousness and can be used to detect such disorders as epilepsy, tumours, or brain damage. *See also* BRAIN DEATH.

**electroendosmosis** *See* ELECTROOSMOSIS.

**electroforming** A method of forming intricate metal articles or parts by *electrodeposition of the metal on a removable conductive mould.

**electroluminescence** *See* LUMINESCENCE.

**electrolysis** The production of a chemical reaction by passing an electric current through an electrolyte. In electrolysis, positive ions migrate to the cathode and negative ions to the anode. The reactions occurring depend on electron transfer at the electrodes and are therefore redox reactions. At the anode, negative ions in solution may lose electrons to form neutral species. Alternatively, atoms of the electrode can lose electrons and go into solution as positive ions. In either case the reaction is an oxidation. At the cathode, positive ions in solution can gain electrons to form neutral species. Thus cathode reactions are reductions.

**(⊕) SEE WEB LINKS**
- An introductory guide to electrolysis from the University of Waterloo, Ontario

**electrolyte** A liquid that conducts electricity as a result of the presence of positive or negative ions. Electrolytes are molten ionic compounds or solutions containing ions, i.e. solutions of ionic salts or of compounds that ionize in solution. Liquid metals, in which the conduction is by free electrons, are not usually regarded as electrolytes. Solid conductors of ions, as in the sodium–sulphur cell, are also known as electrolytes.

**electrolytic capacitor** *See* CAPACITOR.

**electrolytic cell** A cell in which electrolysis occurs; i.e. one in which current is passed through the electrolyte from an external source.

**electrolytic corrosion** Corrosion that occurs through an electrochemical reaction. *See* RUSTING.

**electrolytic gas (detonating gas)** The highly explosive gas formed by the electrolysis of water. It consists of two parts hydrogen and one part oxygen by volume.

**electrolytic rectifier** A *rectifier consisting of two dissimilar electrodes immersed in an electrolyte. By suitable choice of electrodes and electrolyte the cell can be made to pass current easily in one direction but hardly at all in the other. Examples include a lead–aluminium cell with ammonium phosphate(V) electrolyte and a tantalum–lead cell with sulphuric acid as the electrolyte.

**electrolytic refining** The purification of metals by electrolysis. It is commonly applied to copper. A large piece of impure copper is used as the anode with a thin strip of pure copper as the cathode. Copper(II) sulphate solution is the electrolyte. Copper dissolves at the anode: $Cu \rightarrow Cu^{2+} + 2e$, and is deposited at the cathode. The net result is transfer of pure copper from anode to cathode. Gold and silver in the impure copper form a so-called **anode sludge** at the bottom of the cell, which is recovered.

**electrolytic separation** A method of separating isotopes by exploiting the different rates at which they are released in electrolysis. It was formerly used for separating deuterium and hydrogen. On electrolysis of water, hydrogen is formed at the cathode more readily than deuterium, thus the water becomes enriched with deuterium oxide.

**electromagnet** A magnet consisting of a soft ferromagnetic core with a coil of insulated wire wound round it. When a current flows through the wire the core becomes magnetized; when the current ceases to flow the core loses its magnetization. Electromagnets are used in switches, solenoids, electric bells, metal-lifting cranes, and many other applications.

**electromagnetic induction** The production of an electromotive force in a conductor when there is a change of magnetic flux linkage with the conductor or when there is relative motion of the conductor across a magnetic field. The magnitude of the e.m.f. is proportional (and in modern systems of units equal) to the rate of change of the flux linkage or the rate of cutting flux $d\Phi/dt$; the sense of the induced e.m.f. is such that any induced current opposes the change causing the induction, i.e. $E = -d\Phi/dt$. See FARADAY'S LAWS; LENZ'S LAW; NEUMANN'S LAW; INDUCTANCE.

**electromagnetic interaction** See FUNDAMENTAL INTERACTIONS.

**electromagnetic pump** A pump used for moving liquid metals, such as the liquid-sodium coolant in a fast nuclear reactor. The liquid is passed through a flattened pipe over two electrodes between which a direct current flows. A magnetic field at right angles to the current causes a force to be created directly on the liquid, along the axis of the tube. The pump has no moving parts and is therefore safe and trouble free.

**electromagnetic radiation** Energy resulting from the acceleration of electric charge and the associated electric fields and magnetic fields. The energy can be regarded as waves propagated through space (requiring no supporting medium) involving oscillating electric and magnetic fields at right angles to each other and to the direction of propagation. In a vacuum the waves travel with a constant speed (the speed of light) of $2.9979 \times 10^8$ metres per second; if material is present they are slower. Alternatively, the energy can be regarded as a stream of *photons travelling at the speed of light, each photon having an energy $hc/\lambda$, where $h$ is the Planck constant, $c$ is the speed of light, and $\lambda$ is the wavelength of the associated wave. A fusion of these apparently conflicting concepts is possible using the methods of *quantum mechanics or *wave mechanics. The characteristics of the radiation depend on its wavelength. See ELECTROMAGNETIC SPECTRUM.

**electromagnetic spectrum** The range of wavelengths over which *electromagnetic radiation extends. The longest waves ($10^5$–$10^{-3}$ metres) are radio waves, the next longest ($10^{-3}$–$10^{-6}$ m) are infrared waves, then comes the narrow band ($4$–$7 \times 10^{-7}$ m) of visible light, followed by ultraviolet waves ($10^{-7}$–$10^{-9}$ m), X-rays ($10^{-9}$–$10^{-11}$ m), and gamma rays ($10^{-11}$–$10^{-14}$ m).

**electromagnetic units (e.m.u.)** A system of electrical units formerly used in the *c.g.s. system. The e.m.u. of electric current is the **abampere** (all e.m.u. have the prefix

*ab-* attached to the names of practical units). The abampere is the current that, flowing in an arc of a circle (1 centimetre in diameter), exerts a force of 1 dyne on unit magnetic pole at the centre of the circle. In e.m.u. the magnetic constant is of unit magnitude. The system has now been replaced by *SI units for most purposes. *Compare* ELECTROSTATIC UNITS; GAUSSIAN UNITS; HEAVISIDE–LORENTZ UNITS.

**electromagnetic wave** See ELECTROMAGNETIC RADIATION; WAVE.

**electrometallurgy** The uses of electrical processes in the separation of metals from their ores, the refining of metals, or the forming or plating of metals.

**electrometer** A measuring instrument for determining a voltage difference without drawing an appreciable current from the source. Originally electrostatic instruments based on the electroscope, they are now usually based on operational amplifiers, solid-state devices with high input impedances. Electrometers are also used to measure low currents (nanoamperes), by passing the current through a high resistance.

**electromotive force (e.m.f.)** The greatest potential difference that can be generated by a particular source of electric current. In practice this may be observable only when the source is not supplying current, because of its *internal resistance.

**electromotive series (electrochemical series)** A series of chemical elements arranged in order of their *electrode potentials. The hydrogen electrode ($H^+ + e \rightarrow \frac{1}{2}H_2$) is taken as having zero electrode potential. Elements that have a greater tendency than hydrogen to lose electrons to their solution are taken as **electropositive**; those that gain electrons from their solution are below hydrogen in the series and are called **electronegative**. The series shows the order in which metals replace one another from their salts; electropositive metals will replace hydrogen from acids. The chief metals and hydrogen, placed in order in the series, are: potassium, calcium, sodium, magnesium, aluminium, zinc, cadmium, iron, nickel, tin, lead, hydrogen, copper, mercury, silver, platinum, gold. This type of series is sometimes referred to as an **activity series**.

**electron** An *elementary particle, classed as a *lepton, with a rest mass (symbol $m_e$) of $9.109\ 3897(54) \times 10^{-31}$ kg and a negative charge of 1.602 177 33(49) × $10^{-19}$ coulomb. Electrons are present in all atoms in groupings called shells around the nucleus; when they are detached from the atom they are called **free electrons**. The antiparticle of the electron is the **positron**.

**electron affinity** Symbol $A$. The energy change occurring when an atom or molecule gains an electron to form a negative ion. For an atom or molecule X, it is the energy released for the electron-attachment reaction

$$X(g) + e \rightarrow X^-(g)$$

Often this is measured in electronvolts. Alternatively, the molar enthalpy change, $\Delta H$, can be used.

**electron biprism** An arrangement of fields that splits a beam of electrons or other charged particles in an analogous way to an optical biprism.

**electron capture** 1. The formation of a negative ion by an atom or molecule when it acquires an extra free electron. 2. A radioactive transformation in which a nucleus acquires an electron from an inner orbit of the atom, thereby transforming, initially, into a nucleus with the same mass number but an atomic number one less than that of the original nucleus (capture of the electron transforms a proton into a neutron). This type of capture is accompanied by emission of an X-ray photon or Auger electron as the vacancy in the inner orbit is filled by an outer electron.

**electron counting rules** Rules for counting the numbers of electrons that give rise to stable structures in atoms, molecules, and solids. For example, the result that the number of electrons in a full electronic shell in an atom is $n^2$, where $n$ is the principal quantum number associated with the shell, is an example of an electron counting rule. Various rules exist for predicting stable structures in chemical compounds, especially in certain types of cluster compound.

**electron cyclotron maser (ECM)** A type of *maser in which the operation is based on the stimulated emission of radiation caused by the motion of high-energy electrons moving as in a *cyclotron. A full understanding of this type of maser requires the special theory of relativity. There are many practical applications of the device, including plasma heating in fusion research and telecommunications. This mechanism of maser emission may account for short intense pulses in solar microwave radiation and for certain ionospheric radio emissions.

**electron-deficient compound** A compound in which there are fewer electrons forming the chemical bonds than required in normal electron-pair bonds. Such compounds use *multicentre bonds. *See* BORANE.

**electron diffraction** *Diffraction of a beam of electrons by atoms or molecules. The fact that electrons can be diffracted in a similar way to light and X-rays shows that particles can act as waves (*see* DE BROGLIE WAVELENGTH). An electron (mass $m$, charge $e$) accelerated through a potential difference $V$ acquires a kinetic energy $mv^2/2 = eV$, where $v$ is the velocity of the electron. The (nonrelativistic) momentum ($p$) of the electron is $\sqrt{(2eVm)}$. The de Broglie wavelength ($\lambda$) of an electron is given by $h/p$, where $h$ is the Planck constant, thus $\lambda = h/\sqrt{(2eVm)}$. For an accelerating voltage of 3600 V, the wavelength of the electron beam is 0.02 nanometre, some $3 \times 10^4$ times shorter than visible radiation.

Electrons then, like X-rays, show diffraction effects with molecules and crystals in which the interatomic spacing is comparable to the wavelength of the beam. They have the advantage that their wavelength can be set by adjusting the voltage. Unlike X-rays they have very low penetrating power. The first observation of electron diffraction was by George Thomson (1892–1975) in 1927, in an experiment in which he passed a beam of electrons in a vacuum through a very thin gold foil onto a photographic plate. Concentric circles were produced by diffraction of electrons by the lattice. The same year Clinton J. Davisson (1881–1958) and Lester Germer (1896–1971) performed a classic experiment in which they obtained diffraction patterns by glancing an electron beam off the surface of a nickel crystal. Both experiments were important verifications of de Broglie's theory and the new quantum theory.

Electron diffraction, because of the low penetration, cannot easily be used to investigate crystal structure. It is, however, employed to measure bond lengths and angles of molecules in gases. Moreover, it is extensively used in the study of solid surfaces and absorption. The main techniques are low-energy electron diffraction (**LEED**) in which the electron beam is reflected onto a fluorescent screen, and high-energy electron diffraction (**HEED**) used either with reflection or transmission in investigating thin films.

**electronegative** Describing elements that tend to gain electrons and form negative ions. The halogens are typical electronegative elements. For example, in hydrogen chloride, the chlorine atom is more electronegative than the hydrogen and the molecule is polar, with negative charge on the chlorine atom. There are various ways of assigning values for the **electronegativity** of an element. **Mulliken electronegativities** are calculated from $E = (I + A)/2$, where $I$ is ionization potential and $A$ is electron affinity. More commonly, **Pauling electronegativities** are used. These are based on bond dissociation energies using a scale in which fluorine, the most electronegative element, has a value 4. Some other values on this scale are B 2, C 2.5, N 3.0, O 3.5, Si 1.8, P 2.1, S 2.5, Cl 3.0, Br 2.8.

**electron flow** The transfer of electrons along a series of carrier molecules in the *electron transport chain.

**electron gun** A device used in *cathode-ray tubes (including television tubes), electron microscopes, etc., to produce a steady narrow beam of electrons. It usually consists of a heated cathode, control grid, and two or more annular anodes inserted in an evacuated tube. The electrons emitted by the cathode are attracted to the final anode, through which they pass. The intensity of the beam is regulated by the control grid and potential differences between the anodes create electric fields that focus the diverging electrons into a narrow beam.

**electronic mail (e-mail)** Messages, documents, etc., sent between users of computer systems, the computer systems being used to transport and hold the e-mail. The service itself is also referred to as electronic mail. The sender and recipient(s) need not be at their computers at the same time to communicate, and the computer systems may be situated worldwide. The sender creates an e-mail by means of a mail-sending computer program, and a mail transport system then takes responsibility for delivering the e-mail to the indicated address(es).

**electronics** The study and design of control, communication, and computing devices that rely on the movement of electrons in circuits containing semiconductors,

thermionic valves, resistors, capacitors, and inductors. See Chronology.

**electron lens** A device used to focus an electron beam. It is analogous to an optical lens but instead of using a refracting material, such as glass, it uses a coil or coils to produce a magnetic field or an arrangement of electrodes between which an electric field is created. Electron lenses are used in *electron microscopes and *cathode-ray tubes.

**electron microscope** A form of microscope that uses a beam of electrons instead of a beam of light (as in the optical microscope) to form a large image of a very small object. In optical microscopes the resolution is limited by the wavelength of the light. High-energy electrons, however, can be associated with a considerably shorter wavelength than light; for example, electrons accelerated to an energy of $10^5$ electronvolts have a wavelength of 0.004 nanometre (*see* DE BROGLIE WAVELENGTH) enabling a resolution of 0.2–0.5 nm to be achieved. The **transmission electron microscope** (see illustration) has an electron beam, sharply focused by *electron lenses, passing through a very thin metallized specimen (less than 50 nanometres thick) onto a fluorescent screen, where a visual image is formed. This image can be photographed. The **scanning elec-**

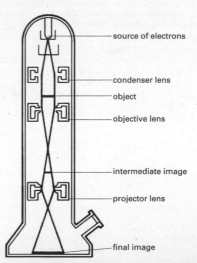

source of electrons

condenser lens

object

objective lens

intermediate image

projector lens

final image

**Electron microscope.** Principle of the transmission electron microscope.

# ELECTRONICS

| | |
|---|---|
| 1887 | Radio waves are discovered by Heinrich Hertz. |
| 1894 | Oliver Lodge invents the 'coherer' for detecting radio waves. Marconi develops radio telegraphy. |
| 1897 | J. J. Thomson discovers the electron. |
| 1902 | US engineer Reginald Fessenden (1866–1932) develops radio telephony. |
| 1903 | Danish engineer Valdemar Poulsen (1869–1942) invents the arc transmitter for radio telegraphy. |
| 1904 | British engineer Ambrose Fleming (1849–1945) invents the diode thermionic valve. |
| 1906 | US engineer Lee De Forest (1873–1961) invents the triode thermionic valve. US electrical engineer Greenleaf Pickard (1877–1956) patents the crystal detector for radios. Fessenden introduces amplitude modulation in radio broadcasting. |
| 1911 | German physicist Karl Braun (1850–1918) invents cathode-ray tube scanning. |
| 1912 | Fessenden develops the heterodyne radio receiver. |
| 1919 | US electrical engineer Edwin Armstrong (1890–1954) develops the superheterodyne radio receiver. |
| 1921 | US physicist Albert Hull (1880–1966) invents the magnetron microwave-generating valve. |
| 1923 | Russian-born US engineer Vladimir Zworykin (1889–1982) invents the iconoscope television camera-tube. |
| 1928 | Scottish inventor John Logie Baird (1888–1946) and Vladimir Zworykin independently develop television. |
| 1930 | Swedish-born US electronics engineer Ernst Alexanderson (1878–1975) invents an all-electronic television system. |
| 1933 | US electrical engineer Edwin Armstrong (1890–1954) develops frequency modulation radio broadcasting. |
| 1947 | US physicists John Bardeen, Walter Brattain (1902–87), and William Shockley (1910–89) invent the point-contact transistor. |
| 1950 | US engineers develop the Videcon television camera tube. |
| 1953 | Chinese-born US computer engineer An Wang (1920–90) invents the magnetic core computer memory. |
| 1954 | US physicist Charles Townes (1915–  ) and Soviet physicists Nikolai Basov and Aleksandr Prokhorov (1916–2002) independently develop the maser. |
| 1958 | US electronics engineers Jack Kilby and Robert Noyce (1927–90) develop integrated circuits. |
| 1960 | US physicist Theodore Maiman (1927–2007) invents the ruby laser. |
| 1961 | US electronics engineer Steven Hofstein develops the field-effect transistor. |
| 1971 | US electronics engineer Marcian Edward Hoff (1937–  ) designs the first microprocessor (Intel 4004). |
| 1977 | US engineers transmit television signals along optical fibres. |

**tron microscope** can be used with thicker specimens and forms a perspective image, although the resolution and magnification are lower. In this type of instrument a beam of primary electrons scans the specimen and those that are reflected, together with any secondary electrons emitted, are collected. This current is used to modulate a separate electron beam in a TV monitor, which scans the screen at the same frequency, consequently building up a picture of the specimen. The resolution is limited to about 10–20 nm. *See also* FIELD-EMISSION MICROSCOPE; FIELD-IONIZATION MICROSCOPE.

**(((()))) SEE WEB LINKS**

• A tutorial featuring virtual scanning electron microscopy, created by the Optical Microscopy Division of the National High Magnetic Field Laboratory

**electron–nuclear double resonance** *See* ENDOR.

**electron optics** The study of the use of *electron lenses in the *electron microscope, *cathode-ray tubes, and other similar devices. The focusing of beams of positive or negative ions also relies on these methods.

**electron paramagnetic resonance (EPR)** A spectroscopic method of locating electrons within the molecules of a paramagnetic substance (*see* MAGNETISM) in order to provide information regarding its bonds and structure. The spin of an unpaired electron is associated with a *magnetic moment that is able to align itself in one of two ways with an applied external magnetic field. These two alignments correspond to different *energy levels, with a statistical probability, at normal temperatures, that there will be slightly more in the lower state than in the higher. By applying microwave radiation to the sample a transition to the higher state can be achieved. The precise energy difference between the two states of an electron depends on the surrounding electrons in the atom or molecule. In this way the position of unpaired electrons can be investigated. The technique is used particularly in studying free radicals and paramagnetic substances such as inorganic complexes. It is also called **electron-spin resonance (ESR)**. *See also* NUCLEAR MAGNETIC RESONANCE.

**electron probe microanalysis (EPM)** A method of analysing a very small quantity of a substance (as little as $10^{-13}$ gram). The

method consists of directing a very finely focused beam of electrons on to the sample to produce the characteristic X-ray spectrum of the elements present. It can be used quantitatively for elements with atomic numbers in excess of 11.

**electron spectroscopy** Any of a number of techniques in which information is obtained by analysing the energy spectrum of electrons. *Photoelectron spectroscopy is a typical example.

**electron-spin resonance** *See* ELECTRON PARAMAGNETIC RESONANCE.

**electron-transfer reaction** A chemical reaction that involves the transfer, addition, or removal of electrons. Electron-transfer reactions often involve complexes of transition metals. In such complexes one general mechanism for electron transfer is the **inner-sphere mechanism**, in which two complexes form an intermediate, with ligand bridges enabling electrons to be transferred from one complex to another complex. The other main mechanism is the **outer-sphere mechanism**, in which two complexes retain all their ligands, with electrons passing from one complex to the other. The rates of electron-transfer reactions vary enormously. These rates can be explained in terms of the way in which molecules of the solvent solvating the reactants rearrange so as to solvate the products in the case of the outer-sphere mechanism. In the case of the inner-sphere (ligand-bridged) reactions the rate of the reaction depends on the intermediate and the way in which the electron is transferred.

**electron transport chain (electron transport system)** A sequence of biochemical reduction–oxidation reactions that effects the transfer of electrons through a series of carriers. An electron transport chain, also known as the **respiratory chain**, forms the final stage of *aerobic respiration. It results in the transfer of electrons or hydrogen atoms derived from the *Krebs cycle to molecular oxygen, with the formation of water. At the same time it conserves energy from food or light in the form of *ATP. The chain comprises a series of *carrier molecules that undergo reversible reduction–oxidation reactions, accepting electrons and then donating them to the next carrier in the chain – a process known as **electron flow**. In the mitochondria, NADH and FADH$_2$, generated by the Krebs cycle, transfer their electrons to a chain comprising flavin mononucleotide (FMN), *ubiquinone, and a series of *cytochromes. This process is coupled to the formation of ATP at three sites along the chain (see OXIDATIVE PHOSPHORYLATION). The ATP is then carried across the mitochondrial membrane in exchange for ADP. An electron transport chain also occurs in *photosynthesis.

**electronvolt** Symbol eV. A unit of energy equal to the work done on an electron in moving it through a potential difference of one volt. It is used as a measure of particle energies although it is not an *SI unit. 1 eV = $1.602 \times 10^{-19}$ joule.

**electroorganic reaction** An organic reaction produced in an electrolytic cell. Electroorganic reactions are used to synthesize compounds that are difficult to produce by conventional techniques. An example of an electroorganic reaction is *Kolbe's method of synthesizing alkanes.

**electroosmosis** The movement of a polar liquid through a membrane under the influence of an applied electric field. The linear velocity of flow divided by the field strength is the **electroosmotic mobility**. Electroosmosis was formerly called **electroendosmosis**.

**electroosmotic mobility** See ELECTROOSMOSIS.

**electropherogram** See CAPILLARY ELECTROPHORESIS.

**electrophile** An ion or molecule that is electron deficient and can accept electrons. Electrophiles are often reducing agents and Lewis *acids. They are either positive ions (e.g. $NO_2^+$) or molecules that have a positive charge on a particular atom (e.g. $SO_3$, which has an electron-deficient sulphur atom). In organic reactions they tend to attack negatively charged parts of a molecule. Compare NUCLEOPHILE.

**electrophilic addition** An *addition reaction in which the first step is attack by an electrophile (e.g. a positive ion) on an electron-rich part of the molecule. An example is addition to the double bonds in alkenes.

**electrophilic substitution** A *substitution reaction in which the first step is attack by an electrophile. Electrophilic substitution is a feature of reactions of benzene (and its compounds) in which a positive ion approaches the delocalized pi electrons on the benzene ring.

**electrophoresis (cataphoresis)** A technique for the analysis and separation of colloids, based on the movement of charged colloidal particles in an electric field. There are various experimental methods. In the simplest, the sample is placed in a U-tube and a buffer solution added to each arm, so that there are sharp boundaries between buffer and sample. An electrode is placed in each arm, a voltage applied, and the motion of the boundaries under the influence of the field is observed. The rate of migration of the particles depends on the field, the charge on the particles, and on other factors, such as the size and shape of the particles. Electrophoresis can also be carried out using an adsorbent, such as a strip of filter paper, soaked in a buffer with two electrodes making contact. The sample is placed between the electrodes and a voltage applied. Different components of the mixture migrate at different rates, so the sample separates into zones. The components can be identified by the rate at which they move. In **gel electrophoresis** the medium is a gel, typically made of polyacrylamide, agarose, or starch. In modern automated DNA sequencers, electrophoresis is carried out in capillary tubes, less than 0.5mm in diameter and about 48cm long, containing the gel.

Electrophoresis, which has also been called **electrochromatography**, is used extensively in studying mixtures of proteins, nucleic acids, carbohydrates, enzymes, etc. In clinical medicine it is used for determining the protein content of body fluids.

**electrophoretic deposition** A technique for coating a material making it an electrode in a bath containing a colloidal suspension of charged particles. Under suitable conditions, the particles are attracted to, and deposited on, the electrode. Electrophoretic deposition is used extensively in industry; for example, in applying paint to metal components.

**electrophoretic effect** The effect in which the mobility of ions in solution moving under the influence of an applied electric field is affected by the flow of ions of opposite charge in the opposite direction.

**electrophorus** An early form of *electrostatic generator. It consists of a flat dielectric plate and a metal plate with an insulated handle. The dielectric plate is charged by friction and the metal plate is placed on it and momentarily earthed, which leaves the metal plate with an induced charge of opposite polarity to that of the dielectric plate. The process can be repeated until all of the original charge has leaked away.

**electroplating** A method of plating one metal with another by *electrodeposition. The articles to be plated are made the cathode of an electrolytic cell and a rod or bar of the plating metal is made the anode. Electroplating is used for covering metal with a decorative, more expensive, or corrosion-resistant layer of another metal.

**electropositive** Describing elements that tend to lose electrons and form positive ions. The alkali metals are typical electropositive elements.

**electroscope** A device for detecting electric charge and for identifying its polarity. In the **gold-leaf electroscope** two rectangular gold leaves are attached to the end of a conducting rod held in an insulated frame. When a charge is applied to a plate attached to the other end of the conducting rod, the leaves move apart owing to the mutual repulsion of the like charges they have received.

**electrospray ionization (ESI)** A technique for producing ions for mass spectrometry, used especially for obtaining ions from large molecules. The sample is dissolved in a volatile solvent, which may also contain volatile acids or bases so that the sample exists in an ionic form. The solution is forced through a charged metal capillary tube and forms an aerosol. Evaporation of the solvent results in single ions of the sample, which are analysed by the mass spectrometer.

**electrostatic field** The *electric field that surrounds a stationary charged body.

**electrostatic generator** A device used to build up electric charge to an extreme potential usually for experimental purposes. The *electrophorus and the *Wimshurst machine were early examples; a more usual device now is the *Van de Graaff generator.

**electrostatic precipitation** A method of removing solid and liquid particles from suspension in a gas. The gas is exposed to an electric field so that the particles are attracted to and deposited on a suitably placed electrode. Electrostatic precipitation is widely used to remove dust and other pollutants from waste gases and from air. *See also* COTTRELL PRECIPITATOR.

**electrostatics** The study of electric charges at rest, the forces between them (*see* COULOMB'S LAW), and the electric fields associated with them. *Compare* ELECTRODYNAMICS.

**electrostatic units (e.s.u.)** A system of electrical units in the *c.g.s. system. The e.s.u. of electric charge is the **statcoulomb** (all e.s.u. have the prefix **stat-** attached to the names of practical units). The statcoulomb is the quantity of electric charge that will repel an equal quantity 1 centimetre distant with a force of 1 dyne. In e.s.u. the electric constant is of unit magnitude. The system has now been replaced for most purposes by *SI units. *Compare* ELECTROMAGNETIC UNITS; GAUSSIAN UNITS; HEAVISIDE–LORENTZ UNITS.

**electrostriction** A change in the dimensions of a body as a result of reorientation of its molecules when it is placed in an electric field. If the field is not homogeneous the body will tend to move; if its relative permittivity is higher than that of its surroundings it will tend to move into a region of higher field strength. *Compare* MAGNETOSTRICTION.

**electrovalent bond** *See* CHEMICAL BOND.

**electroweak theory** A *gauge theory (sometimes called **quantum flavourdynamics**, or **QFD**) that gives a unified description of the electromagnetic and weak interactions (*see* FUNDAMENTAL INTERACTIONS). A successful electroweak theory was proposed in 1967 by Steven Weinberg and Abdus Salam, known as the **Weinberg–Salam model** or **WS model**. Because early developments of these ideas were put forward by Sheldon Glashow, it is sometimes known as the **Glashow–Weinberg–Salam model** or **GWS model**. In this electroweak theory the gauge group is non-Abelian and the gauge symmetry is a *broken symmetry. The electroweak interaction is mediated by photons and by intermediate vector bosons, called the *W boson and the *Z boson. The observation of these particles in 1983–84, with their predicted energies, was a major success of the theory. The theory successfully accounts for existing data for electroweak processes and also predicts the existence of a heavy particle with spin 0, the *Higgs boson.

**electrum** 1. An alloy of gold and silver containing 55–88% of gold. 2. A *German silver alloy containing 52% copper, 26% nickel, and 22% zinc.

**element** A substance that cannot be decomposed into simpler substances. In an element, all the atoms have the same number of protons or electrons, although the number of neutrons may vary. There are 92 naturally occurring elements. *See also* PERIODIC TABLE; TRANSURANIC ELEMENTS; TRANSACTINIDE ELEMENTS.

**elementary particles** The fundamental constituents of all the matter in the universe. By the beginning of the 20th century, the electron and the proton had been discovered, but it was not until 1932 that the existence of the neutron was definitely established. Since 1932, it had been known that atomic nuclei consist of both protons and neutrons (except hydrogen, whose nucleus consists of a lone proton). Between 1900 and 1930, *quantum mechanics was also making progress in the understanding of physics on the atomic scale. Non-relativistic quantum theory was completed in an astonishingly brief period (1923–26), but it was the relativistic version that made the greatest impact on our understanding of elementary particles. Dirac's discovery in 1928 of the equation that bears his name led to the discovery of the positive electron or *positron. The mass of the positron is equal to that of the negative electron while its charge is equal in magnitude but opposite in sign. Pairs of particles related to each other in this way are said to be antiparticles of each other. Positrons have only a transitory existence; that is, they do not form part of ordinary matter. Positrons and electrons are produced simultaneously in high-energy collisions of charged particles or gamma rays with matter in a process called *pair production.

The union of *relativity and quantum mechanics therefore led to speculation as early as 1932 that there might also be antiprotons and antineutrons, bearing a similar relationship to their respective ordinary particles as the positron does to the electron. However, it was not until 1955 that particle beams were made sufficiently energetic to enable these antimatter particles to be observed. It is now understood that all known particles have antimatter equivalents, which are predicted by relativistic quantum equations.

By the mid-1930s the list of known and theoretically postulated particles was still small but steadily growing. At this time the Japanese physicist Hideki Yukawa (1907–81) was studying the possible *fundamental interactions that could hold the nucleus together. Since the nucleus is a closely packed collection of positively charged protons and neutral neutrons, clearly it could not be held together by an electromagnetic force; there had to be a different and very large force capable of holding proton charges together at such close proximity. This force would necessarily be restricted to the short range of nuclear dimensions, because evidence of its existence only arose after the discovery of the constituents of the atomic nucleus. Guided by the properties required of this new force, Yukawa proposed the existence of a particle called the *meson, which was responsible for transmitting nuclear forces. He suggested that protons and neutrons in the nucleus could interact by emitting and absorbing mesons. For this reason this new type of force was called an *exchange force. Yukawa was even able to predict the mass of his meson (meaning 'middle weight'), which turned out to be intermediate between the proton and the electron.

Only a year after Yukawa had made this suggestion, a particle of intermediate mass was discovered in *cosmic radiation. This particle was named the **µ-meson** or **muon**. The µ⁻ has a charge equal to the electron, and its antiparticle µ⁺ has a positive charge of equal magnitude. However, physicists soon discovered that muons do not interact with nuclear particles sufficiently strongly to be Yukawa's meson. It was not until 1947 that a family of mesons with the appropriate properties was discovered. These were the **π-mesons** or **pions**, which occur in three types: positive, negative, and neutral. Pions, which interact strongly with nuclei, have in fact turned out to be the particles predicted by Yukawa in the 1930s. The nuclear force between protons and neutrons was given the name 'strong interaction' (*see* FUNDAMENTAL INTERACTIONS) and until the 1960s it was thought to be an exchange force as proposed by Yukawa.

A theory of the weak interaction was also in its infancy in the 1930s. The weak interaction is responsible for *beta decay, in which a radioactive nucleus is transformed into a slightly lighter nucleus with the emission of an electron. However, beta decays posed a problem because they appeared not to conserve energy and momentum. In 1931 *Pauli proposed the existence of a neutral particle that might be able to carry off the missing energy and momentum in a beta decay and escape undetected. Three years later, *Fermi included Pauli's particle in a comprehensive theory of beta decay, which seemed to explain many experimentally observed results. Fermi called this new particle the *neutrino, the existence of which was finally established in the 1950s.

A plethora of experiments involving the neutrino revealed some remarkable properties for this new particle. The neutrino was found to have an intimate connection with the electron and muon, and indeed never appeared without the simultaneous appearance of one or other of these particles. A conservation law was postulated to explain this observation. Numbers were assigned to the electron, muon, and neutrino, so that during interactions these numbers were conserved; i.e. their algebraic sums before and after these interactions were equal. Since these particles were among the lightest known at the time, these assigned numbers became known as **lepton numbers** (lepton: 'light ones'). In order to make the assignments of lepton number agree with experiment, it is necessary to postulate the existence of two types of neutrino. Each of these types is associated with either the electron or muon; there are thus muon neutrinos and electron neutrinos. In 1978 the **tau particle** or **tauon** was discovered and was added to the list of particles with assigned lepton numbers. The conservation of lepton number in the various interactions involving the tau requires the existence of an equivalent tau neutrino. The six particles with assigned lepton numbers are now known as *leptons.

Neutrinos have zero charge and were originally thought to have zero rest mass, but there has been increasing indirect experimental evidence to the contrary. In 1985 a Soviet team reported a measurement, for the first time, of a non-zero neutrino mass. The

mass measured was extremely small (10 000 times less than the mass of the electron), but subsequent attempts independently to reproduce these results did not succeed. More recently (1998–99), Japanese and US groups have put forward theories and corroborating experimental evidence to suggest, indirectly, that neutrinos do have mass. In these experiments neutrinos are found to apparently 'disappear'. Since it is unlikely that momentum and energy are actually vanishing from the universe, a more plausible explanation is that the types of neutrinos detected are changing into types that cannot be detected. Present theoretical considerations imply that the masses of neutrinos involved cannot be equal to one another, and therefore they cannot all be zero. This speculative work has not yet yielded estimates of the neutrino masses, which is indicated by the use of asterisks in the table.

In the 1960s, the development of high-energy accelerators and more sophisticated detection systems led to the discovery of many new and exotic particles. They were all unstable and existed for only small fractions of a second; nevertheless they set into motion a search for a theoretical description that could account for them all. The large number of these apparently fundamental particles suggested strongly that they do not, in fact, represent the most fundamental level of the structure of matter. Physicists found themselves in a position similar to Mendeleev when the *periodic table was being developed. Mendeleev realized that there had to be a level of structure below the elements themselves, which explained the chemical properties and the interrelations between elements.

Murray Gell-Mann and his collaborators proposed the particle-physics equivalent of the periodic table in 1961. In this structure, leptons were indeed regarded as fundamental particles, but the short-lived particles dis-

| Name | Symbol | Charge (electron charges) | Rest mass (MeV/$c^2$) |
|------|--------|---------------------------|----------------------|
| electron | $e^-$ | −1 | 0.511 |
| electron neutrino | $v_e$ | 0 | * |
| muon | $\mu^-$ | −1 | 105.7 |
| muon neutrino | $v_\mu$ | 0 | * |
| tauon | $\tau^-$ | −1 | 1784 |
| tau neutrino | $v_\tau$ | 0 | * |

**Elementary particles.** Table of leptons.

covered in the 1960s were not. These particles were found to undergo strong interactions, which did not seem to affect the leptons. Gell-Mann called these strongly interacting particles the *hadrons and proposed that they occurred in two different types: baryons and mesons. These two different types corresponded to the two different ways of constructing hadrons from constituent particles, which Gell-Mann called **quarks**. These quarks came in three **flavours**, up (u), down (d), and strange (s). These three quarks were thought to be the fundamental constituents of hadrons, i.e. matter that undergoes strong interactions: baryons are composed of three quarks (u, d, or s) or three antiquarks (ū, d̄, or s̄); mesons are composed of (u, d, or s) quark–antiquark pairs.

No other combinations seemed to be necessary to describe the full variation of the observed hadrons. This scheme even led to the prediction of other particles that were not known to exist in 1961. For example, in 1961 Gell-Mann not only predicted the Ω⁻ (omega-minus) particle, but more importantly told experimentalists exactly how to produce it. The Ω⁻ particle was finally discovered in 1964.

Gell-Mann called his scheme 'the eightfold way', after the similarly named Buddhist principle. The scheme requires that quarks have properties not previously allowed for fundamental particles. For example, quarks have fractional electric charges, i.e. charges of 1/3 and 2/3 of the electron charge. Quarks also have a strong affinity for each other through a new kind of charge known as **colour charge**. Thus colour charge is responsible for strong interactions, and the force is known as the colour force. This is a revision of Yukawa's proposal in 1930. Yukawa's strong force was mediated by π-mesons. The strong force is now thought to be mediated by exchange of particles carrying colour charge called **gluons**. The theory governing these colour charge combinations is modelled on *quantum electrodynamics and is known as *quantum chromodynamics.

In November 1974 the discovery of the ψ (psi) particle initiated what later came to be known as 'the November revolution'. At the time, any known hadron could be described as some combination of u, d, or s quarks. These hadrons were very short-lived with lifetimes of about $10^{-23}$ s. The ψ particle, however, had a lifetime of $10^{-20}$ s; i.e. a thousand times longer. This suggested a com-

pletely different species of particle. It is now universally accepted that the ψ represents a meson-bound state of a new fourth quark, the **charm** (c) quark and its antiquark. In 1977 the list of quarks once again increased with the discovery of a new even heavier meson, called the Y (**upsilon**) meson. This meson was found to have an even longer lifetime than the ψ, and was quickly identified as the carrier of a fifth quark, **bottom** (b).

Thus, by the end of 1977, five flavours of quark (u, d, s, c, b) were known to exist together with six flavours of lepton (e, μ, τ, $v_e$, $τ_μ$, $v_τ$). Assuming that quarks and leptons are the fundamental constituents of matter, many of the strong and weak interactions of hadrons and the weak interactions of leptons can be explained. However, anticipating a symmetry in nature's building blocks, it was expected that a sixth quark would eventually reveal itself. This quark, labelled **top** (t), would be the 2/3 electronic charge partner to the b quark (see table).

| Quark symbol | Name | Charge |
|---|---|---|
| u | up | –2/3 |
| d | down | –1/3 |
| c | charm | –2/3 |
| s | strange | –1/3 |
| t | top | –2/3 |
| b | bottom | –1/3 |

**Elementary particles.** Table of quarks (mass is not shown because quarks are never observed alone).

In 1998 the top quark was found at CERN in Geneva and the symmetry of six quarks with six leptons was finally verified.

In 1978 the **standard model** was proposed as the definitive theory of the fundamental constituents of matter. In the current view, all matter consists of three kinds of particles: leptons, quarks, and mediators (see table overleaf). The mediators are the particles by which the four fundamental interactions are mediated. In the standard model, each of these interactions has a particle mediator. For the electromagnetic reaction it is the *photon.

For weak interactions the force is mediated by three particles called W⁺, W⁻, and Z° *bosons; for the strong force it is the gluon. Current theories of quantum gravity propose the *graviton as the mediator for the gravitational interaction, but this work is highly

| Interaction | Mediator (exchange particle) | Rest mass $(GeV/c^2)$ | Charge |
|---|---|---|---|
| strong | gluon | 0 | 0 |
| electromagnetic | photon | 0 | 0 |
| weak | $W^+$, $W^-$, $Z^\circ$ | 81,81,93 | +1,-1,0 |
| gravitational | graviton | 0 | 0 |

**Elementary particles.** Table of mediators.

speculative and the graviton has never been detected.

**elements of an orbit** Six parameters used to define the path of a celestial body. The shape of the orbit is defined by its eccentricity (*see* CONIC) and semimajor axis. The orientation of the orbit is specified by the *inclination of the orbital plane to the reference plane (usually the *ecliptic) and by the longitude of the ascending *node (the angular distance from the vernal equinox to the ascending node). The position of the body in its orbit is defined by its eccentric *anomaly and the position as a function of time is calculated from the periapsis passage (*see* APSIDES).

**elevation of boiling point** An increase in the boiling point of a liquid when a solid is dissolved in it. The elevation is proportional to the number of particles dissolved (molecules or ions) and is given by $\Delta t = k_B C$, where $C$ is the molal concentration of solute. The constant $k_B$ is the **ebullioscopic constant** of the solvent and if this is known, the molecular weight of the solute can be calculated from the measured value of $\Delta t$. The elevation is measured by a Beckmann thermometer. *See also* COLLIGATIVE PROPERTIES.

**elimination reaction** A reaction in which one molecule decomposes into two, one much smaller than the other.

**Ellnvar** Trade name for a nickel–chromium steel containing about 36% nickel, 12% chromium, and smaller proportions of tungsten and manganese. Its elasticity does not vary with temperature.

**ELISA (enzyme-linked immunosorbent assay)** A sensitive technique (*see* IMMUNOASSAY) for accurately determining the amount of protein or other antigen in a given sample by means of an enzyme-catalysed colour change. Antibody specific to the test

protein is adsorbed onto a solid substrate, such as a PVC sheet, and a measured amount of the sample is added; all molecules of the test protein in the sample are bound by the antibody. A second antibody specific for a second site on the test protein is added; this is conjugated with an enzyme, which catalyses a colour change in the fourth reagent, added finally to the sheet. The colour change can be measured photometrically and compared against a standard curve to give the concentration of protein in the sample. ELISA is widely used for diagnostic and other purposes.

**Ellingham diagram** A diagram used to show the conditions under which a metal oxide can be reduced to a metal. The standard Gibbs free energy of formation of the oxide is considered, for example,

$$M + \tfrac{1}{2}O_2 \rightarrow MO$$

This value, $\Delta G^\ominus$, is plotted against temperature. In general, the result is a straight line. In some cases, there is an abrupt change in the line's slope at a point because of a phase change. The value of $\Delta G^\ominus$ for the reducing agent is also plotted. For example, if the reducing agent is carbon, forming carbon dioxide, it is $\Delta G^\ominus$ for the reaction

$$C + O_2 \rightarrow CO_2$$

Reduction can occur in the range of temperatures in which the carbon curve is lower than the metal curve. The diagram was devised by the physical chemist H. J. T. Ellingham.

**ellipse** A *conic formed by the intersection of a plane with a right circular cone, so that the plane is inclined to the axis of the cone at an angle in excess of half the apex angle of the cone. The ellipse has two vertices, which are joined by a line called the **major axis**. The centre of the ellipse falls on this line, midway between the vertices. The **minor axis** is the line perpendicular to the major axis that passes through the centre and joins two points on the ellipse. The **foci** of an el-

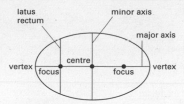

**An ellipse.**

lipse are two points on the major axis so placed that for any point on the ellipse the sum of the distances from that point to each focus is constant. (See illustration.) The area of an ellipse is $\pi ab$, where $a$ and $b$ are half the major and minor axes, respectively. For an ellipse centred at the origin, the equation in Cartesian coordinates is $x^2/a^2 + y^2/b^2 = 1$. The foci are at $(ea, 0)$ and $(-ea, 0)$, where $e$ is the eccentricity. Each of the two chords of the ellipse passing through a focus and parallel to the minor axis is called a **latus rectum** and has a length equal to $2b^2/a$.

**ellipsoid** A solid body formed when an *el lipse is rotated about an axis. If it is rotated about its major axis it is a **prolate ellipsoid**; if it is rotated about its minor axis it is an **oblate ellipsoid**. For an ellipsoid centred at the origin the equation in Cartesian coordinates is:

$$x^2/a^2 + y^2/b^2 + z^2/c^2 = 1.$$

**elliptical galaxy** *See* GALAXY.

**elliptical polarization** *See* POLARIZATION OF LIGHT.

**El Niño** A surge of warm ocean water (the Peru current) that occurs every 5 to 8 years off the eastern coast of South America. See Feature overleaf.

**Elton, Charles Sutherland** (1900–91) British zoologist and ecologist, who founded the Bureau of Animal Population at Oxford in 1932 and the same year became editor of the new *Journal of Animal Ecology*. The first zoologist to study animals in relation to their environment, he explored the nature of food chains and studied population fluctuations.

**eluate** *See* CHROMATOGRAPHY; ELUTION.

**eluent** *See* CHROMATOGRAPHY; ELUTION.

**elution** The process of removing an adsorbed material (**adsorbate**) from an adsorbent by washing it in a liquid (**eluent**). The solution consisting of the adsorbate dissolved in the eluent is the **eluate**. Elution is the process used to wash components of a mixture through a *chromatography column.

**elutriation** The process of suspending finely divided particles in an upward flowing stream of air or water to wash and separate them into sized fractions.

**elytra** The thickened horny forewings of the *Coleoptera (beetles), which cover and protect the membranous hindwings when the insect is at rest.

**e-mail** *See* ELECTRONIC MAIL.

**emanation** The former name for the gas radon, of which there are three isotopes: Rn–222 (radium emanation), Rn–220 (thoron emanation), and Rn–219 (actinium emanation).

**emasculation** The removal of the anthers of a flower in order to prevent self-pollination or the undesirable pollination of neighbouring plants.

**EMBASE (Excerpta Medica Database)** A bibliographic database containing citations, abstracts, indexing terms, and codes covering over 5000 biological and medical journals from more than 70 countries, especially Europe. It contains over 11 million records, with some half a million new ones added each year, and has a particular emphasis on drug-related fields.

**Embden–Meyerhof pathway** *See* GLYCOLYSIS.

**embryo** 1. An animal in the earliest stages of its development, from the time when the fertilized ovum starts to divide (*see* CLEAVAGE), while it is contained within the egg or reproductive organs of the mother, until hatching or birth. A human embryo (see illustration p. 286) is called a *fetus after the first eight weeks of pregnancy. 2. The structure in plants that develops from the zygote prior to germination. In seed plants the zygote is situated in the *embryo sac of the ovule. It divides by mitosis to form the embryonic cell and a structure called the **suspensor**, which embeds the embryo in the surrounding nutritive tissue. The embryonic cell divides continuously and eventually gives rise to the *radicle (young root), *plumule (young shoot), and one or two *cotyledons (seed leaves). Changes also take place in the surrounding tissues of the ovule,

## EL NIÑO

A phenomenon reoccurring every few years in the equatorial part of the Pacific ocean, characterized by movement of a mass of warm water eastward towards the west coast of South America. This change in ocean conditions has long been recognized in Peru, where sailors noticed that an unusual counter-current appeared in certain years around the area of the port of Paita. They named this current El Niño – 'the Christ Child' – because it usually appeared immediately after Christmas. It was also known that the appearance of this current coincided with different weather conditions, particularly increased rainfall and sometimes flooding.

The phenomenon has dramatic effects on the climate and ecology of this part of South America. In particular, it interrupts the Humbolf current, which is a cold ocean current carrying plankton from Arctic regions. El Niño occurs every 4–7 years and the effects last for about 8 months. In the 1960s interest developed in the phenomenon and it is now recognized that El Niño has effects on climate much wider than those observed on the west coast of South America. In extreme cases, as in 1986–87 and 1997–98, it can cause tropical cyclones over the whole Pacific area, drought in southeastern Asia and Australia, and increased rainfall and flooding in parts of North America.

### The mechanism of El Niño

There is no definite agreement about what induces an El Niño event, but the physical mechanism of how it occurs is fairly well understood. The large-scale movement of water in the world's oceans is influenced by, and in turn influences, the prevailing wind patterns. In tropical regions there are persistent **trade winds** flowing from east to west. In the Pacific these push large amounts of water westward towards the coasts of Indonesia. This causes a significant difference in sea level between opposite sides of the Pacific. For instance, the sea level in the Philippines in the west is around 60 cm higher than that on the coast of Panama in the eastern Pacific. The water in the west is also much warmer. This mass of warm ocean in the western Pacific gives Indonesia its high rainfall under normal climatic conditions.

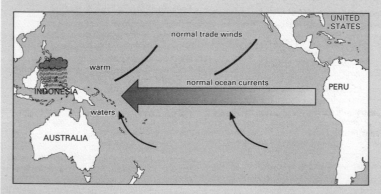

In normal years the trade winds blow from east to west across the Pacific. This causes a difference in sea level and a mass of warm water builds up in the western Pacific, creating a warm area of ocean off the east coast of Indonesia. This area has some of the highest rainfall in the world.

Every few years there is a significant change in the prevailing wind pattern, involving a fall in the intensity of the trade winds. In extreme cases, a reversal of direction of the winds may occur. As a result, the warm water that is piled up in the west flows back eastwards across the Pacific towards the west coast of South America. The event lasts until the normal conditions and wind patterns are re-established.

## The Southern Oscillation and La Niña

El Niño events are connected with another periodic phenomenon occurring not in the ocean but in the atmosphere. The meteorologist Sir Gilbert Walker noted in 1923 that when pressure was high in the Pacific it tends to be low in the Indian ocean, and vice versa. The extent of this is now measured by taking the difference between the surface atmospheric pressure at Darwin in Australia and at Tahiti in the south Pacific. A high pressure at one site is usually accompanied by a low pressure at the other and every few years the pattern reverses. There is a large mass of air slowly oscillating (with a period of a few years) across tropical regions. Walker called this the **Southern Oscillation**.

The Southern Oscillation is part of a large general cycle of coupled air and water flow known as the **El Niño–Southern Oscillation** (ENSO). El Niño is the warm phase of this cycle. In some years, as part of the ENSO cycle, a cold region develops in the eastern tropical Pacific. This is known as **La Niña** ('the little girl').

The Southern oscillation is not the only atmospheric oscillation that occurs. For example there is a **North Atlantic Oscillation** (NAO) measured by the pressure difference between Iceland and the Azores, which is thought to have a major influence on climatic conditions and on the ecosystems of this part of the globe. There is however no Atlantic analogue of El Niño.

## The causes of El Niño

Although the flows of air and water are understood, there is no consensus about why an El Niño forms. Some workers have suggested that the frequency and intensities of El Niño events may be increasing because of global warming. However, there is no direct evidence for this .

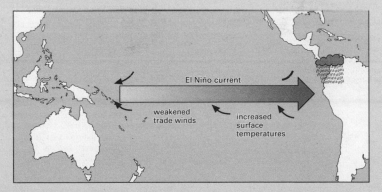

In certain years there is a reduction in the intensity of the trade winds across the Pacific. This allows the mass of warm water in the west to flow across to the east, creating a warm area of ocean off the west coast of South America. This brings rain to Peru but causes drought in southeast Asia and Australia.

chorionic villi | uterus wall
fallopian tube — placenta
— blood vessel
yolk sac — umbilical cord
— amnion
— embryo
— chorion
cervix — amniotic fluid
vagina

**Embryo.** A developing human embryo.

which becomes the *seed enclosing the embryo plant.

**embryology** The study of the development of animals from the fertilized egg to the new adult organism. It is sometimes limited to the period between fertilization of the egg and hatching or birth (*see* EMBRYO).

**(((●))) SEE WEB LINKS**

• Hosted by the University of New South Wales, this dynamic site explores human embryology and development

**embryonic stem cell** *See* STEM CELL.

**embryophyte** A true plant, i.e. one that develops from an embryo and therefore is necessarily multicellular. The term underlines the distinction between plants and algae, which lack embryos.

**embryo sac** A large cell that develops in

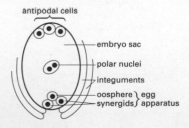

antipodal cells
— embryo sac
— polar nuclei
— integuments
— oosphere ⎱ egg
— synergids ⎰ apparatus

**Embryo sac.**

the *ovule of flowering plants. It is equivalent to the female *gametophyte of lower plants, although it is very much reduced. Typically, it contains eight nuclei formed by division of the original female gamete (see illustration). One forms the *oosphere (egg cell), which is fertilized by a male nucleus and becomes the *embryo. The two polar nuclei fuse with a second male nucleus to form a triploid nucleus that gives rise to the *endosperm. The three remaining nuclei form the antipodal cells.

**emerald** The green gem variety of *beryl: one of the most highly prized gemstones. The finest specimens occur in the Muzo mines, Colombia. Other occurrences include the Ural Mountains, the Transvaal in South Africa, and Kaligunan in India. Emeralds can also be successfully synthesized.

**emergence** A key concept in *complexity theory in which certain features of a complex system occur as a result of collective behaviour of the system. For example, in the kinetic theory of gases the concept of temperature emerges from the average kinetic energy of the very large number of particles in the system. It has been suggested that space and time in physics should ultimately be emergent quantities.

**emergency action code** *See* HAZCHEM CODE.

**emery** A rock composed of corundum

(natural aluminium oxide, $Al_2O_3$) with magnetite, haematite, or spinel. It occurs on the island of Naxos (Greece) and in Turkey. Emery is used as an abrasive and polishing material and in the manufacture of certain concrete floors.

**e.m.f.** *See* ELECTROMOTIVE FORCE.

**emission spectrum** *See* SPECTRUM.

**emissivity** Symbol $\varepsilon$. The ratio of the power per unit area radiated by a surface to that radiated by a *black body at the same temperature. A black body therefore has an emissivity of 1 and a perfect reflector has an emissivity of 0. The emissivity of a surface is equal to its *absorptance.

**emittance** *See* EXITANCE.

**emitter** *See* TRANSISTOR.

**emitter-coupled logic (ECL)** A set of integrated *logic circuits. The input part of an ECL consists of an emitter-coupled *transistor pair which is a very good differential amplifier. The output is through an *emitter follower. ECL circuits are very rapid logic circuits.

**emitter follower** An amplifying circuit using a bipolar junction *transistor with a *common-collector connection. The output is taken from the emitter.

**empirical** Denoting a result that is obtained by experiment or observation rather than from theory.

**empirical formula** *See* FORMULA.

**emulsification** (in digestion) The breakdown of fat globules in the duodenum into tiny droplets, which provides a larger surface area on which the enzyme pancreatic *lipase can act to digest the fats into fatty acids and glycerol. Emulsification is assisted by the action of the bile salts (*see* BILE).

**emulsion** A *colloid in which small particles of one liquid are dispersed in another liquid. Usually emulsions involve a dispersion of water in an oil or a dispersion of oil in water, and are stabilized by an **emulsifier**. Commonly emulsifiers are substances, such as *detergents, that have lyophobic and lyophilic parts in their molecules. Dietary fats are reduced to an emulsion in the duodenum to facilitate their subsequent digestion (*see* EMULSIFICATION).

**en** The symbol for ethylene diamine (1,2-diaminoethane) functioning as a bidentate ligand, used in formulae.

**enamel** The material that forms a covering over the crown of a *tooth (i.e. the part that projects above the gum). Enamel is smooth, white, and extremely hard, being rich in minerals containing calcium, especially *apatite. It is produced by certain cells (**ameloblasts**) of the oral epithelium and protects the underlying dentine of the tooth. Enamel may also be found in the placoid *scales of certain fish, which demonstrates the common developmental origin of scales and teeth.

**enamine** A type of compound with the general formula $R^1R^2C=C(R^3)-NR^4R^5$, where R is a hydrocarbon group or hydrogen. Enamines can be produced by condensation of an aldehyde or ketone with a secondary amine.

**enantiomers** *See* OPTICAL ACTIVITY.

**enantiomorphism** *See* OPTICAL ACTIVITY.

**enantiotropy** *See* ALLOTROPY.

**encephalin** *See* ENKEPHALIN.

**endangered species** A plant or animal species defined by the IUCN (International Union for the Conservation of Nature and Natural Resources) as being in immediate danger of *extinction because its population numbers have reached a critical level or its habitats have been drastically reduced. If these causal factors continue the species is unlikely to survive. A list of endangered species is published by the IUCN, which also defines other categories of threatened species.

(((●))) **SEE WEB LINKS**
• Official website of the IUCN

**endemic** 1. Describing a plant or animal species that is restricted to one or a few localities in its distribution. Endemic species are usually confined to islands and are vulnerable to extinction. 2. Describing a disease or a pest that is always present in an area. For example, malaria is endemic in parts of Africa.

**endergonic reaction** A chemical reaction in which energy is absorbed. *Compare* EXERGONIC REACTION.

**endo-** (in chemistry) Prefix used to designate a bridged ring molecule with a substituent on the ring that is on the same side as the bridge. If the substituent is on the opposite side the compound is designated **exo-**.

**endocannabinoids** *See* CANNABINOIDS.

**endocarp** *See* PERICARP.

**endocrine gland (ductless gland)** Any gland in an animal that manufactures *hormones and secretes them directly into the bloodstream to act at distant sites in the body (known as **target organs** or **cells**). Endocrine glands tend to control slow long-term activities in the body, such as growth and sexual development. In mammals they include the *pituitary, *adrenal, *thyroid, and *parathyroid glands, the *ovary and *testis, the *placenta, and part of the pancreas (*see* ISLETS OF LANGERHANS). The activity of endocrine glands is controlled by negative feedback, i.e. a rise in output of hormone inhibits a further increase in its production, either directly or indirectly via the target organ or cell. *See also* NEUROENDOCRINE SYSTEM. *Compare* EXOCRINE GLAND.

**endocrinology** The study of the structure and functions of the *endocrine glands and of the *hormones they produce.

**endocytosis** The process by which materials enter a cell without passing through the plasma membrane. The membrane folds around material outside the cell, resulting in the formation of a saclike vesicle into which the material is incorporated. This vesicle is then pinched off from the cell surface so that it lies within the cell. Both *phagocytosis and *pinocytosis are forms of endocytosis. *Compare* EXOCYTOSIS.

**endoderm (entoderm)** The internal layer of cells of the *gastrula, which will develop into the alimentary canal (gut) and digestive glands of the adult. *See also* GERM LAYERS.

**endodermis** The innermost layer of the root *cortex of a plant, lying immediately outside the vascular tissue. Various modifications of the endodermal cell walls enable them to regulate the passage of materials both into and out of the vascular system. An endodermis may also be seen in the stems of some plants.

**endoergic** Denoting a nuclear process that absorbs energy. *Compare* EXOERGIC.

**endogamy** The fusion of reproductive cells from closely related parents, i.e. *inbreeding. *Compare* EXOGAMY.

**endogenous** Describing a substance, stimulus, organ, etc., that originates from within an organism. For example, growth rhythms not directed by environmental stimuli are termed endogenous rhythms. Lateral roots, which always grow from inside the main root rather than from its surface, are said to arise endogenously. *Compare* EXOGENOUS.

**endolymph** The fluid that fills the membranous labyrinth of the vertebrate *inner ear. *See* COCHLEA; SEMICIRCULAR CANALS. *Compare* PERILYMPH.

**endometrium** The mucous membrane that lines the *uterus of mammals. It is comprised of an upper mucus-secreting layer, which is shed during menstruation, and a basal layer, which proliferates to form the upper layer. *See also* MENSTRUAL CYCLE.

**endonuclease** An enzyme that catalyses the internal cleavage of nucleic acids. *See also* RESTRICTION ENZYME. *Compare* EXONUCLEASE.

**endoparasite** A parasite that lives inside its host's body. *See* PARASITISM.

**endopeptidase** A protein-digesting enzyme that cleaves a polypeptide chain at specific sites between amino acids. For example, *chymotrypsin cleaves the chain next to aromatic amino acids, such as phenylalanine; *trypsin cleaves the chain next to basic amino acids, such as lysine or arginine; and *pepsin cleaves the chain next to tyrosine and phenylalanine. *Compare* EXOPEPTIDASE.

**endoplasm** *See* CYTOPLASM.

**endoplasmic reticulum (ER)** A system of membranes within the cytoplasm of plant and animal *cells. It forms a link between the plasma and nuclear membranes and is the site of protein synthesis. It is also concerned with the transport of proteins and lipids within the cell. **Rough ER** has *ribosomes attached to its surface; proteins synthesized on the ribosomes are enclosed in vesicles and transported to the *Golgi apparatus. **Smooth ER** lacks ribosomes; it is the site of important metabolic reactions, including phospholipid and fatty-acid synthesis.

**ENDOR** Electron–nuclear double resonance. A magnetic resonance technique involving exitation of both electron spins and nuclear spins. Two sources of radiation are used. One is a fixed source at microwave frequency, which partially saturates the electron spins. The other is a variable radiofrequency source, which excites the atomic nuclear spins. Excitation of the nuclear spins affects the electron spins by hyperfire coupling, increasing the relaxation time of the excited electron spins and increasing the signal strength. The technique, which is usually done at low temperatures, is used to investigate paramagnetic molecules. *See also* ELECTRON PARAMAGNETIC RESONANCE.

**end organ** The structure at the end of a peripheral nerve. Examples of end organs are the muscle *end plate at the end of a motor neuron and the *receptor at the end of a sensory neuron.

**endorphin** Any of a class of three endogenous *opioids – α-, β-, and γ- endorphins – found naturally in brain and other tissues that have pain-relieving effects similar to those of morphine. They are all peptides or polypeptides derived from the precursor pro-opiomelanocortin; for example, β-endorphin is a 31-amino-acid peptide. The endorphins mediate their analgesic effects by binding to opioid receptors. *See also* ENKEPHALIN.

**endoskeleton** A supporting framework that lies entirely within the body of an animal, such as the bony *skeleton of vertebrates or the spicules of a sponge. The function of an endoskeleton is to support the body and in vertebrates it also protects the organs and provides a system of levers on which the muscles can act to produce movement. *Compare* EXOSKELETON.

**endosperm** A nutritive tissue, characteristic of flowering plants, that surrounds the developing embryo in a seed. It develops from nuclei in the *embryo sac and its cells are triploid. In **endospermic** seeds it remains and increases in size; in **nonendospermic** seeds it disappears as the food is absorbed by the embryo, particularly the *cotyledons. Many plants with endospermic seeds, such as cereals and oil crops, are cultivated for the rich food reserves in the endosperm.

**endospore** The resting stage of certain bacteria, formed in response to adverse conditions. The bacterial cell becomes enclosed in a tough resistant protein coat. On return to favourable conditions the spore germinates and reverts to the normal vegetative form of the organism. Endospores can remain viable for long periods, perhaps several thousands of years.

**endosymbiont theory** A theory, devised principally by US biologist Lynn Margulis (1938– ), that eukaryotic organisms evolved from symbiotic associations between prokaryotic ancestors. Free-living aerobic bacteria and chloroxybacteria (*see* CYANOBACTERIA) became incorporated inside larger nucleated prokaryotic cells, where they acted as forerunners of the mitochondria and chloroplasts seen in modern eukaryotes. Such events are held to have occurred on several occasions, producing various lineages of both heterotrophic and phototrophic protoctists, from which evolved ancestors of animals, plants, and fungi. There is strong evidence for the theory, particularly the finding that mitochondria and chloroplasts have DNA similar in form to that of eubacteria, and that they contain prokaryotic-type ribosomes.

**endothelium** A single layer of thin platelike cells that line the inner surfaces of blood and lymph vessels and the heart. Endothelium is derived from the *mesoderm. *Compare* EPITHELIUM; MESOTHELIUM.

**endotherm (homoiotherm)** An animal that can generate and maintain heat within its body independently of the environmental temperature. Mammals and birds are endotherms; they are often described as being **warm-blooded**. *See* HOMOIOTHERMY. *Compare* ECTOTHERM.

**endothermic** Denoting a chemical reaction that takes heat from its surroundings. *Compare* EXOTHERMIC.

**endotoxin** *See* TOXIN.

**end plate** The area of the plasma membrane of a muscle cell that lies immediately beneath a motor nerve ending at a *neuromuscular junction. Release of a *neurotransmitter at the end plate induces contraction of the muscle fibre.

**end point** The point in a titration at which reaction is complete as shown by the *indicator.

**energy** A measure of a system's ability to do work. Like work itself, it is measured in joules. Energy is conveniently classified into two forms: **potential energy** is the energy stored in a body or system as a consequence of its position, shape, or state (this includes gravitational energy, electrical energy, nuclear energy, and chemical energy); **kinetic energy** is energy of motion and is usually defined as the work that will be done by the body possessing the energy when it is brought to rest. For a body of mass $m$ having a speed $v$, the kinetic energy is $mv^2/2$ (classical) or $(m - m_0)c^2$ (relativistic). The rotational kinetic energy of a body having an angular velocity $\omega$ is $I\omega^2/2$, where $I$ is its moment of inertia.

The *internal energy of a body is the sum of the potential energy and the kinetic energy of its component atoms and molecules.

**energy bands** Ranges of energies that electrons can have in a solid. In a single atom, electrons exist in discrete *energy levels. In a crystal, in which large numbers of atoms are held closely together in a lattice, electrons are influenced by a number of adjacent nuclei and the sharply defined levels of the atoms become bands of allowed energy (see illustration); this approach to energy levels in solids is often known as the **band theory**. Each band represents a large number of allowed quantum states. Between the bands are **forbidden bands**. The outermost electrons of the atoms (i.e. the ones responsible for chemical bonding) form the **valence band** of the solid. This is the band, of those occupied, that has the highest energy.

The band structure of solids accounts for their electrical properties. In order to move through the solid, the electrons have to change from one quantum state to another. This can only occur if there are empty quantum states with the same energy. In general, if the valence band is full, electrons cannot change to new quantum states in the same band. For conduction to occur, the electrons have to be in an unfilled band – the **conduc-**

**tion band**. Metals are good conductors either because the valence band and the conduction band are only half-filled or because the conduction band overlaps with the valence band; in either case vacant states are available. In insulators the conduction band and valence band are separated by a wide forbidden band and electrons do not have enough energy to 'jump' from one to the other.

In intrinsic *semiconductors the forbidden gap is narrow and, at normal temperatures, electrons at the top of the valence band can move by thermal agitation into the conduction band (at absolute zero, a semiconductor would act as an insulator). Doped semiconductors have extra bands in the forbidden gap.

**energy flow** (in ecology) The flow of energy that occurs along a *food chain. Energy enters the food chain at the level of the *producers (usually plants) in the form of solar energy. The plants convert solar energy into chemical energy in the process of *photosynthesis. Chemical energy is passed from one trophic level to the next through feeding. Since a large proportion of energy is lost at each trophic level, mostly in the form of heat energy due to respiration, a food chain does not normally consist of more than five trophic levels: the fifth trophic level does not contain enough energy to support further levels. Energy is also lost from the food chain in excretory products and the remains of dead organisms; this is converted into heat energy by the action of *decomposers. *See also* PRODUCTIVITY; PYRAMID OF ENERGY.

**energy landscape** A multidimensional surface in which the energy of a system is

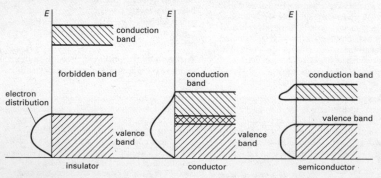

**Energy bands.**

plotted against parameters characterizing that system. For example, in molecular spectroscopy the parameters are the distances between atoms in the molecule. The concept of energy landscape is very useful in many topics in physics and chemistry, including nuclear fission and protein folding. *See also* STRING LANDSCAPE.

**energy level** A definite fixed energy that a system described by *quantum mechanics, such as a molecule, atom, electron, or nucleus, can have. In an atom, for example, the atom has a fixed energy corresponding to the *orbitals in which its electrons move around the nucleus. The atom can accept a quantum of energy to become an excited atom (*see* EXCITATION) if that extra energy will raise an electron to a permitted orbital. Between the **ground state**, which is the lowest possible energy level for a particular system, and the first excited state there are no permissible energy levels. According to the *quantum theory, only certain energy levels are possible. An atom passes from one energy level to the next without passing through fractions of that energy transition. These levels are usually described by the energies associated with the individual electrons in the atoms, which are always lower than an arbitrary level for a free electron. The energy levels of molecules also involve quantized vibrational and rotational motion.

**Engel's salt** *See* POTASSIUM CARBONATE.

**engine** Any device for converting some forms of energy into mechanical work. *See* HEAT ENGINE; CARNOT CYCLE; INTERNAL-COMBUSTION ENGINE; STEAM ENGINE.

**enkephalin (encephalin)** Any of a class of endogenous *opioids consisting of five amino acids and found principally in the central nervous system. They bind to opioid receptors, chiefly in the brain, and their release controls levels of pain and other sensations. *See also* ENDORPHIN.

**enols** Compounds containing the group –CH=C(OH)– in their molecules. *See also* KETO–ENOL TAUTOMERISM.

**enrichment** The process of increasing the abundance of a specified isotope in a mixture of isotopes. It is usually applied to an increase in the proportion of U–235, or the addition of Pu–239 to natural uranium for use in a nuclear reactor or weapon.

**Ensembl** A joint venture between the Eu-ropean Molecular Biology Laboratory (EMBL), the European Bioinformatics Institute (EBI), and the Wellcome Trust's Sanger Institute that provides both sequence data from various eukaryotic genomes and open-source (free) software for researchers handling the data.

**ensemble** A set of systems of particles used in *statistical mechanics to describe a single system. The concept of an ensemble was put forward by the US scientist Josiah Willard Gibbs (1839–1903) in 1902 as a way of calculating the time average of the single system, by averaging over the systems in the ensemble at a fixed time. An ensemble of systems is constructed from knowledge of the single system and can be represented as a set of points in *phase space with each system of the ensemble represented by a point. Ensembles can be constructed both for isolated systems and for open systems.

**entanglement** *See* QUANTUM ENTANGLEMENT.

**enterokinase (enteropeptidase)** An enzyme in the small intestine that activates trypsinogen to *trypsin.

**enteron (coelenteron; gastrovascular cavity)** The body cavity of the coelenterates, which has one opening functioning both as mouth and anus. *See* CNIDARIA.

**enthalpy** Symbol $H$. A thermodynamic property of a system defined by $H = U + pV$, where $H$ is the enthalpy, $U$ is the internal energy of the system, $p$ its pressure, and $V$ its volume. In a chemical reaction carried out in the atmosphere the pressure remains constant and the enthalpy of reaction, $\Delta H$, is equal to $\Delta U + p\Delta V$. For an exothermic reaction $\Delta H$ is taken to be negative.

**entoderm** *See* ENDODERM.

**entomology** The study of insects.

**entomophily** Pollination of a flower in which the pollen is carried on an insect. Entomophilous flowers are usually brightly coloured and scented and often secrete nectar. In some species (e.g. primulas) there are structural differences between the flowers to ensure that cross-pollination occurs. Other examples of entomophilous flowers are orchids and antirrhinums. *Compare* ANEMOPHILY; HYDROPHILY.

**Entrez** An information retrieval service, or browser, provided by the National Center for

Biotechnology Information (NCBI), a division of the US National Library of Medicine. It gives access to a collection of NCBI databases, both bibliographic and biomolecular, including PubMed, *MEDLINE, and OMIM (Online Mendelian Inheritance in Man).

**entropy** Symbol *S*. A measure of the unavailability of a system's energy to do work; in a closed system an increase in entropy is accompanied by a decrease in energy availability. When a system undergoes a reversible change the entropy (*S*) changes by an amount equal to the energy (*Q*) transferred to the system by heat divided by the thermodynamic temperature (*T*) at which this occurs, i.e. $\Delta S = \Delta Q / T$. However, all real processes are to a certain extent irreversible changes and in any closed system an irreversible change is always accompanied by an increase in entropy.

In a wider sense entropy can be interpreted as a measure of disorder; the higher the entropy the greater the disorder (*see* BOLTZMANN FORMULA). As any real change to a closed system tends towards higher entropy, and therefore higher disorder, it follows that the entropy of the universe (if it can be considered a closed system) is increasing and its available energy is decreasing (*see* HEAT DEATH OF THE UNIVERSE). This increase in the entropy of the universe is one way of stating the second law of *thermodynamics.

((●)) **SEE WEB LINKS**

• Translations of papers by Clausius on entropy (1850) and the second law (1865), published in *Annalen der Physik und Chemie*

**environment** (in ecology) The physical, chemical, and biological conditions of the region in which an organism lives. *See also* ECOLOGY; ECOSYSTEM.

**environmental resistance** The sum total of the factors that prevent populations from continually growing and therefore tend to keep populations at constant levels. These factors include predators, disease, and a shortage of any of the various requirements for survival, such as food, water, shelter, and light (which is particularly important for plants). *See also* POPULATION GROWTH.

**enyl complex** A type of complex in which there is a link between the metal atom or ion and the pi electrons of a double bond. *Zeise's salt was the first known example.

**enzyme** A protein that acts as a *catalyst in biochemical reactions. Each enzyme is specific to a particular reaction or group of similar reactions. Many require the association of certain nonprotein *cofactors in order to function. The molecule undergoing reaction (the **substrate**) binds to a specific *active site on the enzyme molecule to form a short-lived intermediate (*see* ENZYME–SUBSTRATE COMPLEX): this greatly increases (by a factor of up to $10^{20}$) the rate at which the reaction proceeds to form the product. Enzyme activity is influenced by substrate concentration and by temperature and pH, which must lie within a certain range. Other molecules may compete for the active site, causing *inhibition of the enzyme or even irreversible destruction of its catalytic properties.

Enzyme production is governed by a cell's genes. Enzyme activity is further controlled by pH changes, alterations in the concentrations of essential cofactors, feedback inhibition by the products of the reaction, and activation by another enzyme, either from a less active form or an inactive precursor (*zymogen). Such changes may themselves be under the control of hormones or the nervous system. *See also* ENZYME KINETICS.

Enzymes are classified into six major groups, according to the type of reaction they catalyse: (1) *oxidoreductases; (2) *transferases; (3) *hydrolases; (4) *lyases; (5) *isomerases; (6) *ligases. The names of most individual enzymes also end in -*ase*, which is added to the names of the substrates on which they act. Thus *lactase is the enzyme that acts to break down lactose; it is classified as a hydrolase.

((●)) **SEE WEB LINKS**

• Information about IUPAC nomenclature

**enzyme inhibition** *See* INHIBITION.

**enzyme kinetics** The study of the rates of enzyme-catalysed reactions. Rates of reaction are usually measured by using the purified enzyme *in vitro* with the substrate and then observing the formation of the product or disappearance of the substrate. As the concentration of the substrate is increased the rate of reaction increases proportionally up to a certain point, after which any further increase in substrate concentration no longer increases the reaction rate (*see* MICHAELIS–MENTEN CURVE). At this point, all active sites of the enzyme are saturated with substrate; any further increase in the rate of reaction will occur only if more enzyme is added. Reaction rates are also affected by the presence of inhibitors (*see* INHIBITION), temperature, and pH (*see* ENZYME).

**enzyme-linked immunosorbent assay** *See* ELISA.

**enzyme–substrate complex** The intermediate formed when a substrate molecule interacts with the *active site of an enzyme. Following the formation of an enzyme–substrate complex, the substrate molecule undergoes a chemical reaction and is converted into a new product. Various mechanisms for the formation of enzyme–substrate complexes have been suggested, including the *lock-and-key mechanism.

**Eocene** The second geological epoch of the *Palaeogene period. It extended from the end of the Palaeocene epoch, about 55 million years ago, to the beginning of the Oligocene epoch, about 34 million years ago. The term was first proposed by Sir Charles Lyell in 1833. Mammals were dominant in the Eocene: rodents, artiodactyls, carnivores, perissodactyls (including early horses), and whales were among the groups to make their first appearance.

**eosin** One of a series of acidic dyes, used in optical microscopy, that colours cytoplasm pink and cellulose red. It is frequently used as a counterstain with *haematoxylin for colouring tissue smears and sections of animal tissue.

**ephedrine** An alkaloid, $C_6H_5CH(OH)CH$-$(CH_3)NHCH_3$, found in plants of the genus *Ephedra*, once used as a bronchodilator in the treatment of asthma. It is also used as a stimulant and appetite suppressant. Structurally, it is a phenylethylamine and is similar to amphetamines, although less active. It is, however, widely used in the illegal synthesis of methamphetamine. The molecule has two chiral centres. If the stereochemical conformations are opposite (i.e. $1R,2S$ or $1S,1R$) the name ephedrine is used. If the conformations are the same ($1R,2R$ or $1S,2S$) then the compound is called **pseudoephedrine**.

**ephemeral** **1.** (in botany) An *annual plant that completes its life cycle in considerably less than one growing season. A number of generations can therefore occur in one year. Many troublesome weeds, such as groundsel and willowherb, are ephemerals. Certain desert plants are also ephemerals, completing their life cycles in a short period

following rain. **2.** (in zoology) A short-lived animal, such as a mayfly.

**ephemeris** A tabulation showing the calculated future positions of the sun, moon, and planets, together with other useful information for astronomers and navigators. It is published at regular intervals.

**ephemeris time (ET)** A time system that has a constant uniform rate as opposed to other systems that depend on the earth's rate of rotation, which has inherent irregularities. It is reckoned from an instant in 1900 (Jan 0d 12h) when the sun's mean longitude was 279.696 677 8°. The unit by which ephemeris time is measured is the tropical year, which contains 31 556 925.9747 **ephemeris seconds**. This fundamental definition of the *second was replaced in 1964 by the caesium second of atomic time.

**epicalyx** A ring of bracts below a flower that resembles a calyx. It is seen, for example, in the strawberry flower.

**epicarp** *See* PERICARP.

**epicentre** The point on the surface of the earth directly above the focus of an earthquake or directly above or below a nuclear explosion.

**epicotyl** The region of a seedling stem above the stalks of the seed leaves (*cotyledons) of an embryo plant. It grows rapidly in seeds showing *hypogeal germination and lifts the stem above the soil surface. *Compare* HYPOCOTYL.

**epicycle** A small circle whose centre rolls around the circumference of a larger fixed circle. The curve traced out by a point on the epicycle is called an **epicycloid**.

**epidemiology** The study of diseases that affect large numbers of people. Traditionally, epidemiologists have been concerned primarily with infectious diseases, such as typhoid and influenza, that arise and spread rapidly among the population as epidemics. However, today the discipline also covers noninfectious disorders, such as diabetes, heart disease, and back pain. Typically the distribution of a disease is charted in order to discover patterns that might yield clues about its mode of transmission or the susceptibility of certain groups of people. This in turn may reveal insights about the causes of the disease and possible preventive measures.

**epidermis** 1. (in zoology) The outermost layer of cells of the body of an animal. In invertebrates the epidermis is normally only one cell thick and is covered by an impermeable *cuticle. In vertebrates the epidermis is the thinner of the two layers of *skin (*compare* DERMIS). It consists of a basal layer of actively dividing cells (*see* MALPIGHIAN LAYER), covered by layers of cells that become impregnated with keratin (*see* KERATINIZATION). The outermost layers of epidermal cells (the *stratum corneum) form a water-resistant protective layer. The epidermis may bear a variety of specialized structures (e.g. *feathers, *hairs). 2. (in botany) The outermost layer of cells covering a plant. It is overlaid by a *cuticle and its functions are principally to protect the plant from injury and to reduce water loss. Some epidermal cells arc modified to form guard cells (*see* STOMA) or hairs of various types (*see* PILIFEROUS LAYER). In woody plants the functions of the shoot epidermis are taken over by the periderm tissues (*see* CORK CAMBIUM) and in mature roots the epidermis is sloughed off and replaced by the *hypodermis.

**epidiascope** An optical instrument used by lecturers, etc., for projecting an enlarged image of either a translucent object (such as a slide or transparency) or an opaque object (such as a diagram or printed page) onto a screen.

**epididymis** A long coiled tube in which spermatozoa are stored in vertebrates. In reptiles, birds, and mammals it is attached at one end to the *testis and opens into the sperm duct (*vas deferens) at the other.

**epigamic** Serving to attract a mate. Epigamic characters include the bright plumage of some male birds.

**epigeal** Describing seed germination in which the seed leaves (cotyledons) emerge from the ground and function as true leaves. Examples of epigeal germination are seen in sycamore and sunflower. *Compare* HYPOGEAL.

**epiglottis** A flexible flap of cartilage in mammals that is attached to the wall of the pharynx near the base of the tongue. During swallowing (*see* DEGLUTITION) it covers the *glottis (the opening to the respiratory tract) and helps to prevent food from entering the trachea (windpipe), although it is not essential for this purpose.

**epilimnion** The upper layer of water in a lake. *Compare* HYPOLIMNION. *See* THERMOCLINE.

**epimerism** A type of optical isomerism in which a molecule has two chiral centres; two optical isomers (**epimers**) differ in the arrangement about one of these centres. *See also* OPTICAL ACTIVITY.

**epinephrine** *See* ADRENALINE.

**epiphysis** The terminal section of a growing bone (especially a long limb bone) in mammals. It is separated from the bone shaft (**diaphysis**) by cartilage. New bone is produced on the side of the cartilage facing the diaphysis, while new cartilage is produced on the other side of the cartilage disc. When the bone reaches adult length the epiphysis merges with the diaphysis.

**epiphyte** A plant that grows upon another plant but is neither parasitic on it nor rooted in the ground. Epiphytes include many mosses and lichens and some tropical orchids.

**episome** A genetic element that can exist and replicate either independently of its host cell's chromosomes or as an integrated part of the chromosomes. Examples include certain bacterial *plasmids.

**epistasis** A gene interaction in which one gene suppresses the effect of another gene that is situated at a different *locus on the chromosome. For example, in guinea pigs the gene that controls the production of melanin is epistatic to the gene that regulates the deposition of melanin. A dominant allele (C) is responsible for the production of melanin, while the amount of melanin deposited is controlled by a second gene, which determines whether the coat colour is black or brown. If an animal is homozygous recessive (cc) for melanin production, the coat colour will be white regardless of the alleles that produce black or brown coloration.

**epitaxy (epitaxial growth)** Growth of a layer of one substance on a single crystal of another, such that the crystal structure in the layer is the same as that in the substrate. It is used in making semiconductor devices.

**epithelium** A tissue in vertebrates consisting of closely packed cells in a sheet with little intercellular material. It covers the outer surfaces of the body and the walls of the internal cavities (coeloms). It also forms glands and parts of sense organs. Its functions are protective, absorptive, secretory, and sensory. The types of cell vary, giving rise to **squamous**, **cuboidal**, **columnar**, and **ciliated epithelia**. **Stratified epithelium** (e.g. in the skin) is made up of several layers of cells. Epithelium is derived from *ectoderm and *endoderm. *Compare* ENDOTHELIUM; MESOTHELIUM.

**epithermal neutron** A neutron with an energy in excess of that associated with a thermal neutron (*see* MODERATOR) but less than that of a *fast neutron, i.e. a neutron having an energy in the range 0.1 to 100 eV.

**EPM** *See* ELECTRON PROBE MICROANALYSIS.

**epoch** **1.** (in astronomy) An arbitrary date used as a fixed reference point, especially for the celestial coordinates of a star or deep-sky object or the orbital elements of a planet. Because the coordinates *right ascension and *declination change over time, the position of an astronomical object given in a catalogue must always be referred to a particular epoch, e.g. the date and time that an observation was made, the start of a year. **2.** The starting date of a calendar or chronological era. **3.** A unit of geological time in which a series of rocks is formed.

**epoetin (EPO)** A drug used medically to treat anaemia, and also used illegally by athletes participating in endurance events. It is a genetically engineered form of the hormone erythropoietin, which regulates the production of red blood cells.

**epoxides** Compounds that contain oxygen atoms in their molecules as part of a three-membered ring (see formula). Epoxides are thus **cyclic ethers**.

**Epoxides.** The functional group in epoxides.

**epoxyethane (ethylene oxide)** A colourless flammable gas, $C_2H_4O$; m.p. $-111°C$; b.p. $13.5°C$. It is a cyclic ether (*see* EPOXIDES) that is made by the catalytic oxidation of ethene. It can be hydrolysed to ethane-1,2-diol and also polymerizes to:

… $-O-C_2H_4-O-C_2H_4-$…,

which is used for lowering the viscosity of water (e.g. in fire fighting).

**epoxy resins** Synthetic resins produced by copolymerizing epoxide compounds with phenols. They contain $-O-$ linkages and epoxide groups and are usually viscous liquids. They can be hardened by addition of agents, such as polyamines, that form cross-linkages. Alternatively, catalysts may be used to induce further polymerization of the resin. Epoxy resins are used in electrical equipment and in the chemical industry (because of resistance to chemical attack). They are also used as adhesives.

**EPR** *See* ELECTRON PARAMAGNETIC RESONANCE.

**epsomite** A mineral form of *magnesium sulphate heptahydrate, $MgSO_4$. $7H_2O$.

**Epsom salt** *See* MAGNESIUM SULPHATE.

**Epstein–Barr virus** *See* HERPESVIRUS.

**equation of motion (kinematic equation)** Any of four equations that apply to bodies moving linearly with uniform acceleration ($a$). The equations, which relate distance covered ($s$) to the time taken ($t$), are:

$$v = u + at$$
$$s = (u + v)t/2$$
$$s = ut + at^2/2$$
$$v^2 = u^2 + 2as,$$

where $u$ is the initial velocity of the body and $v$ is its final velocity.

**equation of state** An equation that relates the pressure $p$, volume $V$, and thermodynamic temperature $T$ of an amount of substance $n$. The simplest is the ideal *gas law:

$$pV = nRT,$$

where $R$ is the universal gas constant. Applying only to ideal gases, this equation takes no account of the volume occupied by the gas molecules (according to this law if the pressure is infinitely great the volume becomes zero), nor does it take into account any forces between molecules. A more accurate equation of state would therefore be

$$(p + k)(V - nb) = nRT,$$

where $k$ is a factor that reflects the decreased pressure on the walls of the container as a result of the attractive forces between particles, and $nb$ is the volume occupied by the particles themselves when the pressure is infinitely high. In the **van der Waals equa-**

**tion of state**, proposed by the Dutch physicist J. D. van der Waals (1837–1923),

$$k = n^2 a / V^2,$$

where $a$ is a constant. This equation more accurately reflects the behaviour of real gases; several others have done better but are more complicated.

**equation of time** The length of time that must be added to the mean solar time, as shown on a clock, to give the apparent solar time, as shown by a sundial. The amount varies during the year, being a minimum of −14.2 minutes in February and a maximum of +16.4 minutes in October. It is zero on four days (April 15/16, June 14/15, Sept. 1/2, Dec. 25/26). The difference arises as a result of two factors: the eccentricity of the earth's orbit and the inclination of the ecliptic to the celestial equator.

**equator 1.** The great circle around the earth that lies in a plane perpendicular to the earth's axis. It is equidistant from the two geographical poles. **2.** The **magnetic equator** is a line of zero magnetic dip (*see* GEOMAGNETISM) that is close to the geographical equator but lies north of it in Africa and south of it in America. **3.** The **celestial equator** is the circle formed on the *celestial sphere by the extension of the earth's equatorial plane. **4.** (in cell biology) *See* SPINDLE.

**equilibrium** A state in which a system has its energy distributed in the statistically most probable manner; a state of a system in which forces, influences, reactions, etc., balance each other out so that there is no net change.

A body is in **static equilibrium** if the resultants of all forces and all couples acting on it are both zero; it may be at rest and will certainly not be accelerated. Such a body at rest is in **stable equilibrium** if after a slight displacement it returns to its original position – for a body whose weight is the only downward force this will be the case if the vertical line through its centre of gravity always passes through its base. If a slight displacement causes the body to move to a new position, then the body is in **unstable equilibrium**.

A body is said to be in **thermal equilibrium** if no net heat exchange is taking place within it or between it and its surroundings. A system is in *chemical equilibrium when a reaction and its reverse are proceeding at equal rates (*see also* EQUILIBRIUM CONSTANT). These are examples of **dynamic equi-**

**librium**, in which activity in one sense or direction is in aggregate balanced by comparable reverse activity.

**equilibrium constant** For a reversible reaction of the type

$$xA + yB \rightleftharpoons zC + wD$$

chemical equilibrium occurs when the rate of the forward reaction equals the rate of the back reaction, so that the concentrations of products and reactants reach steady-state values. It can be shown that at equilibrium the ratio of concentrations

$$[C]^z[D]^w / [A]^x[B]^y$$

is a constant for a given reaction and fixed temperature, called the equilibrium constant $K_c$ (where the $c$ indicates concentrations have been used). Note that, by convention, the products on the right-hand side of the reaction are used on the top line of the expression for equilibrium constant. This form of the equilibrium constant was originally introduced in 1863 by C. M. Guldberg and P. Waage using the law of *mass action. They derived the expression by taking the rate of the forward reaction

$$k_f[A]^x[B]^y$$

and that of the back reaction

$$k_b[C]^z[D]^w$$

Since the two rates are equal at equilibrium, the equilibrium constant $K_c$ is the ratio of the rate constants $k_f / k_b$. The principle that the expression is a constant is known as the **equilibrium law** or **law of chemical equilibrium**.

The equilibrium constant shows the **position** of equilibrium. A low value of $K_c$ indicates that [C] and [D] are small compared to [A] and [B]; i.e. that the back reaction predominates. It also indicates how the equilibrium shifts if concentration changes. For example, if [A] is increased (by adding A) the equilibrium shifts towards the right so that [C] and [D] increase, and $K_c$ remains constant.

For gas reactions, partial pressures are used rather than concentrations. The symbol $K_p$ is then used. Thus, in the example above

$$K_p = p_C^z p_D^w / p_A^x p_B^y$$

It can be shown that, for a given reaction $K_p$ = $K_c(RT)^{\Delta v}$, where $\Delta v$ is the difference in stoichiometric coefficients for the reaction (i.e. $z + w - x - y$). Note that the units of $K_p$ and $K_c$ depend on the numbers of molecules appearing in the stoichiometric equation. The value of the equilibrium constant depends on the temperature. If the forward reaction is exothermic, the equilibrium constant decreases as the temperature rises; if endothermic it increases (*see also* VAN'T HOFF'S ISOCHORE).

The expression for the equilibrium constant can also be obtained by thermodynamics; it can be shown that the standard equilibrium constant $K^\ominus$ is given by $\exp(-\Delta G^\ominus / RT)$, where $\Delta G^\ominus$ is the standard Gibbs free energy change for the complete reaction. Strictly, the expressions above for equilibrium constants are true only for ideal gases (pressure) or infinite dilution (concentration). For accurate work *activities are used.

**equilibrium law**  *See* EQUILIBRIUM CONSTANT.

**equinox**  1. Either of the two points on the *celestial sphere at which the *ecliptic intersects the celestial equator. The sun appears to cross the celestial equator from south to north at the **vernal equinox** and from north to south at the **autumnal equinox**. 2. Either of the two instants at which the centre of the sun appears to cross the celestial equator. In the northern hemisphere the vernal equinox occurs on or about March 21 and the autumnal equinox on or about Sept. 23. In the southern hemisphere the dates are reversed. *See* PRECESSION OF THE EQUINOXES.

**equipartition of energy**  The theory, proposed by Ludwig Boltzmann and given some theoretical support by James Clerk Maxwell, that the energy of gas molecules in a large sample under thermal *equilibrium is equally divided among their available *degrees of freedom, the average energy for each degree of freedom being $kT/2$, where $k$ is the *Boltzmann constant and $T$ is the thermodynamic temperature. The proposition is not generally true if *quantum considerations are important, but is frequently a good approximation.

**equivalence point**  The point in a titration at which reaction is complete. *See* INDICATOR.

**equivalent proportions**  *See* CHEMICAL COMBINATION.

**equivalent weight**  The mass of an element or compound that could combine with or displace one gram of hydrogen (or eight grams of oxygen or 35.5 grams of chlorine) in a chemical reaction. The equivalent weight represents the 'combining power' of

the substance. For an element it is the relative atomic mass divided by the valency. For a compound it depends on the reaction considered.

**erbium**  Symbol Er. A soft silvery metallic element belonging to the *lanthanoids; a.n. 68; r.a.m. 167.26; r.d. 9.006 (20°C); m.p. 1529°C; b.p. 2863°C. It occurs in apatite, gadolinite, and xenotine from certain sources. There are six natural isotopes, which are stable, and twelve artificial isotopes are known. It has been used in alloys for nuclear technology as it is a neutron absorber; it is being investigated for other potential uses. It was discovered by Carl Mosander (1797–1858) in 1843.

(((•))) **SEE WEB LINKS**
• Information from the WebElements site

**erecting prism**  A glass prism used in optical instruments to convert an inverted image into an erect image, as in prismatic binoculars.

**erg**  A unit of work or energy used in the c.g.s. system and defined as the work done by a force of 1 dyne when it acts through a distance of 1 centimetre. $1 \text{ erg} = 10^{-7}$ joule.

**ergocalciferol**  *See* VITAMIN D.

**ergonomics**  The study of the engineering aspects of the relationship between workers and their working environment.

**ergosphere**  The region immediately around a *black hole. The hole's rotation drags the space–time continuum round with it, so that frames of reference are not stationary with reference to the remainder of the universe. The ergosphere's outer boundary is called the stationary limit.

**ergosterol**  A *sterol occurring in fungi, bacteria, algae, and plants. It is converted into vitamin $D_2$ by the action of ultraviolet light.

**Eris**  A *small solar system body made of rock and ice that is the largest *dwarf planet so far known and the ninth largest body orbiting the sun. At an estimated 2400 km or more in diameter, it is at least 5% larger than *Pluto and 27% more massive. Eris was discovered in January 2005 by a team at Mount Palomar led by Michael E. Brown (1965–   ). It is a *scattered disc object, orbiting the sun once every 557.43 years at a mean distance of $10.12 \times 10^9$ km. The eccentricity of its orbit is about 0.44 and when at its farthest point from the sun ($14.6 \times 10^9$ km), it is three times as far away as Neptune. One natural satellite, Dysnomia, orbits Eris at an estimated distance of nearly 37 400 km in a period of 15.77 days.

**Erlangen Programme** The view, put forward by the German mathematician Felix Klein (1849–1925) in 1872 when he became a faculty member of the University of Erlangen, that each type of geometry can be characterized by a group of transformations. There are many physical examples of this, including the Lorentz group of Minkowski space–time and the space groups of crystals. However, not all types of geometry fit into this view readily; in particular, Riemannian geometry, and hence curved space–time, do not fit unless the Erlangen Programme is given a wider interpretation. It has been suggested that such a broader interpretation might lead to important insights into the foundations of fundamental physics.

**erosion** The wearing away of the land surface by natural agents that involves the transport of rock debris. These natural agents include moving waters (e.g. rivers, ocean waves), ice (e.g. glaciers), wind, organisms, and gravity. *See also* SOIL EROSION.

**erratic** A fragment of rock, often unlike the rocks around it, that has been displaced from its original location by the action of a glacier or, more rarely, an iceberg. Erratics may have been moved as little as several metres to more than 800 km. They vary in size from small pebbles to massive boulders, and may be found on the surface or embedded in *boulder clay. They provide geologists with information about the movement of ice sheets.

**erythroblast** Any of the cells in the *myeloid tissue of red bone marrow that develop into erythrocytes (red blood cells). Erythroblasts have a nucleus and are at first colourless, but fill with *haemoglobin as they develop. In mammals the nucleus disappears.

**erythrocyte (red blood cell)** The most numerous type of blood cell, which contains the red pigment *haemoglobin and is responsible for oxygen transport. Mammalian erythrocytes are disc-shaped and lack a nucleus; those of other vertebrates are oval and nucleated. In man the number of erythrocytes in the blood varies between 4.5 and 5.5 million per cubic millimetre. They survive

for about four months and are then destroyed in the spleen and liver. *See also* ERYTHROBLAST. *Compare* LEUCOCYTE.

**Esaki diode** *See* TUNNEL DIODE.

**ESCA** *See* PHOTOELECTRON SPECTROSCOPY.

**escapement** A device in a clock or watch that controls the transmission of power from the spring or falling weight to the hands. It is usually based on a balance wheel or pendulum. It thus allows energy to enter the mechanism in order to move the hands round the face, overcome friction in the gear trains, and maintain the balance wheel or pendulum in continuous motion.

**escape velocity** The minimum speed needed by a space vehicle, rocket, etc., to escape from the gravitational field of the earth, moon, or other celestial body. The gravitational force between a rocket of mass $m$ and a celestial body of mass $M$ and radius $r$ is $MmG/r^2$ (*see* NEWTON'S LAW OF GRAVITATION). Therefore the gravitational potential energy of the rocket with respect to its possible position very far from the celestial body on which it is resting can be shown to be $-GmM/r$, assuming (by convention) that the potential energy is zero at an infinite distance from the celestial body. If the rocket is to escape from the gravitational field it must have a kinetic energy that exceeds this potential energy, i.e. the kinetic energy $mv^2/2$ must be greater than $MmG/r$, or $v > \sqrt{(2MG/r)}$. This is the value of the escape velocity. Inserting numerical values for the earth and moon into this relationship gives an escape velocity from the earth of 11 200 m s$^{-1}$ and from the moon of 2370 m s$^{-1}$.

***Escherichia coli (E. coli)*** A species of Gram-negative aerobic bacteria that is found in the intestine and is also widely used in microbiological and genetics research. The motile rod-shaped cells ferment lactose and are usually harmless commensals, although certain strains are pathogenic and can cause a severe form of food poisoning. Studies of *E. coli* laboratory cultures have revealed much about the genetics of prokaryotes; the species is also frequently used in genetic engineering, particularly as a host for *gene cloning and the expression of recombinant foreign genes in culture.

**ESI** *See* ELECTROSPRAY IONIZATION.

**esker** An elongated, steep-sided ridge of debris left behind by meltwater streams flowing in or under a slow-moving glacier. The ridge may be straight or, more often, sinuous in shape. It may be up to 50 m high and 700 m wide. In very cold regions, some eskers have ice cores. The debris is stratified in layers and generally consists of rounded particles of gravel and sand, with some fine-grain deposits. *See also* KAME.

**ESR** *See* ELECTRON-PARAMAGNETIC RESONANCE.

**essential amino acid** An *amino acid that an organism is unable to synthesize in sufficient quantities. It must therefore be present in the diet. In humans the essential amino acids are histidine, lysine, threonine, methionine, isoleucine, leucine, valine, phenylalanine, and tryptophan. These are required for protein synthesis and deficiency leads to retarded growth and other symptoms. Most of the amino acids required by man are also essential for all other multicellular animals and for most protozoans.

**essential element** Any of a number of elements required by living organisms to ensure normal growth, development, and maintenance. Apart from the elements found in organic compounds (i.e. carbon, hydrogen, oxygen, and nitrogen), plants, animals, and microorganisms all require a range of elements in inorganic forms in varying amounts, depending on the type of organism. The **major elements**, present in tissues in relatively large amounts (greater than 0.005%), are calcium, phosphorus, potassium, sodium, chlorine, sulphur, and magnesium (*see also* MACRONUTRIENT). The **trace elements** occur at much lower concentrations and thus requirements are much less. The most important are iron, manganese, zinc, copper, iodine, cobalt, selenium, molybdenum, chromium, and silicon (*see also* MICRONUTRIENT). Each element may fulfil one or more of a variety of metabolic roles. Sodium, potassium, and chloride ions are the chief electrolytic components of cells and body fluids and thus determine their electrical and osmotic status. Calcium, phosphorus, and magnesium are all present in bone. Calcium is also essential to nerve and muscle activity, while phosphorus is a key constituent of the chemical energy carriers (e.g. *ATP) and the nucleic acids. Sulphur is needed primarily for amino acid synthesis (in plants and microorganisms). The trace elements may serve as *cofactors or as constituents of complex molecules, e.g. iron in

haem and cobalt in vitamin $B_{12}$. *See also* MINERAL DEFICIENCY.

**essential fatty acids (omega fatty acids)** *Fatty acids that must normally be present in the diet of certain animals, including humans. Essential fatty acids belong to the $n$-3 (or omega-3) and $n$-6 (or omega-6) classes of polyunsaturated fatty acids and include *arachidonic (omega-6), *linoleic (omega-6), and *linolenic (omega-3) acids; all possess double bonds at the same two positions along their hydrocarbon chain and so can act as precursors of *prostaglandins. Deficiency can cause dermatosis, weight loss, irregular oestrus, etc. Both omega-3 and omega-6 acids should be consumed in the diet, ideally in roughly the same amounts.

**essential oil** A natural oil with a distinctive scent secreted by the glands of certain aromatic plants. *Terpenes are the main constituents. Essential oils are extracted from plants by steam distillation, extraction with cold neutral fats or solvents (e.g. alcohol), or pressing and used in perfumes, flavourings, and medicine. Examples are citrus oils, flower oils (e.g. rose, jasmine), and oil of cloves.

**esterification** A reaction of an alcohol with an acid to produce an ester and water; e.g.

$$CH_3OH + C_6H_5COOH \rightleftharpoons CH_3OOCC_6H_5 + H_2O$$

The reaction is an equilibrium and is normally slow, but can be speeded up by addition of a strong acid catalyst. The ester can often be distilled off so that the reaction can proceed to completion. The reverse reaction is ester hydrolysis or *saponification. *See also* LABELLING.

**esters** Organic compounds formed by reaction between alcohols and acids (see illustration overleaf). Esters formed from carboxylic acids have the general formula RCOOR'. Examples are ethyl ethanoate, $CH_3COOC_2H_5$, and methyl propanoate, $C_2H_5COOCH_3$. Esters containing simple hydrocarbon groups are volatile fragrant substances used as flavourings in the food industry. Triesters, molecules containing three ester groups, occur in nature as oils and fats. *See also* GLYCERIDE.

(((🌐))) **SEE WEB LINKS**
• Information about IUPAC nomenclature

**etaerio** A cluster of fruits formed from the

$$CH_3 - O \overset{\displaystyle \cdots}{\underset{\displaystyle \cdots}{H}} \quad \overset{\displaystyle \cdots}{\underset{\displaystyle \cdots}{H - O}} \overset{O}{\underset{}{C}} - C_2H_5 \quad \rightleftharpoons \quad CH_3 \overset{O}{\underset{O}{C}} - C_2H_5 \; + \; H_2O$$

methanol           propanoic acid         methyl propanoate    water

**Ester formation.**

unfused carpels of a single flower. For example, the anemone has an etaerio of *achenes, larkspur an etaerio of *follicles, and blackberry an etaerio of *drupes.

**ethanal (acetaldehyde)** A colourless highly flammable liquid aldehyde, $CH_3CHO$; r.d. 0.78; m.p. −121°C; b.p. 20.8°C. It is made from ethene by the *Wacker process and used as a starting material for making many organic compounds. The compound polymerizes if dilute acid is added to give **ethanal trimer** (or **paraldehyde**), which contains a six-membered ring of alternating carbon and oxygen atoms with a hydrogen atom and a methyl group attached to each carbon atom. It is used as a drug for inducing sleep. Addition of dilute acid below 0°C gives **ethanal tetramer** (or **metaldehyde**), which has a similar structure to the trimer but with an eight-membered ring. It is used as a solid fuel in portable stoves and in slug pellets.

**ethanamide (acetamide)** A colourless solid crystallizing in the form of long white crystals with a characteristic smell of mice, $CH_3CONH_2$; r.d. 1.159; m.p. 82.3°C; b.p. 221.25°C. It is made by the dehydration of ammonium ethanoate or by the action of ammonia on ethanoyl chloride, ethanoic anhydride, or ethyl ethanoate.

**ethane** A colourless flammable gaseous hydrocarbon, $C_2H_6$; m.p. −183°C; b.p. −89°C. It is the second member of the *alkane series of hydrocarbons and occurs in natural gas.

**ethanedioic acid** See OXALIC ACID.

**ethane-1,2-diol (ethylene glycol; glycol)** A colourless viscous hygroscopic liquid, $CH_2OHCH_2OH$; m.p. −11.5°C; b.p. 198°C. It is made by hydrolysis of epoxyethane (from ethene) and used as an antifreeze and a raw material for making *polyesters (e.g. Terylene).

**ethanoate (acetate)** A salt or ester of ethanoic acid (acetic acid).

**ethanoic acid (acetic acid)** A clear viscous liquid or glassy solid *carboxylic acid, $CH_3COOH$, with a characteristically sharp odour of vinegar; r.d. 1.049; m.p. 16.6°C; b.p. 117.9°C. The pure compound is called **glacial ethanoic acid**. It is manufactured by the oxidation of ethanol or by the oxidation of butane in the presence of dissolved manganese(II) or cobalt(II) ethanoates at 200°C, and is used in making ethanoic anhydride for producing cellulose ethanoates. It is also used in making ethenyl ethanoate (for polyvinylacetate). The compound is formed by the fermentation of alcohol and is present in vinegar, which is made by fermenting beer or wine. 'Vinegar' made from ethanoic acid with added colouring matter is called 'non-brewed condiment'. In living organisms it combines with *coenzyme A to form acetyl coenzyme A, which plays a crucial role in energy metabolism.

**ethanoic anhydride (acetic anhydride)** A pungent-smelling colourless liquid, $(CH_3CO)_2O$, b.p. 139.5°C. It is used in organic synthesis as an *ethanoylating agent (attacking an −OH or −NH group) and in the manufacture of aspirin and cellulose plastics. It hydrolyses in water to give ethanoic acid.

**ethanol (ethyl alcohol)** A colourless water-soluble *alcohol, $C_2H_5OH$; r.d. 0.789 (0°C); m.p. −117.3°C; b.p. 78.3°C. It is the active principle in intoxicating drinks, in which it is produced by fermentation of sugar using yeast

$$C_6H_{12}O_6 \rightarrow 2C_2H_5OH + 2CO_2$$

The ethanol produced kills the yeast and fermentation alone cannot produce ethanol solutions containing more than 15% ethanol by volume. Distillation can produce a constant-boiling mixture containing 95.6% ethanol and 4.4% water. Pure ethanol (**absolute alcohol**) is made by removing this water by means of drying agents.

The main industrial use of ethanol is as a solvent although at one time it was a major starting point for making other chemicals. For this it was produced by fermentation of molasses. Now ethene has replaced ethanol as a raw material and industrial ethanol is made by hydrolysis of ethene.

**ethanoylating agent (acetylating agent)** A chemical reagent used to introduce an ethanoyl group ($-COCH_3$) instead of hydrogen in an organic compound. Examples include *ethanoic anhydride and ethanoyl chloride (acetyl chloride, $CH_3COCl$).

**ethanoyl chloride (acetyl chloride)** A colourless liquid acyl chloride (*see* ACYL HALIDES), $CH_3COCl$, with a pungent smell; r.d. 1.105; m.p. $-112.15°C$; b.p. $50.9°C$. It is made by reacting ethanoic acid with a halogenating agent such as phosphorus(III) chloride, phosphorus(V) chloride, or sulphur dichloride oxide and is used to introduce ethanoyl groups into organic compounds containing $-OH$, $-NH_2$, and $-SH$ groups. *See* ACYLATION.

**ethanoyl group (acetyl group)** The organic group $CH_3CO-$.

**ethene (ethylene)** A colourless flammable gaseous hydrocarbon, $C_2H_4$; m.p. $-169°C$; b.p. $-103.7°C$. It is the first member of the *alkene series of hydrocarbons. It is made by cracking hydrocarbons from petroleum and is now a major raw material for making other organic chemicals (e.g. ethanal, ethanol, ethane-1,2-diol). It can be polymerized to *polyethene. It occurs naturally in plants, in which it acts as a growth substance (*see* PLANT HORMONE)· its best-known effect is the stimulation of fruit ripening.

**ethenone** *See* KETENE.

**ethenyl ethanoate (vinyl acetate)** An unsaturated organic ester, $CH_2:CHOOCCH_3$; r.d. 0.9; m.p. $-100°C$; b.p. $73°C$. It is made by catalytic reaction of ethanoic acid and ethene and used to make polyvinylacetate.

**ether 1. (aether)** A hypothetical medium once believed to be necessary to support the propagation of electromagnetic radiation. It is now regarded as unnecessary and in modern theory electromagnetic radiation can be propagated through empty space. The existence of the ether was first called in question as a result of the *Michelson–Morley experiment. **2.** *See* ETHOXYETHANE; ETHERS.

**ethers** Organic compounds containing the group $-O-$ in their molecules. Examples are dimethyl ether, $CH_3OCH_3$, and diethyl ether, $C_2H_5OC_2H_5$ (*see* ETHOXYETHANE). They are

volatile highly flammable compounds made by dehydrating alcohols using sulphuric acid.

**((())) SEE WEB LINKS**
• Information about IUPAC nomenclature

**ethology** The study of the biology of *animal behaviour. Central to the ethologist's approach is the principle that animal behaviour (like physical characteristics) is subject to evolution through natural selection. Ethologists therefore seek to explain how the behaviour of animals, both as individuals and social groups, in their natural environment may contribute to the survival of the maximum number of its relatives and offspring. This involves recognizing the stimuli that are important in nature (*see* SIGN STIMULUS) and how innate predispositions interact with *learning in the development of behaviour (*see* INSTINCT). Studies of this sort were pioneered by Konrad Lorenz and Niko Tinbergen and have led to the modern disciplines of sociobiology and behavioural ecology.

**ethoxyethane (diethyl ether; ether)** A colourless flammable volatile *ether, $C_2H_5OC_2H_5$; r.d. 0.71; m.p. $-116°C$; b.p. 34.5°C. It can be made by *Williamson's synthesis. It is an anaesthetic and useful organic solvent.

**ethyl 3-oxobutanoate (ethyl acetoacetate)** A colourless liquid ester with a pleasant odour, $CH_3COCH_2COOC_2H_5$; r.d. 1.03; m.p. $-80°C$; b.p. 180.4°C. It can be prepared by reacting ethyl ethanoate ($CH_3COOC_2H_5$) with sodium or sodium ethoxide. The compound shows keto–enol *tautomerism and contains about 7% of the enol form, $CH_3C(OH):CHCOOC_2H_5$, under normal conditions. Sometimes known as **acetoacetic ester**, it is used in organic synthesis.

**ethyl acetate** *See* ETHYL ETHANOATE.

**ethyl acetoacetate** *See* ETHYL 3-OXOBUTANOATE.

**ethyl alcohol** *See* ETHANOL.

**ethylamine** A colourless flammable volatile liquid, $C_2H_5NH_2$; r.d. 0.69; m.p. $-81°C$; b.p. 16.6°C. It is a primary amine made by reacting chloroethane with ammonia and used in making dyes.

**ethylbenzene** A colourless flammable liquid, $C_6H_5C_2H_5$; r.d. 0.867; m.p. $-95°C$; b.p.

136°C. It is made from ethene and ethybenzene by a *Friedel–Crafts reaction and is used in making phenylethene (for polystyrene).

**ethyl bromide** See BROMOETHANE.

**ethylene** See ETHENE.

**ethylene glycol** See ETHANE-1,2-DIOL.

**ethylene oxide** See EPOXYETHANE.

**ethyl ethanoate (ethyl acetate)** A colourless flammable liquid ester, $C_2H_5OOCCH_3$; r.d. 0.9; m.p. –83.6°C; b.p. 77.06°C. It is used as a solvent and in flavourings and perfumery.

**ethyl group** The organic group $CH_3CH_2$–.

**ethyl iodide** See IODOETHANE.

**ethyne (acetylene)** A colourless unstable gas, $C_2H_2$, with a characteristic sweet odour; r.d. 0.618; m.p. –80.8°C; b.p. –84.0°C. It is the simplest member of the *alkyne series of unsaturated hydrocarbons, and is prepared by the action of water on calcium dicarbide or by adding alcoholic potassium hydroxide to 1,2-dibromoethane. It can be manufactured by heating methane to 1500°C in the presence of a catalyst. It is used in oxyacetylene welding and in the manufacture of ethanal and ethanoic acid. Ethyne can be polymerized easily at high temperatures to give a range of products. The inorganic saltlike dicarbides contain the ion $C_2^{2-}$, although ethyne itself is a neutral compound (i.e. not a protonic acid).

**etiolation** The abnormal form of growth observed when plants grow in darkness or severely reduced light. Such plants characteristically have blanched leaves and shoots, excessively long shoots, and reduced leaves and root systems.

**Eubacteria** See BACTERIA.

**eucaryote** See EUKARYOTE.

**euchromatin** See CHROMATIN.

**Euclid** (c. 330–c. 260 BC) Greek mathematician, who worked at the academy in Alexandria. He is best known for his 13-volume work the *Elements*, which deals mainly with geometry and established the axiomatic method of reasoning.

**Euclidean geometry** The form of geometry set forth in the 13 volumes, called the *Elements*, by Euclid. It is based on a series of axioms and applies to plane figures. Compare NON-EUCLIDEAN GEOMETRY.

**eudicot** Any dicotyledonous flowering plant whose pollen has three apertures (i.e. **triaperturate pollen**), through one of which the pollen tube emerges during pollination. Eudicots contrast with the so-called 'primitive' dicots, such as the magnolia family, which have **uniaperturate pollen** (i.e. with a single aperture). The eudicots form a large clade, but their phylogenetic relationships to other dicots and to monocots remain unclear.

**eudiometer** An apparatus for measuring changes in volume of gases during chemical reactions. A simple example is a graduated glass tube sealed at one end and inverted in mercury. Wires passing into the tube allow the gas mixture to be sparked to initiate the reaction between gases in the tube.

**eugenics** The study of methods of improving the quality of human populations by the application of genetic principles. **Positive eugenics** would seek to do this by selective breeding programmes. **Negative eugenics** aims to eliminate harmful genes (e.g. those causing haemophilia and colour blindness) by counselling any prospective parents who are likely to be *carriers.

**Euglenida** A class of mostly unicellular protists (including *Euglena*) that move by means of undulipodia (flagella). Most euglenids are photosynthetic and inhabit fresh water, and in some earlier classification systems were regarded as green algae (phylum Chlorophyta). However, they lack a cell wall, being covered with a proteinaceous *pellicle, and some forms are colourless and thus ingest food, since they cannot photosynthesize. Euglenids are sometimes classified as a phylum, Euglenophyta, but they are now more usually included in a larger phylum, Discomitochondria, together with these other groups of organisms, on the basis of a common mitochondrial structure (characterized by disc-shaped cristae) and the absence of sexual reproduction.

**Eukarya** See DOMAIN; EUKARYOTE.

**eukaryote (eucaryote)** An organism consisting of cells in which the genetic material is contained within a distinct nucleus. All organisms except bacteria are eukaryotes, which are classified in the **Eukarya**, one of the three *domains of life. See CELL. Compare PROKARYOTE.

**Eulerian fluid dynamics** A formulation of fluid mechanics in which the velocity of the fluid at fixed postions is analysed, i.e. a formulation that describes changes at fixed positions in the fluid, in contrast to *Lagrangian fluid dynamics. It is named after the Swiss mathematician and physicist Leonhard Euler (1707–83).

**Eumetazoa** See METAZOA.

**euphotic zone (epipelagic zone; photic zone)** The topmost layer of a lake or sea in which there is sufficient light for net primary production, i.e. where the energy fixed by photosynthesis exceeds that lost by respiration. The depth varies, depending on such factors as turbidity, supply of nutrients in the water, tidal turbulence, and temperature. It typically ranges from 1 m to about 30 m in lakes and coastal waters, and rarely reaches depths of more than 200 m in the open ocean. Compare APHOTIC ZONE.

**europium** Symbol Eu. A soft silvery metallic element belonging to the *lanthanoids; a.n. 63; r.a.m. 151.96; r.d. 5.245 (20°C); m.p. 822°C; b.p. 1597°C. It occurs in small quantities in bastanite and monazite. Two stable isotopes occur naturally: europium–151 and europium–153, both of which are neutron absorbers. Experimental europium alloys have been tried for nuclear-reactor parts but until recently the metal has not been available in sufficient quantities. It is widely used in the form of the oxide in phosphors for television screens. It was discovered by Sir William Crookes in 1889.

( ((⊕)) SEE WEB LINKS )
• Information from the WebElements site

**Eustachian tube** The tube that connects the *middle ear to the back of the throat (pharynx) in vertebrates. It is normally closed, but during swallowing it opens to allow air into the middle ear, which equalizes the pressure on each side of the *tympanum (eardrum). It was named after the Italian anatomist Bartolomeo Eustachio (?1520–74).

**eustasy** Fluctuations in sea level on a global scale. One cause is the growth and decay of ice masses, such as glaciers and ice sheets (**glacio-eustasy**); since the last ice age sea levels have gradually risen. Tectonic movements can also change the volume of ocean basins (e.g. through sea-floor spreading). Global average sea level rose between 0.1 and 0.2 m during the 20th century.

**eutectic mixture** A solid solution consisting of two or more substances and having the lowest freezing point of any possible mixture of these components. The minimum freezing point for a set of components is called the **eutectic point**. Low-melting-point alloys are usually eutectic mixtures.

**euthanasia** The act of ending the life of a person or animal in order to prevent further suffering, e.g. from an incurable and painful disease. This can be achieved by administering a lethal drug or by withholding vital treatment. In human medicine euthanasia is fraught with ethical and legal problems, and is illegal in most countries. Where it is practised, strict safeguards are enforced to ensure that the patient's wishes are determined and adhered to. Euthanasia is widely performed in veterinary medicine.

**Eutheria (Placentalia)** An infraclass of mammals in which the embryos are retained in a uterus in the mother's body and nourished by a *placenta. The young are thus fully protected during their embryonic development and kept at a constant temperature. Placental mammals evolved during the Cretaceous period (about 125 million years ago). Modern placentals are a highly diverse group that occupy all types of habitat in all parts of the world. They include the orders *Artiodactyla, *Carnivora, *Cetacea, *Chiroptera, *Insectivora, *Perissodactyla, *Primates, *Proboscidea, and *Rodentia. Compare METATHERIA; PROTOTHERIA.

**eutrophic** Describing a body of water (e.g. a lake) with an abundant supply of nutrients and a high rate of formation of organic matter by photosynthesis. Pollution of a lake by *sewage or *fertilizers renders it eutrophic (a process called **eutrophication**). This stimulates excessive growth of algae (see ALGAL BLOOM); the death and subsequent decomposition of these increases the *biochemical oxygen demand and thus depletes the oxygen content of the lake, resulting in the death of the lake's fish and other animals. Compare DYSTROPHIC; OLIGOTROPHIC.

**Evans balance** See GOUY BALANCE.

**evaporation** The change of state of a liquid into a vapour at a temperature below the boiling point of the liquid. Evaporation occurs at the surface of a liquid, some of those molecules with the highest kinetic energies escaping into the gas phase. The result is a fall in the average kinetic energy of the molecules of the liquid and consequently a fall in its temperature.

**evaporative cooling** Cooling of a substance as a result of evaporation. *See also* LASER COOLING.

**even–even nucleus** An atomic nucleus containing an even number of protons and an even number of neutrons. Such nuclei are very common because they are very stable.

**even–odd nucleus** An atomic nucleus containing an even number of protons and an odd number of neutrons.

**event horizon** *See* BLACK HOLE.

**evergreen** (Describing) a plant that bears leaves throughout the year, each leaf being shed independently of the others after two or three years. The leaves of evergreens are often reduced or adapted in some way to prevent excessive water loss; examples are the needles of conifers and the leathery waxy leaves of holly. *Compare* DECIDUOUS.

**evocation** The ability of experimental stimuli (e.g. chemicals or tissue implants) to cause unspecialized embryonic tissue to develop into specialized tissue.

**evo-devo** Short for evolution of development – a field of study concerned with the interplay of developmental genetics and evolution. Advances in molecular biology and genetics have revealed how evolution of diversity in form and function can arise from mutations that alter the course or timing of development. One is alteration of the modular organization of body plans, such as repeated limbs or body segments. Duplication of such a module may lead to its further adaptation into a novel structure, such as pincers or mouthparts. Another area involves changes in the timing, location, or level of gene expression or alterations in the relative growth rates of different parts of the body. In this way relatively few mutations affecting development can have profound effects on the morphology of the mature organism.

**evolute** The locus of the centres of curvature of all the points on a given curve (called the **involute**).

**evolution** The gradual process by which the present diversity of plant and animal life arose from the earliest and most primitive organisms, which is believed to have been continuing for at least the past 3000 million

years. Until the middle of the 18th century it was generally believed that each species was divinely created and fixed in its form throughout its existence (*see* SPECIAL CREATION). Lamarck was the first biologist to publish a theory to explain how one species could have evolved into another (*see* LAMARCKISM), but it was not until the publication of Darwin's *On the Origin of Species* in 1859 that special creation was seriously challenged. Unlike Lamarck, Darwin proposed a feasible mechanism for evolution and backed it up with evidence from the fossil record and studies of comparative anatomy and embryology (*see* DARWINISM; NATURAL SELECTION). The modern version of Darwinism, which incorporates discoveries in genetics made since Darwin's time, remains the most acceptable theory of species evolution (*see also* PUNCTUATED EQUILIBRIUM). More controversial, however, and still to be firmly clarified, are the relationships and evolution of groups above the species level.

( 🌐 ) **SEE WEB LINKS**

• Understanding Evolution, created by the University of California Museum of Paleontology, covers the evidence for and mechanisms of evolution

**exa-** Symbol E. A prefix used in the metric system to denote $10^{18}$ times. For example, $10^{18}$ metres = 1 exametre (Em).

**exbi-** *See* BINARY PREFIXES.

**excess electron** An electron in a *semiconductor that is not required in the bonding system of the crystal lattice and has been donated by an impurity atom. It is available for conduction (**excess conduction**).

**exchange force 1.** A force resulting from the continued interchange of particles in a manner that bonds their hosts together. Examples are the covalent bond involving electrons, and the strong interaction (*see* FUNDAMENTAL INTERACTIONS) in which mesons are exchanged between nucleons or gluons are exchanged between quarks (*see* ELEMENTARY PARTICLES). **2.** *See* MAGNETISM.

**excimer** *See* EXCIPLEX.

**exciplex** A combination of two different atoms that exists only in an excited state. When an exciplex emits a photon of electromagnetic radiation, it immediately dissociates into the atoms, rather than reverting to the ground state. A similar transient excited association of two atoms of the same kind is an **excimer**. An example of an exciplex is the species XeCl* (the asterisk indicates an excited state), which can be formed by an electric discharge in xenon and chlorine. This is used in the **exciplex laser**, in which a population inversion is produced by an electrical discharge.

**excitation 1.** A process in which a nucleus, electron, atom, ion, or molecule acquires energy that raises it to a quantum state (**excited state**) higher than that of its *ground state. The difference between the energy in the ground state and that in the excited state is called the **excitation energy**. *See* COLLECTIVE EXCITATION; ENERGY LEVEL; QUASIPARTICLE. **2.** The process of applying current to the winding of an electromagnet, as in an electric motor. **3.** The process of applying a signal to the base of a transistor or the control electrode of a thermionic valve.

**excitatory postsynaptic potential (EPSP)** The electric potential that is generated in a postsynaptic neuron during the transmission of a nerve impulse (*see* SYNAPSE). It is caused by *depolarization of the postsynaptic membrane when a *neurotransmitter (such as acetylcholine), released from the presynaptic membrane, binds to the postsynaptic membrane. This will induce an *action potential in the receiving neuron if the EPSP is large enough. *Compare* INHIBITORY POSTSYNAPTIC POTENTIAL.

**exciton** An electron–hole pair in a crystal that is bound in a manner analogous to the electron and proton of a hydrogen atom. It behaves like an atomic excitation that passes from one atom to another and may be long-lived. Exciton behaviour in *semiconductors is important.

**exclusion principle** *See* PAULI EXCLUSION PRINCIPLE.

**excretion** The elimination by an organism of the waste products that arise as a result of metabolic activity. These products include water, carbon dioxide, and nitrogenous compounds. Excretion plays an important role in maintaining the constancy of an organism's *internal environment (*see* HOMEOSTASIS). In plants and simple animals waste products are excreted by simple diffusion from the body, but higher animals have specialized organs and organ systems devoted to this function. Examples of excretory organs in vertebrates are the lungs (for carbon dioxide and water), and the *kidneys (for nitrogenous compounds (urea) and water). In addition, mammals excrete small amounts of urea, salts, and water from the skin in sweat.

**exercise** Increased muscular activity, which results in an increase in metabolic rate, heart rate, and oxygen uptake. Exercise also causes an increase in *anaerobic respiration in order to compensate for the *oxygen debt, which results in a build-up of lactic acid in the tissues.

**exergonic reaction** A chemical reaction in which energy is released (*compare* ENDERGONIC REACTION). An *exothermic reaction is an exergonic reaction in which energy is released in the form of heat.

**exfoliation** A type of rock erosion that results from weathering. Layers or shells of rock are gradually removed from a massive outcrop, causing so-called onion-skin weathering. Commonly affecting basalt and granite, it is thought to be caused mainly by variations in temperature between day and night. Pressure changes in newly exposed rocks may also cause large-scale exfoliation.

**exhalation** *See* EXPIRATION.

**exitance** Symbol $M$. The radiant or luminous flux emitted per unit area of a surface. The **radiant exitance** ($M_e$) is measured in watts per square metre (W m$^{-2}$), while the **luminous exitance** ($M_v$) is measured in lumens per square metre (lm m$^{-2}$). Exitance was formerly called **emittance**.

**exo-** *See* ENDO-.

**exocarp** *See* PERICARP.

**exocrine gland** A gland that discharges its secretion into a body cavity (such as the gut) or onto the body surface. Examples are the *sebaceous and *sweat glands, the *mammary glands, and part of the pancreas. Exocrine glands are formed in the embryo from the invagination of epithelial cells. Their secretions pass initially into a cavity (an **alveolus** or **acinus**) and then out through a duct or duct network, along which the secretion may become modified by exchange with the blood across the duct epithelium.

**exocytosis** The passage of material from the inside of the cell to the cell surface within membrane-bound vesicles. The membranes of the vesicles fuse with the plasma membrane, releasing their contents to the exterior. Exocytosis is used both for the removal of waste material from the cell and for secretion; for example of mucus by *goblet cells. *Compare* ENDOCYTOSIS.

**exodermis** *See* HYPODERMIS.

**exoergic** Denoting a nuclear process that gives out energy. *Compare* ENDOERGIC.

**exogamy** The fusion of reproductive cells from distantly related or unrelated organisms, i.e. *outbreeding. *Compare* ENDOGAMY.

**exogenous** Describing substances, stimuli, etc., that originate outside an organism. For example, vitamins that cannot be synthesized by an animal are said to be supplied exogenously in the diet. *Compare* ENDOGENOUS.

**exon** A nucleotide sequence in a gene that codes for part or all of the gene product and is therefore expressed in mature messenger RNA, ribosomal RNA, or transfer RNA. In eukaryotes, exons are separated by noncoding sequences called *introns.

**exonuclease** An enzyme that catalyses the cleavage of nucleotides from the end of a nucleic acid molecule. *Compare* ENDONUCLEASE.

**exopeptidase** A protein-digesting enzyme that cleaves amino acids from the ends of a polypeptide chain. *Carboxypeptidase, which breaks down proteins in the small intestine, is an example of an exopeptidase. *Compare* ENDOPEPTIDASE.

**exoplanet** *See* EXTRASOLAR PLANET.

**exoskeleton** A rigid external covering for the body in certain animals, such as the hard chitinous cuticle of arthropods. An exoskeleton protects and supports the body and provides points of attachment for muscles. The cuticle of arthropods must be shed at intervals to allow growth to occur (*see* ECDYSIS). Other examples of exoskeletons are the shells of molluscs and the bony plates of tortoises and armadillos. *Compare* ENDOSKELETON.

**exosphere** *See* EARTH'S ATMOSPHERE.

**exothermic** Denoting a chemical reaction that releases heat into its surroundings. *Compare* ENDOTHERMIC.

**exotic atom** **1.** An atom in which an electron has been replaced by another negatively charged particle, such as a muon or *meson. In this case the negative particle eventually collides with the nucleus with the emission of X-ray photons. **2.** A system in which the nucleus of an atom has been replaced by a positively charged meson. Such exotic atoms have to be created artifically and are unstable.

**exotoxin** *See* TOXIN.

**expanded plastic** (**cellular plastic**) A plastic in the form of a rigid solid foam. Polystyrene, used as a packing material, is a common example.

**expansion** The writing of a function or quantity as a *series of terms. The series may be finite or infinite. *See* BINOMIAL THEOREM; TAYLOR SERIES.

**expansion of the universe** The hypothesis, based on the evidence of the *redshift, that the distance between the galaxies is continuously increasing. The original theory, which was proposed in 1929 by Edwin Hubble, assumes that the galaxies are flying apart as a consequence of the big bang with which the universe originated. Several variants have since been proposed. *See also* BIG-BANG THEORY; HUBBLE CONSTANT.

**expansivity** (**thermal expansion**) **1. Linear expansivity** is the fractional increase in length of a specimen of a solid, per unit rise in temperature. If a specimen increases in length from $l_1$ to $l_2$ when its temperature is raised $\theta°$, then the expansivity ($\alpha$) is given by:

$$l_2 = l_1(1 + \alpha\theta).$$

This relationship assumes that $\alpha$ is independent of temperature. This is not, in general, the case and a more accurate relationship is:

$$l_2 = l_1(1 + a\theta + b\theta^2 + c\theta^3\ldots),$$

where $a$, $b$, and $c$ are constants. **2. Superficial expansivity** is the fractional increase in area of a solid surface caused by unit rise in temperature, i.e.

$$A_2 = A_1(1 + \beta\theta),$$

where $\beta$ is the superficial expansivity. To a good approximation $\beta = 2\alpha$. **3. Volume expansivity** is the fractional increase in volume of a solid, liquid, or gas per unit rise in temperature, i.e.

$$V_2 = V_1(1 + \gamma\theta),$$

where $\gamma$ is the cubic expansivity and $\gamma = 3\alpha$. For liquids, the expansivity observed directly is called the **apparent expansivity** as the container will also have expanded with the rise in temperature. The **absolute expansivity** is the apparent expansivity plus the volume expansivity of the container. For the expansion of gases, *see* CHARLES' LAW.

**( ⊕ SEE WEB LINKS )**

• Values of the expansivity of selected liquids and solids at the NPL website

**experiment** A process or trial designed to test a scientific theory.

**expert system** A computer program that, using stored data, can reach conclusions in a particular field of knowledge; in effect, it makes decisions. Thus unlike a *database, it processes the data before presenting the result. The user can usually question the system's reasoning, often via a language processor. Such systems are used for troubleshooting in industry, for diagnosing medical disorders, for determining the structures of molecules, etc.

**expiration (exhalation)** The process by which gas is expelled from the lungs (*see* RESPIRATORY MOVEMENT). In mammals, the volume of the thoracic cavity is reduced by contraction of the internal *intercostal muscles and relaxation of the muscles of the diaphragm, assisted by upward pressure of the abdominal organs. As a result, pressure in the lungs exceeds atmospheric pressure and gas flows out of the lungs, allowing the pressures to equalize. *Compare* INSPIRATION.

**expiratory centre** *See* VENTILATION CENTRE.

**explantation** The removal of cells, tissues, or organs of animals and plants for observation of their growth and development in appropriate culture media. *See also* TISSUE CULTURE; ORGAN CULTURE.

**explosive** A compound or mixture that, when ignited or detonated, undergoes a rapid violent chemical reaction that produces large amounts of gas and heat, accompanied by light, sound and a high-pressure shock wave. **Low explosives** burn comparatively slowly when ignited, and are employed as propellants in firearms and guns; they are also used in blasting. Examples include *gunpowder and various smokeless propellants, such as *cordite. **High explosives** decompose very rapidly to produce an uncontrollable blast. Examples of this type include *dynamite, *nitroglycerine, and *trinitrotoluene (TNT); they are exploded using a detonator. Other high-power explosives include pentaerythritol tetranitrate (PETN) and ammonium nitride/fuel oil mix-

ture (ANFO). Cyclonite (RDX) is a military high explosive; mixed with oils and waxes, it forms a plastic explosive (such as Semtex). See also Chronology.

**exponent** A number or symbol that indicates the power to which another number or expression is raised. For example, $(x + y)^n$ indicates that the expression $(x + y)$ is raised to the $n$th power; $n$ is the exponent. Any number or expression in which the exponent is zero is equal to 1, i.e. $x^0 = 1$.

**exponential** A function that varies as the power of another quantity. If $y = a^x$, $y$ varies exponentially with $x$. The function $e^x$, also written as $\exp(x)$, is called the **exponential function** (*see* E). It is equal to the sum of the **exponential series**, i.e.

$$e^x = 1 + x + x^2/2! + x^3/3! + \dots + x^n/n! + \dots$$

**exponential growth** A form of *population growth in which the rate of growth is related to the number of individuals present. Increase is slow when numbers are low but rises sharply as numbers increase. If population number is plotted against time on a graph a characteristic J-shaped curve results (see graph). In animal and plant populations, such factors as overcrowding, lack of nutrients, and disease limit population increase beyond a certain point and the J-shaped exponential curve tails off giving an S-shaped (sigmoid) curve.

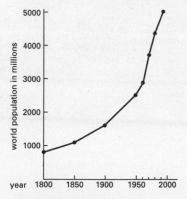

**Exponential growth.** Graph showing exponential growth of the human population.

**exposure meter** A photocell that operates a meter to indicate the correct exposure for a specified film in photography. It en-

## EXPLOSIVES

| | |
|---|---|
| 900–1000 | Gunpowder developed in China. |
| 1242 | English monk Roger Bacon (1220–92) describes the preparation of gunpowder. |
| c.1250 | German alchemist Berthold Schwarz claims to have reinvented gunpowder. |
| 1771 | French chemist Pierre Woulfe discovers picric acid (originally used as a yellow dye). |
| 1807 | Scottish cleric Alexander Forsyth (1767–1843) discovers mercury fulminate. |
| 1833 | French chemist Henri Braconnot (1781–1855) nitrates starch, making a highly flammable compound (crude nitrocellulose). |
| 1838 | French chemist Théophile Pelouze (1807–67) nitrates paper, making crude nitrocellulose. |
| 1845 | German chemist Christian Schönbein (1799–1868) nitrates cotton, making nitrocellulose. |
| 1846 | Italian chemist Ascania Sobrero (1812–88) discovers nitroglycerine. |
| 1863 | Swedish chemist J. Wilbrand discovers trinitrotoluene (TNT). Swedish chemist Alfred Nobel (1833–96) invents a detonating cap based on mercury fulminate. |
| 1867 | Alfred Nobel invents dynamite by mixing nitroglycerine and kieselguhr. |
| 1871 | German chemist Hermann Sprengel shows that picric acid can be used as an explosive. |
| 1875 | Alfred Nobel invents blasting gelatin (nitroglycerine mixed with nitrocellulose). |
| 1885 | French chemist Eugène Turpin discovers ammonium picrate (Mélinite). |
| 1888 | Alfred Nobel invents a propellant from nitroglycerine and nitrocellulose (Ballistite). |
| 1889 | British scientists Frederick Abel (1826–1902) and James Dewar invent a propellant (Cordite) similar to Ballistite. |
| 1891 | German chemist Bernhard Tollens (1841–1918) discovers pentaerythritol tetranitrate (PETN). |
| 1899 | Henning discovers cyclotrimethylenetrinitramine (RDX or cyclonite). |
| 1905 | US army officer B. W. Dunn (1860–1936) invents ammonium picrate explosive (Dunnite). |
| 1915 | British scientists invent amatol (TNT + ammonium nitrate). |
| 1955 | US scientists develop ammonium nitrate–fuel oil mixtures (ANFO) as industrial explosives. |

ables the correct shutter speed and aperture to be chosen for any photographic circumstances. Some cameras have a built-in exposure meter that automatically sets the aperture according to the amount of light available and the chosen shutter speed.

**extended ASCII** A set of characters with *ASCII values between 128 and 255. These characters may include special symbols, graphics characters, and accented characters. The assignment of extended ASCII characters is not standard. It depends on the particular computer system and may also depend on the font being used.

**extended phenotype** The concept, advanced by British biologist Richard Dawkins in his 1982 book of the same title, that the phenotype of an organism extends beyond its body to encompass the organism's behaviour and the consequences of that behaviour. Dawkins cites a beaver's lake as an example. This manifestation of the beaver's instinctive dam-building activities is, he argues, an evolutionary adaptation just as much as, say, the beaver's coat, and is likewise subject to natural selection. Other instances include birds' nests, termite mounds, and spiders' webs.

**extender** An inert substance added to a product (paint, rubber, washing powder, etc.) to dilute it (for economy) or to modify its physical properties.

**extensive variable** A quantity in a *macroscopic system that is proportional to the size of the system. Examples of extensive variables include the volume, mass, and total energy. If an extensive variable is divided by an arbitrary extensive variable, such as the volume, an *intensive variable results. A macroscopic system can be described by one extensive variable and a set of intensive variables.

**extensometer** Any device for measuring the extension of a specimen of a material under longitudinal stress. A common method is to make the specimen form part of a capacitor, the capacitance of which will change with a change in the specimen's dimensions.

**extensor** Any muscle that causes a limb to extend. *See* VOLUNTARY MUSCLE. *Compare* FLEXOR.

**external conversion** A process in which molecules in electronically excited states pass to a lower electronic state (which is frequently the ground state) by colliding with other molecules. In this process the electronic energy is eventually converted into heat. Since this process involves collisions, the rate at which it occurs depends on how frequently collisons occur. As a result, this process occurs much faster in liquids than in gases. It is sometimes called **collision quenching**.

**exteroceptor** Any *receptor that detects external stimuli. Examples of exteroceptors

are the thermoreceptors in the skin, which monitor the temperature of the external environment. *Compare* INTEROCEPTOR.

**extinction** **1.** (in biology) The irreversible condition of a species or other group of organisms of having no living representatives in the wild, which follows the death of the last surviving individual of that species or group. Extinction may occur on a local or global level; it can result from various human activities, including the destruction of habitats or the overexploitation of species that are hunted or harvested as a resource. Species at the top of a *food chain (e.g. large birds of prey) will be more prone to extinction since they exist in relatively small numbers and will be affected by a deleterious change at any of the levels in the food chain. *See also* MASS EXTINCTION. **2.** (in behaviour) The termination of a behaviour pattern that is no longer appropriate. For example, dogs can be conditioned to salivate when they hear a bell ring in the absence of a food stimulus (*see* CONDITIONING). However, if the bell continues to be rung in the absence of food the dogs will gradually stop salivating on hearing the bell. **3.** (in physics) A reduction in the intensity of radiation as a result of absorption or scattering as it passes through matter.

**extinction coefficient** A measure of the extent by which the intensity of a beam of light is reduced by passing through a distance $d$ of a solution having a molar concentration $c$ of the dissolved substance. If the intensity of the light is reduced from $I_1$ to $I_2$, the extinction coefficient is $[\log(I_1/I_2)]/cd$.

**extracellular** Located or occurring outside the cell. *Cuticularization is an example of an extracellular process.

**extracellular matrix (ECM)** The viscous watery fluid that surrounds cells in animal tissues. Secreted by the cells themselves, it is the medium through which they receive materials (e.g. nutrients, hormones) from elsewhere in the body and via which they communicate with other cells. The ECM is the environment in which cells migrate during tissue development and it contains constituents that bind cells together to maintain tissue integrity. It consists of glycoproteins, *collagens, and other structural components. The ECM is especially prominent in connective tissues, such as bone, cartilage, and adipose tissue, in which it is sometimes called **ground substance**.

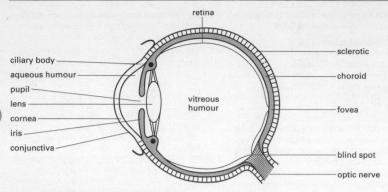

**Eye.** Structure of the vertebrate eye.

**extraction** **1.** The process of obtaining a metal from its ore. **2.** The separation of a component from a mixture by selective solubility. *See* PARTITION.

**extractive distillation** A distillation technique in which a solvent is added to the mixture in order to separate two closely boiling components. The added solvent is usually nonvolatile and is selected for its ability to have different effects on the volatilities of the components.

**extraembryonic membranes (embryonic membranes)** The tissues produced by an animal *embryo for protection and nutrition but otherwise taking no part in its development. The four membranes, which are called **fetal membranes** in man, are the *chorion, *amnion, *allantois, and *yolk sac.

**extraordinary ray** *See* DOUBLE REFRACTION.

**extrapolation** An *approximation technique for finding the value of a function or measurement beyond the values already known. If the values $f(x_0), f(x_1),...,f(x_n)$ of a function of a variable $x$ are known in the interval $[x_0,x_n]$, the value of $f(x)$ for a value of $x$ outside the interval $[x_0,x_n]$ can be found by extrapolation. The techniques used in extrapolation are usually not as good as those used in *interpolation.

**extrasolar planet (exoplanet)** A planet in a solar system beyond our own. As of mid-2009, over 350 extrasolar planets had been discovered since the first confirmed observations were made in the early 1990s. Most ex-trasolar planets have been detected through their effects on the stars that they orbit. The majority of those found so far have been giants, typically much larger than Jupiter. Many occupy orbits very close to their parent star and have been dubbed 'hot Jupiters'.

**extremely high frequency (EHF)** A radio frequency between 30 000 megahertz and 300 gigahertz.

**extremophile** A bacterium that thrives under extreme conditions, e.g. at very high or very low temperatures, or in very salty or acidic environments. For example, certain archaebacteria (*see* ARCHAEA), termed **hyperthermophiles**, live in hot springs at temperatures near or even above 100°C. The enzymes of such organisms exhibit great stability and have been extracted for use in laboratory and commercial processes.

**extrinsic semiconductor** *See* SEMICONDUCTOR.

**eye** The organ of sight. The most primitive eyes are the *eyespots of some unicellular organisms. More advanced eyes are the *ocelli and *compound eyes of arthropods (e.g. insects). The cephalopod molluscs (e.g. the octopus and squid) and vertebrates possess the most highly developed eyes (see illustration). These normally occur in pairs, are nearly spherical, and filled with fluid. Light is refracted by the *cornea through the pupil in the *iris and onto the *lens, which focuses images onto the retina. These images are received by light-sensitive cells in the retina (*see* CONE; ROD), which transmit impulses to the brain via the optic nerve.

**SEE WEB LINKS**

• Gross anatomy of the eye, from emedicine

**eyepiece (ocular)** The lens or system of lenses in an optical instrument that is nearest to the eye. It usually produces a magnified image of the previous image formed by the instrument.

**eyespot (stigma) 1.** A structure found in some free-swimming unicellular algae and in plant reproductive cells that contains orange or red pigments (carotenoids) and is sensitive to light. It enables the cell to move in relation to a light source (*see* PHOTOTAXIS). **2.** A spot of pigment found in some lower animals, e.g. jellyfish.

**eye tooth** A *canine tooth in the upper jaw.

**E–Z convention** A convention for the description of a molecule showing cis–trans isomerism (*see* ISOMERISM). In a molecule ABC=CDE, where A, B, D, and E are substituent groups, the sequence rule (*see* CIP SYSTEM) is applied to the pair A and B to find which has priority and similarly to the pair C and D. If the two groups of highest priority are on the same side of the bond then the isomer is designated $Z$ (from German *zusammen*, together). If they are on opposite sides the isomer is designated $E$ (German *entgegen*, opposite). The letters are used in the names of compounds; for example ($E$)-butenedioic acid (fumaric acid) and ($Z$)-butenedioic acid (maleic acid). In compounds containing two (or more) double bonds numbers are used to designate the bonds (e.g. ($2E$, $4Z$)-2,4-hexadienoic acid). The system is less ambiguous than the cis/trans system of describing isomers.

e

**F₁ (first filial generation)** The first generation of offspring resulting from an arranged cross between *homozygous parents in breeding experiments. *See* MONOHYBRID CROSS.

**F₂ (second filial generation)** The second generation of offspring in breeding experiments, obtained by crosses between individuals of the *F₁ generation. *See* MONOHYBRID CROSS.

**Fabry–Pérot interferometer** A type of *interferometer in which monochromatic light is passed through a pair of parallel half-silvered glass plates producing circular interference fringes. One of the glass plates is adjustable, enabling the separation of the plates to be varied. The wavelength of the light can be determined by observing the fringes while adjusting the separation. This type of instrument is used in spectroscopy.

**face-centred cubic (f.c.c.)** *See* CUBIC CRYSTAL.

**facilitated diffusion** The transport of molecules across the plasma membrane of a living cell by a process that involves a specific transmembrane carrier (*see* TRANSPORT PROTEIN) located within the plasma membrane but does not require expenditure of energy by the cell. The carrier combines with a molecule at one face of the membrane, then changes shape so the molecule is moved through the membrane and released at the opposite face. It enables the diffusion through the membrane of molecules that otherwise could not pass through. *Compare* ACTIVE TRANSPORT.

**fac-isomer** *See* ISOMERISM.

**factorial** The product of a given number and all the whole numbers below it. It is usually writen $n!$, e.g. factorial $4 = 4! = 4 \times 3 \times 2 \times 1 = 24$. Factorial 0 is defined as 1.

**Factor VIII (antihaemophilic factor)** One of the blood *clotting factors. Factor VIII is a soluble protein that stimulates the activation of Factor X by Factor IXa, which in turn converts *prothrombin to thrombin, thus causing the fibrin matrix of a blood clot to form.

*Haemophilia is due to a deficiency or defect of Factor VIII and is treated by administration of blood plasma or plasma concentrate containing the factor. Factor VIII can now be obtained from genetically engineered cell cultures, which avoids the risk of contamination with viruses, notably HIV (the AIDS virus).

**FAD (flavin adenine dinucleotide)** A *coenzyme important in various biochemical reactions. It comprises a phosphorylated vitamin B₂ (riboflavin) molecule linked to the nucleotide adenine monophosphate (AMP). FAD is usually tightly bound to the enzyme forming a **flavoprotein**. It functions as a hydrogen acceptor in dehydrogenation reactions, being reduced to FADH₂. This in turn is oxidized to FAD by the *electron transport chain, thereby generating ATP (two molecules of ATP per molecule of FADH₂).

**faeces** Waste material that is eliminated from the alimentary canal through the *anus. Faeces consist of the indigestible residue of food that remains after the processes of digestion and absorption of nutrients and water have taken place, together with bacteria and dead cells shed from the gut lining.

**Fahrenheit, Gabriel Daniel** (1686–1736) German physicist, who became an instrument maker in Amsterdam. In 1714 he developed the mercury-in-glass thermometer, and devised a temperature scale to go with it (*see* FAHRENHEIT SCALE).

**Fahrenheit scale** A temperature scale in which (by modern definition) the temperature of boiling water is taken as 212 degrees and the temperature of melting ice as 32 degrees. It was invented in 1714 by Gabriel Fahrenheit, who set the zero at the lowest temperature he knew how to obtain in the laboratory (by mixing ice and common salt) and took his own body temperature as 96°F. The scale is no longer in scientific use. To convert to the *Celsius scale the formula is $C = 5(F - 32)/9$.

**Fajans' method**  *See* ADSORPTION INDICATOR.

**Fajans' rules**  Rules indicating the extent to which an ionic bond has covalent character caused by polarization of the ions. Covalent character is more likely if:
(1) the charge of the ions is high;
(2) the positive ion is small or the negative ion is large;
(3) the positive ion has an outer electron configuration that is not a noble-gas configuration.
   The rules were introduced by the Polish–American chemist Kasimir Fajans (1887–1975).

**fallopian tube (oviduct)**  The tube that carries egg cells from the *ovary to the womb in mammals. The eggs are carried by the action of muscles and cilia. It was named after Gabriel Fallopius.

**Fallopius, Gabriel**  (1523–62) Italian anatomist, who was professor of anatomy at Pisa (from 1548) and Padua (from 1551). Best known for his discoveries about the human skeletal and reproductive systems, he identified the oviducts, which are named after him (*fallopian tubes).

**fall-out**  **1. (radioactive fall-out)**  Radioactive particles deposited from the atmosphere either from a nuclear explosion or from a nuclear accident. **Local fall-out**, within 250 km of an explosion, falls within a few hours of the explosion. **Tropospheric fall-out** consists of fine particles deposited all round the earth in the approximate latitude of the explosion within about one week. **Stratospheric fall-out** may fall anywhere on earth over a period of years. The most dangerous radioactive isotopes in fall-out are the fission fragments iodine–131 and strontium–90. Both can be taken up by grazing animals and passed on to human populations in milk, milk products, and meat. Iodine–131 accumulates in the thyroid gland and strontium–90 accumulates in bones.
**2. (chemical fall-out)**  Hazardous chemicals discharged into and subsequently released from the atmosphere, especially by factory chimneys.

**false fruit**  *See* PSEUDOCARP.

**family**  **1.** (in taxonomy) A category used in the *classification of organisms that consists of one or several similar or closely related genera. Similar families are grouped into an order. Family names end in -*aceae* or -*ae* in botany (e.g. Cactaceae) and -*idae* in zoology (e.g. Equidae). The names are usually derived from a type genus (*Cactus* and *Equus* in the examples above) that is characteristic of the whole family (*see* TYPE SPECIMEN). In botany, families are sometimes called **natural orders**. **2.** (in molecular biology) A group of proteins with shared similarities in their amino-acid sequence, and often similarities in function, due to evolutionary divergence from a putative common ancestral protein. For example, the various types and subtypes of *adrenoceptors can be considered as a protein family.

**farad**  Symbol F. The SI unit of capacitance, being the capacitance of a capacitor that, if charged with one coulomb, has a potential difference of one volt between its plates. 1 F = 1 C V$^{-1}$. The farad is too large for most applications; the practical unit is the microfarad ($10^{-6}$ F). The unit is named after Michael Faraday.

**Faraday, Michael**  (1791–1867) British chemist and physicist, who received little formal education. He started to experiment on electricity and in 1812 attended lectures by Sir Humphry *Davy at the Royal Institution; a year later he became Davy's assistant. He remained at the Institution until 1861. Faraday's chemical discoveries include the liquefaction of chlorine (1823) and benzene (1825) as well as the laws of electrolysis (*see* FARADAY'S LAWS). He is probably best remembered for his work in physics: in 1821 he demonstrated electromagnetic rotation (the principle of the *electric motor) and discovered *electromagnetic induction (the principle of the dynamo). In 1845 he discovered the *Faraday effect.

**Faraday cage**  An earthed screen made of metal wire that surrounds an electric device in order to shield it from external electrical fields.

**Faraday constant**  Symbol *F*. The electric charge carried by one mole of electrons or singly ionized ions, i.e. the product of the *Avogadro constant and the charge on an electron (disregarding sign). It has the value 9.648 5309(29) × $10^4$ coulombs per mole. This number of coulombs is sometimes treated as a unit of electric charge called the **faraday**.

**Faraday disc**  *See* HOMOPOLAR GENERATOR.

**Faraday effect**  The rotation of the plane of polarization of electromagnetic radiation on passing through an isotropic medium exposed to a magnetic field. The angle of rotation is proportional to $Bl$, where $l$ is the length of the path of the radiation in the medium and $B$ is the magnetic flux density.

**Faraday's laws** Two laws describing electrolysis:

(1) The amount of chemical change during electrolysis is proportional to the charge passed.

(2) The charge required to deposit or liberate a mass $m$ is given by $Q = Fmz/M$, where $F$ is the Faraday constant, $z$ the charge of the ion, and $M$ the relative ionic mass.

These are the modern forms of the laws. Originally, they were stated by Faraday (1934) in a different form:

(1) The amount of chemical change produced is proportional to the quantity of electricity passed.

(2) The amount of chemical change produced in different substances by a fixed quantity of electricity is proportional to the electrochemical equivalent of the substance.

**Faraday's laws of electromagnetic induction** (1) An e.m.f. is induced in a conductor when the magnetic field surrounding it changes. (2) The magnitude of the e.m.f. is proportional to the rate of change of the field. (3) The sense of the induced e.m.f. depends on the direction of the rate of change of the field.

**farming** *See* AGRICULTURE.

**fascia** A sheet of fibrous connective tissue occurring beneath the skin and also enveloping glands, vessels, nerves, and forming muscle and tendon sheaths.

**fascicle** 1. A small bundle of nerve or muscle fibres. 2. *See* VASCULAR BUNDLE.

**fast green** A green dye used in optical microscopy that stains cellulose, cytoplasm, collagen, and mucus green. It is frequently used to stain plant tissues, with *safranin as a counterstain. Unlike **light green**, a similar dye, it does not fade easily.

**fast-ion conductor (superionic conductor)** A solid conductor of electricity in which the moving particles are ions, which transport electric charge by moving rapidly between vacancies in the lattice of a crystal. Substances of this type are sometimes called **solid electrolytes**. They are used in batteries and fuel cells.

**fast neutron** A neutron resulting from

nuclear fission that has an energy in excess of 0.1 MeV ($1.6 \times 10^{-14}$ J), having lost little of its energy by collision. In some contexts **fast fission** is defined as fission brought about by fast neutrons, i.e. neutrons having energies in excess of 1.5 MeV ($2.4 \times 10^{-13}$ J), the fission threshold of uranium–238. *See also* NUCLEAR REACTOR; SLOW NEUTRON.

**fast reactor** *See* NUCLEAR REACTOR.

**fat** A mixture of lipids, chiefly *triglycerides, that is solid at normal body temperatures. Fats occur widely in plants and animals as a means of storing food energy, having twice the calorific value of carbohydrates. In mammals, fat is deposited in a layer beneath the skin (subcutaneous fat) and deep within the body as a specialized *adipose tissue (*see also* BROWN FAT). The insulating properties of fat are also important, especially in animals lacking fur and those inhabiting cold climates (e.g. seals and whales).

Fats derived from plants and fish generally have a greater proportion of unsaturated *fatty acids than those from mammals. Their melting points thus tend to be lower, causing a softer consistency at room temperatures. Highly unsaturated fats are liquid at room temperatures and are therefore more properly called *oils.

**fat body** 1. An abdominal organ in amphibians attached to the anterior of each kidney. It contains a reserve of fat that nourishes the gonads during the winter hibernation in readiness for the spring breeding season. 2. A mass of fatty tissue spreading throughout the body cavity of insects in which fats, proteins, and glycogen are stored as a reserve for hibernation or pupation.

**fat cell** Any of the cells of *adipose tissue, in which fats (triglycerides) are stored. Fat cells contain enzymes (lipases) that can break down fat into glycerol and fatty acids, which can be transported in the blood to the liver, where they are used in *fatty-acid oxidation.

**fathom** A unit used to describe a depth of water. It is equal to 6 feet (1.83 m).

**fatigue** 1. A decline in the level of response of tissues (such as muscle), cells, etc., to nervous stimulation, which occurs after prolonged and continued stimulation of these structures. 2. *See* METAL FATIGUE.

**fatty acid** An organic compound consisting of a hydrocarbon chain and a terminal carboxyl group (*see* CARBOXYLIC ACIDS). Chain length ranges from one hydrogen atom (methanoic, or formic, acid, HCOOH) to nearly 30 carbon atoms. Ethanoic (acetic), propanoic (propionic), and butanoic (butyric) acids are important in metabolism. Long-chain fatty acids (more than 8–10 carbon atoms) most commonly occur as constituents of certain lipids, notably glycerides, phospholipids, sterols, and waxes, in which they are esterified with alcohols. These long-chain fatty acids generally have an even number of carbon atoms; unbranched chains predominate over branched chains. They may be **saturated** (e.g. *palmitic (hexadecanoic) acid and *stearic (octadecanoic) acid) or **unsaturated**, with one double bond (e.g. *oleic (cis-octodec-9-enoic) acid) or two or more double bonds, in which case they are called **polyunsaturated fatty acids** (e.g. *linoleic acid and *linolenic acid). *See also* ESSENTIAL FATTY ACIDS.

The physical properties of fatty acids are determined by chain length, degree of unsaturation, and chain branching. Short-chain acids are pungent liquids, soluble in water. As chain length increases, melting points are raised and water-solubility decreases. Unsaturation and chain branching tend to lower melting points.

**fatty-acid oxidation (β-oxidation)** The metabolic pathway in which fats are metabolized to release energy. Fatty-acid oxidation occurs continually but does not become a major source of energy until the animal's carbohydrate resources are exhausted, for example during starvation. Fatty-acid oxidation occurs chiefly in mitochondria in animal cells, and in *peroxisomes in plant cells. A series of reactions cleave off two carbon atoms at a time from the hydrocarbon chain of the fatty acid. These two-carbon fragments are combined with *coenzyme A to form *acetyl coenzyme A (acetyl CoA), which then enters the *Krebs cycle. The formation of acetyl CoA occurs repeatedly until all the hydrocarbon chain has been used up.

**fault** A fracture in the earth's crust along which there has been displacement of rock on one side relative to the other. The displacement ranges from a few centimetres to a few kilometres and may occur in a horizontal, oblique, or vertical direction. The extent of vertical displacement of the strata is the **throw**; the horizontal displacement is the **heave**. The side of the fault on which the strata have moved relatively downward is known as the downthrow side; the other is the upthrow side.

**fauna** All the animal life normally present in a given habitat at a given time. *See also* MACROFAUNA; MICROFAUNA. *Compare* FLORA.

**f-block elements** The block of elements in the *periodic table consisting of the lanthanoid series (from cerium to lutetium) and the actinoid series (from thorium to lawrencium). They are characterized by having two $s$-electrons in their outer shell ($n$) and $f$-electrons in their inner ($n$–1) shell.

**f.c.c.** Face-centred cubic. *See* CUBIC CRYSTAL.

**F-centre** *See* COLOUR CENTRE.

**feathers** The body covering of birds, formed as outgrowths of the epidermis and composed of the protein *keratin. Feathers provide heat insulation, they give the body its streamlined shape, and those of the wings and tail are important in flight. Basically a feather consists of a **quill**, which is embedded in the skin attached to a feather follicle and is continuous with the shaft (**rachis**) of the feather, which carries the *barbs. This basic structure is modified depending on the type of feather (*see* CONTOUR FEATHERS; DOWN FEATHERS; FILOPLUMES).

**fecundity** The number of offspring produced by an organism (in higher animals, generally the female of the species) in a given time. Normally all organisms, assum-

(a) original block   (b) normal fault   (c) reverse fault   (d) strike-slip fault

**Faults.**

ing they reach reproductive age, are sufficiently fecund to replace themselves several times over. Darwin noted this, together with the fact that population numbers nevertheless tended to remain fairly constant: these observations led him to formulate his theory of evolution by *natural selection. *Compare* FERTILITY.

**feedback** The use of part of the output of a system to control its performance. In **positive feedback**, the output is used to enhance the input; an example is an electronic oscillator, or the howl produced by a loudspeaker that is placed too close to a microphone in the same circuit. A small random noise picked up by the microphone is amplified and reproduced by the loudspeaker. The microphone now picks it up again; it is further amplified, and fed from the speaker to microphone once again. This continues until the system is overloaded. In **negative feedback**, the output is used to reduce the input. In electronic amplifiers, stability is achieved, and distortion reduced, by using a system in which the input is decremented in proportion as the output increases. A similar negative feedback is used in *governors that reduce the fuel supply to an engine as its speed increases.

Many biological processes rely on negative feedback. As the population of a species expands, so its food supply per individual is diminished; the result is that the population then begins to fall. Many biochemical processes are controlled by feedback *inhibition. Feedback mechanisms play an important role in maintaining a state of equilibrium within an organism (*see* HOMEOSTASIS).

**feeding** *See* INGESTION.

**Fehling's test** A chemical test to detect *reducing sugars and aldehydes in solution, devised by the German chemist Hermann von Fehling (1812–85). **Fehling's solution** consists of Fehling's A (copper(II) sulphate solution) and Fehling's B (alkaline 2,3-dihydroxybutanedioate (sodium tartrate) solution), equal amounts of which are added to the test solution. After boiling, a positive result is indicated by the formation of a brick-red precipitate of copper(I) oxide. Methanal, being a strong reducing agent, also produces copper metal; ketones do not react. The test is now little used, having been replaced by *Benedict's test.

**feldspars** A group of silicate minerals, the most abundant minerals in the earth's crust. They have a structure in which $(Si,Al)O_4$ tetrahedra are linked together with potassium, sodium, and calcium and very occasionally barium ions occupying the large spaces in the framework. The chemical composition of feldspars may be expressed as combinations of the four components:
**anorthite** (An), $CaAl_2Si_2O_8$;
**albite** (Ab), $NaAlSi_3O_8$;
**orthoclase** (Or), $KAlSi_3O_8$;
**celsian** (Ce), $BaAl_2Si_2O_8$.
The feldspars are subdivided into two groups: the **alkali feldspars** (including microcline, orthoclase, and sanidine), in which potassium is dominant with a smaller proportion of sodium and negligible calcium; and the **plagioclase feldspars**, which vary in composition in a series that ranges from pure sodium feldspar (albite) through to pure calcium feldspar (anorthite) with negligible potassium. Feldspars form colourless, white, or pink crystals with a hardness of 6 on the Mohs' scale.

**feldspathoids** A group of alkali aluminosilicate minerals that are similar in chemical composition to the *feldspars but are relatively deficient in silica and richer in alkalis. The structure consists of a framework of $(Si,Al)O_4$ tetrahedra with aluminium and silicon atoms at their centres. The feldspathoids occur chiefly with feldspars but do not coexist with free quartz $(SiO_2)$ as they react with silica to yield feldspars. The chief varieties of feldspathoids are:
nepheline, $KNa_3(AlSiO_4)_4$;
leucite, $KAlSi_2O_6$;
analcime, $NaAlSi_2O_6.H_2O$;
cancrinite, $Na_8(AlSiO_4)_6(HCO_3)_2$;
and the sodalite subgroup comprising
sodalite, $3(NaAlSiO_4).NaCl$;
nosean, $3(NaAlSiO_4).Na_2SO_4$;
haüyne, $3(NaAlSiO_4).CaSO_4$;
lazurite $(Na,Ca)_8(Al,Si)_{12}O_{24}(S,SO_4)$ (*see* LAPIS LAZULI).

**felsic** (from *fel*dspar plus *sili*ca) Denoting any light-coloured silicate material or a rock in which such minerals predominate. Felsic minerals include feldspar, feldspathoids, and quartz. *See also* MAFIC.

**FEM** *See* FIELD-EMISSION MICROSCOPE.

**female 1.** Denoting the gamete (sex cell) that, during *sexual reproduction, fuses with a *male gamete in the process of fertilization. Female gametes are generally larger than the male gametes and are usually immotile (*see* OOSPHERE; OVUM). **2.** (Denoting) an individual organism whose reproductive organs produce only female gametes. *Compare* HERMAPHRODITE.

**femoral** Of or relating to the thigh or the femur (thigh bone). For example, the **femoral artery** runs down the front of the thigh.

**femto-** Symbol f. A prefix used in the metric system to denote $10^{-15}$. For example, $10^{-15}$ second = 1 femtosecond (fs).

**femtochemistry** The study of chemical reactions using lasers that produce very short-duration pulses of light (typically 10–100 femtoseconds, $10^{-14}$–$10^{-13}$ s). The pulses are produced using dye lasers or using solid-state lasers made from titanium-doped sapphire. Femtosecond lasers can be used to study the breaking and formation of individual chemical bonds in compounds.

**femur** **1.** The thigh bone of terrestrial vertebrates. It articulates at one end with the pelvic girdle at the hip joint and at the other (via two *condyles) with the *tibia. **2.** The third segment of an insect's leg, attached to the *trochanter. *See also* COXA.

**fenestra** Either of the two delicate membranes between the *middle ear and the *inner ear. The upper membrane is the **fenestra ovalis** (*see* OVAL WINDOW); the lower membrane is the **fenestra rotunda** (*see* ROUND WINDOW).

**Fenton's reagent** A mixture of hydrogen peroxide and iron(II) sulphate used to produce free radicals by reactions of the type

$$Fe^{2+} + H_2O_2 \rightarrow Fe^{3+} + \cdot OH + OH^-$$

$$Fe^{3+} + H_2O_2 \rightarrow Fe^{2+} + \cdot OOH + H^+$$

It is used in water treatment and as a reagent in organic synthesis to introduce an –OH group into an aromatic drug.

**Fermat's principle** The path taken by a ray of light between any two points in a system is always the path that takes the least time. This principle leads to the law of the rectilinear propagation of light and the laws of reflection and refraction. It was discovered by the French mathematician, Pierre de Fermat (1601–65).

**fermentation** A form of *anaerobic respiration occurring in certain microorganisms, e.g. yeasts. **Alcoholic fermentation** comprises a series of biochemical reactions by which pyruvate (the end product of *glycoly-sis) is converted to ethanol and carbon dioxide. Fermentation is the basis of the baking, wine, and beer industries.

**fermi** A unit of length formerly used in nuclear physics. It is equal to $10^{-15}$ metre. In SI units this is equal to 1 femtometre (fm). It was named after Enrico Fermi.

**Fermi, Enrico** (1901–54) Italian-born US physicist. He became a professor at Rome University, where in 1934 he discovered how to produce slow (thermal) neutrons. He used these to create new radioisotopes, for which he was awarded the 1938 Nobel Prize. In 1938 he and his Jewish wife emigrated to the USA. In 1942 he led the team that built the first atomic pile (nuclear reactor) in Chicago. Fermi was an influential theoretical physicist who, independently of Paul Dirac, discovered Fermi–Dirac statistics. He also proposed the first proper theory of weak interactions.

**Fermi–Dirac statistics** *See* QUANTUM STATISTICS.

**Fermi level** The energy in a solid at which the average number of particles per quantum state is ½; i.e. one half of the quantum states are occupied. The Fermi level in conductors lies in the conduction band (*see* ENERGY BANDS), in insulators it lies in the valence band, and in semiconductors it falls in the gap between the conduction band and the valence band. At absolute zero all the electrons would occupy energy levels up to the Fermi level and no higher levels would be occupied. It is named after Enrico Fermi.

**fermion** An *elementary particle (or bound state of an elementary particle, e.g. an atomic nucleus or an atom) with half-integral spin; i.e. a particle that conforms to Fermi–Dirac statistics (*see* QUANTUM STATISTICS). *Compare* BOSON.

**fermium** Symbol Fm. A radioactive metallic transuranic element belonging to the *actinoids; a.n. 100; mass number of the most stable isotope 257 (half-life 10 days). Ten isotopes are known. The element was first identified by Albert Ghiorso and associates in debris from the first hydrogen-bomb explosion in 1952.

(((•))) **SEE WEB LINKS**
• Information from the WebElements site

**ferns** *See* FILICINOPHYTA.

**ferrate** An iron-containing anion, $FeO_4^{2-}$.

It exists only in strong alkaline solutions, in which it forms purple solutions.

**Ferrel's law** A law stating that a body moving across the surface of the earth will tend to be deflected to the right in the northern hemisphere and to the left in the southern hemisphere as a result of the earth's rotation. It was proposed in 1858 by the US meteorologist William Ferrel (1817–91).

**ferric alum** One of the *alums, $K_2SO_4.Fe_2(SO_4)_3.24H_2O$, in which the aluminium ion $Al^{3+}$ is replaced by the iron(III) (ferric) ion $Fe^{3+}$.

**ferric chloride test** A *presumptive test for morphine. The reagent is a 10% solution of ferric chloride (iron(III) chloride, $Fe Cl_3$) in water. With morphine a blue-green coloration occurs, changing to green.

**ferric compounds** Compounds of iron in its +3 oxidation state; e.g. ferric chloride is iron(III) chloride, $FeCl_3$.

**ferricyanide** A compound containing the complex ion $[Fe(CN)_6]^{3-}$, i.e. the hexacyanoferrate(III) ion.

**ferrimagnetism** See MAGNETISM.

**ferrite** 1. A member of a class of mixed oxides $MO.Fe_2O_3$, where M is a metal such as cobalt, manganese, nickel, or zinc. The ferrites are ceramic materials that show either ferrimagnetism or ferromagnetism, but are not electrical conductors. For this reason they are used in high-frequency circuits as magnetic cores. 2. See STEEL.

**ferroalloys** Alloys of iron with other elements made by smelting mixtures of iron ore and the metal ore; e.g. ferrochromium, ferrovanadium, ferromanganese, ferrosilicon, etc. They are used in making alloy *steels.

**ferrocene** An orange-red crystalline solid, $Fe(C_5H_5)_2$; m.p. 173°C. It can be made by adding the ionic compound $Na^+C_5H_5^-$ (cyclopentadienyl sodium, made from sodium and cyclopentadiene) to iron(III) chloride. In ferrocene, the two rings are parallel, with the iron ion sandwiched between them (hence the name **sandwich compound**: see formula). The bonding is between pi orbitals on the rings and *d*-orbitals on the $Fe^{2+}$ ion. The compound can undergo electrophilic substitution on the $C_5H_5$ rings (they have some aromatic character). It can also be oxidized to the blue ion $(C_5H_5)_2Fe^+$. Ferrocene is the

first of a class of similar complexes called **metallocenes**. Its systematic name is **di-π-cyclopentadienyl iron(II)**.

**Ferrocene.**

**ferrocyanide** A compound containing the complex ion $[Fe(CN)_6]^{4-}$, i.e. the hexacyanoferrate(II) ion.

**ferroelectric materials** Ceramic dielectrics, such as Rochelle salt and barium titanate, that have a domain structure making them analogous to ferromagnetic materials (see MAGNETISM). They exhibit hysteresis and usually the *piezoelectric effect.

**ferromagnetism** See MAGNETISM.

**ferrosoferric oxide** See TRIIRON TETROXIDE.

**ferrous compounds** Compounds of iron in its +2 oxidation state; e.g. ferrous chloride is iron(II) chloride, $FeCl_2$.

**fertile material** A nuclide that can absorb a neutron to form a *fissile material. Uranium–238, for example, absorbs a neutron to form uranium–239, which decays to plutonium–239. This is the type of conversion that occurs in a breeder reactor (see NUCLEAR REACTOR).

**fertility** 1. The potential capability of an organism to reproduce itself. In sexually reproducing plants and animals it is the number of fertilized eggs produced in a given time. For practical purposes this usually cannot be measured, and the only reliable indicators are the numbers of mature seeds produced, eggs laid, or live offspring delivered. However, these measures are strictly referred to as *fecundity, since they exclude fertilized embryos that have failed to develop. 2. The relative ability of a soil to support plant growth. It consists of both physical factors, e.g. particle size and moisture content, and chemical factors, e.g. concentration and availability of nutrients.

**fertilization (syngamy)** The union of

male and female gametes (reproductive cells) during the process of sexual reproduction to form a **zygote**. It involves the fusion of the gametic nuclei (**karyogamy**) and cytoplasm (**plasmogamy**). As each gamete contains only half the correct number of chromosomes, fertilization and zygote formation results in a cell with the full complement of chromosomes, half of which are derived from each of the parents. In animals the process involves fusion of the nuclei of a spermatozoon and an ovum. In most aquatic animals (e.g. fish) this takes place in the surrounding water, into which the gametes are shed. Among most terrestrial animals (e.g. insects, many mammals) fertilization occurs in the body of the female, into which the sperms are introduced. In flowering plants, after *pollination, the male gamete (pollen) produces a *pollen tube, which grows down into the female reproductive organ (carpel) to enable a pollen nucleus to fuse with the egg nucleus.

In **self-fertilization** the male and female gametes are derived from the same individual. Among plants, self-fertilization (also called **autogamy**) is common in many cultivated species, e.g. wheat and oats. However, self-fertilization is a form of *inbreeding and does not allow for the mixing of genetic material; if it occurs over a number of generations it will result in offspring being less vigorous and productive than those resulting from cross-fertilization. In **cross-fertilization** (also called **allogamy** in plants) the gametes are derived from different individuals. In plants the pollen comes either from another flower of the same plant or from a different plant (*see also* INCOMPATIBILITY).

**fertilizer** Any substance that is added to soil in order to increase its productivity. Fertilizers can be of natural origin, such as *composts, or they can be made up of synthetic chemicals, particularly nitrates and phosphates. Synthetic fertilizers can increase crop yields dramatically, but when leached from the soil by rain, which runs into lakes, they also increase the process of eutrophication (*see* ALGAL BLOOM; EUTROPHIC). Bacteria that can fix nitrogen are sometimes added to the soil to increase its fertility; for example, in tropical countries the blue-green bacterium *Anabaena* is added to rice paddies to increase soil fertility.

**FET** *See* TRANSISTOR.

**fetal membranes** *See* EXTRAEMBRYONIC MEMBRANES.

**fetus (foetus)** The *embryo of a mammal, especially a human, when development has reached a stage at which the main features of the adult form are recognizable. In humans the embryo from eight weeks to birth is called a fetus.

**Feulgen's test** A histochemical test in which the distribution of DNA in the chromosomes of dividing cell nuclei can be observed. It was devised by the German chemist R. Feulgen (1884–1955). A tissue section is first treated with dilute hydrochloric acid to remove the purine bases of the DNA, thus exposing the aldehyde groups of the sugar deoxyribose. The section is then immersed in *Schiff's reagent, which combines with the aldehyde groups to form a magenta-coloured compound.

**Feynman diagram** *See* QUANTUM ELECTRODYNAMICS.

**Fibonacci number** Any number in the sequence 0, 1, 1, 2, 3, 5, 8, 13, 21,…, in which each term is the sum of the two preceding terms. The ratio of two consecutive terms tends to $\frac{1}{2}(1 + \sqrt{5})$, which is equal to the *golden ratio. It was named after the Italian mathematician Leonardo Fibonacci (*c.* 1170–1250).

**fibre** **1.** An elongated plant cell whose walls are extensively (usually completely) thickened with lignin. Fibres are found in the vascular tissue, usually in the xylem, where they provide structural support. The term is often used loosely to mean any kind of xylem element. The fibres of many species, e.g. flax, are of commercial importance. **2.** Any of various threadlike structures in the animal body, such as a muscle fibre, a nerve fibre, or a collagen fibre. **3. (dietary fibre; roughage)** The part of food that cannot be digested and absorbed to produce energy. Dietary fibre falls into four groups: **cellulose**, **hemicelluloses**, **lignins**, and **pectins**. Highly refined foods, such as sucrose, contain no dietary fibre. Foods with a high fibre content include wholemeal cereals and flour, root vegetables, nuts, and fruit. In human nutrition a distinction is made between soluble and insoluble fibre. **Soluble fibre** (e.g. in oats, pulses, fruit, vegetables) is broken down by bacteria in the large intestine to yield short-chain fatty acids, some of which can be absorbed and metabolized by the liver. **Insoluble fibre** (e.g. in wholegrain cereals) is resistant to bacterial attack and forms a bulky water-retaining mass that promotes peristalsis and accelerates the passage of faeces. Dietary fibre is considered by some to be helpful in the prevention of many of the diseases of Western civilization, such as diverticulosis, constipation, appendicitis, obesity, and diabetes mellitus.

**fibre optics** The transmission of light along *optical fibres.

(⊕) SEE WEB LINKS
• A comprehensive tutorial from Arc Electronics

**fibrin** The insoluble protein that forms fibres at the site of an injury and is the foundation of a blood clot. *See* BLOOD CLOTTING.

**fibrinogen** The protein dissolved in the blood plasma that, when suitably activated, is converted to insoluble *fibrin fibres. *See* BLOOD CLOTTING.

**fibrinolysis** The breakdown of the protein *fibrin by the enzyme *plasmin (**fibrinase** or **fibrinolysin**), which occurs when blood clots are removed from the circulation.

**fibroblast** A cell that secretes fibres in the intercellular substance of *connective tissue. The cells are long, flat, and star-shaped and lie close to collagen fibres. Fibroblasts are often grown in cell cultures.

**fibrocartilage** *See* CARTILAGE.

**fibrous protein** *See* PROTEIN.

**fibula** The smaller and outer of the two bones between the knee and the ankle in terrestrial vertebrates. *Compare* TIBIA.

**Fick's law** A law describing the diffusion that occurs when solutions of different concentrations come into contact, with molecules moving from regions of higher concentration to regions of lower concentration. Fick's law states that the rate of diffusion d$n$/d$t$, called the **diffusive flux** and denoted $J$, across an area $A$ is given by: d$n$/d$t$ = $J$ = $-DA\partial c/\partial x$, where $D$ is a constant called the **diffusion constant**, $\partial c/\partial x$ is the concentration gradient of the solute, and d$n$/d$t$ is the amount of solute crossing the area $A$ per unit time. $D$ is constant for a specific solute and solvent at a specific temperature. Fick's law was formulated by the German physiologist Adolf Eugen Fick (1829–1901) in 1855.

**field** A region in which a body experiences a *force as the result of the presence of some other body or bodies. A field is thus a method of representing the way in which bodies are able to influence each other. For example, a body that has mass is surrounded by a region in which another body that has mass experiences a force tending to draw the two bodies together. This is the gravitational field (*see* NEWTON'S LAW OF GRAVITATION). The other three *fundamental interactions can also be represented by means of fields of force. However in the case of the *magnetic field and *electric field that together create the electromagnetic interaction, the force can vary in direction according to the character of the field. For example, in the field surrounding a negatively charged body, a positively charged body will experience a force of attraction, while another negatively charged body is repelled.

The strength of any field can be described as the ratio of the force experienced by a small appropriate specimen to the relevant property of that specimen, e.g. force/mass for the gravitational field. *See also* QUANTUM FIELD THEORY.

**field capacity** The amount of water that remains in a soil when excess has drained away. It is held by capillary forces of the soil pores and reflects the physical nature of the soil.

**field coil** The coil in an electrical machine that produces the magnetic field.

**field-effect transistor** (FET) *See* TRANSISTOR.

**field-emission microscope** (FEM) A type of electron microscope in which a high negative voltage is applied to a metal tip placed in an evacuated vessel some distance from a glass screen with a fluorescent coating. The tip produces electrons by **field emission**, i.e. the emission of electrons from an unheated sharp metal part as a result of a high electric field. The emitted electrons form an enlarged pattern on the fluorescent screen, related to the individual exposed planes of atoms. As the resolution of the instrument is limited by the vibrations of the metal atoms, it is helpful to cool the tip in liquid helium. Although the individual atoms forming the point are not displayed, individual adsorbed atoms of other substances can be, and their activity is observable.

**field-ionization microscope** (field-ion microscope; FIM) A type of electron microscope that is similar in principle to the *field-emission microscope, except that a high positive voltage is applied to the metal tip, which is surrounded by low-pressure gas (usually helium) rather than a vacuum. The image is formed in this case by **field ionization**: ionization at the surface of an unheated solid as a result of a strong electric field creating positive ions by electron transfer from surrounding atoms or molecules. The image is formed by ions striking the fluorescent screen. Individual atoms on the surface of the tip can be resolved and, in certain cases, adsorbed atoms may be detected.

**field lens** The lens in the compound eyepiece of an optical instrument that is furthest from the eye. Its function is to increase the field of view by refracting towards the main eye lens rays that would otherwise miss it.

**field magnet** The magnet that provides the magnetic field in an electrical machine. In some small dynamos and motors it is a permanent magnet but in most machines it is an electromagnet.

**filament** 1. (in zoology) A long slender hairlike structure, such as any of the *barbs of a bird's feather. 2. (in botany) The stalk of the *stamen in a flower. It bears the anther and consists mainly of conducting tissue. 3. (in cell biology) Any of the microscopic protein fibres that form part of the *cytoskeleton. They include intermediate filaments and microfilaments (*see* ACTIN). 4. (in physics) A thin wire, often of tungsten, that is heated by an electric current to incandescence, in light bulbs and thermionic valves.

**file** A collection of data stored in a computer. It may consist of program instructions or numerical, textual, or graphical information. It usually consists of a set of similar or related records.

**file transfer protocol** *See* FTP.

**Filicinophyta (Pterophyta)** A phylum of mainly terrestrial vascular plants (*see* TRACHEOPHYTE) – the ferns. Ferns are perennial plants bearing large conspicuous leaves (**fronds**: *see* MEGAPHYLL) usually arising from either a rhizome or a short erect stem. Bracken is a common example. Only the tree ferns have stems that reach an appreciable height. There is a characteristic uncurling of the young leaves as they expand into the adult form. Reproduction is by means of spores borne on the underside of specialized leaves (*sporophylls).

**filler** A solid inert material added to a synthetic resin or rubber, either to change its physical properties or simply to dilute it for economy.

**film badge** A lapel badge containing masked photographic film worn by personnel who could be exposed to ionizing radiation. The film is developed to indicate the extent that the wearer has been exposed to harmful radiation. Typically, the badges contain a number of filters of different materials and thickness, thus allowing an estimation of the type of radiation received.

**filoplumes** Minute hairlike *feathers consisting of a shaft (**rachis**) bearing a few unattached barbs. They are found between the contour feathers.

**filter** 1. (in chemistry) A device for separating solid particles from a liquid or gas. The simplest laboratory filter for liquids is a funnel in which a cone of paper (**filter paper**) is placed. Special containers with a porous base of sintered glass are also used. *See also* GOOCH CRUCIBLE. 2. (in physics) A device placed in the path of a beam of radiation to alter its frequency distribution. For example, a plane pigmented piece of glass may be placed over a camera lens to alter the relative intensity of the component wavelengths of the beam entering the camera. 3. (in electronics) An electrical network that transmits signals within a certain frequency range but attenuates other frequencies.

**filter feeding** A method of feeding in which tiny food particles are strained from the surrounding water by various mechanisms. It is used by many aquatic invertebrates, especially members of the plankton, and by some vertebrates, notably baleen whales. *See also* CILIARY FEEDING; WHALEBONE.

**filter pump** A simple laboratory vacuum pump in which air is removed from a system by a jet of water forced through a narrow nozzle. The lowest pressure possible is the vapour pressure of water.

**filtrate** The clear liquid obtained by filtration.

**filtration** The process of separating solid particles using a filter. In **vacuum filtration**, the liquid is drawn through the filter by a vacuum pump. *Ultrafiltration is filtration under pressure; for example, ultrafiltration of the blood occurs in the *nephrons of the vertebrate kidney.

**FIM** *See* FIELD-IONIZATION MICROSCOPE.

**finder** A small low-powered astronomical telescope, with a wide field of view, that is fixed to a large astronomical telescope so that the large telescope can be pointed in the correct direction to observe a particular celestial body.

**fine chemicals** Chemicals produced industrially in relatively small quantities and with a high purity; e.g. dyes and drugs.

**fineness of gold** A measure of the purity of a gold alloy, defined as the parts of gold in 1000 parts of the alloy by mass. Gold with a fineness of 750 contains 75% gold, i.e. 18 *carat gold.

**fine structure** Closely spaced optical spectral lines arising from *transitions between energy levels that are split by the vibrational or rotational motion of a molecule or by electron spin. They are visible only at high resolution. **Hyperfine structure**, visible only at very high resolution, results from the influence of the atomic nucleus on the allowed energy levels of the atom.

**fine structure constant** Symbol $\alpha$. The dimensionless constant, with a value of about 1/137, that characterizes quantum electrodynamics. It is given by $\alpha = e^2/\hbar c$, where $e$ is the charge on an electron, $\hbar$ is the Dirac constant, and $c$ is the speed of light in free space.

**finite series** *See* SERIES.

**fins** The locomotory organs of aquatic vertebrates. In fish there are typically one or more **dorsal** and **ventral fins** (sometimes continuous), whose function is balance; a **caudal fin** around the tail, which is the main propulsive organ; and two paired fins: the **pectoral fins** attached to the pectoral (shoulder) girdle and the **pelvic fins** attached to the pelvic (hip) girdle, which are used in steering. These paired fins are homologous with the limbs of tetrapods. Fins are strengthened by a number of flexible fin rays, which may be cartilaginous, bony and jointed, horny, or fibrous and jointed.

**firedamp** Methane formed in coal mines.

**first convoluted tubule** *See* PROXIMAL CONVOLUTED TUBULE.

**first-order reaction** *See* ORDER.

**Fischer–Tropsch process** An industrial method of making hydrocarbon fuels from carbon monoxide and hydrogen. The process was invented in 1933 and used by Germany in World War II to produce motor fuel. Hydrogen and carbon monoxide are mixed in the ratio 2:1 (water gas was used with added hydrogen) and passed at 200°C over a nickel or cobalt catalyst. The resulting hydrocarbon mixture can be separated into a higher-boiling fraction for Diesel engines and a lower-boiling gasoline fraction. The gasoline fraction contains a high proportion of straight-chain hydrocarbons and has to be reformed for use in motor fuel. Alcohols, aldehydes, and ketones are also present. The process is also used in the manufacture of SNG from coal. It is named after the German chemist Franz Fischer (1852–1932) and the Czech Hans Tropsch (1839–1935).

**fish** *See* CHONDRICHTHYES (cartilaginous fish); OSTEICHTHYES (bony fish); PISCES.

**FISH** *See* FLUORESCENCE IN SITU HYBRIDIZATION.

**fissile material** A nuclide of an element that undergoes nuclear fission, either spontaneously or when irradiated by neutrons. Fissile nuclides, such as uranium–235 and plutonium–239, are used in *nuclear reactors and nuclear weapons. *Compare* FERTILE MATERIAL.

**fission** 1. (in biology) A type of asexual reproduction occurring in some unicellular organisms, e.g. diatoms, protozoans, and bacteria, in which the parent cell divides to form two (**binary fission**) or more (**multiple fission**) similar daughter cells. 2. (in physics) *See* NUCLEAR FISSION.

**fission products** *See* NUCLEAR FISSION.

**fission-track dating** A method of estimating the age of glass and other mineral objects by observing the tracks made in them by the fission fragments of the uranium nuclei that they contain. By irradiating the objects with neutrons to induce fission and comparing the density and number of the tracks before and after irradiation it is possible to estimate the time that has elapsed since the object solidified.

**fitness** (in genetics) A measure of the relative breeding success of an individual in a given population in a given time. The fittest individuals are those that contribute the most offspring to the next generation. Fitness therefore reflects how well an organism is adapted to its environment, which determines its survival. *See also* INCLUSIVE FITNESS.

**Fittig reaction** *See* WURTZ REACTION.

**Fitzgerald contraction** *See* LORENTZ–FITZGERALD CONTRACTION.

**fixation** 1. The first stage in the preparation of a specimen for microscopical examination, in which the tissue is killed and preserved in as natural a state as possible by immersion in a chemical **fixative**. The fixative prevents the distortion of cell components by denaturing its constituent protein. Some commonly used fixatives are formaldehyde, ethanol, and Bouin's fluid (for light microscopy), and osmium tetroxide and gluteraldehyde (for electron microscopy). Fixation may also be brought about by heat. 2. *See* NITROGEN FIXATION.

**fixed action pattern** *See* INSTINCT.

**fixed point** A temperature that can be accurately reproduced to enable it to be used as the basis of a *temperature scale.

**fixed star** Any of the innumerable heavenly bodies that do not appear to alter their position on the *celestial sphere. They were so called by the ancients to distinguish them from the planets, which were once known as **wandering stars** because they appeared to move relative to the background of fixed stars. The discovery of the *proper motion of stars in the 18th century established that stars are not fixed in the sky although, because of their immense distances from the solar system, they may appear to be so.

**Fizeau, Armand Hippolyte Louis** (1819–96) French physicist. In 1845 he and Léon Foucault (1819–68) took the first photographs of the sun. In 1849 he measured the speed of light (*see* FIZEAU'S METHOD); he also analysed the *Doppler effect for light.

**Fizeau's method** A method of measuring the speed of light, invented by Armand Fizeau in 1849. A cogwheel rotating at high speed enables a series of flashes to be transmitted to a distant mirror. The light reflected back to the cogwheel is observed and the speed of light calculated from the rates of rotation of the wheel required to produce an eclipse of the returning light.

**flaccid** (in botany) Describing plant tissue that has become soft and less rigid than normal because the cytoplasm within its cells has shrunk and contracted away from the cell walls through loss of water (*see* PLASMOLYSIS).

**flagellum** (*pl.* **flagella**) 1. (in prokaryotes) A long slender threadlike structure that protrudes from the cell surface of a bacterium. It rotates from its base and propels the bacterium along. Up to several micrometres in length, a flagellum is constructed of numerous subunits of the protein flagellin. 2. (in eukaryotes) *See* UNDULIPODIUM.

**flame** A hot luminous mixture of gases undergoing combustion. The chemical reactions in a flame are mainly free-radical chain reactions and the light comes from fluorescence of excited molecules or ions or from incandescence of small solid particles (e.g. carbon).

**flame test** A simple test for metals, in which a small amount of the sample (usually moistened with hydrochloric acid) is placed on the end of a platinum wire and held in a Bunsen flame. Certain metals can be detected by the colour produced: barium (green), calcium (brick red), lithium (crimson), potassium (lilac), sodium (yellow), strontium (red).

**flare star** A type of cool *red dwarf whose brightness changes unpredictably and rapidly, probably because of the release of intense amounts of energy brought about by magnetic fields in the star's photosphere. The high-energy flares make the red star appear much brighter, though often for only a few minutes. *See also* SOLAR FLARE.

**flash memory** A form of storage in which the data may be altered electrically. The device does not need refreshing to maintain the data, which is stored even when power is removed. Flash memory finds application in computers, in digital cameras, and in portable storage devices that emulate hard disks (e.g. *USB drives).

**flash photolysis** A technique for studying free-radical reactions in gases. The apparatus used typically consists of a long glass or quartz tube holding the gas, with a lamp outside the tube suitable for producing an intense flash of light. This dissociates molecules in the sample creating free radicals, which can be detected spectroscopically by a beam of light passed down the axis of the tube. It is possible to focus the spectrometer on an absorption line for a particular product and measure its change in intensity with time using an oscilloscope. In this way the kinetics of very fast free-radical gas reactions can be studied.

**flash point** The temperature at which the vapour above a volatile liquid forms a combustible mixture with air. At the flash point the application of a naked flame gives a momentary flash rather than sustained combustion, for which the temperature is too low.

**flatworms** *See* PLATYHELMINTHES.

**flavin adenine dinucleotide** *See* FAD.

**flavones** A group of *flavonoid compounds found in many plants.

**flavonoid** One of a group of naturally occurring phenolic compounds many of which are plant pigments. They include the *anthocyanins, **flavonols**, and **flavones**. Plant-derived flavonoids in foods are strong antioxidants in their natural state but are poorly absorbed from the intestine and rapidly metabolized and excreted. However, they might be beneficial to health by stimulating the body to produce enzymes that help to eliminate mutagens and carcinogens.

**flavoprotein** *See* FAD.

**flavour** *See* ELEMENTARY PARTICLES.

**fleas** *See* SIPHONAPTERA.

**Fleming, Sir Alexander** (1881–1955) British bacteriologist, born in Scotland. He studied medicine at St Mary's Hospital, London, where he remained all his life. In 1922 he identified *lysozyme, an enzyme that destroys bacteria, and in 1928 discovered the antibiotic *penicillin. He shared the 1945 Nobel Prize for physiology or medicine with *Florey and *Chain, who first isolated the drug.

**Fleming's rules** Rules to assist in remembering the relative directions of the field, current, and force in electrical machines. The left hand refers to motors, the right hand to generators. If the forefinger, second finger, and thumb of the left hand are extended at right angles to each other, the forefinger indicates the direction of the field, the second finger the direction of the current, and the thumb the direction of the force. If the right hand is used the digits indicate these directions in a generator. The mnemonic was invented by Sir John Ambrose Fleming (1849–1945).

**flexor** A muscle that causes a limb to bend by bringing the two parts of the limb together. An example is the *biceps. Flexors work antagonistically with *extensors. *See* VOLUNTARY MUSCLE.

**flies** *See* DIPTERA.

**flight** 1. Any form of *locomotion in air, which can be active or passive (**gliding**). Mechanisms of flight have evolved mainly in birds, bats, and insects: these animals are adapted for flight by the presence of **wings**, which increases the ratio of surface area to body weight. Birds possess powerful flight muscles: the **depressor** muscle runs from the underside of the humerus to the sternum and is responsible for the downstroke of the wing; the **levator** muscle works antagonistically, producing the upstroke. Flight in insects works in a similar fashion but the muscles that control the wing movement are attached to the thorax. A few species of mammals, reptiles, and fish have developed flight to a lesser extent. For example, flying squirrels (order Dermoptera) possess a membrane attached to the limbs that can open and function as a parachute, allowing the animals to glide. 2. Part of a survival mechanism in an animal that is generated in response to a threatening situation. A potentially dangerous situation can induce the release of *adrenaline, which prepares the animal for 'fight or flight' by increasing the blood pressure and heart rate and diverting the blood flow to the muscles and heart. 3. *See* AERODYNAMICS.

**flint (chert)** Very hard dense nodules of microcrystalline quartz and chalcedony found in chalk and limestone.

**flip-flop (bistable circuit)** An electronic circuit that has two stable states. It is switched from one stable state to the other by means of a triggering pulse. They are extensively used as *logic circuits in computers.

**flocculation** The process in which particles in a colloid aggregate into larger clumps. Often, the term is used for a reversible aggregation of particles in which the forces holding the particles together are weak and the colloid can be redispersed by agitation. The stability of a lyophobic colloidal dispersion depends on the existence of a layer of electric charge on the surface of the particles. Around this are attracted electrolyte ions of opposite charge, which form a mobile ionic 'atmosphere'. The result is an electrical double layer on the particle, consisting of an inner shell of fixed charges with an outer mobile atmosphere. The potential energy between two particles depends on a repulsive interaction between double layers on adjacent particles and an attractive interaction due to *van der Waals' forces between the particles.

At large separations, the repulsive forces dominate, and this accounts for the overall stability of the colloid. As the particles become closer together, the potential energy increases to a maximum and then falls sharply at very close separations, where the van der Waals forces dominate. This potential-energy minimum corresponds to *coagulation and is irreversible. If the *ionic strength of the solution is high, the ionic atmosphere around the particles is dense and the potential-energy curve shows a shallow minimum at larger separation of particles. This corresponds to flocculation of the particles. Ions with a high charge are particularly effective for causing flocculation and coagulation.

**flocculent** Aggregated in woolly masses; used to describe precipitates.

**floppy disk (diskette)** A flexible plastic disk with a magnetic coating encased in a stiff envelope. It is used to store information in a small computer system. *See* MAGNETIC DISK.

**flora** All the plant life normally present in a given habitat at a given time. *See also* MICROFLORA. *Compare* FAUNA.

**Florey, Howard Walter, Baron** (1898–1968) Australian pathologist, who moved to Oxford in 1922. After working in Cambridge and Sheffield (studying *lysozyme), he returned to Oxford in 1935. There he teamed up with Ernst *Chain and by 1939 they succeeded in isolating and purifying *penicillin. They also developed a method of producing the drug in large quantities and carried out its first clinical trials. The two men shared the 1945 Nobel Prize for physiology or medicine with penicillin's discoverer, Alexander *Fleming.

**florigen** A hypothetical plant growth substance that is postulated to transmit the stimulus for flowering, which is a response to photoperiod (*see* PHOTOPERIODISM), from the leaves to the apex of the plant. However, florigen has never been isolated and some plant physiologists question its existence.

**flower** The structure in angiosperms (flowering plants) that bears the organs for sexual reproduction. Flowers are very variable in form, ranging from the small green insignificant wind-pollinated flowers of many grasses to spectacular brightly coloured insect-pollinated flowers. Flowers are often grouped together into *inflorescences, some of which (e.g. that of dandelion) are so compacted as to resemble a single flower. Typically flowers consist of a receptacle that bears sepals, petals, stamens,

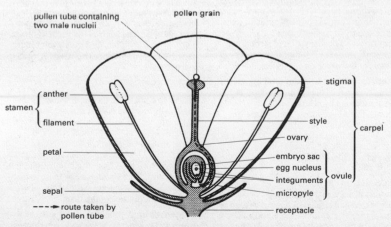

**Flower.** Section through a monocarpellary flower at the time of pollination.

and carpels (see illustration). The flower parts are adapted to bring about pollination and fertilization resulting in the formation of seeds and fruits. The sepals are usually green and leaflike and protect the flower bud. The petals of insect-pollinated flowers are adapted in many ingenious ways to attract insects and, in some instances, other animals. For example, some flowers are adapted to attract short-tongued insects by having an open shallow *corolla tube and nectar situated in an exposed position. Flowers adapted for pollination by long-tongued insects have a long corolla tube of fused petals with nectar in a concealed position. The tongue of the insect brushes against the anthers and stigma before reaching the nectar. Wind-pollinated flowers, in contrast, are inconspicuous. The anthers dangle outside the corolla and the stigmas have a feathery surface to catch the pollen grains.

Some species are adapted for self-pollination and have small flowers, no nectar, and stamens and carpels that mature simultaneously.

**flowering plants** See ANTHOPHYTA.

**fluctuations** Random deviations in the value of a quantity about some average value. In all systems described by *quantum mechanics fluctuations, called **quantum fluctuations**, occur – even at the *absolute zero of thermodynamic temperature as a result, ultimately, of the Heisenberg *uncertainty principle. In any system above absolute zero, fluctuations, called **thermal fluctuations**, occur. It is necessary to take fluctuations into account to obtain a quantitative theory of *phase transitions in three dimensions. The formation of structure in the *early universe is thought to be a result of quantum fluctuations.

**fluidics** The use of jets of fluid in pipes to perform many of the control functions usually performed by electronic devices. Being about one million times slower than electronic devices, fluidic systems are useful where delay lines are required. They are also less sensitive to high temperatures, strong magnetic fields, and ionizing radiation than electronic devices.

**fluidization** A technique used in some industrial processes in which solid particles suspended in a stream of gas are treated as if they were in the liquid state. Fluidization is useful for transporting powders, such as coal dust. **Fluidized beds**, in which solid particles

are suspended in an upward stream, are extensively used in the chemical industry, particularly in catalytic reactions where the powdered catalyst has a high surface area. They are also used in furnaces, being formed by burning coal in a hot turbulent bed of sand or ash through which air is passed. The bed behaves like a fluid, enabling the combustion temperature to be reduced so that the production of polluting oxides of nitrogen is diminished. By adding limestone to the bed with the fuel, the emission of sulphur dioxide is reduced.

High-pressure fluidized beds are also used in power-station furnaces in a **combined cycle** in which the products of combustion from the fluidized bed are used to drive a gas turbine, while a steam-tube boiler in the fluid bed raises steam to drive a steam turbine. This system both increases the efficiency of the combustion process and reduces pollution.

**fluid mechanics** The study of fluids at rest and in motion. **Fluid statics** is concerned with the pressures and forces exerted on liquids and gases at rest. *Hydrostatics is specifically concerned with the behaviour of liquids at rest. In **fluid dynamics** the forces exerted on fluids, and the motion that results from these forces, are examined. It can be divided into *hydrodynamics: the motion of liquids (not only water); and aerodynamics: the motion of gases.

Fluid dynamics is an important science used to solve many of the problems arising in aeronautical, chemical, mechanical, and civil engineering. It also enables many natural phenomena, such as the flight of birds, the swimming of fish, and the development of weather conditions, to be studied scientifically.

**flukes** See TREMATODA.

**flunitrazepam** A *benzodiazepine used medically in some countries as a powerful hypnotic, sedative, and muscle relaxant. It was marketed in the US under the tradename **Rohypnol**. Flunitrazepam has become notorious as a so-called 'date rape' drug. It is quickly eliminated from the body, difficult to detect, and causes amnesia, so that victims cannot remember events that occur when under the influence of the drug.

**fluorescein** A yellowish-red dye that produces yellow solutions with a green fluorescence. It is used in tracing water flow and as an absorption indicator (*see* ADSORPTION IN-DICATOR).

**fluorescence** *See* LUMINESCENCE.

**fluorescence in situ hybridization (FISH)** A technique in which a DNA probe, labelled with a fluorescent dye, base-pairs (hybridizes) with the complementary base sequence of a target nucleotide. It is used in genetic mapping for locating specific genes within a chromosome set. Another application is for locating particular messenger RNAs (mRNAs) within cells.

**fluorescent light** *See* ELECTRIC LIGHTING.

**fluoridation** The process of adding very small amounts of fluorine salts (e.g. sodium fluoride, NaF) to drinking water to prevent tooth decay. The fluoride becomes incorporated into the fluoroapatite (*see* APATITE) of the growing teeth and reduces the incidence of *dental caries.

**fluoride** *See* HALIDE.

**fluorination** A chemical reaction in which a fluorine atom is introduced into a molecule. *See* HALOGENATION.

**fluorine** Symbol F. A poisonous pale yellow gaseous element belonging to group 17 (formerly VIIB) of the periodic table (the *halogens); a.n. 9; r.a.m. 18.9984; d. 1.7 g dm$^{-3}$; m.p. $-219.62°C$; b.p. $-188.1°C$. The main mineral sources are *fluorite ($CaF_2$) and *cryolite ($Na_3AlF$). The element is obtained by electrolysis of a molten mixture of potassium fluoride and hydrogen fluoride. It is used in the synthesis of organic fluorine compounds. Chemically, it is the most reactive and electronegative of all elements. It is a highly dangerous element, causing severe chemical burns on contact with the skin. The element was identified by Scheele in 1771 and first isolated by Moissan in 1886.

(((●))) **SEE WEB LINKS**
• Information from the WebElements site

**fluorite (fluorspar)** A mineral form of calcium fluoride, $CaF_2$, crystallizing in the cubic system. It is variable in colour; the most common fluorites are green and purple (blue john), but other forms are white, yellow, or brown. Fluorite is used chiefly as a flux material in the smelting of iron and steel; it is also used as a source of fluorine and hydrofluoric acid and in the ceramic and optical-glass industries.

**fluorite structure** *See* CALCIUM FLUORIDE.

**fluorocarbons** Compounds obtained by replacing the hydrogen atoms of hydrocarbons by fluorine atoms. Their inertness and high stability to temperature make them suitable for a variety of uses as oils, polymers, etc. *See also* CHLOROFLUOROCARBON; HALON.

**5-fluorouracil (5-FU)** A fluorine derivative of the pyrimidine uracil. It is used in chemotherapy where it inhibits the cell's ability to synthesize DNA. It is often used in a treatment regime along with cisplatin.

**flux** 1. A substance applied to the surfaces of metals to be soldered to inhibit oxidation. 2. A substance used in the smelting of metals to assist in the removal of impurities as slag. 3. The number of particles flowing per unit area of cross section in a beam of particles. 4. *See* LUMINOUS FLUX. 5. *See* MAGNETIC FLUX. 6. *See* ELECTRIC FLUX.

**flux density** 1. *See* MAGNETIC FIELD. 2. *See* ELECTRIC DISPLACEMENT.

**fluxional molecule** A molecule that undergoes alternate very rapid rearrangements of its atoms and thus only has a specific structure for a very short period of time. For example, the molecule $ClF_3$ has a T-shape at low temperatures ($-60°C$); at room temperature the fluorine atoms change position very rapidly and appear to have identical positions.

**fluxmeter** An instrument used to measure *magnetic flux. It is used in conjunction with a coil (the **search coil**) and resembles a moving-coil galvanometer except that there are no restoring springs. A change in the magnetic flux induces a momentary current in the search coil and in the coil of the meter, which turns in proportion and stays in the deflected position. This type of instrument has been largely superseded by a type using the Hall probe (*see* HALL EFFECT).

**flyby** A close approach made by a spacecraft to a planet, satellite, or asteroid without entering orbit or landing, mainly for the purpose of taking photographs or collecting data but also for carrying out a manoeuvre called a gravity assist, by which the spacecraft takes up a tiny fraction of a planet's rotational energy in order to change direction or boost its velocity.

**FM (frequency modulation)** *See* MODULATION.

**fMRI** Functional magnetic resonance imaging. *See* NUCLEAR MAGNETIC RESONANCE.

**f-number** *See* APERTURE.

**foam** A dispersion of bubbles in a liquid. Foams can be stabilized by *surfactants. Solid foams (e.g. expanded polystyrene or foam rubber) are made by foaming the liquid and allowing it to set. *See also* COLLOIDS.

**foaming agent (blowing agent)** A substance used to produce a liquid or solid foam (e.g. an expanded plastic). Physical agents are compressed gases; chemical foaming agents are substances that release gas under certain conditions (e.g. sodium hydrogencarbonate).

**focal length** The distance between the *optical centre of a lens or pole of a spherical mirror and its *principal focus.

**focal point** *See* FOCUS.

**focal ratio** *See* APERTURE.

**Fock degeneracy** A 'hidden' degeneracy that occurs in the spectrum of the hydrogen atom as a result of the rotational invariance in four dimensions associated with the Coulomb interation between the proton and the electron. It was discovered by the Soviet physicist Vladimir Fock (1898–1974) in 1935.

**focus** 1. (in optics) Any point in an optical system through or towards which rays of light are converged. It is sometimes called the **focal point** and sometimes loosely used to mean *principal focus or (particularly by photographers) *focal length. 2. (in mathematics) *See* CONIC; ELLIPSE.

**focusing** (in animal physiology) The process of directing and concentrating light from a source onto the *retina of the eye, by means of the lens, in order to obtain a clear image of objects at a range of distances. *See* ACCOMMODATION.

**foetus** *See* FETUS.

**folacin** *See* FOLIC ACID.

**fold** A wavelike form in layered sedimentary rock strata that results from deformational processes in the earth's crust. Basin-shaped folds in which the beds of rock dip towards each other are known as **synclines**; those in which the beds of rock are folded into an arch shape are known as **anticlines**. More complex folds result where the rock strata are subjected to intense horizontal pressures. See illustration.

**folic acid (folacin)** A vitamin of the *vitamin B complex. In its active form, tetrahydrofolic acid, it is a *coenzyme in various reactions involved in the metabolism of amino acids, purines, and pyrimidines. It is synthesized by intestinal bacteria and is especially abundant in green leafy vegetables. Deficiency causes poor growth and nutritional anaemia.

**follicle** 1. (in animal anatomy) Any enclosing cluster of cells that protects and nourishes a cell or structure within. For example, follicles in the *ovary contain developing egg cells, while *hair follicles envelop the roots of hairs. 2. (in botany) A dry fruit that, when ripe, splits along one side to release its seeds. It is formed from a single carpel containing one or more seeds. Follicles do not occur singly but are grouped to form clusters (**etaerios**). Examples include larkspur, columbine, and monk's hood.

**follicle-stimulating hormone (FSH)** A hormone, secreted by the anterior pituitary gland in mammals, that stimulates, in fe-

monoclinal fold

anticlinal fold

synclinal fold

direction of pressure →

overturned fold

direction of pressure →

recumbent fold

**Folds.**

male mammals, ripening of specialized structures in the ovary (*Graafian follicles) that produce ova and, in males, the formation of sperm in the testis. It is a major constituent of fertility drugs, used to treat failure of ovulation and decreased sperm production. *See also* GONADOTROPHIN.

**food** Any material containing *nutrients, such as carbohydrates, proteins and fats, which are required by living organisms in order to obtain energy for growth and maintenance. Heterotrophic organisms, such as animals, ingest their food (*see also* DIET); autotrophic organisms, such as plants, manufacture their food materials.

**food chain** The transfer of energy from green plants (the primary producers) through a sequence of organisms in which each eats the one below it in the chain and is eaten by the one above. Thus plants are eaten by herbivores, which are then eaten by carnivores. These may in turn be eaten by different carnivores. The position an organism occupies in a food chain is known as its *trophic level. In practice, many animals feed at several different trophic levels, resulting in a more complex set of feeding relationships known as a *food web. *See* BIOENERGETICS; CONSUMER; PRODUCER; PYRAMID OF BIOMASS; PYRAMID OF ENERGY; PYRAMID OF NUMBERS.

**food poisoning** An acute illness caused by food that may be naturally poisonous or contaminated by certain types of pathogenic microorganisms. The most common type of food poisoning in the UK is that caused by the bacteria belonging to the genus *Salmonella, which inhabit the alimentary canal of livestock. Freezing and other types of *food preservation can prevent the growth of the bacteria and thorough cooking will kill the microorganisms before the meat is eaten. However, food poisoning can result if frozen meat is not completely thawed at its centre before cooking, as it may not reach sufficiently high temperatures to kill the bacteria during cooking. Another type of food poisoning, known as **botulism**, is caused by toxins produced by the bacterium *Clostridium botulinum*, which can grow in badly preserved canned foods. Other bacteria causing food poisoning include *Staphylococcus aureus*, *Campylobacter jejuni*, *Listeria monocytogenes*, and pathogenic *Escherichia coli*.

**food preservation** Prevention of the spoilage of food, which is achieved by a variety of techniques. These aim to prevent bacterial and fungal decay and contamination of food, which can cause *food poisoning. For example, dehydration removes the water from food, which prevents microorganisms from growing. Treating food with salt (salting) causes the microorganisms to lose water due to osmosis. Pickling involves treatment with vinegar (ethanoic acid), which reduces the pH and prevents bacteria from growing. Heating food (**blanching**) to temperatures of 90°C denatures the enzymes that cause the breakdown of food and kills many bacteria. The food is then packed in air-tight containers, such as cans or bottles. Heating milk to high temperatures to kill the bacteria is the basis of *pasteurization. Freezing food prevents the growth of bacteria but does not necessarily kill them; thorough cooking is therefore essential. In *freeze drying, food is rapidly frozen and then dehydrated, usually in a vacuum. Preprepared food can be preserved by the addition of chemicals (**food additives**), such as *sodium benzenecarboxylate, proprionates, and sulphur dioxide, but some of these may have adverse side-effects. **Irradiation** is a method of food preservation in which the bacteria are killed by irradiating the food with gamma rays.

**food production** *See* AGRICULTURE.

**food reserves** Reserves of fat, carbohydrate, or (rarely) protein in cells and tissues that function as an important store of energy that can be released and used in ATP production when required by the organism. For example, in animals *fat is stored in adipose tissue, and carbohydrate – in the form of the **storage compound** *glycogen – is stored in liver and muscle cells. In plants *starch is a major storage compound, being found in perennating organs (*see* PERENNATION) and seeds (in which it is mobilized at germination), and oils are important storage materials in some species (e.g. in the seeds of the castor-oil plant).

**food supply** (in human ecology) The production of food for human consumption. *See* AGRICULTURE.

**food web** A system of *food chains that are linked with one another. In a food web a particular organism may feed at more than one trophic level. For example, in a pond food web a freshwater mussel may feed directly on green algae, in which case it is a primary consumer. However, it can also feed on protozoa, which are themselves primary consumers, in which case the mussel is the secondary consumer. A food web does not usually include the decomposers, but these organisms are very important in the flow of energy through a food web (*see* ENERGY FLOW).

**fool's gold** *See* PYRITE.

**foot** The unit of length in *f.p.s. units. It is equal to one-third of a yard and is now therefore defined as 0.3048 metre. Several units based on the foot were formerly used in science, including the units of work, the **foot-pound-force** and the **foot-poundal**, and the illumination units, the **foot-candle** and the **foot-lambert**. These have all been replaced by SI units.

**foramen** An aperture in an animal part or organ, especially one in a bone or cartilage. For example, the **foramen magnum** is the opening at the base of the skull through which the *spinal cord passes.

**forbidden band** *See* ENERGY BANDS.

**forbidden transitions** Transitions between energy levels in a quantum-mechanical system that are not allowed to take place because of *selection rules. In practice, forbidden transitions can occur, but they do so with much lower probability than allowed transitions. There are three reasons why forbidden transitions may occur:
(1) the selection rule that is violated is only an approximate rule. An example is provided by those selection rules that are only exact in the absence of *spin–orbit coupling. When spin–orbit coupling is taken into account, the forbidden transitions become allowed – their strength increasing with the size of the spin–orbit coupling;
(2) the selection rule is valid for dipole radiation, i.e. in the interaction between a quantum-mechanical system, such as an atom, and an electromagnetic field, only the (variable) electric dipole moment is considered. Actual transitions may involve magnetic dipole radiation or quadrupole radiation;
(3) the selection rule only applies for an atom, molecule, etc., in isolation and does not necessarily apply if external fields, collisions, etc., are taken into account.

**force** Symbol *F*. The agency that tends to change the momentum of a massive body, defined as being proportional to the rate of increase of momentum. For a body of mass *m* travelling at a velocity *v*, the momentum is *mv*. In any coherent system of units the force

is therefore given by $F = \mathrm{d}(mv)/\mathrm{d}t$. If the mass is constant $F = m\mathrm{d}v/\mathrm{d}t = ma$, where *a* is the acceleration (*see* NEWTON'S LAWS OF MOTION). The SI unit of force is the newton. Forces occur always in equal and opposite action–reaction pairs between bodies, though it is often convenient to think of one body being in a force *field.

**forced convection** *See* CONVECTION.

**force ratio (mechanical advantage)** The ratio of the output force (load) of a machine to the input force (effort).

**forebrain (prosencephalon)** One of the three sections of the brain of a vertebrate embryo. The forebrain develops to form the *cerebrum, *hypothalamus, and *thalamus in the adult. *Compare* HINDBRAIN; MIDBRAIN.

**foregut 1.** The anterior region of the alimentary canal of vertebrates, up to the anterior part of the duodenum. **2.** The anterior part of the alimentary canal of arthropods. *See also* HINDGUT; MIDGUT.

**forest** An area of vegetation in which the dominant plants are trees; forests constitute major *biomes. Temperate forests have adequate or abundant rainfall and moderate temperatures. They may be dominated by deciduous trees (such as oak, ash, elm, beech, or maple), often growing together to form mixed deciduous forest, as in temperate regions of Europe, Asia, and North America; or by broad-leaved evergreens (such as southern beech, *Nothofagus*), as in Chile. Cold forests, of northern regions, are dominated by evergreen conifers (*see* TAIGA). Tropical forests include *rainforest; monsoon forest, found in SE Asia and having heavy rainfall interspersed with periods of drought; and thorn forest, as in SW North America, SW Africa, and parts of Central and South America and Australia, which has sparse rainfall, is dominated by small thorny trees, and grades into savanna woodland and semidesert. *See also* DEFORESTATION.

**form 1.** A category used in the *classification of organisms into which different types of a variety may be placed. **2.** Any distinct variant within a species. Seasonal variants, e.g. the tawny brown (summer) and blue-white (winter) forms of the blue hare, may be called forms. *See also* POLYMORPHISM.

**formaldehyde** *See* METHANAL.

**formalin** A colourless solution of methanal (formaldehyde) in water with methanol as a stabilizer; r.d. 1.075–1.085. When kept at temperatures below 25°C a white polymer of methanal separates out. It is used as a disinfectant and preservative for biological specimens.

**formate** *See* METHANOATE.

**formic acid** *See* METHANOIC ACID.

**formula** **1.** (in chemistry) A way of representing a chemical compound using symbols for the atoms present. Subscripts are used for the numbers of atoms. The **molecular formula** simply gives the types and numbers of atoms present. For example, the molecular formula of ethanoic acid is $C_2H_4O_2$. The **empirical formula** gives the atoms in their simplest ratio; for ethanoic acid it is $CH_2O$. The **structural formula** gives an indication of the way the atoms are arranged. Commonly, this is done by dividing the formula into groups; ethanoic acid can be written $CH_3.CO.OH$ (or more usually simply $CH_3COOH$). Structural formulae can also show the arrangement of atoms or groups in space. **2.** (in mathematics and physics) A rule or law expressed in algebraic symbols.

**formula weight** The relative molecular mass of a compound as calculated from its molecular formula.

**formylation** A chemical reaction that introduces a formyl group (methanoyl, –CHO) into an organic molecule.

**formyl group** The group HCO .

**Fortin barometer** *See* BAROMETER.

**forward genetics** The traditional approach to genetic investigation, in which the aim is to identify the gene that governs a particular known function (identified by the effect of a mutation of that gene). *Compare* REVERSE GENETICS.

**fossil** The remains or traces of any organism that lived in the geological past. In general only the hard parts of organisms become fossilized (e.g. bones, teeth, shells, and wood) but under certain circumstances the entire organism is preserved. For example, virtually unaltered fossils of extinct mammals, such as the woolly mammoth and woolly rhinoceros, have been found preserved in ice in the Arctic. Small organisms or parts of organisms (e.g. insects, leaves, flowers) have been preserved in *amber.

In the majority of fossils the organism has

been turned to stone – a process known as **petrification**. This may take one of three forms. In **permineralization**, solutions originating underground fill the microscopic cavities in the organism. Minerals in these solutions (e.g. silica or calcite) may actually replace the original material of the organism so that even microscopic structures may be preserved; this process is known as **replacement** (or **mineralization**). A third form of petrification – **carbonization** (or **distillation**) – occurs in certain soft tissues that are composed chiefly of compounds of carbon, hydrogen, and oxygen (e.g. cellulose). After the organism has been buried, and in the absence of oxygen, carbon dioxide and water are liberated until only free carbon remains. This forms a black carbon film in the rock outlining the original organism. **Moulds** are formed when the original fossil is dissolved away leaving a mould of its outline in the solid rock. The deposition of mineral matter from underground solutions in a mould forms a **cast**. Palaeontologists often produce casts from moulds using such substances as dental wax. Moulds of thin organisms (e.g. leaves) are commonly known as **imprints**. **Trace fossils** are the fossilized remnants of the evidence of animal life, such as tracks, trails, footprints, burrows, and **coprolites** (fossilized faeces).

The ideal conditions for the formation of fossils occur in areas of rapid sedimentation, especially those parts of the seabed that lie below the zone of wave disturbance. *See also* CHEMICAL FOSSIL; INDEX FOSSIL; MICROFOSSIL.

**fossil fuel** Coal, oil, and natural gas, the fuels used by humans as a source of energy. They are formed from the remains of living organisms and all have a high carbon or hydrogen content. Their value as fuels relies on the exothermic oxidation of carbon to form carbon dioxide ($C + O_2 \rightarrow CO_2$) and the oxidation of hydrogen to form water ($H_2 + \frac{1}{2}O_2 \rightarrow H_2O$). Fossil fuels are a major source of the greenhouse gas carbon dioxide; as such, their use contributes to the *greenhouse effect and global warming.

**fossil hominid** *See* HOMINID.

**Foucault pendulum** A simple pendulum in which a heavy bob attached to a long wire is free to swing in any direction. As a result of the earth's rotation, the plane of the pendulum's swing slowly turns (at the poles of the earth it makes one complete revolution in 24 hours). It was devised by the French physicist Jean Bernard Léon Foucault (1819–68) in 1851, when it was used to demonstrate the earth's rotation.

**Fourier analysis** The representation of a function $f(x)$, which is periodic in $x$, as an infinite series of sine and cosine functions,

$$f(x) = a_0/2 + \sum_{n=1}^{\infty} (a_n \cos nx + b_n \sin nx)$$

A series of this type is called a *Fourier series. If the function is periodic with a period $2\pi$, the coefficients $a_0$, $a_n$, $b_n$ are:

$$a_0 = \int_{-\pi}^{+\pi} f(x) dx,$$

$$a_n = \int_{-\pi}^{+\pi} f(x) \cos nx dx \ (n = 1,2,3,\ldots),$$

$$b_n = 1/\pi \int_{-\pi}^{+\pi} f(x) \sin nx dx \ (n = 1,2,3,\ldots).$$

Fourier analysis and Fourier series are named after the French mathematician and engineer Joseph Fourier (1768–1830). Fourier series have many important applications in mathematics, science, and engineering, having been invented by Fourier in the first quarter of the 19th century in his analysis of the problem of heat *conduction.

**Fourier series** An expansion of a periodic function as a series of trigonometric functions. Thus,

$$f(x) = a_0 + (a_1 \cos x + b_1 \sin x) + (a_2 \cos 2x + b_2 \sin 2x) + \ldots,$$

where $a_0$, $a_1$, $b_1$, $b_2$, etc., are constants, called **Fourier coefficients**. The series is used in *Fourier analysis.

**Fourier transform** An integral transform of the type:

$$F(y) = \int_{-\infty}^{\infty} f(x) e^{-xy} dy.$$

The inverse is:

$$f(x) = (1/2\pi) \int_{-\infty}^{\infty} F(y) e^{ixy} dy.$$

Fourier transform techniques are used in obtaining information from spectra, especially in NMR and infrared spectroscopy (*see* FOURIER-TRANSFORM INFRARED).

**Fourier-transform infrared (FT-IR)** Infrared spectroscopy in which computers are part of the spectroscopic apparatus and use *Fourier transforms to enable the curve of intensity against wave number to be plotted with very high sensitivity. This has allowed spectra to be obtained in the far infrared region; previously it was difficult to attain spectra in this region as the resolution was obscured by the signal-to-noise ratio being too high to resolve the vibrational and/or rotational spectra of small molecules in their gas phase. FT-IR has been used in research on the atmosphere. Another application of this technique is the detection of impurities in samples of condensed matter.

**four-level laser** A laser in which four energy levels are involved. The disadvantage of a three-level laser is that it is difficult to attain population inversion because many molecules have to be raised from their ground state to an excited state by pumping. In a four-level laser, the laser transition finishes in an initially unoccupied state F, having started in a state I, which is not the ground state. As the state F is initially unoccupied, any population in I constitutes population inversion. Thus laser action is possible if I is sufficiently metastable. If transitions from F to the ground state G are rapid, population inversion is maintained since this lowers the population in F caused by the transition in the laser action.

**fourth dimension** *See* SPACE–TIME.

**four-wave mixing** In nonlinear optics, the production of a photon in a medium as a result of the interaction of three photons. In four-wave mixing, the photon produced has a wavelength (frequency) that is different from the wavelength (frequencies) of any of the three original photons. The effect is important in optical fibres; it can lead to loss of signal when a number of wavelengths are transmitted along the same fibre.

**fovea (fovea centralis)** A shallow depression in the *retina of the eye, opposite the lens, that is present in some vertebrates. This area contains a large concentration of *cones with only a thin layer of overlying nerves. It is therefore specialized for the perception of colour and sharp intense images. The clarity is enhanced in animals with binocular vision.

**f.p.s. units** The British system of units based on the foot, pound, and second. It has now been replaced for all scientific purposes by SI units.

**fractal** A curve or surface generated by a process involving successive subdivision. For example, a **snowflake curve** can be produced by starting with an equilateral triangle and dividing each side into three segments. The middle segments are then replaced by two equal segments, which would form the sides of a smaller equilateral triangle. This gives a 12-sided star-shaped figure. The next stage is to subdivide each of the sides of this figure in the same way, and so on. The result is a developing figure that resembles a snowflake. In the limit, this figure has 'fractional dimension' – i.e. a dimension between that of a line (1) and a surface (2); the dimension of the snowflake curve is 1.26. The study of this type of 'self-similar' figure is used in certain branches of physics – for example, crystal growth. Fractals are also important in *chaos theory and in computer graphics. *See also* MANDELBROT SET.

**fraction** *See* FRACTIONAL DISTILLATION.

**fractional crystallization** A method of separating a mixture of soluble solids by dissolving them in a suitable hot solvent and then lowering the temperature slowly. The least soluble component will crystallize out first, leaving the other components in solution. By controlling the temperature, it is sometimes possible to remove each component in turn.

**fractional distillation (fractionation)** The separation of a mixture of liquids by distillation. Effective separation can be achieved by using a long vertical column (**fractionating column**) attached to the distillation vessel and filled with glass beads. Vapour from the liquid rises up the column until it condenses and runs back into the vessel. The rising vapour in the column flows over the descending liquid, and eventually a steady state is reached in which there is a decreasing temperature gradient up the column. The vapour in the column has more volatile components towards the top and less volatile components at the bottom. Various **fractions** of the mixture can be drawn off at points on the column. Industrially, fractional distillation is performed in large towers containing many perforated trays. It is used extensively in petroleum refining.

**fractionating column** *See* FRACTIONAL DISTILLATION.

**fractionation** *See* FRACTIONAL DISTILLATION.

**fragmentation** A method of asexual reproduction, occurring in some invertebrates, in which parts of the organism break off and subsequently differentiate and develop into new individuals. It occurs especially in certain cnidarians and annelids. Regeneration may occur before separation, producing chains of individuals budding from the parent.

**frame dragging** *See* LENSE–THIRRING EFFECT.

**frame of reference** A set of axes, taken as being for practical purposes at rest, that enables the position of a point (or body) in space to be defined at any instant of time. In a four-dimensional continuum (*see* SPACE–TIME) a frame of reference consists of a set of four coordinate axes, three spatial and one of time.

**francium** Symbol Fr. A radioactive element belonging to group 1 of the periodic table; a.n. 87; r.d. 2.4; m.p. 27±1°C; b.p. 677±1°C. The element is found in uranium and thorium ores. All 22 known isotopes are radioactive, the most stable being francium–223. The existence of francium was confirmed by Marguerite Perey in 1939.

**((●)) SEE WEB LINKS**
• Information from the WebElements site

**Frank, Ilya** *See* CERENKOV, PAVEL ALEKSEYEVICH.

**Frank–Kasper phase** A phase that occurs in certain complex alloys involving both icosahedra and tetrahedra packed closely together. *See also* QUASICRYSTAL.

**Franklin, Benjamin** (1706–90) American scientist and statesman who held various government posts. As an amateur scientist he experimented with electricity, introducing the concepts of 'positive' and 'negative'. In 1752 he carried out the extremely dangerous experiment of flying a kite during a thunderstorm and proved the electrical nature of lightning. He also invented the lightning conductor.

**Franklin, Rosalind** *See* CRICK, FRANCIS HARRY COMPTON.

**Frasch process** A method of obtaining sulphur from underground deposits using a tube consisting of three concentric pipes. Superheated steam is passed down the outer pipe to melt the sulphur, which is forced up through the middle pipe by compressed air fed through the inner tube. The steam in the outer casing keeps the sulphur molten in the pipe. It was named after the German-born US chemist Hermann Frasch (1851–1914).

**fraternal twins (dizygotic twins)** Two individuals that result from a single pregnancy, each having developed from a separate fertilized egg. The two egg cells contain different combinations of *alleles as do the sperm that fertilize them. The twins therefore have no more genetic similarity than brothers or sisters from single births. *Compare* IDENTICAL TWINS.

**Fraunhofer, Josef von** (1787–1826) German physicist, who trained as an optician. In 1814 he observed dark lines in the spectrum of the sun (*see* FRAUNHOFER LINES). He also studied *Fraunhofer diffraction.

**Fraunhofer diffraction** A form of *diffraction in which the light source and the receiving screen are in effect at infinite distances from the diffracting object, so that the wave fronts can be treated as planar rather than spherical. In practice it involves parallel beams of light. It can be regarded as an extreme case of *Fresnel diffraction but is of more practical use in explaining single and multiple slit patterns.

**Fraunhofer lines** Dark lines in the solar spectrum that result from the absorption by elements in the solar chromosphere of some of the wavelengths of the visible radiation emitted by the hot interior of the sun.

**free electron** *See* ELECTRON.

**free-electron laser** A type of *laser that works by the stimulated emission of electromagnetic radiation produced by the motion of electrons moving rapidly in external magnetic fields. A beam of electrons at relativistic speed is produced in an accelerator and passed into a region in which the electrons are forced to follow a sinusoidal path by an array of magnets (the 'undulator'). The associated *synchrotron radiation is highly coherent as in a conventional laser. Free-electron lasers can be tuned to a narrow frequency range and radiation can be produced over a wide range, from microwaves to soft X-rays. The free-electron-laser mechanism might account for some aspects of pulsar radiation.

**free energy** A measure of a system's ability to do work. The **Gibbs free energy** (or **Gibbs function**), $G$, is defined by $G = H − TS$, where $G$ is the energy liberated or absorbed in a reversible process at constant pressure and constant temperature ($T$), $H$ is the *enthalpy and $S$ the *entropy of the system. Changes in Gibbs free energy, $\Delta G$, are useful in indicating the conditions under which a chemical reaction will occur. If $\Delta G$ is positive the reaction will only occur if energy is supplied to force it away from the equilibrium position (i.e. when $\Delta G = 0$). If $\Delta G$ is negative

the reaction will proceed spontaneously to equilibrium.

The **Helmholtz free energy** (or **Helmholtz function**), $F$, is defined by $F = U − TS$, where $U$ is the *internal energy. For a reversible isothermal process, $\Delta F$ represents the useful work available.

**free fall** Motion resulting from a gravitational field that is unimpeded by a medium that would provide a frictional retarding force or buoyancy. In the earth's gravitational field, free fall takes place at a constant acceleration, known as the *acceleration of free fall.

**free radical** An atom or group of atoms with an unpaired valence electron. Free radicals can be produced by photolysis or pyrolysis in which a bond is broken without forming ions (*see* HOMOLYTIC FISSION). Because of their unpaired valence electron, most free radicals are extremely reactive. *See also* CHAIN REACTION.

**free space** A region in which there is no matter and no electromagnetic or gravitational fields. It has a temperature of absolute zero, unit refractive index, and the speed of light is its maximum value. The electric constant (*see* PERMITTIVITY) and the magnetic constant (*see* PERMEABILITY) are defined for free space.

**freeze drying** A process used in dehydrating food, blood plasma, and other heat-sensitive substances. The product is deep-frozen and the ice trapped in it is removed by reducing the pressure and causing it to sublime. The water vapour is then removed, leaving an undamaged dry product.

**freeze fracture** A method of preparing material for electron microscopy that allows the visualization of the interior of plasma membranes and organelles. Cells are frozen at −196°C and cracked so that the plane of fracture runs through the middle of *lipid bilayers, separating the two halves. The exposed surfaces are then coated with carbon and platinum and the organic material is digested with enzymes (**freeze etching**), leaving a carbon–platinum replica of the fractured surface, which can be examined using the microscope.

**freezing mixture** A mixture of components that produces a low temperature. For example, a mixture of ice and sodium chloride gives a temperature of −20°C.

**freezing-point depression** *See* DEPRESSION OF FREEZING POINT.

**Frenkel defect** *See* CRYSTAL DEFECT (Feature).

**freon** *See* CHLOROFLUOROCARBON.

**frequency** Symbol $f$ or $v$. The rate of repetition of a regular event. The number of cycles of a wave, or some other oscillation or vibration, per second is expressed in *hertz. The frequency ($f$) of a wave motion is given by $f = c/\lambda$, where $c$ is the velocity of propagation and $\lambda$ is the wavelength. The frequency associated with a quantum of electromagnetic energy is given by $f = E/h$, where $E$ is the quantum's energy and $h$ is the Planck constant.

**frequency modulation (FM)** *See* MODULATION; RADIO.

**fresnel** A unit of frequency equal to $10^{12}$ hertz. In SI units this is equal to 1 terahertz (THz). It was named after the French physicist A. J. Fresnel (1788–1827).

**Fresnel diffraction** A form of *diffraction in which the light source or the receiving screen, or both, are at finite distances from the diffracting object, so that the wavefronts are not plane, as in *Fraunhofer diffraction. It was studied by A. J. Fresnel.

**Fresnel lens** A lens with one face cut into a series of steps. It enables a relatively light and robust lens of short focal length though poor optical quality to be used in projectors (as condenser lenses), searchlights, spot lights, car headlights, and lighthouses. Such lenses have been made fine enough to serve as magnifiers for reading small print.

**friction** The force that resists the motion of one surface relative to another with which it is in contact. For a body resting on a horizontal surface there is a normal contact force, $R$, between the body and surface, acting perpendicularly to the surface. If a horizontal force $B$ is applied to the body with the intention of moving it to the right, there will be an equal horizontal friction force, $F$, to the left, resisting the motion (see illustration). If $B$ is increased until the body just moves, the value of $F$ will also increase until it reaches the **limiting frictional force** ($F_L$), which is the maximum value of $F$. $F_L$ is then equal to $\mu_s R$, where $\mu_s$ is the **coefficient of static friction**, the value of which depends on the nature of the surfaces. Once the body is moving with constant velocity, the value of

$F$ falls to a value $F_k$, which is equal to $\mu_k R$, where $\mu_k$ is the **coefficient of kinetic friction**. Both $\mu_s$ and $\mu_k$ are independent of the surface area of the body unless this is very small and $\mu_k$ is almost independent of the relative velocity of the body and surface.

Friction occurs because surfaces, however smooth they may look to the eye, have many microscopic humps and crests. Therefore the actual area of contact is very small, and the consequent very high pressure leads to local pressure welding of the surfaces. During motion the welds are broken and remade continually. *See also* ROLLING FRICTION.

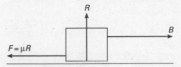

**Friction.**

**Friedel–Crafts reaction** A type of reaction in which an alkyl group (from a haloalkane) or an acyl group (from an acyl halide) is substituted on a benzene ring (see illustration overleaf). The product is an alkylbenzene (for alkyl substitution) or an alkyl aryl ketone (for acyl substitution). The reactions occur at high temperature (about 100°C) with an aluminium chloride catalyst. The catalyst acts as an electron acceptor for a lone pair on the halide atom. This polarizes the haloalkane or acyl halide, producing a positive charge on the alkyl or acyl group. The mechanism is then electrophilic substitution. Alcohols and alkenes can also undergo Friedel–Crafts reactions. The reaction is named after the French chemist Charles Friedel (1832–99) and the US chemist James M. Crafts (1839–1917).

**Frisch, Karl von** *See* LORENZ, KONRAD ZACHARIAS.

**Froehde's test** A *presumptive test for opioids. **Froehde's reagent** consists of 0.5 gram of sodium molybdate ($Na_2MoO_4$) dissolved in 100 ml of concentrated sulphuric acid. Opioids give a pink to purple colour.

**frogs** *See* AMPHIBIA.

**frond** *See* MEGAPHYLL.

**front** (in meteorology) The sloping interface between two *air masses of different origins and thermal characteristics (e.g. temperature and humidity). Where the air benzene + CH₃Cl ⟶ methylbenzene (toluene)

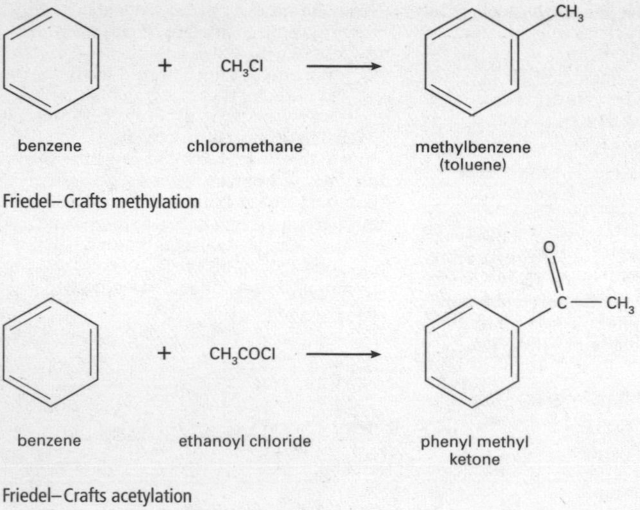

Friedel–Crafts methylation

benzene + CH₃COCl ⟶ phenyl methyl ketone

Friedel–Crafts acetylation

**Friedel-Crafts reaction.**

masses come together the warm air, being lighter, rises and slopes over the colder air. Distinctive weather phenomena are associated with fronts. At a **warm front** (the boundary zone at the front of the warm sector of a *depression), warm air is overtaking cold air and rising above it with a slope of about 1:150. This is associated with cirrus clouds followed by cirrostratus, altostratus, then nimbostratus clouds, as the warm front passes. At a **cold front**, which is usually found to the rear of a depression, warm air is forced to rise by an advancing undercutting wedge of cold air. This is accompanied by a sudden veering of the wind, a fall in temperature, and heavy rainfall. An **occluded front** (or **occlusion**) occurs when a cold front overtakes a warm front in an atmospheric depression.

**frontal lobe** The anterior part of each cerebral hemisphere, which is associated with the higher mental functions, such as abstract thought.

**frontier-orbital theory** A theory of the reactions of molecules that emphasizes the energies and symmetries of **frontier orbitals**: the two orbitals in a molecule that are the occupied orbital of highest energy and the unoccupied orbital of lowest energy. Frontier orbital theory was developed by the

Japanese chemist Kenichi Fukui (1919–98) in the 1950s and is an alternative approach to the *Woodward–Hoffmann rules. It has been very successful in explaining such reactions as the *Diels–Alder reaction.

**Frost diagram** A graph showing how standard electrode potentials vary with oxidation state for different oxidation states of an element. Frost diagrams can be constructed from *Latimer diagrams. For an element M, the standard electrode potential $E^{\ominus}$ is calculated for the reaction.

$$M(N) + Ne^- \rightarrow M(0)$$

where $M(N)$ indicates the species in oxidation state $N$ and $M(0)$ indicates the zero oxidation state. The Frost diagram is then obtained by plotting $NE^{\ominus}$ against $N$ for the different species. $NE^{\ominus}$ is proportional to the standard Gibbs free energy of the particular half reaction.

In a Frost diagram, the lowest point corresponds to the most stable oxidation state of the element. Also, the slope of a line between two points is the standard potential of the couple represented by the points. Like Latimer diagrams, Frost diagrams depend on pH.

**froth flotation** A method of separating mixtures of solids, used industrially for separating ores from the unwanted gangue. The mixture is ground to a powder and water and a frothing agent added. Air is blown through the water. With a suitable frothing agent, the bubbles adhere only to particles of ore and carry them to the surface, leaving the gangue particles at the bottom.

**fructification** *See* SPOROPHORE.

**fructose (fruit sugar; laevulose)** A simple sugar, $C_6H_{12}O_6$, stereoisomeric with glucose (*see* MONOSACCHARIDE). (Although natural fructose is the D-form, it is in fact laevorotatory.) Fructose occurs in green plants, fruits, and honey and tastes sweeter than sucrose (cane sugar), of which it is a constituent. Derivatives of fructose are important in the en-

ergy metabolism of living organisms. Fructose derived from corn starch is now used extensively in food manufacturing as a sucrose substitute.

**fruit** The structure formed from the ovary of a flower, usually after the ovules have been fertilized. It consists of the **fruit wall** (*see* PERICARP) enclosing the seed(s). Other parts of the flower, such as the receptacle, may develop and contribute to the structure, resulting in a **false fruit** (*see* PSEUDOCARP). The fruit may retain the seeds and be dispersed whole (an **indehiscent fruit**), or it may open (dehisce) to release the seeds (a **dehiscent fruit**). Fruits are divided into two main groups depending on whether the

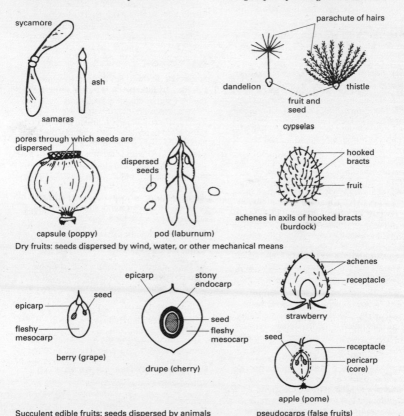

Dry fruits: seeds dispersed by wind, water, or other mechanical means

Succulent edible fruits: seeds dispersed by animals    pseudocarps (false fruits)

**Fruit.**

ovary wall remains dry or becomes fleshy (**succulent**). Succulent fruits are generally dispersed by animals and dry fruits by wind, water, or by some mechanical means. See illustration. *See also* COMPOSITE FRUIT.

**fruit fly** *See* DROSOPHILA.

**fruit sugar** *See* FRUCTOSE.

**frustration** A situation in which there are competing interactions in a system. For example, in a *spin glass the magnetic atoms are subject to both ferromagnetic and antiferromagnetic interactions. Frustration also occurs in the folding of large polymer molecules due to parts of the molecule getting in each other's way. It is thought that proteins fold readily to specific shapes because they have evolved to minimize this type of steric frustration.

**frustum** A solid figure produced when two parallel planes cut a larger solid or when one plane parallel to the base cuts it.

**FSH** *See* FOLLICLE-STIMULATING HORMONE.

**FTP (file transfer protocol)** The protocol used on the *Internet for transferring files (downloading or uploading) from one computer to another.

**5-FU** *See* 5-FLUOROURACIL.

**fucoxanthin** The major *carotenoid pigment present, with chlorophyll, in the brown algae (*see* PHAEOPHYTA).

**fuel** A substance that is oxidized or otherwise changed in a furnace or heat engine to release useful heat or energy. For this purpose wood, vegetable oil, and animal products have largely been replaced by *fossil fuels since the 18th century.

The limited supply of fossil fuels has encouraged the development of *renewable energy sources.

**fuel cell** A cell in which the chemical energy of a fuel is converted directly into electrical energy. The simplest fuel cell is one in which hydrogen is oxidized to form water over porous sintered nickel electrodes. A supply of gaseous hydrogen is fed to a compartment containing the porous anode and a supply of oxygen is fed to a compartment containing the porous cathode; the electrodes are separated by a third compartment containing a hot alkaline electrolyte, such as potassium hydroxide. The electrodes are porous to enable the gases to react with the electrolyte, with the nickel in the electrodes

acting as a catalyst. At the anode the hydrogen reacts with the hydroxide ions in the electrolyte to form water, with the release of two electrons per hydrogen molecule:

$$H_2 + 2OH^- \rightarrow 2H_2O + 2e^-$$

At the cathode, the oxygen reacts with the water, taking up electrons, to form hydroxide ions:

$$\frac{1}{2}O_2 + H_2O + 2e^- \rightarrow 2OH^-$$

The electrons flow through an external circuit as an electric current. The device is a more efficient converter of electric energy than a heat engine, but it is bulky and requires a continuous supply of gaseous fuels. Their use to power electric vehicles is being actively explored. The second generation of fuel cells uses molten salts, particularly lithium or potassium carbonate, as electrolytes. The third generation uses solid conducting ionic oxides. *See also* BIOCHEMICAL FUEL CELL.

**fuel element** *See* NUCLEAR REACTOR.

**fugacity** Symbol $f$. A thermodynamic function used in place of partial pressure in reactions involving real gases and mixtures. For a component of a mixture, $d(\ln f) = d\mu/RT$, where $\mu$ is the chemical potential. It has the same units as pressure and the fugacity of a gas is equal to the pressure if the gas is ideal. The fugacity of a liquid or solid is the fugacity of the vapour with which it is in equilibrium. The ratio of the fugacity to the fugacity in some standard state is the *activity. For a gas, the standard state is chosen to be the state at which the fugacity is 1. The activity then equals the fugacity.

**fullerene** *See* BUCKMINSTERFULLERENE.

**fullerite** *See* BUCKMINSTERFULLERENE.

**fuller's earth** A naturally occurring clay material (chiefly montmorillonite) that decolorizes oil and grease. In the past raw wool was cleaned and whitened by kneading it in water with fuller's earth; a process known as **fulling**. Fuller's earth is now widely used to decolorize fats and oils and also as an insecticide carrier and drilling mud. The largest deposits occur in the USA, UK, and Japan.

**full-wave rectifier** *See* RECTIFIER.

**fulminate** *See* CYANIC ACID.

**fulminic acid** *See* CYANIC ACID.

**fumaric acid** *See* BUTENEDIOIC ACID.

**function** Any operation or procedure that

relates one variable to one or more other variables. If $y$ is a function of $x$, written $y = f(x)$, a change in $x$ produces a change in $y$, and if $x$ is known, $y$ can be determined. $x$ is the **independent variable** and $y$ is the **dependent variable**.

**functional** A function of a function. Functionals are used very extensively in the quantum-mechanical many-body problem, statistical mechanics, and quantum field theory.

**functional group** The group of atoms responsible for the characteristic reactions of a compound. The functional group is –OH for alcohols, –CHO for aldehydes, –COOH for carboxylic acids, etc.

**fundamental** *See* HARMONIC.

**fundamental constants (universal constants)** Those parameters that do not change throughout the universe. The charge on an electron, the speed of light in free space, the Planck constant, the gravitational constant, the electric constant, and the magnetic constant are all thought to be examples. It has been suggested that some fundamental constants might change with time, although there is no conclusive evidence that this occurs.

(((●))) **SEE WEB LINKS**

• Values of fundamental physical constants from NIST

**fundamental interactions** The four different types of interaction that can occur between bodies. These interactions can take place even when the bodies are not in physical contact and together they account for all the observed forces that occur in the universe. While the unification of these four types of interaction into a single theory has long been the aim of physicists, this has not yet been fully achieved. *See also* ELEMENTARY PARTICLES; GAUGE THEORY; UNIFIED-FIELD THEORY.

The **gravitational interaction**, some $10^{40}$ times weaker than the electromagnetic interaction, is the weakest of all. The force that it generates acts between all bodies that have mass and the force is always attractive. The interaction can be visualized in terms of a classical *field of force in which the strength of the force falls off with the square of the distance between the interacting bodies (*see* NEWTON'S LAW OF GRAVITATION). The hypothetical gravitational quantum, the **graviton**, is also a useful concept in some contexts. On the atomic scale the gravitational force is negligibly weak, but on the cosmological scale, where masses are enormous, it is immensely important in holding the components of the universe together. Because gravitational interactions are long-ranged, there is a well-defined macroscopic theory in general relativity. At present, there is no satisfactory quantum theory of gravitational interaction.

The **weak interaction**, some $10^{10}$ times weaker than the electromagnetic interaction, occurs between *leptons and in the decay of hadrons. It is responsible for the *beta decay of particles and nuclei. In the current model, the weak interaction is visualized as a force mediated by the exchange of virtual particles, called intermediate vector bosons. The weak interactions are described by *electroweak theory, which unifies them with the electromagnetic interactions.

The **electromagnetic interaction** is responsible for the forces that control atomic structure, chemical reactions, and all electromagnetic phenomena. It accounts for the forces between charged particles, but unlike the gravitational interaction, can be either attractive or repulsive. Some neutral particles decay by electromagnetic interaction. The interaction is either visualized as a classical field of force (*see* COULOMB'S LAW) or as an exchange of virtual *photons. As with gravitational interactions, the fact that electromagnetic interactions are long-ranged means that they have a well-defined classical theory given by *Maxwell's equations. The quantum theory of electromagnetic interactions is described by *quantum electrodynamics.

The **strong interaction**, some $10^2$ times stronger than the electromagnetic interaction, functions only between *hadrons and is responsible for the force between nucleons that gives the atomic nucleus its great stability. It operates at very short range inside the nucleus ($10^{-15}$ metre) and is visualized as an exchange of virtual mesons. The strong interactions are described by *quantum chromodynamics.

**fundamental units** A set of independently defined *units of measurement that forms the basis of a system of units. Such a set requires three mechanical units (usually of length, mass, and time) and, in some systems, one electrical unit; it has also been found convenient to treat certain other quantities as fundamental, even though they are not strictly independent. In the metric system the centimetre–gram–second (c.g.s.) system was replaced by the metre–kilogram–second (m.k.s.) system; the latter now forms the basis for *SI units. In British Imperial units the foot–pound–second (f.p.s.) system was formerly used.

**fungi** A group of organisms formerly classified in the kingdom Fungi. Molecular studies have shown that fungi are more closely related to animals than plants, and fungi and animals are now classified in the assemblage opisthokonts. Fungi can either exist as single cells or make up a multicellular body called a *mycelium, which consists of filaments known as *hyphae. Most fungal cells are multinucleate and have cell walls composed chiefly of *chitin. Fungi exist primarily in damp situations on land and, because of the absence of chlorophyll, are either parasites or saprotrophs on other organisms. The principal criteria used in classification are the nature of the spores produced and the presence or absence of cross walls within the hyphae (*see* Ascomycota; Basidiomycota; Zygomycota). *See also* LICHENS.

 SEE WEB LINKS
- Tree of Life survey of fungi, including phylogeny plus many links to other sites

**fungicide** *See* PESTICIDE.

**Fungi Imperfecti** *See* DEUTEROMYCOTA.

**funicle** The stalk that attaches an ovule to the placenta in the ovary of a flowering plant. It contains a strand of conducting tissue leading from the placenta into the chalaza.

**furan** A colourless liquid compound, $C_4H_4O$; r.d. 0.94; m.p. –86°C; b.p. 31.4°C. It has a five-membered ring consisting of four $CH_2$ groups and one oxygen atom.

**furanose** A *sugar having a five-membered ring containing four carbon atoms and one oxygen atom.

**furfural** A colourless liquid, $C_5H_4O_2$, b.p.

162°C, which darkens on standing in air. It is the aldehyde derivative of *furan and occurs in various essential oils and in *fusel oil. It is used as a solvent for extracting mineral oils and natural resins and itself forms resins with some aromatic compounds.

**fuse** A length of thin wire made of tinned copper or a metal alloy of low melting point that is designed to melt at a specified current loading in order to protect an electrical device or circuit from overloading. The wire is often enclosed in a small glass or ceramic cartridge with metal ends.

**fused ring** *See* RING.

**fusel oil** A mixture of high-molecular weight *alcohols containing also esters and fatty acids, sometimes formed as a toxic impurity in the distillation products of alcoholic fermentation. It is used as a source of higher alcohols and in making paints and plastics.

**fusible alloys** Alloys that melt at low temperature (around 100°C). They have a number of uses, including constant-temperature baths, pipe bending, and automatic sprinklers to prevent fires from spreading. Fusible alloys are usually *eutectic mixtures of bismuth, lead, tin, and cadmium. Wood's metal and Lipowitz's alloy are examples of alloys that melt at about 70°C.

**fusion** 1. Melting. 2. *See* NUCLEAR FUSION. 3. The combining together of cells, nuclei, or cytoplasm. *See* CELL FUSION; FERTILIZATION.

**fusion reactor** *See* THERMONUCLEAR REACTOR.

**fuzzy logic** A form of logic that allows for degrees of imprecision, used in *artificial intelligence studies. More traditional logics deal with two truth values: 'true' and 'false'. Fuzzy logics are multivalued dealing with concepts such as 'fairly true' and 'more or less true'. These can be represented by numbers within a range [0,1], with the number representing the degree of truth. **Fuzzy control** applies fuzzy logic to the computer-control of processes.

**GABA** *See* GAMMA-AMINOBUTYRIC ACID.

**gabbro** A coarse-grained basic intrusive igneous rock with a similar composition to *basalt, i.e. mainly plagioclase feldspar, and pyroxine, with some olivine. Accessory minerals include apatite, hornblende, ilmenite, and magnetite.

**Gabor, Dennis** (1900–79) Hungarian-born British physicist, who worked as a research engineer from 1927 until 1933, when he joined the British Thomson-Houston company. In 1948 he joined the staff of Imperial College, London. In that same year, while working on electron microscopes, he invented *holography, for which he was awarded the 1971 Nobel Prize for physics.

**Gabriel reaction** A method of making a primary *amine (free from any secondary or tertiary amine impurities) from a haloalkane (alkyl halide) using potassium phthalimide. It is named after Siegmund Gabriel (1851–1924).

**gadolinium** Symbol Gd. A soft silvery metallic element belonging to the *lanthanoids; a.n. 64; r.a.m. 157.25; r.d. 7.901 (20°C); m.p. 1313°C; b.p. 3266°C. It occurs in gadolinite, xonotime, monazite, and residues from uranium ores. There are seven stable natural isotopes and eleven artificial isotopes are known. Two of the natural isotopes, gadolinium–155 and gadolinium–157, are the best neutron absorbers of all the elements. The metal has found limited applications in nuclear technology and in ferromagnetic alloys (with cobalt, copper, iron, and cerium). Gadolinium compounds are used in electronic components. The element was discovered by Jean de Marignac (1817–94) in 1880.

**(⊕) SEE WEB LINKS**
• Information from the WebElements site

**Gaia hypothesis** The theory, based on an idea put forward by the British scientist James Ephraim Lovelock (1919–   ), that the whole earth, including both its biotic (living) and abiotic (nonliving) components, functions as a single self-regulating system. Named after the Greek earth goddess, it proposes that the responses of living organisms to environmental conditions ultimately bring about changes that make the earth better adapted to support life; the system would rid itself of any species that adversely affects the environment. The theory has found favour with many conservationists.

**gain** *See* AMPLIFIER.

**galactic centre** The region at the centre of a galaxy. In the Milky Way (our Galaxy) it corresponds to a radio source in the direction of Sagittarius. This source, called Sagittarius A, may correspond to a massive *black hole.

**galactic merger** The joining of two galaxies that approach within each other's gravitational fields. They spiral together, forming one galaxy and producing a starburst, in which many new stars are created by the collapse of interstellar clouds. Streams of stars, some large enough to be dwarf galaxies, form tails behind the new galaxy.

**galactic nucleus** A bulge of older stars (population II, *see* POPULATION TYPE) that surrounds the centre of a galaxy. The spiral arms of a spiral galaxy (such as the Milky Way) originate at the galactic nucleus.

**galactose** A simple sugar, $C_6H_{12}O_6$, stereoisomeric with glucose, that occurs naturally as one of the products of the enzymic digestion of milk sugar (lactose) and as a constituent of gum arabic.

**galactosidase** *See* LACTASE.

**galaxy** A vast collection of stars, dust, and gas held together by the mutual gravitational attraction between its components. Galaxies are usually classified as elliptical, spiral, or irregular in shape. **Elliptical galaxies** appear like ellipsoidal clouds of stars, with very little internal structure apart from (in some cases) a denser nucleus. **Spiral galaxies** are flat disc-shaped collections of stars with prominent spiral arms. **Irregular galaxies** have no apparent structure or shape.

The sun belongs to a spiral galaxy known as the **Galaxy** (with a capital G) or the **Milky**

**Way System**. There are some $10^{11}$ stars in the system, which is about 30 000 parsecs across with a maximum thickness at the centre of about 4000 parsecs. The sun is about 10 000 parsecs from the centre of the Galaxy.

The galaxies are separated from each other by enormous distances, the nearest large galaxy to our own (the Andromeda galaxy) being about $6.7 \times 10^5$ parsecs away.

**galaxy cluster** A group of *galaxies containing many hundreds of members extending over a radius of up to a few megaparsecs (there also exist small groups of galaxies, such as the *Local Group, with a few tens of members). The richest and most regular clusters, such as the **Coma cluster**, with thousands of members, are gravitationally bound systems; it is not certain whether other less regular and less concentrated clusters are also bound. As well as galaxies, the clusters contain hot **intracluster gas**, at temperatures between $10^7$ and $10^8$ K; this can be detected by its X-ray emission. On a scale larger than clusters there are also **superclusters**, with extents of the order of a hundred megaparsecs, containing about a hundred galaxies. It is not known whether superclusters are gravitationally bound. *See also* MISSING MASS.

**Galen** (*c*. 130–*c*. 200) Greek physician, who studied in Pergamum, Corinth, and Alexandria. He practised in Pergamum before moving to Rome. Galen's writings from this time became the basis of medical teaching and practice for 1500 years.

**galena** A mineral form of lead(II) sulphide, PbS, crystallizing in the cubic system; the chief ore of lead. It usually occurs as grey metallic cubes, frequently in association with silver, arsenic, copper, zinc, and antimony. Important deposits occur in Australia (at Broken Hill), Germany, the USA (especially in Missouri, Kansas, and Oklahoma), and the UK.

**Galilean satellites** The four largest satellites (moons) of Jupiter, so-called because they were discovered and described by Galileo in 1610. They are, in order of increasing distance from the planet, Io, Europa, Ganymede and Callisto. The largest is Ganymede, which, with a diameter of 5262 kilometres, is about 7% larger than the planet Mercury.

**Galilean telescope** *See* TELESCOPE.

**Galilean transformations** A set of equations for transforming the position and motion parameters from a frame of reference with origin at O and coordinates $(x, y, z)$ to a frame with origin at O′ and coordinates at $(x′, y′, z′)$. They are:

$$x′ = x - vt$$
$$y′ = y$$
$$z′ = z$$
$$t′ = t$$

The equations conform to Newtonian mechanics. *Compare* LORENTZ TRANSFORMATIONS.

**Galileo Galilei** (1564–1642) Italian astronomer and physicist. In 1583 he noticed that the time of swing of a *pendulum is independent of its amplitude, and three years later invented a hydrostatic balance for measuring *relative densities. He became a professor in Padua in 1592 and it was there (in 1610) that he made his first astronomical telescope. With it he discovered four satellites of Jupiter, mountains on the moon, and sunspots. Returning to Pisa, his birthplace, he studied motion, demonstrating that the speed of a falling body is independent of its weight. He also gave open support to the sun-centred theory of the universe advocated by *Copernicus, a stand that brought him into conflict with the church. He was summoned to Rome, forced to retract before the Inquisition, and banished under house arrest.

**gall** An abnormal growth of a plant tissue or organ elicited by a foreign organism. Galls most frequently occur as swellings or pits in stems, roots, leaves, and buds. Organisms responsible for their formation include bacteria, viruses, fungi, nematodes, mites, and insects. The gall structure is typically very distinct from surrounding normal tissue and often is characteristic of the eliciting organism. The mechanisms underlying gall formation are known in only a few cases. The bacterium *Agrobacterium tumefaciens*, which is responsible for crown galls, induces a genetic change in infected host tissue by transfer of a plasmid bearing tumour-forming genes.

**gall bladder** A small pouch attached to the *bile duct, present in most vertebrates. *Bile, produced in the *liver, is stored in the gall bladder and released when food (especially fatty substances) enters the duodenum.

**gallium** Symbol Ga. A soft silvery metallic element belonging to group 13 (formerly IIIB) of the periodic table; a.n. 31; r.a.m. 69.72; r.d. 5.90 (20°C); m.p. 29.78°C; b.p. 2403°C. It occurs in zinc blende, bauxite, and kaolin, from which it can be extracted by fractional electrolysis. It also occurs in gallite, $CuGaS_2$, to an extent of 1%; although bauxite only contains 0.01% this is the only commercial source. The two stable isotopes are gallium–69 and gallium–71; there are eight radioactive isotopes, all with short half-lives. The metal has only a few minor uses (e.g. as an activator in luminous paints), but gallium arsenide is extensively used as a semiconductor in many applications. Gallium corrodes most other metals because it rapidly diffuses into their lattices. Most gallium(I) and some gallium(II) compounds are unstable. The element was first identified by Paul Lecoq de Boisbaudran (1838–1912) in 1875.

**((()) SEE WEB LINKS**

• Information from the WebElements site

**gallon 1. (Imperial gallon)** The volume occupied by exactly ten pounds of distilled water of density 0.998 859 gram per millilitre in air of density 0.001 217 gram per millilitre. 1 gallon = 4.546 09 litres (cubic decimetres). **2.** A unit of volume in the US Customary system equal to 0.832 68 Imperial gallon, i.e. 3.785 44 litres.

**GALP** See GLYCERALDEHYDE 3-PHOSPHATE.

**Galvani, Luigi** (1737–98) Italian physiologist. In the late 1770s he observed that the muscles of a dead frog twitched when touched by two different metals. He concluded that the muscle was producing electricity, later disproved by *Volta (who showed that the two metals and body fluids formed a battery). Galvani invented *galvanized iron and the *galvanometer.

**galvanic cell** See VOLTAIC CELL.

**galvanized iron** Iron or steel that has been coated with a layer of zinc to protect it from corrosion. Corrugated mild-steel sheets for roofing and mild-steel sheets for dustbins, etc., are usually galvanized by dipping them in molten zinc. The formation of a brittle zinc–iron alloy is prevented by the addition of small quantities of aluminium or magnesium. Wire is often galvanized by a cold electrolytic process as no alloy forms in this process. Galvanizing is an effective method of protecting steel because even if the surface is scratched, the zinc still protects the underlying metal. See SACRIFICIAL PROTECTION.

**galvanometer** An instrument for detecting and measuring small electric currents. In the moving-coil instrument a pivoted coil of fine insulated copper wire surrounds a fixed soft-iron core between the poles of a permanent magnet. The interaction between the field of the permanent magnet and the sides of the coil, produced when a current flows through it, causes a torque on the coil. The moving coil carries either a pointer or a mirror that deflects a light beam when it moves; the extent of the deflection is a measure of the strength of the current. The galvanometer can be converted into an *ammeter or a *voltmeter. Digital electronic instruments are increasingly replacing the moving-coil type. See also BALLISTIC GALVANOMETER.

**gametangium** An organ that produces gametes. The term is usually restricted to the sex organs of algae, fungi, mosses, and ferns. See ANTHERIDIUM; ARCHEGONIUM; OOGONIUM.

**gamete** A reproductive cell that fuses with another gamete to form a zygote. Examples of gametes are ova and spermatozoa. Gametes are *haploid, i.e. they contain half the normal (diploid) number of chromosomes; thus when two fuse, the diploid number is restored (*see* FERTILIZATION). Gametes are formed by *meiosis. See also SEXUAL REPRODUCTION.

**gametogenesis** The processes involved in the formation of gametes. Gametes are normally formed by *meiosis but sometimes by *mitosis (as in the gametophyte generation of the ferns). In mammals gametogenesis in the female is known as *oogenesis and occurs in the ovaries; in the male it is known as *spermatogenesis and occurs in the testes.

**gametophyte** The generation in the life cycle of a plant that bears the gamete-producing sex organs. The gametophyte is *haploid. It is the dominant phase in the life cycle of mosses and liverworts, the *sporophyte generation depending on it either partially or completely. In clubmosses, horsetails, and ferns it is the *prothallus. In seed plants it is very much reduced. For example, in angiosperms the pollen grain is the male gametophyte and the embryo sac is the female gametophyte. See also ALTERNATION OF GENERATIONS.

**gamma-aminobutyric acid (GABA)** An

inhibitory *neurotransmitter in the central nervous system (principally the brain) that is capable of increasing the permeability of *postsynaptic membranes. GABA is synthesized by *decarboxylation of the amino acid glutamate.

**gamma globulin** *See* GLOBULIN.

**gammahydroxybutyric acid** *See* 4-HYDROXYBUTANOIC ACID.

**gamma-iron** *See* IRON.

**gamma radiation** Electromagnetic radiation emitted by excited atomic nuclei during the process of passing to a lower excitation state. Gamma radiation ranges in energy from about $10^{-15}$ to $10^{-10}$ joule (10 keV to 10 MeV) corresponding to a wavelength range of about $10^{-10}$ to $10^{-14}$ metre. A common source of gamma radiation is cobalt–60, the

$$^{60}_{27}Co \xrightarrow{\beta} {}^{60}_{28}Ni \xrightarrow{\gamma} {}^{60}_{28}Ni$$

The de-excitation of nickel–60 is accompanied by the emission of gamma-ray photons having energies 1.17 MeV and 1.33 MeV.

**gamma-ray astronomy** *Astronomy involving gamma ray photons (with energies in excess of 100 MeV). The cosmic radiation with the highest energy can be detected by electron–photon cascades, which take place in the atmosphere. Gamma rays having lower energies can only be detected above the atmosphere. Many high-energy processes in *astrophysics are responsible for the production of gamma rays; one example is the decay of neutral *pions.

An interesting phenomenon is the **gamma-ray burst**. These events last for a few seconds, during which they are the strongest source of gamma rays in the sky. It is thought that they may be the result of the formation of a *black hole, either when a large star collapses or when two neutron stars collide.

**ganglion** A mass of nervous tissue containing many *cell bodies and *synapses, usually enclosed in a connective-tissue sheath. In vertebrates most ganglia occur outside the central nervous system; exceptions are the *basal ganglia in the brain. In invertebrates ganglia occur along the nerve cords and the most anterior pair (**cerebral ganglia**) are analogous to the vertebrate brain; invertebrate ganglia constitute a part of the central nervous system.

**gangue** Rock and other waste material present in an ore.

**ganoid scale** *See* SCALES.

**garnet** Any of a group of silicate minerals that conform to the general formula $A_3B_2(SiO_4)_3$. The elements representing A may include magnesium, calcium, manganese, and iron(II); those representing B may include aluminium, iron(III), chromium, or titanium. Six varieties of garnet are generally recognized:
pyrope, $Mg_3Al_2Si_3O_{12}$;
almandine, $Fe_3^{2+}Al_2Si_3O_{12}$;
spessartite, $Mn_3Al_2Si_3O_{12}$;
grossularite, $Ca_3Al_2Si_3O_{12}$;
andradite, $Ca_3(Fe^{3+},Ti)_2Si_3O_{12}$;
uvarovite, $Ca_3Cr_2Si_3O_{12}$.
Varieties of garnet are used as gemstones and abrasives.

**gas** A state of matter in which the matter concerned occupies the whole of its container irrespective of its quantity. In an *ideal gas, which obeys the *gas laws exactly, the molecules themselves would have a negligible volume and negligible forces between them, and collisions between molecules would be perfectly elastic. In practice, however, the behaviour of real gases deviates from the gas laws because their molecules occupy a finite volume, there are small forces between molecules, and in polyatomic gases collisions are to a certain extent inelastic (*see* EQUATION OF STATE).

**gas chromatography** A technique for separating or analysing mixtures of gases by *chromatography. The apparatus consists of a very long tube containing the stationary phase. This may be a solid, such as kieselguhr (**gas–solid chromatography**, or **GSC**), or a nonvolatile liquid, such as a hydrocarbon oil coated on a solid support (**gas–liquid chromatography**, or **GLC**). The sample is often a volatile liquid mixture, which is vaporized and swept through the column by a carrier gas (e.g. hydrogen). The components of the mixture pass through the column at different rates because they adsorb to different extents on the stationary phase. They are detected as they leave, either by measuring the thermal conductivity of the gas or by a flame detector.

Gas chromatography is usually used for analysis; components can be identified by the time they take to pass through the column. It is sometimes also used for separating mixtures.

Gas chromatography is often used to separate a mixture into its components, which are then directly injected into a mass spectrometer. This technique is known as **gas chromatography–mass spectroscopy** or **GCMS**.

**gas constant (universal molar gas constant)** Symbol $R$. The constant that appears in the **universal gas equation** (see GAS LAWS). It has the value 8.314 510(70) J $K^{-1}$ $mol^{-1}$.

**gas-cooled reactor** See NUCLEAR REACTOR.

**gaseous exchange** The transfer of gases between an organism and the external environment in either direction. It occurs by diffusion across a *concentration gradient and includes the exchange of oxygen and carbon dioxide in respiration and photosynthesis. Successful gaseous exchange requires a large surface area, as is provided by the alveoli of the lungs and the leaves of plants.

**gas equation** See GAS LAWS.

**gas giant** Any of the four outer planets of the solar system – *Jupiter, *Saturn, *Uranus, or *Neptune – or a planet with similar physical characteristics found in another solar system.

**gas laws** Laws relating the temperature, pressure, and volume of an *ideal gas. *Boyle's law states that the pressure ($p$) of a specimen is inversely proportional to the volume ($V$) at constant temperature ($pV =$ constant). The modern equivalent of *Charles' law states that the volume is directly proportional to the thermodynamic temperature ($T$) at constant pressure ($V/T =$ constant); originally this law stated the constant expansivity of a gas kept at constant pressure. The pressure law states that the pressure is directly proportional to the thermodynamic temperature for a specimen kept at constant volume. The three laws can be combined in the **universal gas equation**, $pV = nRT$, where $n$ is the amount of gas in the specimen and $R$ is the *gas constant. The gas laws were first established experimentally for real gases, although they are obeyed by real gases to only a limited extent; they are obeyed best at high temperatures and low pressures. See also EQUATION OF STATE.

**gasohol** A mixture of petrol (gasoline) and alcohol (i.e. typically ethanol at 10%, or methanol at 3%), used as an alternative fuel for cars and other vehicles in many countries. The ethanol is obtained as a *biofuel by fermentation of agricultural crops or crop

residues, for example sugar cane waste. The percentage of ethanol in the fuel is denoted by the **E number**. Many cars can use a mixture of 85% ethanol and 15% petrol, called E85. Ethanol-based gasohol has a higher octane rating and burns more completely than conventional petrol, thus lowering some emissions. However, the ethanol can damage certain engine components, such as rubber seals. Methanol-based gasohol is more toxic and corrosive, and its emissions include formaldehyde, a known carcinogen. It is not widely used.

**gas oil** A high-density petroleum fraction (between kerosene and lubricating oil), whose molecules have up to 25 carbon atoms. It is used as a domestic and industrial heating fuel.

**gasoline** See PETROLEUM.

**gas-phase electrophoresis** See ION-MOBILITY SPECTROMETRY.

**gas thermometer** A device for measuring temperature in which the working fluid is a gas. It provides the most accurate method of measuring temperatures in the range 2.5 to 1337 K. Using a fixed mass of gas a **constant-volume thermometer** measures the pressure of a fixed volume of gas at relevant temperatures, usually by means of a mercury *manometer and a *barometer.

**gastric** Of or relating to the stomach.

**gastric juice** An acidic mixture of inorganic salts, hydrochloric acid (see OXYNTIC CELL), mucus, and *pepsinogens secreted by **gastric glands** in the stomach lining.

**gastric mill (proventriculus)** A type of *gizzard occurring in many crustaceans. It is situated in the anterior region of the stomach and consists of a set of bones (ossicles) and muscles that grind food particles. The food particles are then filtered by bristles in the posterior section of the stomach.

**gastrin** A hormone, produced in the stomach, that controls the release of gastric juice. The secretion of gastrin is stimulated by the presence of food in the stomach. It is one of the hormones that integrates and controls digestive processes (see also SECRETIN).

**Gastropoda** A class of molluscs that includes the snails, whelks, limpets, land and sea slugs, and conches. Molluscs have a well-developed head with tentacles, a large flattened foot, and a coiled twisted shell. They occupy marine, freshwater, and terrestrial habitats; in the terrestrial and some freshwater gastropods the *mantle cavity acts as a lung instead of enclosing gills.

**gastrula** The stage in the *development of an animal embryo that succeeds the *blastula. It begins with the production of the primary *germ layers and the embryo becomes converted to a cup-shaped structure containing a cavity (the *archenteron).

**gas turbine** An internal-combustion engine in which the products of combustion of a fuel burnt in compressed air are expanded through a turbine. Atmospheric air is compressed by a rotary compressor driven by the turbine, fed into a combustion chamber, and mixed with the fuel (kerosene, natural gas, etc.); the expanding gases drive the turbine and power is taken from the unit by means of rotation of the turbine shaft (as in locomotives) or thrust from a jet (as in aircraft).

**gate** **1.** An electronic circuit with a single output and one or more inputs; the output is a function of the input or inputs. In the **transmission gate** the output waveform is a replica of a selected input during a specific interval. In the **switching gate** a constant output is obtained for a specified combination of inputs. These gates are the basic components of digital computers. *See* LOGIC CIRCUITS. **2.** The electrode in a field-effect *transistor that controls the current through the channel.

**Gattermann reaction** A variation of the *Sandmeyer reaction for preparing chloro- or bromoarenes by reaction of the diazonium compound. In the Gattermann reaction the aromatic amine is added to sodium nitrite and the halogen acid (10°C), then fresh copper powder (e.g. from $Zn + CuSO_4$) is added and the solution warmed. The diazonium salt then forms the haloarene, e.g.

$$C_6H_5N_2{}^+Cl^- \rightarrow C_6H_5Cl + N_2$$

The copper acts as a catalyst. The reaction is easier to perform than the Sandmeyer reaction and takes place at lower temperature, but generally gives lower yields. It was discovered in 1890 by the German chemist Ludwig Gattermann (1860–1920).

**gauche conformation** *See* CONFORMATION.

**gauge boson** A spin-one vector boson that mediates interactions governed by *gauge theories. Examples of gauge bosons

are photons in *quantum electrodynamics, gluons in *quantum chromodynamics, and W and Z bosons that mediate the interactions in the Weinberg–Salam model unifying electromagnetic and weak interactions (*see* ELECTROWEAK THEORY). If the *gauge symmetry of the theory is unbroken, the gauge boson is massless. Examples of massless gauge bosons include the photon and gluon. If the gauge symmetry of the theory is a *broken symmetry, the gauge boson has a nonzero mass, examples being the W and Z bosons. Treating gravity, as described by the general theory of *relativity, as a gauge theory, the gauge boson is the massless spin-two *graviton.

**gauge theory** Any of a number of *quantum field theories put forward to explain fundamental interactions. A gauge theory involves a symmetry *group for the fields and potentials (the **gauge group**). In the case of electrodynamics, the group is Abelian whereas the gauge theories for strong and weak interactions use non-Abelian groups. Non-Abelian gauge theories are known as **Yang–Mills theories**. This difference explains why *quantum electrodynamics is a much simpler theory than *quantum chromodynamics, which describes the strong interactions, and *electroweak theory, which is the unified theory of the weak and electromagnetic interactions. In the case of quantum gravity, the gauge group is even more complicated than the gauge groups for either the strong or weak interactions.

In gauge theories the interactions between particles can be explained by the exchange of particles (intermediate vector bosons, or *gauge bosons), such as gluons, photons, and W and Z bosons.

**gauss** Symbol G. The c.g.s. unit of magnetic flux density. It is equal to $10^{-4}$ tesla.

**Gauss, Karl Friedrich** (1777–1855) German mathematician and physicist, who became director of Göttingen Observatory in 1806. One of the greatest mathematicians of all time, he contributed to the theory of numbers and proved the fundamental theorem of algebra. His collaboration with Wilhelm *Weber on electromagnetism led to the invention of an electric telegraph, and he worked out the relationship between electric flux and electric field (*see* GAUSS' LAW).

**Gaussian distribution** *See* NORMAL DISTRIBUTION.

**Gaussian units** A system of units for electric and magnetic quantities based upon c.g.s. electrostatic and electromagnetic units. Although replaced by *SI units in most branches of science, they are, like Heaviside–Lorentz units, still used in relativity theory and in particle physics. In Gaussian units, the electric and magnetic constants are both equal to unity.

**Gauss' law** The total electric flux normal to a closed surface in an electric field is proportional to the algebraic sum of the electric charges within the surface. A similar law applies to surfaces drawn in a magnetic field and the law can be generalized for any vector field through a closed surface. It was first stated by Karl Gauss.

**gaussmeter** A *magnetometer, especially one calibrated in gauss.

**Gauss' theorem** See DIVERGENCE THEOREM.

**Gay-Lussac's law** **1.** When gases combine chemically the volumes of the reactants and the volume of the product, if it is gaseous, bear simple relationships to each other when measured under the same conditions of temperature and pressure. The law was first stated in 1808 by Joseph Gay-Lussac (1778–1850) and led to *Avogadro's law. **2.** See CHARLES' LAW.

**gaylussite** A mineral consisting of a hydrated mixed carbonate of sodium and calcium, $Na_2CO_3.CaCO_3.5H_2O$.

**GCMS** See GAS CHROMATOGRAPHY.

**gebi-** See BINARY PREFIXES.

**Gegenschein** (German: counterglow) A faint elliptical patch of light visible on a moonless night on the ecliptic at a point 180° from the position of the sun. It is caused by the reflection of sunlight by meteoric particles (see also ZODIACAL LIGHT).

**Geiger, Hans Wilhelm** (1882–1945) German physicist, who carried out research with *Rutherford at Manchester University before returning to Germany in 1912. In 1908 he and Rutherford produced the *Geiger counter, improved in 1928 as the Geiger–Müller counter. In 1909 his scattering experiments with alpha particles led to Rutherford's nuclear theory of the atom.

**Geiger counter (Geiger–Müller counter)** A device used to detect and measure *ionizing radiation. It consists of a tube containing a low-pressure gas (usually a mixture of methane with argon or neon) and a cylindrical hollow cathode through the centre of which runs a fine-wire anode (see illustration). A potential difference of about 1000 volts is maintained between the electrodes. An ionizing particle or photon passing through a window into the tube will cause an ion to be produced and the high p.d. will accelerate it towards its appropriate electrode, causing an avalanche of further ionizations by collision. The consequent current pulses can be counted in electronic circuits or simply amplified to work a small loudspeaker in

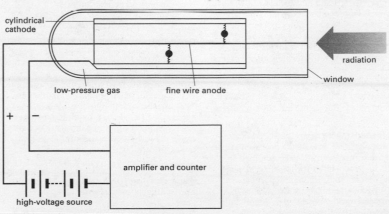

**Geiger counter.**

the instrument. It was first devised in 1908 by Hans Geiger. Geiger and W. Müller produced an improved design in 1928.

**Geissler tube** An early form of gas-discharge tube designed to demonstrate the luminous effects of an electric discharge passing through a low-pressure gas between two electrodes. Modified forms are used in spectroscopy as a source of light. It was invented in 1858 by Heinrich Geissler (1814–79).

**gel** A lyophilic *colloid that has coagulated to a rigid or jelly-like solid. In a gel, the disperse medium has formed a loosely-held network of linked molecules through the dispersion medium. Examples of gels are silica gel and gelatin.

**gelatin(e)** A colourless or pale yellow water-soluble protein obtained by boiling collagen with water and evaporating the solution. It swells when water is added and dissolves in hot water to form a solution that sets to a gel on cooling. It is used in photographic emulsions and adhesives, in bacteriology for preparing culture media, in pharmacy for preparing capsules and suppositories, and in jellies and other foodstuffs.

**gel electrophoresis** *See* ELECTRO-PHORESIS.

**gel filtration** A type of column *chromatography in which a mixture of liquids is passed down a column containing a gel. Small molecules in the mixture can enter pores in the gel and move slowly down the column; large molecules, which cannot enter the pores, move more quickly. Thus, mixtures of molecules can be separated on the basis of their size. The technique is used particularly for separating proteins but it can also be applied to other polymers and to cell nuclei, viruses, etc.

**gelignite** A high explosive made from nitroglycerin, cellulose nitrate, sodium nitrate, and wood pulp.

**Gell-Mann, Murray** (1929– ) US theoretical physicist, who held a professorship at the California Institute of Technology. In 1955 he proposed the property of *strangeness for certain fundamental particles. In 1961 he and Yuval Ne'eman (1925–2006) proposed the eightfold way to define the structure of particles. This led to Gell-Mann's postulate of the quark (*see* ELEMEN-TARY PARTICLES). In 1969 he was awarded the Nobel Prize for physics.

**gem** Designating molecules in which two functional groups are attached to the same atom in a molecule. For example, the compound 1,1-dichloroethane ($CH_3CHCl_2$) is a gem dihalide and can be named *gem*-dichloroethane. *Compare* VICINAL.

**gemmation** A type of *vegetative propagation in which small clumps of undifferentiated cells (**gemmae**) develop on the surface of a plant. These are shed and dispersed to other areas, where they grow to produce new individuals. Gemmation is found only in certain lower plants, such as mosses and liverworts.

**gene** A unit of heredity composed of DNA. In classical genetics (*see* MENDELISM; MENDEL'S LAWS) a gene is visualized as a discrete particle, forming part of a *chromosome, that determines a particular characteristic. It can exist in different forms called *alleles, which determine which aspect of the characteristic is shown (e.g. tallness or shortness for the characteristic of height).

A gene occupies a specific position (*locus) on a chromosome. In view of the discoveries of molecular genetics, it may be defined as the sequence of nucleotides of DNA (or RNA) concerned with a specific function, such as the synthesis of a single polypeptide chain or of a messenger RNA molecule, corresponding to a particular sequence of the *genetic code. One or more of these **structural genes**, coding for protein, may be associated with other genes controlling their expression (*see* OPERON).

**gene amplification** The multiple replication of a section of the *genome, which occurs during a single cell cycle and results in the production of many copies of a specific sequence of the DNA molecule. For example, in the oocytes of amphibians and other animals, in which large numbers of ribosomes are needed, the genes encoding ribosomal RNA are greatly amplified. Viral genes that cause the formation of tumours (*see* ONCO-GENE) are amplified in tumour cells.

**gene bank** *See* DNA LIBRARY.

**gene cloning (DNA cloning)** The production of exact copies (**clones**) of a particular gene or genes using genetic engineering techniques. The DNA containing the target gene(s) is split into fragments using *restriction enzymes. Alternatively, complementary DNA (cDNA) can be made from messenger RNA by the enzyme reverse transcriptase. These fragments are then inserted into cloning *vectors, such as bacterial plasmids or bacteriophages, which transfer the recombinant DNA to suitable host cells, such as the bacterium *E. coli*.

Inside the host cell the recombinant DNA undergoes replication; thus, a bacterial host will give rise to a colony of cells each containing identical cloned DNA fragments. Various screening methods may be used to identify such colonies, enabling them to be selected and cultured. The collection of bacterial colonies containing the cloned fragments represents a *DNA library. Gene cloning facilitates *DNA sequencing; it also enables large quantities of a desired protein product to be produced: human insulin, for example, is now produced by bacteria containing the cloned insulin gene.

**gene expression** The manifestation of the effects of a gene by the production of the particular protein, polypeptide, or type of RNA whose synthesis it controls. Individual genes can be 'switched on' (exert their effects) or 'switched off' according to the needs and circumstances of the cell at a particular time. A number of mechanisms are thought to be responsible for the control of gene expression; the *Jacob–Monod hypothesis postulates the mechanism operating in prokaryotes (*see* OPERON). Control of gene expression is more complicated in eukaryotes, which possess various control mechanisms not seen in prokaryotes. For example, the methylation of cytosine bases of specific genes in eukaryotic DNA (**DNA methylation**) is observed in cells in which the gene is not expressed; if DNA methylation is prevented by the use of inhibitory chemicals, this can cause certain genes to be expressed.

**gene imprinting (molecular imprinting)** The differential expression of a gene according to whether it is derived from the mother or father. This leads to unequal genetic contributions from the mother and father to their offspring and is essential for normal development. For example, loss of maternal imprinting in mice results in an abnormally large fetus, whereas loss of paternal imprinting leads to a small fetus. Several human diseases are associated with failures of imprinting, including Angelman syndrome and Prader–Willi syndrome.

**gene knockout** *See* KNOCKOUT.

**gene library** *See* DNA LIBRARY.

**gene manipulation** *See* BIOTECHNOLOGY; GENETIC ENGINEERING.

**gene mutation** *See* POINT MUTATION.

**Gene Ontology** *See* ONTOLOGY.

**gene pool** All the *genes and their different alleles that are present in a population of a particular species of organism. *See also* POPULATION GENETICS.

**gene probe** A single-stranded DNA or RNA fragment used in genetic engineering to search for a particular gene or other DNA sequence. The probe has a base sequence complementary to the target sequence and will thus attach to it by *base pairing. By labelling the probe with a radioactive isotope or fluorescent label it can be identified on subsequent separation and purification. Probes of varying lengths, up to about 100 nucleotides, can be constructed in the laboratory. They are used in the *Southern blotting technique to identify particular DNA fragments, for instance in conjunction with *restriction mapping to diagnose gene abnormalities or to map certain sequences.

**general circulation of the atmosphere** *See* WIND.

**general theory of relativity** *See* RELATIVITY.

**generation** A group of organisms of approximately the same age within a population. Organisms that are crossed to produce offspring in a genetics study are referred to as the **parental generation** and their offspring as the first **filial generation**. *See also* $F_1$; $F_2$; P.

**generation time** 1. (in physics) The average time that elapses between the creation of a neutron by fission in a nuclear reactor and a fission produced by that neutron. 2. (in biology) The interval between the beginnings of consecutive cell divisions. It may be as short as 20 minutes in bacteria. *See also* INTERPHASE.

**generative nucleus** One of the two male gametes in the *pollen tube of angiosperms.

**generator** Any machine that converts mechanical power into electrical power. Electromagnetic generators are the main source of electricity and may be driven by steam turbines, water turbines, internal-combustion engines, windmills, or by some moving part of any other machine. In power stations, generators produce alternating current and are often called **alternators**.

**gene sequencing** See DNA SEQUENCING.

**gene silencing** See KNOCKOUT; RNA IN-TERFERENCE.

**gene splicing** A stage in the processing of messenger *RNA (see TRANSCRIPTION), occurring only in eukaryote cells, in which non-coding *introns are removed from the primary mRNA transcript and the coding *exons are spliced together to form the functional mRNA molecule. Splicing is catalysed by a complex of small RNA molecules and proteins called a **spliceosome**. In some organisms, self-splicing of mRNA is known to occur.

**gene therapy** The application of genetic engineering techniques to alter or replace defective genes. Techniques currently being investigated involve the transfer of normal genes into the genetic material of the cell to replace the defective gene and the use of *knockout techniques to inactivate defective genes in certain tissues. *Retroviruses are often used as *vectors for transferring genes into cells as part of the natural retrovirus life cycle involves the insertion of their own genetic material into the chromosomes of their host. Alternatively *liposomes may be used. Gene therapy is being developed in an attempt to cure and prevent such single-gene diseases as cystic fibrosis.

**genetically modified organisms (GMOs)** Organisms whose genomes incorporate and express genes from another species. Genetically modified (or **transgenic**) individuals are created by genetic engineering, using suitable *vectors to insert the desired foreign gene into the fertilized egg or early embryo of the host. GMOs offer considerable commercial potential. See Feature (pp 352–353).

**genetic code** The means by which genetic information in *DNA controls the manufacture of specific proteins by the cell. The code takes the form of a series of triplets of bases in DNA, from which is transcribed a complementary sequence of *codons in messenger *RNA (see TRANSCRIPTION). The sequence of these codons determines the sequence of amino acids during *protein synthesis. There are 64 possible codes from the combinations of the four bases present in DNA and mes-

senger RNA and 20 amino acids present in body proteins: some of the amino acids are coded by more than one codon, and some codons have other functions (see START CODON; STOP CODON). See illustration.

**genetic engineering (recombinant DNA technology)** The techniques involved in altering the characters of an organism by inserting genes from another organism into its DNA. This altered DNA (known as **recombinant DNA**) is usually produced by *gene cloning. Genetic engineering has many applications, ranging from the commercial production of hormones, vaccines, etc., to the creation of genetically modified crop plants in agriculture. See GENETICALLY MODIFIED ORGANISMS (Feature). See also DNA LIBRARY; GENE PROBE.

**genetic fingerprinting** See DNA PROFILING.

**genetic mapping** See CHROMOSOME MAP; LINKAGE MAP; PHYSICAL MAP; RESTRICTION MAPPING.

**genetic marker** See MARKER GENE; MOLECULAR MARKER.

**genetics** The branch of biology concerned with the study of heredity and variation. **Classical genetics** is based on the work of Gregor Mendel (see MENDELISM). During the 20th century genetics expanded to overlap with the fields of ecology and animal behaviour (see BEHAVIOURAL GENETICS; POPULATION GENETICS), and important advances in biochemistry and microbiology led to clarification of the chemical nature of *genes and the ways in which they can replicate and be transmitted, creating the field of **molecular genetics**. Since the 1980s, automated DNA sequencing techniques coupled with advances in computerized data handling have enabled rapid determination of the nucleotide sequences of entire genomes. The *bioinformatics revolution has allowed evolutionary relationships to be traced at the genome level and gene function to be analysed at the cellular level. See also GENETIC ENGINEERING.

**((())) SEE WEB LINKS**

- Essential genetic topics described using entertaining animations, produced by the Genetic Science Learning Center of Utah State University

**genetic screening** The process by which the genome of a human or other organism is analysed for genetic markers that indicate

| First base in codon | | Second base in codon | | | | Third base in codon |
|---|---|---|---|---|---|---|
| | **U** | **C** | **A** | **G** | | |
| **U** | UUU Phe | UCU Ser | UAU Tyr | UGU Cys | | U |
| | UUC Phe | UCC Ser | UAC Tyr | UGC Cys | | C |
| | UUA Leu | UCA Ser | UAA (stop codon) | UGA (stop codon)* | | A |
| | UUG Leu | UCG Ser | UAG (stop codon) | UGG Trp | | G |
| **C** | CUU Leu | CCU Pro | CAU His | CGU Arg | | U |
| | CUC Leu | CCC Pro | CAC His | CGC Arg | | C |
| | CUA Leu | CCA Pro | CAA Gln | CGA Arg | | A |
| | CUG Leu | CCG Pro | CAG Gln | CGG Arg | | G |
| **A** | AUU Ile | ACU Thr | AAU Asn | AGU Ser | | U |
| | AUC Ile | ACC Thr | AAC Asn | AGC Ser | | C |
| | AUA Ile | ACA Thr | AAA Lys | AGA Arg† | | A |
| | AUG Met (start codon) | ACG Thr | AAG Lys | AGG Arg | | G |
| **G** | GUU Val | GCU Ala | GAU Asp | GGU Gly | | U |
| | GUC Val | GCC Ala | GAC Asp | GGC Gly | | C |
| | GUA Val | GCA Ala | GAA Glu | GGA Gly | | A |
| | GUG Val | GCG Ala | GAG Glu | GGG Gly | | G |

\*UGA encodes tryptophan in mitochondrial DNA
†AGA is a stop codon in mitochondrial DNA

**Genetic code.**

the presence of particular genes, especially ones that cause or predispose to certain diseases. Increased knowledge of the human genome (*see* HUMAN GENOME PROJECT) and technological advances have simplified genetic screening in persons with a family history of certain inherited diseases, e.g. certain forms of breast cancer. Clinical gene testing is now used routinely to screen for many different genes, either to assess the risk of disease in susceptible individuals or their offspring or to confirm a diagnosis of inherited disease. Commercial gene test kits are also available to the general population, although claims that these can determine the risk of healthy individuals developing, say, heart disease or cancer should be treated with caution. Such tests have major implications for the insurance industry as well as for medicine. For example, some healthy individuals may be expected to pay a higher premium for life insurance because genetic screening reveals the presence of such genes. *See also* PREIMPLANTATION GENETIC DIAGNOSIS.

**genetic variation** *See* VARIATION.

**gene tracking** A method for determining the inheritance of a particular gene in a family. It is used in the diagnosis of genetic diseases, such as cystic fibrosis and Huntington's disease. Molecular markers, such as \*single nucleotide polymorphisms or \*restriction fragment length polymorphisms (RFLPs) situated in or near the locus of interest, are identified using \*gene probes, and suitable markers selected. These can then be traced through members of the family and used to detect the presence or absence of the

# GENETICALLY MODIFIED ORGANISMS

Since the early 1980s developments in genetic engineering have made it possible to produce genetically modified organisms. A gene from one organism is isolated and transferred to cells of another organism, where it is incorporated into the recipient's chromosomes and expressed. Such transgenic organisms can exhibit quite novel characteristics. Since the 1990s there has been a dramatic growth in the commercial applications of this new technology, ranging from the production of human hormones in bacteria and vaccines in yeasts to the development of genetically modified (GM) crop plants.

## Techniques

Various methods are used to introduce novel genes, depending on the nature of the recipient organism. Much of the work with genetic modification of plants involves *protoplasts, cultured spherical cells from which the cell walls have been removed. The Ti plasmid (see illustration) of *Agrobacterium tumefaciens*, the bacterium that is responsible for the tumorous growths of crown-gall disease in plants, has been used successfully as a *vector with certain dicotyledons, including tobacco, tomato, potato, soyabean, and cotton. It works much less well with grasses, cereals, and other monocots. In these plants various other techniques are available, including:

- electroporation – treatment of cells by exposure to an electric field that renders them transiently permeable to DNA fragments;
- microinjection – injection of DNA directly into the cell nucleus;
- biolistics – 'shooting' a cell with a DNA-coated tungsten microprojectile, which penetrates the cell wall with minimal damage.

To produce a transgenic animal the novel genes are inserted at a very early stage of development, e.g. the early embryo or the pronucleus of a fertilized egg, typically using microinjection. The recombinant embryos are then transferred to the uterus of a foster mother where they complete their development.

## Applications

*Plants*

- tolerance to herbicides
- improved insect resistance
- 'vaccination' against specific diseases
- longer 'shelf life' for fruit

*Animals*

- production of therapeutic proteins in milk
- potential for improved growth rates and milk yields
- potential for production of organs for human transplants

## Risks

The use of GM organisms in the environment poses certain potential problems. For example, genes for herbicide or insect resistance may spread from crop plants to wild plants, with possible serious consequences for both agriculture and natural ecosystems. Farmers may be faced with new 'superweeds', while insect populations could decline. Moreover, the products of GM crops have to be fully evaluated to ensure that they are safe to eat. Genetic modification of animals often has unforeseen side-effects and raises ethical issues about such treatments.

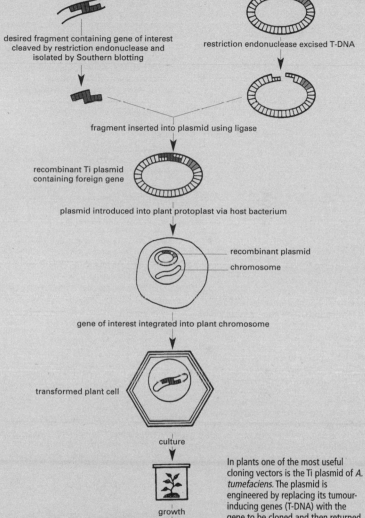

desired fragment containing gene of interest cleaved by restriction endonuclease and isolated by Southern blotting

restriction endonuclease excised T-DNA

g

fragment inserted into plasmid using ligase

recombinant Ti plasmid containing foreign gene

plasmid introduced into plant protoplast via host bacterium

recombinant plasmid

chromosome

gene of interest integrated into plant chromosome

transformed plant cell

culture

growth

transformed plant

In plants one of the most useful cloning vectors is the Ti plasmid of *A. tumefaciens*. The plasmid is engineered by replacing its tumour-inducing genes (T-DNA) with the gene to be cloned and then returned to the bacterium, which is allowed to infect cultured plant protoplasts. Transformed cells are selected by the presence of marker genes and propagated with plant growth substances so they develop into complete plants.

disease locus prenatally in future at-risk pregnancies.

**genome** All the genes contained in a single set of chromosomes, i.e. in a *haploid nucleus. Each parent, through its reproductive cells, contributes its genome to its offspring.

**genome project** Any undertaking, whether by a single organization or a consortium of scientific institutions, to map and sequence the entire genome of an organism. The bacterium *Haemophilus influenzae* was the first organism to be sequenced; the first eukaryotic genome to be sequenced was that of the budding yeast *Saccharomyces cerevisiae*. A massive international collaboration resulted in the sequencing of the human genome, completed in 2003 (*see* HUMAN GENOME PROJECT). According to the Genomes OnLine Database (GOLD), by early 2009 there were nearly 4500 genome projects, either completed or ongoing, involving species drawn from all groups of organisms.

**genomics** The branch of genetics concerned with the study of genomes. It has developed since the 1980s, exploiting automated techniques and computer-based systems to collect and analyse vast amounts of data on nucleotide and amino-acid sequences of various organisms, generated by projects such as the *Human Genome Project. There are several distinct but overlapping areas of genomics. **Structural genomics** is essentially about mapping the genome, and ultimately producing a complete DNA sequence for any particular organism, but is often extended to include determination of three-dimensional molecular structures of nucleic acids and proteins (*see* PROTEOMICS). **Functional genomics** deals with gene expression and how gene products work. This highly complex area, which involves analysis of transcripts of sets of genes (*see* TRANSCRIPTOMICS), seeks to understand how gene expression is controlled and integrated and how gene functions change under different conditions, such as disease states. **Comparative genomics** identifies regions of sequence similarity between genomes of different species. Knowledge of the functional significance of a particular DNA sequence in one species allows predictions about functions of closely matching sequences in other species. In addition, such comparisons permit inferences about mechanisms of gene evolution and give insights into the evolutionary relationships of different organisms. *See also* BIOINFORMATICS; METABOLOMICS.

**genotoxicity** The condition resulting from the interaction of toxic agents (**genotoxins**) with DNA molecules in genes. Since the genes are passed down to the next generation, the toxicity induced by genotoxins is heritable. Genotoxins can induce mutations in chromosomes (**clastogenesis**) or in a small number of base pairs (**mutagenesis**). Genotoxic agents include X-rays, natural *carcinogens, some man-made products (e.g. acridine and vinyl chloride), and viruses.

**genotype** The genetic composition of an organism, i.e. the combination of *alleles it possesses. *Compare* PHENOTYPE.

**genus** (*pl.* **genera**) A category used in the *classification of organisms that consists of a number of similar or closely related species. The common name of an organism (especially a plant) is sometimes similar or identical to that of the genus, e.g. *Lilium* (lily), *Crocus*, *Antirrhinum*. Similar genera are grouped into families. *See also* BINOMIAL NOMENCLATURE.

**geocentric universe** A conception of the universe in which the earth is regarded as being at its centre. It was espoused by many ancient Greek philosophers and reached its most comprehensive form in the Ptolemaic system (*see* PTOLEMAIC ASTRONOMY), which held sway throughout the Middle Ages. In the 16th century, Copernicus revived the idea that the earth travels around the sun, which had been a minority view in antiquity. A century later Galileo and Kepler finally established the validity of the *heliocentric universe and *Copernican astronomy.

**geochemistry** The scientific study of the chemical composition of the earth. It includes the study of the abundance of the earth's elements and their isotopes and the distribution of the elements in environments of the earth (lithosphere, atmosphere, biosphere, and hydrosphere).

**geochronology** *See* VARVE DATING.

**geode** A globular hollow piece of rock, ranging in size from 2–30 cm across, found in lavas and limestone. The outer shell is made of chalcedony (a type of silica) and the interior hollow is lined with well-formed crystals of quartz, aligned towards the centre.

**geodesic (geodesic line)** The shortest distance between two points on a curved surface.

**geodesy** The science concerned with surveying and mapping the earth's surface to determine, for example, its exact size, shape, and gravitational field. The information supplied by geodesy in the form of locations, distances, directions, elevations, and gravity information is of use in civil engineering, navigation, geophysics, and geography.

**geodetic effect** The effect, predicted by the general theory of *relativity, in which a mass (e.g. the earth) distorts space–time.

**(((⊕))) SEE WEB LINKS**
- The home page of NASA's Gravity Probe B experiment to investigate the geodetic effect

**geodynamics** The study of the motions of the earth; it includes those of the crust, mantle, and core, and the earth's rotation.

**geographical information system (GIS)** A computer-based system for the capture, storage, retrieval, manipulation, analysis, and display of spatial data. Geographical information systems have wide applications, for example, in forestry management, estate management, town planning, public utility management, insurance, transportation, and distribution.

**geography** The study of the features of the earth's surface, their distribution and interaction, and the interaction of man with them. It is divided between the physical and social sciences. **Human geography** includes economic geography, political geography, historical geography, and urban geography; **physical geography** encompasses geomorphology, biogeography, climatology, meteorology, pedology, and hydrology. The methods used by geographers include mapmaking (cartography), remote sensing techniques (e.g. aerial photography and satellite imagery), surveying techniques, statistical analysis, and *geographical information systems.

**geological time scale** A time scale that covers the earth's history from its origin, estimated to be about 4600 million years ago, to the present. The chronology is divided into a hierarchy of time intervals: eons, eras, periods, epochs, ages, and chrons (see Appendix).

**geology** The study of the origin, structure, and composition of the earth. It is commonly subdivided into **historical geology**, which includes stratigraphy, palaeontology, and geochronology; and **physical geology**, which includes geomorphology, geophysics, geochemistry, mineralogy, petrology, crystallography, and economic geology.

**geomagnetism** The science concerned with the earth's magnetic field. If a bar magnet is suspended at any point on the earth's surface so that it can move freely in all planes, the north-seeking end of the magnet (N-pole) will point in a broadly northerly direction. The angle ($D$) between the horizontal direction in which it points and the geographic meridian at that point is called the **magnetic declination**. This is taken to be positive to the east of geographic north and negative to the west. The needle will not, however, be horizontal except on the **magnetic equator**. In all other positions it will make an angle ($I$) with the horizontal, called the **inclination** (or **magnetic dip**). At the **magnetic poles** $I = 90°$ (+90° at the N-pole, –90° at the S-pole) and the needle will be vertical. The positions of the poles, which vary with time, were in the 1970s approximately 76.1°N, 100°W (N) and 65.8°S, 139°E (S). The vector intensity $F$ of the geomagnetic field is specified by $I$, $D$, and $F$, where $F$ is the local magnetic intensity of the field measured in gauss or tesla (1 gauss = $10^{-4}$ tesla). $F$, $I$, and $D$, together with the horizontal and vertical components of $F$, and its north and east components, are called the **magnetic elements**. The value of $F$ varies from about 0.2 gauss to 0.6 gauss, in general being higher in the region of the poles than at the equator, but values vary irregularly over the earth's surface with no correlation with surface features. There is also a slow unpredictable change in the local values of the magnetic elements called the **secular magnetic variation**. For example, in London between 1576 and 1800 $D$ changed from +11° to –24° and $I$ varied between 74° and 67°. The study of *palaeomagnetism has extended knowledge of the secular magnetic variation into the geological past and it is clear that the direction of the geomagnetic field has reversed many times. The source of the field and the cause of the variations are not known with any certainty but the source is believed to be associated with *dynamo action in the earth's liquid core.

**(((⊕))) SEE WEB LINKS**
- The website of the USGS National Geomagnetism Program
- Geomagnetic data at the NPL website

**geometrical isomerism** *See* ISOMERISM.

**geometrical optics** *See* OPTICS.

**geometric distribution** The distribution of the number of independent Bernoulli trials before a successful result is obtained; for example, the distribution of the number of times a coin has to be tossed before a head comes up. The probability that the number of trials ($x$) is $k$ is

$P(x=k) = q^{k-1}p$

The mean and variance are $1/p$ and $q/p^2$ respectively.

**geometric mean** *See* MEAN.

**geometric series** A series of numbers or terms in which the ratio of any term to the subsequent term is constant. For example, 1, 4, 16, 64, 256,... has a **common ratio** of 4. In general, a geometric series can be written:

$a + ar + ar^2 ... + ar^{n-1}$

and the sum of $n$ terms is:

$a(r^n - 1)/(r - 1)$.

**geometrized units** A system of units, used principally in general relativity, in which all quantities that have dimensions involving length, mass, and time are given dimensions of a power of length only. This is equivalent to setting the gravitational constant and the speed of light both equal to unity. *See also* GAUSSIAN UNITS; HEAVISIDE–LORENTZ UNITS; NATURAL UNITS; PLANCK UNITS.

**geomorphology** The study of the origin and development of landforms, excluding the major forms of the earth's surface (e.g. mountain chains and ocean basins).

**geo-neutrinos** Neutrinos that are emitted as a result of radioactivity inside the earth. Small numbers of geo-neutrinos have been detected in experiments. It is hoped that eventually they can be used to investigate the internal structure of the earth and to elucidate how radioactivity slows down the cooling of the earth.

**geophysics** The branch of science in which the principles of mathematics and physics are applied to the study of the earth's crust and interior. It includes the study of earthquake waves, geomagnetism, gravitational fields, and electrical conductivity using precise quantitative principles. In applied geophysics the techniques are applied to the discovery and location of eco-

nomic minerals (e.g. petroleum). Meteorology and physical oceanography can also be considered as geophysical sciences.

**geosphere** The nonliving part the earth, in contrast to the living biosphere; it includes the *lithosphere, *hydrosphere, and the *earth's atmosphere. The term is also used synonymously with lithosphere.

**geostationary orbit** *See* SYNCHRONOUS ORBIT.

**geosynchronous orbit** *See* SYNCHRONOUS ORBIT.

**geothermal energy** Heat within the earth's interior that is a potential source of energy. Volcanoes, geysers, hot springs, and fumaroles are all sources of geothermal energy. The main areas of the world in which these energy sources are used to generate power include Larderello (Italy), Wairakei (New Zealand), Geysers (California, USA), and Reykjavik (Iceland). High-temperature porous rock also occurs in the top few kilometres of the earth's crust. Thermal energy from these reservoirs can be tapped by drilling into them and extracting their thermal energy by conduction to a fluid. The hot fluid can then be used for direct heating or to raise steam to drive a turbogenerator.

**geotropism (gravitropism)** The growth of plant organs in response to gravity. A main root is positively geotropic and a main stem negatively geotropic, growing downwards and upwards respectively, irrespective of the positions in which they are placed. For example, if a stem is placed in a horizontal position it will still grow upwards. *Auxins are thought to play a role in geotropism. *See* TROPISM.

**geraniol** An alcohol, $C_9H_{15}CH_2OH$, present in a number of essential oils.

**germanium** Symbol Ge. A lustrous hard metalloid element belonging to group 14 (formerly IVB) of the periodic table; a.n. 32; r.a.m. 72.59; r.d. 5.36; m.p. 937°C; b.p. 2830°C. It is found in zinc sulphide and in certain other sulphide ores, and is mainly obtained as a by-product of zinc smelting. It is also present in some coal (up to 1.6%). Small amounts are used in specialized alloys but the main use depends on its semiconductor properties. Chemically, it forms compounds in the +2 and +4 oxidation states, the germanium(IV) compounds being the more stable. The element also forms a large number of organometallic compounds. Predicted in 1871 as eka-silicon by Dmitri Mendeleev, it was discovered by Winkler in 1886.

**(⊕) SEE WEB LINKS**
• Information from the WebElements site

**German silver (nickel silver)** An alloy of copper, zinc, and nickel, often in the proportions 5:2:2. It resembles silver in appearance and is used in cheap jewellery and cutlery and as a base for silver-plated wire. *See also* ELECTRUM.

**germ cell** Any cell in the series of cells (the **germ line**) that eventually produces *gametes. In mammals the germinal epithelium of the ovaries and testes contain the germ cells.

**germinal epithelium** **1.** A layer of epithelial cells on the surface of the ovary that are continuous with the *mesothelium. These cells do not give rise to the ova (they were formerly thought to do this). **2.** The layer of epithelial cells lining the seminiferous tubules of the testis, which gives rise to spermatogonia (*see* SPERMATOGENESIS).

**germination** **1.** The initial stages in the growth of a seed to form a seedling. The embryonic shoot (plumule) and embryonic root (radicle) emerge and grow upwards and downwards respectively. Food reserves for germination come from *endosperm tissue within the seed and/or from the seed leaves (cotyledons). *See also* EPIGEAL; HYPOGEAL. **2.** The first signs of growth of spores and pollen grains.

**germ layers (primary germ layers)** The layers of cells in an animal embryo at the *gastrula stage, from which are derived the various organs of the animal's body. There are two or three germ layers: an outer layer (*see* ECTODERM), an inner layer (*see* ENDODERM), and in most animal groups a middle layer (*see* MESODERM). *See also* DEVELOPMENT.

**germ plasm** *See* WEISMANNISM.

**gestation** The period in animals bearing live young (especially mammals) from the fertilization of the egg to birth of the young (parturition). In humans gestation is known as **pregnancy** and takes about nine months (40 weeks).

**getter** A substance used to remove small amounts of other substances from a system by chemical combination. For example, a metal such as magnesium may be used to remove the last traces of air when achieving a high vacuum. Various getters are also employed to remove impurities from semiconductors.

**GeV** Gigaelectronvolt, i.e. $10^9$ eV. In the USA this is often written BeV, billion-electronvolt.

**geyser** A hot spring that regularly throws up jets of hot water and steam. The steam is formed in the geyser tube underground where groundwater comes into contact with hot rock (magma). Increased pressure may raise the boiling temperature of the water, accounting for the intermittent nature of the eruption. Steam formed below a column of water forces the water out of the vent before the steam can escape.

**GFP** *See* GREEN FLUORESCENT PROTEIN.

**GHB** Gammahydroxybutyric acid. *See* 4-HYDROXYBUTANOIC ACID.

**ghrelin** A peptide hormone that is secreted chiefly by cells in the stomach lining and increases hunger by stimulating the release of *neuropeptide Y from hypothalamus. It also stimulates release of growth hormone by binding to receptors in the anterior pituitary. The concentration of ghrelin in blood rises during the fasting period before a meal, thereby promoting appetite. Paradoxically, it also appears to suppress the mobilization of fat reserves in adipose tissue. Its role in regulating the body's energy balance has prompted interest in ghrelin as a potential target for anti-obesity treatments.

**giant fibre** A nerve fibre with a very large diameter, found in many types of invertebrate (e.g. earthworms and squids). Its function is to allow extremely rapid transmission of nervous impulses and hence rapid escape movements in emergencies.

**giant impact hypothesis** A hypothesis concerning the origin of the earth's moon, according to which a Mars-sized planet nicknamed Theia, having formed in the same orbital zone as the earth, was pulled toward it and eventually collided with it; Theia's iron core sank into that of the still-evolving earth, and the bulk of Theia's mantle, along with part of the earth's mantle and crust, was ejected into earth orbit. The debris formed a ring around the earth but soon coalesced to form the moon.

**giant magnetoresistance** A type of *magnetoresistance that occurs in thin films consisting of alternating layers of ferromagnetic and nonmagnetic metals. In the presence of an external magnetic field there is a substantial decrease in the electrical resistance due to quantum-mechanical effects associated with the spins of the electrons of the nonmagnetic metal. Giant magnetoresistance is made use of in computer magnetic disk drives. *See also* COLOSSAL MAGNETORESISTANCE; MAGNETORESISTANCE.

**giant molecular cloud (GMC)** A huge area in space several hundred light-years across, consisting of molecular hydrogen. There are as many as 5000 in the Milky Way, and two of them occur in the constellation Orion. They are the largest objects in the Galaxy, each containing up to 10 million times as much material as the sun. Parts of them have collapsed under gravity and formed star-containing emission nebulae.

**giant star** A very large star that is highly luminous. Lying above the main sequence on a *Hertzsprung–Russell diagram, giant stars represent a late stage in *stellar evolution. *See also* RED GIANT; SUPERGIANT.

**gibberellic acid (GA₃)** A plant hormone that is extracted from fungal cultures and is one of the most important commercially available *gibberellins. It was discovered in 1954.

**gibberellin** Any of a group of plant hormones chemically related to terpenes. Gibberellins promote stem elongation and the mobilization of food reserves in germinating seeds and have a role in inducing flowering and fruit development. Commercially available gibberellins, such as *gibberellic acid, are used to manipulate the onset of sexual maturity in various species, e.g. to induce cone bearing in young conifers.

**gibbous** *See* PHASES OF THE MOON.

**Gibbs, Josiah Willard** (1839–1903) US mathematician and physicist, who spent his entire academic career at Yale University. During the 1870s he developed the theory of chemical thermodynamics, devising functions such as Gibbs *free energy; he also derived the *phase rule and was one of the founders of *statistical mechanics. In mathematics he introduced *vector notation.

**Gibbs free energy (Gibbs function)** *See* FREE ENERGY.

**gibbsite** A mineral form of hydrated *aluminium hydroxide ($Al(OH)_3$). It is named after the US mineralogist George Gibbs (d. 1833).

**gibi-** *See* BINARY PREFIXES.

**giga-** Symbol G. A prefix used in the metric system to denote one thousand million times. For example, $10^9$ joules = 1 gigajoule (GJ).

**gilbert** Symbol Gb. The c.g.s. unit of *magnetomotive force equal to $10/4\pi$ (= 0.795 77) ampere-turn. It is named after William Gilbert.

**Gilbert, William** (1544–1603) English physician and physicist. He was physician to Queen Elizabeth I, and in 1600 published his famous book about magnetism, in which he likened the earth to a huge bar magnet. He was the first to use the terms 'magnetic pole' and 'electricity'.

**gill** **1.** (in zoology) A respiratory organ used by aquatic animals to obtain oxygen from the surrounding water. A gill consists essentially of a membrane or outgrowth from the body, with a large surface area and a plentiful blood supply, through which diffusion of oxygen and carbon dioxide between the water and blood occurs. Fishes have **internal gills**, formed as outgrowths from the pharynx wall and contained within *gill slits. Water entering the mouth is pumped out through these slits and over the gills. The gills of most aquatic invertebrates and amphibian larvae are **external gills**, which project from the body so that water passes over them as the animal moves. **2.** (in botany) One of the ridges of tissue that radiate from the centre of the underside of the cap of mushrooms. The spores are produced on these gills.

**gill bar** A cartilaginous support for the tissue between the gill slits in lower chordates, such as lancelets.

**gill slit** An opening leading from the pharynx to the exterior in aquatic vertebrates and lancelets. In lancelets they function in *filter feeding. In fish they contain the *gills and are usually in the form of a series of long slits. They are absent in adult tetrapod vertebrates (except for some amphibians) but their presence in some form in the embryos of all vertebrates is a characteristic of the phylum *Chordata.

**gimbal** A type of mount for an instrument (such as a *gyroscope or compass) in which the instrument is free to rotate about two perpendicular axes.

**gingiva (gum)** The part of the epithelial tissue lining the mouth that covers the jaw bones. It is continuous with the sockets surrounding the roots of the teeth.

**Giorgi units** See M.K.S. UNITS.

**GIS** See GEOGRAPHICAL INFORMATION SYSTEM.

**gizzard** A muscular compartment of the alimentary canal of many animals that is specialized for breaking up food. In birds the gizzard lies between the *proventriculus and the duodenum and contains small stones and grit, which assist in breaking up the food when the gizzard contracts. See also GASTRIC MILL.

**glacial ethanoic acid** See ETHANOIC ACID.

**gland** A group of cells or a single cell in animals or plants that is specialized to secrete a specific substance. In animals there are two types of glands, both of which synthesize their secretions. *Endocrine glands discharge their products directly into the blood vessels; *exocrine glands secrete through a duct or network of ducts into a body cavity or onto the body surface. Secretory cells are characterized by having droplets (**vesicles**) containing their products. See also SECRETION.

In plants glands are specialized to secrete certain substances produced by the plant. The secretions may be retained within a single cell, secreted into a special cavity or duct, or secreted to the outside. Examples are the water glands (*hydathodes) of certain leaves, nectaries (see NECTAR), and the digestive glands of certain carnivorous plants.

**Glaser, Donald Arthur** (1926– ) US physicist. In 1952, at the University of Michigan, he devised the *bubble chamber for detecting ionizing radiation. For this work he was awarded the 1960 Nobel Prize for physics.

**Glashow–Weinberg–Salam model (GWS model)** See ELECTROWEAK THEORY.

**glass** Any noncrystalline solid; i.e. a solid in which the atoms are random and have no long-range ordered pattern. Glasses are often regarded as supercooled liquids. Characteristically they have no definite melting point, but soften over a range of temperatures.

The common glass used in windows, bottles, etc., is **soda glass**, which is made by heating a mixture of lime (calcium oxide), soda (sodium carbonate), and sand (silicon(IV) oxide). It is a form of calcium silicate. Borosilicate glasses (e.g. **Pyrex**) are made by incorporating some boron oxide, so that silicon atoms are replaced by boron atoms. They are tougher than soda glass and more resistant to temperature changes, hence their use in cooking utensils and laboratory apparatus. Glasses for special purposes (e.g. optical glass) have other elements added (e.g. barium, lead).

**glass electrode** A type of *half cell having a glass bulb containing an acidic solution of fixed pH, into which dips a platinum wire. The glass bulb is thin enough for hydrogen ions to diffuse through. If the bulb is placed in a solution containing hydrogen ions, the electrode potential depends on the hydrogen-ion concentration. Glass electrodes are used in pH measurement.

**glass fibres** Melted glass drawn into thin fibres some 0.005 mm–0.01 mm in diameter. The fibres may be spun into threads and woven into fabrics, which are then impregnated with resins to give a material that is both strong and corrosion resistant for use in car bodies and boat building.

**glauberite** A mineral consisting of a mixed sulphate of sodium and calcium, $Na_2SO_4.CaSO_4$.

**Glauber's salt** *Sodium sulphate decahydrate, $Na_2SO_4.10H_2O$, used as a laxative. It is named after Johann Glauber (1604–68).

**GLC (gas–liquid chromatography)** See GAS CHROMATOGRAPHY.

**glenoid cavity** The socket-shaped cavity in the *scapula (shoulder blade) that holds the head of the *humerus in a ball-and-socket joint.

**glia (glial cells; neuroglia)** Cells of the nervous system that support the neurons. There are four classes of glial cells: **astrocytes**, **oligodendrocytes**, **ependymal cells**, and **microglia**. Oligodendrocytes form insulating sheaths of *myelin round neurons in the central nervous system, preventing impulses from travelling between adjacent neurons. Other functions of glial cells include providing nutrients for neurons and controlling the biochemical composition of the fluid surrounding the neurons.

**glide** A symmetry element in a crystal lattice that consists of a combination of a translation with a reflection about a plane. *See also* SCREW.

**global positioning system (GPS)** A satellite-based navigational system that, with the use of a GPS receiver, can determine any point on or above the earth's surface with a high degree of accuracy. The system uses a network of 24 satellites, designed and controlled by the US Department of Defense, originally for military use. Uses include marine and terrestrial navigation systems (e.g. satellite navigation systems in vehicles), surveying, and mapping.

(●) SEE WEB LINKS
• FAQ covering the global positioning system provided by the US Federal Aviation Administration

**global warming** An increase over time of the average air temperature of the earth. Global average surface temperature increased over the 20th century by about 0.6°C, and is predicted to increase by between 1.4°C and 5.8°C during the period 1990–2100. The increase in temperature has been largely attributed to human activity in the form of increased emissions of greenhouse gases (*see* GREENHOUSE EFFECT), especially carbon dioxide, to the atmosphere and the consequent greenhouse effect. It has serious implications, for example, for changes in global climate patterns and in the melting of ice masses, such as the polar ice caps, with consequent raising of sea levels (during the 20th century global average sea level rose between 0.1 and 0.2 m). At an international level, global warming is studied by the Intergovernmental Panel on Climate Change (IPCC). International efforts to control global warming led to the **Kyoto Protocol** (an amendment to the United Nations Framework Convention on Climate Change, which was signed in 1992). The treaty was drawn up in Kyoto, Japan, in 1997, to reduce emissions of greenhouse gases; the treaty took effect in 2005, following ratification by Russia, and will require all ratifying nations to achieve individual emission reduction targets. Among countries not ratifying the treaty, most notable is the USA, the world's largest emitter of greenhouse gases, which withdrew from the Kyoto Protocol in 2001. The term *climate change is used synonymously for global warming in some areas.

**globular cluster** *See* STAR CLUSTER.

**globular protein** *See* PROTEIN.

**globulin** Any of a group of globular proteins that are generally insoluble in water and present in blood, eggs, milk, and as a reserve protein in seeds. Blood serum globulins comprise four types: $\alpha_1$-, $\alpha_2$-, and $\beta$-globulins, which serve as carrier proteins; and $\gamma$-globulins (**gamma globulins**), which include the *immunoglobulins responsible for immune responses.

**glomerular filtrate** The fluid in the lumen of the Bowman's capsule of the *nephron that has been filtered from the capillaries of the glomerulus (*see* ULTRAFILTRATION). The glomerular filtrate has the same composition as the plasma except that it does not contain any of the larger components, such as plasma proteins or cells.

**glomerulus** A tangled mass of blood capillaries enclosed by the cup-shaped end (*Bowman's capsule) of a kidney tubule (*see* NEPHRON). Fluid is filtered from these capillaries into the Bowman's capsule and down the nephron (*see* GLOMERULAR FILTRATE).

**glottis** The opening from the pharynx to the trachea (windpipe). In mammals it also serves as the space for the *vocal cords. *See also* EPIGLOTTIS; LARYNX.

**glove box** A metal box that has gloves fitted to ports in its walls. It is used to manipulate mildly radioactive materials and in laboratory techniques in which an inert, sterile, dry, or dust-free atmosphere has to be maintained.

**glow discharge** An electrical discharge that passes through a gas at low pressure and causes the gas to become luminous. The glow is produced by the decay of excited atoms and molecules.

**glucagon** A hormone, secreted by the *islets of Langerhans in the pancreas, that increases the concentration of glucose in the blood by stimulating the metabolic breakdown of glycogen. It thus antagonizes the effects of *insulin (*see* ANTAGONISM).

**glucan** Any *polysaccharide composed only of glucose residues, e.g. starch and glycogen.

**glucocorticoid** *See* CORTICOSTEROID.

**gluconeogenesis** The synthesis of glucose from noncarbohydrate sources, such as fat and protein. This occurs when the glycogen supplies in the liver are exhausted. The pathway is essentially a reversal of *glycolysis from pyruvate to glucose and it can utilize many sources, including amino acids, glycerol, and *Krebs cycle intermediates. Large-scale protein and fat catabolism normally occurs only in those suffering from starvation or certain endocrine disorders.

**gluconic acid** An optically active hydroxycarboxylic acid,

$$CH_2(OH)(CHOH)_4COOH.$$

It is the carboxylic acid corresponding to the aldose sugar glucose, and can be made by the action of certain moulds.

**glucosamine** See AMINO SUGAR.

**glucose (dextrose; grape sugar)** A white crystalline sugar, $C_6H_{12}O_6$, occurring widely in nature. Like other *monosaccharides, glucose is optically active: most naturally occurring glucose is dextrorotatory. Glucose and its derivatives are crucially important in the energy metabolism of living organisms. It is a major energy source, being transported around the body in blood, lymph, and cerebrospinal fluid to the cells, where energy is released in the process of *glycolysis. Glucose is present in the sap of plants, in fruits, and in honey and is also a constituent of many polysaccharides, most notably of starch and cellulose. These yield glucose when broken down, for example by enzymes during digestion.

**glucuronic acid** A compound, $OC_6H_9O_6$, derived from the oxidation of glucose. It is an important constituent of *gums and *mucilages. Glucuronic acid can combine with hydroxyl (–OH), carboxyl (–COOH), or amino (–NH$_2$) groups to form a **glucuronide**. The addition of a glucuronide group to a molecule (**glucuronidation**) generally increases the solubility of a compound; hence glucuronidation plays an important role in the excretion of foreign substances (see PHASE II METABOLISM).

**glucuronide** See GLUCURONIC ACID.

**glueball** A hypothetical bound state consisting of two or more gluons (see ELEMENTARY PARTICLES). Glueballs are thought to be unstable and decay rapidly into *hadrons. There is some indirect experimental evidence for glueballs.

**gluino** See SUPERSYMMETRY.

**gluon** See ELEMENTARY PARTICLES.

**glutamate** The anion of the amino acid glutamic acid. It functions as a neurotransmitter at excitatory synapses in the verte-brate central nervous system and at excitatory neuromuscular junctions in insects and crustaceans.

**glutamic acid** See AMINO ACID; GLUTAMATE.

**glutamine** See AMINO ACID.

**glutaric acid** See PENTANEDIOIC ACID.

**glutathione** A *peptide comprising the amino acids glutamic acid, cysteine, and glycine. It occurs widely in plants, animals, and microorganisms, serving chiefly as an antioxidant. Reduced glutathione reacts with potentially harmful oxidizing agents and is itself oxidized. This is important in ensuring the proper functioning of proteins, haemoglobin, membrane lipids, etc. Glutathione is also involved in amino acid transport across plasma membranes.

**gluten** A mixture of two proteins, gliadin and glutenin, occurring in the endosperm of wheat grain. Their amino acid composition varies but glutamic acid (33%) and proline (12%) predominate. The composition of wheat glutens determines the 'strength' of the flour and whether or not it is suitable for biscuit or bread making. Sensitivity of the lining of the intestine to gluten occurs in **coeliac disease**, a condition that must be treated by a gluten-free diet.

**glycaemic index (GI)** An indication of how eating a particular food will affect the concentration of glucose in the blood. GI is determined relative to the effects of a standard test meal of (usually) 50 g of glucose in a fasting individual, which is given the value 100. Foods with a high GI (typically >85), such as bread, rice, and potatoes, cause blood glucose to rise rapidly following ingestion, whereas the increase is lower and more prolonged in foods with a low GI (< 60), such as apples, beans, and yoghurt. Lower GI foods are recommended to optimize control of blood glucose for diabetics.

**glycan** See POLYSACCHARIDE.

**glyceraldehyde 3-phosphate (GALP)** A triose phosphate, $CHOH(OH)CH_2OPO_3H_2$, that is an intermediate in the *Calvin cycle (see also PHOTOSYNTHESIS) and glycolysis.

**glycerate 3-phosphate** A phosphorylated three-carbon monosaccharide that is an intermediate in the *Calvin cycle of photosynthesis and also in *glycolysis. It was formerly known as **3-phosphoglycerate** or **phosphoglyceric acid** (**PGA**).

**glyceride (acylglycerol)** A fatty-acid ester of glycerol. Esterification can occur at one, two, or all three hydroxyl groups of the glycerol molecule producing mono-, di-, and triglycerides respectively. *Triglycerides are the major constituent of fats and oils found in living organisms. Alternatively, one of the hydroxyl groups may be esterified with a phosphate group forming a phosphoglyceride (*see* PHOSPHOLIPID) or to a sugar forming a **glycolipid**.

**glycerine** *See* GLYCEROL.

**glycerol (glycerine; propane-1,2,3,-triol)** A trihydric alcohol, $HOCH_2CH(OH)CH_2OH$. Glycerol is a colourless sweet-tasting viscous liquid, miscible with water but insoluble in ether. It is widely distributed in all living organisms as a constituent of the *glycerides, which yield glycerol when hydrolysed. Glycerol itself is used as an *antifreeze molecule by certain organisms.

**glycerophospholipid** *See* PHOSPHOLIPID.

**glycine** A sweet-tasting *amino acid that, besides being a component of proteins, is the main inhibitory neurotransmitter for fast synapses in the spinal cord of vertebrates.

**glycobiology** The study of carbohydrates and carbohydrate complexes, especially *glycoproteins.

**glycogen (animal starch)** A *polysaccharide consisting of a highly branched polymer of glucose occurring in animal tissues, especially in liver and muscle cells. It is the major store of carbohydrate energy in animal cells and is present as granular clusters of minute particles.

**glycogenesis** The conversion of glucose to glycogen, which is stimulated by insulin from the pancreas. Glycogenesis occurs in skeletal muscles and to a lesser extent in the liver. Glucose that is taken up by cells is phosphorylated to glucose 6-phosphate; this is converted successively to glucose 1-phosphate, uridine diphosphate glucose, and finally to glycogen. *See also* GLUCONEOGENESIS. *Compare* GLYCOGENOLYSIS.

**glycogenolysis** The conversion of glycogen to glucose, which occurs in the liver and is stimulated by glucagon from the pancreas and adrenaline from the adrenal medulla. These hormones activate an enzyme that phosphorylates glucose molecules in the glycogen chain to form glucose 1-phosphate, which is converted to glucose 6-phosphate. This is then converted to glucose by a phosphatase enzyme. In skeletal muscle glycogen is degraded to glucose 6-phosphate, which is then converted into pyruvate and used in ATP production during glycolysis and the Krebs cycle. However, pyruvate can also be converted, in the liver, to glucose; thus muscle glycogen is indirectly a source of blood glucose. *Compare* GLYCOGENESIS.

**glycol** *See* ETHANE-1,2-DIOL.

**glycolipid** *See* GLYCERIDE.

**glycolysis (Embden–Meyerhof pathway)** The series of biochemical reactions in which

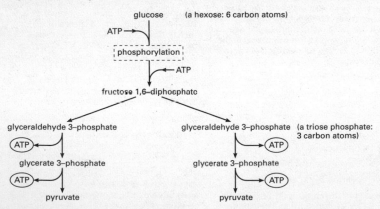

**Glycolysis.** The principal stages.

glucose is broken down to pyruvate with the release of usable energy in the form of *ATP (see illustration). One molecule of glucose undergoes two phosphorylation reactions and is then split to form two triose-phosphate molecules. Each of these is converted to pyruvate. The net energy yield is two ATP molecules per glucose molecule. In *aerobic respiration pyruvate then enters the *Krebs cycle. Alternatively, when oxygen is in short supply or absent, the pyruvate is converted to various products by *anaerobic respiration. Other simple sugars, e.g. fructose and galactose, and glycerol (from fats) enter the glycolysis pathway at intermediate stages. *Compare* GLUCONEOGENESIS.

**glycoprotein** A carbohydrate linked covalently to a protein. Formed in the Golgi apparatus in the process of *glycosylation, glycoproteins are important components of plasma membranes, in which they extend throughout the *lipid bilayer. They are also constituents of body fluids, such as mucus, that are involved in lubrication. Many of the hormone receptors on the surfaces of cells are glycoproteins. Glycoproteins produced by viruses attach themselves to the surface of the host cell, where they act as markers for the receptors of leucocytes. Viral glycoproteins can also act as target molecules and help viruses to detect certain types of host cell; for example, a glycoprotein on the surface of *HIV enables the virus to find and infect white blood cells.

**glycosaminoglycan** Any one of a group of polysaccharides that contain *amino sugars (such as glucosamine). Formerly known as **mucopolysaccharides**, they include *hyaluronic acid and chondroitin (*see* CARTILAGE), which provide lubrication in joints and form part of the matrix of cartilage. The three-dimensional structure of these molecules enables them to trap water, which forms a gel and gives glycosaminoglycans their elastic properties.

**glycoside** Any one of a group of compounds consisting of a pyranose sugar residue, such as glucose, linked to a noncarbohydrate residue (R) by a *glycosidic bond: the hydroxyl group (–OH) on carbon-1 of the sugar is replaced by –OR. Glycosides are widely distributed in plants; examples are the *anthocyanin pigments and the **cardiac glycosides**, such as digoxin and ouabain, which are used medicinally for their stimulant effects on the heart.

**glycosidic bond (glycosidic link)** The type of chemical linkage between the monosaccharide units of disaccharides, oligosaccharides, and polysaccharides, which is formed by the removal of a molecule of water (i.e. a *condensation reaction). The bond is normally formed between the carbon-1 on one sugar and the carbon-4 on the other (see illustration). An α-glycosidic bond is formed when the –OH group on carbon-1 is below the plane of the glucose ring and a β-glycosidic bond is formed when it is above the plane. Cellulose is formed of glucose molecules linked by 1-4 β-glycosidic bonds, whereas starch is composed of 1-4 α-glycosidic bonds.

**glycosylation** The process in which a carbohydrate is joined to another molecule, such as a protein to form a *glycoprotein or to a lipid to form a glycolipid (*see* GLYCERIDE). Glycosylation occurs in the rough endoplasmic reticulum and the *Golgi apparatus of cells.

**glyphosate** *N*-(phosphonomethyl)-glycine: a herbicide, marketed as Roundup, that kills a wide range of plants but shows little persistence in soil and has low toxicity to animals. If applied to the leaves it is rapidly translocated to the rest of the plant,

**Glycosidic bond formation.**

and hence can penetrate the roots of even hardy perennials. It works by blocking the synthesis of aromatic amino acids, so that treated plants are unable to manufacture proteins and other key metabolites. Certain crops, notably soya bean, have been genetically engineered to give them resistance to glyphosate. These 'Roundup-ready' crops, which can be sprayed with the herbicide without being affected, are now widely grown in North America and elsewhere.

**Gnathostomata** A subphylum or super-class of chordates consisting of all vertebrates that possess jaws. It contains six extant classes: *Chondrichthyes (cartilaginous fishes), *Osteichthyes (bony fishes), *Amphibia, *Reptilia, *Aves (birds), and *Mammalia. *See also* CRANIATA.

**gneiss** A coarse-grained rock that is characterized by compositional banding of metamorphic origin. It consists chiefly of irregular granular bands of quartz and feldspar alternating with thin undulating bands of micas and amphiboles. Gneisses are formed during high-grade regional metamorphism; those derived from sedimentary origins are known as **paragneisses** and those from igneous origins as **orthogneisses**.

**gnotobiotic** Designating germ-free conditions, especially those in which experimental animals are inoculated with known strains of microorganisms.

**goblet cell** A goblet-shaped cell, found in the epithelium of the intestine and respiratory system in mammals and in the epidermis of fish, that secretes *mucus. Goblet cells have a wide top and constricted base and possess glycoprotein-containing vesicles.

**Gödel's theorem** A fundamental result in mathematics stating that, in a given mathematical structure, some propositions cannot be proved to be true or false, i.e. the propositions are undecidable, using only the axioms of that mathematical structure. The theorem was proved by the Austrian mathematician Kurt Gödel (1906–78) in 1931. The relevance of this result to fundamental physics has been extensively debated.

**goethite** A yellow-brown mineral, FeO.OH, crystallizing in the orthorhombic system. It is formed as a result of the oxidation and hydration of iron minerals or as a direct precipitate from marine or fresh water (e.g. in swamps and bogs). Most *limonite is

composed largely of cryptocrystalline goethite. Goethite is mined as an ore of iron.

**gold** Symbol Au. A soft yellow malleable metallic *transition element; a.n. 79; r.a.m. 196.967; r.d. 19.32; m.p. 1064.43°C; b.p. 2807±2°C. Gold has a face-centred-cubic crystal structure. It is found as the free metal in gravel or in quartz veins, and is also present in some lead and copper sulphide ores. It also occurs combined with silver in the telluride sylvanite, $(Ag,Au)Te_2$. It is used in jewellery, dentistry, and electronic devices. Chemically, it is unreactive, being unaffected by oxygen. It reacts with chlorine at 200°C to form gold(III) chloride. It forms a number of complexes with gold in the +1 and +3 oxidation states.

**((⊕)) SEE WEB LINKS**
• Information from the WebElements site

**Gold, Thomas** *See* HOYLE, SIR FRED.

**golden ratio** A visually satisfying proportion found by dividing a line in such a way that the ratio of the smaller part to the larger part is the same as that of the larger part to the whole. It is equal to about 8:13 or about 1:1.618034, and can be expressed exactly as 1:½(1 + √5). *See also* FIBONACCI NUMBER.

**Goldschmidt process** A method of extracting metals by reducing the oxide with aluminium powder, e.g.

$$Cr_2O_3 + 2Al \rightarrow 2Cr + Al_2O_3$$

The reaction can also be used to produce molten iron (*see* THERMITE). It was discovered by the German chemist Hans Goldschmidt (1861–1923).

**Goldstone's theorem** The theorem in relativistic quantum field theory that if there is an exact continuous symmetry of the *Hamiltonian or *Lagrangian defining the system, and this is not a symmetry of the *vacuum state (i.e. there is a *broken symmetry), then there must be at least one spin-zero massless particle called a **Goldstone boson**. In the quantum theory of many-body systems Goldstone bosons are *collective excitations. An important exception to Goldstone's theorem is provided in *gauge theories with the Higgs mechanism, whereby the Goldstone bosons gain mass and become *Higgs bosons. The theorem is named after Jeffrey Goldstone.

**Golgi, Camillo** (1843–1926) Italian cytologist, who experimented with cells and tissues while working as a physician. He later became a professor at Pavia University. He devised a method of staining cells using silver salts, which enabled him to study nerve cells. He is best known for his discovery of the cell organelle now called the *Golgi apparatus. For his work on the structure of the nervous system he shared the 1906 Nobel Prize for physiology or medicine with Santiago Ramón y Cajal (1852–1934).

**Golgi apparatus (Golgi complex)** An assembly of vesicles and folded membranes within the cytoplasm of eukaryotic *cells that modifies proteins and packages them and other materials (e.g. polysaccharides) for delivery to the plasma membrane for secretion or to destinations within the cell. Proteins arrive in vesicles following their assembly in the *endoplasmic reticulum; after processing in the Golgi apparatus, they are sorted into **Golgi vesicles**, for secretion, storage, or transport to lysosomes. The apparatus is named after its discoverer, Camillo Golgi.

**gonad** Any of the usually paired organs in animals that produce reproductive cells (gametes). The most important gonads are the male *testis, which produces spermatozoa, and the female *ovary, which produces ova (egg cells). The gonads also produce hormones that control secondary sexual characteristics.

**gonadotrophin (gonadotrophic hormone)** Any of several hormones, secreted by the mammalian anterior *pituitary gland, that stimulate reproductive activity of the testes or ovaries (the gonads). Pituitary gonadotrophins include *follicle-stimulating hormone and *luteinizing hormone. **Chorionic gonadotrophin** is a hormone produced by the placenta of higher mammals that maintains the *corpus luteum. The presence of large amounts of **human chorionic gonadotrophin (hCG)** in the urine of women is an indication of pregnancy.

**Gondwanaland** *See* CONTINENTAL DRIFT.

**Gooch crucible** A porcelain dish with a perforated base over which a layer of asbestos is placed, used for filtration in gravimetric analysis. It is named after US chemist Frank Gooch (1852–1929).

**Gopher** A computer program used on a computer connected to the *Internet that carries out routine tasks of collecting information for the user from services attached to the Internet. Gopher presents the user with a directory of material accessible at a particu-

lar point, a set of documents that can be searched using keywords, or a document containing text or other forms of material that the computer can display. Gopher is simpler but rather less flexible than the *World Wide Web.

**Gouy balance** A method of measuring magnetic susceptibility. The sample is suspended from a balance, with the bottom part of the sample between the poles of an electromagnet. When the magnetic field is switched on, the sample experiences a field gradient which causes an apparent change in weight. In particular, paramagnetic substances show an increase in weight, which, after correction for a smaller diamagnetic contribution, can be used to calculate the paramagnetic part of the susceptibility. This can be used to calculate the number of unimpaired electrons in the sample. Magnetic measurements of this type are widely used to investigate the electronic structures of metal complexes. The **Evans balance** is a portable version of the Gouy balance using permanent magnets and giving a direct readout. Other methods of measuring magnetic susceptibility include magnetic resonance techniques and the use of a SQID (superconducting quantum interference device). The Gouy balance is named after the French scientist Louis Georges Gouy (1854–1926).

**governor** A device that maintains a motor or engine at a constant speed despite variations in the load, using the principle of negative feedback. A common method uses a set of flying balls that reduce the fuel intake as the speed increases. The balls, attached by flexible steel strips to a collar capable of moving vertically up and down a rotating shaft, move outwards as the speed increases. The collar rises as the balls fly out and is coupled to a lever that controls the fuel intake.

**G protein** Any one of a group of proteins that relay signals in mammalian cells. They occur on the inner surface of the plasma membrane and transmit signals from receptors on the outer surface of the cell to intracellular components. G proteins are activated by binding to GTP and become inactive when they bind to GDP (*see* GUANOSINE). The cholera toxin exerts its effects by changing the G protein in the epithelial cells of the intestine so that it is continually activated, which causes an abnormal increase in cellular adenylate cyclase levels. One consequence of this is that sodium ions are ac-

**g**

tively pumped into the intestine, causing water to follow by osmosis: the result is diarrhoea and dehydration.

**G-protein-coupled receptor (GPCR)**
Any of a superfamily of proteins that are located in cell membranes and act as receptors by relaying signals from the exterior to the interior of the cell via associated G proteins. The latter then activate second messengers, which regulate the activity of the cell. A wide range of substances, including peptide hormones, acetylcholine, noradrenaline, gamma-aminobutyric acid, and glutamate, can bind as ligands to the various types of GPCR, at binding sites in the extracellular region of the receptor molecule. Ligand binding changes the shape of the receptor allowing the G protein to bind to it, which activates the G protein by causing GTP to replace bound GDP.

**GPS** *See* GLOBAL POSITIONING SYSTEM.

**Graafian follicle (ovarian follicle)** The fluid-filled cavity that surrounds and protects the developing egg cell in the ovary of a mammal. After the release of the ovum it develops into a *corpus luteum. It is named after the Dutch anatomist Reinier de Graaf (1641–73).

**graben** A long elongated block of rock or crust that sinks between a pair of parallel near-vertical faults, when they move slightly outwards. The higher bordering areas are called horsts.

**grad** *See* GRADIENT OPERATOR.

**gradient** 1. The slope of a line. In Cartesian coordinates, a straight line $y = mx + c$, has a gradient $m$. For a curve, $y = f(x)$, the gradient at a point is the derivative $dy/dx$ at that point, i.e. the slope of the tangent to the curve at that point. 2. *See* GRADIENT OPERATOR.

**gradient operator (grad)** The *operator

$$\nabla = \boldsymbol{i}\,\partial/\partial x + \boldsymbol{j}\,\partial/\partial y + \boldsymbol{k}\,\partial/\partial z,$$

where $\boldsymbol{i}$, $\boldsymbol{j}$, and $\boldsymbol{k}$ are unit vectors in the $x$, $y$, and $z$ directions. Given a scalar function f and a unit vector $\boldsymbol{n}$, the *scalar product $\boldsymbol{n}.\nabla f$ is the rate of change of f in the direction of $\boldsymbol{n}$. *See also* CURL; DIVERGENCE.

**graft** An isolated portion of living tissue that is joined to another tissue, either in the same or a different organism, the consequent growth resulting in fusion of the tissues. (The word is also used for the process of joining the tissues.) Grafting of plant tissues is a horticultural practice used to propagate plants, especially certain bushes and fruit trees, artificially. A shoot or bud of the desired variety (the **scion**) is grafted onto a rootstock of either a common or a wild related species (the **stock**). The scion retains its desirable characteristics (e.g. flower form or fruit yield) and supplies the stock with food made by photosynthesis. The stock supplies the scion with water and mineral salts and affects only the size and vigour of the scion. Animal and human grafts are used to replace faulty or damaged parts of the body. An **autograft** is taken from one part of the body and transferred to another part of the same individual, e.g. a skin graft used for severe burns. An **allograft** (**homograft**) is taken from one individual (the **donor**) and implanted in another of the same species (the **recipient**), the process being known as **transplantation**, e.g. a heart or kidney transplant. In such cases the graft may be regarded by the body as foreign (a state of **incompatibility**): an *immune response follows and the graft is rejected (*see also* HISTO-COMPATIBILITY).

**graft copolymer** *See* POLYMER.

**graft hybrid** A type of plant *chimaera that may be produced when a part of one plant (the scion) is grafted onto another plant of a different genetic constitution (the stock). Shoots growing from the point of union of the graft contain tissues from both the stock and the scion.

**Graham, Thomas** (1805–69) Scottish chemist, who became professor of chemistry at Glasgow University in 1830, moving to University College, London, in 1837. His 1829 paper on gaseous diffusion introduced *Graham's law. He went on to study diffusion in liquids, leading in 1861 to the definition of *colloids.

**Graham's law** The rates at which gases diffuse are inversely proportional to the square roots of their densities. This principle is made use of in the diffusion method of separating isotopes. The law was formulated in 1829 by Thomas Graham.

**gram** Symbol g. One thousandth of a kilogram. The gram is the fundamental unit of mass in *c.g.s. units and was formerly used in such units as the **gram-atom**, **gram-molecule**, and **gram-equivalent**, which have now been replaced by the *mole.

**Gram's stain** A staining method used to differentiate bacteria. The bacterial sample is smeared on a microscope slide, stained with a violet dye, treated with acetone-alcohol (a decolourizer), and finally counter-stained with a red dye. **Gram-positive** bacteria retain the first dye, appearing blue-black under the microscope. In **Gram-negative** bacteria, the acetone-alcohol washes out the violet dye and the counterstain is taken up, the cells appearing red. It is named after the Danish bacteriologist Hans Gram (1853–1938), who first described the technique (since modified) in 1884.

**grand unified theory (GUT)** A theory that attempts to combine the strong, weak, and electromagnetic interactions into a single *gauge theory with a single symmetry group. There are a number of different theories, most of which postulate that the interactions merge at high energies into a single interaction (the standard model emerges from the GUT as a result of *broken symmetry). The energy above which the interactions are the same is around $10^{15}$ GeV, which is much higher than those obtainable with existing accelerators.

One prediction of GUTs is the occurrence of *proton decay. Some also predict that the neutrino has nonzero mass. There is no evidence for proton decay at present, although there is some evidence that neutrinos have very small nonzero masses. *See also* SUPER-STRING THEORY.

**granite** An extremely hard light-coloured acid igneous rock consisting mainly of plagioclase feldspar and quartz (average 25%), with some biotite (mica), hornblende, or other coloured mineral. It generally results from the slow solidification of molten *magma, giving it a coarse grain size. Some granites result from metamorphism of pre-existing rocks.

**granulocyte** Any white blood cell (*see* LEUCOCYTE) that contains granular material (**secretory vessels**) and *lysosomes in its cytoplasm. *Neutrophils and *basophils are examples of granulocytes. *Compare* AGRANU-LOCYTE.

**granum** (*pl.* **grana**) A stack of platelike bodies (**thylakoids**), many of which are found in plant *chloroplasts (each chloroplast contains about 50 grana). Grana bear the light-receptive pigment chlorophyll and contain the enzymes responsible for the light-dependent reactions of *photosynthesis.

**grape sugar** *See* GLUCOSE.

**graph** A diagram that illustrates the relationship between two variables. It usually consists of two perpendicular axes, calibrated in the units of the variables and crossing at a point called the **origin**. Points are plotted in the spaces between the axes and the points are joined to form a curve. *See also* CARTESIAN COORDINATES; POLAR COORDINATES.

**graphene** A single sheet of carbon atoms arranged in hexagons. Graphene can be regarded as a single plane of graphite and is of interest because it has some unusual and potentially useful electrical properties.

**graphite** *See* CARBON.

**graphite-moderated reactor** *See* NUCLEAR REACTOR.

**graph theory** The area of mathematics that deals with *graphs and their properties. It has important applications in *topology and in the construction of certain types of *algorithms, as used in some computer programs.

**graptolites** A group of extinct marine colonial animals that were common in the Palaeozoic era. Graptolites are generally regarded as being related to colonial soft-bodied marine invertebrates called pterobranchs. They had chitinous outer skeletons in the form of simple or branched stems, the individual polyps occupying minute cups (**thecae**) along these stems. Fossils of these skeletons are found in Palaeozoic rocks of all continents; they are particularly abundant in Ordovician and Silurian rock strata, for which they are used as *index fossils. At the end of the Silurian many graptolites became extinct but a few groups continued into the early Carboniferous.

**grass-green bacteria** *See* CYANO-BACTERIA.

**grassland** A major terrestrial *biome in which the dominant plants are species of grass; the rainfall is insufficient to support extensive growth of trees, which are also suppressed by grazing animals. Tropical grassland (**savanna**), which covers much of Africa south of the Sahara, has widely spaced trees, such as acacias and baobabs, and supports large herds of grazing animals and their predators. Temperate grasslands, such as the **steppes** of Asia, the **prairies** of North America, and the **pampas** of South America, have few trees and are largely used for agriculture.

**graticule** (in optics) A network of fine wires or a scale in the eyepiece of a telescope or microscope or on the stage of a microscope, or on the screen of a cathode-ray oscilloscope for measuring purposes.

**grating** See DIFFRACTION GRATING.

**gravimetric analysis** A type of quantitative analysis that depends on weighing. For instance, the amount of silver in a solution of silver salts could be measured by adding excess hydrochloric acid to precipitate silver chloride, filtering the precipitate, washing, drying, and weighing.

**gravitation** See NEWTON'S LAW OF GRAVITATION.

**gravitational collapse** A phenomenon predicted by the general theory of *relativity in which matter collapses as a consequence of gravitational attraction until it becomes a compact object such as a *white dwarf, *neutron star, or *black hole. The type of object depends on the initial mass. The process of gravitational collapse is important in *astrophysics as it gives rise to such phenomena as *supernova explosions and gamma-ray bursts (see GAMMA-RAY ASTRONOMY).

**gravitational constant** Symbol $G$. The constant that appears in *Newton's law of gravitation; it has the value $6.672\ 59(85) \times 10^{-11}$ N m$^2$ kg$^{-2}$. $G$ is usually regarded as a universal constant although, in some models of the universe, it is proposed that it decreases with time as the universe expands.

**gravitational field** The region of space surrounding a body that has the property of *mass. In this region any other body that has mass will experience a force of attraction. The ratio of the force to the mass of the second body is the **gravitational field strength**.

**gravitational interaction** See FUNDAMENTAL INTERACTIONS.

**gravitational lens** An object that deflects light by gravitation as described by the general theory of *relativity; it is analogous to a lens in *optics. The prediction of a gravitational lensing effect in general relativity theory has been confirmed in observations on *quasars. In 1979 a 'double' quasar was

discovered, due to the multiple image of a single quasar caused by gravitational lensing by a galaxy, or cluster of galaxies, along the line of sight between the observer and the quasar. The images obtained by gravitational lensing can be used to obtain information about the mass distribution of the galaxy or cluster of galaxies.

**gravitational mass** See MASS.

**gravitational shift** See REDSHIFT.

**gravitational waves 1.** (in physics) Waves propagated through a *gravitational field. The prediction that an accelerating mass will radiate gravitational waves (and lose energy) comes from the general theory of *relativity. Many attempts have been made to detect waves from space directly using large metal detectors. The theory suggests that a pulse of gravitational radiation (as from a supernova explosion or *black hole) causes the detector to vibrate, and the disturbance is detected by a transducer. The interaction is very weak and extreme care is required to avoid external disturbances and the effects of thermal noise in the detecting system. So far, no accepted direct observations have been made. However, indirect evidence of gravitational waves has come from observations of a pulsar in a binary system with another star. **2.** (in oceanography) Water surface waves transmitted primarily because of the weight of the water in the crests, which causes them to collapse. Ocean waves are of this type.

**(((⊕))) SEE WEB LINKS**

• The home page for the Laser Interferometer Gravitational-Wave Observatory (LIGO)

**graviton** A hypothetical particle or quantum of energy exchanged in a gravitational interaction (see FUNDAMENTAL INTERACTIONS). Such a particle has not been observed but is postulated to make the gravitational interaction consistent with quantum mechanics. It would be expected to travel at the speed of light and have zero rest mass and charge, and spin 2.

**gravitropism** See GEOTROPISM.

**gravity** The phenomenon associated with the gravitational force acting on any object that has mass and is situated within the earth's *gravitational field. The weight of a body (see MASS) is equal to the force of gravity acting on the body. According to Newton's second law of motion $F = ma$, where $F$

is the force producing an acceleration $a$ on a body of mass $m$. The weight of a body is therefore equal to the product of its mass and the acceleration due to gravity ($g$), which is now called the *acceleration of free fall. By combining the second law of motion with *Newton's law of gravitation ($F = GM_1M_2/d^2$) it follows that: $g = GM/d^2$, where $G$ is the *gravitational constant, $M$ is the mass of the earth, and $d$ is the distance of the body from the centre of the earth. For a body on the earth's surface $g = 9.806\ 65$ m s$^{-2}$.

A force of gravity also exists on other planets, moons, etc., but because it depends on the mass of the planet and its diameter, the strength of the force is not the same as it is on earth. If $F_e$ is the force acting on a given mass on earth, the force $F_p$ acting on the same mass on another planet will be given by:

$$F_p = F_e d_e^2 M_p / M_e d_p^2,$$

where $M_p$ and $d_p$ are the mass and diameter of the planet, respectively. Substituting values of $M_p$ and $d_p$ for the moon shows that the force of gravity on the moon is only $1/6$ of the value on earth.

**(∰) SEE WEB LINKS**
- The home page of the NASA/JPL Gravity Recovery and Climate experiment to make precise measurements of the earth's gravitational field

**gray** Symbol Gy. The derived SI unit of absorbed *dose of ionizing radiation (*see* RADIATION UNITS). It is named after the British radiobiologist L. H. Gray (1905–65).

**grazing** The consumption of vegetation, usually on *grassland, by animals, particularly cattle and sheep. Overgrazing can lead to *desertification.

**grazing incidence telescope** A type of astronomical telescope designed to work at X-ray and gamma-ray wavelengths. Its mirror consists of a paraboloidal annulus of metal, which deflects incoming photons striking it at an acute angle. The deflected high-energy photons are detected by a spark chamber.

**Great Attractor** A huge concentration of mass, equivalent to about a million galaxies, beyond the Hydra and Centaurus constellations. Our own Galaxy and others near to it are heading towards the Great Attractor at a rate of about 600 kilometres per second.

**great circle** Any circle on a sphere formed by a plane that passes through the centre of the sphere. The equator and the meridians

of longitude are all great circles on the earth's surface.

**green algae** *See* CHLOROPHYTA.

**green fluorescent protein** (GFP) A naturally fluorescent protein obtained from the jellyfish *Aequorea victoria* and used as a marker for identifying cells containing recombinant DNA or for localizing specific proteins in cells. It absorbs blue light and emits a green fluorescence, and hence the abundance and location of GFP in cells can be visualized microscopically under ultraviolet light. Several colour variants of GFP are now available.

**greenhouse effect** An effect occurring in the atmosphere because of the presence of certain gases (**greenhouse gases**) that absorb infrared radiation. Light and ultraviolet radiation from the sun are able to penetrate the atmosphere and warm the earth's surface. This energy is re-radiated as infrared radiation, which, because of its longer wavelength, is absorbed by such substances as carbon dioxide. The greenhouse effect is a natural phenomenon, without which the earth's climate would be much more hostile to life. However, emissions of carbon dioxide from human activities (e.g. farming, industry, and transport) have increased markedly in the last 150 years or so. The overall effect is that the average temperature of the earth and its atmosphere is increasing (*see* GLOBAL WARMING). The effect is similar to that occurring in a greenhouse, where light and long-wavelength ultraviolet radiation can pass through the glass into the greenhouse but the infrared radiation is absorbed by the glass and part of it is re-radiated into the greenhouse. The greenhouse effect is seen as a major environmental hazard. Average increases in temperature are likely to change weather patterns and agricultural output. It is already causing the polar ice caps to melt, with a corresponding rise in sea level. Carbon dioxide, from coal-fired power stations and car exhausts, is the main greenhouse gas. Other contributory pollutants are nitrogen oxides, ozone, methane, and *chlorofluorocarbons. Many countries have now agreed targets to limit emissions of greenhouse gases, e.g. by switching to renewable energy sources. *See also* POLLUTION.

**greenhouse gas** *See* GREENHOUSE EFFECT.

**greenockite** A mineral form of cadmium sulphide, CdS.

**Green's function** One of a set of functions that are used for solving *differential equations with *boundary conditions. It was named after the British mathematician George Green (1793–1841).

**green vitriol** *See* IRON(II) SULPHATE.

**Gregorian telescope** *See* TELESCOPE.

**Greisen–Zatsepin–Kuzmin limit** *See* GZK LIMIT.

**grey matter** Part of the tissue that makes up the central nervous system of vertebrates. It is brown-grey in colour, consisting largely of nerve *cell bodies, *synapses, and *dendrites. The grey matter is the site of coordination between nerves of the central nervous system. *Compare* WHITE MATTER.

**grid** **1.** (in electricity) The system of overhead wires or underground cables by which electrical power is distributed from power stations to users. The grid is at a high voltage, up to 750 kV in some countries. **2.** (in electronics) *See* CONTROL GRID. **3.** (in cartography) A network of horizontal and vertical lines on a map that provide a means of locating a specific point.

**Griess test** A test for nitrates or nitrites, once widely used to detect the possible existence of gunshot residue.

**Grignard reagents** A class of organometallic compounds of magnesium, with the general formula RMgX, where R is an organic group and X a halogen atom (e.g. $CH_3MgCl$, $C_2H_5MgBr$, etc.). They actually have the structure $R_2Mg.MgCl_2$, and can be made by reacting a haloalkane with magnesium in ether; they are rarely isolated but are extensively used in organic synthesis, when they are made in one reaction mixture. Grignard reagents have a number of reactions that make them useful in organic synthesis. With methanal they give a primary alcohol

$$CH_3MgCl + HCHO \rightarrow CH_3CH_2OH$$

Other aldehydes give a secondary alcohol

$$CH_3CHO + CH_3MgCl \rightarrow (CH_3)_2CHOH$$

With alcohols, hydrocarbons are formed

$$CH_3MgCl + C_2H_5OH \rightarrow C_2H_5CH_3$$

Water also gives a hydrocarbon

$$CH_3MgCl + H_2O \rightarrow CH_4$$

The compounds are named after their discoverer, the French chemist Victor Grignard (1871–1935).

**grooming** The actions of an animal of rearranging fur or feathers and cleaning the body surface by biting, scratching, licking, etc., which is important for removing parasites and spreading oils over the body surface. In many mammals, especially primates, grooming between individuals (**allogrooming**) has an important role in maintaining social cohesion.

**ground state** The lowest stable energy state of a system, such as a molecule, atom, or nucleus. *See* ENERGY LEVEL.

**ground substance** The matrix of connective tissue, in which various cells and fibres are embedded. The ground substance of cartilage consists of *chondrin. *See* EXTRACELLULAR MATRIX.

**ground tissues** All the plant tissues formed by the *apical meristems except the epidermis and vascular tissue. The principal ground tissues are the *cortex, *pith, and primary *medullary rays, and they consist chiefly of *parenchyma.

**Collenchyma** is a form of ground tissue less frequently observed. It consists of living cells with additional cellulose thickening in the walls, giving them additional strength, and is most commonly found in the stem cortex.

**groundwater** **1. (subterranean water)** Water that occurs below the surface of the ground, as opposed to that at the surface. **2. (phreatic water)** Water that occurs below the surface in soil or rocks that are saturated, either in cavities and pores (through which it can flow) or below the water table. Groundwater has an erosive action in permeable rocks, such as limestone, in which it can form underground rivers and caves.

**ground wave** A radio wave that travels in approximately a straight line between points on the earth's surface. For transmission over longer distances sky waves have to be involved. *See* RADIO TRANSMISSION.

**group** **1.** (in physics) A set of elements *A*, *B*, *C*, etc., for which there exists a law of composition, referred to as 'multiplication'. Any two elements can be combined to give a 'product' *AB*.
(1) Every product of two elements is an element of the set.

(2) The operation is associative, i.e. $A(BC) = (AB)C$.

(3) The set has an element $I$, called the **identity element**, such that $IA = AI = A$ for all $A$ in the set.

(4) Each element of the set has an **inverse** $A^{-1}$ belonging to the set such that $AA^{-1} = A^{-1}A = I$.

Although the law of combination is called 'multiplication' this does not necessarily have its usual meaning. For example, the set of integers forms a group if the law of composition is addition.

Two elements $A$, $B$ of a group **commute** if $AB = BA$. If all the elements of a group commute with each other the group is said to be **Abelian**. If this is not the case the group is said to be **non-Abelian**. The distinction between Abelian and non-Abelian groups is of fundamental importance in *gauge theories.

The interest of group theory in physics and chemistry is in analysing symmetry. **Discrete groups** have a finite number of elements, such as the symmetries involved in rotations and reflections of molecules, which give rise to **point groups**. **Continuous groups** have an infinite number of elements where the elements are continuous. An example of a continuous group is the set of rotations about a fixed axis. The **rotation group** thus formed underlies the *quantum theory of *angular momentum, which has many applications to *atoms and *nuclei. More abstract and more general continuous groups describe fundamental interactions by gauge theories.

**2.** (in chemistry) *See* PERIODIC TABLE.

**group 0 elements** *See* NOBLE GASES.

**group 1 elements** A group of elements in the *periodic table: lithium (Li), sodium (Na), potassium (K), rubidium (Rb), caesium (Cs), and francium (Fr). They are known as the *alkali metals. Formerly, they were classified in group I, which consisted of two subgroups: group IA (the main group) and group IB. Group IB consisted of the *coinage metals, copper, silver, and gold, which comprise group 11 and are usually considered with the *transition elements.

**group 2 elements** A group of elements in the *periodic table: beryllium (Be), magnesium (Hg), calcium (Ca), strontium (Sr), barium (Ba), and radium (Ra). They are known as the *alkaline-earth metals. Formerly, they were classified in group II, which consisted of two subgroups: group IIA (the main group, *see* ALKALINE-EARTH METALS) and

group IIB. Group IIB consisted of the three metals zinc (Zn), cadmium (Cd), and mercury (Hg), which have two $s$-electrons outside filled $d$-subshells. Moreover, none of their compounds have unfilled $d$-levels, and the metals are regarded as nontransition elements. They now form group 12 and are sometimes called the **zinc group**. Zinc and cadmium are relatively electropositive metals, forming compounds containing divalent ions $Zn^{2+}$ or $Cd^{2+}$. Mercury is more unreactive and also unusual in forming mercury(I) compounds, which contain the ion $Hg_2^{2+}$.

**groups 3–12** *See* TRANSITION ELEMENTS.

**group 13 elements** A group of elements in the *periodic table: boron (B), aluminium (Al), gallium (Ga), indium (In), and thallium (Tl), which all have outer electronic configurations $ns^2np^1$ with no partly filled inner levels. They are the first members of the $p$-block. The group differs from the alkali metals and alkaline-earth metals in displaying a considerable variation in properties as the group is descended. Formerly, they were classified in group III, which consisted of two subgroups: group IIIB (the main group) and group IIIA. Group IIIA consisted of scandium (Sc), yttrium (Yt), and lanthanum (La), which are generally considered with the *lanthanoids, and actinium (Ac), which is classified with the *actinoids. Scandium and yttrium now belong to group 3 (along with lutetium and lawrencium).

Boron has a small atomic radius and a relatively high ionization energy. In consequence its chemistry is largely covalent and it is generally classed as a metalloid. It forms a large number of volatile hydrides, some of which have the uncommon bonding characteristic of *electron-deficient compounds. It also forms a weakly acidic oxide. In some ways, boron resembles silicon (*see* DIAGONAL RELATIONSHIP).

As the group is descended, atomic radii increase and ionization energies are all lower than for boron. There is an increase in polar interactions and the formation of distinct $M^{3+}$ ions. This increase in metallic character is clearly illustrated by the increasing basic character of the hydroxides: boron hydroxide is acidic, aluminium and gallium hydroxides are amphoteric, indium hydroxide is basic, and thallium forms only the oxide. As the elements of group 13 have a vacant $p$-orbital they display many electron-acceptor properties. For example, many boron compounds form adducts with donors such as ammonia and organic amines (acting as Lewis acids). A large number of complexes of the type $[BF_4]^-$, $[AlCl_4]^-$, $[InCl_4]^-$, $[TlI_4]^-$ are known and the heavier members can expand their coordination numbers to six as in $[AlF_6]^{3-}$ and $[TlCl_6]^3$. This acceptor property is also seen in bridged dimers of the type $Al_2Cl_6$. Another feature of group 13 is the increasing stability of the monovalent state down the group. The electron configuration $ns^2np^1$ suggests that only one electron could be lost or shared in forming compounds. In fact, for the lighter members of the group the energy required to promote an electron from the $s$-subshell to a vacant $p$-subshell is small. It is more than compensated for by the resulting energy gain in forming three bonds rather than one. This energy gain is less important for the heavier members of the group. Thus, aluminium forms compounds of the type AlCl in the gas phase at high temperatures. Gallium similarly forms such compounds and gallium(I) oxide ($Ga_2O$) can be isolated. Indium has a number of known indium(I) compounds (e.g. InCl, $In_2O$, $In_3^I[In^{III}Cl_6]$). Thallium has stable monovalent compounds. In aqueous solution, thallium(I) compounds are more stable than the corresponding thallium(III) compounds. *See* INERT-PAIR EFFECT.

**group 14 elements** A group of elements in the *periodic table: carbon (C), silicon (Si), germanium (Ge), tin (Sn), and lead (Pb), which all have outer electronic configurations $ns^2np^2$ with no partly filled inner levels. Formerly, they were classified in group IV, which consisted of two subgroups: IVB (the main group) and group IVA. Group IVA consisted of titanium (Ti), zirconium (Zr), and hafnium (Hf), which now form group 4 and are generally considered with the *transition elements.

The main valency of the elements is 4, and the members of the group show a variation from nonmetallic to metallic behaviour in moving down the group. Thus, carbon is a nonmetal and forms an acidic oxide ($CO_2$) and a neutral oxide. Carbon compounds are mostly covalent. One allotrope (diamond) is an insulator, although graphite is a fairly good conductor. Silicon and germanium are metalloids, having semiconductor properties. Tin is a metal, but does have a nonmetallic allotrope (grey tin). Lead is definitely a metal. Another feature of the group is the tendency to form divalent compounds as the size of the atom increases.

Thus carbon has only the highly reactive carbenes. Silicon forms analogous silylenes. Germanium has an unstable hydroxide ($Ge(OH)_2$), a sulphide (GeS), and halides. The sulphide and halides disproportionate to germanium and the germanium(IV) compound. Tin has a number of tin(II) compounds, which are moderately reducing, being oxidized to the tin(IV) compound. Lead has a stable lead(II) state. *See* INERT-PAIR EFFECT.

In general, the reactivity of the elements increases down the group from carbon to lead. All react with oxygen on heating. The first four form the dioxide; lead forms the monoxide (i.e. lead(II) oxide, PbO). Similarly, all will react with chlorine to form the tetrachloride (in the case of the first four) or the dichloride (for lead). Carbon is the only one capable of reacting directly with hydrogen. The hydrides all exist from the stable methane ($CH_4$) to the unstable plumbane ($PbH_4$).

**group 15 elements** A group of elements in the *periodic table: nitrogen (N), phosphorus (P), arsenic (As), antimony (Sb), and bismuth (Bi), which all have outer electronic configurations $ns^2np^3$ with no partly filled inner levels. Formerly, they were classified in group V, which consisted of two subgroups: group VB (the main group) and group VA. Group VA consisted of vanadium (V), niobium (Nb), and tantalum (Ta), which are generally considered with the *transition elements:

The lighter elements (N and P) are nonmetals; the heavier elements are metalloids. The lighter elements are electronegative in character and have fairly large ionization energies. Nitrogen has a valency of 3 and tends to form covalent compounds. The other elements have available $d$-sublevels and can promote an $s$-electron into one of these to form compounds with the V oxidation state. Thus, they have two oxides $P_2O_3$, $P_2O_5$, $Sb_2O_3$, $Sb_2O_5$, etc. In the case of bismuth, the pentoxide $Bi_2O_5$ is difficult to prepare and unstable – an example of the increasing stability of the III oxidation state in going from phosphorus to bismuth. The oxides also show how there is increasing metallic (electropositive) character down the group. Nitrogen and phosphorus have oxides that are either neutral ($N_2O$, NO) or acidic. Bismuth trioxide ($Bi_2O_3$) is basic. Bismuth is the only member of the group that forms a well-characterized positive ion $Bi^{3+}$.

**group 16 elements** A group of elements in the *periodic table: oxygen (O), sulphur (S), selenium (Se), tellurium (Te), and polonium (Po), which all have outer electronic configurations $ns^2np^4$ with no partly filled inner levels. They are also called the **chalcogens**. Formerly, they were classified in group VI, which consisted of two subgroups: group VIB (the main group) and group VIA. Group VIA consisted of chromium (Cr), molybdenum (Mo), and tungsten (W), which now form group 6 are generally classified with the *transition elements.

The configurations are just two electrons short of the configuration of a noble gas and the elements are characteristically electronegative and almost entirely nonmetallic. Ionization energies are high, (O 1314 to Po 813 kJ mol$^{-1}$) and monatomic cations are not known. Polyatomic cations do exist, e.g. $O_2^+$, $S_8^{2+}$, $Se_8^{2+}$, $Te_4^{2+}$. Electronegativity decreases down the group but the nearest approach to metallic character is the occurrence of 'metallic' allotropes of selenium, tellurium, and polonium along with some metalloid properties, in particular, marked photoconductivity. The elements of group 16 combine with a wide range of other elements and the bonding is largely covalent. The elements all form hydrides of the type $XH_2$. Apart from water, these materials are all toxic foul-smelling gases; they show decreasing thermal stability with increasing relative atomic mass of X. The hydrides dissolve in water to give very weak acids (acidity increases down the group). Oxygen forms the additional hydride $H_2O_2$ (hydrogen peroxide), but sulphur forms a range of sulphanes, such as $H_2S_2$, $H_2S_4$, $H_2S_6$.

Oxygen forms the fluorides $O_2F_2$ and $OF_2$, both powerful fluorinating agents; sulphur forms analogous fluorides along with some higher fluorides, $S_2F_2$, $SF_2$, $SF_4$, $SF_6$, $S_2F_{10}$. Selenium and tellurium form only the higher fluorides $MF_4$ and $MF_6$; this is in contrast to the formation of lower valence states by heavier elements observed in groups 13, 14, and 15. The chlorides are limited to $M_2Cl_2$ and $MCl_4$; the bromides are similar except that sulphur only forms $S_2Br_2$. All metallic elements form oxides and sulphides and many form selenides.

**group 17 elements** A group of elements in the *periodic table: fluorine (F), chlorine (Cl), bromine (Br), iodine (I), and astatine (At). They are known as the *halogens. Formerly, they were classified in group VII,

which consisted of two subgroups: group VIIB (the main group) and group VIIA. Group VIIA consisted of the elements manganese (Mn), technetium (Te), and rhenium (Re), which now form group 7 and are usually considered with the transition elements.

**group 18 elements** A group of elements in the *periodic table: helium (He), neon (Ne), argon (Ar), krypton (Kr), xenon (Xe), and radon (Rn). Formerly classified as **group 0 elements**, they are usually referred to as the *noble gases.

**Grover's algorithm** An algorithm, invented in 1996 by the Indian–American computer scientist Lov Grover (1961–  ), for searching an unsorted database on a quantum computer. It is a more efficient method of searching than a traditional algorithm, i.e. one for a conventional computer. *See also* SHOR'S ALGORITHM.

**growth** An increase in the dry weight or volume of an organism through cell division and cell enlargement. Growth may continue throughout the life of the organism, as occurs in woody plants, or it may cease at maturity, as in humans and other mammals. *See also* ALLOMETRIC GROWTH; EXPONENTIAL GROWTH.

**growth factor** Any of various chemicals, particularly polypeptides, that have a variety of important roles in the stimulation of new cell growth and cell maintenance. They bind to the cell surface on receptors. Specific growth factors can cause new cell proliferation (**epidermal growth factor**, **insulin-like growth factor (IGF)**, **haemopoietic growth factor** – *see* HAEMOPOIETIC TISSUE) and the migration of cells (**fibroblast growth factor**) and play a role in wound healing (**platelet-derived growth factor; PDGF**). Some growth factors act in the embryonic stage of development; for example, nerve growth factor (*see* NEUROTROPHIN) stimulates the growth of axons and dendrites from developing sensory and sympathetic neurons. Some growth factors or their receptors are involved in the abnormal regulation of growth seen in cancer when produced in excessive amounts or permanently activated.

**growth hormone (GH; somatotrophin)** A hormone, secreted by the mammalian pituitary gland, that stimulates protein synthesis and growth of the long bones in the legs and arms. It also promotes the breakdown and use of fats as an energy source, rather than glucose. Production of growth hormone is greatest during early life. Its secretion is controlled by the opposing actions of two hormones from the hypothalamus: **growth hormone releasing hormone** (**somato-liberin**), which promotes its release; and *somatostatin, which inhibits it. Overproduction of **human growth hormone** (**hGH**) results in gigantism in childhood and *acromegaly in adults; underproduction results in dwarfism. **Bovine somatotrophin** (**BST**) has been used to increase milk and meat production in cattle.

**growth ring** (**annual ring**) Any of the rings that can be seen in a cross-section of a woody stem (e.g. a tree trunk). It represents the *xylem formed in one year as a result of fluctuating activity of the vascular *cambium. In temperate climates pale soft **spring wood**, characterized by large xylem vessels, is formed in spring and early summer. Growth slows down in late summer and a darker dense **autumn wood** with smaller xylem vessels is formed (see illustration). The age of a tree can be determined by counting the rings. Under certain circumstances two or more growth rings may form in one year, giving rise to false annual rings.

**growth substance** *See* PLANT HORMONE.

**GSC** (**gas–solid chromatography**) *See* GAS CHROMATOGRAPHY.

**G-series** *See* NERVE AGENTS.

**GTP** (**guanosine triphosphate**) *See* GUANOSINE.

**guanidine** A crystalline basic compound $HN:C(NH_2)_2$, related to urea.

**guanine** A *purine derivative. It is one of the major component bases of *nucleotides and the nucleic acids *DNA and *RNA.

**guano** An accumulation of the droppings of birds, bats, or seals, usually formed by a long-established colony of animals. It is rich in plant nutrients, and some deposits are extracted for use as fertilizer.

**guanosine** A nucleoside consisting of one guanine molecule linked to a D-ribose sugar molecule. The derived nucleotides, guanosine mono-, di-, and triphosphate (GMP, GDP, and GTP, respectively), participate in various metabolic reactions.

**guard cell** *See* STOMA.

**guild** (in ecology) A group of species within a community that exploit the same resources in a similar way. Thus, for example, different species of snakes and lizards may belong to a guild by virtue of occupying the same types of underground shelters, and different species of seed-eating birds may constitute a guild.

**gullet** *See* OESOPHAGUS.

**gum 1.** Any of a variety of substances obtained from plants. Typically they are insoluble in organic solvents but form gelatinous or sticky solutions with water. Most gums are complex polysaccharides. Commercially im-

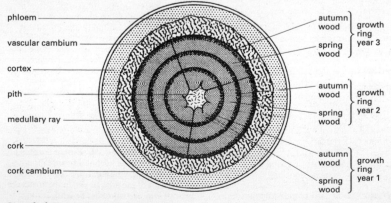

**Growth ring.** Transverse section through a three-year-old woody stem to show the growth rings.

portant examples are gum arabic and gum tragacanth. **Gum arabic** (or **gum acacia**) is obtained from various acacia trees; it is used in the manufacture of confectionery, cosmetics, linctuses and other medicinal products, and gummed labels. **Gum tragacanth**, extracted from trees of the genus *Astragalus*, forms a thick *mucilage in water; it is used in the manufacture of pills and confectionery and as a sauce thickener. Gum resins are mixtures of gums and natural resins.

Gums are produced by the young xylem vessels of some plants (mainly trees) in response to wounding or pruning. The exudate hardens when it reaches the plant surface and thus provides a temporary protective seal while the cells below divide to form a permanent repair. Excessive gum formation is a symptom of some plant diseases. **2.** *See* GINGIVA.

**guncotton** *See* CELLULOSE NITRATE.

**gun metal** A type of bronze usually containing 88–90% copper, 8–10% tin, and 2–4% zinc. Admiralty gunmetal, which is used in shipbuilding, contains 88% copper, 10% tin, and 2% zinc. Because it was easy to cast it was originally used to make cannons; it is still used for bearings and other parts that require high resistance to wear and corrosion.

**gunpowder** An explosive consisting of a mixture of potassium nitrate, sulphur, and charcoal. It was invented by the Chinese, probably in the 10th century, although the English monk Roger Bacon (1214–92) is often credited with its discovery. For many centuries it was the explosive used in firearms; it is no longer used for this purpose, although it is the basis of many fireworks.

**gut** *See* ALIMENTARY CANAL.

**GUT** *See* GRAND UNIFIED THEORY.

**guttation** *See* HYDATHODE.

**guyot (tablemount)** A flat-topped circular underwater mountain. Guyots occur in deep water (usually below 2000 m) and are thought to be conical volcanic peaks levelled off by the action of water currents. They consist mainly of *basalt and tend to occur in groups, similar to island arcs.

**GWS model** Glashow–Weinberg–Salam model. *See* ELECTROWEAK THEORY.

**gymnosperm** Any plant whose ovules and the seeds into which they develop are borne unprotected, rather than enclosed in ovaries, as are those of the flowering plants (the term gymnosperm means naked seed). In traditional systems of classification such plants were classified as the Gymnospermae, a class of the Spermatophyta, but they are now divided into separate phyla: *Coniferophyta (conifers), *Cycadophyta (cycads), Ginkgophyta (ginkgo), and Gnetophyta (e.g. *Welwitschia*).

**gynoecium (gynaecium)** The female sex organs (*carpels) of a flower. *Compare* ANDROECIUM.

**gypsum** A monoclinic mineral form of hydrated *calcium sulphate, $CaSO_4.2H_2O$. It occurs in five varieties: **rock gypsum**, which is often red stained and granular; **gypsite**, an impure earthy form occurring as a surface deposit; **alabaster**, a pure fine-grained translucent form; **satin spar**, which is fibrous and silky; and **selenite**, which occurs as transparent crystals in muds and clays. It is used in the building industry and in the manufacture of cement, rubber, paper, and plaster of Paris.

**gyrocompass** A *gyroscope that is driven continuously so that it can be used as a nonmagnetic compass. When the earth rotates the gyroscope experiences no torque if its spin axis is parallel to the earth's axis; if these axes are not parallel, however, the gyroscope experiences a sequence of restoring torques that tend to make it align itself with the earth's axis. The gyrocompass is therefore an accurate north-seeking device that is uninfluenced by metallic or magnetic objects and it is also more consistent than the magnetic compass. It is therefore widely used on ships, aircraft, missiles, etc.

**gyromagnetic ratio** Symbol γ. The ratio of the angular momentum of an atomic system to its magnetic moment. The inverse of the gyromagnetic ratio is called the **magnetomechanical ratio**.

**gyroscope** A disk with a heavy rim mounted in a double *gimbal so that its axis can adopt any orientation in space. When the disk is set spinning the whole contrivance has two useful properties: (1) Gyroscopic inertia, i.e. the direction of the axis of spin resists change so that if the gimbals are turned the spinning disk maintains the same orientation in space. This property forms the basis of the *gyrocompass and other navigational devices. (2) Precession, i.e. when a gyroscope is subjected to a torque that tends to alter the direction of its axis, the gyroscope turns about an axis at right angles both to the axis about which the torque was applied and to its main axis of spin. This is a consequence of the need to conserve *angular momentum.

In the gyrostabilizer for stabilizing a ship, aircraft, or platform, three gyroscopes are kept spinning about mutually perpendicular axes so that any torque tending to alter the orientation of the whole device affects one of the gyroscopes and thereby activates a servomechanism that restores the original orientation.

**GZK limit (Greisen–Zatsepin–Kuzmin limit)** A limit on the energy of cosmic rays imposed by the interaction between the cosmic rays and the cosmic microwave background radiation. This limit, found by Kenneth Greisen, Vadim Kuzmin, and Georgiy Zatsepin in 1966 using the theory of the *inverse Compton effect, is about $6 \times 10^{19}$ eV. Cosmic rays with higher energies have been detected but it has been established that these come from relatively near sources and so have not had a chance to be slowed down by interaction with the background radiation.

**Haber, Fritz** (1868–1934) German chemist who worked at the Karlsruhe Technical Institute, where he developed the *Haber process in 1908. As a Jew, he left Germany in 1933 to go into exile in Britain, working in Cambridge at the Cavendish Laboratory. For his Haber process, he was awarded the 1918 Nobel Prize for chemistry.

**Haber process** An industrial process for producing ammonia by reaction of nitrogen with hydrogen:

$$N_2 + 3H_2 \rightleftharpoons 2NH_3$$

The reaction is reversible and exothermic, so that a high yield of ammonia is favoured by low temperature (*see* LE CHATELIER'S PRINCIPLE). However, the rate of reaction would be too slow for equilibrium to be reached at normal temperatures, so an optimum temperature of about 450°C is used, with a catalyst of iron containing potassium and aluminium oxide promoters. The higher the pressure the greater the yield, although there are technical difficulties in using very high pressures. A pressure of about 250 atmospheres is commonly employed.

The process is of immense importance for the fixation of nitrogen for fertilizers. It was developed in 1908 by Fritz Haber and was developed for industrial use by Carl Bosch (1874–1940), hence the alternative name **Haber–Bosch process**. The nitrogen is obtained from liquid air. Formerly, the hydrogen was from *water gas and the water-gas shift reaction (the **Bosch process**) but now the raw material (called **synthesis gas**) is obtained by steam *reforming natural gas.

**habit** *See* CRYSTAL.

**habitat** The place in which an organism lives, which is characterized by its physical features or by the dominant plant types. Freshwater habitats, for example, include streams, ponds, rivers, and lakes. *See also* MICROHABITAT.

**habituation** 1. A simple type of learning consisting of a gradual waning in the response of an animal to a continuous or repeated stimulus that is not associated with *reinforcement. 2. The condition of being psychologically, but not physically, dependent on a drug.

**Hadean** The earliest eon in the history of the earth, from the time of the accretion of planetary material, around 4600 million years ago, to the date of the oldest known rocks – and hence the beginning of the geological record – about 3900 million years ago. The young earth was probably a rocky planet with a hot interior and a moist surface with oceans of liquid water. No evidence of life has been found. *Compare* ARCHAEAN.

**hadron** Any of a class of subatomic particles that interact by the strong interaction (*see* FUNDAMENTAL INTERACTIONS). The class includes protons, neutrons, and pions. Hadrons are believed to have an internal structure and to consist of quarks; they are therefore not truly elementary. Hadrons are either *baryons, which decay into protons and are believed to consist of three quarks, or *mesons, which decay into *leptons and photons or into proton pairs and are believed to consist of a quark and an antiquark. *See* ELEMENTARY PARTICLES.

**haem (heme)** An iron-containing molecule that binds with proteins as a *cofactor or *prosthetic group to form the **haemoproteins**. These are *haemoglobin, *myoglobin, and the *cytochromes. Essentially, haem comprises a *porphyrin with its four nitrogen atoms holding the iron(II) atom as a chelate. This iron can reversibly bind oxygen (as in haemoglobin and myoglobin) or (as in the cytochromes) conduct electrons by conversion between iron(II) and iron(III) species.

**haemagglutination** *See* AGGLUTINATION.

**haematin test (Teichmann test)** A test for blood using the presence of characteristic haematin crystals. It was introduced in 1853 by Ludwig Teichmann.

**haematite** A mineral form of iron(III) oxide, $Fe_2O_3$. It is the most important ore of iron and usually occurs in two main forms: as a massive red kidney-shaped ore (**kidney ore**) and as grey to black metallic crystals known as **specular iron ore**. Haematite is the major red colouring agent in rocks; the largest deposits are of sedimentary origin. In industry haematite is also used as a polishing agent (jeweller's rouge) and in paints.

**haematoxylin** A compound used in its oxidized form (**haematein**) as a blue dye in optical microscopy, particularly for staining smears and sections of animal tissue. It stains nuclei blue and is frequently used with *eosin as a counterstain for cytoplasm. Haematoxylin requires a mordant, such as iron alum, which links the dye to the tissue. Different types of haematoxylin can be made up depending on the mordant used, the method of oxidation, and the pH. Examples are **Delafield's haematoxylin** and **Ehrlich's haematoxylin**.

**haemochromogen test (Takayama test)** A test used to confirm the presence of blood. It is a microcrystal test exploiting the characteristic appearance of haemochromogen crystals observed under a microscope. The test was introduced in Japan in 1912 by Masao Takayama.

**haemocoel** The body cavity of arthropods and molluscs, which is filled with blood. The haemocoel is an enlarged blastocoel (*see* BLASTULA) which greatly reduces the coelom (this is restricted to the cavities of the gonads and excretory organs). The haemocoel can act as a *hydrostatic skeleton.

**haemocyanin (hemocyanin)** Any of a group of copper-containing respiratory proteins found in solution in the blood of certain arthropods and molluscs. Haemocyanins contain two copper atoms that reversibly bind oxygen, changing between the colourless deoxygenated form (CuI) and the blue oxygenated form (CuII). In some species, haemocyanin molecules form giant polymers with molecular weights of several million.

**haemoglobin** One of a group of globular proteins occurring widely in animals as oxygen carriers in blood. Vertebrate haemoglobin is contained in the red blood cells (erythrocytes). It comprises two pairs of polypeptide chains, known as α-chains and β-chains (forming the **globin** protein), with each chain folded to provide a binding site for a *haem group. Each of the four haem groups binds one oxygen molecule to form **oxyhaemoglobin**. Dissociation occurs in

oxygen-depleted tissues: oxygen is released and haemoglobin is reformed (*see* OXYGEN DISSOCIATION CURVE). The haem groups also bind other inorganic molecules, including carbon monoxide (to form *carboxyhaemoglobin).

**haemolysis** The breakdown of red blood cells. It can be due to the action of disease-causing microorganisms, poisons, antibodies in mismatched blood transfusions, or certain allergic reactions. It produces anaemia.

**haemophilia** A hereditary sex-linked disease (*see* SEX LINKAGE) in which there is a deficiency or defect of *Factor VIII, causing the blood to clot very slowly. There may be prolonged bleeding following injury and, in severe cases, spontaneous bleeding into the joints and muscles. The disorder is due to the recessive allele of the Factor VIII gene, which is located on the X chromosome. Female carriers of the defective allele are unaffected, whereas all males who inherit a defective allele exhibit the disease.

**haemopoietic tissue** The tissue that gives rise to blood cells in the process of **haemopoiesis**. The haemopoietic tissue of the embryo and fetal stage of vertebrates is the bone marrow, lymph nodes, yolk sac, liver, spleen, and thymus but after birth haemopoiesis occurs in the red bone marrow (*see* MYELOID TISSUE). The different types of *stem cells in haemopoietic tissue that give rise to erythrocytes and leucocytes are all originally derived from **haemopoietic stem cells** (or **haemocytoblasts**). The formation of the different types of blood cell is under the control of **haemopoietic growth factors**, which include hormones and *cytokines.

**hafnium** Symbol Hf. A silvery lustrous metallic *transition element; a.n. 72; r.a.m. 178.49; r.d. 13.3; m.p. 2227±20°C; b.p. 4602°C. The element is found with zirconium and is extracted by formation of the chloride and reduction by the Kroll process. It is used in tungsten alloys in filaments and electrodes and as a neutron absorber. The metal forms a passive oxide layer in air. Most of its compounds are hafnium(IV) complexes; less stable hafnium(III) complexes also exist. The element was first reported by Urbain in 1911, and its existence was finally established by

Dirk Coster (1889–1950) and George de Hevesey (1885–1966) in 1923.

🌐 **SEE WEB LINKS**

• Information from the WebElements site

**Hahn, Otto** (1879–1968) German chemist, who studied in London (with William *Ramsay) and Canada (with Ernest *Rutherford) before returning to Germany in 1907. In 1917, together with Lise *Meitner, he discovered protactinium. In the late 1930s he collaborated with Fritz Strassmann (1902–  ) and in 1938 bombarded uranium with slow neutrons. Among the products was barium, but it was Meitner (now in Sweden) who the next year interpreted the process as *nuclear fission. In 1944 Hahn received the Nobel Prize for chemistry.

**hahnium** *See* TRANSACTINIDE ELEMENTS.

**hair** 1. A multicellular threadlike structure, consisting of many dead keratinized cells, that is produced by the epidermis in mammalian *skin. The section of a hair below the skin surface (the **root**) is contained within a *hair follicle, the base of which produces the hair cells. Hair assists in maintaining body temperature by reducing heat loss from the skin. Bristles and whiskers are specialized types of hair. 2. Any of various threadlike structures on plants, such as a *trichome.

**hair follicle** A narrow tubular depression in mammalian skin containing the root of a *hair. It is lined with epidermal cells and extends down through the epidermis and dermis to its base in the subcutaneous tissue. The ducts of *sebaceous glands empty into hair follicles.

**half cell** An electrode in contact with a solution of ions, forming part of a *cell. Various types of half cell exist, the simplest consisting of a metal electrode immersed in a solution of metal ions. Gas half cells have a gold or platinum plate in a solution with gas bubbled over the metal plate. The commonest is the *hydrogen half cell. Half cells can also be formed by a metal in contact with an insoluble salt or oxide and a solution. The *calomel half cell is an example of this. Half cells are commonly referred to as **electrodes**.

**half-life** *See* DECAY; THERAPEUTIC HALF-LIFE.

**half sandwich** *See* SANDWICH COMPOUND.

**half-thickness** The thickness of a specified material that reduces the intensity

of a beam of radiation to half its original value.

**half-wave plate** *See* RETARDATION PLATE.

**half-wave rectifier** *See* RECTIFIER.

**half-width** The width of a spectral line measured at half its height. In some contexts, the term is used for half the width of the line measured at half its height.

**halide** A compound of a halogen with another element or group. The halides of typical metals are ionic (e.g. sodium fluoride, $Na^+F^-$). Metals can also form halides in which the bonding is largely covalent (e.g. aluminium chloride, $AlCl_3$). Organic compounds are also sometimes referred to as halides; e.g. the alkyl halides (*see* HALOALKANES) and the *acyl halides. Halides are named **fluorides**, **chlorides**, **bromides**, or **iodides**.

**halite** (rock salt) Naturally occurring *sodium chloride (common salt, NaCl), crystallizing in the cubic system. It is chiefly colourless or white (sometimes blue) when pure but the presence of impurities may colour it grey, pink, red, or brown. Halite often occurs in association with anhydrite and gypsum.

**Hall effect** The production of an e.m.f. within a conductor or semiconductor through which a current is flowing when there is a strong transverse magnetic field. The potential difference develops at right angles to both the current and the field. It is caused by the deflection of charge carriers by the field and was first discovered by Edwin Hall (1855–1938). The strength of the electric field $E_H$ produced is given by the relationship $E_H = R_H jB$, where $j$ is the current density, $B$ is the magnetic flux density, and $R_H$ is a constant called the **Hall coefficient**. The value of $R_H$ can be shown to be $1/ne$, where $n$ is the number of charge carriers per unit volume and $e$ is the electronic charge. The effect is used to investigate the nature of charge carriers in metals and semiconductors, in the **Hall probe** for the measurement of magnetic fields, and in magnetically operated switching devices. *See also* QUANTUM HALL EFFECT.

**Halley, Edmund** (1656–1742) British astronomer and mathematician, who published a catalogue of southern stars in 1679, made improvements to barometers, and investigated the optics of *rainbows. In 1705 he calculated the orbit of *Halley's comet and in 1718 discovered the *proper motion of the stars.

**Halley's comet** A bright *comet with a period of 75–76 years. Its last visit to the inner solar system was in 1986. This comet was the first short-period comet to be recognized. Its orbit was first calculated in 1705 by Edmund Halley, after whom it is named. The comet moves around the sun in the opposite direction to the planets and is associated with two meteor showers, the Eta Aquarids (May) and the Orionids (October).

**Hall–Heroult cell** An electrolytic cell used industrially for the extraction of aluminium from bauxite. The bauxite is first purified by dissolving it in sodium hydroxide and filtering off insoluble constituents. Aluminium hydroxide is then precipitated (by adding $CO_2$) and this is decomposed by heating to obtain pure $Al_2O_3$. In the Hall–Heroult cell, the oxide is mixed with cryolite (to lower its melting point) and the molten mixture electrolysed using graphite anodes. The cathode is the lining of the cell, also of graphite. The electrolyte is kept in a molten state (about 850°C) by the current. Molten aluminium collects at the bottom of the cell and can be tapped off. Oxygen forms at the anode, and gradually oxidizes it away. The cell is named after the US chemist Charles Martin Hall (1863–1914), who discovered the process in 1886, and the French chemist Paul Heroult (1863–1914), who discovered it independently in the same year.

**hallucinogen** A drug or chemical that causes alterations in perception (usually visual), mood, and thought. Common hallucinogenic drugs include *lysergic acid diethylamide (LSD) and mescaline. There is no common mechanism of action for this class of compounds although many hallucinogens are structurally similar to *neurotransmitters in the central nervous system, such as serotonin and the catecholamines.

**hallux** The innermost digit on the hindlimb of a tetrapod vertebrate. In man it is the big toe and contains two phalanges. The hallux is absent in some mammals and in many birds it is directed backwards as an adaptation to perching. *Compare* POLLEX.

**halo** 1. A luminous ring that sometimes can be observed around the sun or the moon. It is caused by diffraction of their light by particles in the earth's atmosphere; the radius of the ring is inversely proportional to the predominant particle radius. **2. (galactic halo)** A region of a galaxy, especially a spiral galaxy, forming a spheroidal extension beyond the main aggregation of stars close to the galactic plane or the galactic bulge.

**haloalkanes (alkyl halides)** Organic compounds in which one or more hydrogen atoms of an alkane have been substituted by halogen atoms. Examples are chloromethane, $CH_3Cl$, dibromoethane, $CH_2BrCH_2Br$, etc. Haloalkanes can be formed by direct reaction between alkanes and halogens using ultraviolet radiation. They are usually made by reaction of an alcohol with a halogenating agent.

**halocarbons** Compounds that contain carbon and halogen atoms and (sometimes) hydrogen. The simplest are compounds such as tetrachloromethane ($CCl_4$), tetrabromomethane ($CBr_4$), etc. The *haloforms are also simple halocarbons. The *chlorofluorocarbons (CFCs) contain carbon, chlorine, and fluorine. Similar to these are **hydrochlorofluorocarbons** (HCFCs), which contain carbon, chlorine, fluorine, and hydrogen, and the **hydrofluorocarbons** (HFCs), which contain carbon, fluorine, and hydrogen. The *halons are a class of halocarbons that contain bromine.

**haloform reaction** A reaction for producing haloforms from methyl ketones. An example is the production of chloroform from propanone using sodium chlorate(I) (or bleaching powder):

$$CH_3COCH_3 + 3NaOCl \rightarrow CH_3COCl_3 + 3NaOH$$

The substituted ketone then reacts to give chloroform (trichloromethane):

$$CH_3COCCl_3 + NaOH \rightarrow NaOCOCH_3 + CHCl_3$$

The reaction can also be used for making carboxylic acids, since $RCOCH_3$ gives the product NaOCOR. It is particularly useful for aromatic acids as the starting ketone can be made by a Friedel–Crafts acylation.

The reaction of methyl ketones with sodium iodate(I) gives iodoform (triiodomethane), which is a yellow solid with a characteristic smell. This reaction is used in the **iodoform test** to identify methyl ketones. It also gives a positive result with a secondary alcohol of the formula $RCH(OH)CH_3$ (which is first oxidized to a methylketone) or with ethanol (oxidized to ethanal, which also undergoes the reaction).

**haloforms** The four compounds with formula $CHX_3$, where X is a halogen atom. They are **chloroform** ($CHCl_3$), and, by analogy, **fluoroform** ($CHF_3$), **bromoform** ($CHBr_3$), and **iodoform** ($CHI_3$). The systematic names are trichloromethane, trifluoromethane, etc.

**halogenating agent** *See* HALOGENATION.

**halogenation** A chemical reaction in which a halogen atom is introduced into a compound. Halogenations are described as **chlorination**, **fluorination**, **bromination**, etc., according to the halogen involved. Halogenation reactions may take place by direct reaction with the halogen. This occurs with alkanes, where the reaction involves free radicals and requires high temperature, ultraviolet radiation, or a chemical initiator; e.g.

$$C_2H_6 + Br_2 \rightarrow C_2H_5Br + HBr$$

Halogenation of aromatic compounds can be effected by electrophilic substitution using an aluminium chloride catalyst:

$$C_6H_6 + Cl_2 \rightarrow C_6H_5Cl + HCl$$

Halogenation can also be carried out using compounds, such as phosphorus halides (e.g. $PCl_3$) or sulphur dihalide oxides (e.g. $SOCl_2$), which react with –OH groups. Such compounds are called **halogenating agents**. Addition reactions are also referred to as halogenations; e.g.

$$C_2H_4 + Br_2 \rightarrow CH_2BrCH_2Br$$

**halogens (group 17 elements)** A group of elements in the *periodic table (formerly group VIIB): fluorine (F), chlorine (Cl), bromine (Br), iodine (I), and astatine (At). All have a characteristic electron configuration of noble gases but with outer $ns^2np^5$ electrons. The outer shell is thus one electron short of a noble-gas configuration. Consequently, the halogens are typical nonmetals; they have high electronegativities – high electron affinities and high ionization energies. They form compounds by gaining an electron to complete the stable configuration; i.e. they are good oxidizing agents. Alternatively, they share their outer electrons to form covalent compounds, with single bonds.

All are reactive elements with the reactivity decreasing down the group. The electron affinity decreases down the group and other properties also show a change from fluorine to astatine. Thus, the melting and boiling points increase; at 20°C, fluorine and chlorine are gases, bromine a liquid, and iodine and astatine are solids. All exist as diatomic molecules.

The name 'halogen' comes from the Greek 'salt-producer', and the elements react with metals to form ionic halide salts. They also combine with nonmetals, the activity decreasing down the group: fluorine reacts with all nonmetals except nitrogen and the noble gases helium, neon, and argon; iodine does not react with any noble gas, nor with carbon, nitrogen, oxygen, or sulphur. The elements fluorine to iodine all react with hydrogen to give the acid, with the activity being greatest for fluorine, which reacts explosively. Chlorine and hydrogen react slowly at room temperature in the dark (sunlight causes a free-radical chain reaction). Bromine and hydrogen react if heated in the presence of a catalyst. Iodine and hydrogen react only slowly and the reaction is not complete. There is a decrease in oxidizing ability down the group from fluorine to iodine. As a consequence, each halogen will displace any halogen below it from a solution of its salt, for example:

$$Cl_2 + 2Br^- \rightarrow Br_2 + 2Cl^-$$

The halogens also form a wide variety of organic compounds in which the halogen atom is linked to carbon. In general, the aryl compounds are more stable than the alkyl compounds and there is decreasing resistance to chemical attack down the group from the fluoride to the iodide.

Fluorine has a valency of only 1, although the other halogens can have higher oxidation states using their vacant $d$-electron levels. There is also evidence for increasing metallic behaviour down the group. Chlorine and bromine form compounds with oxygen in which the halogen atom is assigned a positive oxidation state. Only iodine, however, forms positive ions, as in $I^+NO_3^-$.

**halon** A compound obtained by replacing the hydrogen atoms of a hydrocarbon by bromine along with other halogen atoms (*see* HALOCARBONS), for instance halon 1211 is bromochlorodifluoromethane ($CF_2BrCl$) and halon 1301 is bromotrifluoromethane ($CF_3Br$). Halons are very stable and unreactive and are widely used in fire extinguishers. There is concern that they are being broken down in the atmosphere to bromine, which reacts with ozone, leading to depletion of the *ozone layer, and their use is being curtailed.

Although more *chlorofluorocarbons are present in the atmosphere, halons are between three and ten times more destructive of ozone.

**halo nucleus** A type of nucleus in which there are many more neutrons (or, more rarely, more protons) than are present in stable isotopes of that element. Sometimes, a few of the extra neutrons are only weakly bound to the rest of the nucleus and are relatively far from the centre of the nucleus. Halo nuclei are highly unstable; examples include beryllium–11 and carbon–19.

**halophyte** A plant that can tolerate a high concentration of salt in the soil. Such conditions occur in salt marshes and mudflats. Halophytes possess some of the structural modifications of *xerophytes; for example, many of them are *succulents. In addition, they are physiologically adapted to withstand the high salinity of the soil water: their root cells have a higher than normal concentration of solutes, which enables them to take up water by osmosis from the surrounding soil. Examples of halophytes are mangrove trees (*see* MANGROVE SWAMP), thrift (*Armeria*), sea lavender (*Limonium*), and rice grass (*Spartina*). *Compare* HYDROPHYTE; MESOPHYTE.

**Hamilton, Sir William Rowan** (1805–65) Irish mathematician and physicist. Hamilton invented what is now known as the *Hamiltonian formulation of Newtonian mechanics. In doing so, he discovered that there is a very close analogy between Newtonian mechanics and geometrical (ray) optics. He also invented *quaternions in mathematics and found a number of applications of quaternions in physics.

**Hamiltonian** Symbol $H$. A function used to express the energy of a system in terms of its momentum and positional coordinates. In simple cases this is the sum of its kinetic and potential energies. In **Hamiltonian equations**, the usual equations used in mechanics (based on forces) are replaced by equations expressed in terms of momenta. This method of formulating mechanics (**Hamiltonian mechanics**) was first introduced by Sir William Rowan *Hamilton.

**haploid** Describing a nucleus, cell, or organism with a single set of unpaired chromosomes. The haploid number is designated as $n$. Reproductive cells, formed as a result of *meiosis, are haploid. Fusion of two such cells (*see* FERTILIZATION) restores the normal (*diploid) number.

**haplotype** 1. (in genetics) a. A set of linked genes or other genetic markers that are generally inherited together as a unit. This occurs because during meiosis there is little or no *recombination with the corresponding region on the homologous chromosome, and hence shuffling of alleles between the homologous regions is rare. The stretch of DNA containing a haplotype is called a **haplotype block**. b. The entire set of genes occurring on a single chromosome, or haploid set of chromosomes. Hence an individual has two haplotypes for each chromosome, one derived from its mother and one from its father. 2. (in medicine) a. The antigenic constitution of an individual (i.e. antigenic phenotype) resulting from the inheritance of a particular haploid combination of histocompatibility alleles at the HLA locus. b. The set of phenotypic features associated with either the paternal or maternal alleles inherited by an individual.

**hapticity** Symbol η. The number of electrons in a ligand that are directly coordinated to a metal.

**haptotropism** *See* THIGMOTROPISM.

**hard acid** *See* HSAB PRINCIPLE.

**hard base** *See* HSAB PRINCIPLE.

**hardening of oils** The process of converting unsaturated esters of *fatty acids into (more solid) saturated esters by hydrogenation using a nickel catalyst. It is used in the manufacture of margarine from vegetable oils.

**hard ferromagnetic materials** *See* SOFT IRON.

**hardness of water** The presence in water of dissolved calcium or magnesium ions, which form a scum with soap and prevent the formation of a lather. The main cause of hard water is dissolved calcium hydrogencarbonate ($Ca(HCO_3)_2$), which is formed in limestone or chalk regions by the action of dissolved carbon dioxide on calcium carbonate. This type is known as **temporary hardness** because it is removed by boiling:

$$Ca(HCO_3)_2(aq) \rightarrow CaCO_3(s) + H_2O(l) + CO_2(g)$$

The precipitated calcium carbonate is the 'fur' (or 'scale') formed in kettles, boilers, pipes, etc. In some areas, hardness also results from dissolved calcium sulphate ($CaSO_4$) or calcium fluoride ($CaF_2$). These compounds can not be removed by boiling (**permanent hardness**).

Hard water causes problems in washing and by reducing the efficiency of boilers, heating systems, and certain industrial processes. Various methods of **water softening** are used. In public supplies, temporary hardness can be removed by adding lime (calcium hydroxide), which precipitates calcium carbonate

$$Ca(OH)_2(aq) + Ca(HCO_3)_2(aq) \rightarrow$$
$$2CaCO_3(s) + 2H_2O(l)$$

This is known as the **Clark process** (or as **clarking**). It does not remove permanent hardness. Both temporary and permanent hardness can be treated by precipitating calcium carbonate by added sodium carbonate – hence its use as washing soda and in bath salts. Calcium (and other) ions can also be removed from water by ion-exchange using zeolites (e.g. **Permutit**). This method is used in small domestic water-softeners. Another technique is to complex the $Ca^{2+}$ ions and prevent them reacting further. For domestic use polyphosphates (containing the ion $P_6O_{18}^{6-}$, e.g. **Calgon**) are added. Other sequestering agents are also used for industrial water. *See also* SEQUESTRATION.

**hard radiation** Ionizing radiation of high penetrating power, usually gamma rays or short-wavelength X-rays. *Compare* SOFT RADIATION.

**hardware** *See* COMPUTER.

**hardwood** *See* WOOD.

**Hargreaves process** *See* POTASSIUM SULPHATE.

**harmonic** An oscillation having a frequency that is a simple multiple of a **fundamental** sinusoidal oscillation. The fundamental frequency of a sinusoidal oscillation is usually called the **first harmonic**. The **second harmonic** has a frequency twice that of the fundamental and so on (see illustration). A taut string or column of air, as in a violin or organ, will sound upper harmonics at the same time as the fundamental sounds. This is because the string or column of air divides itself into sections, each section then vibrating as if it were a whole. The upper harmonics are also called **overtones**, but the second harmonic is the first overtone, and so on. Musicians, however, often regard harmonic and overtone as synonymous, not counting the fundamental as a harmonic.

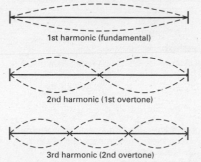

1st harmonic (fundamental)

2nd harmonic (1st overtone)

3rd harmonic (2nd overtone)

**Harmonics.**

**harmonic mean** *See* MEAN.

**harmonic motion** *See* SIMPLE HARMONIC MOTION.

**harmonic oscillator** A system (in either *classical physics or *quantum mechanics) that oscillates with *simple harmonic motion. The harmonic oscillator is exactly soluble in both classical mechanics and quantum mechanics. Many systems exist for which harmonic oscillators provide very good approximations. An example in classical mechanics is a simple *pendulum, while at low temperatures atoms vibrating about their mean positions in molecules or crystal lattices can be regarded as good approximations to harmonic oscillators in quantum mechanics. Even if a system is not exactly a harmonic oscillator the solution of the harmonic oscillator is frequently a useful starting point for solving such systems using *perturbation theory. *Compare* ANHARMONIC OSCILLATOR.

**harmonic series (harmonic progression)** A series or progression in which the reciprocals of the terms have a constant difference between them, e.g. $1 + 1/2 + 1/3 + 1/4 \ldots + 1/n$.

**Harvard classification** *See* SPECTRAL CLASS.

**harvesting** 1. The processes involved in gathering in ripened crops (*see* AGRICULTURE). 2. The collection of cells from cell cultures or of organs from donors for the purpose of transplantation (*see* GRAFT).

**Harvey, William** (1578–1657) English physician, who worked at St Bartholomew's Hospital, London, from 1609 and from 1618 was court physician. He is best known for discovering the *circulation of the blood, which he announced in 1628.

**hashish** *See* CANNABIS.

**hassium** Symbol Hs. A radioactive transactinide element; a.n. 108. It was first made in 1984 by Peter Armbruster and a team in Darmstadt, Germany. It can be produced by bombarding lead-208 nuclei with iron-58 nuclei. Only a few atoms have ever been produced. The name comes from the Latinized form of Hesse, the German state where it was first synthesized.

(((●))) SEE WEB LINKS
• Information from the WebElements site

**haustorium** A specialized structure of certain parasitic plants and fungi that penetrates the cells of the host plant to absorb nutrients. In parasitic fungi haustoria are formed from enlarged hyphae and in parasitic flowering plants, such as the dodder (*Cuscuta*), they are outgrowths of the stem.

**Haversian canals** Narrow tubes within compact *bone containing blood vessels and nerves. They generally run parallel to the bone surface. Each canal surrounded by a series of rings of bone (**lamellae**) is known as a **Haversian system**. Haversian systems are joined to each other by bone material. They are named after Clopton Havers (1650–1702).

**Hawking, Stephen William** (1942– ) British cosmologist and physicist, who in 1979 became the Lucasian Professor of mathematics at Cambridge University. Working with Roger Penrose (1931– ), who had shown how a singularity results from a *black hole, Hawking postulated that the original big bang must have come from a singularity (*see* BIG-BANG THEORY). He also showed how black holes can emit particles by the *Hawking process. Hawking is the author of such popular science works as *A Brief History of Time* (1988) and *The Universe in a Nutshell* (2001).

**Hawking process** Emission of particles by a *black hole as a result of quantum-mechanical effects. The process was first suggested by Stephen Hawking. The gravitational field of the black hole causes production of particle–antiparticle pairs in the

vicinity of the event horizon (the process is analogous to that of pair production). One member of each pair (either the particle or the antiparticle) falls into the black hole, while the other escapes. To an external observer it appears that the black hole is emitting radiation (**Hawking radiation**). The energy of the particles that fall in is negative and exactly balances the (positive) energy of the escaping particles. This negative energy reduces the mass of the black hole and the net result of the process is that the emitted particle flux appears to carry off the black-hole mass. It can be shown that the black hole radiates like a *black body, with the energy distribution of the particles obeying *Planck's radiation law for a temperature that is inversely proportional to the mass of the hole. For a black hole of the mass of the sun, this temperature is only about $10^{-7}$ K, so the energy loss is negligible. However, for a 'mini' black hole, such as might have been formed in the early universe, with a mass of order $10^{12}$ kg (and a radius of order $10^{-15}$ m), the temperature would be of order $10^{11}$ K and the hole would radiate copiously (at a rate of about $6 \times 10^9$ W) a flux of gamma rays, neutrinos, and electron–positron pairs. (The observed levels of cosmic gamma rays put strong constraints on the number of such 'mini' black holes, suggesting that there are too few of them to solve the *missing-mass problem.)

**hazchem code** (emergency action code; **EAC**) A code designed to be displayed when hazardous chemicals are transported or stored in bulk. It is used to help the emergency services to take action quickly in any accident. The code consists of a number followed by one or two letters. The number indicates the type of substance to be used in treating the accident (e.g. stream of water, fine spray, foam, dry agent). The first letter indicates the type of protective clothing needed along with information about the possibility of violent reaction on whether the substance should be contained or diluted. The second letter, where it exists, is letter E, indicating that people have to be evacuated from the neighbourhood of the incident. In the UK, the code is usually displayed as part of a panel, which includes an international UN number for the substance, a telephone number for specialist advice, the company name, and a symbol indicating the danger (e.g. a skull and crossbones for toxic substances).

**HCFC (hydrochlorofluorocarbon)** *See* HALOCARBONS.

**h.c.p.** Hexagonal close packing. *See* CLOSE PACKING.

**headspace** The space above a sample held in a sealed container. Headspace analysis is used in forensic science to investigate the volatile constituents of a sample.

**health physics** The branch of medical physics concerned with the protection of medical, scientific, and industrial workers from the hazards of ionizing radiation and other dangers associated with atomic physics. Establishing the maximum permissible *dose of radiation, the disposal of radioactive waste, and the shielding of dangerous equipment are the principal activities in this field.

**hearing** The sense by which sound is de-tected. In vertebrates the organ of hearing is the *ear. In higher vertebrates variation in air pressure caused by sound waves are am-plified in the outer and middle ears and transmitted to the inner ear, where sensory cells in the *cochlea register the vibrations. The resulting information is transmitted to the brain via the auditory nerve. The ear can distinguish between sounds of different in-tensity (*loudness) and frequency (*pitch).

**heart** A hollow muscular organ that, by means of regular contractions, pumps blood through the circulatory system. The verte-brate heart is composed of a specialized muscle (*see* CARDIAC MUSCLE). Mammals have a four-chambered heart consisting of two atria and two ventricles; the right and left sides are completely separate from each other so there is no mixing of oxygenated and deoxygenated blood (see illustration). Oxygenated blood from the pulmonary veins

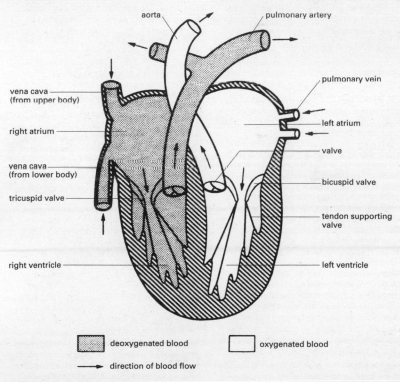

**Heart.** Structure of the mammalian heart.

enters the heart through the left atrium, passes to the left ventricle, and leaves the heart through the *aorta. Deoxygenated blood from the *venae cavae enters the right atrium and is pumped through the right ventricle to the pulmonary artery, which conveys it to the lungs for oxygenation. The tricuspid and bicuspid valves ensure that there is no backflow of blood. The contractions of the heart are initiated and controlled by the sinoatrial node (*see* PACEMAKER); an average adult human heart contracts about 70 times per minute. *See also* CARDIAC CYCLE; CARDIAC OUTPUT.

The hearts of other vertebrates are similar except in the number of atria and ventricles (there may be one or two) and in the degree of separation of oxygenated and deoxygenated blood. Invertebrates, however, show great variation in the form and functioning of the heart.

**heartwood (duramen)** The wood at the centre of a tree trunk or branch. It consists of dead *xylem cells heavily thickened with lignin and provides structural support. Many heartwood cells contain oils, gums, and resins, which darken the wood. *Compare* SAPWOOD.

**heat** The process of energy transfer from one body or system to another as a result of a difference in temperature. The energy in the body or system before or after transfer is also sometimes called heat.

A body in equilibrium with its surroundings contains energy (the kinetic and potential energies of its atoms and molecules) but this is called **internal energy**, $U$, rather than heat. When such a body changes its temperature or phase there is a change in internal energy, $\Delta U$, which (according to the first law of *thermodynamics) is given by $\Delta U = Q - W$, where $Q$ is the heat absorbed by the body from the surroundings and $W$ is the work done simultaneously on the surroundings. To use the word 'heat' for both $U$ and $Q$ is clearly confusing. Note also that certain physical quantities are described as *heat of atomization, *heat of combustion, etc. What is usually used is a standard molar *enthalpy change for the process under consideration. The units are kJ mol$^{-1}$; a negative value indicates that energy is liberated. *See also* HEAT CAPACITY; HEAT TRANSFER; LATENT HEAT.

**heat balance** **1.** A balance sheet showing all the heat inputs to a system (such as a chemical process, furnace, etc.) and all the heat outputs. **2.** The equilibrium that exists on the average between the radiation received from the sun by the earth and its atmosphere and that reradiated or reflected by the earth and the atmosphere. In general, the regions of the earth nearer the equator than about 35°N or S receive more energy from the sun than they are able to reradiate, whereas those regions polewards of 35°N or S receive less energy than they lose. The excess of heat received by the low latitudes is carried to the higher latitudes by atmospheric and oceanic circulations.

**heat capacity (thermal capacity)** The ratio of the heat supplied to an object or specimen to its consequent rise in temperature. The **specific heat capacity** is the ratio of the heat supplied to unit mass of a substance to its consequent rise in temperature. The **molar heat capacity** is the ratio of the heat supplied to unit amount of a substance to its consequent rise in temperature. In practice, heat capacity ($C$) is measured in joules per kelvin, specific heat capacity ($c$) in J K$^{-1}$ kg$^{-1}$, and molar heat capacity ($C_m$) in J K$^{-1}$ mol$^{-1}$. For a gas, the values of $c$ and $C_m$ are commonly given either at **constant volume**, when only its *internal energy is increased, or at **constant pressure**, which requires a greater input of heat as the gas is allowed to expand and do work against the surroundings. The symbols for the specific and molar heat capacities at constant volume are $c_v$ and $C_v$, respectively; those for the specific and molar heat capacities at constant pressure are $c_p$ and $C_p$.

**heat death of the universe** The condition of the universe when *entropy is maximized and all large-scale samples of matter are at a uniform temperature. In this condition no energy is available for doing work and the universe is finally unwound. The condition was predicted by Rudolf Clausius, who introduced the concept of entropy. Clausius's dictum that "the energy of the universe is constant, its entropy tends to a maximum" is a statement of the first two laws of thermodynamics. These laws apply in this sense only to closed systems and, for the predicted heat death to occur, the universe must be a closed system.

**heat engine** A device for converting heat into work. The heat is derived from the combustion of a fuel. In an *internal-combustion engine the fuel is burnt inside the engine, whereas in a *steam engine or steam turbine, examples of external-combustion engines, the fuel is used to raise steam outside the engine and then some of the steam's internal energy is used to do work inside the engine. Engines usually work on cycles of operation, the most efficient of which would be the *Carnot cycle. This cannot be realized in practice, but the *Rankine cycle is approximated by some engines.

**heat exchanger** A device for transferring heat from one fluid to another without permitting the two fluids to contact each other. A simple industrial heat exchanger consists of a bundle of parallel tubes, through which one fluid flows, enclosed in a container, through which the other fluid flows in the opposite direction (a **counter-current heat exchanger**).

**heat of atomization** The energy required to dissociate one mole of a given substance into atoms.

**heat of combustion** The energy liberated when one mole of a given substance is completely oxidized.

**heat of crystallization** The energy liberated when one mole of a given substance crystallizes from a saturated solution of the same substance.

**heat of dissociation** The energy absorbed when one mole of a given substance is dissociated into its constituent elements.

**heat of formation** The energy liberated or absorbed when one mole of a compound is formed in their *standard states from its constituent elements.

**heat of neutralization** The energy liberated in neutralizing one mole of an acid or base.

**heat of reaction** The energy liberated or absorbed as a result of the complete chemical reaction of molar amounts of the reactants.

**heat of solution** The energy liberated or absorbed when one mole of a given substance is completely dissolved in a large volume of solvent (strictly, to infinite dilution).

**heat pump** A device for transferring heat from a low temperature source to a high temperature region by doing work. It is essentially a refrigerator with a different emphasis. The working fluid is at one stage a vapour, which is compressed by a pump adi-

abatically so that its temperature rises. It is then passed to the radiator, where heat is given out to the surroundings (the space to be heated) and the fluid condenses to a liquid. It is then expanded into an evaporator where it takes up heat from its surroundings and becomes a vapour again. The cycle is completed by returning the vapour to the compressor. Heat pumps are sometimes adapted as dual-purpose space-heating-in-winter and air-conditioning-in-summer devices.

**heat radiation (radiant heat)** Energy in the form of electromagnetic waves emitted by a solid, liquid, or gas as a result of its temperature. It can be transmitted through space; if there is a material medium this is not warmed by the radiation except to the extent that it is absorbed. Although it covers the whole electromagnetic spectrum, the highest proportion of this radiation lies in the infrared portion of the spectrum at normal temperatures. *See* BLACK BODY.

**heat shield** A specially prepared surface that prevents a spacecraft or capsule from overheating as it re-enters the earth's atmosphere. The surface is coated with a plastic impregnated with quartz fibres, which is heated by friction with air molecules as the craft enters the atmosphere, causing the outer layer to vaporize. In this way about 80% of the energy is reradiated and the craft is safeguarded from an excessive rise in temperature.

**heat-shock protein (HSP)** *See* MOLECULAR CHAPERONE.

**heat transfer** The transfer of energy from one body or system to another as a result of a difference in temperature. The heat is transferred by *conduction (*see also* CONDUCTIVITY), *convection, and radiation (*see* HEAT RADIATION).

**Heaviside–Kennelly layer** *See* EARTH'S ATMOSPHERE.

**Heaviside–Lorentz units** A system of units for electric and magnetic quantities based upon c.g.s. electrostatic and electromagnetic units. They are the rationalized forms of *Gaussian units and, like the latter, are widely used in particle physics and relativity in preference to the *SI units now employed for general purposes in physics. They are named after Oliver Heaviside (1850–1925) and Hendrik Lorentz (1853–1928).

**heavy-fermion system** A solid in which the electrons have a very high effective mass; i.e. they act as if they had masses several hundred times the normal mass of the electron. An example of a heavy-fermion system is the cerium–copper–silicon compound $CeCuSi_2$. The electrons with high effective mass are $f$-electrons in narrow energy bands associated with strong many-body effects. Substances containing such electrons have unusual thermodynamic, magnetic, and superconducting properties, which are still not completely understood. The *superconductivity of such materials has a more complicated mechanism than that for metals described by the BCS theory, since the Cooper pairs are formed from *quasiparticles with very high effective masses rather than from electrons.

**heavy hydrogen** See DEUTERIUM.

**heavy metal** A metal with a high relative atomic mass. The term is usually applied to common transition metals, such as copper, lead, and zinc. These metals are a cause of environmental *pollution (**heavy-metal pollution**) from a number of sources, including lead in petrol, industrial effluents, and leaching of metal ions from the soil into lakes and rivers by acid rain.

**heavy spar** A mineral form of *barium sulphate, $BaSO_4$.

**heavy water (deuterium oxide)** Water in which hydrogen atoms, $^1H$, are replaced by the heavier isotope deuterium, $^2H$ (symbol D). It is a colourless liquid, which forms hexagonal crystals on freezing. Its physical properties differ from those of 'normal' water; r.d. 1.105; m.p. 3.8°C; b.p. 101.4°C. Deuterium oxide, $D_2O$, occurs to a small extent (about 0.003% by weight) in natural water, from which it can be separated by fractional distillation or by electrolysis. It is useful in the nuclear industry because of its ability to reduce the energies of fast neutrons to thermal energies (see MODERATOR) and because its absorption cross-section is lower than that of hydrogen and consequently it does not appreciably reduce the neutron flux. In the laboratory it is used for *labelling other molecules for studies of reaction mechanisms. Water also contains the compound HDO.

**hecto-** Symbol h. A prefix used in the metric system to denote 100 times. For example, 100 coulombs = 1 hectocoulomb (hC).

**Heisenberg, Werner Karl** (1901–76) German physicist, who became a professor at the University of Leipzig and, after World War II, at the Kaiser Wilhelm Institute in Göttingen. In 1923 he was awarded the Nobel Prize for physics for his work on *matrix mechanics. But he is best known for his 1927 discovery of the *uncertainty principle.

**Heisenberg uncertainty principle** See UNCERTAINTY PRINCIPLE.

**heliacal rising** The first appearance of a star, a constellation, a planet, or the moon above the eastern horizon just before dawn following a period of invisibility during which the object had been rising in daylight. Heliacal risings of the same star or constellation happen one year apart.

**helicate** A type of inorganic molecule containing a double helix of bipyridyl-derived molecules formed around a chain of up to five copper(I) ions. See also SUPRAMOLECULAR CHEMISTRY.

**heliocentric universe** A conception of the universe in which the sun is taken to be at its centre. Aristarchus of Samos (310–230 BC) was the first formally to propose the heliocentric model, but it failed to predict what was actually observed and thus made little headway against the rival, more intuitive *geocentric universe, which held sway throughout the Middle Ages. Nicolaus Copernicus revived an essentially heliocentric view, which was upheld by Galileo against strong opposition from the church on the grounds that if the earth was not at the centre of the universe man's position in God's creation was diminished. In the modern view the sun is at the centre of the *solar system, but the sun, with its family of planets and other bodies, is just one of an enormous number of stars in the Galaxy, which is itself one of an enormous number of *galaxies.

**helioseismology** The study of waves in the sun. The waves studied in helioseismology are not seismic waves, as in geophysics, but pressure waves generated by turbulence in the solar convection layer. These cause slight changes in the pattern of light from the sun and the pattern of pressure waves can be inferred from these changes, leading to information about conditions within the sun.

**heliosphere** The vast bubble-like region enveloping the *solar system and carried with it through space, within which the *solar wind plays a dominant role.

**heliotropism** See PHOTOTROPISM.

**helium** Symbol He. A colourless odourless gaseous nonmetallic element belonging to group 18 of the periodic table (*see* NOBLE GASES); a.n. 2; r.a.m. 4.0026; d. 0.178 g dm$^{-3}$; m.p. $-272.2°C$ (at 20 atm.); b.p. $-268.93°C$. The element has the lowest boiling point of all substances and can be solidified only under pressure. Natural helium is mostly helium–4, with a small amount of helium–3. There are also two short-lived radioactive isotopes: helium–5 and –6. It occurs in ores of uranium and thorium and in some natural-gas deposits. It has a variety of uses, including the provision of inert atmospheres for welding and semiconductor manufacture, as a refrigerant for superconductors, and as a diluent in breathing apparatus. It is also used in filling balloons. Chemically it is totally inert and has no known compounds. It was discovered in the solar spectrum in 1868 by Joseph Lockyer (1836–1920).

( ) SEE WEB LINKS
• Information from the WebElements site

**Helmholtz, Hermann Ludwig Ferdinand von** (1821–94) German physiologist and physicist. In 1850 he measured the speed of a nerve impulse and in 1851 invented the ophthalmoscope. In physics, he discovered the conservation of energy (1847), introduced the concept of *free energy, and invented *vortex dynamics.

**Helmholtz coils** Two coaxial parallel flat coils with the same radius placed a distance apart that is equal to the radius. If the same current is flowing in both coils then the value of the magnetic field strength is approximately uniform between the coils. Coils of this type are used to create fields and, in some cases, are employed to counter the effect of the earth's magnetic field. Helmholtz coils are also used in magnetic measurement, with the coils connected to a *fluxmeter. If a small magnet is placed between the coils and then removed, the integrated signal on the fluxmeter is proportional to the magnet's magnetic moment.

**Helmholtz free energy** See FREE ENERGY.

**helper T cell** See T CELL.

**heme** See HAEM.

**hemiacetals** See ACETALS.

**hemicellulose** A *polysaccharide found in the cell walls of plants. The branched chains

of this molecule bind to cellulose microfibrils, together with pectins, forming a network of cross-linked fibres.

**hemihedral form** The form of a crystal in which only half the number of faces required for the symmetry are present. *Compare* HOLOHEDRAL FORM.

**hemihydrate** A crystalline hydrate containing two molecules of compound per molecule of water (e.g. $2CaSO_4.H_2O$).

**hemiketals** See KETALS.

**Hemiptera** An order of insects comprising the true bugs. Hemipterans typically have oval flattened bodies with two pairs of wings, which are folded back across the abdomen at rest. The forewings are hardened, either at their bases only (in the suborder Heteroptera) or uniformly (in the suborder Homoptera). The mouthparts are modified for piercing and sucking, with long slender stylets forming a double tube. Many bugs feed on plant sap and are serious agricultural pests, including aphids, leaf-hoppers, scale insects, mealy bugs, etc. Others are carnivorous, and the order contains many aquatic species, such as the water boatmen, which have legs adapted for swimming and the exchange of respiratory gases.

**henry** Symbol H. The *SI unit of inductance equal to the inductance of a closed circuit in which an e.m.f. of one volt is produced when the electric current in the circuit varies uniformly at a rate of one ampere per second. It is named after Joseph Henry.

**Henry, Joseph** (1797–1878) US physicist, who became professor of natural philosophy at Princeton in 1832. In 1829 he made an *electric motor, and used insulated windings to produce a powerful *electromagnet. A year later he discovered *electromagnetic induction (independently of *Faraday), and in 1832 he discovered self-induction (*see* INDUCTANCE). In 1835 he invented the electric *relay.

**Henry's law** At a constant temperature the mass of gas dissolved in a liquid at equilibrium is proportional to the partial pressure of the gas. The law, discovered in 1801 by the British chemist and physician William Henry (1775–1836), is a special case of the partition law. It applies only to gases that do not react with the solvent.

**heparin** A glycosaminoglycan (mucopolysaccharide) with *anticoagulant properties, occurring in vertebrate tissues, especially the lungs and blood vessels. Pharmaceutical preparations of heparin are administered to prevent or dissolve blood clots.

**hepatic** Of or relating to the liver. For example, the **hepatic portal vein** (see HEPATIC PORTAL SYSTEM) and the **hepatic artery** supply blood to the liver, and the **hepatic vein** carries blood away from the liver.

**Hepaticae** *See* HEPATOPHYTA.

**hepatic portal system** The vein (**hepatic portal vein**) or veins that transport blood containing the absorbed products of digestion from the intestine directly to the liver.

**Hepatophyta (Marchantiophyta)** A phylum comprising the liverworts – simple plants that lack vascular tissue and possess rudimentary rootlike organs (rhizoids). Liverworts occur in moist situations (including fresh water) and as epiphytes on other plants. Like the mosses (see BRYOPHYTA), liverworts show marked alternation of generations between haploid gamete-bearing forms (gametophytes) and diploid spore-bearing forms (sporophytes), the latter being dependent on the former for nutrients, etc. The plant body (gametophyte) may be a thallus, growing closely pressed to the ground (thallose liverworts, e.g. *Pellia*), or it may bear many leaflike lobes (leafy liverworts). It gives rise to leafless stalks bearing capsules (sporophytes). Spores formed in the capsules are released and grow to produce new plants.

Liverworts were formerly placed in the class Hepaticae in the phylum Bryophyta, which now contains only the mosses.

**heptadecanoic acid (margaric acid)** A white crystalline carboxylic acid with a linear chain of carbon atoms, $C_{16}H_{33}COOH$; m.p. 59–61°C; b.p. 227°C (100 mm Hg). It is present in certain natural fats.

**heptahydrate** A crystalline hydrate that has seven moles of water per mole of compound.

**heptane** A liquid straight-chain alkane obtained from petroleum, $C_7H_{16}$; r.d. 0.684; m.p. –90.6°C; b.p. 98.4°C. In standardizing *octane numbers, heptane is given a value zero.

**heptaoxodiphosphoric(V) acid** *See* PHOSPHORIC(V) ACID.

**heptavalent (septivalent)** Having a valency of seven.

**herb** 1. A *herbaceous plant, i.e. a seed-bearing plant that does not form hard woody tissue. 2. A plant with medicinal or culinary uses. Culinary herbs are usually plants whose leaves are used for flavouring food.

**herbaceous** Describing a plant that contains little permanent woody tissue. The aerial parts of the plant die back after the growing season. In *annuals the whole plant dies; in *biennials and herbaceous *perennials the plant has organs (e.g. bulbs or corms) that survive beneath the soil in unfavourable conditions.

**herbal cannabinoids** *See* CANNABINOIDS.

**herbicide** *See* PESTICIDE.

**Herbig–Haro object (HH object)** A strange bright region in interstellar space. It contains gas and dust that is excited to emit electromagnetic radiation by a jet of gas, which may be the *bipolar outflow from a young *T Tauri star or other protostar.

**herbivore** An animal that eats vegetation, especially any of the plant-eating mammals, such as ungulates (cows, horses, etc.). Herbivores are characterized by having teeth adapted for grinding plants and alimentary canals specialized for digesting cellulose (see CAECUM).

**heredity** The transmission of characteristics from parents to offspring via the chromosomes. The study of heredity (*genetics) was first undertaken by Gregor Mendel, who derived a series of laws that govern heredity (see MENDEL'S LAWS).

**Hermann–Mauguin system (international system)** A notation used to describe the symmetry of point groups. In contrast to the *Schoenflies system, which is used for isolated molecules (e.g. in spectroscopy), the Hermann–Mauguin system is used in *crystallography. Some of the categories are the same as the Schoenflies system. $n$ is the same group as $C_n$. $nmm$ is the same group as $C_{nv}$. There are two $m$s because of two distinct types of mirror plane containing the $n$-fold axis. $n22$ is the same group as $D_n$. The other categories do not coincide with the Schoenflies system. $\bar{n}$ is a group with an $n$-fold rotation–inversion axis and includes $C_{3h}$ as $\bar{6}$, $S_4$ as $\bar{4}$, $S_6$ as $\bar{3}$, and $S_2$ as $\bar{1}$. $n/m$ is the same group as $C_{nh}$ except that $C_{3h}$ is regarded as $\bar{6}$. $n2m$ is the same group as $D_{nd}$, except that $D_{3h}$ is regarded as $62m$. $n/m\,2/m\,2/m$, abbreviated to $n/mmm$, is the same group as $D_{nh}$, except that $D_{3h}$ is regarded as $62m$. (Unlike the Schoenflies system, the Hermann–Mauguin system regards the three-fold axis as a special case.) As regards the cubic groups, $O_h$ is denoted $m3m$ (or $4/m\,\overline{3}\,2/m$), $O$ is denoted $432$, $T_h$ is denoted $m3$ (or $2/m\,\overline{3}$), $T_d$ is denoted $\overline{4}3m$, and $T$ is denoted $23$. In the Hermann–Mauguin system all the cubic groups have 3 as the second number because of the three-fold axis that occurs in all cubic groups. It is named after the German crystallographer Carl Hermann (1898–1961) and the French minerologist Charles-Victor Mauguin (1878–1958).

**hermaphrodite (bisexual) 1.** An animal, such as the earthworm, that has both male and female reproductive organs. **2.** A plant whose flowers contain both stamens and carpels. This is the usual arrangement in most plants. *Compare* MONOECIOUS; DIOECIOUS.

**heroin (diacetylmorphine)** A *narcotic compound that is a synthetic derivative of morphine (*see* OPIATE). The compound is easily absorbed by the brain and is used as a powerful *analgesic. Highly addictive, it is abused by drug users.

**herpesvirus** One of a group of complex DNA-containing viruses causing infections in man and most other vertebrates that tend to recur. The group includes **herpes simplex**, the agent of cold sores; **herpes varicella/zoster**, the virus causing chickenpox and shingles; **Epstein–Barr (EB) virus**, the causal agent of glandular fever and also implicated in the cancer Burkitt's lymphoma; and the *cytomegalovirus.

**hertz** Symbol Hz. The *SI unit of frequency equal to one cycle per second. It is named after Heinrich Hertz.

**Hertz, Heinrich Rudolf** (1857–94) German physicist, who worked as an engineer before attending Berlin University. He is best known for his 1888 discovery of *radio waves, as predicted by James Clerk *Maxwell. The SI unit of frequency is named after him.

**Hertzsprung–Russell diagram (H–R diagram)** A graphical representation of the absolute magnitude of stars (usually along the $y$-axis) plotted against the spectral class or colour index ($x$-axis): see illustration. The $y$-axis then represents the energy output of the star and the $x$-axis its surface temperature. The majority of stars on such a diagram fall on a band running from the top left to the bottom right of the graph. These are called **main-sequence stars** (the sun falls into this class). The few stars falling in the lower left portion are called *white dwarfs. The *giant

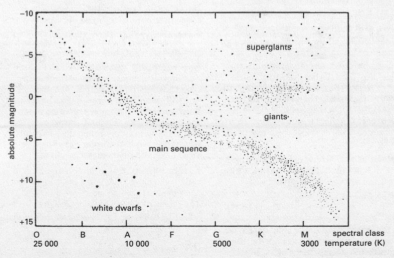

**Hertzsprung–Russell diagram.**

stars fall in a cluster above the main sequence and the *supergiants are above them. The diagram, which was first devised in 1911 by Ejnar Hertzsprung (1873–1969) and in 1913 by Henry Russell (1897–1957), forms the basis of the theory of *stellar evolution.

**hesperidium** *See* BERRY.

**Hess's law** If reactants can be converted into products by a series of reactions, the sum of the heats of these reactions (with due regard to their sign) is equal to the heat of reaction for direct conversion from reactants to products. More generally, the overall energy change in going from reactants to products does not depend on the route taken. The law can be used to obtain thermodynamic data that cannot be measured directly. For example, the heat of formation of ethane can be found by considering the reactions:

$$2C(s) + 3H_2(g) + 3\frac{1}{2}O_2(g) \rightarrow 2CO_2(g) + 3H_2O(l)$$

The heat of this reaction is $2\Delta H_C + 3\Delta H_H$, where $\Delta H_C$ and $\Delta H_H$ are the heats of combustion of carbon and hydrogen respectively, which can be measured. By Hess' law, this is equal to the sum of the energies for two stages:

$$2C(s) + 3H_2(g) \rightarrow C_2H_6(g)$$

(the heat of formation of ethane, $\Delta H_f$) and

$$C_2H_6(g) + 3\frac{1}{2}O_2 \rightarrow 2CO_2(g) + 3H_2O(l)$$

(the heat of combustion of ethane, $\Delta H_E$). As $\Delta H_E$ can be measured and as

$$\Delta H_f + \Delta H_E = 2\Delta H_c + 3\Delta H_H$$

$\Delta H_f$ can be found. Another example is the use of the *Born–Haber cycle to obtain lattice energies. The law was first put forward in 1840 by the Russian chemist Germain Henri Hess (1802–50). It is sometimes called the **law of constant heat summation** and is a consequence of the law of conservation of energy.

**hetero atom** An odd atom in the ring of a heterocyclic compound. For instance, nitrogen is the hetero atom in pyridine.

**heterochromatin** *See* CHROMATIN.

**heterocyclic** *See* CYCLIC.

**heterodont** Describing animals that possess teeth of more than one type (i.e. *incisors, *canine teeth, *premolars, and *molars), each with a particular function.

Most mammals are heterodont. *See* DENTITION. *Compare* HOMODONT.

**heterodyne** Denoting a device or method of radio reception in which *beats are produced by superimposing a locally generated radio wave on an incoming wave. In the *superheterodyne receiver the intermediate frequency is amplified and demodulated. In the **heterodyne wavemeter**, a variable-frequency local oscillator is adjusted to give a predetermined beat frequency with the incoming wave, enabling the frequency of the incoming wave to be determined.

**heterogametic sex** The sex that is determined by possession of two dissimilar *sex chromosomes (e.g. XY). In humans and many other mammals this is the male sex. The heterogametic sex produces reproductive cells (gametes) of two kinds, half containing an X chromosome and half a Y chromosome. *Compare* HOMOGAMETIC SEX.

**heterogeneous** Relating to two or more phases, e.g. a heterogeneous *catalyst. *Compare* HOMOGENEOUS.

**heterolytic fission** The breaking of a bond in a compound in which the two fragments are oppositely charged ions. For example, HCl → H+ + Cl−. *Compare* HOMOLYTIC FISSION.

**heteropolar bond** *See* CHEMICAL BOND.

**heteropoly compound** *See* CLUSTER COMPOUND.

**heteropolymer** *See* POLYMER.

**heterosis** *See* HYBRID VIGOUR.

**heterotrophic nutrition** A type of nutrition in which energy is derived from the intake and digestion of organic substances, normally plant or animal tissues. The breakdown products of digestion are used to synthesize the organic materials required by the organism. All animals obtain their food this way: they are **heterotrophs**. *See also* INGESTION. *Compare* AUTOTROPHIC NUTRITION.

**heterozygous** Describing an organism or cell in which the *alleles at a given locus on *homologous chromosomes are different. The aspect of the feature displayed by the organism will be that determined by the *dominant allele. Heterozygous organisms, called **heterozygotes**, do not breed true. *Compare* HOMOZYGOUS.

**heuristic** Denoting a method of solving a

problem for which no *algorithm exists. It involves trial and error, as in *iteration.

**Heusler alloys** Ferromagnetic alloys containing no ferromagnetic elements. The original alloys contained copper, manganese, and tin and were first made by Conrad Heusler (19th-century mining engineer).

**Hewish, Antony** *See* RYLE, SIR MARTIN.

**hexacyanoferrate(II) (ferrocyanide)** A complex iron-containing anion, $[Fe(CN)_6]^{4-}$, used as a solution of its potassium salt as a test for ferric iron (iron(III)), with which it forms a dark blue precipitate of Prussian blue. The sodium salt is used as an anticaking agent in common salt.

**hexacyanoferrate(III) (ferricyanide)** A complex iron-containing anion, $[Fe(CN)_6]^{3-}$, used as a solution of its potassium salt as a test for ferrous iron (iron(II)), with which it forms a dark blue precipitate of Prussian blue.

**hexadecane (cetane)** A colourless liquid straight-chain alkane hydrocarbon, $C_{16}H_{34}$, used in standardizing cetane ratings for Diesel fuel.

**hexadecanoate** *See* PALMITATE.

**hexadecanoic acid** *See* PALMITIC ACID.

**hexagonal close packing** *See* CLOSE PACKING.

**hexagonal crystal** *See* CRYSTAL SYSTEM.

**hexahydrate** A crystalline compound that has six moles of water per mole of compound.

**hexamine (hexamethylene tetramine)** A white crystalline compound, $C_6H_{12}N_4$, made by the condensation of methanal with ammonia. It has been used as a solid fuel for camping stoves and as an antiseptic for treating urinary infections in medicine. It is used in the production of *cyclonite.

**hexanedioate (adipate)** A salt or ester of hexanedioic acid.

**hexanedioic acid (adipic acid)** A carboxylic acid, $(CH_2)_4(COOH)_2$; r.d. 1.36; m.p. 153°C; b.p. 265°C (100 mmHg). It is used in the manufacture of *nylon 6,6. *See also* POLYMERIZATION.

**6-hexanelactam** *See* CAPROLACTAM.

**hexanitrohexaazaisowurtzitane** *See* HNIW.

**hexanoate (caproate)** A salt or ester of hexanoic acid.

**hexanoic acid (caproic acid)** A liquid fatty acid, $CH_3(CH_2)_4COOH$; r.d. 0.93; m.p. –2 to –1.5°C; b.p. 205°C. Glycerides of the acid occur naturally in cow and goat milk and in some vegetable oils.

**Hexapoda (Insecta)** A subphylum (or class) of arthropods comprising about a million known species (many more are thought to exist). They are distributed worldwide in nearly all terrestrial habitats. Ranging in length from 0.5 to over 300 mm, an insect's body consists of a head, a thorax of three segments and usually bearing three pairs of legs and one or two pairs of wings, and an abdomen of eleven segments. The head possesses a pair of sensory *antennae and a pair of large *compound eyes, between which are three simple eyes (*ocelli). The *mouthparts are variously adapted for either chewing or sucking, enabling insects to feed on a wide range of plant and animal material. Insects owe much of their success to having a highly waterproof *cuticle (to resist desiccation) and wings – outgrowths of the body wall that confer the greater mobility of flight. Breathing occurs through a network of tubes (*see* TRACHEA).

Most insect species have separate sexes and undergo sexual reproduction. In some, this may alternate with asexual *parthenogenesis and in a few, males are unknown and reproduction is entirely asexual. The newly hatched young grow by undergoing a series of moults. In the more primitive groups, e.g. *Dermaptera, *Orthoptera, *Dictyoptera, and *Hemiptera, the young (called a *nymph) resembles the adult. More advanced groups, e.g. *Coleoptera, *Diptera, *Lepidoptera, and *Hymenoptera, undergo *metamorphosis, in which the young (called a *larva) is transformed into a quiescent *pupa from which the fully formed adult emerges. Insects are of vital importance in many ecosystems and many are of economic significance – as pests or disease vectors or beneficially as crop pollinators or producers of silk, honey, etc.

(())) **SEE WEB LINKS**
- This section of the Natural History Museum's website highlights various insect groups and other aspects of entomology

**hexose** A *monosaccharide that has six carbon atoms in its molecules.

**hexyl group (hexyl radical)** The organic group $CH_3CH_2CH_2CH_2CH_2CH_2-$, derived from hexane.

**HFC (hydrofluorocarbon)** *See* HALOCARBONS.

**hibernation** A sleeplike state in which some animals pass the winter months as a way of surviving food scarcity and cold weather. Various physiological changes occur, such as lowering of the body temperature and slowing of the pulse rate and other vital processes, and the animal lives on its reserve of body fat. Animals that hibernate include bats, hedgehogs, and many fish, amphibians, and reptiles. *See also* DORMANCY. *Compare* AESTIVATION.

**hidden matter** *See* MISSING MASS.

**hidden-variables theory** A theory that denies that the specification of a physical system given by a state described by *quantum mechanics is a complete specification. A successful hidden-variables theory has never been constructed. There are important considerations preventing the construction of a simple hidden-variables theory, notably *Bell's theorem.

The only type of hidden-variables theories that appear not to have been ruled out are **non-local hidden-variables theories**, i.e. theories in which hidden parameters can affect parts of the system in arbitrarily distant regions simultaneously. A hidden-variables theory that does not satisfy this definition is called a **local hidden-variables theory**.

**hierarchy** (in biology) A type of social organization in which individuals are ranked according to their status or dominance relative to other group members. This affects their behaviour in various ways, e.g. by determining their access to food or to mates. Many vertebrate animals and some invertebrates live in hierarchical social groups.

**Higgs boson** A spin-zero particle with a nonzero mass, predicted by Peter Higgs (1929– ) to exist in certain *gauge theories, in particular in *electroweak theory (the Weinberg–Salam model). The Higgs boson has not yet been found but it is thought likely that it will be found by larger *accelerators in the next few years, especially since other associated features of the theory, including W and Z bosons, have been found.

**Higgs field** The symmetry-breaking field associated with a *Higgs boson. The Higgs field can either be an elementary *scalar quantity or the field associated with a *bound state of two *fermions. In the Weinberg–Salam model (*see* ELECTROWEAK THEORY), the Higgs field is taken to be a scalar field. It is not known whether this assumption is correct, although attempts to construct electroweak theory involving bound states for the Higgs field, known as **technicolour theory**, have not been successful. Higgs fields also occur in many-body systems, which can be expressed in terms of *quantum field theory with Higgs bosons; an example is the BCS theory of *superconductivity, in which the Higgs field is associated with a *Cooper pair, rather than an elementary scalar field.

**higher dimensions** Dimensions of space or time that exist in addition to the three space dimensions and one time dimension. Higher dimensions can be very small, as in *Kaluza–Klein theory, or large, as in the *brane world. There is no observational evidence for the existence of higher dimensions but they do appear to be a necessary part of certain fundamental theories.

**high frequency (HF)** A radio frequency in the range 3–30 megahertz; i.e. having a wavelength in the range 10–100 metres.

**high-performance liquid chromatography (HPLC)** A sensitive technique for separating or analysing mixtures, in which the sample is forced through the chromatography column under pressure.

**high-speed steel** A steel that will remain hard at dull red heat and can therefore be used in cutting tools for high-speed lathes. It usually contains 12–22% tungsten, up to 5% chromium, and 0.4–0.7% carbon. It may also contain small amounts of vanadium, molybdenum, and other metals.

**high-temperature superconductivity** *See* SUPERCONDUCTIVITY.

**high tension (HT)** A high potential difference, usually one of several hundred volts or more. Batteries formerly used in the anode circuits of radio devices using valves were usually called **high-tension batteries** to distinguish them from the batteries supplying the heating filaments.

**high-velocity star** An old population II star (*see* POPULATION TYPE) that does not orbit the galactic centre but travels far away from the plane of the Galaxy. Such stars move no faster than the stars near them but appear to do so because of their eccentric orbits.

**Hilbert space** A linear *vector space that can have an infinite number of dimensions. The concept is of interest in physics because the state of a system in *quantum mechanics is represented by a vector in Hilbert space. The dimension of the Hilbert space has nothing to do with the physical dimension of the system. The Hilbert space formulation of quantum mechanics was put forward by the Hungarian-born US mathematician John von Neumann (1903–57) in 1927. Other formulations of quantum mechanics, such as *matrix mechanics and *wave mechanics, can be deduced from the Hilbert space formulation. Hilbert space is named after the German mathematician David Hilbert (1862–1943), who invented the concept early in the 20th century.

**hilum** A scar on the seed coat of a plant marking the point at which the seed was attached to the fruit wall by the *funicle. It is a feature that distinguishes seeds from fruits.

**hindbrain (rhombencephalon)** One of the three sections of the brain of a vertebrate embryo. It develops to form the *cerebellum, *pons, and *medulla oblongata, which control and coordinate fundamental physiological processes (including respiration and circulation of blood). *Compare* FOREBRAIN; MIDBRAIN.

**hindgut** 1. The posterior part of the alimentary canal of vertebrates, comprising the posterior section of the colon. 2. The posterior section of the alimentary canal of arthropods. *See also* FOREGUT; MIDGUT.

**hip girdle** *See* PELVIC GIRDLE.

**hippocampus** A part of the vertebrate brain consisting of two ridges, one over each of the two lateral *ventricles. It is highly developed in advanced mammals (primates and whales) and its function seems to be concerned with forming new memories – by encoding information that can be stored as long-term memory elsewhere in the brain.

**Hirudinea** A class of freshwater and terrestrial annelid worms that comprises the leeches. They have suckers at both anterior and posterior ends but no bristles. Some are blood-sucking parasites of vertebrates and invertebrates but the majority are predators.

**histamine** A substance that is released during allergic reactions, e.g. hay fever. Formed from the amino acid histidine, histamine can occur in various tissues but is concentrated in connective tissue. It is one of the inflammatory mediators released from mast cells in response to antigen binding to IgE antibodies on the surface of the mast cell. It causes dilation and increased permeability of small blood vessels, which results in such symptoms as localized swelling, itching, sneezing, and runny eyes and nose. The effects of histamine can be countered by the administration of *antihistamine drugs.

**histidine** *See* AMINO ACID.

**histiocyte** *See* MACROPHAGE.

**histochemistry** The study of the distribution of the chemical constituents of tissues by means of their chemical reactions. It utilizes such techniques as *staining, light and electron microscopy, *autoradiography, and *chromatography.

**histocompatibility** The degree to which tissue from one organism will be tolerated by the immune system of another organism. For any animal, it is essential that its immune system can distinguish its own tissues from foreign cells or tissues, so that only the latter are attacked. This self-recognition is achieved principally by a set of marker molecules, called **histocompatibility proteins** (or **histocompatibility antigens**), which occur on the surfaces of cells. These proteins (in humans also called **human leucocyte antigens**, or **HLAs**) are encoded in vertebrates by a cluster of genes called the *major histocompatibility complex (MHC), which in humans includes the *HLA system. Each species has a unique set of histocompatibility proteins, and there is also wide variation within any given species. This explains why in human transplantation it is very difficult to match donor and recipient tissue exactly. MHC proteins also play a vital role in the immune responses of lymphocytes, notably by enabling *T cells to identify foreign antigens.

**histocompatibility protein** *See* HISTOCOMPATIBILITY; HLA SYSTEM.

**histology** The microscopic study of the tissues of living organisms. The study of cells, a specialized branch of histology, is known as *cytology.

**histone** Any of a group of water-soluble proteins found in association with the *DNA of plant and animal chromosomes. They contain a large proportion of the basic (positively charged) amino acids lysine, arginine, and histidine. They are involved in the condensation and coiling of chromosomes during cell division, and chemical modification of histones is a key aspect of suppressing or activating gene activity. Histones do not occur in vertebrate sperm cells (*see* PROTAMINE) or in bacteria.

**HIV (human immunodeficiency virus)** The *retrovirus that causes *AIDS in humans. It has a specific affinity for the helper *T cells of its host, binding to *CD4 antigens on the cell surface and thereby disabling these cells. The membrane envelope glycoproteins encasing the virus show great variability in their amino-acid sequences, hence the difficulty of preparing an effective AIDS vaccine. Two varieties (serovars) are known: HIV-1 and HIV-2. The latter, which is less virulent, is found chiefly in Africa. HIV is thought to have originated from chimpanzees in central Africa.

**HLA system (human leucocyte antigen system)** A series of gene loci, forming part of the *major histocompatibility complex in humans, that encode a group of proteins that act as antigens and are important in determining the acceptance or rejection by the body of a tissue or organ transplant (*see* GRAFT). These antigens are one group of the so-called **histocompatibility proteins** or **antigens** (*see* HISTOCOMPATIBILITY). Two individuals with identical HLA types are said to be **histocompatible**. Successful transplantation requires a minimum number of HLA differences between the donor's and recipient's tissues.

**HMX (octogen; cyclotetramethylenetetranitramine)** A colourless crystalline compound, $C_4H_8N_8O_8$; r.d. 1.9; m.p. 276–286°C. It has a structure similar to that of *cyclonite (RDX), but with an eight-membered ring rather than a six-membered ring. An extremely powerful explosive, it is used mainly for military purposes and as a rocket propellant. The name comes from 'High Molecular Weight RDX'.

**HNIW (hexanitrohexaazaisowurtzitane)** A powerful explosive, $C_6N_{12}O_{12}$. It has a three-dimensional bridged structure containing six $N-NO_2$ groups. Also known as **CL20**, it is extremely sensitive and so far has not been produced in bulk.

**hnRNP** *See* RIBONUCLEOPROTEIN.

**hoar frost** A deposit of interlocking ice crystals formed on objects (such as branches of trees and hedgerows) that are exposed to air in which there is *supersaturation with water vapour. The formation of hoar frost is similar to the formation of dew with the difference that the temperature of the object on which the hoar frost forms is below freezing temperature, whereas this is not the case with dew.

**Hodgkin, Sir Alan** *See* ECCLES, SIR JOHN CAREW.

**HOE** *See* HOLOGRAPHIC OPTICAL ELEMENT.

**Hogness box** *See* TATA BOX.

**hole** A vacant electron position in the lattice structure of a solid that behaves like a mobile positive *charge carrier with a negative *rest energy. *See* SEMICONDUCTOR.

**holmium** Symbol Ho. A soft silvery metallic element belonging to the *lanthanoids; a.n. 67; r.a.m. 164.93; r.d. 8.795 (20°C); m.p. 1474°C; b.p. 2695°C. It occurs in apatite, xenotime, and some other rare-earth minerals. There is one natural isotope, holmium–165; eighteen artificial isotopes have been produced. There are no uses for the element, which was discovered by Per Cleve (1840–1905) and J. L. Soret in 1879.

**(⊕) SEE WEB LINKS**
• Information from the WebElements site

**Holocene (Recent)** The most recent geological epoch of the *Neogene period, comprising roughly the past 11 500 years since the end of the *Pleistocene up to the present. It follows the final glacial of the Pleistocene and thus is sometimes known as the **Postglacial** epoch. Some geologists consider the Holocene to be an interglacial phase of the Pleistocene that will be followed by another glacial.

**holocrine secretion** *See* SECRETION.

**holoenzyme** A complex comprising an enzyme molecule and its *cofactor. Only in this state is an enzyme catalytically active. *Compare* APOENZYME.

**holographic hypothesis** A key principle of *quantum gravity postulating that it is possible to obtain information about the bulk of a system, such as a black hole, from the surface of that system. This hypothesis was suggested as a general feature of quantum gravity in the mid-1990s and found to hold in superstring theory a few years later.

**holographic optical element (HOE)** An optical device consisting of a hologram, used to focus or deflect electromagnetic radiation. Holographic optical elements are relatively easy to make and can function as lenses, mirrors, diffraction gratings, beam splitters, etc., in place of more traditional optical components.

**holography** A method of recording and displaying a three-dimensional image of an object, usually using *coherent radiation from a *laser and photographic plates (see illustration). The light from a laser is divided so that some of it (the reference beam) falls directly on a photographic plate. The other part illuminates the object, which reflects it back onto the photographic plate. The two beams form interference patterns on the plate, which when developed is called the **hologram**. To reproduce the image of the object, the hologram is illuminated by coherent light, ideally the original reference beam. The hologram produces two sets of diffracted waves; one set forms a virtual image coinciding with the original object position and the other forms a real image on the other side of the plate. Both are three-dimensional. The method was invented by Dennis Gabor in 1948. More recent techniques can produce holograms visible in white light.

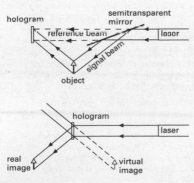

**Holography.**

**holohedral form** The form of a crystal in which the full number of faces required for the symmetry are present. *Compare* HOLO-HEDRAL FORM.

**holophytic** Describing organisms that feed like plants, i.e. that are photoautotrophic. *See* AUTOTROPHIC NUTRITION.

**holotype** *See* TYPE SPECIMEN.

**holozoic** Describing organisms that feed by ingesting complex organic matter, which is subsequently digested and absorbed. *See* HETEROTROPHIC NUTRITION; INGESTION.

**homeobox** A nucleotide sequence containing about 180 base pairs, which are identical or very similar in many eukaryotic organisms, that encodes a series of amino acids known as a **homeodomain**. Present in many eukaryotic regulatory proteins, this sequence is an important region involved in the binding of regulatory proteins to the DNA molecule. A homeobox was first identified in the *homeotic genes of *Drosophila* and has since been found in the homeotic genes of many animals (including humans) and in plants.

**homeostasis** The regulation by an organism of the chemical composition of its body fluids and other aspects of its *internal environment so that physiological processes can proceed at optimum rates. It involves monitoring changes in the external and internal environment by means of *receptors and adjusting the composition of the body fluids accordingly; *excretion and *osmoregulation are important in this process. Examples of homeostatic regulation are the maintenance of the *acid–base balance and body temperature (*see* HOMOIOTHERMY; POIKILOTHERMY).

**homeotic genes** A class of genes that play a central role in controlling the early development and differentiation of embryonic tissues in eukaryotic organisms. They include the **Hox genes**, which control the development of structures along the head-to-tail (anteroposterior) axis of a wide range of animals. Homeotic genes code for proteins that bind DNA and regulate the expression of a wide range of other genes. This binding capability resides in a structural domain of the protein called a **homeodomain**, encoded by a nucleotide sequence (*homeobox) that is characteristic of homeotic genes. These genes were first identified in *Drosophila* fruit flies, through the occurrence of mutations that cause the transformation of one organ into another – the phenomenon of **homeosis**. In vertebrates there are four clusters of homeotic genes located on separate chromosomes.

**hominid** Any member of the primate family Hominidae, which includes humans and their fossil ancestors (**fossil hominids**) in the genus *Homo. This family is now widely regarded as also including the extant great apes (chimpanzees, gorillas, orang-utan, formerly constituting the family Pongidae) and extinct groups, such as *Australopithecus and *Dryopithecus.

**SEE WEB LINKS**

- Interactive documentary narrated by palaeontologist Donald Johanson and presented by the Institute of Human Origins, Arizona State University

*Homo* The genus of primates that includes modern humans (*H. sapiens*, the only living representative) and various extinct species. The oldest *Homo* fossils are those of *H. habilis* and *H. rudolfensis*, which first appeared in Africa 2.2–2.4 million years ago. Both species used simple stone tools. *H. habilis* appears to have been 1–1.5 m tall and had more human-like features and a larger brain than *Australopithecus*. *H. erectus* diverged from *H. ergaster* in Africa and subsequently spread to Asia between 1.8 and 1.5 million years ago. Fossils of *H. erectus*, which was formerly called *Pithecanthropus* (ape man), include Java man and Peking man. They are similar to present-day humans except that there was a prominent ridge above the eyes and no forehead or chin. They used crude stone tools and fire. *H. ergaster* may also have given rise to *H. heidelbergensis* (represented by Heidelberg man and Boxgrove man). This species now contains all hominid specimens with a mixture of 'erectus-like' and 'modern' characters, dating from some 800 000 years ago to the emergence of *H. sapiens* between 130 000 and 90 000 years ago. Among them are the ancestors of both *H. neanderthalensis* (*Neanderthal man) and *H. sapiens*. See also CROMAGNON MAN.

**homocyclic** See CYCLIC.

**homodont** Describing animals whose teeth are all of the same type. Most vertebrates except mammals are homodont. *Compare* HETERODONT.

**homogametic sex** The sex that is determined by possession of two similar *sex chromosomes (i.e. XX). In humans and many other mammals this is the female sex. All the reproductive cells (gametes) produced by the homogametic sex have the same kind of sex chromosome (i.e. an X chromosome). *Compare* HETEROGAMETIC SEX.

**homogamy** The condition in a flower in which the male and female reproductive organs mature at the same time, allowing self-fertilization. *Compare* DICHOGAMY.

**homogeneous** Relating to only one phase, e.g. a homogeneous mixture, a homogeneous *catalyst. *Compare* HETEROGENEOUS.

**homoiothermy** The maintenance by an animal of its internal body temperature at a relatively constant value by using metabolic processes to counteract fluctuations in the temperature of the environment. Homoiothermy occurs in birds and mammals, which are described as *endotherms. The heat produced by their tissue metabolism and the heat lost to the environment are balanced by various means to keep body temperature constant: 36–38°C in mammals and 38–40°C in birds. The *hypothalamus in the brain monitors blood temperature and controls thermoregulation by both nervous and hormonal means. This produces both short-term responses, such as shivering or sweating, and long-term adjustments to metabolism according to seasonal changes in climate (acclimatization). Endotherms generally possess insulating feathers or fur. Their relatively high internal temperature permits fast action of muscles and nerves and enables them to lead highly active lives even in cold climates. However, in certain animals, homoiothermy is abandoned during periods of *hibernation. *Compare* POIKILOTHERMY.

**homoleptic compound** A chemical complex with only one type of ligand, as in nickel carbonyl or lead tetraethyl.

**homologous** (in biology) Describing a character that is shared by a group of species because it is inherited from a common ancestor. Such characters, called **homologies**, are used in *cladistics to determine the evolutionary relationships of species or higher taxa. They are divided into two types: a **shared derived homology** is unique to a particular group and may be used to define a *monophyletic group; a **shared ancestral homology** is not unique to the group, or may not be exhibited by all descendants of the ancestor in which it arose. Even though homologous features share the same evolutionary origin, they may have developed different functions. For example the wings of a bat, the flippers of a dolphin, and the arms of a human are homologous organs, having evolved from the paired pectoral fins of a fish ancestor. *Compare* ANALOGOUS.

**homologous chromosomes** Chromosomes having the same pattern of genes along their lengths although the nature of the genes may differ (*see* ALLELE). In *diploid nuclei, pairs of homologous chromosomes form bivalents at the start of *meiosis (*see* PAIRING). One member of each pair comes from the female parent and the other from the male.

**homologous series** A series of related chemical compounds that have the same functional group(s) but differ in formula by a fixed group of atoms. For instance, the simple carboxylic acids; methanoic (HCOOH), ethanoic ($CH_3COOH$), propanoic ($C_2H_5COOH$), etc., form a homologous series in which each member differs from the next by $CH_2$. Successive members of such a series are called **homologues**.

**homolytic fission** The breaking of a bond in a compound in which the fragments are uncharged free radicals. For example, $Cl_2 \rightarrow Cl\cdot + Cl\cdot$. *Compare* HETEROLYTIC FISSION.

**homoplasy** The similarity of a particular character in two different, yet often related, groups of organisms that is not the result of common ancestry. Such a similarity may arise due to convergent or parallel evolution, and is therefore potentially misleading when examining shared characters in constructing phylogenetic trees (*see* CLADISTICS). For example, wings in bats and birds are a convergent, and therefore homoplasic, character.

**homopolar bond** *See* CHEMICAL BOND.

**homopolar generator** A simple electric generator consisting of a metal disc rotating between two poles of a magnet. Contacts are made to the axle and the rim of the disc. A radial e.m.f. is produced. At constant rotational speed, the device produces a steady direct current and generators of this type are used in certain specialized applications. It can also be used as a simple motor if a direct current is supplied. A device of this type (known as the **Barlow wheel**) was invented in 1822 by the physicist Peter Barlow (1776–1862). This had a star-shaped wheel with the points of the star dipping into a pool of mercury to give the electrical con-

tact. A generator with a disc was used by Michael Faraday in his experiments, and the device is sometimes known as the **Faraday disc**.

**homopolymer** *See* POLYMER.

**homozygous** Describing an organism or cell in which the *alleles at a given locus on homologous chromosomes are identical (they may be either dominant or recessive). Homozygous organisms, which are called **homozygotes**, breed true when crossed with genetically identical organisms. *Compare* HETEROZYGOUS.

**Hooke, Robert** (1635–1703) English physicist, who worked at Oxford University, where he assisted Robert *Boyle. Among his many achievements were the law of elasticity (*see* HOOKE'S LAW), the watch balance wheel, and the compound *microscope. In 1665, using his microscope to study vegetable tissues, he saw 'little boxes', which he named 'cells'.

**Hooke's law** The *stress applied to any solid is proportional to the *strain it produces within the elastic limit for that solid. The ratio of longitudinal stress to strain is equal to the Young modulus of elasticity (*see* ELASTIC MODULUS). The law was first stated by Robert Hooke in the form "Ut tensio, sic vis."

**hormone** 1. A substance that is manufactured and secreted in very small quantities into the bloodstream by an *endocrine gland or a specialized nerve cell (*see* NEUROHORMONE) and regulates the growth or functioning of a specific tissue or organ in a distant part of the body. For example, the hormone *insulin controls the rate and manner in which glucose is used by the body. Other hormones include the *sex hormones, *corticosteroids, *adrenaline, *thyroxine, and *growth hormone. 2. *See* PLANT HORMONE.

**hornblende** Any of a group of common rock-forming minerals of the amphibole group with the generalized formula:

$$(Ca,Na)_2(Mg,Fe,Al)_5(Al,Si)_8O_{22}(OH,F)_2$$

Hornblendes consist mainly of calcium, iron, and magnesium silicate.

**horsepower (hp)** An imperial unit of power originally defined as 550 foot-pound force per second; it is equal to 745.7 watts.

**horsetails** *See* SPHENOPHYTA.

**horst** *See* GRABEN.

**host** An organism whose body provides nourishment and shelter for a parasite (*see* PARASITISM). A **definitive** (or **primary**) **host** is one in which an animal parasite becomes sexually mature; an **intermediate** (or **secondary**) **host** is one in which the parasite passes the larval or asexual stages of its life cycle.

**host–guest chemistry** *See* SUPRAMOLECULAR CHEMISTRY.

**hot-wire instrument** An electrical measuring instrument (basically an ammeter) in which the current to be measured is passed through a thin wire and causes its temperature to rise. The temperature rise, which is proportional to the square of the current, is measured by the expansion of the wire. Such instruments can be used for either direct current or alternating current.

**Hox genes** *See* HOMEOTIC GENES.

**Hoyle, Sir Fred** (1915–2001) British astronomer, who in 1958 became a professor of astronomy at Cambridge University. He is best known for his proposal in 1948, with Hermann Bondi (1919–2005) and Thomas Gold (1920–2004), of the *steady-state theory of the universe. He also carried out theoretical work on the formation of elements in stars.

**HPLC** *See* HIGH-PERFORMANCE LIQUID CHROMATOGRAPHY.

**HSAB principle** A method of classifying Lewis acids and bases (*see* ACID) developed by Ralph Pearson in the 1960s. The acronym stands for 'hard and soft acids and bases'. It is based in empirical measurements of stability of compounds with certain ligands. **Hard acids** tend to complex with halide ions in the order

$$F^- > Cl^- > Br^- > I^-$$

**Soft acids** complex in the opposite order. Compounds that complex with hard acids are **hard bases**; ones that more readily form complexes with soft acids are called **soft bases**. In general, soft acids and bases are more easily polarized than hard acids and bases and consequently have more covalent character in the bond. The idea is an extension of the *type A and B metals concept to compounds other than metal complexes.

**HTML (hypertext markup language)** The computer data format used to transfer and display data on the *World Wide Web. It involves a set of rules for codes inserted into text files to indicate such things as special typefaces, paragraph breaks, illustration positions, and links to other *hypertext pages.

**(((●))) SEE WEB LINKS**
• The HTML (version 4.01) specification

**HTTP** Hypertext transport protocol: an application-level protocol with the lightness and speed necessary for distributed collaborative hypermedia information systems. It is generic, stateless, and object-oriented, with typing and negotiation of data representation, allowing systems to be built independently of the data being transferred. By extension of its request methods (commands), it can be used for many tasks, such as name servers and distributed object-management systems. HTTP has been in use by the *World Wide Web since 1990.

**(((●))) SEE WEB LINKS**
• The HTTP (version 1.1) specification

**Hubble, Edwin Powell** (1889–1953) US astronomer, who worked at both the Yerkes Observatory and the Mount Wilson Observatory. Most of his studies involved nebulae and galaxies, which he classified in 1926. In 1929 he established the *Hubble constant, which enabled him to estimate the age of the universe. The *Hubble space telescope is named after him.

**Hubble constant** The rate at which the velocity of recession of the galaxies increases with distance as determined by the *redshift. The value is not agreed upon but current measurements indicate that it lies between 49 and 95 km s$^{-1}$ per megaparsec. The reciprocal of the Hubble constant, the **Hubble time**, is a measure of the age of the universe, assuming that the expansion rate has remained constant. In fact, it is necessary to take account of the fact that the expansion of the universe is accelerating to get an accurate determination of its age.

**Hubble space telescope** An orbiting telescope managed by the US National Aeronautics and Space Administration (NASA) and the European Space Agency (ESA) and launched in April 1990, operating in the ultraviolet, visible, and near-infrared regions of the spectrum at high resolution free from the distorting effects of the earth's atmosphere. After five servicing missions (1993–2009), the telescope continues to add to the wealth of spectacular images and other data that it had already provided, expanding our understanding of the universe, especially its early phases.

**Hückel rule** *See* AROMATIC COMPOUND.

**hue** *See* COLOUR.

**Hulse–Taylor pulsar** A binary system discovered in 1974 consisting of a pulsar orbiting with a companion star. It was used to demonstrate, by measurements of a period of time, that the system was emitting gravitational waves. It is named after the US physicists Russell Hulse (1950– ) and Joseph Taylor (1941– ).

**human chorionic gonadotrophin (hCG)** *See* GONADOTROPHIN.

**Human Genome Project** A coordinated international project, begun in 1988, to map the entire human *genome so that the genes could be isolated and sequenced (*see* DNA SEQUENCING). It involved the production of a *DNA library. The haploid human genome contains about $3 \times 10^9$ nucleotide base pairs, but only 22 000 – 25 000 genes. The full draft sequence was completed in 2000 and published in February 2001, and the high-quality finished sequence was completed in April 2003, two years ahead of schedule.

**human growth hormone (hGH)** *See* GROWTH HORMONE.

**human immunodeficiency virus** *See* HIV.

**Hume-Rothery rules** A set of empirical rules put forward in 1926 by the British metallurgist William Hume-Rothery (1899–1968) to describe how one metal dissolves in another metal. He found that one metallic element will not dissolve in another if the difference in their atomic radii is greater than 15 per cent or if the electronegativities of the two elements differ substantially. Also, a metal with a lower valency is more likely to dissolve in one with a higher valency, than vice versa. These rules were subsequently justified by the quantum theory of electrons in solids.

**humerus** The long bone of the upper arm of tetrapod vertebrates. It articulates with the *scapula (shoulder blade) at the *glenoid cavity and with the *ulna and *radius (via a *condyle) at the elbow.

**humidity** The concentration of water vapour in the atmosphere. The **absolute humidity** is the mass of water vapour per unit volume of air, usually expressed in kg m$^{-3}$. **Relative humidity** is the ratio, expressed as a percentage, of the moisture in the air to the moisture it would contain if it were saturated at the same temperature and pressure. **Specific humidity** is also sometimes used: this is the mass of water vapour in the atmosphere per unit mass of air. *See also* HYGROMETER.

**humoral** Relating to the blood or other body fluids. For example, **humoral immunity** is immunity conferred by the antibodies present in the blood, lymph, and tissue fluids (*see* IMMUNITY).

**humus** The dark-coloured amorphous colloidal material that constitutes the organic component of soil. It is formed by the decomposition of plant and animal remains and excrement (*see* LITTER) and has a complex and variable chemical composition. Being a colloid, it can hold water and therefore improves the water-retaining properties of soil; it also enhances soil fertility and workability. Acidic humus (**mor**) is found in regions of coniferous forest, where the decay is brought about mainly by fungi. Alkaline humus (**mull**) is typically found in grassland and deciduous forest: it supports an abundance of microorganisms and small animals (e.g. earthworms).

**Hund coupling cases** *See* COUPLING.

**Hund's rules** Empirical rules in atomic *spectra that determine the lowest energy level for a configuration of two equivalent electrons (i.e. electrons with the same $n$ and $l$ quantum numbers), in a many electron atom. (1) The lowest energy state has the maximum *multiplicity consistent with the *Pauli exclusion principle. (2) The lowest energy state has the maximum total electron *orbital angular momentum quantum number, consistent with rule (1). These rules were put forward by the German physicist Friedrich Hund (1896–1997) in 1925.

**hunting** The oscillation of a gauge needle or engine speed about a mean value. In a rotating mechanism, set to operate at a constant speed, pulsation above and below the set speed can occur, especially if the speed is controlled by a governor. It can be corrected using a damping device.

**hurricane** 1. A *tropical cyclone with surface wind speeds in excess of 64 knots (117 km/h) that occurs in the North Atlantic Ocean, Caribbean Sea, Gulf of Mexico, and eastern and central North Pacific Ocean (east of the dateline). (Tropical cyclones that occur in the North Pacific Ocean are known as **typhoons**.) Hurricanes cause widespread damage over land areas and are a considerable hazard to shipping. In order for a hurricane to form, an extensive area of ocean (at least 5° of latitude from the equator) must have a surface temperature in excess of 27°C. At the earth's surface a hurricane appears as a roughly circular vortex with the **eye** at the centre, in which pressure is very low. Winds spiral around this core at great speed in an anticlockwise direction in the northern hemisphere. A hurricane is usually accompanied by deep clouds and torrential rain over a wide area. **2.** *See* Beaufort wind scale.

**Hutton, James** (1726–97) British geologist, born in Scotland. He studied law, medicine, and industrial chemistry before returning to Edinburgh in 1768 to devote himself to geology. He advocated the theory that contemporary geological processes are the same as those that have always prevailed. This theory was not generally accepted until popularized by John Playfair (1748–1819) and championed by Charles *Lyell.

**Huxley, Sir Andrew** *See* Eccles, Sir John Carew.

**Huygens, Christiaan** (1629–95) Dutch astronomer and physicist. Born in the Hague, he made several visits to Paris between 1655 and 1681 and in 1666 became a founding member of the French Academy of Sciences. In 1657 he designed the first pendulum clock and wrote a description of it. He had already made improvements to astronomical telescopes, with which he discovered Saturn's satellite Titan in 1655 and the true shape of Saturn's rings in 1659. His greatest achievement, however, was the wave theory of light, announced in 1690 (*see* Huygens' construction).

**Huygens' construction (Huygens' principle)** Every point on a wavefront may itself be regarded as a source of secondary waves. Thus, if the position of a wavefront at any instant is known, a simple construction enables its position to be drawn at any subsequent time. The construction was first used by Christiaan Huygens.

**hyaline cartilage** *See* cartilage.

**hyaluronic acid** A *glycosaminoglycan (mucopolysaccharide) that is part of the *ground substance of connective tissue. Hyaluronic acid is a major component of articular cartilage, helping in tissue repair and the lubrication of joints. It also plays a role in cell movement and proliferation.

**hyaluronidase** Any of a family of enzymes that break down *hyaluronic acid, thereby decreasing its viscosity and increasing the permeability of connective tissue. They are used in medicine to increase the absorption and diffusion of drugs administered by injection or application.

**hybrid** The offspring of a mating in which the parents differ in at least one characteristic. The term is usually used of offspring of widely different parents, e.g. different varieties or species. Hybrids between different animal species are usually sterile. *See also* hybrid vigour.

**hybridization** **1.** The production of one or more *hybrid organisms by the mating of genetically different parents. **2.** The production of hybrid cells. *See* cell fusion (somatic cell hybridization). **3.** *See* DNA hybridization.

**hybridoma** A type of hybrid cell that is produced by the fusion of a tumour cell (a myeloma cell) with a normal antibody-producing *B cell (*see* cell fusion). The resulting hybrid cell line is able to produce large amounts of normal antibody, which is described as monoclonal (*see* monoclonal antibody) as it results from a cloned cell line.

**hybrid orbital** *See* orbital.

**hybrid vigour (heterosis)** The increased vigour displayed by the offspring from a cross between genetically different parents. Hybrids from crosses between different crop varieties ($F_1$ hybrids) are often stronger and produce better yields than the original varieties. Mules, the offspring of mares crossed with donkeys, have greater strength and resistance to disease and a longer lifespan than either parent.

**hydathode** A pore found in the *epidermis of the leaves of certain plants. Like *stomata, hydathodes are surrounded by two crescent-shaped cells but these, unlike guard cells, do not regulate the size of the aperture. Hydathodes secrete water under conditions in which *transpiration is inhibited; for example, when the atmosphere is very humid. This process of water loss is called **guttation**.

**hydracid**  *See* BINARY ACID.

**hydrate**  A substance formed by combination of a compound with water. *See* WATER OF CRYSTALLIZATION.

**hydrated alumina**  *See* ALUMINIUM HYDROXIDE.

**hydrated aluminium hydroxide**  *See* ALUMINIUM HYDROXIDE.

**hydration**  *See* SOLVATION.

**hydraulic press**  A device in which a force ($F_1$) applied to a small piston ($A_1$) creates a pressure ($p$), which is transmitted through a fluid to a larger piston ($A_2$), where it gives rise to a larger force ($F_2$): see illustration. This depends on Pascal's principle that the pressure applied anywhere in an enclosed fluid is transmitted equally in all directions. The principle of the hydraulic press is widely used in jacks, vehicle brakes, presses, and earth-moving machinery, usually with oil as the working fluid.

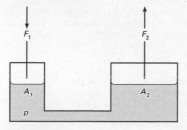

**Hydraulic press.**

**hydraulics**  The study of water or other fluids at rest or in motion, particularly with respect to their engineering uses. The study is based on the principles of *hydrostatics and *hydrodynamics.

**hydrazine**  A colourless liquid or white crystalline solid, $N_2H_4$; r.d. 1.01 (liquid); m.p. 1.4°C; b.p. 113.5°C. It is very soluble in water and soluble in ethanol. Hydrazine is prepared by the **Raschig synthesis** in which ammonia reacts with sodium(I) chlorate (sodium hypochlorite) to give $NH_2Cl$, which then undergoes further reaction with ammonia to give $N_2H_4$. Industrial production must be carefully controlled to avoid a side reaction leading to $NH_4Cl$. The compound is a weak base giving rise to two series of salts, those based on $N_2H_5^+$, which are stable in water (sometimes written in the form $N_2H_4.HCl$ rather than $N_2H_5^+Cl^-$), and a less stable and extensively hydrolysed series based on $N_2H_6^{2+}$. Hydrazine is a powerful reducing agent and reacts violently with many oxidizing agents, hence its use as a rocket propellant. It reacts with aldehydes and ketones to give *hydrazones.

**hydrazoic acid**  *See* HYDROGEN AZIDE.

**hydrazones**  Organic compounds containing the group =C:NNH$_2$, formed by condensation of substituted hydrazines with aldehydes and ketones (see illustration). **Phenylhydrazones** contain the group =C:NNHC$_6$H$_5$.

(🌐) **SEE WEB LINKS**
• Information about IUPAC nomenclature

**hydride**  A chemical compound of hydrogen and another element or elements. Nonmetallic hydrides (e.g. ammonia, methane, water) are covalently bonded. The alkali metals and alkaline earths (*s-block elements) form salt-like hydrides containing the hydride ion H$^-$, which produce hydrogen on reacting with water. Hydride-forming *transition elements form interstitial hydrides, with the hydrogen atoms 'trapped' within the gaps in the lattice of metal atoms. Complex hydrides, such as *lithium tetrahydroaluminate(III), have hydride ions as *ligands; many are powerful reducing agents.

**hydriodic acid**  An acid made by dissolving *hydrogen iodide in water. On standing in air it is slowly oxidized to iodine (making the solution brown); it finds use as a reducing agent.

**hydrobromic acid**  *See* HYDROGEN BROMIDE.

**Hydrazones.** Formation of a hydrazone from a ketone; the same reaction occurs with an aldehyde (R′ = H). If R″ = C$_6$H$_5$, the product is phenylhydrazone.

**hydrocarbons** Chemical compounds that contain only carbon and hydrogen. A vast number of hydrocarbon compounds exist, the main types being the *alkanes, *alkenes, *alkynes, and *arenes.

**hydrochloric acid** See HYDROGEN CHLORIDE.

**hydrochloride** See AMINE SALTS.

**hydrochlorofluorocarbon (HCFC)** See HALOCARBONS.

**hydrocortisone** See CORTISOL.

**hydrocyanic acid** See HYDROGEN CYANIDE.

**hydrodynamics** The study of the motion of incompressible fluids and the interaction of such fluids with their boundaries.

**hydroelectric power** Electric power generated by a flow of water. A natural waterfall provides a source of energy, in the form of falling water, which can be used to drive a water *turbine. This turbine can be coupled to a generator to provide electrical energy. Hydroelectric generators can be arranged to work in reverse so that during periods of low power demand current can be fed to the generator, which acts as a motor. This motor drives the turbine, which then acts as a pump. The pump then raises water to an elevated reservoir so that it can be used to provide extra power at peak-load periods.

**hydrofluoric acid** See HYDROGEN FLUORIDE.

**hydrofluorocarbon (HFC)** See HALOCARBONS.

**hydrogen** Symbol H. A colourless odourless gaseous chemical element; a.n. 1; r.a.m. 1.008; d. 0.0899 g dm$^{-3}$; m.p. −259.14°C; b.p. −252.87°C. It is the lightest element and the most abundant in the universe. It is present in water and in all organic compounds. There are three isotopes: naturally occurring hydrogen consists of the two stable isotopes hydrogen–1 (99.985%) and *deuterium. The radioactive *tritium is made artificially. The gas is diatomic and has two forms: **orthohydrogen**, in which the nuclear spins are parallel, and **parahydrogen**, in which they are antiparallel. At normal temperatures the gas is 25% parahydrogen. In the liquid it is 99.8% parahydrogen. The main source of hydrogen is steam *reforming of natural gas. It can also be made by the Bosch process (see HABER PROCESS) and by electrolysis of water. The main use is in the Haber process for making

ammonia. Hydrogen is also used in various other industrial processes, such as the reduction of oxide ores, the refining of petroleum, the production of hydrocarbons from coal, and the hydrogenation of vegetable oils. Considerable interest has also been shown in its potential use in a 'hydrogen fuel economy' in which primary energy sources not based on fossil fuels (e.g. nuclear, solar, or geothermal energy) are used to produce electricity, which is employed in electrolysing water. The hydrogen formed is stored as liquid hydrogen or as metal *hydrides. Chemically, hydrogen reacts with most elements. It was discovered by Henry Cavendish in 1776.

**((()) SEE WEB LINKS**

• Information from the WebElements site

**hydrogen acceptor** See HYDROGEN CARRIER.

**hydrogenation** 1. A chemical reaction with hydrogen; in particular, an addition reaction in which hydrogen adds to an unsaturated compound. Nickel is a good catalyst for such reactions. 2. The process of converting coal to oil by making the carbon in the coal combine with hydrogen to form hydrocarbons. See FISCHER–TROPSCH PROCESS; BERGIUS PROCESS.

**hydrogen azide (hydrazoic acid; azoimide)** A colourless liquid, HN$_3$; r.d. 1.09; m.p. −80°C; b.p. 37°C. It is highly toxic and a powerful reductant, which explodes in the presence of oxygen and other oxidizing agents. It is prepared by the reaction of sodium amide and sodium nitrate at 175°C followed by distillation of a mixture of the resulting sodium azide and a dilute acid. See also AZIDES.

**hydrogen bomb** See NUCLEAR WEAPONS.

**hydrogen bond** A type of electrostatic interaction between electronegative atoms (fluorine, nitrogen, or oxygen) in one molecule and hydrogen atoms bound to electronegative atoms in another molecule. It is a strong dipole–dipole attraction caused by the electron-withdrawing properties of the electronegative atom. Thus, in the water molecule the oxygen atom attracts the electrons in the O–H bonds. The hydrogen atom has no inner shells of electrons to shield the nucleus, and there is an electrostatic interaction between the hydrogen proton and a lone pair of electrons on an oxygen atom in a neighbouring molecule. Each oxygen atom has two lone pairs and can make hydrogen bonds to two different hydrogen atoms. The strengths of hydrogen bonds are about one tenth of the strengths of normal covalent bonds. Hydrogen bonding does, however, have significant effects on physical properties. Thus it accounts for the unusual properties of *water and for the relatively high boiling points of $H_2O$, HF, and $NH_3$ (compared with $H_2S$, HCl, and $PH_3$). It is also of great importance in living organisms. Hydrogen bonding occurs between bases in the chains of DNA (*see* BASE PAIRING). It also occurs between the C=O and N–H groups in proteins, and is responsible for maintaining the secondary structure.

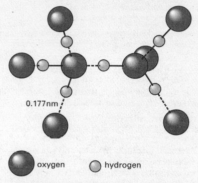

0.177 nm

● oxygen    ○ hydrogen

**Hydrogen bonds.** Shown as dotted lines between water molecules.

**hydrogen bromide** A colourless gas, HBr; m.p. –88.5°C; b.p. –67°C. It can be made by direct combination of the elements using a platinum catalyst. It is a strong acid dissociating extensively in solution (**hydrobromic acid**).

**hydrogencarbonate (bicarbonate)** A salt of *carbonic acid in which one hydrogen atom has been replaced; it thus contains the hydrogencarbonate ion $HCO_3^-$.

**hydrogen carrier (hydrogen acceptor)** A molecule that accepts hydrogen atoms or ions, becoming reduced in the process (*see* OXIDATION–REDUCTION). The *electron transport system, whose function is to generate energy in the form of ATP during respiration, involves a series of hydrogen carriers, including *NAD and *FAD, which pass on the hydrogen (derived from the breakdown of glucose) to the next carrier in the chain.

**hydrogen chloride** A colourless fuming gas, HCl; m.p. –114.8°C; b.p. –85°C. It can be prepared in the laboratory by heating sodium chloride with concentrated sulphuric acid (hence the former name **spirits of salt**). Industrially it is made directly from the elements at high temperature and used in the manufacture of PVC and other chloro compounds. It is a strong acid and dissociates fully in solution (**hydrochloric acid**).

**hydrogen cyanide (hydrocyanic acid; prussic acid)** A colourless liquid or gas, HCN, with a characteristic odour of almonds; r.d. 0.699 (liquid at 22°C); m.p. –14°C; b.p. 26°C. It is an extremely poisonous substance formed by the action of acids on metal cyanides. Industrially, it is made by catalytic oxidation of ammonia and methane with air and is used in producing acrylate plastics. Hydrogen cyanide is a weak acid ($K_a = 2.1 \times 10^{-9}$ mol $dm^{-3}$). With organic carbonyl compounds it forms *cyanohydrins.

**hydrogen electrode** *See* HYDROGEN HALF CELL.

**hydrogen fluoride** A colourless liquid, HF; r.d. 0.99; m.p. –83°C; b.p. 19.5°C. It can be made by the action of sulphuric acid on calcium fluoride. The compound is an extremely corrosive fluorinating agent, which attacks glass. It is unlike the other hydrogen halides in being a liquid (a result of *hydrogen-bond formation). It is also a weaker acid than the others because the small size of the fluorine atom means that the H–F bond is shorter and stronger. Solutions of hydrogen fluoride in water are known as **hydrofluoric acid**.

**hydrogen half cell (hydrogen electrode)** A type of *half cell in which a metal foil is immersed in a solution of hydrogen ions and hydrogen gas is bubbled over the foil. The standard hydrogen electrode, used in measuring standard *electrode potentials, uses a platinum foil with a 1.0 M solution of hydrogen ions, the gas at 1 atmosphere pressure, and a temperature of 25°C. It is written $Pt(s)|H_2(g)$, $H^+(aq)$, the effective reaction being $H_2 \rightarrow 2H^+ + 2e$.

**hydrogenic** Describing an atom or ion that has only one electron; for example, H, $He^+$, $Li^{2+}$, $C^{5+}$. Hydrogenic atoms (or ions) do not involve electron–electron interactions and are easier to treat theoretically.

**hydrogen iodide** A colourless gas, HI;

m.p. –51°C; b.p. –35.38°C. It can be made by direct combination of the elements using a platinum catalyst. It is a strong acid, dissociating extensively in solution (**hydroiodic acid**), and a reducing agent.

**hydrogen ion** *See* ACID; pH SCALE.

**hydrogen molecule ion** The simplest type of molecule ($H_2^+$), consisting of two hydrogen nuclei and one electron. In the *Born–Oppenheimer approximation, in which the nuclei are regarded as being fixed, the *Schrödinger equation for the hydrogen molecule ion can be solved exactly. This enables ideas and approximation techniques concerned with chemical bonding to be tested quantitatively.

**hydrogen peroxide** A colourless or pale blue viscous unstable liquid, $H_2O_2$; r.d. 1.44; m.p. –0.41°C; b.p. 150.2°C. As with water, there is considerable hydrogen bonding in the liquid, which has a high dielectric constant. It can be made in the laboratory by adding dilute acid to barium peroxide at 0°C. Large quantities are made commercially by electrolysis of $KHSO_4.H_2SO_4$ solutions. Another industrial process involves catalytic oxidation (using nickel, palladium, or platinum with an anthraquinone) of hydrogen and water in the presence of oxygen. Hydrogen peroxide readily decomposes in light or in the presence of metal ions to give water and oxygen. It is usually supplied in solutions designated by volume strength. For example, 20-volume hydrogen peroxide would yield 20 volumes of oxygen per volume of solution. Although the *peroxides are formally salts of $H_2O_2$, the compound is essentially neutral. Thus, the acidity constant of the ionization

$$H_2O_2 + H_2O \rightleftharpoons H_3O^+ + HO_2^-$$

is $1.5 \times 10^{-12}$ mol dm$^{-3}$. It is a strong oxidizing agent, hence its use as a mild antiseptic and as a bleaching agent for cloth, hair, etc. It has also been used as an oxidant in rocket fuels.

**hydrogen spectrum** The atomic spectrum of hydrogen is characterized by lines corresponding to radiation quanta of sharply defined energy. A graph of the frequencies at which these lines occur against the ordinal number that characterizes their position in the series of lines, produces a smooth curve indicating that they obey a formal law. In 1885 Johann Balmer (1825–98) discovered the law having the form:

$$1/\lambda = R(1/n_1^2 - 1/n_2^2)$$

This law gives the so-called **Balmer series** of lines in the visible spectrum in which $n_1 = 2$ and $n_2 = 3,4,5,...$, $\lambda$ is the wavelength associated with the lines, and $R$ is the *Rydberg constant.

In the **Lyman series**, discovered by Theodore Lyman (1874–1954), $n_1 = 1$ and the lines fall in the ultraviolet. The Lyman series is the strongest feature of the solar spectrum as observed by rockets and satellites above the earth's atmosphere. In the **Paschen series**, discovered by Louis Paschen (1865–1947), $n_1 = 3$ and the lines occur in the far infrared. The **Brackett series** ($n_1 = 4$), **Pfund series** ($n_1 = 5$), and **Humphreys series** ($n_1 = 6$) also occur in the far infrared.

(((⬡))) SEE WEB LINKS

• A translation of Balmer's 1885 paper on the spectral lines of hydrogen in *Annalen der Physik und Chemie*

**hydrogensulphate (bisulphate)** A salt containing the ion $HSO_4^-$ or an ester of the type $RHSO_4$, where R is an organic group. It was formerly called **hydrosulphate**.

**hydrogen sulphide (sulphuretted hydrogen)** A gas, $H_2S$, with an odour of rotten eggs; r.d. 1.54 (liquid); m.p. –85.5°C; b.p. –60.7°C. It is soluble in water and ethanol and may be prepared by the action of mineral acids on metal sulphides, typically hydrochloric acid on iron(II) sulphide (*see* KIPP'S APPARATUS). Solutions in water (known as **hydrosulphuric acid**) contain the anions HS$^-$ and minute traces of S$^{2-}$ and are weakly acidic. Acid salts (those containing the HS$^-$ ion) are known as **hydrogensulphides** (formerly **hydrosulphides**). In acid solution hydrogen sulphide is a mild reducing agent. Hydrogen sulphide has an important role in traditional qualitative chemical analysis, where it precipitates metals with insoluble sulphides (in acid solution: Cu, Pb, Hg, Cd, Bi, As, Sb, Sn). The formation of a black precipitate with alkaline solutions of lead salts may be used as a test for hydrogen sulphide but the characteristic smell is usually sufficient. Hydrogen sulphide is exceedingly poisonous (more toxic than hydrogen cyanide).

The compound burns in air with a blue flame to form sulphur(IV) oxide ($SO_2$); solutions of hydrogen sulphide exposed to the air undergo oxidation but in this case only to elemental sulphur. North Sea gas contains some hydrogen sulphide (from S-proteins in plants) as do volcanic emissions. *See also* CLAUS PROCESS.

**hydrogensulphite (bisulphite)** A salt containing the ion $^-HSO_3$ or an ester of the type $RHSO_3$, where R is an organic group.

**hydrography** The study of the waters of the earth's surface, including the oceans, seas, lakes, and rivers. It involves the measurement of these features and the presentation of the information on hydrographic charts.

**hydroiodic acid** *See* HYDROGEN IODIDE.

**hydrolase** Any of a class of enzymes that catalyse the addition of water to, or the removal of water from, a molecule. Hydrolases play an important role in the construction and breakdown of storage materials, such as starch.

**hydrological cycle (water cycle)** The circulation of water between the atmosphere, land, and oceans on the earth (see illustration). Water evaporates from water bodies on earth to form water vapour in the atmosphere. This may condense to form clouds and be returned to the earth's surface as rainfall, hail, snow, etc. Some of this precipitation is returned to the atmosphere directly through evaporation or transpiration by plants; some flows off the land surface as overland flow, eventually to be returned to the oceans via rivers; and some infiltrates the ground to flow underground forming groundwater storage.

**hydrology** The scientific study of terrestrial water, in particular inland water before its discharge into the oceans or evaporation into the atmosphere. It includes the study of the occurrence and movement of water and ice on or under the earth's surface. The science has many important applications, for example in flood control, irrigation, domestic and industrial uses, and hydroelectric power.

**hydrolysis** A chemical reaction of a compound with water. For instance, salts of weak acids or bases hydrolyse in aqueous solution, as in

$$Na^+CH_3COO^- + H_2O \rightleftharpoons Na^+ + OH^- + CH_3COOH$$

The reverse reaction of *esterification is another example. *See also* SOLVOLYSIS.

**hydromagnesite** A mineral form of basic *magnesium carbonate, $3MgCO_3.Mg(OH)_2.3H_2O$.

**hydrometer** An instrument for measuring the density or relative density of liquids. It usually consists of a glass tube with a long bulb at one end. The bulb is weighted so that the device floats vertically in the liquid, the relative density being read off its calibrated stem by the depth of immersion.

**hydronium ion** *See* OXONIUM ION.

**hydrophilic** Having an affinity for water. *See* LYOPHILIC.

**hydrophily** A rare form of pollination in which pollen is carried to a flower by water. It occurs by one of two methods. In Canadian pondweed (*Elodea canadensis*) the male flowers break off and float downstream until they contact the female flowers. In *Zostera*, a

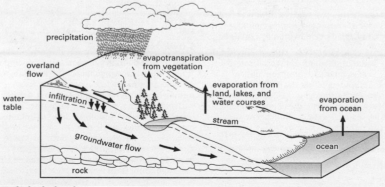

**Hydrological cycle.**

marine species, the filamentous pollen grains are themselves carried in the water. *Compare* ANEMOPHILY; ENTOMOPHILY.

**hydrophobic** Lacking affinity for water. *See* LYOPHOBIC.

**hydrophyte** Any plant that lives either in very wet soil or completely or partially submerged in water. Structural modifications of hydrophytes include the reduction of mechanical and supporting tissues and vascular tissue, the absence or reduction of a root system, and specialized leaves that may be either floating or finely divided, with little or no cuticle. Examples of hydrophytes are waterlilies and certain pondweeds. *Compare* HALOPHYTE; MESOPHYTE; XEROPHYTE.

**hydroponics** A commercial technique for growing certain crop plants in culture solutions rather than in soil. The roots are immersed in an aerated solution containing the correct proportions of essential mineral salts. The technique is based on various water culture methods used in the laboratory to assess the effects of the absence of certain mineral elements on plant growth.

**hydroquinone** *See* BENZENE-1,4-DIOL.

**hydrosol** A sol in which the continuous phase is water. *See* COLLOIDS.

**hydrosphere** The water on the surface of the earth. Some 74% of the earth's surface is covered with water, 97% (or some $10^{21}$ kilograms) of which is in the oceans. Icecaps and glaciers contain about $3 \times 10^{19}$ kg, rivers about $10^{15}$ kg, lakes and inland seas about $2 \times 10^{17}$ kg, and groundwater (down to 4000 metres) about $8 \times 10^{19}$ kg. Water in the atmosphere contains only about $10^{16}$ kg.

**hydrostatics** The study of liquids at rest, with special reference to storage tanks, dams, bulkheads, and hydraulic machinery.

**hydrostatic skeleton** The system of support found in soft-bodied invertebrates, which relies on the incompressibility of fluids contained within the body cavity. For example, in earthworms the coelomic fluid is under pressure within the coelom and therefore provides support for internal organs.

**hydrosulphate** *See* HYDROGENSULPHATE.

**hydrosulphide** *See* HYDROGEN SULPHIDE.

**hydrosulphuric acid** *See* HYDROGEN SULPHIDE.

**hydrotropism** The growth of a plant part

in response to water. Roots, for example, grow towards water in the soil. *See* TROPISM.

**hydroxide** A metallic compound containing the ion OH⁻ (**hydroxide ion**) or containing the group –OH (hydroxyl group) bound to a metal atom. Hydroxides of typical metals are basic; those of *metalloids are amphoteric.

**hydroxoacid** A type of acid in which the acidic hydrogen is on a hydroxyl group attached to an atom that is not attached to an oxo (=O) group. An example is

$$Si(OH)_4 + H_2O \rightarrow Si(OH)_3(O)^- + H_3O^+.$$

*Compare* OXOACID.

**hydroxonium ion** *See* OXONIUM ION.

**4-hydroxybutanoic acid (gammahydroxybutyric acid; GHB)** A naturally occurring carboxylic acid, $HO(CH_2)_3COOH$, found in small amounts in most living things. It is used as a therapeutic drug to treat insomnia, depression, and alcoholism. GHB, as it is commonly known, is also used as an illegal club drug and as a date-rape drug.

**hydroxycerussite** *See* LEAD(II) CARBONATE HYDROXIDE.

**hydroxylamine** A colourless solid, $NH_2OH$, m.p. 33°C. It explodes on heating and may be employed as an oxidizing agent or reducing agent. Hydroxylamine is made by the reduction of nitrates or nitrites, and is used in the manufacture of nylon. With aldehydes and ketones it forms *oximes.

**(((⊕))) SEE WEB LINKS**
• Information about IUPAC nomenclature

**hydroxyl group** The group –OH in a chemical compound.

**2-hydroxypropanoic acid** *See* LACTIC ACID.

**5-hydroxytryptamine** *See* SEROTONIN.

**hygrometer** An instrument for measuring *humidity in the atmosphere. The mechanical type uses an organic material, such as human hair, which expands and contracts with changes in atmospheric humidity. The expansion and contraction is used to operate a needle. In the electric type, the change in resistance of a hygroscopic substance is used as an indication of humidity. In **dew-point hygrometers** a polished surface is reduced in temperature until water vapour from the atmosphere forms on it. The temperature of this dew point enables the relative humidity of the atmosphere to be calculated. In the **wet-and-dry bulb hygrometer**, two thermometers are mounted side by side, the bulb of one being surrounded by moistened muslin. The thermometer with the wet bulb will register a lower temperature than that with a dry bulb owing to the cooling effect of the evaporating water. The temperature difference enables the relative humidity to be calculated. Only the dew-point hygrometer can be operated as an absolute instrument; all the others must ultimately be calibrated against this.

**hygroscopic** Describing a substance that can take up water from the atmosphere. *See also* DELIQUESCENCE.

**hymen** A fold of mucous membrane that covers the opening of the vagina at birth. It normally perforates at puberty, to allow the flow of menstrual blood, but if the opening is small it may be ruptured during the first occasion of sexual intercourse.

**Hymenoptera** An order of insects that includes the ants, bees, wasps, ichneumon flies, and sawflies. Hymenopterans generally have a narrow waist between thorax and abdomen. The smaller hindwings are interlocked with the larger forewings by a row of tiny hooks on the leading edges of the hindwings. Some species are wingless. The mouthparts are typically adapted for biting, although some advanced forms (e.g. bees) possess a tubelike proboscis for sucking liquid food, such as nectar. The long slender *ovipositor can serve for sawing, piercing, or stinging. Metamorphosis occurs via a pupal stage to the adult form. *Parthenogenesis is common in the group.

Ants and some bees and wasps live in colonies, often comprising numerous individuals divided into *castes and organized into a coordinated and complex society. The colony of the honeybee (*Apis mellifera*), for example, consists of workers (sterile females), *drones (fertile males), and usually a single fertile female – the queen. The sole concern of the queen is egg laying. She determines the gender of the egg by either withholding or releasing stored sperm. Unfertilized eggs become males; fertilized eggs become females. The workers fulfil a variety of tasks, including nursing the developing larvae, building the wax cells (combs) of the hive, guarding the colony, and foraging for nectar and pollen. The single function of the

larger drones is to mate with the young queen on her nuptial flight.

**hyoid arch** The second of seven bony V-shaped arches that support the gills of primitive vertebrates. In advanced vertebrates, part of the hyoid arch has evolved to form the stapes (one of the *ear ossicles). The rest of it forms the **hyoid bone**, which supports the tongue.

**hyper-** A prefix denoting over, above, high; e.g. hypersonic, hyperpolarization, hypertonic.

**hyperbola** A *conic with eccentricity $e > 1$. It has two branches (see graph). For a hyperbola centred at the origin, the **transverse axis** runs along the $x$-axis between the vertices and has length $2a$. The **conjugate axis** runs along the $y$-axis and has length $2b$. There are two **foci** on the $x$-axis at $(ae, 0)$ and $(-ae, 0)$. The **latus rectum**, the chords through the foci perpendicular to the transverse axis, have length $2b^2/a$. The equation of the hyperbola is:

$$x^2/a^2 - y^2/b^2 = 1,$$

and the asymptotes are $y = \pm bx/a$.

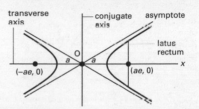

**Hyperbola.**

**hyperbolic functions** A set of functions, **sinh**, **cosh**, and **tanh**, that have similar properties to *trigonometric functions but are related to the hyperbola in the manner that trigonometric functions are related to the circle. The hyperbolic sine (sinh) of the angle $x$ is defined by:

$$\sinh x = \tfrac{1}{2}(e^x - e^{-x}).$$

Similarly,

$$\cosh x = \tfrac{1}{2}(e^x + e^{-x})$$

$$\tanh x = (e^x - e^{-x})/(e^x + e^{-x})$$

Hyperbolic secant (sech), cosecant (cosech), and cotangent (coth) are the reciprocals of cosh, sinh, and tanh, respectively.

**hypercharge** A quantized property of *baryons (*see* ELEMENTARY PARTICLES) that provides a formal method of accounting for the nonoccurrence of certain expected decays by means of the strong interaction (*see* FUNDAMENTAL INTERACTIONS). Hypercharge is in some respects analogous to electric charge but it is not conserved in weak interactions. Nucleons have a hypercharge of +1, and the *pion has a value of 0.

**hyperfine structure** *See* FINE STRUCTURE.

**hypermetropia (hyperopia)** Long-sightedness, in which the lens of the eye is unable to accommodate sufficiently to throw the image of near objects onto the retina. It is caused usually by shortness of the eyeball rather than any fault in the lens system. Spectacles with converging lenses are required to focus the image onto the surface of the retina.

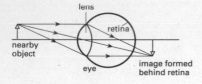

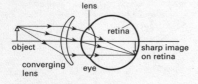

**Hypermetropia.**

**hypernetted chain approximation** An approximation used in the theory of fluids relating functions that give the correlation between two particles to radial distribution functions.

**hypernova** An explosive event in which a large star collapses with the formation of a *black hole. An event of this type would be even more violent than a *supernova (in which a star collapses to form a neutron star). Hypernovae may be the cause of gamma-ray bursts.

**hyperon** A shortlived *elementary particle; it is classified as a *baryon and has a nonzero *strangeness.

**hyperplasia** Increase in the size of a tissue or organ due to an increase in the number of its component cells. *Compare* HYPERTROPHY.

**hypersensitivity** Increased or abnormal sensitivity to compounds, which can elicit a specific immune response accompanied by tissue damage. Hypersensitivity reactions include *allergies and *anaphylaxis.

**hypersonic** Denoting a velocity in excess of Mach 5 (*see* MACH NUMBER). **Hypersonic flight** is flight at hypersonic speeds in the earth's atmosphere.

**hypertension** *See* BLOOD PRESSURE.

**hypertext** A technique by which textual documents can be created and viewed on a computer screen so that one or more documents can be browsed in any order by the selection of key words or phrases by the user. The selected text leads (by underlying searches through associated files, indexes, etc.) to the display of another part of the document, or of some other document. **Hypermedia** is an extension of this technique enabling links to be made between text, images, sounds, etc. *See also* WORLD WIDE WEB.

**hypertext markup language** *See* HTML.

**hypertonic solution** A solution that has a higher osmotic pressure than some other solution. *Compare* HYPOTONIC SOLUTION.

**hypertrophy** An increase in the size of a tissue or organ due to an increase in the size of its component cells. Hypertrophy often occurs in response to an increased workload in an organ, which may result from malfunction or disease. *Compare* HYPERPLASIA.

**hyperventilation** An increase in the amount of air taken into the lungs caused by an increase in the depth or rate of breathing. *See also* VENTILATION.

**hypha** A delicate filament in fungi many of which may form either a loose network (*mycelium) or a tightly packed interwoven mass of *pseudoparenchyma, as in the fruiting body of mushrooms. Hyphae may be branched or unbranched and may or may not possess cross walls. The cell wall consists of either fungal cellulose or *chitin. The cell wall is lined with cytoplasm, which often contains oil globules and glycogen, and there is a central vacuole. The hyphae produce enzymes that in parasitic fungi digest the host tissue, and in saprotrophic fungi digest dead organic matter.

**hypo-** A prefix denoting under, below, low; e.g. hypogyny, hyponasty, hypotonic.

**hypochlorite** *See* CHLORATES.

**hypochlorous acid** *See* CHLORIC(I) ACID.

**hypocotyl** The region of the stem beneath the stalks of the seed leaves (*cotyledons) and directly above the young root of an embryo plant. It grows rapidly in seedlings showing *epigeal germination and lifts the cotyledons above the soil surface. In this region (the **transition zone**) the arrangement of vascular bundles in the root changes to that of the stem. *Compare* EPICOTYL.

**hypodermis (exodermis)** The outermost layer of cells in the plant *cortex, lying immediately below the epidermis. These cells are sometimes modified to give additional structural support or to store food materials or water. After the loss of the *piliferous layer of the root the hypodermis takes over the protective functions of the epidermis.

**hypogeal** 1. Describing seed germination in which the seed leaves (cotyledons) remain below ground. Examples of hypogeal germination are seen in oak and runner bean. *Compare* EPIGEAL. 2. Describing fruiting bodies that develop underground, such as truffles and peanuts.

**hypolimnion** The lower layer of water in a lake. *See* THERMOCLINE. *Compare* EPILIMNION.

**hypophosphorus acid** *See* PHOSPHINIC ACID.

**hypophysis** *See* PITUITARY GLAND.

**hyposulphite** *See* SULPHINATE.

**hyposulphurous acid** *See* SULPHINIC ACID.

**hypotenuse** The longest side of a right-angle triangle.

**hypothalamus** Part of the vertebrate brain that is derived from the *forebrain and located on the ventral surface below the *thalamus and the *cerebrum. The hypothalamus regulates a wide variety of physiological processes, including maintenance of body temperature, water balance, sleeping, and feeding, via both the *autonomic nervous system (which it controls) and the *neuroendocrine system. Its endocrine functions are largely mediated by the *pituitary gland. The pituitary responds to releasing hormones produced by the hypothalamus, which in this way indirectly controls hormone production in other glands.

**hypothesis** *See* LAWS, THEORIES, AND HYPOTHESES.

**hypothesis test** In *statistics, a method of assessing how plausible a null hypothesis or statement is by comparing it with a test sample. For example, a null hypothesis will be accepted in a test carried out at the 10% significance level if the value of the sample is what can be expected from 90% of all random samples.

**hypotonic solution** A solution that has a lower osmotic pressure than some other solution. *Compare* HYPERTONIC SOLUTION.

**hypsometer** A device for calibrating thermometers at the boiling point of water. As the boiling point depends on the atmospheric pressure, which in turn depends on the height above sea level, the apparatus can be used to measure height above sea level.

**hysteresis** A phenomenon in which two physical quantities are related in a manner that depends on whether one is increasing or decreasing in relation to the other. The repeated measurement of *stress against *strain, with the stress first increasing and then decreasing, will produce for some specimens a graph that has the shape of a closed loop. This is known as a **hysteresis cycle**. The most familiar hysteresis cycle, however, is produced by plotting the magnetic flux density (*B*) within a ferromagnetic material against the applied magnetic field strength (*H*).

If the material is initially unmagnetized at O it will reach saturation at P as *H* is increased. As the field is reduced and again increased the loop PQRSTP is formed (see graph). The area of this loop is proportional to the energy loss (**hysteresis loss**) occurring

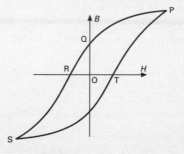

**Hysteresis.**

during the cycle. The value of $B$ equal to OQ is called the **remanance** (or retentivity) and is the magnetic flux density remaining in the material after the saturating field has been reduced to zero. This is a measure of the tendency of the magnetic domain patterns (*see* MAGNETISM) to remain distorted even after

the distorting field has been removed. The value of $H$ equal to OR is called the **coercive force** (or coercivity) and is the field strength required to reduce the remaining flux density to zero. It is a measure of the difficulty of restoring the symmetry of the domain patterns.

**IAA (indoleacetic acid)** *See* AUXIN.

**ice** *See* WATER.

**ice age** A period in the earth's history during which ice advanced towards the equator and a general lowering of temperatures occurred. The last major ice age, that of the Pleistocene epoch (sometimes known as the **Ice Age**), ended about 10 000 years ago. At least four major ice advances (glacials) occurred during the Pleistocene; these were separated by interglacials during which the ice retreated and temperatures rose. At present it is not known if the earth is between ice ages or is in an interglacial of the Pleistocene Ice Age. It has been established that ice ages also occurred during the Precambrian (over 500 million years ago) and during the Permo-Carboniferous (about 250 million years ago).

**ice point** The temperature at which there is equilibrium between ice and water at standard atmospheric pressure (i.e. the freezing or melting point under standard conditions). It was used as a fixed point (0°) on the Celsius scale, but the kelvin and the International Practical Temperature Scale are based on the *triple point of water.

**ichthyosaur** Any extinct aquatic reptile belonging to the order Ichthyosauria, which lived during the Mesozoic era (250–90 million years ago). The earliest forms resembled lizards with flippers and probably swam in an eel-like fashion. Their successors evolved to become progressively more fishlike, with thickened bodies, finlike fore and hind limbs, and crescent-shaped tails. Fossilized skeletal remains show that the eyes were unusually large and adapted for hunting in poor visibility. Ichthyosaurs typically lived on a diet of squid, and some species attained lengths of over 15 m.

**iconoscope** A form of television camera tube (*see* CAMERA) in which the beam of light from the scene is focused on to a thin mica plate. One side of the plate is faced with a thin metallic electrode, the other side being coated with a mosaic of small globules of a photoemissive material. The light beam falling on the mosaic causes photoemission of electrons, creating a pattern of positive charges in what is effectively an array of tiny capacitors. A high-velocity electron beam scans the mosaic, discharging each capacitor in turn through the metallic electrode. The resulting current is fed to amplification circuits, the current from a particular section of the mosaic depending on the illumination it has received. In this way the optical information in the light beam is converted into an electrical signal.

**ICSH** *See* LUTEINIZING HORMONE.

**ideal crystal** A single crystal with a perfectly regular lattice that contains no impurities, imperfections, or other defects.

**ideal gas (perfect gas)** A hypothetical gas that obeys the *gas laws exactly. An ideal gas would consist of molecules that occupy negligible space and have negligible forces between them. All collisions made between molecules and the walls of the container or between molecules and other gas molecules would be perfectly elastic, because the molecules would have no means of storing energy except as translational kinetic energy.

**ideal solution** *See* RAOULT'S LAW.

**identical twins (monozygotic twins)** Two individuals that develop from a single fertilized egg cell by its division into two genetically identical parts. Each part eventually gives rise to a separate individual and these twins have identical DNA sequences. However, differences in patterns of methylation and acetylation of their DNA can result in differences in gene expression, which may explain observed differences in looks, etc. *Compare* FRATERNAL TWINS.

**identity** Symbol ≡. A statement of equality that applies for all values of the unknown quantity. For example, $5y \equiv 2y + 3y$.

**idiogram** *See* KARYOGRAM.

**idiosyncrasy** An abnormal reaction to a drug or other foreign substance shown by an individual, which is usually genetically determined. An individual that shows immunological idiosyncrasy is said to be **hypersensitive** to a particular substance, agent, etc.

**IE** Ionization energy. *See* IONIZATION POTENTIAL.

**Ig** *See* IMMUNOGLOBULIN.

**IGF (insulin-like growth factor)** *See* GROWTH FACTOR.

**igneous rocks** A group of rocks formed from the crystallization of magma (molten silicate liquid). Igneous rocks form one of the three major rock categories (*see also* METAMORPHIC ROCKS; SEDIMENTARY ROCKS). According to the depth at which the magma solidifies, igneous rocks may be classified as plutonic, hypabyssal, or volcanic. Plutonic rocks solidify slowly at great depth, typically forming large intrusive masses (e.g. batholiths and stocks), and generally have the coarsest texture. Examples of plutonic rocks are granite, syenite, diorite, and gabbro. Volcanic (extrusive) rocks are formed from magma that has been poured out onto the earth's surface; these rocks (e.g. basalt, andesite) are characteristically fine-grained or glassy as a result of their rapid cooling. Hypabyssal rocks (e.g. diorite), which cool at shallower depths than the plutonic, are intermediate in character and medium-grained. They commonly occur in the form of small intrusions, such as dykes and sills. Igneous rocks may also be classified chemically according to their silica content as acidic (over 66% silica), intermediate (55–66%), basic (45–55%), or ultrabasic (under 45%).

**ignition temperature 1.** The temperature to which a substance must be heated before it will burn in air. **2.** The temperature to which a *plasma has to be raised in order that nuclear fusion will occur.

**ileum** The portion of the mammalian *small intestine that follows the *jejunum and precedes the *large intestine. It is a site of digestion and absorption. The internal lining of the ileum bears numerous small outgrowths (*see* VILLUS), which increase its absorptive surface area.

**ilium** The largest of the three bones that make up each half of the *pelvic girdle. The ilium bears a flattened wing of bone that is attached by ligaments to the sacrum (*see* SACRAL VERTEBRAE). *See also* ISCHIUM; PUBIS.

**illuminance (illumination)** Symbol *E*. The energy in the form of visible radiation reaching a surface per unit area in unit time; i.e. the luminous flux per unit time. It is measured in *lux (lumens per square metre).

**image** A representation of a physical object formed by a lens, mirror, or other optical instrument. If the rays of light actually pass through the image, it is called a **real image**. If a screen is placed in the plane of a real image it will generally become visible. If the image is seen at a point from which the rays appear to come to the observer, but do not actually do so, the image is called a **virtual image**. No image will be formed on a screen placed at this point. Images may be **upright** or **inverted** and they may be **magnified** or **diminished**.

**image converter** An electronic device in which an image formed by invisible radiation (usually gamma rays, X-rays, ultraviolet, or infrared) is converted into a visible image. Commonly the invisible radiation is focused on to a photocathode, which emits electrons when it is exposed to the radiation. These electrons fall on a fluorescent anode screen, after acceleration and focusing by a system of electron lenses. The fluorescent screen produces a visible image. The device is used in fluoroscopes, infrared telescopes, ultraviolet microscopes, and other devices.

**imaginary number** A number that is a multiple of $\sqrt{-1}$, which is denoted by i; for example $\sqrt{-3} = i\sqrt{3}$. *See also* COMPLEX NUMBER.

**imago** The adult sexually mature stage in the life cycle of an insect after metamorphosis.

**imbibition** The uptake of water by substances that do not dissolve in water, so that the process results in swelling of the substance. Imbibition is a property of many biological substances, including cellulose (and other constituents of plant cell walls), starch, and some proteins. It occurs in dry seeds before they germinate and – together with osmosis – is responsible for the uptake of water by growing plant cells.

**imides** Organic compounds containing the group –CO.NH.CO.– (the **imido group**).

**SEE WEB LINKS**
• Information about IUPAC nomenclature

**imido group** *See* IMIDES.

**imines** Compounds containing the group –NH– in which the nitrogen atom is part of a ring structure, or the group =NH, in which the nitrogen atom is linked to a carbon atom by a double bond. In either case, the group is referred to as an **imino group**.

**(⊕) SEE WEB LINKS**
• Information about IUPAC nomenclature

**imino group** *See* IMINES.

**immersion objective** An optical microscope objective in which the front surface of the lens is immersed in a liquid on the cover slip of the microscope specimen slide. Cedar-wood oil (for an **oil-immersion lens**) or sugar solution is frequently used. It has the same refractive index as the glass of the cover slip, so that the object is effectively immersed in it. The presence of the liquid increases the effective aperture of the objective, thus increasing the resolution.

**immune response** The reaction of the body to foreign or potentially dangerous substances (*antigens), particularly disease-producing microorganisms. *See* IMMUNITY.

**immunity** The ability of an animal to resist infection or to counter the harmful effects of toxins produced by infecting organisms. Immunity depends on the presence in the body of a range of defensive cells and substances, notably *antibodies and white blood cells (*lymphocytes), which produce an immune response. **Innate** (**inherited** or **natural**) **immunity** is the body's first line of defence and plays a vital role in controlling invading organisms during the early stages of an infection. Phagocytic macrophages can engulf bacteria and other microorganisms and induce them to secrete an array of *cytokines. These substances attract additional immune cells to the site and initiate the process of *inflammation. The *complement system of defensive proteins is also activated. The innate response plays a crucial role in promoting **adaptive** (or **acquired**) **immunity**. Dendritic cells ingest foreign material, such as bacteria and virus particles (*see* ANTIGEN), and present it to T cells as part of the immune response; adaptive immunity can persist throughout the lifetime of the individual. **Active immunity** arises when the body produces antibodies against an invading foreign substance (antigen), either through infection or *immunization. **Humoral immunity** is when B lymphocytes produce free antibodies that circulate in the bloodstream (*see* B CELL); **cell-mediated immunity** is caused by the action of T lymphocytes (*see* T CELL).

**Passive immunity** is induced by injection of serum taken from an individual already immune to a particular antigen; it can also be acquired by the transfer of maternal antibodies to offspring via the placenta or breast milk (*see* COLOSTRUM). Active immunity tends to be long-lasting; passive immunity is short-lived. *See also* AUTOIMMUNITY.

**(⊕) SEE WEB LINKS**
• A series of colour slides illustrating the working of the immune system, produced by the US National Cancer Institute

**immunization** The production of *immunity in an individual by artificial means. **Active immunization** (**vaccination**) involves the introduction, either orally or by injection (**inoculation**), of specially treated bacteria, viruses, or their toxins to stimulate the production of *antibodies (*see* VACCINE). **Passive immunization** is induced by the injection of preformed antibodies.

**immunoassay** Any of various techniques that measure the amount of a particular substance by virtue of its binding antigenically to a specific antibody. In **solid-phase immunoassay** the specific antibody is attached to a solid supporting medium, such as a PVC sheet. The sample is added and any test antigens will bind to the antibody. A second antibody, specific for a different site on the antigen, is added. This carries a radioactive or fluorescent label, enabling its concentration, and thus that of the test antigen, to be determined by comparison with known standards. Variations on this technique include *ELISA and *Western blotting.

**immunoglobulin (Ig)** One of a group of proteins (*globulins) in the body that act as *antibodies. They are produced by specialized white blood cells called *B cells and are present in blood serum and other body fluids. There are several classes (e.g. IgE, IgG, and IgM) having different functions.

**immunosuppression** The suppression of an *immune response. Immunosuppression is necessary following organ transplants in order to prevent the host rejecting the grafted organ (*see* GRAFT); it is artificially induced by radiation or chemical agents that inhibit cell division of *lymphocytes. Immunosuppression occurs naturally in certain diseases, notably *AIDS.

**impact printer** *See* PRINTER.

**impedance** Symbol $Z$. The quantity that

measures the opposition of a circuit to the passage of a current and therefore determines the amplitude of the current. In a d.c. circuit this is the resistance ($R$) alone. In an a.c. circuit, however, the *reactance ($X$) also has to be taken into account, according to the equation: $Z^2 = R^2 + X^2$, where $Z$ is the impedance. The **complex impedance** is given by $Z = R + iX$, where $i = \sqrt{-1}$. The real part of the complex impedance, the resistance, represents the loss of power according to *Joule's law. The ratio of the imaginary part, the reactance, to the real part is an indication of the difference in phase between the voltage and the current.

**imperfect fungi**  *See* DEUTEROMYCOTA.

**Imperial units**  The British system of units based on the pound and the yard. The former f.p.s. system was used in engineering and was loosely based on Imperial units; for all scientific purposes *SI units are now used. Imperial units are also being replaced for general purposes by metric units.

**implant**  Any substance, device, or tissue that is inserted into the body. For example, drug implants and heart pacemakers are typically inserted under the skin.

**implantation (nidation)**  (in embryology) The embedding of a fertilized mammalian egg into the wall of the uterus (womb) where it will continue its development. After fertilization in the fallopian tube the egg passes into the womb in the form of a ball of cells (**blastocyst**). Its outer cells destroy cells of the uterine wall, forming a cavity into which the blastocyst sinks.

**implosion**  An inward collapse of a vessel, especially as a result of evacuation.

**imprinting**  1. (in behaviour) A specialized form of learning in which young animals, during a particularly sensitive period in their early development, learn to recognize and approach some large moving object nearby. In nature this is usually the mother, though simple models or individuals of a different species (including humans) may suffice. Imprinting was first described by Konrad *Lorenz, working with young ducks and geese. *See* LEARNING (Feature). 2. (in genetics) *See* GENE IMPRINTING.

**impulse**  1. (in physiology) (**nerve impulse**) The signal that travels along the length of a *nerve fibre and is the means by which information is transmitted through the nervous system. It is marked by the flow of ions across the membrane of the *axon caused by changes in the permeability of the membrane, producing a reduction in potential difference that can be detected as the *action potential. The strength of the impulse produced in any nerve fibre is constant (*see* ALL-OR-NONE RESPONSE). 2. (in physics) Symbol $J$. The product of a force $F$ and the time $t$ for which it acts. If the force is variable, the impulse is the integral of $F dt$ from $t_0$ to $t_1$. The impulse of a force acting for a given time interval is equal to the change in momentum produced over that interval, i.e. $J = m(v_1 - v_0)$, assuming that the mass ($m$) remains constant while the velocity changes from $v_0$ to $v_1$.

**IMS**  *See* ION-MOBILITY SPECTROMETRY.

**inbreeding**  Mating between closely related individuals, the extreme condition being self-fertilization, which occurs in many plants and some primitive animals. A population of inbreeding individuals generally shows less variation than an *outbreeding population. Continued inbreeding among a normally outbreeding population leads to **inbreeding depression** (the opposite of *hybrid vigour) and an increased incidence of harmful characteristics. For example, in humans, certain mental and other defects tend to occur more often in families with a history of cousin marriages.

**incandescence**  The emission of light by a substance as a result of raising it to a high temperature. An **incandescent lamp** is one in which light is emitted by an electrically heated filament. *See* ELECTRIC LIGHTING.

**incisor**  A sharp flattened chisel-shaped *tooth in mammals that is adapted for biting food and – in rodents – for gnawing. In humans there are normally two pairs of incisors (central and lateral) in each jaw. *See* PERMANENT TEETH.

**inclination**  1. *See* GEOMAGNETISM. 2. The angle between the orbital plane of a planet, satellite, or comet and the plane of the earth's *ecliptic.

**inclusive fitness**  The quality that organisms attempt (unconsciously) to maximize as the result of natural selection acting on genes that are influential in controlling their behaviour and physiology. It includes the individual's own **reproductive success** (usually taken as the number of its offspring that survive to adulthood) and also the effects of the individual's actions on the reproductive success of its relatives, because relatives have a higher probability of sharing some identical genes with the individual than do other members of the population. When interactions between relatives are likely to occur (which happens during the lives of many animals and plants) *kin selection will operate.

**incoherent scattering** *See* COHERENT SCATTERING.

**incompatibility** 1. The condition that exists when foreign grafts or blood transfusions evoke a marked *immune response and are rejected. 2. The phenomenon in which pollen from one flower fails to fertilize other flowers on the same plant (**self-incompatibility**) or on other genetically similar plants. This genetically determined mechanism prevents self-fertilization (breeding between likes) and promotes cross-fertilization (breeding between individuals with different genetic compositions). *See also* ALLOGAMY; FERTILIZATION; POLLINATION.

**incomplete dominance** The condition that arises when neither *allele controlling a characteristic is dominant and the aspect displayed by the organism results from the partial influence of both alleles. For example, a snapdragon plant with alleles for red and for white flowers produces pink flowers. *Compare* CODOMINANCE.

**incubation** 1. The process of maintaining the fertilized eggs of birds and of some reptiles and egg-laying mammals at the optimum temperature for the successful development of the embryos. A period of incubation follows the laying of the eggs and precedes their hatching. 2. The process of maintaining a *culture of bacteria or other microorganisms at the optimum temperature for growth of the culture. 3. The phase in the development of an infectious disease between initial infection and the appearance of the first symptoms.

**incus (anvil)** The middle of the three *ear ossicles of the mammalian middle ear.

**indefinite inflorescence** *See* RACEMOSE INFLORESCENCE.

**indefinite integral** *See* INTEGRATION.

**indehiscent** Describing a fruit or fruiting body that does not open to release its seeds or spores when ripe. Instead, release occurs

when the fruit wall decays or, if eaten by an animal, is digested. *Compare* DEHISCENCE.

**indene** A colourless flammable hydrocarbon, $C_9H_8$; r.d. 0.996; m.p. $-1.8°C$; b.p. $182.6°C$. Indene is an aromatic hydrocarbon with a five-membered ring fused to a benzene ring. It is present in coal tar and is used as a solvent and raw material for making other organic compounds.

**independent assortment** The separation of the alleles of one gene into the reproductive cells (gametes) independently of the way in which the alleles of other genes have segregated. By this process all possible combinations of alleles should occur equally frequently in the gametes. In practice this does not happen because alleles situated on the same chromosome tend to be inherited together. However, if the allele pairs $Aa$ and $Bb$ are on different chromosomes, the combinations $AB$, $Ab$, $aB$, and $ab$ will normally be equally likely to occur in the gametes. *See* MEIOSIS; MENDEL'S LAWS.

**indeterminacy** *See* UNCERTAINTY PRINCIPLE.

**index fossil (zone fossil)** An animal *fossil of a group that existed continuously during a particular span of geological time and can therefore be used to date the rock in which it is found. Index fossils are found chiefly in sedimentary rocks. They are an essential tool in stratigraphy for comparing the geological ages of sedimentary rock formations. Examples are *ammonites and *graptolites.

**indicator** A substance used to show the presence of a chemical substance or ion by its colour. **Acid–base indicators** are compounds, such as phenolphthalein and methyl orange, that change colour reversibly, depending on whether the solution is acidic or basic. They are usually weak acids in which the un-ionized form HA has a different colour from the negative ion $A^-$. In solution the indicator dissociates slightly

$$HA \rightleftharpoons H^+ + A^-$$

In acid solution the concentration of $H^+$ is high, and the indicator is largely undissociated HA; in alkaline solutions the equilibrium is displaced to the right and $A^-$ is formed. Useful acid–base indicators show a sharp colour change over a range of about 2 pH units. In titration, the point at which the reaction is complete is the **equivalence point** (i.e. the point at which equivalent quantities of acid and base are added). The **end point** is the point at which the indicator just changes colour. For accuracy, the two must be the same. During a titration the pH changes sharply close to the equivalence point, and the indicator used must change colour over the same range.

Other types of indicator can be used for other reactions. Starch, for example, is used in iodine titrations because of the deep blue complex it forms. **Oxidation–reduction indicators** are substances that show a reversible colour change between oxidized and reduced forms. *See also* ADSORPTION INDICATOR.

**indicator species** A plant or animal species that is very sensitive to a particular environmental factor, so that its presence (or absence) in an area can provide information about the levels of that factor. For example, some lichens are very sensitive to the concentration of sulphur dioxide (a major pollutant) in the atmosphere. Examination of the lichens present in an area can provide a good indication of the prevailing levels of sulphur dioxide.

**indigenous** Describing a species that occurs naturally in a certain area, as distinct from one introduced by humans; native.

**indigo** A blue vat dye, $C_{16}H_{10}N_2O_2$. It occurs as the glucoside **indican** in the leaves of plants of the genus *Indigofera*, from which it was formerly extracted. It is now made synthetically.

**indium** Symbol In. A soft silvery element belonging to group 13 (formerly IIIB) of the periodic table; a.n. 49; r.a.m. 114.82; r.d. 7.31 (20°C); m.p. 156.6°C; b.p. 2080±2°C. It occurs in zinc blende and some iron ores and is obtained from zinc flue dust in total quantities of about 40 tonnes per annum. Naturally occurring indium consists of 4.23% indium–113 (stable) and 95.77% indium–115 (half-life $6 \times 10^{14}$ years). There are a further five short-lived radioisotopes. The uses of the metal are small – some special-purpose electroplates and some special fusible alloys. Several semiconductor compounds are used, such as InAs, InP, and InSb. With only three electrons in its valency shell, indium is an electron acceptor and is used to dope pure germanium and silicon; it forms stable indium(I), indium(II), and indium(III) compounds. The element was discovered in 1863 by Ferdinand Reich (1799–1882) and Hieronymus Richter (1824–90).

**SEE WEB LINKS**
• Information from the WebElements site

**indole** A yellow solid, $C_8H_7N$, m.p. 52°C. Its molecules consist of a benzene ring fused to a nitrogen-containing five-membered ring. It occurs in some plants and in coal tar, and is produced in faeces by bacterial action. It is used in making perfumes. Indole has the nitrogen atom positioned next to the fused benzene ring. An isomer with the nitrogen two atoms away from the fused ring is called **isoindole**.

**indoleacetic acid (IAA)** *See* AUXIN.

**induced emission (stimulated emission)** The emission of a photon by an atom in the presence of *electromagnetic radiation. The atom can become excited by the absorption of a photon of the right energy and, having become excited, the atom can emit a photon. The rate of absorption is equal to the rate of induced emission, both rates being proportional to the density of photons of the electromagnetic radiation. The relation between induced emission and *spontaneous emission is given by the *Einstein coefficients. The process of induced emission is essential for the operation of *lasers and *masers. *See also* QUANTUM THEORY OF RADIATION.

**induced fission** *See* NUCLEAR FISSION.

**inducer** *See* OPERON.

**inductance** The property of an electric circuit or component that causes an e.m.f. to be generated in it as a result of a change in the current flowing through the circuit (**self-inductance**) or of a change in the current flowing through a neighbouring circuit with which it is magnetically linked (**mutual inductance**). In both cases the changing current is associated with a changing magnetic field, the linkage with which in turn induces the e.m.f. In the case of self inductance, $L$, the e.m.f., $E$, generated is given by $E = -L.dI/dt$, where $I$ is the instantaneous current and the minus sign indicates that the e.m.f. induced is in opposition to the change of current. In the case of mutual inductance, $M$, the e.m.f., $E_1$, induced in one circuit is given by $E_1 = -M.dI_2/dt$, where $I_2$ is the instantaneous current in the other circuit.

**induction** **1.** (in embryology) The ability of natural stimuli to cause unspecialized embryonic tissue to develop into specialized tissue. **2.** (in obstetrics) The initiation of childbirth by artificial means; for example, by injection of the hormone *oxytocin. **3.** (in physics) A change in the state of a body produced by a field. *See* ELECTROMAGNETIC INDUCTION; INDUCTANCE.

**induction coil** A type of *transformer used to produce a high-voltage alternating current or pulses of high-voltage current from a low-voltage direct-current source. The induction coil is widely used in spark-ignition *internal-combustion engines to produce the spark in the sparking plugs. In such an engine the battery is connected to the primary winding of the coil through a circuit-breaking device driven by the engine and the e.m.f. generated in the secondary winding of the coil is led to the sparking plugs through the distributor. The primary coil consists of relatively few turns, whereas the secondary consists of many turns of fine wire.

**induction heating** The heating of an electrically conducting material by *eddy currents induced by a varying electromagnetic field. Induction heating may be an undesirable effect leading to power loss in transformers and other electrical devices. It is, however, useful for melting and heat-treating and in forging and rolling metals, as well as for welding, brazing, and soldering. The material to be heated is inserted into a coil through which an alternating current flows and acts as the short-circuited secondary of a *transformer. Eddy currents induced in the material within the coil cause the temperature of the material to rise.

**induction motor** *See* ELECTRIC MOTOR.

**inductive effect** The effect of a group or atom of a compound in pulling electrons towards itself or in pushing them away. Inductive effects can be used to explain some aspects of organic reactions. For instance, electron-withdrawing groups, such as $-NO_2$, $-CN$, $-CHO$, $-COOH$, and the halogens substituted on a benzene ring, reduce the electron density on the ring and decrease its susceptibility to further (electrophilic) substitution. Electron-releasing groups, such as $-OH$, $-NH_2$, $-OCH_3$, and $-CH_3$, have the opposite effect.

**indusium** The kidney-shaped covering of the *sorus of certain ferns that protects the developing sporangia. It withers when the sorus ripens to expose the sporangia.

**industrial melanism** The increase of melanic (dark) forms of an animal in areas darkened by industrial pollution. The example most often quoted is that of the peppered moth (*Biston betularia*), melanic forms of which markedly increased in the industrial north of England during the 19th century. Experiments have shown that the dark forms increase in polluted regions because they are less easily seen by birds against a dark background; conversely the paler forms survive better in unpolluted areas.

**inelastic collision** A collision in which some of the kinetic energy of the colliding bodies is converted into internal energy in one body so that kinetic energy is not conserved. In collisions of macroscopic bodies some kinetic energy is turned into vibrational energy of the atoms, causing a heating effect. Collisions between molecules of a gas or liquid may also be inelastic as they cause changes in vibrational and rotational *energy levels. In nuclear physics, an inelastic collision is one in which the incoming particle causes the nucleus it strikes to become excited or to break up. **Deep inelastic scattering** is a method of probing the structure of subatomic particles in much the same way as Rutherford probed the inside of the atom (*see* RUTHERFORD SCATTERING). Such experiments were performed on protons in the late 1960s using high-energy electrons at the Stanford Linear Accelerator Centre (SLAC). As in Rutherford scattering, deep inelastic scattering of electrons by proton targets revealed that most of the incident electrons interacted very little and pass straight through, with only a small number bouncing back. This indicates that the charge in the proton is concentrated in small lumps, reminiscent of Rutherford's discovery that the positive charge in an atom is concentrated at the nucleus. However, in the case of the proton, the evidence suggested three distinct concentrations of charge and not one.

**inequality** A relationship between two quantities in which one of the quantities is not equal to (or not necessarily equal to) the other quantity. If the quantities are $a$ and $b$, two inequalities exist: $a$ is greater than $b$, written $a > b$, and $a$ is less than $b$, i.e. $a < b$. Similar statements can take the form: $a$ is greater than or equal to $b$, written $a \geq b$, and $a$ is less than or equal to $b$, which is denoted $a \leq b$. There are many applications of inequalities in physical science, an example being the Heisenberg *uncertainty principle.

**inert gases** *See* NOBLE GASES.

**inertia** The property of matter that causes it to resist any change in its motion. Thus, a body at rest remains at rest unless it is acted upon by an external force and a body in motion continues to move at constant speed in a straight line unless acted upon by an external force. This is a statement of Newton's first law of motion. The *mass of a body is a measure of its inertia. *See* MACH'S PRINCIPLE; INERTIAL FRAME.

**inertial frame** A *frame of reference in which bodies move in straight lines with constant speeds unless acted upon by external forces, i.e. a frame of reference in which free bodies are not accelerated. Newton's laws of motion are valid in an inertial system but not in a system that is itself accelerated with respect to such a frame.

**inertial mass** *See* MASS.

**inert-pair effect** An effect seen especially in groups 13 and 14 of the periodic table, in which the heavier elements in the group tend to form compounds with a valency two lower than the expected group valency. It is used to account for the existence of thallium(I) compounds in group 13 and lead(II) in group 14. In forming compounds, elements in these groups promote an electron from a filled *s*-level state to an empty *p*-level. The energy required for this is more than compensated for by the extra energy gain in forming two more bonds. For the heavier elements, the bond strengths or lattice energies in the compounds are lower than those of the lighter elements. Consequently the energy compensation is less important and the lower valence states become favoured.

**infection** The invasion of any living organism by disease-causing microorganisms (*see* PATHOGEN), which proceed to establish themselves, multiply, and produce various symptoms in their host. Pathogens may invade via a wound or (in animals) through the mucous membranes lining the alimentary, respiratory, and reproductive tracts, and may be transmitted by an infected individual, a *carrier, or an arthropod *vector. Symptoms in animals appear after an initial symptomless **incubation period** and typically consist of localized *inflammation, often with pain and fever. Infections are combatted by the body's natural defences

(*see* IMMUNITY). Treatment with drugs (*see* ANTIBIOTICS; ANTISEPTIC) is effective against most bacterial, fungal, and protozoan infections; some viral infections respond to *antiviral drugs. *See also* IMMUNIZATION.

**inferior** Describing a structure that is positioned below or lower than another structure in the body. For example, in flowering plants the ovary is described as inferior when it is located below the other organs of the flower. *Compare* SUPERIOR.

**infinite series** *See* SERIES.

**infinitesimal** Vanishingly small but not zero. Infinitesimal changes are notionally made in the *calculus, which is sometimes called the **infinitesimal calculus**.

**infinity** Symbol ∞. A quantity having a value that is greater than any assignable value. Minus infinity, −∞, is a quantity having a value that is less than any assignable value.

**inflammation** The defence reaction of tissue to injury, infection, or irritation by chemicals or physical agents. Activated macrophages and other cells in the affected tissue release various substances, including *histamine, *serotonin, *kinins, and *prostaglandins. These cause localized dilatation of blood vessels so that fluid leaks out and blood flow is increased. They also attract white blood cells (neutrophils and monocytes) to the site. Overall, these responses lead to swelling, redness, heat, and often pain. White blood cells, particularly *phagocytes, enter the tissue and an immune response is stimulated (*see* IMMUNITY). A gradual healing process usually follows.

**inflation** *See* EARLY UNIVERSE.

**inflationary universe** *See* EARLY UNIVERSE.

**inflection** A point on a curve at which the tangent changes from rotation in one direction to rotation in the opposite direction. If the curve $y = f(x)$ has a stationary point $dy/dx = 0$, there is either a maximum, minimum, or inflection at this point. If $d^2y/dx^2 = 0$, the stationary point is a point of inflection.

**inflorescence** A particular arrangement of flowers on a single main stalk of a plant. There are many different types of inflorescence, which are classified into two main groups depending on whether the tip of the flower axis goes on producing new flower buds during growth (*see* RACEMOSE INFLORES-

CENCE) or loses this ability (*see* CYMOSE INFLO-RESCENCE).

**information technology**  *See* IT.

**information theory**  The branch of mathematics that analyses information mathematically. Several branches of physics have been related to information theory. For example, an increase in *entropy has been expressed as a decrease in information. It has been suggested that it may be possible to express the basic laws of physics using information theory. *See also* LANDAUER'S PRINCIPLE; ZEILINGER'S PRINCIPLE.

**infradian rhythm**  *See* BIORHYTHM.

**infrared astronomy**  The study of radiation from space in the infrared region of the spectrum (*see* INFRARED RADIATION). Some infrared radiation is absorbed by water and carbon dioxide molecules in the atmosphere but there are several narrow atmospheric *windows in the near-infrared (1.15–1.3 μm, 1.5–1.75 μm, 2–2.4 μm, 3.4–4.2 μm, 4.6–4.8 μm, 8–13 μm, and 16–18 μm). Longer wavelength observations must be made from balloons, rockets, or satellites. Infrared sources are either thermal, i.e. emitted by the atoms or molecules of gases or dust particles in the temperature range 100–3000 K, or electronic, i.e. emitted by high-energy electrons interacting with magnetic fields as in *synchrotron radiation. Detectors are either modified reflecting *telescopes or solid-state photon detectors, usually incorporating photovoltaic devices (*see* PHOTOELECTRIC EFFECT).

**infrared radiation (IR)**  Electromagnetic radiation with wavelengths longer than that of red light but shorter than radiowaves, i.e. radiation in the wavelength range 0.7 micrometre to 1 millimetre. It was discovered in 1800 by William Herschel (1738–1822) in the sun's spectrum. The natural vibrational frequencies of atoms and molecules and the rotational frequencies of some gaseous molecules fall in the infrared region of the electromagnetic spectrum. The infrared absorption spectrum of a molecule is highly characteristic of it and the spectrum can therefore be used for molecular identification. Glass is opaque to infrared radiation of wavelength greater than 2 micrometres and other materials, such as germanium, quartz, and polyethene, have to be used to make lenses and prisms. Photographic film can be made sensitive to infrared up to about 1.2 μm.

**infrared spectroscopy (IR spectroscopy)**  A technique for chemical analysis and the determination of structure. It is based on the principles that molecular vibrations occur in the infrared region of the electromagnetic spectrum and functional groups have characteristic absorption frequencies. The frequencies of most interest range from 2.5 to 16 μm; however, in IR spectroscopy it is common to use the reciprocal of the wavelength, and thus this range becomes 4000–625 cm$^{-1}$. Examples of typical vibrations are centred on 2900 cm$^{-1}$ for C–H stretching in alkanes, 1600 cm$^{-1}$ for N–H stretching in amino groups, and 2200 cm$^{-1}$ for C≡C stretching in alkynes. In an IR spectrometer there is a source of IR light, covering the whole frequency range of the instrument, which is split into two beams of equal intensity. One beam is passed through the sample and the other is used as a reference against which the first is then compared. The spectrum is usually obtained as a chart showing absorption peaks, plotted against wavelength or frequency. The sample can be a gas, liquid, or solid.

**infrasound**  Soundlike waves with frequencies below the audible limit of about 20 hertz.

**ingestion (feeding)**  A method of *heterotrophic nutrition in which food is taken into an organism and subsequently digested (*see* DIGESTION). Ingestion is the principal mechanism of animal nutrition. *See also* MACROPHAGOUS; MICROPHAGOUS.

**inhalation**  *See* INSPIRATION.

**inheritance**  The transmission of particular characteristics from generation to generation by means of the *genetic code, which is transferred to offspring in the gametes. *See also* MENDEL'S LAWS.

**inhibition**  **1.** (in chemistry) A reduction in the rate of a catalysed reaction by substances called **inhibitors**. Inhibitors may work by poisoning catalysts for the reaction or by removing free radicals in a chain reaction. **Enzyme inhibition** affects biochemical reactions, in which the catalysts are *enzymes. **Competitive inhibition** occurs when the inhibitor molecules resemble the substrate molecules and bind to the *active site of the enzyme, so preventing normal enzymatic activity. Competitive inhibition can be reversed by increasing the concentration of the substrate. In **noncompetitive inhibition**

the inhibitor binds to a part of the enzyme or *enzyme–substrate complex other than the active site, known as an **allosteric site**. This deforms the active site so that the enzyme cannot catalyse the reaction. Noncompetitive inhibition cannot be reversed by increasing the concentration of the substrate. The toxic effects of many substances are produced in this way. Inhibition by reaction products (**feedback inhibition**) is important in the control of enzyme activity. *See also* ALLOSTERIC ENZYME. **2.** (in physiology) The prevention or reduction of the activity of effectors (such as muscles) by means of certain nerve impulses. Inhibitory activity often provides a balance to stimulation of a process; for example, the impulse to stimulate contraction of a voluntary muscle may be accompanied by an inhibitory impulse to prevent contraction of its antagonist.

**inhibitory postsynaptic potential (IPSP)** The electric potential that is generated in a postsynaptic neuron when an inhibitory neurotransmitter (such as gamma-aminobutyric acid) is released into the synapse and causes a slight increase in the potential difference across the postsynaptic membrane. This makes the neuron less likely to transmit an impulse. *Compare* EXCITATORY POSTSYNAPTIC POTENTIAL.

**initial** One of a group of cells (or, in lower plants, a single cell) that divides to produce the cells of a plant tissue or organ. The cells of the apical meristem, cambium, and cork cambium are initials.

**initiation codon** *See* START CODON.

**innate behaviour** An inherited pattern of behaviour that appears in a similar form in all normally reared individuals of the same sex and species. *See* INSTINCT.

**inner** Describing a chemical compound formed by reaction of one part of a molecule with another part of the same molecule. Thus, a lactam is an inner amide; a lactone is an inner ester.

**inner ear** The structure in vertebrates, surrounded by the temporal bone of the skull, that contains the organs of balance and hearing. It consists of soft hollow sensory structures (the **membranous labyrinth**), containing fluid (**endolymph**), surrounded by fluid (**perilymph**), and encased in a bony cavity (the **bony labyrinth**). It consists of two chambers, the *sacculus and *utriculus, from

which arise the *cochlea and *semicircular canals respectively.

**inner transition series** *See* TRANSITION ELEMENTS.

**innervation** The supply of nerve fibres to and from an organ.

**innominate artery** A short artery that branches from the aorta to divide into the *subclavian artery (the main artery to the arm) and the right *carotid artery (which supplies blood to the head).

**innominate bone** One of the two bones that form each half of the *pelvic girdle in adult vertebrates. This bone is formed by the fusion of the *ilium, *ischium, and *pubis.

**inoculation** **1.** *See* VACCINE. **2.** The placing of a small sample of microorganisms or any other type of cell into a *culture medium so that the cells can grow and proliferate.

**inoculum** A small amount of material containing bacteria, viruses, or other microorganisms that is used to start a culture.

**inorganic chemistry** The branch of chemistry concerned with compounds of elements other than carbon. Certain simple carbon compounds, such as $CO$, $CO_2$, $CS_2$, and carbonates and cyanides, are usually treated in inorganic chemistry.

**inositol** A cyclic alcohol, $C_6H_{12}O_6$, that is a constituent of certain cell phosphoglycerides. It is sometimes classified as a member of the vitamin B complex but it can be synthesized by many animals and it is not regarded as an essential nutrient in humans. **Phosphatidyl inositol**, a constituent of plasma membranes, is a precursor of the intracellular messenger molecules, **inositol 1,4,5-trisphosphate (IP$_3$)** and **diacylglycerol**; these are produced in response to the binding of substances, e.g. serotonin, to their receptors on the cell surface. These pathways mediate such cellular events as smooth muscle contraction, adrenaline secretion, and histamine secretion.

**Insecta** *See* HEXAPODA.

**insecticide** *See* PESTICIDE.

**Insectivora** An order of small, mainly nocturnal, mammals that includes the hedgehogs, moles, and shrews. They have long snouts covered with stiff tactile hairs and their teeth are specialized for seizing and crushing insects and other small prey. The insectivores have changed very little since they evolved in the Cretaceous period, 130 million years ago.

**insectivore** An animal that eats insects, especially a mammal of the order Insectivora (hedgehogs, shrews, etc.).

**insectivorous plant** *See* CARNIVOROUS PLANT.

**insertion** 1. (of muscles) *See* VOLUNTARY MUSCLE. 2. (in genetics) A *point mutation in which an extra nucleotide base is added to the DNA sequence. This results in the misreading of the base sequence during the *translation stage of protein synthesis.

**insertion sequence** *See* TRANSPOSON.

**insight learning** A form of learning in which an animal responds to new situations by adapting experiences gained in other contexts. Insight learning requires an animal to solve problems by viewing a situation as a whole instead of relying wholly on trial-and-error learning. Chimpanzees are capable of insight learning. *See* LEARNING (Feature).

**insolation** (from *in*coming *sol*ar radi*ation*) The solar radiation that is received at the earth's surface per unit area. It is related to the *solar constant, the duration of daylight, the altitude of the sun, and the latitude of the receiving surface. It is measured in MJ m$^{-2}$.

**inspiration (inhalation)** The process by which gas is drawn into the lungs through the trachea (*see* RESPIRATORY MOVEMENT). In mammals the rib cage is raised by contraction of the external *intercostal muscles and the muscles of the diaphragm. These actions enlarge the thorax, so that pressure in the lung cavity is reduced below atmospheric pressure, which causes an influx of air until the pressures are equalized. *Compare* EXPIRATION.

**inspiratory centre** *See* VENTILATION CENTRE.

**instantaneous value** The value of any varying quantity at a specified instant.

**instar** A stage in the larval development of an insect between two moults (ecdyses). There are usually a number of larval instars before the pupal stage and metamorphosis.

**instinct** An innate tendency to behave in a particular way, which does not depend critically on particular learning experiences for its development and therefore is seen in a

similar form in all normally reared individuals of the same sex and species. Much instinctive behaviour takes the form of **fixed action patterns**. These are movements that – once started – are performed in a stereotyped way unaffected by external stimuli. For example, a frog's prey-catching tongue flick is performed in the same way whether or not anything is caught. Some complex instinctive behaviour, however, requires some learning by the animal before it is perfected. Birdsong, for example, consists of an innate component that is modified and made more complex by the influence of other birds, the habitat, etc.

**insulator** A substance that is a poor conductor of heat and electricity. Both properties usually occur as a consequence of a lack of mobile electrons. *See* ENERGY BANDS.

**insulin** A protein hormone, secreted by the β (or B) cells of the *islets of Langerhans in the pancreas, that promotes the uptake of glucose by body cells, particularly in the liver and muscles, and thereby controls its concentration in the blood. Insulin was the first protein whose amino-acid sequence was fully determined (in 1955). Underproduction of insulin results in the accumulation of large amounts of glucose in the blood and its subsequent excretion in the urine. This condition, known as **diabetes mellitus**, can be treated successfully by insulin injections.

**insulin-like growth factor (IGF)** *See* GROWTH FACTOR.

**Integer** Any one of the positive or negative whole numbers.

**integral calculus** *See* CALCULUS.

**integrand** *See* INTEGRATION.

**integrated circuit** A miniature electronic circuit produced within a single crystal of a *semiconductor, such as silicon. They range from simple logic circuits, little more than 1 mm square, to large-scale circuits measuring up to 8 mm square and containing a million or so transistors (active components) and resistors or capacitors (passive components). They are widely used in memory circuits, microcomputers, pocket calculators, and electronic watches on account of their low cost and bulk, reliability, and high speed. They are made by introducing impurities into specific regions of the semiconductor crystal by a variety of techniques.

**integration** 1. (in physiology) The coordination within the brain of separate but related nervous processes. For example, sensory information from the inner ear and the eye are both necessary for the sense of balance. These stimuli must be integrated by the brain not only with each other but also with various motor nerves, which coordinate the muscles that control posture. **2.** (in mathematics) The process of continuously summing changes in a function f($x$). It is the basis of the integral *calculus and the opposite process to *differentiation. The function to be integrated is called the **integrand** and the result of integration on the integrand is called the **integral**. For example, the integration of f($x$) is written ∫f($x$)d$x$, the differential d$x$ being added to indicate that f($x$) must be integrated with respect to $x$. To complete the integration, a **constant of integration**, $C$, must be added where no interval over which the integration takes place is given. This is called an **indefinite integral**. If the interval is specified, e.g.

$$\int_a^b (x)\mathrm{d}x,$$

no constant of integration is required and the result is a **definite integral**. This means that f($x$) is to be integrated between the values $x = +r$ and $x = -r$.

**integument 1.** The outermost body layer of an animal, characteristically comprising a layer of living cells – the *epidermis – together with a superficial protective coat, which may be a secreted hardened *cuticle, as in arthropods, or dead keratinized cells, as in vertebrates (*see* SKIN). **2.** The outer protective covering of a plant *ovule. It is perforated by a small pore, the *micropyle. Usually two integuments are present in angiosperms and one in gymnosperms. After fertilization the integuments form the *testa of the seed.

**intelligence** The coordination of *memory, *learning, and reasoning in animals. Intelligence has also been defined as the ability of an animal to form associative links between events or objects of which it has had no previous experience (*see* INSIGHT LEARNING). In humans intelligence is generally expressed as an **intelligence quotient** (**IQ**): the mental age of the subject (as measured by standard tests) divided by his or her real age × 100.

**intensity 1.** The rate at which radiant energy is transferred per unit area. *See* RADIANT INTENSITY. **2.** The rate at which sound energy is transferred as measured relative to

some reference value. *See* DECIBEL. **3.** Magnetic intensity. *See* MAGNETIC FIELD. **4.** Electric intensity. *See* ELECTRIC FIELD. **5.** *See* LUMINOUS INTENSITY.

**intensive variable** A quantity in a *macroscopic system that has a well defined value at every point inside the system and that remains (nearly) constant when the size of the system is increased. Examples of intensive variables are the pressure, temperature, density, specific heat capacity at constant volume, and viscosity. An intensive variable results when any *extensive variable is divided by an arbitrary extensive variable such as the volume. A macroscopic system can be described by one extensive variable and a set of intensive variables.

**interaction** An effect involving a number of bodies, particles, or systems as a result of which some physical or chemical change takes place to one or more of them. *See also* FUNDAMENTAL INTERACTIONS.

**interactome** All the interactions that occur between the various molecules produced by an organism, in all its cells and tissues and at all stages of its life. Of central significance are the interactions of proteins – the protein interactome – because these are the molecules encoded by the organism's genes and are fundamental to all other cellular processes. Determining which proteins form complexes, or bind together in some way, enables researchers to identify the components of the complex pathways and networks that govern different aspects of cellular functioning and potentially play a role in health and disease.

**intercalation cell** A type of secondary cell in which layered electrodes, usually made of metal oxides or graphite, store positive ions between the crystal layers of an electrode. In one type, lithium ions form an intercalation compound with a graphite electrode when the cell is charged. During discharge, the ions move through an electrolyte to the other electrode, made of manganese oxide, where they are more tightly bound. When the cell is being charged, the ions move back to their positions in the graphite. This backwards and forwards motion of the ions has led to the name **rocking-chair cell** for this type of system. Such cells have the advantage that only minor physical changes occur to the electrodes during the charging and discharging processes and the electrolyte is not decomposed but simply serves as a conductor of ions. Consequently, such cells can be recharged many more times than, say, a lead-acid accumulator, which eventually suffers from degeneration of the electrodes. **Lithium cells**, based on this principle, have been used in portable electronic equipment, such as camcorders. They have also been considered for use in electric vehicles.

**intercalation compound** A type of compound in which atoms, ions, or molecules are trapped between layers in a crystal lattice. There is no formal chemical bonding between the host crystal and the trapped molecules (*see also* CLATHRATE). Such compounds are formed by *lamellar solids and are often nonstoichiometric; examples are graphitic oxide (graphite–oxygen) and the mineral *muscovite.

**intercellular** (in biology) Located or occurring between cells. *Compare* INTRACELLULAR.

**intercostal muscles** The muscles located between the *ribs, surrounding the lungs. Comprising the superficial **external intercostal muscles** and the deep **internal intercostal muscles**, they play an essential role in breathing (*see* EXPIRATION; INSPIRATION).

**interference** The interaction of two or more wave motions affecting the same part of a medium so that the instantaneous disturbances in the resultant wave are the vector sum of the instantaneous disturbances in the interfering waves.

The phenomenon was first described by Thomas Young in 1801 in light waves; it provided strong evidence for the wave theory of light. In the apparatus known as **Young's slits**, light is passed from a small source through a slit in a screen and the light emerging from this slit is used to illuminate two adjacent slits on a second screen. By allowing the light from these two slits to fall on a third screen, a series of parallel interference fringes is formed. Where the maximum values of the two waves from the slits coincide a bright fringe occurs (**constructive interference**) and where the maxima of one wave coincide with the minima of the other dark fringes are produced (**destructive interference**). (In Young's original experiment, two pinholes were used rather than slits). *Newton's rings are also an interference effect. Because *lasers produce *coherent radiation they are also used to produce interference effects, one application of their use being *holography. *See also* INTERFEROMETER.

**interferometer** An instrument designed to produce optical *interference fringes for measuring wavelengths, testing flat surfaces, measuring small distances, etc. *See also* ECHELON; FABRY–PÉROT INTERFEROMETER; MICHELSON–MORLEY EXPERIMENT. In astronomy, radio interferometers are one of the two basic types of *radio telescopes.

**interferon (IFN)** Any of a number of proteins (*see* CYTOKINE) that increase the resistance of cells to attack by viruses. In humans, three groups of interferons have been discovered: α-interferons from white blood cells; β-interferons from connective tissue fibroblasts; and γ-interferons from T cells and *natural killer cells (NK cells). α- and β-interferons induce intrinsic resistance to viral infection in all cells by triggering the expression of genes that encode antiviral proteins. Moreover, they activate NK cells, which selectively kill virus-infected cells, and promote synthesis of MHC proteins by all cell types (*see* HISTOCOMPATIBILITY), thereby protecting uninfected cells from attack by the NK cells. The actions of γ-interferon include macrophage activation, increasing the expression of MHC molecules, and suppression of helper *T cells. It is produced by cytotoxic T cells. Interferons are produced commercially for therapeutic purposes using genetically engineered bacteria or human tissue culture.

**interhalogen** A chemical compound formed between two *halogens. Interhalogens are highly reactive and volatile, made by direct combination of the elements. They include compounds with two atoms (ClF, IBr, etc.), four atoms ($ClF_3$, $IF_3$, etc.), six atoms ($BrF_5$, $IF_5$, etc.) and $IF_7$ with eight atoms.

**interleukin** Any of several *cytokines that are produced by leucocytes. **Interleukin-1 (IL-1)** is secreted by antigen-activated macrophages and activates *T cells. **Interleukin-2 (IL-2)** stimulates the proliferation of T cells, which also secrete it. **Interleukin-3** is a growth factor for haemopoietic cells, and **interleukin-4** induces B cells to proliferate and produce antibodies. Nearly 30 interleukins are now known to exist, and some are manufactured using recombinant DNA technology, for use as therapeutic agents.

**intermediate bond** *See* CHEMICAL BOND.

**intermediate coupling** *See* J-J COUPLING.

**intermediate frequency** *See* HETERO-DYNE; SUPERHETERODYNE RECEIVER.

**intermediate neutron** A *neutron with kinetic energy in the range $10^2$–$10^5$ electron-volts ($1.6 \times 10^{-17}$ – $1.6 \times 10^{-14}$ joule).

**intermediate vector boson** *See* W BOSON; Z BOSON.

**intermetallic compound** A compound consisting of two or more metallic elements present in definite proportions in an alloy.

**intermolecular forces** Weak forces occurring between molecules. *See* VAN DER WAALS' FORCE; HYDROGEN BOND.

**internal-combustion engine** A *heat engine in which fuel is burned in combustion chambers within the engine rather than in a separate furnace (as with the steam engine). The first working engine was the four-stroke **Otto engine** produced in 1876 by Nikolaus Otto (1832–91). In this type of engine a piston descends in a cylinder, drawing in a charge of fuel and air through an inlet valve; after reaching the bottom of its stroke the piston rises in the cylinder with the valves closed and compresses the charge; at or near the top of its stroke the charge is ignited by a spark and the resulting increase in pressure from the explosion forces the piston down again; on the subsequent upstroke the exhaust valve opens and the burnt gases are pushed out of the combustion chamber. The cycle is then repeated. Otto's engine used gas as a fuel; however, the invention of the carburettor and the development of the oil industry at the end of the 19th century enabled the Otto engine to become the source of power for the emerging motor car. A variation of the Otto four-stroke engine is the two-stroke engine that has no complicated valve system, the explosive charge entering and leaving the cylinder through ports in the cylinder that are covered and uncovered by the moving piston.

An alternative to the Otto engine, especially for heavy vehicles where weight is not a problem, is the compression-ignition **Diesel engine** invented by Rudolf Diesel (1858–1913) in about 1896. In this type of engine there are no sparking plugs; instead air is compressed in the cylinder, causing its temperature to rise to about 550°C. Oil is then sprayed into the combustion chamber and ignites on contact with the hot air. While the spark-ignition petrol engine typically works on a *compression ratio of 8 or 9 to 1, the Diesel engine has to have a compression ratio of between 15 and 25 to 1. This requires a much heavier, and therefore more expensive, engine. *See also* GAS TURBINE.

**internal conversion** A process in which an excited atomic nucleus (*see* EXCITATION) decays to the *ground state and the energy released is transferred by electromagnetic coupling to one of the bound electrons of that atom rather than being released as a photon. The coupling is usually with an electron in the K-, L-, or M-shell of the atom, and this **conversion electron** is ejected from the atom with a kinetic energy equal to the difference between the nuclear transition energy and the binding energy of the electron. The resulting ion is itself in an excited state and usually subsequently emits an Auger electron (*see* AUGER EFFECT) or an X-ray photon.

**internal energy** Symbol $U$. The total of the kinetic energies of the atoms and molecules of which a system consists and the potential energies associated with their mutual interactions. It does not include the kinetic and potential energies of the system as a whole nor their nuclear energies or other intra-atomic energies. The value of the absolute internal energy of a system in any particular state cannot be measured; the significant quantity is the change in internal energy, $\Delta U$. For a closed system (i.e. one that is not being replenished from outside its boundaries) the change in internal energy is equal to the heat absorbed by the system ($Q$) from its surroundings, less the work done ($W$) by the system on its surroundings, i.e. $\Delta U = Q - W$. *See also* ENERGY; HEAT; THERMO-DYNAMICS.

**internal environment** The conditions that prevail within the body of an organism, particularly with respect to the composition of the *tissue fluid. The concept of an internal environment was first proposed by the French physiologist Claude Bernard (1813–78), who stated that maintenance of a constant internal environment was necessary for the survival of an organism in a varying external environment. Selective absorption of materials across plasma membranes plays a large part in controlling the internal environment of both animals and plants. Animals in addition can regulate their body fluids by the action of hormones and the nervous system. *See* HOMEOSTASIS.

**internal resistance** The resistance within a source of electric current, such as a cell or generator. It can be calculated as the difference between the e.m.f. ($E$) and the potential difference ($V$) between the terminals divided by the current being supplied ($I$), i.e. $r = (E - V)/I$, where $r$ is the internal resistance.

**international candle** A former unit of *luminous intensity. It has now been replaced by the *candela, to which it is approximately equal.

**international date line** An imaginary line on the earth's surface that joins the north and south poles and approximately follows the 180° meridian through the Pacific Ocean. This line has been agreed internationally to mark the beginning and end of a day. A traveller moving towards the east, against the sun's apparent movement, gains 1 hour for every 15° of longitude; westward he loses time at the same rate. In crossing the dateline therefore he is deemed to compensate for this by losing or gaining (respectively) one day. The 180° meridian was chosen as the date line by the International Meridian Conference in 1884.

**International Nucleotide Sequence Database Collaboration** (INSDC) A collaborative venture that effectively divides up the task of collecting, updating, and storing the nucleotide sequence data reported by researchers throughout the world. It comprises three databases, which exchange information on a daily basis; these are GenBank, run by the US National Institutes of Health, the DNA Data Bank of Japan (DDBJ), and the Nucleotide Sequence Database of the European Molecular Biology Laboratory (EMBL). Information from the entire collection can be accessed via any one of the partner organizations.

**International Practical Temperature Scale** *See* TEMPERATURE SCALES.

**International Space Station** A space-based scientific research facility constructed in earth orbit at an altitude of 350 km from prefabricated modules and components. It is a joint project involving the space agencies of the United States, Russia, Japan, Canada, and the ten member nations of the European Space Agency, with additional contributions from Brazil and Italy. Assembly of the station began in 1998 and was scheduled to be completed by 2011. It has been crewed continuously since November 2000.

**international system** *See* HERMANN–MAUGUIN SYSTEM.

**Internet (Net)** The global network that links most of the world's computer networks. It does not offer services to users, but serves primarily to interconnect other networks on which services are located. These include basic services for *electronic mail, the transfer of computer files, and remote log-in, and high-level services including the *World Wide Web. The Internet is informal, with a minimal level of administration by governing bodies.

**internode** 1. (in botany) The part of a plant stem between two *nodes. 2. (in neurology) The myelinated region of a nerve fibre between two nodes of Ranvier. *See* MYELIN SHEATH.

**interoceptor** A *receptor that detects stimuli from the internal environment of an organism. *Chemoreceptors that detect changes in the levels of oxygen concentration in the blood are examples. *Compare* EXTEROCEPTOR.

**interphase** The period following the completion of *cell division, when the nucleus is not dividing. During this period changes in both the nucleus and the cytoplasm result in the complete development of the daughter cells. *See* CELL CYCLE.

**interplanetary space (interplanetary medium)** The space occupied by the sun and the planets, dwarf planets, asteroids, comets, *trans-Neptunian objects, and all other bodies within the *solar system. Apart from these, the **interplanetary matter** that saturates this region of space consists mostly of the energetic particles that make up the *solar wind. The solar wind consists primarily of protons emerging from the sun at a rate of about $10^9$ kilograms per second. At the earth's distance from the sun the particle density is only a few particles per $cm^3$. Apart from this very tenuous gas, there are also dust particles in interplanetary space, largely believed to originate in the belt of asteroids. Particles weighing about 1 g produce visible meteors in the earth's atmosphere; micrometeorites as small as 1 nanogram can be detected by their impact on spacecraft.

**interpolation** An *approximation technique for finding the value of a function or a measurement that lies within known values. If the values $f(x_0)$, $f(x_1)$, ..., $f(x_n)$ of a function f of a variable $x$ are known in the interval

$[x_0, x_n]$, the value of f($x$) for a value of $x$ inside the interval $[x_0, x_n]$ can be found by interpolation. One method of interpolation, called **linear interpolation** for $x_0 < x < x_1$, gives:

$$f(x) \cong f(x_0) + [f(x_1) - f(x_0)] \, (x - x_0)/(x_1 - x_0),$$

which is derived using the assumption that between the points $x_0$ and $x_1$, the graph of the function f($x$) can be regarded as a straight line. More complicated methods of interpolation exist, using more than two values for the function. The techniques used for interpolation are usually much better than the techniques used in *extrapolation.

**intersex** An organism displaying characteristics that are intermediate between those of the typical male and typical female of its species. For example, a human intersex may have testes that fail to develop, so that although he is technically a man he has the external appearance of a woman. Intersexes may be produced in various ways; for example, by malfunctioning of the sex hormones. *See also* HERMAPHRODITE.

**interspecific competition** *See* COMPETITION.

**interstellar molecules** *See* ASTROCHEMISTRY.

**interstellar space (interstellar medium)** The space between the stars. The **interstellar matter** that occupies this space constitutes several percent of the Galaxy's total mass and it is from this matter that new stars are formed. The matter is primarily hydrogen, in which a number of other molecules and radicals have been detected, together with small solid dust grains. On average the density of matter in interstellar space is about $10^6$ hydrogen atoms per cubic metre, but the gas is not uniformly distributed, being clumped into **interstellar clouds** of various sizes and densities.

**interstitial** *See* CRYSTAL DEFECT (Feature).

**Interstitial cell** A cell that forms part of the connective tissue (the **interstitium**) between other tissues and structures, especially any of the cells of the *testis that lie between the seminiferous tubules and secrete androgens in response to stimulation by interstitial-cell-stimulating hormone (*see* LUTEINIZING HORMONE).

**interstitial-cell-stimulating hormone** *See* LUTEINIZING HORMONE.

**interstitial compound** A compound in which ions or atoms of a nonmetal occupy interstitial positions in a metal lattice. Such compounds often have metallic properties. Examples are found in the *carbides, *borides, and *silicides.

**intervertebral disc** Any of the discs of cartilage that separate the bones of the *vertebral column. The intervertebral discs allow the vertebral column a certain degree of flexibility and they also absorb shock.

**intestinal juice (succus entericus)** A slightly alkaline liquid containing mucus that is secreted into the lumen of the small intestine from the cells that line the *crypts of Lieberkühn. Together with pancreatic juice, the intestinal juice provides an alkaline environment that helps in the absorption of digested food molecules entering the small intestine in chyme from the stomach.

**intestine** The portion of the *alimentary canal posterior to the stomach. Its major functions are the final digestion of food matter from the stomach, the absorption of soluble food matter, the absorption of water, and the production of *faeces. *See* LARGE INTESTINE; SMALL INTESTINE.

**intracellular** (in biology) Located or occurring within cells. *Compare* INTERCELLULAR.

**intraspecific competition** *See* COMPETITION.

**intrinsic factor** *See* VITAMIN B COMPLEX.

**intrinsic semiconductor** *See* SEMICONDUCTOR.

**intron (intervening sequence)** A nucleotide sequence in a gene that does not code for the gene product (*compare* EXON). Introns, which occur principally in eukaryotes, are transcribed into messenger *RNA but are subsequently removed from the transcript before translation (*see* GENE SPLICING). Their functional significance is still subject to debate.

**intrusion** An upwelling of *magma or other molten rock into an existing rock. The intrusion may force its way through or follow such weaknesses as joints and bedding planes. The heat of the molten intrusion may bring about changes in the composition of the country rock it invades. There are various kinds of igneous intrusions, including *batholiths, *dykes, laccoliths, sills, and *xenoliths.

**inulin** A polysaccharide, made up from fructose molecules, that is stored as a food reserve in the roots or tubers of many plants, such as the dahlia.

**Invar** A trade name for an alloy of iron (63.8%), nickel (36%), and carbon (0.2%) that has a very low *expansivity over a a restricted temperature range. It is used in watches and other instruments to reduce their sensitivity to changes in temperature.

**inverse Compton effect** The gain in energy of low-energy photons when they are scattered by free electrons of much higher energy. As a consequence, the electrons lose energy. The effect is thought to be important in certain astrophysical processes. *See also* COMPTON EFFECT; GZK LIMIT.

**inverse functions** If $y = f(x)$ and a function can be found so that $x = g(y)$, then $g(y)$ is said to be the inverse function of $f(x)$. If $y$ is a trigonometrical function of the angle $x$, say $y = \sin x$, then $x$ is the **inverse trigonometrical function** of $y$, written $x = \arcsin y$ or $\sin^{-1}y$. Similarly, the other trigonometrical functions form the inverse trigonometrical functions $\cos^{-1}y$, $\tan^{-1}y$, $\cot^{-1}y$, $\sec^{-1}y$, and $\csc^{-1}y$. **Inverse hyperbolic functions** are also formed in this way, e.g. $\operatorname{arcsinh}y$ or $\sinh^{-1}y$, $\cosh^{-1}y$, and $\tanh^{-1}y$.

**inverse-square law** A law in which the magnitude of a physical quantity is proportional to the reciprocal of the square of the distance from the source of that property. *Newton's law of gravitation and *Coulomb's law are both examples.

**inversion** 1. (in chemistry) A chemical reaction involving a change from one optically active configuration to the opposite configuration. The Walden inversion is an example. *See* NUCLEOPHILIC SUBSTITUTION. 2. (in genetics) A *chromosome mutation caused by reversal of part of a chromosome, so that the genes within that part are in inverse order. Inversion mutations usually occur during *crossing over in meiosis. 3. (in genetics) A *point mutation caused by the reversal of two or more bases in the DNA sequence within a gene.

**inversion layer** *See* TRANSISTOR.

**inversion temperature** *See* JOULE–THOMSON EFFECT.

**invertebrate** Any animal that lacks a vertebral column (backbone). Invertebrates include all nonchordate animals as well as the more primitive chordates (*see* CHORDATA).

***in vitro*** Describing biological processes that are made to occur outside the living body, in laboratory apparatus (literally 'in glass', i.e. in a test tube). In *in vitro* fertilization, mature egg cells are removed from the ovary of a woman unable to conceive normally and fertilized externally; the resultant blastocyst is implanted into her uterus. *Compare* IN VIVO.

***in vivo*** Describing biological processes as they are observed to occur in their natural environment, i.e. within living organisms. *Compare* IN VITRO.

**involucre** A protective structure in some flowering plants and bryophytes. In flowering plants it consists of a ring of *bracts arising beneath the flower cluster of those species with a *capitulum (i.e. members of the dandelion family) or an *umbel (i.e. members of the carrot family). In mosses and liverworts the involucre is a projection of tissue from the thallus that arches over the developing *archegonium.

**involuntary** (in biology) Not under the control of the will of an individual. Involuntary responses by muscles, glands, etc., occur automatically when required; many such responses, such as gland secretion, heartbeat, and peristalsis, are controlled by the *autonomic nervous system and effected by *involuntary muscle.

**involuntary muscle (smooth muscle)** Muscle whose activity is not under the control of the will; it is supplied by the *autonomic nervous system. Involuntary muscle comprises long spindle-shaped cells without striations. These cells occur singly, in groups, or as sheets in the skin, around hair follicles, and in the digestive tract, respiratory tract, urinogenital tract, and the circulatory system. The cells contract slowly in spontaneous rhythms or when stretched; they may show sustained contraction (tonus) for long periods without fatigue. *Compare* CARDIAC MUSCLE; VOLUNTARY MUSCLE.

**involute** *See* EVOLUTE.

**involution** 1. A decrease in the size of an organ or the body. It may be associated with functional decline, as occurs in the ageing process, or follow enlargement, as when the uterus returns to its normal size after pregnancy. **2.** The turning or rolling inwards of cells that occurs during the development of some vertebrate embryos.

**iodic acid** Any of various oxoacids of iodine, such as iodic(V) acid and iodic(VII) acid. When used without an oxidation state specified, the term usually refers to iodic(V) acid ($HIO_3$).

**iodic(V) acid** A colourless or very pale yellow solid, $HIO_3$; r.d. 4.63; decomposes at 110°C. It is soluble in water but insoluble in pure ethanol and other organic solvents. The compound is obtained by oxidizing iodine with concentrated nitric acid, hydrogen peroxide, or ozone. It is a strong acid and a powerful oxidizing agent.

**iodic(VII) acid (periodic acid)** A hygroscopic white solid, $H_5IO_6$, which decomposes at 140°C and is very soluble in water, ethanol, and ethoxyethane. Iodic(VII) acid may be prepared by electrolytic oxidation of concentrated solutions of iodic(V) acid at low temperatures. It is a weak acid but a strong oxidizing agent.

**iodide** *See* HALIDE.

**iodine** Symbol I. A dark violet nonmetallic element belonging to group 17 of the periodic table (*see* HALOGENS); a.n. 53; r.a.m. 126.9045; r.d. 4.94; m.p. 113.5°C; b.p. 184.35°C. The element is insoluble in water but soluble in ethanol and other organic solvents. When heated it gives a violet vapour that sublimes. Iodine is required as a trace element (*see* ESSENTIAL ELEMENT) by living organisms; in animals it is concentrated in the thyroid gland as a constituent of thyroid hormones. The element is present in sea water and was formerly extracted from seaweed. It is now obtained from oil-well brines (displacement by chlorine). There is one stable isotope, iodine–127, and fourteen radioactive isotopes. It is used in medicine as a mild antiseptic (dissolved in ethanol as **tincture of iodine**), and in the manufacture of iodine compounds. Chemically, it is less reactive than the other halogens and the most electropositive (metallic) halogen. It was discovered in 1812 by Bernard Courtois (1777–1838).

(((())) **SEE WEB LINKS**
• Information from the WebElements site

**iodine(V) oxide (iodine pentoxide)** A white solid, $I_2O_5$; r.d. 4.799; decomposes at 300–350°C. It dissolves in water to give iodic(V) acid and also acts as an oxidizing agent.

**iodine value** A measure of the amount of unsaturation in a fat or vegetable oil (i.e. the number of double bonds). It is obtained by finding the percentage of iodine by weight absorbed by the sample in a given time under standard conditions.

**iodoethane (ethyl iodide)** A colourless liquid *haloalkane, $C_2H_5I$; r.d. 1.9; m.p. –108°C; b.p. 72°C. It is made by reacting ethanol with a mixture of iodine and red phosphorus.

**iodoform** *See* TRIIODOMETHANE.

**iodoform test** *See* HALOFORM REACTION.

**iodomethane (methyl iodide)** A colourless liquid haloalkane, $CH_3I$; r.d. 2.28; m.p. –66.45°C; b.p. 42.4°C. It can be made by reacting methanol with a mixture of iodine and red phosphorus.

**ion** An atom or group of atoms that has either lost one or more electrons, making it positively charged (a cation), or gained one or more electrons, making it negatively charged (an anion). *See also* IONIZATION.

**ion channel** A protein that spans a cell membrane to form a water-filled pore through which ions can pass in or out of the cell or cell compartment. Ion channels are found in the plasma membrane and in certain internal cell membranes. They vary in how they open and close and in their selectivity to different ions: some may be specific for one particular ion, whereas others may admit two or more similar ions (e.g. $K^+$ and $Na^+$). The electrical and chemical environment inside cells, including the resting potential, is determined largely by the numbers, types, and activity of the cell's ion channels; they play a crucial role in the excitability of nerve and muscle cells.

**ion engine** A type of jet-propulsion engine that may become important for propelling or controlling spacecraft. It consists of a unit producing a beam of ions, which are accelerated by an electric or electromagnetic field. Reaction forces from the high-speed ions causes propulsion in much the same way as that caused by exhaust gas of a rocket. However, a separate beam of electrons or ions of opposite polarity to the propelling beam must also be ejected from the engine to enable recombination to take place behind the vehicle (to avoid the vehicle becoming charged). Ion engines provide high *specific impulse and therefore low propellant consumption. The three main components of an ion engine are the power generator, the propellant feed, and the thruster. The power generator may be a nuclear reactor or a solar-energy collector. If it is the former, a gas turbine is coupled to the reactor and the turbine drives an electric generator. A solar-energy unit provides electricity direct. The propellant chosen needs to have an ion of medium mass (low mass for high specific impulse, high mass for high thrust) and a low first *ionization potential. Caesium and mercury are materials currently envisaged as suitable propellants. The thruster consists of an ionizer to produce the ions, an accelerator to provide and shape the accelerating field, and a neutralizer (usually an electron emitter) to neutralize the fast-moving ion beam after ejection.

**ion exchange** The exchange of ions of the same charge between a solution (usually aqueous) and a solid in contact with it. The process occurs widely in nature, especially in the absorption and retention of water-soluble fertilizers by soil. For example, if a potassium salt is dissolved in water and applied to soil, potassium ions are absorbed by the soil and sodium and calcium ions are released from it.

The soil, in this case, is acting as an ion exchanger. Synthetic **ion-exchange resins** consist of various copolymers having a cross-linked three-dimensional structure to which ionic groups have been attached. An **anionic resin** has negative ions built into its structure and therefore exchanges positive ions. A **cationic resin** has positive ions built in and exchanges negative ions. Ion-exchange resins, which are used in sugar refining to remove salts, are synthetic organic polymers containing side groups that can be ionized. In anion exchange, the side groups are ionized basic groups, such as $-NH_3^+$ to which anions $X^-$ are attached. The exchange reaction is one in which different anions in the solution displace the $X^-$ from the solid. Similarly, cation exchange occurs with resins that have ionized acidic side groups such as $-COO^-$ or $-SO_2O^-$, with positive ions $M^+$ attached.

Ion exchange also occurs with inorganic polymers such as *zeolites, in which positive ions are held at sites in the silicate lattice. These are used for water-softening, in which $Ca^{2+}$ ions in solution displace $Na^+$ ions in the

zeolite. The zeolite can be regenerated with sodium chloride solution. **Ion-exchange membranes** are used as separators in electrolytic cells to remove salts from sea water (*see also* DESALINATION) and in producing deionized water. Ion-exchange resins are also used as the stationary phase in **ion-exchange chromatography**.

**ionic bond** *See* CHEMICAL BOND.

**ionic crystal** *See* CRYSTAL.

**ionic product** The product of the concentrations of ions present in a given solution taking the stoichiometry into account. For a sodium chloride solution the ionic product is $[Na^+][Cl^-]$; for a calcium chloride solution it is $[Ca^{2+}][Cl^-]^2$. In pure water, there is an equilibrium with a small amount of self-ionization:

$$H_2O \rightleftharpoons H^+ + OH^-$$

The equilibrium constant of this dissociation is given by

$$K_W = [H^+][OH^-]$$

since the concentration $[H_2O]$ can be taken as constant. $K_W$ is referred to as the ionic product of water. It has the value $10^{-14}$ $mol^2$ $dm^{-6}$ at 25°C. In pure water (i.e. no added acid or added alkali) $[H^+] = [OH^-] = 10^{-7}$ $mol$ $dm^{-3}$. *See also* SOLUBILITY PRODUCT; pH SCALE.

**ionic radius** A value assigned to the radius of an ion in a crystalline solid, based on the assumption that the ions are spherical with a definite size. X-ray diffraction can be used to measure the internuclear distance in crystalline solids. For example, in NaF the Na – F distance is 0.231 nm, and this is assumed to be the sum of the $Na^+$ and $F^-$ radii. By making certain assumptions about the shielding effect that the inner electrons have on the outer electrons, it is possible to assign individual values to the ionic radii – $Na^+$ 0.096 nm; $F^-$ 0.135 nm. In general, negative ions have larger ionic radii than positive ions. The larger the negative charge, the larger the ion; the larger the positive charge, the smaller the ion.

**ionic strength** Symbol $I$. A function expressing the effect of the charge of the ions in a solution, equal to the sum of the molality of each type of ion present multiplied by the square of its charge:

$$I = \Sigma m_i z_i^2.$$

**ion implantation** The technique of implanting ions in the lattice of a semiconductor crystal in order to modify its electronic properties. It is used as an alternative to diffusion, or in conjunction with it, in the manufacture of integrated circuits and solid-state components.

**ionization** The process of producing *ions. Certain molecules (*see* ELECTROLYTE) ionize in solution; for example, *acids ionize when dissolved in water (*see also* SOLVATION):

$$HCl \rightarrow H^+ + Cl^-$$

Electron transfer also causes ionization in certain reactions; for example, sodium and chlorine react by the transfer of a valence electron from the sodium atom to the chlorine atom to form the ions that constitute a sodium chloride crystal:

$$Na + Cl \rightarrow Na^+Cl^-$$

Ions may also be formed when an atom or molecule loses one or more electrons as a result of energy gained in a collision with another particle or a quantum of radiation (*see* PHOTOIONIZATION). This may occur as a result of the impact of *ionizing radiation or of thermal ionization and the reaction takes the form

$$A \rightarrow A^+ + e$$

Alternatively, ions can be formed by electron capture, i.e.

$$A + e \rightarrow A^-$$

**ionization chamber** An instrument for detecting *ionizing radiation. It consists of two electrodes contained in a gas-filled chamber with a potential difference maintained between them. Ionizing radiation entering the chamber ionizes gas atoms, creating electrons and positive ions. The electric field between the electrodes drives the electrons to the anode and the positive ions to the cathode. This current is, in suitable conditions, proportional to the intensity of the radiation. *See also* GEIGER COUNTER.

**ionization energy (IE)** *See* IONIZATION PO TENTIAL.

**ionization gauge** A vacuum gauge consisting of a three-electrode system inserted into the container in which the pressure is to be measured. Electrons from the cathode are attracted to the grid, which is positively biased. Some pass through the grid but do not reach the anode, as it is maintained at a negative potential. Some of these electrons do, however, collide with gas molecules, ionizing them and converting them to positive

ions. These ions are attracted to the anode; the resulting anode current can be used as a measure of the number of gas molecules present. Pressure as low as $10^{-6}$ pascal can be measured in this way.

**ionization potential (IP)** Symbol *I*. The minimum energy required to remove an electron from a specified atom or molecule to such a distance that there is no electrostatic interaction between ion and electron. Originally defined as the minimum potential through which an electron would have to fall to ionize an atom, the ionization potential was measured in volts. It is now, however, defined as the energy to effect an ionization and is conveniently measured in electron-volts (although this is not an SI unit).

The energy to remove the least strongly bound electron is the **first ionization potential**. Second, third, and higher ionization potentials can also be measured, although there is some ambiguity in terminology. Thus, in chemistry the second ionization potential is often taken to be the minimum energy required to remove an electron from the singly charged ion; the second IP of lithium would be the energy for the process

$$Li^+ \rightarrow Li^{2+} + e$$

In physics, the second ionization potential is the energy required to remove an electron from the next to highest energy level in the neutral atom or molecule; e.g.

$$Li \rightarrow Li^{*+} + e,$$

where $Li^{*+}$ is an excited singly charged ion produced by removing an electron from the K-shell.

**((⊕)) SEE WEB LINKS**

- Values for ionization energies of neutral atoms at the NIST website

**ionizing radiation** Radiation of sufficiently high energy to cause *ionization in the medium through which it passes. It may consist of a stream of high-energy particles (e.g. electrons, protons, alpha-particles) or short-wavelength electromagnetic radiation (ultraviolet, X-rays, gamma-rays). This type of radiation can cause extensive damage to the molecular structure of a substance either as a result of the direct transfer of energy to its atoms or molecules or as a result of the secondary electrons released by ionization (*see* SECONDARY EMISSION). In biological tissue the effect of ionizing radiation can be very serious, usually as a consequence of the ejection of an electron from a water molecule and the oxidizing or reducing effects of the resulting highly reactive species:

$$2H_2O \rightarrow e^- + H_2O^* + H_2O^+$$

$$H_2O^* \rightarrow .OH + .H$$

$$H_2O^+ + H_2O \rightarrow .OH + H_3O^+$$

where the dot before a radical indicates an unpaired electron and * denotes an excited species.

**ion-microprobe analysis** A technique for analysing the surface composition of solids. The sample is bombarded with a narrow beam (as small as 2 μm diameter) of high-energy ions. Ions ejected from the surface by sputtering are detected by mass spectrometry. The technique allows quantitative analysis of both chemical and isotopic composition for concentrations as low as a few parts per million.

**ion-mobility spectrometry (IMS)** A technique for detecting low concentrations of specific compounds, based on the rate at which their ions migrate through an electric field. The instrument operates in the gas phase at atmospheric pressure. The sample vapour enters an ionizing region, where ions can be produced by a variety of methods. In compact instruments the source is usually a small amount of radioactive material. The ions are allowed in pulses into a drift tube, where they move to a detector under the influence of a homogeneous electric field. The rate of movement depends on the way the ions interact with neutral molecules in the tube and this depends on the ion's size and shape. The spectrum is a plot of detector signal against time, and is characteristic of the sample being ionized. Ion-mobility spectrometers are compact, sensitive, and fast-acting. They are widely used in screening for drugs and explosives at airports, border crossings, etc. often the technique is to wipe a swab over luggage and place it in the instrument. More sophisticated instruments combine IMS with gas chromatography or mass spectrometry. The technique is sometimes referred to as **gas-phase electrophoresis**.

**ionomer** A thermoplastic resin with ionic bonds between the polymer chains.

**ionosphere** *See* EARTH'S ATMOSPHERE; RADIO TRANSMISSION.

**ionospheric wave** *See* RADIO TRANSMISSION.

**ion pair** A pair of oppositely charged ions produced as a result of a single ionization; e.g.

$$HCl \rightarrow H^+ + Cl^-.$$

Sometimes a positive ion and an electron are referred to as an ion pair, as in

$$A \rightarrow A^+ + e^-.$$

**ion pump** A type of *vacuum pump that can reduce the pressure in a container to about 1 nanopascal by passing a beam of electrons through the residual gas. The gas is ionized and the positive ions formed are attracted to a cathode within the container where they remain trapped. The pump is only useful at very low pressures, i.e. below about 1 micropascal. The pump has a limited capacity because the absorbed ions eventually saturate the surface of the cathode. A more effective pump can be made by simultaneously *sputtering a film of metal, so that fresh surface is continuously produced. The device is then known as a **sputter-ion pump**.

**ion trap** A device used to trap ions by electrical or magnetic fields (or a combination of both). There are a number of types. The **Paul trap** has a ring electrode with a hyperbolic section and two hyperbolic end caps. Ions are trapped in an oscillating field produced by applying an oscillating voltage (about 1 MHz) between the ring electrode and the end caps. The device was invented in the 1950s by the German physicist Wolfgang Paul (1913–93). The **Penning trap** has a similar geometry and operation, but there is also a positive direct voltage (about 100 V) on the cap electrodes with respect to the ring electrode, and an axial magnetic field (about 5 tesla) to confine the particles. This type of trap was developed in 1959 by the German–American physicist Hans Dehmelt (1922–   ). He named it after Frans Penning, who had invented the *Penning gauge, which also uses a magnetic field. Ion traps can be used for storing and investigating the properties of ions and other charged particles and can also be used in *mass spectrometry.

**IP** *See* IONIZATION POTENTIAL.

**IP₃** *See* INOSITOL.

**IR** *See* INFRARED RADIATION.

**iridium** Symbol Ir. A silvery metallic *transition element (*see also* PLATINUM METALS); a.n. 77; r.a.m. 192.20; r.d. 22.42; m.p. 2410°C; b.p. 4130°C. It occurs with platinum and is mainly used in alloys with platinum and osmium. The element forms a range of iridium(III) and iridium(IV) complexes. It was discovered in 1804 by Smithson Tennant (1761–1815).

• Information from the WebElements site

**iridium anomaly** The occurrence of unusually high concentrations of the relatively scarce metal iridium at the boundaries of certain geological strata. Two such layers have been discovered, one at the end of the Cretaceous, 65 million years ago, and the second at the end of the Eocene, 34 million years ago. One theory to account for these suggests that on each occasion a huge iridium-containing meteorite may have collided with the earth, producing a cloud of dust that settled out to form an iridium-rich layer. The environmental consequences of such an impact, notably in causing a general warming of the earth by the *greenhouse effect, may have led to the extinction of the dinosaurs at the end of the Cretaceous and the extinction of many radiolarians at the end of the Eocene. See ALVAREZ EVENT.

**iris 1.** (in anatomy) The pigmented ring of muscular tissue, lying between the cornea and the lens, in the eyes of vertebrates and some cephalopod molluscs. It has a central hole (the **pupil**) through which light enters the eye and it contains both circular and radial muscles. Reflex contraction of the former occurs in bright light to reduce the diameter of the pupil; contraction of the radial muscles in dim light increases the pupil diameter and therefore the amount of light entering the eye. Colour is determined by the amount of the pigment melanin in the iris. Blue eyes result from relatively little melanin; grey and brown eyes from increasingly larger amounts. **2.** (in physics) See DIAPHRAGM.

**iron** Symbol Fe. A silvery malleable and ductile metallic *transition element; a.n. 26; r.a.m. 55.847; r.d. 7.87; m.p. 1535°C; b.p. 2750°C. The main sources are the ores *haematite ($Fe_2O_3$), *magnetite ($Fe_3O_4$), limonite ($FeO(OH)_nH_2O$), ilmenite ($FeTiO_3$), siderite ($FeCO_3$), and pyrite ($FeS_2$). The metal is smelted in a *blast furnace to give impure *pig iron, which is further processed to give *cast iron, *wrought iron, and various types of *steel. The pure element has three crystal forms: **alpha-iron**, stable below 906°C with a body-centred-cubic structure; **gamma-iron**, stable between 906°C and 1403°C with a nonmagnetic face-centred-cubic structure; and **delta-iron**, which is the body-centred-cubic form above 1403°C. Alpha-iron is ferromagnetic up to its Curie point (768°C). The element has nine isotopes (mass numbers 52–60), and is the fourth most abundant in the earth's crust. It is required as a trace element (see ESSENTIAL ELEMENT) by living organisms. Iron is quite reactive, being oxidized by moist air, displacing hydrogen from dilute acids, and combining with nonmetallic elements. It forms ionic salts and numerous complexes with the metal in the +2 or +3 oxidation states. Iron(VI) also exists in the ferrate ion $FeO_4^{2-}$, and the element also forms complexes in which its oxidation number is zero (e.g. $Fe(CO)_5$).

• Information from the WebElements site

**iron(II) chloride** A green-yellow deliquescent compound, $FeCl_2$; hexagonal; r.d. 3.16; m.p. 670°C. It also exists in hydrated forms: $FeCl_2.2H_2O$ (green monoclinic; r.d. 2.36) and $FeCl_2.4H_2O$ (blue-green monoclinic deliquescent; r.d. 1.93). Anhydrous iron(II) chloride can be made by passing a stream of dry hydrogen chloride over the heated metal; the hydrated forms can be made using dilute hydrochloric acid or by recrystallizing with water. It is converted into iron(III) chloride by the action of chlorine.

**iron(III) chloride** A black-brown solid, $FeCl_3$; hexagonal; r.d. 2.9; m.p. 306°C; decomposes at 315°C. It also exists as the hexahydrate $FeCl_3.6H_2O$, a brown-yellow deliquescent crystalline substance (m.p. 37°C; b.p. 280–285°C). Iron(III) chloride is prepared by passing dry chlorine over iron wire or steel wool. The reaction proceeds with incandescence when started and iron(III) chloride sublimes as almost black iridescent scales. The compound is rapidly hydrolysed in moist air. In solution it is partly hydrolysed; hydrolysis can be suppressed by the addition of hydrochloric acid. The compound dissolves in many organic solvents, forming solutions of low electrical conductivity: in ethanol, ethoxyethane, and pyridine the molecular weight corresponds to $FeCl_3$ but is higher in other solvents corresponding to $Fe_2Cl_6$. The vapour is also dimerized. In many ways the compound resembles aluminium chloride, which it may replace in Friedel–Crafts reactions.

**iron(II) oxide** A black solid, FeO; cubic; r.d. 5.7; m.p. 1420°C. It can be obtained by heating iron(II) oxalate; the carbon monoxide formed produces a reducing atmosphere thus preventing oxidation to iron(III) oxide. The compound has the sodium chloride structure, indicating its ionic nature, but the crystal lattice is deficient in iron(II) ions and it is nonstoichiometric. Iron(II) oxide dissolves readily in dilute acids.

**iron(III) oxide** A red-brown to black insoluble solid, $Fe_2O_3$; trigonal; r.d. 5.24; m.p. 1565°C. There is also a hydrated form, $Fe_2O_3.xH_2O$, which is a red-brown powder; r.d. 2.44–3.60. (*See* RUSTING.)

Iron(III) oxide occurs naturally as *haematite and can be prepared by heating iron(III) hydroxide or iron(II) sulphate. It is readily reduced on heating in a stream of carbon monoxide or hydrogen.

**iron pyrites** *See* PYRITE.

**ironstone** A sedimentary rock so-called because of its content of iron, usually in the form of the minerals haematite, limonite, or pyrite, which sometimes give the surface of the rock a typical rusty colour. It is found in beds, layers, or nodules. The iron minerals often take the form of small spherical oolites embedded in the rock.

**iron(II) sulphate** An off-white solid, $FeSO_4.H_2O$; monoclinic; r.d. 2.970. There is also a heptahydrate form, $FeSO_4.7H_2O$; blue-green monoclinic; r.d. 1.898; m.p. 64°C. The heptahydrate is the best known iron(II) salt and is sometimes called **green vitriol** or **copperas**. It is obtained by the action of dilute sulphuric acid on iron in a reducing atmosphere. The anhydrous compound is very hygroscopic. It decomposes at red heat to give iron(III) oxide, sulphur trioxide, and sulphur dioxide. A solution of iron(II) sulphate is gradually oxidized on exposure to air, a basic iron(III) sulphate being deposited.

**iron(III) sulphate** A yellow hygroscopic compound, $Fe_2(SO_4)_3$; rhombic; r.d. 3.097; decomposes above 480°C. It is obtained by heating an aqueous acidified solution of iron(II) sulphate with hydrogen peroxide:

$$2FeSO_4 + H_2SO_4 + H_2O_2 \rightarrow Fe_2(SO_4)_3 + 2H_2O$$

On crystallizing, the hydrate $Fe_2(SO_4)_3.9H_2O$ is formed. The acid sulphate $Fe_2(SO_4)_3.H_2SO_4.8H_2O$ is deposited from solutions containing a sufficient excess of sulphuric acid.

**irradiance** Symbol *E*. The *radiant flux per unit area reaching a surface; in SI units it is measured in watts per square metre ($W\,m^{-2}$). Irradiance refers to electromagnetic radiation of all kinds, whereas *illuminance refers only to visible radiation.

**irradiation** Exposure to any form of radiation; often exposure to *ionizing radiation is implied. *See also* FOOD PRESERVATION.

**irrational number** A number that cannot be expressed as the ratio of two integers. An irrational number may be a *surd, such as $\sqrt{2}$ or $\sqrt{3}$, which can be expressed to any desired degree of accuracy but cannot be assigned an exact value. Alternatively, it may be a *transcendental number, such as $\pi$ or e. *Compare* RATIONAL NUMBER.

**irreversibility** The property of a system that precludes a change to the system from being a *reversible process. The paradox that although the equations describing the bodies in a system, such as Newton's laws of motion, Maxwell's equation, or Schrödinger's equation are invariant under *time reversal, events involving systems made up from large numbers of these bodies are not reversible. The process of scrambling an egg is an example. The resolution of this paradox requires the concept of *entropy using *statistical mechanics. Irreversibility occurs in the transition from an ordered arrangement to a disordered arrangement, which is a natural trend, since changes in a closed system occur in the direction of increasing entropy. Irreversibility also occurs in processes that violate T symmetry. According to the *CPT theorem, processes that violate CP also violate T and hence are irreversible. This has been observed in some weak interactions.

**irreversible reaction** *See* CHEMICAL REACTION.

**irrigation** The provision of water for crops by artificial methods; for example by constructing ditches, pipe systems, and canals. Irrigation can lead to problems when the water leaches trace elements from the soil; selenium, for example, can be toxic to both local fauna and flora. Irrigation can also increase the salinity of the soil, if diverted rivers are used to provide the water. Evaporation of surface water leaves a crust of salt, which can drain down to deeper layers of the soil.

**irritability** *See* SENSITIVITY.

**IR spectroscopy** *See* INFRARED SPECTROS-COPY.

**ischium** The most posterior of the three bones that make up each half of the \*pelvic girdle. *See also* ILIUM; PUBIS.

**isentropic process** Any process that takes place without a change of \*entropy. The quantity of heat transferred, $\delta Q$, in a reversible process is proportional to the change in entropy, $\delta S$, i.e. $\delta Q = T\delta S$, where $T$ is the thermodynamic temperature. Therefore, a reversible \*adiabatic process is isentropic, i.e. when $\delta Q$ equals zero, $\delta S$ also equals zero.

**Ising model** A simplified model of a magnetic system consisting of an array of magnetic spins. Spins may have one of two values and interactions occur with nearest neighbours. There are also random thermal fluctuations depending on the temperature of the system. At low temperatures there is a net magnetization as a result of alignment of spins. At high temperature there is no net magnetization. The model was first proposed by the German physicist Ernst Ising (1900–98), who studied the one-dimensional case in 1924. The two-dimensional case for a square lattice was solved exactly by Lars Onsager in 1944. Only approximate solutions have been found for three-dimensional models. The Ising model is very important in statistical mechanics and can be used to investigate other types of phase transition.

**ISIS/Draw** A commonly used chemical drawing program for 2D and 3D structures, copyright of MDL Information Systems, Inc. The program had certain additional features including calculation of molecular weight, calculation of percentages of elements present, IUPAC name generation, and viewing in RasMol. The final version of ISIS/Draw was released in 2002; it has been replaced by Accelrys Draw, the latest version of which was released in 2011.

**islets of Langerhans** Small groups of cells in the pancreas that function as an endocrine gland. The alpha (or A) cells secrete the hormone \*glucagon, the beta (or B) cells secrete \*insulin, and the D cells secrete \*somatostatin. The islets are named after their discoverer, the German anatomist and microscopist Paul Langerhans (1847–88).

**iso-** Prefix denoting that a compound is an isomer, e.g. isopentane ($CH_3CH(CH_3)C_2H_5$,

2-methylbutane) is an isomer of pentane. *See* ISOMERISM.

**isobar** **1.** A line on a map or chart that joins points or places that have the same atmospheric pressure. **2.** A curve on a graph representing readings taken at constant pressure. **3.** One of two or more nuclides that have the same number of nucleons but different \*atomic numbers. Radium–88, actinium–89, and thorium–90 are isobars as each has a \*nucleon number of 228.

**isobaric spin** *See* ISOTOPIC SPIN.

**isocline** A line on a map or chart joining points or places of equal magnetic dip (*see* GEOMAGNETISM).

**isocyanate** *See* CYANIC ACID.

**isocyanic acid** *See* CYANIC ACID.

**isocyanide** *See* ISONITRILE.

**isocyanide test** A test for primary amines by reaction with an alcoholic solution of potassium hydroxide and trichloromethane.

$$RNH_2 + 3KOH + CHCl_3 \rightarrow RNC + 3KCl + 3H_2O$$

The isocyanide RNC is recognized by its unpleasant smell. This reaction of primary amines is called the **carbylamine reaction**.

**isodiaphere** One of two or more nuclides in which the difference between the number of neutrons and the number of protons is the same. A nuclide and its product after losing an \*alpha particle are isodiapheres.

**isodynamic line** A line on a map or chart joining points or places at which the total strengths of the earth's magnetic field are equal (*see* GEOMAGNETISM).

**isoelectric point** The pH of a medium at which a protein carries no net charge and therefore will not migrate in an electric field. Proteins precipitate most readily at their isoelectric points; this property can be utilized to separate mixtures of proteins or amino acids.

**isoelectronic** Describing compounds that have the same numbers of valence electrons. For example, nitrogen ($N_2$) and carbon monoxide (CO) are isoelectronic molecules.

**isoenzyme** *See* ISOZYME.

**isogamy** Sexual reproduction involving the production and fusion of gametes that are similar in size and structure. It occurs in some protoctists, e.g. certain protozoans and algae. *Compare* ANISOGAMY.

**isogonal line**  A line on a map or chart joining points or places of equal magnetic declination (*see* GEOMAGNETISM).

**isoindole**  *See* INDOLE.

**isolating mechanism**  Any of the biological properties of organisms that prevent interbreeding (and therefore exchange of genetic material) between members of different species that inhabit the same geographical area. These mechanisms include **seasonal isolation**, in which the *breeding seasons of the different populations do not overlap; and **behavioural isolation**, in which different *courtship behaviour in the populations ensures that mating takes place only between members of the same species. Both these are examples of **premating mechanisms. Postmating mechanisms** include hybrid infertility and inviability.

**isoleptic complex**  A metal complex in which all the ligands are the same.

**isoleucine**  *See* AMINO ACID.

**isomerase**  Any of a class of *enzymes that catalyse the rearrangement of the atoms within a molecule, thereby converting one isomer into another.

**isomerism**  1. (in chemistry) The existence of chemical compounds (**isomers**) that have the same molecular formulae but different molecular structures or different arrangements of atoms in space. In **structural isomerism** the molecules have different molecular structures: i.e. they may be different types of compound or they may simply differ in the position of the functional group in the molecule. Structural isomers generally have different physical and chemical properties. In **stereoisomerism**, the isomers have the same formula and functional groups, but differ in the arrangement of groups in space. Optical isomerism is one form of this (*see* OPTICAL ACTIVITY). Another type is **cis–trans isomerism** (formerly **geometrical isomerism**), in which the isomers have different positions of groups with respect to a double bond or central atom (see illustration overleaf).

Octahedral complexes can display cis–trans isomerism if they have formulae of the type $MX_2Y_4$. Octahedral complexes with formulae of the type $MX_3Y_3$ can display a different type of isomerism. If the three X ligands are in a plane that includes the metal atom and the three Y ligands are in a different plane at right angles, then the structure is a **mer-isomer** (meridional). If the three X ligands are all on one face of the octahedron and the three Y ligands are on an opposite face, then it is a **fac-isomer** (facial). See illustration overleaf. *See also* AMBIDENTATE; E–Z CONVENTION. 2. (in physics) The existence of atomic nuclei that have the same atomic number and the same mass number but different energy states.

**isomers**  *See* ISOMERISM.

**isometric**  1. (in technical drawing) Denoting a projection in which the three axes are equally inclined to the surface of the drawing and lines are drawn to scale. 2. (in crystallography) Denoting a system in which the axes are perpendicular to each other, as in cubic crystals. 3. (in physics) Denoting a line on a graph illustrating the way in which temperature and pressure are interrelated at constant volume.

**isomorphism**  The existence of two or more substances (**isomorphs**) that have the same crystal structure, so that they are able to form *solid solutions.

**isonitrile (isocyanide; carbylamine)**  An organic compound containing the group –NC, in which the bonding is to the nitrogen atom.

**(())) SEE WEB LINKS**
• Information about IUPAC nomenclature

**iso-octane**  *See* OCTANE; OCTANE NUMBER.

**isopoly compound**  *See* CLUSTER COMPOUND.

**isoprene**  A colourless liquid diene, $CH_2$:$C(CH_3)CH$:$CH_2$. The systematic name is **2-methylbuta-1,3-diene**. It is the structural unit in *terpenes and natural *rubber, and is used in making synthetic rubbers.

**isospin**  *See* ISOTOPIC SPIN.

**isostasy**  The theoretical equilibrium that tends to exist in the earth's crust. If this equilibrium is disturbed, for example as a result of erosion or deposition, compensatory movements in the earth's crust occur: areas of deposition sink, whereas areas of erosion rise. Continental ice sheets have been an important cause of isostatic movements of the earth's crust. The growth of ice sheets and the resulting additional mass of the ice is compensated by the downward

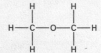

1-chloropropane       2-chloropropane

structural isomers in which the functional group has different positions

methoxymethane       ethanol

structural isomers in which the functional groups are different

*trans*-but-2-ene       *cis*-but-2-ene

cis–trans isomers in which the groups are distributed on a double bond

cis–trans isomers in a square-planar complex

keto form       enol form

keto–enol tautomerism

**Isomerism.**

deflection of the lithospheric plate and the displacement of asthenospheric material (*see* ASTHENOSPHERE) beneath it. When the ice melts, the displaced material flows back causing the overlying lithospheric plate to rise.

**isotactic polymer** *See* POLYMER.

**isotherm** **1.** A line on a map or chart joining points or places of equal temperature. **2.** A curve on a graph representing readings taken at constant temperature (e.g. the relationship between the pressure and volume of a gas at constant temperature).

**isothermal process** Any process that

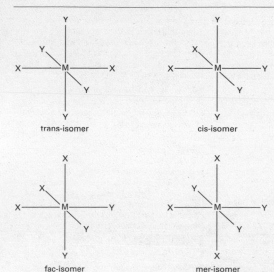

trans-isomer         cis-isomer

fac-isomer         mer-isomer

**Isomerism.**

takes place at constant temperature. In such a process heat is, if necessary, supplied or removed from the system at just the right rate to maintain constant temperature. *Compare* ADIABATIC PROCESS.

**isotone** One of two or more nuclides that contain the same number of neutrons but different numbers of protons. The naturally occurring isotones, for example, strontium–88 and yttrium–89 (both with 50 neutrons), give an indication of the stability of certain nuclear configurations.

**isotonic** Describing solutions that have the same osmotic pressure.

**isotope** One of two or more atoms of the same element that have the same number of protons in their nucleus but different numbers of neutrons. Hydrogen (1 proton, no neutrons), deuterium (1 proton, 1 neutron), and tritium (1 proton, 2 neutrons) are isotopes of hydrogen. Most elements in nature consist of a mixture of isotopes. *See* ISOTOPE SEPARATION.

**(((●))) SEE WEB LINKS**
• Isotopic compositions at the NIST website

**isotope separation** The separation of the *isotopes of an element from each other on the basis of slight differences in their

physical properties. For laboratory quantities the most suitable device is often the mass spectrometer. On a larger scale the methods used include gaseous diffusion (widely used for separating isotopes of uranium in the form of the gas uranium hexafluoride), distillation (formerly used to produce heavy water), electrolysis (requiring cheap electrical power), thermal diffusion (formerly used to separate uranium isotopes, but now considered uneconomic), centrifuging, and laser methods (involving the excitation of one isotope and its subsequent separation by electromagnetic means).

**isotopic number (neutron excess)** The difference between the number of neutrons in an isotope and the number of protons.

**isotopic signature** The relative proportions of certain isotopes in a sample of tissue, water, air or other material. In biology the proportions of stable isotopes of common elements in organic matter, especially carbon, nitrogen, and oxygen, can provide useful clues about growing conditions, lifestyle, and metabolism of the organism from which the material was derived. Differences in the relative tissue abundance of the natural stable isotopes of carbon (i.e. $^{12}C$ and $^{13}C$), oxygen ($^{16}O$ and $^{18}O$) and nitrogen

($^{14}$N and $^{15}$N) reflect their differing availability in the environment or in the diet, or some form of selective metabolism by the organism. So, for example, human hair samples with differing isotopic signatures can indicate the dietary preferences of the individuals.

**isotopic spin** (isospin; isobaric spin) A quantum number applied to hadrons (*see* ELEMENTARY PARTICLES) to distinguish between members of a set of particles that differ in their electromagnetic properties but are otherwise apparently identical. For example if electromagnetic interactions and weak interactions are ignored, the proton cannot be distinguished from the neutron in their strong interactions: isotopic spin was introduced to make a distinction between them. The use of the word 'spin' implies only an analogy to angular momentum, to which isotopic spin has a formal resemblance.

**isotropic** Denoting a medium whose physical properties are independent of direction. *Compare* ANISOTROPIC.

**isozyme** (isoenzyme) One of several forms of an enzyme in an individual or population that catalyse the same reaction but differ from each other in such properties as substrate affinity and maximum rates of enzyme–substrate reaction (*see* MICHAELIS–MENTEN CURVE).

**IT** (information technology) The use of computers and telecommunications equipment (with their associated microelectronics) to send, receive, store and manipulate data. The data may be textual, numerical, audio or video, or any combination of these. *See also* WORLD WIDE WEB.

**iteration** The process of successive approximations used as a technique for solving a mathematical problem. The technique can be used manually but is widely used by computers.

**IUPAC** International Union of Pure and Applied Chemistry. An international non-governmental body formed in 1919 to foster worldwide communications in chemical science, both academic and industrial. IUPAC is the international authority defining recommended terminology, atomic weights, isotopic abundances, and other data.

(((●))) **SEE WEB LINKS**
• The IUPAC home page

**IUPAP** International Union of Pure and Applied Physics. An international nongovernmental body formed in 1922 to stimulate and promote international cooperation in physics. IUPAP fosters the preparation and publication of tables of physical constants and promotes international agreements on the use of symbols, units, nomenclature, and standards.

(((●))) **SEE WEB LINKS**
• The IUPAP home page

**jacinth** *See* ZIRCON.

**Jacob–Monod hypothesis** The theory postulated by the French biologists François Jacob (1920– ) and Jacques Monod (1910–76) in 1961 to explain the control of *gene expression in bacteria (*see* OPERON). Jacob and Monod investigated the expression of the gene that codes for the enzyme β-galactosidase, which breaks down lactose; the operon that regulates lactose metabolism is called the *lac* operon.

**jade** A hard semiprecious stone consisting either of jadeite or nephrite. **Jadeite**, the most valued of the two, is a sodium aluminium pyroxene, $NaAlSi_2O_6$. It is prized for its intense translucent green colour but white, green and white, brown, and orange varieties also occur. The only important source of jadeite is in the Mogaung region of upper Burma. **Nephrite** is one of the amphibole group of rock-forming minerals. It occurs in a variety of colours, including green, yellow, white, and black. Important sources include Siberia, Turkistan, New Zealand, Alaska, China, and W USA.

**jadeite** *See* JADE.

**Jahn–Teller effect** If a likely structure of a nonlinear molecule or ion would have degenerate orbitals (i.e. two molecular orbitals with the same energy levels) the actual structure of the molecule or ion is distorted so as to split the energy levels ('raise' the degeneracy). The effect is observed in inorganic complexes. For example, the ion $[Cu(H_2O)_6]^{2+}$ is octahedral and the six ligands might be expected to occupy equidistant positions at the corners of a regular octahedron. In fact, the octahedron is distorted, with four ligands in a square and two opposite ligands further away. If the 'original' structure has a centre of symmetry, the distorted structure must also have a centre of symmetry. **Jahn–Tellar splitting** is the splitting of spectral lines as a result of this effect. H. A. Jahn and Edward Teller predicted this effect in 1937 using group theory.

**jargoon** *See* ZIRCON.

**jasper** An impure variety of *chalcedony. It is associated with iron ores and as a result contains iron oxide impurities that give the mineral its characteristic red or reddish-brown colour. Jasper is used as a gemstone.

**JavaScript** A scripting language designed to add features to web pages. JavaScript code is embedded in the HTML code and is run by the web browser. JavaScript is loosely based on the Java programming language and was produced by Netscape and Sun Microsystems. The language has been standardized by ECMA International, and its official name is **ECMAScript**.

(( )) SEE WEB LINKS
• Mozilla's documentation for JavaScript

**jaw** The part of the vertebrate skeleton that provides a support for the mouth and holds the teeth. It consists of two bones, the upper jaw (maxilla) and the lower jaw (mandible). Members of the Agnatha lack jaws.

**Jaynes–Cummings model** A model used in *quantum optics and atomic physics to describe the interactions between an atom with two energy levels and a quantized mode of an electromagnetic field. This model, which was put forward by the American physicists Edwin Jaynes (1922–98) and Fred Cummings in 1963, has proved to be very useful in establishing which aspects of quantum optics are purely quantum mechanical and which can be dealt with by using quantum mechanics for the two-level atom and classical electrodynamics for the electromagnetic field.

**Jeans instability** Instability in a cloud of gas in space due to fluctuations in the density of the gas, causing the matter in the cloud to clump together and lead to gravitational collapse. The conditions under which this occurs were worked out by Sir James Hopwood Jeans (1877–1946) in terms of Newtonian gravity. The analogous analysis of this problem using general relativity theory is the basis of the theory of *structure formation.

**jejunum** The portion of the mammalian

*small intestine that follows the *duodenum and precedes the *ileum. The surface area of the lining of the jejunum is greatly increased by numerous small outgrowths (see VILLUS). This facilitates the absorption of digested material, which is the prime function of the jejunum.

**jellyfish** See CNIDARIA.

**Jenner, Edward** (1749–1823) British physician, who is best known for introducing smallpox vaccination to Britain in 1796 (announced two years later), using a vaccine made from cowpox.

**jet** A variety of *coal that can be cut and polished and is used for jewellery, ornaments, etc.

**jet propulsion (reaction propulsion)** The propulsion of a body by means of a force produced by discharging a fluid in the form of a jet. The backward-moving jet of fluid reacts on the body in which it was produced, in accordance with Newton's third law of motion, to create a reactive force that drives the body forward. Jet propulsion occurs in nature, the squid using a form of it to propel itself through water. Although jet-propelled boats and cars have been developed, the main use of jet propulsion is in aircraft and spacecraft. Jet propulsion is the only known method of propulsion in space. In the atmosphere, jet propulsion becomes more efficient at higher altitudes, as efficiency is inversely proportional to the density of the medium through which a body is flying. The three principal means of providing jet propulsion are the turbojet, the ramjet, and the rocket. The **turbojet** is an air-breathing *heat engine based on the *gas turbine, used to power jet aircraft. The **ramjet** is also an air-breathing engine, but compression of the oxidant is achieved by the forward motion of the device through the atmosphere. This enables the compressor and turbine of the gas turbine to be dispensed with and the remaining system consists simply of an inlet diffuser, a combustion chamber in which fuel is burnt, and a jet nozzle through which the products of combustion are discharged. Used in guided missiles, the ramjet must be accelerated to its operating velocity before it can fly (see also PULSE JET). These two forms of jet propulsion, being air-breathing engines, can only be used in the earth's atmosphere. The *rocket, however, carries its own oxidant and can thus be used in space. See also ION ENGINE.

**jet stream** A narrow wind current that occurs in the earth's atmosphere above the lower troposphere (see EARTH'S ATMOSPHERE). The wind flows towards the east at speeds of between 60 km/h (summer) and 125 km/h (winter).

**jeweller's rouge** Red powdered haematite, iron(III) oxide, $Fe_2O_3$. It is a mild abrasive used in metal cleaners and polishes.

**j-j coupling** A type of *coupling in many-fermion systems, such as electrons in atoms and nucleons in nuclei, in which the energies associated with the spin–orbit interactions are much higher than the energies associated with electrostatic repulsion. *Multiplets of many-electron atoms having a large atomic number are characterized by j-j coupling. Multiplets in the *shell model of nuclei characterized by j-j coupling are invoked to explain the *magic numbers of nuclei. The multiplets of many atoms and nuclei are intermediate between j-j coupling and *Russell–Saunders coupling (**intermediate coupling**).

**JMol** A commonly used molecular viewing program similar to RasMol. It can be used as an applet in a web page.

(⊕) SEE WEB LINKS

• Details and a download for JMol from Source-Forge

**joint** The point of contact between two (or more) bones, together with the tissues that surround it. Joints fall into three classes that differ in the degree of freedom of movement they allow: (1) **immovable joints**, e.g. the *sutures between the bones that form the cranium; (2) **slightly movable joints**, e.g. the *symphyses between the vertebrae of the spinal column; and (3) **freely movable** or **synovial joints**, e.g. those that occur between the limb bones. Synovial joints include the **ball-and-socket joints** (between the limbs and the hip and shoulder girdles), which allow movement in all directions; and the **hinge joints** (e.g. at the knee and elbow), which allow movement in one plane only. A synovial joint is bound by ligaments and lined with *synovial membrane.

**Joliot-Curie, Irène** (1897–1956) French physicist, daughter of Marie and Pierre *Curie, who was educated by her mother and her scientist associates. In 1921 she began work at the Radium Institute, becoming director in 1946. In 1926 she married **Frédéric Joliot** (1900–58). They shared the 1935 Nobel

Prize for chemistry for their discovery of artificial radioactivity the previous year.

**joliotium**  *See* TRANSACTINIDE ELEMENTS.

**Joly's steam calorimeter**  An apparatus invented by John Joly (1857–1933) to measure the specific heat capacity of a gas at constant volume. Two equal spherical containers are suspended from the opposite ends of a balance arm. One sphere is evacuated and the other contains the sample gas. The whole apparatus is enclosed in a steam bath, the specific heat capacity of the sample gas being calculated from the difference between the masses of the water that condenses on each sphere.

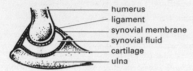

- humerus
- ligament
- synovial membrane
- synovial fluid
- cartilage
- ulna

a hinge joint (the elbow)

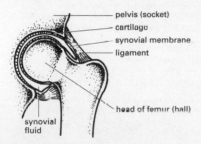

- pelvis (socket)
- cartilage
- synovial membrane
- ligament

- head of femur (ball)

synovial fluid

a ball-and-socket joint (the hip)

**Joint.** Types of freely movable joint.

**Josephson effects**  Electrical effects observed when two superconducting materials (at low temperature) are separated by a thin layer of insulating material (typically a layer of oxide less than $10^{-8}$ m thick). If normal metallic conductors are separated by such a barrier it is possible for a small current to flow between the conductors by the *tunnel effect. If the materials are superconductors (*see* SUPERCONDUCTIVITY), several unusual phenomena occur:
(1) A supercurrent can flow through the barrier; i.e. it has zero resistance.

(2) If this current exceeds a critical value, this conductivity is lost; the barrier then only passes the 'normal' low tunnelling current and a voltage develops across the junction.
(3) If a magnetic field is applied below the critical current value, the current density changes regularly with distance across the junction. The net current through the barrier depends on the magnetic field applied. As the field is increased the net current increases from zero to a maximum, decreases to zero, increases again to a (lower) maximum, decreases, and so on. If the field exceeds a critical value the superconductivity in the barrier vanishes and a potential difference develops across the junction.
(4) If a potential difference is applied across the junction, a high-frequency alternating current flows through the junction. The frequency of this current depends on the size of the potential difference.

A junction of this type is called a **Josephson junction**; two or more junctions joined by superconducting paths form a **Josephson interferometer**. Such junctions can be used in measuring fundamental constants, in defining a voltage standard, and in the highly accurate measurement of magnetic fields. An important potential use is in logic components in high-speed computers. Josephson junctions can switch states very quickly (as low as 6 picoseconds). Moreover they have very low power consumption and can be packed closely without generating too much heat. It is possible that computers based on such devices could operate 50 times faster than the best existing machines. The effects are named after Brian Josephson (1940– ), who predicted them theoretically in 1962.

**joule**  Symbol J. The *SI unit of work and energy equal to the work done when the point of application of a force of one newton moves, in the direction of the force, a distance of one metre. 1 joule = $10^7$ ergs = 0.2388 calorie. It is named after James Joule.

**Joule, James Prescott**  (1818–89) British physicist. In 1840 he discovered the relationship between electric current, resistance, and the amount of heat produced. In 1849 he gave an account of the *kinetic theory of gases, and a year later announced his best-known finding, the *mechanical equivalent of heat. Later, with William Thomson (Lord *Kelvin), he discovered the *Joule–Thomson effect.

**Joule heating** The production of heat in a conductor as a result of the passage of an electric current through the conductor. The quantity of heat produced is given by *Joule's law (def. 1).

**Joule's laws** **1.** The heat ($Q$) produced when an electric current ($I$) flows through a resistance ($R$) for a time ($t$) is given by $Q = I^2Rt$. **2.** The *internal energy of a given mass of gas is independent of its volume and pressure, being a function of temperature alone. This law applies only to *ideal gases (for which it provides a definition of thermodynamic temperature) as in a real gas intermolecular forces would cause changes in the internal energy should a change of volume occur. *See also* JOULE–THOMSON EFFECT.

**Joule–Thomson effect (Joule–Kelvin effect)** The change in temperature that occurs when a gas expands through a porous plug into a region of lower pressure. For most real gases the temperature falls under these circumstances as the gas has to do internal work in overcoming the intermolecular forces to enable the expansion to take place. This is a deviation from *Joule's law (def. 2). There is usually also a deviation from *Boyle's law, which can cause either a rise or a fall in temperature since any increase in the product of pressure and volume is a measure of external work done. At a given pressure, there is a particular temperature, called the **inversion temperature** of the gas, at which the rise in temperature from the Boyle's law deviation is balanced by the fall from the Joule's law deviation. There is then no temperature change. Above the inversion temperature the gas is heated by expansion; below it, it is cooled. The effect was discovered by James Joule working in collaboration with William Thomson (later Lord Kelvin).

**Jovian** Relating to the planet Jupiter.

**JUGFET** *See* TRANSISTOR.

**jugular vein** A paired vein in the neck of mammals that returns blood from the head to the heart. It joins the subclavian vein at the base of the neck.

**jumping gene** *See* TRANSPOSON.

**junction detector (solid-state detector)** A sensitive detector of *ionizing radiation in which the output is a current pulse proportional to the energy falling in or near the depletion region of a reverse-biased *semiconductor junction. The first types

were made by evaporating a thin layer of gold on to a polished wafer of $n$-type germanium; however, gold–silicon devices can be operated at room temperature and these have superseded the germanium type, which have to be operated at the temperature of liquid nitrogen to reduce noise. When the gold–silicon junction is reverse-biased a depletion region, devoid of charge carriers (electrons and holes), forms in the silicon. Incoming ionizing radiation falling in this depletion region creates pairs of electrons and holes, which both have to be collected in order to give an output pulse proportional to the energy of the detected particle.

Junction detectors are used in medicine and biology as well as in space systems.

**junction transistor** *See* TRANSISTOR.

**Jupiter** The largest planet in the *solar system and the fifth in order from the sun ($778.54 \times 10^6$ km distant). Its equatorial diameter is 142 985 km and its *sidereal period is 11.86 years. Jupiter is a *gas giant. Its mass has been calculated as $1.899 \times 10^{27}$ kg or 318 times that of the earth. Viewed through an optical telescope it appears as an elliptical disc crossed by a number of bands. These are zones and belts of high and low atmospheric pressure respectively in the planet's outermost cloud layers. A peculiar feature of the Jovian disc is the **Great Red Spot**, an anticyclonic vortex, roughly oval in shape and brick-red in colour, that is located in the planet's southern hemisphere. It rotates anticlockwise once every 6 earth days. With a *radio telescope thermal radiation can be detected from the Jovian stratosphere of mainly hydrogen and helium, indicating a temperature between 100 and 130 K close to the outermost layer. Jupiter puts out more heat than it receives from the sun, so it must have an internal heat source. Microwave nonthermal emission can also be detected from the Jovian *Van Allen belts. The planet has at least 63 satellites, the largest of which are known as the *Galilean satellites. A planetary ring system was discovered in 1979 by the US Voyager I probe.

(((•))) **SEE WEB LINKS**

- Transcript of 2007 podcast discussion between Fraser Cain and Dr Pamela Gay, astronomer, writer, and teacher

**Jurassic** The second geological period of the Mesozoic era. It followed the Triassic, which ended about 200 million years ago, and extended until the beginning of the Cretaceous period, about 145 million years ago. It was named in 1829 by A. Brongniart after the Jura Mountains on the borders of France and Switzerland. Jurassic rocks include clays and limestones in which fossil flora and fauna are abundant. Plants included ferns, cycads, ginkgos, rushes, and conifers. Important invertebrates included *ammonites (on which the Jurassic is zoned), corals, bra-chiopods, bivalves, and echinoids. Reptiles dominated the vertebrates; the first flying reptiles – the pterosaurs – and the first primitive bird, *Archaeopteryx*, appeared.

**juvenile hormone** A hormone secreted by insects from a pair of endocrine glands (**corpora allata**) close to the brain. It inhibits metamorphosis and maintains the presence of larval features.

j

**kainite** A naturally occurring double salt of magnesium sulphate and potassium chloride, $MgSO_4.KCl.3H_2O$.

**Kainozoic** *See* CENOZOIC.

**kalinite** A mineral form of *aluminium potassium sulphate $(Al_2(SO_4)_3.K_2SO_4.24H_2O)$.

**kallidin** *See* KININ.

**Kaluza–Klein theory** A type of *unified-field theory that postulates a generalization of the general theory of relativity to higher than four space–time dimensions. In five space–time dimensions this gives general relativity and electromagnetic interactions. In higher space–time dimensions Kaluza–Klein theories give general relativity and more general *gauge theories. A combination of Kaluza–Klein theory and *supersymmetry gives rise to *supergravity, which needs eleven space–time dimensions. In these theories it is proposed that the higher dimensions are 'rolled up' to become microscopically small (a process known as **spontaneous compactification**) with four macroscopic space–time dimensions remaining. It is named after the German mathematician Theodor Kaluza (1885–1954) and the Swedish physicist Oscar Klein (1894–1977).

**kame** An isolated mound of rock particles, originally formed at the lower end or side of a slow-moving glacier. The mound consists of layers of gravel and sand, which were transported by meltwater and left behind when the ice melted. *See also* ESKER.

**kaolin (china clay)** A soft white clay that is composed chiefly of the mineral kaolinite (*see* CLAY MINERALS). It is formed during the weathering and hydrothermal alteration of other clays or feldspar. Kaolin is mined in the UK, France, the Czech Republic, and USA. Besides its vital importance in the ceramics industry it is also used extensively as a filler in the manufacture of rubber, paper, paint, and textiles and as a constituent of medicines.

**kaon** A K-meson. *See* MESON.

**karst** A type of broken limestone terrain, characterized by fissures (grikes) and depressions (dolines). A typical karst landscape, lacking vegetation, resembles an area paved with large slabs of limestone. Rainwater percolates through the fissures into the rock beneath, where it can follow bedding planes and form underground streams, which carve out caves. Surface streams may disappear down *sink holes.

**karyogamy** The fusion of nuclei or nuclear material that occurs during sexual reproduction. *See* FERTILIZATION.

**karyogram (idiogram)** A diagram representing the characteristic features of the *chromosomes of a species.

**karyokinesis** The division of a cell nucleus. *See* MEIOSIS; MITOSIS.

**karyotype** The number and structure of the *chromosomes in the nucleus of a cell. The karyotype is identical in all the *diploid cells of an organism.

**Kastle Meyer test (phenolphthalein test)** A presumptive test used to indicate blood. Phenolphthalein and hydrogen peroxide are used; reaction with haemoglobin in the blood gives a pink colour.

**katal** Symbol kat. A non-SI unit of enzyme activity defined as the catalytic activity of an enzyme that increases the rate of conversion of a specified chemical reaction by 1 mol $s^{-1}$ under specified assay conditions.

**Kater's pendulum** A complex *pendulum designed by Henry Kater (1777–1835) to measure the acceleration of free fall. It consists of a metal bar with knife edges attached near the ends and two weights that can slide between the knife edges. The bar is pivoted from each knife edge in turn and the positions of the weights are adjusted so that the period of the pendulum is the same with both pivots. The period is then given by the formula for a simple pendulum, which enables $g$ to be calculated.

**katharometer** An instrument for comparing the thermal conductivities of two gases by comparing the rate of loss of heat from two heating coils surrounded by the gases. The instrument can be used to detect the presence of a small amount of an impurity in air and is also used as a detector in gas chromatography.

**kb** *See* KILOBASE.

**KDa** *See* KILODALTON.

**keel (carina)** The projection of bone from the sternum (breastbone) of a bird or bat, to which the powerful flight muscles are attached. The sterna of flightless birds (e.g. ostrich and emu) lack keels.

**keeper** A piece of soft iron used to bridge the poles of a permanent magnet when it is not in use. It reduces the leakage field and thus preserves the magnetization.

**Kekulé, Friedrich August von Stradonitz** (1829–96) German chemist, who became professor at Ghent (1858) and later at Bonn (1867). He studied the structures of organic molecules and is best remembered for his structure for *benzene, which he correctly interpreted as having a symmetrical ring of six carbon atoms.

**Kekulé structure** A proposed structure of *benzene in which the molecule has a hexagonal ring of carbon atoms linked by alternating double and single bonds. Kekulé structures contribute to the resonance hybrid of benzene. The structure was suggested in 1865 by Friedrich August Kekulé.

**kelp** Any large brown seaweed (*see* PHAEO-PHYTA) or its ash, used as a source of iodine.

**kelvin** Symbol K. The *SI unit of thermodynamic *temperature equal to the fraction 1/273.16 of the thermodynamic temperature of the *triple point of water. The magnitude of the kelvin is equal to that of the degree celsius (centigrade), but a temperature expressed in degrees celsius is numerically equal to the temperature in kelvins less 273.15 (i.e. °C = K – 273.15). The *absolute zero of temperature has a temperature of 0 K (–273.15°C). The former name **degree kelvin** (symbol °K) became obsolete by international agreement in 1967. The unit is named after Lord Kelvin.

**Kelvin, Baron** (William Thomson; 1824–1907) British physicist, born in Belfast, who became professor of natural philosophy at Glasgow University in 1846. He carried out important experimental work on electro-

magnetism, inventing the mirror *galvanometer and contributing to the development of telegraphy. He also worked with James *Joule on the *Joule–Thomson (or Joule–Kelvin) effect. His main theoretical work was in *thermodynamics, in which he stressed the importance of the conservation of energy (*see* CONSERVATION LAW). He also introduced the concept of *absolute zero and the Kelvin temperature scale based on it; the unit of thermodynamic temperature is named after him.

**Kelvin effect** *See* THOMSON EFFECT.

**Kepler, Johannes** (1571–1630) German astronomer, whose achievements helped vindicate the heliocentric theory of *Copernicus. In 1600 Kepler went to Prague to work for Tycho *Brahe, whom he succeeded as imperial mathematician to the court of the Holy Roman Emperor Rudolf II. During his time in Prague, Kepler used the decades of accurate and systematic observational data collected by Tycho to formulate his three laws of planetary motion (*see* KEPLER'S LAWS) and to publish the *Rudolphine Tables* (1627), a set of mathematical tables providing instructions for finding the positions of the planets together with a comprehensive star catalogue. Kepler also did pioneering work in optics, making an improved refracting telescope (the **Keplerian telescope**).

**Kepler's laws** Three laws of planetary motion formulated by Johannes Kepler on the basis of observations made by Tycho Brahe. Kepler published the first and second laws in 1609 and the third in 1619. The laws state that: (1) the orbits of the planets are elliptical with the sun at one *focus of the ellipse; (2) each planet revolves around the sun so that an imaginary line (the **radius vector**) connecting the planet to the sun sweeps out equal areas in equal time periods; (3) the ratio of the square of each planet's *sidereal period to the cube of its distance from the sun is a constant for all the planets.

**keratin** Any of a group of fibrous *proteins occurring in hair, feathers, hooves, and horns. Keratins have coiled polypeptide chains that combine to form supercoils of several polypeptides linked by disulphide bonds between adjacent cysteine amino acids. Aggregates of these supercoils form microfibrils, which are embedded in a protein matrix. This produces a strong but elastic structure.

**keratinization (cornification)** The process in which the cytoplasm of the outermost cells of the mammalian *epidermis is replaced by *keratin. Keratinization occurs in the *stratum corneum, feathers, hair, claws, nails, hooves, and horns.

**kerosine** See PETROLEUM.

**Kerr effect** The ability of certain substances to refract differently light waves whose vibrations are in two directions (see DOUBLE REFRACTION) when the substance is placed in an electric field. The effect, discovered in 1875 by John Kerr (1824–1907), is caused by the fact that certain molecules have electric *dipoles, which tend to be orientated by the applied field; the normal random motions of the molecules tends to destroy this orientation and the balance is struck by the relative magnitudes of the field strength, the temperature, and the magnitudes of the dipole moments.

The Kerr effect is observed in a **Kerr cell**, which consists of a glass cell containing the liquid or gaseous substance; two capacitor plates are inserted into the cell and light is passed through it at right angles to the electric field. There are two principal indexes of refraction: $n_o$ (the ordinary index) and $n_e$ (the extraordinary index). The difference in the velocity of propagation in the cell causes a phase difference, $\delta$, between the two waves formed from a beam of monochromatic light, wavelength $\lambda$, such that

$$\delta = (n_o - n_e)x/\lambda,$$

where $x$ is the length of the light path in the cell. Kerr also showed empirically that the ratio

$$(n_o - n_e)\lambda = BE^2,$$

where $E$ is the field strength and $B$ is a constant, called the **Kerr constant**, which is characteristic of the substance and approximately inversely proportional to the thermodynamic temperature.

The **Kerr shutter** consists of a Kerr cell filled with a liquid, such as nitrobenzene, placed between two crossed polarizers; the electric field is arranged to be perpendicular to the axis of the light beam and at 45° to the axis of the polarizers. In the absence of a field there is no optical path through the device. When the field is switched on the nitrobenzene becomes doubly refracting and a path opens between the crossed polarizers.

**ketals** Organic compounds, similar to *acetals, formed by addition of an alcohol to a ketone. If one molecule of ketone (RR'CO) reacts with one molecule of alcohol R"OH, then a **hemiketal** is formed. The rings of ketose sugars are hemiketals. Further reaction produces a full ketal (RR'C(OR")$_2$).

**(())) SEE WEB LINKS**
• Information about IUPAC nomenclature

**ketamine** A veterinary anaesthetic that is used illegally as a club drug. It is a class A drug in the UK.

**ketene** 1. The compound $CH_2$=C=O (**ethenone**). 2. Any of a class of compounds of the type $R_1R_2$=C=O, where $R_1$ and $R_2$ are organic groups. Ketenes are reactive compounds and are often generated in a reaction medium for organic synthesis.

**(())) SEE WEB LINKS**
• Information about IUPAC nomenclature

**keto–enol tautomerism** A form of tautomerism in which a compound containing a –$CH_2$–CO– group (the **keto form** of the molecule) is in equilibrium with one containing the –CH=C(OH)– group (the **enol**). It occurs by migration of a hydrogen atom between a carbon atom and the oxygen on an adjacent carbon. See ISOMERISM.

**keto form** See KETO–ENOL TAUTOMERISM.

**ketohexose** See MONOSACCHARIDE.

**ketone body** Any of three compounds, acetoacetic acid (3-oxobutanoic acid, $CH_3COCH_2COOH$), β-hydroxybutyric acid (3-hydroxybutanoic acid, $CH_3CH(OH)$-$CH_2COOH$), and acetone (propanone, $CH_3COCH_3$), all of which are produced by the liver as a result of the metabolism of body fat deposits. Ketone bodies are normally used as energy sources by peripheral tissues. However, if carbohydrate supply is limited (e.g. during starvation or in diabetics), the blood level of ketone bodies rises and they may be present in urine, giving it a characteristic 'pear drops' odour. This condition is called **ketosis**.

**ketones** Organic compounds that contain the carbonyl group (>C=O) linked to two hydrocarbon groups. The **ketone group** is a carbonyl group with two single bonds to other carbon atoms. In systematic chemical nomenclature, ketone names end with the suffix -*one*. Examples are propanone (acetone), $CH_3COCH_3$, and butanone (methyl ethyl ketone), $CH_3COC_2H_5$. Ketones can be made by oxidizing secondary alcohols to convert the C–OH group to C=O. Certain ketones form addition compounds with sodium hydrogensulphate(IV) (sodium hydrogensulphite). They also form addition compounds with hydrogen cyanide to give *cyanohydrins and with alcohols to give *ketals. They undergo condensation reactions to yield *oximes, *hydrazones, phenylhydrazones, and *semicarbazones. These are reactions that they share with aldehydes. Unlike aldehydes, they do not affect Fehling's solution or Tollen's reagent and do not easily oxidize. Strong oxidizing agents produce a mixture of carboxylic acids; butanone, for example, gives ethanoic and propanoic acids.

🌐 **SEE WEB LINKS**
• Information about IUPAC nomenclature

**ketopentose** *See* MONOSACCHARIDE.

**ketose** *See* MONOSACCHARIDE.

**keystone species** A species whose impact on its community is disproportionately large relative to its abundance. This is generally because it alone fulfils some crucial functional role in the community, the continuation of which is essential for the survival of numerous other species. Classic examples are the beaver (*Castor* spp.), whose dam building creates the unique beaver ponds on which many other species depend, and the bison (*Bos bison*), responsible for a mosaic-like grazing pattern that underpinned the biodiversity of the grasslands of North America.

**khat** A plant, *Catha edilis*, found in East Africa and the Arabian peninsular. The leaves are chewed to give a mildly stimulating and euphoric effect. Its activity comes from two alkaloids: *cathine and, in fresh leaves, the more potent *cathinone. It is a controlled substance in many countries including the US. In the UK it is not a controlled substance and is used by certain Somalian and Yemeni ethnic groups.

**kibi-** *See* BINARY PREFIXES.

**kidney** The main organ of *excretion of vertebrates, through which nitrogenous waste material (usually in the form of *urine) is eliminated from the body. In mammals there is a pair of kidneys situated in the abdomen (see illustration). Each has an outer **cortex** and an inner **medulla** and is made up of tubular units called *nephrons, through which nitrogenous waste is filtered from the

blood, with the formation of urine. The nephrons drain into a basin-like cavity in the kidney (the **renal pelvis**), which leads to the *ureter and *bladder.

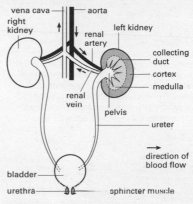

**Kidney.** The kidneys of a mammal.

**kieselguhr** A soft fine-grained deposit consisting of the siliceous skeletal remains of diatoms, formed in lakes and ponds. Kieselguhr is used as an absorbent, filtering material, filler, and insulator.

**kieserite** A mineral form of *magnesium sulphate monohydrate, $MgSO_4.H_2O$.

**killer cell** *See* NATURAL KILLER CELL; T CELL.

**kilo** Symbol k. A prefix used in the metric system to denote 1000 times. For example, 1000 volts = 1 kilovolt (kV).

**kilobase** Symbol kb. A unit used at the molecular level for measuring distances along nucleic acids, chromosomes, or genes, equal to 1000 bases (equivalent to 1000 nucleotides or base pairs). *See also* BASE PAIR.

**kilodalton** Symbol kDa. A non-SI unit of mass used to express molecular mass, especially for large molecules, such as proteins and polysaccharides. It is equal to 1000 daltons (*see* ATOMIC MASS UNIT).

**kilogram** Symbol kg. The *SI unit of mass defined as a mass equal to that of the international platinum–iridium prototype kept by the International Bureau of Weights and Measures at Sèvres, near Paris.

**kiloton weapon** A nuclear weapon with an explosive power equivalent to one thousand tons of TNT. *Compare* MEGATON WEAPON.

**kilowatt-hour** Symbol kWh. The commercial unit of electrical energy. It is equivalent to a power consumption of 1000 watts for 1 hour. 1 kWh = $3.6 \times 10^6$ joules.

**kimberlite** A rare igneous rock that often contains diamonds. It occurs as narrow pipe intrusions but is often altered and fragmented. It consists of olivine and phlogopite mica, usually with calcite, serpentine, and other minerals. The chief occurrences of kimberlite are in South Africa, especially at Kimberley (after which the rock is named), and in the Yakutia area of Siberia.

**kinase** An enzyme that can transfer a phosphate group from a high-energy phosphate, such as ATP, to an organic molecule. *Phosphorylation is normally required to activate the molecule, which is often an enzyme. For example, kinases activate the precursors of enzymes secreted in pancreatic juice (*see* CHYMOTRYPSIN; TRYPSIN).

**kinematic equation** *See* EQUATION OF MOTION.

**kinematics** The branch of mechanics concerned with the motions of objects without being concerned with the forces that cause the motion. In this latter respect it differs from *dynamics, which is concerned with the forces that affect motion. *See also* EQUATION OF MOTION.

**kinematic viscosity** Symbol ν. The ratio of the *viscosity of a liquid to its density. The SI unit is $m^2 s^{-1}$.

**kinesis** The movement of a cell or organism in response to a stimulus in which the rate of movement depends on the intensity (rather than the direction) of the stimulus. For example, a woodlouse moves slowly in a damp atmosphere and quickly in a dry one.

**kinetic effect** A chemical effect that depends on reaction rate rather than on thermodynamics. For example, diamond is thermodynamically less stable than graphite; its apparent stability depends on the vanishingly slow rate at which it is converted. *Overpotential in electrolytic cells is another example of a kinetic effect. **Kinetic isotope effects** are changes in reaction rates produced by isotope substitution. For example, if the slow step in a chemical reaction is the breaking of a C–H bond, the rate for the deuterated compound would be slightly

lower because of the lower vibrational frequency of the C–D bond. Such effects are used in investigating the mechanisms of chemical reactions.

**kinetic energy** *See* ENERGY.

**kinetic equations** Equations used in *kinetic theory. The *Boltzmann equation is an example of a kinetic equation. An important application of kinetic equations is to calculate *transport coefficients (and inverse transport coefficients), such as *conductivity and *viscosity in *non-equilibrium statistical mechanics. In general, kinetic equations do not have exact solutions for interacting systems. If the system is near to *equilibrium an approximation technique can be used by regarding the deviation from equilibrium as a *perturbation.

**kinetic isotope effect** *See* KINETIC EFFECT.

**kinetics** The branch of physical chemistry concerned with measuring and studying the rates of chemical reactions. The main aim of chemical kinetics is to determine the mechanism of reactions by studying the rate under different conditions (temperature, pressure, etc.).

**kinetic theory** A theory, largely the work of Count Rumford, James Joule, and James Clerk Maxwell, that explains the physical properties of matter in terms of the motions of its constituent particles. In a gas, for example, the pressure is due to the incessant impacts of the gas molecules on the walls of the container. If it is assumed that the molecules occupy negligible space, exert negligible forces on each other except during collisions, are perfectly elastic, and make only brief collisions with each other, it can be shown that the pressure *p* exerted by one mole of gas containing *n* molecules each of mass *m* in a container of volume *V*, will be given by:

$$p = nm(\bar{c})^2/3V,$$

where $\bar{c}^2$ is the mean square speed of the molecules. As according to the *gas laws for one mole of gas: $pV = RT$, where *T* is the thermodynamic temperature, and *R* is the molar *gas constant, it follows that:

$$RT = nm(\bar{c})^2/3$$

Thus, the thermodynamic temperature of a gas is proportional to the mean square speed of its molecules. As the average kinetic *energy of translation of the molecules is $m(\bar{c})^2/2$, the temperature is given by:

$$T = (m(\bar{c})^2/2)(2n/3R)$$

The number of molecules in one mole of any gas is the *Avogadro constant, $N_A$; therefore in this equation $n = N_A$. The ratio $R/N_A$ is a constant called the *Boltzmann constant ($k$). The average kinetic energy of translation of the molecules of one mole of any gas is therefore $3kT/2$. For monatomic gases this is proportional to the *internal energy ($U$) of the gas, i.e.

$$U = N_A 3kT/2$$

and as $k = R/N_A$

$$U = 3RT/2$$

For diatomic and polyatomic gases the rotational and vibrational energies also have to be taken into account (*see* DEGREES OF FREEDOM).

In liquids, according to the kinetic theory, the atoms and molecules still move around at random, the temperature being proportional to their average kinetic energy. However, they are sufficiently close to each other for the attractive forces between molecules to be important. A molecule that approaches the surface will experience a resultant force tending to keep it within the liquid. It is, therefore, only some of the fastest moving molecules that escape; as a result the average kinetic energy of those that fail to escape is reduced. In this way evaporation from the surface of a liquid causes its temperature to fall.

In a crystalline solid the atoms, ions, and molecules are able only to vibrate about the fixed positions of a *crystal lattice; the attractive forces are so strong at this range that no free movement is possible.

**kinetochore** A platelike structure by which the microtubules of the *spindle attach to the *centromere of a chromosome during nuclear division. It acts as a motor, pulling the centromere along the attached microtubules towards the spindle pole.

**kinetosome (basal body)** *See* UNDULIPODIUM.

**kingdom** In traditional classification systems, the highest category into which organisms are classified. The original two kingdoms, Plantae (*see* PLANT) and Animalia (*see* ANIMAL), were over time increased to five: Bacteria (or Prokaryotae; *see* BACTERIA), Protoctista (including protozoa and algae), Fungi (*see* FUNGI), Plantae, and Animalia.

However, the discovery of the archaebacteria (*see* ARCHAEA) led taxonomists to suggest a superordinate category in the taxonomic hierarchy – the *domain. According to modern molecular systematics, there are three domains, but the number of kingdoms is much harder to determine.

**kinin** 1. One of a group of peptides, occurring in blood, that are involved in inflammation. Kinins are formed in response to blood-vessel injury by the enzymatic splitting of blood plasma globulins (**kininogens**) at the site of inflammation. Kinins include **bradykinin** and **kallidin**. They cause local increases in the permeability of small blood vessels. 2. *See* CYTOKININ.

**kinomere** *See* CENTROMERE.

**kin selection** Natural selection of genes that tend to cause the individuals bearing them to be altruistic to close relatives. These relatives therefore have a higher probability of bearing identical copies of those same genes than do other members of the population. Thus kin selection for a gene that tends to cause an animal to share food with a close relative will result in the gene being spread through the population because it (unconsciously) benefits itself. The more closely two animals are related, the higher the probability that they share some identical genes and therefore the more closely their interests coincide. Parental care is a special case of kin selection. *See* INCLUSIVE FITNESS.

**Kipp's apparatus** A laboratory apparatus for making a gas by the reaction of a solid with a liquid (e.g. the reaction of hydrochloric acid with iron sulphide to give hydrogen sulphide). It consists of three interconnected glass globes arranged vertically, with the solid in the middle globe. The upper and lower globes are connected by a tube and contain the liquid. The middle globe has a tube with a tap for drawing off gas. When the tap is closed, pressure of gas forces the liquid down in the bottom reservoir and up into the top, and reaction does not occur. When the tap is opened, the release in pressure allows the liquid to rise into the middle globe, where it reacts with the solid. It is named after Petrus Kipp (1808–64).

**Kirchhoff, Gustav Robert** (1824–87) German physicist, who in 1850 became a professor at Breslau and four years later joined Robert *Bunsen at Heidelberg. In 1845, while still a student, he formulated

*Kirchhoff's laws concerning electric circuits. With Bunsen he worked on spectroscopy, a technique that led them to discover the elements *caesium (1860) and *rubidium (1861).

**Kirchhoff's law of radiation** A law stating that the emissivity of a body is equal to its absorptance at the same temperature.

**Kirchhoff's laws** Two laws relating to electric circuits, first formulated by Gustav Kirchhoff. (a) The current law states that the algebraic sum of the currents flowing through all the wires in a network that meet at a point is zero. (b) The voltage law states that the algebraic sum of the e.m.f.s within any closed circuit is equal to the sum of the products of the currents and the resistances in the various portions of the circuit.

**Kirkwood gap** Any of several spaces in the distribution of *asteroids in the main belt that correspond to locations of *orbital resonance with Jupiter. Consequently any asteroids found there have long ago been perturbed by Jupiter's enormous gravitational influence into more eccentric orbits. The gaps were discovered by the American astronomer Daniel Kirkwood (1814–95).

**Kjeldahl's method** A method for measuring the percentage of nitrogen in an organic compound. The compound is boiled with concentrated sulphuric acid and copper(II) sulphate catalyst to convert any nitrogen to ammonium sulphate. Alkali is added and the mixture heated to distil off ammonia. This is passed into a standard acid solution, and the amount of ammonia can then be found by estimating the amount of unreacted acid by titration. The amount of nitrogen in the original specimen can then be calculated. The method was developed by the Danish chemist Johan Kjeldahl (1849–1900).

**klinostat** A device used in experiments to test the influence of gravity on the growth movements of plants (*see* GEOTROPISM). It consists of a motor that slowly rotates a drum inside which seedlings are attached. This prevents any single part of the seedlings from receiving uninterrupted gravitational stimulation and results in horizontal growth of the seedlings.

**klystron** An electron tube that generates or amplifies microwaves by **velocity modulation**. Several types are used; in the simple two-cavity klystron a beam of high-energy electrons from an electron gun is passed

through a *resonant cavity, where it interacts with high-frequency radio waves. This microwave energy modulates the velocities of the electrons in the beam, which then enters a drift space where the faster electrons overtake the slower ones to form bunches. The bunched beam now has an alternating component, which is transferred to an output cavity and thence to an output waveguide.

**knee-jerk reflex** *See* STRETCH REFLEX.

**knockin** A technique related to gene *knockout in which a gene is inserted into the genome of a cell, cell line, or organism. Such an insertion can be targeted to a particular site in the genome and may replace or supplement existing genes. Moreover, it can be induced to switch on only in certain tissues, or at certain stages of development, thus mimicking the normal behaviour of genes.

**knocking** The metallic sound produced by a spark-ignition petrol engine under certain conditions. It is caused by rapid combustion of the unburnt explosive mixture in the combustion chambers ahead of the flame front. As the flame travels from the sparking plug towards the piston it compresses and heats the unburnt gases ahead of it. If the flame front moves fast enough, normal combustion occurs and the explosive mixture is ignited progressively by the flame. If it moves too slowly, ignition of the last part of the unburnt gas can occur very rapidly before the flame reaches it, producing a shock wave that travels back and forth across the combustion chamber. The result is overheating, possible damage to the plugs, an undesirable noise, and loss of power (probably due to preignition caused by overheated plugs). Knocking can be avoided by an engine design that increases turbulence in the combustion chamber and thereby increases flame speed. It also can be avoided by reducing the compression ratio, but this involves loss of efficiency. The most effective method is to use high-octane fuel (*see* OCTANE NUMBER), which has a longer self-ignition delay than low-octane fuels. This can be achieved by the addition of an **antiknock agent**, such as lead(IV) tetraethyl, to the fuel, which retards the combustion chain reactions. However, lead-free petrol is now preferred to petrol containing lead tetraethyl owing to environmental dangers arising from lead in the atmosphere. In the USA the addition of lead compounds is now forbidden. New formulae for petrol are designed to raise the octane number without polluting the atmosphere. These new formulae include increasing the content of aromatics and oxygenates (oxygen-containing compounds, such as alcohols). However, it is claimed that the presence in the atmosphere of incompletely burnt aromatics constitutes a cancer risk.

**knockout** Inactivation of a particular gene or genes within an organism or cell in order to assess the impact of this defect on the organism. One technique involves genetically engineering laboratory animals (especially mice) so that a normal gene is replaced with a defective homologous gene. Experiments with mice treated in this way reveal how defects in particular genes can affect the development and life of the animal. A less laborious technique uses the phenomenon of *RNA interference to suppress the expression of specific genes in cultured tissue cells – so-called **gene silencing**.

**knotane** *See* MOLECULAR KNOT.

**Knudsen flow** *See* MOLECULAR FLOW.

**Koch, Robert** *See* EHRLICH, PAUL.

**Kohlrausch's law** If a salt is dissolved in water, the conductivity of the (dilute) solution is the sum of two values – one depending on the positive ions and the other on the negative ions. The law, which depends on the independent migration of ions, was deduced experimentally by the German chemist Friedrich Kohlrausch (1840–1910).

**Kolbe's method** A method of making alkanes by electrolysing a solution of a carboxylic acid salt. For a salt $Na^+RCOO^-$, the carboxylate ions lose electrons at the cathode to give radicals:

$$RCOO^- - e \rightarrow RCOO\cdot$$

These decompose to give alkyl radicals

$$RCOO\cdot \rightarrow R\cdot + CO_2$$

Two alkyl radicals couple to give an alkane

$$R\cdot + R\cdot \rightarrow RR$$

The method can only be used for hydrocarbons with an even number of carbon atoms, although mixtures of two salts can be electrolysed to give a mixture of three products. The method was discovered by the German chemist Herman Kolbe (1818–84), who electrolysed pentanoic acid ($C_4H_9COOH$) in 1849 and obtained a hydrocarbon, which he assumed was the substance 'butyl' $C_4H_9$ (actually octane, $C_8H_{18}$).

**Kovar** A trade name for an alloy of iron, cobalt, and nickel with an *expansivity similar to that of glass. It is therefore used in making glass-to-metal seals, especially in circumstances in which a temperature variation can be expected.

**Krebs, Sir Hans Adolf** (1900–81) German-born British biochemist, who emigrated to Britain in 1933, working at Sheffield University before moving to Oxford in 1954. Krebs is best known for the *Krebs cycle, the basis of which he discovered in 1937. Details were later added by Fritz Lipmann (1899–1986), with whom Krebs shared the 1953 Nobel Prize for physiology or medicine.

**Krebs cycle (citric acid cycle; tricarboxylic acid cycle; TCA cycle)** A cyclical series of biochemical reactions that is fundamental to the metabolism of aerobic organisms, i.e. animals, plants, and many microorganisms (see illustration overleaf). The enzymes of the Krebs cycle are located in the *mitochondria and are in close association with the components of the *electron transport chain. The two-carbon *acetyl coenzyme A (acetyl CoA) reacts with the four-carbon oxaloacetate to form the six-carbon citrate. In a series of seven reactions, this is reconverted to oxaloacetate and produces two molecules of carbon dioxide. Most importantly, the cycle generates one molecule of guanosine triphosphate (GTP – equivalent to 1 ATP) and reduces three molecules of the coenzyme *NAD to NADH and one molecule of the coenzyme *FAD to $FADH_2$. NADH and $FADH_2$ are then oxidized by the electron transport chain to generate three and two molecules of ATP respectively. This gives a net yield of 12 molecules of ATP per molecule of acetyl CoA.

Acetyl CoA can be derived from carbohydrates (via *glycolysis), fats, or certain amino acids. (Other amino acids may enter the cycle at different stages.) Thus the Krebs cycle is the central 'crossroads' in the complex system of metabolic pathways and is involved not only in degradation and energy production but also in the synthesis of biomolecules. It is named after its principal discoverer, Sir Hans Krebs.

**Kroll process** A process for producing certain metals by reducing the chloride with magnesium metal, e.g.

$$TiCl_4 + 2Mg \rightarrow Ti + 2MgCl_2$$

**krypton** Symbol Kr. A colourless gaseous

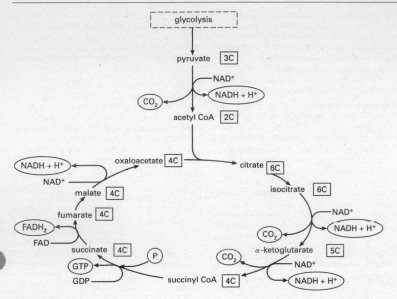

**Krebs cycle.**

element belonging to group 0 (the *noble gases) of the periodic table; a.n. 36; r.a.m. 83.80; d. 3.73 g m$^{-3}$; m.p. $-156.6°C$; b.p. $-152.3°C$. Krypton occurs in the air (0.0001% by volume) from which it can be extracted by fractional distillation of liquid air. Usually, the element is not isolated but is used with other inert gases in fluorescent lamps, etc. The element has five natural isotopes (mass numbers 78, 80, 82, 83, 84) and there are five radioactive isotopes (76, 77, 79, 81, 85). Krypton–85 (half-life 10.76 years) is produced in fission reactors and it has been suggested that an equilibrium amount will eventually occur in the atmosphere. The element is practically inert and forms very few compounds (certain fluorides, such as KrF$_2$, have been reported).

**⊕ SEE WEB LINKS**
- Information from the WebElements site

***K* selection** A type of selection that favours organisms with a low rate of reproduction but whose populations expand to the maximum number of individuals that the habitat can support (the *carrying capacity of the habitat). *K*-selected species (or *K* **strategists**) tend to be highly adapted to

their environment and are able to compete successfully for food and other resources. They also tend to inhabit stable environments and have relatively long life spans. *Compare* R SELECTION.

**K–T boundary** *See* ALVAREZ EVENT.

**Kuiper belt** A large wide ring of small celestial bodies thought to be composed mostly of ice that orbit the sun beyond the planet Neptune. Short-period *comets are thought to originate either in this region or from further out in the *scattered disc. Kuiper belt objects (KBOs) orbit close to the plane of the solar system in a region that extends from Neptune's orbital path 30 astronomical units (AU) out from the sun to a distance of about 55 AU. Some astronomers believe that there could be as many as 70 000 KBOs in excess of 100 km across. *See* SOLAR SYSTEM (Feature).

**⊕ SEE WEB LINKS**
- The Kuiper Belt Home Page, maintained by David Jewett, one of the co-discoverers of the first known Kuiper belt object in 1992

**Kundt's tube** An apparatus designed by August Kundt (1839–94) in 1866 to measure the speed of sound in various fluids. It consists of a closed glass tube into which a dry powder (such as lycopodium) has been sprinkled. The source of sound in the original device was a metal rod clamped at its centre with a piston at one end, which is inserted into the tube. When the rod is stroked, sound waves generated by the piston enter the tube. If the position of the piston in the tube is adjusted so that the gas column is a whole number of half wavelengths long, the dust will be disturbed by the resulting *stationary waves forming a series of striations,

enabling distances between *nodes to be measured. The vibrating rod can be replaced by a small loudspeaker fed by an oscillator.

**Kupfer nickel**  A naturally occurring form of nickel arsenide, NiAs; an important ore of nickel.

**kurchatovium**  *See* TRANSACTINIDE ELEMENTS.

**kwashiorkor**  *See* MALNUTRITION.

**Kyoto Protocol**  *See* GLOBAL WARMING.

**labelling 1. (isotopic labelling)** The process of replacing a stable atom in a compound with a radioisotope of the same element to enable its path through a biological or mechanical system to be traced by the radiation it emits. In some cases a different stable isotope is used and the path is detected by means of a mass spectrometer. A compound containing either a radioactive or stable isotope is called a **labelled compound** and the atom used is a **label**. If a hydrogen atom in each molecule of the compound has been replaced by a tritium atom, the compound is called a **tritiated compound**. A radioactive labelled compound will behave chemically and physically in the same way as an otherwise identical stable compound, and its presence can easily be detected using a *Geiger counter. This process of **radioactive tracing** is widely used in chemistry, biology, medicine, and engineering. For example, it can be used to follow the course of the reaction of a carboxylic acid with an alcohol to give an ester, e.g.

$$CH_3COOH + C_2H_5OH \rightarrow C_2H_5COOCH_3 + H_2O$$

To determine whether the noncarbonyl oxygen in the ester comes from the acid or the alcohol, the reaction is performed with the labelled compound $CH_3CO^{18}OH$, in which the oxygen in the hydroxyl group of the acid has been 'labelled' by using the $^{18}O$ isotope. It is then found that the water product is $H_2^{18}O$; i.e. the oxygen in the ester comes from the alcohol, not the acid. **2.** The addition to a target substance of a readily identifiable marker, such as a fluorescent dye. This enables the presence and in some cases the amount of target molecule to be determined by (for example) a fluorescence detector. Markers such as fluorescence are now widely used in automated assay techniques.

**labia** *See* LABIUM.

**labile** Describing a chemical compound in which certain atoms or groups can easily be replaced by other atoms or groups. The term is applied to coordination complexes in which ligands can easily be replaced by other ligands in an equilibrium reaction.

**labium 1.** The lower lip in the *mouthparts of an insect, which is used in feeding and is formed by the fusion of a pair of appendages (the second *maxillae). **2.** Either member of two pairs of fleshy folds that form part of the *vulva. The outer and larger pair, the **labia majora**, are covered by pubic hair and contain adipose tissue; the smaller **labia minora** lack adipose tissue and pubic hair. Both pairs of labia contain sebaceous glands.

**labrum** The upper lip in the *mouthparts of an insect. It is formed from a plate of cuticle hinged to the head above the mouth and is used in feeding.

**labyrinth** The system of cavities and tubes that comprises the *inner ear of vertebrates. It consists of a system of membranous structures (**membranous labyrinth**) housed in a similar shaped bony cavity (**bony labyrinth**).

***lac* operon** The *operon that regulates lactose metabolism in the bacterium *Escherichia coli*. Its form was first postulated in 1961 by François Jacob (1920– ) and Jacques Monod (1910–76) to explain control of β-galactosidase synthesis, and it is used as a model for the structure and regulation of genes in prokaryotes. *See* JACOB–MONOD HYPOTHESIS.

**lacrimal gland (lachrymal gland)** The tear gland, present in the eyelids of some vertebrates. The fluid (tears) produced by this gland cleanses and lubricates the exposed surface of the eye; it drains into the nose through the lacrimal duct.

**lactams** Organic compounds containing a ring of atoms in which the group –NH.CO.– forms part of the ring. Lactams can be formed by reaction of an –NH$_2$ group in one part of a molecule with a –COOH group in the other to give a cyclic amide (see illustration). They can exist in an alternative tautomeric form, the **lactim** form, in which the hydrogen atom on the nitrogen has migrated to the oxygen of the carbonyl to give amino acid    lactam

**Lactam formation.**

–N=C(OH)–. The pyrimidine base uracil is an example of a lactam.

**SEE WEB LINKS**
• Information about IUPAC nomenclature of lactams
• Information about IUPAC nomenclature of lactims

**lactase (galactosidase)** The enzyme that breaks down the milk sugar, lactose, to glucose and galactose.

**lactate** A salt or ester of lactic acid (i.e. a 2-hydroxypropanoate).

**lactation** The discharge of milk from the *mammary glands. This generally only occurs after birth of the young and is stimulated by the sucking action of the infants. Lactation is under the control of hormones, notably *prolactin and *oxytocin.

**lacteal** A minute blind-ended lymph vessel that occurs in each *villus of the small intestine. Digested fats are absorbed into the lacteals (*see* CHYLE) and transported to the bloodstream through the *thoracic duct.

**lactic acid (2-hydroxypropanoic acid)** A clear odourless hygroscopic syrupy liquid, $CH_3CH(OH)COOH$, with a sour taste; r.d. 1.206; m.p. 18°C; b.p. 122°C. It is prepared by the hydrolysis of ethanal cyanohydrin or the oxidation of propan-1,2-diol using dilute nitric acid. Lactic acid is manufactured by the fermentation of lactose (from milk) and used in the dyeing and tanning industries. It is an alpha hydroxy *carboxylic acid. *See also* OPTICAL ACTIVITY.

Lactic acid is produced from pyruvic acid in active muscle tissue when oxygen is limited and subsequently removed for conversion to glucose by the liver. During strenuous exercise it may build up in the muscles, causing cramplike pains. It is also produced by fermentation in certain bacteria and is characteristic of sour milk.

**lactims** *See* LACTAMS.

**lactogenic hormone** *See* PROLACTIN.

**lactones** Organic compounds containing a ring of atoms in which the group –CO.O– forms part of the ring. Lactones can be formed (or regarded as formed) by reaction of an –OH group in one part of a molecule with a –COOH group in the other to give a cyclic ester (see illustration). This type of reaction occurs with γ-hydroxy carboxylic acids such as $CH_2(OH)CH_2CH_2COOH$ (in which the hydroxyl group is on the third carbon from the carboxyl group). The resulting γ-lactone has a five-membered ring. Similarly, δ-lactones have six-membered rings. β-lactones, with a four-membered ring, are not produced directly from β-hydroxy acids, but can be synthesized by other means.

**SEE WEB LINKS**
• Information about IUPAC nomenclature

hydroxy carboxylic acid    lactone

**Lactone formation.**

**lactose (milk sugar)** A sugar comprising one glucose molecule linked to a galactose molecule. Lactose is manufactured by the mammary gland and occurs only in milk. For example, cows' milk contains about 4.7% lactose. It is less sweet than sucrose (cane sugar).

**lacuna** A gap or cavity in the tissues of an organism; for example, the hollow centre of certain plant stems or any of the small cavities in bone in which the bone-forming cells are found.

**Ladenburg benzene** An (erroneous) structure for *benzene proposed by Albert Ladenburg (1842–1911), in which the six carbon atoms were arranged at the corners of a triangular prism and linked by single bonds to each other and to the six hydrogen atoms.

**laevorotatory** Designating a chemical compound that rotates the plane of plane-polarized light to the left (anticlockwise for someone facing the oncoming radiation). *See* OPTICAL ACTIVITY.

**laevulose** *See* FRUCTOSE.

**lag** *See* PHASE ANGLE.

**lag phase** *See* BACTERIAL GROWTH CURVE.

**Lagrangian** Symbol $L$. A function used to define a dynamical system in terms of functions of coordinates, velocities, and times given by:

$$L = T - V$$

where $T$ is the kinetic energy of the system and $V$ is the potential energy of the system. The Lagrangian formulation of dynamics has the advantage that it does not deal with many vector quantities, such as forces and accelerations, but only with two scalar functions, $T$ and $V$. This leads to great simplifications in calculations. **Lagrangian dynamics** was formulated by Joseph Louis Lagrange (1736–1813).

**Lagrangian coherent structure** A boundary that separates a moving fluid into a region in which there is turbulence and a region in which there is not turbulence. This boundary can move with time. The concept is useful in fluid mechanics.

**Lagrangian fluid dynamics** A formulation of fluid mechanics in which the trajectories of small regions of the moving fluid are analysed, in contrast to *Eulerian fluid dynamics.

**Lagrangian point** One of two points, in the plane of two large objects orbiting each other, at which a third much smaller object can remain in stable equilibrium. There are three other theoretical Lagrangian points but they are unstable. Each stable Lagrangian point occurs on the orbit of the smaller of the two large objects, at the apex of an equilateral triangle that has as its base a line joining the two large objects. The Trojan *asteroids occupy such Lagrangian points on the orbit of Jupiter around the sun. They are named after Joseph Louis Lagrange (1736–1813).

**LAH** Lithium aluminium hydride; *see* LITHIUM TETRAHYDROALUMINATE(III).

**lake** A pigment made by combining an organic dyestuff with an inorganic compound (usually an oxide, hydroxide, or salt). Absorption of the organic compound on the inorganic substrate yields a coloured complex, as in the combination of a dyestuff with a *mordant. Lakes are used in paints and printing inks.

**Lamarck, Jean-Baptiste de Monet, Chevalier de** (1744–1829) French natural historian. In 1778 he published a flora of France, which included a dichotomous identification key, and later worked on the classification of invertebrates, published in a seven-volume natural history (1815–22). In 1809 he put forward a theory of *evolution that has become known as *Lamarckism (later rejected in favour of Darwinism).

**Lamarckism** One of the earliest superficially plausible theories of *evolution, proposed by Jean-Baptiste de Lamarck in 1809. He suggested that changes in an individual are acquired during its lifetime, chiefly by increased use or disuse of organs in response to "a need that continues to make itself felt", and that these changes are inherited by its offspring. Thus the long neck and limbs of a giraffe are explained as having evolved by the animal stretching its neck to browse on the foliage of trees. This so-called inheritance of acquired characteristics has never unquestionably been demonstrated to occur and the theory was largely displaced by *Darwinism. Lamarckism is also incompatible with the *Central Dogma of molecular biology. *See also* LYSENKOISM.

**lambda particle** A spin $-\frac{1}{2}$ electrically neutral *baryon made up of one up quark, one down quark, and one strange quark. The mass of the lambda particle is 1115.60 MeV and its average lifetime is $2.6 \times 10^{-10}$ s.

**lambda phage** A temperate *bacteriophage that infects cells of the bacterium *Escherichia coli*, where it can either exist as a quiescent prophage (in a state called **lysogeny**) or undergo replication leading to lysis of the host cell and release of new phage particles. Lambda phage has been intensively studied as a model of viral infection and replication and is much used in genetic research and in genetic engineering. Modified lambda phages are used as *vectors in gene cloning, especially for packaging relatively large amounts of foreign DNA.

**lambda point** Symbol $\lambda$. The temperature of 2.186 K below which helium–4 becomes a superfluid. The name derives from the shape of the curve of specific heat capacity against temperature, which is shaped like a Greek letter lambda ($\lambda$) at this point. *See* SUPERFLUIDITY.

**lambert** A former unit of *luminance equal to the luminance of a uniformly diffusing surface that emits or reflects one lumen per square centimetre. It is approximately equal to $3.18 \times 10^3$ Cd m$^{-2}$. It is named after Johann H. Lambert (1728–77).

**Lambert's laws** (1) The *illuminance of a surface illuminated by light falling on it perpendicularly from a point source is inversely proportional to the square of the distance between the surface and the source. (2) If the rays make an angle $\theta$ with the normal to the surface, the illuminance is proportional to $\cos\theta$. (3) **(Bouguer's law)** The *luminous intensity ($I$) of light (or other electromagnetic radiation) decreases exponentially with the distance $d$ that it enters an absorbing medium, i.e.

$$I = I_0 \exp(-\alpha d)$$

where $I_0$ is the intensity of the radiation that enters the medium and $\alpha$ is its **linear absorption coefficient**. These laws were first stated (for light) by Johann H. Lambert (1728–77).

**Lamb shift** A small energy difference between two levels ($^2S_{1/2}$ and $^2P_{1/2}$) in the *hydrogen spectrum. The shift results from the quantum interaction between the atomic electron and the electromagnetic radiation. It was first explained by Willis Eugene Lamb (1913–2008).

**lamella** 1. (in botany) **a.** Any of the paired folds of membranes seen between the *grana in a plant chloroplast. **b.** Any of the spore-bearing gills on the underside of the cap of many mushrooms and toadstools. *See also* MIDDLE LAMELLA. 2. (in zoology) Any of various thin layers of membranes, especially any of the thin layers of tissue of which compact bone is formed.

**lamellar solids** Solid substances in which the crystal structure has distinct layers (i.e. has a layer lattice). The *micas are an example of this type of compound. *Intercalation compounds are lamellar compounds formed by interposition of atoms, ions, etc., between the layers of an existing element or compound. For example, graphite is a lamellar solid. With strong oxidizing agents (e.g. a mixture of concentrated sulphuric and nitric acids) it forms a nonstoichiometric 'graphitic oxide', which is an intercalation compound having oxygen atoms between the layers of carbon atoms.

**Lamellibranchia** *See* BIVALVIA.

**lamina** 1. The thin and usually flattened blade of a leaf, in which photosynthesis and transpiration occurs. The bulk of the lamina is made up of *mesophyll cells interspersed by a network of veins (*vascular bundles). The mesophyll is enclosed by a protective epidermis that produces a waxy cuticle. 2. The leaflike part of the thallus of certain algae, notably kelps. *See also* STIPE.

**laminar flow** *Streamline flow of a fluid in which the fluid moves in layers without fluctuations or turbulence so that successive particles passing the same point have the same velocity. It occurs at low *Reynolds numbers, i.e. low velocities, high viscosities, low densities or small dimensions. The flow of lubricating oil in bearings is normally laminar because of the thinness of the lubricant layer.

**laminated core** A core for a transformer or other electrical machine in which the ferromagnetic alloy is made into thin sheets (laminations), which are oxidized or varnished to provide a relatively high resistance between them. This has the effect of reducing *eddy currents, which occur when alternating currents are used.

**lamp black** A finely divided (microcrystalline) form of carbon made by burning organic compounds in insufficient oxygen. It is used as a black pigment and filler.

**lancelet** *See* CHORDATA.

**Landauer's principle** The principle put forward by Rolf Landauer in the 1960s that energy has to be expended to erase information. This principle links thermodynamics and information theory.

**Landau level** A quantized energy level that occurs when an electrically charged particle is moving in an external magnetic field. The existence of such levels was predicted by Lev Landau in 1930. The concept of Landau levels is important in the theory of the *quantum Hall effect.

**Landé interval rule** A rule in atomic spectra stating that if the *spin–orbit coupling is weak in a given multiplet, the energy differences between two successive $J$ levels (where $J$ is the total resultant angular momentum of the coupled electrons) are proportional to the larger of the two values of $J$. The rule was stated by the German-born US physicist Alfred Landé (1888–1975) in 1923. It can be deduced from the quantum theory of angular momentum. In addition to assuming *Russell–Saunders coupling, the Landé interval rule assumes that the interactions between spin magnetic moments can be ignored, an assumption that is not correct for very light atoms, such as helium. Thus the Landé interval rule is best obeyed by atoms with medium atomic numbers.

**Langlands program** A set of conjectures put forward in 1967 by the Canadian mathematician Robert Langlands (1936– ), relating different branches of mathematics. Some of these conjectures have physical realizations in *quantum field theory and *superstring theory.

**Langmuir adsorption isotherm** An equation used to describe the amount of gas adsorbed on a plane surface, as a function of the pressure of the gas in equilibrium with the surface. The Langmuir adsorption isotherm can be written:

$$\theta = bp/(1 + bp),$$

where $\theta$ is the fraction of the surface covered by the adsorbate, $p$ is the pressure of the gas, and $b$ is a constant called the **adsorption coefficient**, which is the equilibrium constant for the process of adsorption. The Langmuir adsorption isotherm was derived by the US chemist Irving Langmuir (1881–1957), using the *kinetic theory of gases and making the assumptions that:
(1) the adsorption consists entirely of a monolayer at the surface;
(2) there is no interaction between molecules on different sites and each site can hold only one adsorbed molecule;
(3) the heat of adsorption does not depend on the number of sites and is equal for all sites.
The Langmuir adsorption isotherm is of limited application since for real surfaces the energy is not the same for all sites and interactions between adsorbed molecules cannot be ignored.

**Langmuir–Blodgett film** A film of molecules on a surface that can contain multiple layers of film. A Langmuir–Blodgett film with multiple layers can be made by dipping a plate into a liquid so that it is covered by a monolayer and then repeating the process. This process, called the **Langmuir–Blodgett technique**, enables a multilayer to be built up, one monolayer at a time. Langmuir–Blodgett films have many potential practical applications, including insulation for optical and semiconductor devices and selective membranes in biotechnology.

**lanolin** An emulsion of purified wool fat in water, containing cholesterol and certain terpene alcohols and esters. It is used in cosmetics.

**lansfordite** A mineral form of *magnesium carbonate pentahydrate, $MgCO_3.5H_2O$.

**lanthanides** See LANTHANOIDS.

**lanthanoid contraction** See LANTHANOIDS.

**lanthanoids (lanthanides; lanthanons; rare-earth elements)** A series of elements in the *periodic table, generally considered to range in proton number from cerium (58) to lutetium (71) inclusive. The lanthanoids all have two outer $s$-electrons (a $6s^2$ configuration), follow lanthanum, and are classified together because an increasing proton number corresponds to increase in number of $4f$ electrons. In fact, the $4f$ and $5d$ levels are close in energy and the filling is not smooth. The outer electron configurations are as follows:
57 lanthanum (La) $5d^16s^2$
58 cerium (Ce) $4f5d^16s^2$ (or $4f^26s^2$)
59 praseodymium (Pr) $4f^36s^2$
60 neodymium (Nd) $4f^46s^2$
61 promethium (Pm) $4f^56s^2$
62 samarium (Sm) $4f^66s^2$
63 europium (Eu) $4f^76s^2$
64 gadolinium (Gd) $4f^75d^16s^2$
65 terbium (Tb) $4f^96s^2$
66 dysprosium (Dy) $4f^{10}6s^2$
67 holmium (Ho) $4f^{11}6s^2$
68 erbium (Er) $4f^{12}6s^2$
69 thulium (Tm) $4f^{13}6s^2$
70 ytterbium (Yb) $4f^{14}6s^2$
71 lutetium (Lu) $4f^{14}5d^16s^2$
Note that lanthanum itself does not have a $4f$ electron but it is generally classified with the lanthanoids because of its chemical similarities, as are yttrium (Yt) and scandium (Sc). Scandium, yttrium, and lanthanum are $d$-block elements; the lanthanoids and *actinoids make up the $f$-block.
The lanthanoids are sometimes simply called the **rare earths**, although strictly the 'earths' are their oxides. Nor are they particularly rare: they occur widely, usually together. All are silvery very reactive metals. The $f$-electrons do not penetrate to the outer part of the atom and there is no $f$-orbital participation in bonding (unlike the $d$-orbitals of the main *transition elements) and the elements form few coordination compounds. The main compounds contain $M^{3+}$ ions. Cerium also has the highly oxidizing $Ce^{4+}$

state and europium and ytterbium have a $M^{2+}$ state.

The 4*f* orbitals in the atoms are not very effective in shielding the outer electrons from the nuclear charge. In going across the series the increasing nuclear charge causes a contraction in the radius of the $M^{3+}$ ion – from 0.1061 nm in lanthanum to 0.0848 nm in lutetium. This effect, the **lanthanoid contraction**, accounts for the similarity between the transition elements zirconium and hafnium.

**lanthanons**  See LANTHANOIDS.

**lanthanum**  Symbol La. A silvery metallic element belonging to group 3 (formerly IIIA) of the periodic table and often considered to be one of the *lanthanoids; a.n. 57; r.a.m. 138.91; r.d. 6.146 (20°C); m.p. 921°C; b.p. 3457°C. Its principal ore is bastnasite, from which it is separated by an ion-exchange process. There are two natural isotopes, lanthanum–139 (stable) and lanthanum–138 (half-life $10^{10}$–$10^{15}$ years). The metal, being pyrophoric, is used in alloys for lighter flints and the oxide is used in some optical glasses. The largest use of lanthanum, however, is as a catalyst in cracking crude oil. Its chemistry resembles that of the lanthanoids. The element was discovered by Carl Mosander (1797–1858) in 1839.

(((⊕))) SEE WEB LINKS
• Information from the WebElements site

**lapis lazuli**  A blue rock that is widely used as a semiprecious stone and for ornamental purposes. It is composed chiefly of the deep blue mineral **lazurite** embedded in a matrix of white calcite and usually also contains small specks of pyrite. It occurs in only a few places in crystalline limestones as a contact metamorphic mineral. The chief source is Afghanistan; lapis lazuli also occurs near Lake Baikal in Siberia and in Chile. It was formerly used to make the artists' pigment ultramarine.

**Laplace equation**  The partial differential equation:

$$\partial^2 u/\partial x^2 + \partial^2 u/\partial y^2 + \partial^2 u/\partial z^2 = 0$$

It may also be written in the form $\nabla^2 u = 0$, where $\nabla^2$ is called the **Laplace operator**. It was formulated by the French mathematician Pierre Laplace (1749–1827).

**Large Electron-Positron Collider**  See CERN.

**Large Hadron Collider (LHC)**  A proton collider at *CERN, which started operating in 2008. It accelerates two beams of protons to very high energies and then causes the beams to collide with each other, giving effective energies of about 14 TeV. It is hoped that the LHC will discover the *Higgs boson and provide information on physics beyond the standard model, such as *supersymmetry.

(((⊕))) SEE WEB LINKS
• The Large Hadron Collider home page at CERN

**large intestine**  The portion of the alimentary canal of vertebrates between the *small intestine and the *anus. It consists of the *caecum, *colon, and *rectum and its principal function is the absorption of water and formation of faeces.

**large-scale structure**  The structure of the distribution of visible matter in the universe at very large scales. This structure includes galaxies, clusters of galaxies, superclusters, and voids. See also STRUCTURE FORMATION.

**Larmor precession**  A precession of the motion of charged particles in a magnetic field. It was first deduced in 1897 by Sir Joseph Larmor (1857–1942). Applied to the orbital motion of an electron around the nucleus of an atom in a magnetic field of flux density $B$, the frequency of precession is given by $eB/4\pi m\mu$, where $e$ and $m$ are the electronic charge and mass respectively, $\mu$ is the permeability, and $v$ is the velocity of the electron. This is known as the **Larmor frequency**.

**larva**  The juvenile stage in the life cycle of most invertebrates, amphibians, and fish, which hatches from the egg, is unlike the adult in form, and is normally incapable of sexual reproduction (see PAEDOGENESIS). It develops into the adult by undergoing *metamorphosis. Larvae can feed themselves and are otherwise self-supporting. -- Examples are the tadpoles of frogs, the caterpillars of butterflies, and the ciliated planktonic larvae of many marine animals. Compare NYMPH.

**larynx**  The anterior portion of the *trachea (windpipe) of tetrapod vertebrates, which in amphibians, reptiles, and mammals contains the *vocal cords. Movement of the cartilage in the walls of the larynx (by means of the laryngeal muscles) alters the tension of the vocal cords. This changes the pitch of the sound emitted by the vocal cords when they vibrate. The final voiced sound is modified by resonance within the oral and nasal cavities.

**laser** (*l*ight *a*mplification by *s*timulated *e*mission of *r*adiation) A light amplifier usually used to produce monochromatic coherent radiation in the infrared, visible, and ultraviolet regions of the *electromagnetic spectrum. Lasers that operate in the X-ray region of the spectrum are also being developed.

Nonlaser light sources emit radiation in all directions as a result of the spontaneous emission of photons by thermally excited solids (filament lamps) or electronically excited atoms, ions, or molecules (fluorescent lamps, etc.). The emission accompanies the spontaneous return of the excited species to the *ground state and occurs randomly, i.e. the radiation is not coherent. In a laser, the atoms, ions, or molecules are first 'pumped' to an excited state and then stimulated to emit photons by collision of a photon of the same energy. This is called **stimulated emission**. In order to use it, it is first necessary to create a condition in the amplifying medium, called **population inversion**, in which the majority of the relevant entities are excited. Random emission from one entity can then trigger coherent emission from the others that it passes. In this way amplification is achieved.

The laser amplifier is converted to an oscillator by enclosing the amplifying medium within a resonator. Radiation then introduced along the axis of the resonator is reflected back and forth along its path by a mirror at one end and by a partially transmitting mirror at the other end. Between the mirrors the waves are amplified by stimulated emission. The radiation emerges through the semitransparent mirror at one end as a powerful coherent monochromatic parallel beam of light. The emitted beam is uniquely parallel because waves that do not bounce back and forth between the mirrors quickly escape through the sides of the oscillating medium without amplification.

Some lasers are solid, others are liquid or gas devices. Population inversion can be achieved by **optical pumping** with flashlights or with other lasers. It can also be achieved by such methods as chemical reactions, discharges in gases, and recombination emission in semiconducting materials (*see* RECOMBINATION PROCESS).

Lasers have found many uses since their invention in 1960, including laser welding, surgery, *holography, printing, optical communications, and the reading of digital information. In chemistry, their main use has been in the study of photochemical reactions and in the spectroscopic investigation of molecules. *See also* DYE LASER.

**laser cooling** A technique for producing extremely low temperatures using lasers to slow down and trap atoms. The basic method is to direct a set of crossed laser beams at a sample of gas, with the wavelength set so that photons are absorbed by the atoms. One atom moving towards the photon beam will lose momentum on absorbing a photon and be cooled. An atom moving away from the incident photons will gain energy on absorption. Atoms moving towards the incident photons 'see' the incident photons as having a slightly different frequency than those moving away because of the *Doppler effect, and it is possible to adjust the incident laser frequency by a small amount so that atoms are more likely to absorb when they are moving towards the oncoming photons. This results in a net cooling effect, a technique known as **Doppler cooling**. It produces a region of slow moving atoms at the intersection of the laser beams – a state of matter sometimes called **optical molasses**. Further cooling, to temperatures below the theoretical limit for Doppler cooling, can be obtained by a mechanism known as **Sisyphus cooling**. Here the

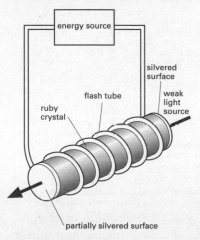

**Laser.** A simple ruby laser.

atom moves through a standing wave created by the laser. As it moves to the top of each 'hill' it loses energy and at the top it is optically pumped to a state at the bottom of the 'valley'. Consequently, the effect is of an atom always moving up a potential gradient and losing energy. The name comes from the character Sisyphus in Greek mythology, who was condemned by the Gods continuously to push a boulder to the top of a hill, only for it to roll back down again when he reached the summit.

Work on laser cooling has also involved methods of trapping atoms. The **magneto-optical trap** (**MOT**) uses six crossed laser beams together with an applied magnetic field to keep cooled atoms together. This allows a further method of cooling in which the height of the trap is lowered so as to let the more energetic atoms escape (a method known as **evaporative cooling**). Techniques of this type have led to temperatures less than $10^{-6}$ K and to the discovery of the Bose–Einstein condensate (*see* BOSE–EINSTEIN CONDENSATION).

**laser printer** *See* PRINTER.

**latent heat** Symbol $L$. The quantity of heat absorbed or released when a substance changes its physical phase at constant temperature (e.g. from solid to liquid or from liquid to gas at the melting point or from liquid to gas at the boiling point). For example, the latent heat of vaporization is the energy a substance absorbs from its surroundings in order to overcome the attractive forces between its molecules as it changes from a liquid to a gas and in order to do work against the external atmosphere as it expands. In thermodynamic terms the latent heat is the *enthalpy of evaporation ($\Delta H$), i.e. $L = \Delta H = \Delta U + p\Delta V$, where $\Delta U$ is the change in the internal energy, $p$ is the pressure, and $\Delta V$ is the change in volume.

The **specific latent heat** (symbol $l$) is the heat absorbed or released per unit mass of a substance in the course of its isothermal change of phase. The **molar latent heat** is the heat absorbed or released per unit amount of substance during an isothermal change of state.

**latent learning** A form of *learning in which there is apparently no immediate reward for the animal, and what is learnt remains 'latent'. The prime example is an animal exploring its surroundings. Learning about the geography of its home area may

bring an animal no immediate benefits, but can prove vital in the future when fleeing a predator or searching for food. Many insects learn the details of landmarks near their nest by making orientation flights. This process enables them to locate the nest when returning from distant sites.

**latent period** The short time that elapses between the reception of a stimulus and the start of the response. For a contracting muscle the latent period lasts about 0.02 seconds.

**lateral inversion (perversion)** The type of reversal that occurs with an image formed by a plane mirror. A person with a mole on his left cheek sees an image in a plane mirror of a person with a mole on his right cheek. Since, however, that is (correctly) to the observer's left, the real reversal is of front and back; the image is 'turned through' itself to face the object – hence the alternative name.

**lateral velocity** The component of a celestial body's velocity that is at 90° to its *line-of-sight velocity.

**laterite** A layer of deposits composed largely of hydroxides of iron and aluminium formed from the weathering of rocks in humid tropical and subtropical climates. Laterites range from soft earthy materials to hard dense rocks. It hardens on exposure to the atmosphere and is used as a building material. In the weathering process silica, alkalis, and alkaline earths are removed leaving behind concentrations of iron and aluminium oxides; this process is known as **laterization**.

**latex** A milky fluid of mixed composition found in some herbaceous plants and trees. Its function is not clear but it may assist in protecting wounds (*compare* GUM) and it may be involved in the nutrition of the plant. The latex of some species, notably rubber trees, is collected for commercial purposes.

**Latimer diagram** A simple diagram summarizing the standard potentials for an element in different oxidation states. The different species are written in a horizontal line in order of decreasing oxidation state, with the most highly oxidized species on the left. Arrows are written between adjacent species with the standard potential in volts indicated above the arrow. Often the oxidation number is included in the diagram. The standard electrode potential for nonadjacent species can be calculated, but values for

**latitude and longitude**  1. (in geography) Imaginary lines on the earth's surface, enabling any point to be defined in terms of two angles subtended at its centre (see illustration). **Parallels of latitude** are circles drawn round the earth parallel to the equator; their diameters diminish as they approach the poles. These parallels are specified by the angle subtended at the centre of the earth by the arc formed between a point on the parallel and the equator. All points on the equator therefore have a latitude of 0°, while the north pole has a latitude of 90°N and the south pole of 90°S. Parallels of latitude 1° apart are separated on the earth's surface by about 100 km.

**Meridians of longitude** are half *great circles passing through both poles; they cross parallels of latitude at right angles. In 1884 the meridian through Greenwich, near London, was selected as the prime meridian and designated as 0°. Other meridians are defined by the angle between the plane of the meridian and the plane of the prime meridian specifying whether it is E or W of the prime meridian. At the equator meridians 1° apart are separated by about 112 km.
**2.** (in astronomy) The **celestial latitude** of a star, or other celestial body, is its angular distance north (taken as positive) or south (taken as negative) of the ecliptic measured along the great circle through the body and the poles of the ecliptic. The **celestial longitude** is the angular distance from the vernal equinox measured eastwards along the ecliptic to the intersection of the body's circle of longitude; it is measured in the same direction as the sun's apparent annual motion.

**lattice**  The regular arrangement of atoms, ions, or molecules in a crystalline solid. *See* CRYSTAL LATTICE.

**lattice energy**  A measure of the stability of a *crystal lattice, given by the energy that would be released per mole if atoms, ions, or molecules of the crystal were brought together from infinite distances apart to form the lattice. *See* BORN–HABER CYCLE.

**lattice vibrations**  The periodic vibrations of the atoms, ions, or molecules in a *crystal lattice about their mean positions. On heating, the amplitude of the vibrations increases until they are so energetic that the lattice breaks down. The temperature at which this happens is the melting point of the solid and the substance becomes a liquid. On cooling, the amplitude of the vibrations diminishes. At *absolute zero a residual vibration persists, associated with the *zero-point energy of the substance. The increase in the electrical resistance of a conductor is due to increased scattering of the free conduction electrons by the vibrating lattice particles.

**latus rectum**  *See* ELLIPSE; HYPERBOLA; PARABOLA.

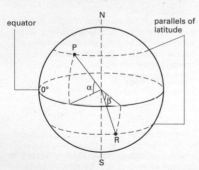

The latitude of P is given by the angle α. In this case it would be α°N. The latitude of R is β°S.

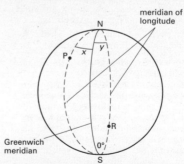

The longitude of P is given by the angle x. In this case it would be x°W. R has a longitude of y°E.

**Latitude and longitude.**

**Laue, Max Theodor Felix von** (1879–1960) German physicist, who became a professor at Berlin in 1919, moving in 1943 to the Max Planck Institute at Göttingen. He is best known for his discovery in 1912 of *X-ray diffraction, for which he was awarded the 1914 Nobel Prize for physics.

**laughing gas** See DINITROGEN OXIDE.

**launch vehicle** A rocket used to launch a satellite, spaceprobe, space station, etc. Multistage rockets are usually used, the empty tanks and engine of the first two stages being jettisoned before the desired orbit is reached. The **launch window** is the time interval during which the vehicle must be launched so that it can achieve its correct orbit or the trajectory it must take in order to reach its destination.

**Laurasia** See CONTINENTAL DRIFT.

**lauric acid** See DODECANOIC ACID.

**lava** Molten rock (magma) that rises to the surface from below ground, usually through a volcano. A typical free-flowing lava consists of a basic rock, such as *basalt; acidic lavas are more viscous. There are various types: **aa** solidifies as rough blocks; **pahoehoe** takes the form of ropy strands; **pillow lava**, named after its shape, occurs where molten lava flows into the sea or erupts from an underwater volcano. Lava that contains volatile materials solidifies as *pumice.

**Laves phase** A phase that occurs for alloys of the type $AB_2$, in which the atoms of the element A are arranged in a diamond-like structure and the atoms of the element B are arranged in tetrahedra around the atoms of A. The phases are named after the German crystallographer Fritz Laves (1906–1978).

**Lavoisier, Antoine Laurent** (1743–1794) French chemist, who collected taxes for the government in Paris. In the 1770s he discovered oxygen and nitrogen in air and demolished the *phlogiston theory of combustion by demonstrating the role of oxygen in the process. In 1783 he made water by burning hydrogen in oxygen (see CAVENDISH, HENRY). He also devised a rational nomenclature for chemical compounds. In 1794 he was tried by the Jacobins as an opponent of the Revolution (because of his tax-gathering), found guilty, and guillotined.

**law of chemical equilibrium** See EQUILIBRIUM CONSTANT.

**law of conservation of energy** See CONSERVATION LAW.

**law of conservation of mass** See CONSERVATION LAW.

**law of constant composition** See CHEMICAL COMBINATION.

**law of definite proportions** See CHEMICAL COMBINATION.

**law of mass action** See MASS ACTION.

**law of multiple proportions** See CHEMICAL COMBINATION.

**law of octaves (Newlands' law)** An attempt at classifying elements made by John Newlands (1837–98) in 1863. He arranged 56 elements in order of increasing atomic mass in groups of eight, pointing out that each element resembled the element eight places from it in the list. He drew an analogy with the notes of a musical scale. **Newlands' octaves** were groups of similar elements distinguished in this way: e.g. oxygen and sulphur; nitrogen and phosphorus; and fluorine, chlorine, bromine, and iodine. In some cases it was necessary to put two elements in the same position. The proposal was rejected at the time. See PERIODIC TABLE.

(⊕)) SEE WEB LINKS
• John Newlands' paper

**law of reciprocal proportions** See CHEMICAL COMBINATION.

**lawrencium** Symbol Lr. A radioactive metallic transuranic element belonging to the *actinoids; a.n. 103; mass number of the first discovered isotope 257 (half-life 8 seconds). A number of very short-lived isotopes have now been synthesized. The element was identified by Albert Ghiorso and associates in 1961. The alternative name **unnil-trium** has been proposed.

(⊕)) SEE WEB LINKS
• Information from the WebElements site

**laws of chemical combination** See CHEMICAL COMBINATION.

**Lawson criterion** A condition for the release of energy from a *thermonuclear reactor first laid down by J. D. Lawson in 1957. It is usually stated as the minimum value for the product of the density ($n_G$) of the fusion-fuel particles and the *containment time ($\tau$) for energy breakeven, i.e. it is a measure of the density of the reacting particles required and the time for which they need to react in order to produce more energy than was used in raising the temperature of the reacting particles to the *ignition temperature. For a 50:50 mixture of deuterium and tritium at the ignition temperature, the value of $n_G\tau$ is between $10^{14}$ and $10^{15}$ cm$^{-3}$ s.

**laws, theories, and hypotheses** In science, a law is a descriptive principle of nature that holds in all circumstances covered by the wording of the law. There are no loopholes in the laws of nature and any exceptional event that did not comply with the law would require the existing law to be discarded or would have to be described as a miracle. Eponymous laws are named after their discoverers (e.g. *Boyle's law); some laws, however, are known by their subject matter (e.g. the law of conservation of mass), while other laws use both the name of the discoverer and the subject matter to describe them (e.g. *Newton's law of gravitation).

A description of nature that encompasses more than one law but has not achieved the uncontrovertible status of a law is sometimes called a **theory**. Theories are often both eponymous and descriptive of the subject matter (e.g. Einstein's theory of relativity and Darwin's theory of evolution).

A **hypothesis** is a theory or law that retains the suggestion that it may not be universally true. However, some hypotheses about which no doubt still lingers have remained hypotheses (e.g. Avogadro's hypothesis), for no clear reason. Clearly there is a degree of overlap between the three concepts.

**layer lattice** A crystal structure in which the atoms are chemically bonded in plane layers, with relatively weak forces between atoms in adjacent layers. Graphite and micas are examples of substances having layer lattices (i.e. they are *lamellar solids).

**lazurite** *See* LAPIS LAZULI.

**LCD** Liquid-crystal display; a flat-panel display that is used with many computers. LCDs are also used in other digital instruments and in flat-screen televisions. Early LCDs suffered from poor contrast between light and dark combined with narrow viewing angles. Several different forms of construction now offer improved viewing characteristics.

LCD technology is based on **liquid crystals**. These are common organic compounds that, between specific temperature limits, change their crystal structure to allow them to flow like a liquid. **Supertwisted nematic**

**displays** use rod-shaped (nematic) crystals. The crystals are organized between two transparent polarized layers with 90° between the directions of polarization. The crystals form a spiral between the two layers so that light can be rotated and passed through the material unchanged. When an electric field is applied, the orientation of the crystals is disturbed thus stopping the light passing. Controlling the electric field applied to each pixel results in an image. It is possible to switch modes up to 120 hertz. In consequence, by shuttering white light through coloured dye filters it is possible to turn a monochrome display into a colour one. The same shuttering system can be used to generate stereo images.

Supertwisted nematic displays may be **passive-matrix LCDs**, containing no active (switching) electronic components. Nowadays much higher performance, especially for colour displays, is obtained from **active-matrix LCDs**. In this construction, a thin-film transistor is added to each pixel to ensure an adequate and constant drive is maintained between refresh cycles. This gives a more uniform display and wider viewing angle.

**LCP** *See* LIQUID-CRYSTAL POLYMER.

**LD$_{50}$** Lethal dose 50, or median lethal dose: the amount of a pharmacological or toxic substance (such as ionizing radiation) that causes death in 50% of a group of experimental animals. For each LD$_{50}$ the species and weight of the animal and the route of administration of the substance is specified. LD$_{50}$s are used both in toxicology and in the *bioassay of therapeutic compounds.

**L-dopa** *See* DOPA.

**L–D process** *See* BASIC-OXYGEN PROCESS.

**leaching** Extraction of soluble components of a solid mixture by percolating a solvent through it.

**lead** (in physics) *See* PHASE ANGLE.

**lead** (in chemistry) Symbol Pb. A heavy dull grey soft ductile metallic element belonging to *group 14 (formerly IVB) of the periodic table; a.n. 82; r.a.m. 207.19; r.d. 11.35; m.p. 327.5°C; b.p. 1740°C. The main ore is the sulphide galena (PbS); other minor sources include anglesite (PbSO$_4$), cerussite (PbCO$_3$), and litharge (PbO). The metal is extracted by roasting the ore to give the oxide, followed by reduction with carbon. Silver is also recovered from the ores. Lead has a variety of uses including building construction, lead-plate accumulators, bullets, and shot, and is a constituent of such alloys as solder, pewter, bearing metals, type metals, and fusible alloys. Chemically, it forms compounds with the +2 and +4 oxidation states, the lead(II) state being the more stable.

**(((⊕))) SEE WEB LINKS**
• Information from the WebElements site

**lead(II) acetate** *See* LEAD(II) ETHANOATE.

**lead–acid accumulator** An accumulator in which the electrodes are made of lead and the electrolyte consists of dilute sulphuric acid. The electrodes are usually cast from a lead alloy containing 7–12% of antimony (to give increased hardness and corrosion resistance) and a small amount of tin (for better casting properties). The electrodes are coated with a paste of lead(II) oxide (PbO) and finely divided lead; after insertion into the electrolyte a 'forming' current is passed through the cell to convert the PbO on the negative plate into a sponge of finely divided lead. On the positive plate the PbO is converted to lead(IV) oxide ($PbO_2$). The equation for the overall reaction during discharge is:

$$PbO_2 + 2H_2SO_4 + Pb \rightarrow 2PbSO_4 + 2H_2O$$

The reaction is reversed during charging. Each cell gives an e.m.f. of about 2 volts and in motor vehicles a 12-volt battery of six cells is usually used. The lead–acid battery produces 80–120 kJ per kilogram. *Compare* NICKEL–IRON ACCUMULATOR.

**lead(II) carbonate** A white solid, $PbCO_3$, insoluble in water; rhombic; r.d. 6.6. It occurs as the mineral *cerussite, which is isomorphous with aragonite and may be prepared in the laboratory by the addition of cold ammonium carbonate solution to a cold solution of a lead(II) salt (acetate or nitrate). It decomposes at 315°C to lead(II) oxide and carbon dioxide.

**lead(II) carbonate hydroxide (white lead; basic lead carbonate)** A powder, $2PbCO_3.Pb(OH)_2$, insoluble in water, slightly soluble in aqueous carbonate solutions; r.d. 6.14; decomposes at 400°C. Lead(II) carbonate hydroxide occurs as the mineral **hydroxycerussite** (of variable composition). It was previously manufactured from lead in processes using spent tanning bark or horse manure, which released carbon dioxide. It is currently made by electrolysis of mixed solutions (e.g. ammonium nitrate, nitric acid,

sulphuric acid, and acetic acid) using lead anodes. For the highest grade product the lead must be exceptionally pure (known in the trade as 'corroding lead') as small amounts of metallic impurity impart grey or pink discolorations. The material was used widely in paints, both for art work and for commerce, but it has the disadvantage of reacting with hydrogen sulphide in industrial atmospheres and producing black lead sulphide. The poisonous nature of lead compounds has also contributed to the declining importance of this material.

**lead-chamber process** An obsolete method of making sulphuric acid by the catalytic oxidation of sulphur dioxide with air using a potassium nitrate catalyst in water. The process was carried out in lead containers (which was expensive) and only produced dilute acid. It was replaced in 1876 by the *contact process.

**lead dioxide** *See* LEAD(IV) OXIDE.

**lead equivalent** A measure of the absorbing power of a radiation screen, expressed as the thickness of a lead screen in millimetres that would afford the same protection as the material being considered.

**lead(II) ethanoate (lead(II) acetate)** A white crystalline solid, $Pb(CH_3COO)_2$, soluble in water and slightly soluble in ethanol. It exists as the anhydrous compound (r.d. 3.25; m.p. 280°C), as a trihydrate, $Pb(CH_3COO)_2.3H_2O$ (monoclinic; r.d. 2.55; loses water at 75°C), and as a decahydrate, $Pb(CH_3COO)_2.10H_2O$ (rhombic; r.d. 1.69). The common form is the trihydrate. Its chief interest stems from the fact that it is soluble in water and it also forms a variety of complexes in solution. It was once known as **sugar of lead** because of its sweet taste.

**lead(IV) ethanoate (lead tetra-acetate)** A colourless solid, $Pb(CH_3COO)_4$, which decomposes in water and is soluble in pure ethanoic acid; monoclinic; r.d. 2.228; m.p. 175°C. It may be prepared by dissolving dilead(II) lead(IV) oxide in warm ethanoic acid. In solution it behaves essentially as a covalent compound (no measurable conductivity) in contrast to the lead(II) salt, which is a weak electrolyte.

**lead(IV) hydride** *See* PLUMBANE.

**lead monoxide** *See* LEAD(II) OXIDE.

**lead(II) oxide (lead monoxide)** A solid yellow compound, PbO, which is insoluble in water; m.p. 886°C. It exists in two crystalline forms: **litharge** (tetrahedral; r.d. 9.53) and **massicot** (rhombic; r.d. 8.0). It can be prepared by heating the nitrate, and is manufactured by heating molten lead in air. If the temperature used is lower than the melting point of the oxide, the product is massicot; above this, litharge is formed. Variations in the temperature and in the rate of cooling give rise to crystal vacancies and red, orange, and brown forms of litharge can be produced. The oxide is amphoteric, dissolving in acids to give lead(II) salts and in alkalis to give *plumbates.

**lead(IV) oxide (lead dioxide)** A dark brown or black solid with a rutile lattice, $PbO_2$, which is insoluble in water and slightly soluble in concentrated sulphuric and nitric acids; r.d. 9.375; decomposes at 290°C. Lead(IV) oxide may be prepared by the oxidation of lead(II) oxide by heating with alkaline chlorates or nitrates, or by anodic oxidation of lead(II) solutions. It is an oxidizing agent and readily reverts to the lead(II) oxidation state, as illustrated by its conversion to $Pb_3O_4$ and PbO on heating. It reacts with hydrochloric acid to evolve chlorine. Lead(IV) oxide has been used in the manufacture of safety matches and was widely used until the mid-1970s as an adsorbent for sulphur dioxide in pollution monitoring.

**lead(II) sulphate** A white crystalline solid, $PbSO_4$, which is virtually insoluble in water and soluble in solutions of ammonium salts; r.d. 6.2; m.p. 1170°C. It occurs as the mineral **anglesite**; it may be prepared in the laboratory by adding any solution containing sulphate ions to solutions of lead(II) ethanoate. The material known as **basic lead(II) sulphate** may be made by shaking together lead(II) sulphate and lead(II) hydroxide in water. This material has been used in white paint in preference to lead(II) carbonate hydroxide, as it is not so susceptible to discoloration through reaction with hydrogen sulphide. The toxicity of lead compounds has led to a decline in the use of these compounds.

**lead(II) sulphide** A black crystalline solid, PbS, which is insoluble in water; r.d. 7.5; m.p. 1114°C. It occurs naturally as the metallic-looking mineral *galena (the principal ore of lead). It may be prepared in the laboratory by the reaction of hydrogen sulphide with

soluble lead(II) salts. Lead(II) sulphide has been used as an electrical rectifier.

**lead tetra-acetate** *See* LEAD(IV) ETHANOATE.

**lead(IV) tetraethyl (tetraethyl lead)** A colourless liquid, $Pb(C_2H_5)_4$, insoluble in water, soluble in benzene, ethanol, ether, and petroleum; r.d. 1.659; m.p. –137°C; b.p. 200°C. It may be prepared by the reaction of hydrogen and ethene with lead but a more convenient laboratory and industrial method is the reaction of a sodium–lead alloy with chloroethane. A more recent industrial process is the electrolysis of ethylmagnesium chloride (the Grignard reagent) using a lead anode and slowly running additional chloroethane onto the cathode. Lead tetraethyl is used in fuel for internal-combustion engines (along with 1,2-dibromoethane) to increase the *octane number and reduce preignition. However, its use in petrol results in the emission of hazardous lead compounds into the atmosphere. Pressure from environmental groups has encouraged a reduction in the use of lead(IV) tetraethyl and an increasing use of lead-free petrol. *See* KNOCKING.

**leaf** A flattened structure that develops from a superficial group of tissues, the leaf buttress, on the side of the stem apex. Each leaf has a lateral bud in its axil. Leaves are arranged in a definite pattern (*see* PHYLLOTAXIS) and usually show limited growth. Each consists of a broad flat *lamina (leaf blade) and a leaf base, which attaches the leaf to the stem; a leaf stalk (**petiole**) may also be present. The leaves of bryophytes are simple appendages, which are not homologous with the leaves of vascular plants as they develop on the gametophyte generation.

Leaves show considerable variation in size, shape, arrangement of veins, type of attachment to the stem, and texture. They may be **simple** or divided into **leaflets**, i.e. **compound** (see illustration). Types of leaf include: *cotyledons (seed leaves); **scale leaves**, which lack chlorophyll and develop on rhizomes or protect the inner leaves of a bud; **foliage leaves**, which are the main organs for photosynthesis and transpiration; and *bracts and **floral leaves**, such as sepals, petals, stamens, and carpels, which are specialized for reproduction.

Leaves may be modified for special purposes. For example the leaf bases of bulbs

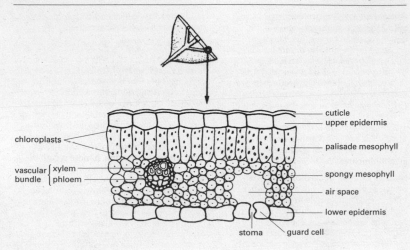

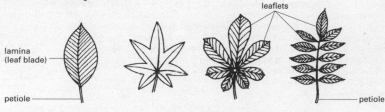

Transverse section through a leaf blade

Simple leaves

Compound leaves

**Leaf.**

arc swollen with food to survive the winter. In some plants leaves are reduced to spines for protection and their photosynthetic function is carried out by another organ, such as a *cladode.

**leaf buttress** *See* PRIMORDIUM.

**leaf litter** *See* LITTER.

**learning** A process by which an animal's response to a particular situation may be permanently altered, usually in a beneficial way, as a result of its experience. Learning allows an animal to respond more flexibly to the situations it encounters: learning abilities in different species vary widely and are adapted to the species' environment. *See also* CONDITIONING; HABITUATION; IMPRINTING; INSIGHT LEARNING; LATENT LEARNING. See Feature overleaf.

**least-squares method** A method of fitting a curve (or line) to points on a *graph. The best fit occurs when the sum of the squares of the distances from the curve to the points is a minimum. It assumes that random measurement errors follow a *normal distribution.

**Leblanc process** An obsolete process for manufacturing sodium carbonate. The raw materials were sodium chloride, sulphuric acid, coke, and limestone (calcium carbonate), and the process involved two stages. First the sodium chloride was heated with sulphuric acid to give sodium sulphate:

$$2NaCl(s) + H_2SO_4(l) \rightarrow Na_2SO_4(s) + 2HCl(g)$$

The sodium sulphate was then heated with coke and limestone:

## LEARNING IN ANIMALS

An animal's survival prospects are greatly improved if the animal alters its behaviour according to its experience. Learning increases its chances of obtaining food, avoiding predators, and adjusting to other changes in its environment. The importance of learning in the development of behaviour was stressed particularly by US experimental psychologists, such as John B. Watson (1878–1958) and B. F. Skinner (1904–90), who studied animals under carefully controlled laboratory conditions. They demonstrated how rats and pigeons could be trained, or 'conditioned', by exposing them to stimuli in the form of food rewards or electric shocks. This work was criticized by others, notably the ethologists, who preferred to observe animals in their natural surroundings and who stressed the importance of inborn mechanisms, such as instinct, in behavioural development. A synthesis between these two once-conflicting approaches has now been achieved: learning is regarded as a vital aspect of an animal's development, occurring in response to stimuli in the animal's environment but within constraints set by the animal's genes. Hence young animals are receptive to a wide range of stimuli but are genetically predisposed to respond to those that are most significant, such as those from their mother.

### Conditioning

The classical demonstration of conditioning was undertaken by Ivan *Pavlov in the early 1900s. He showed how dogs could learn to associate the ringing of a bell with the presentation of food, and after a while would salivate at the sound of the bell alone. He measured the amount of saliva produced by a dog, and showed that this increased as the animal learnt to associate the sound of the bell with presentation of food. The dog became conditioned to respond to the sound of the bell. Such learning is widespread among animals. Pavlov's experiment involved positive conditioning, but negative conditioning can also occur. For example, a young bird quickly learns to associate the black-and-orange markings of the cinnabar moth's caterpillars with their unpleasant taste, and to avoid eating such caterpillars in future.

### Trial-and-error learning

This occurs when the spontaneous behaviour of an animal produces a reward. For example, a hungry cat placed in a box is required to pull a string loop to open the door and gain access to food. After various scratching and reaching movements, it accidentally pulls the loop and is released from the box. Its behaviour is instrumental in securing a reward. On subsequent occasions, the cat's attention becomes increasingly focused on the loop, until eventually it pulls the loop straightaway on entering the box.

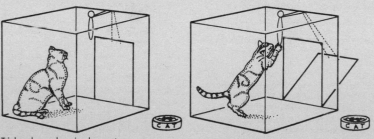

Trial-and-error learning by a cat

## Insight learning

Chimpanzees can learn to stack crates or boxes to form a platform or to manipulate poles in order to reach an otherwise inaccessible bunch of bananas. A chimp may apparently solve such a problem suddenly, as if gaining insight after mental consideration of the problem. Such complex learning benefits from previous experience, in this instance by simply 'playing' with crates, boxes, or poles.

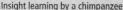
Insight learning by a chimpanzee

## Imprinting

This is a form of learning found in young animals, especially young birds, in which they form an attachment to their mother in early life, thereby ensuring that they are taken care of and do not wander off. For example, chicks or ducklings follow the first large moving object that they encounter after hatching. This is normally their mother, but artificially incubated youngsters can become imprinted on a wooden decoy, as illustrated here, or even on a human being – as originally demonstrated in goslings and ducklings by Konrad *Lorenz. Imprinting occurs during a particularly sensitive period of development: the attachment formed by an animal to an imprinted individual or object lasts well into its adult life.

Imprinting in ducklings

$$Na_2SO_4 + 2C + CaCO_3 \rightarrow Na_2CO_3 + CaS + 2CO_2$$

Calcium sulphide was a by-product, the sodium carbonate being extracted by crystallization. The process, invented in 1783 by the French chemist Nicolas Leblanc (1742–1806), was the first for producing sodium carbonate synthetically (earlier methods were from wood ash and other vegetable sources). By the end of the 19th century it had been largely replaced by the *Solvay process.

**lechatelierite** A mineral form of *silicon(IV) oxide, $SiO_2$.

**Le Chatelier's principle** If a system is in equilibrium, any change imposed on the system tends to shift the equilibrium to nullify the effect of the applied change. The principle, which is a consequence of the law of conservation of energy, was first stated in 1888 by Henri Le Chatelier (1850–1936). It is applied to chemical equilibria. For example, in the gas reaction

$$2SO_2 + O_2 \rightleftharpoons 2SO_3$$

an increase in pressure on the reaction mixture displaces the equilibrium to the right, since this reduces the total number of molecules present and thus decreases the pressure. The standard enthalpy change for the forward reaction is negative (i.e. the reaction is exothermic). Thus, an increase in temperature displaces the equilibrium to the left since this tends to reduce the temperature. The *equilibrium constant thus falls with increasing temperature.

**(🌐) SEE WEB LINKS**
• Le Chatelier's original paper

**lecithin (phosphatidylcholine)** A phosphoglyceride (*see* PHOSPHOLIPID) containing the amino alcohol *choline esterified to the phosphate group. It is the most abundant animal phospholipid (being a component of plasma membranes) and also occurs in higher plants, but rarely in microorganisms.

**Leclanché cell** A primary *voltaic cell consisting of a carbon rod (the anode) and a zinc rod (the cathode) dipping into an electrolyte of a 10–20% solution of ammonium chloride. *Polarization is prevented by using a mixture of manganese dioxide mixed with crushed carbon, held in contact with the anode by means of a porous bag or pot; this reacts with the hydrogen produced. This wet form of the cell, devised in 1867 by Georges Leclanché (1839–82), has an e.m.f. of about 1.5 volts. The *dry cell based on it is widely used in torches, radios, and calculators.

**lectin** Any of a group of proteins, found in a variety of organisms, that bind to specific carbohydrate groups. Lectins derived from plant seeds, such as **concanavalin A**, can cause cells to clump together by forming cross links between the oligosaccharide groups on cell surfaces. Lectins are widely used for diagnosis and experimental purposes, e.g. to identify mutant cells in cell cultures, to determine blood groups by triggering *agglutination of red blood cells, or in mapping the surface of plasma membranes. In legumes lectins take part in the recognition of suitable bacterial partners for the plant in establishing root nodule symbioses.

**LED** *See* LIGHT-EMITTING DIODE.

**leeches** *See* HIRUDINEA.

**LEED** Low-energy electron diffraction. *See* ELECTRON DIFFRACTION.

**Leeuwenhoek, Anton van** (1632–1723) Dutch microscopist, who had little formal education. He is known for accurately grinding small lenses to make simple microscopes, with which he made the first observations of red blood cells, protozoa, and spermatozoa. He communicated regularly with the Royal Society in London, which published many of his findings in its *Philosophical Transactions*.

**Leggett's theorem** A result in *quantum mechanics that is a generalization of *Bell's theorem. This theorem, which was put forward in 2003 by the British physicist Anthony Leggett (1938– ), states that nonlocal *hidden-variables theories make predictions that contradict the predictions of quantum mechanics. It was subsequently found that experiments agree with the predictions of quantum mechanics rather than of nonlocal hidden-variables theories.

**legume (pod)** A dry fruit formed from a single carpel and containing one or more seeds, which are shed when mature. It is the characteristic fruit of the Leguminosae (Fabaceae; pea family). It splits, often explosively, along both sides and the two halves of the fruit move apart to expose the seeds. A special form of the legume is the *lomentum.

**Leishman's stain** A neutral stain for blood smears devised by the British surgeon Sir William Boog Leishman (1865–1926). It consists of a mixture of *eosin (an acidic stain), and *methylene blue (a basic stain) in alcohol and is usually diluted and buffered before use. It stains the different components of blood in a range of shades between red and blue. The similar **Wright's stain** is favoured by American workers.

### Lemaître, Georges Edouard

(1894–1966) Belgian astronomer, who was ordained as a priest in 1923. He went to work at Louvain University in 1925, becoming professor of astronomy two years later. He is best known for his *big-bang theory of the origin of the universe.

**lens** 1. (in physics) A curved, ground, and polished piece of glass, moulded plastic, or other transparent material used for the refraction of light. A **converging lens** is one that brings the rays of a parallel beam of light to a real *principal focus. They include biconvex, planoconvex, and converging meniscus lenses. **Diverging lenses** cause the rays of a parallel beam to diverge as if from a virtual principal focus; these include the biconcave, planoconcave, and diverging meniscus lenses. See illustrations.

The **centre of curvature** of a lens face is the centre of the sphere of which the surface of the lens is a part. The **optical axis** is the line joining the two centres of curvature of a lens or, in the case of a lens with one plane surface, the line through one centre of curvature that is normal to the plane surface. The **optical centre** of a lens is the point within a lens on the optical axis through which any rays entering the lens pass without deviation. The distance between the optical centre and the principal focus of a lens is called the **focal length** ($f$). The distance ($v$) between the lens and the image it forms is related to the distance ($u$) between the lens and the object by:

$$1/v + 1/u = 1/f,$$

provided that the *real-is-positive convention is used. This takes distances to real objects, images, and foci as positive; those to virtual objects, images, and foci as negative. The equation does not always apply if the alternative New Cartesian convention (see SIGN CONVENTION) is used.

2. (in anatomy) A transparent biconvex structure in the eyes or analogous organs of many animals, responsible for directing light onto light-sensitive cells. In vertebrates it is a flexible structure centred behind the iris and attached by **suspensory ligaments** to the *ciliary body. In terrestrial species its main function is to focus images onto the retina. To focus on near objects, the circular muscles in the ciliary body contract and the lens becomes more convex; contraction of the radial muscles in the ciliary body flattens the lens for focusing on distant objects (see also ACCOMMODATION).

**Lense–Thirring effect** An effect predicted to occur in general relativity theory by J. Lense and Hans Thirring in 1918 in which a compact rotating body causes the space near it to rotate in the same direction. The phenomenon is also known as **frame dragging**. It has been reported in observations of neutron stars and black holes. Measurements have also been made using shifts in the orbits of satellites around the earth.

### (((∰))) SEE WEB LINKS

• The home page of NASA's Gravity Probe B experiment to investigate frame dragging

**lenticel** Any of the raised pores in the stems of woody plants that allow gas exchange between the atmosphere and the internal tissues. The pore is formed by the *cork cambium, which, at certain points, produces a loose bulky form of cork that

converging lenses

biconvex          planoconvex          converging
                                       meniscus

diverging lenses

biconcave         planoconcave          diverging
                                        meniscus

radius of curvature
centre of curvature
optical axis
optical centre

**Lenses.**

pushes through the outer tissues to create the lenticel.

**Lenz's law** An induced electric current always flows in such a direction that it opposes the change producing it. This law, first stated by Heinrich Lenz (1804–65) in 1835, is a particular example of the law of conservation of energy.

**LEP** Large Electron-Positron Collider. *See* CERN.

**Lepidoptera** An order of insects comprising the butterflies and moths, found mainly in tropical regions. Adults possess two pairs of membranous wings, often brightly coloured and usually coupled together. The wings, body, and legs are covered with minute scales. Adult mouthparts are generally modified to form a long proboscis for sucking nectar, fruit juices, etc. Butterflies are typically small-bodied, active during daylight, and rest with their wings folded vertically; moths have larger bodies, are nocturnal, and rest with their wings in various positions. The larvae (caterpillars) have a prominent head and a segmented wormlike body, most segments bearing a pair of legs. They chew leaves and stems, sometimes causing considerable damage to crop plants. The larvae undergo metamorphosis via a *pupa (chrysalis) to the adult form. In some groups, the pupa is enclosed in a cocoon of silk derived from silk glands (modified salivary glands); others use leaves, etc. to build a cocoon.

**leptin** A protein hormone, comprising 167 amino acids in humans, that is secreted by adipose tissue and regulates adipose tissue mass and energy balance. It acts on leptin receptors in the hypothalamus and inhibits expression of *neuropeptide Y, thereby countering the appetite-stimulating effects of the latter and inhibiting food intake. It also promotes synthesis of the appetite suppressant *melanocyte-stimulating hormone. Deficiency of leptin or its receptors leads to severe obesity.

**lepton** Any of a class of *elementary particles that consists of the *electron, muon, tau particle, and three types of *neutrino (one associated with each of the other types of lepton). For each lepton there is an equivalent antiparticle. The antileptons have a charge opposite that of the leptons; the antineutrinos, like the neutrinos, have no charge. The electron, muon, and tau particle

all have a charge of −1. These three particles differ from each other only in mass: the muon is 200 times more massive than the electron and the tau particle is 3500 times more massive than the electron. Leptons interact by the electromagnetic interaction and the weak interaction (*see* FUNDAMENTAL INTERACTIONS).

**lepton number** *See* ELEMENTARY PARTICLES.

**leptotene** The beginning of the first prophase of *meiosis, when the chromatids can be seen and *pairing begins.

**Leslie's cube** A metal box in the shape of a cube in which each of the four vertical sides have different surface finishes. When hot water is placed in the cube, the emissivity of the finishes can be compared. The device was first used by Sir John Leslie (1766–1832).

**lethal allele (lethal gene)** A mutant form of a gene that eventually results in the death of an organism if expressed in the phenotype. Most lethal genes are recessive; for example, sickle-cell anaemia (*see* POLYMORPHISM) results from a recessive lethal gene that causes the production of abnormal and inefficient haemoglobin.

**lethal dose 50** *See* $LD_{50}$.

**leucine** *See* AMINO ACID.

**leucocyte (white blood cell)** A colourless cell with a nucleus, found in blood and lymph. Leucocytes are formed in lymph nodes and red bone marrow and are capable of amoeboid movement. They can produce *antibodies and move through the walls of vessels to migrate to the sites of injuries, where they surround and isolate dead tissue, foreign bodies, and bacteria. There are two major types: those without granules in the cytoplasm, such as *lymphocytes and *monocytes (*see* AGRANULOCYTE), and those with granular cytoplasm (*granulocytes), which include *basophils and *neutrophils.

**leuco form** *See* DYES.

**leucomalachite green test** A *presumptive test for blood. The reagent is the dye leucomalachite green dissolved in water along with sodium perborate ($NaBO_3$). A blue-green colour indicates a positive result.

**leucoplast** Any *plastid in plant cells that contains no pigment and is therefore colourless. Leucoplasts are usually found in tissues not normally exposed to light and frequently contain reserves of starch, protein, or oil. *Compare* CHROMOPLAST.

**leukaemia** *See* CANCER.

**level** An instrument used in *surveying to determine heights. It usually consists of a telescope and attached spirit level mounted on a tripod. The level is set up between a point of known height and a point for which the height is required. Before use it is adjusted until the line of sight is exactly horizontal. Sightings are then made onto a graduated levelling staff at the two points. The difference in elevation between the two points can then be calculated from the readings taken at these points.

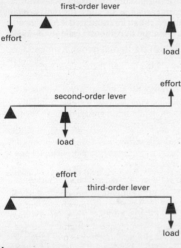

**Lever.**

**lever** A simple machine consisting of a rigid bar pivoted about a fulcrum. The mechanical advantage or *force ratio of a lever (the ratio of load to effort) is equal to the ratio of the perpendicular distance of the line of action of the effort from the fulcrum to the perpendicular distance of the line of action of the load from the fulcrum. In a first-order lever the fulcrum comes between load and effort. In a second-order lever the load comes between the fulcrum and the effort. In a third-order lever the effort comes between the fulcrum and the load. See illustrations.

**Lewis acid and base** *See* ACID.

**Leyden jar** An early form of *capacitor consisting of a glass jar with a layer of metal foil on the outside and a similar layer on the inside. Contact to the inner foil is by means of a loose chain hanging inside the jar. It was invented in the Dutch town of Leyden in about 1745.

**LF** *See* LOW FREQUENCY.

**LH** *See* LUTEINIZING HORMONE.

**LHC** *See* LARGE HADRON COLLIDER.

**libration** The phenomenon that enables 59% of the moon's surface to be observed from earth over a 30-year period, in spite of its *synchronous rotation. **Physical libration** arises from slight variations in the rotation of the moon on its axis, caused by minor distortions in its physical shape. **Geometric librations** are apparent oscillations arising from the fact that the moon is observed from slightly different directions at different times. The geometric **libration in longitude** results from the nonuniform orbital motion of the moon. The geometric **libration in latitude** arises because the moon's axis of rotation is not perpendicular to its orbital plane; it enables more of the lunar polar regions to be observed.

**lice** *See* MALLOPHAGA (bird lice); SIPHUNCU-LATA (sucking lice).

**lichens** A group of organisms that are symbiotic associations (*see* SYMBIOSIS) between a fungus (usually one of the *Ascomycota) and a green alga or a cyanobacterium. The fungus usually makes up most of the plant body and the cells of the alga or bacterium are distributed within it. The alga or bacterium photosynthesizes and passes most of its food to the fungus and the fungus protects its partner's cells. The lichen reproduces by means of fungal spores, which must find a suitable partner on germination. Lichens are slow growing but can live in regions that are too cold or exposed for other plants. They may form a flattened crust or be erect and branching. Many grow as *epiphytes, especially on tree trunks. Some species are very sensitive to air pollution and have been used as *indicator species. Lichens are classified as fungi, usually being placed in the taxon of the fungal partner; some authorities group them together in the phylum Mycophycophyta.

**Liebermann test** A *presumptive test sometimes used for cocaine and morphine.

The **Liebermann reagent** is a solution of potassium nitrite ($KNO_2$) in sulphuric acid. With morphine a black colour is produced; cocaine gives a yellow colour.

**Liebig condenser** A laboratory condenser having a straight glass tube surrounded by a coaxial glass jacket through which cooling water is passed. It is named after the German organic chemist Justus von Liebig (1803–73).

**life cycle** The complete sequence of events undergone by organisms of a particular species from the fusion of gametes in one generation to the same stage in the following generation. In most animals gametes are formed by *meiosis of germ cells in the reproductive organs of the parents. The zygote, formed by the fusion of two gametes, eventually develops into an organism essentially similar to the parents. In plants, however, the products of meiosis are spores, which develop into plants (the *gametophyte generation) often very different in form from the spore-forming (*sporophyte) generation. The sporophyte generation is restored when gametes, formed by the gametophyte generation, fuse. See ALTERNATION OF GENERATIONS.

**ligament** A resilient but flexible band of tissue (chiefly *collagen) that holds two or more bones together at a movable *joint. Ligaments restrain the movement of bones at a joint and are therefore important in preventing dislocation.

**ligand** 1. (in chemistry) An ion or molecule that donates a pair of electrons to a metal atom or ion in forming a coordination *complex. Molecules that function as ligands are acting as Lewis bases (see ACID). For example, in the complex hexaquocopper(II) ion $[Cu(H_2O)_6]^{2+}$ six water molecules coordinate to a central $Cu^{2+}$ ion. In the tetrachloroplatinate(II) ion $[PtCl_4]^{2-}$, four $Cl^-$ ions are coordinated to a central $Pt^{2+}$ ion. A feature of such ligands is that they have lone pairs of electrons, which they donate to empty metal orbitals. A certain class of ligands also have empty $p$- or $d$-orbitals in addition to their lone pair of electrons and can produce complexes in which the metal has low oxidation state. A double bond is formed between the metal and the ligand: a sigma bond by donation of the lone pair from ligand to metal, and a pi bond by **back donation** of electrons on the metal to empty $d$-orbitals on the ligand. Carbon monoxide is the most important such ligand, forming metal carbonyls (e.g. $Ni(CO)_4$).

The examples given above are examples of **monodentate** ligands (literally: 'having one tooth'), in which there is only one point on each ligand at which coordination can occur. Some ligands are **polydentate**; i.e. they have two or more possible coordination points. For instance, 1,2-diaminoethane, $H_2NC_2H_4NH_2$, is a **bidentate** ligand, having two coordination points. Certain polydentate ligands can form *chelates.
2. (in cell biology) A molecule that binds to a protein with a high degree of specificity. Examples are the substrate of an enzyme and a hormone binding to a cell receptor.

**ligand-field theory** An extension of *crystal-field theory describing the properties of compounds of transition-metal ions or rare-earth ions in which covalent bonding between the surrounding molecules (see LIGAND) and the transition-metal ions is taken into account. This may involve using valence-bond theory or molecular-orbital theory. Ligand-field theory was developed extensively in the 1930s. As with crystal-field theory, ligand-field theory indicates that energy levels of the transition-metal ions are split by the surrounding ligands, as determined by *group theory. The theory has been very successful in explaining the optical, spectroscopic, and magnetic properties of the compounds of transition-metal and lanthanide ions.

**ligase** Any of a class of enzymes that catalyse the formation of covalent bonds using the energy released by the cleavage of ATP. Ligases are important in the synthesis and repair of many biological molecules, including DNA (see DNA LIGASE), and are used in genetic engineering to insert foreign DNA into cloning *vectors.

**light** The form of *electromagnetic radiation to which the human eye is sensitive and on which our visual awareness of the universe and its contents relies (see COLOUR).

The finite velocity of light was suspected by many early experimenters in optics, but it was not established until 1676 when Ole Rømer (1644–1710) measured it. Sir Isaac Newton investigated the optical *spectrum and used existing knowledge to establish a primarily **corpuscular theory** of light, in which it was regarded as a stream of particles that set up disturbances in the 'aether' of space. His successors adopted the corpuscles but ignored the wavelike disturbances until Thomas Young rediscovered the *interference of light in 1801 and showed that a **wave theory** was essential to interpret this type of phenomenon. This view was accepted for most of the 19th century and it enabled James Clerk Maxwell to show that light forms part of the *electromagnetic spectrum. He believed that waves of electromagnetic radiation required a special medium to travel through, and revived the name 'luminiferous ether' for such a medium. The *Michelson–Morley experiment in 1887 showed that, if the medium existed, it could not be detected; it is now generally accepted that the ether is an unnecessary hypothesis. In 1905 Albert Einstein showed that the *photoelectric effect could only be explained on the assumption that light consists of a stream of discrete *photons of electromagnetic energy. This renewed conflict between the corpuscular and wave theories has gradually been resolved by the evolution of the *quantum theory and *wave mechanics. While it is not easy to construct a model that has both wave and particle characteristics, it is accepted, according to Bohr's theory of *complementarity, that in some experiments light will appear wavelike, while in others it will appear to be corpuscular. During the course of the evolution of wave mechanics it has also become evident that electrons and other elementary particles have dual wave and particle properties.

**light bulb** *See* ELECTRIC LIGHTING.

**light-dependent reaction** *See* PHOTOSYNTHESIS.

**light-emitting diode (LED)** A *semiconductor device that converts electrical energy into light or infrared radiation in the range 550 nm (green light) to 1300 nm (infrared radiation). The most commonly used LED (see illustration) emits red light and consists of gallium arsenide–phosphide on a gallium arsenide substrate, light being emitted at a *p–n* junction, when electrons and holes recombine (*see* RECOMBINATION PROCESS). LEDs are extensively used for displaying letters and numbers in digital instruments in which a self-luminous display is required.

**light green** *See* FAST GREEN.

**light-independent reaction** *See* PHOTOSYNTHESIS.

**lightning** A high-energy luminous electrical discharge that passes between a charged cloud and a point on the surface of the earth, between two charged clouds, or between oppositely charged layers of the same cloud. In general, the upper parts of clouds are positively charged and the lower parts are negatively charged; the reasons for this separation of charge are complex.

Lightning usually occurs in the form of a downward step leader followed by an intensely luminous return stroke, which can produce instantaneous temperatures as high as 30 000°C. In the typical step leader a surge of electrons descends in approximately 50-metre steps with about 50-microsecond pauses between steps. When this leader reaches the earth a surge of charge returns up the preionized path taken by the leader. Cloud-to-cloud strokes also involve a leader and return stroke. The average current in a lightning stroke is about 10 000 amperes, but maximum currents in the return stroke can reach 20 000 A. *See also* BALL LIGHTNING.

**light year** A unit of distance used in astronomy; the distance travelled by light in a

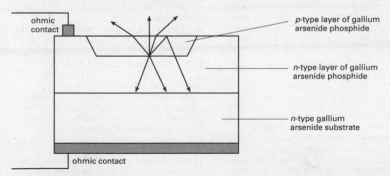

ohmic contact

*p*-type layer of gallium arsenide phosphide

*n*-type layer of gallium arsenide phosphide

*n*-type gallium arsenide substrate

ohmic contact

**Light-emitting diode.**

vacuum during one year. It is equal to $9.4650 \times 10^{15}$ metres or $5.8785 \times 10^{12}$ miles.

**lignin** A complex organic polymer that is deposited within the cellulose of plant cell walls during secondary thickening. Lignification makes the walls woody and therefore rigid. See SCLERENCHYMA.

**lignite** See COAL.

**ligule** 1. A membranous scalelike outgrowth from the leaves of certain flowering plants. Many grasses have a ligule at the base of the leaf blade. 2. A small membranous structure that develops on the upper surface of a young leaf base in certain clubmosses, for example *Selaginella*. It withers as the plant matures. 3. A strap-shaped extension from the corolla tube in certain florets of a *capitulum, termed **ligulate** (or **ray**) **florets**.

**limb** 1. An appendage of a vertebrate animal, such as the leg or arm of a mammal or the wing of a bird. See also PENTADACTYL LIMB. 2. The expanded upper part of a sepal, petal, or leaf. 3. The widened upper section of a gamopetalous *corolla.

**lime** See CALCIUM OXIDE.

**limestone** A sedimentary rock that is composed largely of carbonate minerals, especially carbonates of calcium and magnesium. *Calcite and *aragonite are the chief minerals; *dolomite is also present in the dolomitic limestones. There are many varieties of limestones but most are deposited in shallow water. **Organic limestones** (e.g. *chalk) are formed from the calcareous skeletons of organisms; **precipitated limestones** include oolite, which is composed of ooliths – spherical bodies formed by the precipitation of carbonate around a nucleus; and **clastic limestones** are derived from fragments of pre-existing calcareous rocks.

**limewater** A saturated solution of *calcium hydroxide in water. When carbon dioxide gas is bubbled through limewater, a 'milky' precipitate of calcium carbonate is formed:

$$Ca(OH)_2(aq) + CO_2(g) \rightarrow CaCO_3(s) + H_2O(l)$$

If the carbon dioxide continues to be bubbled through, the calcium carbonate eventually redissolves to form a clear solution of calcium hydrogencarbonate:

$$CaCO_3(s) + CO_2(g) + H_2O(g) \rightarrow Ca(HCO_3)_2(aq)$$

If cold limewater is used the original calcium carbonate precipitated has a calcite structure; hot limewater yields an aragonite structure.

**liming** The application of lime (calcium hydroxide) to soils to increase levels of calcium and decrease acidity.

**limit** The value that a function approaches as the independent variable approaches a specified value.

**limit cycle** See ATTRACTOR.

**limiting factor** Any environmental factor that – by its decrease, increase, absence, or presence – limits the growth, metabolic processes, or distribution of organisms or populations. In a desert ecosystem, for example, low rainfall and high temperature will be factors limiting colonization. When a metabolic process is affected by more than one factor, the **law of limiting factors** states that its rate is limited by the factor that is nearest its minimum value. For example, photosynthesis is affected by many factors, such as light, temperature, and carbon dioxide concentration, but on a warm sunny day carbon dioxide concentration will be the limiting factor as light and temperature will be at optimum levels.

**limiting friction** The friction force that just balances a moving force applied to a solid body resting on a solid surface when the body fails to move. If the moving force exceeds the limiting friction, the body will begin to move.

**limnology** The study of the physical and biological characteristics of lakes and other bodies of fresh water.

**limonite** A generic term for a group of hydrous iron oxides, mostly amorphous. *Goethite and *haematite are important constituents, together with colloidal silica, clays, and manganese oxides. Limonite is formed by direct precipitation from marine or fresh water in shallow seas, lagoons, and bogs (thus it is often called **bog iron ore**) and by oxidation of iron-rich minerals. It is used as an ore of iron and as a pigment.

**linac** See LINEAR ACCELERATOR.

**Linde process** A process for the *liquefaction of gases by the Joule–Thomson effect. In this process, devised by Carl von Linde

(1842–1934) for liquefying air, the air is freed of carbon dioxide and water and compressed to 150 atmospheres. The compressed gas is passed through a copper coil to an expansion nozzle within a Dewar flask. The emerging air is cooled by the Joule–Thomson effect as it expands and then passes back within a second copper coil that surrounds the first coil. Thus the expanded gas cools the incoming gas in a process that is said to be **regenerative**. Eventually the air is reduced to its *critical temperature and, at the pressure of 150 atmospheres (well above its critical pressure), liquefies. The process is also used for other gases, especially hydrogen and helium. Hydrogen has first to be cooled below its inversion temperature (*see* JOULE–THOMSON EFFECT) using liquid air; helium has first to be cooled below its inversion temperature using liquid hydrogen.

**linear absorption coefficient** *See* LAMBERT'S LAWS.

**linear accelerator (linac)** A type of particle *accelerator in which charged particles are accelerated in a straight line, either by a steady electric field or by means of radiofrequency electric fields.
**Van de Graaff accelerator**. This device accelerates charged particles by applying a high electrical potential difference generated by a *Van de Graaff generator. The potential difference can be kept steady to within one part in a thousand, forming a beam of accelerated particles of uniform energy. The maximum electrical potential attainable is typically about 10 MV and depends on the insulating properties of the gas around the Van de Graaff sphere. It is increased by enclosing the whole generator in a pressure vessel containing an inert gas at a pressure of about 20 atmospheres. A source, at the same potential as the sphere, produces charged particles, which enter a column of cylindrical electrodes, each of which is at a lower potential than the one above it. The ions are accelerated as they pass through the gaps between the cylinders. The nonuniform electric fields between the gaps have the effect of focusing the beam of charged particles.
**Drift-tube accelerator**. In this device charged particles are accelerated inside a line of hollow metal cylinders called **drift tubes**. The cylinders are connected alternately to opposite terminals of an alternating potential difference produced by either a *magnetron or a *klystron. The arrangement ensures that adjacent cylinders are always at

opposite electrical potentials. For example, a proton beam may be injected into the first of the line of drift tubes from a Van de Graaff accelerator. Protons reaching the gap between the first two tubes will be accelerated into the second tube, when the alternating potential makes the first tube positive and the second tube negative. This enables the protons emerging into the gap between two cylinders to be accelerated into the next cylinder. All parts of a particular tube are at the same potential, since the metal acts as an equipotential surface. Therefore within a cylinder the particles travel at a constant speed (hence 'drift tube'). It follows that the energy of the beam is increased every time the protons cross between drift tubes, and therefore a device with a large number of gaps can produce extremely high-energy beams using only moderate supply voltages. The Berkeley proton accelerator has a drift-tube arrangement of 47 cylinders, 12 miles long, and accelerates protons up to 31.5 GeV.
**Travelling-wave accelerator**. This apparatus uses radio-frequency electromagnetic waves to accelerate charged particles. Charged particles are fed into the travelling-wave accelerator at close to the speed of light and are carried down a *wave guide by the electric field component of a radio wave. The very high initial speeds for charged particles are needed to match the phase velocity of radio signals propagating down the wave guide. However, this means that travelling-wave accelerators are suitable only for accelerating light particles, such as electrons. The electrons can be accelerated to initial speeds of 98% of the speed of light by a Van de Graaff accelerator. At such high initial speeds, there is little scope for further acceleration and any increase in electron energy provided by the accelerator results from the relativistic increase in mass. The Stanford linear accelerator (SLAC) uses the travelling-wave principle. SLAC is capable of accelerating electrons and positrons to 50 GeV in a tube two miles long.

**linear energy transfer (LET)** The energy transferred per unit path length by a moving high-energy charged particle (such as an electron or a proton) to the atoms and molecules along its path. It is of particular importance when the particles pass through living tissue as the LET modifies the effect of a specific dose of radiation. LET is proportional to the square of the charge on the particle and increases as the velocity of the particle decreases.

**linear equation** An equation between two variables that gives a straight line when plotted on a graph. It has the general form $y = mx + c$, where $m$ is the gradient of the line and $c$ is the intercept of the line on the $y$-axis (in Cartesian coordinates).

**linear expansivity** See EXPANSIVITY.

**linear molecule** A molecule in which the atoms are in a straight line, as in carbon dioxide, $O=C=O$.

**linear momentum** See MOMENTUM.

**linear motor** A form of induction motor in which the stator and armature are linear and parallel, rather than cylindrical and coaxial. In some experimental trains the magnetic force between the primary winding in the vehicle and the secondary winding on the ground support the vehicle on an air cushion thus eliminating track friction. However, because of the high cost of the installation and the low efficiency the device has not yet found commercial application.

**line defect** See CRYSTAL DEFECT (Feature).

**line notation** A notation system for writing the structure of a chemical compound as a string of letters, numbers, and symbols. Examples of line notation are Wiswesser line notation (WLN), SMILES, SYBYL line notation (SLN), ROSDAL, and InChI.

**line-of-sight velocity (radial velocity)** The component of a celestial body's velocity along the line of sight of the observer. It is usually given in relation to the sun to avoid complications arising from the earth's orbital motion. Line-of-sight velocity is normally calculated from the *Doppler effect on the body's spectrum, a *redshift indicating a receding body (taken as a positive velocity) and a blueshift indicating an approaching body (taken as negative).

**line printer** See PRINTER.

**lines of force** Imaginary lines in a *field of force that enable the direction and strength of the field to be visualized. They are used primarily in electric and magnetic fields; in electric fields they are sometimes called **tubes of force**, to express their characteristic of being perpendicular to a conducting surface. The tangent to a line of force at any point gives the direction of the field at that point and the number of lines per unit area

perpendicular to the force represents the *intensity of the field.

**line spectrum** See SPECTRUM.

**linkage** The tendency for two different genes on the same chromosome to remain together during the separation of *homologous chromosomes at meiosis. Linkage can be broken by *crossing over or by a *chromosome mutation, when sections of chromosomes are exchanged and new combinations of genes are produced. See also SEX LINKAGE.

**linkage map** A *chromosome map showing the relative positions of *genes along the length of the chromosomes of an organism. It is constructed by making crosses and observing whether certain characteristics tend to be inherited together. The closer together two allele pairs are situated on *homologous chromosomes, the less often will they be separated and rearranged as the reproductive cells are formed (see CHIASMA; CROSSING OVER). The proportion of offspring that show *recombination of the alleles concerned thus reflects their spacing and is used as a unit of length in mapping chromosomes. The information obtained from such a **classical linkage map** can be combined with a restriction map, which is a linkage map of sites cleaved by restriction enzymes (see RESTRICTION MAPPING), providing a huge number of potential marker sites for genes of interest. Linkage maps provide valuable frameworks for constructing detailed *physical maps giving the base sequence of the chromosomal DNA.

**Linnaean system** See BINOMIAL NOMENCLATURE.

**Linnaeus, Carolus** (Carl Linné; 1707–78) Swedish botanist. He travelled round Europe and by 1735 had described more than 100 new species of plants. In 1749 he announced his system of *binomial nomenclature, which, with modification, has been used ever since for all organisms.

**linoleic acid** A liquid polyunsaturated *fatty acid with two double bonds, $CH_3(CH_2)_4CH:CHCH_2CH:CH(CH_2)_7COOH$. Linoleic acid is abundant in many plant fats and oils, e.g. linseed oil, groundnut oil, and soya-bean oil. It is an *essential fatty acid.

**linolenic acid** A liquid polyunsaturated *fatty acid with three double bonds in its structure: $CH_3CH_2CH:CHCH_2CH:CHCH_2-CH:CH(CH_2)_7COOH$. Linolenic acid occurs in certain plant oils, e.g. linseed and soya-bean oil, and in algae. It is one of the *essential fatty acids.

**linseed oil** A pale yellow oil pressed from flax seed. It contains a mixture of glycerides of fatty acids, including linoleic acid and linolenic acid. It is a *drying oil, used in oil paints, varnishes, linoleum, etc.

**Linz–Donawitz process** See BASIC-OXYGEN PROCESS.

**lipase** An enzyme secreted by the pancreas that catalyses the breakdown of fats into fatty acids and glycerol in the small intestine.

**lipid** Any of a diverse group of organic compounds, occurring in living organisms, that are insoluble in water but soluble in organic solvents, such as chloroform, benzene, etc. Lipids are broadly classified into two categories: **complex lipids**, which are esters of long-chain fatty acids and include the *glycerides (which constitute the *fats and *oils of animals and plants), glycolipids, *phospholipids, and *waxes; and **simple lipids**, which do not contain fatty acids and include the *steroids and *terpenes.

Lipids have a variety of functions in living organisms. Fats and oils are a convenient and concentrated means of storing food energy in plants and animals. Phospholipids and *sterols, such as cholesterol, are major components of cell membranes (see LIPID BILAYER). Waxes provide vital waterproofing for body surfaces. Terpenes include vitamins A, E, and K, and phytol (a component of chlorophyll) and occur in essential oils, such as menthol and camphor. Steroids include the adrenal hormones, sex hormones, and bile acids.

Lipids can combine with proteins to form *lipoproteins, e.g. in plasma membranes. In

bacterial cell walls, lipids may associate with polysaccharides to form **lipopolysaccharides**.

**lipid bilayer** The arrangement of lipid molecules in biological *membranes, which takes the form of a double sheet. Each lipid molecule comprises a hydrophilic 'head' (having a high affinity for water) and a hydrophobic 'tail' (having a low affinity for water). In the lipid bilayer the molecules are aligned so that their hydrophilic heads face outwards, forming the outer and inner surfaces of the membrane, while the hydrophobic tails face inwards, away from the external aqueous environment. See illustration.

**Lipmann, Fritz** See KREBS, SIR HANS ADOLF.

**lipoic acid** A vitamin of the *vitamin B complex. It is one of the *coenzymes involved in the decarboxylation of pyruvate by the enzyme pyruvate dehydrogenase. This reaction has to take place before carbohydrates can enter the *Krebs cycle during aerobic respiration. Good sources of lipoic acid include liver and yeast.

**lipolysis** The breakdown of storage lipids in living organisms. Most long-term energy reserves are in the form of triglycerides in fats and oils. When these are needed, e.g. during starvation, lipase enzymes convert the triglycerides into glycerol and the component fatty acids. These are then transported to tissues and oxidized to provide energy.

**lipoprotein** One of a group of compounds consisting of a lipid combined with a protein. Lipoproteins are the main structural

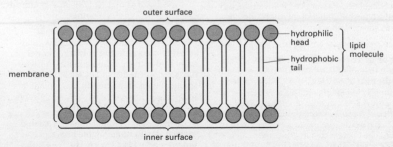

**Lipid bilayer.**

materials of the membranes of cells and cell organelles. They also occur in blood and lymph, being the form in which lipids are transported in these media. *Cholesterol is transported in the bloodstream mainly in the form of **low-density lipoproteins** (**LDLs**) and is removed by means of LDL receptors in cell membranes; the LDLs are bound to the receptors, which are then taken into the cells. Lack of LDL receptors, occurring as a genetic defect in some individuals, is believed to be a cause of high levels of cholesterol in the blood, predisposing to atherosclerosis. **Very low-density lipoproteins** (**VLDLs**) are formed in the liver and are the precursors of LDLs, while **high-density lipoproteins** (**HDLs**), the smallest of all lipoproteins, transport cholesterol from tissues to the liver.

**liposome** A microscopic spherical membrane-enclosed vesicle or sac (20–30 nm in diameter) made artificially in the laboratory by the addition of an aqueous solution to a phospholipid gel. The membrane resembles a cell membrane and the whole vesicle is similar to a cell organelle. Liposomes can be incorporated into living cells and are used to transport relatively toxic drugs into diseased cells, where they can exert their maximum effects. For example, liposomes containing the drug methotrexate, used in the treatment of cancer, can be injected into the patient's blood. The cancerous organ is at a higher temperature than normal body temperature, so that when the liposome passes through its blood vessels, the membrane melts and the drug is released. The study of the behaviour of liposome membranes is used in research into membrane function, particularly to observe the behaviour of membranes during anaesthesia with respect to permeability changes.

**lipotropin** Either of two peptide hormones produced in the anterior pituitary gland that trigger the mobilization of fat deposits and the transfer of lipid components to the bloodstream. β-lipotropin is formed by cleavage of the precursor pro-opiomelanocortin, and is itself cleaved to form γ-lipotropin and other peptides, including endorphins.

**liquation** The separation of mixtures of solids by heating to a temperature at which lower-melting components liquefy.

**liquefaction of gases** The conversion of a gaseous substance into a liquid. This is usually achieved by one of four methods or by a combination of two of them:
(1) by vapour compression, provided that the substance is below its *critical temperature;
(2) by refrigeration at constant pressure, typically by cooling it with a colder fluid in a countercurrent heat exchanger;
(3) by making it perform work adiabatically against the atmosphere in a reversible cycle;
(4) by the *Joule–Thomson effect (*see also* LINDE PROCESS).

Large quantities of liquefied gases are now used commercially, especially *liquefied petroleum gas and liquefied natural gas.

**liquefied natural gas (LNG)** *See* LIQUEFIED PETROLEUM GAS.

**liquefied petroleum gas (LPG)** Various petroleum gases, principally propane and butane, stored as a liquid under pressure. It is used as an engine fuel and has the advantage of causing very little cylinder-head deposits.

**Liquefied natural gas** (**LNG**) is a similar product and consists mainly of methane. However, it cannot be liquefied simply by pressure as it has a low critical temperature of 190 K and must therefore be cooled to below this temperature before it will liquefy. Once liquefied it has to be stored in well-insulated containers. It provides a convenient form in which to ship natural gas in bulk from oil wells or gas-only wells to users. It is also used as an engine fuel.

**liquid** A phase of matter between that of a crystalline solid and a *gas. In a liquid, the large-scale three-dimensional atomic (or ionic or molecular) regularity of the solid is absent but, on the other hand, so is the total disorganization of the gas. Although liquids have been studied for many years there is still no simple comprehensive theory of the liquid state. It is clear, however, from diffraction studies that there is a short-range structural regularity extending over several molecular diameters. These bundles of ordered atoms, molecules, or ions move about in relation to each other, enabling liquids to have almost fixed volumes, which adopt the shape of their containers.

**liquid crystal** A substance that flows like a liquid but has some order in its arrangement of molecules. **Nematic crystals** have long molecules all aligned in the same direction, but otherwise randomly arranged. **Cholesteric** and **smectic** liquid crystals also have aligned molecules, which are arranged in distinct layers. In cholesteric crystals, the axes of the molecules are parallel to the plane of the layers; in smectic crystals they are perpendicular. Concepts from the theory of *phase transitions, such as *order parameters and *broken symmetry, have proved useful in analysing the properties of liquid crystals.

**liquid-crystal display** See LCD.

**liquid-crystal polymer** A polymer with a liquid-crystal structure, this being the most thermodynamically stable. Liquid-crystal polymers contain long rigid chains and combine strength with lightness. They are, however, difficult to produce commercially.

**liquid-drop model** A model of the atomic nucleus in which the nucleons are regarded as being analogous to the molecules in a liquid, the interactions between which maintain the droplet shape by surface tension. The model has been useful in the theory of nuclear fission.

**liquidus** A line on a phase diagram above which a substance is liquid.

***l*-isomer** See OPTICAL ACTIVITY.

**L-isomer** See ABSOLUTE CONFIGURATION.

**Lissajous figures** A curve in one plane traced by a point moving under the influence of two independent harmonic motions. In the common case the harmonic motions are simple, perpendicular to each other, and have a simple frequency ratio. They can be displayed by applying sinusoidal alternating potentials to the X- and Y-inputs of a *cathode-ray oscilloscope. They are named after Jules Lissajous (1822–80).

**(●) SEE WEB LINKS**
• An interactive simulation of an oscilloscope producing Lissajou's figures from the University of Florida

***Listeria*** A genus of rod-shaped aerobic motile Gram-positive bacteria. Only one species, *L. monocytogenes*, causes disease (**listeriosis**). It is resistant to physical and chemical treatments and can occur as a contaminant in certain foods, in faeces, etc. Listeriosis can take various forms, depending on the site of infection: localization in the central nervous system causes meningoencephalitis, while uterine infection can result in abortion or congenital handicap in the fetus.

**litharge** See LEAD(II) OXIDE.

**lithia** See LITHIUM OXIDE.

**lithium** Symbol Li. A soft silvery metal, the first member of group 1 (formerly IA) of the periodic table (see ALKALI METALS); a.n. 3; r.a.m. 6.939; r.d. 0.534; m.p. 180.54°C; b.p. 1347°C. It is a rare element found in spodumene (LiAlSi$_2$O$_6$), petalite (LiAlSi$_4$O$_{10}$), the mica lepidolite, and certain brines. It is usually extracted by treatment with sulphuric acid to give the sulphate, which is converted to the chloride. This is mixed with a small amount of potassium chloride, melted, and electrolysed. The stable isotopes are lithium–6 and lithium–7. Lithium–5 and lithium–8 are short-lived radioisotopes. The metal is used to remove oxygen in metallurgy and as a constituent of some Al and Mg alloys. It is also used in batteries and is a potential tritium source for fusion research. Lithium salts are used in psychomedicine. The element reacts with oxygen and water; on heating it also reacts with nitrogen and hydrogen. Its chemistry differs somewhat from that of the other group 1 elements because of the small size of the Li$^+$ ion.

**(●) SEE WEB LINKS**
• Information from the WebElements site

**lithium aluminium hydride** See LITHIUM TETRAHYDROALUMINATE(III).

**lithium battery** A type of voltaic cell containing lithium or lithium compounds. The most commonly used has a metallic lithium anode and a manganese dioxide (MnO$_2$) cathode, the electrolyte being a solution of lithium salts in an organic solvent. Batteries of this type have an output of about 3 volts. They are more expensive than alkaline batteries, but last longer. Li–MnO$_2$ batteries are also produced in a flat disk form for use in digital watches and other small portable devices. A number of other more specialized lithium primary batteries are available but are not in general use. See also LITHIUM-ION BATTERY.

**lithium carbonate** A white solid, Li$_2$CO$_3$; r.d. 2.11; m.p. 723°C; decomposes above 1310°C. It is produced commercially by treating the ore with sulphuric acid at 250°C and leaching the product to give a solution of lithium sulphate. The carbonate is then obtained by precipitation with sodium carbonate solution. Lithium carbonate is used in the prevention and treatment of manic-depressive (bipolar) disorders. It is also used industrially in ceramic glazes.

**lithium deuteride** *See* LITHIUM HYDRIDE.

**lithium hydride** A white solid, LiH; cubic; r.d. 0.82; m.p. 680°C; decomposes at about 850°C. It is produced by direct combination of the elements at temperatures above 500°C. The bonding in lithium hydride is believed to be largely ionic; i.e. $Li^+H^-$ as supported by the fact that hydrogen is released from the anode on electrolysis of the molten salt. The compound reacts violently and exothermically with water to yield hydrogen and lithium hydroxide. It is used as a reducing agent to prepare other hydrides and the $^2H$ isotopic compound, **lithium deuteride**, is particularly valuable for deuterating a range of organic compounds. Lithium hydride has also been used as a shielding material for thermal neutrons.

**lithium hydrogencarbonate** A compound, $LiHCO_3$, formed by the reaction of carbon dioxide with aqueous lithium bicarbonate and known only in solution. It has found medicinal uses similar to those of lithium carbonate and is sometimes included in proprietary mineral waters.

**lithium hydroxide** A white crystalline solid, LiOH, soluble in water, slightly soluble in ethanol and insoluble in ether. It is known as the monohydrate (monoclinic; r.d. 1.51) and in the anhydrous form (tetragonal, r.d. 1.46; m.p. 450°C; decomposes at 924°C). The compound is made by reacting lime with lithium salts or lithium ores. Lithium hydroxide is basic but has a closer resemblance to group 2 hydroxides than to the other group 1 hydroxides (an example of the first member of a periodic group having atypical properties).

**lithium-ion battery** A type of rechargeable cell in which the anode is carbon and the cathode is a metal oxide (e.g. cobalt(IV) oxide, $CoO_2$). The electrolyte is a lithium salt such as the borate ($LiBO_4$) or chlorate ($LiClO_4$) in an organic solvent. The action of the cell depends on movement of Li ions between anode and cathode with oxidation of the cobalt ions during charging and reduction during discharge. Lithium-ion batteries are light and have a low self-discharge rate, although the capacity does deteriorate with age. They are extensively used in mobile phones, laptops, camcorders, and similar devices, as well as electric cars. *See also* LITHIUM BATTERY.

**lithium oxide (lithia)** A white crystalline compound, $Li_2O$; cubic; r.d. 2.01; m.p. 1700°C. It can be obtained from a number of lithium ores; the main uses are in lubricating greases, ceramics, glass and refractories, and as a flux in brazing and welding.

**lithium sulphate** A white or colourless crystalline material, $Li_2SO_4$, soluble in water and insoluble in ethanol. It forms a monohydrate (monoclinic; r.d. 1.88) and an anhydrous form, which exists in α- (monoclinic), β- (hexagonal) and γ- (cubic) forms; r.d. 2.23. The compound is prepared by the reaction of the hydroxide or carbonate with sulphuric acid. It is not isomorphous with other group 1 sulphates and does not form alums.

**lithium tetrahydroaluminate(III) (lithium aluminium hydride; LAH)** A white or light grey powder, $LiAlH_4$; r.d. 0.917; decomposes at 125°C. It is prepared by the reaction of excess lithium hydride with aluminium chloride. The compound is soluble in ethoxyethane, reacts violently with water to release hydrogen, and is widely used as a powerful reducing agent in organic chemistry. It should always be treated as a serious fire risk in storage.

**lithosphere** The earth's crust (*see* EARTH). Sometimes the lithosphere is also understood to include the mantle and sometimes the mantle and the core.

**litmus** A water-soluble dye extracted from certain lichens. It turns red under acid conditions and blue under alkaline conditions, the colour change occurring over the pH range 4.5–8.3 (at 25°C). It is not suitable for titrations because of the wide range over which the colour changes, but is used as a rough *indicator of acidity or alkalinity, both in solution and as litmus paper (absorbent paper soaked in litmus solution).

**litre** Symbol l or L. A unit of volume in the metric system regarded as a special name for the cubic decimetre. It was formerly defined as the volume of 1 kilogram of pure water at 4°C at standard pressure, which is equivalent to 1.000 028 $dm^3$.

**litter** Dead organic matter in the soil that has not yet decomposed. It consists of fallen leaves and other plant remains (**leaf litter**), animal excrement, etc. After decomposition by *decomposers and *detritivores litter becomes *humus.

**littoral** Designating or occurring in the marginal shallow-water zone of a sea or lake, especially (in the sea) between high and low tide lines. In this zone enough light penetrates to the bottom to support rooted aquatic plants. *Compare* PROFUNDAL; SUBLITTORAL.

**liver** A large lobed organ in the abdomen of vertebrates that plays an essential role in many metabolic processes by regulating the composition and concentration of nutrients and toxic materials in the blood. It receives the products of digestion dissolved in the blood via the *hepatic portal vein and its most important functions are to convert excess glucose to the storage product *glycogen, which serves as a food reserve; to break down excess amino acids to ammonia, which is converted to *urea or *uric acid and excreted via the kidneys; and to store and break down fats (*see* LIPOLYSIS). Other functions of the liver are (1) the production of *bile; (2) the breakdown (*detoxification) of poisonous substances in the blood; (3) the removal of damaged red blood cells; (4) the synthesis of vitamin A and the blood-clotting substances prothrombin and fibrinogen; and (5) the storage of iron.

**(⊕) SEE WEB LINKS**
• Summary of liver anatomy, functions, and diseases, compiled by the British Liver Trust

**liverworts** *See* HEPATOPHYTA.

**living fossil** Any organism whose closest relatives are extinct and that was once itself thought to be extinct. An example is the coelacanth, a primitive fish that was common in the Devonian era, the first recent living specimen of which was discovered in 1938.

**lixiviation** The separation of mixtures by dissolving soluble constituents in water.

**lizards** *See* SQUAMATA.

**Lloyd's mirror** An optical arrangement for producing interference fringes. A slit is illuminated by monochromatic light and placed close to a plane mirror. Interference occurs between direct light from the slit and light reflected from the mirror. It was first used by Humphrey Lloyd (1800–81) in 1834.

**LNG** *See* LIQUEFIED PETROLEUM GAS.

**loaded concrete** Concrete containing elements (such as iron or lead) with a high mass number; it is used in making the radiation shield around nuclear reactors.

**loam** A fertile *soil that is made up of organic matter mixed with clay, sand, and silt. Loams differ in their ratios of clay, sand, and silt, which influences which types of plants they can support.

**Local Group** The group of *galaxies of which our own Galaxy is a member. It consists of some 30–40 known members, the most massive of which are the Galaxy and the Andromeda galaxy.

**localization** 1. The confinement of electrons to a particular atom in a molecule or to a particular chemical bond. 2. In the theory of *disordered solids, the concept that an electron is concentrated around a specific site and cannot contribute to the solid's electrical conductivity (at *absolute zero) by moving through the system. In one dimension any amount of disorder makes all electron states localized. In three dimensions a small amount of disorder makes electron states near the top and the bottom of the *energy bands localized; states in the centre of the bands are called **extended** states because they can propagate through the system and hence contribute to electrical conductivity. The dividing energies between localized and extended states are called **mobility edges**. Given sufficient disorder all states become localized. In two dimensions all electron states in disordered solids are thought to be localized, with some states being strongly localized around specific sites while other states are weakly localized around specific sites. Localization also occurs in disordered solids for other *excitations, such as *phonons and *spin waves.

**localized bond** A *chemical bond in which the electrons forming the bond remain between (or close to) the linked atoms. *Compare* DELOCALIZATION.

**local oscillator** An *oscillator in a *heterodyne or *superheterodyne radio receiver. It supplies the radio-frequency signal that beats with the incoming signal to produce the intermediate frequency.

**Local Supercluster** The flattened *galaxy cluster of which the *Local Group is a member. It is about 100 million light-years across, with the Virgo cluster at its centre.

**lock-and-key mechanism** A mechanism

proposed in 1890 by Emil Fischer (1852–1919) to explain binding between the active site of an enzyme and a substrate molecule. The active site was thought to have a fixed structure (the lock), which exactly matched the structure of a specific substrate (the key). Thus the enzyme and substrate interact to form an *enzyme–substrate complex. The substrate is converted to products that no longer fit the active site and are therefore released, liberating the enzyme. Observations made by X-ray diffraction studies have shown that the active site of an enzyme is more flexible than the lock-and-key theory would suggest.

**locomotion** The ability of an organism to move in a particular direction in its environment, which requires a propulsive force acting against a supporting structure. Most animals and many single-celled organisms have powers of locomotion. Some protists possess contractile fibres that exert force on the plasma membrane to change the shape of the cell; this may be combined with *cytoplasmic streaming to bring about locomotion (*see* AMOEBOID MOVEMENT). In many other protists and bacteria the propulsive force is provided by the action of *undulipodia or *flagella. In animals the force required to initiate locomotion is generated by *muscles, which act against a supporting framework provided by a *skeleton. *See also* FINS; FLIGHT.

**locule (loculus)** A small cavity in a plant or animal body. In plants the locule of the ovary is the cavity containing the ovules and the locules of the anther contain the developing pollen grains.

**locus 1.** (in mathematics) A set of points whose location is specified by an equation. For example, if a point moves so that the sum of its distances from two fixed points is constant, the locus of the point is an *ellipse. **2.** (in genetics) The position of a gene on a chromosome or within a DNA (or RNA) molecule. The alleles of a gene occupy the same locus on *homologous chromosomes.

**lodestone** *See* MAGNETITE.

**Lodge, Sir Oliver Joseph** (1851–1940) British physicist, who became principal of the new Birmingham University in 1900. His best-known work was in *radio, particularly his invention in 1894 of the 'coherer', used as a detector in early radio receivers (*see* DEMODULATION). After 1910 he became increasingly interested in spiritualism and reconciling science and religion.

**logarithm** The power to which a number, called the **base**, has to be raised to give another number. Any number $y$ can be written in the form $y = x^n$. $n$ is then the logarithm to the base $x$ of $y$, i.e. $n = \log_x y$. If the base is 10, the logarithms are called **common logarithms**. **Natural** (or **Napierian**) **logarithms** (named after John Napier) are to the base e = 2.718 28…, written $\log_e y$ or $\ln y$. Logarithms were formerly used to facilitate calculations, before the advent of electronic calculators.

A logarithm contains two parts, an integer and a decimal. The integer is called the **characteristic**, and the decimal is called the **mantissa**. For example, the logarithm to the base 10 of 210 is 2.3222, where 2 is the characteristic and 0.3222 is the mantissa.

**logarithmic scale 1.** A scale of measurement in which an increase or decrease of one unit represents a tenfold increase or decrease in the quantity measured. Decibels and pH measurements are common examples of logarithmic scales of measurement. **2.** A scale on the axis of a graph in which an increase of one unit represents a tenfold increase in the variable quantity. If a curve $y = x^n$ is plotted on graph paper with logarithmic scales on both axes, the result is a straight line of slope $n$, i.e. $\log y = n \log x$, which enables $n$ to be determined.

**logarithmic series** The expansion of a logarithmic function, such as $\log_e(1 + x)$, i.e. $x - x^2/2 + x^3/3 - \ldots + (-1)^n x^n/n$, or $\log_e(1 - x)$, i.e. $-x - x^2/2 - x^3/3 \ldots - x^n/n$.

**logic circuits** The basic switching circuits or *gates used in digital computers and other digital electronic devices. The output signal, using a *binary notation, is controlled by the logic circuit in accordance with the input system. The three basic logic circuits are the **AND**, **OR**, and **NOT circuits**. The **AND** circuit gives a binary 1 output if a binary 1 is present on each input circuit; otherwise the output is a binary 0. The **OR** circuit gives a binary 1 output if a binary 1 is present on at least one input circuit; otherwise the output is binary 0. The **NOT** circuit inverts the input signal, giving a binary 1 output for a binary 0 input or a 0 output for a 1 input.

Often these basic logic circuits are used in combination, e.g. a **NAND circuit** consists of **NOT + AND** circuits. In terms of electronic equipment, logic circuits are now almost exclusively embodied into *integrated circuits.

**log phase** *See* BACTERIAL GROWTH CURVE.

**lomentum** A type of dry dehiscent fruit formed from a single carpel but divided into one-seeded compartments by constrictions between the seeds. *Legumes (e.g. those of *Acacia*) and *siliquas (e.g. those of wild radish) can be divided in this way.

**lone pair** A pair of electrons having opposite spin in an orbital of an atom. For instance, in ammonia the nitrogen atom has five electrons, three of which are used in forming single bonds with hydrogen atoms. The other two occupy a filled atomic orbital and constitute a lone pair (see illustration). The orbital containing these electrons is equivalent to a single bond (sigma orbital) in spatial orientation, accounting for the pyramidal shape of the molecule. In the water molecule, there are two lone pairs on the oxygen atom. In considering the shapes of molecules, repulsions between bonds and lone pairs can be taken into account:

lone pair–lone pair > lone pair–bond > bond–bond.

**Lone pair.** Lone pair of electrons in ammonia.

**long-day plant** A plant in which flowering can be induced or enhanced by long days, usually of more than 12 hours of daylight. Examples are spinach and spring barley. *See* PHOTOPERIODISM. *Compare* DAY-NEUTRAL PLANT; SHORT-DAY PLANT.

**longitude** *See* LATITUDE AND LONGITUDE.

**longitudinal wave** *See* WAVE.

**long period** *See* PERIODIC TABLE.

**long-sightedness** *See* HYPERMETROPIA.

**loop of Henle** The hairpin-shaped section of a kidney tubule situated between the proximal and distal tubules in the *nephron. The loop of Henle extends from the cortex into the medulla; it consists of a thin descending limb, which is permeable to water, and a thick ascending limb, which is impermeable to water. Complex movements of

ions and water across the walls of the loop results in the production of concentrated urine in the *collecting duct. It is named after Friedrich Henle (1809–85).

**loop variable** A quantity that is used to characterize field theories as an alternative to characterizing them in terms of fields and potentials. A loop variable is like a line of force or flux line that is closed to form a loop. Loop variables have been used with great success in *quantum electrodynamics, *quantum chromodynamics, and *quantum gravity.

**Lorentz–Fitzgerald contraction (Fitzgerald contraction)** The contraction of a moving body in the direction of its motion. It was proposed independently by Hendrik Lorentz (1853–1928) and George Fitzgerald (1851–1901) in 1892 to account for the null result of the *Michelson–Morley experiment. The contraction was given a theoretical background in Einstein's special theory of *relativity. In this theory, an object of length $l_0$ at rest in one *frame of reference will appear, to an observer in another frame moving with relative velocity $v$ with respect to the first, to have length

$$l_0\sqrt{(1 - v^2/c^2)},$$

where $c$ is the speed of light. The original hypothesis regarded this contraction as a real one accompanying the absolute motion of the body. The contraction is in any case negligible unless $v$ is of the same order as $c$.

**Lorentz–Lorenz equation** A relation between the *polarizability $\alpha$ of a molecule and the *refractive index $n$ of a substance made up of molecules with this polarizability. The Lorentz–Lorenz equation can be written in the form $\alpha = (3/4\pi N)\,[(n^2-1)/(n^2 + 2)]$, where $N$ is the number of molecules per unit volume. The equation provides a link between a microscopic quantity (the polarizability) and a macroscopic quantity (the refractive index). It was derived using macroscopic electrostatics in 1880 by Hendrik Lorentz and independently by the Danish physicist Ludwig Valentin Lorenz also in 1880. *Compare* CLAUSIUS–MOSSOTTI EQUATION.

**Lorentz transformations** A set of equations for transforming the position and motion parameters from a frame of reference with origin at O and coordinates $(x,y,z)$ to a frame moving relative to it with origin at O' and coordinates $(x',y',z')$. They replace the *Galilean transformations used in *Newtonian mechanics and are used in relativistic mechanics. They are:

$$x' = \beta(x - vt)$$
$$y' = y$$
$$z' = z$$
$$t' = \beta(t - vx/c^2),$$

where $v$ is the relative velocity of separation of O and O', $c$ the speed of light, and $\beta = 1/\sqrt{(1 - v^2/c^2)}$. The above equations apply for constant $v$ in the $xx'$ direction with O and O' coinciding at $t = t' = 0$.

**Lorenz, Konrad Zacharias** (1903–89) Austrian ethologist who studied medicine, becoming a lecturer at Vienna in 1937. Watching the behaviour of birds on his private estate, he made his studies of *imprinting. For this work he shared the 1973 Nobel Prize for physiology or medicine with Karl von Frisch (1886–1982) and Niko *Tinbergen.

**Loschmidt's constant (Loschmidt number)** Symbol $N_L$. The number of particles per unit volume of an *ideal gas at STP. It has the value $2.686\,763(23) \times 10^{25}$ m$^{-3}$ and was first worked out by the Bohemian physical chemist Joseph Loschmidt (1821–95).

**Lotka–Volterra mechanism** A simple chemical reaction mechanism proposed as a possible mechanism of *oscillating reactions. The process involves a conversion of a reactant R into a product P. The reactant flows into the reaction chamber at a constant rate and the product is removed at a constant rate, i.e. the reaction is in a steady state (but not in chemical equilibrium). The mechanism involves three steps:

$$R + X \rightarrow 2X$$
$$X + Y \rightarrow 2Y$$
$$Y \rightarrow P$$

The first two steps involve *autocatalysis: the first step is catalysed by the reactant X and the second by the reactant Y. The kinetics of such a reaction can be calculated numerically, showing that the concentrations of both X and Y increase and decrease periodically with time. This results from the autocatalytic action. Initially, the concentration of X is small, but, as it increases, there is a rapid increase in the rate of the first reaction because of the autocatalytic action of X. As the concentration of X builds up, the rate of the second reaction also increases. Initially, the concentration of Y is low but there is a sudden surge in the rate of step 2, resulting from the autocatalytic action of Y. This lowers the concentration of X and slows

down step 1, so the concentration of X falls. Less X is now available for the second step and the concentration of Y also starts to fall. With this fall in the amount of Y, less X is removed, and the first reaction again begins to increase. These processes are repeated, leading to repeated rises and falls in the concentrations of both X and Y. The cycles are not in phase, peaks in the concentration of Y occurring later than peaks in X.

In fact, known oscillating chemical reactions have different mechanisms to the above, but the scheme illustrates how oscillation may occur. This type of process is found in fields other than chemistry; they were investigated by the Italian mathematician Vito Volterra (1860–1940) in models of biological systems (e.g. predator–prey relationships).

**loudness** The physiological perception of sound intensity. As the ear responds differently to different frequencies, for a given intensity loudness is dependent on frequency. Sounds with frequencies between 1000 hertz and 5000 Hz are louder than sounds of the same intensity at higher or lower frequencies. Duration is also a factor in loudness, long bursts of sound being louder than short bursts. Loudness increases up to a duration of about 0.2 second; above this limit loudness does not increase with duration.

Relative loudness is usually measured on the assumption of proportionality to the logarithm of the intensity (for a given frequency), i.e. proportionality to the relative intensity on the *decibel scale. A subjective judgment is made of the relative intensity above threshold that a note of 1000 Hz must have to match the specimen sound; the loudness of this, in *phons, is then equal to that relative intensity in decibels.

**loudspeaker** A transducer for converting an electrical signal into an acoustic signal. Usually it is important to preserve as many characteristics of the electrical waveform as possible. The device must be capable of reproducing frequencies in the range 150–8000 hertz for speech and 20–20 000 Hz for music.

The most common loudspeaker consists of a moving-coil device. In this a cone-shaped diaphragm is attached to a coil of wire and made to vibrate in accordance with the electrical signal by the interaction between the current passing through the coil and a steady magnetic field from a permanent magnet surrounding it.

**low** See DEPRESSION.

**low-dimensional system** A *condensed-matter system in which the spatial dimension is less than three. In practice, a **two-dimensional system** is a thin film or layer and a **one-dimensional system** is a thin wire. Two-dimensional systems have applications to *semiconductor technology, in devices such as MOSFETs (see TRANSISTOR). The behaviour of low-dimensional systems is of interest because the problems for low-dimensional systems (particularly one-dimensional systems) are much easier to solve than the corresponding problems in three dimensions. Clusters of atoms and very small crystals can be considered as **zero-dimensional systems**.

**lowering of vapour pressure** A reduction in the saturated vapour pressure of a pure liquid when a solute is introduced. If the solute is a solid of low vapour pressure, the decrease in vapour pressure of the liquid is proportional to the concentration of particles of solute; i.e. to the number of dissolved molecules or ions per unit volume. To a first approximation, it does not depend on the nature of the particles. See COLLIGATIVE PROPERTIES; RAOULT'S LAW.

**low frequency (LF)** A radio frequency in the range 30–300 kilohertz; i.e. having a wavelength in the range 1–10 kilometre.

**Lowry–Brønsted theory** See ACID.

**low-temperature physics** Physics at low temperatures, especially at temperatures close to absolute zero.

**LSD** See LYSERGIC ACID DIETHYLAMIDE.

**L-series** See ABSOLUTE CONFIGURATION.

**lubrication** The use of a substance to prevent contact between solid surfaces in relative motion in order to reduce friction, wear, overheating, and rusting. Liquid hydrocarbons (oils), either derived from petroleum or made synthetically, are the most widely used lubricants as they are relatively inexpensive, are good coolants, provide the appropriate range of viscosities, and are thermally stable. Additives include polymeric substances that maintain the desired viscosity as the temperature increases, antioxidants that prevent the formation of a sludge, and alkaline-earth phenates that neutralize acids and reduce wear.

At high temperatures, solid lubricants, such as graphite or molybdenum disulphide, are often used. Semifluid lubricants (greases) are used to provide a seal against moisture and dirt and to remain attached to vertical surfaces. They are made by adding gelling agents, such as metallic soaps, to liquid lubricants.

Recent technology has made increasing use of gases as lubricants, usually in air bearings. Their very low viscosities minimize energy losses at the bearings but necessitate some system for pumping the gas continuously to the bearings. The principle is that of the hovercraft.

**lumbar vertebrae** The *vertebrae in the region of the lower back. They occur below the *thoracic vertebrae and above the *sacral vertebrae. In mammals they bear processes for the attachment of back muscles.

**lumen** 1. The space enclosed by a vessel, duct, or other tubular or saclike organ. The central cavity of blood vessels and of the digestive tract are examples. 2. Symbol lm. The SI unit of *luminous flux equal to the flux emitted by a uniform point source of 1 candela in a solid angle of 1 steradian.

**luminance (photometric brightness)** Symbol $L$. The *luminous intensity of any surface in a given direction per unit projected area of the surface, viewed from that direction. It is given by the equation $L = dI/(dA\cos\theta)$, where $I$ is the luminous intensity and $\theta$ is the angle between the line of sight and the normal to the surface area $A$ being considered. It is measured in candela per square metre.

**luminescence** The emission of light by a substance for any reason other than a rise in its temperature. In general, atoms of substances emit *photons of electromagnetic energy when they return to the *ground state after having been in an excited state (see EXCITATION). The causes of the excitation are various. If the exciting cause is a photon, the process is called **photoluminescence**; if it is an electron it is called **electroluminescence**. **Chemiluminescence** is luminescence resulting from a chemical reaction (such as the slow oxidation of phosphorus); *bioluminescence is the luminescence produced by a living organism (such as a firefly). If the luminescence persists significantly after the exciting cause is removed it is called **phosphorescence**; if it does not it is called **fluorescence**. This distinction is arbitrary since there must always be some delay; in some definitions a persistence of more than

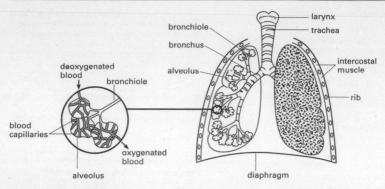

**Lung.** The lungs and air passages of a mammal (right lung cut open to show internal structure).

10 nanoseconds ($10^{-8}$ s) is treated as phosphorescence.

**luminol test** A *presumptive test for blood. The reagent is a mixture of 3-aminophthalhydrazide, sodium carbonate, and sodium perborate. When sprayed with the reagent, traces of blood (even old blood) emit a faint chemoluminescence.

**luminosity** **1.** *Luminous intensity in a particular direction; the apparent brightness of an image. **2.** The brightness of a star defined as the total energy radiated in unit time. It is related to the surface area ($A$) and the **effective temperature** ($T_e$; the temperature of a black body having the same radius as the star and radiating the same amount of energy per unit area in one second) by a form of *Stefan's law, i.e.

$$L = A\sigma T_e^4$$

where σ is the Stefan constant and $L$ is the luminosity.

**luminous exitance** *See* EXITANCE.

**luminous flux** Symbol $\Phi_v$. A measure of the rate of flow of light, i.e. the radiant flux in the wavelength range 380–760 nanometres, corrected for the dependence on wavelength of the sensitivity of the human eye. It is measured by reference to emission from a standard source, usually in lumens.

**luminous intensity** Symbol $I_v$. A measure of the light-emitting ability of a light source, either generally or in a particular direction. It is measured in candelas.

**lunar eclipse** *See* ECLIPSE.

**lunation** *See* SYNODIC MONTH.

**lung** The *respiratory organ of air-breathing vertebrates. A pair of lungs is situated in the thorax, within the ribcage. Each consists essentially of a thin moist membrane that is folded to increase its surface area. Exchange of oxygen and carbon dioxide takes place between blood capillaries on one side of the membrane and air on the other. The lung is supplied with air through a *bronchus. In mammals and reptiles the membrane of the lung takes the form of numerous sacs (*see* ALVEOLUS) that are connected to the bronchus via *bronchioles (see illustration). The lungs themselves contain no muscular tissue and are ventilated by *respiratory movements, the mechanisms of which vary with the species.

**lungfish** *See* DIPNOI.

**luteinizing hormone (LH; interstitial-cell-stimulating hormone; ICSH)** A hormone, secreted by the anterior pituitary gland in mammals, that stimulates in males the production of sex hormones (*androgens) by the *interstitial cells of the testes and in females ovulation, *progesterone synthesis, and *corpus luteum formation.

**luteotrophic hormone** *See* PROLACTIN.

**lutetium** Symbol Lu. A silvery metallic element belonging to the *lanthanoids; a.n. 71; r.a.m. 174.97; r.d. 9.8404 (20°C); m.p. 1663°C; b.p. 3402°C. Lutetium is the least abundant of the elements and the little quantities that are available have been obtained by processing other metals. There are two natural isotopes, lutetium–175 (stable) and lutetium–176 (half-life $2.2 \times 10^{10}$ years). The element is used as a catalyst. It was first identified by Gerges Urbain (1872–1938) in 1907.

 **SEE WEB LINKS**
• Information from the WebElements site

**Luttinger liquid (Tomonaga–Luttinger liquid)** A model of interacting fermions in one dimension proposed by the Japanese physicist Sin-Itiro Tomonaga (1906–1979) in 1950 and analysed further by the American physicist Joaquin Luttinger (1923–97) in 1963. A Luttinger liquid is a *non-Fermi liquid and hence has different properties from, for example, the properties of electrons in ordinary metals. The electrons in one-dimensional chains of certain organic molecules form a Luttinger liquid.

**lux** Symbol lx. The SI unit of *illuminance equal to the illumination produced by a *luminous flux of 1 lumen distributed uniformly over an area of 1 square metre.

**lyase** Any of a class of enzymes that catalyse either the cleavage of a double bond and the addition of new groups to a substrate, or the formation of a double bond.

**Lycophyta (Lycopodophyta)** A phylum of *tracheophyte plants containing the clubmosses (genus *Lycopodium*) and related genera (including *Selaginella*) as well as numerous extinct forms, which reached their peak in the Carboniferous period with giant coal-forming tree species. Lycophytes have roots and their stems are covered with numerous small leaves. Reproduction is by means of spores; the sporangia are usually grouped into cones.

**lye** See POTASSIUM HYDROXIDE.

**Lyell, Sir Charles** (1797–1875) British geologist, born in Scotland. Poor eyesight made him change from his legal studies to geology, which resulted in his theory that rocks are formed by a slow continuous process. He is also known for his three-volume *The Principles of Geology* (1831–33), which was to become the standard textbook for generations.

**Lyman series** See HYDROGEN SPECTRUM.

**lymph** The colourless liquid found within the *lymphatic system, into which it drains from the spaces between the cells. Lymph (called **tissue fluid** in the intercellular spaces) resembles *blood plasma, consisting mostly of water with dissolved salts and proteins. Fats are found in suspension and their presence varies with food intake. The lymph eventually enters the bloodstream near the heart.

**lymphatic system** The network of vessels that conveys *lymph from the tissue fluids to the bloodstream. Tiny *lacteals (in the small intestine) and **lymph capillaries** (in other tissues) drain into larger tubular vessels that converge to form the right lymphatic duct and the *thoracic duct, which connect with the venous blood supply to the heart. Associated with the lymphatic vessels at intervals along the system are the *lymph nodes. The lymph capillary walls are very permeable, so lymph bathing the body's tissues can drain away molecules that are too large to pass through blood capillary walls. Lymph is pumped by cycles of contraction and relaxation of the lymphatic vessels and also by the action of adjoining muscles.

**lymph capillary** See LYMPHATIC SYSTEM.

**lymph node** A mass of *lymphoid tissue, many of which occur at intervals along the *lymphatic system. Lymph in the lymphatic vessels flows through the lymph nodes, which filter out bacteria and other foreign particles, so preventing them from entering the bloodstream and causing infection. The lymph nodes also produce *lymphocytes. In humans, major lymph nodes occur in the neck, under the arms, and in the groin.

**lymphocyte** A type of white blood cell (*leucocyte) that has a large nucleus and little cytoplasm. Lymphocytes are formed in the *lymph nodes and provide about a quarter of all leucocytes. They are important in the body's defence and are responsible for immune reactions as the presence of *antigens stimulates them to produce *antibodies. There are two principal populations of lymphocytes: **B lymphocytes** (*see* B CELL), which produce circulating antibodies and are responsible for humoral *immunity; and **T lymphocytes** (*see* T CELL), which are responsible for cell-mediated immunity.

**lymphoid tissue** The type of tissue found in the *lymph nodes, *tonsils, *spleen, and *thymus. It is responsible for producing lymphocytes and therefore contributes to the body's defence against infection.

**lymphokine** See CYTOKINE.

**lymphoma** See CANCER.

**lyophilic** Having an affinity for a solvent

('solvent-loving'; if the solvent is water the term **hydrophilic** is used). *See* COLLOIDS.

**lyophobic** Lacking any affinity for a solvent ('solvent-hating'; if the solvent is water the term **hydrophobic** is used). *See* COLLOIDS.

**Lysenkoism** The official Soviet science policy governing the work of geneticists in the USSR from about 1940 to 1960. It was named after its chief promoter, the agriculturalist Trofim Lysenko (1898–1976). Lysenkoism dismissed all the advances that had been made in classical genetics, denying the existence of genes, and held that the variability of organisms was produced solely by environmental changes. There was also a return to a belief in the inheritance of acquired characteristics (*see* LAMARCKISM). This state of affairs continued, despite overwhelming conflicting evidence from Western scientists, because it provided support for communist theory.

**lysergic acid diethylamide (LSD)** A chemical derivative of lysergic acid that has potent hallucinogenic properties (*see* HALLUCINOGEN). It occurs in the cereal-fungus ergot and was first synthesized in 1943. LSD acts as an *antagonist at *serotonin receptors.

**lysigeny** The localized disruption of plant cells to form a cavity (surrounded by remnants of the broken cells) in which secretions accumulate. Examples are the oil cavities in the leaves of citrus trees. *Compare* SCHIZOGENY.

**lysine** *See* AMINO ACID.

**lysis** The destruction of a living cell. This may be effected by *lysosomes or *lymphocytes, either as part of the normal metabolic process (as when cells are damaged or worn out) or as a reaction against invading cells (e.g. bacteria). *Bacteriophages eventually cause lysis of their host cells.

**lysogeny** *See* LAMBDA PHAGE; PROPHAGE.

**lysosome** A membrane-bound sac (organelle) found in animal cells and in single-celled eukaryotes. It contains hydrolytic enzymes that degrade aged or defective cell components or material taken in by the cell from its environment, such as food particles or bacteria. In plant cells, the *vacuole contains hydrolytic enzymes equivalent to those in the lysosome and can degrade materials in a manner similar to a lysosome.

**lysozyme** An antibacterial enzyme widely distributed in body fluids and secretions, including tears and saliva. It disrupts the polysaccharide components of bacterial cell walls, leaving them susceptible to destruction.

**machine** A device capable of making the performance of mechanical work easier, usually by overcoming a force of resistance (the load) at one point by the application of a more convenient force (the effort) at some other point. In physics, the six so-called **simple machines** are the lever, wedge, inclined plane, screw, pulley, and wheel and axle.

**Mach number** The ratio of the relative speeds of a fluid and a rigid body to the speed of sound in that fluid under the same conditions of temperature and pressure. If the Mach number exceeds 1 the fluid or body is moving at a **supersonic speed**. If the Mach number exceeds 5 it is said to be **hypersonic**. The number is named after Ernst Mach (1838–1916).

**Mach's principle** The *inertia of any particular piece of matter is attributable to the interaction between that piece of matter and the rest of the universe. A body in isolation would have zero inertia. This principle was stated by Ernst Mach in the 1870s and was made use of by Einstein in his general theory of *relativity. The significance of Mach's principle in general relativity theory is still a contentious issue.

**Maclaurin's series** *See* TAYLOR SERIES.

**macrofauna** The larger animals, collectively, which can be observed without the aid of a microscope (*compare* MICROFAUNA). The macrofauna sometimes includes small soil-dwelling invertebrates, such as annelids and nematodes, but these may be separated into an intermediate category, the **mesofauna**.

**macromolecular crystal** A crystalline solid in which the atoms are all linked together by covalent bonds. Carbon (in diamond), boron nitride, and silicon carbide are examples of substances that have macromolecular crystals. In effect, the crystal is a large molecule (hence the alternative description **giant-molecular**), which accounts for the hardness and high melting point of such materials.

**macromolecule** A very large molecule.

Natural and synthetic polymers have macromolecules, as do many proteins and nucleic acids. *See also* COLLOIDS.

**macronutrient** A chemical element required by plants in relatively large amounts. Macronutrients include carbon, hydrogen, oxygen, nitrogen, phosphorus, potassium, sulphur, magnesium, calcium, and iron. *See also* ESSENTIAL ELEMENT. *Compare* MICRONUTRIENT.

**macrophage** A large phagocytic cell (*see* PHAGOCYTE) that can ingest pathogenic microorganisms (e.g. bacteria, protozoa) or cell debris and forms part of the body's immune system. Macrophages develop from precursor cells (promonocytes) in bone marrow, become wandering *monocytes in the bloodstream, and then settle as mature macrophages in various tissues, including lymph nodes, connective tissues (as **histiocytes**), lungs, the linings of liver sinusoids and the spleen, skin, and nervous tissues (**microglia**). Collectively the macrophages make up the **mononuclear phagocyte system**.

**macrophagous** Describing a method of feeding in heterotrophic organisms in which food is ingested in the form of relatively large chunks. *Compare* MICROPHAGOUS.

**macrophyll** *See* MEGAPHYLL.

**macroscopic** Designating a size scale very much larger than that of atoms and molecules. Macroscopic objects and systems are described by *classical physics although *quantum mechanics can have macroscopic consequences. *Compare* MESOSCOPIC; MICROSCOPIC.

**macula** 1. A patch of sensory hair cells in the *utriculus and *sacculus of the inner ear that provides information about the position of the body in relation to gravity. The hairs of the cells are embedded in an **otolith**, a gelatinous cap containing particles of calcium carbonate. Movement of the particles in response to gravity pulls the gelatinous mass downwards, which bends the hairs and triggers a nerve impulse to the brain. 2. An area of the *retina of the vertebrate eye with increased *visual acuity. Maculae occur in some animals that lack *foveae and often surround foveae in those animals that possess them.

**Madelung constant** A constant arising in calculations of the cohesion of ionic crystals. The electrostatic interaction per ion pair, $U$, is given by $U(r) = -\alpha e^2/r$, where $\alpha$ is the Madelung constant and $e^2/r$ is the Coulomb interaction between the ions, with $r$ being the lattice constant. The value of $\alpha$ depends on the type of lattice. For the sodium chloride lattice, $\alpha$ has a value of about 1.75. A more realistic calculation of cohesion is obtained if short-range repulsions with an inverse power law are included, i.e.

$$U(r) = \alpha e^2/r - C/r^n,$$

where $C$ and $n$ are constants. The value of $\alpha$ can be used in calculations to determine $C$ and $n$. It was first introduced by the German physicist Erwin Madelung (1881–1972) in 1918.

**Madelung's rule** An empirical rule for determining the order in which atomic orbitals are filled. This rule, which was proposed in 1936 by Erwin Madelung, comes in two parts. The first part states that the order of filling is the order of increasing $n + l$, where $n$ is the principal quantum number and $l$ is the azimuthal quantum number. The second part states that for two orbitals with an equal value of $n + l$ the order of filling is the order of increasing $n$. The first part arises because increasing $n + l$ increases the number of nodes of the wavefunction, and hence increases the energy. The second part arises because the approximate *Fock degeneracy of many-electron atoms means that the value of $n$ has a stronger effect on the energy than the value of $l$.

**mafic** (from *ma*gnesium + *fer*ric) Denoting any dark-coloured ferromagnesian mineral or a rock in which such minerals predominate. Mafic minerals incude amphibole, olivine, and pyroxine. *See also* FELSIC.

**Magellanic clouds** Two small galaxies that are members of the *Local Group and are situated close to the Milky Way. They are visible from the southern hemisphere of the earth and were first recorded by Ferdinand Magellan (1480–1521) in 1519.

**magic numbers** Numbers of neutrons or protons that occur in atomic nuclei to produce very stable structures. The magic numbers for both protons and neutrons are 2, 8, 20, 28, 50, and 82. For neutrons 126 and 184 are also magic numbers and for protons 114 is a magic number. The relationship between stability and magic numbers led to a nuclear *shell model in analogy to the electron shell model of the atom.

**magma** Hot molten material that originates within the earth's crust or mantle and when cooled and solidified forms igneous rock. Most magmas are composed largely of silicates with suspended crystals and dissolved gases. Magma is extruded as *lava onto the surface of the earth as a result of volcanic activity; magma that cools and solidifies within the earth's crust may form either plutonic (at great depths) or hypabyssal (at intermediate depths) rocks.

**Magnadur** A trade name for a ceramic material used to make permanent magnets. It consists of sintered iron oxide and barium oxide.

**Magnalium** A trade name for an aluminium-based alloy of high reflectivity for light and ultraviolet radiation that contains 1–2% of copper and between 5% and 30% of magnesium. Strong and light, these alloys also sometimes contain other elements, such as tin, lead, and nickel.

**magnesia** *See* MAGNESIUM OXIDE.

**magnesite** A white, colourless, or grey mineral form of *magnesium carbonate, $MgCO_3$, crystallizing in the trigonal system. It is formed as a replacement mineral of magnesium-rich rocks when carbon dioxide is available. Magnesite is mined both as an ore for magnesium and as a source of magnesium carbonate. It occurs in Austria, USA, Greece, Norway, India, Australia, and South Africa.

**magnesium** Symbol Mg. A silvery metallic element belonging to group 2 (formerly IIA) of the periodic table (*see* ALKALINE-EARTH METALS); a.n. 12; r.a.m. 24.312; r.d. 1.74; m.p. 648.8°C; b.p. 1090°C. The element is found in a number of minerals, including magnesite ($MgCO_3$), dolomite ($MgCO_3.CaCO_3$), and carnallite ($MgCl_2.KCl.6H_2O$). It is also present in sea water, and it is an *essential element for living organisms. Extraction is by electrolysis of the fused chloride. The element is used in a number of light alloys (e.g. for aircraft). Chemically, it is very reactive. In air it forms a protective oxide coating but when ignited it burns with an intense white flame. It also reacts with the halogens, sulphur, and nitrogen. Magnesium was first isolated by Sir Humphry Davy in 1808.

**(⊕) SEE WEB LINKS**

• Information from the WebElements site

**magnesium bicarbonate** *See* MAGNE-SIUM HYDROGENCARBONATE.

**magnesium carbonate** A white compound, $MgCO_3$, existing in anhydrous and hydrated forms. The anhydrous material (trigonal; r.d. 2.96) is found in the mineral *magnesite. There is also a trihydrate, $MgCO_3.3H_2O$ (rhombic; r.d. 1.85), which occurs naturally as **nesquehonite**, and a pentahydrate, $MgCO_3.5H_2O$ (monoclinic; r.d. 1.73), which occurs as **lansfordite**. Magnesium carbonate also occurs in the mixed salt *dolomite ($CaCO_3.MgCO_3$) and as **basic magnesium carbonate** in the two minerals **artinite** ($MgCO_3.Mg(OH)_2.3H_2O$) and **hydromagnesite** ($3MgCO_3.Mg(OH)_2.3H_2O$). The anhydrous salt can be formed by heating magnesium oxide in a stream of carbon dioxide:

$$MgO(s) + CO_2(g) \rightarrow MgCO_3(s)$$

Above 350°C, the reverse reaction predominates and the carbonate decomposes. Magnesium carbonate is used in making magnesium oxide and is a drying agent (e.g. in table salt). It is also used as a medical antacid and laxative (the basic carbonate is used) and is a component of certain inks and glasses.

**magnesium chloride** A white solid compound, $MgCl_2$. The anhydrous salt (hexagonal; r.d. 2.32; m.p. 714°C; b.p. 1412°C) can be prepared by the direct combination of dry chlorine with magnesium:

$$Mg(s) + Cl_2(g) \rightarrow MgCl_2(s)$$

The compound also occurs naturally as a constituent of carnallite ($KCl.MgCl_2$). It is a deliquescent compound that commonly forms the hexahydrate, $MgCl_2.6H_2O$ (monoclinic; r.d. 1.57). When heated, this hydrolyses to give magnesium oxide and hydrogen chloride gas. The fused chloride is electrolysed to produce magnesium and it is also used for fireproofing wood, in magnesia cements and artificial leather, and as a laxative.

**magnesium hydrogencarbonate (magnesium bicarbonate)** A compound, $Mg(HCO_3)_2$, that is stable only in solution. It is formed by the action of carbon dioxide on a suspension of magnesium carbonate in water:

$$MgCO_3(s) + CO_2(g) + H_2O(l) \rightarrow$$
$$Mg(HCO_3)_2(aq)$$

On heating, this process is reversed. Magnesium hydrogencarbonate is one of the compounds responsible for temporary *hardness of water.

**magnesium hydroxide** A white solid compound, $Mg(OH)_2$; trigonal; r.d. 2.36; decomposes at 350°C. Magnesium hydroxide occurs naturally as the mineral **brucite** and can be prepared by reacting magnesium sulphate or chloride with sodium hydroxide solution. It is used in the refining of sugar and in the processing of uranium. Medicinally it is important as an antacid (**milk of magnesia**) and as a laxative.

**magnesium oxide (magnesia)** A white compound, $MgO$; cubic; r.d. 3.58; m.p. 2800°C. It occurs naturally as the mineral **periclase** and is prepared commercially by thermally decomposing the mineral *magnesite:

$$MgCO_3(s) \rightarrow MgO(s) + CO_2(g)$$

It has a wide range of uses, including reflective coatings on optical instruments and aircraft windscreens and in semiconductors. Its high melting point makes it useful as a refractory lining in metal and glass furnaces.

**magnesium peroxide** A white solid, $MgO_2$. It decomposes at 100°C to release oxygen and also releases oxygen on reaction with water:

$$2MgO_2(s) + 2H_2O \rightarrow 2Mg(OH)_2 + O_2$$

The compound is prepared by reacting sodium peroxide with magnesium sulphate solution and is used as a bleach for cotton and silk.

**magnesium sulphate** A white soluble compound, $MgSO_4$, existing as the anhydrous compound (rhombic; r.d. 2.66; decomposes at 1124°C) and in hydrated crystalline forms. The monohydrate $MgSO_4.H_2O$ (monoclinic; r.d. 2.45) occurs naturally as the mineral **kieserite**. The commonest hydrate is the heptahydrate, $MgSO_4.7H_2O$ (rhombic; r.d. 1.68), which is called **Epsom salt(s)**, and occurs naturally as the mineral **epsomite**. This is a white powder with a bitter saline taste, which loses $6H_2O$ at 150°C and $7H_2O$ at 200°C. It is used in sizing and fireproofing cotton and silk, in tanning leather, and in the manufacture of fertilizers, explosives, and matches. In medicine, it is used as a laxative. It is also used in veterinary medicine for treatment of local inflammations and infected wounds.

**magnet** A piece of magnetic material (*see* MAGNETISM) that has been magnetized and is therefore surrounded by a *magnetic field. A magnet, often in the shape of a bar or horseshoe, that retains appreciable magnetization indefinitely (provided it is not heated, beaten, or exposed to extraneous magnetic fields) is called a **permanent magnet**. *See also* ELECTROMAGNET.

**magnetic bottle** A nonuniform *magnetic field used to contain the plasma in a *thermonuclear reactor. At the temperature of a thermonuclear reaction ($10^8$ K) any known substance would vaporize and the plasma has therefore to be contained in such a way that it does not come into contact with a material surface. The magnetic bottle provides a means of achieving this, by deflecting away from its boundaries the moving charged particles that make up the plasma.

**magnetic bubble memory** A form of computer memory in which a small magnetized region of a substance is used to store information. Bubble memories consist of materials, such as magnetic garnets, that are easily magnetized in one direction but hard to magnetize in the perpendicular direction. A thin film of these materials deposited on a nonmagnetic substrate constitutes a bubble-memory chip. When a magnetic field is applied to such a chip, by placing it between two permanent magnets, cylindrical domains (called magnetic bubbles) are formed. These bubbles constitute a magnetic region of one polarity surrounded by a magnetic region of the opposite polarity. Information is represented as the presence or absence of a bubble at a specified storage location and is retrieved by means of a rotating magnetic field. Typically a chip measures 15 mm$^2$, or 25 mm$^2$ enclosed in two permanent magnets and two rotating field coils; each chip can store up to one million bits.

**magnetic circuit** A closed path containing a *magnetic flux. The path is clearly delimited only if it consists mainly or wholly of ferromagnetic or other good magnetic materials; examples include transformer cores and iron parts in electrical machines. The design of these parts can often be assisted by analogy with electrical circuits, treating the *magnetomotive force as the analogue of

e.m.f., the magnetic flux as current, and the *reluctance as resistance. There is, however, no actual flow around a magnetic circuit.

**magnetic compass** *See* COMPASS.

**magnetic constant** *See* PERMEABILITY.

**magnetic declination** *See* GEOMAGNETISM.

**magnetic dip** *See* GEOMAGNETISM.

**magnetic disk** A smooth aluminium disk, usually 35.6 cm in diameter, both surfaces of which are coated with magnetic iron oxide. The disks are used as a recording medium in computers. Data is recorded in concentric tracks on both surfaces with up to 236 tracks per centimetre. The disks rotate at 3600 revolutions per minute, information being put onto the disk and removed from it by a record-playback head. *See also* FLOPPY DISK.

**magnetic domain** *See* MAGNETISM.

**magnetic elements** *See* GEOMAGNETISM.

**magnetic equator** *See* EQUATOR; GEOMAGNETISM.

**magnetic field** A *field of force that exists around a magnetic body (*see* MAGNETISM) or a current-carrying conductor. Within a magnetic field a magnetic dipole may experience a torque and a moving charge may experience a force. The strength and direction of the field can be given in terms of the **magnetic flux density** (or **magnetic induction**), symbol $B$; it can also be given in terms of the **magnetic field strength** (**magnetizing force** or **magnetic intensity**), symbol $H$.

The magnetic flux density is a vector quantity and is the *magnetic flux per unit area of a magnetic field at right angles to the magnetic force. It can be defined in terms of the effects the field has, for example by $B = F/qv\sin\theta$, where $F$ is the force a moving charge $q$ would experience if it was travelling at a velocity $v$ in a direction making an angle $\theta$ with that of the field. The *SI unit is the tesla.

The magnetic field strength is also a vector quantity and is related to $B$ by: $H = B/\mu$, where $\mu$ is the *permeability of the medium. The SI unit of field strength is the ampere per metre (A m$^{-1}$).

**magnetic field strength** *See* MAGNETIC FIELD.

**magnetic flux** Symbol $\Phi$. A measure of quantity of magnetism, taking account of the strength and the extent of a *magnetic field. The flux $d\Phi$ through an element of area $dA$ perpendicular to $B$ is given by $d\Phi = BdA$. The *SI unit of magnetic flux is the weber.

**magnetic flux density**  *See* MAGNETIC FIELD.

**magnetic force**  The attractive or repulsive force exerted on a *magnetic pole or a moving electric charge in a *magnetic field.

**magnetic induction**  *See* MAGNETIC FIELD.

**magnetic intensity**  *See* MAGNETIC FIELD.

**magnetic meridian**  *See* MERIDIAN.

**magnetic mirror**  A device used to contain *plasma in thermonuclear experimental devices. It consists of a region of high magnetic field strength at the end of a containment tube. Ions entering the region reverse their motion and return to the plasma from which they have emerged. *See also* MAGNETIC BOTTLE.

**magnetic moment**  The ratio between the maximum torque ($T_{max}$) exerted on a magnet, current-carrying coil, or moving charge situated in a *magnetic field and the strength of that field. It is thus a measure of the strength of a magnet or current-carrying coil. In the Sommerfeld approach this quantity (also called **electromagnetic moment** or **magnetic area moment**) is $T_{max}/B$. In the Kennelly approach the quantity (also called **magnetic dipole moment**) is $T_{max}/H$.

In the case of a magnet placed in a magnetic field of field strength $H$, the maximum torque $T_{max}$ occurs when the axis of the magnet is perpendicular to the field. In the case of a coil of $N$ turns and area $A$ carrying a current $I$, the magnetic moment can be shown to be $m = T/B = NIA$ or $m = T/H = \mu NIA$. Magnetic moments are measured in *SI units in A $m^2$.

An orbital electron has an orbital magnetic moment $IA$, where $I$ is the equivalent current as the electron moves round its orbit. It is given by $I = q\omega/2\pi$, where $q$ is the electronic charge and $\omega$ is its angular velocity. The orbital magnetic moment is therefore $IA = q\omega A/2\pi$, where $A$ is the orbital area. If the electron is spinning there is also a spin magnetic moment (*see* SPIN); atomic nuclei also have magnetic moments (*see* NUCLEAR MOMENT).

**magnetic monopole**  A hypothetical magnetic entity consisting of an isolated elementary north or south pole. It has been postulated as a source of a *magnetic field by analogy with the way in which an electrically charged particle produces an electric field. Numerous ingenious experiments have been designed to detect the monopole but so far none has produced an unequivocal result. Magnetic monopoles are predicted to exist in certain *gauge theories with *Higgs bosons. In particular, some *grand unified theories predict very heavy monopoles (with mass of order $10^{16}$ GeV). Magnetic monopoles are also predicted to exist in *Kaluza–Klein theories and *superstring theory.

**magnetic permeability**  *See* PERMEABILITY.

**magnetic poles**  1. *See* GEOMAGNETISM. 2. The regions of a *magnet from which the magnetic forces appear to originate. A magnetized bar has a pole at each end; if it is freely suspended in the earth's magnetic field (*see* GEOMAGNETISM) it will rotate so that one end points approximately towards the earth's geographical north pole. This end is called the north-seeking end or the north pole of the magnet. The other end is accordingly called the south-seeking end or south pole. In the obsolete theory associated with the *c.g.s. system of units, a **unit magnetic pole** was treated as one of a pair, which repelled each other with a force of 1 dyne when separated by 1 cm in space.

**magnetic potential**  *See* MAGNETOMOTIVE FORCE.

**magnetic quantum number**  *See* ATOM.

**magnetic resonance imaging (MRI)** *See* NUCLEAR MAGNETIC RESONANCE.

**magnetic susceptibility**  *See* SUSCEPTIBILITY.

**magnetic tape**  A plastic tape coated with a ferromagnetic material – iron oxide powder, chromium dioxide, or, for the best results, particles of pure iron. The tape is used for recording data in tape recorders and computers. To record, the tape is passed over a recording head containing a gap in a magnetic circuit whose magnetization is modulated by the information to be recorded; the information is imprinted on the tape in the form of the direction of magnetization of the individual particles of iron oxide. The playback procedure is the reverse of recording; the tape containing its orientation of tiny magnets is fed over the gap of the same (now the playback) head, in whose coil corresponding e.m.f.s are generated by induction.

**magnetic trap** A device for trapping electrically neutral particles that have magnetic moments using nonuniform magnetic fields. Magnetic traps have been used in research on very low temperatures and the formation of Bose–Einstein condensates.

**magnetic variation (secular magnetic variation)** *See* GEOMAGNETISM.

**magnetism** A group of phenomena associated with *magnetic fields. Whenever an electric current flows a magnetic field is produced; as the orbital motion and the *spin of atomic electrons are equivalent to tiny current loops, individual atoms create magnetic fields around them, when their orbital electrons have a net *magnetic moment as a result of their angular momentum. The magnetic moment of an atom is the vector sum of the magnetic moments of the orbital motions and the spins of all the electrons in the atom. The macroscopic magnetic properties of a substance arise from the magnetic moments of its component atoms and molecules. Different materials have different characteristics in an applied magnetic field; there are four main types of magnetic behaviour:

(a) In **diamagnetism** the magnetization is in the opposite direction to that of the applied field, i.e. the *susceptibility is negative. Although all substances are diamagnetic, it is a weak form of magnetism and may be masked by other, stronger, forms. It results from changes induced in the orbits of electrons in the atoms of a substance by the applied field, the direction of the change (in accordance with *Lenz's law) opposing the applied flux. There is thus a weak negative susceptibility and a relative permeability that is slightly less than one.

(b) In **paramagnetism** the atoms or molecules of the substance have net orbital or spin magnetic moments that are capable of being aligned in the direction of the applied field. They therefore have a positive (but small) susceptibility and a relative permeability slightly in excess of one. Paramagnetism occurs in all atoms and molecules with unpaired electrons; e.g. free atoms, free radicals, and compounds of transition metals containing ions with unfilled electron shells. It also occurs in metals as a result of the magnetic moments associated with the spins of the conducting electrons.

(c) In **ferromagnetic** substances, within a certain temperature range, there are net atomic magnetic moments, which line up in such a way that magnetization persists after the removal of the applied field. Below a certain temperature, called the *Curie point (or Curie temperature) an increasing magnetic field applied to a ferromagnetic substance will cause increasing magnetization to a high value, called the **saturation magnetization**. This is because a ferromagnetic substance consists of small (1–0.1 mm across) magnetized regions called **domains**. The total magnetic moment of a sample of the substance is the vector sum of the magnetic moments of the component domains. Within each domain the individual atomic magnetic moments are spontaneously aligned by **exchange forces**, related to whether or not the atomic electron spins are parallel or antiparallel. However, in an unmagnetized piece of ferromagnetic material the magnetic moments of the domains themselves are not aligned; when an external field is applied those domains that are aligned with the field increase in size at the expense of the others. In a very strong field all the domains are lined up in the direction of the field and provide the high observed magnetization. Iron, nickel, cobalt, and their alloys are ferromagnetic. Above the Curie point, ferromagnetic materials become paramagnetic. A variant of exchange is **superexchange**, i.e. a magnetic interaction that can occur when two magnetic ions are separated by a nonmagnetic ion, with the interaction being mediated by the electrons in the nonmagnetic ion. Superexchange is important in magnetic insulators.

(d) Some metals, alloys, and transition-element salts exhibit another form of magnetism called **antiferromagnetism**. This occurs below a certain temperature, called the *Néel temperature, when an ordered array of atomic magnetic moments spontaneously forms in which alternate moments have opposite directions. There is therefore no net resultant magnetic moment in the absence of an applied field. In manganese fluoride, for example, this antiparallel arrangement occurs below a Néel temperature of 72 K. Below this temperature the spontaneous ordering opposes the normal tendency of the magnetic moments to align with the applied field. Above the Néel temperature the substance is paramagnetic.

A special form of antiferromagnetism is **ferrimagnetism**, a type of magnetism exhibited by the *ferrites. In these materials the magnetic moments of adjacent ions are antiparallel and of unequal strength, or the number of magnetic moments in one direction is greater than those in the opposite direction. By suitable choice of rare-earth ions in the ferrite lattices it is possible to design ferrimagnetic substances with specific magnetizations for use in electronic components. *See also* GEOMAGNETISM.

**SEE WEB LINKS**
• Values for the magnetic properties of materials at the NPL website

**magnetite** A black mineral form of iron oxide crystallizing in the cubic system. It is a mixed iron(II)-iron(III) oxide, $Fe_3O_4$, and is one of the major ores of iron. It is strongly magnetic and some varieties, known as **lodestone**, are natural magnets; these were used as compasses in the ancient world. Magnetite is widely distributed and occurs as an accessory mineral in almost all igneous and metamorphic rocks. The largest deposits of the mineral occur in N Sweden.

**magneto** An alternating-current generator used as a high-tension source in the ignition systems of petrol engines in which there are no batteries, e.g. in some tractor, marine, and aviation engines. Most modern magnetos consist of a permanent-magnet rotor revolving within a primary (low-voltage) winding around which a secondary winding is placed in which to induce the high voltage

needed to produce the spark across the points of the plugs. Magnetos are geared to the engine shaft, the speed depending on the number of poles of the magneto and the number of engine cylinders. A make-and-break device is incorporated in the primary winding; when the primary current stops the change of flux within the secondary induces in it a large e.m.f.

**magnetobremsstrahlung** *See* SYNCHROTRON RADIATION.

**magnetocaloric effect** A reversible change of temperature resulting from a change in the magnetization of a ferromagnetic or paramagnetic substance (*see* MAGNETISM). The change in temperature $\Delta T$, accompanying an adiabatic change of magnetic field $\Delta H$, is:

$$\Delta T/\Delta H = -T/C_H(\partial M/\partial T)_H$$

$C_H$ is the specific heat capacity per unit volume at constant $H$ and $M$ is the magnetization.

**magnetochemistry** The branch of physical chemistry concerned with measuring and investigating the magnetic properties of compounds. It is used particularly for studying transition-metal complexes, many of which are paramagnetic because they have unpaired electrons. Measurement of the magnetic susceptibility allows the magnetic moment of the metal atom to be calculated, and this gives information about the bonding in the complex.

**magnetohydrodynamics (MHD)** The

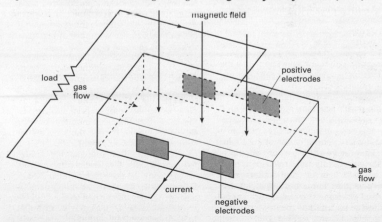

**Magnetohydrodynamics.** Magnetohydrodynamic generator.

study of the interactions between a conducting fluid and a *magnetic field. MHD is important in the study of controlled thermonuclear reactions in which the conducting fluid is a *plasma confined by a magnetic field. Other important applications include the **magnetohydrodynamic power generator** (see illustration). In the open-cycle MHD generator a fossil fuel, burnt in oxygen or preheated compressed air, is seeded with an amount of low *ionization potential (such as potassium or caesium). This element is thermally ionized at the combustion temperature (usually over 2500 K) producing sufficient free electrons (e.g. $K \rightarrow K^+ + e$) to provide adequate electrical conductivity. The interaction between the moving conducting fluid and the strong applied magnetic field across it generates an e.m.f. on the Faraday principle, except that the solid conductor of the conventional generator is replaced by a fluid conductor. The power output per unit fluid volume ($W$) is given by $W = k\sigma v^2 B^2$, where $\sigma$ is the conductivity of the fluid, $v$ is its velocity, $B$ is the magnetic flux density, and $K$ is a constant. Devices of this kind are in use in some power stations, where they are suitable for helping to meet high short-term demands and have the ability of increasing the thermal efficiency of a steam-turbine generator from about 40% to 50%. In experimental closed-cycle systems the fluid is continuously recirculated through a compressor; the fluid consists of a heated and seeded noble gas or a liquid metal.

**magnetomechanical ratio** *See* GYROMAGNETIC RATIO.

**magnetometer** An instrument for measuring the magnitude, and sometimes the direction, of a magnetic field. **Absolute magnetometers** measure the field without reference to a standard magnetic instrument. The most widely used are the **vibration magnetometer**, the **deflection galvanometer**, and the more modern **nuclear magnetometer**. The vibration instrument was devised by Gauss in 1832 and depends on the rate of oscillation of a small bar magnet suspended in a horizontal plane. The same magnet is then used as a fixed deflector to deflect a second similarly suspended magnet. The deflection galvanometer uses a Helmholtz coil system of known dimensions with a small magnet suspended at its centre. The deflected magnet comes to rest at a position controlled by the earth's

magnetic field, the coil's magnetic field, and the angle through which the coil must be turned to keep the magnet and the coil in alignment. The sensitive nuclear magnetometers are based on measuring the audiofrequency voltage induced in a coil by the precessing protons in a sample of water. Various **relative magnetometers** are also in use, especially for measuring the earth's magnetic field and in calibrating other equipment.

**magnetomotive force (m.m.f.)** The analogue of *electromotive force in a *magnetic circuit. Mathematically, it is the circular integral of $H\cos\theta \, ds$, where $H\cos\theta$ is the component of the *magnetic field strength in the direction of a path of length d$s$. The m.m.f. is measured in *SI units in ampere-turns. It was formerly called the **magnetic potential**.

**magneton** A unit for measuring *magnetic moments of nuclear, atomic, or molecular magnets. The **Bohr magneton**, $\mu_B$, has the value of the classical magnetic moment of an electron, given by

$$\mu_B = eh/4\pi m_e = 9.274 \times 10^{-24} \text{ A m}^2,$$

where $e$ and $m_e$ are the charge and mass of the electron and $h$ is the Planck constant. The **nuclear magneton**, $\mu_N$, is obtained by replacing the mass of the electron by the mass of the proton and is therefore given by

$$\mu_N = \mu_B m_e / m_p = 5.05 \times 10^{-27} \text{ A m}^2.$$

**magneto-optical effects** Effects resulting from the influence of a *magnetic field upon matter that is in the process of emitting or absorbing light. Examples are the *Faraday effect and the *Zeeman effect.

**magneto-optical trap** *See* LASER COOLING.

**magnetoresistance** An increase in the resistance of a metal due to the presence of a magnetic field, which alters the paths of the electrons. At normal temperatures the change in resistance resulting from the magnetic field is small but at very low temperatures the increase is considerable. The theory of magnetoresistance is too complicated to be explained quantitatively by the simple model of electrical *conductivity in metals, which assumes that it results from the movement of free electrons. To obtain a quantitative explanation, it is necessary to take into account the *energy-band structure of metals. Usually magnetoresistance is a

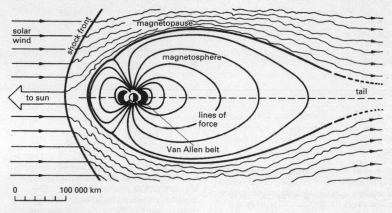

**Magnetosphere.**

fairly small effect (about 5%). However, under certain circumstances much larger effects are possible.

**magnetosphere** A comet-shaped region surrounding the earth and other magnetic planets in which the charged particles of the *solar wind are controlled by the planet's magnetic field rather than the sun's magnetic field. It extends for some 60 000 km on the side facing the sun but on the opposite side it extends to a much greater extent. The boundary of the magnetosphere is known as the **magnetopause** (see illustration). The magnetosphere of the earth includes the *Van Allen belts.

**magnetostriction** The change in length of a ferromagnetic material (*see* MAGNETISM) when it is magnetized. It results from changes in the boundaries of the domains. A ferromagnetic rod exposed to an alternating field will vibrate along its length. This appears to be a major source of transformer hum, which can be removed by using a magnetic steel containing 6.5% silicon. Magnetostriction of a nickel transducer is used to generate and receive ultrasonic waves.

**magnetron** A microwave generator in which electrons, generated by a heated cathode, move under the combined force of an electric field and a magnetic field. The cathode consists of a central hollow cylinder, the outer surface of which carries the barium and strontium oxide electron emitters. The anode is also a cylinder, arranged concentrically around the cathode, and it contains a

series of quarter-wavelength *resonant cavities arranged around its inner surface. The electric field is applied radially between anode and cathode, the magnetic field is coaxial with the cathode. The whole device is maintained in a vacuum enclosure. The magnetron is extensively used as a generator for radar installations and can produce microsecond pulses of up to 10 MW.

**magnification** A measure of the extent to which an optical system enlarges or reduces an image. The **linear magnification**, $m$, is the ratio of the image height to the object height. If this ratio is greater than one the system is enlarging, if it is less than one, it is reducing. The **angular magnification**, $M$ or $\gamma$, is the ratio of the angles formed by the final image and the object (when viewed directly, in the most favourable position available) at the eye. This is also sometimes called the **magnifying power** of an optical system.

**magnifying power** *See* MAGNIFICATION.

**magnitude** A measure of the relative brightness of a star or other celestial object. The **apparent magnitude** depends on the star's *luminosity, its distance, and the absorption of light between the object and the earth. In 1856 the astronomer N. R. Pogson devised a scale in which a difference of five magnitudes corresponds to a brightness ratio of 100 to 1. Two stars that differ by one magnitude therefore have a brightness ratio of $(100)^{0.2}:1 = 2.512$, known as the **Pogson ratio**. This scale is now universally adopted. Apparent magnitudes are not a measure of luminosity, which is defined in terms of the **absolute magnitude**. This is the apparent magnitude of a body if it was situated at a standard distance of 10 parsecs.

**Magnoliophyta** See ANTHOPHYTA.

**magnon** See SPIN WAVE.

**Magnox** A group of magnesium alloys used to enclose uranium fuel elements in *nuclear reactors. They usually contain some aluminium as well as other elements, such as beryllium.

**mainframe computer** See COMPUTER.

**main-sequence stars** See HERTZSPRUNG–RUSSELL DIAGRAM.

**major histocompatibility complex (MHC)** A large gene cluster that encodes various components of the immune system, including the histocompatibility antigens and components of the *complement system. In humans the MHC contains over 200 genes, is located on chromosome 6, and includes the *HLA system. Other vertebrate species have similar MHC regions. Certain MHC genes can have many variant alleles; this produces an enormous diversity of antigens in a population, each individual possessing a unique set.

**majority carrier** See SEMICONDUCTOR.

**Maksutov telescope** See TELESCOPE.

**malachite** A secondary mineral form of copper carbonate–hydroxide, $CuCO_3$. $Cu(OH)_2$. It is bright green and crystallizes in the monoclinic system but usually occurs as aggregates of fibres or in massive form. It is generally found with *azurite in association with the more important copper ores and is itself mined as an ore of copper (e.g. in the Democratic Republic of Congo). It is also used as an ornamental stone and as a gemstone.

**MALDI** Matrix absorption laser desorption ionization. A technique for producing ions for mass spectroscopy, used especially for large biological compounds. The sample is absorbed on an inert matrix from which ions are desorbed by a laser.

**male 1.** Denoting the gamete (sex cell) that, during *sexual reproduction, fuses with a *female gamete in the process of fertilization. Male gametes are generally smaller than the female gametes and are usually motile (see SPERMATOZOON). **2.** (Denoting)

an individual whose reproductive organs produce only male gametes. *Compare* HERMAPHRODITE.

**maleic acid** See BUTENEDIOIC ACID.

**maleic anhydride** A colourless solid, $C_4H_2O_3$, m.p. 53°C, the anhydride of *cis*-butenedioic acid (maleic acid). It is a cyclic compound with a ring containing four carbon atoms and one oxygen atom, made by the catalytic oxidation of benzene or its derivatives at high temperatures. It is used mainly in the manufacture of alkyd and polyester resins and copolymers.

**malic acid (2-hydroxybutanedioic acid)** A white crystalline solid, HOOCCH-(OH)CH$_2$COOH. L-malic acid occurs in living organisms as an intermediate metabolite in the *Krebs cycle and also (in certain plants) in photosynthesis. It is found especially in the juice of unripe fruits, e.g. green apples, and contributes to their sour taste.

**malleus (hammer)** The first of the three *ear ossicles of the mammalian *middle ear.

**Mallophaga** An order of wingless insects comprising the bird lice. Bird lice are minute with dorsoventrally flattened ovoid bodies, reduced eyes, and biting mouthparts. They are ectoparasites of birds, feeding on particles of dead skin, feather fragments, and sometimes blood. The eggs hatch to form nymphs resembling the adults.

**malnutrition** The condition arising due to the lack of one or more of the *nutrients that are required in the *diet to maintain health. Malnutrition can result from a reduced intake of nutrients (**undernourishment**), an inability to use absorbed nutrients, failure to meet a required increase in nutrient intake, or nutrient losses. There are three stages in the process of malnutrition: first, the carbohydrate stores in the body are depleted; secondly, the fat reserves are metabolized (*see* FATTY-ACID OXIDATION); and finally, proteins are broken down to provide energy. Death may result after protein levels have been reduced to half their normal value. **Kwashiorkor** is a type of malnutrition that develops when the diet lacks proteins and hence *essential amino acids. Malnutrition due to reduced absorption of nutrients in the intestine can develop with a cereal-based diet, due to sensitivity of the intestinal lining to gluten, a protein found in cereals. *See also* MINERAL DEFICIENCY.

**malonic acid** *See* PROPANEDIOIC ACID.

**Malpighian body (Malpighian corpuscle)** The part of a *nephron in the kidney that consists of its cup-shaped end together with the *glomerulus that it encloses. It is named after its discoverer, the Italian anatomist Marcello Malpighi (1628–94).

**Malpighian layer (stratum germinativum)** The innermost layer of the *epidermis of mammalian *skin, separated from the underlying dermis by a fibrous *basement membrane. It is only in this layer of the epidermis that active cell division (*mitosis) occurs. As the cells produced by these divisions age and mature, they migrate upwards through the layers of the epidermis to replace the cells being continuously worn away at the surface.

**maltase** A membrane-bound enzyme in the small intestine that hydrolyses the disaccharide maltose into glucose.

**maltose (malt sugar)** A sugar consisting of two linked glucose molecules that results from the action of the enzyme *amylase on starch. Maltose occurs in high concentrations in germinating seeds; malt, used in the manufacture of beer and malt whisky, is produced by allowing barley seeds to germinate and then slowly drying them.

**malt sugar** *See* MALTOSE.

**Mammalia** A class of vertebrates containing some 4250 species. Mammals are warm-blooded animals (*see* HOMOIOTHERMY), typically having sweat glands whose secretion cools the skin and an insulating body covering of hair. All female mammals have *mammary glands, which secrete milk to nourish the young. Mammalian teeth are differentiated into incisors, canines, premolars, and molars and the middle ear contains three sound-conducting *ear ossicles. The four-chambered heart enables complete separation of oxygenated and deoxygenated blood and a muscular *diaphragm takes part in breathing movements, both of which ensure that the tissues are well supplied with oxygen. This, together with well-developed sense organs and brain, have enabled mammals to pursue an active life and to colonize a wide variety of habitats.

Mammals evolved from carnivorous reptiles in the Triassic period about 225 million years ago. There are two subclasses: the primitive egg-laying *Prototheria (monotremes) and the Theria, which includes all other mammals and consists of the infraclasses *Metatheria (marsupials) and *Eutheria (placental mammals).

(⊕) **SEE WEB LINKS**
- The Life of Mammals website from the BBC

**mammary glands** The milk-producing organs (possibly modified sweat glands) of female mammals, which provide food for the young (*see* MILK; COLOSTRUM). Their number (2 to 20) and position (on the chest or abdomen) vary according to the species. In most mammals the gland openings project as a **nipple** or **teat**. Nipples have a number of milk-duct openings; teats have one duct leading from a storage cavity.

**Mandelbrot set** A *fractal that produces complex self-similar patterns. In mathematical terms, it is the set of values of $c$ that make the series $z_n + 1 = (z_n)^2 + c$ converge, where $c$ and $z$ are complex numbers and $z$ begins at the origin $(0,0)$. It was discovered by and named after the Polish-born French mathematician Benoît Mandelbrot (1924–2010).

**Mandelin test** A *presumptive test for amphetamines and alkaloids. The **Mandelin reagent** is a 1% solution of ammonium vanadate ($NH_4VO_3$) in concentrated sulphuric acid. Different substances give different colours. Mescaline, for example, produces an orange colour, heroin a brown colour, and amphetamine a blue-green colour.

**mandible** 1. One of a pair of horny *mouthparts in insects, crustaceans, centipedes, and millipedes. The mandibles lie in front of the weaker *maxillae and their lateral movements assist in biting and crushing the food. 2. The lower jaw of vertebrates. 3. Either of the two parts of a bird's beak.

**Mandibulata** *See* ARTHROPOD.

**manganate(VI)** A salt containing the ion $MnO_4^{2-}$. Manganate(VI) ions are dark green; they are produced by manganate(VII) ions in basic solution.

**manganate(VII) (permanganate)** A salt containing the ion $MnO_4^-$. Manganate(VII) ions are dark purple and strong oxidizing agents.

**manganese** Symbol Mn. A grey brittle metallic *transition element, a.n. 25; r.a.m. 54.94; r.d. 7.2; m.p. 1244°C; b.p. 1962°C. The main sources are pyrolusite ($MnO_2$) and rhodochrosite ($MnCO_3$). The metal can be extracted by reduction of the oxide using magnesium (*Kroll process) or aluminium (*Goldschmidt process). Often the ore is mixed with iron ore and reduced in an electric furnace to produce ferromanganese for use in alloy steels. The element is fairly electropositive; it combines with oxygen, nitrogen, and other nonmetals when heated (but not with hydrogen). Salts of manganese contain the element in the +2 and +3 oxidation states. Manganese(II) salts are the more stable. It also forms compounds in higher oxidation states, such as manganese(IV) oxide and manganate(VI) and manganate(VII) salts. The element was discovered in 1774 by Karl Scheele.

**(((⊕))) SEE WEB LINKS**
• Information from the WebElements site

**manganese nodule** An irregular lump of rock containing manganese, found on the deep ocean floor, particularly the north Pacific. The nodules range in size from 0.5 to 25 cm across and have a banded structure, built up on a central particle, such as a pebble or even a shark's tooth. They contain up to 24% manganese, with some iron (14%), nickel (1%), and copper (0.5%), and sometimes cobalt (0.5%). They form when metal-bearing solutions well up from the ocean floor. Various methods have been tried to 'mine' them, although none has yet been adopted commercially.

**manganese(IV) oxide (manganese dioxide)** A black oxide made by heating manganese(II) nitrate. The compound also occurs naturally as pyrolusite. It is a strong oxidizing agent, used as a depolarizing agent in voltaic cells.

**manganic compounds** Compounds of manganese in its +3 oxidation state; e.g. manganic oxide is manganese(III) oxide, $Mn_2O_3$.

**manganin** A copper alloy containing 13–18% of manganese and 1–4% of nickel. It has a high electrical resistance, which is relatively insensitive to temperature changes. It is therefore suitable for use in resistance wire.

**manganous compounds** Compounds of manganese in its +2 oxidation state; e.g. manganous oxide is manganese(II) oxide, MnO.

**mangrove swamp** A region of vegetation, found along tropical coasts, in which mangrove trees (*Rhizophora* species) pre-dominate. The waterlogged soil is highly saline, and – like other *halophytes – mangroves are adapted to withstand these conditions; they also possess aerial roots (**pneumatophores**) through which gaseous exchange occurs, to counteract effects of the badly aerated soil. Mangroves play a vital role in protecting coastal regions from the effects of tropical storms and high tides.

**Mannich reaction** A reaction in which a primary or secondary amine reacts with methanal (formaldehyde) and a carbonyl compound to produce an amino-carbonyl compound. It takes place in two stages. First the amine reacts with methanol to form a Schiff base:

$$R_2N + H_2CO \rightarrow R_2N^+{=}CH_2.$$

This then reacts with the carbonyl compound:

$$R_2N^+{=}CH_2 + R^1R^2CHCOR^3 \rightarrow$$
$$R_2N{-}CH_2{-}C(R^1R^2)COR^3.$$

The reaction was first reported by Carl Mannich in 1912.

**mannitol** A polyhydric alcohol, $CH_2OH$-$(CHOH)_4CH_2OH$, derived from mannose or fructose. It is the main soluble sugar in fungi and an important carbohydrate reserve in brown algae. Mannitol is used as a sweetener in certain foodstuffs.

**mannose** A *monosaccharide hexose, $C_6H_{12}O_6$, stereoisomeric with glucose, that occurs naturally only in polymerized forms called **mannans**. These are found in plants, fungi, and bacteria, serving as food energy stores.

**manometer** A device for measuring pressure differences, usually by the difference in height of two liquid columns. The simplest type is the U-tube manometer, which consists of a glass tube bent into the shape of a U. If a pressure to be measured is fed to one side of the U-tube and the other is open to the atmosphere, the difference in level of the liquid in the two limbs gives a measure of the unknown pressure.

**mantissa** *See* LOGARITHM.

**mantle** 1. (in zoology) The fold of skin covering the dorsal surface of the body of molluscs, which extends into lateral flaps that protect the gills in the **mantle cavity** (the space between the body and mantle). The outer surface of the mantle secretes the shell (in species that have shells). 2. (in geology) *See* EARTH.

**many-body problem** The problem that it is very difficult to obtain exact solutions to systems involving interactions between more than two bodies – using either classical mechanics or quantum mechanics. To understand the physics of many-body systems it is necessary to make use of *approximation techniques or *model systems that capture the essential physics of the problem. For some problems, such as the **three-body problem** in classical mechanics, it is possible to obtain qualitative information about the system. Useful concepts in the quantum theory of many-body systems are *quasiparticles and *collective excitations. If there are a great many bodies interacting, such as the molecules in a gas, the problem can be analysed using the techniques of *statistical mechanics.

**map projections** The methods used to represent the spherical surface of the earth on a plane surface. The circles of latitude and longitude are represented by a network or graticule of lines. Directions, areas, distances, and shape can never all be recreated accurately and the map projection chosen for a particular area will thus depend on the purpose for which the map is to be used and on the part of the world represented. There are three main groups of projections:
(1) **Cylindrical projections** are obtained by projecting the globe onto a cylinder that intersects the earth. If the axis of the cylinder is parallel to the axis of the earth the meridians and parallels will appear as straight lines. The three basic types of cylindrical projection are the simple cylindrical, the equal-area, and Mercator's projections. A modified equal-area form of cylindrical projection was invented in 1973 by the German historian Arno Peters – and is known as the **Peters' projection**. It draws attention to Third World countries, which are prominently placed in the centre of the map. Landforms close to the equator are elongated, while those in high lattitudes are compressed.
(2) **Conic (conical) projections** result from the projection of the meridians and parallels onto a cone. In conic projections the axis of the cone is usually parallel to the earth's axis; as a result the meridians appear as radiating straight lines and the parallels are shown as concentric arcs. Scale is correct only along the standard parallel (i.e. the parallel along which the cone intersects the globe).
(3) **Azimuthal (zenithal) projections** are constructed as if a plane is placed at a tan-

gent to the earth's surface and the portion of the earth covered is transferred onto the plane. On this projection all great circles that pass through the centre of the projection appear as straight lines; all points have their true compass bearings. Examples of azimuthal projections include the polar azimuthal and Lambert's azimuthal.

**marble** A metamorphic rock composed of recrystallized *calcite or *dolomite. Pure marbles are white but such impurities as silica or clay minerals result in variations of colour. Marble is extensively used for building purposes and ornamental use; the pure white marble from Carrara in Italy is especially prized by sculptors. The term is applied commercially to any limestone or dolomite that can be cut and polished.

**Marchantiophyta** See HEPATOPHYTA.

**Marconi, Guglielmo** (1874–1937) Italian electrical engineer, who in 1894 began experimenting with Hertzian waves (see HERTZ, HEINRICH RUDOLF), making the first *radio transmissions. Moving to London in 1896, for the next few years he worked on improving the range and reliability of his equipment. This enabled him in late 1901 to transmit Morse signals across the Atlantic Ocean, establishing radio telegraphy and more importantly the use of radio waves as a communications medium. In 1909 he and Karl *Braun were awarded the Nobel Prize for physics.

**margaric acid** See HEPTADECANOIC ACID.

**marijuana** See CANNABIS.

**marker gene** A gene used to identify a particular bacterial colony or bacteriophage plaque. Such genes are incorporated into cloning *vectors to enable the isolation and replication of colonies containing a desired vector. Typically, marker genes confer resistance to specific antibiotics or produce colour changes (**genetic marker**). A gene that acts as a tag for another, closely linked, gene. Such markers are used in mapping the order of genes along chromosomes and in following the inheritance of particular genes: genes closely linked to the marker will generally be inherited with it. Markers must be readily identifiable in the phenotype, for instance by controlling an easily observable feature (such as eye colour). See also MOLECULAR MARKER.

**Markov chain** In *statistics, a series of

random events or states, chosen from a specific collection, in which the probability of each event is determined only by its predecessor. It was named after the Russian mathematician Andrei Markov (1856–1922).

**Markovnikoff's rule** When an acid HA adds to an alkene, a mixture of products can be formed if the alkene is not symmetrical. For instance, the reaction between $C_2H_5CH:CH_2$ and HCl can give $C_2H_5CH_2CH_2Cl$ or $C_2H_5CHClCH_3$. In general, a mixture of products occurs in which one predominates over the other. In 1870, Vladimir Markovnikoff (1837–1904) proposed the rule that the main product would be the one in which the hydrogen atom adds to the carbon having the larger number of hydrogen atoms (the latter product above). This occurs when the mechanism is *electrophilic addition, in which the first step is addition of $H^+$. The electron-releasing effect of the alkyl group ($C_2H_5$) distorts the electron-distribution in the double bond, making the carbon atom furthest from the alkyl group negative. This is the atom attacked by $H^+$ giving the carbonium ion $C_2H_5C^+HCH_3$, which further reacts with the negative ion $Cl^-$.

Under certain circumstances **anti-Markovnikoff** behaviour occurs, in which the opposite effect is found. This happens when the mechanism involves free radicals and is common in addition of hydrogen bromide when peroxides are present.

**Marquis test** A widely used *presumptive test that gives a variety of colour changes with a range of compounds. It is particularly useful for detecting opiate alkaloids and for amphetamines and methamphetamine. **Marquis reagent** is a mixture of methanal (formaldehyde) solution in water with sulphuric acid. Mescaline gives an orange coloration. With morphine, a violet colour is produced. Amphetamines give an orange-red colour and methamphetamine gives an orange colour (the two can be distinguished by the Simon test). The mechanism involves attack of the aldehyde and a substituted aromatic ring to form a carbocation. Further reaction forms a coloured dimer of the original molecule.

**Mars** The seventh largest *planet in the *solar system and the fourth in order from the *sun. Its mean distance from the sun is $227.94 \times 10^6$ km, its mass is $6.4219 \times 10^{23}$ kg (about 11% that of earth), and its mean diameter is 6795 km; it has a *sidereal period of 686.98 days. Its period of axial rotation is known as a **sol** and is equal to 24h 37.4m. The bulk of our knowledge about Mars has come from the orbiters, landers, and rovers that have visited the planet since 1965. They reveal it as a barren, rocky world with a thin atmosphere (less than 1% of the pressure at the earth's surface) consisting of about 95% carbon dioxide with the remaining 5% made up of nitrogen, argon, oxygen, and water vapour. The surface material is largely basalt and covered with a thin layer of reddish iron-rich claylike soil and light dust. The terrain includes volcanic plateaux and calderas as well as vast and deep impact basins and lunar-like craters. Olympus Mons, the highest mountain in the solar system, rises more than 26 km above the Martian surface, and Valles Marineris is the largest known canyon system, at 4000 km long, 200 km wide, and up to 7 km deep. Mars's *axial tilt of 25.19° means that it experiences seasons similar to earth. Like earth, Mars has polar ice caps. Because Mars is more than 37 million km closer to the sun at *perihelion than at *aphelion, climatic fluctuations can be extreme. Owing to the thinness of the atmosphere, temperatures range between 133 K and 293 K. Mars has two small satellites, Phobos (approximately $20 \times 23 \times 28$ km) and Deimos ($10 \times 12 \times 16$ km), neither of which is sufficiently massive to have contracted to a sphere.

(((•))) SEE WEB LINKS
- A comprehensive guide to Mars and how to observe it
- Google Mars – interactive images of the planet

**marsh gas** Methane formed by rotting vegetation in marshes.

**Marsh's test** A chemical test for arsenic in which hydrochloric acid and zinc are added to the sample, arsine being produced by the nascent hydrogen generated. Gas from the sample is led through a heated glass tube and, if arsine is present, it decomposes to give a brown deposit of arsenic metal. The arsenic is distinguished from antimony (which gives a similar result) by the fact that antimony does not dissolve in sodium chlorate(I) (hypochlorite). The test was devised in 1836 by the British chemist James Marsh (1789–1846).

**marsupials** See METATHERIA.

**martensite** A solid solution of carbon in

alpha-iron (*see* IRON) formed when *steel is cooled too rapidly for pearlite to form from austenite. It is responsible for the hardness of quenched steel.

**mascagnite** A mineral form of *ammonium sulphate, $(NH_4)_2SO_4$.

**mascon** A gravitational anomaly on the surface of the moon resulting from a concentration of mass below the lunar surface. They occur in circular lunar maria and were caused either by the mare basalt as it flooded the basins or by uplift of high-density mantle material when the basins were formed.

**maser** (*m*icrowave *a*mplification by *s*timulated *e*mission of *r*adiation) A device for amplifying or generating *microwaves by means of stimulated emission (*see* LASER). As oscillators, masers are used in *atomic clocks, while they are used as amplifiers in *radio astronomy, being especially suitable for amplifying feeble signals from space.

In the **ammonia gas maser** (devised in 1954) a molecular beam of ammonia passes through a small orifice into a vacuum chamber, where it is subjected to a nonuniform electric field. This field deflects ground-state ammonia molecules, shaped like a pyramid with the three hydrogen atoms forming the plane of the base and the single nitrogen atom forming the apex. The ground-state molecule has a dipole moment on account of its lack of symmetry and it is for this reason that it suffers deflection. Excited molecules, in which the nitrogen atom vibrates back and forth through the plane of the hydrogen atoms, have no resultant dipole moment and are not deflected. The beam, now consisting predominantly of excited molecules, is passed to a resonant cavity fed with the microwave radiation corresponding to the energy difference between the excited and the ground states. This causes stimulated emission as the excited molecules fall to the ground state and the input microwave radiation is amplified coherently. This arrangement can also be made to oscillate and in this form is the basis of the *ammonia clock.

In the more versatile **solid-state maser** a magnetic field is applied to the electrons of paramagnetic (*see* MAGNETISM) atoms or molecules. The energy of these electrons is quantized into two levels, depending on whether or not their spins are parallel to the magnetic field. The situation in which there are more parallel magnetic moments than

antiparallel can be reversed by sudden changes in the magnetic field. This electron-spin resonance in paramagnetic materials allows amplification over broader bandwidths than gas masers.

**mass** A measure of a body's *inertia, i.e. its resistance to acceleration. According to Newton's laws of motion, if two unequal masses, $m_1$ and $m_2$, are allowed to collide, in the absence of any other forces both will experience the same force of collision. If the two bodies acquire accelerations $a_1$ and $a_2$ as a result of the collision, then $m_1a_1 = m_2a_2$. This equation enables two masses to be compared. If one of the masses is regarded as a standard of mass, the mass of all other masses can be measured in terms of this standard. The body used for this purpose is a 1-kg cylinder of platinum–iridium alloy, called the international standard of mass. Mass defined in this way is called the **inertial mass** of the body.

Mass can also be defined in terms of the gravitational force it produces. Thus, according to Newton's law of gravitation, $m_g = Fd^2/MG$, where $M$ is the mass of a standard body situated a distance $d$ from the body of mass $m_g$; $F$ is the gravitational force between them and $G$ is the *gravitational constant. The mass defined in this way is the **gravitational mass**. In the 19th century Lóránd Eötvös (1848–1919) showed experimentally that gravitational and inertial mass are indistinguishable, i.e. $m_i = m_g$. Experiments performed in the 20th century confirmed this to greater accuracy.

Although mass is formally defined in terms of its inertia, it is usually measured by gravitation. The **weight** ($W$) of a body is the force by which a body is gravitationally attracted to the earth corrected for the effect of rotation and equals the product of the mass of the body and the *acceleration of free fall ($g$), i.e. $W = mg$. In the general language, weight and mass are often used synonymously; however, for scientific purposes they are different. Mass is measured in kilograms; weight, being a force, is measured in newtons. Weight, moreover, depends on where it is measured, because the value of $g$ varies at different localities on the earth's surface. Mass, on the other hand, is constant wherever it is measured, subject to the special theory of *relativity. According to this theory, announced by Albert Einstein in 1905, the mass of a body is a measure of its total energy content. Thus, if the energy of a body increases, for example by an increase in kinetic energy or temperature, then its mass will increase. According to this law an increase in energy $\Delta E$ is accompanied by an increase in mass $\Delta m$, according to the **mass–energy equation** $\Delta m = \Delta E / c^2$, where $c$ is the speed of light. Thus, if 1 kg of water is raised in temperature by 100 K, its internal energy will increase by $4 \times 10^{-12}$ kg. This is, of course, a negligible increase and the mass–energy equation is only significant for extremely high energies. For example, the mass of an electron is increased sevenfold if it moves relative to the observer at 99% of the speed of light.

The origin of mass is not yet fully understood.

**mass action** The law of mass action states that the rate at which a chemical reaction takes place at a given temperature is proportional to the product of the **active masses** of the reactants. The active mass of a reactant is taken to be its molar concentration. For example, for a reaction

$$xA + yB \rightarrow \text{products}$$

the rate is given by

$$R = k[A]^x[B]^y$$

where $k$ is the *rate constant. The principle was introduced by C. M. Guldberg and P. Waage in 1863. It is strictly correct only for ideal gases. In real cases *activities can be used. *See also* EQUILIBRIUM CONSTANT.

**mass concentration** *See* CONCENTRATION.

**mass decrement** *See* MASS DEFECT.

**mass defect** 1. The difference between the rest mass of an atomic nucleus and the sum of the rest masses of its individual nucleons in the unbound state. It is thus the mass equivalent of the *binding energy on the basis of the mass–energy equation (*see* MASS; RELATIVITY). 2. (mass decrement) The difference between the rest mass of a radioactive nucleus before decay and the total rest mass of the decay products.

**mass–energy equation** *See* MASS; RELATIVITY.

**mass extinction** The extinction of a large number of species within a relatively short interval of the geological time scale. The fossil record provides evidence for several mass extinctions, perhaps as many as 20, since the start of the Phanerozoic eon about 570 million years ago. Such extinctions cause radical changes in the characteristic fossil assem-

blages of rock, which have been reflected in the naming of strata by geologists. Hence, mass extinctions often mark the boundaries between geological strata and between the corresponding geological time intervals. The biggest mass extinctions occurred at the end of the Permian period (about 245 million years ago), when over 80% of all marine invertebrate genera disappeared (including the trilobites), and at the end of the Cretaceous (*see* ALVAREZ EVENT).

**mass flow** A hypothesis to explain the movement of sugars in the phloem tissue of plants. At a **source** (site of production) sugars are loaded into *companion cells and thence into the *sieve elements, causing water to follow by osmosis. The pressure of water in the tubes (the hydrostatic pressure) causes it to move along the tubes to a **sink** (site of utilization), where the reverse process occurs. Here sugars diffuse or are actively transported from the sieve elements into the companion cells and then into the surrounding tissues, establishing a concentration gradient from source to sink. However, the mass flow hypothesis does not explain how different solutes can be transported in the phloem in different directions at the same time.

**massicot** *See* LEAD(II) OXIDE.

**mass number** *See* NUCLEON NUMBER.

**mass spectrometry** (mass spectroscopy) A technique used to determine relative atomic masses and the relative abundance of isotopes, and for chemical analysis and the study of ion reactions. In a **mass spectrometer** a sample (usually gaseous) is ionized and the positive ions produced are accelerated into a high-vacuum region containing electric and magnetic fields. These fields deflect and focus the ions onto a detector. The fields can be varied in a controlled way so that ions of different types can impinge on the detector. A **mass spectrum** is thus obtained consisting of a series of peaks of variable intensity to which mass/charge ($m/e$) values can be assigned. The original ions are usually produced by electron impact, although ion impact, photoionization, and field ionization are also used. For organic molecules, the mass spectrum consists of a series of peaks, one corresponding to the parent ion and the others to fragment ions produced by the ionization process. Different molecules can be identified by their characteristic pattern of lines. Analysis of mixtures can be done by gas chromatography–mass spectroscopy (*see* GAS CHROMATOGRAPHY). Other types of mass spectrometer exist. In a **quadrupole mass spectrometer** the ions pass along a region surrounded by four parallel rods. Variable voltages applied to the rods produce an oscillating electric field. Varying the frequency of oscillation allows different ions to pass through to a detector. In a **time-of-flight mass spectrometer** the ions are accelerated by an electric field and then enter a drift tube through which they pass to a detector. Different types of ion are distinguished by their time of flight in the drift tube.

**mass spectrum** *See* SPECTRUM.

**mast cell** A large cell with densely granular cytoplasm that is found in connective tissues, for example around blood vessels and in the skin. Mast-cell granules contain mediators of inflammation, such as *histamine and prostaglandin D$_2$, as well as various *cytokines, which cause a local increase in blood flow. The granule contents are released from the cell in response to tissue injury or as part of an allergic response. Release is triggered by binding of antigen to a type of antibody (IgE) that is bound to the mast cell. The cell also releases *heparin.

**mastication** The process of chewing food, which involves movements of the jaws and teeth. Mastication breaks up the food into small particles, which provides a greater surface area for digestion and enables the formation of a *bolus, which is small enough to pass through the oesophagus.

**mastoid process** An outgrowth from the temporal bone of the skull containing air cavities that communicate with the cavity of the middle ear. In humans it is a route through which infection may spread from the middle ear.

**masurium** A former name for *technetium.

**maternal effect genes** Genes expressed in maternal follicle cells whose products (messenger RNAs and proteins) diffuse into the egg cell to influence its early development. Gradients of the products are established in the egg cytoplasm; following fertilization and subsequent cell division of the zygote, these gradients influence zygotic gene expression and cause regional differentiation of the embryo. For example, in many types of embryo, maternal effect genes are responsible for determining polarity, i.e. which end is the 'head' and which is the 'tail'.

**mating** *See* SEXUAL INTERCOURSE.

**matrix** (*pl.* **matrices**) **1.** (in chemistry) A continuous solid phase in which particles (atoms, ions, etc.) are embedded. Unstable species, such as free radicals, can be trapped in an unreactive substrate, such as solid argon, and studied by spectroscopy. The species under investigation are separated by the matrix, hence the term **matrix isolation** for this technique. **2.** (in geology) The fine-grained material of rock in which the coarser-grained material is embedded. **3.** (in mathematics) A set of quantities in a rectangular array, used in certain mathematical operations. The array is usually enclosed in large parentheses or in square brackets. **4.** (in histology) The component of tissues (e.g. bone and cartilage) in which the cells of the tissue are embedded. *See also* EXTRACELLULAR MATRIX.

**matrix mechanics** A formulation of *quantum mechanics using matrices (*see* MATRIX) to represent states and operators. Matrix mechanics was the first formulation of quantum mechanics to be stated (by Werner Heisenberg in 1925) and was developed by Heisenberg and Max Born (1882–1970) and the German physicist Pascual Jordan (1902–80). It was shown by Erwin Schrödinger in 1926 to be equivalent to the *wave mechanics formulation of quantum mechanics.

**Matura diamond** *See* ZIRCON.

**maturity** **1.** The stage in a life cycle that is reached when a developing organism has taken on the appearance of the adult form and is capable of reproduction. **2.** The stage reached in the formation of gametes (*gametogenesis) following meiotic division of precursor cells and their development into fully functional gametes. **Maturation** comprises the divisions and other processes leading to the formation of gametes.

**maxilla** **1.** One of a pair of *mouthparts in insects, crustaceans, centipedes, and millipedes. They lie behind the *mandibles and their lateral movements assist in feeding. Crustaceans have two pairs of maxillae but in insects the second pair are fused together forming the *labium. **2.** One of a pair of large tooth-bearing bones in the upper jaw of vertebrates. In mammals they carry all the upper teeth except the incisors.

**maximum and minimum thermometer** A thermometer designed to record both the maximum and minimum temperatures that have occurred over a given time period. It usually consists of a graduated capillary tube at the base of which is a bulb containing ethanol. The capillary contains a thin thread of mercury with a steel index at each end. As the temperature rises the index is pushed up the tube, where it remains in position to show the maximum temperature reached; as the temperature falls the lower index is pushed down the tube and similarly remains in position at the lowest temperature. The indexes are reset by means of a permanent magnet.

**maximum permissible dose** *See* DOSE.

**maxwell** A unit of magnetic flux in the *c.g.s.* system, equal to the flux through 1 square centimetre perpendicular to a magnetic field of 1 gauss. 1 maxwell is equal to $10^{-8}$ weber. It is named after James Clerk Maxwell.

**Maxwell, James Clerk** (1831–79) British physicist, born in Edinburgh, who held academic posts at Aberdeen, London, and Cambridge. In the 1860s he was one of the founders of the *kinetic theory of gases, but his best-known work was a mathematical analysis of electricity, magnetism, and *electromagnetic radiation, published in 1865.

**Maxwell–Boltzmann distribution** A law describing the distribution of speeds among the molecules of a gas. In a system consisting of $N$ molecules that are independent of each other except that they exchange energy on collision, it is clearly impossible to say what velocity any particular molecule will have. However, statistical statements regarding certain functions of the molecules were worked out by James Clerk Maxwell and Ludwig Boltzmann. One form of their law states that $n = N\exp(-E/RT)$, where $n$ is the number of molecules with energy in excess of $E$, $T$ is the thermodynamic temperature, and $R$ is the *gas constant.

**Maxwell's equations** A set of differential equations describing the space and time dependence of the electromagnetic field and forming the basis of classical electrodynamics. In *SI units the equations are:

(1) $\text{div}\boldsymbol{D} = \rho$

(2) $\text{curl}\boldsymbol{E} = -\partial\boldsymbol{B}/\partial t$

(3) $\text{div}\boldsymbol{B} = 0$

(4) $\text{curl}\boldsymbol{H} = \partial\boldsymbol{D}/\partial t + \boldsymbol{J}$

where $\boldsymbol{D}$ is the electric displacement, $\boldsymbol{E}$ is the electric field strength, $\boldsymbol{B}$ is the magnetic flux density, $\boldsymbol{H}$ is the magnetic field strength, $\rho$ is the volume charge density, and $\boldsymbol{J}$ is the electric current density. Note that in relativity and particle physics it is common to use *Gaussian or *Heaviside–Lorentz units, in which case Maxwell's equations include $4\pi$ and the speed of light $c$. Maxwell's equations have the following interpretation. Equation (1) represents *Coulomb's law; equation (2) represents *Faraday's laws of electromagnetic induction; equation (3) represents the absence of *magnetic monopoles; equation (4) represents a generalization of *Ampère's law.

**Mayer's test** A general *presumptive test for cocaine, morphine, heroin, and other alkaloids. **Mayer's reagent** is a solution of potassium mercury iodide in water. A positive result is indicated by a cream precipitate.

**MDA (methylenedioxyamphetamine)** A hallucinogenic drug, $C_{10}H_{13}CO_2$, originally designed for medical use but now extensively used as a club drug its effects are similar to those of MDMA (*see* ECSTASY).

**MDMA** Methylenedioxymethamphetamine. *See* ECSTASY.

**McClintock, Barbara** (1902–92) US botanist and geneticist, who joined the Cold Spring Harbor Laboratory of the Carnegie Institute. She is best known for her discovery of 'jumping genes' (*see* TRANSPOSON), which move along a chromosome and exert control over other genes. She carried out her work with maize plants, but such controlling elements were later found in bacteria and other organisms. For this work she was awarded the 1983 Nobel Prize for physiology or medicine.

**McLeod gauge** A vacuum pressure gauge in which a relatively large volume of a low-pressure gas is compressed to a small volume in a glass apparatus (see illustration). The volume is reduced to an extent that causes the pressure to rise sufficiently to support a column of fluid high enough to read. This simple device, which relies on *Boyle's law, is suitable for measuring pressures in the range $10^3$ to $10^{-3}$ pascal.

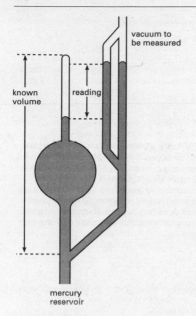

vacuum to
be measured

known
volume

reading

mercury
reservoir

**McLeod gauge.**

**mean** A representative or expected value for a set of numbers.
**1.** The **arithmetic mean** or **average** (usually just called the **mean**) of $n$ values $x_1, x_2, ..., x_n$ is given by:

$$(x_1 + x_2 + x_3 + ... + x_n)/n$$

If $x_1, x_2, ..., x_k$ occur with frequencies (weights) $w_1, w_2, ..., w_k$ then the **weighted mean** is:

$$(w_1x_1 + w_2x_2 + ... + w_kx_k)/(w_1 + w_2 + ... + w_k)$$

**2.** The **harmonic mean** $H$ is given by:

$$n/[(1/x_1) + (1/x_2) + ... + (1/x_n)]$$

**3.** The **geometric mean** $G$ is given by:

$$(x_1, x_2, ..., x_n)^{1/n}$$

**mean deviation** *See* DEVIATION.

**mean free path** The average distance travelled between collisions by the molecules in a gas, the electrons in a metallic crystal, the neutrons in a moderator, etc. According to the *kinetic theory the mean free path between elastic collisions of gas molecules of diameter $d$ (assuming they are rigid spheres) is $1/\sqrt{2}n\pi d^2$, where $n$ is the number of molecules per unit volume in the gas. As $n$

is proportional to the pressure of the gas, the mean free path is inversely proportional to the pressure.

**mean free time** The average time that elapses between the collisions of the molecules in a gas, the electrons in a crystal, the neutrons in a moderator, etc. *See* MEAN FREE PATH.

**mean life** *See* DECAY.

**mean solar day** *See* DAY.

**mean-spherical approximation** An approximation used in the theory of liquids that relates pair distribution functions to correlation functions. The mean-spherical approximation enables exact solutions to be found for several potentials describing the interactions between molecules in a liquid.

**measurements of central tendency** The general name given for the types of average used in statistics, i.e. the *mean, the *median, and the *mode.

**meatus** A small canal or passage in the body. An example is the **external auditory meatus** of the *outer ear in mammals, which connects the exterior opening to the eardrum.

**mebi-** *See* BINARY PREFIXES.

**mechanical advantage** *See* FORCE RATIO.

**mechanical bonding** Bonding that involves a mechanical constraint preventing two parts of a molecule separating, rather than a chemical linkage based on transfer or sharing of electrons. It is found in *rotaxanes, *catenanes, and *molecular knots.

**mechanical equivalent of heat** Symbol $J$. The ratio of a unit of mechanical energy to the equivalent unit of thermal energy, when a system of units is used in which they differ. $J$ has the value $4.1868 \times 10^7$ ergs per calorie. The concept loses its usefulness in *SI units in which all forms of energy are expressed in joules and $J$ therefore has a value of 1.

**mechanics** The study of the interactions between matter and the forces acting on it. *Statics is broadly concerned with the action of forces when no change of momentum is concerned, while *dynamics deals with cases in which there is a change of momentum. *Kinematics is the study of the motion of bodies without reference to the forces affecting the motion. These classical sciences are concerned with macroscopic bodies in the solid state, while *fluid mechanics is the science of the interaction between forces and fluids.

**mechanism** (in chemistry) The way in which a particular chemical reaction occurs, described in terms of the steps involved. For example, the hydrolysis of an alkyl chloride proceeds by the $S_N1$ mechanism (*see* NUCLEOPHILIC SUBSTITUTION).

**mechanoreceptor** A *receptor that responds to such mechanical stimuli as touch, sound, and pressure. The skin is rich in mechanoreceptors.

**Mecke's test** A *presumptive test for amphetamines, methamphetamines, and heroin. **Mecke's reagent** consists of 1 gram of selenious acid in 100 ml of concentrated sulphuric acid. Different substances give different results. Ecstasy, for example, gives a light blue colour, turning to turquoise, and then dark blue. Heroin gives a yellow colour changing to green. LSD gives an olive-green colour, changing to black. Mescaline gives a brownish-orange colour.

**Medawar, Sir Peter Brian** (1915–87) British immunologist. Born in Brazil and educated at Oxford, he held posts in zoology there and at Birmingham and London. He turned to medical biology and studied rejection in tissue grafts, experimenting with mouse embryos and demonstrating the phenomenon of acquired immunological tolerance – the failure of the immune response to particular antigens when these are injected before birth. For this work he shared the 1960 Nobel Prize for physiology or medicine with Sir Macfarlane *Burnet.

**median** 1. The middle number of a set of numbers arranged in order. When there is an even number of numbers the median is the average of the middle two. For example, the median of 1,7,21,33,37 is 21, and of 1,7,21,33,37,54 is $(21 + 33)/2 = 27$. **2.** A straight line in a triangle that joins the vertex to the mid-point of the base.

**median eye (pineal eye)** An eyelike structure, with a lens and retina, found on the top of the head of some lizards, *Sphenodon,* and the Cyclostomata (lampreys) as well as in many fossil vertebrates. It corresponds to the *pineal gland of other vertebrates and is thought to act as a photoreceptor.

**median lethal dose** *See* $LD_{50}$.

**mediastinum** 1. A membrane in the midline of the *thorax of mammals that separates the lungs. 2. The space between the two lungs, which is occupied by the heart and oesophagus.

**medium frequency (MF)** A radio frequency in the range 0.3–3 megahertz; i.e. having a wavelength in the range 100–1000 metres.

**MEDLINE (Medical Literature Analysis and Retrieval System Online)** A bibliographic database administered by the US National Library of Medicine (NLM) that contains over 16 million references to journal articles in life sciences and medicine. Coverage extends to more than 5000 journals, with particular emphasis on US journals, and over half a million new records are added annually. The articles are indexed using the NLM's Medical Subject Heading (MeSH) thesaurus, which permits searching using both alphabetical and hierarchical approaches. MEDLINE is the major component of the NLM's **PubMed** database and can be accessed via *Entrez.

**medulla** 1. (in zoology) The central tissue of various organs, including the adrenal glands (**adrenal medulla**) and kidneys (**renal medulla**). 2. (in botany) *See* PITH.

**medulla oblongata** Part of the vertebrate *brainstem, derived from the *hindbrain, that is continuous with the spinal cord. Its function is to regulate the autonomic pathways controlling respiration, heart beat, blood pressure, and other involuntary processes. It relays nerve signals between the brain and spinal cord and gives rise to many of the *cranial nerves.

**medullary ray (ray)** Any of the vertical plates of *parenchyma cells running radially through the cylinder of vascular tissue in the stems and roots of plants. Each may be one to many cells in width. **Primary medullary rays** occur in young plants and in those not showing secondary thickening; they pass from the cortex through to the pith. **Secondary medullary rays** are produced by the vascular *cambium and terminate in xylem and phloem tissues. Medullary rays store and transport food materials.

**medullated nerve fibre** A nerve fibre that is characterized by a *myelin sheath, which insulates the axon.

**medusa** The free-swimming stage in the life cycle of the *Cnidaria. Medusae are umbrella-shaped, with tentacles round the edge and the mouth in the centre underneath. They swim by pulsations of the body and reproduce sexually. In the Hydrozoa (e.g. *Hydra*) they alternate in the life cycle with *polyps, from which they are produced by budding. In the Scyphozoa, which includes all the common jellyfish, the medusa is the dominant form and the polyp is reduced or absent.

**mega- 1.** Symbol M. A prefix used in the metric system to denote one million times. For example, $10^6$ volts = 1 megavolt (MV). **2.** A prefix denoting large size; e.g. meganucleus, megasporangium.

**megaphyll** A type of foliage leaf in ferns and seed plants that has branched or parallel vascular bundles running through the lamina. The megaphylls of ferns are large pinnate leaves called **fronds**. A megaphyll was formerly called a **macrophyll**. *Compare* MICROPHYLL.

**megaspore** *See* SPOROPHYLL.

**megasporophyll** *See* SPOROPHYLL.

**megaton weapon** A nuclear weapon with an explosive power equivalent to one million tons of TNT. *Compare* KILOTON WEAPON.

**meiosis (reduction division)** A type of cell division that gives rise to four reproductive cells (gametes) each with half the chromosome number of the parent cell. Two consecutive divisions occur (see illustration overleaf). In the first, *homologous chromosomes become paired and may exchange genetic material (*see* CROSSING OVER) before moving away from each other into separate daughter nuclei. This is the actual reduction division because each of the two nuclei so formed contains only half of the original chromosomes. The daughter nuclei then divide by mitosis and four *haploid cells are produced. *See also* PROPHASE; METAPHASE; ANAPHASE; TELOPHASE.

**Meissner effect** The falling off of the magnetic flux within a superconducting metal when it is cooled to a temperature below the critical temperature in a magnetic field. It was discovered by Walther Meissner (1882–1974) in 1933 when he observed that the earth's magnetic field was expelled from the interior of tin crystals below 3.72 K, indicating that as *superconductivity appeared the material became perfectly diamagnetic. *See* MAGNETISM.

**Meitner, Lise** (1878–1968) Austrian-born Swedish physicist, who went to Berlin to study *radioactivity with Otto *Hahn. In 1917 they discovered *protactinium. After World War I Meitner and Hahn returned to Berlin, where in 1935 they bombarded uranium with neutrons. In 1938 she left Germany, with other Jewish scientists, and went to the Nobel Institute in Stockholm. In 1939 she and Otto Frisch (1904–79) explained Hahn's results in terms of *nuclear fission.

**meitnerium** Symbol Mt. A radioactive *transactinide element; a.n. 109. It was first made in 1982 by Peter Armbruster and a team in Darmstadt, Germany, by bombarding bismuth–209 nuclei with iron–58 nuclei. Only a few atoms have ever been detected.

**(((●))) SEE WEB LINKS**

• Information from the WebElements site

**melamine** A white crystalline compound, $C_3N_6H_6$. Melamine is a cyclic compound having a six-membered ring of alternating C and N atoms, with three $NH_2$ groups. It can be copolymerized with methanal to give thermosetting **melamine resins**, which are used particularly for laminated coatings.

**melanin** Any of a group of polymers, derived from the amino acid tyrosine, that cause pigmentation of eyes, skin, and hair in vertebrates. Melanins are produced by specialized epidermal cells called **melanophores** (or **melanocytes**); their dispersion in these cells is controlled by *melanocyte-stimulating hormone and *melatonin. Certain invertebrates, fungi, and microorganisms also produce melanin pigments. The 'ink' of the octopus and squid is a notable example. Hereditary *albinism is caused by the absence of the enzyme tyrosinase, which is necessary for melanin production.

**melanism** Black coloration of the body caused by overproduction of the pigment melanin, often as a reaction to the environment. There are several species of melanic moths in industrially polluted areas (*see* INDUSTRIAL MELANISM) and the panther is a melanic form of leopard.

**melanocyte-stimulating hormone (MSH)** Any of several related peptide hormones that are produced from the precursor

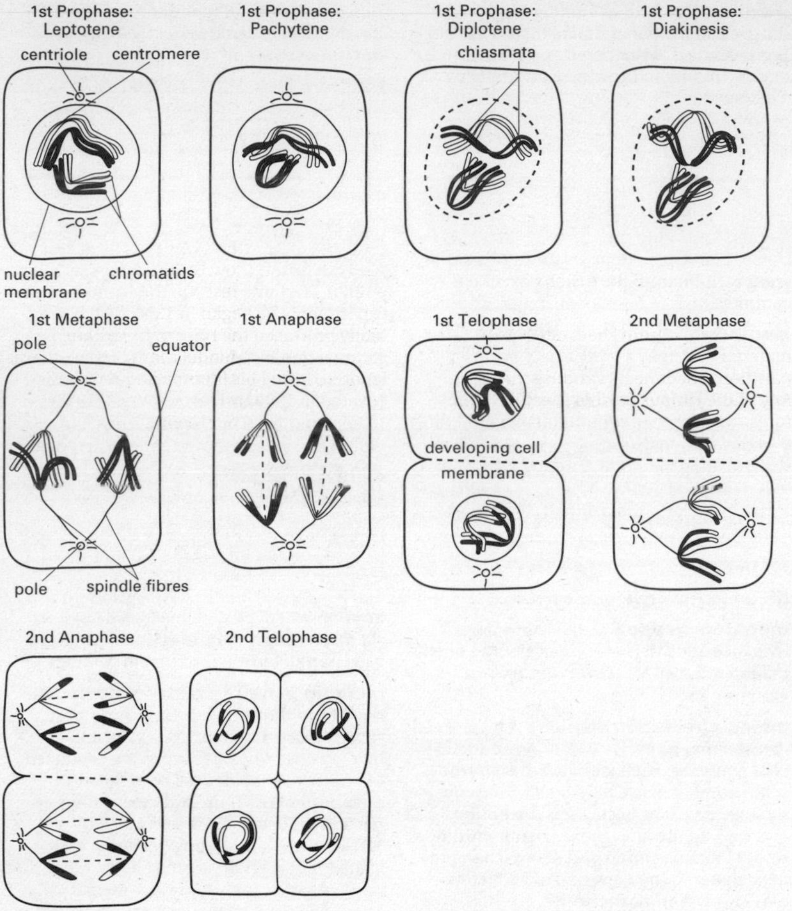

**1st Prophase: Leptotene**

centriole  centromere

nuclear membrane  chromatids

**1st Prophase: Pachytene**

**1st Prophase: Diplotene**

chiasmata

**1st Prophase: Diakinesis**

**1st Metaphase**

pole  equator

pole  spindle fibres

**1st Anaphase**

**1st Telophase**

developing cell membrane

**2nd Metaphase**

**2nd Anaphase**

**2nd Telophase**

**Meiosis.** The stages of meiosis in a cell containing two pairs of homologous chromosomes.

pro-opiomelanocortin (POMC) and secreted by the anterior or intermediate lobe of the pituitary gland. MSH stimulates dispersal of melanin in the *chromatophores of amphibian skin, causing the skin to darken; in humans it stimulates production and dispersal of melanin in the pigment cells (melanocytes) of the skin. Certain neurons in the hypothalamus release α-MSH, which is a potent suppressor of appetite and has a key role in regulating energy balance; it has potential as an antiobesity agent. α-MSH also stimulates sexual activity and is involved in regulation of heart rate and blood pressure.

**melatonin** A hormone derived from *serotonin and secreted by the pineal gland and retinas of vertebrates. Melatonin secretion by the pineal is linked to the dark–light cycle of the organism's environment, being greatest at night and lowest by day. It is used as a drug to treat sleep disorders and symptoms of jet lag. The hormone is involved in regulating certain diurnal and seasonal changes in the body, such as the reproductive cycle in seasonally breeding animals. Melatonin also controls pigmentation changes; it triggers aggregation of the pigment *melanin into melanophores in the skin, causing the skin to turn pale.

**melting point (m.p.)** The temperature at which a solid changes into a liquid. A pure substance under standard conditions of pressure (usually 1 atmosphere) has a single reproducible melting point. If heat is gradually and uniformly supplied to a solid the consequent rise in temperature stops at the melting point until the fusion process is complete.

**membrane** A thin sheet of tissue or other material that lines a body cavity, forms a partition, or connects various structures. Any of the various flexible sheetlike structures, composed predominantly of lipids and proteins, that occur in living cells, such as the *plasma membrane forming the cell boundary. See CELL MEMBRANE.

**membrane bone (dermal bone)** *Bone formed directly in connective tissue, rather than by replacing cartilage (*compare* CARTILAGE BONE). Some face bones, skull bones, and part of the clavicle are membrane bones. Small areas of membrane become jelly-like and attract calcium salts. Bone-forming cells break down these areas forming a bone lattice, which eventually fills in.

**membranous labyrinth** The soft tubular sensory structures that form the *inner ear of vertebrates and are housed within the bony labyrinth.

**meme** A self-replicating unit of cultural inheritance analogous to a gene. The term was introduced by British biologist Richard Dawkins in his book *The Selfish Gene* (1976) to denote a cultural entity, such as a song, a method for making paper aeroplanes, a religion, or a recipe, that is transmitted between individuals and across generations, so that it is inherited and (potentially) can change over time.

**memory 1.** (in biology) The means by which information is stored in the brain. The exact mechanism of processing and storing information is not known but is thought to involve the construction of circuits of *neurons, which are strengthened by repeated use. Memory is essential to the processes of *learning and recognition of individuals and objects. **2.** (in computing) The part of a computer in which data is stored while it is

being worked on. A typical microcomputer, for example, has a comparatively small amount of read-only memory (*see* ROM) and a large amount of random-access memory (*see* RAM). Only data in ROM is preserved when the machine is switched off; any data in RAM must be saved to disk if it is wanted again.

**memory cell** *See* B CELL.

**Mendel, Johann Gregor** (1822–84) Austrian geneticist, who from 1843 lived as a monk in Brünn (now Brno, in the Czech Republic). His fame rests on the plant-breeding experiments he began in 1856, which eventually produced the rules of inheritance summarized in *Mendel's laws. His work was ignored during his lifetime and only rediscovered in 1900 by Hugo de Vries (1848–1935) and others. *See* MENDELISM.

**Mendeleev, Dmitri Ivanovich** (1834–1907) Russian chemist, who became professor of chemistry at St Petersburg in 1866. His most famous work, published in 1869, was the compilation of the *periodic table of the elements, based on the *periodic law.

**Mendeleev's law** *See* PERIODIC LAW.

**mendelevium** Symbol Md. A radioactive metallic transuranic element belonging to the *actinoids; a.n. 101; mass number of the first discovered nuclide 256 (half-life 1.3 hours). Several short-lived isotopes have now been synthesized. The element was first identified by Albert Ghiorso, Glenn Seaborg (1912–99) and associates in 1955. The alternative name **unnilunium** has been proposed.

(⊕) SEE WEB LINKS
• Information from the WebElements site

**Mendelism** The theory of heredity that forms the basis of classical *genetics, proposed by Gregor Mendel in 1866 and formulated in two laws (*see* MENDEL'S LAWS; PARTICULATE INHERITANCE). Mendel suggested that individual characteristics were determined by inherited 'factors', and when improved microscopes revealed details of cell structure the behaviour of Mendel's factors could be related to the behaviour of chromosomes during *meiosis.

**Mendel's laws** Two laws summarizing Gregor Mendel's theory of inheritance (*see also* MENDELISM). The **Law of Segregation** states that each hereditary characteristic is controlled by two 'factors' (now called *alleles), which segregate (separate) and pass into separate germ (reproductive) cells. The **Law of Independent Assortment** states that pairs of 'factors' segregate independently of each other when germ cells are formed (*see also* INDEPENDENT ASSORTMENT). These laws are the foundation of genetics.

**Mendius reaction** A reaction in which an organic nitrile is reduced by nascent hydrogen (e.g. from sodium in ethanol) to a primary amine:

$$RCN + 2H_2 \rightarrow RCH_2NH_2$$

**meninges** The three membranes that surround the brain and spinal cord of vertebrates: the *pia mater, the *arachnoid membrane, and the outer *dura mater. The pia and arachnoid are separated by the **subarachnoid space**, which contains *cerebrospinal fluid.

**meniscus** **1.** A concave or convex upper surface that forms on a liquid in a tube as a result of *surface tension. **2.** *See* CONCAVE.

**menopause** The time in a woman's life when ovulation and menstruation cease (*see* MENSTRUAL CYCLE). It normally occurs between the ages of 45 and 55. The effects of the gonadotrophic hormones, *follicle-stimulating hormone and *luteinizing hormone, in the ovaries decrease so that the follicles do not develop properly. There is a change in the balance of the hormones oestrogen and progesterone, secreted by the ovaries, which may be associated with certain physical symptoms, such as weight gain and 'hot flushes', and there may also be mood changes. These symptoms can be treated by hormone replacement therapy (HRT) with oestrogens and progestogens.

**menstrual cycle** The approximately monthly cycle of events associated with *ovulation that replaces the *oestrous cycle in most primates (including humans). The lining of the uterus becomes progressively thicker with more blood vessels in preparation for the *implantation of a fertilized egg cell (blastocyst). Ovulation occurs during the middle of the cycle (the fertile period). If fertilization does not occur the uterine lining breaks down and is discharged from the body (**menstruation**); the discharge is known as a 'period'. In women the fertile period is 11–15 days after the end of the last menstruation.

**menstruation** *See* MENSTRUAL CYCLE.

**menthol** A white crystalline terpene alcohol, $C_{10}H_{19}OH$; r.d. 0.89; m.p. 42°C; b.p. 103–104°C. It has a minty taste and is found in certain essential oils (e.g. peppermint) and used as a flavouring.

**mercaptans** *See* THIOLS.

**mercapto group** *See* THIOLS.

**mercuric compounds** Compounds of mercury in its +2 oxidation state; e.g. mercuric chloride is mercury(II) chloride, $HgCl_2$.

**mercurous compounds** Compounds of mercury in its +1 oxidation state; e.g. mercury(I) chloride is mercurous chloride, $HgCl$.

**mercury** Symbol Hg. A heavy silvery liquid metallic element belonging to the *zinc group; a.n. 80; r.a.m. 200.59; r.d. 13.55; m.p. –38.87°C; b.p. 356.58°C. The main ore is the sulphide cinnabar (HgS), which can be decomposed to the elements. Mercury is used in thermometers, barometers, and other scientific apparatus, and in dental amalgams. The element is less reactive than zinc and cadmium and will not displace hydrogen from acids. It is also unusual in forming mercury(I) compounds containing the $Hg_2^{2+}$ ion, as well as mercury(II) compounds containing $Hg^{2+}$ ions. It also forms a number of complexes and organomercury compounds.

**((()) SEE WEB LINKS**

• Information from the WebElements site

**Mercury** The smallest *planet in the *solar system and the closest to the sun. Its mean distance from the *sun is $57.91 \times 10^6$ km, its mass is $3.3022 \times 10^{23}$ kg (about 5.5% that of earth), and its mean equatorial diameter is 4879.4 km; it has a *sidereal period of 87.97 days. Mercury's period of axial rotation of 58.65 days is two-thirds of its sidereal period. Flybys in 1974–75 by the US probe Mariner 10 revealed Mercury to be a rocky world with surface features similar to those on the earth's *moon, with many impact craters and plains similar to the lunar maria. It has a negligible atmosphere consisting mostly of traces of hydrogen and helium. Its relative density, at 5.427, puts it second only to the earth and, given its small size, implies that it has a large, iron-rich core, which may be surrounded by a mantle up to 700 km thick and a crust 100–300 km thick. The temperature at the equator soars to 700 K at perihelion rapidly plunging to 110 K at night. In 2004 the US Mercury Surface, Space Environment, Geochemistry, and Ranging (MES-SENGER) probe was launched to examine Mercury's physical characteristics and environment from orbit. MESSENGER is scheduled to enter orbit around Mercury in 2011 for a year-long mission.

**mercury cell** A primary *voltaic cell consisting of a zinc anode and a cathode of mercury(II) oxide (HgO) mixed with graphite. The electrolyte is potassium hydroxide (KOH) saturated with zinc oxide, the overall reaction being:

$$Zn + HgO \rightarrow ZnO + Hg$$

The e.m.f. is 1.35 volts and the cell will deliver about 0.3 ampere-hour per $cm^3$.

**mercury(I) chloride** A white salt, $Hg_2Cl_2$; r.d. 7.15; sublimes at 400°C. It is made by heating mercury(II) chloride with mercury and is used in calomel cells (so called because the salt was formerly called **calomel**) and as a fungicide.

**mercury(II) chloride** A white salt, $HgCl_2$; r.d. 5.4; m.p. 276°C; b.p. 302°C. It is made by reacting mercury with chlorine and used in making other mercury compounds.

**mercury(II) fulminate** A grey crystalline solid, $Hg(CNO)_2.\frac{1}{2}H_2O$, made by the action of nitric acid on mercury and treating the solution formed with ethanol. It is used as a detonator for cartridges and can be handled safely only under cold water.

**mercury(II) oxide** A yellow or red oxide of mercury, HgO. The red form is made by heating mercury in oxygen at 350°C; the yellow form, which differs from the red in particle size, is precipitated when sodium hydroxide solution is added to a solution of mercury(II) nitrate. Both forms decompose to the elements at high temperature. The black precipitate formed when sodium hydroxide is added to mercury(I) nitrate solution is sometimes referred to as mercury(I) oxide ($Hg_2O$) but is probably a mixture of HgO and free mercury.

**mercury(II) sulphide** A red or black compound, HgS, occurring naturally as the minerals cinnabar (red) and metacinnabar (black). It can be obtained as a black precipitate by bubbling hydrogen sulphide through a solution of mercury(II) nitrate. The red form is obtained by sublimation. The compound is also called **vermilion** (used as a pigment).

**mercury-vapour lamp** A type of discharge tube in which a glow discharge takes place in mercury vapour. The discharge takes place in a transparent tube of fused silica or quartz into the ends of which molybdenum and tungsten electrodes are sealed; this tube contains argon and a small amount of pure mercury. A small arc is struck between a starter electrode and one of the main electrodes causing local ionization of some argon atoms. The ionized atoms diffuse through the tube causing the main discharge to strike; the heat from this vaporizes the mercury droplets, which become ionized current carriers. Radiation is confined to four visible wavelengths in the visible spectrum and several strong ultraviolet lines. The light is bluish but can be changed by the use of *phosphors on an outer tube. The outer tube is also usually used to filter out excessive ultraviolet radiation. The lamp is widely used for street lighting on account of its low cost and great reliability and as a source of ultraviolet radiation.

**mericarp** *See* SCHIZOCARP.

**meridian 1.** *See* LATITUDE AND LONGITUDE. **2. (magnetic meridian)** An imaginary great circle on the earth's surface that passes through the north and south magnetic poles. A compass needle on the earth's surface influenced only by the earth's magnetic field (*see* GEOMAGNETISM) comes to rest along a magnetic meridian. **3. (celestial meridian)** A great circle of the *celestial sphere that passes through the zenith and the celestial poles. It meets the horizon at the north and south points.

**mer-isomer** *See* ISOMERISM.

**meristem** A plant tissue consisting of actively dividing cells that give rise to cells that differentiate into new tissues of the plant. The most important meristems are those occurring at the tip of the shoot and root (*see* APICAL MERISTEM) and the lateral meristems in the older parts of the plant (*see* CAMBIUM; CORK CAMBIUM).

**merocrine secretion** *See* SECRETION.

**meromictic lake** *See* DIMICTIC LAKE.

**mescaline** A powerful hallucinogenic compound obtained from **peyote** – the flowering head of a type of Mexican cactus. Mescaline is a class A drug in the UK. It can be detected by the Mecke test.

**mesencephalon** *See* MIDBRAIN.

**mesentery** A thin sheet of tissue, bounded on each side by \*peritoneum, that supports the gut and other organs in the body cavities of animals. Vertebrates have a well-developed dorsal mesentery that anchors the stomach and intestine and contains blood vessels and nerves supplying the gut. The reproductive organs and their ducts are also supported by mesenteries.

**mesocarp** *See* PERICARP.

**mesoderm** The layer of cells in the \*gastrula that lies between the \*ectoderm and \*endoderm. It develops into the muscles, circulatory system, and sex organs and in vertebrates also into the excretory system and skeleton. *See also* GERM LAYERS.

**mesoglea** The gelatinous noncellular layer between the endoderm and ectoderm in the body wall of coelenterates. It may be thin, as in *Hydra*, or tough and fibrous, as in the larger jellyfish and sea anemones. It often contains cells that have migrated from the two body layers but these do not form tissues and organs and the mesoglea is not homologous with the mesoderm of \*triploblastic animals.

**meso-isomer** *See* OPTICAL ACTIVITY.

**mesomerism** A former name for \*resonance in molecules.

**meson** Any of a class of \*elementary particles that are a subclass of the \*hadrons. According to current quark theory mesons consist of quark–antiquark pairs. They exist with positive, negative, and zero charges, but when charged the charge has the same magnitude as that of the electron. They include the **kaon**, **pion**, and **psi** particles. Mesons are believed to participate in the forces that hold nucleons together in the nucleus. The muon, originally called a mu-meson, was thought to be a meson but is now recognized as a \*lepton.

**meson-catalysed fusion** *See* NUCLEAR FUSION.

**mesophyll** The internal tissue of a leaf blade (lamina), consisting of \*parenchyma cells. There are two distinct forms. **Palisade mesophyll** lies just beneath the upper epidermis and consists of cells elongated at right angles to the leaf surface. They contain a large number of \*chloroplasts and their principal function is photosynthesis. **Spongy mesophyll** occupies most of the remainder of the lamina. It consists of spherical loosely arranged cells containing fewer chloroplasts than the palisade mesophyll. Between these cells are air spaces leading to the \*stomata.

**mesophyte** Any plant adapted to grow in soil that is well supplied with water and mineral salts. Such plants wilt easily when exposed to drought conditions as they are not adapted to conserve water. The majority of flowering plants are mesophytes. *Compare* HALOPHYTE; HYDROPHYTE; XEROPHYTE.

**mesoscopic** Designating a size scale intermediate between those of the \*microscopic and the \*macroscopic states. Mesoscopic objects and systems require \*quantum mechanics to describe them. Many devices in \*electronics are mesoscopic.

**mesothelium** A single layer of thin platelike cells covering the surface of the inside of the abdominal cavity and thorax and surrounding the heart, forming part of the \*peritoneum and \*pleura (*see* SEROUS MEMBRANE). It is derived from the \*mesoderm. *Compare* ENDOTHELIUM; EPITHELIUM.

**mesotrophic** Describing a body of water, such as a lake, that is intermediate between a \*eutrophic lake and an \*oligotrophic lake in the amount of nutrients contained within it.

**Mesozoic** The geological era that extended from the end of the \*Palaeozoic era, about 251 million years ago, to the beginning of the \*Cenozoic era, about 65 million years ago. It comprises the \*Triassic, \*Jurassic, and \*Cretaceous periods. The Mesozoic era is often known as the **Age of Reptiles** as these animals, which included the dinosaurs, pterosaurs, and ichthyosaurs, became the dominant lifeform; most became extinct before the end of the era.

**messenger RNA** *See* RNA.

**Messier Catalogue** A list of nebulae, galaxies, and star clusters, originally published (with 45 entries) in 1774. Such objects are referred to by their Messier numbers; e.g. the Andromeda galaxy is M31. It is named after its originator, Charles Messier (1730–1817).

**meta-** **1.** Prefix designating a benzene compound in which two substituents are in the 1,3 positions on the benzene ring. The abbreviation *m*- is used; for example, *m*-xylene is 1,3-dimethylbenzene. *Compare* ORTHO-; PARA-. **2.** Prefix designating a lower oxo acid, e.g. metaphosphoric acid. *Compare* ORTHO-.

**metabolic pathway** *See* METABOLISM.

**metabolic rate** A measure of the energy used by an animal in a given time period. The metabolic rate of an animal is affected by several interacting factors, including temperature and the level of activity. The metabolic rate of a resting animal is known as the *basal metabolic rate (BMR).

**metabolic waste** The *waste products, collectively, of metabolism.

**metabolism** The sum of the chemical reactions that occur within living organisms. The various compounds that take part in or are formed by these reactions are called **metabolites**. In animals many metabolites are obtained by the digestion of food, whereas in plants only the basic starting materials (carbon dioxide, water, and minerals) are externally derived. The synthesis (*anabolism) and breakdown (*catabolism) of most compounds occurs by a number of reaction steps, the reaction sequence being termed a **metabolic pathway**. Some pathways (e.g. *glycolysis) are linear; others (e.g. the *Krebs cycle) are cyclic. The changes at each step in a pathway are usually small and are promoted by efficient biological catalysts – the enzymes. In this way the amounts of energy required or released at any given stage are minimal, which helps in maintaining a constant *internal environment. Various *feedback mechanisms exist to govern *metabolic rates.

**metabolite** *See* METABOLISM.

**metabolome** The entire complement of metabolites found within a cell under defined conditions, such as a particular physiological or developmental state. The metabolome is determined using various forms of high-throughput mass spectroscopy. It excludes nucleic acids and other large molecules, giving a 'snapshot' of the cell's metabolic state. *See* METABOLOMICS.

**metabolomics** The study of how the pool of metabolites (*see* METABOLOME) of cells changes under various physiological or developmental conditions or in response to genetic modification (e.g. mutation).

**metaboric acid** *See* BORIC ACID.

**metacarpal** One of the bones in the *metacarpus.

**metacarpus** The hand (or corresponding part of the forelimb) in terrestrial verte-brates, consisting of a number of rod-shaped bones (**metacarpals**) that articulate with the bones of the wrist (*see* CARPUS) and those of the fingers (*see* PHALANGES). The number of metacarpals varies between species: in the basic *pentadactyl limb there are five, but this number is reduced in many species.

**metal** Any of a class of chemical elements that are typically lustrous solids that are good conductors of heat and electricity. Not all metals have all these properties (e.g. mercury is a liquid). In chemistry, metals fall into two distinct types. Those of the *s*- and *p*-blocks (e.g. sodium and aluminium) are generally soft silvery reactive elements. They tend to form positive ions and so are described as electropositive. This is contrasted with typical nonmetallic behaviour of forming negative ions. The *transition elements (e.g. iron and copper) are harder substances and generally less reactive. They form coordination complexes. All metals have oxides that are basic, although some, such as aluminium, have *amphoteric properties.

**metaldehyde** A solid compound, $C_4O_4H_4(CH_3)_4$, formed by polymerization of ethanal (acetaldehyde) in dilute acid solutions below 0°C. The compound, a tetramer of ethanal, is used in slug pellets and as a fuel for portable stoves.

**metal fatigue** A cumulative effect causing a metal to fail after repeated applications of *stress, none of which exceeds the ultimate *tensile strength. The **fatigue strength** (or **fatigue limit**) is the stress that will cause failure after a specified number (usually $10^7$) of cycles. The number of cycles required to produce failure decreases as the level of stress or strain increases. Other factors, such as corrosion, also reduce the fatigue life.

**metallic bond** A chemical bond of the type holding together the atoms in a solid metal or alloy. In such solids, the atoms are considered to be ionized, with the positive ions occupying lattice positions. The valence electrons are able to move freely (or almost freely) through the lattice, forming an 'electron gas'. The bonding force is electrostatic attraction between the positive metal ions and the electrons. The existence of free electrons accounts for the good electrical and thermal conductivities of metals. *See also* ENERGY BANDS.

**metallic crystal** A crystalline solid in which the atoms are held together by *metallic bonds. Metallic crystals are found in some *interstitial compounds as well as in metals and alloys.

**metallized dye** *See* DYES.

**metallocene** A type of organometallic complex in which one or more aromatic rings (e.g. $C_5H_5^-$ or $C_6H_6$) coordinate to a metal ion or atom by the pi electrons of the ring. *Ferrocene was the first such compound to be discovered.

**metallography** The microscopic study of the structure of metals and their alloys. Both optical *microscopes and *electron microscopes are used in this work.

**metalloid (semimetal)** Any of a class of chemical elements intermediate in properties between metals and nonmetals. The classification is not clear cut, but typical metalloids are boron, silicon, germanium, arsenic, and tellurium. They are electrical semiconductors and their oxides are amphoteric.

**metallurgy** The branch of applied science concerned with the production of metals from their ores, the purification of metals, the manufacture of alloys, and the use and performance of metals in engineering practice. **Process metallurgy** is concerned with the extraction and production of metals, while **physical metallurgy** concerns the mechanical behaviour of metals.

**metamagnet** A material that is an antiferromagnet in the absence of an external magnetic field but undergoes a first-order transition to a phase in which there is a nonzero ferromagnetic moment when the external magnetic field becomes sufficiently large. Iron(II) chloride is an example of a metamagnet.

**metamaterial** A type of synthetic composite material with a complex nanostructure, constructed so as to have unusual properties that do not occur naturally. A particular type consists of materials that have a negative refractive index. There has been a considerable amount of research into using these as 'invisibility cloaks' for microwaves and possibly visible radiation.

**metameric segmentation (metamerism; segmentation)** The division of an animal's body (except at the head region – *see* CEPHALIZATION) into a number of compartments (**segments** or **metameres**) each containing the same organs. Metameric seg-

mentation is most strongly marked in annelid worms (e.g. earthworms), in which the muscles, blood vessels, nerves, etc. are repeated in each segment. In these animals the segmentation is obvious both externally and internally. It also occurs internally in arthropods and in the embryonic development of all vertebrates, in which it is confined to parts of the muscular, skeletal, and nervous systems and does not show externally.

**metamict state** The amorphous state of a substance that has lost its crystalline structure as a result of the radioactivity of uranium or thorium. **Metamict minerals** are minerals whose structure has been disrupted by this process. The metamictization is caused by alpha particles and the recoil nuclei from radioactive disintegration.

**metamorphic rocks** One of the three major rock categories (*see also* IGNEOUS ROCKS; SEDIMENTARY ROCKS). Metamorphic rock is formed when pre-existing rock is subjected to either chemical or physical alteration by heat, pressure, or chemically active fluids. It involves three main processes of formation. **Contact metamorphism** results from the intrusion of a mass of molten rock into sedimentary rock. Heat from the intrusion spreads into the surrounding sediments causing mineralogical changes to take place. **Regional metamorphism** is developed over large areas and is associated with mountain building. Sediments collect in large depressions, known as geosynclines, in the earth's crust. As successive layers accumulate the lower layers subside into the crust and are subjected to increasing heat and pressure, causing the rocks to be metamorphosed. Eventually these rocks may also be uplifted and folded to form mountain chains. **Dislocation metamorphism** results from the more localized mechanical shearing and crushing of rocks, for example along fault planes.

Metamorphic rocks are characteristically resistant and tend to form upland areas. Examples of metamorphic rocks include marble (metamorphosed limestone) and slate (metamorphosed shale).

**metamorphosis** The rapid transformation from the larval to the adult form that occurs in the life cycle of many invertebrates and amphibians. Examples are the changes from a tadpole to an adult frog and from a pupa to an adult insect. Metamorphosis often involves considerable destruction of larval tissues by lysosomes, and in both insects and amphibians it is controlled by hormones.

**metaphase** The stage of cell division during which the membrane around the nucleus breaks down, the *spindle forms, and centromeres attach the chromosomes to the equator of the spindle. In the first metaphase of *meiosis pairs of chromosomes (bivalents) are attached, while in *mitosis and the second metaphase of meiosis, individual chromosomes are attached.

**metaphosphoric acid** *See* PHOSPHORIC(V) ACID.

**metaplasia** The transformation of a tissue into a different type. This is an abnormal process; for example, metaplasia of the epithelium of the bronchi may be an early sign of cancer.

**metaplumbate** *See* PLUMBATE.

**metastable state** A condition of a system in which it has a precarious stability that can easily be disturbed. It is unlike a state of stable equilibrium in that a minor disturbance will cause a system in a metastable state to fall to a lower energy level. A book lying on a table is in a state of stable equilibrium; a thin book standing on edge is in metastable equilibrium. Supercooled water is also in a metastable state. It is liquid below 0°C; a grain of dust or ice introduced into it will cause it to freeze. An excited state of an atom or nucleus that has an appreciable lifetime is also metastable.

**metastannate** *See* STANNATE.

**metastasis** *See* CANCER.

**metatarsal** One of the bones in the *metatarsus.

**metatarsus** The foot (or corresponding part of the hindlimb) in terrestrial vertebrates, consisting of a number of rod-shaped bones (**metatarsals**) that articulate with the bones of the ankle (*see* TARSUS) and those of the toes (*see* PHALANGES). The number of metatarsals varies between species: in the basic *pentadactyl limb there are five, but this number is reduced in some species.

**Metatheria** An infraclass of mammals containing the marsupials. The female bears an abdominal pouch (**marsupium**) into which the newly born young, which are in a very immature state, move to complete their development. They obtain nourishment from the mother's mammary teats. Modern marsupials are restricted to Australasia (where they include the kangaroos, koala bears, phalangers, and bandicoots) and America (the opossums). Marsupials evolved during the early to mid- Cretaceous period, 140–125 million years ago. In Australia, where the marsupials have been isolated for millions of years, they show the greatest diversity of form, having undergone *adaptive radiation to many of the niches occupied by placental mammals elsewhere. *Compare* EUTHERIA; PROTOTHERIA.

**metathesis** A type of reaction in which radicals are exchanged. In inorganic chemistry, it is also called **double decomposition**. A simple example is

$$KCL + A_gNO_3 \rightarrow KNO_3 + A_gCl.$$

Metathesis of alkenes is an important type of reaction in synthetic organic chemistry. It involves exchange of groups. For example

$$RHC=CH_2 + RHC=CH_2 \rightarrow RHC=CHR + H_2C=CH_2.$$

Reactions of this type are catalysed by metal alkylides (containing an $M=CR_2$ grouping) and the intermediate is a four-membered ring containing the metal ion. The catalysts most often used are the **Schrock catalysts** based on molybdenum and the **Grubbs catalysts** based on ruthenium. The American chemists Richard Schrock and Robert Grubbs shared the Nobel prize for chemistry in 2005 for work in this field.

**Metazoa (Eumetazoa)** A subkingdom comprising all multicellular animals. In some classifications it also includes the *Porifera (sponges) and *Placozoa, which some authorities place in a separate subkingdom, Parazoa.

**meteor** A streak of light observable in the sky when a particle of matter enters the earth's atmosphere and becomes incandescent as a result of friction with atmospheric atoms and molecules. These particles of matter are known collectively as **meteoroids**. Meteoroids that survive their passage through the atmosphere and strike the earth's surface are known as **meteorites**. Only some 2500 meteorites are known, excluding the **micrometeorites** (bodies less than 1 mm in diameter). Meteorites consist mainly of silicate materials (stony meteorites) or iron (iron meteorites). It is estimated that the earth collects over $10^8$ kg of meteoritic material every year, mostly in the form of micrometeorites. Micrometeorites survive atmospheric friction because their small size enables them to radiate away the heat generated by friction before they vaporize.

**meteorite** *See* METEOR.

**meteoroid** A particle of dust (from a comet) or rock (from an asteroid) in space on a collision course with the earth. When it enters the earth's atmosphere, it burns up as a *meteor or hits the ground as a meteorite.

**meteorology** The study of the physical phenomena and processes taking place in the atmosphere and its interactions with the ground surface. This knowledge is applied to weather forecasting. The chief branches of meteorology are *dynamical meteorology, *micrometeorology, and *synoptic meteorology.

**methacrylate** A salt or ester of methacrylic acid (2-methylpropenoic acid).

**methacrylate resins** *Acrylic resins obtained by polymerizing 2-methylpropenoic acid or its esters.

**methacrylic acid** *See* 2-METHYLPROPENOIC ACID.

**methadone** A synthetic opioid, $C_{21}H_{27}NO$, used medically as an analgesic for chronic pain and also as a substitute for heroin in the treatment of addiction. Methadone is itself addictive and considerable quantities of 'street' methadone are used in the UK.

**methamphetamine** *See* AMPHETAMINE.

**methanal (formaldehyde)** A colourless gas, HCHO; r.d. 0.815 (at –20°C); m.p. –92°C; b.p. –21°C. It is the simplest *aldehyde, made by the catalytic oxidation of methanol (500°C; silver catalyst) by air. It forms two polymers: *methanal trimer and polymethanal. *See also* FORMALIN.

**methanal trimer** A cyclic trimer of methanal, $C_3O_3H_6$, obtained by distillation of an acidic solution of methanal. It has a six-membered ring of alternating –O– and –$CH_2$– groups.

**methane** A colourless odourless gas, $CH_4$; m.p. –182.5°C; b.p. –164°C. Methane is the simplest hydrocarbon, being the first member of the *alkane series. It is the main constituent of natural gas (~99%) and as such is an important raw material for producing

other organic compounds. It can be converted into methanol by catalytic oxidation.

**methanide** *See* CARBIDE.

**methanoate (formate)** A salt or ester of methanoic acid.

**methanogen** Any of various archaebacteria (*see* BACTERIA) that produce methane; they include such genera as *Methanobacillus* and *Methanothrix*. Methanogens are obligate anaerobes (*see* ANAEROBIC RESPIRATION) found in oxygen-deficient environments, such as marshes, swamps, sludge (formed during *sewage treatment), and the digestive systems of ruminants. They mostly obtain their energy by reducing carbon dioxide and oxidizing hydrogen, with the production of methane:

$$CO_2 + 4H_2 \rightarrow CH_4 + 2H_2O.$$

Formate, methanol, or acetate may also be used as substrates by certain methanogens. Methanogenic bacteria are important in the production of *biogas.

**methanoic acid (formic acid)** A colourless pungent liquid, HCOOH; r.d. 1.2; m.p. 8°C; b.p. 101°C. It can be made by the action of concentrated sulphuric acid on the sodium salt (sodium methanoate), and occurs naturally in ants and stinging nettles. Methanoic acid is the simplest of the *carboxylic acids.

**methanol (methyl alcohol)** A colourless liquid, $CH_3OH$; r.d. 0.79; m.p. –93.9°C; b.p. 64.96°C. It is made by catalytic oxidation of methane (from natural gas) using air. Methanol is used as a solvent (*see* METHYLATED SPIRITS) and as a raw material for making methanal (mainly for urea–formaldehyde resins). It was formerly made by the dry distillation of wood (hence the name **wood alcohol**).

**methionine** *See* AMINO ACID.

**method of mixtures** A method of determining the specific heat capacities of liquids or a liquid and a solid by mixing known masses of the substances at different temperatures and measuring the final temperature of the mixture.

**methoxy group** The organic group $CH_3O–$.

**methyl acetate** *See* METHYL ETHANOATE.

**methyl alcohol** *See* METHANOL.

**methylamine** A colourless flammable gas,

$CH_3NH_2$; m.p. $-93.5°C$; b.p. $-6.3°C$. It can be made by a catalytic reaction between methanol and ammonia and is used in the manufacture of other organic chemicals.

**methylated spirits** A mixture consisting mainly of ethanol with added methanol (~9.5%), pyridine (~0.5%), and blue dye. The additives are included to make the ethanol undrinkable so that it can be sold without excise duty for use as a solvent and a fuel (for small spirit stoves).

**methylation** A chemical reaction in which a methyl group ($CH_3–$) is introduced in a molecule. A particular example is the replacement of a hydrogen atom by a methyl group, as in a *Friedel–Crafts reaction.

**methylbenzene (toluene)** A colourless liquid, $CH_3C_6H_5$; r.d. 0.9; m.p. $-95°C$; b.p. $111°C$. Methylbenzene is derived from benzene by replacement of a hydrogen atom by a methyl group. It can be obtained from coal tar or made from methylcyclohexane (extracted from crude oil) by catalytic dehydrogenation. Its main uses are as a solvent and as a raw material for producing TNT.

**methyl bromide** *See* BROMOMETHANE.

**2-methylbuta-1,3-diene** *See* ISOPRENE.

**methyl chloride** *See* CHLOROMETHANE.

**methylene** The highly reactive *carbene, $:CH_2$. The divalent $CH_2$ group in a compound is the **methylene group**.

**methylene blue** A blue dye used in optical microscopy to stain nuclei of animal tissues. It is also suitable as a vital stain and a bacterial stain.

**methylenedioxymethamphetamine (MDMA)** *See* ECSTASY.

**methyl ethanoate (methyl acetate)** A colourless volatile fragrant liquid, $CH_3$-$COOCH_3$; r.d. 0.92; m.p. $-98°C$; b.p. $54°C$. A typical *ester, it can be made from methanol and methanoic acid and is used mainly as a solvent.

**methyl ethyl ketone** *See* BUTANONE.

**methyl group (methyl radical)** The organic group $CH_3–$.

**methyl methacrylate** An ester of methacrylic acid (2-methylpropenoic acid), $CH_2$:$C(CH_3)COOCH_3$, used in making *methacrylate resins.

**methyl orange** An organic dye used as an acid–base *indicator. It changes from red below pH 3.1 to yellow above pH 4.4 (at 25°C) and is used for titrations involving weak bases.

**methylphenols (cresols)** Organic compounds having a methyl group and a hydroxyl group bound directly to a benzene ring. There are three isomeric methylphenols with the formula $CH_3C_6H_4OH$, differing in the relative positions of the methyl and hydroxyl groups. A mixture of the three can be obtained by distilling coal tar and is used as a germicide and antiseptic.

**2-methylpropenoic acid (methacrylic acid)** A white crystalline unsaturated soluble carboxylic acid, $CH_2$:$C(CH_3)COOH$, used in making *methacrylate resins.

**methyl red** An organic dye similar in structure and use to methyl orange. It changes from red below pH 4.4 to yellow above pH 6.0 (at 25°C).

**methylxanthines** Derivatives of xanthine in which one or more hydrogen atoms have been substituted by methyl groups. The common ones are the trimethylxanthine *caffeine and the dimethylxanthines *theophylline and *theobromine.

**metre** Symbol m. The SI unit of length, being the length of the path travelled by light in vacuum during a time interval of 1/299 792 458 of a second. This definition, adopted by the General Conference on Weights and Measures in October 1983, replaced the 1960 definition based on the krypton lamp, i.e. 1 650 763.73 wavelengths in a vacuum of the radiation corresponding to the transition between the levels $2p^{10}$ and $5d^5$ of the nuclide krypton–86. This definition replaced the older (1927) definition of a metre based on a platinum–iridium bar of standard length. When the *metric system was introduced in 1791 in France, the metre was intended to be one ten-millionth of the earth's meridian quadrant passing through Paris. However, the original geodetic surveys proved the impractibility of such a standard and the original platinum metre bar, the *mètre des archives*, was constructed in 1793.

**metre bridge** *See* WHEATSTONE BRIDGE.

**metric system** A decimal system of units originally devised by a committee of the French Academy, which included J. L. Lagrange and P. S. Laplace, in 1791. It was based on the *metre, the gram defined in terms of the mass of a cubic centimetre of water, and the second. This centimetre-gram-second system (*see* C.G.S. UNITS) later gave way for scientific work to the metre-kilogram-second system (*see* M.K.S. UNITS) on which *SI units are based.

**metric ton (tonne)** A unit of mass equal to 1000 kg or 2204.61 lb. 1 tonne = 0.9842 ton.

**metrology** The scientific study of measurement, especially the definition and standardization of the units of measurement used in science.

**MHC** *See* MAJOR HISTOCOMPATIBILITY COMPLEX.

**MHD** *See* MAGNETOHYDRODYNAMICS.

**mho** A reciprocal ohm, the former name of the unit of electrical *conductance now known as the siemens.

**mica** Any of a group of silicate minerals with a layered structure. Micas are composed of linked $SiO_4$ tetrahedra with cations and hydroxyl groupings between the layers. The general formula of the micas is $X_2Y_{4-6}Z_8O_{20}(OH,F)_4$, where X = K,Na,Ca; Y = Al,Mg,Fe,Li; and Z = Si,Al. The three main mica minerals are:
*muscovite, $K_2Al_4(Si_6Al_2O_{20})(OH,F)_4$;
*biotite, $K_2(Mg,Fe^{2+})_{6-4}(Fe^{3+},Al,Ti)_{0-2}$- $(Si_{6-5}Al_{2-3}O_{20})(OH,F)_4$;
lepidolite, $K_2(Li,Al)_{5-6}(Si_{6-7}Al_{2-1}O_{20})(OH,F)_4$. Micas have perfect basal cleavage and the thin cleavage flakes are flexible and elastic. Flakes of mica are used as electrical insulators and as the dielectric in capacitors.

**micelle** An aggregate of molecules in a *colloid. For example, when soap or other *detergents dissolve in water they do so as micelles – small clusters of molecules in which the nonpolar hydrocarbon groups are in the centre and the hydrophilic polar groups are on the outside solvated by the water molecules. Phospholipids in aqueous solution also form micelles. The products of fat digestion are dispersed into micelles by the action of bile salts, which facilitates their absorption in the small intestine.

**Michaelis–Menten curve** A graph that shows the relationship between the concentration of a substrate and the rate of the corresponding enzyme-controlled reaction. It is named after Leonor Michaelis (1875–1949) and L. M. Menten. The curve only applies to enzyme reactions involving a single sub-

strate. The graph can be used to calculate the **Michaelis constant** ($K_m$), which is the concentration of a substrate required in order for an enzyme to act at half of its maximum velocity ($V_{max}$). The Michaelis constant is a measure of the affinity of an enzyme for a substrate. A low value corresponds to a high affinity, and vice versa. *See also* ENZYME KINETICS.

**((⊕)) SEE WEB LINKS**
• Original paper on Michaelis–Menten kinetics

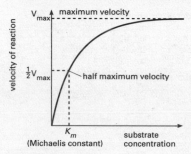

**Michaelis–Menten curve.**

**Michelson–Morley experiment** An experiment, conducted in 1887 by the US physicists Albert Michelson (1852–1931) and Edward Morley (1838–1923), that attempted to measure the velocity of the earth through the *ether. Using a modified **Michelson interferometer** (see illustration) they expected to observe a shift in the interference fringes formed when the instrument was rotated through 90°, showing that the speed of light measured in the direction of the earth's rota-

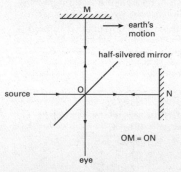

**Michelson-Morley experiment.** Michelson interferometer.

tion, or orbital motion, is not identical to its speed at right angles to this direction. No shift was observed. An explanation was finally provided by the *Lorentz–Fitzgerald contraction, which provided an important step in the formulation of Einstein's special theory of *relativity and the abandonment of the ether concept.

**((⊕)) SEE WEB LINKS**

• The original 1887 paper in *The American Journal of Science*

**micro-** **1.** A prefix denoting very small size; e.g. microgamete, micronucleus. **2.** Symbol μ. A prefix used in the metric system to denote one millionth. For example, $10^{-6}$ metre = 1 micrometre ($\mu m$).

**microarray** A glass slide or bead on which are deposited biomolecules or other material in a regular micro-scale pattern to enable automated simultaneous multiple assays of target substances or activities. Microarrays are powerful analytical tools with wide-ranging applications. They can be designed to carry small DNA molecules (*see* DNA MICRO-ARRAY), proteins (e.g. antibodies or antigens), carbohydrates or other organic molecules, or even individual living cells. These reagents are applied to the glass substrate in a regular microscopic grid pattern, each being identified by its unique coordinate, or address, on the grid. Interaction of a target substance (e.g. an antibody or a complementary nucleic acid) with a particular address on the microarray activates or attaches a label (e.g. a fluorescent dye). The microarray can then be 'read' by a scanner, which automatically assesses the amount of label at each address, and hence the amount of target substance. Even smaller-scale **nanoarrays** are already being developed, to increase further the scope and speed of this technology.

**microbalance** A sensitive *balance capable of weighing masses of the order $10^{-6}$ to $10^{-9}$ kg.

**microbiology** The scientific study of microorganisms (e.g. bacteria, viruses, and fungi). Originally this was directed towards their effects (e.g. in causing disease and decay), but during the 20th century the emphasis shifted to their physiology, biochemistry, and genetics. Microbes are now recognized as important vehicles for the study of biochemical and genetic processes common to all living organisms, and their rapid growth enables their laboratory culture in large numbers for studies in genetics.

**((⊕)) SEE WEB LINKS**

• Home page of the Society for General Microbiology

**microcavity** A cavity with reflecting faces that is so small that specifically quantum mechanical aspects of quantum electrodynamics, such as spontaneous emission, can be investigated. Microcavities range in size from nanometres to micrometres.

**microclimate** The local climate of a small area or of a particular *habitat, which is different from the **macroclimate** of the larger surrounding geographical area.

**microcomputer** A *computer in which the central processing unit is implemented by means of a semiconductor chip or chip set, known as a **microprocessor**. The power of a microcomputer is determined not only by the speed and power of the processor but also by features of the other components in the computer, such as the storage capacity of the main (*RAM) memory and the disks used as *backing store, as well as the operating system and other software used. Microcomputers are used in a wide variety of forms, including *personal computers, electronic point-of-sales terminals, and cash-dispensing automated teller machines.

**microcrystal test** *See* CRYSTAL TEST.

**microdissection (micromanipulation)** A technique used for the dissection of living cells under the high power of an optical microscope. It utilizes minute mechanically manipulated instruments, such as needles, scalpels, *micropipettes, and lasers. For example, the instruments may be used to remove a single nucleus from one cell and to implant it in another (*see* NUCLEAR TRANS-FER).

**microelectronics** The techniques of designing and making electronic circuits of very small size. As a result of these techniques a single *silicon chip measuring less than a centimetre in either direction can contain many thousands of transistors and may constitute the central processing unit of a microcomputer. In addition to an enormous drop in size, compared to an equivalent valve-operated device, these microelectronic circuits are some 100 000 times more reliable than their thermionic predecessors.

**microfauna** **1.** Animals that cannot be seen with the naked eye. They are normally observed with the aid of a microscope. *Compare* MACROFAUNA. **2.** The animals that live in a particular *microhabitat.

**microfibril** A microscopic fibre. Plant cell walls contain microfibrils, about 5 nm in diameter, each consisting of parallel cellulose chains that are associated together to form a rod or a flat ribbon. Cellulose microfibrils are arranged in layers at right angles to each other.

**microflora** **1.** Plants and algae that cannot be seen with the naked eye. They are normally observed with the aid of a microscope. **2.** The plants and algae that live in a particular *microhabitat.

**microfossil** A *fossil that is so small that it can only be studied under a microscope. Microfossils include bacteria, diatoms, and protozoa and parts of organisms, such as plant pollen and skeletal fragments. Microfossils are important in the correlation of rocks where only small samples are available. The study of microfossils, particularly pollen, is known as *palynology.

**microglia** *See* GLIA; MACROPHAGE.

**microhabitat** The local habitat of a particular organism or microorganism. There are normally a number of different microhabitats within a large *habitat (**macrohabitat**), each with its distinct set of environmental conditions. For example, in a stream macrohabitat there will exist different microhabitats, depending on oxygen content, pH, speed of water flow, and other factors in localized areas of the stream.

**micromanipulation** *See* MICRODISSECTION.

**micrometeorite** *See* METEOR.

**micrometeorology** The branch of meteorology concerned with small-scale processes at work within the lowest layers of the atmosphere, including the interaction of the atmosphere with the ground surface. Examples of the processes studied include mountain and valley winds and land and sea breezes.

**micrometer** A gauge for measuring small diameters, thicknesses, etc., accurately. It consists of a G-shaped device in which the gap between the measuring faces is adjusted by means of an accurately calibrated screw, the end of which forms one of the measuring faces.

**micron** The former name for the *SI unit now called the micrometre, i.e. $10^{-6}$ m.

**micronutrient** A chemical element required by plants in relatively small quantities. Micronutrients are typically found in cofactors and coenzymes. They include copper, zinc, molybdenum, manganese, cobalt, and boron. *See* ESSENTIAL ELEMENT. *Compare* MACRONUTRIENT.

**microorganism (microbe)** Any organism that can be observed only with the aid of a microscope. Microorganisms include bacteria, viruses, protozoans, and some algae and fungi. *See* MICROBIOLOGY.

**microphagous** Describing the method of feeding of those heterotrophic organisms that take in their food in the form of tiny particles. *Filter feeding and *ciliary feeding are examples of this type of feeding. *Compare* MACROPHAGOUS.

**microphone** A *transducer in which sound waves are converted into corresponding variations in an electrical signal for amplification, transmission to a distant point, or recording. Various types of device are used. In the **dynamic microphone** the sound waves impinge on a conductor of low mass supported in a magnetic field and cause it to oscillate at the frequency of the sound waves. These movements induce an e.m.f. in the conductor that is proportional to its velocity. The moving conductor consists of a metal ribbon, a wire, or a coil of wire. In the **moving-iron microphone**, sound waves cause a light armature to oscillate so that it varies the reluctance of a magnetic circuit. In a coil surrounding this path the varying reluctance is experienced as a variation in the magnetic flux within it, which induces a corresponding e.m.f. In the **carbon microphone**, widely used in telephones, a diaphragm constitutes a movable electrode in contact with carbon granules, which are also in contact with a fixed electrode. The movement of the diaphragm, in response to the sound waves, varies the resistance of the path through the granules to the fixed electrode. *See also* CAPACITOR MICROPHONE; CRYSTAL MICROPHONE.

**microphyll** A type of foliage leaf in clubmosses and horsetails that has a single unbranched midrib. Such leaves are generally no more than a few millimetres long. *Compare* MEGAPHYLL.

**micropipette** A glass pipette with an ultrafine tip, typically less than 1 μm in diameter. It can be inserted into single cells or other microscopic structures and used, for example, to inject materials. The micropipette is usually held by a **micromanipulator**, a mechanical device that allows precise movement of the tip.

**microprocessor** *See* COMPUTER.

**micropropagation** The *in vitro* propagation of plants by cloning (*see* CLONE). Typically, this involves culturing excised meristematic tissue on a special medium that encourages axillary bud development. The new shoots are then separated and cultures, and the cycle is repeated until finally the shoots are transferred to a medium that promotes root development, to produce plantlets. Micropropagation is used in agriculture, horticulture, and forestry as special genotypes can be bred and maintained, the process is rapid, and plants can be kept disease-free.

**micropyle** 1. A small opening in the surface of a plant ovule through which the pollen tube passes prior to fertilization. It results from the incomplete covering of the nucellus by the integuments. It remains as an opening in the testa of most seeds through which water is absorbed. 2. A small pore in some animal cells or tissues; for example, in insect eggs (*see* CHORION).

**microRNA (miRNA)** A small RNA molecule that is encoded by a cell and can 'silence' the expression of a particular target gene within the cell (*see* RNA INTERFERENCE). miRNAs bind to target messenger RNA (mRNA) molecules and suppress translation of the mRNA into protein. They regulate expression of perhaps a third of all protein-coding genes and are involved in many aspects of embryological development, cell differentiation, cell death, and cancer.

**microsatellite DNA** *See* REPETITIVE DNA.

**microscope** Any device for forming a magnified image of a small object. The **simple microscope** consists of a biconvex magnifying glass or an equivalent system of lenses, either hand-held or in a simple frame. The **compound microscope** (see illustration) uses two lenses or systems of lenses, the second magnifying the real image formed by the first. The lenses are usually mounted at the opposite ends of a tube that has mechanical controls to move it in relation to the object. An optical condenser and mirror, often with a separate light source, provide illumination of the object. The widely used **binocular microscope** consists of two separate instruments fastened together so that one eye looks through one while the other eye looks through the other. This gives stereoscopic vision and reduces eye strain. *See also* ATOMIC FORCE MICROSCOPY; ELECTRON MICROSCOPE; FIELD-EMISSION MICROSCOPE; FIELD-IONIZATION MICROSCOPE; PHASE-CONTRAST MICROSCOPE; SCANNING TUNNELLING MICROSCOPY; ULTRAVIOLET MICROSCOPE. See also Chronology: Microscopy.

**(⊕) SEE WEB LINKS**
- Webpages from Botany online, describing the basic microscopical techniques used in biology

**microscopic** Designating a size scale comparable to the subatomic particles, atoms, and molecules. Microscopic objects and systems are described by *quantum mechanics. *Compare* MACROSCOPIC; MESOSCOPIC.

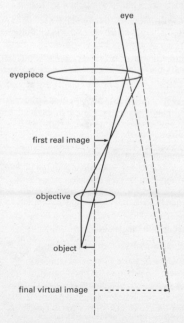

**Microscope.** Compound microscope.

## MICROSCOPY

c.1590   Dutch spectacle-makers Hans and Zacharias Janssen invent the compound microscope.

1610   German astronomer Johannes Kepler (1571–1630) invents the modern compound microscope.

1675   Anton van Leeuwenhoek invents the simple microscope.

1826   British biologist Dames Smith (d. 1870) constructs a microscope with much reduced chromatic and spherical aberrations.

1827   Italian scientist Giovanni Amici (1786–1863) invents the reflecting achromatic microscope.

1861   British chemist Joseph Reade (1801–70) invents the kettledrum microscope condenser.

1912   British microscopist Joseph Barnard (1870–1949) invents the ultramicroscope.

1932   Dutch physicist Frits Zernike (1888–1966) invents the phase-contrast microscope.

1936   German-born US physicist Erwin Mueller (1911–77) invents the field-emission microscope.

1938   German engineer Ernst Ruska (1906–88) develops the electron microscope.

1940   Canadian scientist James Hillier (1915–2007) makes a practical electron microscope.

1951   Erwin Mueller invents the field-ionization microscope.

1978   US scientists of the Hughes Research Laboratory invent the scanning ion microscope.

1981   Swiss physicists Gerd Binning (1947– ) and Heinrich Rohrer (1933– ) invent the scanning tunnelling microscope.

1985   Gerd Binning invents the atomic force microscope.

1987   James van House and Arthur Rich invent the positron microscope.

**microsome** A fragment of *endoplasmic reticulum formed when cells or tissues are disrupted. Microsomes can be isolated by centrifugation and are commonly used to investigate the functional properties of endoplasmic reticulum, such as enzymic activity and protein synthesis.

**microspore** *See* SPOROPHYLL.

**microsporophyll** *See* SPOROPHYLL.

**microtome** A machine used to cut thin sections (3–5 μm thick) of plant or animal tissue for microscopical observation. There are various designs of microtome, each basically consisting of a steel knife, a block for supporting the specimen, and a device for moving the specimen towards the knife. The

specimen is usually supported by being embedded in wax; if a **freezing microtome** is used, the specimen is frozen. An **ultramicrotome** is used to cut much thinner sections (20–100 nm thick) for electron microscopy. The biological material is embedded in plastic or resin, sectioned with a glass or diamond knife, and the cut sections are allowed to float on the surface of water in an adjacent water bath.

**microtubule** A microscopic tubular structure in eukaryotic cells that is composed of the protein **tubulin** and occurs singly or in pairs, triplets, or bundles. Microtubules help cells to maintain their shape (*see* CYTOSKELETON); they also occur in cilia and flagella (*see* UNDULIPODIUM) and in the *centrioles and they form the *spindle during nuclear division.

**microvillus** One of a number of minute finger-like projections on the free surfaces of epithelial cells. Microvilli are covered with plasma membrane and their cytoplasm is continuous with the main cell cytoplasm. Their purpose is to increase the absorptive or secretory surface area of the cell, and they are abundant on the villi of the intestine, where they form a *brush border.

**microwave background radiation** A cosmic background of radiation in the frequency range $3 \times 10^{11}$ hertz to $3 \times 10^{8}$ hertz discovered in 1965. Believed to have emanated from the primordial fireball of the big bang with which the universe is thought to have originated (*see* BIG-BANG THEORY), the radiation has an energy density in intergalactic space of some $4 \times 10^{-14}$ J m$^{-3}$. *See also* COBE; WMAP.

**microwave optics** The study of the behaviour of microwaves by analogy with the behaviour of light waves. On the large scale microwaves are propagated in straight lines and, like light waves, they undergo reflection, refraction, diffraction, and polarization.

**microwaves** Electromagnetic waves with wavelengths in the range $10^{-3}$ to 0.03 m.

**microwave spectroscopy** A sensitive technique for chemical analysis and the determination of molecular structure (bond lengths, bond angles, and dipole moments), and also relative atomic masses. It is based on the principle that microwave radiation (*see* MICROWAVES) causes changes in the rotational energy levels of molecules and absorption consequently occurs at characteristic frequencies. In a microwave spectrometer a microwave source, usually a klystron valve, produces a beam that is passed through a gaseous sample. The beam then impinges on the detector, usually a crystal detector, and the signal (wavelength against intensity) is displayed, either as a printed plot or on an oscilloscope. As microwaves are absorbed by air the instrument is evacuated.

**midbrain (mesencephalon)** One of the three sections of the brain of a vertebrate embryo. Unlike the *forebrain and the *hindbrain, the midbrain does not undergo further subdivision to form additional zones. In mammals it becomes part of the *brainstem, but in amphibians, reptiles, and birds the roof of the midbrain becomes enlarged as the **tectum**, a dominant centre for integration, and may include a pair of **optic lobes**.

**middle ear (tympanic cavity)** The air-filled cavity within the skull of vertebrates that lies between the *outer ear and the *inner ear. It is linked to the pharynx (and therefore to outside air) via the *Eustachian tube and in mammals contains the three *ear ossicles, which transmit auditory vibrations from the outer ear (via the *tympanum) to the inner ear (via the oval window).

**middle lamella** A thin layer of material, consisting mainly of pectins, that binds together the walls of adjacent plant cells.

**midgut 1.** The middle section of the alimentary canal of vertebrates, which is concerned with digestion and absorption. It comprises most of the small intestine. **2.** The middle section of the alimentary canal of arthropods. *See also* FOREGUT; HINDGUT.

**mid-ocean ridge** A long chain of underwater mountains, several thousand metres high, that runs for a total of about 50 000 km across the floors of the major oceans. The ridge corresponds to tectonic plate margins (*see* PLATE TECTONICS), at which upwelling magma breaches the comparatively thin oceanic crust. The sea floor spreads as the plates gradually move apart. Underwater volcanoes are a feature of some ridges.

**migration 1.** (in chemistry) The movement of a group, atom, or double bond from one part of a molecule to another. **2.** (in physics) The movement of ions under the influence of an electric field. **3.** (in biology) The seasonal movement of complete populations of animals to a more favourable environment. It is usually a response to lower temperatures resulting in a reduced food supply, and is often triggered by a change in day length (*see* PHOTOPERIODISM). Migration is common in mammals (e.g. porpoises), fish (e.g. eels and salmon), and some insects but is most marked in birds. The Arctic tern, for example, migrates annually from its breeding ground in the Arctic circle to the Antarctic – a distance of some 17 600 km. Migrating animals possess considerable powers of orientation; birds seem to possess a compass sense, using the sun, pole stars, and (in cloud) the earth's magnetic lines of force as reference points (*see* NAVIGATION).

**milk**　The fluid secreted by the *mammary glands of mammals. It provides a balanced and highly nutritious food for offspring. Cows' milk comprises about 87% water, 3.6% lipids (triglycerides, phospholipids, cholesterol, etc.), 3.3% protein (largely casein), 4.7% lactose (milk sugar), and, in much smaller amounts, vitamins (especially vitamin A and many B vitamins) and minerals (notably calcium, phosphorus, sodium, potassium, magnesium, and chlorine). Composition varies among species; human milk contains less protein and more lactose.

**milk of magnesia**　See MAGNESIUM HYDROXIDE.

**milk sugar**　See LACTOSE.

**milk teeth**　See DECIDUOUS TEETH.

**Milky Way**　See GALAXY.

**Miller indices**　A set of three numbers that characterize a face of a crystal. The French mineralogist René Just Haüy (1743–1822) proposed the **law of rational intercepts**, which states that there is always a set of axes, known as **crystal axes**, that allows a crystal face to be characterized in terms of intercepts of the face with these axes. The reciprocals of these intercepts are small rational numbers. When the fractions are cleared there is a set of three integers. These integers are known as the Miller indices of the crystal face after the British mineralogist William Hallowes Miller (1810–80), who pointed out that crystal faces could be characterized by these indices. If a plane is parallel to one of the crystal axes then its intercept is at infinity and hence its reciprocal is 0. If a face cuts a crystal axis on the negative side of the origin then the intercept, and hence its reciprocal, i.e. the Miller index for that axis, are negative. This is indicated by a bar over the Miller index. For example, the Miller indices for the eight faces of an octahedron are (III), ($\bar{\text{I}}$II), ($\text{I}\bar{\text{I}}$I), (II$\bar{\text{I}}$), ($\bar{\text{I}}\bar{\text{I}}$I), ($\bar{\text{I}}$I$\bar{\text{I}}$), (I$\bar{\text{I}}\bar{\text{I}}$), and ($\bar{\text{I}}\bar{\text{I}}\bar{\text{I}}$).

**milli-**　Symbol m. A prefix used in the metric system to denote one thousandth. For example, 0.001 volt = 1 millivolt (mV).

**millibar**　See BAR.

**Millikan, Robert Andrews**　(1868–1953) US physicist, who after more than 20 years at the University of Chicago went to the California Institute of Technology in 1921. His best-known work, begun in 1909, was to determine the charge on the *electron in his oil-drop experiment, which led to the award of the 1923 Nobel Prize for physics. He then went on to do important work on *cosmic radiation.

(((●))) SEE WEB LINKS

• One of Millikan's original papers (1911) in *The Physical Review*

**millipedes**　See DIPLOPODA.

**mimicry**　The resemblance of one animal to another, which has evolved as a means of protection. In one form of mimicry the markings of certain harmless insects closely resemble the *warning coloration of another insect (the **model**). Predators that have learnt to avoid the model will also avoid good mimics of it. This phenomenon is often found among butterflies. A second form of mimicry involves the mutual resemblance of a group of animals, all harmful, such as the wasp, bee, and hornet, so that a predator, having experienced one, will subsequently avoid them all.

**mineral**　A naturally occurring substance that has a characteristic chemical composition and, in general, a crystalline structure. The term is also often applied generally to organic substances that are obtained by mining (e.g. coal, petroleum, and natural gas) but strictly speaking these are not minerals, being complex mixtures without definite chemical formulas. Rocks are composed of mixtures of minerals. Minerals may be identified by the properties of their crystal system, hardness (measured on the Mohs' scale), relative density, lustre, colour, cleavage, and fracture. Many names of minerals end in -*ite*.

**mineral acid**　A common inorganic acid, such as hydrochloric acid, sulphuric acid, or nitric acid.

**mineral deficiency**　Lack of any essential mineral nutrient, such as nitrogen, phosphorus, or potassium, in living organisms, which can result in mineral deficiency diseases. In humans, for example, lack of calcium causes poor bone development, and lack of nitrogen can cause the disease kwashiorkor, due to a deficiency in protein intake (*see* MALNUTRITION). In plants mineral deficiency results in stunted growth and *chlorosis. A deficiency of trace elements (*see* ESSENTIAL ELEMENT) also leads to diseases; for example, a deficiency of iron can cause anaemia in humans and chlorosis in plants.

**mineralocorticoid**　See CORTICOSTEROID.

**mineralogy** The branch of geology concerned with the study of *minerals.

**mineral salts** Inorganic salts that need to be ingested or absorbed by living organisms for healthy growth and maintenance. They comprise the salts of the trace elements in animals (*see* ESSENTIAL ELEMENT) and the *micronutrients of plants.

**minicomputer** A *computer that is intermediate between a mainframe and a *microcomputer in processing power, and can be used by several people at once.

**minimal supersymmetric standard model (MSSM)** The smallest possible model that combines the standard model of elementary particle theory with *supersymmetry. It predicts that all the fermions and gauge bosons of the standard model should have supersymmetric partners and that there should be five *Higgs bosons.

**minisatellite DNA** *See* REPETITIVE DNA.

**minority carrier** *See* SEMICONDUCTOR.

**minor planets** Small solar system bodies directly orbiting the sun that include *dwarf planets, *asteroids, and *centaurs, but not comets and meteoroids. The term 'minor planet' was originally an alternative name for asteroid, but the discovery in the late 20th century of anomalous objects such as (2060) Chiron (now classified as centaurs) and the first *trans-Neptunian objects led to its broader use.

**minute** 1. One sixtieth of an hour. 2. One sixtieth of a degree (angle).

**Miocene** The first epoch of the *Neogene period, stretching from the end of the Oligocene, about 23 million years ago, to the start of the Pliocene, roughly 5 million years ago. It saw the radiation of several modern mammal groups, including the ruminants, certain rodents (beavers, porcupines, and cavies), and the apes. Cooling of the climate during the Oligocene resulted in a continuous shift to deciduous hardwood species, such as oak and maple, at the expense of conifers during the Miocene.

**mirabilite** A mineral form of *sodium sulphate, $Na_2SO_4.10H_2O$.

**mirage** An optical phenomenon that occurs as a result of the bending of light rays through layers of air having very large temperature gradients. An **inferior mirage** occurs when the ground surface is strongly

heated and the air near the ground is much warmer that the air above. Light rays from the sky are strongly refracted upwards near the surface giving the appearance of a pool of water. A **superior mirage** occurs if the air close to the ground surface is much colder than the air above. Light is bent downwards from the object towards the viewer so that it appears to be elevated or floating in the air.

**Mira-type variable star (Mira Ceti variable)** A star in the red giant or red supergiant category whose radiated energy varies regularly over a quite long period. Pulsation of the surface layers is thought to be the cause of the variation, which recurs every 2 to 30 months. The stars are named after their prototype, Mira Ceti.

**miRNA** *See* MICRORNA.

**mirror** A surface that reflects most of the light falling on it. A **plane mirror** is a flat surface that produces an erect virtual *image of a real object, in which front and back are reversed. **Spherical mirrors** are formed from the surfaces of spheres and form images of real objects in much the same way as lenses.

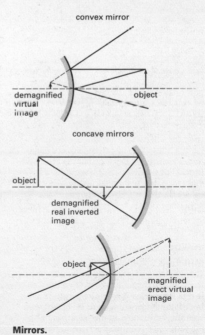

convex mirror

demagnified virtual image

object

concave mirrors

object

demagnified real inverted image

object

magnified erect virtual image

**Mirrors.**

A convex mirror forms erect virtual images. They are commonly used as rear-view mirrors in road vehicles, and give a diminished wide-angle image. A concave mirror can form either inverted real images or erect virtual images. (See illustrations.) Spherical mirrors obey the *lens equation (using the real-positive sign convention) and are subject to some *aberrations similar to those of lenses.

**misch metal** An alloy of cerium (50%), lanthanum (25%), neodymium (18%), praseodymium (5%), and other rare earths. It is used alloyed with iron (up to 30%) in lighter flints, and in small quantities to improve the malleability of iron. It is also added to copper alloys to make them harder, to aluminium alloys to make them stronger, to magnesium alloys to reduce creep, and to nickel alloys to reduce oxidation.

**missing mass** The mass of matter in the universe that cannot be observed by direct observations of its emitted or absorbed electromagnetic radiation. There are a number of astrophysical observations that suggest that the actual mass of the universe is much greater than that estimated by observations using optical telescopes, radio telescopes, etc. It is thought that there is a considerable amount of **dark matter** (or **hidden matter**) causing this discrepancy. Various explanations have been put forward for the missing mass, including black holes, brown dwarfs, cosmic strings, axions, neutrinos, monopoles, and various exotic particles, such as **weakly interacting massive particles (WIMPS)**, which are predicted to exist by supersymmetry. The universe contains far more missing matter than directly observable matter.

**mitochondrial DNA (mtDNA)** A circular ring of DNA found in mitochondria. In mammals mtDNA makes up less than 1% of the total cellular DNA, but in plants the amount is variable. It codes for ribosomal and transfer RNA but only some mitochondrial proteins (up to 30 proteins in animals), the nuclear DNA being required for encoding most of these. Human mtDNA contains 37 genes encoding 13 proteins and some RNAs and is generally inherited only via the female line. *See also* MITOCHONDRIAL EVE.

**mitochondrial Eve** The hypothetical female claimed by some biologists to be the ancestor of all humankind. Analysis of *mitochondrial DNA (mtDNA) from groups of

people throughout the world suggests that mitochondrial Eve lived around 140 000 years ago, probably in Africa (hence she is also known as 'African Eve'). Mitochondrial DNA is particularly useful for investigating recent genetic history as it mutates quickly (ten times more rapidly than nuclear DNA) and in humans is inherited solely through the female line (therefore it does not undergo recombination by *crossing over). The uniformity of the different samples of mtDNA indicates that modern humans evolved relatively recently from a single region in Africa. This view has been reinforced by studies of Y chromosomes from different groups around the world, which are transmitted only through the male line.

**mitochondrion** A structure within the cytoplasm of eukaryotic *cells that carry out aerobic respiration: it is the site of the *Krebs cycle and *electron transport chain, and therefore the cell's energy production. Mitochondria vary greatly in shape, size, and number but are typically oval or sausage-shaped and bounded by two membranes, the inner one being folded into finger-like projections (**cristae**); they contain their own DNA (*see* MITOCHONDRIAL DNA). They are most numerous in cells with a high level of metabolic activity.

**mitosis** The division of a cell to form two daughter cells each having a nucleus containing the same number and kind of chromosomes as the mother cell. The changes during divisions are clearly visible with a light microscope. Each chromosome divides lengthwise into two *chromatids, which separate and form the chromosomes of the resulting daughter nuclei. The process is divided into four stages, *prophase, *metaphase, *anaphase, and *telophase, which merge into each other (see illustration). Mitotic divisions ensure that all the cells of an individual are genetically identical to each other and to the original fertilized egg. *See also* CELL CYCLE.

**mitral valve** *See* BICUSPID VALVE.

**Mitscherlich's law (law of isomorphism)** Substances that have the same crystal structure have similar chemical formulae. The law can be used to determine the formula of an unknown compound if it is isomorphous with a compound of known formula. It is named after Eilhard Mitscherlich (1794–1863).

**mixed function oxidase** (mixed function oxygenase) *See* MONOOXYGENASE.

**mixed-valence compounds** (**intermediate-valence compounds**) Compounds in which energy levels of *f* electrons coexist with energy bands of *s* and/or *d* electrons. Such compounds have unusual thermal and magnetic properties.

**mixture** A system of two or more distinct chemical substances. **Homogeneous mixtures** are those in which the atoms or molecules are interspersed, as in a mixture of gases or in a solution. **Heterogeneous mixtures** have distinguishable phases, e.g. a mixture of iron filings and sulphur. In a mixture there is no redistribution of valence electrons, and the components retain their individual chemical properties. Unlike compounds, mixtures can be separated by physical means (distillation, crystallization, etc.).

**m.k.s. units** A *metric system of units devised by A. Giorgi (and sometimes known as **Giorgi units**) in 1901. It is based on the metre, kilogram, and second and grew from the earlier *c.g.s. units. The electrical unit chosen to augment these three basic units

was the ampere and the *permeability of space (magnetic constant) was taken as $10^{-7}$ H m$^{-1}$. To simplify electromagnetic calculations the magnetic constant was later changed to $4\pi \times 10^{-7}$ H m$^{-1}$ to give the **rationalized MKSA system**. This system, with some modifications, formed the basis of *SI units, now used in most scientific work.

**m.m.f.** *See* MAGNETOMOTIVE FORCE.

**mmHg** A unit of pressure equal to that exerted under standard gravity by a height of one millimetre of mercury, or 133.322 pascals.

**mobility** (of an ion) Symbol *u*. The terminal speed of an ion in an electric field divided by the field strength.

**mobility edge** *See* LOCALIZATION.

**mode** The number that occurs most frequently in a set of numbers. For example, the mode (modal value) of {7, 6, 2, 1, 2, 1, 2, 4} is 2. If a continuous random variable has probability density function f(*x*), the mode is the value of *x* for which f(*x*) is a maximum. If such a variable has a frequency curve that is

Prophase

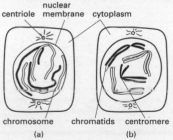

centriole   nuclear membrane   cytoplasm

chromosome   chromatids   centromere
(a)                    (b)

Metaphase            Anaphase            Telophase
pole of spindle      equator of spindle

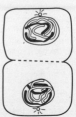

pole of spindle      spindle fibres

**Mitosis.** The stages of mitosis in a cell containing two pairs of homologous chromosomes.

approximately symmetrical and has only one mode, then

(mean − mode) = 3(mean − median).

**model** A simplified description of a physical system intended to capture the essential aspects of the system in a sufficiently simple form to enable the mathematics to be solved. In practice some models require *approximation techniques to be used, rather than being exactly soluble. When exact solutions are available they can be used to examine the validity of the approximation techniques.

**modem** (derived from modulator/demodulator) A device that can convert a digital signal (consisting of a stream of *bits) into an analogue (smoothly varying) signal, and vice versa. Modems are therefore required to link digital devices, such as computers, over an analogue telephone line.

**moderator** A substance that slows down free neutrons in a *nuclear reactor, making them more likely to cause fissions of atoms of uranium–235 and less likely to be absorbed by atoms of uranium–238. Moderators are light elements, such as deuterium (in heavy water), graphite, and beryllium, to which neutrons can impart some of their kinetic energy on collision without being captured. Neutrons that have had their energies reduced in this way (to about 0.025 eV, equivalent to a speed of 2200 m s$^{-1}$) are said to have been **thermalized** or to have become **thermal neutrons**.

**modern synthesis** *See* NEO-DARWINISM.

**modified Newtonian dynamics (MOND)** A modification of the Newtonian theory of gravity that has been used as an alternative to the idea of dark matter (*see* MISSING MASS) to explain why the motion of stars in galaxies is not in accord with the expectations of standard Newtonian theory. It is very difficult to provide a theorical justification for MOND by deriving it from a more general theory of gravity than Newtonian gravity. It is generally, but not universally, thought that MOND is less good at describing observations on the motions of stars in galaxies than are theories involving the existence of dark matter.

**modifier gene** A gene that influences the expression of another gene. For example, one gene controls whether eye colour is blue or brown but other (modifier) genes can also influence the colour by affecting the amount or distribution of pigment in the iris.

**modulation** The process of changing an electrical signal. In radio transmission, it is the process of superimposing the characteristics of a periodic signal onto a *carrier wave so that the information contained in the signal can be transmitted by the wave. The simplest form of modulation is **amplitude modulation (AM)**, in which the amplitude of the carrier is increased or diminished as the signal amplitude increases and diminishes. The modulated wave is composed of the carrier wave plus upper and lower sidebands. In **single-sideband modulation (SSB)** the carrier and one of the sidebands of an amplitude-modulated waveform are suppressed. This saves on bandwidth occupancy and signal power. In **frequency modulation (FM)**, the frequency of the carrier is increased or diminished as the signal amplitude increases and diminishes but the carrier amplitude remains constant. In **phase modulation**, the relative phase of the carrier is varied in accordance with the signal amplitude. (See illustrations.) Both frequency modulation and phase modulation are forms of **angle modulation**.

In **pulse modulation** the information is

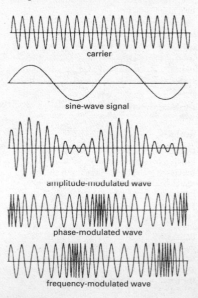

carrier

sine-wave signal

amplitude-modulated wave

phase-modulated wave

frequency-modulated wave

**Modulation.**

transmitted by controlling the amplitude, duration, position, or presence of a series of pulses. Morse code is a simple form of a pulse modulation.

**modulus** *See* ABSOLUTE VALUE.

**modulus of elasticity** *See* ELASTIC MODULUS.

**Moho (Mohorovičić discontinuity)** A discontinuity within the *earth that marks the junction between the crust and the underlying mantle. Below the discontinuity earthquake seismic waves undergo a sudden increase in velocity, a feature that was first observed in 1909 by the Croatian geophysicist Andrija Mohorovičić (1857–1936), after whom the discontinuity was named. The Moho lies at a depth of about 10–12 km below the oceans and about 33–35 km below the continents.

**Mohs' scale** A hardness scale in which a series of ten minerals are arranged in order, each mineral listed being scratched by and therefore softer than those below it. The minerals are: (1) talc; (2) gypsum; (3) calcite; (4) fluorite; (5) apatite; (6) orthoclase; (7) quartz; (8) topaz; (9) corundum; (10) diamond. As a rough guide a mineral with a value up to 2.5 on this scale can be scratched by a fingernail, up to 4 can be scratched by a coin, and up to 6 by a knife. The scale was devised by Friedrich Mohs (1773–1839).

**moissanite** A clear form of silicon carbide used as an inexpensive substitute for diamond. It is named after the French chemist Henri Moissan (1852–1907), who discovered it in 1893.

**molal concentration** *See* CONCENTRATION.

**molality** *See* CONCENTRATION.

**molar** 1. (in physics and chemistry) Denoting that an extensive physical property is being expressed per *amount of substance, usually per mole. For example, the molar heat capacity of a compound is the heat capacity of that compound per unit amount of substance; in SI units it would be expressed in $J K^{-1} mol^{-1}$. **2.** (in chemistry) Having a concentration of one mole per $dm^3$. **3.** (in anatomy) A broad ridged tooth in the adult dentition of mammals (*see* PERMANENT TEETH), found at the back of the jaws behind the premolars. There are two or more molars on each side of both jaws; their surfaces are raised into ridges or *cusps for grinding food

during chewing. In man the third (and most posterior) molar does not appear until young adulthood: these molars are known as **wisdom teeth**.

**molar conductivity** Symbol $\Lambda$. The conductivity of that volume of an electrolyte that contains one mole of solution between electrodes placed one metre apart.

**molar heat capacity** *See* HEAT CAPACITY.

**molarity** *See* CONCENTRATION.

**molar latent heat** *See* LATENT HEAT.

**molar volume (molecular volume)** The volume occupied by a substance per unit amount of substance.

**mole** Symbol mol. The SI unit of *amount of substance. It is equal to the amount of substance that contains as many elementary units as there are atoms in 0.012 kg of carbon–12. The elementary units may be atoms, molecules, ions, radicals, electrons, etc., and must be specified. 1 mole of a compound has a mass equal to its *relative molecular mass expressed in grams.

**molecular beam** A beam of atoms, ions, or molecules at low pressure, in which all the particles are travelling in the same direction and there are few collisions between them. They are formed by allowing a gas or vapour to pass through an aperture into an enclosure, which acts as a collimator by containing several additional apertures and vacuum pumps to remove any particles that do not pass through the apertures. Molecular beams are used in studies of surfaces and chemical reactions and in spectroscopy.

**molecular biology** The study of the structure and function of large molecules associated with living organisms, in particular proteins and the nucleic acids *DNA and *RNA. **Molecular genetics** is a specialized branch, concerned with the analysis of genes (*see* DNA SEQUENCING).

**(((⊕))) SEE WEB LINKS**
• Webpage from the Stanford Encyclopedia of Philosophy summarizing the history, concepts and experimental approaches of molecular biology

**molecular chaperone** Any of a group of proteins in living cells that assist newly synthesized or denatured proteins to fold to their functional three-dimensional structures. The chaperones bind to the protein and prevent improper interactions within

m

the polypeptide chain, so that it assumes the correct folded orientation. This process may require energy in the form of ATP. There are several families of chaperones, including five classes of **heat-shock proteins**. These are manufactured in response to raised temperature or other forms of stress, and are presumed to help protect the cell from damage by refolding proteins that have been partially denatured by heat.

**molecular clock** The concept that during evolution the number of substitutions in the nucleotides of nucleic acids (DNA or RNA), and hence in the proteins encoded by the nucleic acids, is proportional to time. Hence, by comparing the DNA or proteins of species that diverged a known length of time ago (e.g. determined from fossil evidence), it is possible to calculate the average substitution rate, thereby calibrating the 'molecular clock'. Comparative studies of different proteins in various groups of organisms tend to show that the average number of amino-acid substitutions per site per year is typically around $10^{-9}$. These results indicate a fairly constant rate of molecular evolution in comparable sequences of macromolecules in different organisms.

**molecular distillation** Distillation in high vacuum (about 0.1 pascal) with the condensing surface so close to the surface of the evaporating liquid that the molecules of the liquid travel to the condensing surface without collisions. This technique enables very much lower temperatures to be used than are used with distillation at atmospheric pressure and therefore heat-sensitive substances can be distilled. Oxidation of the distillate is also eliminated as there is no oxygen present.

**molecular flow (Knudsen flow)** The flow of a gas through a pipe in which the mean free path of gas molecules is large compared to the dimensions of the pipe. This occurs at low pressures; because most collisions are with the walls of the pipe rather than other gas molecules, the flow characteristics depend on the relative molecular mass of the gas rather than its viscosity. The effect was studied by M. H. C. Knudsen (1871–1949).

**molecular formula** *See* FORMULA.

**molecular imprinting** *See* GENE IMPRINTING.

**molecularity** The number of molecules involved in forming the activated complex in a step of a chemical reaction. Reactions are said to be **unimolecular**, **bimolecular**, or **trimolecular** according to whether 1, 2, or 3 molecules are involved.

**molecular knot (knotane)** A type of compound in which one or more chains of atoms in the molecule are looped in the configuration of a knot. The molecule may have only one closed chain forming the knot. If the knot is a trefoil knot the compound is chiral. Alternatively, molecular knots may have two or more separate loops tied together in a knot. In such cases there is no formal chemical bonding between the rings and they are held by *mechanical bonding. Molecular knots can be produced synthetically and also occur naturally in certain proteins.

**molecular marker** Any site (locus) in the genome of an organism at which the DNA base sequence varies among the different individuals of a population. Such markers generally have no apparent effect on the phenotype of the individual, but they can be determined by biochemical analysis of the DNA and are used for a variety of purposes, including chromosome mapping, DNA profiling, and genetic screening. Genetic tools, such as restriction, enzymes and the polymerase chain reaction plus the growing abundance of DNA sequence data, coupled with automated high-throughput assays, have revealed several classes of molecular markers, including *restriction fragment length polymorphisms (RFLPs), minisatellite and microsatellite DNA (*see* REPETITIVE DNA), and *single nucleotide polymorphisms (SNPs).

**molecular orbital** *See* ORBITAL.

**molecular sieve** Porous crystalline substances, especially aluminosilicates (*see* ZEOLITE), that can be dehydrated with little change in crystal structure. As they form regularly spaced cavities, they provide a high surface area for the adsorption of smaller molecules.

The general formula of these substances is $M_nO.Al_2O_3.xSiO_2.yH_2O$, where $M$ is a metal ion and $n$ is twice the reciprocal of its valency. Molecular sieves are used as drying agents and in the separation and purification of fluids. They can also be loaded with chemical substances, which remain separated from any reaction that is taking place around them, until they are released by heating or by displacement with a more strongly adsorbed substance. They can thus be used as cation exchange mediums and as catalysts and catalyst supports. They are also used as the stationary phase in certain types of *chromatography (**molecular-sieve chromatography**).

## molecular systematics (biochemical taxonomy)

The use of amino-acid or nucleotide-sequence data in determining the evolutionary relationships of different organisms. Essentially it involves comparing the sequences of functionally homologous molecules from each organism being studied, and determining the number of differences between them. The greater the number of differences, the more distantly related the organisms are likely to be. Moreover, since the number of nucleotide substitutions, and hence substitutions of corresponding amino acids, is generally proportional to time, some indication of the time scale involved can be obtained (*see* MOLECULAR CLOCK). This information has proved particularly useful where there are gaps in the fossil record and can be combined with other evidence from morphology, physiology, and embryology to produce more accurate phylogenetic trees. In microbiology molecular systematics has transformed bacterial phylogeny, in particular establishing the view that there are two quite distinct lineages of *bacteria, the archaebacteria and eubacteria. There has been an equally radical reassessment of the classification of eukaryotes, which on current molecular evidence form an unrooted phylogenetic tree of eight or nine assemblages.

## molecular volume

*See* MOLAR VOLUME.

## molecular weight

*See* RELATIVE MOLECULAR MASS.

## molecule

One of the fundamental units forming a chemical compound; the smallest part of a chemical compound that can take part in a chemical reaction. In most covalent compounds, molecules consist of groups of atoms held together by covalent or coordinate bonds. Covalent substances that form *macromolecular crystals have no discrete molecules (in a sense, the whole crystal is a molecule). Similarly, ionic compounds do not have single molecules, being collections of oppositely charged ions.

## mole fraction

Symbol $X$. A measure of the amount of a component in a mixture. The mole fraction of component A is given by $X_A$ = $n_A/N$, where $n_A$ is the amount of substance of A (for a given entity) and $N$ is the total amount of substance of the mixture (for the same entity).

## Molisch's test

*See* ALPHA-NAPHTHOL TEST.

## Mollusca

A phylum of soft-bodied invertebrates characterized by an unsegmented body differentiated into a **head**, a ventral muscular **foot** used in locomotion, and a dorsal **visceral hump** covered by a fold of skin – the *mantle – which secretes a protective shell in many species. Respiration is by means of gills or a lunglike organ and the feeding organ is a *radula. Molluscs occur in marine, freshwater, and terrestrial habitats and there are six classes, including the *Gastropoda (snails, slugs, limpets, etc.), *Bivalvia (e.g. mussels, oysters), and *Cephalopoda (squids and octopuses).

**(())) SEE WEB LINKS**
• Website of the Malacological Society of London, dedicated to the advancement of knowledge and research on molluscs

## molybdenum

Symbol Mo. A silvery hard metallic *transition element; a.n. 42; r.a.m. 95.94; r.d. 10.22; m.p. 2617°C; b.p. 4612°C. It is found in molybdenite ($MoS_2$), the metal being extracted by roasting to give the oxide, followed by reduction with hydrogen. The element is used in alloy steels. Molybdenum(IV) sulphide ($MoS_2$) is used as a lubricant. Chemically, it is unreactive, being unaffected by most acids. It oxidizes at high temperatures and can be dissolved in molten alkali to give a range of molybdates and polymolybdates. Molybdenum was discovered in 1778 by Karl Scheele.

**(())) SEE WEB LINKS**
• Information from the WebElements site

## moment of a force

A measure of the turning effect produced by a force about an axis. The magnitude of the moment is the product of the force and the perpendicular distance from the axis to the line of action of the force. An object will be in rotational equilibrium if the algebraic sum of all the

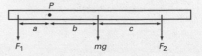

**Moment of a force**. For equilibrium $mgb + F_2(b + c) = F_1a$, where $mg$ is the weight of the beam acting through its centre of mass.

moments of the forces on it about any axis is zero. See illustration.

**moment of inertia** Symbol $I$. The moment of inertia of a massive body about an axis is the sum of all the products formed by multiplying the magnitude of each element of mass ($\delta m$) by the square of its distance ($r$) from the line, i.e. $I_m = \Sigma r^2 \delta m$. It is the analogue in rotational dynamics of mass in linear dynamics. The basic equation is $T = I\alpha$, where $T$ is the torque causing angular acceleration $\alpha$ about the specified axis.

**momentum** The **linear momentum** ($p$) of a body is the product of its mass ($m$) and its velocity ($v$), i.e. $p = mv$. *See also* ANGULAR MOMENTUM.

**monatomic molecule** A 'molecule' consisting of only one atom (e.g. Ar or He), distinguished from diatomic and polyatomic molecules.

**MOND** *See* MODIFIED NEWTONIAN DYNAMICS.

**Mond process** A method of obtaining pure nickel by heating the impure metal in a stream of carbon monoxide at 50–60°C. Volatile nickel carbonyl ($Ni(CO)_4$) is formed, and this can be decomposed at higher temperatures (180°C) to give pure nickel. The method was invented by the German–British chemist Ludwig Mond (1839–1909).

**Monel metal** An alloy of nickel (60–70%), copper (25–35%), and small quantities of iron, manganese, silicon, and carbon. It is used to make acid-resisting equipment in the chemical industry.

**monoamine oxidase (MAO)** An enzyme that breaks down monoamines (e.g. *adrenaline and *noradrenaline) in the body by oxidation. Drugs that inhibit this enzyme are used to treat forms of depression.

**monobasic acid** An *acid that has only one acidic hydrogen atom in its molecules. Hydrochloric ($HCl$) and nitric ($HNO_3$) acids are common examples.

**monochasium** *See* CYMOSE INFLORESCENCE.

**monochromatic radiation** Electromagnetic radiation, especially visible radiation, of only one frequency or wavelength. Completely monochromatic radiation cannot be produced, but *lasers produce radiation within a very narrow frequency band. *Compare* POLYCHROMATIC RADIATION.

**monochromator** A device that provides

*monochromatic radiation from a polychromatic source. In the case of visible radiation, for example, a prism can be used together with slits to select a small range of wavelengths.

**monoclinic** *See* CRYSTAL SYSTEM.

**monoclonal antibody** A specific *antibody produced by one of numerous identical cells derived from a single parent cell. (The population of these cells comprises a *clone and each cell is said to be **monoclonal**.) The parent cell is obtained by the fusion of a normal antibody-producing cell (a lymphocyte) with a cell derived from a malignant tumour of *lymphoid tissue of a mouse. This hybrid cell then multiplies rapidly and yields large amounts of antibody. Monoclonal antibodies are used to identify a particular antigen within a mixture and can therefore be used for identifying blood groups; they also enable the production of highly specific, and therefore effective, *vaccines.

**Monocotyledoneae** A class of flowering plants (*see* ANTHOPHYTA), distinguished by having one seed leaf (*cotyledon) within the seed. The monocotyledons generally have parallel leaf veins, scattered vascular bundles within the stems, and flower parts in threes or multiples of three. Monocotyledon species include some crop plants (e.g. cereals, onions, fodder grasses), ornamentals (e.g. tulips, orchids, lilies), and a very limited number of trees (e.g. the palms). *Compare* DICOTYLEDONEAE; EUDICOT.

**monoculture** *See* AGRICULTURE.

**monocyte** The largest form of white blood cell (*leucocyte) in vertebrates. Monocytes have a kidney-shaped nucleus, are actively phagocytic, and are the precursors of macrophages.

**monoecious** Describing plant species that have separate male and female flowers on the same plant. Examples of monoecious plants are maize and birch. *Compare* DIOECIOUS.

**monohybrid cross** A genetic cross between parents that differ in the alleles they possess for one particular gene, one parent having two dominant alleles and the other two recessives. All the offspring (called **monohybrids**) have one dominant and one recessive allele for that gene (i.e. they are hybrid at that one locus). Crossing between

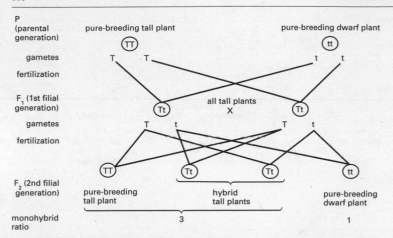

P
(parental generation)          pure-breeding tall plant                    pure-breeding dwarf plant
                                      TT                                           tt

gametes                        T        T                                      t        t

fertilization

$F_1$ (1st filial generation)              Tt        all tall plants        Tt
                                                          X

gametes                        T        t                              T        t

fertilization

$F_2$ (2nd filial generation)      TT          Tt              Tt          tt

                               pure-breeding        hybrid            pure-breeding
                               tall plant        tall plants          dwarf plant

monohybrid ratio                              3                              1

**Monohybrid cross.** The inheritance of stem lengths in garden peas.

these offspring yields a characteristic 3:1 (monohybrid) ratio in the following generation of dominant:recessive phenotypes (see illustration). *Compare* DIHYBRID CROSS.

**monohydrate** A crystalline compound having one mole of water per mole of compound.

**monomer** A molecule (or compound) that joins with others in forming a dimer, trimer, or polymer.

**mononuclear phagocyte system (reticuloendothelial system)** *See* MACROPHAGE.

**monooxygenase (mixed function oxidase; mixed function oxygenase)** Any of a large group of enzymes that perform oxidation–reduction reactions in which one atom of the oxygen molecule is incorporated into the chemical donor substrate and the other oxygen atom is combined with hydrogen ions to form water. Such enzymes are commonly involved in the detoxification of harmful substances by tissues, being located in the vertebrate liver. *Cytochrome oxidase is an example.

**monophyletic** Denoting any group of organisms that are assumed to have originated from the same ancestor, i.e. any family, class, etc., of a natural classification. Sometimes the term has a more limited meaning and designates only those groups that include *all* the descendants of a common ancestor. In this restricted sense the birds are considered monophyletic because they are the sole descendants of a group of arboreal Triassic reptiles but the modern reptiles are not, because their common amphibian ancestor also gave rise to the birds and mammals. Such groups as the reptiles are described as **paraphyletic**. *Compare* POLYPHYLETIC.

**monophyodont** Describing a type of dentition that consists of a single set of teeth that last for the entire lifespan of an animal. *Compare* DIPHYODONT; POLYPHYODONT.

**monopodium** The primary axis of growth in such plants as pine trees. It consists of a single main stem that continues to grow from the tip and gives rise to lateral branches. *Compare* SYMPODIUM.

**monosaccharide (simple sugar)** A carbohydrate that cannot be split into smaller units by the action of dilute acids. Monosaccharides are classified according to the number of carbon atoms they possess: **trioses** have three carbon atoms; **tetroses**, four; **pentoses**, five; **hexoses**, six; etc. Each of these is further divided into **aldoses** and **ketoses**, depending on whether the molecule contains an aldehyde group (–CHO) or a ketone group (–CO–). For example glucose, having six carbon atoms and an aldehyde group, is an **aldohexose** whereas fructose is a **ketohexose**. These aldehyde and ketone groups confer reducing properties on monosaccharides: they can be oxidized to yield sugar acids. They also react with phosphoric acid to produce phosphate esters (e.g. in *ATP), which are important in cell metabo-

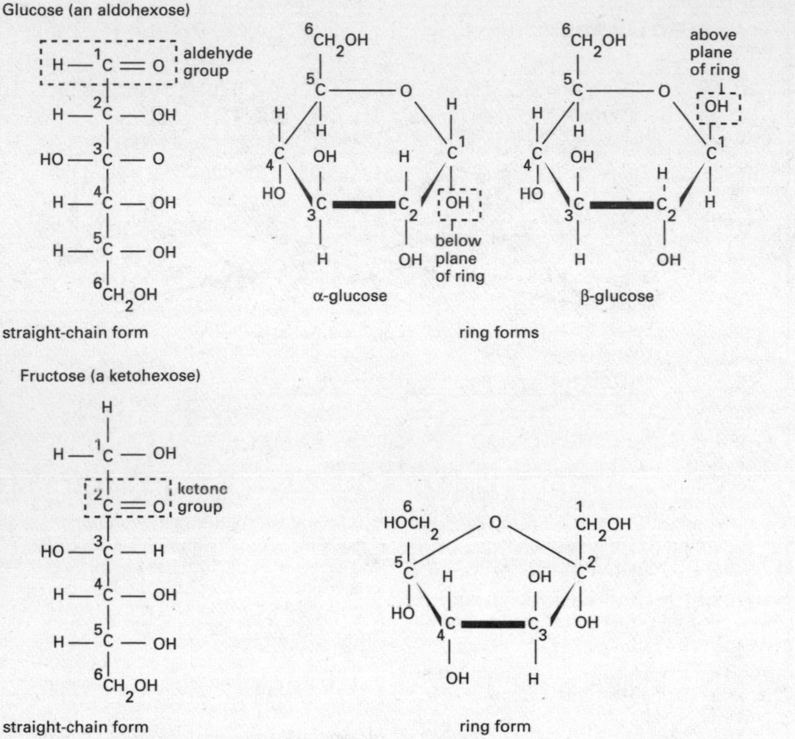

Glucose (an aldohexose)

straight-chain form     ring forms

α-glucose     β-glucose

Fructose (a ketohexose)

straight-chain form     ring form

**Monosaccharides.**

lism. Monosaccharides can exist as either straight-chain or ring-shaped molecules (see illustration). They also exhibit *optical activity, giving rise to both dextrorotatory and laevorotatory forms.

**monosodium glutamate (MSG)** A white solid, $C_8H_8NNaO_4.H_2O$, used extensively as a flavour enhancer, especially in convenience foods. It is a salt of glutamic acid (an *amino acid), from which it is prepared. It can cause an allergic reaction in some susceptible people who consume it.

**monotremes** *See* PROTOTHERIA.

**monotropy** *See* ALLOTROPY.

**monovalent (univalent)** Having a valency of one.

**monozygotic twins** *See* IDENTICAL TWINS.

**Monte Carlo simulation** A method that involves random sampling for the mathematical simulation of physical systems. Monte Carlo calculations are applied to problems that can be formulated in terms of probability and are usually carried out by computer. Such calculations have been performed for nuclei, atoms, molecules, solids, liquids, and nuclear reactors. The technique is named after the gambling centre in Monaco, renowned for its casino.

**moon** The earth's only natural satellite, and the only body outside the earth to have been reached by human beings (1969). It is located at a mean distance of 384 400 km and orbits around the common centre of mass that it shares with our planet once every 27.322 days. It has a diameter of 3476 km. Locked by the earth into *synchronous rotation, the moon always presents the same face to an earth-based observer. The moon is

## MOON EXPLORATION

| | |
|---|---|
| c.150 BC | Greek astronomer Hipparcus of Nicaea (c. 190–c. 120 BC) determines the distance to the moon. |
| 1610 | Galileo uses a telescope to observe the surface features of the moon. |
| 1647 | German astronomer Johannes Hevelius (1611–87) draws first map of the moon. |
| 1757 | French mathematician Alexis Clairaut (1713–65) calculates the mass of the moon. |
| 1840 | British-born US chemist William Draper (1811–82) takes photographs (daguerrotypes) of the moon. |
| 1959 | Soviet space probe Lunik I flies past the moon; Lunik II crashes on the moon; Lunik III bypasses the moon and returns pictures of the far side. |
| 1962 | US Ranger 4 space probe hits the moon. |
| 1964 | US Ranger 7 photographs the moon before crash-landing. |
| 1965 | US Ranger 8 returns TV pictures from the surface of the moon. |
| 1966 | US Lunar Orbiters 1 and 2 orbit the moon, returning photographs; US Surveyors 5 and 6 make soft landings. Soviet Luna 9 and 13 make soft landings; Luna 10 and 11 orbit the moon. |
| 1967 | US Surveyors 3, 5, and 6 make soft landings; Lunar Orbiters 3 and 4 photograph the surface. |
| 1968 | US Surveyor 7 lands on the moon. Soviet Zond 5 and 6 orbit the moon and return to earth. |
| 1969 | Astronauts from Apollo 11 and 12 land on the moon. |
| 1970 | Soviet Luna 16 lands on the moon and releases Lunakod I robot vehicle. |
| 1971 | Astronauts from Apollo 14 and 15 land on the moon. |
| 1972 | Astronauts from Apollo 16 and 17 land on the moon. |
| 1973 | Soviet Luna 21 lands on the moon and releases Lunakod II robot vehicle. |
| 1990 | Japanese scientists launch satellite Hagoromo (from another space probe) to orbit the moon. |
| 1994 | US orbiting probe Clementine maps the moon. |
| 1998 | US Lunar Prospector orbits the moon. |

visible because it reflects light from the sun. As it orbits the earth, increasing and then decreasing portions of its surface are illuminated producing apparent changes of shape called *phases of the moon. On average, the moon completes one **lunation** (cycle of phases) every 29.530 59 days. Its gravitational influence on the earth plays the main role in the production of *tides.

The moon has no atmosphere or surface water. Its surface temperature varies between 80 K (night minimum) and 400 K (noon at the equator). The moon's surface consists of craters, mountains, and dark lava plains called **maria** (from the Latin *mare*, 'sea' and is scarred by long channels called **rilles**. In addition, almost all the surface is covered with a loose layer of soil and dust known as the **regolith**. See Chronology: Moon Exploration.

**((⊕)) SEE WEB LINKS**

- Lunar Navigator Full moon atlas
- Exploring the moon: a site devoted to the US Mercury, Gemini and Apollo missions

**Moore's law** The statement that the number of transistors that can be placed on an integrated circuit doubles every two years.

This statement was first made by Gordon Moore (1929– ), the president of Intel, in 1965 and it has remained valid for the first fifty years of the existence of integrated circuits. However, there are various reasons for thinking that this will come to an end in the future. For example, as circuits become smaller, the quantum effects associated with individual atoms and electrons become more significant.

**moraine** A deposit of rock debris scoured from a valley floor by a glacier and left behind when the ice melts. The pieces of rock vary in size from boulders to fine sand, typically resembling *boulder clay. There are various types: a ground moraine forms underneath the glacier; a lateral moraine forms at the sides; a medial moraine occurs where two lateral moraines, from different glaciers, meet; and a terminal moraine is deposited at the lower end or toe of the glacier, usually indicating the farthest point reached by the ice. *See also* ESKER; KAME.

**mordant** A substance used in certain dyeing processes. Mordants are often inorganic oxides or salts, which are absorbed on the fabric. The dyestuff then forms a coloured complex with the mordant, the colour depending on the mordant used as well as the dyestuff. *See also* LAKE.

**Morgan, Thomas Hunt** (1866–1945) US geneticist, who held professorships at Bryn Mawr College (1891–1904), Columbia University (1904–28), and the California Institute of Technology (1928–45). He is best known for his discovery of *crossing over during *meiosis, so modifying Mendel's law of *independent assortment. For this work Morgan was awarded the 1933 Nobel Prize for physiology or medicine.

**morph** Any of the distinct common forms found in a population displaying *polymorphism.

**morphine** An opiate that is the main active constituent of opium. It is used medically in the relief of severe pain, and can be acetylated to produce heroin. Morphine can be detected by the *Marquis test.

**morphogenesis** The development, through growth and differentiation, of form and structure in an organism.

**morphology** The study of the form and structure of organisms, especially their external form. *Compare* ANATOMY.

**mortality** *See* DEATH RATE.

**mosaic evolution** The evolution of different parts of an organism at different rates. For example, many aspects of the human phenotype have evolved relatively slowly or not at all since the hominids diverged from their primate ancestors, one notable exception being the nervous system, which has given humans their overwhelming selective advantage. This high degree of evolutionary independence among different aspects of the phenotype permits flexibility; for example, when a population is faced with new selection pressures in a changing environment, only the most crucial components need to evolve, not the entire phenotype.

**mosaic gold** *See* TIN(IV) SULPHIDE.

**Moseley's law** The frequencies of the lines in the *X-ray spectra of the elements are related to the atomic numbers of the elements. If the square roots of the frequencies of corresponding lines of a set of elements are plotted against the atomic numbers a straight line is obtained. The law was discovered by Henry Moseley (1887–1915).

(((●))) SEE WEB LINKS

• Original paper on the law

**moss agate** *See* AGATE.

**Mössbauer effect** An effect occurring when certain nuclides decay with emission of gamma radiation. For an isolated nucleus, the gamma radiation would usually have a spread of energies because the energy of the process is partitioned between the gamma-ray photon and the recoil energy of the nucleus. In 1957 Rudolph Mössbauer (1929– ) found that in certain solids, in which the emitting nucleus is held by strong forces in the lattice, the recoil energy is taken up by the whole lattice. Since this may typically contain $10^{10}$–$10^{20}$ atoms, the recoil energy is negligible and the energy of the emitted photon is sharply defined in a very narrow energy spread.

The effect is exploited in **Mössbauer spectroscopy** in which a gamma-ray source is mounted on a moving platform and a similar sample is mounted nearby. A detector measures gamma rays scattered by the sample. The source is moved slowly towards the sample at a varying speed, so as to continuously change the frequency of the emitted gamma radiation by the Doppler effect. A sharp decrease in the signal from the detector at a particular speed (i.e. frequency) indicates resonance absorption in the sample nuclei. The effect is used to investigate nuclear energy levels. In chemistry, Mössbauer spectroscopy can also give information about the bonding and structure of compounds because **chemical shifts** in the resonance energy are produced by the presence of surrounding atoms.

**mosses** *See* BRYOPHYTA.

**MOT** *See* LASER COOLING.

**moths** *See* LEPIDOPTERA.

**motion** A change in the position of a body or system with respect to time, as measured by a particular observer in a particular *frame of reference. Only relative motion can be measured; absolute motion is meaningless. *See also* EQUATION OF MOTION; NEWTON'S LAWS OF MOTION.

**motivation** The internal conditions responsible for temporary reversible changes in the responsiveness of an animal to external stimulation. Thus an animal that has been deprived of food will accept less palatable food than one that has not been deprived: the difference is attributed to a change in feeding motivation. Changes in responsiveness due to maturation, *learning, or injury are not usually readily reversible and are therefore not considered to be due to changes in motivation. Early attempts to describe motivation in terms of a number of separate 'drives' (e.g. food drive, sex drive) have not found general favour, partly because 'drives' interact with one another; for example, water deprivation often affects an animal's willingness to feed.

**motor** Any device for converting chemical energy or electrical energy into mechanical energy. *See* ELECTRIC MOTOR; INTERNAL-COMBUSTION ENGINE; LINEAR MOTOR.

**motor generator** An electric motor mechanically coupled to an electric generator. The motor is driven by a supply of specified voltage, frequency, or number of phases and the generator provides an output in which one or more of these parameters is different to suit a particular purpose.

**motor neuron** A *neuron that transmits nerve impulses from the central nervous system to an effector organ (such as a muscle or gland) and thereby initiates a physiological response (e.g. muscle contraction).

**Mott insulator** A substance that is an insulator because of electron correlation and in which the highest occupied energy band is not necessarily full. Certain transition metal oxides in which there are narrow bands are Mott insulators. The concept was put forward and developed by the British physicist Sir Nevill Francis Mott (1905–1996), starting in 1949.

**moulting** **1.** The seasonal loss of hair, fur, or feathers that occurs in mammals and birds. **2.** The periodic loss of the integument of arthropods and reptiles. *See* ECDYSIS.

**mouse** A simple device that is connected to a personal computer by cable and can be moved by hand over a flat surface, its movements being sensed by the rotation of a ball in its base. These movements are communicated to the computer and cause corresponding movements of the cursor on the screen; the cursor indicates the active position on the screen. One or more buttons on the mouse can be pressed to initiate an action, for example to indicate a desired cursor position for typing or to select an item from a menu of options.

**mouth** The opening of the *alimentary canal, which in most animals is used for the *ingestion of food. It leads to the *buccal cavity (mouth cavity).

**mouth cavity** *See* BUCCAL CAVITY.

**mouthparts** Modified paired appendages on the head segments of arthropods, used for feeding. A typical insect has a *labium (lower lip), one pair each of *mandibles and *maxillae, and a *labrum (upper lip), although in many the mouthparts are modified to form piercing stylets or a sucking proboscis. Crustaceans, centipedes, and millipedes have one pair of mandibles and two pairs of maxillae used for cutting and holding the food. Crustaceans also have several pairs of **maxillipedes**.

**moving-coil instrument** A measuring instrument in which current or voltage is determined by the couple on a small coil pivoted between the poles of a magnet with curved poles, giving a radial magnetic field. When a current flows through the coil it turns against a return spring. If the angle through which it turns is $\alpha$, the current $I$ is given by $I = k\alpha / BAN$, where $B$ is the magnetic flux density, $A$ is the area of the coil, and $N$ is its number of turns; $k$ is a constant depending on the strength of the return spring. The instrument is suitable for measuring d.c. but can be converted for a.c. by means of a rectifier network. It is usually made as a *galvanometer and converted to an ammeter or voltmeter by means of a *shunt or a *multiplier.

**moving-iron instrument** A measuring instrument in which current or voltage is determined by the force of attraction on a bar of soft iron pivoted within the magnetic field of a fixed coil or by the repulsion between the poles induced in two soft iron rods within the coil. As the deflection caused by the passage of a current through the coil does not depend on the direction of the current, moving-iron instruments can be used with either d.c. or a.c. without a rectifier. They are, however, less sensitive than *moving-coil instruments.

**moving-iron microphone** See MICROPHONE.

**mRNA** See RNA.

**MRSA** See STAPHYLOCOCCUS.

**MSG** See MONOSODIUM GLUTAMATE.

**MSH** See MELANOCYTE-STIMULATING HORMONE.

**MSSM** See MINIMAL SUPERSYMMETRIC STANDARD MODEL.

**mtDNA** See MITOCHONDRIAL DNA.

**mucilage** Any of a large group of complex polysaccharides frequently present in the cell walls of aquatic plants and in the seed coats of certain other species. Mucilages are hard when dry and slimy when wet. Like *gums they probably have a general protective function or serve to anchor the plant.

**mucopolysaccharide** See GLYCOSAMINOGLYCAN.

**mucous membrane (mucosa)** A layer of tissue comprising an epithelium supported on connective tissue. Within the epithelium are special **goblet cells**, which secrete *mucus onto the surface, and the epithelium often bears cilia. Mucous membranes line body cavities communicating with the exterior, including the alimentary and respiratory tracts. Compare SEROUS MEMBRANE.

**mucus** The slimy substance secreted by goblet cells onto the surface of a *mucous membrane to protect and lubricate it and to trap bacteria, dust particles, etc. Mucus consists of water, various **mucoproteins** (glycoproteins), cells, and salts. The mucoprotein

chains, in which the carbohydrate component is a large polysaccharide, are joined by disulphide bridges to form long **mucin** strands, which readily form gels with water.

**muffle furnace** An insulated furnace, usually electrically heated, used for producing controlled high temperatures. In the laboratory, muffle furnaces are used for drying solids, sintering, and studying high-temperature reactions. They typically operate in the range 100–1200°C.

**multiaccess system** A system allowing several users of a computer, at different terminals, to make apparently simultaneous use of the computer without being aware of each other.

**multicellular** Describing tissues, organs, or organisms that are composed of a number of cells. Compare UNICELLULAR.

**multicentre bond** A bond formed between three, and sometimes more, atoms that contains only a single pair of electrons. The structure of *boranes can be explained by considering them to be *electron-deficient compounds containing multi-centre bonds.

**multidecker sandwich** See SANDWICH COMPOUND.

**multidimensional spectroscopy** A type of spectroscopy used to study complex systems and large molecules such as proteins, for which traditional spectroscopy does not give clear results. The basic technique is to excite all possible resonances in the system and analyse the radiation produced using Fourier analysis. Originally used with *nuclear magnetic resonance, versions of multidimensional spectroscopy have been developed for optical and infrared regions of the spectrum.

**multifactorial inheritance** See QUANTITATIVE INHERITANCE.

**multimedia** A combination of various media, such as text, sound, and moving and still images, now often held on *CD-ROM. The user can make use of the different media in an integrated way.

**multimeter** An electrical measuring instrument designed to measure potential differences or currents over a number of ranges. It also usually has an internal dry cell enabling resistances to be measured. Most multimeters are moving-coil instruments with a switch to enable series resistors or parallel resistors to be incorporated into the circuit.

**multiple alleles** Three or more alternative forms of a gene (*alleles) that can occupy the same *locus. However, only two of the alleles can be present in a single organism. For example, the *ABO system of blood groups is controlled by three alleles, only two of which are present in an individual.

**multiple bond** A bond between two atoms that contains more than one pair of electrons. Such bonds primarily involve sigma bonding with secondary contribution from pi bonding (or, sometimes, delta bonding). *See* ORBITAL.

**multiple proportions** *See* CHEMICAL COMBINATION.

**multiplet** **1.** A spectral line formed by more than two (*see* DOUBLET) closely spaced lines. **2.** A group of *elementary particles that are identical in all respects except that of electric charge.

**multiplication factor** Symbol $k$. The ratio of the average number of neutrons produced in a *nuclear reactor per unit time to the number of neutrons lost by absorption or leakage in the same time. If $k = 1$, the reactor is said to be **critical**. If $k > 1$ it is **supercritical** and if $k < 1$ it is **subcritical**. *See also* CRITICAL REACTION.

**multiplicity** A quantity used in atomic *spectra to describe the energy levels of many-electron atoms characterized by *Russell–Saunders coupling given by $2S + 1$, where $S$ is the total electron *spin quantum number. The multiplicity of an energy level is indicated by a left superscript to the value of $L$, where $L$ is the resultant electron *orbital angular momentum of the individual electron orbital angular momenta $l$.

**multiplier** A fixed resistance used with a voltmeter, usually a *moving-coil instrument, to vary its range. Many voltmeters are provided with a series of multipliers from which the appropriate value can be selected. If the original instrument requires $i$ amperes for full-scale deflection and the resistance of the moving coil is $r$ ohms, the value $R$ of the resistance of the multiplier required to give a full-scale deflection when a voltage $V$ is applied across the terminals is given by $R = V/i - r$.

**multivibrator** An electronic *oscillator consisting of two active devices, usually transistors, interconnected in an electrical network. The purpose of the device is to generate a continuous square wave with which to store information in binary form in a logic circuit. This is achieved by applying a portion of the output voltage or current of each active device to the input of the other with the appropriate magnitude and polarity, so that the devices are conducting alternately for controllable periods.

**mu-mesic atom** *See* MUONIC ATOM.

**Mumetal** The original trade name for a ferromagnetic alloy, containing 78% nickel, 17% iron, and 5% copper, that had a high *permeability and a low *coercive force. More modern versions also contain chromium and molybdenum. These alloys are used in some transformer cores and for shielding various devices from external magnetic fields.

**Muntz metal** A form of *brass containing 60% copper, 39% zinc, and small amounts of lead and iron. Stronger than alpha-brass, it is used for hot forgings, brazing rods, and large nuts and bolts. It is named after G. F. Muntz (1794–1857).

**muon** *See* LEPTON; ELEMENTARY PARTICLES.

**muonic atom (mu-mesic atom)** An *exotic atom in which one of the electrons has been replaced by a muon. Since the mass of a muon is 207 times that of an electron, the average radius of the orbit of a muon is much smaller than that of a corresponding electron. Muonic atoms provide tests for quantum electrodynamics. They are also used in research into muon-catalysed fusion (*see* NUCLEAR FUSION).

**Musci** *See* BRYOPHYTA.

**muscle** A tissue consisting of sheets or bundles of cells (**muscle fibres**) that are capable of contracting, so producing movement or tension in the body. There are three types of muscle. *Voluntary muscle produces voluntary movement (e.g. at joints); *involuntary muscle mainly effects the movements of hollow organs (e.g. intestine and bladder); and *cardiac muscle occurs only in the heart.

**(((⊕))) SEE WEB LINKS**

- Interactive tutorial identifying the principal muscles of the human body

**muscle spindle** A receptor in vertebrate muscle that is sensitive to stretch (*see* PROPRIOCEPTOR). Muscle spindles run parallel to normal muscle fibres; each consists of a capsule containing small striated muscle fibres (**intrafusal fibres**). Muscle spindles are responsible for the adjustment of muscle tone and play an important role in the subconscious maintenance of posture and movement. *See also* STRETCH REFLEX.

**muscovite (white mica; potash mica)** A mineral consisting of potassium aluminosilicate, $K_2Al_4(Si_6Al_2)O_{20}(OH,F)_4$; one of the most important members of the *mica group of minerals. It is chemically complex and has a sheetlike crystal structure (*see* INTERCALATION COMPOUND). It is usually silvery-grey in colour, sometimes tinted with green, brown, or pink. Muscovite is a common constituent of certain granites and pegmatites. It is also common in metamorphic and sedimentary rocks. It is widely used in industry, for example in the manufacture of electrical equipment and as a filler in roofing materials, wallpapers, and paint.

**mustard gas** *See* SULPHUR MUSTARD; NITROGEN MUSTARDS.

**mutagen** An agent that causes an increase in the number of mutants (*see* MUTATION) in a population. Mutagens operate either by causing changes in the DNA of the *genes, so interfering with the coding system, or by causing chromosome damage. Various chemicals (e.g. *colchicine) and forms of radiation (e.g. X-rays) have been identified as mutagens.

**mutant** (Denoting) a gene or an organism that has undergone a heritable change, especially one with visible effects (i.e. the change in *genotype is associated with a change in *phenotype). *See* MUTATION.

**mutarotation** Change of optical activity with time as a result of spontaneous chemical reaction.

**mutation** A sudden random change in the genetic material of a cell that potentially can cause it and all cells derived from it to differ in appearance or behaviour from the normal type. An organism affected by a mutation (especially one with visible effects) is described as a **mutant**. **Somatic mutations** affect the nonreproductive cells and are therefore restricted to the tissues of a single organism but **germ-line mutations**, which occur in the reproductive cells or their precursors, may be transmitted to the organism's descendants and cause abnormal development.

Mutations occur naturally at a low rate but this may be increased by radiation and by some chemicals (*see* MUTAGEN). Most are *point mutations, which consist of invisible changes in the DNA of the chromosomes, but some (the *chromosome mutations) affect the appearance or the number of the chromosomes. An example of a chromosome mutation is that giving rise to *Down's syndrome.

Mutations that alter phenotypes are generally harmful, but a very small proportion may increase an organism's *fitness; these spread through the population over successive generations by natural selection. Mutation is therefore essential for evolution, being the ultimate source of genetic variation.

(((●))) **SEE WEB LINKS**

• First of two topics on mutation from the online resource DNA From The Beginning

**mutual inductance** *See* INDUCTANCE.

**mutualism** An interaction between two species in which both species benefit. (The term *symbiosis is often used synonymously with mutualism.) A well-known example of mutualism is the association between termites and the specialized protozoans that inhabit their guts. The protozoans, unlike the termites, are able to digest the cellulose of the wood that the termites eat and release sugars that the termites absorb. The termites benefit by being able to use wood as a foodstuff, while the protozoans are supplied with food and a suitable environment. *See also* MYCORRHIZA.

**mycelium** A network of *hyphae that forms the body of a fungus. It consists of feeding hyphae together with reproductive hyphae, which produce *sporangia and *gametangia.

**mycology** The scientific study of *fungi.

**Mycophycophyta** *See* LICHENS.

**mycoplasmas** A group of bacteria that lack a rigid cell wall and are among the smallest living cells (diameter 0.1 μm–0.8 μm). They are either saprotrophic or parasitic and are found on animal mucous and synovial membranes, in insects, and in plants (in which they seem to inhabit sieve tubes). They cause a variety of diseases, including pleuropneumonia in cattle – hence they were formerly also known as **pleuropneumonia-like organisms** (**PPLO**). Due to their small size and flexible cell wall they can pass through a 0.2-μm-diameter filter and they represent a major contaminant of biotechnological products, such as monoclonal antibodies and vaccines, and of other cell cultures, in which they may exist symbiotically with the cells. Eight genera have been described (including *Mycoplasma*) with over 120 species.

**mycorrhiza** The mutually beneficial association (*see* MUTUALISM) formed between fungi and the roots of plants. This is a very common form of mutualism; the absorption of mineral ions by the plant roots is enhanced by the presence of the fungus, which benefits by obtaining soluble organic nutrients from the root cells. **Ectotrophic mycorrhizas** form a network of hyphae around the root and grow into the air spaces between the cells of the root. The hyphae of **endotrophic mycorrhizas** enter the cortex cells of the roots.

**Mycota** In older classification systems, a kingdom comprising the *fungi.

**myelin** A *phospholipid produced by the *Schwann cells of the nervous system. Myelin forms an insulating layer around the nerve fibres (*see* MYELIN SHEATH).

**myelin sheath (medullary sheath)** The layer of fatty material that surrounds and electrically insulates the axons of most vertebrate and some invertebrate neurons. The myelin sheath enables a more rapid transmission of nerve impulses (at speeds up to 120 m s⁻¹). It consists of layers of membrane derived from *Schwann cells. The sheath is interrupted at intervals along the axon by **nodes of Ranvier**; myelinated sections of axon are called **internodes**.

**myeloid tissue** Tissue within red *bone marrow that produces the blood cells. It is found around the blood vessels and contains various cells that are precursors of the blood cells. *See* HAEMOPOIETIC TISSUE.

**myeloma** *See* CANCER.

**myofibril** *See* VOLUNTARY MUSCLE.

**myogenic** Originating in or produced by muscle cells. The contractions of *cardiac muscle fibres are described as myogenic, since they are produced spontaneously, without requiring stimulation from nerve cells (*see* PACEMAKER).

**myoglobin** A globular protein occurring widely in muscle tissue as an oxygen carrier. It comprises a single polypeptide chain and a *haem group, which reversibly binds a molecule of oxygen. This is only relinquished at relatively low external oxygen concentrations, e.g. during strenuous exercise when muscle oxygen demand outpaces supply from the blood. Myoglobin thus acts as an emergency oxygen store.

**myopia** Short-sightedness. It results from the lens of the eye refracting the parallel rays of light entering it to a focus in front of the retina generally because of an abnormally long eyeball. The condition is corrected by using diverging spectacle lenses to move the image back to the retina.

**myosin** A contractile protein that interacts with *actin to bring about contraction of muscle or cell movement. The type of myosin molecule found in muscle fibres consists of a tail, by which it aggregates with other myosin molecules to form so-called 'thick filaments'; and a globular head, which has sites for the attachment of actin and ATP molecules. *See* SARCOMERE.

**myotatic reflex** *See* STRETCH REFLEX.

**myotome** One of a series of segmented muscle blocks found in fishes and lancelets. Myotomes are arranged in pairs on either side of the body that work antagonistically (*see* ANTAGONISM) against the backbone (or notochord), providing a means of locomotion by causing the tail to sweep from side to side.

**Myriapoda** In some classifications, a subphylum of arthropods that comprises the classes *Chilopoda (centipedes), *Diplopoda (millipedes), Pauropoda (pauropods), and Symphyla (symphilids). In other classifications the Myriapoda is a class containing only the centipedes and millipedes.

**myristic acid** *See* TETRADECANOIC ACID.

**Myxomycota** *See* SLIME MOULDS.

**myxovirus** One of a group of RNA-containing viruses associated with various diseases of humans and other vertebrates. **Orthomyxoviruses** produce diseases of the respiratory tract, e.g. influenza; **paramyxoviruses** include the causal agents of mumps, measles, and fowl pest.

**NAD (nicotinamide adenine dinucleotide)**
A *coenzyme, derived from the B vitamin
*nicotinic acid, that participates in many
biological dehydrogenation reactions. NAD
is characteristically loosely bound to the enzymes concerned. It normally carries a positive charge and can accept one hydrogen
atom and two electrons to become the reduced form, NADH. NADH is generated during the oxidation of food, especially by the
reactions of the *Krebs cycle. It then gives up
its two electrons (and single proton) to the
*electron transport chain, thereby reverting
to NAD+ and generating three molecules of
ATP per molecule of NADH.

NADP (**nicotinamide adenine dinucleotide phosphate**) differs from NAD only
in possessing an additional phosphate
group. It functions in the same way as NAD
although anabolic reactions (*see* ANABOLISM)
generally use NADPH (reduced NADP) as a
hydrogen donor rather than NADH. Enzymes tend to be specific for either NAD or
NADP as coenzyme.

**nadir** The point diametrically opposite the
*zenith on the *celestial sphere.

**NAND circuit** *See* LOGIC CIRCUITS.

**nano-** Symbol n. A prefix used in the metric
system to denote $10^{-9}$. For example, $10^{-9}$
metre = 1 nanometre (nm).

**nanoarray** *See* MICROARRAY.

**nanotechnology** The development and
use of devices that have a size of only a few
nanometres. Research has been carried out
into very small components, which depend
on electronic effects and may involve movement of a countable number of electrons in
their action. Such devices would act much
faster than larger components. Considerable
interest has been shown in the production of
structures on a molecular level by suitable
sequences of chemical reactions. It is also
possible to manipulate individual atoms on
surfaces using *atomic force microscopy.

**nanotube** *See* BUCKMINSTERFULLERENE.

**napalm** A substance used in incendiary
bombs and flame throwers, made by forming a gel of petrol with aluminium soaps
(aluminium salts of long-chain carboxylic
acids, such as palmitic acid).

**naphtha** Any liquid hydrocarbon or mixture obtained by the fractional distillation of
petroleum. It is generally applied to higher
*alkane fractions with nine or ten carbon
atoms. Naphtha is used as a solvent and as a
starting material for *cracking into more
volatile products, such as petrol.

**naphthalene** A white volatile solid, $C_{10}H_8$
(see formula); r.d. 1.025; m.p. 80.55°C; b.p.
218°C. Naphthalene is an aromatic hydrocarbon with an odour of mothballs and is
obtained from crude oil. It is a raw material
for making certain synthetic resins.

**Naphthalene.**

**naphthols** Two phenols derived from
naphthalene with the formula $C_{10}H_7OH$,
differing in the position of the –OH group.
The most important is naphthalen-2-ol (β-
naphthol), with the –OH in the 2-position.
It is a white solid (r.d. 1.28; m.p. 123–124°C;
b.p. 295°C) used in rubber as an antioxidant.
Naphthalen-2-ol will couple with diazonium
salts at the 1-position to form red *azo compounds, a reaction used in testing for the
presence of primary amines (by making the
diazonium salt and adding naphthalen-2-ol).

**naphthyl group** The group $C_{10}H_7$– obtained by removing a hydrogen atom from
naphthalene. There are two forms depending on whether the hydrogen is removed
from the 1- or 2-position.

**Napier, John** (1550–1617) Scottish mathematician, who is best known for devising
*logarithms, announced in 1614. His tables,

which used the base e, were later modified by Henry Briggs (1561–1630) to base 10.

**Napierian logarithm** *See* LOGARITHM.

**narcotic** Any drug that induces stupor and relieves pain, especially morphine and other *opiates. Such narcotics are addictive and cause dependence, and their medical use is strictly controlled.

**nares (nostrils)** The paired openings of the *nasal cavity in vertebrates. All vertebrates have **external nares**, which open to the exterior; in some species these are situated on a *nose. **Internal nares** (or **choanae**) are present only in air-breathing vertebrates (including lungfish) and open into the mouth cavity. In mammals they open posteriorly, beyond the secondary *palate.

**nasal cavity** The cavity in the head of a vertebrate that is lined by a membrane rich in sensitive olfactory receptors (*see* OLFACTION). It is connected to the exterior by external nostrils and (in air-breathing vertebrates) to the respiratory system by internal *nares.

**nascent hydrogen** A reactive form of hydrogen generated *in situ* in the reaction mixture (e.g. by the action of acid on zinc). Nascent hydrogen can reduce elements and compounds that do not readily react with 'normal' hydrogen. It was once thought that the hydrogen was present as atoms, but this is not the case. Probably hydrogen molecules are formed in an excited state and react before they revert to the ground state.

**nastic movements** Movements of plant organs in response to external stimuli that are independent of the direction of the stimuli. Examples are the opening of crocus and tulip flowers in response to a rise in temperature (**thermonasty**), the opening of evening primrose flowers at night (**photonasty**), and the folding up and drooping of leaves of the sensitive plant (*Mimosa pudica*) when lightly touched (**haptonasty**). *Compare* TROPISM. *See also* NYCTINASTY.

**natality** *See* BIRTH RATE.

**natron** A mineral form of hydrated sodium carbonate, $Na_2CO_3.H_2O$.

**Natta process** *See* ZIEGLER PROCESS.

**natural abundance** *See* ABUNDANCE.

**natural frequency** 1. The frequency of the free oscillation of a system. 2. The frequency at which resonance occurs in an electrical circuit.

**natural gas** A naturally occurring mixture of gaseous hydrocarbons that is found in porous sedimentary rocks in the earth's crust, usually in association with *petroleum deposits. It consists chiefly of methane (about 85%), ethane (up to about 10%), propane (about 3%), and butane. Carbon dioxide, nitrogen, oxygen, hydrogen sulphide, and sometimes helium may also be present. Natural gas, like petroleum, originates in the decomposition of organic matter. It is widely used as a fuel and also to produce carbon black and some organic chemicals. Natural gas occurs on every continent, the major reserves occurring in the USA, Russia, Kazakhstan, Turkmenistan, Ukraine, Algeria, Canada, and the Middle East. *See also* LIQUEFIED PETROLEUM GAS.

**natural group** A group of organisms of any taxonomic rank that are believed to be descended from a common ancestor (*see* MONOPHYLETIC). In an ideal natural classification all taxa should be natural groups. *See also* CLADISTICS.

**natural history** 1. The study of living organisms in their natural habitats. 2. The study of all natural phenomena.

**natural killer cell (NK cell)** A lymphoid cell that recognizes and destroys tissue cells infected with pathogenic organisms. NK cells are important as an early line of defence against infection, and they play a significant role in combating infections by, for example, herpesviruses. They become activated in response to interferons or the release of cytokines by macrophages, bind to target cells, and release cytotoxic granules onto the surface of their target. The toxic effector molecules penetrate the target's plasma membrane and induce programmed cell death (*see* APOPTOSIS).

**natural logarithm** *See* LOGARITHM.

**natural selection** The process that, according to *Darwinism, brings about the evolution of new species of animals and plants. Darwin noted that the size of any population tends to remain constant despite the fact that more offspring are produced than are needed to maintain it. He also saw that variations existed between individuals of the population and concluded that disease, competition, and other forces acting on the population eliminated those individuals less well adapted to their environment. The survivors would pass on any heritable advantageous characteristics (i.e. characteristics with survival value) to their offspring and in time the composition of the population would change in adaptation to a changing environment. Over a long period of time this process could give rise to organisms so different from the original population that new species are formed. *See also* ADAPTIVE RADIATION. *Compare* PUNCTUATED EQUILIBRIUM.

**natural units**  A system of units, used principally in particle physics, in which all quantities that have dimensions involving length, mass, and time are given dimensions of a power of energy (usually expressed in electronvolts). This is equivalent to setting the rationalized *Planck constant and the speed of light both equal to unity. *See also* GAUSSIAN UNITS; GEOMETRIZED UNITS; HEAVISIDE–LORENTZ UNITS; PLANCK UNITS.

**nature and nurture**  The combined effects of inherited factors (nature) and environmental factors (nurture) on the development of an organism. The genetic potential of an organism will only be realized under appropriate environmental conditions. *See also* PHENOTYPE.

**nautical mile**  A measure of distance used at sea. In the UK it is defined as 6080 feet but the international definition is 1852 metres. 1 international nautical mile is therefore equivalent to 1.15078 land (statute) miles.

**navigation**  (in biology) The complex process that enables animals to travel along a particular course in order to reach a specific destination. Navigation is an important aspect of behaviour in many animals, particularly those, such as birds, fish, and some insects, that undergo *migrations. Landmarks, such as coastlines and mountain ranges, are important reference points for navigation but many animals can navigate successfully without the aid of these, by using the sun, stars, magnetic fields, odours, and polarized light. For example, birds use the sun and stars as landmarks and are sensitive to the earth's magnetic fields, while salmon can identify the unique odour of their home river. *See also* DANCE OF THE BEES.

**NBS**  *See* N-BROMOSUCCINIMIDE.

**Neanderthal man**  A form of fossil human that lived in Europe and western Asia between about 200 000 and 28 000 years ago, when the climate was much colder than today. Neanderthals were thought to be a subspecies of *Homo sapiens* but are now generally regarded as a distinct species, *H. neanderthalensis*. The fossil remains indicate that Neanderthals were fairly short, strongly built, and had low brows but that the brain size was the same as, or larger than, modern humans'. They were nomadic cave dwellers who buried their dead. Neanderthals became extinct abruptly; they may have been exterminated by incoming modern humans, with their more advanced stone tool technology. The name is derived from the site in the Neander valley, Germany, where fossils were found in 1856.

**near point**  The nearest point at which the human eye can focus an object. As the lens becomes harder with age, the extent to which accommodation can bring a near object into focus decreases. Therefore with advancing age the near point recedes – a condition known as *presbyopia.

**nebula**  A cloud of interstellar dust and/or gas. There are two classes: **bright nebulae**, which are visible as misty patches of light; and **dark nebulae**, which can only be seen if they happen to lie in front of something bright, such as a bright nebula. **Emission nebulae** are bright nebulae in which the gas atoms have been ionized by ultraviolet radiation from nearby stars and light is emitted as these ions interact with free electrons in the gas. **Reflection nebulae** are dark dust clouds that become visible by reflecting light from nearby stars. Some nebulae are the cradles of star formation. *See also* GIANT MOLECULAR CLOUD.

**nebular hypothesis**  The theory that the solar system evolved from a vast rotating cloud of interstellar gas and dust particles (*see* NEBULA), which collapsed under its own gravitational attraction to form a flattened spinning disc called the **solar nebula** from which the sun and the other bodies evolved. The nebular hypothesis was advanced in 1796 by the French astronomer Pierre-Simon de Laplace (1749–1827). Laplace's model formed the starting point for the currently accepted **solar nebular disc model**. *See* SOLAR SYSTEM (Feature).

**nectar**  A sugary liquid produced in plants by **nectaries**, regions of secretory cells on the receptacle or other parts of a flower. It attracts pollinating insects or other animals.

**neon**

**Néel temperature** The temperature above which an antiferromagnetic substance becomes paramagnetic (*see* MAGNETISM). The *susceptibility increases with temperature, reaching a maximum at the Néel temperature, after which it abruptly declines. The phenomenon was discovered around 1930 by the French physicist Louis Néel (1904–2000).

**Ne'eman, Yuval** *See* GELL-MANN, MURRAY.

**negative binomial distribution** *See* PASCAL'S DISTRIBUTION.

**negative charge** *See* CHARGE.

**negative feedback** *See* FEEDBACK.

**nekton** *Pelagic organisms that actively swim through the water. Examples are fish, jellyfish, turtles, and whales. *Compare* PLANKTON.

**nematic crystal** *See* LIQUID CRYSTAL.

**nematoblast** *See* THREAD CELL.

**Nematoda** A phylum of invertebrates comprising the roundworms. They are characterized by a smooth narrow cylindrical unsegmented body tapered at both ends. They shed their tough outer cuticle four times during life to allow growth. The microscopic free-living forms are found in all parts of the world, where they play an important role in the destruction and recycling of organic matter. The many parasitic nematodes are much larger; they include the filaria (*Wuchereria*) and Guinea worm (*Dracunculus*), which cause serious diseases in humans.

**(()) SEE WEB LINKS**
• Website of the Society of Nematologists. Education Committee link accesses many images of nematodes

**neo-Darwinism (modern synthesis)** The current theory of the process of *evolution, formulated between about 1920 and 1950, that combines evidence from classical genetics with the Darwinian theory of evolution by *natural selection (*see* DARWINISM). It makes use of modern knowledge of genes and chromosomes to explain the source of the genetic variation upon which selection works. This aspect was unexplained by traditional Darwinism.

**neodymium** Symbol Nd. A soft silvery metallic element belonging to the *lan-thanoids; a.n. 60; r.a.m. 144.24; r.d. 7.004 (20°); m.p. 1021°C; b.p. 3068°C. It occurs in bastnasite and monazite, from which it is recovered by an ion-exchange process. There are seven naturally occurring isotopes, all of which are stable, except neodymium–144, which is slightly radioactive (half-life $10^{10}$–$10^{15}$ years). Seven artificial radioisotopes have been produced. The metal is used to colour glass violet-purple and to make it dichroic. It is also used in misch metal (18% neodymium) and in neodymium–iron–boron alloys for magnets. It was discovered by Carl von Welsbach (1856–1929) in 1885.

**(()) SEE WEB LINKS**
• Information from the WebElements site

**Neogene** The current geological period of the *Cenozoic era, consisting of the *Miocene, *Pliocene, *Pleistocene, and *Holocene epochs. The Neogene began 23 million years ago, following the *Palaeogene period, and corresponds to the latter third of the *Tertiary period and the entire *Quaternary period, both of which are no longer officially recognized divisions.

**neo-Lamarckism** Any of the comparatively modern theories of evolution based on Lamarck's theory of the inheritance of acquired characteristics (*see* LAMARCKISM). These include the unfounded dogma of *Lysenkoism and the controversial experiments on the inheritance of acquired immunological tolerance in mice.

**Neolithic** The New Stone Age: a stage of human cultural and technological evolution that began in the Middle East in approximately 8500 BC. It was characterized by farming using wild and domesticated crops and herding of livestock and (in the mid- and late Neolithic) by the making of pottery. Grinding and polishing of stone tools was also practised. The Bronze Age superseded Neolithic cultures from around 3500 BC in the Middle East.

**neon** Symbol Ne. A colourless gaseous element belonging to group 18 (formerly group 0) of the periodic table (the *noble gases); a.n. 10; r.a.m. 20.179; d. 0.9 g dm$^{-3}$; m.p. –248.67°C; b.p. –246.05°C. Neon occurs in air (0.0018% by volume) and is obtained by fractional distillation of liquid air. It is used in discharge tubes and neon lamps, in which it has a characteristic red glow. It forms hardly any compounds (neon fluorides have been reported). The element was discovered in 1898 by Sir William Ramsey and M. W. Travers.

**⊕ SEE WEB LINKS**
• Information from the WebElements site

**neon lamp** A small lamp consisting of a pair of electrodes, treated to emit electrons freely, sealed in a glass bulb containing neon gas at low pressure. When a minimum voltage of between 60 and 90 volts is applied across the electrodes, the kinetic energy of the electrons is sufficient to ionize the neon atoms around the cathode, causing the emission of a reddish light. With d.c. the glow is restricted to the cathode; with a.c. both electrodes act alternately as cathodes and a glow appears to emanate from both electrodes. The device consumes a very low power and is widely used as an indicator light showing that a circuit is live.

**neoplasm (tumour)** Any new abnormal growth of cells, forming either a harmless (benign) tumour or a malignant one (*see* CANCER).

**neoprene** A synthetic rubber made by polymerizing the compound 2-chlorobuta-1,2-diene. Neoprene is often used in place of natural rubber in applications requiring resistance to chemical attack.

**neoteny** The retention of the juvenile body form, or particular features of it, in a mature animal. For example, the axolotl, a salamander, retains the gills of the larva in the adult. Neoteny is thought to have been an important mechanism in the evolution of certain groups, such as humans, who are believed to have developed from juvenile forms of apes.

**neper** A unit used to express a ratio of powers, currents, etc., used especially in telecommunications to denote the attenuation of an amplitude $A_1$ to an amplitude $A_2$ as $N$ nepers, where

$$N = \log_e (A_2/A_1).$$

1 neper = 8.686 decibels. The unit is named after John Napier.

**nephelometry** The measurement of the turbidity (cloudiness) of a liquid. Nephelometers generally have a light source (often a laser) and a detector to measure the amount of light scattered by suspended particles. They are used in a number of fields including water quality control and blood analysis (for protein content).

**nephrite** *See* JADE.

**nephron** The excretory unit of the vertebrate *kidney (see illustration). Many constituents of the blood are filtered from the glomerulus into the Bowman's capsule at one end of the nephron. The *glomerular filtrate passes along the length of the nephron and some of its water, plus some salts, glucose, and amino acids, are reabsorbed into the surrounding blood capillaries (*see* PROXIMAL CONVOLUTED TUBULE; LOOP OF HENLE; DISTAL CONVOLUTED TUBULE). More water is reabsorbed in the *collecting duct, and the resulting concentrated solution of nitrogen-containing waste matter (*urea) plus inorganic salts drains from the collecting ducts of the nephrons and is discharged as urine into the ureter.

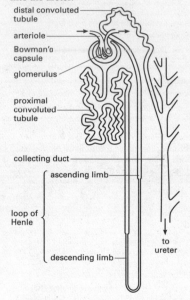

distal convoluted tubule
arteriole
Bowman's capsule
glomerulus
proximal convoluted tubule
collecting duct
ascending limb
loop of Henle
descending limb
to ureter

**A nephron.**

**Neptune** The fourth largest *planet in the *solar system and the eighth in order from the *sun. Its mean distance from the sun is $4452.94 \times 10^6$ km, its mass is $1.0243 \times 10^{26}$ kg (more than 17 times that of earth), and its mean diameter is 49 528 km; it has a *sidereal period of 164.79 years. Its period of axial rotation is 16h 6.6m. Neptune was discovered in 1846 by Johann Galle (1812–1910) on the basis of predictions made by John Couch Adams and Urbain Leverrier (1811–77), who had separately observed perturbations in the orbit of *Uranus. Neptune is a *gas giant; traces of methane in its outer atmosphere absorb red and infrared light, giving it its blue colour. Inside, Neptune is composed of ice and rock. At the planet's cloud tops, the temperature is only 55 K. Neptune has a faint fragmentary ring system. **Triton**, the largest of the 13 Neptunian satellites, measures 2707 km in diameter.

**SEE WEB LINKS**
• NASA's introduction to the planet and its moons

**neptunium** Symbol Np. A radioactive metallic transuranic element belonging to the *actinoids; a.n. 93; r.a.m. 237.0482. The most stable isotope, neptunium–237, has a half-life of $2.2 \times 10^6$ years and is produced in small quantities as a by-product by nuclear reactors. Other isotopes have mass numbers 229–236 and 238–241. The only other relatively long-lived isotope is neptunium–236 (half-life $5 \times 10^3$ years). The element was first produced by Edwin McMillan (1907–91) and Philip Abelson (1913–2004) in 1940.

**SEE WEB LINKS**
• Information from the WebElements site

**neptunium series** See RADIOACTIVE SERIES.

**neritic zone** The region of the sea over the continental shelf, which is less than 200 metres deep (approximately the maximum depth for organisms carrying out photosynthesis). Compare OCEANIC ZONE.

**Nernst effect** An effect in which a temperature gradient along an electric conductor or semiconductor placed in a perpendicular magnetic field, causes a potential difference to develop in the third perpendicular direction between opposite edges of the conductor. This effect, an analogue of the *Hall effect, was discovered in 1886 by the German physicist Walter Nernst (1864–1941).

**Nernst heat theorem** A statement of the third law of *thermodynamics in a restricted form: if a chemical change takes place between pure crystalline solids at *absolute zero there is no change of entropy.

**nerve** A strand of tissue comprising many *nerve fibres plus supporting tissues (see GLIA), enclosed in a connective-tissue sheath. Nerves connect the central nervous system with the organs and tissues of the body. A nerve may carry only motor nerve fibres (**motor nerve**) or only sensory fibres (**sensory nerve**) or it may be mixed and carry both types (**mixed nerve**). Although the nerve fibres are in close proximity within the nerve, their physiological responses are independent of each other.

**nerve agents** A class of highly toxic compounds that act by affecting the action of acetylcholine, a neurotransmitter regulated by the enzyme acetylcholinesterase (AChE). The nerve agents, which are organophosphates, bind to AChE and block its action. Consequently there is nothing to stop the build-up of acetylcholine, and exposure to small amounts of nerve gas can kill in minutes. There are two series of agents. The **G-series** was produced by German scientists in the 1930s and 1940s (the G stands for German). The main ones are *tabun (GA), *sarin (GB), *soman (GD), and *cyclosarin (GF). A further series of nerve agents is the **V-series**. The most common of these is *VX, which was discovered at Porton Down in the UK in 1952. Nerve agents are classified as weapons of mass destruction by the UN.

**nerve cell** See NEURON.

**nerve cord** A large bundle of nerve fibres, running down the longitudinal axis of the body, that forms an important part of the *central nervous system. Most invertebrates have a pair of solid nerve cords, situated ventrally and bearing segmentally arranged *ganglia. All animals of the phylum *Chordata have a dorsal hollow nerve cord; in vertebrates this is the *spinal cord.

**nerve fibre** The *axon of a *neuron together with the tissues associated with it (such as a *myelin sheath). The length and diameter of nerve fibres are very variable, even within the same organism. See also GIANT FIBRE.

**nerve growth factor (NGF)** See NEUROTROPHIN.

**nerve impulse** See IMPULSE.

**nerve net** A network of nerve cells connected with each other by synapses or fusion. The nervous system of certain invertebrates (e.g. coelenterates and echinoderms) consists exclusively of a nerve net in the body wall.

**nervous system** The system of cells and tissues in multicellular animals by which information is conveyed between sensory cells and organs and effectors (such as muscles and glands). It consists of the *central nervous system (in vertebrates the *brain and *spinal cord; in invertebrates the *nerve cord and *ganglia) and the *peripheral nervous system. Its function is to receive, transmit, and interpret information and then to formulate appropriate responses for the effector organs. It also serves to coordinate responses that require more than one physiological process. Nervous tissue consists of *neurons, which convey the information in the form of *impulses, and supporting tissue.

**(⊕) SEE WEB LINKS**

- Basic interactive guide to the anatomy of the human nervous system; from Human Anatomy Online

**nesquehonite** A mineral form of *magnesium carbonate trihydrate, $MgCO_3.3H_2O$.

**Nessler's reagent** A solution of mercury(II) iodide ($HgI_2$) in potassium iodide and potassium hydroxide. It is used in testing for ammonia, with which it forms a brown coloration or precipitate.

**Net** See INTERNET.

**Neumann's law** The magnitude of an electromagnetically induced e.m.f. ($E$) is given by $E = -d\Phi/dt$, where $\Phi$ is the magnetic flux. This is a quantitative statement of *Faraday's second law of electromagnetic induction and is sometimes known as the Faraday–Neumann law.

**neural network** A network of processors designed to mimic the transmission of impulses in the human brain. Neural networks are either electronic constructions or, often, computer-simulated structures. Each processor ('neurone') multiplies its input signal by a weighting factor and the final output signal depends on these factors, which can be adjusted. Such networks can be 'taught' to recognize patterns in large amounts of data. They are used in research into artificial intelligence and have also been applied in predicting financial market trends.

**neural tube** A hollow tube of tissue in the early embryo of vertebrates that subsequently develops into the brain and spinal cord. It forms by folding of the ectodermal **neural plate**, and has a central canal running through it. Sometimes the folds of the neural plate fail to close properly, resulting in **neural tube defects** (such as spina bifida) in the fetus.

**neuroendocrine system** Any of the systems of dual control of certain activities in the body of some higher animals by nervous and hormonal stimulation. For example, the posterior *pituitary gland and the medulla of the *adrenal gland receive direct nervous stimulation to secrete their hormones, whereas the anterior pituitary gland is stimulated by *releasing hormones from the hypothalamus.

**neurohormone** Any hormone that is produced not by an endocrine gland but by a specialized nerve cell and is secreted from nerve endings into the bloodstream or directly to the tissue or organ whose growth or function it controls (see NEUROSECRETION). Examples of neurohormones are *noradrenaline, *antidiuretic hormone, and hormones associated with metamorphosis and moulting in insects (see ECDYSONE; JUVENILE HORMONE). *Compare* NEUROPEPTIDE.

**neuromuscular junction** The point where a muscle fibre comes into contact with a motor neuron carrying nerve impulses from the central nervous system. The impulses travel from the neuron to the muscle fibre by means of a neurotransmitter, in a similar way to the transmission of impulses across a *synapse between two neurons. The neurotransmitter is released from vesicles at the end of the motor neuron into a small gap (the cleft), where it diffuses to the *end plate of the muscle fibre and depolarizes the membrane. When depolarization has reached a certain threshold an action potential is triggered in the muscle fibre.

**neuron (neurone; nerve cell)** An elongated branched cell that is the fundamental unit of the *nervous system, being specialized for the conduction of *impulses. A neuron consists of a *cell body, containing the nucleus and *Nissl granules; *dendrites, which receive incoming impulses and pass them towards the cell body; and an *axon, which conducts impulses away from the cell body, sometimes over long distances. Impulses are passed from one neuron to the next via *synapses. *Sensory neurons transmit information from receptors to the central nervous system. *Motor neurons conduct information from the central nervous system to *effectors (e.g. muscles). See illustration.

**neuropeptide** Any of numerous peptides

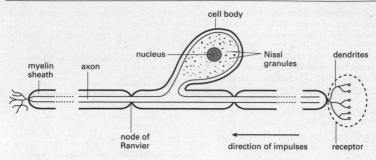

Sensory neuron

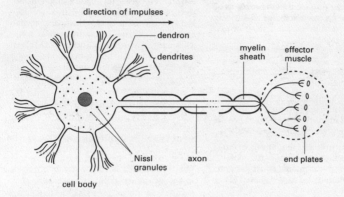

Motor neuron

**Neuron.**

that influence the activity of neurons. Examples include the hypothalamic *releasing hormones, *antidiuretic hormone, and the gastric peptides (e.g. *VIP) released from cells in the duodenal wall. Neuropeptides may act as neurotransmitters, as cotransmitters to modify the action of neurotransmitters, and/or as *neurohormones.

**neuropeptide Y (NPY)** A 36-amino acid peptide that has key roles in energy metabolism and in regulating the activity of the heart and blood vessels. It is a potent stimulant of appetite, being released from cells in the hypothalamus, where it activates NPY receptors, which are *G-protein-coupled receptors. It also regulates secretion of gonadotrophin-releasing hormone from the hypothalamus. Its release is triggered by the gut hormone *ghrelin and inhibited by the

hormone *leptin. NPY is widely distributed in the brain and sympathetic nervous system and commonly occurs in nerve cells that secrete the neurotransmitter noradrenaline. In general it modulates the effects of adrenaline and noradrenaline, for example by decreasing contraction of heart muscle and reducing blood flow through coronary vessels. It is also significant in the reduction of anxiety and control of blood pressure.

**neurosecretion** The secretion of *neurohormones by **neurosecretory cells**, which possess characteristics of both nerve cells and endocrine cells. They are found, for example, in the hypothalamus, where they receive nerve impulses from other parts of the brain but transmit these signals to the pituitary gland by neurohormones that are released into the blood.

**neurotransmitter** A chemical that mediates the transmission of a nerve impulse across a *synapse. Examples are *adrenaline and *noradrenaline (in adrenergic nerves) and *acetylcholine (in cholinergic nerves). The neurotransmitter is released at the synaptic knob at the tip of the axon into the synaptic cleft. It diffuses across to the opposite membrane (the postsynaptic membrane), where it initiates the propagation of a nerve impulse in the next neuron.

**neurotrophin (NT)** Any of several growth factors that promote the development, maintenance, and repair of neurons. Each consists of two identical polypeptides held together by disulphide bonds. The first to be discovered, **nerve growth factor** (NGF), is essential for the normal growth and survival of neurons of the sympathetic nervous system and also sensory neurons. It is produced by the neurons themselves and by astrocytes. Schwann cells, fibroblasts, and certain other cells.

**neuter** An organism that does not possess either male or female reproductive organs. Cultivated ornamental flowers that have neither pistils nor stamens are called neuters.

**neutral** Describing a compound or solution that is neither acidic nor basic. A neutral solution is one that contains equal numbers of both protonated and deprotonated forms of the solvent.

**neutralization** The process in which an acid reacts with a base to form a salt and water.

**neutrino** A *lepton (*see also* ELEMENTARY PARTICLES) that exists in three forms, one in association with the electron, one with the muon, and one with the tau particle. Each form has its own antiparticle. The neutrino, which was postulated in 1930 to account for the 'missing' energy in *beta decay, was identified tentatively in 1953 and, more definitely, in 1956. Neutrinos have no charge and travel at speeds very close to the speed of light. In some *grand unified theories they are predicted to have nonzero mass and there is now a large amount of indirect evidence for this, although the values of neutrino masses have not been determined.

**neutrino astronomy** A branch of astronomy that gives information about objects by studying the neutrinos they emit, using detectors based at earth observatories. Since neutrinos are very difficult to detect, the only bodies that have been studied in this way are the sun and the supernova SN1987A, which exploded in 1987. It is hoped that in the future it will be possible to investigate other aspects of neutrino astronomy, including the *cosmic neutrino background.

**(((∰))) SEE WEB LINKS**
- Official homepage of the Super-Kamiokande neutrino observatory

**neutron** A neutral hadron (*see* ELEMENTARY PARTICLES) that is stable in the atomic nucleus but decays into a proton, an electron, and an antineutrino with a mean life of 12 minutes outside the nucleus. Its rest mass (symbol $m_n$) is slightly greater than that of the proton, being $1.674\,9286(10) \times 10^{-27}$ kg. Neutrons occur in all atomic nuclei except normal hydrogen. The neutron was first reported in 1932 by James Chadwick (1891–1974).

**(((∰))) SEE WEB LINKS**
- Chadwick's 1932 letter to *Nature* on the discovery of the neutron
- A review of the background to Chadwick's discovery of the neutron, from the American Institute of Physics

**neutron activation analysis** *See* ACTIVATION ANALYSIS.

**neutron bomb** *See* NUCLEAR WEAPONS.

**neutron diffraction** The scattering of neutrons by atoms in solids, liquids, or gases. This process has given rise to a technique, analogous to *X-ray diffraction techniques, using a flux of thermal neutrons from a nuclear reactor to study solid-state structure and phenomena. Thermal neutrons have average kinetic energies of about 0.025 eV ($4 \times 10^{-21}$ J) giving them an equivalent wavelength of about 0.1 nanometre, which is suitable for the study of interatomic interference. There are two types of interaction in the scattering of neutrons by atoms: one is the interaction between the neutrons and the atomic nucleus, the other is the interaction between the *magnetic moments of the neutrons and the spin and orbital magnetic moments of the atoms. The latter interaction has provided valuable information on antiferromagnetic and ferrimagnetic materials (*see* MAGNETISM). Interaction with the atomic nucleus gives diffraction patterns that complement those from X-rays. X-rays, which interact with the extranuclear electrons, are not suitable for investigating light elements (e.g. hydrogen), whereas neutrons do give diffraction patterns from such atoms because they interact with nuclei.

**neutron drip** *See* NEUTRON STAR.

**neutron excess** *See* ISOTOPIC NUMBER.

**neutron interferometer** A type of interferometer that uses the interference caused by the wave nature of neutrons to investigate many phenomena described by quantum mechanics. Neutron interferometers use large single crystals of silicon. They have been used to demonstrate the *spinor nature of wavefunctions for fermions. Neutron interferometers have also been used to investigate the structure of matter.

**neutron number** Symbol N. The number of neutrons in an atomic nucleus of a particular nuclide. It is equal to the difference between the *nucleon number and the *atomic number.

**neutron star** A compact stellar object that is supported against collapse under self-gravity by the *degeneracy pressure of the neutrons of which it is primarily composed. Neutron stars are believed to be formed as the end products of the evolution of stars of mass greater than a few (4–10) solar masses. The core of the evolved star collapses and (assuming that its mass is greater than the *Chandrasekhar limit for a *white dwarf), at the very high densities involved (about $10^{14}$ kg m$^{-3}$), electrons react with protons in atomic nuclei to produce neutrons. The neutron-rich nuclei thus formed release free neutrons in a process known as **neutron drip**. The density increases to about $10^{17}$ kg m$^{-3}$, at which most of the electrons and protons have been converted to a *degenerate gas of neutrons and the atomic nuclei have lost their separate identities. If the mass of the core exceeds the Oppenheimer–Volkoff limit for a neutron star, then further collapse will occur, leading to the formation of a *black hole.

*Pulsars are believed to be rapidly rotating magnetized neutron stars and many X-ray sources are thought to be neutron stars in binary systems with another star, from which material is drawn into an accretion disc. This material, heated to a very high temperature, emits radiation in the X-ray region.

**neutron temperature** A concept used to express the energies of neutrons that are in thermal equilibrium with their surroundings, assuming that they behave like a monatomic gas. The neutron temperature T,

on the Kelvin scale, is given by $T = 2E/3k$, where $E$ is average neutron energy and $k$ the *Boltzmann constant.

**neutrophil** A type of white blood cell (*leucocyte) that has a lobed nucleus and granular cytoplasm (*see* GRANULOCYTE). Neutrophils engulf bacteria (*see* PHAGOCYTOSIS) and release various substances, such as *lysozyme and oxidizing agents.

**New General Catalogue (NGC)** A list of nonstellar objects, originally published (with 7840 entries) in 1888 by J. L. E. Dreyer. Such objects are referred to by their New General Catalogue numbers; e.g. the Orion nebula is NGC1976.

(⊕) **SEE WEB LINKS**
• The interactive NGC Catalogue online, hosted by the Students for the Exploration and Development of Space [SEDS]

**Newlands' law** *See* LAW OF OCTAVES.

**newton** Symbol N. The *SI unit of force, being the force required to give a mass of one kilogram an acceleration of 1 m s$^{-2}$. It is named after Sir Isaac Newton.

**Newton, Sir Isaac** (1642–1727) English mathematician and physicist, one of the world's greatest scientists. He went to Cambridge University in 1661 and stayed for nearly 40 years except for 1665–67, when he returned to his home at Woolsthorpe in Lincolnshire (because of the Plague), where some of his best work was done. In 1699 he was made Master of the Royal Mint. He was reluctant to publish his work and his great mathematical masterpiece, the *Principia*, did not appear until 1687. In it he introduced *calculus and formulated *Newton's laws of motion. In 1665 he derived *Newton's law of gravitation, and in optics he produced *Newton's formula for a lens and, in 1672, his theories about *light and the spectrum (*see also* NEWTON'S RINGS); these were summarized in his *Opticks* of 1704. Also in the late 1660s he constructed a reflecting *telescope. The SI unit of force is named after him.

**Newtonian fluid** A fluid in which the velocity gradient is directly proportional to the shear stress. If two flat plates of area $A$ are separated by a layer of fluid of thickness $d$ and move relative to each other at a velocity $v$, then the rate of shear is $v/d$ and the shear stress is $F/A$, where $F$ is the force applied to each (see illustration). For a Newtonian fluid $F/A = \mu v/d$, where $\mu$ is the constant of proportionality and is called the Newtonian *viscosity. Many liquids are Newtonian fluids over a wide range of temperatures and pressures. However, some are not; these are called **non-Newtonian fluids**. In such fluids there is a departure from the simple Newtonian relationships. For example, in some liquids the viscosity increases as the velocity gradient increases, i.e. the faster the liquid moves the more viscous it becomes. Such liquids are said to be **dilatant** and the phenomenon they exhibit is called **dilatancy**. It occurs in some pastes and suspensions. More common, however, is the opposite effect in which the viscosity depends not only on the velocity gradient but also on the time for which it has been applied. These liquids are said to exhibit **thixotropy**. The faster a **thixotropic liquid** moves the less viscous it becomes. This property is used in nondrip paints (which are more viscous on the brush than on the wall) and in lubricating oils (which become thinner when the parts they are lubricating start to move). Another example is the non-Newtonian flow of macromolecules in solution or in polymer melts. In this case the shearing force $F$ is not parallel to the shear planes and the linear relationship does not apply. In general, the many types of non-Newtonian fluid are somewhat complicated and no theory has been developed to accommodate them fully.

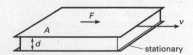

**Newtonian fluid.**

**Newtonian mechanics** The system of *mechanics that relies on *Newton's laws of motion. Newtonian mechanics is applicable to bodies moving at speeds relative to the observer that are small compared to the speed of light. Bodies moving at speeds comparable to the speed of light require an approach based on *relativistic mechanics, in which the mass of a body changes with its speed.

**Newtonian telescope** See TELESCOPE.

**Newton's formula** For a lens, the distances $p$ and $q$ between two conjugate points and their respective foci is given by $pq = f^2$, where $f$ is the focal length of the lens.

**Newton's law of cooling** The rate at which a body loses heat is proportional to the difference in temperature between the body and the surroundings. It is an empirical law that is only true for substantial temperature differences if the heat loss is by forced convection or conduction.

**Newton's law of gravitation** There is a force of attraction between any two massive particles in the universe. For any two point masses $m_1$ and $m_2$, separated by a distance $d$, the force of attraction $F$ is given by $F = m_1 m_2 G / d^2$, where $G$ is the *gravitational constant. Real bodies having spherical symmetry act as point masses positioned at their centres of mass.

**Newton's laws of motion** The three laws of motion on which *Newtonian mechanics is based. (1) A body continues in a state of rest or uniform motion in a straight line unless it is acted upon by external forces. (2) The rate of change of momentum of a moving body is proportional to and in the same direction as the force acting on it, i.e. $F = \mathrm{d}(mv)/\mathrm{d}t$, where $F$ is the applied force, $v$ is the velocity of the body, and $m$ its mass. If the mass remains constant, $F = m\mathrm{d}v/\mathrm{d}t$ or $F = ma$, where $a$ is the acceleration. (3) If one body exerts a force on another, there is an equal and opposite force, called a **reaction**, exerted on the first body by the second.

**Newton's rings** **1.** (in optics) *Interference fringes formed by placing a slightly convex lens on a flat glass plate. If monochromatic light is reflected by the two close surfaces into the observer's eye at a suitable angle, the point of contact of the lens is seen as a dark spot surrounded by a series of bright and dark rings. The radius of the $n$th dark ring is given by $r_n = \sqrt{nR\lambda}$, where $\lambda$ is the wavelength and $R$ is the radius of curvature of the lens. The phenomenon is used in the quality testing of lens surfaces. With white light, coloured rings are formed. **2.** (in photography) The irregular patterns produced by thin film interference between a projected transparency and its cover glass.

**niacin** See NICOTINIC ACID.

**niche** See ECOLOGICAL NICHE.

**Nichrome** Trade name for a group of nickel–chromium alloys used for wire in heating elements as they possess good resistance to oxidation and have a high resistivity. Typical is Nichrome V containing 80% nickel and 19.5% chromium, the balance consisting of manganese, silicon, and carbon.

**nickel** Symbol Ni. A malleable ductile silvery metallic *transition element; a.n. 28; r.a.m. 58.70; r.d. 8.9; m.p. 1450°C; b.p. 2732°C. It is found in the minerals pentlandite (NiS), pyrrhoite ((Fe,Ni)S), and garnierite ((Ni,Mg)$_6$(OH)$_6$Si$_4$O$_{11}$.H$_2$O). Nickel is also present in certain iron meteorites (up to 20%). The metal is extracted by roasting the ore to give the oxide, followed by reduction with carbon monoxide and purification by the *Mond process. Alternatively electrolysis is used. Nickel metal is used in special steels, in Invar, and, being ferromagnetic, in magnetic alloys, such as *Mumetal. It is also an effective catalyst, particularly for hydrogenation reactions (*see also* RANEY NICKEL). The main compounds are formed with nickel in the +2 oxidation state; the +3 state also exists (e.g. the black oxide, Ni$_2$O$_3$). Nickel was discovered by Axel Cronstedt (1722–65) in 1751.

**(⊕) SEE WEB LINKS**
• Information from the WebElements site

**nickel arsenide structure** A type of ionic crystal structure in which the anions have a distorted hexagonal close packed arrangement with the cations occupying the octahedral holes. Each type of ion has a coordination number of 6. Examples of compounds with this structure are NiAs, NiS, FeS, and CoS.

**(⊕) SEE WEB LINKS**
• An interactive version of the structure

**nickel–cadmium cell** *See* NICKEL–IRON ACCUMULATOR.

**nickel carbonyl** A colourless volatile liquid, Ni(CO)$_4$; m.p. –25°C; b.p. 43°C. It is formed by direct combination of nickel metal with carbon monoxide at 50–60°C. The reaction is reversed at higher temperatures, and the reactions are the basis of the *Mond process for purifying nickel. The nickel in the compound has an oxidation state of zero, and the compound is a typical example of a complex with pi-bonding *ligands, in which filled *d*-orbitals on the nickel overlap with empty *p*-orbitals on the carbon.

**nickelic compounds** Compounds of nickel in its +3 oxidation state; e.g. nickelic oxide is nickel(III) oxide (Ni$_2$O$_3$).

**nickel–iron accumulator (Edison cell; NIFE cell)** A *secondary cell devised by Thomas Edison (1847–1931) having a positive plate of nickel oxide and a negative plate of iron both immersed in an electrolyte of potassium hydroxide. The reaction on discharge is

$$2NiOOH.H_2O + Fe \rightarrow 2Ni(OH)_2 + Fe(OH)_2,$$

the reverse occurring during charging. Each cell gives an e.m.f. of about 1.2 volts and produces about 100 kJ per kilogram during each discharge. The **nickel–cadmium cell** is a similar device with a negative cadmium electrode. It is often used as a *dry cell. *Compare* LEAD–ACID ACCUMULATOR.

**nickelous compounds** Compounds of nickel in its +2 oxidation state; e.g. nickelous oxide is nickel(II) oxide (NiO).

**nickel(II) oxide** A green powder, NiO; r.d. 6.6. It can be made by heating nickel(II) nitrate or carbonate with air excluded.

**nickel(III) oxide (nickel peroxide; nickel sesquioxide)** A black or grey powder, Ni$_2$O$_3$; r.d. 4.8. It is made by heating nickel(II) oxide in air and used in *nickel–iron accumulators.

**nickel silver** *See* GERMAN SILVER.

**Nicol prism** A device for producing plane-polarized light (*see* POLARIZER). It consists of two pieces of calcite cut with a 68° angle and stuck together with Canada balsam. The extraordinary ray (*see* DOUBLE REFRACTION) passes through the prism while the ordinary ray suffers total internal reflection at the interface between the two crystals, as the refractive index of the calcite is 1.66 for the ordinary ray and that of the Canada balsam is 1.53. Modifications of the prism using different shapes and cements are used for special purposes. It was devised in 1828 by William Nicol (1768–1851).

**nicotinamide** *See* NICOTINIC ACID.

**nicotinamide adenine dinucleotide** *See* NAD.

**nicotinamide adenine dinucleotide phosphate (NADP)** *See* NAD.

**nicotine** A colourless poisonous *alkaloid present in tobacco. It is used as an insecticide.

**nicotinic acid (niacin)** A vitamin of the *vitamin B complex. It can be manufactured by plants and animals from the amino acid tryptophan. The amide derivative, **nicotinamide**, is a component of the coenzymes *NAD and NADP. These take part in many metabolic reactions as hydrogen acceptors.

Deficiency of nicotinic acid causes the disease *pellagra in humans. Apart from tryptophan-rich protein, good sources are liver and groundnut and sunflower meals.

**nictitating membrane** A clear membrane forming a third eyelid in amphibians, reptiles, birds, and some mammals (but not humans). It can be drawn across the cornea independently of the other eyelids, thus clearing the eye surface and giving added protection without interrupting the continuity of vision.

**nidation** See IMPLANTATION.

**nido-structure** See BORANE.

**nielsbohrium** See TRANSACTINIDE ELEMENTS.

**NIFE cell** See NICKEL–IRON ACCUMULATOR.

**ninhydrin** A brown crystalline solid, $C_9H_4O_3.H_2O$, which decomposes at 242°C. It is used as a test for amino acids, peptides, and proteins, with which it gives a deep blue colour. For this reason it has been used as a spray reagent for 'developing' paper chromatograms. It is also used in forensic science to develop latent fingerprints.

**niobium** Symbol Nb. A soft ductile grey-blue metallic transition element; a.n. 41; r.a.m. 92.91; r.d. 8.57; m.p. 2468°C; b.p. 4742°C. It occurs in several minerals, including niobite ($Fe(NbO_3)_2$), and is extracted by several methods including reduction of the complex fluoride $K_2NbF_7$ using sodium. It is used in special steels and in welded joints (to increase strength). Niobium–zirconium alloys are used in superconductors. Chemically, the element combines with the halogens and oxidizes in air at 200°C. It forms a number of compounds and complexes with the metal in oxidation states 2, 3, or 5. The element was discovered by Charles Hatchett (c. 1765–1847) in 1801 and first isolated by Christian Blomstrand (1826–97) in 1864. Formerly, it was called **columbium**.

**(((⊕))) SEE WEB LINKS**
• Information from the WebElements site

**nipple** See MAMMARY GLANDS.

**Nissl granules (Nissl bodies)** Particles seen within the cell bodies of *neurons. They are rich in RNA and stain strongly with basic dyes. They are named after F. Nissl (1860–1919), the German neurologist who discovered them.

**nit** A unit of *luminance equal to one *candela per square metre.

**nitrate** A salt or ester of nitric acid.

**nitrating mixture** A mixture of concentrated sulphuric and nitric acids, used to introduce a nitro group ($-NO_2$) into an organic compound. Its action depends on the presence of the nitronium ion, $NO_2^+$. It is mainly used to introduce groups into the molecules of *aromatic compounds (the nitro group can subsequently be converted into or replaced by others) and to make commercial *nitro compounds, such as the explosives cellulose trinitrate (nitrocellulose), glyceryl trinitrate (nitroglycerine), trinitrotoluene (TNT), and picric acid (trinitrophenol).

**nitration** A type of chemical reaction in which a nitro group ($-NO_2$) is added to or substituted in a molecule. Nitration can be carried out by a mixture of concentrated nitric and sulphuric acids. An example is electrophilic substitution of benzene (and benzene compounds), where the electrophile is the nitryl ion $NO_2^+$.

**nitre (saltpetre)** Commercial *potassium nitrate; the name was formerly applied to natural crustlike efflorescences, occurring in some arid regions.

**nitre cake** See SODIUM HYDROGEN-SULPHATE.

**nitric acid** A colourless corrosive poisonous liquid, $HNO_3$; r.d. 1.50; m.p. –42°C; b.p. 83°C. Nitric acid may be prepared in the laboratory by the distillation of a mixture of an alkali-metal nitrate and concentrated sulphuric acid. The industrial production is by the oxidation of ammonia to nitrogen monoxide, the oxidation of this to nitrogen dioxide, and the reaction of nitrogen dioxide with water to form nitric acid and nitrogen monoxide (which is recycled). The first reaction ($NH_3$ to NO) is catalysed by platinum or platinum/rhodium in the form of fine wire gauze. The oxidation of NO and the absorption of $NO_2$ to form the product are noncatalytic and proceed with high yields but both reactions are second-order and slow. Increases in pressure reduce the selectivity of the reaction and therefore rather large gas absorption towers are required. In practice the absorbing acid is refrigerated to around 2°C and a commercial 'concentrated nitric acid' at about 67% is produced.

Nitric acid is a strong acid (highly dissociated in aqueous solution) and dilute solutions behave much like other mineral acids. Concentrated nitric acid is a strong oxidizing agent. Most metals dissolve to form nitrates but with the evolution of nitrogen oxides. Concentrated nitric acid also reacts with several nonmetals to give the oxo acid or oxide. Nitric acid is generally stored in dark brown bottles because of the photolytic decomposition to dinitrogen tetroxide. *See also* NITRATION.

**nitric oxide** *See* NITROGEN MONOXIDE.

**nitrides** Compounds of nitrogen with a more electropositive element. Boron nitride is a covalent compound having macromolecular crystals. Certain electropositive elements, such as lithium, magnesium, and calcium, react directly with nitrogen to form ionic nitrides containing the $N^{3-}$ ion. Transition elements form a range of interstitial nitrides (e.g. $Mn_4N$, $W_2N$), which can be produced by heating the metal in ammonia.

**nitriding** The process of hardening the surface of steel by producing a layer of iron nitride. One technique is to heat the metal in ammonia gas. Another is to dip the hot metal in a bath of molten sodium cyanide.

**nitrification** A chemical process in which nitrogen (mostly in the form of ammonia) in plant and animal wastes and dead remains is oxidized at first to nitrites and then to nitrates. These reactions are effected mainly by the nitrifying bacteria *Nitrosomonas* and *Nitrobacter* respectively. Unlike ammonia, nitrates are readily taken up by plant roots; nitrification is therefore a crucial part of the *nitrogen cycle. Nitrogen-containing compounds are often applied to soils deficient in this element, as fertilizer. *Compare* DENITRIFICATION.

**nitrile rubber** A copolymer of buta-1,3-diene and propenonitrile. Nitrile rubbers are commercially important synthetic rubbers because of their resistance to oil and many solvents.

**nitriles (cyanides)** Organic compounds containing the group –CN bound to an organic group. Nitriles are made by reaction between potassium cyanide and haloalkanes in alcoholic solution, e.g.

$$KCN + CH_3Cl \rightarrow CH_3CN + KCl$$

An alternative method is dehydration of amides

$$CH_3CONH_2 - H_2O \rightarrow CH_3CN$$

They can be hydrolysed to amides and carboxylic acids and can be reduced to amines.

**(((●))) SEE WEB LINKS**
• Information about IUPAC nomenclature

**nitrite** A salt or ester of nitrous acid. The salts contain the dioxonitrate (III) ion, $NO_2^-$, which has a bond angle of 115°.

**nitrobenzene** A yellow oily liquid, $C_6H_5NO_2$; r.d. 1.2; m.p. 6°C; b.p. 211°C. It is made by the *nitration of benzene using a mixture of nitric and sulphuric acids.

**nitrocellulose** *See* CELLULOSE NITRATE.

**nitro compounds** Organic compounds containing the group –$NO_2$ (the **nitro group**) bound to a carbon atom. Nitro compounds are made by *nitration reactions. They can be reduced to aromatic amines (e.g. nitrobenzene can be reduced to phenylamine). *See also* EXPLOSIVE.

**nitrogen** Symbol N. A colourless gaseous element belonging to *group 15 (formerly VB) of the periodic table; a.n. 7; r.a.m. 14.0067; d. 1.2506 g dm$^{-3}$; m.p. –209.86°C; b.p. –195.8°C. It occurs in air (about 78% by volume) and is an essential constituent of proteins and nucleic acids in living organisms (*see* NITROGEN CYCLE). Nitrogen is obtained for industrial purposes by fractional distillation of liquid air. Pure nitrogen can be obtained in the laboratory by heating a metal azide. There are two natural isotopes: nitrogen–14 and nitrogen–15 (about 3%). The element is used in the *Haber process for making ammonia and is also used to provide an inert atmosphere in welding and metallurgy. The gas is diatomic and relatively inert – it reacts with hydrogen at high temperatures and with oxygen in electric discharges. It also forms *nitrides with certain metals. Nitrogen was discovered in 1772 by Daniel Rutherford (1749–1819).

**(((●))) SEE WEB LINKS**
• Information from the WebElements site

**nitrogenase** An important enzyme complex that is present in those microorganisms that are capable of fixing atmospheric nitrogen (*see* NITROGEN FIXATION). Nitrogenase catalyses the conversion of atmospheric nitrogen into ammonia, which can then be used to synthesize nitrites, nitrates, or amino acids. The two main enzymes within the nitrogenase complex are **dinitrogenase reductase** and **dinitrogenase**.

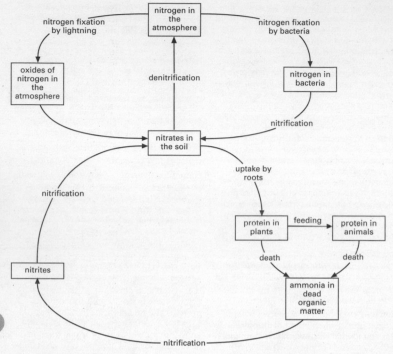

**The nitrogen cycle.**

**nitrogen cycle** One of the major cycles of chemical elements in the environment (*see* BIOGEOCHEMICAL CYCLE). Nitrates in the soil are taken up by plant roots and may then pass along *food chains into animals. Decomposing bacteria convert nitrogen-containing compounds (especially ammonia) in plant and animal wastes and dead remains back into nitrates, which are released into the soil and can again be taken up by plants (*see* NITRIFICATION). Though nitrogen is essential to all forms of life, the huge amount present in the atmosphere is not directly available to most organisms (*compare* CARBON CYCLE). It can, however, be assimilated by some specialized bacteria (*see* NITROGEN FIXATION) and is thus made available to other organisms indirectly. Lightning flashes also make some nitrogen available to plants by causing the combination of atmospheric nitrogen and oxygen to form oxides of nitrogen, which enter the soil and form nitrates. Some nitrogen is returned from the soil to the atmosphere by denitrifying bacteria (*see* DENITRIFICATION). See illustration.

**nitrogen dioxide** *See* DINITROGEN TETROXIDE.

**nitrogen fixation** A chemical process in which atmospheric nitrogen is assimilated into organic compounds in living organisms and hence into the *nitrogen cycle. The ability to fix nitrogen is limited to certain bacteria (e.g. *Azotobacter*, *Anabaena*). Some bacteria (e.g. *Rhizobium*) are able to fix nitrogen in association with cells in the roots of leguminous plants, such as peas and beans, in which they form characteristic *root nodules; cultivation of legumes is therefore one way of increasing soil nitrogen. Various chemical processes are used to fix atmospheric nitrogen in the manufacture of *fertilizers. These include the *Birkeland–Eyde process, the cyanamide process (*see* CALCIUM DICARBIDE), and the *Haber process.

**nitrogen monoxide (nitric oxide)** A colourless gas, NO; m.p. −163.6°C; b.p. −151.8°C. It is soluble in water, ethanol, and ether. In the liquid state nitrogen monoxide is blue in colour (r.d. 1.26). It is formed in many reactions involving the reduction of nitric acid, but more convenient reactions for the preparation of reasonably pure NO are reactions of sodium nitrite, sulphuric acid, and either sodium iodide or iron(II) sulphate. Nitrogen monoxide reacts readily with oxygen to give nitrogen dioxide and with the halogens to give the nitrosyl halides XNO (X = F,Cl,Br). It is oxidized to nitric acid by strong oxidizing agents and reduced to dinitrogen oxide by reducing agents. The molecule has one unpaired electron, which accounts for its paramagnetism and for the blue colour in the liquid state. This electron is relatively easily removed to give the **nitrosyl ion** NO⁺, which is the ion present in such compounds as NOClO₄, NOBF₄, NOFeCl₄, (NO)₂PtCl₆ and a ligand in complexes, such as Co(CO)₃NO.

In mammals and other vertebrates, nitrogen monoxide plays several important roles. For example, it acts as a gaseous mediator in producing such responses as dilation of blood vessels, relaxation of smooth muscle, and inhibition of platelet aggregation, and it acts as a neurotransmitter in the central nervous systems. In certain cells of the immune system it is converted to the peroxynitrite ion (⁻O–O–N=O), which has activity against pathogens.

**nitrogen mustards** A group of nitrogen compounds similar to *sulphur mustard. They were used as chemotherapy agents in cancer treatment. Like sulphur mustard, they are powerful blistering agents. Large quantities were made during World War II, although none were used in combat.

**nitrogenous base** A basic compound containing nitrogen. The term is used especially of organic ring compounds, such as adenine, guanine, cytosine, and thymine, which are constituents of nucleic acids. *See* AMINE SALTS.

**nitrogen oxides** Oxides of nitrogen (NOₓ), such as nitrogen monoxide (NO) and dinitrogen oxide (N₂O), many of which are pollutants contributing to *acid rain. Nitrogen oxides are expelled in the emissions from car exhausts, aircraft, and factories. *See also* AIR POLLUTION.

**nitroglycerine** An explosive made by reacting 1,2,3-trihydroxypropane (glycerol) with a mixture of concentrated sulphuric and nitric acids. Despite its name and method of preparation, it is not a nitro compound, but an ester of nitric acid, CH₂(NO₃)CH(NO₃)CH₂(NO₃). It is used in dynamites.

**nitro group** *See* NITRO COMPOUNDS.

**nitronium ion** *See* NITRYL ION.

**nitrosonium ion** The positive ion NO⁺, present in certain salts such as the chlorate (NO⁺ClO₄) and the borofluoride (NO⁺BF₄).

**nitrosyl ion** The ion NO⁺. *See* NITROGEN MONOXIDE.

**nitrous acid** A weak acid, HNO₂, known only in solution and in the gas phase. It is prepared by the action of acids upon nitrites, preferably using a combination that removes the salt as an insoluble precipitate (e.g. Ba(NO₂)₂ and H₂SO₄). The solutions are unstable and decompose on heating to give nitric acid and nitrogen monoxide. Nitrous acid can function both as an oxidizing agent (forms NO) with I⁻ and Fe²⁺, or as a reducing agent (forms NO₃⁻) with, for example, Cu²⁺; the latter is most common. It is widely used (prepared *in situ*) for the preparation of diazonium compounds in organic chemistry. The full systematic name is **dioxonitric(III) acid**.

**nitrous oxide** *See* DINITROGEN OXIDE.

**nitryl ion (nitronium ion)** The ion NO₂⁺, found in mixtures of nitric acid and sulphuric acid and solutions of nitrogen oxides in nitric acid. Nitryl salts, such as NO₂⁺ClO₄⁻, can be isolated but are extremely reactive. Nitryl ions generated *in situ* are used for *nitration in organic chemistry.

**NMR** *See* NUCLEAR MAGNETIC RESONANCE.

**nobelium** Symbol No. A radioactive metallic transuranic element belonging to the *actinoids; a.n. 102; mass number of most stable element 254 (half-life 55 seconds). Seven isotopes are known. The element was first identified with certainty by Albert Ghiorso and Glenn Seaborg (1912–99) in 1966.

(((●))) SEE WEB LINKS
• Information from the WebElements site

**noble gases (inert gases; rare gases; group 18 elements)** A group of monatomic gaseous elements forming group 18 (formerly group 0) of the *periodic table: helium (He), neon (Ne), argon (Ar), krypton (Kr), xenon (Xe), and radon (Rn). The electron configuration of helium is $1s^2$. The configurations of the others terminate in $ns^2np^6$ and all inner shells are fully occupied. The elements thus represent the termination of a period and have closed-shell configuration and associated high ionization energies (He 2370 to Rn 1040 kJ $mol^{-1}$) and lack of chemical reactivity. Being monatomic the noble gases are spherically symmetrical and have very weak interatomic interactions and consequent low enthalpies of vaporization. The behaviour of the lighter members approaches that of an ideal gas at normal temperatures; with the heavier members increasing polarizability and dispersion forces lead to easier liquefaction under pressure. Four types of 'compound' have been described for the noble gases but of these only one can be correctly described as compounds in the normal sense. One type consists of such species as $HHe^+$, $He_2^+$, $Ar_2^+$, $HeLi^+$, which form under highly energetic conditions, such as those in arcs and sparks. They are short-lived and only detected spectroscopically. A second group of materials described as inert-gas–metal compounds do not have defined compositions and are simply noble gases adsorbed onto the surface of dispersed metal. The third type, previously described as 'hydrates' are in fact clathrate compounds with the noble gas molecule trapped in a water lattice. True compounds of the noble gases were first described in 1962 and several fluorides, oxyfluorides, fluoroplatinates, and fluoroantimonates of *xenon are known. A few krypton fluorides and a radon fluoride are also known although the short half-life of radon and its intense alpha activity restrict the availability of information. Apart from argon, the noble gases are present in the atmosphere at only trace levels. Helium may be found along with natural gas (up to 7%), arising from the radioactive decay of heavier elements (via alpha particles).

**noble metal** A metal characterized by it lack of chemical reactivity, particularly to acids and atmospheric corrosion. Examples include gold, palladium, platinum, and rhodium.

**no-cloning theorem** A result stating that it is not possible to copy quantum information perfectly. This is the case because the Heisenberg uncertainty principle means that it is not possible to obtain complete information about a quantum state and because examining a quantum state alters that state. This theorem was proved by William Wooters and Wojciech Zurek in 1982.

**nodal points** Two points on the axis of a system of lenses; if the incident ray passes through one, the emergent ray will pass through the other.

**node** 1. (in botany) The part of a plant stem from which one or more leaves arise. The nodes at the stem apex are very close together and remain so in species of monocotyledons that form bulbs. In older regions of the stem they are separated by areas of stem called **internodes**. 2. (in anatomy) A natural thickening or bulge in an organ or part of the body. Examples are the **sinoatrial node** that controls the heartbeat (see PACE-MAKER) and the *lymph nodes. 3. (in physics) A point of minimum disturbance in a *stationary-wave system. 4. (in astronomy) Either of two points at which the orbit of a celestial body intersects a reference plane, usually the plane of the *ecliptic or the celestial equator (see CELESTIAL SPHERE).

**node of Ranvier** See MYELIN SHEATH.

**nodule** (in botany) See ROOT NODULE.

**no-hair theorem** See BLACK HOLE.

**noise** 1. Any undesired sound. It is measured on a *decibel scale ranging from the threshold of hearing (0 dB) to the threshold of pain (130 dB). Between these limits a whisper registers about 20 dB, heavy urban traffic about 90 dB, and a heavy hammer on steel plate about 110 dB. A high noise level (industrial or from overamplified music, for example) can cause permanent hearing impairment. 2. Any unwanted disturbance within a useful frequency band in a communication channel.

**nomad** (in cytology) A cell that migrates or wanders from its site of formation. Certain types of *phagocytes are nomads.

**Nomarski microscope** A type of light microscope that is useful for viewing live transparent unstained specimens, such as cells or microscopic organisms. The shadow-cast images give the illusion of depth to the outlines and surface features of organelles or other structures. An incident beam of plane-polarized light is split into parallel beams by a prism (**Nomarski prism**) so that different parts of the beam pass through closely adjacent areas of the specimen. Slight differences in thickness and refractive index within the specimen cause interference between the beams as they exit the specimen and are recombined by a second prism: parts of the beam that are in phase will reinforce each other and produce a bright image, whereas parts that are out of phase will cancel each other out and produce a dark image. It is named after the Polish-born physicist Georges Nomarski (1919–97).

(🌐) SEE WEB LINKS

• Overview of Nomarski microscopy with interactive tutorials, from Olympus Microscopy Resource Center

**nomogram** A graph consisting of three lines, each with its own scale, each line representing the values of a variable over a specified range. A ruler laid between two points on two of the lines enables the value of the third variable to be read off the third line.

**nonahydrate** A crystalline compound that has nine moles of water per mole of compound.

**nonanoic acid (perargonic acid)** A clear oily liquid carboxylic acid, $CH_3(CH_2)_7COOH$; r.d. 0.9; m.p. 12.5°C; b.p. 254°C. It is found as esters in oil of pelargonium and certain esters are used as flavourings.

**nonbenzenoid aromatics** Aromatic compounds that have rings other than benzene rings. Examples are the cyclopentadienyl anion, $C_5H_5^-$, and the tropylium cation, $C_7H_7^+$. See also ANNULENES.

**noncompetitive inhibition** See INHIBITION.

**noncyclic phosphorylation (noncyclic photophosphorylation)** See PHOTOPHOSPHORYLATION.

**non-equilibrium statistical mechanics** The statistical mechanics of systems not in thermal equilibrium. One of the main purposes of non-equilibrium statistical mechanics is to calculate *transport coefficients and inverse transport coefficients, such as *conductivity and *viscosity, from first principles and to provide a basis for *transport theory. The non-equilibrium systems easiest to understand are those near thermal equilibrium. For systems far from equilibrium, such possibilities as *chaos, *turbulence, and *self-organization can arise due to nonlinearity.

**non-equilibrium thermodynamics** The thermodynamics of systems not in thermal *equilibrium. The non-equilibrium systems easiest to understand are those near thermal equilibrium. For systems far from equilibrium, more complicated patterns, such as *chaos and *self-organization, can arise due to nonlinearity. Which behaviour is observed depends on the value of certain parameters in the system.

**non-Euclidean geometry** A type of geometry that does not comply with the basic postulates of *Euclidean geometry, particularly a form of geometry that does not accept Euclid's postulate that only one straight line can be drawn through a point in space parallel to a given straight line. Several types of non-Euclidean geometry exist.

**non-Fermi liquid** A many-fermion system that, unlike the electrons in ordinary metals, cannot readily be decribed in terms of *quasiparticles. Examples of non-Fermi liquids include *Luttinger liquids and high-temperature *superconductivity. The theoretical analysis of non-Fermi liquids is more difficult than for ordinary many-fermion systems.

**nonferrous metal** Any metal other than iron or any alloy that does not contain iron. In commercial terms this usually means aluminium, copper, lead, nickel, tin, zinc, or their alloys.

**nonlinear optics** A branch of optics concerned with the optical properties of matter subjected to intense electromagnetic fields. For nonlinearity to manifest itself, the external field should not be negligible compared to the internal fields of the atoms and molecules of which the matter consists. *Lasers are capable of generating external fields sufficiently intense for nonlinearity to occur. Indeed, the subject of nonlinear optics has been largely developed as a result of the invention of the laser. In nonlinear optics the induced electric polarization (see DIELECTRIC) of the medium is not a linear function of the strength of the external *electromagnetic radiation. This leads to more complicated phenomena than can occur in **linear optics**, in which the induced polarization is proportional to the strength of the external electromagnetic radiation.

**nonmetal** An element that is not a *metal. Nonmetals can either be *insulators or *semiconductors. At low temperatures nonmetals are poor conductors of both electricity and heat as few free electrons move through the material. If the conduction band is near to the valence band (see ENERGY BANDS) it is possible for nonmetals to conduct electricity at high temperatures but, in contrast to metals, the conductivity increases with increasing temperature. Nonmetals are electronegative elements, such as carbon, nitrogen, oxygen, phosphorus, sulphur, and the halogens. They form compounds that contain negative ions or covalent bonds. Their oxides are either neutral or acidic.

**non-Newtonian fluid** See NEWTONIAN FLUID.

**nonpolar compound** A compound that has covalent molecules with no permanent dipole moment. Examples of nonpolar compounds are methane and benzene.

**nonpolar solvent** See SOLVENT.

**nonreducing sugar** A sugar that cannot donate electrons to other molecules and therefore cannot act as a reducing agent. Sucrose is the most common nonreducing sugar. The linkage between the glucose and fructose units in sucrose, which involves aldehyde and ketone groups, is responsible for the inability of sucrose to act as a *reducing sugar.

**nonrelativistic quantum theory** See QUANTUM THEORY.

**nonrenewable energy sources** See RENEWABLE ENERGY SOURCES.

**nonsense mutation** A mutation in one of the nucleotides in a DNA sequence that generates a *stop codon, resulting in the premature termination of synthesis of a protein.

**nonstoichiometric compound (Berthollide compound)** A chemical compound in which the elements do not combine in simple ratios. For example, rutile (titanium(IV) oxide) is often deficient in oxygen, typically having a formula $TiO_{1.8}$.

**noradrenaline (norepinephrine)** A hormone produced by the *adrenal glands and also secreted from nerve endings in the *sympathetic nervous system as a chemical transmitter of nerve impulses (see NEUROTRANSMITTER). Many of its general actions are similar to those of *adrenaline, but it is more concerned with maintaining normal body activity than with preparing the body for emergencies.

**Nordhausen sulphuric acid** See DISULPHURIC(VI) ACID.

**norepinephrine** See NORADRENALINE.

**normal** 1. (in mathematics) A line drawn at right angles to a surface. 2. (in chemistry) Having a concentration of one gram equivalent per $dm^3$.

**normal distribution (Gaussian distribution)** The type of statistical distribution followed by, for example, the same measurement taken several times, where the variation of a quantity ($x$) about its mean value ($\mu$) is entirely random. A normal distribution has the probability density function

$$f(x) = \exp[-(x-\mu)^2/2\sigma^2]/\sigma\sqrt{2\pi}$$

where $\sigma$ is known as the **standard deviation**. The distribution is written $N(\mu,\sigma^2)$. The graph of $f(x)$ is bell-shaped and symmetrical about $x = \mu$. The standard normal distribution has $\mu = 0$ and $\sigma^2 = 1$. $x$ can be standardized by letting $z = (x-\mu)/\sigma$. The values $z_\alpha$, for which the area under the curve from $-\infty$ to $z_\alpha$ are $\alpha$, are tabulated; i.e. $z$ is such that

$$P(z \le z_\alpha) = \alpha$$

Hence

$$P(a<x\le b) = P(a-\mu)/\sigma < z \le (b-\mu)/\sigma$$

can be found. See also POISSON DISTRIBUTION; T-DISTRIBUTION.

**normalizing** The process of heating steel to above an appropriate critical temperature followed by cooling in still air. The process promotes the formation of a uniform internal structure and the elimination of internal stress.

**Northern blotting** See SOUTHERN BLOTTING.

**nose** The protuberance on the face of some vertebrates that contains the nostrils (see NARES) and part of the *nasal cavity. It therefore forms part of the olfactory system (see OLFACTION) and the external opening of the respiratory system.

**nostrils** See NARES.

**NOT circuit** See LOGIC CIRCUITS.

**note** 1. A musical sound of specified pitch. 2. A representation of such a sound in a musical score. Such a representation has a specified duration as well as a specified pitch.

**notochord** An elastic skeletal rod lying lengthwise beneath the nerve cord and above the alimentary canal in the embryos or adults of all chordate animals (*see* CHORDATA). Its function is to strengthen and support the body and act as a protagonist for the muscles. It is found in both adult and larval lancelets but in adult vertebrates it is largely replaced by the *vertebral column.

**nova** A star that, over a period of only a few days, becomes $10^3$–$10^4$ times brighter than it was. Some 10–15 such events occur in the Milky Way each year. Novae are believed to be close *binaries, one component of which is usually a *white dwarf and the other a *red giant. Matter is transferred from the red giant to the white dwarf, on whose surface it accumulates, eventually leading to a thermonuclear explosion. *See also* SUPERNOVA.

**NSOM** Near-field scanning optical microscopy. A form of scanning probe microscopy in which a probe with a very small aperture is used, giving high resolving power.

**⊕ SEE WEB LINKS**
• An account of the technique

**N.T.P.** *See* S.T.P.

***n*-type conductivity** *See* SEMICONDUCTOR; TRANSISTOR.

**nucellus** The tissue that makes up the greater part of the ovule of seed plants. It contains the *embryo sac and nutritive tissue. It is enclosed by the integuments except for a small gap, the *micropyle. In certain flowering plants it may persist after fertilization and provide nutrients for the embryo.

**nuclear battery** A single cell, or battery of cells, in which the energy of particles emitted from the atomic nucleus is converted internally into electrical energy. In the high-voltage type, a beta-emitter, such as strontium–90, krypton–85, or tritium, is sealed into a shielded glass vessel, the electrons being collected on an electrode that is separated from the emitter by a vacuum or by a solid dielectric. A typical cell delivers some 160 picoamperes at a voltage proportional to the load resistance. It can be used to maintain the voltage of a charged capacitor. Of greater use, especially in space technology, are the various types of low-voltage nuclear batteries. Typical is the gas-ionization device in which a beta-emitter ionizes a gas in an electric field. Each beta-particle produces about 200 ions, thus multi-plying the current. The electric field is obtained by the contact potential difference between two electrodes, such as lead dioxide and magnesium. Such a cell, containing argon and tritium, gives about 1.6 nanoamperes at 1.5 volts. Other types use light from a phosphor receiving the beta-particles to operate photocells or heat from the nuclear reaction to operate a thermopile.

**nuclear–cytoplasmic ratio** A measure of the size of a cell nucleus in relation to the cytoplasm. The nuclear–cytoplasmic ratio is often used as an index in the comparison of cells from normal and abnormal tissues. For example, cultured cancer cells show an increase in the nuclear–cytoplasmic ratio.

**nuclear energy** Energy obtained as a result of *nuclear fission or *nuclear fusion. The nuclear fission of one uranium atom yields about $3.2 \times 10^{-11}$ joule, whereas the combustion of one carbon atom yields about $6.4 \times 10^{-19}$ joule. Mass for mass, uranium yields about 2 500 000 times more energy by fission than carbon does by combustion. The nuclear fusion of deuterium to form helium releases about 400 times as much energy as the fission of uranium (on a mass basis).

**nuclear fission** A nuclear reaction in which a heavy nucleus (such as uranium) splits into two parts (**fission products**), which subsequently emit either two or three neutrons, releasing a quantity of energy equivalent to the difference between the rest mass of the neutrons and the fission products and that of the original nucleus. Fission may occur without external influence (**spontaneous fission**) or as a result of irradiation by neutrons (**induced fission**). For example, the fission of a uranium–235 nucleus by a *slow neutron may proceed thus:

$$^{235}U + n \rightarrow {}^{148}La + {}^{85}Br + 3n$$

The energy released is approximately $3 \times 10^{-11}$ J per $^{235}U$ nucleus. For 1 kg of $^{235}U$ this is equivalent to 20 000 megawatt-hours – the amount of energy produced by the combustion of $3 \times 10^6$ tonnes of coal. Nuclear fission is the process used in *nuclear reactors and atom bombs (*see* NUCLEAR WEAPONS).

**⊕ SEE WEB LINKS**
• Background to Lise Meitner's codiscovery of nuclear fission

**nuclear force** A strong attractive force (resulting from the strong interaction) between *nucleons in the atomic nucleus that holds the nucleus together. At close range (up to about $2 \times 10^{-15}$ metre) these forces are some 100 times stronger than electromagnetic forces. *See* FUNDAMENTAL INTERACTIONS.

**nuclear fuel** A substance that will sustain a fission chain reaction so that it can be used as a source of *nuclear energy. The fissile isotopes are uranium–235, uranium–233, plutonium–241, and plutonium–239 (*see* FISSILE MATERIAL). The first occurs in nature as 1 part in 140 of natural uranium, the others have to be made artificially. $^{233}$U is produced when thorium–232 captures a neutron and $^{239}$Pu is produced by neutron capture in $^{238}$U. $^{232}$Th and $^{238}$U are called fertile isotopes (*see* FERTILE MATERIAL).

**nuclear fusion** A type of *nuclear reaction in which atomic nuclei of low atomic number fuse to form a heavier nucleus with the release of large amounts of energy. In *nuclear fission reactions a neutron is used to break up a large nucleus, but in nuclear fusion the two reacting nuclei themselves have to be brought into collision. As both nuclei are positively charged there is a strong repulsive force between them, which can only be overcome if the reacting nuclei have very high kinetic energies. These high kinetic energies imply temperatures of the order of $10^8$ K. As the kinetic energy required increases with the nuclear charge (i.e. atomic number), reactions involving low atomic-number nuclei are the easiest to produce. At these elevated temperatures, however, fusion reactions are self-sustaining; the reactants at these temperatures are in the form of a *plasma (i.e. nuclei and free electrons) with the nuclei possessing sufficient energy to overcome electrostatic repulsion forces. The fusion bomb (*see* NUCLEAR WEAPONS) and the stars generate energy in this way. It is hoped that the method will be harnessed in the *thermonuclear reactor as a source of energy for man's use.

Typical fusion reactions with the energy release in joules are:

$$D + D - T + p + 6.4 \times 10^{-13} \text{ J}$$

$$T + D = {}^4He + n + 28.2 \times 10^{-13} \text{ J}$$

$$^6Li + D = 2{}^4He + 35.8 \times 10^{-13} \text{ J}$$

By comparison the formation of a water molecule from atoms of hydrogen and oxygen is accompanied by the release of $1.5 \times 10^{-19}$ J.

A large amount of work has been done on **cold fusion**; i.e. fusion that can occur at lower temperatures than those necessary to overcome the electrostatic repulsion between nuclei. The most productive approach is **meson-catalysed fusion**, in which the deuterium atoms have their electrons replaced by negative muons to give 'muonic atoms' of deuterium. The muon is 207 times heavier than the electron, so the muonic deuterium atom is much smaller and is able to approach another deuterium atom more closely, allowing nuclear fusion to occur. The muon is released to form another muonic atom, and the process continues. The limiting factor is the short lifetime of the muon, which restricts the number of fusion reactions it can catalyse. The term 'cold fusion' is also applied to the technique of producing new *transactinide elements by bombarding nuclei of one element with nuclei of another at an energy precisely chosen to allow the fusion reaction to occur.

(((●))) **SEE WEB LINKS**

• Introduction to nuclear fusion from Worsley School, Alberta

**nuclear isomerism** A condition in which atomic nuclei with the same number of neutrons and protons have different lifetimes. This occurs when nuclei exist in different unstable quantum states, from which they decay to lower excited states or to the ground state, with the emission of gamma-ray photons. If the lifetime of a particular excited state is unusually long it is said to be isomeric, although there is no fixed limit separating isomeric decays from normal decays.

**nuclear magnetic resonance (NMR)** The absorption of electromagnetic radiation at a suitable precise frequency by a nucleus with a nonzero magnetic moment in an external magnetic field. The phenomenon occurs if the nucleus has nonzero *spin, in which case it behaves as a small magnet. In an external magnetic field, the nucleus's magnetic moment vector precesses about the field direction but only certain orientations are allowed by quantum rules. Thus, for hydrogen (spin of ½) there are two possible states in the presence of a field, each with a slightly different energy. Nuclear magnetic resonance is the absorption of radiation at a photon energy equal to the difference between these levels, causing a transition from a lower to a higher energy state. For practical purposes, the difference in energy levels is small and the radiation is in the radiofrequency region of the electromagnetic spectrum. It depends on the field strength.

## MAGNETIC RESONANCE IMAGING

A diagnostic imaging technique based on the phenomenon of *nuclear magnetic resonance (NMR). NMR is a process in which *protons interact with a strong magnetic field and with radio waves to generate electrical pulses that can be processed in a similar way to computerized *tomography. The medical application of NMR, began in the 1950s, but the first images of live patients were not produced until the late 1970s. Images produced by MRI are similar to those produced by computerized tomography using X-rays, but without the radiation hazard.

A major factor in the high costs of MRI is the need for a *superconducting magnet to produce the very strong magnetic fields (0.1 – 2 tesla). A niobium–titanium alloy, which becomes superconducting at −269°C, is used to construct the field coils. These need to be immersed in liquid helium. Superimposed on this large magnetic field are smaller fields, with known gradients in two directions. These gradient fields produce a unique value of the magnetic field strength at each point within the instrument (see illustration).

Some nuclei in the atoms of a patient's tissues have a *spin, which makes them behave as tiny nuclear magnets. The purpose of the large magnetic field is to align these nuclear magnets. Having achieved this alignment, the area under xamination is subjected to pulses of radiofrequency (RF) radiation. At a resonant frequency of RF pulses the nuclei under examination undergo *Larmor precession. This phenomenon may be thought of as a 'tipping' of the nuclear magnets away from the strong field alignment. The nuclear magnets then precess or 'wobble', about the axis of the main field as the nuclei regain their alignment with that field.

The speed at which the nuclei return to the steady stage gives rise to two parameters, known as **relaxation times**. Because these relaxation times for nuclei depend on their atomic environment, they may be used to identify nuclei. Small changes in the magnetic field produced as the nuclei precess induce currents in a receiving coil. These signals are digitized before being stored in the memory of a computer.

The resulting set of RF pulse sizes and sequences identify a variety of resonance situations. By analysing these sequences and knowing the unique value of magnetic field strength within the volume under investigation, the resonance signals may be decoded to give estimates of the compositions of the patient's tissues. A three dimensional map of the composition can then be produced, using colour to indicate contrast between differing tissue compositions.

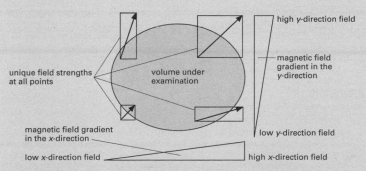

**MRI.** The way unique field strengths are produced at different points in a specimen

NMR can be used for the accurate determination of nuclear moments. It can also be used in a sensitive form of magnetometer to measure magnetic fields. In medicine, **magnetic resonance imaging** (MRI) has been developed, in which images of tissue are produced by magnetic-resonance techniques. **Functional MRI** (fMRI) can detect changes in blood flow that accompany neural activity and is therefore used to study brain function. See Feature.

The main application of NMR is as a technique for chemical analysis and structure determination, known as **NMR spectroscopy**. It depends on the fact that the electrons in a molecule shield the nucleus to some extent from the field, causing different atoms to absorb at slightly different frequencies (or at slightly different fields for a fixed frequency). Such effects are known as **chemical shifts**. There are two methods of NMR spectroscopy. In **continuous wave** (**CW**) **NMR**, the sample is subjected to a strong field, which can be varied in a controlled way over a small region. It is irradiated with radiation at a fixed frequency, and a detector monitors the field at the sample. As the field changes, absorption corresponding to transitions occurs at certain values, and this causes oscillations in the field, which induce a signal in the detector. **Fourier transform** (**FT**) **NMR** uses a fixed magnetic field and the sample is subjected to a high-intensity pulse of radiation covering a range of frequencies. The signal produced is analysed mathematically to give the NMR spectrum. The most common nucleus studied is $^1$H. For instance, an NMR spectrum of ethanol ($CH_3CH_2OH$) has three peaks in the ratio 3:2:1, corresponding to the three different hydrogen-atom environments. The peaks also have a fine structure caused by interaction between spins in the molecule. Other nuclei can also be used for NMR spectroscopy (e.g. $^{13}$C, $^{14}$N, $^{19}$F) although these generally have lower magnetic moment and natural abundance than hydrogen. *See also* ELECTRON PARAMAGNETIC RESONANCE.

**(((⊕))) SEE WEB LINKS**

• A comprehensive online textbook covering all aspects of magnetic resonance imaging

**nuclear moment** A property of atomic nuclei in which lack of spherical symmetry of the nuclear charge gives rise to electric moments and the intrinsic spin and rotational motion of the component nucleons give rise to magnetic moments.

**nuclear physics** The physics of atomic nuclei and their interactions, with particular reference to the generation of *nuclear energy.

**nuclear power** Electric power or motive power generated by a *nuclear reactor.

**nuclear reaction** Any reaction in which there is a change to an atomic nucleus. This may be a natural spontaneous disintegration or an artificial bombardment of a nucleus with an energetic particle, as in a *nuclear reactor. Nuclear reactions are commonly represented by enclosing within a bracket the symbols for the incoming and outgoing particles; the initial nuclide is shown before the bracket and the final nuclide after it. For example, the reaction:

$$^{12}C + {}^2H \rightarrow {}^{13}N + {}^1n$$

is shown as $^{12}C(d,n)^{13}N$, where $d$ is the symbol for a deuteron.

**nuclear reactor** A device in which a *nuclear fission *chain reaction is sustained and controlled in order to produce *nuclear energy, radioisotopes, or new nuclides. The fuels available for use in a fission reactor are uranium–235, uranium–233, and plutonium–239; only the first occurs in nature (as 1 part in 140 of natural uranium), the others have to be produced artificially (*see* NUCLEAR FUEL). When a uranium–235 nucleus is made to undergo fission by the impact of a neutron it breaks into two roughly equal fragments, which release either two or three very high-energy neutrons. These *fast neutrons need to be slowed down to increase the probability that they will cause further fissions of $^{235}$U nuclei and thus sustain the chain reaction. This slowing down process occurs naturally to a certain extent when the neutrons collide with other nuclei; unfortunately, however, the predominant uranium isotope, $^{238}$U, absorbs fast neutrons to such an extent that in natural uranium the fission reaction is not self-sustaining. In order to create a controlled self-sustaining chain reaction it is necessary either to slow down the neutrons (using a *moderator in a **thermal reactor**) to greatly reduce the number absorbed by $^{238}$U, or to reduce the predominance of $^{238}$U in natural uranium by enriching it with more $^{235}$U than it normally contains. In a **fast reactor** the fuel used is enriched uranium and no moderator is employed.

In thermal reactors, neutrons are slowed down by collisions with light moderator atoms (such as graphite, deuterium, or beryllium); they are then in thermal equilibrium with the surrounding material and are known as **thermal neutrons**. In a **heterogeneous thermal reactor** the fuel and moderator are in separate solid and liquid phases (e.g. solid uranium fuel and a heavy water moderator). In the **homogeneous thermal reactor** the fuel and moderator are mixed together, for example in a solution, molten dispersion, slurry, or suspension.

In the reactor **core** the **fuel elements** encase the fuel; in a heterogeneous reactor the fuel elements may fit into a lattice that also contains the moderator. The progress of the reaction is controlled by *control rods, which when lowered into the core absorb neutrons and so slow down or stop the chain reaction. The heat produced by the nuclear reaction in the core is used to generate electricity by the same means as in a conventional power station, i.e. by raising steam to drive a steam turbine that turns a generator. The heat is transferred to the steam-raising boiler or heat-exchanger by the *coolant. Water is frequently used as the coolant; in the case of the **boiling-water reactor** (BWR) and the **pressurized-water reactor** (PWR) water is both coolant and moderator. In the BWR the primary coolant drives the turbine; in the PWR the primary coolant raises steam in a secondary circuit for driving the turbine. In the **gas-cooled reactor** the coolant is a gas, usually carbon dioxide with an outlet temperature of about 350°C, or 600°C in the case of the **advanced gas-cooled reactor** (AGR).

In fast reactors, in which there is no moderator, the temperature is higher and a liquid-metal coolant is used, usually liquid sodium. Some fast reactors are used as converters or breeders. A **converter reactor** is one that converts *fertile material (such as $^{238}$U) into *fissile material (such as $^{239}$Pu). A **breeder reactor** produces the same fissile material as it uses. For example, a **fast breeder reactor** using uranium enriched with $^{239}$Pu as the fuel can produce more $^{239}$Pu than it uses by converting $^{238}$U to $^{239}$Pu. *See also* THERMONUCLEAR REACTOR.

(🌐) **SEE WEB LINKS**
• The website of the World Nuclear Association, an association of companies in the nuclear industry

**nuclear transfer** A technique used in cloning animals in which a nucleus from a donor cell (adult or embryo) is injected into an unfertilized egg cell from which the chromosomes have been removed by micropipette; this is then stimulated by electrical pulses to begin dividing and develop as an embryo (*see* CLONE). The technique has been used successfully with various mammal species, most famously producing Dolly the sheep in 1997. Dolly was the first mammal to be cloned from a fully differentiated adult body cell. The donor cells were taken from a culture of sheep udder cells and starved into a state of quiescence in a low-nutrient medium. This was done to switch off all but essential genes and better mimic a natural fertilization. There are several advantages in using adult body cells: cultures are easier to obtain and maintain, and there is greater scope for genetically engineering such cells and screening

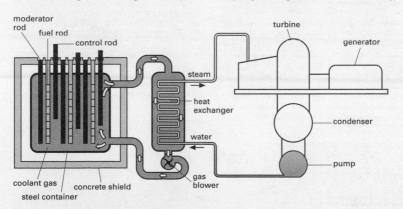

moderator
rod
    fuel rod
        control rod                                    turbine
                                                                    generator


                                    steam


                                    heat
                                    exchanger                      condenser

                                    water
                                                                   pump
coolant gas    concrete shield      gas
    steel container                 blower

**Nuclear reactor.** A schematic diagram of a gas-cooled reactor.

them to select successfully modified cells. Nuclear transfer is now used increasingly to replicate elite animals in the livestock industry, to produce genetically engineered mammals for commercial use (e.g. goats that secrete human proteins in their milk), and to replicate endangered species. However, the failure rate is generally high, and even the few live clones produced often have congenital defects that shorten their lives. This shows that 'reprogramming' differentiated body cells poses formidable technical obstacles.

**nuclear waste** *See* RADIOACTIVE WASTE.

**nuclear weapons** Weapons in which an explosion is caused by *nuclear fission, *nuclear fusion, or a combination of both. In the fission bomb (**atomic bomb** or **A-bomb**) two subcritical masses (*see* CRITICAL MASS) of a *fissile material (uranium–235 or plutonium–239) are brought together by a chemical explosion to produce one supercritical mass. The resulting nuclear explosion is typically in the *kiloton range with temperatures of the order $10^8$ K being reached. The fusion bomb (**thermonuclear weapon**, **hydrogen bomb**, or **H-bomb**) relies on a nuclear-fusion reaction, which becomes self-sustaining at a critical temperature of about $35 \times 10^6$ K. Hydrogen bombs consist of either two-phase fission-fusion devices in which an inner fission bomb is surrounded by a hydrogenous material, such as heavy hydrogen (deuterium) or lithium deuteride, or a three-phase fission-fusion-fission device, which is even more powerful. The *megaton explosion produced by such a thermonuclear reaction has not yet been used in war. A special type of fission-fusion bomb is called a **neutron bomb**, in which most of the energy is released as high-energy neutrons. This neutron radiation destroys people but provides less of the shock waves and blast that destroy buildings.

**nuclease** Any enzyme that breaks down nucleic acids to nucleotides. Nucleases are found in the small intestine. *See also* DNASE; ENDONUCLEASE; EXONUCLEASE.

**nucleic acid** A complex organic compound in living cells that consists of a chain of *nucleotides. There are two types: *DNA (deoxyribonucleic acid) and *RNA (ribonucleic acid).

 SEE WEB LINKS
• Information about IUPAC nomenclature

**nucleoid (nuclear region)** The part of a cell of a bacterium (i.e. a prokaryotic *cell) that contains the genetic material *DNA and therefore controls the activity of the cell. It corresponds to the nucleus of the more advanced eukaryotic cells but is not bounded by a membrane.

**nucleolus** A small dense round body within the nondividing *nucleus of eukaryotic cells that consists of protein, DNA, and ribosomal *RNA. It plays an important role in *ribosome manufacture (and therefore protein synthesis).

**nucleon** A *proton or a *neutron.

**nucleon emission** A decay mechanism in which a particularly unstable *nuclide regains some stability by the emission of a nucleon, i.e. a proton or neutron. Proton emitters have fewer neutrons than their stable isotopes. Proton emitters are therefore found below the *Segrè plot stability line. For example, $^{17}$Ne has three fewer neutrons than its most abundant stable isotope, $^{20}$Ne. There are no naturally occurring proton emitters. Neutron emitters have many more neutrons than their stable isotopes. For this reason, neutron emitters are found above the stability line on the Segrè plot and in most cases can also decay by negative *beta decay. There are no naturally occurring neutron emitters. They are usually produced in nuclear reactors by the negative beta decay of fission products. An example is $^{99}$Y, which has 10 more neutrons than the stable isotope $^{89}$Y.

**nucleonics** The technological aspects of *nuclear physics, including the design of nuclear reactors, devices to produce and detect radiation, and nuclear transport systems. It is also concerned with the technology of *radioactive waste disposal and with radioisotopes.

**nucleon number (mass number)** Symbol A. The number of *nucleons in an atomic nucleus of a particular nuclide.

**nucleophile** An ion or molecule that can donate electrons. Nucleophiles are often oxidizing agents and Lewis bases. They are either negative ions (e.g. $Cl^-$) or molecules that have electron pairs (e.g. $NH_3$). In organic reactions they tend to attack positively charged parts of a molecule. *Compare* ELECTROPHILE.

**nucleophilic addition** A type of addition

reaction in which the first step is attachment of a *nucleophile to a positive (electron-deficient) part of the molecule. *Aldehydes and *ketones undergo reactions of this type because of polarization of the carbonyl group (carbon positive).

**nucleophilic substitution** A type of substitution reaction in which a *nucleophile displaces another group or atom from a compound. For example, in

$$CR_3Cl + OH^- \rightarrow CR_3OH + Cl^-$$

the nucleophile is the $OH^-$ ion. There are two possible mechanisms of nucleophilic substitution. In $S_N1$ **reactions**, a positive carbonium ion is first formed:

$$CR_3Cl \rightarrow CR_3^+ + Cl^-$$

This then reacts with the nucleophile

$$CR_3^+ + OH^- \rightarrow CR_3OH$$

The $CR_3^+$ ion is planar and the $OH^-$ ion can attack from either side. Consequently, if the original molecule is optically active (the three R groups are different) then a racemic mixture of products results.

The alternative mechanism, the $S_N2$ **reaction**, is a concerted reaction in which the nucleophile approaches from the side of the R groups as the other group (Cl in the example) leaves. In this case the configuration of the molecule is inverted. If the original molecule is optically active, the product has the opposite activity, an effect known as **Walden inversion**. The notations $S_N1$ and $S_N2$ refer to the kinetics of the reactions. In the $S_N1$ mechanism, the slow step is the first one, which is unimolecular (and first order in $CR_3Cl$). In the $S_N2$ reaction, the process is bimolecular (and second order overall).

**nucleoplasm (karyoplasm)** The material contained within the *nucleus of a cell. The nucleoplasm is bound by the nuclear envelope, which separates it from the cytoplasm.

**nucleoprotein** Any compound present in cells of organisms that consists of a nucleic acid (DNA or RNA) combined with a protein. Chromosomes consist of DNA and proteins, mostly histones, as do ribosomes (see RIBONUCLEOPROTEIN).

**nucleoside** An organic compound consisting of a nitrogen-containing *purine or *pyrimidine base linked to a sugar (ribose or deoxyribose). An example is *adenosine. Compare NUCLEOTIDE.

**nucleosynthesis** The synthesis of chemical elements by nuclear processes. There are several ways in which nucleosynthesis can take place. **Primordial nucleosynthesis** took place very soon after the *big bang, when the universe was extremely hot. This process was responsible for the cosmic abundances observed for light elements, such as helium. Explosive nucleosynthesis can also occur during the explosion of a *supernova. However, **stellar nucleosynthesis**, which takes place in the centre of stars at very high temperatures, is now the principal form of nucleosynthesis. The exact process occurring in stellar nucleosynthesis depends on the temperature, density, and chemical composition of the star. The synthesis of helium from protons and of carbon from helium can both occur in stellar nucleosynthesis. Some light elements, such as boron, are made by *spallation. See also CARBON CYCLE; EARLY UNIVERSE; STELLAR EVOLUTION.

**nucleotide** An organic compound consisting of a nitrogen-containing *purine or *pyrimidine base linked to a sugar (ribose or deoxyribose) and a phosphate group. *DNA and *RNA are made up of long chains of nucleotides (i.e. **polynucleotides**). Compare NUCLEOSIDE.

**nucleus** (of atom) The central core of an atom that contains most of its mass. It is positively charged and consists of one or more nucleons (protons or neutrons). The positive charge of the nucleus is determined by the number of protons it contains (see ATOMIC NUMBER) and in the neutral atom this is balanced by an equal number of electrons, which move around the nucleus. The simplest nucleus is the hydrogen nucleus, consisting of one proton only. All other nuclei also contain one or more neutrons. The neutrons contribute to the atomic mass (see NUCLEON NUMBER) but not to the nuclear charge. The most massive nucleus that occurs in nature is uranium–238, containing 92 protons and 146 neutrons. The symbol used for this *nuclide is $^{238}_{92}U$, the upper figure being the nucleon number and the lower figure the atomic number. In all nuclei the nucleon number ($A$) is equal to the sum of the atomic number ($Z$) and the neutron number ($N$), i.e. $A = Z + N$.

**nucleus** (of cell) The large body embedded in the cytoplasm of all plant and animal *cells (but not bacterial cells) that contains the genetic material *DNA organized into *chromosomes. The nucleus functions as the control centre of the cell. It is bounded by a double membrane (the **nuclear envelope**), which is perforated by many **nuclear pores** for the selective transfer of water-soluble molecules between the nucleus and cytoplasm. When the cell is not dividing, a *nucleolus is present in the nucleus and the chromosomal material (*chromatin) is dispersed throughout the nucleus. In dividing cells the chromosomes become much shorter and thicker and the nucleolus disappears. The contents of the nucleus constitute the **nucleoplasm**. In certain protozoans there are two nuclei per cell, a **macronucleus** (or **meganucleus**) concerned with vegetative functions and a **micronucleus** involved in sexual reproduction.

**nuclide** A type of atom as characterized by its *atomic number and its *neutron number. An *isotope is a member of a series of different atoms that have the same atomic number but different neutron numbers (e.g. uranium–238 and uranium–235 are isotopes of uranium); a nuclide refers only to a particular nuclear species (e.g. the nuclides uranium–235 and plutonium–239 are fissile). The term is also used for a type of nucleus.

**(((•))) SEE WEB LINKS**

• An interactive chart giving detailed information on all nuclides, from the National Nuclear Data Center, Brookhaven National Laboratory

**null method** A method of making a measurement in which the quantity to be measured is balanced by another similar reading by adjusting the instrument to read zero (*see* WHEATSTONE BRIDGE).

**numerical analysis** The analysis of problems by means of calculations involving numbers rather than analytical formulae. Numerical analysis is used extensively for problems too complicated to be solved analytically (either exactly or approximately). Numerical analysis can be performed using either electronic calculators or computers.

**numerical taxonomy** *See* TAXONOMY.

**nut** A dry single-seeded fruit that develops from more than one carpel and does not shed its seed when ripe. The fruit wall is woody or leathery. Many nuts are enclosed in a hard or membranous cup-shaped structure, the *cupule. The term nut is often loosely used of any hard fruit. For example,

the walnut is in fact a *drupe and the Brazil nut is a seed.

**nutation 1.** (in astronomy) An irregular periodic oscillation of the earth's poles. It causes an irregularity of the precessional circle traced by the celestial poles and results from the varying distances and relative directions of the sun and the moon. **2.** (in botany) The spiral movement of a plant organ during growth, also known as **circumnutation**. It is seen in climbing plants and helps the plant find a suitable support to twine around. Examples are the coiling movements of the shoot tips of runner beans and of the tendrils of sweet peas.

**nutrient** Any substance that is required for the nourishment of an organism, providing a source of energy or structural components. In animals nutrients form part of the *diet and include the **major nutrients**, i.e. carbohydrates, proteins (*see also* ESSENTIAL AMINO ACID), and lipids (*see also* ESSENTIAL FATTY ACIDS), as well as vitamins and certain minerals (*see* ESSENTIAL ELEMENT). Plant nutrients, derived from carbon dioxide in the atmosphere and water (containing minerals) absorbed from the soil by the roots, are *macronutrients or *micronutrients.

**nutrition** The process by which organisms obtain energy (in the form of food) for growth, maintenance, and repair. There are two main types of nutrition: *heterotrophic nutrition, employed by animals, fungi, and certain bacteria; and *autotrophic nutrition, found in most plants and bacteria.

**nyctinasty (sleep movements)** *Nastic movements of plant organs in response to the changes in light and temperature that occur between day and night (and vice versa). Examples are the opening and closing of many flowers and the folding together of the leaflets of clover and other plants at night.

**nylon** Any of various synthetic polyamide fibres having a protein-like structure formed by the condensation between an amino group of one molecule and a carboxylic acid group of another. There are three main nylon fibres, nylon 6, nylon 6,6, and nylon 6,10. Nylon 6, for example Enkalon and Celon, is formed by the self-condensation of 6-aminohexanoic acid. Nylon 6,6, for example Bri nylon, is made by polycondensation of hexanedioic acid (adipic acid) and 1,6-diaminohexane (hexamethylenediamine) having an average formula weight between 12 000 and 15 000. Nylon 6,10 is made by polymerizing decanedioic acid and 1,6-diaminohexane.

**nymph** The juvenile stage of certain insects, such as dragonflies, grasshoppers, and earwigs, which resembles the adult except that the wings and reproductive organs are undeveloped. There is no pupal stage, and the nymph develops directly into the adult. *Compare* LARVA.

**OB association** An area in space in which young high-mass O and B type stars predominate (*see* SPECTRAL CLASS). The stars emit strong ultraviolet radiation, which ionizes the surrounding hydrogen and forms emission nebulae. Shock waves compress dust and gas, which may collapse under the force of gravity and begin the formation of more stars.

**objective** The *lens or system of lenses nearest to the object being examined through an optical instrument.

**occipital condyle** A single or paired bony knob that protrudes from the occipital bone of the skull and articulates with the first cervical vertebra (the *atlas). In humans there is a pair of occipital condyles, one on each side of the *foramen magnum. Occipital condyles are absent in most fish, which cannot move their heads.

**occluded front** *See* FRONT.

**occlusion** **1.** The trapping of small pockets of liquid in a crystal during crystallization. **2.** The absorption of a gas by a solid such that atoms or molecules of the gas occupy interstitial positions in the solid lattice. Palladium, for example, can occlude hydrogen.

**occultation** The obscuring of a star, planet, or other celestial body by the moon or another planet. A solar *eclipse is a form of occultation.

**oceanic zone** The region of the open sea beyond the edge of the continental shelf, where the depth is greater than 200 metres. *Compare* NERITIC ZONE.

**oceanography** The study of the oceans. It includes the origin, structure, and form of the oceans, the nature of the seafloor and its sediments, the characteristics of the ocean waters (e.g. tides, salinity, and currents), and the types of flora and fauna living within the oceans. The effects of human intervention also form an important aspect of oceanography.

**ocean trench** A deep narrow depression in the ocean floor, often thousands of metres deep (the Mariana Trench in the Pacific Ocean is 10 850 m deep). Trenches usually form near the edge of a continent where the tectonic plate carrying the ocean is being subducted beneath the continental plate (*see* PLATE TECTONICS). They are often associated with earthquakes and island arcs.

**ocellus** A simple eye occurring in insects and other invertebrates. It typically consists of light-sensitive cells and a single cuticular lens.

**ochre** A yellow or red mineral form of iron(III) oxide, $Fe_2O_3$, used as a pigment.

**octadecanoate** *See* STEARATE.

**octadecanoic acid** *See* STEARIC ACID.

**octadecenoic acid** A straight-chain unsaturated fatty acid with the formula $C_{17}H_{33}COOH$. **Cis-octadec-9-enoic acid** (*see* OLEIC ACID) has the formula $CH_3(CH_2)_7CH:CH(CH_2)_7COOH$. The glycerides of this acid are found in many natural fats and oils.

**octahedral** *See* COMPLEX.

**octahydrate** A crystalline hydrate that has eight moles of water per mole of compound.

**octane** A straight-chain liquid *alkane, $C_8H_{18}$; r.d. 0.7; m.p. –56.79°C; b.p. 125.66°C. It is present in petroleum. The compound is isomeric with 2,2,4-trimethylpentane, $(CH_3)_3CCH_2CH(CH_3)_2$, **iso-octane**). *See* OCTANE NUMBER.

**octane number** A number that provides a measure of the ability of a fuel to resist *knocking when it is burnt in a spark-ignition engine. It is the percentage by volume of iso-octane ($C_8H_{18}$; 2,2,4-trimethylpentane) in a blend with normal heptane ($C_7H_{16}$) that matches the knocking behaviour of the fuel being tested in a single cylinder four-stroke engine of standard design. *Compare* CETANE NUMBER.

**octanitrocubane** *See* CUBANE.

**octanoic acid (caprylic acid)** A colourless liquid straight-chain saturated *carboxylic acid, $CH_3(CH_2)_6COOH$; b.p. 239.3°C.

**octavalent** Having a valency of eight.

**octave** **1.** The interval between two musical notes that have fundamental frequencies in the ratio 2:1; the word describes the interval in terms of the light notes of the diatonic scale. **2.** *See* LAW OF OCTAVES.

**octet** A stable group of eight electrons in the outer shell of an atom (as in an atom of a noble gas).

**octogen** *See* HMX.

**ocular** *See* EYEPIECE.

**odd–even nucleus** An atomic nucleus containing an odd number of protons and an even number of neutrons.

**odd–odd nucleus** An atomic nucleus containing an odd number of protons and an odd number of neutrons. There are very few stable odd–odd nuclei.

**Odonata** An order of insects containing the dragonflies and damselflies, most of which occur in tropical regions. Adult dragonflies have a pair of prominent *compound eyes, a compact thorax bearing two pairs of delicate membranous wings, and a long slender abdomen. They are strong fliers and prey on other insects, either in flight or at rest. The eggs are laid near or in water, and the newly hatched nymphs are aquatic and resemble the adults, with rudimentary wings. They breathe through gills and feed on small aquatic animals. The nymph leaves the water for its final moult into the terrestrial adult.

**odontoblast** A cell that is responsible for producing the *dentine of vertebrate teeth. Odontoblasts are found around the lining of the *pulp cavity and have processes that extend into the dentine.

**oersted** Symbol Oe. The unit of magnetic field strength in the *c.g.s. system. A field has a strength of one oersted if it exerts a force of one dyne on a unit magnetic pole placed in it. It is equivalent to $10^3/4\pi$ A m$^{-1}$. The unit was named after Hans Oersted.

**Oersted, Hans Christian** (1777–1851) Danish physicist, who became a professor at Copenhagen in 1806. His best-known discovery came during a lecture in 1820, when he observed the deflection of a compass needle near a wire carrying an electric current. He had discovered electromagnetism.

**oesophagus (gullet)** The section of the *alimentary canal that lies between the *pharynx and the stomach. It is a muscular tube whose function is to transfer food to the stomach by means of wavelike contractions (*peristalsis) along its length.

**oestrogen** One of a group of female sex hormones, produced principally by the ovaries, that promote the onset of *secondary sexual characteristics (such as breast enlargement and development in women) and control the *oestrous cycle (*menstrual cycle in humans). **Oestradiol** is the most important. Oestrogens are secreted at particularly high levels during ovulation, stimulating the uterus to prepare for pregnancy. They are used in *oral contraceptives (with *progestogens) and as treatment for various disorders of the female reproductive organs. Small amounts of oestrogens are produced by the adrenal glands and testes.

**oestrous cycle** The cycle of reproductive activity shown by most sexually mature nonpregnant female mammals except most primates (*compare* MENSTRUAL CYCLE). There are four phases:
(1) **pro-oestrus** – *Graafian follicles develop in the ovary and secrete oestrogens;
(2) **oestrus** (**heat**) – ovulation normally occurs, the female is ready to mate and becomes sexually attractive to the male;
(3) **metoestrus** – *corpus luteum develops from ruptured follicle;
(4) **dioestrus** – *progesterone secreted by corpus luteum prepares uterus for implantation.

The length of the cycle depends on the species: larger mammals typically have a single annual cycle with a well-defined breeding season (they are described as **monoestrous**). The males have a similar cycle of sexual activity. Other species may have many cycles per year (i.e. they are **polyoestrous**) and the male may be sexually active all the time.

**oestrus (heat)** *See* OESTROUS CYCLE.

**offset** *See* RUNNER.

**offspring (progeny)** New individual organisms that result from the process of sexual or asexual reproduction. *See also* $F_1$; $F_2$.

**ohm** Symbol $\Omega$. The derived *SI unit of electrical resistance, being the resistance between two points on a conductor when a constant potential difference of one volt, applied between these points, produces a current of one ampere in the conductor. The former **international ohm** (sometimes called the 'mercury ohm') was defined in terms of the resistance of a column of mercury. The unit is named after Georg Ohm.

**Ohm, Georg Simon** (1787–1854) German physicist, who taught in Cologne, Berlin, Nuremberg, and finally (1849) Munich. He is best known for formulating *Ohm's law in 1827. The unit of electrical resistance is named after him.

**ohmmeter** Any direct-reading instrument for measuring the value of a resistance in ohms. The instrument commonly used is a *multimeter capable of measuring also both currents and voltages. To measure resistance a dry cell and resistor are switched in series with the moving coil *galvanometer and the unknown resistance is connected across the instrument's terminals. The value of the resistance is then read off an ohms scale. Such instruments are increasingly being replaced by electronic digital multimeters.

**Ohm's law** The ratio of the potential difference between the ends of a conductor to the current flowing through it is constant. This constant is the *resistance of the conductor, i.e. $V = IR$, where $V$ is the potential difference in volts, $I$ is the current in amperes, and $R$ is the resistance in ohms. The law was discovered in 1827 by Georg Ohm. Most materials do not obey this simple linear law; those that do are said to be **ohmic** but remain so only if physical conditions, such as temperature, remain constant. Metals are the most accurately ohmic conductors.

**oil** Any of various viscous liquids that are generally immiscible with water. Natural plant and animal oils are either volatile mixtures of terpenes and simple esters (e.g. *essential oils) or are *glycerides of fatty acids. Mineral oils are mixtures of hydrocarbons (e.g. *petroleum).

**oil-immersion lens** *See* IMMERSION OBJECTIVE.

**oil of vitriol** *See* SULPHURIC ACID.

**oil of wintergreen** Methyl salicylate (methyl 2-hydroxybenzoate, $C_8H_8O_3$), a colourless aromatic liquid ester, b.p. 223°C. It occurs in the essential oils of some plants, and is manufactured from salicylic acid. It is easily absorbed through the skin and used in medicine for treating muscular and sciatic

pain. Because of its attractive smell it is also used in perfumes and food flavourings.

**oil sand (tar sand; bituminous sand)** A sandstone or porous carbonate rock that is impregnated with hydrocarbons. The largest deposit of oil sand occurs in Alberta, Canada (the Athabasca tar sands); there are also deposits in the Orinoco Basin of Venezuela, Russia, USA, Madagascar, Albania, Trinidad, and Romania.

**oil shale** A fine-grained carbonaceous sedimentary rock from which oil can be extracted. The rock contains organic matter – **kerogen** – which decomposes to yield oil when heated. Deposits of oil shale occur on every continent, the largest known reserves occurring in Colorado, Utah, and Wyoming in the USA. Commercial production of oil from oil shale is generally considered to be uneconomic unless the price of petroleum rises above the recovery costs for oil from oil shale. However, threats of declining conventional oil resources have resulted in considerable interest and developments in recovery techniques.

**Oklo reactors** Naturally occurring nuclear fission reactors that are believed to have existed in uranium deposits at Oklo in Gabon, West Africa, about 2000 million years ago. In 1972, French scientists noticed a slight difference in the normal $^{235}U/^{238}U$ ratio in uranium ore from Oklo. Further detailed investigations showed that there had been 15 natural reactors in the ore deposits at Oklo, operating intermittently for about 1 million years. It is thought that the geology of the mine was an important factor in the creation of these reactors, in particular, the seepage of water through overlying rock, which functioned as a moderator. A similar natural reactor has been found at Bangombe, some miles south of Oklo, but no other comparable reactors have been found anywhere in the world. The Oklo reactors are of considerable interest. They involve basic nuclear processes occurring occurring 2000 million years ago and might give insights into the time dependence of *fundamental constants. More practically, Oklo can be regarded as a 2000-million-year experiment in the containment of nuclear waste. The reactors shut down naturally when the proportion of $^{235}U$ decreased, and – for the same reason – natural reactors of this type could not occur today. The products of the reactor have, however, been localized because of the geology of the region, in particular, beds of granite underlying the ore deposits.

**Olbers' paradox** If the universe is infinite, uniform, and unchanging the sky at night would be bright, as in whatever direction one looked one would eventually see a star. The number of stars would increase in proportion to the square of the distance from the earth; the intensity of light reaching the earth from a given star is inversely proportional to the square of the distance. Consequently, the whole sky should be about as bright as the sun. The paradox, that this is not the case, was stated by Heinrich Olbers (1758–1840) in 1826. (It had been discussed earlier, in 1744, by J. P. L. Chesaux.) The paradox is resolved by the fact that, according to the *big-bang theory, the universe is not infinite, not uniform, and not unchanging. For instance, light from the most distant galaxies displays an extreme *redshift and ceases to be visible.

**oleate** A salt or ester of *oleic acid.

**olefines** *See* ALKENES.

**oleic acid** An unsaturated *fatty acid with one double bond, $CH_3(CH_2)_7$-$CH:CH(CH_2)_7COOH$; r.d. 0.9; m.p. 13°C. Oleic acid is one of the most abundant constituent fatty acids of animal and plant fats, occurring in butterfat, lard, tallow, groundnut oil, soya-bean oil, etc. Its systematic chemical name is **cis-octadec-9-enoic acid**.

**oleum** *See* DISULPHURIC(VI) ACID.

**olfaction** The sense of smell or the process of detecting smells. This is achieved by receptors in **olfactory organs** (such as the *nose) that are sensitive to air- or waterborne chemicals. Stimulation of these receptors results in the transmission of information to the brain via the **olfactory nerve**.

(((●))) SEE WEB LINKS
• Entertaining tutorial on the sense of smell compiled by Tim Jacob of Cardiff University

**Oligocene** The third geological epoch of the *Palaeogene period. It began about 34 million years ago, following the Eocene epoch, and extended for about 11 million years to the beginning of the Miocene epoch. The epoch was characterized by the continued rise of mammals; the first pigs, rhinoceroses, and tapirs made their appearance.

**Oligochaeta** A class of hermaphrodite annelid worms that bear only a few bristles (*chaetae). Oligochaetes are very abundant in freshwater and terrestrial habitats. The most familiar members of the class are the earthworm (*Lumbricus*) and the freshwater bloodworm (*Tubifex*).

**oligopeptide** *See* PEPTIDE.

**oligosaccharide** A carbohydrate (a type of *sugar) whose molecules contain a chain of up to 20 united monosaccharides. Oligosaccharides are formed as intermediates during the digestion of *polysaccharides, such as cellulose and starch.

**oligotrophic** Describing a body of water (e.g. a lake) with a poor supply of nutrients and a low rate of formation of organic matter by photosynthesis. *Compare* DYSTROPHIC; EUTROPHIC.

**olivine** An important group of rock-forming silicate minerals crystallizing in the orthorhombic system. Olivine conforms to the general formula $(Mg,Fe)_2SiO_4$ and comprises a complete series from pure magnesium silicate (forsterite, $Mg_2SiO_4$) to pure iron silicate (fayalite, $Fe_2SiO_4$). It is green, brown-green, or yellow-green in colour.

**omasum** The third of four chambers that form the stomach of ruminants. *See* RUMINANTIA.

**omega fatty acid** *See* ESSENTIAL FATTY ACIDS.

**omega-minus particle** A spin 3/2 *baryon made up three strange quarks (*see* STRANGENESS). The existence of the omega-minus particle, as well as its properties, were predicted by Murray Gell-Mann in 1962 as part of a scheme to classify baryons, called the **eightfold way**. The omega-minus particle was subsequently discovered experimentally, thus demonstrating the validity of the eightfold way. This discovery was historically very important in the theoretical understanding of the strong interactions (*see* FUNDAMENTAL INTERACTIONS). The mass of the omega mass particle is 1672.5 MeV and its average lifetime is $0.8 \times 10^{-10}$ s. The omega minus particle has an electric charge of –1 (its *antiparticle has a charge of +1). *See* ELEMENTARY PARTICLES.

(((●))) SEE WEB LINKS
• A bubble-chamber photograph of the discovery of the omega-minus particle

**ommatidium** *See* COMPOUND EYE.

**omnivore** An animal that eats both animal and vegetable matter. Pigs, for example, are omnivorous. *Compare* CARNIVORE; HERBIVORE.

**oncogene** A dominant mutant allele of a cellular gene (a **proto-oncogene**) that disrupts cell growth and division and is capable of transforming a normal cell into a cancerous cell. Mutations in proto-oncogenes tend to relax mechanisms that control the cell cycle and accelerate cell division, leading to the cell proliferation that is characteristic of cancer. Some oncogenic mutations cause inhibition of programmed cell death (*apoptosis), so that cancerous cells are less likely to be destroyed by the body's defences. Certain oncogenes of vertebrates are derived from viruses (*see* ONCOGENIC).

**oncogenic** Describing a chemical, organism, or environmental factor that causes the development of cancer. Some viruses are oncogenic to vertebrates, notably the *retroviruses (including the Rous sarcoma virus of chickens), human papillomavirus, and human adenovirus. These viruses contain genes (known as *oncogenes) that become integrated into the host cell's DNA and are responsible for the transformation of a normal cell into a cancerous cell. *See also* GROWTH FACTOR.

**one gene–one polypeptide hypothesis** The theory that each *gene is responsible for the synthesis of a single *polypeptide. It was originally stated as the **one gene–one enzyme hypothesis** by the US geneticist George Beadle (1903–89) in 1945 but later modified when it was realized that genes also encoded nonenzyme proteins and individual polypeptide chains. It is now known that some genes code for various types of RNA involved in protein synthesis.

**one-pot synthesis** A method of synthesizing organic compounds in which the materials used are mixed together in a single vessel and allowed to react, rather than conducting the reaction in a sequence of separate stages.

**onium ion** An ion formed by adding a proton to a neutral molecule, e.g. the hydroxonium ion ($H_3O^+$) or the ammonium ion ($NH_4^+$).

**ontogeny** The developmental course of an organism from the fertilized egg through to maturity. It has been suggested that "ontogeny recapitulates *phylogeny", i.e. the stages of development, especially of the embryo, reflect the evolutionary history of the organism. This idea is now discredited.

**ontology** A specification of the assumptions, terms, or concepts underlying a particular field of knowledge. For example, the **Gene Ontology** (GO) project is an international collaboration between various databases in the field of genomics to standardize terminology used by researchers. Standardization of terms in this way is vital for efficient searching of databases, particularly for devising and using automated search programs.

**oocyte** *See* OOGENESIS.

**oogamy** Sexual reproduction involving the formation and subsequent fusion of a large, usually stationary, female gamete and a small motile male gamete. The female gamete may contain nourishment for the development of the embryo, which is often retained and protected by the parent organism.

**oogenesis** The production and growth of the ova (egg cells) in the animal ovary. Special cells (**oogonia**) within the ovary divide repeatedly by mitosis to produce large numbers of prospective egg cells (**oocytes**). When mature, these undergo meiosis, which halves the number of chromosomes. During the first meiotic division a **polar body** and a secondary oocyte are produced. At the second meiotic division the secondary oocyte produces an ovum and a second polar body. Oocytes may be present in the ovaries at birth and may represent the total number of eggs to be produced.

**oogonium** 1. The female sex organ (*gametangium) of algae and fungi. 2. Any of the immature sex cells in the animal ovary that give rise to oocytes by mitotic divisions (*see* OOGENESIS).

**Oort cloud (Öpik-Oort cloud)** A spherical cloud of *comets that is believed to surround the entire solar system and provide a reservoir of long-period comets. It is estimated to contain up to $10^{12}$ comets and to extend from between 2000 and 5000 *astronomical units (AU) to 50 000 AU from the sun. Disturbances caused by a passing star push comets into eccentric solar orbits that may bring them to the inner solar system or else eject them from the solar system altogether. It is named after Ernst Öpik (1893–1985) and Jan Oort (1899–1971).

**oosphere (ovum; egg cell)** The nonmotile female gamete in plants and some algae. In angiosperms (flowering plants) it is a cell in the *embryo sac of the ovule. In other plants it is situated in an *archegonium. In algae, such as *Fucus*, the oosphere is protected by an **oogonium** until it is shed into the water prior to fertilization. Many oospheres store food in the form of starch or oil droplets.

**oospore** A zygote that is produced as a result of *oogamy in certain algae and fungi. It contains food reserves, develops a protective outer covering, and enters a resting phase before germination. *Compare* ZYGOSPORE.

**opacity** The extent to which a medium is opaque to electromagnetic radiation, especially to light. It is the reciprocal of the *transmittance. A medium that is opaque to X-rays and gamma rays is said to be **radiopaque**.

**opal** A hydrous amorphous form of silica. Many varieties of opal occur, some being prized as gemstones. Common opal is usually milk white but the presence of impurities may colour it yellow, green, or red. Precious opals, which are used as gemstones, display the property of **opalescence** – a characteristic internal play of colours resulting from the interference of light rays within the stone. Black opal has a black background against which the colours are displayed. The chief sources of precious opals are Australia and Mexico. Geyserite is a variety deposited by geysers or hot springs. Another variety, diatomite, is made up of the skeletons of diatoms.

**open chain** *See* CHAIN.

**open cluster (galactic cluster)** *See* STAR CLUSTER.

**open-hearth process** A traditional but now obsolete method for manufacturing steel by heating together scrap, pig iron, etc., in a refractory-lined shallow open furnace heated by burning producer gas in air. It has been replaced by the *basic-oxygen process.

**open reading frame** *See* READING FRAME.

**opera glasses** *See* BINOCULARS.

**operator** A mathematical symbol indicating that a specified operation should be carried out. For example, the operator √ in √$x$ indicates that the square root of $x$ should be

taken; the operator d/d$x$ in d$y$/d$x$ indicates that $y$ should be differentiated with respect to $x$, etc.

**operculum** 1. (in zoology) A lid or flap of skin covering an aperture, such as the gill slit cover of fish and larval amphibians and the horny calcareous operculum secreted by many gastropod molluscs, which closes the opening of the shell when the animal is inside. 2. (in botany) The cone-shaped lid of the *capsule of mosses, which is forcibly detached to release the spores.

**operon** A functionally integrated genetic unit for the control of gene expression in bacteria, as proposed in the *Jacob–Monod hypothesis. Typically it comprises a closely linked group of **structural genes**, coding for protein, and adjacent loci controlling their expression – an **operator site** and a **promoter site**. The structural genes tend to encode enzymes concerned with a particular biochemical pathway. *Transcription of the structural genes is prevented by binding of a **repressor** molecule to the operator site. Another molecule, the **inducer**, can bind to the repressor molecule, preventing it from binding to the operator and thus allowing the promoter site to bind the enzyme RNA polymerase, thereby initiating transcription. The repressor molecule is encoded by a **regulator gene**, which may be close to or distant from the operon. Some operons also have an **attenuator region** preceding the first structural gene, where transcription may either stall or proceed according to the amount of end product in the cell. *See also* LAC OPERON.

**opiate** One of a group of drugs derived from **opium**, an extract of the poppy plant *Papaver somniferum* that depresses brain function (a **narcotic** action). Opiates include *morphine and its synthetic derivatives, such as *heroin and codeine. They are used in medicine chiefly to relieve pain, but the use of morphine and heroin is strictly controlled since they can cause drug dependence and tolerance.

**opioid** Any one of a group of substances that produce pharmacological and physiological effects similar to those of morphine. The **endogenous opioids**, which occur naturally in the body, include the *endorphins and *enkephalins.

**Oppenheimer–Volkoff limit** The maximum mass a neutron star can have before it undergoes gravitational collapse to a *black hole. It is more difficult to estimate this limit than the analogous *Chandrasekhar limit for white dwarf stars. It is thought that the Oppenheimer–Volkoff limit is between two and three times the mass of the sun. It was first calculated by Robert Oppenheimer and George Volkoff in 1939.

**opposition** The point at which a planet having its orbit outside that of the earth is in a line with the earth and the sun. When a planet is in opposition it can be observed during the night and is near to its closest point to the earth; it is therefore a favourable opportunity for observation.

**opsin** The lipoprotein component of *rhodopsin, the light-sensitive pigment that occurs in the rod cells of the retina.

**optical activity** The ability of certain substances to rotate the plane of plane-polarized light as it passes through a crystal, liquid, or solution. It occurs when the molecules of the substance are asymmetric, so that they can exist in two different structural forms each being a mirror image of the other (see CHIRALITY ELEMENT). The two forms are **optical isomers** or **enantiomers**. The existence of such forms is also known as **enantiomorphism** (the mirror images being **enantiomorphs**). One form will rotate the light in one direction and the other will rotate it by an equal amount in the other. The two possible forms are described as *dextrorotatory or *laevorotatory according to the direction of rotation. An equimolar mixture of the two forms is not optically active. It is called a **racemic mixture** (or **racemate**). Prefixes are used to designate the isomer: (+)- (dextrorotatory), (–)- (laevorotatory), and (±)- (racemic mixture) are now preferred to, and increasingly used for, the former *d*-, *l*-, and *dl*-, respectively. In addition, certain molecules can have a **meso-isomer** in which one part of the molecule is a mirror image of the other. Such molecules are not optically active.

**Optical activity.**

Molecules that show optical activity have no plane of symmetry. The commonest case of this is in organic compounds in which a carbon atom is linked to four different groups. An atom of this type is said to be a **chiral centre**. Asymmetric molecules showing optical activity can also occur in inorganic compounds. For example, an octahedral complex in which the central ion coordinates to six different ligands would be optically active. Many naturally occurring compounds show optical isomerism and usually only one isomer occurs naturally. For instance, glucose is found in the dextro-rotatory form. The other isomer, (–)- or *l*-glucose, can be synthesized in the laboratory, but cannot be synthesized by living organisms. See also ABSOLUTE CONFIGURATION.

**optical axis (principal axis; optic axis)** The line passing through the *optical centre and the centre of a curvature of a *lens or spherical *mirror.

**optical brightener** See BRIGHTENERS.

**optical centre** The point at the geometrical centre of a *lens through which a ray of light entering the lens passes without deviation.

**optical fibre** A *wave guide through which light can be transmitted with very little leakage through the sidewalls. In the **step-index fibre** a pure glass core, with a diameter between 6 and 250 micrometres, is surrounded by a coaxial glass or plastic cladding of lower refractive index. The cladding is usually between 10 and 150 micrometres thick. The interface between core and cladding acts as a cylindrical mirror at which *total internal reflection of the transmitted light takes place. This structure enables a beam of light to travel through many kilometres of fibre. In the **graded-index fibre**, each layer of glass, from the fibre axis to its outer wall, has a slightly lower refractive index than the layer inside it. This arrangement also prevents light from escaping through the fibre walls by a combination of refraction and total internal reflection, and can be made to give the same transit time for rays at different angles.

**Fibre-optic systems** use optical fibres to transmit information, in the form of coded pulses or fragmented images (using bundles of fibres), from a source to a receiver. They are also used in medical instruments (**fibrescopes**) to examine internal body cavities, such as the stomach and bladder.

**optical flat** A flat glass disc having very accurately polished surfaces so that the deviation from perfect flatness does not exceed (usually) 50 nanometres. It is used to test the flatness of such plane surfaces as gauge anvils by means of the *interference patterns formed when parallel beams of light pass through the flat and are reflected by the surface being inspected.

Surfaces are said to be **optically flat** if the deviation from perfect flatness is smaller than the wavelength of light.

**optical glass** Glass used in the manufacture of lenses, prisms, and other optical parts. It must be homogeneous and free from bubbles and strain. Optical **crown glass** may contain potassium or barium in place of the sodium of ordinary crown glass and has a refractive index in the range 1.51 to 1.54. **Flint glass** contains lead oxide and has a refractive index between 1.58 and 1.72. Higher refractive indexes are obtained by adding lanthanoid oxides to glasses; these are now known as lanthanum crowns and flints.

**optical isomers** *See* OPTICAL ACTIVITY.

**optical lattice** A periodic array produced by interference between laser beams propagating in opposite directions. The interference results in a periodic potential, and it is possible to trap atoms at the minima of this potential. Optical lattices can be formed in two or three dimensions. They have been used to study many aspects of quantum mechanics, particularly the theory of many-body systems, and as *optical traps.

**optical lever** An experimental device used to measure angular rotation (e.g. in a *galvanometer or *torsion balance). Typically, a small mirror is attached to the rotating object, and a beam of light is directed onto the mirror and reflected onto a scale. The angle turned through by the beam is twice the angle turned through by the mirror.

**optical microscope** *See* MICROSCOPE.

**optical molasses** *See* LASER COOLING.

**optical pumping** *See* LASER.

**optical pyrometer** *See* PYROMETRY.

**optical rotary dispersion (ORD)** The effect in which the amount of rotation of plane-polarized light by an optically active compound depends on the wavelength. A graph of rotation against wavelength has a characteristic shape showing peaks or troughs.

**optical rotation** Rotation of plane-polarized light. *See* OPTICAL ACTIVITY.

**optical telescope** *See* TELESCOPE.

**optical temperature** *See* RADIATION TEMPERATURE.

**optical trap** A device for trapping small particles using electromagnetic radiation. Both *optical lattices and *optical tweezers can be used as optical traps.

**optical tweezers** A device that is used to manipulate particles such as atoms or molecules. It works by pushing the particles about by *radiation pressure associated with a laser. Since optical tweezers can be used to manipulate strands of DNA they may have important applications in biotechnology.

**optic axis 1.** The direction in a doubly refracting crystal in which light is transmitted without double refraction. **2.** *See* OPTICAL AXIS.

**optic lobes** *See* MIDBRAIN.

**optic nerve** The second *cranial nerve: a paired sensory nerve that runs from each eye to the brain. It is responsible for conveying visual stimuli received by the rods and cones in the retina to the brain for interpretation.

**optics** The study of *light and the phenomena associated with its generation, transmission, and detection. In a broader sense, optics includes all the phenomena associated with infrared and ultraviolet radiation. **Geometrical optics** assumes that light travels in straight lines and is concerned with the laws controlling the reflection and refraction of rays of light. **Physical optics** deals with phenomena that depend on the wave nature of light, e.g. diffraction, interference, and polarization.

**oral cavity** *See* BUCCAL CAVITY.

**oral contraceptive** Any hormonal preparation taken in the form of a pill to prevent conception (*see* BIRTH CONTROL). The most common form is the combined pill, which contains an *oestrogen and a *progestogen. Both act to suppress ovulation, while the progestogen additionally causes changes in the viscosity of cervical mucus and alters the lining of the womb, both of which decrease the chances of fertilization should ovulation occur. The so-called 'minipill' contains only a progestogen and has fewer side effects than the combined pill. Emergency contraception (the so-called 'morning-after pill'), to prevent pregnancy after unprotected sexual intercourse, consists of two spaced doses of either a combined oestrogen–progestogen preparation or an oestrogen alone, the first dose being taken within 72 hours of intercourse.

**orbit** **1.** (in astronomy) The path through space of one celestial body about another. For one small body moving in the gravitational field of another the orbit is a \*conic section. Most such orbits are elliptical and most planetary orbits in the solar system are nearly circular. The shape and size of an elliptical orbit is specified by its eccentricity, $e$, and the length of its semimajor axis, $a$. **2.** (in physics) The path of an electron as it travels round the nucleus of an atom. *See* ORBITAL. **3.** (in anatomy) Either of the two sockets in the skull of vertebrates that house the eyeballs.

**orbital** A region in which an electron may be found in an atom or molecule. In the original \*Bohr theory of the atom the electrons were assumed to move around the nucleus in circular orbits, but further advances in quantum mechanics led to the view that it is not possible to give a definite path for an electron. According to \*wave mechanics, the electron has a certain probability of being in a given element of space. Thus for a hydrogen atom the electron can be anywhere from close to the nucleus to out in space but the maximum probability in spherical shells of equal thickness occurs in a spherical shell around the nucleus with a radius equal to the Bohr radius of the atom. The probabilities of finding an electron in different regions can be obtained by solving the Schrödinger wave equation to give the wave function $\psi$, and the probability of location per unit volume is then proportional to $|\psi|^2$. Thus the idea of electrons in fixed orbits has been replaced by that of a probability distribution around the nucleus – an **atomic orbital** (see illustration). Alternatively, the orbital can be thought of as an electric charge distribution (averaged over time). In representing orbitals it is convenient to take a surface enclosing the space in which the electron is likely to be found with a high probability.

The possible atomic orbitals correspond to subshells of the atom. Thus there is one $s$-orbital for each shell (orbital quantum number $l = 0$). This is spherical. There are three

$p$-orbitals (corresponding to the three values of $l$) and five $d$-orbitals. The shapes of orbitals depend on the value of $l$. For instance, $p$-orbitals each have two lobes; most $d$-orbitals have four lobes.

In molecules, the valence electrons move under the influence of two nuclei (in a bond involving two atoms) and there are corresponding **molecular orbitals** for electrons (see illustration). It is convenient in considering these to regard them as formed by overlap of atomic orbitals. In a hydrogen molecule the $s$-orbitals on the two atoms overlap and form a molecular orbital between the two nuclei. This is an example of a **sigma orbital**. In a double bond, as in ethene, one bond is produced by overlap along the line of axes to form a sigma orbital. The other is produced by sideways overlap of the lobes of the $p$-orbitals (see illustration). The resulting molecular orbital has two parts, one on each side of the sigma orbital – this is a **pi orbital**. It is also possible for a **delta orbital** to form by lateral overlap of two $d$-orbitals. In fact, the combination of two atomic orbitals produces two molecular orbitals with different energies. The one of lower energy is the **bonding orbital**, holding the atoms together; the other is the **antibonding orbital**, which would tend to push the atoms apart. In the case of valence electrons, only the lower (bonding) orbital is filled.

In considering the formation of molecular orbitals it is often useful to think in terms of **hybrid** atomic orbitals. For instance, carbon has in its outer shell one $s$-orbital and three $p$-orbitals. In forming methane (or other tetrahedral molecules) these can be regarded as combining to give four equivalent $sp^3$ hybrid orbitals, each with a lobe directed to a corner of a tetrahedron. It is these that overlap with the $s$-orbitals on the hydrogen atoms. In ethene, two $p$-orbitals combine with the $s$-orbital to give three $sp^2$ hybrids with lobes in a plane pointing to the corners of an equilateral triangle. These form the sigma orbitals in the C–H and C–C bonds. The remaining $p$-orbitals (one on each carbon) form the pi orbital. In ethyne, $sp^2$ hybridization occurs to give two hybrid orbitals on each atom with lobes pointing along the axis. The two remaining $p$-orbitals on each carbon form two pi orbitals. Hybrid atomic orbitals can also involve $d$-orbitals. For instance, square-planar complexes use $sp^2d$ hybrids; octahedral complexes use $sp^3d^2$.

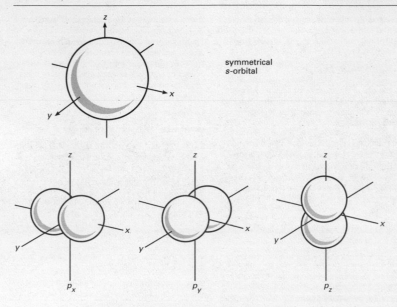

symmetrical
s-orbital

three equivalent *p*-orbitals, each having 2 lobes

atomic orbitals

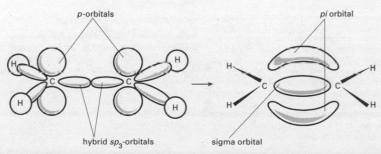

*p*-orbitals

*pi* orbital

hybrid *sp*₃-orbitals          sigma orbital

molecular orbitals: formation of the double bond in ethene

**Orbitals.**

**orbital quantum number** *See* ATOM.

**orbital resonance** An effect in *celestial mechanics resulting from a situation in which two bodies have periods of revolution that are in a simple whole-number ratio, allowing each body to have a regularly recurring gravitational effect upon the other. Depending on the mass of each body con-

cerned, orbital resonance can either stabilize the orbits and protect them from being perturbed, or destabilize the orbit of one of the bodies, ejecting it from its path into a new one or making its path more elliptical. *See also* KIRKWOOD GAP.

**orbital velocity (orbital speed)** The speed of a satellite, spacecraft, or other body travelling in an *orbit around the earth or around some other celestial body. If the orbit is elliptical, the orbital speed, $v$, is given by:

$$v = \sqrt{[gR^2(2/r - 1/a)]},$$

where $g$ is the accceleration of free fall, $R$ is the radius of the orbited body, $a$ is the semi-major axis of the orbit, and $r$ is the distance between the orbiting body and the centre of mass of the system. If the orbit is circular, $r = a$ and $v = \sqrt{(gR^2/r)}$.

**OR circuit** See LOGIC CIRCUITS.

**order** 1. (in chemistry) In the expression for the rate of a chemical reaction, the sum of the powers of the concentrations is the overall order of the reaction. For instance, in a reaction

A + B → C

the rate equation may have the form

$$R = k[A][B]^2$$

This reaction would be described as **first order** in A and **second order** in B. The overall order is three. The order of a reaction depends on the mechanism and it is possible for the rate to be independent of concentration (**zero order**) or for the order to be a fraction. See also MOLECULARITY; PSEUDO ORDER.
2. (in mathematics) The number of times a variable is differentiated. $\mathrm{d}y/\mathrm{d}x$ represents a first-order derivative, $\mathrm{d}^2y/\mathrm{d}x^2$ a second-order derivative, etc. In a *differential equation the order of the highest derivative is the order of the equation. $\mathrm{d}^2y/\mathrm{d}x^2 + 2\mathrm{d}y/\mathrm{d}x = 0$ is a second-order equation of the first degree. See also DEGREE.
3. (in taxonomy) A category used in the *classification of organisms that consists of one or several similar or closely related families. Similar orders form a class. Order names typically end in -ales in botany, e.g. Rosales (roses and orchard fruits), and in -a in zoology, e.g. Carnivora (flesh eaters).
4. (in physics) A category of *phase transition.

**order of magnitude** A value expressed to the nearest power of ten.

**order parameter** A quantity that characterizes the order of a phase of a system below its *transition temperature. An order parameter has a non-zero value below the transition temperature and a zero value above the transition temperature. An example of an order parameter is magnetization (see MAGNETISM) in a ferromagnetic system. If the phase transition is continuous (i.e.

there is no *latent heat), the order parameter goes to zero continuously as the transition temperature is approached from below. **Disorder parameters** are quantities that are non-zero above the transition temperature and zero beneath it. Order parameters are associated with the *broken symmetry of a system.

**ordinary ray** . See DOUBLE REFRACTION.

**ordinate** See CARTESIAN COORDINATES.

**Ordovician** The second geological period of the Palaeozoic era, following the Cambrian and preceding the Silurian periods. It began about 488 million years ago and lasted for about 44 million years. The period was named by the British geologist Charles Lapworth (1842–1920) in 1879. *Graptolites, in deep-water deposits, are the dominant fossils. Other fossils include *trilobites, brachiopods, bryozoans, gastropods, bivalves, echinoids, crinoids, nautiloid cephalopods, and the first corals.

**ore** A naturally occurring mineral from which a metal and certain other elements (e.g. phosphorus) can be extracted, usually on a commercial basis. Metals may be present in ores in the native form, but more commonly they occur combined as oxides, sulphides, sulphates, silicates, etc.

**ore dressing** See BENEFICIATION.

**oregonator** A type of chemical reaction mechanism that causes an *oscillating reaction. It is the type of mechanism responsible for the *B–Z reaction, and involves five steps of the form:

A + Y → X

X + Y → C

A + X → 2X + Z

2X → D

Z → Y

Autocatalysis occurs as in the *Lotka–Volterra mechanism and the *brusselator. The mechanism was named after Oregon in America, where the research group that discovered it is based.

**organ** Any distinct part of an organism that is specialized to perform one or a number of functions. Examples are ears, eyes, lungs, and kidneys (in animals) and leaves, roots, and flowers (in plants). A given organ will contain many different *tissues.

**organ culture** The culture of complete living organs (**explants**) of animals and plants outside the body in a suitable culture medium. Animal organs must be small enough to allow the nutrients in the culture medium to penetrate all the cells. Whole plant roots and even root systems can be kept alive in such conditions for a considerable period of time. *See also* EXPLANTATION.

**organelle** A minute structure within a eukaryotic *cell that has a particular function. Examples of organelles are the nucleus, mitochondria, and lysosomes.

**organic chemistry** The branch of chemistry concerned with compounds of carbon.

**organic evolution** The process by which changes in the genetic composition of populations of organisms occur in response to environmental changes. *See* ADAPTATION; EVOLUTION. *Compare* BIOCHEMICAL EVOLUTION.

**organism** An individual living system, such as an animal, plant, or *microorganism, that is capable of reproduction, growth, and maintenance.

**organizer** An area of an animal embryo that causes adjacent areas of the embryo to develop in a certain way. The **primary organizer** (blastopore lip or archenteron roof) directs the overall development of the *gastrula.

**organo-** Prefix used before the name of an element to indicate compounds of the elements containing organic groups (with the element bound to carbon atoms). For example, lead(IV) tetraethyl is an organolead compound.

**organogenesis** The formation of organs during embryonic development. In animals this begins following the rearrangement of the cells at gastrulation, when the three germ layers are fully formed in their correct positions. Dividing cells of the *gastrula begin to differentiate and the rudimentary organs and organ systems begin to form. *See* DIFFERENTIATION; ECTODERM; ENDODERM; MESODERM.

**organometallic compound** A compound in which a metal atom or ion is bound to an organic group. Organometallic compounds may have single metal–carbon bonds, as in the aluminium alkyls (e.g. $Al(CH_3)_3$). In some cases, the bonding is to the pi electrons of a double bond, as in complexes formed between platinum and ethene, or to the pi electrons of a ring, as in *ferrocene.

**SEE WEB LINKS**
- Information about IUPAC nomenclature
- Further information about nomenclature

**orgasm** The climax of sexual excitement in humans, which – in males – coincides with *ejaculation. A sense of physiological and emotional release is accompanied by a feeling of extreme pleasure.

**origin** *See* CARTESIAN COORDINATES.

**origin of life** The process by which living organisms developed from inanimate matter, which is generally thought to have occurred on earth between 3500 and 4000 million years ago. It is supposed that the primordial atmosphere was like a chemical soup containing all the basic constituents of organic matter: ammonia, methane, hydrogen, and water vapour. These underwent a process of chemical evolution using energy from the sun and electric storms to combine into ever more complex molecules, such as amino acids, proteins, and vitamins. Eventually self-replicating nucleic acids, the basis of all life, could have developed. The very first organisms may have consisted of such molecules bounded by a simple membrane. *See* PROTEINOID.

**origin of the elements** The nuclear processes that give rise to chemical elements. There is not one single process that can account for all the elements. The abundance of the chemical elements is determined not just by the stability of the nuclei of the atoms but also how readily the nuclear processes leading to the existence of these atoms occur. Most of the helium in the universe was produced by fusion in the early universe when the temperature and the pressure were very high. Most of the elements between helium and iron were made in nuclear fusion reactions inside stars. Since iron is at the bottom of an energy valley of stability, energy needs to be put into a nucleus heavier than iron for a fusion reaction to occur. Inside stars some heavy elements are built up by the **s-process**, where s stands for slow, in which high-energy neutrons are absorbed by a nucleus, with the resulting nucleus undergoing beta decay to produce a nucleus with a higher atomic number. Other heavy elements are produced by the **r-**

**process**, where r stands for rapid, which occurs in supernova explosions.

**ornithine (Orn)** An *amino acid, $H_2N(CH_2)_3CH(NH_2)COOH$, that is not a constituent of proteins but is important in living organisms as an intermediate in the reactions of the *urea cycle and in arginine synthesis.

**ornithine cycle** *See* UREA CYCLE.

**orogenesis** The process by which major mountain chains are formed. This includes the deformational processes of thrusting, folding, and faulting that result from the collision of two continents. Examples of mountains formed through orogenesis include the Alpine–Himalayan, Appalachian, and Cordilleran orogenic belts.

**orpiment** A natural yellow mineral form of arsenic(III) sulphide, $As_2S_3$. The name is also used for the synthetic compound, which is used as a pigment.

**ortho-** **1.** Prefix indicating that a benzene compound has two substituted groups in the 1,2 positions (i.e. on adjacent carbon atoms). The abbreviation *o-* is used; for example *o*-dichlorobenzene is 1,2-dichlorobenzene. *Compare* META-; PARA-. **2.** Prefix formerly used to indicate the most hydrated form of an acid. For example, phosphoric(V) acid, $H_3PO_4$, was called orthophosphoric acid to distinguish it from the lower metaphosphoric acid, $HPO_3$, which is actually $(HPO_3)_n$.

**orthoboric acid** *See* BORIC ACID.

**orthoclase** *See* FELDSPARS.

**orthogenesis** An early theory of the nature of evolutionary change, which proposed that organisms evolve along particular paths predetermined by some factor in their genetic make-up. More recent understanding of selection pressure and other external forces that can be shown experimentally to affect the survival of organisms has proved the improbability of the theory.

**orthohydrogen** *See* HYDROGEN.

**orthophosphoric acid** *See* PHOSPHORIC(V) ACID.

**orthoplumbate** *See* PLUMBATE.

**orthopositronium** *See* POSITRONIUM.

**Orthoptera** A large order of insects containing the grasshoppers, locusts, crickets, and – in some classification systems – the cockroaches (*see* DICTYOPTERA). They are characterized by enlarged hind legs modified for jumping and biting mouthparts and produce sounds by *stridulation. The crickets and long-horned grasshoppers (e.g. *Gryllus*, *Tettigonia*) have long threadlike antennae and stridulate by rubbing together modified veins on their forewings. The hearing organs are on the front legs. The short-horned grasshoppers and locusts (e.g. *Chorthippus*, *Locusta*) have short antennae and stridulate by rubbing pegs on the hind leg against a hardened vein on the forewing. The hearing organs are on the abdomen.

**orthorhombic** *See* CRYSTAL SYSTEM.

**orthosilicate** *See* SILICATE.

**orthostannate** *See* STANNATE.

**orthotropism** The tendency for a *tropism (growth response of a plant) to be orientated directly in line with the stimulus concerned. An example is the vertical growth of main stems and roots in response to gravity (**orthogeotropism**). *Compare* PLAGIOTROPISM.

**oscillating reaction (clock reaction)** A type of chemical reaction in which the concentrations of the products and reactants change periodically, either with time or with position in the reacting medium. Thus, the concentration of a component may increase with time to a maximum, decrease to a minimum, then increase again, and so on, continuing the oscillation over a period of time. Systems are also known in which spirals and other patterns spread through the reacting medium, demonstrating a periodic spatial variation. Oscillating chemical reactions have certain features in common. They all occur under conditions far from chemical equilibrium and all involve *autocatalysis, i.e. a product of a reaction step acts as a catalyst for that step. This autocatalysis drives the oscillation by a process of positive feedback. Moreover, oscillating chemical reactions are associated with the phenomenon known as **bistability**. In this, a reaction may be in a steady-state condition, with reactants flowing into a reaction zone while products are flowing out of it. Under these conditions, the concentrations in the reaction zone may not change with time, although the reaction is not in a state of chemical equilibrium. Bistable systems have two possible stable steady states. Interaction with an additional substance in the reaction medium causes the system to oscillate between the states as the concentrations change. Oscillating chemical reactions are thought to occur in a number of biochemical processes. For example, they occur in glycolysis, in which ATP is produced by enzyme-catalysed reactions. They are also known to regulate the rhythm of the heartbeat. Most have highly complex reaction mechanisms. *See* LOTKA–VOLTERRA MECHANISM; BRUSSELATOR; OREGONATOR. *See also* CHAOTIC REACTION.

**oscillator** An electronic device that produces an alternating output of known frequency. If the output voltage or current has the form of a sine wave with respect to time, the device is called a **sinusoidal** (or **harmonic**) **oscillator**. If the output voltage changes abruptly from one level to another (as in a *square wave or *sawtooth wave) it is called a **relaxation oscillator**. A harmonic oscillator consists of a frequency-determining circuit or device, such as a *resonant circuit, maintained in oscillation by a source of power that by positive feedback also makes up for the resistive losses. In some relaxation oscillators the circuit is arranged so that in each cycle energy is stored in a reactive element (a capacitor or inductor) and subsequently discharged over a different time interval. *See also* MULTIVIBRATOR.

**oscillatory universe** A cosmological model in which the universe has repeated periods of expansion followed by contraction; i.e. a series of big bangs interspersed with big crunches. It was originally put forward in 1922 by the Russian physicist Alexander Friedman (1888–1925). The theory was criticized because it led eventually to an increasing entropy (i.e. heat death). *See also* EKPYROTIC UNIVERSE.

**oscilloscope** *See* CATHODE-RAY OSCILLO-SCOPE.

**osmiridium** A hard white naturally occurring alloy consisting principally of osmium (17–48%) and iridium (49%). It also contains small quantities of platinum, rhodium, and ruthenium. It is used for making small items subject to wear, e.g. electrical contacts or the tips of pen nibs.

**osmium** Symbol Os. A hard blue-white metallic *transition element; a.n. 76; r.a.m. 190.2; r.d. 22.57; m.p. 3045°C; b.p. 5027°C. It is found associated with platinum and is used in certain alloys with platinum and irid-ium (*see* OSMIRIDIUM). Osmium forms a number of complexes in a range of oxidation states. It was discovered by Smithson Tennant (1761–1815) in 1804.

(🌐) **SEE WEB LINKS**
• Information from the WebElements site

**osmium(IV) oxide (osmium tetroxide)** A yellow solid, $OsO_4$, made by heating osmium in air. It is used as an oxidizing agent in organic chemistry, as a catalyst, and as a fixative in electron microscopy.

**osmometer** *See* OSMOSIS.

**osmoreceptor** A receptor situated in the hypothalamus of the brain that responds to an increase in the concentration of the extracellular fluid. This results in the release of *antidiuretic hormone (ADH) and the subsequent conservation of water, thereby maintaining the *homeostasis of the body fluids.

**osmoregulation** The control of the water content and the concentration of salts in the body of an animal or protist. In freshwater species osmoregulation must counteract the tendency for water to pass into the animal by *osmosis. Various methods have been developed to eliminate the excess, such as *contractile vacuoles in protozoans and *kidneys with well-developed glomeruli in freshwater fish. Marine vertebrates have the opposite problem: they prevent excessive water loss and enhance the excretion of salts by having kidneys with few glomeruli and short tubules. In terrestrial vertebrates the dangers of desiccation are reduced by the presence of long convoluted kidney tubules, which increase the reabsorption of water and salts.

**osmosis** The passage of a solvent through a **semipermeable membrane** separating two solutions of different concentrations. A semipermeable membrane is one through which the molecules of a solvent can pass but the molecules of most solutes cannot. There is a thermodynamic tendency for solutions separated by such a membrane to become equal in concentration, the water (or other solvent) flowing from the weaker to the stronger solution. Osmosis will stop when the two solutions reach equal concentration, and can also be stopped by applying a pressure to the liquid on the stronger-solution side of the membrane. The pressure required to stop the flow from a pure solvent into a solution is a characteristic of the solution, and is called the **osmotic pressure** (symbol $\Pi$). Osmotic pressure depends only on the concentration of particles in the solution, not on their nature (i.e. it is a *colligative property). For a solution of $n$ moles in volume $V$ at thermodynamic temperature $T$, the osmotic pressure is given by $\Pi V = nRT$, where $R$ is the gas constant. Osmotic-pressure measurements are used in finding the relative molecular masses of compounds, particularly macromolecules. A device used to measure osmotic pressure is called an **osmometer**.

The distribution of water in living organisms is dependent to a large extent on osmosis, water entering the cells through their membranes. A cell membrane is not truly semipermeable as it allows the passage of certain solute molecules; it is described as **partially permeable**. Osmosis in plants is now usually described in terms of *water potential: water moves from an area of high (less negative) water potential to an area of low (more negative) water potential (*see also* PLASMOLYSIS; TURGOR; WILTING). Animals have evolved various means to counteract the effects of osmosis (*see* OSMOREGULATION); in water relations in animals solutions are still described in terms of osmotic pressure.

**osmotic pressure** *See* OSMOSIS.

**ossification** The process of *bone formation. It is brought about by the action of special cells called *osteoblasts, which deposit layers of bone in connective tissue. Some bones are formed directly in connective tissue (*see* MEMBRANE BONE); others are formed by the replacement of cartilage (*see* CARTILAGE BONE).

**Osteichthyes** The class of vertebrates comprising the bony fishes – marine and freshwater fish with a bony skeleton. All have gills covered with a bony operculum, and a layer of thin overlapping bony *scales covers the entire body surface. Bony fish have a *swim bladder, which acts as a hydrostatic organ enabling the animal to remain suspended in the water at any depth. In some fish this bladder acts as a lung. *See also* DIPNOI; TELEOSTEI. *Compare* CHONDRICHTHYES.

**osteoblast** Any of the cells, found in *bone, that secrete collagen and other substances that form the matrix of bone. Osteoblasts are derived from **osteoprogenitor cells** in the bone marrow; they eventually become *osteocytes (bone cells). *See also* OSSIFICATION.

**osteoclast** Any of the cells in *bone that

are involved in the breakdown of bone matrix to enable the further development and remodelling of bone during growth and repair.

**osteocyte** Any of the cells, found in bone, that are derived from *osteoblasts and perform activities required for the maintenance of the bone tissue.

**Ostrogradsky's theorem** *See* DIVERGENCE THEOREM.

**Ostwald's dilution law** An expression for the degree of dissociation of a weak electrolyte. For example, if a weak acid dissociates in water

$$HA \rightleftharpoons H^+ + A^-$$

the dissociation constant $K_a$ is given by

$$K_a = \alpha^2 n / (1 - \alpha) V$$

where $\alpha$ is the degree of dissociation, $n$ the initial amount of substance (before dissociation), and $V$ the volume. If $\alpha$ is small compared with 1, then $\alpha^2 = KV/n$; i.e. the degree of dissociation is proportional to the square root of the dilution. The law was first put forward by the German physical chemist Wilhelm Ostwald (1853–1932) to account for electrical conductivities of electrolyte solutions.

**otolith** A gelatinous mass containing a high concentration of particles of calcium carbonate, which forms part of the *macula of the inner ear.

**Otto engine** *See* INTERNAL-COMBUSTION ENGINE.

**ounce** 1. One sixteenth of a pound (avoirdupois), equal to 0.028 349 kg. 2. Eight drachms (Troy), equal to 0.031 103 kg. 3. (fluid ounce) Eight fluid drachms, equal to 0.028 413 dm$^3$.

**outbreeding** Mating between unrelated or distantly related individuals of a species. Outbreeding populations usually show more variation than *inbreeding ones and have a greater potential for adapting to environmental changes. Outbreeding increases the number of *heterozygous individuals, so that disadvantageous recessive characteristics tend to be masked by dominant alleles.

**outer ear (external ear)** The part of the ear external to the *tympanum (eardrum). It is present in mammals, birds, and some reptiles and consists of a tube (the **external auditory meatus**) that directs sound waves onto the tympanum. In mammals it may in-

**oxalate**

clude an external *pinna, which extends beyond the skull.

**oval window (fenestra ovalis)** A membrane-covered opening between the middle ear and the inner ear (see EAR), situated above the *round window. Vibrations of the tympanum are transferred across the middle ear by the *ear ossicles and transmitted to the inner ear by the oval window, which is connected to the third ear ossicle (stapes).

**ovarian follicle** See GRAAFIAN FOLLICLE.

**ovary** **1.** The reproductive organ in female animals in which eggs (ova) are produced. In most vertebrates there are two ovaries (in some fish the ovaries fuse together to form a single structure and in birds the left ovary only is functional). As well as eggs, they produce steroid hormones (see OESTROGEN; PROGESTERONE). In mammals each ovary is situated close to the opening of a *fallopian tube; it contains numerous follicles in which the eggs develop and from which they are released in a regular cycle. See also GRAAFIAN FOLLICLE; MENSTRUAL CYCLE; OOGENESIS; OVULATION; REPRODUCTIVE SYSTEM. **2.** The hollow base of the *carpel of a flower, containing one or more *ovules. After fertilization, the ovary wall develops into the fruit enclosing the seeds. In some species, the carpels are fused together to form a complex ovary.

**overdamped** See DAMPING.

**overpopulation** The situation that arises when rapid growth of a population, usually a human population, results in numbers that cannot be supported by the available resources, such as space and food. This occurs when the birth rate exceeds the death rate, or when immigration exceeds emigration, or when a combination of these factors exists. See POPULATION GROWTH.

**overpotential** A potential that must be applied in an electrolytic cell in addition to the theoretical potential required to liberate a given substance at an electrode. The value depends on the electrode material and on the current density. It is a kinetic effect occurring because of the significant activation energy for electron transfer at the electrodes, and is particularly important for the liberation of such gases as hydrogen and oxygen. For example, in the electrolysis of a solution of zinc ions, hydrogen ($E^{\ominus} = 0.00$ V) would be expected to be liberated at the cathode in preference to zinc ($E^{\ominus} = -0.76$ V). In fact, the high overpotential of hydrogen on zinc

(about 1 V under suitable conditions) means that zinc can be deposited instead.

**overtones** See HARMONIC.

**oviduct** The tube that conveys an animal egg cell from the ovary to other parts of the reproductive system or to the outside. Eggs are passed along the oviduct by the action of muscles and cilia. See FALLOPIAN TUBE.

**oviparity** Reproduction in which fertilized eggs are laid or spawned by the mother and hatch outside her body. It occurs in most animals except marsupial and placental mammals. Compare OVOVIVIPARITY; VIVIPARITY.

**ovipositor** An organ at the hind end of the abdomen of female insects through which eggs are laid. It consists of a pair of modified appendages and is often long and piercing, so that eggs can be laid in otherwise inaccessible places. The sting of bees and wasps is a modified ovipositor.

**ovoviviparity** Reproduction in which fertilized eggs develop and hatch in the oviduct of the mother. It occurs in many invertebrates and in some fish and reptiles (e.g. the viper). Compare OVIPARITY; VIVIPARITY.

**ovulation** The release of an egg cell from the ovary, which in mammals is stimulated by *luteinizing hormone. The developing egg cell within its follicle migrates to the ovary surface; when mature, it is released from the follicle (which breaks open) into the body cavity, from where it passes into the oviduct. See also MENSTRUAL CYCLE.

**ovule** The part of the female reproductive organs of seed plants that consists of the *nucellus, *embryo sac, and *integuments. The ovules of gymnosperms are situated on ovuliferous scales of the female cones while those of angiosperms are enclosed in the carpel. After fertilization, the ovule becomes the seed.

**ovuliferous scale** One of a group of large woody specialized leaves that form the female *cone of conifers and related trees. It bears the ovules, which develop into seeds.

**ovum (egg cell)** (pl. **ova**) **1.** (in zoology) The mature reproductive cell (see GAMETE) of female animals, which is produced by the ovary (see OOGENESIS). It is spherical, has a nucleus, is covered with a vitelline membrane, and is not mobile. **2.** (in botany) The *oosphere of plants.

**oxalate** A salt or ester of *oxalic acid.

**oxalic acid (ethanedioic acid)** A crystalline solid, $(COOH)_2$, that is slightly soluble in water. Oxalic acid is strongly acidic and very poisonous. It occurs in certain plants, e.g. sorrel and the leaf blades of rhubarb.

**oxaloacetic acid** A compound, $HO_2CCH_2COCO_2H$, that plays an integral role in the *Krebs cycle. The anion, oxaloacetate, reacts with the acetyl group from acetyl coenzyme A to form citrate.

**oxazole** A heterocyclic compound having a nitrogen atom and an oxygen atom in a five-membered ring, $C_3H_3NO$.

**oxbow lake** A crescent-shaped lake formed when a meander of a slow-flowing river is cut off from the main channel after the river, in flood, crosses the neck of land between two bends. Most oxbow lakes, or cutoffs, soon silt up.

**oxfuel** A liquid fuel containing added alcohols or ethers to act as an additional source of oxygen during combustion of the fuel. It has been claimed that such additives help to lower the concentration of carbon monoxide in engine emissions.

**oxidant** See OXIDIZING AGENT.

**oxidase** Any enzyme that catalyses *oxidation–reduction reactions that involve the transfer of electrons to molecular oxygen.

**oxidation** See OXIDATION–REDUCTION.

**oxidation number (oxidation state)** See OXIDATION–REDUCTION.

**oxidation–reduction (redox)** Originally, **oxidation** was simply regarded as a chemical reaction with oxygen. The reverse process – loss of oxygen – was called **reduction**. Reaction with hydrogen also came to be regarded as reduction. Later, a more general idea of oxidation and reduction was developed in which oxidation was loss of electrons and reduction was gain of electrons. This wider definition covered the original one. For example, in the reaction

$$4Na(s) + O_2(g) \rightarrow 2Na_2O(s)$$

the sodium atoms lose electrons to give $Na^+$ ions and are oxidized. At the same time, the oxygen atoms gain electrons and are reduced. These definitions of oxidation and reduction also apply to reactions that do not involve oxygen. For instance in

$$2Na(s) + Cl_2(g) \rightarrow 2NaCl(s)$$

the sodium is oxidized and the chlorine reduced. Oxidation and reduction also occurs at the electrodes in *cells.

This definition of oxidation and reduction applies only to reactions in which electron transfer occurs – i.e. to reactions involving ions. It can be extended to reactions between covalent compounds by using the concept of **oxidation number** (or **state**). This is a measure of the electron control that an atom has in a compound compared to the atom in the pure element. An oxidation number consists of two parts:
(1) Its sign, which indicates whether the control has increased (negative) or decreased (positive).
(2) Its value, which gives the number of electrons over which control has changed.

The change of electron control may be complete (in ionic compounds) or partial (in covalent compounds). For example, in $SO_2$ the sulphur has an oxidation number +4, having gained partial control over 4 electrons compared to sulphur atoms in pure sulphur. The oxygen has an oxidation number –2, each oxygen having lost partial control over 2 electrons compared to oxygen atoms in gaseous oxygen. Oxidation is a reaction involving an increase in oxidation number and reduction involves a decrease. Thus in

$$2H_2 + O_2 \rightarrow 2H_2O$$

the hydrogen in water is +1 and the oxygen –2. The hydrogen is oxidized and the oxygen is reduced.

The oxidation number is used in naming inorganic compounds. Thus in $H_2SO_4$, sulphuric(VI) acid, the sulphur has an oxidation number of +6. Compounds that tend to undergo reduction readily are *oxidizing agents; those that undergo oxidation are *reducing agents.

**oxidative decarboxylation** The reaction in the *Krebs cycle in which oxygen, derived from two water molecules, is used to oxidize two carbon atoms to two molecules of carbon dioxide. The two carbon atoms result from the *decarboxylation reactions that occur during the Krebs cycle as the six-carbon compound citrate is converted to the four-carbon compound oxaloacetate.

**oxidative phosphorylation** A reaction occurring during the final stages of *aerobic respiration, in which ATP is formed from ADP and phosphate coupled to electron transport in the *electron transport chain. The reaction occurs in the mitochondria and is the cell's principal method of storing the energy released by the oxidation of food. *See also* PHOSPHORYLATION.

**oxides** Binary compounds formed between elements and oxygen. Oxides of nonmetals are covalent compounds having simple molecules (e.g. CO, $CO_2$, $SO_2$) or giant molecular lattices (e.g. $SiO_2$). They are typically acidic or neutral. Oxides of metals are ionic, containing the $O^{2-}$ ion. They are generally basic or *amphoteric. Various other types of ionic oxide exist (*see* OZONIDES; PEROXIDES; SUPEROXIDES).

**oxidizing acid** An acid that can act as a strong oxidizing agent as well as an acid. Nitric acid is a common example. It is able to attack metals, such as copper, that are below hydrogen in the electromotive series, by oxidizing the metal:

$$2HNO_3 + Cu \rightarrow CuO + H_2O + 2NO_2$$

This is followed by reaction between the acid and the oxide:

$$2HNO_3 + CuO \rightarrow Cu(NO_3)_2 + H_2O$$

**oxidizing agent (oxidant)** A substance that brings about oxidation in other substances. It achieves this by being itself reduced. Oxidizing agents contain atoms with high oxidation numbers; that is the atoms have suffered electron loss. In oxidizing other substances these atoms gain electrons.

**oxidoreductase** Any of a class of enzymes that catalyse *oxidation–reduction reactions, i.e. they are involved in the transfer of hydrogen or electrons between molecules. They include the *oxidases and *dehydrogenases.

**oximes** Compounds containing the group C:NOH, formed by reaction of an aldehyde or ketone with hydroxylamine ($H_2NOH$) (see illustration). Ethanal ($CH_3CHO$), for example, forms the oxime $CH_3CH{:}NOH$.

**(⊕) SEE WEB LINKS**
• Information about IUPAC nomenclature

**oxo-** Prefix indicating the presence of oxygen in a chemical compound.

**oxoacid** An acid in which the acidic hydrogen is part of a hydroxyl group bound to an

atom that is bound to an oxo group (=O). Sulphuric acid is an example. *Compare* HYDROXOACID.

**3-oxobutanoic acid (acetoacetic acid)** A colourless syrupy liquid, $CH_3COCH_2COOH$. It is an unstable compound, decomposing into propanone and carbon dioxide. The acid can be prepared from its ester, *ethyl 3-oxobutanoate.

**oxonium ion** An ion of the type $R_3O^+$, in which R indicates hydrogen or an organic group, especially the ion $H_3O^+$, which is formed when *acids dissociate in water. This is also called the **hydroxonium ion** or the **hydronium ion**.

**oxo process** An industrial process for making aldehydes by reaction between alkanes, carbon monoxide, and hydrogen (cobalt catalyst using high pressure and temperature).

**oxyacetylene burner** A welding or cutting torch that burns a mixture of oxygen and acetylene (ethyne) in a specially designed jet. The flame temperature of about 3300°C enables all ferrous metals to be welded. For cutting, the point at which the steel is to be cut is preheated with the oxyacetylene flame and a powerful jet of oxygen is then directed onto the steel. The oxygen reacts with the hot steel to form iron oxide and the heat of this reaction melts more iron, which is blown away by the force of the jet.

**oxycodone** An opioid, $C_{18}H_{21}N_2$, similar in structure to codeine but with a –OH group in codeine replaced by a carbonyl group. It is an analgesic often used for the treatment of chronic pain. It is also used illegally and is a controlled substance in most countries.

**oxygen** Symbol O. A colourless odourless gaseous element belonging to *group 16 (formerly VIB) of the periodic table; a.n. 8; r.a.m. 15.9994; d. 1.429 g $dm^{-3}$; m.p. –218.4°C; b.p. –183°C. It is the most abundant element in the earth's crust (49.2% by weight) and is present in the atmosphere (28% by volume). Atmospheric oxygen is of vital importance for all organisms that carry out *aerobic res-

$$R-C{\overset{O}{\underset{R'}{\Big\langle}}} \quad + \quad {\overset{H}{\underset{H}{\Big\rangle}}}N-O-H \quad \xrightarrow{-H_2O} \quad R-C{\overset{N-O-H}{\underset{R'}{\Big\langle}}}$$

ketone                    hydroxylamine                        oxime

**Oximes.**

piration. For industrial purposes it is obtained by fractional distillation of liquid air. It is used in metallurgical processes, in high-temperature flames (e.g. for welding), and in breathing apparatus. The common form is diatomic (**dioxygen**, $O_2$); there is also a reactive allotrope *ozone ($O_3$). Chemically, oxygen reacts with most other elements forming *oxides. The element was discovered by Joseph Priestley in 1774.

**((⊕)) SEE WEB LINKS**
• Information from the WebElements site

**oxygen cycle** The cycling of oxygen between the biotic and abiotic components of the environment (*see* BIOGEOCHEMICAL CYCLE). The oxygen cycle is closely linked to the *carbon cycle and the water cycle (*see* HYDROLOGICAL CYCLE). In the process of respiration oxygen is taken in by living organisms and released into the atmosphere, combined with carbon, in the form of carbon dioxide. Carbon dioxide enters the carbon cycle or is taken up by plants for *photosynthesis. During photosynthesis oxygen is evolved by the chemical splitting of water and returned to the atmosphere. In the upper atmosphere, *ozone is formed from oxygen and dissociates to release oxygen (*see* OZONE LAYER).

**oxygen debt** The physiological state that exists in a normally aerobic animal when insufficient oxygen is available for metabolic requirements (e.g. during a period of strenuous physical activity). To meet the body's increased demand for energy, pyruvate is converted anaerobically (i.e. in the absence of oxygen) to lactic acid, which requires oxygen for its breakdown and accumulates in the tissues. When oxygen is available again lactic acid is oxidized in the liver, thus repaying the debt.

**oxygen dissociation curve** The S-shaped curve produced when the percentage saturation of haemoglobin with oxygen (i.e. the percentage of binding sites of haemoglobin that are occupied by oxygen molecules) is plotted against the partial pressure of oxygen ($pO_2$), which is a measure of the oxygen concentration in the surrounding medium. The steep rise of the curve indicates the high affinity of haemoglobin for oxygen: a small increase in $pO_2$ results in a relatively sharp increase in the percentage saturation of haemoglobin with oxygen. Therefore in the lungs, where the $pO_2$ is high, the blood is rapidly saturated with oxygen. Conversely, a

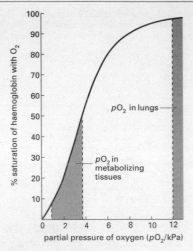

Oxygen dissociation curve.

small drop in $pO_2$ results in a large drop in percentage saturation of haemoglobin. Thus in tissues that utilize oxygen at a high rate, where the $pO_2$ is low, oxygen readily dissociates from haemoglobin and is released for use by the tissues. *See also* BOHR EFFECT.

**oxyhaemoglobin** *See* HAEMOGLOBIN.

**oxyntic cell (parietal cell)** Any of the cells in the wall of the stomach that produce hydrochloric acid, which forms part of the *gastric juice. Hydrochloric acid is required for the conversion of pepsinogen to pepsin in the lumen of the stomach and kills various microorganisms that enter with food. The oxyntic cells also produce intrinsic factor, which is involved in the absorption of vitamin $B_{12}$ in the small intestine (*see* VITAMIN B COMPLEX).

**oxytocin** A hormone, produced by birds and mammals, that in mammals causes both contraction of smooth muscle in the uterus during birth and expulsion of milk from the mammary glands during suckling. Oxytocin is produced in the neurosecretory cells of the hypothalamus (*see* NEUROSECRETION) but is stored and secreted by the posterior pituitary gland.

**ozonation** The formation of *ozone ($O_3$) in the earth's atmosphere. In the upper atmosphere (stratosphere) about 20–50 km above the surface of the earth, oxygen molecules

($O_2$) dissociate into their constituent atoms under the influence of *ultraviolet light of short wavelength (below about 240 nm). These atoms combine with oxygen molecules to form ozone (*see* OZONE LAYER). Ozone is also formed in the lower atmosphere (troposphere) from nitrogen oxides and other pollutants by photochemical reactions (*see* PHOTOCHEMICAL SMOG). Tropospheric ozone is increasingly a cause of impaired plant growth and reduced crop yields.

**ozone (trioxygen)** A colourless gas, $O_3$, soluble in cold water and in alkalis; m.p. $-192.7°C$; b.p. $-111.9°C$. Liquid ozone is dark blue in colour and is diamagnetic (dioxygen, $O_2$, is paramagnetic). The gas is made by passing oxygen through a silent electric discharge and is usually used in mixtures with oxygen. It is produced in the stratosphere by the action of high-energy ultraviolet radiation on oxygen (*see* OZONATION) and its presence there acts as a screen for ultraviolet radiation (*see* OZONE LAYER). Ozone is also one of the greenhouse gases (*see* GREENHOUSE EFFECT). It is a powerful oxidizing agent and is used to form ozonides by reaction with alkenes and subsequently by hydrolysis to carbonyl compounds.

**ozone hole** *See* OZONE LAYER.

**ozone layer (ozonosphere)** A layer of the *earth's atmosphere in which most of the atmosphere's ozone is concentrated. It occurs 15–50 km above the earth's surface and is virtually synonymous with the stratosphere. In this layer most of the sun's *ultraviolet radiation is absorbed by the ozone molecules, causing a rise in the temperature of the stratosphere and preventing vertical mixing so that the stratosphere forms a stable layer. By absorbing most of the solar ultraviolet radiation the ozone layer protects living organisms on earth. The fact that the ozone layer is thinnest at the equator is believed to account for the high equatorial incidence of skin cancer as a result of exposure to unabsorbed solar ultraviolet radiation. In the 1980s it was found that depletion of the ozone layer was occurring over both the poles, creating **ozone holes**. This is thought to have been caused by a series of complex photochemical reactions involving *nitrogen oxides produced from aircraft and, more seriously, *chlorofluorocarbons (CFCs) and halons. CFCs rise to the stratosphere, where they react with ultraviolet light to release chlorine atoms; these atoms, which are highly reactive, catalyse the destruction of ozone. Use of CFCs is now much reduced in an effort to reverse this human-induced damage to the ozone layer. *See also* AIR POLLUTION.

**(⊕) SEE WEB LINKS**

• Description of the ozone layer, its depletion, and steps to protect it, produced by the US Environmental Protection Agency

**ozonides** 1. A group of compounds formed by reaction of ozone with alkali metal hydroxides and formally containing the ion $O_3^-$. 2. Unstable compounds formed by the addition of ozone to the C=C double bond in alkenes. *See* OZONOLYSIS.

**ozonolysis** A reaction of alkenes with ozone to form an ozonide. It was once used to investigate the structure of alkenes by hydrolysing the ozonide to give aldehydes or ketones, for instance:

$$R_2C{:}CHR' \rightarrow R_2CO + R'CHO$$

These could be identified, and the structure of the original alkene determined.

**P (parental generation)** The individuals that are selected to begin a breeding experiment, crosses between which yield the *$F_1$ generation. Only pure-breeding (homozygous) individuals are selected for the P generation.

**pacemaker 1. (sinoatrial node)** A small mass of specialized muscle cells in the mammalian heart, found in the wall of the right atrium near the opening for the vena cava. The cells initiate and maintain the heart beat: by their rhythmic and spontaneous contractions they stimulate contraction of the atria (*see also* ATRIOVENTRICULAR NODE). The cells themselves are controlled by the autonomic nervous system, which determines the heart rate. Similar pacemakers occur in the hearts of other vertebrates.
**2.** An electronic or nuclear battery-charged device that can be implanted surgically into the chest to produce and maintain the heart beat. These devices are used when the heart's own pacemaker is defective or diseased.

**pachytene** The period in the first prophase of *meiosis when paired *homologous chromosomes are fully contracted and twisted around each other.

**packing density 1.** The number of devices (such as *logic circuits) or integrated circuits per unit area of a *silicon chip.
**2.** The quantity of information stored in a specified space of a storage system associated with a computer, e.g. *bits per inch of magnetic tape.

**packing fraction** The algebraic difference between the relative atomic mass of an isotope and its mass number, divided by the mass number.

**paedogenesis** Reproduction by an animal that is still in the larval or pre-adult form. Paedogenesis is a form of *neoteny and is particularly marked in the axolotl, a larval form of the salamander, which retains its larval features owing to a thyroid deficiency but can breed, producing individuals like itself. If the thyroid hormone thyroxine is given, metamorphosis occurs.

**pahoehoe** *See* LAVA.

**pairing (synapsis)** The close association between *homologous chromosomes that develops during the first prophase of *meiosis. The two chromosomes move together and an exact pairing of corresponding points along their lengths occurs as they lie side by side. The resulting structure is called a **bivalent**.

**pair production** The creation of an electron and a positron from a photon in a strong electric field, such as that surrounding an atomic nucleus. The electron and the positron each have a mass of about $9 \times 10^{-31}$ kg, which is equivalent on the basis of the mass–energy equation ($E = mc^2$) to a total of $16 \times 10^{-14}$ J. The frequency, ν, associated with a photon of this energy (according to $E = h\nu$) is $2.5 \times 10^{20}$ Hz. Pair production thus requires photons of high quantum energy (Bremsstrahlung or gamma rays). Any excess energy is taken up as kinetic energy of the products.

**palaeobotany** The branch of *palaeontology concerned with the study of plants through geological time, as revealed by their *fossil remains (*see also* PALYNOLOGY). It overlaps with other aspects of plant study, including anatomy, ecology, evolution, and taxonomy.

**Palaeocene** The earliest geological epoch of the *Palaeogene period. It began about 65 million years ago, following the Cretaceous period, and extended for about 10 million years to the beginning of the *Eocene (the Palaeocene is sometimes included in the Eocene). It was named by the palaeobotanist W. P. Schimper in 1874. A major floral and faunal discontinuity occurred between the end of the Cretaceous and the beginning of the Palaeocene: following the extinction of many reptiles the mammals became abundant on land. By the end of the epoch primates and rodents had evolved.

**palaeoclimatology** The study of climates of earlier geological periods. This is based largely on the study of sediments that were laid down during these periods and of fossils. The changes in the positions of the continents as a result of *continental drift and *plate tectonics complicate the study.

**palaeoecology** The study of the relationships of *fossil organisms to each other and to their environments. It involves the study both of the fossils and of the surrounding rocks in which they are found. Trace fossils may provide information on the behaviour of the organism.

**Palaeogene** The older of the two geological periods of the *Cenozoic era, consisting of the *Palaeocene, *Eocene, and *Oligocene epochs. The Palaeogene began 65 million years ago and lasted until the start of the *Neogene period, 23 million years ago. It thus corresponds approximately to the first two-thirds of the *Tertiary period, a division that is no longer officially recognized.

**Palaeolithic** The Old Stone Age, lasting in Europe from about 2.5 million to 9000 years ago, during which humans used primitive stone tools made by chipping stones and flints.

**palaeomagnetism** The study of magnetism in rocks, which provides information on variations in the direction and intensity of the earth's magnetic field with time. During the formation of an igneous or sedimentary rock containing magnetic minerals the polarity of the earth's magnetic field at that time becomes 'frozen' into the rock. Studies of this fossil magnetism in samples of rocks have enabled the former positions of magnetic poles at various geological times to be located. It has also revealed that periodic reversals in the geomagnetic field have taken place (i.e. the N-pole becomes the S-pole and vice versa). This information has been important in plate tectonics in establishing the movements of lithospheric plates over the earth's surface. The magnetic reversals provided crucial evidence for the sea-floor spreading hypothesis proposed in the early 1960s.

**palaeontology** The study of extinct organisms, including their structure, environment, evolution, and distribution, as revealed by their *fossil remains. Palaeontological work also makes important contributions to geology in revealing stratigraphic relationships between rock strata and determining the physical appearance and climate of the earth during past geological ages. See

*also* PALAEOBOTANY; PALAEOECOLOGY; PALAEOZOOLOGY.

**Palaeozoic** The first era of *Phanerozoic time. It follows the *Precambrian and is subdivided into the Lower Palaeozoic, comprising the *Cambrian, *Ordovician, and *Silurian periods, and the Upper Palaeozoic, comprising the *Devonian, *Carboniferous, and *Permian periods. It extended from about 542 million years ago to about 251 million years ago, when it was succeeded by the *Mesozoic era.

**palaeozoology** The branch of *palaeontology concerned with the study of animals throughout geological time, as revealed by their *fossil remains.

**palate** The roof of the mouth cavity of vertebrates, which separates the *buccal and nasal cavities. In mammals it is divided into two zones, the bony **hard palate** and the **soft palate**, and completely separates the buccal cavity from the air passage to enable simultaneous eating and breathing.

**palisade mesophyll** *See* MESOPHYLL.

**palladium** Symbol Pd. A soft white ductile *transition element (*see also* PLATINUM METALS); a.n. 46; r.a.m. 106.4; r.d. 12.02; m.p. 1552°C; b.p. 3140±1°C. It occurs in some copper and nickel ores and is used in jewellery and as a catalyst for hydrogenation reactions. Chemically, it does not react with oxygen at normal temperatures. It dissolves slowly in hydrochloric acid. Palladium is capable of occluding 900 times its own volume of hydrogen. It forms few simple salts, most compounds being complexes of palladium(II) with some palladium(IV). It was discovered by William Woolaston (1766–1828) in 1803.

**(🌐) SEE WEB LINKS**
• Information from the WebElements site

**pallium** *See* CEREBRAL CORTEX.

**palmitate (hexadecanoate)** A salt or ester of palmitic acid.

**palmitic acid (hexadecanoic acid)** A 16-carbon saturated fatty acid, $CH_3(CH_2)_{14}$-COOH; r.d. 0.85; m.p. 63°C; b.p. 390°C. Glycerides of palmitic acid occur widely in plant and animal oils and fats.

**palp** An elongated sensory organ, usually near the mouth, in many invertebrates. Examples are the tactile head appendages of polychaete worms, the ciliated flap of tissue that produces feeding currents in bivalve molluscs, the distal part of the *mandibles of crustaceans, and the olfactory parts of the first and second *maxillae of some insects.

**palynology (micropalaeontology)** The study of fossil pollen and spores (**pollen analysis**) and various other *microfossils, such as coccoliths and dinoflagellates. Palynology is used in stratigraphy, palaeoclimatology, and archaeology. Pollen and spores are very resistant to decay and therefore their fossils are found in sedimentary rocks. They may be extracted by various methods, including boiling with potassium hydroxide solution, washing with strong oxidizing mixtures, and centrifuging repeatedly. Spores and pollen are classified according to shape, form of aperture, and both internal and external details of the exine (outer coat). They indicate the nature of the dominant flora, and therefore the climate and conditions of the period in which they lived.

**pancreas** A gland in vertebrates lying between the duodenum and the spleen. Under the influence of the hormone *secretin it secretes **pancreatic juice** containing digestive enzymes or their precursors (mainly *trypsin, *chymotrypsin, *amylase, and *lipase) into the duodenum via the pancreatic duct. It also contains groups of cells – the *islets of Langerhans – that function as an *endocrine gland, producing the hormones *insulin and *glucagon, which regulate blood sugar levels.

**pancreozymin** *See* CHOLECYSTOKININ.

**panicle** A type of flowering shoot common in the grass family. The primary axis bears groups of *racemes and is itself racemose, as the youngest groups of flowers are at the top (e.g. oat). The term may be used loosely for any form of branched *racemose inflorescence; for example, the horse chestnut is a raceme of cymes. Both these arrangements are seen in the family Polygonaceae (docks and sorrels).

**pantothenic acid** A vitamin of the *vitamin B complex. It is a constituent of *coenzyme A, which performs a crucial role in the oxidation of fats, carbohydrates, and certain amino acids. Deficiency rarely occurs because the vitamin occurs in many foods, especially cereal grains, peas, egg yolk, liver, and yeast.

**papain** A protein-digesting enzyme (*see* PROTEASE) occurring in the fruit of the West Indian papaya tree (*Carica papaya*). It is used as a digestant and in the manufacture of meat tenderizers.

**paper chromatography** A technique for analysing mixtures by *chromatography, in which the stationary phase is absorbent paper. A spot of the mixture to be investigated is placed near one edge of the paper and the sheet is suspended vertically in a solvent, which rises through the paper by capillary action carrying the components with it. The components move at different rates, partly because they absorb to different extents on the cellulose and partly because of partition between the solvent and the moisture in the paper. The paper is removed and dried, and the different components form a line of spots along the paper. Colourless substances are detected by using ultraviolet radiation or by spraying with a substance that reacts to give a coloured spot (e.g. ninhydrin gives a blue coloration with amino acids). The components can be identified by the distance they move in a given time.

**papilla** Any cone-shaped protuberance projecting from the surface of an organ or organism. Papillae occur, for example, on the tongue, in the kidneys, and, in plants, on the surface of many petals.

**papovavirus** One of a group of DNA-containing viruses that produce tumours in their hosts. **Papillomaviruses** produce non-malignant tumours (such as warts) in all vertebrates and certain cancers (e.g. cervical cancer) in humans. **Polyomaviruses** produce malignant tumours in certain classes of vertebrates (not including humans).

**pappus** A group of modified *sepals, often in the form of a ring of silky hairs. For example, when the fruit of the dandelion matures a pappus of hairs persists at the top of a thin stalk forming a parachute-like structure, which serves to disperse the fruit.

**para-** **1.** Prefix designating a benzene compound in which two substituents are in the 1,4 positions, i.e. directly opposite each other, on the benzene ring. The abbreviation *p-* is used; for example, *p*-xylene is 1,4-dimethylbenzene. *Compare* ORTHO-; META-. **2.** Prefix denoting the form of diatomic molecules in which the nuclei have opposite spins, e.g. parahydrogen. *Compare* ORTHO-.

**parabola** A *conic with eccentricity $e = 1$. It is the locus of a point that moves so that its distance from the **focus** is equal to its per-

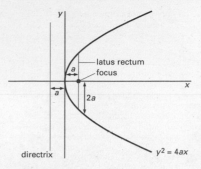

**Parabola.**

pendicular distance from the **directrix**. A chord through the focus, perpendicular to the axis, is called the **latus rectum**. For a parabola with its vertex at the origin, lying symmetrically about the $x$-axis, the equation is $y^2 = 4ax$, where $a$ is the distance from the vertex to the focus. The directrix is the line $x = -a$, and the latus rectum is $4a$. See illustration.

**parabolic reflector (paraboloidal reflector)** A reflector having a section that is a parabola. A concave parabolic reflector will reflect a parallel beam of radiation through its focus and, conversely, will produce a parallel beam if the source of the radiation is placed at its focus. Parabolic mirrors are used in reflecting optical *telescopes to collect the light and in some light sources that require a parallel beam of light. In radio telescopes a dish aerial may also consist of a parabolic reflector.

**paraboloid** A solid formed by rotating a parabola about its axis of symmetry.

**paraffin** See PETROLEUM.

**paraffins** See ALKANES.

**paraffin wax** See PETROLEUM.

**paraformaldehyde** See METHANAL.

**parahydrogen** See HYDROGEN.

**paraldehyde** See ETHANAL.

**parallax** 1. An apparent displacement of a distant object (with respect to a more distant background) when viewed from two different positions. If such an object is viewed from two points at either end of a base line, the angle between the lines joining the object to the ends of the base line is the **angle**

of parallax. If the base line is the distance between the two eyes of an observer the angle is called the **binocular parallax**.
2. The angular displacement in the apparent position of a celestial body when observed from two different points. **Diurnal parallax** results from the earth's daily rotation, the celestial body being viewed from the surface of the earth rather than from its centre. **Annual parallax** is caused by the earth's motion round the sun, the celestial body being viewed from the earth rather than from the centre of the sun. **Secular parallax** is caused by the motion of the solar system relative to the fixed stars.

**parallel circuits** A circuit in which the circuit elements are connected so that the current divides between them. For resistors in parallel, the total resistance, $R$, is given by $1/R = 1/r_1 + 1/r_2 + 1/r_3 \dots$, where $r_1$, $r_2$, and $r_3$ are the resistances of the individual elements. For capacitors in parallel, the total capacitance, $C$, is given by $C = c_1 + c_2 + c_3 \dots$.

**parallelepiped (parallelopiped)** A solid with six faces, all of which are parallelograms.

**parallel evolution** The development of related organisms along similar evolutionary paths due to strong selection pressures acting on all of them in the same way. It is debatable if the phenomenon really exists: many argue that all evolution is ultimately *convergent or divergent (see ADAPTIVE RADIATION).

**parallelogram of forces** See PARALLELOGRAM OF VECTORS.

**parallelogram of vectors** A method of determining the *resultant of two *vector quantities. The two vector quantities are represented by two adjacent sides of a parallelogram and the resultant is then the diagonal through their point of intersection. The magnitude and direction of the resultant is found by scale drawing or by trigonometry. The method is used for such vectors as forces (**parallelogram of forces**) and velocities (**parallelogram of velocities**). See illustration overleaf.

**parallelogram of velocities** See PARALLELOGRAM OF VECTORS.

**parallel processing** A technique that allows more than one process – stream of activity – to be running at any given moment in a computer system, hence processes can be

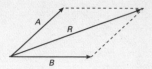

A is the velocity of the boat with respect to the water; B is the velocity of the water with respect to the bank

R is the resultant velocity of the boat with respect to the bank

**Parallelogram of velocities.**

executed in parallel. This means that two or more processors are active among a group of processes at any instant.

**parallel spins** Neighbouring spinning electrons in which the *spins, and hence the magnetic moments, of the electrons are aligned in the same direction. The interaction between the magnetic moments of electrons in atoms is dominated by exchange interactions (see EXCHANGE FORCE). Under some circumstances the exchange interactions between magnetic moments favour parallel spins, while under other conditions they favour *antiparallel spins. The case of ferromagnetism (see MAGNETISM) is an example of a system with parallel spins.

**paramagnetism** See MAGNETISM.

**parametric equation** An equation of a curve expressed in the form of the parameters that locate points on the curve. The parametric equations of a straight line are $x = a + bt$, $y = c + dt$. For a circle, they are $x = a\cos\theta$, $y = a\sin\theta$.

**paraphyletic** See MONOPHYLETIC.

**parapositronium** See POSITRONIUM.

**parasitic capture** The absorption of a neutron by a nuclide that does not result in either fission or a useful artificial element.

**parasitism** An association in which one organism (the **parasite**) lives on (**ectoparasitism**) or in (**endoparasitism**) the body of another (the *host), from which it obtains its nutrients. Some parasites inflict comparatively little damage on their host, but many cause characteristic diseases (these are, however, never immediately fatal, as killing the host would destroy the parasite's source of food). Parasites are usually highly specialized for their way of life, which may involve one host or several (if the *life cycle requires it). They typically produce vast numbers of

eggs, very few of which survive to find their way to another suitable host. **Obligate parasites** can only survive and reproduce as parasites; **facultative parasites** can also live as *saprotrophs. The parasites of humans include fleas and lice (which are ectoparasites), various bacteria, protozoans, and fungi (endoparasites causing characteristic diseases), and tapeworms (e.g. *Taenia solium*, which lives in the gut).

((())) **SEE WEB LINKS**

• Comprehensive descriptions of human parasites feature in this diagnostic website from the US Centers for Disease Control and Prevention

**parasympathetic nervous system** Part of the *autonomic nervous system. Its nerve endings release acetylcholine as a *neurotransmitter and its actions tend to antagonize those of the *sympathetic nervous system. For example, the parasympathetic nervous system increases salivary gland secretion, decreases heart rate, and promotes digestion (by increasing *peristalsis). While the sympathetic nervous system has opposite effects.

**parathyroid glands** Two pairs of *endocrine glands situated behind, or embedded within, the thyroid gland in higher vertebrates. They produce *parathyroid hormone, which controls the amount of calcium in the blood. See also C CELL.

**parathyroid hormone (PTH; parathormone; parathyrin)** A peptide hormone secreted by the *parathyroid gland in response to low levels of calcium in the blood. It acts to maintain normal blood levels of calcium by (1) increasing the number of osteoclasts, cells that break down the bone matrix and release calcium into the blood; (2) increasing the reabsorption of calcium and magnesium ions in the kidney tubules, so that their concentration is maintained in the blood; (3) converting *vitamin D to its active form, which increases calcium absorption in the intestine. Parathyroid hormone acts in opposition to *calcitonin.

**paraxial ray** A ray of light that falls on a reflecting or refracting surface close to and almost parallel to the axis. It is for such rays that simple lens theory can be developed, by means of making small angle approximations.

**parenchyma** 1. A plant tissue consisting of roughly spherical relatively undifferentiated cells, frequently with air spaces between them. The cortex and pith are composed of parenchyma cells (*see* GROUND TISSUES). 2. Loose *connective tissue formed of large cells. Its function is to pack the spaces between organs in some simple acoelomate animals, such as flatworms (Platyhelminthes).

**parent** 1. (in biology) **a.** Either male or female partner that together produce offspring in the process of sexual reproduction. *See also* P. **b.** Denoting an organism or cell that gives rise to new organisms or cells, as by asexual reproduction or cell division. 2. (in physics) *See* DAUGHTER.

**parity** Symbol *P*. The property of a *wave function that determines its behaviour when all its spatial coordinates are reversed in direction, i.e. when $x,y,z$ are replaced by $-x,-y,-z$. If a wave function $\psi$ satisfies the equation $\psi(x,y,z) = \psi(-x,-y,-z)$ it is said to have even parity, if it satisfies $\psi(x,y,z) = -\psi(-x,-y,-z)$ it has odd parity. In general,

$$\psi(x,y,z) = P\psi(-x,-y,-z),$$

where $P$ is a quantum number called parity that can have the value +1 or –1. The principle of **conservation of parity** (or **space-reflection symmetry**) would hold if all physical laws could be stated in a coordinate system independent of left- or right-handedness. If parity was conserved there would therefore be no fundamental way of distinguishing between left and right. In electromagnetic and strong interactions, parity is, in fact, conserved. In 1956, however, it was shown that parity is not conserved in weak interactions. In the beta decay of cobalt–60, for example, the electrons from the decay are emitted preferentially in a direction opposite to that of the cobalt spin. This experiment provides a fundamental distinction between left and right.

**parsec** A unit of length used to express astronomical distance. The distance at which the mean radius of the earth's orbit subtends an angle of one second of arc. One parsec is equal to $3.0857 \times 10^{16}$ metres or 3.2616 light years.

**parsimony** *See* PRINCIPLE OF PARSIMONY.

**parthenocarpy** The formation of fruit without prior fertilization of the flower by pollen. The resulting fruits are seedless and therefore do not contribute to the reproduction of the plant; examples are bananas and pineapples. *Plant hormones may have a role in this phenomenon, which can be induced by auxins in the commercial production of tomatoes and other fruits.

**parthenogenesis** The development of an organism from an unfertilized egg. This occurs sporadically in many plants (e.g. dandelions and hawkweeds) and in a few animals, but in some species it is the main and sometimes only method of reproduction. For example, in some species of aphid, males are absent or very rare. The eggs formed by the females contain the full (diploid) number of chromosomes and are genetically identical. Variation is consequently very limited in species that reproduce parthenogenetically.

**partial** A simple component of a complex tone. When a musical instrument produces a note, say, middle C, it will produce a complex tone in which the fundamental frequency is mixed with a number of partials. Some of these partials, for example, if the note is produced by bowing a taut string, will be *harmonics, i.e. integral multiples of the fundamental. If the string is struck, however, some of the partials can be inexact multiples of the fundamental. Partials are not therefore identical with harmonics.

**partial derivative** The infinitesimal change in a function consisting of two or more variables when one of the variable changes and the others remain constant. If $z$ = f($x,y$), $\partial z/\partial x$ is the partial derivative of $z$ with respect to $x$, while $y$ remains unchanged. A **partial differential equation**, such as the *Laplace equation, is an equation containing partial derivatives of a function.

**partial eclipse** *See* ECLIPSE.

**partial pressure** *See* DALTON'S LAW.

**particle** 1. (in physics) One of the fundamental components of matter. *See* ELEMENTARY PARTICLES. 2. (in mechanics) A hypothetical body that has mass but no physical extension. As it is regarded as having no volume, a particle is incapable of rotation and therefore can only have translational motion. Thus a real body may often, for translational purposes, be regarded as a particle located at the body's centre of mass and having a mass equal to that of the whole body.

**particle physics** The study of *elementary particles.

**particulate inheritance** The transmission from parent to offspring of separate units that determine characteristics. Gregor Mendel observed that *recessive characteristics, absent in the offspring of a cross in which only one parent possessed them, reappeared repeatedly in the progeny of subsequent crosses. This led him to formulate his theory of inherited 'factors' (now called *alleles) that retain their identity through succeeding generations (*see* MENDEL'S LAWS). *Compare* BLENDING INHERITANCE.

**partition** If a substance is in contact with two different phases then, in general, it will have a different affinity for each phase. Part of the substance will be absorbed or dissolved by one and part by the other, the relative amounts depending on the relative affinities. The substance is said to be **partitioned** between the two phases. For example, if two immiscible liquids are taken and a third compound is shaken up with them, then an equilibrium is reached in which the concentration in one solvent differs from that in the other. The ratio of the concentrations is the **partition coefficient** of the system. The **partition law** states that this ratio is a constant for given liquids.

**partition coefficient** *See* PARTITION.

**partition function** The quantity $Z$ defined by $Z = \Sigma \exp(-E_i/kT)$, where the sum is taken over all states $i$ of the system. $E_i$ is the energy of the $i$th state, $k$ is the *Boltzmann constant, and $T$ is the thermodynamic *temperature. $Z$ is a quantity of fundamental importance in equilibrium *statistical mechanics. For a system in which there are non-trivial interactions, it is very difficult to calculate the partition function exactly. For such systems it is necessary to use *approximation techniques and/or *model systems. The partition function links results at the atomic level to *thermodynamics, since $Z$ is related to the Helmholtz *free energy $F$ by $F = kT\ln Z$.

**parton** A pointlike, almost free, particle postulated as a component of nucleons. The parton model enabled the results of very high-energy experiments on nucleons to be understood. *See* QUANTUM CHROMODYNAMICS.

**parturition** The act of giving birth to young at the end of the *gestation period. Fetal hormones cause the process to start.

**pascal** The *SI unit of pressure equal to one newton per square metre.

**Pascal, Blaise** (1623–62) French mathematician and physicist. An infant prodigy, he had already made a mechanical calculating machine by 1642. In physics he formulated *Pascal's law concerning fluid pressure and the principle behind the hydraulic press. The SI unit of pressure is named after him.

**Pascal's distribution (negative binomial distribution)** The distribution of the number of independent *Bernoulli trials performed up to and including the $r^{th}$ success. The probability that the number of trials, $x$, is equal to $k$ is given by

$$P(x=k) = {}^{k-1}C_{r-1}p^r q^{k-r}$$

The mean and variance are $r/p$ and $rq/p^2$ respectively. *See also* GEOMETRIC DISTRIBUTION.

**Pascal's law** In a confined fluid, externally applied pressure is transmitted uniformly in all directions. In a static fluid, force is transmitted at the speed of sound throughout the fluid and acts at right angles to any surface in or bounding the fluid. This principle is made use of in the hydraulic jack, the pneumatic tyre, and similar devices. The law was discovered in 1647 by Blaise Pascal.

**Pascal's triangle** A triangular array of numbers in which each number is the sum of the two numbers immediately above it (except for the 1s):

$$1$$
$$1\ 1$$
$$1\ 2\ 1$$
$$1\ 3\ 3\ 1$$
$$1\ 4\ 6\ 4\ 1$$
$$1\ 5\ 10\ 10\ 5\ 1$$
$$1\ 6\ 15\ 20\ 15\ 6\ 1$$
$$1\ 7\ 21\ 35\ 35\ 21\ 7\ 1$$

and so on. The numbers in each row are the coefficients of the expansion of the binomial $(x + y)^n$ (*see* BINOMIAL THEOREM). It is named after Blaise Pascal.

**Paschen–Back effect** An effect in atomic line *spectra that occurs when the atoms are placed in a strong magnetic field. Spectral lines that give the anomalous *Zeeman effect when the atoms are placed in a weaker magnetic field have a different splitting pattern in a very strong magnetic field in which the spectral lines go back to the pattern of the normal Zeeman effect. The Paschen–Back effect is named after the German physicists Louis Paschen (1865–1947) and Ernest Back (1881–1959), who discovered it in 1912.

**Paschen series**   *See* HYDROGEN SPECTRUM.

**passive**   Describing a solid that has reacted with another substance to form a protective layer, so that further reaction stops. The solid is said to have been 'rendered passive'. For example, aluminium reacts spontaneously with oxygen in air to form a thin layer of *aluminium oxide, which prevents further oxidation. Similarly, pure iron forms a protective oxide layer with concentrated nitric acid and is not dissolved further.

**passive device**   1. An electronic component, such as a capacitor or resistor, that is incapable of amplification. 2. An artificial *satellite that reflects an incoming signal without amplification. 3. A solar-power device that makes use of an existing structure to collect and utilize solar energy without the use of pumps, fans, etc. 4. A radar device that provides information for navigation, guidance, surveillance, etc., by receiving the microwave radiation. Such a passive device emits no microwave energy itself and therefore does not disclose its position. 5. A system that detects an object by the radiation that it emits, rather than by reflecting radiation off it, as in a passive infrared detector (**PIT detector**). *Compare* ACTIVE DEVICE.

**passive immunity**   *See* IMMUNITY.

**passive transport**   *See* DIFFUSION.

**Pasteur, Louis**   (1822–95) French chemist and microbiologist, who held appointments in Strasbourg (1849–54) and Lille (1854–57), before returning to Paris to the Ecole Normale and the Sorbonne. From 1888 to his death he was director of the Pasteur Institute. In 1848 he discovered *optical activity, in 1860 relating it to molecular structure. In 1856 he began work on *fermentation, and by 1862 was able to disprove the existence of *spontaneous generation. He introduced *pasteurization (originally for wine) in 1863. He went on to study disease and developed vaccines against cholera (1880), anthrax (1882), and rabies (1885).

**pasteurization**   The treatment of milk to destroy disease-causing bacteria, such as those of tuberculosis, typhoid, and brucellosis. Milk is heated to 65°C for 30 minutes or to 72°C for 15 minutes followed by rapid cooling to below 10°C. The method was devised by the French microbiologist Louis Pasteur (1822–95).

**patella (kneecap)**   A small rounded movable bone that is situated in a tendon in front of the knee joint in most mammals (including humans). The function of the patella is to protect the knee.

**path integral formulation**   A formulation of quantum mechanics put forward by Richard Feynman in 1942 in which all the possible paths a particle in a quantum mechanical system can take, weighted by the probability of each path occurring, are added up. Path integrals have been used extensively, both in analysing the foundations of quantum mechanics and in solving certain types of problem.

**pathogen**   Any disease-causing microorganism. Pathogens include viruses, rickettsiae, and many bacteria, fungi, and protozoans. *See* INFECTION.

**pathology**   The study of the changes in organs and tissues that are caused by or give rise to disease. This involves the examination of tissue samples, X-ray photographs, or other evidence taken from living patients or from cadavers. **Clinical pathology** applies these findings to clinical cases, particularly in the development of diagnostic tests and treatments. In **experimental pathology**, disease processes are studied using experimental animals, cell cultures, or other means.

**patristic**   Denoting similarity between organisms resulting from common ancestry. *Compare* HOMOPLASY.

**Pauli, Wolfgang Ernst**   (1900–58) Austrian-born Swiss physicist. After studying with Niels *Bohr and Max Born, he taught at Heidelberg and, finally Zurich. His formulation in 1925 of the *Pauli exclusion principle explained the electronic make-up of atoms. For this work he was awarded the 1945 Nobel Prize for physics. In 1930 he predicted the existence of the *neutrino, which was finally discovered in 1956 by Clyde Cowan (1919–74) and Frederick Reines (1918–98).

**Pauli exclusion principle**   The quantum-mechanical principle, applying to fermions but not to bosons, that no two identical particles in a system, such as electrons in an atom or quarks in a hadron, can possess an identical set of quantum numbers. It was first formulated by Wolfgang Pauli in 1925. The origin of the Pauli exclusion principle lies in the *spin–statistics theorem of relativistic quantum field theory.

**Pauling, Linus Carl** (1901–94) US chemist. After spending two years in Europe, he became a professor at the Californian Institute of Technology. His original work was on chemical bonding; in the mid-1930s he turned to the structure of proteins, for which he was awarded the 1954 Nobel Prize for chemistry. He was also an active campaigner against nuclear weapons and in 1962 was awarded the Nobel Peace Prize.

**Paul trap** *See* ION TRAP.

**Pavlov, Ivan Petrovich** (1849–1936) Russian physiologist, who became professor of physiology in St Petersburg in 1886. While working on the physiology of digestion he discovered that the mere sight of food stimulates the production of digestive juices. For this work he was awarded the 1904 Nobel Prize for physiology or medicine. Pavlov went on to demonstrate operant *conditioning in dogs and other animals. *See also* LEARNING (Feature).

**p-block elements** The block of elements in the periodic table consisting of the main groups 13 (B to Tl), 14 (C to Pb), 15 (N to Bi), 16 (O to Po), 17 (F to At) and 18 (He to Rn). The outer electronic configurations of these elements all have the form $ns^2np^x$ where $x$ = 1 to 6. Members at the top and on the right of the p-block are nonmetals (C, N, P, O, F, S, Cl, Br, I, At). Those on the left and at the bottom are metals (Al, Ga, In, Tl, Sn, Pb, Sb, Bi, Po). Between the two, from the top left to bottom right, lie an ill-defined group of metalloid elements (B, Si, Ge, As, Te).

**PC** *See* PERSONAL COMPUTER.

**PCB** *See* POLYCHLORINATED BIPHENYL.

**PCP** *See* PHENCYCLIDINE.

**PCR** *See* POLYMERASE CHAIN REACTION.

**p.d. (potential difference)** *See* ELECTRIC POTENTIAL.

**PDGF** *See* GROWTH FACTOR.

**peacock ore** *See* BORNITE.

**peak oil** The point at which global oil production reaches a maximum and then begins to decline because of diminishing reserves. Thereafter, the oil extracted from new sources cannot keep pace with depletion of existing oilfields, and demand is forecast to exceed supply, so causing the price of oil and its derivatives to rise inexorably. This in turn, it is argued, will have dramatic economic and social consequences. The timing of peak oil is controversial, with many estimates lying between 2010 and 2015. However, this depends on continued high demand for oil, which may ease if alternatives, such as *renewable energy sources, become more widely adopted.

**pearl ash** *See* POTASSIUM CARBONATE.

**pearlite** *See* STEEL.

**Pearson symbol** A symbolic notation devised by W. B. Pearson for indicating the structure of a crystal. There are three parts:
(1) A lower case letter denoting one of six crystal systems: a = anorthic (triclinic); m = monoclinic; o = orthorhombic; t = tetragonal; h = hexagonal or rhombohedral; c = cubic.
(2) An upper-case letter denoting the lattice type: P = primitive; I = body-centred; F = face-centred (full); C = face-centred (side); R = rhombohedral.
(3) A number giving the number of atoms in the unit cell.
For example, sodium chloride has a cubic, face-centred structure with 8 atoms in the unit cell, so has Pearson symbol cF8. The first two letters of a Pearson symbol give the *Bravais lattice.

**peat** A mass of dark-brown or black fibrous plant debris produced by the partial disintegration of vegetation in wet places. It may accumulate in depressions. When subjected to burial and hence pressure and heat it may be converted to *coal. Peat is used to improve soil and as a fuel, especially in Ireland and Sweden.

**pebi-** *See* BINARY PREFIXES.

**peck order** *See* DOMINANT.

**pecten** Any of various comblike structures in animals. The pecten in the eyes of birds consists essentially of a network of blood vessels attached to the optic nerve and projecting into the vitreous humour. Its function is uncertain, but it may be involved in supplying the retina with nutrients and oxygen. A simple form of this structure is found in the eyes of reptiles.

**pectin (pectic substance)** A mixture of polysaccharides made up primarily of a sugar acid (galacturonic acid). Pectin is an important constituent of plant cell walls and the *middle lamella between adjacent cell walls; it is also found in certain plant juices. Normally present in an insoluble form, in ripening fruits and in tissues affected by certain diseases it changes into a soluble form, which is evidenced by softening of the tissues. It is used in making jam as it forms a gel with sucrose.

**pectoral fins** *See* FINS.

**pectoral girdle (shoulder girdle)** The bony or cartilaginous structure in vertebrates to which the anterior limbs (pectoral fins, forelegs, or arms) are attached. In mammals it consists of two dorsal *scapulae (shoulder blades) attached to the backbone and two ventral *clavicles (collar bones) attached to the sternum (breastbone).

**pedicel** The stalk attaching a flower to the main floral axis (*see* PEDUNCLE). Some flowers, described as **sessile**, do not have a pedicel and arise directly from the peduncle.

**pedology** The science of the study of soils, including their origin and characteristics and their utilization.

**peduncle** The main stalk of a plant that bears the flowers, which may be solitary or grouped in an *inflorescence. *Compare* PEDICEL.

**pegmatite** Very coarse-grained igneous rock. Granite pegmatite, the commonest type, consists chiefly of alkali feldspar and quartz; accessory minerals, such as mica, tourmaline, topaz, beryl, fluorite, cassiterite, and garnet, may also be present. Many pegmatites are thus economically important as sources of these minerals. The individual crystals may be extremely large; for example, mica and quartz crystals over 3 m in length have been found.

**pelagic** Describing organisms that swim or drift in a sea or a lake, as distinct from those that live on the bottom (*see* BENTHOS). Pelagic organisms are divided into *plankton and *nekton.

**pelargonic acid** *See* NONANOIC ACID.

**Pelecypoda** *See* BIVALVIA.

**pellagra** A disease resulting from a deficiency of *nicotinic acid, which is characterized by dermatitis and mental disorder.

**pellicle** The thin outer covering, composed of protein, that protects and maintains the shape of certain unicellular organisms, e.g. *Euglena*. It is transparent and in ciliated organisms, e.g. *Paramecium*, contains small pores through which the cilia emerge.

**Peltier effect** The change in temperature produced at a junction between two dissimilar metals or semiconductors when an electric current passes through the junction. The direction of the current determines whether the temperature rises or falls. The first metals to be investigated were bismuth and copper; if the current flows from bismuth to copper the temperature rises. If the current is reversed the temperature falls. The effect was discovered in 1834 by the French physicist Jean Peltier (1785–1845) and has been used recently for small-scale refrigeration. *Compare* SEEBECK EFFECT.

**pelvic fins** *See* FINS.

**pelvic girdle (pelvis; hip girdle)** The bony or cartilaginous structure in vertebrates to which the posterior limbs (pelvic fins or legs) are attached. The pelvic girdle articulates dorsally with the backbone; it is made up of two halves, each produced by the fusion of the *ilium, *ischium, and *pubis.

**pelvis** 1. *See* PELVIC GIRDLE. 2. The lower part of the abdomen in the region of the pelvic girdle. 3. A conical chamber in the *kidney into which urine drains from the kidney tubules before passing to the *ureter.

**pen drive** *See* USB DRIVE.

**pendulum** Any rigid body that swings about a fixed point. The **ideal simple pendulum** consists of a bob of small mass oscillating back and forth through a small angle at the end of a string or wire of negligible mass. Such a device has a period $2\pi\sqrt{(l/g)}$, where $l$ is the length of the string or wire and $g$ is the *acceleration of free fall. This type of pendulum moves with *simple harmonic motion.

The **compound pendulum** consists of a rigid body swinging about a point within it. The period of such a pendulum is given by

$$T = 2\pi\sqrt{[(h^2 + k^2)/hg]},$$

where $k$ is the radius of gyration about an axis through the centre of mass and $h$ is the distance from the pivot to the centre of mass. *See also* KATER'S PENDULUM.

**penicillin** An *antibiotic derived from the mould *Penicillium notatum*; specifically it is known as **penicillin G** (benzylpenicillin) or **penicillin V** (phenoxymethylpenicillin), which belong to a class of similar substances called penicillins. They are all active against a wide variety of bacteria, producing their effects by disrupting synthesis of the bacterial cell wall, and are used to treat a variety of infections caused by these bacteria.

**penis** The male reproductive organ of mammals (and also of some birds and reptiles) used to introduce sperm into the female reproductive tract to ensure internal fertilization. It contains a duct (the *urethra) through which the sperms pass. The penis becomes erect during precopulatory activity, either by filling with blood or haemolymph or by the action of muscles, and can be inserted into the vagina (or cloaca). In mammals the urine also leaves the body through the penis.

**Penning gauge** A type of cold-cathode *ionization gauge in which a discharge is maintained between two electrodes with a potential difference of a few kilovolts. An axial magnetic field is also applied to cause electrons to move in spiral paths and increase the ionization current and sensitivity. This type of gauge is used in the range $10^{-3}$–$10^{-5}$ torr. It was invented in 1936 by the Dutch physicist Frans Penning (1894–1953).

**Penning trap** *See* ION TRAP.

**Penrose, Sir Roger** (1931– ) British mathematician and physicist. Penrose was the first to point out that singularities are inevitable features of the general theory of relativity. He has made other important contributions to the theory of relativity and its quantization, including his work on *twistor theory. He also discovered Penrose patterns, which are a two-dimensional analogue of quasicrystals. *See also* HAWKING, STEPHEN WILLIAM.

**Penrose process** A process by which the rotational energy of a rotating black hole can be extracted. An object close to the event horizon may split into two particles. One, with negative energy, falls into the black hole, causing the rotation rate to decrease. The other, with positive rotation, moves away. The result is that energy is extracted at the expense of rotational energy of the black hole. This process was suggested by Sir Roger Penrose in 1969. *See also* BLANDFORD–ZNAJEK PROCESS.

**pentadactyl limb** A limb with five digits, characteristic of tetrapod vertebrates (amphibians, reptiles, birds, and mammals). It evolved from the paired fins of primitive fish as an adaptation to locomotion on land and is not found in modern fish. The limb has three parts (see illustration): the upper arm or thigh containing one long bone, the forearm or shank containing two long bones, and the hand or foot, which contains a number of small bones. This basic design is modified in many species, according to the function of the limb, particularly by the loss or fusion of the terminal bones.

**pentaerythritol** A white crystalline compound, $C(CH_2OH)_4$; m.p. 260°C; b.p. 276°C (30 mmHg). It is used in making the explosive pentaerythritol trinitrate and in producing resins and other organic products.

**pentaerythritol tetranitrate (PETN)** A powerful high explosive made from pentaerythritol, $C(CH_2ONO_2)_4$.

**pentahydrate** A crystalline hydrate that has five moles of water per mole of compound.

**pentane** A straight-chain alkane hydrocarbon, $C_5H_{12}$; r.d. 0.63; m.p. –129.7°C; b.p. 36.1°C. It is obtained by distillation of petroleum.

**pentanedioic acid (glutaric acid)** A simple dicarboxylic acid, $HOOC(CH_2)_3COOH$; m.p. 96°C; b.p. 200°C. It is used in the production of certain polymers.

**pentanoic acid (valeric acid)** A colourless liquid *carboxylic acid, $CH_3(CH_2)_3COOH$; r.d. 0.9; m.p. –34°C; b.p. 186.05°C. It is used in the perfume industry.

**pentaquark** A long-lived particle consisting of five quarks with a mass of just over 1500 MeV, which has been predicted to exist. There is no evidence for the existence of this particle.

**pentavalent (quinquevalent)** Having a valency of five.

**pentlandite** A mineral consisting of a mixed iron–nickel sulphide, $(Fe,Ni)_9S_8$, crystallizing in the cubic system; the chief ore of nickel. It is yellowish-bronze in colour with a metallic lustre. The chief occurrence of the mineral is at Sudbury in Ontario, Canada.

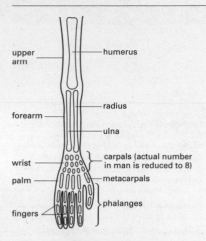

A basic pentadactyl forelimb, as exemplified by the human arm

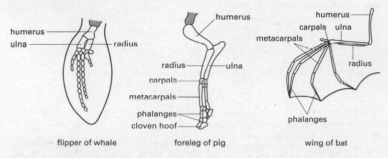

flipper of whale          foreleg of pig          wing of bat

The modified pentadactyl forelimb of various vertebrates

**Pentadactyl limb.**

**pentode** A *thermionic valve with a **suppressor grid** between the anode and the screen grid of a tetrode. Its purpose is to suppress the loss of electrons from the anode as a result of secondary emission. The suppressor grid is maintained at a negative potential relative to the anode and to the screen grid.

**pentose** A sugar that has five carbon atoms per molecule. *See* MONOSACCHARIDE.

**pentose phosphate pathway (pentose shunt)** A series of biochemical reactions that results in the conversion of glucose 6-phosphate to ribose 5-phosphate and generates NADPH, which provides reducing power for other metabolic reactions, such as synthesis of fatty acids. Ribose 5-phosphate and its derivatives are components of such molecules as ATP, coenzyme A, NAD, FAD, DNA, and RNA. In plants the pentose phosphate pathway also plays a role in the synthesis of sugars from carbon dioxide. In animals the pathway occurs at various sites, including the liver and adipose tissue.

**pentyl group (pentyl radical)** The organic group $CH_3CH_2CH_2CH_2CH_2^-$, derived from pentane.

**penumbra** *See* SHADOW.

**pepo** *See* BERRY.

**pepsin** An enzyme that catalyses the

**Peptide.** Formation of a peptide bond.

breakdown of proteins to polypeptides in the vertebrate stomach. It is secreted as an inactive precursor, *pepsinogen.

**pepsinogen** The inactive precursor of the enzyme *pepsin. Pepsinogen is secreted by the lining of the vertebrate stomach into the lumen, where it is converted to pepsin by hydrochloric acid and also by the action of pepsin itself.

**peptidase** *See* ENDOPEPTIDASE; EXOPEPTIDASE; PROTEASE.

**peptide** Any of a group of organic compounds comprising two or more amino acids linked by **peptide bonds**. These bonds are formed by the reaction between adjacent carboxyl (–COOH) and amino (–NH₂) groups with the elimination of water (see illustration). **Dipeptides** contain two amino acids, **tripeptides** three, and so on. *Polypeptides contain more than 20 and usually 100–300. Naturally occurring **oligopeptides** (of less than 20 amino acids) include the tripeptide glutathione and the pituitary hormones antidiuretic hormone and oxytocin, which are octapeptides. Peptides also result from protein breakdown, e.g. during digestion.

((⊕)) **SEE WEB LINKS**

• Information about IUPAC nomenclature of peptides

**per-** Prefix indicating that a chemical compound contains an excess of an element, e.g. a peroxide.

**percentile** For a random variable in *statistics, any of the 99 values that divide its distribution such that an integral percentage of the collection lies below that value. For example, the 85th percentile is the value of a variable that has 85% of the collection below that value. The 25th percentile is called the lower **quartile**, the 50th percentile is the **median**, and the 75th percentile is the upper quartile.

**perchlorate** *See* CHLORATES.

**perchloric acid** *See* CHLORIC(VII) ACID.

**Percus–Yevick approximation** An approximation used in *statistical mechanics to calculate the radial distribution function of a system. This approximation, which was devised by Jerome Percus and George Yevick in 1958, has been used extensively in the theory of liquids.

**perdisulphuric acid** *See* PEROXOSULPHURIC(VI) ACID.

**perennation** The survival of biennial or perennial plants from one year to the next by vegetative means. In biennials and herbaceous perennials the aerial parts of the plant die down and the plants survive by means of underground storage roots (e.g. carrot), *rhizomes (e.g. couch grass, Solomon's seal), *tubers (e.g. dahlia), *bulbs (e.g. daffodil, snowdrop), or *corms (e.g. crocus, gladiolus). These **perennating organs** are also frequently responsible for *vegetative propagation. Woody perennials survive the winter by reducing their metabolic activity (e.g. by leaf loss in deciduous trees and shrubs).

**perennial** A plant that lives for a number of years. Woody perennials (trees and shrubs) have a permanent aerial form, which continues to grow year after year. Herbaceous (i.e. nonwoody) perennials have aerial shoots that die down each autumn and are replaced in spring by new shoots from an underground structure (*see* PERENNATION). Lupin and rhubarb are examples of herbaceous perennials. *Compare* ANNUAL; BIENNIAL; EPHEMERAL.

**perfect gas** *See* IDEAL GAS; GAS.

**perfect pitch** *See* ABSOLUTE PITCH.

**perfect solution** *See* RAOULT'S LAW.

**perianth** The part of a flower situated outside the stamens and carpels. In dicotyledons it consists of two distinct whorls, the outer of sepals (*see* CALYX) and the inner of petals (*see* COROLLA). In monocotyledons the two whorls are similar and often brightly coloured. In wind-pollinated flowers both whorls may be reduced or absent. In many horticultural varieties the number of perianth parts is multiplied, but the resulting 'double' flowers are often sterile.

**periastron** See APASTRON.

**pericardial cavity** The cavity in vertebrates that contains the heart and is bounded by a membrane (the *pericardium). It is part of the *coelom.

**pericardium (pericardial membrane)** The membrane that encloses the pericardial cavity, containing the vertebrate heart. The pericardium holds the heart in position while allowing it to relax and contract. It consists of two main parts: a tough outer fibrous layer (**fibrous pericardium**) and the more delicate **serous pericardium**, which consists of a double layer of *serous membrane, the inner layer being in close contact with the heart.

**pericarp (fruit wall)** The part of a fruit that develops from the ovary wall of a flower. The type of fruit that develops depends on whether the pericarp becomes dry and hard or soft and fleshy. The pericarp can be made up of three layers. The outer skin (**epicarp** or **exocarp**) may be tough and hard; the middle layer (**mesocarp**) may be succulent as in peach, hard as in almond, or fibrous as in coconut; and the inner layer (**endocarp**) may be hard and stony as in many *drupes, membranous as in citrus fruits, or indistinguishable from the mesocarp, as in many *berries.

**pericycle** A plant tissue comprising the outermost layer of the root vascular tissue, lying immediately beneath the *endodermis. Lateral roots originate from the pericycle.

**pericynthion** The point in the orbit around the moon of a satellite launched from the earth that is nearest to the moon. For a satellite launched from the moon the equivalent point is the **perilune**. *Compare* APOCYNTHION.

**periderm** See CORK CAMBIUM.

**perigee** The point in the orbit of the moon or an artificial earth satellite when it is closest to the earth. See APOGEE.

**perihelion** The point in the solar orbit of a planet, comet, or other solar system object, natural or artificial, at which it is nearest to the sun. At the beginning of the 21st century, the earth is at perihelion on or about 3 January. Its distance from the sun at that point is

about 0.9833 astronomical unit. *Compare* APHELION.

**perilymph** The fluid of the *inner ear that fills the space between the bony labyrinth and the membranous labyrinth. *Compare* ENDOLYMPH.

**period** 1. The time taken for one complete cycle of an oscillating system or wave. 2. *See* PERIODIC TABLE. 3. *See* MENSTRUAL CYCLE. 4. *See* GEOLOGICAL TIME SCALE.

**period doubling** A mechanism for describing the transition to *chaos in certain dynamical systems. If the force on a body produces a regular orbit with a specific *period a sudden increase in the force can suddenly double the period of the orbit and the motion becomes more complex. The original simple motion is called a **one-cycle**, while the more complicated motion after the period doubling is called a **two-cycle**. The process of period doubling can continue until a motion called an **n-cycle** is produced. As $n$ increases to infinity the motion becomes non-periodic. The period-doubling route to chaos occurs in many systems involving nonlinearity, including lasers and certain chaotic chemical reactions. The period-doubling route to chaos was postulated and investigated by the US physicist Mitchell Feigenbaum in the early 1980s. Routes to chaos other than period doubling also exist.

**periodic acid** See IODIC(VII) ACID.

**periodic law** The principle that the physical and chemical properties of elements are a periodic function of their proton number. The concept was first proposed in 1869 by Dimitri Mendeleev, using relative atomic mass rather than proton number, as a culmination of efforts to rationalize chemical properties by Johann Döbereiner (1817), John Newlands (1863), and Lothar Meyer (1864). One of the major successes of the periodic law was its ability to predict chemical and physical properties of undiscovered elements and unknown compounds that were later confirmed experimentally. See PERIODIC TABLE.

**periodic motion** Any motion of a system that is continuously and identically repeated. The time $T$ that it takes to complete one cycle of an oscillation or wave motion is called the *period, which is the reciprocal of the *frequency. See PENDULUM; SIMPLE HARMONIC MOTION.

**periodic table** A table of elements arranged in order of increasing proton number to show the similarities of chemical elements with related electronic configurations. (The original form was proposed by Dimitri Mendeleev in 1869 using relative atomic masses.) In the modern **short form**, the *lanthanoids and *actinoids are not shown. The elements fall into vertical columns, known as **groups**. Going down a group, the atoms of the elements all have the same outer shell structure, but an increasing number of inner shells. Traditionally, the alkali metals were shown on the left of the table and the groups were numbered IA to VIIA, IB to VIIB, and 0 (for the noble gases). All the elements in the middle of the table are classified as *transition elements and the nontransition elements are regarded as **main-group** elements. Because of confusion in the past regarding the numbering of groups and the designations of subgroups, modern practice is to number the groups across the table from 1 to 18 (see Appendix). Horizontal rows in the table are **periods**. The first three are called **short periods**; the next four (which include transition elements) are **long periods**. Within a period, the atoms of all the elements have the same number of shells, but with a steadily increasing number of electrons in the outer shell. The periodic table can also be divided into four **blocks** depending on the type of shell being filled: the *s-block, the *p-block, the *d-block, and the *f-block.

There are certain general features of chemical behaviour shown in the periodic table. In moving down a group, there is an increase in metallic character because of the increased size of the atom. In going across a period, there is a change from metallic (electropositive) behaviour to nonmetallic (electronegative) because of the increasing number of electrons in the outer shell. Consequently, metallic elements tend to be those on the left and towards the bottom of the table; nonmetallic elements are towards the top and the right.

There is also a significant difference between the elements of the second short period (lithium to fluorine) and the other elements in their respective groups. This is because the atoms in the second period are smaller and their valence electrons are shielded by a small $1s^2$ inner shell. Atoms in the other periods have inner s- and p-electrons shielding the outer electrons from the nucleus. Moreover, those in the second period only have s- and p-orbitals available for bonding. Heavier atoms can also promote electrons to vacant d-orbitals in their outer shell and use these for bonding. *See also* DIAGONAL RELATIONSHIP; INERT-PAIR EFFECT.

((⊕)) SEE WEB LINKS

- The WebElements table produced by Mark Winter at the University of Sheffield
- Over 50 different forms of the periodic table in the Chemogenesis web book by Mark R. Leach

**periodontal membrane** The membrane of connective tissue that surrounds the root of a *tooth and anchors it to its socket in the jawbone. Fibres of the periodontal membrane pass into the *cement covering the root, which provides a firm attachment.

**periosteum** The outer membrane that surrounds a bone. It contains connective tissue, capillaries, nerves, and a number of types of bone cell. The periosteum plays an important role in bone repair and growth.

**peripheral device** Any device, such as an input or output device, connected to the central processing unit of a *computer. Backing store is also usually regarded as a peripheral.

**peripheral nervous system** All parts of the nervous system excluding the *central nervous system. It consists of all the *cranial and *spinal nerves and their branches, which link the *receptors and *effectors with the central nervous system. *See also* AUTONOMIC NERVOUS SYSTEM.

**periscope** An optical device that enables an observer to see over or around opaque objects. The simplest type consists of a long tube with mirrors at each end set at 45° to the direction to be viewed. A better type uses internally reflecting prisms instead of plane mirrors. Periscopes are used in tanks (to enable the observer to see over obstacles without being shot at) and in submarines (when the vessel is submerged). Such periscopes are usually quite complicated instruments and include telescopes.

**Perissodactyla** An order of mammals having hoofed feet with an odd number of toes. They are all herbivores and include the tapirs, rhinoceros, and horse. The teeth are large and specialized for grinding. Cellulose digestion occurs in the caecum and large intestine. Fossils of the Eocene epoch, 60 million years ago, show that these animals were at that time already distinct from the cloven-hoofed *Artiodactyla.

**peristalsis** Waves of involuntary muscular contraction and relaxation that pass along the alimentary canal, forcing food contents along. It is brought about by contraction of the circular muscles of the gut wall in sequence.

**peristome 1.** A ring of toothlike structures around the opening of a moss *capsule. The teeth tend to bend and twist in dry weather, so opening the mouth of the capsule and allowing the spores to escape. In wet weather they close over the opening of the capsule. **2.** The area around the mouth in many invertebrates and some protists. It sometimes assists in food collecting. Examples are the spirally ciliated groove around the mouth of some ciliate protozoans and the first segment of the earthworm.

**peritoneum** The thin layer of tissue (*see* SEROUS MEMBRANE) that lines the abdominal cavity of vertebrates and covers the abdominal organs. *See also* MESENTERY.

**Perkin, Sir William Henry** (1838–1907) British chemist, who while still a student accidentally produced mauvine, the first aniline dye and the first dyestuff to be synthesized. Perkin built a factory to produce it, and made a fortune.

**permaculture** A permanent agriculture based on cropping from perennial plants and trees and often incorporating livestock. Such systems, inspired by natural ecosystems, are typically designed to be diverse, stable, and resilient, with minimal need for energy or artificial fertilizers and pesticides. However, they are generally relatively small in scale and are geared to fulfilling local needs or self-sufficiency for the farmer, as opposed to the large-scale cash cropping of conventional agriculture.

**permafrost** Permanently frozen soil and subsoil that occurs in arctic, subarctic, and alpine regions. It ranges from 30 cm to over 1000 m thick and covers nearly a fifth of the land surface of the earth. In summer the top few centimetres may thaw, forming pools of meltwater (which cannot drain through the frozen soil beneath).

**Permalloys** A group of alloys of high magnetic permeability consisting of iron and nickel (usually 40–80%) often with small amounts of other elements (e.g. 3–5% molybdenum, copper, chromium, or tungsten). They are used in thin foils in electronic transformers, for magnetic shielding, and in computer memories.

**permanent gas** A gas, such as oxygen or nitrogen, that was formerly thought to be impossible to liquefy. A permanent gas is now regarded as one that cannot be liquefied by pressure alone at normal temperatures (i.e. a gas that has a critical temperature below room temperature).

**permanent hardness** *See* HARDNESS OF WATER.

**permanent magnet** *See* MAGNET.

**permanent teeth** The second and final set of teeth that mammals produce after shedding the *deciduous teeth. An adult

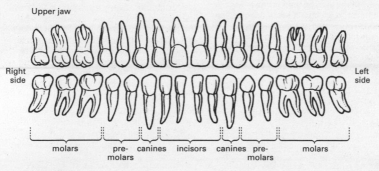

Upper jaw

Right side / Left side

molars | pre-molars | canines | incisors | canines | pre-molars | molars

Lower jaw

**Permanent teeth.**

human normally has 32 permanent teeth, consisting of incisors, canines, molars, and premolars (see illustration). These usually appear between the ages of approximately 6 and 21 years. *See also* DENTAL FORMULA; DIPHYODONT.

**permanganate** *See* MANGANATE(VII).

**permeability (magnetic permeability)** Symbol $\mu$. The ratio of the magnetic flux density, $B$, in a substance to the external field strength, $H$; i.e. $\mu = B/H$. The permeability of free space, $\mu_0$, is also called the **magnetic constant** and has the value $4\pi \times 10^{-7}$ H m$^{-1}$ in *SI units. The relative permeability of a substance, $\mu_r$, is given by $\mu/\mu_0$ and is therefore dimensionless. *See* MAGNETISM.

**Permian** The last geological period in the Palaeozoic era. It extended from the end of the Carboniferous period, about 299 million years ago, to the beginning of the Mesozoic era, about 251 million years ago. It was named by the British geologist Roderick Murchison (1792–1871) in 1841 after the Perm province in Russia. In some areas continental conditions prevailed, which continued into the following period, the Triassic. These conditions resulted in the deposition of the New Red Sandstone. During the period a number of animal groups became extinct, including the trilobites, tabulate and rugose corals, and blastoids (*see* MASS EXTINCTION). Amphibians and reptiles continued to be the dominant land animals and gymnosperms replaced ferns, clubmosses, and horsetails as the dominant plants.

**permittivity** Symbol $\varepsilon$. The ratio of the *electric displacement in a medium to the intensity of the electric field producing it. It is important for electrical insulators used as *dielectrics.

If two charges $Q_1$ and $Q_2$ are separated by a distance $r$ in a vacuum, the force $F$ between the charges is given by:

$$F = Q_1 Q_2 / r^2 4\pi\varepsilon_0$$

In this statement of *Coulomb's law using *SI units, $\varepsilon_0$ is called the absolute permittivity of free space, which is now known as the **electric constant**. It has the value $8.854 \times 10^{-12}$ F m$^{-1}$.

If the medium between the charges is anything other than a vacuum the equation becomes:

$$F = Q_1 Q_2 / r^2 4\pi\varepsilon$$

and the force between the charges is reduced. $\varepsilon$ is the **absolute permittivity** of the

new medium. The **relative permittivity** ($\varepsilon_r$) of a medium, formerly called the **dielectric constant**, is given by $\varepsilon_r = \varepsilon/\varepsilon_0$.

**permonosulphuric(VI) acid** *See* PEROXOSULPHURIC(VI) ACID.

**permutations and combinations** A combination is any subset of a particular set of objects, regardless of the order of selection. If the set consists of $n$ objects, $r$ objects can be selected giving $n!/r!(n-r)!$ different combinations. This can be written $_nC_r$.

A permutation is an ordered subset (i.e. attention is paid to the order of selection or arrangement) of a particular set of objects. If the set consists of $n$ objects, $r$ such objects can be selected to give $n!/(n-r)!$ permutations. This is written $_nP_r$.

**Permutit** Trade name for a *zeolite used for water softening.

**peroxides 1.** A group of inorganic compounds that contain the $O_2^{2-}$ ion. They are notionally derived from hydrogen peroxide, $H_2O_2$, but these ions do not exist in aqueous solution due to extremely rapid hydrolysis to OH$^-$. **2.** A class of organic compounds that contain the –O–O– group. Organic peroxides tend to be unstable and some are explosive. They can be used to initiate free-radical reactions.

**(())) SEE WEB LINKS**
• Information about IUPAC nomenclature

**peroxisome** An organelle, found in the cytoplasm of virtually all plant and animal cells, that contains several enzymes involved in oxidation processes. Peroxisomes are bound by a single membrane; the enzymes they contain include urate oxidase and catalase (which catalyses the breakdown of hydrogen peroxide, produced as a by-product of the activity of the other enzymes). Peroxisomes are active in the process of *detoxification, being particularly abundant in the liver and in the *proximal convoluted tubules of the kidney. In plant cells they are the main site of *fatty-acid oxidation.

**peroxodisulphuric acid** *See* PEROXOSULPHURIC(VI) ACID.

**peroxomonosulphuric(VI) acid** *See* PEROXOSULPHURIC(VI) ACID.

**peroxosulphuric(VI) acid** The term commonly refers to **peroxomonosulphuric(VI) acid**, $H_2SO_5$, which is also called **permonosulphuric(VI) acid** and **Caro's acid**.

It is a crystalline compound made by the action of hydrogen peroxide on concentrated sulphuric acid. It decomposes in water and the crystals decompose, with melting, above 45°C. The compound **peroxodisulphuric acid**, $H_2S_2O_8$, also exists (formerly called **perdisulphuric acid**). It is made by the high-current electrolysis of sulphate solutions. It decomposes at 65°C (with melting) and is hydrolysed in water to give the mono acid and sulphuric acid. Both peroxo acids are very powerful oxidizing agents. *See also* SUL-PHURIC ACID (for structural formulas).

**perpetual motion** 1. Perpetual motion of the first kind. Motion in which a mechanism, once started, would continue indefinitely to perform useful work without being supplied with energy from an outside source. Such a device would contravene the first law of *thermodynamics and is therefore not feasible. Many historical attempts, exercising great ingenuity, were constructed before the concept of energy and its conservation were understood. Some attempts have been made, since the first law of thermodynamics became generally accepted, by inventors seeking to establish loopholes in the laws of nature. 2. Perpetual motion of the second kind. Motion in which a mechanism extracts heat from a source and converts all of it into some other form of energy. An example of such a mechanism would be a ship that utilized the internal energy of the oceans for propulsion. Such a device does not contravene the first law of thermodynamics but it does contravene the second law. In the case of the ship, the sea would have to be at a higher temperature than the ship to establish a useful flow of heat. This could not occur without an external energy source. 3. Perpetual motion of the third kind. A form of motion that continues indefinitely but without doing any useful work. An example is the random molecular motion in a substance. This type postulates the complete elimination of friction. A mechanism consisting of frictionless bearings maintained in a vacuum could turn indefinitely, once started, without contravening the first or second laws of thermodynamics, provided it did no external work. Experience indicates that on the macroscopic scale such a condition cannot be achieved. On the microscopic scale, however, a superconducting ring of wire will apparently sustain a perpetual current flow without the application of an external force. This could be considered a form of

perpetual motion of the third kind, if the energy required to cool the wire to superconducting temperatures is ignored.

**persistent organic pollutant (POP)** A toxic substance that is not biodegradable and persists in the environment. POPs can concentrate as they move up the food chain and are generally deleterious to health. In 2001, an international conference in Sweden resulted in an agreement (the **Stockholm Convention**) to reduce or eliminate the 12 POPs of greatest concern, and to add other chemicals to the list in the future. The 12 chemicals were the pesticides aldrin, chlordane, *DDT, dieldrin, endrin, heptachlor, mirex, and toxaphene, together with the industrial chemical hexachlorobenzene and the groups *polychlorinated biphenyls, *dioxins, and *furans.

• Official Stockholm Convention site

**personal computer** A general-purpose *microcomputer designed for use by one person at a time. The original Personal Computer (or PC) was a highly successful product from IBM. An **IBM-compatible** computer is functionally identical to an IBM PC and able to accept all hardware and software intended for it. The abbreviation 'PC' is now most often used to mean an IBM-compatible computer as opposed to other systems. Personal computers range widely in capability and cost. They may take the form of desktop computers or be portable versions, such as laptop, notebook, or subnotebook computers.

**Perspex** Trade name for a form of *polymethylmethacrylate.

**perturbation** A departure by a celestial body from the trajectory or orbit it would follow if it moved only under the influence of a single central force. According to *Kepler's law, for example, a single planet orbiting the sun would move in an elliptical orbit. In fact, planets are perturbed from elliptical orbits by the gravitational forces exerted on them by other planets. Similarly, the moon's orbit round the earth is perturbed by the gravitational effect of the sun and the trajectories of comets are perturbed when they pass close to planets.

**perturbation theory** A method used in calculations in both classical physics (e.g. planetary orbits) and quantum mechanics (e.g. atomic structure), in which the system is divided into a part that is exactly calculable and a small term, which prevents the whole system from being exactly calculable. The technique of perturbation theory enables the effects of the small term to be calculated by an infinite series (which in general is an asymptotic series). Each term in the series is a 'correction term' to the solutions of the exactly calculable system. In classical physics, perturbation theory can be used for calculating planetary orbits. In quantum mechanics, it can be used to calculate the energy levels in molecules. In the many-body problem in quantum mechanics and in relativistic quantum field theory, the terms in perturbation theory may be represented pictorially by Feynman diagrams (*see* QUANTUM ELECTRODYNAMICS).

**pervasive computing**  *See* UBIQUITOUS COMPUTING.

**perversion**  *See* LATERAL INVERSION.

**pest**  Any of various organisms, such as fungi, insects, rodents, and plants, that harm crops or livestock or otherwise interfere with the wellbeing of human beings. **Weeds** are plant pests that grow where they are not wanted – often on cultivated land, where they compete with crop plants for space, light, nutrients, etc. Pests are controlled by the use of *pesticides and *biological control methods.

**pesticide**  Any chemical compound used to kill pests that destroy agricultural production or are in some way harmful to humans. Pesticides include **herbicides** (such as *2,4-D and Paraquat), which kill unwanted plants or weeds; **insecticides** (such as *pyrethrum), which kill insect pests; **fungicides**, which kill fungi; and **rodenticides** (such as *warfarin), which kill rodents. The problems associated with pesticides are that they are very often nonspecific and may therefore be toxic to organisms that are not pests; they may also be nonbiodegradable, so that they persist in the environment and may accumulate in living organisms (*see* BIOACCUMULATION). Organophosphorus insecticides, such as malathion and parathion, are biodegradable but can also damage the respiratory and nervous systems in humans as well as killing useful insects, such as bees. They act as *anticholinesterases. Organochlorine insecticides, such as dieldrin, aldrin, and *DDT, are very persistent and not easily biodegradable.

**PET**  *See* POSITRON EMISSION TOMOGRAPHY.

**peta-**  Symbol P. A prefix used in the metric system to denote one thousand million million times. For example, $10^{15}$ metres = 1 petametre (Pm).

**petal**  One of the parts of the flower that make up the *corolla. Petals of insect-pollinated plants are usually brightly coloured and often scented. Those of wind-pollinated plants are usually reduced or absent. Petals are considered to be modified leaves but their structure is simpler. Epidermal hairs may be present and the cuticle is often covered by lines or dots known as **honey guides**, which direct insects to the *nectar.

**Peters' projection**  *See* MAP PROJECTIONS.

**petiole**  The stalk that attaches a *leaf blade to the stem. Leaves without petioles are described as **sessile**.

**Petri dish**  A shallow circular flat-bottomed dish made of glass or plastic and having a fitting lid. It is used in laboratories chiefly for culturing bacteria and other microorganisms. It was invented by the German bacteriologist Julius Petri (1852–1921).

**petrification**  *See* FOSSIL.

**petrochemicals**  Organic chemicals obtained from petroleum or natural gas.

**petrolatum**  *See* PETROLEUM JELLY.

**petroleum**  A naturally occurring oil that consists chiefly of hydrocarbons with some other elements, such as sulphur, oxygen, and nitrogen. In its unrefined form petroleum is known as **crude oil** (sometimes **rock oil**). Petroleum is believed to have been formed from the remains of living organisms that were deposited, together with rock particles and biochemical and chemical precipitates, in shallow depressions, chiefly in marine conditions. Under burial and compaction the organic matter went through a series of processes before being transformed into petroleum, which migrated from the source rock to become trapped in large underground reservoirs beneath a layer of impermeable rock. The petroleum often floats above a layer of water and is held under pressure beneath a layer of *natural gas.

Petroleum reservoirs are discovered through geological exploration: commercially important oil reserves are detected by exploratory narrow-bore drilling. The major known reserves of petroleum are in Saudi Arabia, Russia, China, Kuwait, Iran, Iraq,

Mexico, USA, United Arab Emirates, Libya, Venezuela, and beneath the North Sea. The oil is actually obtained by the sinking of an oil well. Before it can be used it is separated by fractional distillation in oil refineries. The main fractions obtained are:

(1) **Refinery gas** A mixture of methane, ethane, butane, and propane used as a fuel and for making other organic chemicals.

(2) **Gasoline** A mixture of hydrocarbons containing 5 to 8 carbon atoms, boiling in the range 40–180°C. It is used for motor fuels and for making other chemicals.

(3) **Kerosine** (or **paraffin oil**) A mixture of hydrocarbons having 11 or 12 carbon atoms, boiling in the range 160–250°C. Kerosine is a fuel for jet aircraft and for oil-fired domestic heating. It is also cracked to produce smaller hydrocarbons for use in motor fuels.

(4) **Diesel oil** (or **gas oil**) A mixture of hydrocarbons having 13 to 25 carbon atoms, boiling in the range 220–350°C. It is a fuel for diesel engines.

The residue is a mixture of higher hydrocarbons. The liquid components are obtained by vacuum distillation and used in lubricating oils. The solid components (**paraffin wax**) are obtained by solvent extraction. The final residue is a black tar containing free carbon (**asphalt** or **bitumen**).

**petroleum ether** A colourless volatile flammable mixture of hydrocarbons (not an ether), mainly pentane and hexane. It boils in the range 30–70°C and is used as a solvent.

**petroleum jelly (petrolatum)** A semi-solid mixture of hydrocarbons extracted from petroleum. It is used in skincare products and cosmetics, and as a lubricant. Petroleum is widely available under the tradename **Vaseline**.

**pewter** An alloy of lead and tin. It usually contains 63% tin; pewter tankards and food containers should have less than 35% of lead so that the lead remains in solid solution with the tin in the presence of weak acids in the food and drink. Copper is sometimes added to increase ductility and antimony is added if a hard alloy is required.

**peyote** *See* MESCALINE.

**Pfund series** *See* HYDROGEN SPECTRUM.

**PGD** *See* PREIMPLANTATION GENETIC DIAGNOSIS.

**pH** *See* pH SCALE.

**PHA** *See* PHYTOHAEMAGGLUTININ.

**Phaeophyta (brown algae)** A phylum of *algae in which the green chlorophyll pigments are usually masked by the brown pigment fucoxanthin. Brown algae are usually marine (being abundant in cold water) and many species, such as the wracks (*Fucus*), inhabit intertidal zones. They vary in size from small branched filaments to ribbon-like bodies (known as kelps) many metres long.

**phage** *See* BACTERIOPHAGE.

**phagocyte** A cell that is able to engulf and break down foreign particles, cell debris, and disease-producing microorganisms (*see* PHAGOCYTOSIS). Some protists and certain mammalian cells (e.g. *macrophages and *monocytes) are phagocytes. Phagocytes are important elements in the natural defence mechanism of most animals.

**phagocytosis** The process by which foreign particles invading the body or minute food particles are engulfed and broken down by certain animal cells (known as *phagocytes). The plasma membrane of the phagocyte invaginates to capture the particle and then closes around it to form a vesicle. This then coalesces with a *lysosome, which contains enzymes that break down the particle. *See* ENDOCYTOSIS. *Compare* PINOCYTOSIS.

**phalanges** The bones that make up the *digits of the hand or foot in vertebrates. They articulate with the *metacarpals of the hand or with the *metatarsals of the foot. In the basic *pentadactyl limb there are two phalanges for the first digit (the thumb or big toe in humans) and three for each of the others.

**phane** *See* CYCLOPHANE.

**Phanerozoic** The most recent eon of geological time, represented by rock strata containing clearly recognizable fossils. It comprises the *Palaeozoic, *Mesozoic, and *Cenozoic eras and has extended for about 542 million years from the beginning of the Cambrian period. *Compare* PROTEROZOIC.

**pharmacogenomics (pharmacogenetics)** The study of how genes affect the actions of drugs. The enormous growth in knowledge about human genetics arising from the *Human Genome Project, coupled with the rapid advance of computer systems to analyse the vast amounts of data, has revolutionized drug discovery and development. This approach, which combines *genomics and pharmacology, improves understanding of drug actions, suggests new potential drug molecules, and enables computer-based searches for likely drug targets. It also raises the prospect of drugs being tailor-made to suit the genetic make-up of particular patients or groups of patients. This more precise targeting of drugs should make drugs more effective, with less risk of adverse side effects.

**pharmacokinetics** The movement of foreign substances, particularly drugs, throughout the body of an animal. Processes that influence the pharmacokinetics of a compound include uptake, distribution throughout the body tissues, the length of time the compound remains in the body, and its rate of clearance (e.g. by metabolism or excretion).

**pharmacology** The study of the properties of drugs and their effects on living organisms. Clinical pharmacology is concerned with the effects of drugs in treating disease.

**pharynx** 1. The cavity in vertebrates between the mouth and the *oesophagus and windpipe (*trachea), which serves for the passage of both food and respiratory gases. The presence of food in the pharynx stimulates swallowing (*see* DEGLUTITION). In fish and aquatic amphibians the pharynx is perforated by *gill slits. 2. The corresponding region in invertebrates.

**phase** 1. A homogeneous part of a heterogeneous system that is separated from other parts by a distinguishable boundary. A mixture of ice and water is a two-phase system. A solution of salt in water is a single-phase system. 2. A description of the stage that a periodic motion has reached, usually by comparison with another such motion of the same frequency. Two varying quantities are said to be **in phase** if their maximum and minimum values occur at the same instants; otherwise, there is said to be a **phase difference**. *See also* PHASE ANGLE. 3. One of the circuits in an electrical system or device in which there are two or more alternating currents that are not in phase with each other. In a three-phase system the displacement between the currents is one third of a period. 4. *See* PHASES OF THE MOON.

**phase angle** The difference in *phase between two sinusoidally varying quantities. The displacement $x_1$ of one quantity at time $t$ is given by $x_1 = a\sin\omega t$, where $\omega$ is the angu-

lar frequency and $a$ is the amplitude. The displacement $x_2$ of a similar wave that reaches the end of its period $T$, a fraction β of the period before the first is said to **lead** the first quantity by a time β$T$; if it reaches the end of its period, a fraction β of the period after the first quantity it **lags** by a time β$T$. The value of $x_2$ is then given by $x_2 = a\sin(\omega t + \phi)$. $\phi$ is called the phase angle and it is equal to $2\pi\beta$.

**phase-contrast microscope** A type of *microscope that is widely used for examining such specimens as biological cells and tissues. It makes visible the changes in phase that occur when nonuniformly transparent specimens are illuminated. In passing through an object the light is slowed down and becomes out of phase with the original light. With transparent specimens having some structure *diffraction occurs, causing a larger phase change in light outside the central maximum of the pattern. The phase-contrast microscope provides a means of combining this light with that of the central maximum by means of an annular diaphragm and a **phase-contrast plate**, which produces a matching phase change in the light of the central maximum only. This gives greater contrast to the final image, due to constructive interference between the two sets of light waves. This is **bright contrast**; in **dark contrast** a different phase-contrast plate is used to make the same structure appear dark, by destructive interference of the same waves.

**phase diagram** A graph showing the relationship between solid, liquid, and gaseous *phases over a range of conditions (e.g. temperature and pressure). *See* STEEL.

**phase I metabolism** The first stage in the conversion of a foreign compound, such as a drug or toxin, into a form that can be eliminated by the body. Common reactions during this phase are oxidation, reduction, and hydrolysis; the resulting metabolites are chemically more reactive than the parent compound, enabling them to undergo the reactions of the second stage (*see* PHASE II METABOLISM).

**phase II metabolism** The second stage in adapting foreign compounds for elimination from the body (*compare* PHASE I METABOLISM). Phase II metabolism involves the addition of chemical groups (e.g. glycine or acetate), which usually makes the compound less toxic to body tissues and easier to excrete.

**phase modulation** *See* MODULATION.

**phase rule** For any system at equilibrium, $P + F = C + 2$, where $P$ is the number of distinct phases, $C$ the number of components, and $F$ the number of degrees of freedom of the system. The relationship, derived by Josiah Willard Gibbs in 1876, is often called the **Gibbs phase rule**.

**phases of the moon** The shapes of the illuminated surface of the moon as seen from the earth. The shape changes as a result of the relative positions of the earth, sun, and moon.

**New moon** occurs when the nearside is totally unilluminated by the sun. As the moon moves eastwards in its orbit the sunrise *terminator crosses the nearside from east to west producing a **crescent moon**. The moon is half illuminated at **first quarter**. When it is more than half-phase but less than full phase it is said to be a **gibbous moon**. When the moon is at *opposition the nearside is fully illuminated producing a **full moon**. The sunset terminator then follows to produce a waning gibbous moon, **last quarter**, a waning crescent moon, and eventually the next new moon.

**phase space** For a system with $n$ degrees of freedom, the $2n$-dimensional space with coordinates $(q_1, q_2, ..., q_n, p_1, p_2, ..., p_n)$, where the $q$s describe the degrees of freedom of the system and the $p$s are the corresponding momenta. Each point represents a state of the system. In a gas of $N$ point particles, each particle has three positional coordinates and three corresponding momentum coordinates, so that the phase space has $6N$-dimensions. If the particles have internal degrees of freedom, such as the vibrations and rotations of molecules, then these must be included in the phase space, which is consequently of higher dimension than that for point particles. As the system changes with time the representative points trace out a curve in phase space known as a **trajectory**. *See also* ATTRACTOR; CONFIGURATION SPACE; STATISTICAL MECHANICS.

**phase speed (phase velocity)** Symbol $V_p$. The speed of propagation of a pure sine wave. $V_p = \lambda f$, where $\lambda$ is the wavelength and $f$ is the frequency. The value of the phase speed depends on the nature of the medium

through which it is travelling and may also depend on the mode of propagation. For electromagnetic waves travelling through space the phase speed $c$ is given by $c^2 = 1/\varepsilon_0\mu_0$, where $\varepsilon_0$ and $\mu_0$ are the electric constant and the magnetic constant respectively.

**phase transition** A change in a feature that characterizes a system. Examples of phase transitions are changes from solid to liquid, liquid to gas, and the reverse changes. Other examples of phase transitions include the transition from a paramagnet to a ferromagnet (*see* MAGNETISM) and the transition from a normally conducting metal to a superconductor. Phase transitions can occur by altering such variables as temperature and pressure.

Phase transitions can be classified by their **order**. If there is non-zero *latent heat at the transition it is said to be a **first-order transition**. If the latent heat is zero it is said to be a **second-order transition**.

Some *models describing phase transitions, particularly in *low-dimensional systems, are amenable to exact mathematical solutions. An effective technique for understanding phase transitions is the *renormalization group since it can deal with problems involving different length-scales, including the feature of **universality**, in which very different physical systems behave in the same way near a phase transition. *See also* ORDER PARAMETER; RENORMALIZATION GROUP; TRANSITION POINT; BROKEN SYMMETRY; EARLY UNIVERSE.

**phasor** A rotating *vector that represents a sinusoidally varying quantity. Its length represents the amplitude of the quantity and it is imagined to rotate with angular velocity equal to the angular frequency of the quantity, so that the instantaneous value of the quantity is represented by its projection upon a fixed axis. The concept is convenient for representing the *phase angle between two quantities; it is shown on a diagram as the angle between their phasors.

**phellem** *See* CORK.

**phelloderm** *See* CORK CAMBIUM.

**phellogen** *See* CORK CAMBIUM.

**phencyclidine (PCP)** A hallucinogenic drug, originally used as a veterinary anaesthetic. It is usually used as a powder (known as 'angel dust').

**phenetic** Describing a system of *classification of organisms based on similarities and differences in as many observable characteristics as possible. A phenetic system does not aim to reflect evolutionary descent, although it may well do so. *Compare* PHYLOGENETIC.

**phenol (carbolic acid)** A white crystalline solid, $C_6H_5OH$; r.d. 1.1; m.p. 43°C; b.p. 182°C. It is made by the *cumene process or by the *Raschig process and is used to make a variety of other organic chemicals. *See also* PHENOLS.

**phenol–formaldehyde resin** A class of resins produced by polymerizing phenols with formaldehyde (methanol) using acid or basic catalysts. They are generally cross-linked thermosetting materials. *Bakelite is the original example.

**phenolic resins** Synthetic resins made by copolymerizing phenols with aldehydes. *Phenol–formaldehyde resins are the commonest type.

**phenolphthalein** A dye used as an acid-base *indicator. It is colourless below pH 8 and red above pH 9.6. It is used in titrations involving weak acids and strong bases. It is also used as a laxative.

**phenolphthalein test** *See* KASTLE MEYER TEST.

**phenols** Organic compounds that contain a hydroxyl group (–OH) bound directly to a carbon atom in a benzene ring. Unlike normal alcohols, phenols are acidic because of the influence of the aromatic ring. Thus, phenol itself ($C_6H_5OH$) ionizes in water:

$$C_6H_5OH \rightarrow C_6H_5O^- + H^+$$

Phenols are made by fusing a sulphonic acid salt with sodium hydroxide to form the sodium salt of the phenol. The free phenol is liberated by adding sulphuric acid.

**phenotype** The observable characteristics of an organism. These are determined by its genes (*see* GENOTYPE), the dominance relationships between the *alleles, and by the interaction of the genes with the environment.

**phenylalanine** *See* AMINO ACID.

**phenylamine (aniline; aminobenzene)** A colourless oily liquid aromatic *amine, $C_6H_5NH_2$, with an 'earthy' smell; r.d. 1.0217; m.p. –6.3°C; b.p. 184.1°C. The compound turns brown on exposure to sunlight. It is basic, forming the **phenylammonium** (or

anilinium) **ion**, $C_6H_5NH_3^+$, with strong acids. It is manufactured by the reduction of nitrobenzene or by the addition of ammonia to chlorobenzene using a copper(II) salt catalyst at 200°C and 55 atm. The compound is used extensively in the rubber industry and in the manufacture of drugs and dyes.

**phenylammonium ion** The ion $C_6H_5NH_3^+$, derived from *phenylamine.

**N-phenylethanamide** *See* ACETANILIDE.

**phenylethene (styrene)** A liquid hydrocarbon, $C_6H_5CH:CH_2$; r.d. 0.9; m.p. –31°C; b.p. 145°C. It can be made by dehydrogenating ethylbenzene and is used in making polystyrene.

**phenyl group** The organic group $C_6H_5-$, present in benzene.

**phenylhydrazine** A toxic colourless dense liquid, $C_6H_8N_2$, b.p. 240°C, which turns brown on exposure to air. It is a powerful reducing agent, made from *diazonium salts of benzene. It is used to identify aldehydes and ketones, with which it forms condensation products called *hydrazones. With glucose and similar sugars it forms osazones. For such tests, the nitro derivative 2,4-dinitrophenylhydrazine (DNP) is often preferred as this generally forms crystalline derivatives that can be identified by their melting points. Phenylhydrazine is also used to make dyes and derivatives of *indole.

**phenylhydrazones** *See* HYDRAZONES.

**phenylketonuria** A genetic disorder in which there is disordered metabolism of the amino acid phenylalanine, leading to severe mental retardation of affected children. The disease is caused by the absence or deficiency of the enzyme phenylalanine hydroxylase, which results in the accumulation of phenylalanine in all body fluids. There are also high levels of the ketone phenylpyruvate in the urine, hence the name of the disease. The disease occurs in individuals who are homozygous for the defective recessive allele on chromosome 12; both parents of such individuals are thus heterozygous carriers of the allele. The advent of *gene probes has greatly aided accurate diagnosis, of both phenylketonurics and carriers.

**phenylmethanol (benzyl alcohol)** A liquid aromatic alcohol, $C_6H_5CH_2OH$; r.d. 1.04; m.p. –15.3°C; b.p. 205.4°C. It is used mainly as a solvent.

**phenylmethylamine** *See* BENZYLAMINE.

**3-phenylpropenoic acid** *See* CINNAMIC ACID.

**pheromone (ectohormone)** A chemical substance emitted by an organism into the environment as a specific signal to another organism, usually of the same species. Pheromones play an important role in the social behaviour of certain animals, especially insects and mammals. They are used to attract mates, to mark trails, and to promote social cohesion and coordination in colonies. Pheromones are usually highly volatile organic acids or alcohols and can be effective at minute concentrations.

**Phillips process** A process for making high-density polyethene by polymerizing ethene at high pressure (30 atmospheres) and 150°C. The catalyst is chromium(III) oxide supported on silica and alumina.

**phloem (bast)** A tissue that conducts food materials in vascular plants from regions where they are produced (notably the leaves) to regions, such as growing points, where they are needed. It consists of hollow tubes (sieve tubes) that run parallel to the long axis of the plant organ and are formed from elongated cells (*sieve elements) joined end to end. The end walls of these cells are broken down to a greater or lesser extent to allow passage of materials. In young plants and in newly formed tissues of mature plants the phloem is formed by the activity of the *apical meristem. In most plants secondary phloem is later differentiated by the vascular *cambium and this replaces the earlier formed phloem in older regions. *See also* COMPANION CELL. *Compare* XYLEM.

**phlogiston theory** A former theory of combustion in which all flammable objects were supposed to contain a substance called **phlogiston**, which was released when the object burned. The existence of this hypothetical substance was proposed in 1669 by Johann Becher, who called it 'combustible earth' (*terra pinguis*: literally 'fat earth'). For example, according to Becher, the conversion of wood to ashes by burning was explained on the assumption that the original wood consisted of ash and *terra pinguis*, which was released on burning. In the early 18th century Georg Stahl renamed the substance phlogiston (from the Greek for 'burned') and extended the theory to include the calcination (and corrosion) of metals.

Thus, metals were thought to be composed of **calx** (a powdery residue) and phlogiston; when a metal was heated, phlogiston was set free and the calx remained. The process could be reversed by heating the metal over charcoal (a substance believed to be rich in phlogiston, because combustion almost totally consumed it). The calx would absorb the phlogiston released by the burning charcoal and become metallic again.

The theory was finally demolished by Antoine Lavoisier, who showed by careful experiments with reactions in closed containers that there was no *absolute* gain in mass – the gain in mass of the substance was matched by a corresponding loss in mass of the air used in combustion. After experiments with Priestley's dephlogisticated air, Lavoisier realized that this gas, which he named oxygen, was taken up to form a calx (now called an oxide). The role of oxygen in the new theory was almost exactly the opposite of phlogiston's role in the old. In combustion and corrosion phlogiston was released; in the modern theory, oxygen is taken up to form an oxide.

**phloroglucinol** A red dye (usually acidified with hydrochloric acid) that stains lignin in plant cells red.

**phon** A unit of loudness of sound that measures the intensity of a sound relative to a reference tone of defined intensity and frequency. The reference tone usually used has a frequency of 1 kilohertz and a root-mean-square sound pressure of $2 \times 10^{-5}$ pascal. The observer listens with both ears to the reference tone and the sound to be measured alternately. The reference tone is then increased until the observer judges it to be of equal intensity to the sound to be measured. If the intensity of the reference tone has been increased by $n$ *decibels to achieve this, the sound being measured is said to have an intensity of $n$ phons. The decibel and phon scales are not identical as the phon scale is subjective and relies on the sensitivity of the ear to detect changes of intensity with frequency.

**phonochemistry** *See* SONOCHEMISTRY.

**phonon** A quantum of *crystal-lattice vibrational energy having a magnitude $hf$, where $h$ is the *Planck constant and $f$ is the frequency of the vibration. Phonons are analogous to the quanta of light, i.e. *photons. The concept of phonons is useful in the treatment of the thermal conductivity of nonmetallic solids and, through consideration of electron–phonon interactions, the temperature dependence of the electrical conductivity of metals.

**phosgene** *See* CARBONYL CHLORIDE.

**phosphagen** A compound in animal tissues that provides a reserve of chemical energy in the form of high-energy phosphate bonds. The most common phosphagens are *creatine phosphate, occurring in vertebrate muscle and nerves, and arginine phosphate, found in most invertebrates. During tissue activity (e.g. in muscle contraction) phosphagens give up their phosphate groups, thereby generating *ATP from ADP. The phosphagens are then reformed when ATP is available.

**phosphatase** An enzyme that catalyses the removal of a phosphate group from an organic compound.

**phosphates** Salts based formally on phosphorus(V) oxoacids and in particular salts of *phosphoric(V) acid, $H_3PO_4$. A large number of polymeric phosphates also exist, containing P–O–P bridges. These are formed by heating the free acid and its salts under a variety of conditions; as well as linear polyphosphates, cyclic polyphosphates and cross-linked polyphosphates or ultraphosphates are known.

**phosphatide** *See* PHOSPHOLIPID.

**phosphatidylcholine** *See* LECITHIN.

**phosphide** A binary compound of phosphorus with a more electropositive element. Phosphides show a wide range of properties. Alkali and alkaline earth metals form ionic phosphides, such as $Na_3P$ and $Ca_3P_2$, which are readily hydrolysed by water. The other transition-metal phosphides are inert metallic-looking solids with high melting points and electrical conductivities.

**phosphine** A colourless highly toxic gas, $PH_3$; m.p. –133°C; b.p. –87.7°C; slightly soluble in water. Phosphine may be prepared by reacting water or dilute acids with calcium phosphide or by reaction between yellow phosphorus and concentrated alkali. Solutions of phosphine are neutral but phosphine does react with some acids to give phosphonium salts containing $PH_4^+$ ions, analogous to the ammonium ions. Phosphine prepared in the laboratory is usually contaminated with diphosphine and is spontaneously flammable but the pure compound is not so. Phosphine can function as a ligand in binding to transition-metal ions. Dilute gas mixtures of very pure phosphine and the rare gases are used for doping semiconductors.

**phosphinic acid (hypophosphorus acid)** A white crystallline solid, $H_3PO_2$; r.d. 1.493; m.p. 26.5°C; decomposes above 130°C. It is soluble in water, ethanol, and ethoxyethane. Salts of phosphinic acid may be prepared by boiling white phosphorus with the hydroxides of group 1 or group 2 metals. The free acid is made by the oxidation of phosphine with iodine. It is a weak monobasic acid in which it is the –O–H group that is ionized to give the ion $H_2PO_2^-$. The acid and its salts are readily oxidized to the orthophosphate and consequently are good reducing agents.

**phosphite** *See* PHOSPHONIC ACID.

**phospholipid (phosphatide)** One of a group of lipids having both a phosphate group and one or more fatty acids. **Glycerophospholipids** (or **phosphoglycerides**) are based on *glycerol; the three hydroxyl groups are esterified with two fatty acids and a phosphate group, which may itself be bound to one of a variety of simple organic groups (e.g. in *lecithin (phosphatidylcholine) it is choline). **Sphingolipids** are based on the alcohol sphingosine and contain only one fatty acid linked to an amino group. With their hydrophilic polar phosphate groups and hydrophobic hydrocarbon 'tails', phospholipids readily form membrane-like structures in water. They are a major component of plasma membranes (*see* LIPID BILAYER).

**phosphonate** *See* PHOSPHONIC ACID.

**phosphonic acid (phosphorous acid; orthophosphorous acid)** A colourless to pale-yellow deliquescent crystalline solid, $H_3PO_3$; r.d. 1.65; m.p. 73.6°C; decomposes at 200°C; very soluble in water and soluble in alcohol. Phosphonic acid may be crystallized from the solution obtained by adding ice-cold water to phosphorus(III) oxide or phosphorus trichloride. The structure of this material is unusual in that it contains one direct P–H bond and is more correctly written $(HO)_2HPO$. The acid is dibasic, giving rise to the ions $H_2PO_3^-$ and $HPO_3^{2-}$ (**phosphonates**; formerly **phosphites**), and has moderate reducing properties. On heating it gives phosphine and phosphoric(V) acid.

**phosphonium ion** The ion $PH_4^+$, or the

corresponding organic derivatives of the type $R_3PH^+$, $RPH_3^+$. The phosphonium ion $PH_4^+$ is formally analogous to the ammonium ion $NH_4^+$ but $PH_3$ has a much lower proton affinity than $NH_3$ and reaction of $PH_3$ with acids is necessary for the production of phosphonium salts.

**phosphor** A substance that is capable of *luminescence (including phosphorescence). Phosphors that release their energy after a short delay of between $10^{-10}$ and $10^{-4}$ second are sometimes called **scintillators**.

**phosphor bronze** An alloy of copper containing 4% to 10% of tin and 0.05% to 1% of phosphorus as a deoxidizing agent. It is used particularly for marine purposes and where it is exposed to heavy wear, as in gear wheels. *See also* BRONZE.

**phosphorescence** *See* LUMINESCENCE.

**phosphoric(V) acid (orthophosphoric acid)** A white rhombic solid, $H_3PO_4$; r.d. 1.834; m.p. 42.35°C; loses water at 213°C; very soluble in water and soluble in ethanol. Phosphoric(V) acid is very deliquescent and is generally supplied as a concentrated aqueous solution. It is the most commercially important derivative of phosphorus, accounting for over 90% of the phosphate rock mined. It is manufactured by two methods; the **wet process**, in which the product contains some of the impurities originally present in the rock and applications are largely in the fertilizer industry, and the **thermal process**, which produces a much purer product suitable for the foodstuffs and detergent industries. In the wet process the phosphate rock, $Ca_3(PO_4)_2$, is treated with sulphuric acid and the calcium sulphate removed either as gypsum or the hemihydrate. In the thermal process, molten phosphorus is sprayed and burned in a mixture of air and steam. Phosphoric(V) acid is a weak tribasic acid, which is best visualized as $(HO)_3PO$. Its full systematic name is **tetraoxo-phosphoric(V) acid**. It gives rise to three series of salts containing **phosphate(V)** ions based on the anions $[(HO)_2PO_2]^-$, $[(HO)PO_3]^{2-}$, and $PO_4^{3-}$. These salts are acidic, neutral, and alkaline in character respectively and phosphate ions often feature in buffer systems. There is also a wide range of higher acids and acid anions in which there is some P–O–P chain formation. The simplest of these is **pyrophosphoric acid** (technically **heptaoxodiphosphoric(V) acid**), $H_4P_2O_7$, produced by heating phos-

phoric(V) acid (solid) and phosphorus(III) chloride oxide. **Metaphosphoric acid** is a glassy polymeric solid $(HPO_2)_x$.

**phosphorous acid** *See* PHOSPHONIC ACID.

**phosphorus** Symbol P. A nonmetallic element belonging to *group 15 (formerly VB) of the periodic table; a.n. 15; r.a.m. 30.9738; r.d. 1.82 (white), 2.34 (red); m.p. 44.1°C (α-white); b.p. 280°C (α-white). It occurs in various phosphate rocks, from which it is extracted by heating with carbon (coke) and silicon(IV) oxide in an electric furnace (1500°C). Calcium silicate and carbon monoxide are also produced. Phosphorus has a number of allotropic forms. The α-white form consists of $P_4$ tetrahedra (there is also a β-white form stable below –77°C). If α-white phosphorus is dissolved in lead and heated at 500°C a violet form is obtained. Red phosphorus, which is a combination of violet and white phosphorus, is obtained by heating α-white phosphorus at 250°C with air excluded. There is also a black allotrope, which has a graphite-like structure, made by heating white phosphorus at 300°C with a mercury catalyst. The element is highly reactive. It forms metal *phosphides and covalently bonded phosphorus(III) and phosphorus(V) compounds. Phosphorus is an *essential element for living organisms. It is an important constituent of tissues (especially bones and teeth) and of cells, being required for the formation of nucleic acids and energy-carrying molecules (e.g. ATP) and also involved in various metabolic reactions. The element was discovered by Hennig Brand (*c.* 1630–9?) in 1669.

(((●))) SEE WEB LINKS

• Information from the WebElements site

**phosphorus(III) bromide (phosphorus tribromide)** A colourless fuming liquid, $PBr_3$; r.d. 2.85; m.p. –40°C; b.p. 173°C. It is prepared by passing bromine vapour over phosphorus but avoiding an excess, which would lead to the phosphorus(V) bromide. Like the other phosphorus(III) halides, $PBr_3$ is pyramidal in the gas phase. In the liquid phase the P–Br bonds are labile; for example, $PBr_3$ will react with $PCl_3$ to give a mixture of products in which the halogen atoms have been redistributed. Phosphorus(III) bromide is rapidly hydrolysed by water to give phosphonic acid and hydrogen bromide. It reacts readily with many organic hydroxyl groups and is used as a reagent for introducing bromine atoms into organic molecules.

**phosphorus(V) bromide (phosphorus pentabromide)** A yellow readily sublimable solid, $PBr_5$, which decomposes below 100°C and is soluble in benzene and carbon tetrachloride (tetrachloromethane). It may be prepared by the reaction of phosphorus(III) bromide with bromine or the direct reaction of phosphorus with excess bromine. It is very readily hydrolysed to give hydrogen bromide and phosphoric(V) acid. An interesting feature of this material is that in the solid state it has the structure $[PBr_4]^+Br^-$. It is used in organic chemistry as a brominating agent.

**phosphorus(III) chloride (phosphorus trichloride)** A colourless fuming liquid, $PCl_3$; r.d. 1.57; m.p. −112°C; b.p. 75.5°C. It is soluble in ether and in carbon tetrachloride but reacts with water and with ethanol. It may be prepared by passing chlorine over excess phosphorus (excess chlorine contaminates the product with phosphorus(V) chloride). The molecule is pyramidal in the gas phase and possesses weak electron-pair donor properties. It is hydrolysed violently by water to phosphonic acid and hydrogen chloride. Phosphorus(III) chloride is an important starting point for the synthesis of a variety of inorganic and organic derivatives of phosphorus.

**phosphorus(V) chloride (phosphorus pentachloride)** A yellow-white rhombic solid, $PCl_5$, which fumes in air; r.d. 4.65; m.p. 166.8°C (under pressure); sublimes at 160–162°C. It is decomposed by water to give hydrogen chloride and phosphoric(V) acid. It is soluble in organic solvents. The compound may be prepared by the reaction of chlorine with phosphorus(III) chloride. Phosphorus(V) chloride is structurally interesting in that in the gas phase it has the expected trigonal bipyramidal form but in the solid phase it consists of the ions $[PCl_4]^+[PCl_6]^-$. The same ions are detected when phosphorus(V) chloride is dissolved in polar solvents. It is used in organic chemisty as a chlorinating agent.

**phosphorus(III) chloride oxide (phosphorus oxychloride; phosphoryl chloride)** A colourless fuming liquid, $POCl_3$; r.d. 1.67; m.p. 2°C; b.p. 105.3°C. It may be prepared by the reaction of phosphorus(III) chloride with oxygen or by the reaction of phosphorus(V) oxide with phosphorus(V) chloride. Its reactions are very similar to those of phosphorus(III) chloride. Hydrolysis with water gives phosphoric(V) acid. Phosphorus(III) chloride

oxide has a distorted tetrahedral shape and can act as a donor towards metal ions, thus giving rise to a series of complexes.

**phosphorus cycle** The cycling of *phosphorus between the biotic and abiotic components of the environment (*see* BIOGEOCHEMICAL CYCLE). Inorganic phosphates ($PO_4^{3-}$, $HPO_4^{2-}$, or $H_2PO_4^-$) are absorbed by plants from the soil and bodies of water and eventually pass into animals through food chains. Within living organisms phosphates are built up into nucleic acids and other organic molecules. When plants and animals die, phosphates are released and returned to the abiotic environment through the action of bacteria. On a geological time scale, phosphates in aquatic environments eventually become incorporated into and form part of rocks; through a gradual process of erosion, these phosphates are returned to the soil, seas, rivers, and lakes. Phosphorus-containing rocks are mined for the manufacture of fertilizers, which provide an additional supply of inorganic phosphate to the abiotic environment.

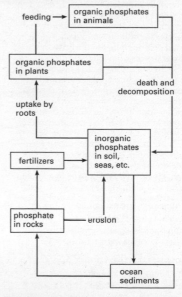

Phosphorus cycle.

**phosphorus(III) oxide (phosphorus trioxide)** A white or colourless waxy solid,

$P_4O_6$; r.d. 2.13; m.p. 23.8°C; b.p. 173.8°C. It is soluble in ether, chloroform, and benzene but reacts with cold water to give phosphonic acid, $H_3PO_3$, and with hot water to give phosphine and phosphoric(V) acid. The compound is formed when phosphorus is burned in an oxygen-deficient atmosphere (about 50% yield). As it is difficult to separate from white phosphorus by distillation, the mixture is irradiated with ultraviolet radiation to convert excess white phosphorus into the red form, after which the oxide can be separated by dissolution in organic solvents. Although called a trioxide for historical reasons, phosphorus(III) oxide consists of $P_4O_6$ molecules of tetrahedral symmetry in which each phosphorus atom is linked to the three others by an oxygen bridge. The chemistry is very complex. Above 210°C it decomposes into red phosphorus and polymeric oxides. It reacts with chlorine and bromine to give oxo-halides and with alkalis to give phosphonates (*see* PHOSPHONIC ACID).

**phosphorus(V) oxide (phosphorus pentoxide; phosphoric anhydride)** A white powdery and extremely deliquescent solid, $P_4O_{10}$; r.d. 2.39; m.p. 580°C (under pressure); sublimes at 300°C. It reacts violently with water to give phosphoric(V) acid. It is prepared by burning elemental phosphorus in a plentiful supply of oxygen, then purified by sublimation. The hexagonal crystalline form consists of $P_4O_{10}$ molecular units; these have the phosphorus atoms arranged tetrahedrally, each P atom linked to three others by oxygen bridges and having in addition one terminal oxygen atom. The compound is used as a drying agent and as a dehydrating agent; for example, amides are converted into nitriles and sulphuric acid is converted to sulphur trioxide.

**phosphorus oxychloride** *See* PHOSPHORUS(III) CHLORIDE OXIDE.

**phosphorus pentabromide** *See* PHOSPHORUS(V) BROMIDE.

**phosphorus pentachloride** *See* PHOSPHORUS(V) CHLORIDE.

**phosphorus tribromide** *See* PHOSPHORUS(III) BROMIDE.

**phosphorus trichloride** *See* PHOSPHORUS(III) CHLORIDE.

**phosphorus trioxide** *See* PHOSPHORUS(III) OXIDE.

**phosphorylase** *See* PHOSPHORYLATION.

**phosphorylation** The introduction of a phosphate group ($PO_4^{3-}$) to a biomolecule in a reaction that is normally controlled by a **phosphorylase** enzyme. Phosphate is able to combine easily with inert organic compounds, making them chemically active. The first stage in many biochemical reactions is phosphorylation. The conversion of AMP and ADP to *ATP occurs by phosphorylation reactions in two main metabolic pathways, *oxidative phosphorylation and *photophosphorylation. The formation of other nucleotides also involves a phosphorylation reaction. The activity of many enzymes is controlled by phosphorylation: certain enzymes are activated when they are phosphorylated (*see* KINASE), while others are deactivated. Phosphorylation of these enzymes is under the control of hormones and other messengers.

**phosphoryl chloride** *See* PHOSPHORUS(III) CHLORIDE OXIDE.

**phot** A unit of illuminance equal to $10^4$ lux or one lumen per square centimetre.

**photic zone** The upper layer of a sea or a lake, in which there is sufficient light for photosynthesis. The limit of the photic zone varies from less than a metre to more than 200 metres, depending on the turbidity of the water.

**photino** *See* SUPERSYMMETRY.

**photoautotroph** An autotrophic organism, such as a green plant or a phototrophic bacterium, that synthesizes its organic materials using energy derived from the sun (solar energy) in the process of photosynthesis. *See* AUTOTROPHIC NUTRITION.

**photocathode** A *cathode that emits electrons when light falls upon it, as a result of the *photoelectric effect. *See* PHOTOELECTRIC CELL.

**photocell** *See* PHOTOELECTRIC CELL.

**photochemical reaction** A chemical reaction caused by light or ultraviolet radiation. The incident photons are absorbed by reactant molecules to give excited molecules or free radicals, which undergo further reaction.

**photochemical smog** A noxious smog produced by the reaction of nitrogen oxides with hydrocarbons in the presence of ultraviolet light from the sun. The reaction is very complex and one of the products is ozone.

**photochemistry** The branch of chemistry concerned with *photochemical reactions.

**photochromism** A change of colour occurring in certain substances when exposed to light. Photochromic materials are used in sunglasses that darken in bright sunlight.

**photoconductive effect** *See* PHOTOELECTRIC EFFECT.

**photodiode** A semiconductor *diode used to detect the presence of light or to measure its intensity. It usually consists of a $p–n$ junction device in a container that focuses any light in the environment close to the junction. The device is usually biased in reverse so that in the dark the current is small; when it is illuminated the current is proportional to the amount of light falling on it. *See* PHOTOELECTRIC EFFECT.

**photodisintegration** The decay of a nuclide as a result of the absorption of a gamma-ray photon.

**photoelasticity** An effect in which certain materials exhibit double refraction when subjected to stress. It is used in a technique for detecting strains in transparent materials (e.g. Perspex, celluloid, and glass). When polarized white light is passed through a stressed sample, the birefringence causes coloured patterns to be seen on the viewing screen of a suitable *polarimeter. If monochromatic polarized light is used, a complex pattern of light and dark fringes is produced.

**photoelectric cell (photocell)** Any of several devices that produce an electric signal in response to exposure to electromagnetic radiation. The original photocells utilized photoemission from a photosensitive cathode (**photocathode**). The electrons emitted are attracted to an anode. A positive potential on the anode enables a current to flow through an external circuit, the current being proportional to the intensity of the illumination on the cathode. The electrodes are enclosed in an evacuated glass tube (*see also* PHOTOMULTIPLIER).

More modern light-sensitive devices utilize the photoconductive effect and the photovoltaic effect (*see* PHOTOELECTRIC EFFECT; PHOTODIODE; PHOTOTRANSISTOR; SOLAR CELL).

**photoelectric effect** The liberation of electrons (*see* PHOTOELECTRON) from a substance exposed to electromagnetic radiation. The number of electrons emitted depends on the intensity of the radiation. The kinetic energy of the electrons emitted depends on the frequency of the radiation. The effect is a quantum process in which the radiation is regarded as a stream of *photons, each having an energy $hf$, where $h$ is the Planck constant and $f$ is the frequency of the radiation. A photon can only eject an electron if the photon energy exceeds the *work function, $\phi$, of the solid, i.e. if $hf_0 = \phi$ an electron will be ejected; $f_0$ is the minimum frequency (or **threshold frequency**) at which ejection will occur. For many solids the photoelectric effect occurs at ultraviolet frequencies or above, but for some materials (having low work functions) it occurs with light. The maximum kinetic energy, $E_m$, of the photoelectron is given by the *Einstein equation: $E_m = hf - \phi$ (*see also* PHOTOIONIZATION).

Apart from the liberation of electrons from atoms, other phenomena are also referred to as photoelectric effects. These are the **photoconductive effect** and the **photovoltaic effect**. In the photoconductive effect, an increase in the electrical conductivity of a semiconductor is caused by radiation as a result of the excitation of additional free charge carriers by the incident photons. **Photoconductive cells**, using such photosensitive materials as cadmium sulphide, are widely used as radiation detectors and light switches (e.g. to switch on street lighting).

In the photovoltaic effect, an e.m.f. is produced between two layers of different materials as a result of irradiation. The effect is made use of in **photovoltaic cells**, most of which consist of $p–n$ *semiconductor junctions (*see also* PHOTODIODE; PHOTOTRANSISTOR). When photons are absorbed near a $p–n$ junction new free charge carriers are produced (as in photoconductivity); however, in the photovoltaic effect the electric field in the junction region causes the new charge carriers to move, creating a flow of current in an external circuit without the need for a battery. *See also* PHOTOELECTRIC CELL.

**photoelectron** An electron emitted from a substance by irradiation as a result of the *photoelectric effect or *photoionization.

**photoelectron spectroscopy** A technique for determining the *ionization potentials of molecules. The sample is a gas or vapour irradiated with a narrow beam of ultraviolet radiation (usually from a helium source at 58.4 nm, 21.21 eV photon energy). The photoelectrons produced in accordance with the *Einstein equation are passed through a slit into a vacuum region, where they are deflected by magnetic or electrostatic fields to give an energy spectrum. The photoelectron spectrum obtained has peaks corresponding to the ionization potentials of the molecule (and hence the orbital energies). The technique also gives information on the vibrational energy levels of the ions formed. **ESCA** (electron spectroscopy for chemical analysis) is a similar analytical technique in which a beam of X-rays is used. In this case, the electrons ejected are from the inner shells of the atoms. Peaks in the electron spectrum for a particular element show characteristic chemical shifts, which depend on the presence of other atoms in the molecule.

**photoemission** The process in which electrons are emitted by a substance as a result of irradiation. *See* PHOTOELECTRIC EFFECT; PHOTOIONIZATION.

**photofission** A *nuclear fission that is caused by a gamma-ray photon.

**photographic density** A measure of the opacity of a photographic emulsion (negative or transparency). *See* DENSITOMETER.

**photography** The process of forming a permanent record of an image. Traditionally photography uses specially treated film or paper. In normal black-and-white photography a *camera is used to expose a film or plate to a focused image of the scene for a specified time. The film or plate is coated with an emulsion containing silver salts and the exposure to light causes the silver salts to break down into silver atoms; where the light is bright dark areas of silver are formed on the film after development (by a mild reducing agent) and fixing. The negative so formed is printed, either by a contact process or by projection. In either case light passing through the negative film falls on a sheet of paper also coated with emulsion. Where the negative is dark, less light passes through and the resulting positive is light in this area, corresponding with a light area in the original scene. As photographic emulsions are sensitive to ultraviolet and X-rays, they are widely used in studies involving these forms of electromagnetic radiation. *See also* COLOUR PHOTOGRAPHY.

**photoionization** The *ionization of an atom or molecule as a result of irradiation by electromagnetic radiation. For a photoionization to occur the incident photon of the radiation must have an energy in excess of the *ionization potential of the species being irradiated. The ejected photoelectron will have an energy, $E$, given by $E = hf - I$, where $h$ is the Planck constant, $f$ is the frequency of the incident radiation, and $I$ is the ionization potential of the irradiated species.

**photolithography** A technique used in the manufacture of semiconductor components, integrated circuits, etc. It depends on the principle of masking selected areas of a surface and exposing the unmasked areas to such processes as the introduction of impurities, deposition of thin films, removal of material by etching, etc. The technique has been developed for use on tiny structures (typically measured in micrometres), which can only be examined by means of an electron microscope.

**photoluminescence** *See* LUMINESCENCE.

**photolysis** A chemical reaction produced by exposure to light or ultraviolet radiation. Photolytic reactions often involve free radicals, the first step being homolytic fission of a chemical bond. (*See* FLASH PHOTOLYSIS.) The photolysis of water, using energy from sunlight absorbed by chlorophyll, produces gaseous oxygen, electrons, and hydrogen ions and is a key reaction in *photosynthesis.

**photometer** An instrument used to measure *luminous intensity, illumination, and other photometric quantities. The older types rely on visual techniques to compare a source of light with a standard source. More modern photometers use *photoelectric cells, of the photoconductive, photoemissive, or photovoltaic types. The photovoltaic types do not require an external power source and are therefore very convenient to use but are relatively insensitive. The photoemissive type usually incorporates a *photomultiplier, especially for use in astronomy and with other weak sources. Photoconductive units require only low-voltage supplies, which makes them convenient for commercial illumination meters and photographers' exposure meters.

**photometric brightness** *See* LUMINANCE.

**photometry** The study of visual radiation, especially the calculations and measurements of *luminous intensity, *luminous flux, etc. In some cases photometric calculations and measurements extend into the near infrared and the near ultraviolet.

In photometry, two types of measurement are used: those that measure **luminous** quantities rely on the use of the human eye (for example, to compare the illuminance of two surfaces); those called **radiant** quantities rely on the use of photoelectric devices to measure electromagnetic energy. *See also* PHOTOMETER.

**(⊕) SEE WEB LINKS**

• Photometric data available at the NPL website

**photomicrography** The use of photography to obtain a permanent record (a **photomicrograph**) of the image of an object as viewed through a microscope.

**photomorphogenesis** The development of plants under the influence of light. All the processes crucial to the growth and development of plants are triggered by light, including seed germination, stem elongation, chloroplast formation, and flowering. These light responses are mediated by various photoreceptor systems, including light-sensitive molecules, principally *phytochrome. The photoreceptors interact with cell signalling networks to regulate the expression of genes involved in development and also influence the production of plant hormone.

**photomultiplier** A sensitive type of *photoelectric cell in which electrons emitted from a photocathode are accelerated to a second electrode where several electrons are liberated by each original photoelectron, as a result of *secondary emission. The whole process is repeated as many times as necessary to produce a useful electric current by secondary emission from the last electrode. A photomultiplier is thus a photocathode with the output amplified by an electron multiplier. The initial photocurrent can be amplified by a factor of $10^8$. Photomultipliers are thus useful when it is necessary to detect low intensities of light, as in stellar photometry, star and planet tracking in guidance systems, and more mundanely in process control.

**photon** A particle with zero rest mass consisting of a *quantum of electromagnetic radiation. The photon may also be regarded as a unit of energy equal to $hf$, where $h$ is the *Planck constant and $f$ is the frequency of the radiation in hertz. Photons travel at the speed of light. They are required to explain the photoelectric effect and other phenomena that require light to have particle character.

**photoneutron** A neutron emitted by an atomic nucleus undergoing a *photonuclear reaction.

**photonics** The study of devices analogous to those used in electronics, but with the electrons replaced by photons. Thus, photonics is concerned with devices involving the transmission, modulation, reflection, refraction, amplification, detection, and guidance of light. Examples are *lasers and *optical fibres. Photonics is used extensively in *telecommunications.

**photonuclear reaction** A *nuclear reaction that is initiated by a (gamma-ray) photon.

**photoperiodism** The response of an organism to changes in day length (**photoperiod**). It enables organisms to exploit favourable conditions associated with seasonal changes in climate and vegetation, for example to produce flowers when pollinating insects are abundant. Generally, the organism must experience a particular ratio of light to dark in the 24-hour period for the physiological switch from say, nonflowering to flowering, to take place. This ratio is the **critical photoperiod** (or **critical day length**), which varies between species: some actions are triggered when the photoperiod falls below the critical threshold (*see* SHORT-DAY PLANT); others are prompted when day length exceeds the critical photoperiod (*see* LONG-DAY PLANT). Plants have various photoreceptor molecules that are sensitive to light, including *phytochrome. In animals, photoperiodic control of breeding is controlled by *melatonin.

**photophosphorylation** The formation of ATP from ADP and inorganic phosphate using light energy in *photosynthesis (*compare* OXIDATIVE PHOSPHORYLATION). There are two pathways: noncyclic photophosphorylation and cyclic photophosphorylation. In **noncyclic photophosphorylation** electrons derived from the *photolysis of water pass from a complex of chlorophyll molecules along a series of carrier molecules to $NADP^+$. As they do so some of their energy is coupled to the formation of ATP from ADP and inorganic phosphate. $NADP^+$ is reduced to NADPH, which provides reducing power for the light-independent reactions of photosynthesis. In **cyclic photophosphorylation** the electrons are recycled through the electron carrier system back to the chlorophyll complex, resulting in further ATP formation (but no NADPH).

**photopic vision** The type of vision that occurs when the cones in the eye are the principal receptors, i.e. when the level of illumination is high. Colours can be identified with photopic vision. *Compare* SCOTOPIC VISION.

**photoprotection** Protection of a plant's photosynthetic apparatus from the harmful effects of light. During periods of peak light intensity plants are able to utilize less than half the incoming energy. The surplus energy poses the risk of photooxidation, and the formation of highly reactive superoxide radicals that can destroy the cell's chlorophyll and many other cellular components. Much of the excess energy is trapped and dissipated as heat by *carotenoids.

**photoreceptor** A sensory cell or group of cells that reacts to the presence of light. It usually contains a pigment that undergoes a chemical change when light is absorbed, thus stimulating a nerve. *See* EYE.

**photosensitive substance** 1. Any substance that when exposed to electromagnetic radiation produces a photoconductive, photoelectric, or photovoltaic effect. 2. Any substance, such as the emulsion of a photo-graphic film, in which electromagnetic radiation produces a chemical change.

**photosphere** The visible surface of the *sun or other star and the source of its continuous spectrum. It is a gaseous layer several hundreds of kilometres thick with an average temperature of 5780 K. Where the photosphere merges with the *chromosphere the temperature is 4000 K.

**photosynthesis** The chemical process by which green plants, algae, and certain bacteria synthesize organic compounds from carbon dioxide and water in the presence of sunlight. It occurs in the *chloroplasts and there are two principal types of reaction. In the **light-dependent reactions**, which require the presence of light, energy from sunlight is absorbed by *photosynthetic pigments (chiefly the green pigment *chlorophyll) and used to bring about the *photolysis of water:

$$H_2O \rightarrow 2H^+ + 2e^- + \tfrac{1}{2}O_2.$$

The electrons released by this reaction pass along a series of electron carriers (*see* ELECTRON TRANSPORT CHAIN); as they do so they lose their energy, which is used to convert ADP to ATP in the process of *photophosphorylation. The electrons and protons pro-

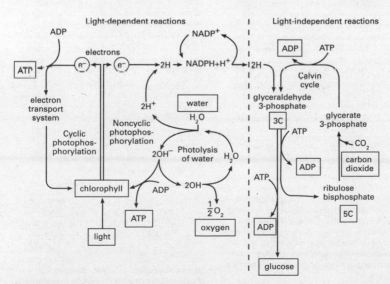

**Photosynthesis.**

duced by the photolysis of water are used to reduce NADP:

$$2H^+ + 2e^- + NADP^+ \rightarrow NADPH + H^+.$$

The ATP and NADPH produced during the light-dependent reactions provide energy and reducing power, respectively, for the ensuing **light-independent reactions** (formerly called the dark reaction), which do not require the presence of light. During these reactions carbon dioxide is reduced to carbohydrate in a metabolic pathway known as the *Calvin cycle. Photosynthesis can be summarized by the equation:

$$CO_2 + 2H_2O \rightarrow [CH_2O] + H_2O + O_2.$$

Since virtually all other forms of life are directly or indirectly dependent on plants for food, photosynthesis is the basis for all life on earth. Furthermore virtually all the atmospheric oxygen has originated from oxygen released during photosynthesis.

(⊕) **SEE WEB LINKS**

• Summary of photosynthesis, with illustrations and animations, compiled by the Royal Society of Chemistry

**photosynthetic pigments** The pigments responsible for the capture of light energy during the light-dependent reactions of *photosynthesis. In plants, algae, and cyanobacteria the green pigment *chlorophyll *a* is the principal light receptor, absorbing blue and red light. However *carotenoids also absorb light energy and pass this on to the chlorophyll molecules.

**phototaxis** The movement of a cell (e.g. a gamete) or a unicellular organism in response to light. For example, certain algae (e.g. *Chlamydomonas*) can perceive light by means of a sensitive eyespot and move to regions of higher light concentration to enhance photosynthesis. *See* TAXIS.

**phototransistor** A junction *transistor that is photosensitive. When radiation falls on the emitter-base junction, new free charge carriers are created in the base region and the collector current is increased. Phototransistors are similar to *photodiodes except that the primary photoelectric current is amplified internally and it is therefore more sensitive to light than the photodiode. Some types can be used as switching or bistable devices, a small intensity of radiation switching them from a low to high current state.

**phototropism (heliotropism)** The growth of plant organs in response to light. Aerial shoots usually grow towards light, while some aerial roots grow away from light. The phototropic response is triggered by photosensitive molecules and brought about by *auxin. *See* TROPISM.

**photovoltaic effect** *See* PHOTOELECTRIC EFFECT.

**pH scale** A logarithmic scale for expressing the acidity or alkalinity of a solution. To a first approximation, the pH of a solution can be defined as $-\log_{10}c$, where $c$ is the concentration of hydrogen ions in moles per cubic decimetre. A neutral solution at 25°C has a hydrogen-ion concentration of $10^{-7}$ mol dm$^{-3}$, so the pH is 7. A pH below 7 indicates an acid solution; one above 7 indicates an alkaline solution. More accurately, the pH depends not on the concentration of hydrogen ions but on their *activity, which cannot be measured experimentally. For practical purposes, the pH scale is defined by using a hydrogen electrode in the solution of interest as one half of a cell, with a reference electrode (e.g. a calomel electrode) as the other half cell. The pH is then given by $(E - E_R)F/2.303RT$, where $E$ is the e.m.f. of the cell and $E_R$ the standard electrode potential of the reference electrode, and $F$ the Faraday constant. In practice, a glass electrode is more convenient than a hydrogen electrode.

pH stands for 'potential of hydrogen'. The scale was introduced by Søren Sørensen (1868–1939) in 1909.

**phthalic acid** A colourless crystalline dicarboxylic acid, $C_6H_4(COOH)_2$; r.d. 1.6; m.p. 207°C. The two –COOH groups are substituted on adjacent carbon atoms of the ring, the technical name being **benzene-1,2-dicarboxylic acid**. The acid is made from **phthalic anhydride** (benzene-1,2-dicarboxylic anhydride, $C_8H_4O_3$), which is made by the catalytic oxidation of naphthalene. The anhydride is used in making plasticizers and polyester resins.

**phthalic anhydride** *See* PHTHALIC ACID.

**phthalocyanine** A synthetic compound having molecules with four isoindole rings linked by –N= bridges. The structure is similar to that of the *porphyrins. It can form complexes with central metal ions. Copper phthalocyanines are used as dyes.

**phycomycetes** In older classification schemes, all primitive *fungi, many of which are found in water (e.g. the water moulds, which may be parasitic on fish) or in damp areas. Many are unicellular but those that form mycelia generally have hyphae lacking cross walls, which distinguishes them from the *Ascomycota and *Basidiomycota. They include the *Zygomycota.

**phyllotaxis (phyllotaxy)** The arrangement of leaves on a plant stem. The leaves may be inserted in whorls or pairs at each node or singly up the stem. When arranged in pairs the two leaves arise on opposite sides of the stem and are usually at right angles to the leaf pairs above and below them. Single leaves may be inserted alternately or in a spiral pattern up the stem. Phyllotaxis generally results in the minimum of shading of leaves by those above them.

**phylogenetic** Describing a system of *classification of organisms that aims to show their evolutionary history. *Compare* PHENETIC.

**phylogenomics** The field of *bioinformatics that integrates knowledge about the evolutionary history of organisms (phylogeny) with structural and functional analysis of their genomes (genomics) and proteins (proteomics). A basic supposition is that well-characterized genes (or proteins) in one organism provide valuable insights into the likely sequence and function of homologous genes or proteins in related organisms. Conversely, comparisons of gene or protein sequences from different species can provide evidence of possible homology, and hence shared evolutionary origins.

**phylogeny** The evolutionary history of an organism or group of related organisms. *Compare* ONTOGENY.

**phylum** A category used in the *classification of organisms that consists of one or several similar or closely related classes. Examples of phyla are the Rhodophyta, Ascomycota, Bryophyta, Annelida, and Chordata. In traditional classification schemes phyla are grouped into kingdoms, and for plants the *division is sometimes used instead of the phylum.

**physical chemistry** The branch of chemistry concerned with the effect of chemical structure on physical properties. It includes chemical thermodynamics and electrochemistry.

**physical map** (in genetics) Any map that shows the arrangement of the material (i.e.

the nucleoprotein) making up a chromosome or segment of a genome (*see* CHROMOSOME MAP). The coarsest physical maps are ones depicting chromosome banding patterns, which are dark and light transverse bands obtained by staining entire chromosomes in mitosis. These **cytological maps** enable characterization of individual chromosomes and can reveal gross anomalies, such as missing or duplicated segments. On a much larger scale are **contig maps**; these show the order of cloned DNA segments taken from a *DNA library and fitted together to form a series of overlapping, or contiguous, segments, called a **contig**. Such segments are roughly on a gene-length scale. Once a contig has been correctly aligned, the base sequence of each component segment can be determined (*see* DNA SEQUENCING), and hence the overall sequence of the chromosomal DNA can be pieced together. *Compare* LINKAGE MAP.

**physics** The study of the laws that determine the structure of the universe with reference to the matter and energy of which it consists. It is concerned not with chemical changes that occur but with the forces that exist between objects and the interrelationship between matter and energy. Traditionally, the study was divided into separate fields: heat, light, sound, electricity and magnetism, and mechanics (*see* CLASSICAL PHYSICS). Since the turn of the century, however, quantum mechanics and relativistic physics have become increasingly important; the growth of modern physics has been accompanied by the studies of atomic physics, nuclear physics, and particle physics. The physics of astronomical bodies and their interactions is known as **astrophysics**, the physics of the earth is known as **geophysics**, and the study of the physical aspects of biology is called **biophysics**. *See also* THEORETICAL PHYSICS.

**physiological saline** A liquid medium in which animal tissues may be kept alive for a few hours during experiments without pathological changes or distortion of the cells taking place. Such fluids are salt solutions that are isotonic with and have the same pH as the body fluids of the animal. A well-known example is **Ringer's solution**, formulated by the British physiologist S. Ringer (1835–1910), which is a mixture of sodium chloride, calcium chloride, sodium bicarbonate, and potassium chloride solutions.

p

**physiological specialization** The occurrence within a species of several forms that are identical in appearance but differ in physiology: these are termed **physiological races**. For example, many pathogenic fungi develop new physiological races in response to the strong selection pressure exerted when disease-resistant crop varieties are sown over large areas.

**physiology** The branch of biology concerned with the vital functions of plants and animals, such as nutrition, respiration, reproduction, and excretion.

**physisorption** See ADSORPTION.

**phyto-** Prefix denoting plants. For example, phytopathology is the study of plant diseases.

**phytochrome** A protein-based plant pigment present in small quantities in many plant organs. It exists in two interconvertible forms: a physiologically active form, which forms when the plant is illuminated with red light or normal daylight; and an inactive form, formed when the plant is exposed to far-red light or darkness. The active form regulates many plant processes, such as seed germination and the initiation of flowering.

**phytogeography** See PLANT GEOGRAPHY.

**phytohaemagglutinin (PHA)** Any of various plant-derived compounds that induce changes in lymphocytes normally associated with antigen challenge. These changes include cell enlargement, increased RNA and DNA synthesis, and, finally, cell division. Response to PHAs is used to test for competence of cell-mediated *immunity, for example in patients suffering chronic virus infections.

**phytohormone** See PLANT HORMONE.

**phytoplankton** The photosynthesizing organisms of *plankton, consisting chiefly of microscopic algae, such as diatoms and dinoflagellates. Near the surface of the sea there may be many millions of such plants per cubic metre. Members of the phytoplankton are of great importance as they form the basis of food for all other forms of aquatic life, being the primary *producers. *Compare* ZOOPLANKTON.

**phytoremediation** The use of plants to decontaminate polluted land, water, or air. Different plant species that can grow on contaminated sites have evolved various ways of countering high concentrations of heavy metals, oil, solvents, or other toxic substances. Some absorb and degrade the toxins internally, whereas others secrete substances that stabilize or neutralize toxins in the soil. Yet others take up and accumulate the toxins in their tissues, so that when the plants are harvested from the site, the toxins are removed from the site.

**pi** Symbol $\pi$. The ratio of the circumference of any circle to its diameter. It is a *transcendental number with the value 3.141 592....

**pia mater** The innermost of the three membranes (*meninges) that surround the brain and spinal cord of vertebrates. The pia mater lies immediately adjacent to the central nervous system, and the *choroid plexus, which secretes cerebrospinal fluid, is an extension of it.

**piano stool** See SANDWICH COMPOUND.

**pi bond** See ORBITAL.

**pico-** Symbol p. A prefix used in the metric system to denote $10^{-12}$. For example, $10^{-12}$ farad = 1 picofarad (pF).

**picornavirus** One of a group of small RNA-containing viruses (*pico* = small; hence pico-RNA-virus) commonly present in the alimentary and respiratory tracts of vertebrates. They cause mild infections of these tracts but the group also includes the polioviruses, which attack the central nervous system causing poliomyelitis; and the causal agent of foot and mouth disease in cattle, sheep, and pigs.

**picrate** A salt or ester of picric acid.

**picric acid (2,4,6-trinitrophenol)** A yellow highly explosive nitro compound, $C_6H_2(NO_2)_3$; r.d. 1.8; m.p. 122°C.

**pie chart** A diagram in which percentages are shown as sectors of a circle. If $x$ percent of the prey of a carnivore comprises species $X$, $y$ percent species $Y$, and $z$ percent species $Z$, a pie chart would show three sectors having central angles $3.6x°$, $3.6y°$, and $3.6z°$.

**pi electron** An electron in a pi orbital. See ORBITAL.

**piezoelectric effect** The generation of a potential difference across the opposite faces of certain nonconducting crystals (**piezoelectric crystals**) as a result of the application of mechanical stress between these faces. The electric polarization produced is proportional to the stress and the direction of the polarization reverses if the stress changes from compression to tension. The **reverse piezoelectric effect** is the opposite phenomenon: if the opposite faces of a piezoelectric crystal are subjected to a potential difference, the crystal changes its shape. Rochelle salt and quartz are the most frequently used piezoelectric materials. While Rochelle salt produces the greater polarization for a given stress, quartz is more widely used as its crystals have greater strength and are stable at temperatures in excess of 100°C.

If a quartz plate is subjected to an alternating electric field, the reverse piezoelectric effect causes it to expand and contract at the field frequency. If this field frequency is made to coincide with the natural elastic frequency of the crystal, the plate resonates; the direct piezoelectric effect then augments the applied electric field. This is the basis of the *crystal oscillator and the *quartz clock. *See also* CRYSTAL MICROPHONE; CRYSTAL PICK-UP.

**pig iron** The impure form of iron produced by a blast furnace, which is cast into pigs (blocks) for converting at a later date into cast iron, steel, etc. The composition depends on the ores used, the smelting procedure, and the use to which the pigs will later be put.

**pigment** (in biology) A compound that gives colour to a tissue. Pigments perform a variety of functions. For example, *haemoglobin in vertebrate erythrocytes gives blood its characteristic red colour and enables the transport of oxygen throughout the body. Other biological pigments include *chlorophyll, a photosynthetic pigment in plants that is responsible for their green coloration; and *melanin, a brown pigment in animals that provides protection from ultraviolet light and can be used in camouflaging colorations.

**pileus** The umbrella-shaped cap of certain fungi, such as mushrooms. Spores are produced from *gills or pores on the lower surface.

**piliferous layer** The part of the root epidermis that bears *root hairs. It extends over a region about 4–10 mm behind the root tip. Beyond this the piliferous layer is sloughed off to reveal the hypodermis.

**pillow lava** *See* LAVA.

**Piltdown man** Fossil remains, purported to have been found by Charles Dawson (1864–1916) at Piltdown, Sussex, in 1912, that were named *Eoanthropus dawsoni* and described as a representative of the true ancestors of modern humans. The skull resembled that of a human but the jaw was apelike. In 1953 dating techniques showed the specimen to be a fraud.

**pi-meson** *See* PION.

**pinacol rearrangement** A reaction in which the diol pinacol, $(CH_3)_2$ COH-$(CH_3)_2COH$, converts to the ketone pinacolone, $CH_3COC(CH_3)_3$, under acid conditions, with loss of a molecule of water. A methyl group moves from one carbon atom to an adjacent one in order to stabilize an intermediate carbocation. The reaction, also called the **pinacol-pinacolone rearrangement**, gives its name to a class of similar rearrangements.

**pinch effect** A magnetic attraction between parallel conductors carrying currents flowing in the same direction. The force was noticed in early induction furnaces. Since the late 1940s it has been widely studied as a means of confining the hot plasma in a *thermonuclear reactor. In an experimental toroidal thermonuclear reactor a large electric current is induced in the plasma by electromagnetic induction; this current both heats the plasma and draws it away from the walls of the tube as a result of the pinch effect.

**pineal eye** *See* MEDIAN EYE.

**pineal gland** An outgrowth of the *forebrain. In humans its functions are obscure, but in other vertebrates it acts as an endocrine gland, secreting the hormone *melatonin.

**pinna (auricle)** The visible part of the *outer ear, present in some mammals. It is made of cartilage and its function is to channel sound waves into the external auditory meatus. In some species the pinna is movable and aids in detecting the direction from which a sound originates.

**pinocytosis** The process by which a living cell engulfs a minute droplet of liquid. It involves a mechanism similar to *phagocytosis. *See* ENDOCYTOSIS.

**pion (pi-meson)** An *elementary particle classified as a *meson. It exists in three forms: neutral, positively charged, and negatively charged. The charged pions decay into

muons and neutrinos; the neutral pion decays into two gamma-ray photons. Pions
consist of a quark and an anti-quark.

**pi orbital** *See* ORBITAL.

**piperidine** A saturated heterocyclic compound having a nitrogen atom in a six-membered ring, $C_5H_{11}N$; r.d. 0.86; m.p. -7°C; b.p.
106°C. The structure is present in many alkaloids.

**pipette** A graduated tube used for transferring measured volumes of liquid.

**Pirani gauge** An instrument used to
measure low pressures ($1-10^{-4}$ torr; 100–0.01
Pa). It consists of an electrically heated filament, which is exposed to the gas whose
pressure is to be measured. The extent to
which heat is conducted away from the filament depends on the gas pressure, which
thus controls its equilibrium temperature.
Since the resistance of the filament is dependent on its temperature, the pressure is
related to the resistance of the filament. The
filament is arranged to be part of a *Wheatstone bridge circuit and the pressure is read
from a microammeter calibrated in pressure
units. As the effect depends on the thermal
conductivity of the gas, the calibration has to
be made each time the pressure of a different gas is measured.

**PIR detector** *See* PASSIVE DEVICE.

**pirssonite** A mineral consisting of a hydrated mixed carbonate of sodium and calcium, $Na_2CO_3.CaCO_3.2H_2O$.

**Pisces** In some classifications, a superclass
of the *Gnathostomata (jawed chordates)
comprising the fishes (*compare* TETRAPODA).
There are two classes of modern fish: *Chondrichthyes (cartilaginous fishes) and *Osteichthyes (bony fishes).

**((⊕)) SEE WEB LINKS**
• The Australian Museum Fish Site, an extensive
  site catering for fish enthusiasts at various levels.

**pistil** The female part of a flower, consisting either of a single *carpel (**simple pistil**)
or a group of carpels (**compound pistil**).

**pitch** 1. (in chemistry) A black or dark-
brown residue resulting from the distillation
of coal tar, wood tar, or petroleum (bitumen). The term is also sometimes used for
the naturally occurring petroleum residue
(asphalt). Pitch is used as a binding agent
(e.g. in road tars), for waterproofing (e.g. in
roofing felts), and as a fuel. 2. (in physics)

The property of a sound that characterizes
its highness or lowness to an observer. It is
related to, but not identical with, frequency.
Below about 1000 Hz the pitch is slightly
higher than the frequency and above 1000
the position is reversed. The loudness of a
sound also affects the pitch. Up to 1000 Hz
an increase in loudness causes a decrease in
pitch. From about 1000 to 3000 Hz the pitch
is independent of loudness, while above
3000 Hz an increase in loudness seems to
cause a raising of pitch. Pitch is usually
measured in mels; a note of 1000 Hz frequency with a loudness of 40 decibels above
the absolute threshold of hearing has a pitch
of 1000 mels. 3. (in mechanics) *See* SCREW.

**pitchblende** *See* URANINITE.

**pitfall trap** A simple trap for small invertebrate animals consisting of a tin that is
placed in the ground with its rim at ground
level. The trap, which contains some kind of
bait, can be covered by a tile suspended
above ground level by stones so that rain
does not enter the tin.

**pith** 1. (medulla) The cylinder of *parenchyma tissue found in the centre of
plant stems to the inside of the vascular tissue. It is light in weight and has been put to
various commercial uses, notably the manufacture of pith helmets. 2. (not in scientific
usage) The white tissue below the rind of
many citrus fruits. 3. To destroy the central
nervous system of an animal, especially a
laboratory animal such as a frog, by severing
its spinal cord.

*Pithecanthropus* *See* HOMO.

**Pitot tube** A device for measuring the
speed of a fluid. It consists of two tubes, one
with an opening facing the moving fluid and
the other with an opening at 90° to the direction of the flow. The two tubes are connected
to the opposite sides of a manometer so that
the difference between the dynamic pressure in the first tube and the static pressure
in the second tube can be measured. The
speed $v$ of the flow of an incompressible
fluid is then given by: $v^2 = 2(P_2 - P_1)/\rho$, where
$P_2$ is the dynamic pressure, $P_1$ is the static
pressure, and $\rho$ is the density of the fluid.
The device has a wide variety of applications.
It was devised by the Italian-born French engineer Henri Pitot (1695–1771).

**pituitary gland (pituitary body; hypoph-
ysis)** A pea-sized endocrine gland attached
by a thin stalk to the *hypothalamus at the

base of the brain. It consists of two main lobes, the anterior and the posterior, separated in nonhumans by an intermediate lobe. The **anterior pituitary** secretes such hormones as *growth hormone, the *gonadotrophins, *prolactin, thyroid-stimulating hormone (*see* THYROID GLAND) and *ACTH. Because these hormones regulate the growth and activity of several other endocrine glands, the anterior pituitary is often referred to as the **master endocrine gland**. Activity of the anterior pituitary itself is regulated by specific *releasing hormones produced by the hypothalamus (*see also* NEUROENDOCRINE SYSTEM). The **posterior pituitary** secretes the hormones *oxytocin and antidiuretic hormone, and the **intermediate lobe** secretes *melanocyte-stimulating hormone.

**pixel** One of the tiny dots that make up an image on the screen of a computer's *visual-display unit (VDU) or on some types of television receiver; it is short for picture element. Screen resolution is determined by the number of pixels (the more pixels, the better the resolution), and each pixel is given a brightness and colour. A typical high-resolution colour VDU screen has a $1024 \times 768$ array of pixels.

**p$K$ value** A measure of the strength of an acid or base on a logarithmic scale. For an acid, the value, denoted p$K_a$, is given by $\log_{10}(1/K_a)$, where $K_a$ is the acid dissociation constant. Similarly, for a base p$K_b$ is given by $\log_{10}(1/K_b)$. p$K_a$ values are often used to compare the strengths of different acids.

**placenta** 1. The organ in mammals and other viviparous animals by means of which the embryo is attached to the wall of the uterus. It is composed of embryonic and maternal tissues: extensions of the *chorion and *allantois grow into the uterine wall so that materials (e.g. oxygen, nutrients) can pass between the blood of the embryo and its mother (there is, however, no direct connection between the maternal and embryonic blood). The placenta is eventually expelled as part of the *afterbirth. 2. A ridge of tissue on the ovary wall of flowering plants to which the ovules are attached. The arrangement of ovules on the placenta (**placentation**) is variable, depending on the number of carpels and whether they are free (*see* APOCARPY) or fused (*see* SYNCARPY).

**Placentalia** *See* EUTHERIA.

**placoid scale (denticle)** *See* SCALES.

**Placozoa** A phylum of simple aquatic animals containing just a single known species, *Trichoplax adhaerens*. This has a transparent round flattened body, between 0.2 and 3 mm in diameter, without head, tail, or appendages. It is covered in cilia, which it uses to crawl over surfaces, and it feeds by secreting enzymes from its ventral surface. An adult comprises a few thousand cells of just four types, whose DNA content is the smallest of any animal. The evolutionary relationships of placozoans with other animals, especially the sponges (*Porifera), remain speculative. Molecular studies of ribosomal RNA sequences have indicated that the Placozoa may be secondarily simplified descendants of more complex ancestors and are not closely related to the sponges.

**plagiotropism** The tendency for a *tropism (growth response of a plant) to be orientated at an angle to the line of action of the stimulus concerned. For example, the growth of lateral branches and lateral roots is at an oblique angle to the stimulus of gravity (**plagiogeotropism**). *Compare* ORTHOTROPISM.

**planarians** *See* TURBELLARIA.

**Planck, Max Carl Ernst Ludwig** (1858–1947) German physicist, who became a professor at Berlin University in 1892. Here he formulated the *quantum theory, which had its basis in a paper of 1900. (See also *Planck constant; *Planck's radiation law). One of the most important scientific discoveries of the century, this theory earned him the 1918 Nobel Prize for physics.

**Planck constant** Symbol $h$. The fundamental constant equal to the ratio of the energy $E$ of a quantum of energy to its frequency $\nu$: $E = h\nu$. It has the value $6.626\ 176 \times 10^{-34}$ J s. It is named after Max Planck. In quantum-mechanical calculations (especially particle physics) the **rationalized Planck constant** (or **Dirac constant**) $\hbar = h/2\pi = 1.054\ 589 \times 10^{-34}$ J s is frequently used.

**Planck length** The length scale at which a classical description of gravity ceases to be valid, and *quantum mechanics must be taken into account. It is given by $L_P = \sqrt{(G\hbar/c^3)}$, where $G$ is the gravitational constant, $\hbar$ is the rationalized Planck constant, and $c$ is the speed of light. The value of the Planck length is of order $10^{-35}$ m (twenty orders of magnitude smaller than the size of a proton, $10^{-15}$ m).

**Planck mass** The mass of a particle whose Compton wavelength is equal to the *Planck length. It is given by $m_P = \sqrt{(\hbar/G)}$, where $\hbar$ is the rationalized Planck constant, $c$ is the speed of light, and $G$ is the gravitational constant. The description of an elementary particle of this mass, or particles interacting with energies per particle equivalent to it (through $E = mc^2$), requires a *quantum theory of gravity. Since the Planck mass is of order $10^{-8}$ kg (equivalent energy $10^{19}$ GeV), and, for example, the proton mass is of order $10^{-27}$ kg and the highest energies attainable in present-day particle accelerators are of order $10^3$ GeV, quantum-gravitational effects do not arise in laboratory particle physics. However, energies equivalent to the Planck mass did occur in the early universe according to *big-bang theory, and a quantum theory of gravity is important for discussing conditions there (see PLANCK TIME).

**Planck's radiation law** A law giving the distribution of energy radiated by a *black body. It introduced into physics the novel concept of energy as a quantity that is radiated by a body in small discrete packets rather than as a continuous emission. These small packets became known as quanta and the law formed the basis of *quantum theory. The **Planck formula** gives the energy radiated per unit time at frequency $\nu$ per unit frequency interval per unit solid angle into an infinitesimal cone from an element of the black-body surface that is of unit area in projection perpendicular to the cone's axis. The expression for this **monochromatic specific intensity** $I_\nu$ is:

$$I_\nu = 2hc^{-2}\nu^3/[\exp(h\nu/kT - 1],$$

where $h$ is the *Planck constant, $c$ is the *speed of light, $k$ is the *Boltzmann constant, and $T$ is the thermodynamic temperature of the black body. $I_\nu$ has units of watts per square metre per steradian per hertz (W m$^{-2}$ sr$^{-1}$ Hz$^{-1}$). The monochromatic specific intensity can also be expressed in terms of the energy radiated at wavelength $\lambda$ per unit wavelength interval; it is then written as $I_\lambda$, and the Planck formula is:

$$I_\lambda = 2hc^2\lambda^{-5}/[\exp(hc/\lambda kT) - 1].$$

There are two important limiting cases of the Planck formula. For low frequencies $\nu \ll kT/h$ (equivalently, long wavelengths $\lambda \gg hc/kT$) the **Rayleigh–Jeans formula** is valid:

$$I_\nu = 2c^{-2}\nu^2 kT,$$

or

$$I_\lambda = 2c\lambda^{-4}kT.$$

Note that these expressions do not involve the Planck constant. They can be derived classically and do not apply at high frequencies, i.e. high energies, when the quantum nature of *photons must be taken into account. The second limiting case is the **Wien formula**, which applies at high frequencies $\nu \gg kT/h$ (equivalently, short wavelengths $\lambda \ll hc/kT$):

$$I_\nu = 2hc^{-2}\nu^3 \exp(-h\nu/kT),$$

or

$$I_\lambda = 2hc^2\lambda^{-5} \exp(-hc/\lambda kT).$$

*See also* WIEN'S DISPLACEMENT LAW.

**Planck time** The time taken for a photon (travelling at the speed of light $c$) to move through a distance equal to the *Planck length. It is given by $t_P = \sqrt{(G\hbar/c^5)}$, where $G$ is the gravitational constant and $\hbar$ is the rationalized Planck constant. The value of the Planck time is of order $10^{-43}$ s. In the *big-bang theory, up until a time $t_P$ after the initial instant, it is necessary to use a *quantum theory of gravity to describe the evolution of the universe.

**Planck units** A system of *units, used principally in discussions of *quantum theories of gravity, in which length, mass, and time are expressed as multiples of the *Planck length, mass, and time respectively. This is equivalent to setting the gravitational constant, the speed of light, and the reduced Planck constant all equal to unity. All quantities that ordinarily have dimensions involving length, mass, and time become dimensionless in Planck units. Since, in the subject area where Planck units are used, it is normal to employ *Gaussian or *Heaviside–Lorentz units for electromagnetic quantities, these then become dimensionless. *See also* GEOMETRIZED UNITS; NATURAL UNITS.

**plane** A flat surface defined by the condition that any two points in the plane are joined by a straight line that lies entirely in the surface.

**plane-polarized light** *See* POLARIZATION OF LIGHT.

**planet (major planet)** A body in orbit around a star that has great enough mass to have contracted to a more or less spherical shape under its own gravitation and to have removed all other bodies, planetesimals, etc., from its orbital zone. The earth is one of eight planets orbiting the sun, the others being Mercury, Venus, Mars, Jupiter, Saturn, Uranus, and Neptune. Pluto, which was thought to be a planet when it was discovered in 1930, is now classified as a *dwarf planet. Within our solar system, Mercury, Venus, the earth, and Mars are the **terrestrial planets**; Jupiter, Saturn, Uranus, and Neptune are *gas giants. Mercury and Venus are also called the **inner** (or **inferior**) **planets** because their orbits lie within that of the earth; Mars, Jupiter, Saturn, Uranus, and Neptune are the **outer** (or **superior**) **planets**. *See also* EXTRASOLAR PLANET; SOLAR SYSTEM.

**(⊕) SEE WEB LINKS**
- The International Astronomical Union's Question and Answer Sheet about planets
- A multimedia tour of the planets, the sun, and the other members of the solar system

**planetesimal** One of the many small objects that underwent the process of *accretion to form the planets of the solar system. Planetesimals probably ranged in size from about 1 millimetre to several kilometres across.

**planimeter** An instrument used to measure the area of a closed curve. The outline of the curve is followed by a pointer on the instrument and the area is given on a graduated disc.

**plankton** Minute *pelagic organisms that drift or float passively with the current in a sea or lake. Plankton includes many microscopic organisms, such as algae, protozoans, various animal larvae, and some worms. It forms an important food source for many other members of the aquatic community and is divided into *zooplankton and *phytoplankton. *Compare* NEKTON.

**plano-concave lens** *See* CONCAVE; LENS.

**plano-convex lens** *See* CONVEX; LENS.

**plant** 1. In traditional classification, any living organism of the kingdom Plantae. Plants are distinguished from other multicellular organisms by their life cycles, in which a haploid generation (*gametophyte) alternates with a diploid generation (*sporophyte). (*See also* ALTERNATION OF GENERATIONS.) Most plants manufacture carbohydrates by *photosynthesis, in which simple inorganic substances are built up into organic compounds. The radiant energy needed for this process is absorbed by

*chlorophyll, a complex pigment not found in animals. Plants also differ from animals in the possession of *cell walls (usually composed of *cellulose). Plants are immobile, as there is no necessity to search for food, and they respond slowly to external stimuli. For a classification of the plant kingdom, see Appendix. **2.** Any member of an assemblage of eukaryotic organisms – the **Archaeplastida** – recognized on the basis of recent molecular data. As well as land plants, it contains green algae, red algae, and glaucophytes (algae whose chloroplasts are similar to cyanobacteria).

**(⊕) SEE WEB LINKS**
- Movable hyperbolic tree of plant phylogeny, devised by 'Deep Green' and hosted by the University of California, Berkeley

**plant geography** (phytogeography) The study of the distribution of vegetation around the world, with particular emphasis on the influence of the environmental factors that determine this distribution.

**plant hormone** (growth substance; phytohormone) Any of a number of organic chemicals that are synthesized by plants and regulate growth and development. They are usually made in a particular region, such as the shoot tip, and transported to other regions, where they take effect. *See* ABSCISIC ACID; AUXIN; CYTOKININ; ETHENE; GIBBERELLIN.

**plantigrade** Describing the gait of many mammals, including humans, in which the whole lower surface of the foot is on the ground. *Compare* DIGITIGRADE; UNGULIGRADE.

**planula** The ciliated free-swimming larva of many cnidarians, consisting of a solid mass of cells. It eventually settles on a suitable surface and develops into a *polyp.

**plaque** 1. A thin layer of organic material covering all or part of the exposed surface of a tooth. It contains dissolved food (mostly sugar) and bacteria. The bacteria in plaque metabolize the sugar and produce acid, which eats into the surface of the enamel of the tooth and eventually causes tooth decay (*dental caries). **2.** A clear area in a bacterial culture grown on an agar plate due to *lysis of the bacteria by a bacteriophage.

**plasma** 1. (in physics) A highly ionized gas in which the number of free electrons is approximately equal to the number of positive ions. Sometimes described as the fourth state of matter, plasmas occur in interstellar space, in the atmospheres of stars (including the sun), in discharge tubes, and in experimental thermonuclear reactors.

Because the particles in a plasma are charged, its behaviour differs in some respects from that of a gas. Plasmas can be created in the laboratory by heating a low-pressure gas until the mean kinetic energy of the gas particles is comparable to the *ionization potential of the gas atoms or molecules. At very high temperatures, from about 50 000 K upwards, collisions between gas particles cause cascading ionization of the gas. However, in some cases, such as a fluorescent lamp, the temperature remains quite low as the plasma particles are continually colliding with the walls of the container, causing cooling and recombination. In such cases ionization is only partial and requires a large energy input. In *thermonuclear reactors an enormous plasma temperature is maintained by confining the plasma away from the container walls using electromagnetic fields (*see* PINCH EFFECT). The study of plasmas is known as **plasma physics**. **2.** (in physiology) *See* BLOOD PLASMA.

**plasma cells** Antibody-producing cells found in the epithelium of the lungs and gut and also in bone-forming tissue. They develop in the lymph nodes, spleen, and bone marrow when antigens stimulate lymphocytes to form the precursor cells that give rise to them (*see* B CELL).

**plasma display** A form of display used in television, in which light output is produced from the interaction between an electric current and an ionized inert gas such as neon. The display consists of a matrix of individual cells, one per pixel. A typical monochrome or grey-scale display generates red or orange light. Colour systems generate ultraviolet radiation, which excites phosphors (red, green, and blue) on the surface of the display; the excited phosphors emit light on return to the ground state. Plasma panels are rugged, largely immune to external fields, and do not suffer from flicker. Fabrication of large displays is possible. The device is essentially bistable so no special circuitry is required to isolate individual cells from their neighbours.

**plasmagel** The specialized outer gel-like *cytoplasm of living cells (such as *Amoeba*) that move by extruding part of the cell (known as a *pseudopodium) in the direc-

tion of motion. A reversible conversion of plasmagel to the more fluid *plasmasol is involved in the continuous flow forward of cytoplasm necessary for forming a pseudopodium. *See also* CYTOPLASMIC STREAMING.

**plasma membrane (plasmalemma; cell membrane)** The partially permeable membrane forming the boundary of a cell. It consists mostly of protein and lipid (*see* LIPID BILAYER) and plays various crucial roles in the cell's activities. A key task is to regulate the flow of materials and signals into and out of the cell; this is accomplished, for example, by membrane proteins that act as *ion channels or as *receptors for signal molecules arriving at the cell surface. The plasma membrane is the site of junctions with neighbouring cells and forms attachments to the *extracellular matrix, thus ensuring tissue integrity. In plants, fungi, bacteria, and many protists, it helps in assembling a cell wall or capsule on its outer surface.

**plasma protein** Any of the protein constituents of blood plasma. They comprise *albumins, *globulins, and *fibrinogen.

**plasmasol** The specialized inner sol-like *cytoplasm of living cells that move by producing *pseudopodia. *Compare* PLASMAGEL.

**plasmid** A structure in bacterial cells consisting of DNA that can exist and replicate independently of the chromosome. Plasmids provide genetic instructions for certain cell activities (e.g. resistance to antibiotic drugs). They can be transferred from cell to cell in a bacterial colony. Plasmids are widely used as *vectors to produce recombinant DNA for *gene cloning. *See also* GENETICALLY MODIFIED ORGANISMS (Feature).

**plasmin (fibrinase; fibrinolysin)** An enzyme, present in blood plasma, that breaks down a blood clot by destroying the fibrin threads of the clot and by inactivating factors involved in blood clotting, such as prothrombin and the *clotting factors. This occurs during *fibrinolysis. Plasmin is derived from an inactive precursor, *plasminogen.

**plasminogen** The inactive precursor of the enzyme *plasmin. Plasminogen is incorporated into blood clots and is converted to plasmin during *fibrinolysis.

**plasmodesmata** (*sing.* **plasmodesma**) Fine cytoplasmic strands that connect the *protoplasts of adjacent plant cells by passing through their cell walls and thus permit the passage of substances between cells. Plasmodesmata are cylindrical in shape (about 20–40 nm in diameter) and are lined by the plasma membrane of the two adjacent cells. The endoplasmic reticula of the two adjacent cells are connected by a narrower structure, the **desmotubule**, which runs through the centre of a plasmodesma.

**plasmolysis** The loss of water by *osmosis from a plant cell to the extent that the cytoplasm shrinks away from the cell wall. This happens when the cell is placed in a solution that has a higher solute concentration than that of the cell sap, i.e. it has a lower (more negative) *water potential, since water always moves from an area of high water potential to an area of low water potential. Compare TURGOR.

**plasmon** A *collective excitation for quantized oscillations of the electrons in a metal.

**plasmonics** The design and study of devices that make use of *plasmons, i.e. plasmonics is to plasmons what *electronics is to electrons and *photonics is to photons. Plasmonics has many appealing possibilities in technology but has the practical difficulty that plasmons are short-lived excitations.

**plaster of Paris** The hemihydrate of *calcium sulphate, $2CaSO_4.H_2O$, prepared by heating the mineral gypsum. When ground to a fine powder and mixed with water, plaster of Paris sets hard, forming interlocking crystals of gypsum. The setting results in an increase in volume and so the plaster fits tightly into a mould. It is used in pottery making, as a cast for setting broken bones, and as a constituent of the plaster used in the building industry.

**plasticity** The property of solids that causes them to change permanently in size or shape as a result of the application of a stress in excess of a certain value, called the **yield point**.

**plasticizer** A substance added to a synthetic resin to make it flexible. See PLASTICS.

**plastics** Materials that can be shaped by applying heat or pressure. Most plastics are made from polymeric synthetic *resins, although a few are based on natural substances (e.g. cellulose derivatives or shellac). They fall into two main classes. **Thermoplastic materials** can be repeatedly softened by heating and hardened again on cooling.

**Thermosetting materials** are initially soft, but change irreversibly to a hard rigid form on heating. Plastics contain the synthetic resin mixed with such additives as pigments, plasticizers (to improve flexibility), antioxidants and other stabilizers, and fillers. See Chronology overleaf.

**plastid** An *organelle within a plant cell, often occurring in large numbers. Apart from the nucleus, plastids are the largest solid inclusions in a plant cell. For convenience they are classified into those containing pigments (*chromoplasts) and those that are colourless (*leucoplasts), although changes from one to the other frequently occur. Plastids develop from **proplastids**, colourless bodies found in meristematic and immature cells; they also arise by division of existing plastids. See also CHLOROPLAST.

**plastoglobulus** (pl. **plastoglobuli**) A densely staining droplet found, often in large numbers, in plastids of plant cells. Plastoglobuli consist of lipid pigment and are especially prominent in coloured plastids (*chromoplasts), for example in ripening fruits. Plastoglobuli also occur in chloroplasts, but are masked by the green chlorophyll. When the chlorophyll breaks down as the leaves start to die in autumn, the pigmented plastoglobuli become apparent as the red or yellow 'fall' colours.

**plastron** See CARAPACE.

**platelet (thrombocyte)** A minute disc-shaped cell fragment in mammalian blood. Platelets are formed as fragments of larger cells (**megakaryocytes**) found in red bone marrow; they have no nucleus. They play an important role in *blood clotting, releasing thromboxane $A_2$ (see PROSTAGLANDIN), serotonin, and other chemicals in response to local tissue damage, which causes a chain of events leading to the formation of a plug at the site of the damage, thus preventing further blood loss. There are about 250 000 platelets per cubic millimetre of blood.

**platelet-derived growth factor** See GROWTH FACTOR.

**plate tectonics** The theory, developed in the early 1960s, that the surface of the earth is made of lithospheric plates, which have moved throughout geological time resulting in the present-day positions of the continents. The theory explains the locations of mountain building as well as earthquakes and volcanoes. The rigid lithospheric plates

## PLASTICS

| | |
|---|---|
| 1839 | US inventor Charles Goodyear (1800–60) developed a process to vulcanize rubber |
| 1855 | British chemist Alexander Parkes (1813–90) patents Parkesine, a plastic made from nitrocellulose, methanol, and wood pulp; it is later called 'celluloid'. |
| 1860 | British chemist Charles Williams (1829–1910) prepares isoprene (synthetic rubber). |
| 1868 | US printer John Hyatt (1837–1920) develops commercial process for making celluloid. |
| 1884 | French chemist Hilaire de Chardonnet (1839–1924) develops process for making rayon. |
| 1892 | British chemists Edward Bevan (1856–1921) and Charles Cross (1855–1935) develop the viscose process for making rayon. |
| 1899 | British chemist Frederick Kipping (1863–1949) discovers silicone plastics. |
| 1901 | German chemists Krische and Spitteler make formaldehyde–casein plastic (Galalith). |
| 1905 | Belgian-born US chemist Leo Baekland (1863–1944) invents Bakelite. |
| 1912 | Swiss chemist Jacques Brandenberger produces Cellophane (viscose cellulose film). |
| 1913 | US Formica Insulation company markets plastic laminate made from formaldehyde resins. |
| 1918 | Hans John prepares urea–formaldehyde resin. |
| 1926 | German chemist Hermann Staudinger (1881–1965) discovers the polymeric nature of plastics. |
| 1930 | US chemist Waldo Semon develops PVC (polyvinyl chloride). |
| 1930 | Canadian chemist William Chalmers discovers polymethylmethacrylate (Perspex and Plexiglass). |
| 1930 | German chemists at IG Farbenindustrie produce polystyrene. |
| 1931 | Wallace Carothers invents nylon. |
| 1938 | US chemist Roy Plunkett produces polytetrafluoroethene (PTFE). |
| 1939 | British company ICI develops commercial process for making polyethene. |
| 1941 | British chemists John Whinfield (1901–66) and J. Dickson develop Terylene (Dacron). |
| 1941 | German company IG Farbenindustrie produces polyurethane. |
| 1943 | US Dow Corning company produces silicone plastics. |
| 1947 | British chemists produce acrylic fibres. |
| 1953 | German chemist Karl Ziegler (1896–1973) discovers catalyst for making high-density polyethene. |
| 1954 | Italian chemist Giulio Natta (1903–79) develops industrial process for making high-density polyethene (using Ziegler catalyst). |
| 1989 | Italian company Ferruzzi produces biodegradable plastic (based on starch). |

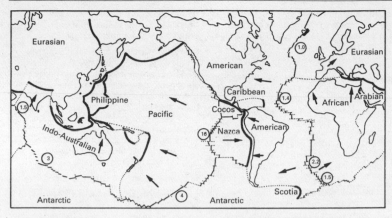

direction of plate movement — constructive plate margins

.......... transform faults ③ spreading rates (centimetres per year)

—— destructive plate margins (subduction zones)

The lithospheric plates

Cross-section showing a constructive plate margin, where two plates are drawing apart along a mid-ocean ridge, and a destructive plate margin, where the oceanic plate is being subducted below the continental plate. Volcanic activity and earthquakes are associated with these margins

## Plate tectonics.

consist of continental and oceanic crust together with the upper mantle, which lie above the weaker plastic asthenosphere. These plates move relative to each other across the earth. Six major plates (Eurasian, American, African, Pacific, Indian, and Antarctic) are recognized, together with a number of smaller ones. The plate margins (boundaries) coincide with zones of seismic and volcanic activity.

A **constructive** (or **divergent**) plate margin occurs when two plates move away from each other. It is marked by a mid-oceanic ridge where basaltic material wells up from the mantle to form new oceanic crust, in a process known as **sea-floor spreading**. The production of new crust at constructive plate

margins is compensated for by the destruction of material along a **destructive** (or **convergent**) plate margin. Along these margins, which are also known as **subduction zones** and marked by an oceanic trench, one plate (usually oceanic) is forced to plunge down beneath the other (which may be continental or oceanic). The crust becomes partially melted and rises to form a chain of volcanoes in the upper plate parallel to the trench. When two continental plates collide the compression results in the formation of mountain chains (e.g. the Himalayas formed by the collision of the Indian and Eurasian plates). A third type of plate margin – the **transform plate margin** – occurs where two plates are slipping past each other.

**platinum** Symbol Pt. A silvery white metallic *transition element (*see also* PLATINUM METALS); a.n. 78; r.a.m. 195.09; r.d. 21.45; m.p. 1772°C; b.p. 3827±100°C. It occurs in some nickel and copper ores and is also found native in some deposits. The main source is the anode sludge obtained in copper–nickel refining. The element is used in jewellery, laboratory apparatus (e.g. thermocouples, electrodes, etc.), electrical contacts, and in certain alloys (e.g. with iridium or rhodium). It is also a hydrogenation catalyst. The element does not oxidize nor dissolve in hydrochloric acid. Most of its compounds are platinum(II) or platinum(IV) complexes.

 **SEE WEB LINKS**
• Information from the WebElements site

**platinum black** Black finely divided platinum metal produced by vacuum evaporation and used as an absorbent and a catalyst.

**platinum metals** The three members of the second and third transition series immediately proceeding silver and gold: ruthenium (Ru), rhodium (Rh), and palladium (Pd); and osmium (Os), iridium (Ir), and platinum (Pt). These elements, together with iron, cobalt, and nickel, were formerly classed as group VIII of the periodic table. The platinum-group metals are relatively hard and resistant to corrosion and are used in jewellery and in some industrial applications (e.g. electrical contacts). They have certain chemical similarities that justify classifying them together. All are resistant to chemical attack. In solution they form a vast range of complex ions. They also form coordination compounds with carbon monoxide and other pi-bonding ligands. A number of complexes can be made in which a hydrogen atom is linked directly to the metal. The metals and their organic compounds have considerable catalytic activity. *See also* TRANSITION ELEMENTS.

**Platonic hydrocarbons** Hydrocarbons that have the same molecular geometry as the five Platonic solids, i.e. the tetrahedron, the cube, the octahedron, the dodecahedron, and the icosahedron. These hydrocarbons have carbon atoms at the vertices of the polyhedra and single bonds along the edges. In fact, there are only two known Platonic hydrocarbons: cubane ($C_8H_8$) and dodecahedrane ($C_{20}H_{20}$). The hydrocarbon based on a tetrahedron ($C_4H_4$) does not exist because of angle strain, but substituted derivatives $C_4X_4$ are known. Angle strain is also the reason for the nonexistence of octahedrane ($C_6H_6$) and icosahedrane ($C_{12}H_{12}$).

**Platyhelminthes** A phylum of invertebrates comprising the flatworms, characterized by a flattened unsegmented body. The simple nervous system shows some concentration of cells at the head end. The mouth leads to a simple branched gut without an anus. Flatworms are hermaphrodite but self-fertilization is unusual. Many species are parasitic. The phylum contains the classes *Turbellaria (planarians), *Trematoda (flukes), and *Cestoda (tapeworms).

**Playfair, John** *See* HUTTON, JAMES.

**pleiomorphism** The existence of distinctly different forms during the life cycle of an individual, e.g. the caterpillar, pupa, and winged adult of a butterfly.

**Pleistocene** The third epoch of the *Neogene period. It extended from the end of the Pliocene, about 1.8 million years ago, to the beginning of the Holocene, (the present epoch), about 11 500 years ago. The Pleistocene is often known as the **Ice Age** as it was characterized by a series of glacials, in which ice margins advanced towards the equator, separated by interglacials when the ice retreated. *See also* ICE AGE.

**pleochroic** Denoting a crystal that appears to be of different colours, depending on the direction from which it is viewed. It is caused by polarization of light as it passes through an anisotropic medium.

**pleura (pleural membrane)** The double membrane that lines the thoracic cavity and covers the exterior surface of the lungs. It is a *serous membrane forming a closed sac, with a small space (the **pleural cavity**) between the two layers.

**plexus** A compact branching network of nerves or blood vessels, such as the **brachial plexus** – a network of spinal nerves that supply branches to the forelimbs in vertebrates. *See also* CHOROID PLEXUS.

**Pliocene** The second epoch of the *Neogene period. Preceded by the Miocene and followed by the Pleistocene, it extended from about 5 million years ago to 1.8 million years ago. Mammals similar to modern forms existed during the epoch and the australopithecines (*see* AUSTRALOPITHECUS), early forerunners of humans, appeared.

**plumbago** See CARBON.

**plumbane (lead(IV) hydride)** An extremely unstable gas, $PbH_4$, said to be formed by the action of acids on magnesium–lead alloys. It was first reported in 1924, although doubts have since been expressed about the existence of the compound. It demonstrates the declining stability of the hydrides in group 14. More stable organic derivatives are known; e.g. trimethyl plumbane, $(CH_3)_3PbH$.

**plumbate** A compound formed by reaction of lead oxides (or hydroxides) with alkali. The oxides of lead are amphoteric (weakly acidic) and react to give plumbate ions. With the lead(IV) oxide, reaction with molten alkali gives the plumbate(IV) ion

$$PbO_2 + 2OH^- \rightarrow PbO_3^{2-} + H_2O$$

In fact, various ions are present in which the lead is bound to hydroxide groups, the principal one being the hexahydroxoplumbate(IV) ion $Pb(OH)_6^{2-}$. This is the negative ion present in crystalline 'trihydrates' of the type $K_2PbO_3.3H_2O$. Lead(II) oxide gives the trihydroxoplumbate(II) ion in alkaline solutions

$$PbO(s) + OH^-(aq) + H_2O(l) \rightarrow$$
$$Pb(OH)_3^{2-}(aq)$$

Plumbate(IV) compounds were formerly referred to as **orthoplumbates** ($PbO_4^{4-}$) or **metaplumbates** ($PbO_3^{2-}$). Plumbate(II) compounds were called **plumbites**.

**plumbic compounds** Compounds of lead in its higher (+4) oxidation state; e.g. plumbic oxide is lead(IV) oxide, $PbO_2$.

**plumbite** See PLUMBATE.

**plumbous compounds** Compounds of lead in its lower (+2) oxidation state; e.g. plumbous oxide is lead(II) oxide, PbO.

**plumule** 1. (in zoology) A *down feather. 2. (in botany) The part of a plant embryo that develops into the shoot system. It consists of the stem apex and first leaves. In seedlings showing *epigeal germination the plumule grows above the soil surface together with the cotyledons; in seeds showing *hypogeal germination, the plumule alone emerges. Compare RADICLE.

**pluripotent** See STEM CELL.

**Pluto** A *trans-Neptunian object that is the second largest *dwarf planet so far known and is the tenth largest body in the *solar system. It was discovered in 1930 by Clyde Tombaugh (1906–97) and for 76 years was regarded as the ninth planet of the solar system. It is now classed as a *Kuiper belt object, the largest known at present. Its official designation is (134340) Pluto. Pluto orbits the sun at a mean distance of $5906.376 \times 10^6$ km, but its orbit is so eccentric that its *perihelion is $4436.825 \times 10^6$ km, well inside Neptune's orbit, and its aphelion is $7375.928 \times 10^6$ km. Pluto's orbital period is 248.09 years. With a diameter of 2300 km and a mass of about $1.3 \times 10^{22}$ kg, Pluto is smaller than the earth's moon. It appears to be half rock and half ice, the ice – frozen nitrogen, methane, and carbon monoxide – surrounding a rocky core. Pluto's average temperature has been measured at 43 K. It has three natural satellites: Charon (discovered in 1978), which at 1200 km in diameter is more than half Pluto's size, and the much smaller Nix and Hydra (both discovered in 2005).

**((()) SEE WEB LINKS**

• NASA's introduction to the planet and its satellites

**plutonium** Symbol Pu. A dense silvery radioactive metallic transuranic element belonging to the *actinoids; a.n. 94; mass number of most stable isotope 244 (half-life $7.6 \times 10^7$ years); r.d. 19.84; m.p. 641°C; b.p. 3232°C. Thirteen isotopes are known, by far the most important being plutonium–239 (half-life $2.44 \times 10^4$ years), which undergoes *nuclear fission with slow neutrons and is therefore a vital power source for *nuclear weapons and some *nuclear reactors. About 20 tonnes of plutonium are produced annually by the world's nuclear reactors, a detailed inventory of every gram of which is kept in order to prevent its military misuse. The element was first produced by Seaborg, McMillan, Kennedy, and Wahl in 1940.

**((()) SEE WEB LINKS**

• Information from the WebElements site

**pnicogens** See PNICTOGENS.

**pnictides** See PNICTOGENS.

**pnictogens (pnicogens)** The elements of group 15 of the periodic table, i.e. nitrogen (N), arsenic (As), antimony (Sb), and bismuth (Bi). The elements As, Sb, and Bi form a range of solid-state ternary compounds with interesting electrical and magnetic properties. Examples are $LaSiP_3$, $K_6Bi_2Sn_{23}$, and $Yb_{14}MnSb_{11}$. Compounds of this type are known as **pnictides**.

**pod**   *See* LEGUME.

**podcast**   A package of multimedia files that can be downloaded from the Internet and played on a computer or a mobile device. The first podcasts were of music tracks that could be played on MP3 players; the technique was subsequently generalized to include other audio content and then additional file types, in particular images and video. New podcasts are generally advertised by web feeds, then downloaded automatically to subscribers' computers and, if appropriate, transferred to portable players. Established broadcasting organizations are now making their television and radio output available as podcasts.

**podzol (podsol)**   A type of soil that forms under cool humid conditions where the vegetation is coniferous or heath. It is characteristic of the taiga zones of Eurasia and North America. The soil has an ashen-coloured upper layer (the A horizon) overlying a layer (the B horizon) containing iron or humus that has been leached down from the A horizon and redeposited.

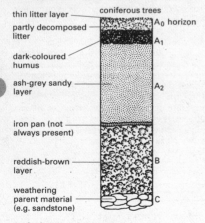

thin litter layer — $A_0$ horizon
coniferous trees
partly decomposed litter — $A_1$
dark-coloured humus
ash-grey sandy layer — $A_2$
iron pan (not always present)
reddish-brown layer — B
weathering parent material (e.g. sandstone) — C

**Podzol.** A podzol soil profile.

**poikilothermy**   The passive variation in the internal body temperature of an animal, which depends on the temperature of the environment. All animals except birds and mammals exhibit poikilothermy and are described as *ectotherms. Although unable to maintain a constant body temperature, they can respond to compensate for very low or very high temperatures. For example, the tis-

sue composition (especially cell osmotic pressure) can change to regulate the blood flow to peripheral tissues (and thus increase heat loss or heat absorption), and the animals can actively seek sun or shade. Seasonal changes in metabolism are usually under hormonal control. In particularly hot climates, ectotherms may undergo *aestivation to escape the heat. *Compare* HOMOIO-THERMY.

**point-contact transistor**   *See* TRANSISTOR.

**point defect**   *See* CRYSTAL DEFECT.

**point discharge**   *See* CORONA.

**point group**   The group formed by all the symmetry operations applied to a pattern arranged around a fixed point. The symmetry operations of molecules form point groups. There are 32 point groups, called **crystallographic point groups**, that have symmetries that are also compatible with the translational symmetries of crystals.

**point mutation (gene mutation)**   A change in the nucleotide sequence of the DNA within a gene; a gene in which such a change has occurred is known as a *mutant gene or allele. The DNA sequence can be altered in several ways; for example by *insertion, *substitution, *deletion, and *inversion. Point mutations can result in a misreading of the genetic code during the translation phase of protein synthesis and can change the order of amino acids making up a protein, which may or may not affect the function of that protein. *Compare* CHROMOSOME MUTATION; SINGLE NUCLEOTIDE POLYMORPHISM. *See also* MUTATION.

**poise**   A *c.g.s. unit of viscosity equal to the tangential force in dynes per square centimetre required to maintain a difference in velocity of one centimetre per second between two parallel planes of a fluid separated by one centimetre. 1 poise is equal to $10^{-1}$ N s m$^{-2}$.

**Poiseuille's equation**   An equation relating the volume flow rate, $V$, of a fluid through a cylindrical tube to the pressure difference, $p$, between the ends of the tube: $V = \pi p r^4/8 l \eta$, where $r$ is the radius and $l$ the length of the tube; $\eta$ is the viscosity of the fluid. It applies if the Reynolds number is less than 2000 and was first stated by Jean Louis Poiseuille (1799–1869).

**poison**   **1.** (in biology) Any substance that is

injurious to the health of a living organism.
**2.** (in chemistry) A substance that prevents
the activity of a catalyst. **3.** (in physics) A
substance that absorbs neutrons in a nuclear
reactor and therefore slows down the reaction. It may be added intentionally for this
purpose or may be formed as a fission product and need to be periodically removed.

**Poisson distribution** A probability distribution for a discrete random variable. It is
defined, for a variable ($r$) that can take values
in the range 0, 1, 2, ..., and has a mean value
μ, as

$$P(r) = e^{-\mu r}/r!$$

A binomial distribution with a small frequency of success $p$ in a large number $n$ of
trials can be approximated by a Poisson distribution with mean $np$. It is named after the
French mathematician and mathematical
physicist Siméon-Denis Poisson (1781–
1840). *See also* NORMAL DISTRIBUTION;
T-DISTRIBUTION.

**Poisson's ratio** The ratio of the lateral
strain to the longitudinal strain in a
stretched rod. If the original diameter of the
rod is $d$ and the contraction of the diameter
under stress is $\Delta d$, the lateral strain $\Delta d/d = s_d$;
if the original length is $l$ and the extension
under stress $\Delta l$, the longitudinal strain is $\Delta l/l$
= $s_l$. Poisson's ratio is then $s_d/s_l$. For steels the
value is between 0.28 and 0.30 and for aluminium alloys it is about 0.33. It was first introduced by Simeon Poisson (1781–1840).

**polar body** *See* OOGENESIS.

**polar compound** A compound that is either ionic (e.g. sodium chloride) or that has
molecules with a large permanent dipole
moment (e.g. water).

**polar coordinates** A system used in analytical geometry to locate a point $P$, with reference to two or three axes. The distance of $P$
from the origin is $r$, and the angle between
the $x$-axis and the **radius vector** $OP$ is θ, thus
in two-dimensional polar coordinates the
coordinates of $P$ are $(r,θ)$. If the *Cartesian
coordinates of $P$ are $(x,y)$ then, $x = r\cosθ$ and
$y = r\sinθ$.

In three dimensions the point $P$ may be
regarded as lying on the surface of a cylinder,
giving **cylindrical polar coordinates**, or on
the surface of a sphere, giving **spherical
polar coordinates**. In the former the coordinates of $P$ would be $(r,θ,z)$; in the latter they
would be $(r,θ,\phi)$ (see illustration).

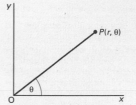

two-dimensional coordinates

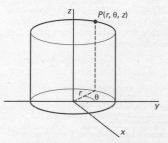

cylindrical polar coordinates

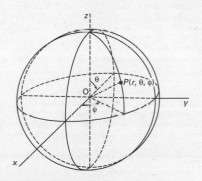

spherical polar coordinates

**Polar coordinates.**

**polarimeter (polariscope)** An instrument
used to determine the angle through which
the plane of polarization of plane-polarized
light is rotated on passing through an optically active substance. Essentially, a polarimeter consists of a light source, a
**polarizer** (e.g. a sheet of Polaroid) for producing plane-polarized light, a transparent
cell containing the sample, and an **analyser**.
The analyser is a polarizing material that can

be rotated. Light from the source is plane-polarized by the polarizer and passes through the sample, then through the analyser into the eye or onto a light-detector. The angle of polarization is determined by rotating the analyser until the maximum transmission of light occurs. The angle of rotation is read off a scale. Simple portable polarimeters are used for estimating the concentrations of sugar solutions in confectionary manufacture. *See also* PHOTOELASTICITY.

**polarity** 1. The property of a cell, tissue, or organism of being structurally and/or functionally different at opposite ends of its long axis. For example, plants consist of roots, which grow in the direction of the force of gravity, and stems, which grow away from the gravitational force (*see* GEOTROPISM). 2. The property of molecules of having an uneven distribution of electrons, so that one part has a positive charge and the other a negative charge.

**polarizability** Symbol $\alpha$. A measure of the response of a molecule to an external electric field. When a molecule is placed in an external electric field, the displacement of electric charge induces a dipole in the molecule (*see* ELECTRIC DISPLACEMENT). If the *electric field strength is denoted $E$ and the electrical dipole moment induced by this electric field $p$, the polarizability $\alpha$ is defined by $p = \alpha E$. To calculate the polarizability from first principles it is necessary to use the quantum mechanics of molecules. However, if regarded as a parameter, the polarizability $\alpha$ provides a link between microscopic and macroscopic theories as in the *Clausius–Mossotti equation and the *Lorentz–Lorenz equation.

**polarization** 1. The process of confining the vibrations of the vector constituting a transverse wave to one direction. In unpolarized radiation the vector oscillates in all directions perpendicular to the direction of propagation. *See* POLARIZATION OF LIGHT. 2. The formation of products of the chemical reaction in a *voltaic cell in the vicinity of the electrodes resulting in increased resistance to current flow and, frequently, to a reduction in the e.m.f. of the cell. *See also* DEPOLARIZATION. 3. The partial separation of electric charges in an insulator subjected to an electric field. 4. The separation of charge in a polar *chemical bond.

**polarization of light** The process of confining the vibrations of the electric vector of light waves to one direction. In unpolarized light the electric field vibrates in all directions perpendicular to the direction of propagation. After reflection or transmission through certain substances (*see* POLAROID) the electric field is confined to one direction and the radiation is said to be **plane-polarized light**. The plane of plane-polarized light can be rotated when it passes through certain substances (*see* OPTICAL ACTIVITY).

In **circularly polarized light**, the tip of the electric vector describes a circular helix about the direction of propagation with a frequency equal to the frequency of the light. The magnitude of the vector remains constant. In **elliptically polarized light**, the vector also rotates about the direction of propagation but the amplitude changes; a projection of the vector on a plane at right angles to the direction of propagation describes an ellipse. Circularly and elliptically polarized light are produced using a *retardation plate.

**polarizer** A device used to plane-polarize light (*see* POLARIZATION OF LIGHT). *Nicol prisms or *Polaroid can be used as polarizers. If a polarizer is placed in front of a source of unpolarized light, the transmitted light is plane-polarized in a specific direction. As the human eye is unable to detect that light is polarized, it is necessary to use an *analyser to detect the direction of polarization. **Crossing** a polarizer and analyser causes extinction of the light, i.e. if the plane of polarization of the polarizer and the plane of the analyser are perpendicular, no light is transmitted when the polarizer and analyser are combined. Both a polarizer and an analyser are components of a *polarimeter.

**polarizing angle** *See* BREWSTER'S LAW.

**polar molecule** A molecule that has a dipole moment; i.e. one in which there is some separation of charge in the *chemical bonds, so that one part of the molecule has a positive charge and the other a negative charge.

**polarography** An analytical technique having an electrochemical basis. A dropping-mercury electrode is used as the cathode along with a large nonpolarizable anode, and a dilute solution of the sample. The **dropping-mercury electrode** (**DME**) consists of a narrow tube through which mercury is slowly passed into the solution so as to form small drops at the end of the tube, which fall away. In this way the cathode can have a low surface area and be kept clean. A variable potential is applied to the cell and a plot of current against potential (a **polarogram**) made. As each chemical species is reduced at the cathode (in order of their electrode potentials) a step-wise increase in current is obtained. The height of each step is proportional to the concentration of the component. It is also possible to use an electrode formed from a small rotating disc in place of the dropping-mercury electrode. The technique is useful for detecting trace amounts of metals and for the investigation of solvated complexes.

**Polaroid** Trade name for a doubly refracting material that plane-polarizes unpolarized light passed through it. It consists of a plastic sheet strained in a manner that makes it birefringent by aligning its molecules. Sunglasses incorporating a Polaroid material absorb light that is vibrating horizontally – produced by reflection from horizontal surfaces – and thus reduce glare.

**polaron** A coupled electron–ion system that arises when an electron is introduced into the conduction band of a perfect ionic crystal and induces lattice polarization around itself.

**polar solvent** See SOLVENT.

**polar vector** A *vector that reverses its sign when the coordinate system is changed to a new system by a reflection in the origin (i.e. $x'_i = -x_i$). Compare AXIAL VECTOR.

**pole** 1. See MAGNETIC POLES; MAGNETIC MONOPOLE. 2. The *optical centre of a curved mirror.

**pollen** The mass of grains containing the male gametes of seed plants, which are produced in large numbers in the *pollen sacs. The pollen grains of insect-pollinated plants may be spiny or pitted and are usually larger than those of wind-pollinated plants, which are usually smooth and light. The pollen grain represents the male *gametophyte generation; it contains two male gamete nuclei. The wall of the mature pollen grain consists of the tough outer wall (**exine**) and the more delicate narrower **intine**. The latter gives rise to the *pollen tube. See also POLLINATION.

**pollen analysis** See PALYNOLOGY.

**pollen sac** The structure in seed plants in which pollen is produced. In angiosperms there are usually four pollen sacs in each *anther. In gymnosperms variable numbers of pollen sacs are borne on the microsporophylls that make up the male *cone.

**pollen tube** An outgrowth of a pollen grain, which transports the male gametes to the ovule. It will only grow if the pollen grain is compatible with the female tissue. In angiosperms, the pollen grain is deposited on the stigma and the pollen tube grows down through the style and into the ovule. In some conifers, e.g. *Pinus* (pines), the pollen tube penetrates the *nucellus but does not develop further until the following year, when the female part of the plant is mature. See also EMBRYO SAC; FERTILIZATION.

**pollex** The innermost digit on the forelimb of a tetrapod vertebrate. It contains two phalanges (see PENTADACTYL LIMB) and in humans and higher primates it is the thumb, which is opposable (i.e. capable of facing and touching the other digits) and gives the hand greater manipulating ability. In some mammals a pollex is absent. Compare HALLUX.

**pollination** The transfer of pollen from an anther (the male reproductive organ) to a stigma (the receptive part of the female reproductive organ), either of the same flower (**self-pollination**) or of a different flower of the same species (**cross-pollination**). Cross-pollination involves the action of a pollinating agent to effect transfer of the pollen (see ANEMOPHILY; ENTOMOPHILY; HYDROPHILY). See also FERTILIZATION; INCOMPATIBILITY.

**pollutant** Any substance, produced and released into the environment as a result of human activities, that has damaging effects on living organisms. Pollutants may be toxic substances (e.g. *pesticides) or natural constituents of the atmosphere (e.g. carbon dioxide) that are present in excessive amounts. See POLLUTION.

**pollution** An undesirable change in the physical, chemical, or biological characteristics of the natural environment, brought about by human activities. It may be harmful to human or nonhuman life. Pollution may affect the soil, rivers, seas, or the atmosphere (see AIR POLLUTION). There are two main classes of *pollutants: those that are **biodegradable** (e.g. sewage), i.e. can be rendered harmless by natural processes and need therefore cause no permanent harm if adequately dispersed or treated; and those that are **nonbiodegradable** (e.g. *heavy metals (such as lead) in industrial effluents and

*DDT and other chlorinated hydrocarbons used as pesticides), which eventually accumulate in the environment and may be concentrated in food chains. Other forms of pollution in the environment include noise (e.g. from jet aircraft, traffic, and industrial processes), thermal pollution (e.g. the release of excessive waste heat into lakes or rivers causing harm to wildlife), and light pollution (from street lights, buildings, etc., which can disorientate wildlife). Recent pollution problems include the disposal of radioactive waste; *acid rain; *photochemical smog; increasing levels of human waste; high levels of carbon dioxide and other greenhouse gases in the atmosphere (*see* GREENHOUSE EFFECT); damage to the *ozone layer by nitrogen oxides, *chlorofluorocarbons (CFCs), and *halons; and pollution of inland waters by agricultural *fertilizers and *sewage effluent, causing eutrophication (*see* EUTROPHIC). Attempts to contain or prevent pollution include strict regulations concerning factory emissions, the use of smokeless fuels, the banning of certain pesticides, the greater use of renewable energy sources, restrictions on the use of chlorofluorocarbons, and the introduction, in some countries, of catalytic converters to cut pollutants in car exhausts.

(((●))) **SEE WEB LINKS**
• The Blacksmith Institute's top ten worst pollution problems facing the world

**polonium** Symbol Po. A rare radioactive metallic element of group 16 (formerly VIB) of the periodic table; a.n. 84; r.a.m. 210; r.d. 9.32; m.p. 254°C; b.p. 962°C. The element occurs in uranium ores to an extent of about 100 micrograms per 1000 kilograms. It has over 30 isotopes, more than any other element. The longest-lived isotope is polonium–209 (half-life 103 years). Polonium has attracted attention as a possible heat source for spacecraft as the energy released as it decays is $1.4 \times 10^5$ J $kg^{-1}$ $s^{-1}$. It was discovered by Marie Curie in 1898 in a sample of pitchblende.

(((●))) **SEE WEB LINKS**
• Information from the WebElements site

**poly-** Prefix indicating a polymer, e.g. polyethene. Sometimes brackets are used in polymer names to indicate the repeated unit, e.g. poly(ethene).

**polyamide** A type of condensation polymer produced by the interaction of an amino group of one molecule and a carboxylic acid group of another molecule to give a protein-like structure. The polyamide chains are linked together by hydrogen bonding.

**polyatomic molecule** A molecule formed from several atoms (e.g. pyridine, $C_5H_5N$, or dinitrogen tetroxide, $N_2O_4$), as distinguished from diatomic and monatomic molecules.

**polybasic acid** An acid with more than one replaceable hydrogen atom. Examples include the dibasic sulphuric acid ($H_2SO_4$) and tribasic phosphoric(V) (orthophosphoric) acid ($H_3PO_4$). Replacement of all the hydrogens by metal atoms forms normal salts. If not all the hydrogens are replaced, *acid salts are formed.

**Polychaeta** A class of annelid worms in which each body segment has a pair of flattened fleshy lobes (**parapodia**) bearing numerous bristles (*chaetae). All polychaetes are aquatic and most of them are marine. They include the fanworms (*Sabella*), which construct tubes of sand, etc., in which they live; and the lugworms (*Arenicola*) and ragworms (*Nereis*), which burrow in sand or mud.

**polychlorinated biphenyl (PCB)** Any of a number of derivatives of biphenyl ($C_6H_5C_6H_5$) in which some of the hydrogen atoms on the benzene rings have been replaced by chlorine atoms. Polychlorinated biphenyls are used in the manufacture of certain polymers as electrical insulators. They are highly toxic and are suspected to be carcinogenic; their increasing use has caused concern because they have been shown to accumulate in the food chain.

**polychloroethene (PVC; polyvinyl chloride)** A tough white solid material, which softens with the application of a plasticizer, manufactured from chloroethene by heating in an inert solvent using benzoyl peroxide as an initiator, or by the free-radical mechanism initiated by heating chloroethene in water with potassium persulphate or hydrogen peroxide. The polymer is used in a variety of ways, being easy to colour and resistant to fire, chemicals, and weather.

**polychromatic radiation** Electromagnetic radiation that consists of a mixture of different wavelengths. This need not refer only to visible radiation. *Compare* MONO-CHROMATIC RADIATION.

**polycyclic** Denoting a compound that has two or more rings in its molecules. Polycyclic compounds may contain single rings (as in phenylbenzene, $C_6H_5.C_6H_5$) or fused rings (as in naphthalene, $C_{10}H_8$).

**polydioxoboric(III) acid** *See* BORIC ACID.

**polyembryony** 1. The formation of more than one embryo in a plant seed. Often one embryo develops from the fertilized egg cell, while the others have formed asexually from other tissues in the ovule. 2. The formation of more than one embryo from a single animal zygote. *Identical twins are produced in this way.

**polyene** An unsaturated hydrocarbon that contains two or more double carbon–carbon bonds in its molecule.

**polyester** A condensation polymer formed by the interaction of polyhydric alcohols and polybasic acids. Linear polyesters are saturated thermoplastics and linked by dipole–dipole attraction as the carbonyl groups are polarized. They are extensively used as fibres (e.g. **Terylene**). Unsaturated polyesters readily copolymerize to give thermosetting products. They are used in the manufacture of glass-fibre products. *See also* ALKYD RESIN.

**polyethene (polyethylene; polythene)** A flexible waxy translucent polyalkene thermoplastic made in a variety of ways producing a polymer of varying characteristics. In the ICI process, ethene containing a trace of oxygen is subjected to a pressure in excess of 1500 atmospheres and a temperature of 200°C. Low-density polyethene (r.d. 0.92) has a formula weight between 50 000 and 300 000, softening at a temperature around 110°C, while the high-density polyethene (r.d. 0.945–0.96) has a formula weight up to 3 000 000, softening around 130°C. The low-density polymer is less crystalline, being more atactic. Polyethene is used as an insulator; it is acid resistant and is easily moulded and blown. *See* PHILLIPS PROCESS; ZIEGLER PROCESS.

**polyethylene** *See* POLYETHENE.

**polygene** Any of a group of genes influencing a quantitative trait, e.g. height in humans. *See* QUANTITATIVE INHERITANCE.

**polygenic inheritance** *See* QUANTITATIVE INHERITANCE.

**polygon** A plane figure with a number of sides. In a **regular polygon** all the sides and internal angles are equal. For such a polygon with $n$ sides, the interior angle is $(180 - 360/n)$ degrees and the sum of the interior angles is $(180n - 360)$ degrees.

**polygon of forces** A polygon in which the sides represent, in magnitude and direction, all forces acting on a rigid body. The side required to close the polygon represents the resultant of a system of forces.

**polyhedron** A solid bounded by polygonal faces. In a **regular polyhedron** all the faces are congruent regular polygons. The cube is one of five possible regular polyhedrons. The others are the **tetrahedron** (four triangular faces), the **octahedron** (eight triangular faces), the **dodecahedron** (twelve pentagonal faces), and the **icosahedron** (twenty triangular faces).

**polyhydric alcohol** An *alcohol that has several hydroxyl groups per molecule.

**polymer** A substance having large molecules consisting of repeated units (the monomers). See Feature overleaf.

**((()) SEE WEB LINKS**
• Information about IUPAC nomenclature

**polymerase** Any enzyme that catalyses the elongation of a polymeric molecule. **RNA polymerases** catalyse the synthesis of RNA using as a template either an existing DNA strand (**DNA-dependent RNA polymerase**) or an RNA strand. **DNA polymerases** catalyse the elongation of a new DNA strand during DNA replication, using an existing DNA strand as template. **RNA-directed DNA polymerase** is more usually known as *reverse transcriptase.

**polymerase chain reaction (PCR)** A technique used to replicate a fragment of DNA so as to produce many copies of a particular DNA sequence. PCR is commonly employed as an alternative to *gene cloning as a means of amplifying genetic material for *DNA sequencing. The technique has also proved invaluable in forensic science, enabling amplification of minute traces of genetic material for *DNA profiling or for detecting microsatellite DNA (*see* REPETITIVE DNA). The two strands of the DNA are separated by heating and short sequences of a single DNA strand (**primers**) are added, together with a supply of free nucleotides and DNA *polymerase obtained from a bacterium that can withstand extreme heat. In a

## POLYMERS

Polymers are substances that have *macromolecules composed of many repeating units (known as 'mers'). A large number of naturally occurring substances are polymers including *rubber and many substances based on glucose, such as the polysaccharides *cellulose and *starch (in plants) and *glycogen (in animals). *Proteins, nucleic acids, and inorganic macromolecular substances, such as *silicates, are other examples.

### Synthetic polymers

One of the unique features of the chemistry of carbon is its ability to form long chains of atoms. This property is the basis of an important area of industrial chemistry concerned with the manufacture of polymeric materials with a variety of properties (*see* PLASTICS). The molecules in these materials are essentially long chains of atoms of various lengths. In some polymers, cross-linkage occurs between the chains. Synthetic polymers are formed by chemical reactions in which individual molecules (monomers) join together to form larger units (*see* POLYMERIZATION). Two types of polymer, **homopolymers** and **heteropolymers**, can be distinguished.

### Homopolymers

These are polymers formed from a single monomer. An example is *polyethene (polyethylene), which is made by polymerization of ethene ($CH_2:CH_2$). Typically such polymers are formed by *addition reactions involving unsaturated molecules. Other similar examples are *polypropene (polypropylene), polystyrene, and *polytetrafluoroethene (PTFE). Homopolymers may also be made by *condensation reactions (as in the case of *polyurethane).

Addition polymerization of ethene to form polyethene: a homopolymer

1,6-diaminohexane          hexanedioic acid

Condensation polymerization to form nylon: a heteropolymer

Formation of polymers

## Heteropolymers

These are also known as **copolymers**. They are made from two (or more) different monomers, which usually undergo a condensation reaction with the elimination of a simple molecule, such as water. A typical example is the condensation of 1,6-diaminohexane (hexamethylenediamine) with hexanedioic acid (adipic acid) to form nylon 6,6. The reaction occurs between the amine groups on the diaminohexane and the carboxyl groups on the hexanedioic acid, with elimination of water molecules (see diagram opposite).

The properties of a polymeric plastic can most easily be modified if it is a copolymer of two or more different monomers. A well-known example is ABS (acrylonitrile–butadiene–styrene) copolymer, commonly used for the body shells of computers and other electronic apparatus. Its properties can be preselected by varying the proportions of the component monomers.

## Stereospecific polymers

In both normal polyethene and nylon the polymer molecules take the form of long chains of various lengths with no regular arrangement of the subunits. Such polymers are said to be **atactic**. If the constituent subunits repeat along the chain in a regular way, a stereospecific polymer may result. The polymer may be **isotactic**, with a particular group always along the same side of the main chain, or **syndiotactic**, with the group alternating from side to side of the chain. Stereospecific polymerization can be performed by use of certain catalytic agents (*see* ZIEGLER PROCESS).

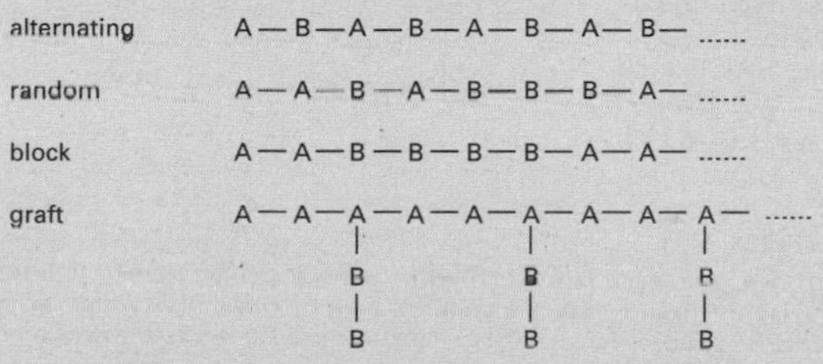

Types of copolymer depending on the arrangement of the monomers A and B

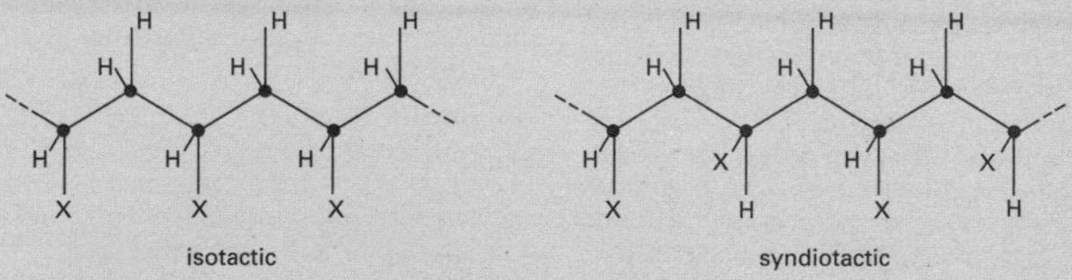

isotactic                     syndiotactic

Types of stereospecific polymer

Structure of polymers

series of heating and cooling cycles, the DNA sequence flanked by the primers doubles with each cycle and is thus rapidly amplified.

**polymerization** A chemical reaction in which molecules join together to form a polymer. If the reaction is an addition reaction, the process is **addition polymerization**; condensation reactions cause **condensation polymerization**, in which a small molecule is eliminated during the reaction. Polymers consisting of a single monomer are **homopolymers**; those formed from two different monomers are **copolymers**.

**polymethanal** A solid polymer of methanal, formed by evaporation of an aqueous solution of methanal.

**polymethylmethacrylate** A clear thermoplastic acrylic material made by polymerizing methyl methacrylate. The technical name is **poly(methyl 2-methylpropenoate)**. It is used in such materials as **Perspex**.

**polymorphism** 1. (in biology) The existence of two or more distinctly different forms (**morphs**) within a plant or animal species. An example is the *caste system of social insects, in which there are workers, drones, and queens. This is an **environmental polymorphism**, i.e. the differences are caused by environmental rather than genetic factors, in this case by the larvae receiving different types of food. There are also **heritable** or **genetic polymorphisms**. An example is the occurrence of sickle-cell disease, a genetic disease that principally affects Black populations of central Africa and is characterized by an abnormal form of the blood pigment haemoglobin and sickle-shaped red blood cells. Three different types of individual occur in such populations: those who have two genes (*AA*) for normal haemoglobin and therefore do not suffer from the disease; those with one normal and one abnormal gene (*AS*), who are described as having the **sickle-cell trait** and generally suffer no symptoms; and those with two abnormal genes (*SS*), who suffer a chronic and eventually fatal form of anaemia. Normally such a harmful gene would have been eliminated from the population by the process of natural selection, but it is maintained in this case because people with the sickle-cell trait are resistant to a severe form of malaria endemic in central Africa. *See also* RESTRICTION FRAGMENT LENGTH POLYMORPHISM; SINGLE NUCLEOTIDE POLYMORPHISM. 2. (in chemistry) The existence of chemical substances

in two (**dimorphism**) or more physical forms. *See* ALLOTROPY.

**polynomial** A mathematical expression containing three or more terms. It has the general form $a_0x^n + a_1x^{n-1} + a_2x^{n-2} + \ldots + a_n$, where $a_0$, $a_1$, etc., are constants and $n$ is the highest power of the variable, called the **degree** of the polynomial.

**polynucleotide** *See* NUCLEOTIDE.

**polyp** The sedentary stage in the life cycle of the *Cnidaria, consisting of a cylindrical body fixed at one end to a firm base and having a mouth surrounded by a ring of tentacles at the other. Some polyps (e.g. *Hydra*) are single; others (e.g. the corals and *Obelia*) form colonies. Polyps typically reproduce asexually by budding to form either new polyps or *medusae. The latter reproduce sexually giving rise to new polyps. Sea anemones are solitary polyps that reproduce sexually to form new polyps.

**polypeptide** A *peptide comprising 20 or more amino acids. Polypeptides that constitute proteins usually contain 100–300 amino acids. Shorter ones include certain antibiotics, e.g. gramicidin, and some hormones, e.g. *ACTH, which has 39 amino acids. The properties of a polypeptide are determined by the type and sequence of its constituent amino acids.

**polyphyletic** Denoting any group of organisms the members of which have originated from several different ancestors. An example is the group including all insectivorous animals. Polyphyletic groups are not natural groups and do not have any place in natural classifications. *Compare* MONOPHYLETIC.

**polyphyodont** Describing a type of dentition in which the teeth are continuously shed and replaced during the lifetime of the animal. Sharks and frogs have a polyphyodont dentition. *Compare* DIPHYODONT; MONOPHYODONT.

**polyploid** Describing a nucleus that contains more than two sets of chromosomes (*see* DIPLOID) or a cell or organism containing such nuclei. For example, *triploid plants have three sets of chromosomes and **tetraploid** plants have four. Polyploidy is far more common in plants than in animals; many crops, in particular, are polyploid (bread wheat, for example, is **hexaploid**, i.e. $6n$). It can be induced chemically with *colchicine. *See also* ALLOPOLYPLOID; AUTOPOLYPLOID.

**polypropene (polypropylene)** An isotactic polymer existing in both low and high formula-weight forms. The lower-formula-weight polymer is made by passing propene at moderate pressure over a heated phosphoric acid catalyst spread on an inert material at 200°C. The reaction yields the trimer and tetramer. The higher-formula-weight polymer is produced by passing propene into an inert solvent, heptane, which contains a trialkyl aluminium and a titanium compound. The product is a mixture of isotactic and atactic polypropene, the former being the major constituent. Polypropene is used as a thermoplastic moulding material.

**polypropylene** *See* POLYPROPENE.

**polyribosome (polysome)** An aggregate of ribosomes in association with a single messenger RNA molecule during the *translation process of protein synthesis. In eukaryotes, polyribosomes are attached to the surface of the rough endoplasmic reticulum and the outer membrane of the nucleus; in bacteria they are found free in the cytoplasm.

**polysaccharide (glycan)** Any of a group of carbohydrates comprising long chains of monosaccharide (simple sugar) molecules. **Homopolysaccharides** consist of only one type of monosaccharide; **heteropolysaccharides** contain two or more different types. Polysaccharides may have molecular weights of up to several million and are often highly branched. Some important examples are starch, glycogen, and cellulose.

**polysome** *See* POLYRIBOSOME.

**polyspermy** The entry of several sperms into the egg during fertilization although only one sperm nucleus actually fuses with the egg nucleus. Polyspermy occurs in animals with yolky eggs (e.g. birds).

**polystyrene** A clear glasslike material manufactured by free-radical polymerization of phenylethene (styrene) using benzoyl peroxide as an initiator. It is used as both a thermal and electrical insulator and for packing and decorative purposes.

**polysulphides** *See* SULPHIDES.

**polytetrafluoroethene (PTFE)** A thermosetting plastic with a high softening point (327°C) prepared by the polymerization of tetrafluoroethene under pressure (45–50 atmospheres). The reaction requires an initiator, ammonium peroxosulphate. The polymer has a low coefficient of friction and its 'anti-stick' properties are probably due to its helical structure with the fluorine atoms on the surface of an inner ring of carbon atoms. It is used for coating cooking utensils and nonlubricated bearings.

**polythene** *See* POLYETHENE.

**polythionate** A salt of a polythionic acid.

**polythionic acids** Oxo acids of sulphur with the general formula $HO.SO_2.S_n.SO_2.OH$, where $n = 0$–4. *See also* SULPHURIC ACID.

**polyurethane** A polymer containing the urethane group $-NH.CO.O-$, prepared by reacting di-isocyanates with appropriate diols or triols. A wide range of polyurethanes can be made, and they are used in adhesives, durable paints and varnishes, plastics, and rubbers. Addition of water to the polyurethane plastics turns them into foams.

**polyvinylacetate (PVA)** A thermoplastic polymer used in adhesives and coatings. It is made by polymerizing vinyl acetate ($CH_2$: $CHCOOCH_3$).

**polyvinyl chloride** *See* POLYCHLOROETHENE.

**polyyne** An unsaturated hydrocarbon that contains two or more triple carbon–carbon bonds in its molecule.

**pome** A type of fruit characteristic of apples and pears. The flesh of the fruit develops from the *receptacle of the flower, which completely encloses the fused carpels. After fertilization the carpels form the 'core' of the fruit, which contains the seeds. *See also* PSEUDOCARP.

**pons (pons Varolii)** A thick tract of nerve fibres in the brain that links the medulla oblongata to the midbrain. Its function is to relay impulses between different parts of the brain. The pons is named after its discoverer, the Italian anatomist C. Varoli (1543–75).

**POP** *See* PERSISTENT ORGANIC POLLUTANT.

**population** (in ecology) **1.** A group of individuals of the same species within a *community. The nature of a population is determined by such factors as density, *sex ratio, birth and death rates, emigration, and immigration. **2.** The total number of indi-

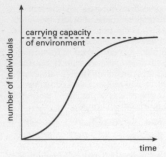 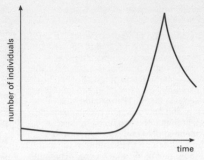

Sigmoid growth curve          J-shaped growth curve

**Population growth curves.**

viduals of a given species or other class of organisms in a defined area, e.g. the population of rodents in Britain.

**population dynamics** The study of the fluctuations that occur in the numbers of individuals in animal and plant populations and the factors controlling these fluctuations. An important distinction is maintained between those factors that are dependent on population density and have a stabilizing effect (e.g. food supply) and those that are independent of population density (e.g. catastrophes, such as flooding).

**population genetics** The study of the distribution of inherited *variation among a group of organisms of the same species. The potential for change depends on the sum total of alleles that are available to the organisms (the **gene pool**), and estimates of changes in allele frequency in a population give an indication of its response to a changing condition.

**population growth** The increase in a population that occurs when the *birth rate is higher than the *death rate, or when immigration exceeds emigration, or when a combination of these factors is present. A growth curve, obtained by plotting population size against time, is typically S-shaped (sigmoid) or J-shaped (see graphs). A sigmoid curve shows an initial phase of *exponential growth. The curve levels off when the environment has reached its *carrying capacity. A J-shaped growth curve shows an initial phase of exponential growth that ceases abruptly, with a sudden decrease in population numbers. This decrease may be caused by a number of factors, such as the

end of the life cycle of the prey or any other factor contributing to *environmental resistance that may suddenly take effect. *See also* BACTERIAL GROWTH CURVE.

**population inversion** *See* LASER.

**population type** A method of classifying stars as either population I or population II bodies, devised in 1944 by Walter Baade (1893–1960). Population I stars are the young metal-rich highly luminous stars found in the spiral arms of galaxies. Population II stars are older metal-deficient stars that occur in the centres of galaxies and in galactic halos.

**Porifera** The phylum of marine and freshwater invertebrates that comprise the sponges, which live permanently attached to rocks or other surfaces. The body of a sponge is hollow and consists basically of an aggregation of cells between which there is little nervous coordination. The body is supported by an internal skeleton of spicules of chalk, silica, or fibrous protein (bath sponges have protein skeletons). Undulipodium-bearing cells (**choanocytes**) cause water to flow in through openings in the body wall and out through openings at the top; food particles are filtered from the water by the choanocytes.

**((()))** SEE WEB LINKS
- Overview of the sponges, including their biology and classification, from the University of Michigan's Museum of Zoology Diversity Web

**porphyrin** Any of a group of organic pigments characterized by the possession of a cyclic group of four linked nitrogen-containing rings (a **tetrapyrrole** nucleus),

**Porphyrin.**

the nitrogen atoms of which are often coordinated to metal ions. Porphyrins differ in the nature of their side-chain groups. They include the *chlorophylls, which contain magnesium; and *haem, which contains iron and forms the *prosthetic group of haemoglobin, myoglobin, and the cytochromes.

**portal vein (portal circulation; portal system)** Any vein that collects blood from one network of capillaries and transports it directly to a second capillary network in another region of the body, without returning to the heart. *See* HEPATIC PORTAL SYSTEM.

**positive charge** *See* CHARGE.

**positive feedback** *See* FEEDBACK.

**positron** The antiparticle of the *electron. *See also* ANNIHILATION; ELEMENTARY PARTICLES; PAIR PRODUCTION.

**positron emission tomography (PET)** A noninvasive imaging technique that can produce three-dimensional pictures of certain biochemical changes within the body, such as areas of increased tissue metabolism. It is used for research in several areas, including neurology and pharmacology, and as a diagnostic technique in medicine, for example in detecting brain tumours. PET is based on the detection of photons produced by the decay of short-lived radioactive isotopes, such as fluorine-18 in labelled fluorodeoxyglucose (*see* LABELLING), injected into the body. The isotopes emit positrons, which almost immediately collide with electrons producing a pair of photons that travel in opposite directions. Simultaneously arriving photon pairs are detected by the scanner on either side of the body as flashes of light, and a computer calculates the point of their mutual origin within the body. By collating many thousands of such coincidence flashes, the computer creates a map showing where

the isotope is concentrated, which corresponds to locally increased blood flow or uptake as a consequence of elevated tissue activity. PET scans are often now integrated with images obtained by computed *tomography or magnetic resonance imaging (*see* NUCLEAR MAGNETIC RESONANCE) so that the biochemical information can be correlated with anatomy of the site.

**positronium** A bound state consisting of an electron and a positron. There are two types of positronium: **orthopositronium**, in which the spins of the two constituents are parallel, and **parapositronium**, in which the spins are anti-parallel. Both forms have brief existences, with orthopositronium decaying into three photons in about $1.5 \times 10^{-7}$ s and parapositronium decaying into two photons in the even shorter time of about $10^{-10}$ s. Positronium has a hydrogen-like spectrum, but with different values of the frequencies since a positron is much lighter than a proton.

**positron microscope** An instrument for studying solids, similar in principle to a scanning electron mircoscope, but using positrons rather than electrons. Scanning positron microscopes are used in investigating surfaces and defects.

**postcaval vein** *See* VENA CAVA.

**posterior 1.** Designating the part of an animal that is to the rear, i.e. that follows when the animal is moving. In humans and bipedal animals (e.g. kangaroos) the posterior surface is equivalent to the *dorsal surface. **2.** Designating the side of a flower or axillary bud that faces towards the flower stalk or main stem, respectively. *Compare* ANTERIOR.

**post-Newtonian approximation** An approximation used in the general theory of relativity in which the effects of relativity are regarded as small corrections to the Newtonian solution to the problem being considered. The post-Newtonian approximation is useful for problems in which the correction is small, as in considering relativistic effects on the motions of planets in the solar system. It cannot be used for systems such as black holes, which involve strong gravitational fields.

**postsynaptic membrane** The membrane at the end of a neuron that receives an impulse at a *synapse.

**potash** Any of a number of potassium compounds, such as the carbonate or the hydroxide.

**potash alum** *See* ALUMINIUM POTASSIUM SULPHATE; ALUMS.

**potash mica** *See* MUSCOVITE.

**potassamide** *See* POTASSIUM MONOXIDE.

**potassium** Symbol K. A soft silvery metallic element belonging to group 1 (formerly IA) of the periodic table (*see* ALKALI METALS); a.n. 19; r.a.m. 39.098; r.d. 0.86; m.p. 63.7°C; b.p. 774°C. The element occurs in seawater and in a number of minerals, such as sylvite (KCl), carnallite ($KCl.MgCl_2.6H_2O$), and kainite ($MgSO_4.KCl.3H_2O$). It is obtained by electrolysis. The metal has few uses but potassium salts are used for a wide range of applications. Potassium is an *essential element for living organisms. The potassium ion, $K^+$, is the most abundant cation in plant tissues, being absorbed through the roots and being used in such processes as protein synthesis. In animals the passage of potassium and sodium ions across the nerve-cell membrane is responsible for the changes of electrical potential that accompany the transmission of impulses. Chemically, it is highly reactive, resembling sodium in its behaviour and compounds. It also forms an orange-coloured superoxide, $KO_2$, which contains the $O_2^-$ ion. Potassium was discovered by Sir Humphry Davy in 1807.

**(⊕) SEE WEB LINKS**
• Information from the WebElements site

**potassium–argon dating** A *dating technique for certain rocks that depends on the decay of the radioisotope potassium–40 to argon–40, a process with a half-life of about $1.27 \times 10^{10}$ years. It assumes that all the argon–40 formed in the potassium-bearing mineral accumulates within it and that all the argon present is formed by the decay of potassium–40. The mass of argon–40 and potassium–40 in the sample is estimated and the sample is then dated from the equation:

$$^{40}Ar = 0.1102 \, ^{40}K(e^{\lambda t} - 1),$$

where $\lambda$ is the decay constant and $t$ is the time in years since the mineral cooled to about 300°C, when the $^{40}Ar$ became trapped in the crystal lattice. The method is effective for micas, feldspar, and some other minerals.

**potassium bicarbonate** *See* POTASSIUM HYDROGENCARBONATE.

**potassium bichromate** *See* POTASSIUM DICHROMATE.

**potassium bromide** A white or colourless crystalline solid, KBr, slightly hygroscopic and soluble in water and very slightly soluble in ethanol; cubic; r.d. 2.75; m.p. 734°C; b.p. 1435°C. Potassium bromide may be prepared by the action of bromine on hot potassium hydroxide solution or by the action of iron(III) bromide or hydrogen bromide on potassium carbonate solution. It is used widely in the photographic industry and is also used as a sedative. Because of its range of transparency to infrared radiation, KBr is used both as a matrix for solid samples and as a prism material in infrared spectroscopy.

**potassium carbonate (pearl ash; potash)** A translucent (granular) or white (powder) deliquescent solid known in the anhydrous and hydrated forms. $K_2CO_3$ (monoclinic; r.d. 2.4; m.p. 891°C) decomposes without boiling. $2K_2CO_3.3H_2O$ (monoclinic; r.d. 2.04) dehydrates to $K_2CO_3.H_2O$ above 100°C and to $K_2CO_3$ above 130°C. It is prepared by the Engel–Precht process in which potassium chloride and magnesium oxide react with carbon dioxide to give the compound **Engel's salt**, $MgCO_3.KHCO_3.4H_2O$. This is decomposed in solution to give the hydrogencarbonate, which can then be calcined to $K_2CO_3$. Potassium carbonate is soluble in water (insoluble in alcohol) with significant hydrolysis to produce basic solutions. Industrial uses include glasses and glazes, the manufacture of soft soaps, and in dyeing and wool finishing. It is used in the laboratory as a drying agent.

**potassium chlorate** A colourless crystalline compound, $KClO_3$, which is soluble in water and moderately soluble in ethanol; monoclinic; r.d. 2.32; m.p. 356°C; decomposes above 400°C giving off oxygen. The industrial route to potassium chlorate involves the fractional crystallization of a solution of potassium chloride and sodium chlorate but it may also be prepared by electrolysis of hot concentrated solutions of potassium chloride. It is a powerful oxidizing agent finding applications in weedkillers and disinfectants and, because of its ability to produce oxygen, it is used in explosives, pyrotechnics, and matches.

**potassium chloride** A white crystalline solid, KCl, which is soluble in water and very slightly soluble in ethanol; cubic; r.d. 1.98;

m.p. 772°C; sublimes at 1500°C. Potassium chloride occurs naturally as the mineral **sylvite** (KCl) and as **carnallite** (KCl.MgCl$_2$. 6H$_2$O); it is produced industrially by fractional crystallization of these deposits or of solutions from lake brines. It has the interesting property of being more soluble than sodium chloride in hot water but less soluble in cold. It is used as a fertilizer, in photography, and as a source of other potassium salts, such as the chlorate and the hydroxide. It has low toxicity.

**potassium chromate** A bright yellow crystalline solid, K$_2$CrO$_4$, soluble in water and insoluble in alcohol; rhombic; r.d. 2.73; m.p. 968.3°C; decomposes without boiling. It is produced industrially by roasting powdered chromite ore with potassium hydroxide and limestone and leaching the resulting cinder with hot potassium sulphate solution. Potassium chromate is used in leather finishing, as a textile mordant, and in enamels and pigments. In the laboratory it is used as an analytical reagent and as an indicator. Like other chromium(III) compounds it is toxic when ingested or inhaled.

**potassium chromium sulphate (chrome alum)** A violet or ruby-red crystalline solid, K$_2$SO$_4$.Cr$_2$(SO$_4$)$_3$.24H$_2$O, that is soluble in water and insoluble in ethanol; cubic or octahedral; r.d. 1.826; m.p. 89°C; loses 10H$_2$O at 100°C, 12H$_2$O at 400°C. Six water molecules surround each of the chromium(III) ions and the remaining ones are hydrogen bonded to the sulphate ions. Like all alums, the compound may be prepared by mixing equimolar quantities of the constituent sulphates. *See* ALUMS.

**potassium cyanide (cyanide)** A white crystalline or granular deliquescent solid, KCN, soluble in water and in ethanol and having a faint characteristic odour of almonds (due to hydrolysis forming hydrogen cyanide at the surface); cubic; r.d. 1.52; m.p. 634°C. It is prepared industrially by the absorption of hydrogen cyanide in potassium hydroxide. The compound is used in the extraction of silver and gold, in some metal-finishing processes and electroplating, as an insecticide and fumigant (source of HCN), and in the preparation of cyanogen derivatives. In the laboratory it is used in analysis, as a reducing agent, and as a stabilizing *lig-and for low oxidation states. The salt itself is highly toxic and aqueous solutions of potassium cyanide are strongly hydrolysed to give

rise to the slow release of equally toxic hydrogen cyanide gas.

**potassium dichromate (potassium bichromate)** An orange-red crystalline solid, K$_2$Cr$_2$O$_7$, soluble in water and insoluble in alcohol; monoclinic or triclinic; r.d. 2.68; monoclinic changes to triclinic at 241.6°C; m.p. 396°C; decomposes above 500°C. It is prepared by acidification of crude potassium chromate solution (the addition of a base to solutions of potassium dichromate reverses this process). The compound is used industrially as an oxidizing agent in the chemical industry and in dyestuffs manufacture, in electroplating, pyrotechnics, glass manufacture, glues, tanning, photography and lithography, and in ceramic products. Laboratory uses include application as an analytical reagent and as an oxidizng agent. Potassium dichromate is toxic and considered a fire risk on account of its oxidizing properties.

**potassium dioxide** *See* POTASSIUM SUPEROXIDE.

**potassium hydride** A white or greyish white crystalline solid, KH; r.d. 1.43–1.47. It is prepared by passing hydrogen over heated potassium and marketed as a light grey powder dispersed in oil. The solid decomposes on heating and in contact with moisture and is an excellent reducing agent. Potassium hydride is a fire hazard because it produces hydrogen on reaction with water.

**potassium hydrogencarbonate (potassium bicarbonate)** A white crystalline solid, KHCO$_3$, soluble in water and insoluble in ethanol; r.d. 2.17; decomposes about 120°C. It occurs naturally as **calcinite** and is prepared by passing carbon dioxide into saturated potassium carbonate solution. It is used in baking, soft-drinks manufacture, and in CO$_2$ fire extinguishers. Because of its buffering capacity, it is added to some detergents and also used as a laboratory reagent.

**potassium hydrogentartrate (cream of tartar)** A white crystalline acid salt, HOOC(CHOH)$_2$COOK. It is obtained from deposits on wine vats (argol) and used in baking powders.

**potassium hydroxide (caustic potash; lye)** A white deliquescent solid, KOH, often sold as pellets, flakes, or sticks, soluble in water and in ethanol and very slightly soluble in ether; rhombic; r.d. 2.044; m.p. 360.4°C; b.p. 1320°C. It is prepared industrially by the electrolysis of concentrated potassium chloride solution but it can also be made by heating potassium carbonate or sulphate with slaked lime, $Ca(OH)_2$. It closely resembles sodium hydroxide but is more soluble and is therefore preferred as an absorber for carbon dioxide and sulphur dioxide. It is also used in the manufacture of soft soap, other potassium salts, and in Ni–Fe and alkaline storage cells. Potassium hydroxide is extremely corrosive to body tissues and especially damaging to the eyes.

**potassium iodate** A white crystalline solid, $KIO_3$, soluble in water and insoluble in ethanol; monoclinic; r.d. 3.9; m.p. 560°C. It may be prepared by the reaction of iodine with hot concentrated potassium hydroxide or by careful electrolysis of potassium iodide solution. It is an oxidizing agent and is used as an analytical reagent. Some potassium iodate is used as a food additive.

**potassium iodide** A white crystalline solid, KI, with a strong bitter taste, soluble in water, ethanol, and acetone; cubic; r.d. 3.13; m.p. 681°C; b.p. 1330°C. It may be prepared by the reaction of iodine with hot potassium hydroxide solution followed by separation from the iodate (which is also formed) by fractional crystallization. In solution it has the interesting property of dissolving iodine to form the triiodide ion $I_3^-$, which is brown. Potassium iodide is widely used as an analytical reagent, in photography, and also as an additive to table salt to prevent goitre and other disorders due to iodine deficiency.

**potassium manganate(VII) (potassium permanganate)** A compound, $KMnO_4$, forming purple crystals with a metallic sheen, soluble in water (intense purple solution), acetone, and methanol, but decomposed by ethanol; r.d. 2.70; decomposition begins slightly above 100°C and is complete at 240°C. The compound is prepared by fusing manganese(IV) oxide with potassium hydroxide to form the manganate and electrolysing the manganate solution using iron electrodes at about 60°C. An alternative route employs production of sodium manganate by a similar fusion process, oxidation with chlorine and sulphuric acid, then treatment with potassium chloride to crystallize the required product.

Potassium manganate(VII) is widely used as an oxidizing agent and as a disinfectant in a variety of applications, and as an analytical reagent.

**potassium monoxide** A grey crystalline solid, $K_2O$; cubic; r.d. 2.32; decomposition occurs at 350°C. It may be prepared by the oxidation of potassium metal with potassium nitrate. It reacts with ethanol to form potassium ethoxide ($KOC_2H_5$), and with liquid ammonia to form potassium hydroxide and **potassamide** ($KNH_2$).

**potassium nitrate (saltpetre)** A colourless rhombohedral or trigonal solid, $KNO_3$, soluble in water, insoluble in alcohol; r.d. 2.109; transition to trigonal form at 129°C; m.p. 334°C; decomposes at 400°C. It occurs naturally as **nitre** and may be prepared by the reaction of sodium nitrate with potassium chloride followed by fractional crystallization. It is a powerful oxidizing agent (releases oxygen on heating) and is used in gunpowder and fertilizers.

**potassium nitrite** A white or slightly yellow deliquescent solid, $KNO_2$, soluble in water and insoluble in ethanol; r.d. 1.91; m.p. 440°C; may explode at 600°C. Potassium nitrite is prepared by the reduction of potassium nitrate. It reacts with cold dilute mineral acids to give nitrous acid and is also able to behave as a reducing agent (if oxidized to the nitrate) or as an oxidizing agent (if reduced to nitrogen). It is used in organic synthesis because of its part in diazotization, and in detecting the presence of the amino groups in organic compounds.

**potassium permanganate** *See* POTASSIUM MANGANATE(VII).

**potassium sulphate** A white crystalline powder, $K_2SO_4$, soluble in water and insoluble in ethanol; rhombic or hexagonal; r.d. 2.66; m.p. 1069°C. It occurs naturally as **schönite** (Strassfurt deposits) and in lake brines, from which it is separated by fractional crystallization. It has also been produced by the **Hargreaves process**, which involves the oxidation of potassium chloride with sulphuric acid. In the laboratory it may be obtained by the reaction of either potassium hydroxide or potassium carbonate with sulphuric acid. Potassium sulphate is used in cements, in glass manufacture, as a food additive, and as a fertilizer (source of $K^+$) for chloride-sensitive plants, such as tobacco and citrus.

**potassium sulphide** A yellow-red or brown-red deliquescent solid, $K_2S$, which is soluble in water and in ethanol but insoluble in diethyl ether; cubic; r.d. 1.80; m.p. 840°C.

It is made industrially by reducing potassium sulphate with carbon at high temperatures in the absence of air. In the laboratory it may be prepared by the reaction of hydrogen sulphide with potassium hydroxide. The pentahydrate is obtained on crystallization. Solutions are strongly alkaline due to hydrolysis. It is used as an analytical reagent and as a depilatory. Potassium sulphide is generally regarded as a hazardous chemical with a fire risk; dusts of $K_2S$ have been known to explode.

**potassium sulphite** A white crystalline solid, $K_2SO_3$, soluble in water and very sparingly soluble in ethanol; r.d. 1.51; decomposes on heating. It is a reducing agent and is used as such in photography and in the food and brewing industries, where it prevents oxidation.

**potassium superoxide (potassium dioxide)** A yellow paramagnetic solid, $KO_2$, produced by burning potassium in an excess of oxygen; it is very soluble (by reaction) in water, soluble in ethanol, and slightly soluble in diethyl ether; m.p. 380°C. When treated with cold water or dilute mineral acids, hydrogen peroxide is obtained. The compound is a powerful oxidizing agent and on strong heating releases oxygen with the formation of the monoxide, $K_2O$.

**potential barrier** A region containing a maximum of potential that prevents a particle on one side of it from passing to the other side. According to classical theory a particle must possess energy in excess of the height of the potential barrier to pass it. However, in quantum theory there is a finite probability that a particle with less energy will pass through the barrier (*see* TUNNEL EFFECT). A potential barrier surrounds the atomic nucleus and is important in nuclear physics; a similar but much lower barrier exists at the interface between *semiconductors and metals and between differently doped semiconductors. These barriers are important in the design of electronic devices.

**potential difference** *See* ELECTRIC POTENTIAL.

**potential divider** *See* VOLTAGE DIVIDER.

**potential energy** *See* ENERGY.

**potentiometer** 1. *See* VOLTAGE DIVIDER. 2. An instrument for measuring, comparing, or dividing small potential differences. A typical example of its use is the measurement of

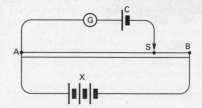

**Potentiometer.**

the e.m.f. ($E_1$) of a cell by comparing it with the e.m.f. ($E_2$) of a standard cell. In this case a circuit is set up as illustrated, in which AB is a wire of uniform resistance and S is a sliding contact onto this wire. An accumulator X maintains a steady current through the wire. To measure the e.m.f. of a cell C, it is connected up as shown in the diagram and the sliding contact moved until the e.m.f. of C exactly balances the p.d. from the accumulator, as indicated by a zero reading on the galvanometer G. If the length AS is then $l_1$, the value of $E_1$ is given by $E_1/E_2 = l_1/l_2$, where $l_2$ is the length AS when the standard cell is used as the cell C.

**potentiometric titration** A titration in which the end point is found by measuring the potential on an electrode immersed in the reaction mixture.

**potometer** An apparatus used to measure the rate of water loss from a shoot (*see* TRANSPIRATION) under natural or artificial conditions.

**pound** The unit of mass in the *f.p.s. system of units defined as 0.453 592 37 kilogram. Before 1963 it was defined in terms of a platinum cylinder called the Imperial Standard Pound.

**poundal** The unit of force in the *f.p.s. system of units equal to the force required to impart to a mass of one pound an acceleration of one foot per second per second.

**Pourbaix diagram** A diagram showing how oxidation–reduction behaviour for compounds of a given element depends on pH.

**powder metallurgy** A process in which powdered metals or alloys are pressed into a variety of shapes at high temperatures. The process started with the pressing of powdered tungsten into incandescent lamp filaments in the first decade of the 20th century and is now widely used for making self-lubricating bearings and cemented tungsten carbide cutting tools.

The powders are produced by atomization of molten metals, chemical decomposition of a compound of the metal, or crushing and grinding of the metal or alloy. The parts are pressed into moulds at pressures ranging from $140 \times 10^6$ Pa to $830 \times 10^6$ Pa after which they are heated in a controlled atmosphere to bond the particles together (*see* SINTERING).

**powder method** *See* DEBYE–SCHERRER METHOD.

**power** **1.** (in physics) Symbol $P$. The rate at which work is done or energy is transferred. In SI units it is measured in watts (joules per second). *See also* HORSEPOWER. **2.** (in mathematics) The number of times a quantity is multiplied; e.g. $x^5$ is the fifth power of $x$. A **power series** is one in which the power of the variable increases with each term, e.g. $a_0 + a_1 x + a_2 x^2 + a_3 x^3 + \ldots + a_n x^n$.

**power factor** *See* ELECTRIC POWER.

**power reactor** A *nuclear reactor designed to produce electrical power.

**poxvirus** One of a group of DNA-containing viruses, often enclosed in an outer membrane, that typically produce skin lesions in vertebrates. They include the viruses causing smallpox (variola), cowpox (vaccinia), and myxomatosis (myxoma). Some poxviruses produce tumours.

**Poynting vector** Symbol $S$. A vector that gives a measure of the flow of energy in an electromagnetic field. It is given by the vector product $S = E \times H$, where $E$ is the electric field strength and $H$ is the magnetic field strength. It was discovered by the British physicist John Henry Poynting (1852–1914) in 1884.

**praseodymium** Symbol Pr. A soft silvery metallic element belonging to the *lanthanoids; a.n. 59; r.a.m. 140.91; r.d. 6.773; m.p. 931°C; b.p. 3512°C. It occurs in bastnasite and monazite, from which it is recovered by an ion-exchange process. The only naturally occurring isotope is praseodymium–141, which is not radioactive; however, fourteen radioisotopes have been produced. It is used in mischmetal, a rare-earth alloy containing 5% praseodymium, for use in lighter flints. Another rare-earth mixture containing 30% praseodymium is used as a catalyst in cracking crude oil. The element was discovered by Carl von Welsbach (1856–1929) in 1885.

**(((•))) SEE WEB LINKS**
- Information from the WebElements site

**preamplifier** An *amplifier in a radio, record player, etc., providing a first stage of amplification. It is usually located close to the signal source (i.e. the aerial or pick-up) and the signal is then transmitted by cable to the main amplifier. Preamplification at this early stage improves the signal-to-noise ratio of the whole system.

**Precambrian** Describing the time from the formation of the earth, nearly 5 billion years ago, to the beginning of the Cambrian period, some 542 million years ago. The term 'Precambrian' is no longer used for a specific geological time interval but remains as a general adjective. Precambrian time is now divided into three eons: *Hadean, *Archaean, and *Proterozoic. Fossils are rare, although *stromatolites indicate that there were flourishing populations of bacteria. However, subsequent metamorphism of Precambrian rocks makes correlation of rocks and events extremely difficult. The largest areas of exposed Precambrian rocks are the shield areas, such as the Canadian (Laurentian) Shield and the Baltic Shield.

**precaval vein** *See* VENA CAVA.

**precessional motion** A form of motion that occurs when a torque is applied to a rotating body in such a way that it tends to change the direction of its axis of rotation. It arises because the resultant of the angular velocity of rotation and the increment of angular velocity produced by the torque is an angular velocity about a new direction; this commonly changes the axis of the applied torque and leads to sustained rotation of the original axis of rotation.

A spinning top, the axis of which is not exactly vertical, has a torque acting on it as a result of gravity. Instead of falling over, the top precesses about a vertical line through the pivot. The earth also experiences a torque and undergoes a slow precession, primarily as a result of the gravitational attraction of the sun and the moon on its equatorial bulge (*see* PRECESSION OF THE EQUINOXES).

**precession of the equinoxes** The slow westward motion of the *equinoxes about the ecliptic as a result of the earth's *precessional motion. The equinoxes move round the ecliptic with a period of 25 800 years.

**precipitate** A suspension of small solid particles produced in a liquid by chemical reaction.

**precipitation** 1. All liquid and solid forms of water that are deposited from the atmosphere; it includes rain, drizzle, snow, hail, dew, and hoar frost. 2. The formation of a precipitate.

**precipitin** Any *antibody that combines with its specific soluble *antigen to form a precipitate. The term is sometimes applied to the precipitate itself. *See also* AGGLUTINATION.

**precursor** A compound that leads to another compound in a series of chemical reactions.

**predation** An interaction between two populations of animals in which one (the **predator**) hunts, captures, and kills the other (the **prey**) for food. Predator–prey relationships form important links in many food chains. They are also important in regulating population sizes of both predator and prey, especially when the predator relies on a single prey species. The term predation is also used, more loosely, for any feeding relationship in which an organism feeds on any other living organism (animal or plant).

**predator** An animal that obtains its food by *predation. All predators are *carnivores, although not all carnivores are predators.

**pregnancy** *See* GESTATION.

**preimplantation genetic diagnosis (PGD)** The screening of early embryos for disease-causing genes to enable the selection of 'healthy' embryos. The technique is used in conjunction with *in vitro* fertilization, which typically yields a number of embryos. A single cell is removed from an eight-stage embryo and subjected to genetic testing; for example, it may be tested for a specific disease allele. If the results are satisfactory, the embryo is implanted in the mother's uterus, and development proceeds. Removal of a single cell at this stage does not affect the embryo's subsequent development. PGD can help especially when couples who are being treated for fertility problems also have a history of genetic disease. However, use of PGD can be extended in nontherapeutic ways, such as choosing a baby's sex or selecting particular desired traits to

produce so-called 'designer babies'. These highly controversial applications of PGD are prohibited in certain countries, including the UK.

**premolar** A broad ridged tooth in mammals that is situated behind the *canine teeth (when present) and in front of the *molars. Premolars are adapted for grinding and chewing food and are present in both the deciduous and permanent dentitions.

**premutation** A gene variant (allele) that produces a normal individual but is predisposed to become a full mutation in subsequent generations. Genetic analysis of human families has revealed the involvement of premutations in several genetic diseases. For example, in the gene for Huntington's disease, normal individuals have a string of 6 to 39 CAG repeats near the start of the coding sequence. In individuals with the disease, this region extends to 36–180 CAG repeats. Individuals with CAG repeats in the low 30s have a premutation for Huntington's disease; this region of the gene is amplified during meiosis to become an abnormal allele, sufficient to produce the disease in that individual's offspring.

**preons** Hypothetical entities that are postulated as being 'building blocks' of quarks and leptons. There is no experimental evidence for preons but the idea has considerable theoretical appeal. It is expected that evidence for preons would only be obtained at much higher energies than are available from present accelerators.

**presbyopia** A loss of accommodation that normally develops in human eyes over the age of 45–50 years. Vision of distant objects remains unchanged but accommodation of the eye to near objects is reduced as a result of loss of elasticity in the lens of the eye. The defect is corrected by reading glasses using weak converging lenses.

**pressure** The force acting normally on unit area of a surface or the ratio of force to area. It is measured in *pascals in SI units. **Absolute pressure** is pressure measured on a gauge that reads zero at zero pressure rather than at atmospheric pressure. **Gauge pressure** is measured on a gauge that reads zero at atmospheric pressure.

**pressure gauge** Any device used to measure pressure. Three basic types are in use: the liquid-column gauge (e.g. the mercury *barometer and the *manometer), the expanding-element gauge (e.g. the *Bourdon gauge and the aneroid *barometer), and the electrical transducer. In the last category the *strain gauge is an example. Capacitor pressure gauges also come into this category. In these devices, the pressure to be measured displaces one plate of a capacitor and thus alters its capacitance.

**pressurized-water reactor** *See* NUCLEAR REACTOR.

**presumptive** Describing embryonic tissue that is not yet *determined but which will eventually develop into a certain kind of tissue by virtue of its position in the embryo.

**presumptive test** A simple test for a given substance using a reagent that changes colour when mixed with the substance under investigation. Presumptive tests are not definitive and further confirmatory tests are always required. They are used extensively in forensic science. Examples are the Duquenois–Levine test for marijuana and Scott's test for cocaine. In general analytical chemistry, presumptive tests are often called **spot tests**.

**presynaptic membrane** The membrane of a neuron that releases neurotransmitter into the synaptic cleft between nerve cells (*see* SYNAPSE).

**Prévost's theory of exchanges** A body emits and absorbs radiant energy at equal rates when it is in equilibrium with its surroundings. Its temperature then remains constant. If the body is not at the same temperature as its surroundings there is a net flow of energy between the surroundings and the body because of unequal emission and absorption. The theory was proposed by Pierre Prévost (1751–1839) in 1791.

**prey** An animal that is a source of food for a predator. *See* PREDATION.

**Pribnow box** A *consensus sequence of nucleotides – TATAAT – occurring in the promoter region of prokaryote genes (*see* OPERON) about 10 nucleotides before the start of transcription. The predominance of adenine and thymine bases means that hydrogen bonding between the two DNA strands in this region is relatively weak, enabling the strands to be separated more easily to permit transcription by RNA polymerase. *See also* TATA BOX.

**prickle** A hard sharp protective outgrowth, many of which may cover the surface of a plant. It contains cortical and vascular tissue and is not regarded as an epidermal outgrowth. *Compare* SPINE; THORN.

**Priestley, Joseph** (1733–1804) British chemist, who in 1755 became a Presbyterian minister. In Leeds, in 1767, he experimented with carbon dioxide ('fixed air') from a nearby brewery; with it he invented soda water. He moved to a ministry in Birmingham in 1780, and in 1791 his revolutionary views caused a mob to burn his house, as a result of which he emigrated to the USA in 1794. In the early 1770s he experimented with combustion and produced the gases hydrogen chloride, sulphur dioxide, and dinitrogen oxide (nitrous oxide). In 1774 he isolated oxygen (*see also* LAVOISIER, ANTOINE LAURENT).

**primary** Any celestial object that has one or more other objects in direct orbit around it. For example, the sun is the primary for the bodies in the *solar system, while the earth is the moon's primary.

**primary alcohol** *See* ALCOHOLS.

**primary amine** *See* AMINES.

**primary cell** A *voltaic cell in which the chemical reaction producing the e.m.f. is not satisfactorily reversible and the cell cannot therefore be recharged by the application of a charging current. *See* DANIELL CELL; LECLANCHÉ CELL; WESTON CELL; MERCURY CELL. *Compare* SECONDARY CELL.

**primary colour** Any one of a set of three coloured lights that can be mixed together to give the sensation of white light as well as approximating all the other colours of the spectrum. An infinite number of such sets exists, the condition being that none of the individual colours of a set should be able to be matched by mixing the other two; however, unless the colours are both intense and very different the range that they can match well will be limited. The set of primary colours most frequently used is red, green, and blue. *See also* COLOUR.

**primary consumer** *See* CONSUMER.

**primary growth** The increase in size of shoots and roots of plants that results from the activity of the *apical (tip) meristems and subsequent expansion of the cells produced. The tissues thus produced are called **primary tissues** and the resultant plant parts constitute the **primary plant body**. *Compare* SECONDARY GROWTH.

**primary producer**  *See* PRODUCER.

**primary structure**  *See* PROTEIN.

**primary winding**  The winding on the input side of a *transformer or *induction coil. *Compare* SECONDARY WINDING.

**Primates**  An order of mammals that includes the monkeys, apes, and humans. Primates evolved from arboreal insectivores 63–70 million years ago. They are characterized by thumbs and big toes that are opposable (i.e. capable of facing and touching the other digits), which permits manual dexterity, and forward-facing eyes allowing *binocular vision. The brain, particularly the cerebrum, is relatively large and well-developed, accounting for the intelligence and quick reactions of these mammals. The young are usually produced singly and undergo a long period of growth and development to the adult form.

(((●))) SEE WEB LINKS
• Wide-ranging account of primate characteristics and taxonomy, compiled by Dennis O'Neil, Palomar College, California

**primordium**  A group of cells that represents the initial stages in development of a plant organ. Root and shoot primordia are present in a young plant embryo while leaf primordia (or **leaf buttresses**) are seen as small bulges just below the shoot apex.

**principal axis**  *See* OPTICAL AXIS.

**principal focus**  A point through which rays close to and parallel to the axis of a lens or spherical mirror pass, or appear to pass, after refraction or reflection. A mirror has one principal focus, a lens has a principal focus on both sides.

**principal plane**  The plane that is perpendicular to the optical axis of a lens and that passes through the optical centre. A thick lens has two principal planes, each passing through a *principal point.

**principal point**  Either of two points on the principal axis of a thick lens from which simply related distances can be measured, as from the optical centre of a thin lens.

**principle of parsimony**  The principle that the most acceptable explanation of an occurrence, phenomenon, or event is the simplest, involving the fewest entities, assumptions, or changes. In phylogenetics, for example, the preferred tree showing evolutionary relationships between species, molecules, or other entities is the one that requires the least amount of evolutionary change, that is, maximum parsimony.

**principle of superposition**  The resultant displacement at any point in a region through which two waves of the same type pass is the algebraic sum of the displacements that the two would separately produce at that point. Both waves leave the region of superposition unaltered.

**printed circuit**  An electronic circuit consisting of a conducting material deposited (printed) onto the surface of an insulating sheet. These devices are now common in all types of electronic equipment, facilitating batch production and eliminating the unreliability of the hand-soldered joint.

**printer**  A device for producing a printed version of text (and, sometimes, pictures) from a computer. There are several types. Impact printers work on the same principle as a typewriter, in which a ribbon is hit by a surface embossed with the character. A **line printer** produces a whole line of text at a time. In this device the characters are held on a row of spinning cylinders. In a **daisy-wheel printer**, the characters are held on the ends of a series of arms radiating from the centre of a wheel. A **dot-matrix printer** forms the image of each character from a rectangular matrix of dots. An **inkjet printer** works by spraying fine jets of quick-drying ink onto the paper. A **laser printer** uses a xerographic technique in which a photosensitive plate is scanned by a low-power laser.

**prion**  An abnormal form of a normal cell protein (PrP) found in the brain of mammals that is believed to be the agent responsible for the diseases scrapie in sheep, *bovine spongiform encephalopathy (BSE) in cattle, and *Creutzfeldt–Jakob disease in humans. Produced by mutation of the normal PrP gene, the abnormal prion protein interacts with normal protein, causing it to accumulate in the brain and progressively damage and destroy brain cells. It can be transmitted to other individuals of the same or closely related species, by injection or ingestion of infected tissue, and some forms appear to be transmissible between species that are not closely related (e.g. between cattle and humans).

**prism**  1. (in mathematics) A polyhedron with two parallel congruent polygons as bases and parallelograms for all other faces.

A **triangular prism** has triangular bases.
**2.** (in optics) A block of glass or other transparent material, usually having triangular bases. Prisms have several uses in optical systems: they can be used to deviate a ray, to disperse white light into the visible spectrum, or to erect an inverted image (*see* BINOCULARS). Prisms of other materials are used for different kinds of radiation. *See also* NICOL PRISM; WOLLASTON PRISM.

**prismane** A saturated hydrocarbon, $C_6H_6$, in which the six carbon atoms are arranged at the corners of a triangular prism. The structure was suggested in 1869 by Albert Ladenburg as a possible structure for benzene (since referred to as **Ladenburg benzene**). The actual compound was synthesized by T. J. Katz and N. Acton in 1973. Hexamethylprismane, in which the carbon atoms are linked to methyl groups rather than hydrogen atoms, was synthesized in 1966.

**private key** *See* PUBLIC KEY ENCRYPTION.

**probability** The likelihood of a particular event occurring. If there are $n$ equally likely outcomes of some experiment, and $a$ ways in which event E could occur, then the probability of event E is $a/n$. For instance, if a die is thrown there are 6 possible outcomes and 3 ways in which an odd number may occur. The probability of throwing an odd number is $3/6 = 1/2$.

**Proboscidea** The order of mammals that comprises the elephants. They are herbivorous, with a muscular trunk (*proboscis) used for drinking, bathing, and collecting food. The tusks are continuously growing upper incisors and the enormous ridged molar teeth are produced in sequence to replace worn teeth throughout life. The order, which evolved in the Eocene epoch, was formerly much larger and more widespread than it is today and included the extinct mammoths. There are only two species of modern elephants: the African and Indian species.

**proboscis** **1.** The trunk of an elephant: a muscular and very flexible elongation of the nose, which has a finger-like extremity and is capable of picking up and moving objects, taking in water, collecting food, etc. **2.** The elongated mouthparts of certain invertebrates, such as the two-winged flies (Diptera).

**procambium** A plant tissue formed by the *apical meristems of shoots and roots. It consists of cells elongated parallel to the long axis of the plant. The procambium subsequently gives rise to the primary *vascular tissue.

**procaryote** *See* PROKARYOTE.

***Proconsul*** A genus of extinct apes known from fossil remains, roughly 23–14 million years old, found in East Africa and assigned to at least three species. These apes were quadrupeds and appear to have been tailless.

**producer** An organism considered as a source of energy for those above it in a *food chain (i.e. at the next *trophic level). Green plants, which convert energy from sunlight into chemical energy, are **primary producers**; herbivores are **secondary producers**, as they utilize energy from green plants and supply energy for carnivores. *Compare* CONSUMER.

**producer gas (air gas)** A mixture of carbon monoxide and nitrogen made by passing air over very hot carbon. Usually some steam is added to the air and the mixture contains hydrogen. The gas is used as a fuel in some industrial processes.

**product** *See* CHEMICAL REACTION.

**productivity (production)** (in ecology) The rate at which an organism, population, or community assimilates energy (**gross productivity**) or makes energy potentially available (as body tissue) to an animal that feeds on it (**net productivity**). The difference between these two rates is due to the rate at which energy is lost through excretion and respiration. Thus **gross primary productivity** is the rate at which plants (or other *producers) assimilate light energy, and **net primary productivity** is the rate at which energy is incorporated as plant tissue. It is measured in kilojoules per square metre per year ($kJ\ m^{-2}\ y^{-1}$). In terrestrial plants, much of the net productivity is not actually available to *consumers, e.g. tree roots are not eaten by herbivores. *See also* ENERGY FLOW.

**profundal** Occurring in or designating the deep-water zone of an inland lake. Light intensity, oxygen concentration, and (during summer and autumn) temperature are markedly lower than in the surface layer. *Compare* LITTORAL; SUBLITTORAL.

**progeny** *See* OFFSPRING.

**progesterone** A hormone, produced primarily by the *corpus luteum of the ovary but also by the placenta, that prepares the inner lining of the uterus for implantation of a fertilized egg cell. If implantation fails, the corpus luteum degenerates and progesterone production ceases accordingly. If implantation occurs, the corpus luteum continues to secrete progesterone, under the influence of *luteinizing hormone and *prolactin, for several months of pregnancy, by which time the placenta has taken over this function. During pregnancy, progesterone maintains the constitution of the uterus and prevents further release of eggs from the ovary. Small amounts of progesterone are produced by the testes. *See also* PROGESTOGEN.

**progestogen** One of a group of naturally occurring or synthetic hormones that maintain the normal course of pregnancy. The best known is *progesterone. In high doses progestogens inhibit secretion of *luteinizing hormone, thereby preventing ovulation, and alter the consistency of mucus in the vagina so that conception tends not to occur. They are therefore used as major constituents of *oral contraceptives.

**program** *See* COMPUTER.

**programmed cell death** *See* APOPTOSIS.

**progressive wave** *See* WAVE.

**projectile** Any body that is thrown or projected. If the projectile is discharged on the surface of the earth at an angle θ to the horizontal it will describe a parabolic flight path (if θ < 90° and the initial velocity < the *escape velocity). Neglecting air resistance, the maximum height of this flight path will be $(v^2\sin^2\theta)/2g$, where $v$ is the velocity of discharge and $g$ is the acceleration of free fall. The horizontal distance covered will be $(v^2\sin2\theta)/g$ and the time of the flight will be $(2v\sin\theta)/g$.

**projective relativity theory** A type of *unified-field theory in which projective geometry is used. This type of unified-field theory has not been successful, but it may be related to *twistor theory.

**projector** An optical device for throwing a large image of a two-dimensional object onto a screen. In an **episcope**, light is reflected from the surface of an opaque two-dimensional object (such as a diagram or photographic print) and an enlarged image

is thrown onto a distant screen by means of a system of mirrors and lenses. The **diascope** passes light through the two-dimensional object (such as a photographic transparency, slide, or film) and uses a converging projection lens to form an enlarged image on a distant screen. An **epidiascope** is a device that can be used as both episcope and diascope. An **overhead projector** is a form of diascope that throws its image on a wall or screen behind and above the operator. In a **motion-picture projector** (or **ciné projector**) the film, consisting of a long sequence of transparent pictures, is driven by a motor past the light source in such a way that each picture comes to rest for a brief period in front of the light source. The illusion of motion is created as each image on the screen is replaced by the next; during the picture change the light is interrupted.

**prokaryote (procaryote)** Any organism in which the genetic material is not enclosed in a cell nucleus. Prokaryotes consist exclusively of bacteria, i.e. archaebacteria and eubacteria, which are now generally classified in separate domains, *Archaea and Eubacteria. *See also* ENDOSYMBIONT THEORY.

**Prokhorov, Aleksandr** *See* BASOV, NIKOLAI GENNEDIYEVITCH.

**prolactin (lactogenic hormone; luteotrophic hormone; luteotrophin)** A hormone produced by the anterior pituitary gland. In mammals it stimulates the mammary glands to secrete milk (*see* LACTATION) and the corpus luteum of the ovary to secrete the hormone *progesterone. Secretion of prolactin is increased by suckling. In birds prolactin stimulates secretion of crop milk by the crop glands.

**proline** *See* AMINO ACID.

**PROM** *See* ROM.

**promethium** Symbol Pm. A soft silvery metallic element belonging to the *lanthanoids; a.n. 61; r.a.m. 145; r.d. 7.26 (20°C); m.p. 1080°C; b.p. 2460°C. The only naturally occurring isotope, promethium–147, has a half-life of only 2.52 years. Eighteen other radioisotopes have been produced, but they have very short half-lives. The only known source of the element is nuclear-waste material. Promethium–147 is of interest as a beta-decay power source but the promethium–146 and –148, which emit penetrating gamma radiation, must first be removed. It was discovered by J. A. Marinsky, L. E. Glendenin, and C. D. Coryell in 1947.

**SEE WEB LINKS**
• Information from the WebElements site

**prominence** A cloud of hot gas from the sun's chromosphere that rises into the lower corona. Cooler than its surroundings, it forms a bright luminous plume lasting for minutes or even for months, and is best seen at the edge of the sun against the blackness of space. Against the sun's disc it appears as a dark feature and is known as a filament. The plumes may be straight (surge and spray prominences), arched (arch prominences) or form a loop (loop prominences). They often form over sunspots, where they are supported by magnetic fields. *See also* SOLAR FLARE.

**promoter** 1. (in chemistry) A substance added to a catalyst to increase its activity. 2. (in protein synthesis) The region of a DNA molecule that signals the start of transcription. *See* OPERON; PRIBNOW BOX; TATA BOX.

**prompt neutrons** The neutrons emitted during a nuclear fission process within less than a microsecond of fission. *Compare* DELAYED NEUTRONS.

**pronation** Rotation of the lower forearm so that the hand faces backwards or downwards with the radius and ulna crossed. *Compare* SUPINATION.

**proof** A measure of the amount of alcohol (ethanol) in drinks. **Proof spirit** contains 49.28% ethanol by weight (about 57% by volume). Degrees of proof express the percentage of proof spirit present, so 70° proof spirit contains $0.7 \times 57\%$ alcohol.

**1,2-propadiene (allene)** A colourless gas, $CH_2CCH_2$; r.d. 1.79; m.p. –136°C; b.p. –34.5°C. Propadiene may be prepared from 1,3-dibromopropane ($CH_2BrCHCH_2Br$) by the action of zinc dust. *See also* ALLENES.

**propagation** 1. (in botany) *See* VEGETATIVE PROPAGATION. 2. (in neurophysiology) The process whereby a nerve *impulse travels along the axon of a neuron. *Compare* TRANSMISSION.

**propagule** Any cellular structure produced by an organism that is capable of dispersing and surviving in the environment before developing into a new individual. Examples are seeds, spores, and cysts.

**propanal (propionaldehyde)** A colourless liquid *aldehyde, $C_2H_5CHO$; m.p. –81°C; b.p. 48.8°C.

**propane** A colourless gaseous hydrocarbon, $C_3H_8$; m.p. –190°C; b.p. –42°C. It is the third member of the *alkane series and is obtained from petroleum. Its main use is as bottled gas for fuel.

**propanedioic acid (malonic acid)** A white crystalline dicarboxylic acid, $HOOCCH_2COOH$; m.p. 132°C. It decomposes above its melting point to ethanoic acid. Propanedioic acid is used in the synthesis of other dicarboxylic acids.

**propanoic acid (propionic acid)** A colourless liquid *carboxylic acid, $CH_3CH_2COOH$; r.d. 0.99; m.p. –20.8°C; b.p. 141°C. It is used to make calcium propanate – an additive in bread.

**propanol** Either of two *alcohols with the formula $C_3H_7OH$. Propan-1-ol is $CH_3CH_2CH_2OH$ and propan-2-ol is $CH_3CH(OH)CH_3$. Both are colourless volatile liquids. Propan-2-ol is used in making propanone (acetone).

**propanone (acetone)** A colourless flammable volatile compound, $CH_3COCH_3$; r.d. 0.79; m.p. –95.4°C; b.p. 56.2°C. The simplest *ketone, propanone is miscible with water. It is made by oxidation of propan-2-ol (*see* PROPANOL) or is obtained as a by-product in the manufacture of phenol from cumene; it is used as a solvent and as a raw material for making plastics.

**propellant** 1. A substance that burns rapidly in a controlled way, used to propel a projectile (e.g. from a gun). In firearms, gunpowder and cordite are common examples. 2. A fuel used in a rocket engine. Usually the propellant is a fuel and an oxidizer; for example kerosene or liquid hydrogen with a liquid oxygen propellant. 3. A substance used to produce the spray in an aerosol can. Aerosol propellants are volatile substances that can be liquefied under pressure and are able to dissolve the working substance. When the pressure is released, the liquid vaporizes, producing the spray. Formerly chlorofluorocarbons were used but their use has been discontinued because of their effect on the ozone layer. Most aerosol cans use liquid hydrocarbon mixtures as the propellant.

**propenal (acrolein)** A colourless pungent liquid unsaturated aldehyde, $CH_2:CHCHO$;

r.d. 0.84; m.p. –87°C; b.p. 53°C. It is made from propene and is used in producing polyester and polyurethane resins.

**propene (propylene)** A colourless gaseous hydrocarbon, $CH_3CH:CH_2$; m.p. –185.25°C; b.p. –47.4°C. It is an *alkene obtained from petroleum by cracking alkanes. Its main use is in the manufacture of polypropene.

**propenoate (acrylate)** A salt or ester of *propenoic acid.

**propenoic acid (acrylic acid)** An unsaturated liquid *carboxylic acid, $CH_2:CHCOOH$; m.p. 13°C; b.p. 141.6°C. It readily polymerizes and it is used in the manufacture of *acrylic resins.

**propenonitrile (acrylonitrile; vinyl cyanide)** A colourless liquid, $H_2C:CHCN$; r.d. 0.81; m.p. –83.5°C. It is an unsaturated nitrile, made from propene and used to make acrylic resins.

**propenyl group (allyl group)** The organic group $H_2C=CHCH_2–$.

**proper motion** The apparent angular motion of a star on the *celestial sphere, expressed in arcseconds per year. This is motion in a direction that is perpendicular to the line of sight. Proper motion is a combination of the star's own actual movement through space and its apparent changes in position caused by the movements of the sun and the earth.

**prophage** The DNA of a temperate *bacteriophage following its incorporation into the host bacterium. The process of incorporation of the viral DNA is known as **lysogeny**.

**prophase** The first stage of cell division, during which chromosomes contract and divide along their length (except for the centromeres) into chromatids. In *mitosis, the chromosomes remain separate from each other. In the first division of *meiosis, homologous chromosomes become paired (*see* PAIRING). By the end of first prophase the two chromosomes begin to move apart.

**propionaldehyde** *See* PROPANAL.

**proplastid** *See* PLASTID.

**proportional counter** A type of detector for *ionizing radiation in which the size of the output pulse is proportional to the number of ions formed in the initial ionizing event. It operates in a voltage region, called the **proportional region**, intermediate between that of an *ionization chamber and a *Geiger counter, avalanche ionization being limited to the immediate vicinity of the primary ionization rather than the entire length of the central wire electrode.

**proportional limit** *See* ELASTICITY.

**proprioceptor** Any *receptor that is sensitive to movement, pressure, or stretching within the body. Proprioceptors occurring in muscles, tendons, and ligaments are important for the coordination of muscular activity and the maintenance of balance and posture. *See also* MUSCLE SPINDLE.

**prop root** Any of the modified roots that arise from the stem of certain plants and provide extra support. Such stems are usually tall and slender and the prop roots develop at successively higher levels as the stem elongates, as in the maize plant. **Buttress roots**, which develop at the base of the trunks of many tropical trees, are similar but tend to have a more flattened appearance. **Stilt roots** are stouter than prop roots. Those formed at the base of the mangrove tree provide firm anchorage in the soft mud of the swamps.

**propylene** *See* PROPENE.

**propyl group** The organic group $CH_3CH_2CH_2–$.

**prosencephalon** *See* FOREBRAIN.

**prostacyclin** *See* PROSTAGLANDIN.

**prostaglandin** Any of a group of lipid-soluble organic compounds synthesized within cell membranes from arachidonic acid and detected in most body tissues. They cause a range of physiological effects in animals, including the contraction of smooth muscle; natural and synthetic prostaglandins are used to induce abortion or labour in humans and domestic animals. Two prostaglandin derivatives have antagonistic effects on blood circulation: **thromboxane $A_2$**, released by platelets activated by local tissue damage, promotes blood clotting and causes constriction of blood vessels, while **prostacyclin** inhibits blood clotting by preventing aggregation of platelets and causes blood vessels to dilate. Prostaglandins are also involved in inflammation, being released from macrophages and mast cells. *See also* ASPIRIN.

**prostate gland** A gland in male mammals that surrounds and opens into the urethra where it leaves the bladder. During ejaculation it secretes a slightly alkaline fluid into the *semen that activates the sperms and prevents them from sticking together.

**prosthetic group** A tightly bound non-peptide inorganic or organic component of a protein. Prosthetic groups may be lipids, carbohydrates, metal ions, phosphate groups, etc. Some *coenzymes are more correctly regarded as prosthetic groups.

**protactinium** Symbol Pa. A radioactive metallic element belonging to the *actinoids; a.n. 91; r.a.m. 231.036; r.d. 15.37 (calculated); m.p. <1600°C (estimated). The most stable isotope, protactinium–231, has a half-life of $3.43 \times 10^4$ years; at least ten other radioisotopes are known. Protactinium–231 occurs in all uranium ores as it is derived from uranium–235. Protactinium has no practical applications; it was discovered by Lise Meitner and Otto Hahn in 1917.

**(((⊕))) SEE WEB LINKS**
• Information from the WebElements site

**protamine** Any of a group of proteins of relatively low molecular weight found in association with the chromosomal *DNA of vertebrate sperm cells. They contain a single polypeptide chain comprising about 67% arginine. Protamines serve in packaging the highly condensed DNA of the germ-cell chromosomes. Protamine sulphate is used therapeutically to treat heparin overdosage.

**protandry** 1. The condition in which the male reproductive organs (stamens) of a flower mature before the female ones (carpels), thereby ensuring that self-fertilization does not occur. Examples of protandrous flowers are ivy and rosebay willowherb. *Compare* PROTOGYNY; HOMOGAMY. *See also* DICHOGAMY. 2. The condition in some hermaphrodite or colonial invertebrates in which the male gonads or individuals are sexually mature before the female ones. *Compare* PROTOGYNY.

**protease (peptidase; proteinase; proteolytic enzyme)** Any enzyme that catalyses the hydrolysis of proteins into smaller *peptide fractions and amino acids, a process known as **proteolysis**. Examples are *pepsin and *trypsin. Several proteases, acting sequentially, are normally required for the complete digestion of a protein to its constituent amino acids.

**protecting group** A group used to protect a certain functional group in a chemical synthesis. For example, a hydroxyl group (–OH) can be converted into an acetyl group (–OOCCH₃) to protect it taking part in a certain step of the synthesis. In this case, the acetyl is the protecting group. Later it can easily be changed back into the original hydroxyl group.

**protein** Any of a large group of organic compounds found in all living organisms. Proteins comprise carbon, hydrogen, oxygen, and nitrogen and most also contain sulphur; molecular weights range from six to several thousand *kilodaltons. Protein molecules consist of one or several long chains (*polypeptides) of *amino acids linked in a characteristic sequence. This sequence is called the **primary structure** of the protein. These polypeptides may undergo coiling or pleating, the nature and extent of which is described as the **secondary structure**. The three-dimensional shape of the coiled or pleated polypeptides is called the **tertiary structure**. **Quaternary structure** specifies the structural relationship of the component polypeptides.

Proteins may be globular or fibrous, with various intermediate forms. **Globular proteins** have compact rounded molecules and are usually water-soluble. Of prime importance are the *enzymes, proteins that catalyse biochemical reactions. Other globular proteins include the *antibodies, which combine with foreign substances in the body; the carrier proteins, such as *haemoglobin; the storage proteins (e.g. *casein in milk and *albumin in egg white), and certain hormones (e.g. *insulin). **Fibrous proteins** are generally insoluble in water and consist of long coiled strands or flat sheets, which confer strength and elasticity. In this category are *keratin and *collagen. Actin and myosin are the principal fibrous proteins of muscle, the interaction of which brings about muscle contraction. *Blood clotting involves the fibrous protein called fibrin.

When heated over 50°C or subjected to strong acids or alkalis, proteins lose their specific tertiary structure and may form insoluble coagulates (e.g. egg white). This usually inactivates their biological properties.

**(((⊕))) SEE WEB LINKS**
• Comprehensive survey of protein biochemistry, from the Virtual Library of Biochemistry, Molecular Biology and Cell Biology

**protein blotting** *See* WESTERN BLOTTING.

**protein engineering** The techniques used to alter the structure of proteins (especially enzymes) in order to improve their use to humans. This involves artificially modifying the DNA sequences that encode them so that, for example, new amino acids are inserted into existing proteins. Synthesized lengths of novel DNA can be used to produce new proteins by cells or other systems containing the necessary factors for *transcription and *translation. Alternatively, new proteins can be synthesized by **solid state synthesis**, in which polypeptide chains are assembled under the control of chemicals. One end of the chain is anchored to a solid support and the chemicals selectively determine which amino acids are added to the free end. The appropriate chemicals can be renewed during the process; when synthesized, the polypeptide is removed and purified. Protein engineering is used to synthesize enzymes (so-called 'designer enzymes') used in biotechnology. The three-dimensional tertiary structure of proteins is crucially important for their function, and this can be investigated using computer-aided modelling.

**protein kinase** An enzyme that catalyses the transfer of a phosphate group from ATP to an intracellular protein, thereby affecting the biological activity of the protein (*see* KINASE). Protein kinases phosphorylate specific amino-acid residues of their target proteins, usually either serine, threonine, or tyrosine. They play an important role in increasing or decreasing enzyme activity and in transmitting signals from receptors on the cell surface. The activity of the protein kinases is itself controlled by cyclic AMP, calcium ions, or other intracellular chemicals. It can be reversed by the action of phosphatase enzymes in the cell.

**proteinoid** A protein-like substance formed by polymerization of amino acids under inorganic conditions, such as heating to over 140°C. In the 1970s it was discovered that proteinoids could also be formed by relatively mild heating (70°C) in the presence of certain inorganic catalysts (e.g. phosphoric acid). In water, proteinoids aggregate to form small round bodies called **proteinoid microspheres**, or 'protocells'. These have certain attributes of living cells, such as a differentially permeable filmlike outer layer, the ability to swell and shrink due to osmotic movements of water, and the capability for budding and binary fission. It has been proposed that such microspheres could have provided a suitable vehicle for the chemical components of life to evolve a primitive form of metabolism and pave the way for the emergence of the first living cells.

**protein sequencing** The process of determining the amino-acid sequence of a protein or its component polypeptides. The technique most commonly used is Edman degradation (devised by Pehr Edman), in which the terminal amino-acid residues are removed sequentially and identified chromatographically. Each step is automated and the whole process can now be performed by a single machine – the sequenator. Large polypeptides must be cleaved into smaller peptides before sequencing.

The results of this chemical sequencing can often be compared with the amino-acid sequence deduced by *DNA sequencing. The gene coding for the protein under investigation may be found by screening a *DNA library, for example by *Western blotting. However, the base sequence of the gene gives only the amino-acid sequence of the nascent protein, i.e. before post-translational modification. The sequence of the functional protein can only be found by chemical analysis.

**protein synthesis** The process by which living cells manufacture proteins from their constituent amino acids, in accordance with the genetic information carried in the DNA of the chromosomes. This information is encoded in messenger *RNA, which is transcribed from DNA in the nucleus of the cell (*see* GENETIC CODE, TRANSCRIPTION). the sequence of amino acids in a particular protein is determined by the sequence of nucleotides in messenger RNA. At the ribosomes the information carried by messenger RNA is translated into the sequence of amino acids of the protein in the process of *translation.

**proteolysis** The enzymic splitting of proteins. *See* PROTEASE.

**proteolytic enzyme** *See* PROTEASE.

**proteome** The entire complement of proteins synthesized by a cell or organism at a given time. This can be determined by analysing protein constituents of cell contents using such techniques as gel electrophoresis, high-throughput liquid chromatography, microarrays, and mass spectroscopy, coupled with automated database searching to identify proteins. Unlike the genome, the proteome is constantly changing due to the influence of intracellular and extracellular factors. *See* PROTEOMICS.

**proteomics** The study of the proteins synthesized by a particular cell or organism (*see* PROTEOME). This vast and rapidly expanding field is of fundamental significance to many areas of biology and medicine. It endeavours to determine what proteins a cell makes, how and when it makes them and in what quantities, how different proteins function, where they function, and how they interact with other cell components, including other proteins. Moreover, proteomics seeks to discover the internal and external factors that influence the proteome, for example during development, disease, or ageing. *See also* TRANSCRIPTOMICS.

**Proterozoic** The eon of geological time extending from the end of the *Archaean, about 2500 million years ago, to the start of the present eon (*see* PHANEROZOIC), about 542 million years ago. Life in the early Proterozoic was dominated by bacteria, which flourished in shallow seas and muds. They depended on a wide variety of metabolic strategies, including photosynthesis, which were crucial in determining the composition of the earth's atmosphere and oceans. The oldest eukaryotic fossils date from after the middle Proterozoic, about 1200 million years ago. These early protists are thought to have arisen through symbiotic associations of various prokaryotes (*see* ENDOSYMBIONT THEORY), probably on several independent occasions.

**prothallus** A small flattened multicellular structure that represents the independent *gametophyte generation of clubmosses, horsetails, and ferns. In some of these plants a single prothallus bears both male and female sex organs. In others there are separate male and female prothalli.

**prothrombin (Factor II)** One of the blood *clotting factors. It is the precursor of the enzyme thrombin, which catalyses the formation of the fibrin matrix of the blood clot. Prothrombin synthesis occurs in the liver and is dependent on adequate supplies of vitamin K. *See also* BLOOD CLOTTING.

**protist** Any eukaryotic organisms that is unicellular or colonial in form and lacks cellular differentiation into tissues. Protists include algae, simple fungi, and protozoa. The kingdom Protista was originally proposed by Ernst Haeckel in 1866 to include the algae, bacteria, fungi, and protozoa; it was later restricted first to unicellular organisms, and then to protozoa, unicellular algae, and organisms then regarded as simple fungi. Molecular studies have revealed that this kingdom is no longer taxonomically valid, and 'protist' is now purely a descriptive term.

**Protoctista** A former kingdom consisting of unicellular or simple multicellular organisms that possess nuclei and cannot be classified as animals, plants, or fungi.

**protogyny** 1. The condition in which the female reproductive organs (carpels) of a flower mature before the male ones (stamens), thereby ensuring that self-fertilization does not occur. Examples of protogynous flowers are plantain and figwort. *Compare* PROTANDRY; HOMOGAMY. *See also* DICHOGAMY. 2. The condition in hermaphrodite or colonial invertebrates in which the female gonads or individuals are sexually mature before the male ones. *Compare* PROTANDRY.

**proton** An *elementary particle that is stable, bears a positive charge equal in magnitude to that of the *electron, and has a mass of $1.672\,614 \times 10^{-27}$ kg, which is 1836.12 times that of the electron. The proton occurs in all atomic nuclei (the hydrogen nucleus consists of a single proton).

**proton decay** A process of the type

$$p \rightarrow e^+ + \pi^0$$

where a proton decays into a positron and a pion, predicted to occur in *grand unified theories (GUTs) because baryon number is no longer conserved. The lifetime depends on the theory used and is typically $10^{35}$ years, but a combination of GUTs and *supersymmetry gives a lifetime of about $10^{45}$ years. Considerable experimental effort has been spent in looking for proton decay, so far with no success.

**protonic acid** An *acid that forms positive hydrogen ions (or, strictly, oxonium ions) in aqueous solution. The term is used to distinguish 'traditional' acids from Lewis acids or from Lowry–Brønsted acids in nonaqueous solvents.

**proton number** *See* ATOMIC NUMBER.

**proto-oncogene** *See* ONCOGENE.

**protoplasm** The material comprising the living contents of a *cell, i.e. all the substances in a cell except large vacuoles and material recently ingested or to be excreted. The term is no longer used; the material of the cell is now referred to as the *nucleoplasm and the *cytoplasm.

**protoplast (energid)** The living unit of a cell, consisting of the nucleus and cytoplasm bounded by the plasma membrane. Protoplasts of bacterial and plant cells can be prepared by removing the cell wall; they are used to study the processes involved in cell metabolism and reproduction.

**protostar** See STELLAR EVOLUTION.

**Prototheria** A subclass of mammals – the monotremes – that lay large yolky eggs. It contains only the duckbilled platypus and the spiny anteater. After hatching, the young feed on milk from simple mammary glands inside a maternal abdominal pouch. In the anteater the eggs are also incubated in this pouch, while the platypus builds an underground nest. Adult monotremes have no true teeth. Their skeleton resembles that of a reptile, and although they are warm-blooded the body temperature is somewhat variable. They are believed to have originated at least 150 million years ago.

**protozoa** A group of unicellular or acellular, usually microscopic, eukaryotic organisms classified in various phyla. They were formerly regarded either as a phylum of simple animals or as members of the kingdom Protista (see PROTIST). They are very widely distributed in marine, freshwater, and moist terrestrial habitats; most protozoans are saprotrophs, but some are parasites, including the agents causing malaria (*Plasmodium*) and sleeping sickness (*Trypanosoma*), and a few contain chlorophyll and carry out photosynthesis, like plants. Protozoan cells may be flexible or rigid, with an outer *pellicle or protective **test**. In some (such as *Paramecium* and *Trypanosoma*) cilia or flagella (see UNDULIPODIUM) are present for locomotion; others (such as *Amoeba*) have *pseudopodia for movement and food capture. *Contractile vacuoles occur in freshwater protozoans. Reproduction is usually asexual, by binary *fission, but some protozoans undergo a form of sexual reproduction (see CONJUGATION).

**Prout's hypothesis** The hypothesis put forward by the British chemist William Prout (1785–1850) in 1815 that all atomic weights are integer multiples of the atomic weight of hydrogen and hence that all atoms are made out of hydrogen. Subsequent work on atomic weights in the 19th century showed that this hypothesis is incorrect (with chlorine having an atomic weight of 35.5 being a glaring example of this). The understanding of atomic structure that emerged in the 20th century, with atomic number being the number of protons in an atom and noninteger atomic weights being due to mixtures of isotopes, has vindicated the spirit of Prout's hypothesis.

(((( ))))  SEE WEB LINKS
• William Prout's original paper

**proventriculus** 1. The anterior part of the stomach of a bird, where digestive enzymes are secreted. Food passes from the proventriculus to the *gizzard. 2. See GASTRIC MILL.

**provirus** The intermediate stage in the infection of a host cell by a virus, e.g. a *retrovirus, in which the viral genome is integrated into the host cell DNA, where it can undergo successive replications before being transcribed to form new RNA viruses. A provirus, notably that of *HIV, can remain dormant for long periods before being transcribed.

**proximal** Denoting the part of an organ that is nearest to the organ's point of attachment. For example, the knuckles are at the proximal end of the fingers. *Compare* DISTAL.

**proximal convoluted tubule (first convoluted tubule)** The section of a *nephron situated between Bowman's capsule and the loop of Henle in the vertebrate kidney. Reabsorption of salt, water, and glucose from the *glomerular filtrate occurs in this tubule; at the same time certain substances, including uric acid and drug metabolites, are actively transferred from the blood capillaries into the tubule. Both activities are facilitated by finger-like projections (see BRUSH BORDER) on the inner surface of the tubule, which increase its effective surface area.

**prussic acid** See HYDROGEN CYANIDE.

**pseudoaromatic (antiaromatic)** A compound that has a ring of atoms containing alternating double and single bonds, yet does not have the characteristic properties of *aromatic compounds. Such compounds do not obey the Hückel rule. Cyclooctatetraene ($C_8H_8$), for instance, has a ring of eight carbon atoms with conjugated double bonds, but the ring is not planar and the compound acts like an alkene, undergoing addition reactions. *See also* ANNULENES.

**pseudocarp (false fruit)** A fruit that incorporates, in addition to the ovary wall, other parts of the flower, such as the *receptacle. For example, the fleshy part of the strawberry is formed from the receptacle and the 'pips' on the surface are the true fruits. *See also* COMPOSITE FRUIT; POME; SOROSIS; SYCONUS.

**pseudoephedrine** *See* EPHEDRINE.

**pseudogene** A sequence of nucleotides in DNA that resembles a functional gene but is not transcribed. Pseudogenes are thought to arise by duplication of an existing gene through unequal crossing-over during meiosis, with accompanying loss of the promoter or other flanking regions required for transcription. For example, the α- and β-globin gene clusters in humans contain several pseudogenes.

**pseudohalogens** A group of compounds, including cyanogen $(CN)_2$ and thiocyanogen $(SCN)_2$, that have some resemblance to the halogens. Thus, they form hydrogen acids (HCN and HSCN) and ionic salts containing such ions as $CN^-$ and $SCN^-$.

**pseudo order** An order of a chemical reaction that appears to be less than the true order because of the experimental conditions used. Pseudo orders occur when one reactant is present in large excess. For example, a reaction of substance A undergoing hydrolysis may appear to be proportional only to [A] because the amount of water present is so large.

**pseudoparenchyma** A tissue that superficially resembles plant parenchyma but is made up of an interwoven mass of hyphae (in fungi) or filaments (in algae). Examples of pseudoparenchymatous structures are the fruiting bodies (mushrooms, toadstools, etc.) of certain fungi and the thalli of certain red and brown algae.

**pseudopodium** A temporary outgrowth of the cell of some protozoans (e.g. *Amoeba*), which serves as a feeding and locomotory organ. Pseudopodia may be blunt or threadlike, form a branching network, or be stiffened with an internal supporting rod. Phagocytic white blood cells also form pseudopodia to engulf invading bacteria.

**pseudopregnancy** A state resembling

pregnancy that may occur in some mammals (e.g. rabbits and rodents) in which many of the phenomena of pregnancy are present but there is no fetus developing in the uterus. It is caused by an extended dioestrus (*see* OESTROUS CYCLE) in the absence of fertilization.

**pseudo-scalar** A *scalar quantity that changes sign when the coordinate system is changed to a new system by a reflection in the origin (i.e. $x'_i = -x_i$). It is the *scalar product of an *axial vector and a *polar vector.

**pseudo-vector** *See* AXIAL VECTOR.

**psilocin** *See* PSILOCYBIN.

**psilocybin** A hallucinogen, similar in effect to *mescaline, found in certain species of mushroom. It is accompanied by a related, more active, compound **psilocin**. Both are classified as class A drugs in the UK. They can be detected by the Marquis test or the Froehde test.

**psi particle (J particle)** A *meson discovered in 1974, which led to the extension of the quark model and the hypothesis that a fourth quark existed with the property of charm (*see* ELEMENTARY PARTICLES). The psi particle is believed to consist of a charmed quark and its antiquark.

**psychrometer** *See* HYGROMETER.

**Pteridophyta** In traditional classification systems, a division of the plant kingdom that included ferns, horsetails, and clubmosses, i.e. the nonseed-bearing tracheophytes. These are now classified as separate phyla: *Filicinophyta (ferns), *Sphenophyta (horsetails), and *Lycophyta (clubmosses).

**Pteridospermales** *See* CYCADOFILICALES.

**pterodactyls** *See* PTEROSAURIA.

**Pterophyta** *See* FILICINOPHYTA.

**Pteropsida** In older classifications, a subdivision of tracheophytes that contained the ferns and seed plants, or a class of the *Pteridophyta containing only the ferns.

**Pterosauria** An extinct order of flying reptiles – the pterodactyls – that lived in the late Triassic, Jurassic, and Cretaceous periods (220–70 million years ago). Pterodactyls had beaked jaws and an elongated fourth finger that supported a membranous wing. They had long jointed tails, no feathers, and could probably only fly by soaring.

**PTFE** *See* POLYTETRAFLUOROETHENE.

**PTH** *See* PARATHYROID HORMONE.

**Ptolemaic astronomy** The system of astronomy embodied in the model of a *geocentric universe developed by the Alexandrian astronomer Claudius Ptolemaeus (Ptolemy) (100–178 AD). His model, the **Ptolemaic system** was a refinement of those developed by Plato, Aristotle, and especially the 3rd-century BC mathematician Apollonius of Perga. In Ptolemaic astronomy, the spherical earth was at rest, lying at the centre of a universe made up of a series of concentric spheres; the sun, the moon, and each of the five known planets moved round the earth in a circular orbit within its own sphere, on a path called the **deferent**. In addition to this motion the orbiting bodies also described **epicycles**, small circles about points on the deferent. Ptolemy published his system together with a star catalogue and mathematical treatise that owed much to the work of the ancient Greek astronomer Hipparchus. It is best known in an Arabic translation, the *Almagest*, or 'Great Book'. The system gave moderately good predictions of phenomena and planetary positions, and won the support of the Christian and Islamic religious authorities because it emphasized circles and uniform perpetual motion, which were symbols of divine perfection and unchanging eternity, and placed the earth and humanity at the very heart of God's creation. For 1600 years the Ptolemaic system remained the accepted model of the universe but in the 17th century it was completely replaced by the model of a heliocentric universe and *Copernican astronomy.

**ptyalin** An enzyme that digests carbohydrates (*see* AMYLASE). It is present in mammalian *saliva and is responsible for the initial stages of starch digestion.

***p*-type conductivity** *See* SEMICONDUCTOR; TRANSISTOR.

**puberty** *See* ADOLESCENCE.

**pubis** One of the three bones that make up each half of the *pelvic girdle. It is the most anterior of the three pelvic bones. In mammals and many reptiles the pubes are united at a slightly movable joint, the **pubic symphysis**. *See also* ILIUM; ISCHIUM.

**public key certificate** *See* CERTIFICATE.

**public key encryption (asymmetric encryption)** In computing, a type of cryptography where one of a pair of keys is used to encrypt a message and the other is used to decrypt it. It does not matter which key is used to encrypt and which to decrypt: both combinations will work; however, attempting to encrypt and decrypt with the same key will not work. The two keys are integers that are related mathematically, but – crucially – it must not be viable to deduce the other key if only one is known. In computational terms this means that it must be proven that there does not exist an algorithm to calculate the second key efficiently from the first, and that the range of possible integers must be large enough to make a 'brute force' attack impractical.

In usage, a given key pair belongs to a specific person or organization. One key of the pair, the **public key**, can be distributed freely; the other, the **private key**, must be kept secret by the owner. The keys can then be used in one of two ways:

(1) Anybody wishing to send confidential data encrypts it with the recipient's public key. It can then only be decrypted by the recipient's private key, which only the recipient possesses.

(2) Anybody wishing to add a digital signature to data encrypts the signature with their private key. If the signature is valid when decrypted with the signer's public key, a user can be confident that the data indeed originated with the signer and has not been altered (*see also* CERTIFICATE).

**PubMed** *See* MEDLINE.

**pulley** A simple machine consisting of a wheel with a flat, crowned, or grooved rim to take a belt, rope, or chain with which a load can be raised (see illustration overleaf).

In fig (a), assuming the system is frictionless, the force $P$ in any part of the rope is constant, therefore $2P = L$, where $L$ is the load. In general, $nP = L$, where $n$ is the number of supporting ropes. In fig (b), the number of supporting ropes is 4. The mechanical advantage of a pulley system is the ratio of the load, $L$, to the effort applied to the free end of the rope, $P$, i.e. mechanical advantage $= L/P = L(L/n)^{-1} = n$. Thus in fig (b) the mechanical advantage is 4. A combination of ropes and pulleys as in fig (b) is called a **block and tackle**.

**pulmonary** Of or relating to the lungs.

**pulmonary artery** The artery that conveys deoxygenated blood from the right ven-

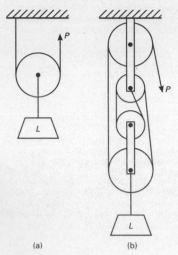

(a)       (b)

**Pulleys.**

tricle of the heart to the lungs, where it receives oxygen.

**pulmonary circulation** The part of the
circulatory system of birds and mammals
that transports deoxygenated blood from the
right side of the heart to the lungs and returns oxygenated blood to the left side of the
heart. *Compare* SYSTEMIC CIRCULATION. *See*
DOUBLE CIRCULATION.

**pulmonary vein** The vein that conveys
oxygenated blood from the lungs to the left
atrium of the heart.

**pulp cavity** The central region of a tooth,
which is connected by a narrow channel at
the tip of the root with the surrounding tissues. The pulp cavity contains the **pulp** –
connective tissue in which blood vessels and
nerve fibres are embedded, and it is lined
with *odontoblasts, which produce the *dentine.

**pulsar** A celestial source of radiation emitted in brief (0.03 second to 4 seconds) regular pulses. First discovered in 1967, pulsars
are believed to be rotating *neutron stars.
The strong magnetic field of the neutron star
concentrates charged particles in two regions and the radiation is emitted in two directional beams. The pulsing effect occurs as
the beams rotate, periodically pointing in
the direction of the earth. Most pulsars are

radio sources (emit electromagnetic radiation of radio frequencies) but a few that emit
light or X-rays have been detected. Over 300
pulsars are now known in our Galaxy.

**pulsatance** *See* ANGULAR FREQUENCY.

**pulse 1.** (in physics) **a.** A brief variation in a
quantity, usually for a finite time, especially
in a quantity that is normally constant. **b.** A
series of such variations having a regular
waveform in which the variable quantity
rises sharply from a base value to a maximum value and then falls back to the base
value in a relatively short time. **2.** (in physiology) A series of waves of dilation that pass
along the arteries, caused by pressure of
blood pumped from the heart through contractions of the left ventricle. In humans it
can be felt easily where arteries pass close to
the skin surface, e.g. at the wrist.

**pulse jet** A type of ramjet (*see* JET PROPUL-
SION) in which a louvred valve at the front of
the projectile is blown open by the ram effect of the moving projectile and remains
open until pressure has built up in the combustion chamber. Fuel is then admitted and
the mixture exploded by spark ignition. This
closes the louvred valve and produces thrust
at the open rear end of the projectile. The
German flying bombs of World War II were
powered by pulse jets.

**pulse modulation** *See* MODULATION.

**pulvinus** A group of cells at the base of a
leaf or leaflet in certain plants that, by
rapidly losing water, brings about changes in
the position of the leaves. In the sensitive
plant (*Mimosa pudica*), the pulvinus is responsible for the folding of the leaves that
occurs at nightfall or when the plant is
touched or injured.

**pumice** A porous volcanic rock that is light
and full of cavities due to expanding gases
that were liberated from solution in *lava
while it solidified. Pumice is often light
enough to float on water. It is usually acid
(siliceous) in composition, and is used as an
abrasive and for polishing.

**pump** A device that imparts energy to a
fluid in order to move it from one place or
level to another or to raise its pressure (*compare* VACUUM PUMP). **Centrifugal pumps** and
*turbines have rotating impellers, which in-
crease the velocity of the fluid, part of the energy so acquired by the fluid then being
converted to pressure energy. Displacement

pumps act directly on the fluid, forcing it to flow against a pressure. They include piston, plunger, gear, screw, and cam pumps. *See also* ELECTROMAGNETIC PUMP.

**punctuated equilibrium** A controversial hypothesis, published in 1972 by N. Eldredge and Stephen J. Gould, proposing that in evolutionary history the development of new species occurs very rapidly in short bursts (lasting typically less than 100 000 years), which are separated by long periods in which little evolutionary change occurs. This hypothesis, which contradicted the orthodox Darwinian view of evolution as a gradual and continuous process, was based on studies of various fossil lineages (e.g. ammonite molluscs) in which forms intermediate between species are absent. Subsequent scrutiny of the evidence supports a pattern of punctuated equilibrium for some, but not all, lineages, so it cannot be regarded as universal. For example, the rodent lineage shows as much morphological change between speciation events as during speciation.

**pupa** The third stage of development in the life cycle of some insects. During the pupal stage locomotion and feeding cease and *metamorphosis from the larva to the adult form takes place. There are three types of pupa. The commonest is the **exarate** or free pupa, in which the wings and other appendages are visible and movable. In the **obtect** type the wings are stuck to the body and immovable, as in the **chrysalis** of a butterfly or moth; and in the **co-arctate** type an exarate pupa develops within a hard barrel-shaped **puparium**, as in the housefly and other Diptera.

**pupil** *See* IRIS.

**pure line** A population of plants or animals all having a particular feature that has been retained unchanged through many generations. The organisms are *homozygous and are said to 'breed true' for the feature concerned.

**purine** An organic nitrogenous base (see formula), sparingly soluble in water, that gives rise to a group of biologically impor-

**Purine.**

tant derivatives, notably *adenine and *guanine, which occur in *nucleotides and nucleic acids (DNA and RNA).

**Purkyne fibres (Purkinje fibres)** Modified fibres in the mammalian heart that originate in the *bundle of His and spread out in a network over the ventricles. Action potentials generated in the sinoatrial node (the *pacemaker of the heart) are conducted extremely rapidly through the ventricles, due to the extensive branching of the Purkyne fibres, causing both ventricles to contract almost simultaneously. They are named after the Czech physiologist Johannes Purkyne (1787–1869).

**putrefaction** The microbial decomposition of organic matter, especially the anaerobic breakdown of proteinaceous material with the production of foul-smelling amines.

**PVA** *See* POLYVINYLACETATE.

**PVC** *See* POLYCHLOROETHENE.

**pyloric sphincter** *See* SPHINCTER; STOMACH.

**pyramid** A solid having a polygonal base with *n* sides, each side forming the base of a triangle. The *n* triangles so formed have a common vertex. The **axis** of the pyramid is a line joining the vertex to the centre of symmetry of the base. If the axis is perpendicular to the base the solid is a **right pyramid**. A **square pyramid** has a square base and a **triangular pyramid** has a triangular base (*see* TETRAHEDRON). The volume of a pyramid is one third of the base area multiplied by the height.

**pyramid of biomass** A diagrammatic representation of the amount of organic material (*see* BIOMASS), measured in grams of dry mass per square metre ($g\ m^{-2}$), found in a particular habitat at ascending *trophic levels of a *food chain. Biomass decreases at each ascending level of the food chain. A pyramid of biomass is a more accurate representation of the flow of energy through a food chain than a *pyramid of numbers, but seasonal variations in the rate of turnover of the organisms at a particular level may result in higher or lower values for the amount of biomass sampled at a particular time than the average amount over the whole year. The best representation of energy flow in a food chain is a *pyramid of energy.

**pyramid of energy** A diagrammatic representation of the amount of energy, meas-

p

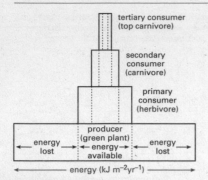

**Pyramid of energy.**

ured in kilojoules per square metre per year ($kJ\ m^{-2}\ yr^{-1}$), available at ascending *trophic levels of a *food chain in a particular habitat. A pyramid of energy is the most accurate representation of the *energy flow through a food chain as it indicates how much energy is lost at each trophic level (through respiration, etc.). *Compare* PYRAMID OF BIOMASS; PYRAMID OF NUMBERS.

**pyramid of numbers** A diagramatic representation of the numbers of animals found in an area at ascending *trophic levels of a *food chain (e.g. a woodland food chain: see illustration). Because only a small proportion of the energy taken in by an organism is converted to tissue and is thus available to consumers at the next trophic level, the number of organisms that can be supported at each level is generally much less than the number at the level that supplies its food (i.e. the level below). *See also* PYRAMID OF BIOMASS; PYRAMID OF ENERGY.

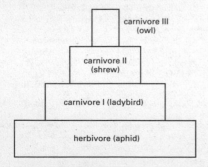

**Pyramid of numbers.**

**pyran** A heterocyclic compound having an oxygen atom and two double bonds in a six-membered ring, $C_5H_6O$. There are two isomers depending on the position of the $CH_2$ group.

**pyranose** A *sugar having a six-membered ring containing five carbon atoms and one oxygen atom.

**pyrazine** An unsaturated heterocyclic compound having two nitrogen atoms in a six-membered ring, $C_4H_4N_2$; r.d. 1.03; m.p. 52°C; b.p. 115°C. Pyrazine is a symmetric diazine.

**pyrazole** An unsaturated heterocyclic compound having two nitrogen atoms in a five-membered ring, $C_3H_4N_2$; m.p. 66–70°C; b.p. 168–188°C. Pyrazine is a symmetric diazine.

**pyrenocarp** *See* DRUPE.

**pyrenoid** A spherical protein body found in the *chloroplasts of many algae. Pyrenoids are associated with the storage of starch: layers of starch are often found around them.

**pyrethrum** 1. Any of several plants of the genus *Chrysanthemum* that contain natural insecticidal compounds (**pyrethrins**). Various synthetic insecticides – the **pyrethroids** – are chemically similar to pyrethrins. Examples include permethrin. The pyrethrins penetrate the insect's cuticle and are fast-acting, nontoxic to many animals and to plants, and readily biodegradable. 2. An insecticidal preparation containing natural pyrethrins.

**pyridine** A colourless liquid with a strong unpleasant smell, $C_5H_5N$ (see formula); r.d. 0.98; m.p. –42°C; b.p. 115°C. Pyridine is an aromatic heterocyclic compound present in coal tar. It is used in making other organic chemicals.

**Pyridine.**

**pyridoxine** *See* VITAMIN B COMPLEX.

**pyrimidine** An organic nitrogenous base (see formula), sparingly soluble in water, that gives rise to a group of biologically important derivatives, notably *uracil,

**Pyrimidine.**

*thymine, and *cytosine, which occur in *nucleotides and nucleic acids (DNA and RNA).

**pyrite (iron pyrites)** A mineral form of iron(II) sulphide, $FeS_2$. Superficially it resembles gold in appearance, hence it is also known as **fool's gold**, but it is harder and more brittle than gold (which may be cut with a knife). Pyrite crystallizes in the cubic system, is brass yellow in colour, has a metallic lustre, and a hardness of 6–6.5 on the Mohs' scale. It is the most common and widespread of the sulphide minerals and is used as a source of sulphur for the production of sulphuric acid. Sources include the Rio Tinto mines in Spain.

**pyro-** Prefix denoting an oxo acid that could be obtained from a lower acid by dehydration of two molecules. For example, pyrosulphuric acid is $H_2S_2O_7$ (i.e. $2H_2SO_4$ minus $H_2O$).

**pyroboric acid** *See* BORIC ACID.

**pyroelectricity** The property of certain crystals, such as tourmaline, of acquiring opposite electrical charges on opposite faces when heated. In tourmaline a rise in temperature of 1 K at room temperature produces a polarization of some $10^{-5}$ C m$^{-2}$.

**pyrogallol** 1,2,3-trihydroxybenzene, $C_6H_3(OH)_3$, a white crystalline solid, m.p. 132°C. Alkaline solutions turn dark brown on exposure to air through reaction with oxygen. It is a powerful reducing agent, employed in photographic developers. It is also used in volumetric gas analysis as an absorber of oxygen.

**pyrolusite** A soft black mineral consisting mainly of manganese dioxide ($MnO_2$); it is the most important ore of manganese. It is generally found as a secondary deposit of masses of fibrous or needle-shaped crystals. As well as an ore, it is used as a decolourizer and oxidizing agent.

**pyrolysis** Chemical decomposition occurring as a result of high temperature.

**pyrometric cones** *See* SEGER CONES.

**pyrometry** The measurement of high temperatures from the amount of radiation emitted, using a **pyrometer**. Modern **narrow-band** or **spectral** pyrometers use infrared-sensitive *photoelectric cells behind filters that exclude visible light. In the **optical pyrometer** (or disappearing filament pyrometer) the image of the incandescent source is focused in the plane of a tungsten filament that is heated electrically. A variable resistor is used to adjust the current through the filament until it blends into the image of the source, when viewed through a red filter and an eyepiece. The temperature is then read from a calibrated ammeter or a calibrated dial on the variable resistor. In the **total-radiation pyrometer** radiation emitted by the source is focused by a concave mirror onto a blackened foil to which a thermopile is attached. From the e.m.f. produced by the thermopile the temperature of the source can be calculated.

**pyrophoric** Igniting spontaneously in air. **Pyrophoric alloys** are alloys that give sparks when struck. *See* MISCH METAL.

**pyrophosphoric acid** *See* PHOSPHORIC(V) ACID.

**pyrosilicate** *See* SILICATE.

**pyrosulphuric acid** *See* DISULPHURIC(VI) ACID.

**pyroxenes** A group of ferromagnesian rock-forming silicate minerals. They are common in basic igneous rocks but may also be developed by metamorphic processes in gneisses, schists, and marbles. Pyroxenes have a complex crystal chemistry; they are composed of continuous chains of silicon and oxygen atoms linked by a variety of other elements. They are related to the *amphiboles, from which they differ in cleavage angles. The general formula is $X_{1-p}Y_{1+p}Z_2O_6$, where $X$ = Ca,Na; $Y$ = Mg,Fe$^{2+}$,Mn,Li,Al,Fe$^{3+}$,Ti; and $Z$ = Si,Al.

Orthorhombic pyroxenes (**orthopyroxenes**), $(Mg,Fe)_2Si_2O_6$, vary in composition between the end-members enstatite ($Mg_2Si_2O_6$) and orthoferrosilite ($Fe_2Si_2O_6$). Monoclinic pyroxenes (**clinopyroxenes**), the larger group, include:
diopside, $CaMgSi_2O_6$;
hedenbergite, $CaFe^{2+}Si_2O_6$;
johannsenite, $CaMnSi_2O_6$;
augite, $(Ca,Mg,Fe,Ti,Al)_2(Si,Al)_2O_6$;
aegirine, $NaFe^{3+}Si_2O_6$;
jadeite (*see* JADE);
pigeonite $(Mg,Fe^{2+},Ca)(Mg,Fe^{2+})Si_2O_6$.

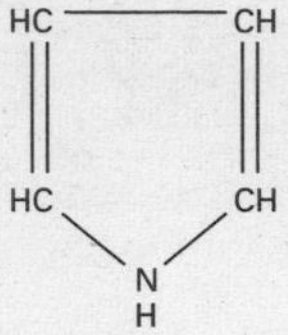

**Pyrrole.**

**pyrrole** An organic nitrogen-containing compound (see formula) that forms part of the structure of *porphyrins.

**pyrrolidine** A saturated heterocyclic compound having one nitrogen atom in a five-membered ring, $C_4H_9N$; r.d. 0.87; m.p. –63°C b.p. 87°C. It is found in certain plants and the ring structure is present in many alkaloids.

**pyruvic acid (2-oxopropanoic acid)** A colourless liquid organic acid, $CH_3COCOOH$.

Pyruvate is an important intermediate compound in metabolism, being produced during *glycolysis and converted to acetyl coenzyme A, required for the *Krebs cycle. Under anaerobic conditions pyruvate is converted to lactate or ethanol.

**Pythagoras of Samos** (*c.* 580–*c.* 500 BC) Greek philosopher and mathematician, who in about 520 BC went to Croton in Italy, where he founded an academy at which numbers and their mystical significance were studied. Pythagoras discovered *irrational numbers and the celebrated *Pythagoras' theorem.

**Pythagoras' theorem** For a right-angled triangle of lengths $h$ (where $h$ is the hypotenuse, the side opposite the right angle), $a$, and $b$ (where $a$ and $b$ are the other two sides), the relationship

$$h^2 = a^2 + b^2.$$

p

**QCD** *See* QUANTUM CHROMODYNAMICS.

**QED** *See* QUANTUM ELECTRODYNAMICS.

**QFD** Quantum flavourdynamics. *See* ELECTROWEAK THEORY.

**QSG** *See* QUASARS.

**QSO** *See* QUASARS.

**QSS** *See* QUASARS.

**quadrat** An ecological sampling unit consisting of a small square area of ground within which all species of interest are noted or measurements taken. Quadrats may be spaced over a larger area to form an overall view when a total survey would be impracticable, or they may be used to sample along a *transect.

**quadrate** A paired bone in the upper jaw of bony fishes, amphibians, reptiles, and birds that articulates with the lower jawbone. It is absent in mammals, being reduced to a small bone (the incus) in the middle ear (*see* EAR OSSICLES).

**quadratic equation** An equation of the second degree having the form $ax^2 + bx + c = 0$. Its roots are:
$$x = [-b \pm \surd(b^2 - 4ac)]/2a.$$

**quadrature** The position of the moon or an outer planet when the line joining it to the earth makes a right angle with a line joining the earth to the sun.

**quadrivalent** Having a valency of four.

**quadrupole mass spectrometer** *See* MASS SPECTROMETRY.

**qualitative analysis** *See* ANALYSIS.

**qualitative variation** *See* DISCONTINUOUS VARIATION.

**quality of sound (timbre)** The quality a musical note has as a result of the presence of *harmonics. A pure note consists only of the fundamental; however, a note from a musical instrument will have several harmonics present, depending on the type of instrument and the way in which it is played. For example, a plucked string (as in a guitar) produces a series of harmonics of diminishing intensity, whereas a struck string (as in a piano) produces a series of harmonics of more nearly equal intensity.

**quanglement** *See* QUANTUM ENTANGLEMENT.

**quantitative analysis** *See* ANALYSIS.

**quantitative inheritance (multifactorial inheritance, polygenic inheritance)** The determination of a particular characteristic, e.g. height or skin colour, by many genes each having a small effect individually (*see* QUANTITATIVE TRAIT). Characteristics controlled in this way show *continuous variation.

**quantitative trait** Any phenotypic trait that shows *continuous variation and can be measured quantitatively in terms of length, weight, concentration, test score, etc. Such traits, which include height, intelligence, and obesity, are determined by the cumulative effects of numerous genes at **quantitative trait loci**.

**quantitative variation** *See* CONTINUOUS VARIATION.

**quantization** The process of constructing a quantum theory for a system, using the original classical theory as a basis. The starting point for such a process is to write the *Lagrangian or *Hamiltonian of the classical system. The formulation of the quantum theory for the system can be performed using a formalism such as *matrix mechanics, or *wave mechanics. The application of these methods leads to the conclusion that energy levels in systems, such as atoms, are discrete (**quantized**) rather than continuous. Before the discovery of quantum mechanics in the mid 1920s, quantization involved a series of ad hoc postulates for atomic systems, such as the *Bohr theory and its extensions.

**quantum** (*pl.* **quanta**) The minimum amount by which certain properties, such as energy or angular momentum, of a system can change. Such properties do not, therefore, vary continuously, but in integral multiples of the relevant quantum, and are described as **quantized**. This concept forms the basis of the *quantum theory. In waves and fields the quantum can be regarded as an excitation, giving a particle-like interpretation to the wave or field. Thus, the quantum of the electromagnetic field is the *photon and the *graviton is the quantum of the gravitational field. *See* QUANTUM ME-CHANICS.

**quantum chaos** The *quantum mechanics of systems for which the corresponding classical system can exhibit *chaos. This subject was initiated by Einstein in 1917, who showed that the quantization conditions associated with the *Bohr theory need to be modified for systems that show chaos in classical mechanics. The subject of quantum chaos is an active field of research in which many basic issues still require clarification. It appears that systems exhibiting chaos in classical mechanics do not necessarily exhibit chaos in quantum mechanics.

**quantum chromodynamics (QCD)** A *gauge theory that describes the strong interaction in terms of quarks and antiquarks and the exchange of massless gluons between them (*see also* ELEMENTARY PARTICLES). Quantum chromodynamics is similar to quantum electrodynamics (QED), with colour being analogous to electric charge and the gluon being the analogue of the photon. The gauge group of QCD is non-Abelian and the theory is much more complicated than quantum electrodynamics; the gauge symmetry in QCD is not a *broken symmetry.

QCD has the important property of *asymptotic freedom – that at very high energies (and, hence, short distances) the interactions between quarks tend to zero as the distance between them tends to zero. Because of asymptotic freedom, perturbation theory may be used to calculate the high energy aspects of strong interactions, such as those described by the *parton model.

**quantum cloning** *See* NO-CLONING THEOREM.

**quantum dot** A quantum-mechanical system, usually made from a semiconductor, in which electrons can be confined into a small region a few nanometres in size containing a few thousand atoms. Such systems act as 'artificial atoms' with their own sets of quantum states for the electrons. Quantum dots have important potential practical applications.

**quantum electrodynamics (QED)** The study of the properties of electromagnetic radiation and the way in which it interacts with charged matter in terms of *quantum mechanics. The collision of a moving electron with a proton, in this theory, can be visualized by a space–time diagram (**Feynman diagram**) in which photons are exchanged (see illustration).

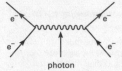

**Quantum electrodynamics.** An example of a Feynman diagram for electron-electron scattering.

*Perturbation-theory calculations using Feynman diagrams enable an agreement between theory and experiment to a greater accuracy than one part in $10^9$ to be obtained. Because of this, QED is the most accurate theory known in physical science. Although many of the effects calculated in QED are very small (about $4 \times 10^{-6}$ eV), such as *energy level splitting in the spectra of *atoms, they are of great significance for demonstrating the physical reality of fluctuations and polarization in the vacuum state.

QED is a *gauge theory for which the gauge *group is Abelian.

**quantum electronics** The application of *quantum optics and the specifically quantum-mechanical properties of electrons to the design of electronic devices.

**quantum entanglement (quanglement)** A phenomenon in quantum mechanics in which a particle or system does not have a definite state but exists as an intermediate form of two 'entangled' states. One of these states is realized when a 'measurement' is made. *See* BELL'S THEOREM.

**quantum field theory** A quantum mechanical theory applied to systems that have an infinite number of degrees of freedom. In quantum field theories, particles are represented by fields that have quantized normal modes of oscillation. For instance, *quantum electrodynamics is a quantum field theory in which the photon is emitted or absorbed by particles; the photon is the quantum of the electromagnetic field. **Relativistic quantum field theories** are used to describe fundamental interactions between elementary particles. They predict the existence of *antiparticles and also show the connection between spin and statistics that leads to the Pauli exclusion principle (*see* SPIN–STATISTICS THEOREM). In spite of their success, it is not clear whether a quantum field theory can give a completely unified description of all interactions (including the gravitational interaction).

**quantum flavourdynamics (QFD)** *See* ELECTROWEAK THEORY.

**quantum gravity** An aspect of *quantum theory that attempts to incorporate the *gravitational field as described by the general theory of *relativity; no such theory has yet been accepted, however. Unlike the *quantum field theories for the other three *fundamental interactions, the procedure of *renormalization does not work for quantum gravity, although there is some evidence that *superstring theory can provide a quantum theory of gravity free of infinities. An approximation to quantum gravity is given by **quantum field theory in curved space–time**, in which the gravitational interactions are treated classically, while all other interactions are treated by *quantum mechanics. An important aspect of quantum field theory in curved space–time is its description of Hawking radiation (*see* HAWKING PROCESS). It is necessary to consider quantum gravity in the very *early universe, just after the *big bang, and the singularities associated with *black holes can also be interpreted as requiring a quantum theory of gravity.

**quantum Hall effect** A quantum mechanical version of the *Hall effect found at very low temperatures, in which the Hall coefficient $R_H$ is proportional to $h/e^2$, where $h$ is the Planck constant and $e$ is the charge of the electron. Thus, the Hall coefficient is quantized. There are two types of quantum Hall effect. The **integer quantum Hall effect** has $R_H$ given as an integer with great precision. It can be used for precision measurements of constants such as $e$ and $h$. In the **fractional quantum Hall effect**, $R_H$ has fractional values.

The integer quantum Hall effect can be understood in terms of noninteracting electrons, whereas the fractional effect is thought to result from many-electron interactions in two-dimensional systems, and to be an example of anyons (*see* QUANTUM STATISTICS).

**quantum interferometry** The study of interference phenomena in which the interference is caused by the quantum-mechanical wave nature of particles such as electrons and neutrons. Fundamental aspects of quantum mechanics can be examined experimentally using such techniques. *See also* NEUTRON INTERFEROMETER.

**quantum jump** A change in a system (e.g. an atom or molecule) from one quantum state to another.

**quantum mechanics** A system of mechanics based on *quantum theory, which arose out of the failure of classical mechanics and electromagnetic theory to provide a consistent explanation of both electromagnetic waves and atomic structure. Many phenomena at the atomic level puzzled physicists at the beginning of the 20th century because there seemed to be no way of explaining them without making use of incompatible concepts. One such phenomenon was the emission of *electrons from the surface of a metal illuminated by light. Einstein realized that the classical description of light as a wave on an electromagnetic field could not explain this *photoelectric effect, as it is called. Experiments showed that electrons would be emitted only if the incident light was of a sufficiently short wavelength, while the intensity of the light appeared not to be relevant. It seemed not to make sense that small ripples of short wavelength could easily knock electrons out of the metal, but a huge tidal wave of long wavelength could not. In 1905 Einstein abandoned classical mechanics and sought an explanation of this photoelectric effect in Planck's work on thermal radiation (*see* PLANCK'S RADIATION LAW). In this work light energy is regarded as being imparted to matter in discrete packets rather than continuously, as one might expect from a wave. Einstein assumed that in the photoelectric effect, light behaves as a shower of particles, each with energy $E$ given by Planck's expression:

$$E = hf,$$

where $f$ is the light's frequency and $h$ is the *Planck constant. Each particle of light, which Einstein called a *photon, would impart its energy to a single electron in the metal. The electron would be liberated only if the photon could impart at least the required amount of energy. However many photons were falling on the surface of the metal, no electrons would be liberated unless individual photons had the required energy ($hf$) to break the attractive forces holding the electrons in the metal. This elegantly quantified reversion to Newton's corpuscular theory of light by Einstein was one of the milestones in the development of quantum mechanics.

Further verification that light could indeed behave as a shower of particles came from the *Compton effect. In Compton scattering, X-radiation is scattered off an electron in a manner that resembles a particle collision. The momentum imparted to the electron can be predicted by assuming that the X-ray possesses the momentum of a photon. An expression for photon momentum is suggested by the classical theory of radiation pressure. It is known that if energy is transported by an electromagnetic wave at a rate $W$ joules per unit area per second, the wave exerts a radiation pressure $W/c$, where $c$ is the speed of light. Planck's expression for the energy of photons therefore led to an equivalent expression for the momentum $p$ of these photons:

$$p = h/\lambda,$$

where $\lambda$ is the wavelength of the light. Experimental studies of the Compton effect produce results in good agreement with this expression.

Both the photoelectric effect and the Compton effect imply that light imparts energy and momentum to matter in the form of packets. It is as if energy and momentum are fundamental 'currencies' of physical interaction, and that these currencies exist in denominations that are all multiples of the Planck constant. These quantities are said to be 'quantized' and a packet of energy or momentum is called a *quantum. Quantum mechanics is essentially concerned with the exchange of these quanta of energy and momentum. For more than a century before the birth of quantum mechanics, experiments had indicated that light behaves as a wave. The successful explanation of the photoelectric and Compton effects demonstrated that in some situations light interacts with matter as if it is a shower of particles. The principle that two models are required to explain the nature of light was called by Niels Bohr *complementarity. This principle was extended by the French aristocrat, Louis de Broglie, who suggested in 1923 that particles of matter might also behave as waves in certain circumstances.

Louis de Broglie received the Nobel Prize for this idea in 1929 after the successful measurement of the *de Broglie wavelength of an electron in 1927 by Clinton Davisson and Lester Germer, who had observed the *diffraction of electrons by single crystals of nickel. The behaviour of individual electrons seemed random and unpredictable, but when a large number had passed through the crystal a typical diffraction pattern emerged. This provided proof that the electron, which until then had been thought of simply as a particle of matter, could under the right circumstances exhibit wavelike properties. Classical mechanics and electromagnetism were based on two kinds of entity: matter and fields. In classical physics, matter consists of particles and waves are oscillations on a field. Quantum mechanics blurs the distinction between matter and field. Modern physicists are forced to concede that the universe is made up of entities that exhibit a *wave–particle duality.

A new representation of matter and field is needed to fully appreciate this wave–particle duality. In quantum mechanics, an electron is represented by a *complex number called a *wave function that depends on time and space coordinates. The wave function behaves like a classical wave displacement on a medium (e.g. on a string), exhibiting wave behaviour, interference, diffraction, etc. However, unlike a classical wave displacement, the wave function is essentially a complex quantity. Since an electron's observable properties do not involve complex numbers, it follows that an electron's wave function cannot itself be identified with a single physical property of the electron. The diffraction of electrons observed by Davisson and Germer revealed that, although the behaviour of the individual electrons is random and unpredictable, when a large number have passed through the apparatus, a diffraction pattern is formed whose intensity distribution is proportional to the intensity of the associated wave function. The intensity of the wave function, $\Psi$, is the square of its *absolute value $|\Psi|^2$. Therefore, although an electron's wave function itself has no physical significance, the square of its absolute value at a point turns out to be proportional to the probability of finding an electron at that point.

The electron wave function must satisfy a wave equation based on the conservation of energy and momentum for the electron. There are two main ways of treating this wave equation: a classical or a relativistic treatment. The resulting wave equations are called **eigenvalue equations** because they have the same form as equations in a branch of mathematics called **eigenvalue problems**, i.e.

$$\Omega\Psi = \omega\Psi,$$

where $\Omega$ is some mathematical operation (multiplication by some number, differentiation, etc.) on the wave function $\Psi$ and $\omega$, called the **eigenvalue** in quantum mechanics, is always a real number. In such an equation the wave function is often called the **eigenfunction**. This form of treatment of quantum mechanics is known as **wave mechanics** (*see also* SCHRÖDINGER EQUATION).

The energy $E$ and momentum $p$ of an electron are associated with the frequency $f$ and wavelength $\lambda$ of the electron's wave using the expressions $E = hf$ and $p = h/\lambda$. While the wave equation expresses the behaviour of the wavelike properties of a particle, it does not define the physical attributes it has as a particle. As a particle, the electron has an easily defined spatial and temporal position, not possessed by an oscillation of some kind of field, whose influence extends over a region of space and time. The incompatibility of these two views of the electron leads to the Heisenberg *uncertainty principle. Heisenberg recognized that if matter had wavelike properties, the physical attributes normally associated with matter (position, momentum, kinetic energy, etc.) would have to be expressed in a statistical, rather than a deterministic, manner. This is illustrated by the electron diffraction patterns of Davisson and Germer. Individual electrons somehow fell onto the apparatus to form a pattern statistically consistent with the intensity of a wave function. It is as if the final wave function were made up of a superposition of all the possible positions that the electrons could fall onto, the waves of electrons constructively and destructively interfering to form the final diffraction pattern.

It is known that a clever superposition of waves of different wavelengths can lead to a construction of a wave packet of finite extension (*see* FOURIER ANALYSIS). However, to produce a packet that exists at a point of zero width requires an infinite number of waves to superimpose. Heisenberg realized that these packets of waves must account for the way in which matter particles, such as electrons, could retain some semblance of their particle-like qualities. However, this must also mean that there is an inherent un-

certainty in position and momentum associated with electrons and indeed all particles of matter (*see* UNCERTAINTY PRINCIPLE). Since waves of different wavelength correspond to different possible momentum values of an electron, a superposition of such waves to form a particle at a point would correspond to an infinite uncertainty in the momentum of the electron. Therefore the more one knows about the position of an electron the less one will know about its momentum and vice versa. A similar uncertainty between the energy of the electron and its temporal position also exists. Quantities that are related by such an uncertainty principle are said to be **incompatible**.

An alternative to the wave mechanical treatment of quantum mechanics is an equivalent formalism called *matrix mechanics, which is based on mathematical operators. *See also* HIDDEN-VARIABLES THEORY; BELL'S THEOREM.

**quantum number** *See* ATOM; SPIN.

**quantum optics** The study of those aspects of light that rely on *quantum mechanics. Quantum optics makes use of the *quantum theory of radiation to describe photons, *coherent radiation, and the interaction of photons and atoms. The study of *lasers is an important branch of quantum optics; other applications include photonics and *quantum electronics.

**quantum phase transition** A phase transition that occurs at absolute zero temperature when some parameter, such as a magnetic field or pressure, is changed. In contrast to ordinary phase transitions, which are associated with thermal fluctuations, quantum phase transitions are associated with quantum fluctuations. There are several systems in condensed matter in which such phase transitions can be observed.

**quantum simulation** The mathematical modelling of systems of large numbers of atoms or molecules by computer studies of relatively small clusters. It is possible to obtain information about solids and liquids in this way and to study surface properties and reactions.

**quantum spin liquid** A quantum-mechanical system in which a large number of electron spins are coupled together to give a distinct quantum system. Quantum spin liquids can be used to investigate aspects of the many-body problem in quantum mechanics and have potentially important applications in technology.

**quantum state** The state of a quantized system as described by its quantum numbers. For instance, the state of a hydrogen *atom is described by the four quantum numbers $n$, $l$, $m_l$, $m_s$. In the ground state they have values 1, 0, 0, and ½ respectively.

**quantum statistics** A statistical description of a system of particles that obeys the rules of *quantum mechanics rather than classical mechanics. In quantum statistics, energy states are considered to be quantized. **Bose–Einstein statistics** apply if any number of particles can occupy a given quantum state. Such particles are called **bosons**. Bosons have an angular momentum $nh/2\pi$, where $n$ is zero or an integer and $h$ is the Planck constant. For identical bosons the *wave function is always symmetric. If only one particle may occupy each quantum state, **Fermi–Dirac statistics** apply and the particles are called **fermions**. Fermions have a total angular momentum $(n + ½)h$ and any wave function that involves identical fermions is always antisymmetric.

The relation between the spin and statistics of particles is given by the *spin–statistics theorem.

In two-space dimensions, it is possible that there are particles (or *quasiparticles) that have statistics intermediate between bosons and fermions. These particles are known as **anyons**; for identical anyons the wave function is not symmetric (a phase sign of +1) or antisymmetric (a phase sign of –1), but interpolates continuously between +1 and –1. Anyons may be involved in the fractional *quantum Hall effect.

**quantum theory** The theory devised by Max Planck in 1900 to account for the emission of the black-body radiation from hot bodies. According to this theory energy is emitted in quanta (*see* QUANTUM), each of which has an energy equal to $hv$, where $h$ is the *Planck constant and $v$ is the frequency of the radiation. This theory led to the modern theory of the interaction between matter and radiation known as *quantum mechanics, which generalizes and replaces classical mechanics and Maxwell's electromagnetic theory. In **nonrelativistic quantum theory** particles are assumed to be neither created nor destroyed, to move slowly relative to the speed of light, and to have a mass that does not change with velocity. These assumptions

apply to atomic and molecular phenomena and to some aspects of nuclear physics. **Relativistic quantum theory** applies to particles that travel at or near the speed of light.

**quantum theory of radiation** The study of the emission and absorption of *photons of electromagnetic radiation by atomic systems using *quantum mechanics. Photons are emitted by atoms when a transition occurs from an excited state to the ground state. If an atom is exposed to external electromagnetic radiation a transition can occur from the ground state to an excited state by absorption of a photon. An excited atom can lose the energy it has gained by *stimulated emission. However, an atom can also emit a photon in the absence of external electromagnetic radiation, a process called *spontaneous emission. The quantum theory of radiation was initiated by Einstein in 1916–17, as an extension of *Planck's radiation law. The theory is quantified by the *Einstein coefficients. The quantum theory of radiation is the basis of the theory behind the operation of *lasers and *masers.

**quantum Turing machine** A mathematical concept which has the same relation to a quantum computer as a *Turing machine does to a traditional computer. This means that a quantum Turing machine is concerned with the problem of whether systematic algorithms exist for problems in quantum computation.

**quantum well** A potential well in which electrons in a three-dimensional system are confined to a plane, i.e. to two dimensions. Quantum wells can be made with a semiconductor that is sandwiched between layers of materials with larger energy band gaps. These wells are used to study two-dimensional systems and also have technological applications including a type of *laser (a **quantum well laser**).

**quantum wire** A linear conductor that is narrow enough for quantum effects to affect the resistance. If the wire diameter is small then the electrons have quantized energy levels for motion perpendicular to the axis. Consequently the resistance of the wire is also quantized. Carbon nanotubes have been investigated as practical implementations of quantum wires.

**quantum Zeno effect** A phenomenon in which constant observation of an unstable particle prevents it from decaying. In general, such a particle would be expected to decay within a time τ. If numerous measurements are made at intervals within time τ, the wavefunction is constantly collapsed, and the lifetime of the particle is prolonged. It can also be shown that if measurements are made at time intervals longer than τ there is an **anti-Zeno effect** in which the decay rate increases. These effects were predicted theoretically in 1977 and have subsequently been observed to occur in a number of experiments. The name is an allusion to a paradox put forward by the Greek philosopher Zeno of Elea (*b. c.* 490 BC). In the 'arrow paradox' he argued that at every moment of time, a moving arrow is occupying a definitive position. Therefore at every moment, the arrow is immobile, and the arrow never moves.

**quarantine** A period of isolation imposed on an animal that moves from an area where particular diseases are prevalent to an area where those diseases are not prevalent. In the UK domestic animals or livestock entering the country are quarantined for the necessary incubation period (*see* INFECTION) in order to prevent the spread of a particular disease.

**quark** *See* ELEMENTARY PARTICLES.

**quark confinement** The hypothesis that free quarks can never be seen in isolation. It is a result of *quantum chromodynamics, in which the property of *asymptotic freedom means that the interactions between quarks get weaker as the distance between them gets smaller, and tends to zero as the distance between them tends to zero. Conversely, the attractive interactions between quarks get stronger as the distance between them gets greater, and the quark-confinement hypothesis is that the quarks cannot escape from one another. It is possible that at very high temperatures, such as those in the *early universe, quarks may become free. The temperature at which this occurs is called the **deconfinement temperature**. The hypothesis of quark confinement has not been proved theoretically in a conclusive way, although there is a lot of evidence for it.

**quark–gluon plasma** A state of matter in which quarks and gluons are not confined into baryons and mesons but exist as a hot plasma. It is thought that this state existed in the early universe until there was a phase transition about $10^{-5}$ s after the big bang when the universe cooled to a sufficiently

low temperature that *quark confinement into hadrons occurred. This state was created in the laboratory in the early years of the 21st century.

**quart** A unit of capacity equal to one quarter of a *gallon.

**quarter-wave plate** *See* RETARDATION PLATE.

**quartile** *See* PERCENTILE.

**quartz** The most abundant and common mineral, consisting of crystalline silica (silicon dioxide, $SiO_2$), crystallizing in the trigonal system. It has a hardness of 7 on the Mohs' scale. Well-formed crystals of quartz are six-sided prisms terminating in six-sided pyramids. Quartz is ordinarily colourless and transparent, in which form it is known as **rock crystal**. Coloured varieties, a number of which are used as gemstones, include *amethyst, citrine quartz (yellow), rose quartz (pink), milk quartz (white), smoky quartz (grey-brown), *chalcedony, *agate, and *jasper. Quartz occurs in many rocks, especially igneous rocks such as granite and quartzite (of which it is the chief constituent), metamorphic rocks such as gneisses and schists, and sedimentary rocks such as sandstone and limestone. The mineral has the property of being piezoelectric and hence is used to make oscillators for clocks (*see* QUARTZ CLOCK), radios, and radar instruments. It is also used in optical instruments and in glass, glaze, and abrasives.

**quartz clock** A clock based on a piezoelectric crystal of quartz. Each quartz crystal has a natural frequency of vibration, which depends on its size and shape. If such a crystal is introduced into an oscillating electronic circuit that resonates at a frequency very close to that of the natural frequency of the crystal, the whole circuit (including the crystal) will oscillate at the crystal's natural frequency and the frequency will remain constant over considerable periods (a good crystal will maintain oscillation for a year with an accumulated error of less than 0.1 second). In a quartz clock or watch the alternating current from the oscillating circuit containing such a crystal is amplified and the frequency subdivided until it is suitable to drive a synchronous motor, which in turn drives a gear train to operate hands. Alternatively it is used to activate a digital display.

**quasars** A class of astronomical objects that appear on optical photographs as starlike but have large *redshifts quite unlike those of stars. They were first observed in 1961 when it was found that strong radio emission was emanating from many of these starlike bodies. Over 200 000 such objects are now known and their redshifts can be as high as 6.4. The redshifts are characteristic of the *expansion of the universe. This **cosmological redshift** is the explanation of the high observed redshifts of quasars favoured by most astronomers. (A few, however, maintain that the redshift could be a local Doppler effect, characteristic of movement relative to the earth and sun of nearby objects in the Galaxy, or a gravitational effect.) If the redshifts are cosmological, quasars are the most distant objects in the universe, some being up to $10^{10}$ light-years away. The exact nature of quasars is unknown but it is believed that they are the nuclei of galaxies in which there is violent activity. The luminosity of the nucleus is so much greater than that of the rest of the galaxy that the source appears pointlike. It has been proposed that the power source in a quasar is a supermassive *black hole accreting material from the stars and gas in the surrounding galaxy.

The name 'quasar' is today regarded as a contraction of quasistellar object (QSO) or quasistellar galaxy (QSG). However, the first quasars to be discovered were radio sources, and the term was coined as a contraction of 'quasi-stellar radio source'.

**quasicrystal** A type of solid structure in which there is long-range order but in which the symmetry of the structure is not allowed in crystallography. AlMn is an example of a substance that forms quasicrystals.

**quasiparticle** A long-lived single-particle *excitation in the quantum theory of many-body systems, in which the excitations of the individual particles are modified by their interactions with the surrounding medium.

**Quaternary** Formerly, the second period of the Cenozoic era, extending from 2 million years ago, following the *Tertiary period, to the present. The beginning of the Quaternary was usually based on the onset of a worldwide cooling. During the period four principal glacial phases occurred in Europe and North America, in which ice advanced towards the equator, separated by interglacials during which conditions became warmer and the ice sheets and glaciers re-

treated. The last glacial ended about 10 000 years ago. The Quaternary has been replaced by the *Neogene period.

**quaternary ammonium compounds** *See* AMINE SALTS.

**quaternary structure** *See* PROTEIN.

**quaternion** Symbol $Q$. A mathematical quantity defined by $Q = a + bi + cj + dk$, where $a$, $b$, $c$, and $d$ are real numbers and $i$, $j$, and $k$ are unit vectors in the $x$, $y$, and $z$ directions respectively. Quaternions were invented by Sir William *Hamilton in 1843 and have been used extensively in physics, especially for problems involving rotation. Quaternions have close connections with the way in which *spin is analysed in quantum mechanics.

**quenching** **1.** (in metallurgy) The rapid cooling of a metal by immersing it in a bath of liquid in order to improve its properties. Steels are quenched to make them harder but some nonferrous metals are quenched for other reasons (copper, for example, is made softer by quenching). **2.** (in physics) The process of inhibiting a continuous discharge in a *Geiger counter so that the incidence of further ionizing radiation can cause a new discharge. This is achieved by introducing a quenching vapour, such as methane mixed with argon or with neon, into the tube.

**quicklime** *See* CALCIUM OXIDE.

**quinhydrone electrode** A *half cell consisting of a platinum electrode in an equimolar solution of quinone (cyclohexadiene-1,4-dione) and hydroquinone (benzene-1,4-diol). It depends on the oxidation–reduction reaction

$$C_6H_4(OH)_2 \rightleftharpoons C_6H_4O_2 + 2H^+ + 2e$$

**quinine** A white solid, $C_{20}H_{24}N_2O_2.3H_2O$, m.p. 57°C. It is a poisonous alkaloid occurring in the bark of the South American cinchona tree, although it is now usually produced synthetically. It forms salts and is toxic to the malarial parasite, and so quinine and its salts are used to treat malaria; in small doses it may be prescribed for colds and influenza. In dilute solutions it has a pleasant astringent taste and is added to some types of tonic water.

**quinol** *See* BENZENE-1,4-DIOL.

**quinone** **1.** *See* CYCLOHEXADIENE-1,4-DIONE.

**2.** Any similar compound containing C=O groups in an unsaturated ring.

**quintessence** A form of energy postulated to be causing the expansion of the universe to accelerate; i.e. the *dark energy in the universe. Theories of quintessence do not assume that dark energy is described by a *cosmological constant. No convincing theory of quintessence has yet emerged.

q

**Rabi model** A quantum mechanical model describing the interaction between a two-level system and an electromagnetic field. The model was originally proposed by the Galician-born American physicist Isidor Rabi (1898–1988) in the context of *nuclear magnetic resonance, with the two levels being spin-up and spin-down and the field being an external magnetic field.

**race** 1. (in biology) A category used in the *classification of organisms that consists of a group of individuals within a species that are geographically, ecologically, physiologically, or chromosomally distinct from other members of the species. The term is frequently used in the same sense as *subspecies. **Physiological races**, for example, are identical in appearance but differ in function. They include strains of fungi adapted to infect different varieties of the same crop species.

spike
(e.g. plantain)

raceme
(e.g. lupin)

corymb
(e.g. candytuft)

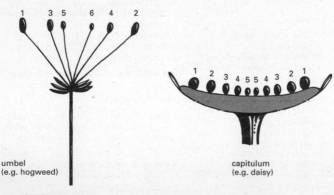

umbel
(e.g. hogweed)

capitulum
(e.g. daisy)

1 = oldest flower

**Racemose inflorescences.**

**2.** (in anthropology) A distinct human type possessing several characteristics that are genetically inherited. The major races are Mongolian, Caucasian, and Ethiopian.

**racemate** *See* RACEMIC MIXTURE.

**raceme** A type of *racemose inflorescence in which the main flower stalk is elongated and bears stalked flowers. An example is the lupin. *See also* PANICLE.

**racemic mixture (racemate)** A mixture of equal quantities of the (+)- or $d$- and (−)- or $l$- forms of an optically active compound. Racemic mixtures are denoted by the prefix (±)- or $dl$-. A racemic mixture shows no *optical activity.

**racemization** A chemical reaction in which an optically active compound is converted into a *racemic mixture.

**racemose inflorescence (indefinite inflorescence)** A type of flowering shoot (*see* INFLORESCENCE) in which the growing region at the tip of the flower stalk continues to produce new flower buds during growth. As a result, the youngest flowers are at the top and the oldest flowers are at the base of the stalk. In a flattened inflorescence, the youngest flowers are in the centre and the oldest flowers are on the outside. Types of racemose inflorescence include the *capitulum, *catkin, *corymb, *raceme, *spadix, *spike, and *umbel (see illustration). *Compare* CYMOSE INFLORESCENCE.

**rachis (rhachis) 1.** The main axis of a compound leaf or an inflorescence. **2.** The shaft of a *feather. **3.** The backbone.

**rad** *See* RADIATION UNITS.

**radar (radio detection and ranging)** A method of detecting the presence, position, and direction of motion of distant objects (such as ships and aircraft) by means of their ability to reflect a beam of electromagnetic radiation of centimetric wavelengths. It is also used for navigation and guidance. It consists of a transmitter producing radio-frequency radiation, often pulsed, which is fed to a movable aerial from which it is transmitted as a beam. If the beam is interrupted by a solid object, a part of the energy of the radiation is reflected back to the aerial. Signals received by the aerial are passed to the receiver, where they are amplified and detected. An echo from a reflection of a solid object is indicated by a sudden rise in the detector output. The time taken for a pulse

to reach the object and be reflected back ($t$) enables the distance away ($d$) of the target to be calculated from the equation $d = ct/2$, where $c$ is the speed of light. In some systems the speed of the object can be measured using the *Doppler effect. The output of the detector is usually displayed on a cathode-ray tube in a variety of different formats (see illustration).

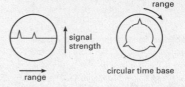

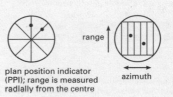

**Radar.** Types of cathode-ray tube radar display.

**radial symmetry** The arrangement of parts in an organ or organism such that cutting through the centre of the structure in any direction produces two halves that are mirror images of each other. The stems and roots of plants usually show radial symmetry, while all animals belonging to the Cnidaria (e.g. jellyfish) and Echinodermata (e.g. starfish) are radially symmetrical – and typically sessile – in their adult form. The term **actinomorphy** is used to describe radial symmetry in flowers (e.g. a buttercup flower). *Compare* BILATERAL SYMMETRY.

**radial velocity** *See* LINE-OF-SIGHT VELOCITY.

**radian** *See* CIRCULAR MEASURE.

**radiance** Symbol $L_e$. The radiant intensity per unit transverse area, in a given direction, of a source of radiation. It is measured in $W\,sr^{-1}\,m^{-2}$.

**radiant energy** Energy transmitted as electromagnetic radiation.

**radiant exitance** *See* EXITANCE.

**radiant flux** Symbol $\Phi_e$. The total power emitted, received, or passing in the form of electromagnetic radiation. It is measured in watts.

**radiant intensity** Symbol $I_e$. The *radiant flux per unit solid angle emitted by a point source. It is measured in watts per steradian.

**radiation** 1. Energy travelling in the form of electromagnetic waves or photons. 2. A stream of particles, especially alpha- or beta-particles from a radioactive source or neutrons from a nuclear reactor. 3. *See* ADAPTIVE RADIATION.

**radiation belts** *See* VAN ALLEN BELTS.

**radiation damage** Harmful changes that occur to inanimate materials and living organisms as a result of exposure to energetic electrons, nucleons, fission fragments, or high-energy electromagnetic radiation. In inanimate materials the damage may be caused by electronic excitation, ionization, transmutation, or displacement of atoms. In organisms, these mechanisms can cause changes to cells that alter their genetic structure, interfere with their division, or kill them. In humans, these changes can lead to **radiation sickness**, **radiation burns** (from large doses of radiation), or to long-term damage of several kinds, the most serious of which result in various forms of cancer (especially leukaemia).

**radiationless decay** Decay of an atom or molecule from an excited state to a lower energy state without the emission of electromagnetic radiation. A common example of a radiationless process is the *Auger effect, in which an electron rather than a photon is emitted as a result of decay.

**radiation pressure** The pressure exerted on a surface by electromagnetic radiation. As radiation carries momentum as well as energy it exerts a force when it meets a surface, i.e. the *photons transfer momentum when they strike the surface. The pressure is usually negligible on large bodies, for example, the pressure of radiation from the sun on the surface of the earth is of the order of $10^{-5}$ pascal, but on small bodies it can have a considerable effect, driving them away from the radiation source. Radiation pressure is also important in the interiors of stars of very high mass.

**radiation temperature** The surface temperature of a celestial body as calculated by *Stefan's law, assuming that the body behaves as a *black body. The radiation temperature is usually measured over a narrow portion of the electromagnetic spectrum, such as the visible range (which gives the **optical temperature**).

**radiation units** Units of measurement used to express the *activity of a radionuclide and the *dose of ionizing radiation. The units **curie**, **roentgen**, **rad**, and **rem** are not coherent with SI units but their temporary use with SI units has been approved while the derived SI units **becquerel**, **gray**, and **sievert** become familiar.

The becquerel (Bq), the SI unit of activity, is the activity of a radionuclide decaying at a rate, on average, of one spontaneous nuclear transition per second. Thus $1$ Bq $= 1$ s$^{-1}$. The former unit, the curie (Ci), is equal to $3.7 \times 10^{10}$ Bq. The curie was originally chosen to approximate the activity of 1 gram of radium–226.

The gray (Gy), the SI unit of absorbed dose, is the absorbed dose when the energy per unit mass imparted to matter by ionizing radiation is 1 joule per kilogram. The former unit, the rad (rd), is equal to $10^{-2}$ Gy.

The sievert (Sv), the SI unit of dose equivalent, is the dose equivalent when the absorbed dose of ionizing radiation multiplied by the stipulated dimensionless factors is $1$ J kg$^{-1}$. As different types of radiation cause different effects in biological tissue a weighted absorbed dose, called the **dose equivalent**, is used in which the absorbed dose is modified by multiplying it by dimensionless factors stipulated by the International Commission on Radiological Protection. The former unit of dose equivalent, the rem (originally an acronym for *r*oentgen *e*quivalent *m*an), is equal to $10^{-2}$ Sv.

In SI units, exposure to ionizing radiation is expressed in coulombs per kilogram, the quantity of X- or gamma-radiation that produces ion pairs carrying 1 coulomb of charge in 1 kilogram of pure dry air. The former unit, the roentgen (R), is equal to $2.58 \times 10^{-4}$ C kg$^{-1}$.

**radiative collision** A collision between charged particles in which part of the kinetic energy is radiated in the form of photons.

**radiative forcing** Any perturbation in the balance of incoming and outgoing radiation (i.e. the net irradiance) at the upper surface of the earth's atmosphere, the troposphere. Such a perturbation forces a change in the earth's climate system until a new irradiance balance is attained. Positive radiative forcing has a warming effect, whereas negative radiative forcing causes the system to cool. Various factors, both natural and anthropogenic (human-derived), can lead to radiative forcing, notably changes in incoming solar radiation or atmospheric reflectivity (albedo). For example, the albedo is increased by a rise in concentration of tropospheric sulphate aerosols derived from volcanic eruptions, producing a cooling effect. Conversely, greenhouse gases, such as carbon dioxide and methane, absorb outgoing infrared radiation, causing positive forcing and hence warming.

**radical** 1. (in chemistry) A group of atoms, either in a compound or existing alone. *See* FREE RADICAL; FUNCTIONAL GROUP. 2. (in mathematics) A root of a number or a quantity. The symbol √ is called the **radical sign**.

(((•))) SEE WEB LINKS

• Information about IUPAC nomenclature
• List of radical names

**radicle** The part of a plant embryo that develops into the root system. The tip of the radicle is protected by a root cap and points towards the micropyle. On germination it breaks through the testa and grows down into the soil. *Compare* PLUMULE.

**radio** A means of transmitting information in which the transmission medium consists of electromagnetic radiation. Information is transmitted by means of the *modulation of a *carrier wave in a transmitter; the modulated carrier wave is fed to a transmitting aerial from which it is broadcast through the atmosphere or through space. A receiving aerial forms part of a *resonant circuit, which can be tuned to the frequency of the carrier wave, enabling the receiver that it feeds selectively to amplify and then to demodulate the transmitted signal. A replica of the original information is thus produced by the receiver. *See also* RADIO TRANSMISSION.

**radioactive age** The age of an archaeological or geological specimen as determined by a process that depends on a radioactive decay. *See* CARBON DATING; FISSION-TRACK DATING; POTASSIUM–ARGON DATING; RUBIDIUM–STRONTIUM DATING; URANIUM–LEAD DATING.

**radioactive dating** The determination of the age of a geological or archaeological specimen by a process that depends on radioactive decay. The age so determined is known as the **radioactive age**. *See* CARBON DATING; FISSION-TRACK DATING; POTASSIUM–ARGON DATING; RUBIDIUM–STRONTIUM DATING; URANIUM–LEAD DATING.

**radioactive series** A series of radioactive nuclides in which each member of the series is formed by the decay of the nuclide before it. The series ends with a stable nuclide. Three radioactive series occur naturally, those headed by thorium–232 (**thorium series**), uranium–235 (**actinium series**), and uranium–238 (**uranium series**). All three series end with an isotope of lead. The neptunium series starts with the artificial isotope plutonium–241, which decays to neptunium–237, and ends with bismuth–209.

**radioactive tracing** *See* LABELLING.

**radioactive waste (nuclear waste)** Any solid, liquid, or gaseous waste material that contains *radionuclides. These wastes are produced in the mining and processing of radioactive ores, the normal running of nuclear power stations and other reactors, the manufacture of nuclear weapons, and in hospitals and research laboratories. Because high-level radioactive wastes can be extremely dangerous to all living matter and because they may contain radionuclides having half-lives of many thousands of years, their disposal has to be controlled with great stringency. High-level waste (e.g. spent nuclear fuel) requires to be cooled artificially and is therefore stored for several decades by its producers before it can be disposed of. Intermediate-level waste (e.g. processing plant sludge and reactor components) is solidified, mixed with concrete, packed in steel drums, and stored in special sites at power stations before being buried in concrete chambers in deep mines or below the seabed. Low-level waste (e.g. solids or liquids lightly contaminated by radioactive substances) is disposed of in steel drums in special sites in concrete-lined trenches.

**radioactivity** The spontaneous disintegration of certain atomic nuclei accompanied by the emission of alpha particles (helium nuclei), beta particles (electrons or positrons), or gamma radiation (short-wavelength electromagnetic waves).

 **Natural radioactivity** is the result of the spontaneous disintegration of naturally occurring radioisotopes. Many radioisotopes can be arranged in three *radioactive series. The rate of disintegration is uninfluenced by chemical changes or any normal changes in their environment. However, radioactivity can be induced in many nuclides by bombarding them with neutrons or other particles. *See also* DECAY; IONIZING RADIATION; RADIATION UNITS.

**radio astronomy** The study of the radio-frequency radiation emitted by celestial objects, from planets and stars to nebulae, galaxies, and quasars. This branch of astronomy began in 1932 when a US engineer, Karl Jansky (1905–40), first detected radio waves from outside the earth's atmosphere by pointing a radio antenna at the centre of the Galaxy in Sagittarius. *See* RADIO SOURCE; RADIO TELESCOPE; RADIO WINDOW.

**radiobiology** The branch of biology concerned with the effects of radioactive substances on living organisms and the use of radioactive tracers to study metabolic processes (*see* LABELLING).

**radiocarbon dating** *See* CARBON DATING.

**radiochemistry** The branch of chemistry concerned with radioactive compounds and with ionization. It includes the study of compounds of radioactive elements and the preparation and use of compounds containing radioactive atoms. *See* LABELLING; RADIOLYSIS.

**radio frequencies** The range of frequencies, between about 3 kilohertz and 300 gigahertz, over which electromagnetic radiation is used in radio transmission. It is subdivided into eight equal bands: *very low frequency, *low frequency, *medium frequency, *high frequency, *very high frequency, *ultra high frequency, *super high frequency, and *extremely high frequency.

**radio galaxies** A *radio source outside the Galaxy that has been identified with an optically visible galaxy. These radio galaxies are distinguished from normal galaxies by having a radio power output some $10^6$ times greater (i.e. up to $10^{38}$ watts rather than $10^{32}$ W). The source of the radio-frequency energy is associated with violent activity, involving the ejection of relativistic jets of particles from the nucleus of the galaxy. It has been suggested that the radio sources are powered by supermassive *black holes in the nucleus.

**radiogenic** Resulting from radioactive decay.

**radiography** The process or technique of producing images of an opaque object on photographic film or on a fluorescent screen by means of radiation (either particles or electromagnetic waves of short wavelength, such as X-rays and gamma-rays). The photograph produced is called a **radiograph**. The process is widely used in diagnostic *radiology, using X-rays, and in flaw detection in industrial products, using high-energy X-rays, gamma-radiation, neutron beams, and (more recently) beams of charged particles. *See also* AUTORADIOGRAPHY.

**radio interferometry** *See* RADIO TELESCOPE.

**radioisotope (radioactive isotope)** An isotope of an element that is radioactive. *See* LABELLING.

**radiolocation** The location of distant objects by means of *radar.

**radiology** The study and use of X-rays, radioactive materials, and other ionizing radiations for medical purposes, especially for diagnosis (**diagnostic radiology**) and the treatment of cancer and allied diseases (**therapeutic radiology** or **radiotherapy**).

**radiolysis** The use of ionizing radiation to produce chemical reactions. The radiation used includes alpha particles, electrons, neutrons, X-rays, and gamma rays from radioactive materials or from accelerators. Energy transfer produces ions and excited species, which undergo further reaction. A particular feature of radiolysis is the formation of short-lived solvated electrons in water and other polar solvents.

**radiometric dating (radioactive dating)** *See* DATING TECHNIQUES; RADIOACTIVE AGE.

**radionuclide (radioactive nuclide)** A *nuclide that is radioactive.

**radiopaque (radio-opaque)** Describing a medium that is opaque to X-rays and gamma rays. Examples are barium salts, used in diagnostic radiology of the digestive tract. *See also* OPACITY.

**radiopharmaceuticals** Compounds used in medicine that have a radioactive atom in the molecule. Radiopharmaceuticals are used both for diagnostic purposes (as in radionuclide imaging) or for therapy (e.g. certain cancer treatments).

**radiosonde** A meteorological instrument that measures temperature, pressure, humidity, and winds in the upper atmosphere. It consists of a package of instruments and a radio transmitter attached to a balloon. The data is relayed back to earth by the transmitter. The position of the balloon can be found by radar and from its changes in position the wind velocities can be calculated.

**radio source** An astronomical object that has been observed with a *radio telescope to emit radio-frequency electromagnetic radiation. Radio sources within the Galaxy include Jupiter, the sun, pulsars, star-forming regions, supernova remnants, and background radiation arising from *synchrotron radiation. Sources outside the Galaxy include galaxies, *radio galaxies, and *quasars. Radio sources were formerly known as **radio stars**.

**radio star** *See* RADIO SOURCE.

**radio telescope** An instrument for detecting and measuring electromagnetic radiation of radio frequencies that have passed through the *radio window in the earth's atmosphere and reached the surface of the earth. There are a great diversity of *radio sources within the universe and radio telescopes are required to detect both continuous emissions and specific spectral lines. They therefore require the highest possible angular resolution so that the details of radio sources can be studied and they should be able to pick up weak signals. The simplest radio telescope consists of a paraboloidal steerable-dish aerial together with ancillary amplifiers. The paraboloidal dish surface reflects the incoming signal to the principal

focus of the reflector. At this point the radio-frequency signals are amplified up to 1000 times and converted to a lower, intermediate, frequency before transmission by cable to the control building. Here the intermediate frequency is amplified again and passed to the detector and display unit. As the radio waves arriving from the surface of the reflector at the focus must be in phase, the surface of the dish must be very accurately constructed; for example, a 100-metre-diameter dish must be accurate to the nearest millimetre, when receiving radiation of 1 cm wavelength. To overcome the problem of constructing large dishes to such a high accuracy, the technique of **radio interferometry** has been developed. In this technique an array of small aerials connected by cable is used to simulate a large dish aerial. In earth-rotation **aperture synthesis**, the individual positions and displacements of an array of only a few such small aerials can be made to simulate an enormous dish aerial as the earth rotates. All but the smallest steerable dishes are constructed from metal mesh so that wind can pass through them. A few very large fixed dishes have been built into the earth's surface.

**radiotherapy** *See* RADIOLOGY.

**radio transmission** The transmission of radio waves from a transmitting aerial to a receiving aerial. The radiation may take several paths (see illustration). The sum of the line-of-sight ground wave, the reflected ground wave, and the surface wave is called

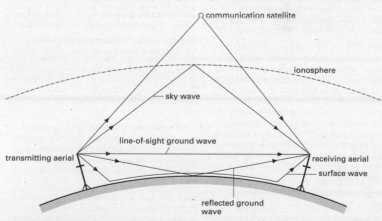

**Radio transmission.**

the **ground wave** (or **tropospheric wave**). **Sky waves** (or **ionospheric waves**) are reflected by the ionosphere and enable long-distance transmissions to be made. The ionization of atoms and molecules in the ionosphere is caused largely by solar ultraviolet and X-radiation and therefore conditions differ between night and day. Ionization in the lower E-region of the ionosphere falls off at night in the absence of sunlight and ions and electrons tend to recombine. However, in the less dense (higher) F-region there are fewer collisions between ions and electrons and therefore fewer recombinations at night. The F-region is therefore a more effective reflector at night.

The UHF and VHF waves used in television broadcasting penetrate the ionosphere with little reflection. Therefore TV broadcasts can only be made over long distances by means of artificial satellites. *See also* RADIO.

**radiotransparent** Transparent to radiation, especially to X-rays and gamma-rays.

**radio window** A region of the electromagnetic spectrum in the radio-frequency band within which radio waves can be transmitted through the earth's atmosphere without significant reflection or attenuation by constituents of the atmosphere. It extends from about 10 megahertz to 100 gigahertz and enables radiation in this range from celestial radio sources to be picked up by *radio telescopes on the earth's surface. Below 100 MHz incoming radio waves are reflected by the ionosphere and those above 100 GHz are increasingly affected by molecular absorption.

**radium** Symbol Ra. A radioactive metallic element belonging to *group 2 (formerly IIA) of the periodic table; a.n. 88; r.a.m. 226.0254; r.d. ~5; m.p. 700°C; b.p. 1140°C. It occurs in uranium ores (e.g. pitchblende). The most stable isotope is radium–226 (half-life 1602 years), which decays to radon. It is used as a radioactive source in research and, to some extent, in radiotherapy. The element was isolated from pitchblende in 1898 by Marie and Pierre Curie.

**(((⊕))) SEE WEB LINKS**
• Information from the WebElements site

**radius 1.** (in anatomy) The smaller of the two bones in the lower section of the forelimb of a tetrapod vertebrate (*compare*

ULNA). The radius articulates with some carpal bones and the ulna at the wrist and with the *humerus at the elbow. This sophisticated articulation of the radius enables humans (and some other animals) to twist the forearm (*see* PRONATION; SUPINATION). **2.** (in mathematics) *See* CIRCLE.

**radius of curvature** *See* CENTRE OF CURVATURE.

**radius of gyration** Symbol $k$. The square root of the ratio of the *moment of inertia of a rigid body about an axis to the body's mass, i.e. $k^2 = I/m$, where $I$ is the body's moment of inertia and $m$ is its mass. If a rigid body has a moment of inertia $I$ about an axis and mass $m$ it behaves as if all its mass is rotating at a distance $k$ from the axis.

**radius vector** *See* POLAR COORDINATES.

**radon** Symbol Rn. A colourless radioactive gaseous element belonging to group 18 of the periodic table (the *noble gases); a.n. 86; r.a.m. 222; d. 9.73 g dm$^{-3}$; m.p. –71°C; b.p. –61.8°C. At least 20 isotopes are known, the most stable being radon–222 (half-life 3.8 days). It is formed by decay of radium–226 and undergoes alpha decay. It is used in radiotherapy. Radon occurs naturally, particularly in areas underlain by granite, where it is thought to be a health hazard. As a noble gas, radon is practically inert, although a few compounds, e.g. radon fluoride, can be made. It was first isolated by William Ramsey and Robert Whytlaw-Gray (1877–1958) in 1908.

**(((⊕))) SEE WEB LINKS**
• Information from the WebElements site

**radula** A tonguelike organ of molluscs, consisting of a horny strip whose surface is studded with rows of horny teeth for rasping food. In some species it is modified for scraping or boring.

**raffinate** A liquid purified by solvent extraction.

**raffinose** A white solid carbohydrate, $C_{18}H_{32}O_{16}$, m.p. 80°C. It is a trisaccharide (a type of *sugar) consisting of fructose, galactose and glucose. It occurs naturally in sugar-beet and cotton-seed residues.

**rainbow** An optical phenomenon that appears as an arc of the colours of the spectrum across the sky when falling water droplets are illuminated by sunlight from behind the observer. The colours are produced by the refraction and internal reflection of the sunlight by the water drops. Two bows may be visible, the inner ring being known as the primary bow and the outer, in which the colours are reversed, as the secondary bow.

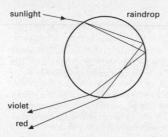

Primary rainbow (one internal reflection)

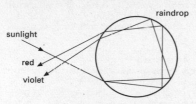

Secondary rainbow (two internal reflections)

**Rainbow.**

**rainfall** *See* PRECIPITATION.

**rainforest** Any major terrestrial *biome in which trees form the dominant plants and annual rainfall is high (over 200 cm). Tropical rainforest is restricted to equatorial regions, such as the Amazon basin, central west Africa, and SE Asia. It is dominated by broadleaved evergreens and shows a very rich species diversity (*see* BIODIVERSITY). The leafy crowns of the trees typically form three layers of canopy, since the trees grow to different heights, which prevents much sunlight from reaching ground level. This limits the number of herbaceous plants and small shrubs that grow on the forest floor, but *epiphytes, vines, and creepers are abundant. The average temperature is about 27°C, which – together with the high humidity – encourages rapid decomposition of leaf litter, releasing minerals that replace those leached from the soil by the heavy rain. If the forest canopy is removed, the soil is destroyed rapidly due to leaching by rain. The

soil of a rainforest, known as **latosol**, is acidic and typically red, due to the oxidation of iron oxide ($Fe_2O_3$) in the topsoil. Rainforests are thought to contain many undiscovered plant species that could be of benefit in the fields of medicine and biotechnology. The continued destruction of rainforest in many parts of the world, particularly in South America and SE Asia (*see* DEFORESTATION), will not only result in the loss of these and other species but also contributes to the *greenhouse effect.

(((●))) **SEE WEB LINKS**

• Website for Rainforest Action Network, an international organization that campaigns to protect the rainforests

**RAM** Random-access memory. The main memory of a computer, fabricated from *integrated circuits, in which data can only be stored temporarily – until the power supply is turned off. RAM consists of arrays of 'cells', each capable of holding one *bit of data. Cells are completely independent so that the access time to any location is fixed (and extremely rapid) hence the term random access.

**r.a.m.** *See* RELATIVE ATOMIC MASS.

**Raman effect** A type of inelastic *scattering of light and ultraviolet radiation discovered in 1928 by Chandrasekara Raman (1880–1970). If a beam of monochromatic light is passed through a transparent substance some of the radiation will be scattered. Although most of the scattered radiation will be the same as the incident frequency, some will have frequencies above (anti-Stokes radiation) and below (Stokes radiation) that of the incident beam. This effect is known as **Raman scattering** and is due to inelastic collisions between photons and molecules leading to changes in the vibrational and rotational energy levels of the molecules. An increase in frequency represents a loss in molecular energy; a decrease represents a gain in molecular energy.

This effect is used in **Raman spectroscopy** for investigating the vibrational and rotational energy levels of molecules. Because the scattering intensity is low, a laser source is used.

(((●))) **SEE WEB LINKS**

• The original paper in *Nature* (1928) by Raman and Krishnan

***Ramapithecus*** *See* SIVAPITHECUS.

**ramjet** *See* JET PROPULSION.

**Ramón y Cajal, Santiago** See GOLGI, CAMILLO.

**Ramsay, Sir William** (1852–1916) British chemist, born in Glasgow. After working under Robert *Bunsen, he returned to Glasgow before taking up professorships at Bristol (1880–87) and London (1887–1912). In the early 1890s he worked with Lord *Rayleigh on the gases in air and in 1894 they discovered *argon. In 1898, with Morris Travers (1872–1961), he discovered *neon, *krypton, and *xenon. Six years later he discovered the last of the noble gases, *radon. He was awarded the Nobel Prize for chemistry in 1904, the year in which Rayleigh received the physics prize.

**Ramsden eyepiece** An eyepiece for optical instruments consisting of two identical plano-convex lenses with their convex faces pointing towards each other. They are separated by a distance of two thirds of the focal length of either lens.

**Randall–Sundrum scenario** A variant of the *brane world picture in which gravity is trapped by warping of the higher dimensions. This scenario, put forward by the American physicist Lisa Randall (1962–  ) and Raman Sundrum in 1999, has attracted a great deal of interest because it provides a possible explanation of why gravity is weaker than the other forces.

**random alloy** See DISORDERED SOLID.

**random walk** The problem of determining the distance from a starting position made by a walker, who can either move forward (toward $+x$) or backwards (toward $-x$) with the choice being made randomly (e.g. by tossing a coin). The progress of the walker is characterized by the net distance $D_N$ travelled in $N$ steps. After $N$ steps the *root mean square value $D_{rms}$, which is the average distance away from the starting position, is given by $D_{rms} = \sqrt{N}$. Physical applications of the random walk include diffusion and the related problem of Brownian motion as well as problems involving the structures of polymers and disordered solids.

**Raney nickel** A spongy form of nickel made by the action of sodium hydroxide on a nickel–aluminium alloy. The sodium hydroxide dissolves the aluminium leaving a highly active form of nickel with a large surface area. The material is a black pyrophoric powder saturated with hydrogen. It is an extremely efficient catalyst, especially for hy-

drogenation reactions at room temperature. It was discovered in 1927 by the American chemist M. Raney.

**rank (category)** (in biology) The position or status of a *taxon in a *classification hierarchy. Examples of ranks are the family, genus, and species.

**Rankine cycle** A cycle of operations in a heat engine. The Rankine cycle more closely approximates to the cycle of a real steam engine that does the *Carnot cycle. It therefore predicts a lower ideal thermal efficiency than the Carnot cycle. In the Rankine cycle (see illustration), heat is added at constant pressure $p_1$, at which water is converted in a boiler to superheated steam; the steam expands at constant entropy to a pressure $p_2$ in the cylinder; heat is rejected at constant pressure $p_2$ in a condenser; the water so formed is compressed at constant entropy to pressure $p_1$ by a feed pump. The cycle was devised by William Rankine (1820–70).

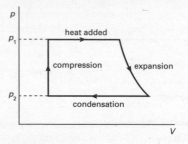

**Rankine cycle.**

**Rankine temperature scale** An absolute temperature scale based on the Fahrenheit scale. Absolute zero on this scale, $0°R$, is equivalent to $-459.67°F$ and the melting point of ice ($32°F$) is therefore $(459.67 + 32 = 491.67)°R$. The scale was devised by William Rankine.

**ranksite** A mineral consisting of a mixed sodium carbonate, sodium sulphate, and potassium chloride, $2Na_2CO_3.9Na_2SO_4.KCl$.

**Raoult's law** The partial vapour pressure of a solvent is proportional to its mole fraction. If $p$ is the vapour pressure of the solvent (with a substance dissolved in it) and $X$ the mole fraction of solvent (number of moles of solvent divided by total number of moles) then $p = p_0X$, where $p_0$ is the vapour pressure of the pure solvent. A solution that obeys

Raoult's law is said to be an **ideal solution**. In general the law holds only for dilute solutions, although some mixtures of liquids obey it over a whole range of concentrations. Such solutions are **perfect solutions** and occur when the intermolecular forces between molecules of the pure substances are similar to the forces between molecules of one and molecules of the other. Deviations in Raoult's law for mixtures of liquids cause the formation of *azeotropes. The law was discovered by the French chemist François Raoult (1830–1901).

**(((⊕))) SEE WEB LINKS**

• Raoult's original paper in *Comptes Rendus* (1887)

**rapeseed methyl ester** *See* BIOFUEL.

**rare-earth elements** *See* LANTHANOIDS.

**rarefaction** A reduction in the pressure of a fluid and therefore of its density.

**rare gases** *See* NOBLE GASES.

**RAS** *See* REFLECTION ANISOTROPY SPECTROSCOPY.

**Raschig process** An industrial process for making chlorobenzene (and phenol) by a gas-phase reaction between benzene vapour, hydrogen chloride, and oxygen (air) at 230°C

$$2C_6H_6 + 2HCl + O_2 \rightarrow 2H_2O + 2C_6H_5Cl$$

The catalyst is copper(II) chloride. The chlorobenzene is mainly used for making phenol by the reaction

$$C_6H_5Cl + H_2O \rightarrow HCl + C_6H_5OH$$

This reaction proceeds at 430°C with a silicon catalyst. The process was invented by the German chemist Fritz Raschig (1863–1928).

**Raschig synthesis** *See* HYDRAZINE.

**RasMol** A commonly used program for visualizing molecules. It can be used with a range of file formats and the display can be choosen (e.g. wireframe, ball-and-stick, space filling, etc.). The original version was written by Roger Sayle of Glaxo around 1990. It is freely available.

**(((⊕))) SEE WEB LINKS**

• A downloadable version of RasMol

**raster** The pattern of scanning lines on the screen of the cathode-ray tube in a television receiver or other device that provides a visual display.

**rate constant (velocity constant)** Symbol $k$. The constant in an expression for the rate of a chemical reaction in terms of concentrations (or activities). For instance, in a simple unimolecular reaction

$$A \rightarrow B$$

the rate is proportional to the concentration of A, i.e.

$$rate = k[A]$$

where $k$ is the rate constant, which depends on the temperature. The equation is the **rate equation** of the reaction, and its form depends on the reaction mechanism.

**rate-determining step** The slowest step in a chemical reaction that involves a number of steps. In such reactions, there is often a single step that is appreciably slower than the other steps, and the rate of this determines the overall rate of the reaction.

**rationalized Planck constant** *See* PLANCK CONSTANT.

**rationalized units** A system of units in which the defining equations have been made to conform to the geometry of the system in a logical way. Thus equations that involve circular symmetry contain the factor $2\pi$, while those involving spherical symmetry contain the factor $4\pi$. *SI and *Heaviside–Lorentz units are rationalized; *Gaussian units are unrationalized.

**rational number** Any number that can be expressed as the ratio of two integers. For example, 0.3333... is rational because it can be written as 1/3. $\sqrt{2}$, however, is *irrational.

**Ratitae (Palaeognathae)** A group comprising the flightless birds, including the ostrich, kiwi, and emu. They have long legs, heavy bones, small wings, a flat breastbone, and curly feathers. These birds are thought to have descended from a variety of flying birds and are not representatives of a single homologous group.

**ray** **1.** (in optics) A narrow beam of radiation or an idealized representation of such a beam on a **ray diagram**, which can be used to indicate the positions of the object and image in a system of lenses or mirrors. **2.** (in botany) *See* MEDULLARY RAY.

**Rayleigh, Baron** (John William Strutt; 1842–1919) British physicist, who built a private laboratory after working at Cambridge

University. His work in this laboratory included the discovery of Rayleigh *scattering. He also worked in acoustics, electricity, and optics, as well as collaborating with William *Ramsay on the discovery of *argon. He was awarded the 1904 Nobel Prize for physics.

**Rayleigh criterion** *See* RESOLVING POWER.

**Rayleigh–Jeans formula** *See* PLANCK'S RADIATION LAW.

**Rayleigh scattering** *See* SCATTERING OF ELECTROMAGNETIC RADIATION.

**rayon** A textile made from cellulose. There are two types, both made from wood pulp. In the viscose process, the pulp is dissolved in carbon disulphide and sodium hydroxide to give a thick brown liquid containing cellulose xanthate. The liquid is then forced through fine nozzles into acid, where the xanthate is decomposed and a cellulose filament is produced. The product is **viscose rayon**. In the acetate process cellulose acetate is made and dissolved in a solvent. The solution is forced through nozzles into air, where the solvent quickly evaporates leaving a filament of **acetate rayon**.

**RBS** *See* RUTHERFORD BACKSCATTERING SPECTROMETRY.

**RDX** *See* CYCLONITE.

**reactance** Symbol $X$. A property of a circuit containing inductance or capacitance that together with any resistance makes up its *impedance. The impedance $Z$ is given by $Z^2 = R^2 + X^2$, where $R$ is the resistance. For a pure capacitance $C$, the reactance is given by $X_C = 1/2\pi fC$, where $f$ is the frequency of the *alternating current; for a pure inductance $L$, $X_L = 2\pi fL$. If the resistance, inductance, and capacitance are in series the impedance $Z = \sqrt{[R^2 + (X_L - X_C)^2]}$. Reactance is measured in ohms.

**reactant** *See* CHEMICAL REACTION.

**reaction** 1. A force that is equal in magnitude but opposite in direction to some other force, in accordance with Newton's third law of motion. If a body A exerts a force on body B, then B exerts an equal and opposite force on A. Thus, every force could be described as 'a reaction', and the term is better avoided, although it is still used in such terms as 're-action propulsion'. **2.** *See* CHEMICAL REACTION.

**reaction propulsion** *See* JET PROPULSION.

**reaction time** (latent period) The period of time between the detection of a stimulus at a sensory receptor and the performance of the appropriate response by the effector organ. This delay is caused by the time taken for the impulse to travel across the synapses of adjacent neurons. The reaction time for a *reflex response, involving only a single linking synapse, is very short.

**reactive dye** *See* DYES.

**reactive oxygen species** (ROS) Any of various chemical species that contain highly active oxygen, particularly the *free radical superoxide anion ($O_2^-$) and its derivatives, including hydrogen peroxide ($H_2O_2$), singlet oxygen (a metastable high-energy form of molecular oxygen), hydroxyl radical (OH·), and hypohalite ions (e.g. hypochlorite, OCl⁻). Superoxide is produced normally as a by-product of aerobic respiration inside mitochondria; both it and other ROS are potentially harmful to living cells because of their highly reactive nature. They can damage DNA and other cell components and are implicated in various diseases (e.g. cancer, heart disease), hence cells have an array of mechanisms to remove them, including the enzyme superoxide dismutase and various *antioxidants. However, certain cells and tissues in both plants and animals produce ROS in order to destroy invading pathogenic organisms, such as fungi and bacteria.

**reactor** 1. *See* NUCLEAR REACTOR. **2.** Any device, such as an inductor or capacitor, that introduces *reactance into an electrical circuit.

**reading frame** A sequence of bases in messenger RNA (or deduced from DNA) that encodes for a polypeptide. Since each coding unit (*codon) of the genetic code consists of three consecutive bases, the reading frame is established according to precisely where translation starts. The hallmark of a functional gene is that it is transcribed to produce an **open reading frame** (ORF), consisting of a *start codon to pinpoint where translation should start, a *stop codon to signal termination of translation, and typically a long sequence of codons that specify the constituent amino acids of the polypeptide (as well as *introns in most eukaryote genes).

**reagent** A substance reacting with another substance. Laboratory reagents are compounds, such as sulphuric acid, hydrochloric acid, sodium hydroxide, etc., used in chemical analysis or experiments.

**realgar** A red mineral form of arsenic(II) sulphide, $As_2S_2$.

**real gas** A gas that does not have the properties assigned to an *ideal gas. Its molecules have a finite size and there are forces between them (*see* EQUATION OF STATE).

**real image** *See* IMAGE.

**real-is-positive convention (real-positive convention)** A convention used in optical formulae relating to lenses and mirrors. In this convention, distances from optical components to real objects, images, and foci are taken as positive, whereas distances to virtual points are taken as negative.

**rearrangement** A type of chemical reaction in which the atoms in a molecule rearrange to form a new molecule.

**Réaumur temperature scale** A temperature scale in which the melting point of ice is taken as 0°R and the boiling point of water as 80°R. It was devised by René Antoine Réaumur (1683–1757).

**recalescence** A phenomenon that occurs during the cooling of iron and other ferromagnetic metals (*see* MAGNETISM) after they have been heated. When iron is heated to white heat and then allowed to cool, it abruptly evolves heat at a certain temperature. This evolution of heat, which slows down the cooling process and can lead to a brief reheating, is caused by an *exothermic reaction when the structure of the crystal changes. The temperature at which this occurs is called the **recalescence point**. For pure iron there are two recalescence points: at 780°C and 880°C. A reverse phenomenon causing cooling, called **decalescence**, occurs when ferromagnetic metals are heated.

**Recent** *See* HOLOCENE.

**receptacle 1. (thalamus; torus)** The tip of a flower stalk, which bears the petals, sepals, stamens, and carpels. The way the receptacle develops determines the position of the flower parts. It can be dilated and dome-shaped, saucer-shaped, or hollow and enclosing the gynoecium. In some plants it may become part of the fruit (*see* PSEUDO-CARP). **2.** A swollen part of the thallus of some algae, e.g. *Fucus*, that bears the conceptacles in which the sex organs are situated.

**receptor** (in biology) **1.** A cell or group of cells specialized to detect a particular stimulus and to initiate the transmission of impulses via the sensory nerves. The eyes, ears, nose, skin, and other sense organs all contain specific receptors responding to external stimuli (*see* EXTEROCEPTOR); other receptors are sensitive to changes within the body (*see* INTEROCEPTOR). *See also* BARO-RECEPTOR; CHEMORECEPTOR; MECHANO-RECEPTOR; OSMORECEPTOR; PROPRIOCEPTOR. **2.** An area of a plasma membrane, consisting of a specially adapted membrane protein, that can bind with a specific hormone, neurotransmitter, drug, or other chemical, thereby initiating a change within the cell.

**recessive** The *allele that is not expressed in the *phenotype when two different alleles are present in the cells of an organism. The aspect of a characteristic controlled by a recessive allele only appears when two such alleles are present, i.e. in the **double recessive** condition. *Compare* DOMINANT.

**recipient** An individual who receives tissues or organs of the body from another (the *donor).

**reciprocal cross** A *cross reversing the roles of males and females to confirm the results obtained from an earlier cross. For example, if the pollen (male) from tall plants is transferred to the stigmas (female) of dwarf plants in one cross, the reciprocal cross would use the pollen of dwarf plants to pollinate the stigmas of tall plants.

**reciprocal proportions** *See* CHEMICAL COMBINATION.

**recombinant DNA** DNA that contains genes from different sources that have been combined by the techniques of *genetic engineering rather than by breeding experiments. Genetic engineering is therefore also known as **recombinant DNA technology**. Recombinant DNA is formed during *gene cloning and in the creation of *genetically modified organisms.

**recombination** The rearrangement of genes that occurs when reproductive cells (gametes) are formed. It results from the *independent assortment of parental sets of chromosomes and exchange of chromosomal material (*see* CROSSING OVER) that occur during *meiosis. Recombination results in offspring that have a combination of characteristics different from that of their parents.

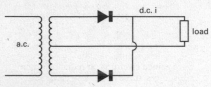

Half-wave rectification

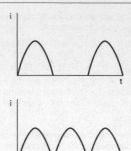

Full-wave rectification

**Rectifier.**

Recombination can also be induced artificially by \*genetic engineering techniques.

**recombination era** The period after the big bang when the universe was sufficiently cool for electrons and nuclei to form atoms for the first time. The recombination occurred about 300 000 years after the big bang. Up to this time, photons in the universe were scattered by free electrons and it is sometimes said that the universe was 'opaque'. With recombination, photons were able to travel without impediment, and the universe became 'transparent'. *See also* MICROWAVE BACKGROUND RADIATION.

**recombination process** The process in which a neutral atom or molecule is formed by the combination of a positive ion and a negative ion or electron; i.e. a process of the type:

$$A^+ + B^- \rightarrow AB$$

or

$$A^+ + e^- \rightarrow A$$

In recombination, the neutral species formed is usually in an excited state, from which it can decay with emission of light or other electromagnetic radiation.

**recreational drug** A drug used for recreation, as opposed to medical use. Strictly, recreational drugs include alcohol and nicotine, but the term is often understood to mean substances such as marijuana, cocaine, and amphetamines.

**recrystallization** A process of repeated crystallization in order to purify a substance

or to obtain more regular crystals of a purified substance.

**rectification** 1. (in physics) The process of obtaining a direct current from an alternating electrical supply. *See* RECTIFIER. 2. (in chemistry) The process of purifying a liquid by \*distillation. *See* FRACTIONAL DISTILLATION.

**rectified spirit** A constant-boiling mixture of \*ethanol (95.6°) and water; it is obtained by distillation.

**rectifier** An electrical device that allows more current to flow in one direction than the other, thus enabling alternating e.m.f.s to drive only direct current. The device most commonly used for rectification is the semiconductor \*diode. In **half-wave rectification**, achieved with one diode, a pulsating current is produced. In **full-wave rectification** two diodes are used, one pair conducting during the first half cycle and the other conducting during the second half (see illustration). The full-wave rectified signal can be smoothed using a capacitor or an inductor.

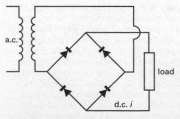

**Rectifier.** Bridge rectifier.

The **bridge rectifier** illustrated also gives full-wave rectification.

**rectum** The portion of the *alimentary canal between the *colon and the *anus. Its main function is the storage of *faeces prior to elimination.

**recycling** 1. The recovery and processing of materials after they have been used, which enables them to be reused. For example, used paper, cans, and glass can be broken down into their constituents, which form the raw materials for the manufacture of new products. **2.** The continual movement of *essential elements between the biotic (living) and abiotic (nonliving) components of the environment. *See* CARBON CYCLE; NITROGEN CYCLE; OXYGEN CYCLE; PHOSPHORUS CYCLE; SULPHUR CYCLE.

**red algae** *See* RHODOPHYTA.

**red blood cell** *See* ERYTHROCYTE.

**red dwarf** A *dwarf star on the main sequence (*see* HERTZSPRUNG–RUSSELL DIAGRAM) that is much cooler, smaller, and less massive than the sun. These stars display a number of magnetic phenomena. *See also* FLARE STAR.

**red giant** A *giant star thought to be in the later stages of *stellar evolution. It has a surface temperature in the range 2000–3000 K and a diameter 10–100 times that of the sun. *See also* HERTZSPRUNG–RUSSELL DIAGRAM.

**red lead** *See* DILEAD(II) LEAD(IV) OXIDE.

**redox** *See* OXIDATION–REDUCTION.

**redshift** 1. **(Doppler redshift)** A displacement of the lines in the spectra of certain galaxies towards the red end of the visible spectrum (i.e. towards longer wavelengths). It is usually interpreted as a *Doppler effect resulting from the recession of the galaxies along the line of sight (*see* EXPANSION OF THE UNIVERSE). The redshift is usually expressed as $\Delta\lambda/\lambda$, where $\Delta\lambda$ is the shift in wavelength of radiation of wavelength $\lambda$. For relatively low velocities of recession this is equivalent to $v/c$, where $v$ is the relative velocity of recession and $c$ is the speed of light. If very high velocities of recession are involved, a relativistic version of $v/c$ is required (*see* RELATIVISTIC SPEED). The redshift of spectral lines occurs in all regions of the electromagnetic spectrum; ultraviolet can be shifted into the visible region and visible radiation can be shifted into the infrared region.

**2. (gravitational** *or* **Einstein redshift)** A similar displacement of spectral lines towards the red caused not by a Doppler effect but by a high gravitational field. This type of redshift was predicted by Einstein and some astronomers believe that this is the cause of the large redshifts of *quasars, which can be as high as 6.4.

**reducing agent (reductant)** A substance that brings about reduction in other substances. It achieves this by being itself oxidized. Reducing agents contain atoms with low oxidation numbers; that is the atoms have gained electrons. In reducing other substances, these atoms lose electrons.

**reducing sugar** A monosaccharide or disaccharide sugar that can donate electrons to other molecules and can therefore act as a reducing agent. The possession of a free ketone ($-CO-$) or aldehyde ($-CHO$) group enables most monosaccharides and disaccharides to act as reducing sugars. Reducing sugars can be detected by *Benedict's test. *Compare* NONREDUCING SUGAR.

**reductant** *See* REDUCING AGENT.

**reduction** *See* OXIDATION–REDUCTION.

**reduction division** *See* MEIOSIS.

**reef** An outcrop of rock that projects above the surface in shallow seas, particularly one constructed by calcareous animals, such as *corals, shellfish, and some algae. Reefs may be mounds or ridges, often with very steep sides, or even a chain of islands. A fringing reef is attached to the coast and acts as a breakwater; a barrier reef runs parallel to the coast, usually with a deeper lagoon on the landward side. Reefs provide rich habitats for fish and invertebrates. *See also* ATOLL.

**refinery gas** *See* PETROLEUM.

**refining** The process of purifying substances or extracting substances from mixtures.

**reflectance** The ratio of the radiant flux reflected by a surface to that falling on it. This quantity is also known as the **radiant reflectance**. The radiant reflectance measured for a specified wavelength of the incident radiant flux is called the **spectral reflectance**.

**reflecting telescope (reflector)** *See* TELESCOPE.

**reflection** The return of all or part of a

beam of particles or waves when it encounters the boundary between two media. The **laws of reflection** state: (1) that the incident ray, the reflected ray, and the normal to the reflecting interface at the point of incidence are all in the same plane; (2) that the angle of incidence equals the angle of reflection. *See also* MIRROR; REFLECTANCE; TOTAL INTERNAL REFLECTION.

**reflection anisotropy spectroscopy (RAS)** A spectroscopic technique for studying surfaces by reflection of light or ultraviolet radiation. A beam of plane-polarized monochromatic radiation is directed at right angles onto a single-crystal surface and the reflected radiation along the surface is measured in two mutually perpendicular directions. If a suitable surface plane is chosen, there is a difference in reflectivity in these directions – a reflection anisotropy – and for a cubic crystal this is a property of the surface rather than the bulk. The change in this effect with wavelength gives a reflection anisotropy spectrum, which gives information about electronic surface states. Surface reactions can be investigated by the changes they cause in the spectrum.

**reflex** An automatic and innate response to a particular stimulus. A reflex response is extremely rapid. This is because it is mediated by a simple nervous circuit called a **reflex arc**, which at its simplest involves only a receptor linked to a sensory neuron, which synapses with a motor neuron (supplying the effector) in the spinal cord. Such reflexes are known as **spinal reflexes**; examples are the **withdrawal reflex** of the hand from a painful stimulus (such as fire) and the *stretch reflex. **Cranial reflexes** are mediated by pathways in the cranial nerves and brain; examples are the blinking and swallowing reflexes. *See also* CONDITIONING.

**refluxing** A laboratory technique in which a liquid is boiled in a container attached to a condenser (**reflux condenser**), so that the liquid continuously flows back into the container. It is used for carrying out reactions over long periods in organic synthesis.

**reforestation** The replanting of trees on areas of land where forests have been cleared by felling or burning (*see* DEFORESTATION) or by natural means. Reforestation is particularly important in countries, such as Brazil, where large areas of forest have been destroyed by deforestation, although planted forest has much less species diversity (*see*

BIODIVERSITY) than the original forest. It also helps to counteract global emissions of carbon dioxide, by fixing the gas as plant material. Hence reforestation can play a part in slowing *global warming.

**reforming** The conversion of straight-chain alkanes into branched-chain alkanes by *cracking or by catalytic reaction. It is used in petroleum refining to produce hydrocarbons suitable for use in gasoline. Benzene is also manufactured from alkane hydrocarbons by catalytic reforming. **Steam reforming** is a process used to convert methane (from natural gas) into a mixture of carbon monoxide and hydrogen, which is used to synthesize organic chemicals. The reaction

$$CH_4 + H_2O \rightarrow CO + 3H_2$$

occurs at about 900°C using a nickel catalyst.

**refracting telescope (refractor)** *See* TELESCOPE.

**refraction** The change of direction suffered by wavefront as it passes obliquely from one medium to another in which its speed of propagation is altered. The phenomenon occurs with all types of waves, but is most familiar with light waves. In optics the direction is changed in accordance with **Snell's law**, i.e. $n_1 \sin i = n_2 \sin r$, where $i$ and $r$ are the angles made by the incident beam of radiation and the refracted beam to the normal (an imaginary line perpendicular to the interface between the two media); $n_1$ and $n_2$ are the *refractive indices of the two media. This law is also known as one of the **laws of refraction**. The other law of refraction is that the incident ray, the refracted ray, and the normal at the point of incidence lie in the same plane. The change of direction results from a change in the speed of propagation

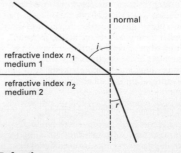

**Refraction.**

and the consequent change in wavelength (see illustration).

**refractive index (refractive constant)** Symbol $n$. The **absolute refractive index** of a medium is the ratio of the speed of electromagnetic radiation in free space to the speed of the radiation in that medium. As the refractive index varies with wavelength, the wavelength should be specified. It is usually given for yellow light (sodium D-lines; wavelength 589.3 nm). The **relative refractive index** is the ratio of the speed of light in one medium to that in an adjacent medium. *See also* REFRACTION.

**refractivity** A measure of the extent to which a medium will deviate a ray of light entering its surface. In some contexts it is equal to $(n-1)$, where $n$ is the *refractive index.

**refractometer** Any of various instruments for measuring the *refractive index of a substance or medium. An example is the **Pulfrich refractometer**, which is a glass block with a polished top, with a small cell on top of the block for liquid samples. A telescope rotating on a vertical circular scale is used to find the angle ($\alpha$) between the top of the block and the direction in which the limiting ray (from incident light parallel to the top face) leaves the side of the block. If the refractive index of the block ($n_g$) is known, that of the liquid can be calculated using $n = \sqrt{(n_g^2 - \sin^2\alpha)}$.

**refractory 1.** Having a high melting point. Metal oxides, carbides, and silicides tend to be refractory, and are extensively used for lining furnaces. **2.** A refractory material.

**refractory period** The period after the transmission of an impulse in a nerve or muscle in which the membrane of the axon or muscle fibre regains its ability to transmit impulses (*see* ACTION POTENTIAL). This period lasts approximately 3 milliseconds and is divided into an **absolute refractory period**, during which a second impulse may not be generated; and a **relative refractory period**, during which it is possible to generate an impulse only if there is an abnormally strong stimulus.

**refrigerant** *See* REFRIGERATION.

**refrigeration** The process of cooling a substance and of maintaining it at a temperature below that of its surroundings. See Feature overleaf.

**regeneration** The growth of new tissues or organs to replace those lost or damaged by injury. Many plants can regenerate a complete plant from a shoot segment or a single leaf, this being the basis of many horticultural propagation methods (*see* CUTTING). The capacity for regeneration in animals is less marked. Some planarians and sponges can regenerate whole organisms from small pieces, and crustaceans (e.g. crabs), echinoderms (e.g. brittlestars), and some reptiles and amphibians can grow new limbs or tails (*see* AUTOTOMY), but in mammals regeneration is largely restricted to wound healing and, in certain cases, regrowth of damaged nerve fibres.

**regioselectivity** The effect in which certain positions in the molecule are favoured over others in a reaction. An example is the way in which certain groups on the benzene ring direct further substituents to the meta position or to the ortho and para positions. *Markovnikoff's rule for electrophilic additions to alkenes is another example.

**regma** A dry fruit that is characteristic of the geranium family. It is similar to the *carcerulus but breaks up into one-seeded parts, each of which splits open to release a seed.

**Regnault's method** A technique for measuring gas density by evacuating and weighing a glass bulb of known volume, admitting gas at known pressure, and reweighing. The determination must be carried out at constant known temperature and the result corrected to standard temperature and pressure. The method is named after the French chemist Henri Victor Regnault (1810–78).

**regolith** A layer of weathered rock fragments that lies below the soil and above the bedrock below. In the tropics it may be up to 60 m thick. Fine regolith differs from soil in that it has no organic content (humus) and cannot therefore support plant life, although the roots of trees and other plants may penetrate it for water.

**regulator gene** *See* OPERON.

**regulatory genes** Genes that control development by regulating the expression of structural genes responsible for the formation of body components. They encode *transcription factors, which interact with regulatory sites of other genes causing activation or repression of developmental path-

## REFRIGERATION

In the domestic refrigerator a cycle of operations equivalent to those of a *heat pump is used.

### The vapour-compression cycle

In this cycle (see illustration) a volatile liquid refrigerant, such as a fluorocarbon, is pumped through the cooling coils of the evaporator to the ice-making compartment of the refrigerator. Inside these coils the refrigerant evaporates, taking the latent heat required to effect the change of state from the inside of the ice-making compartment, causing the temperature in this compartment to fall. The vapour is then passed to an electrically driven compressor before entering the condenser, where the high-pressure gas is converted back to a liquid. The heat produced by this second change of state is given out, usually at the back of the refrigerator. The liquid refrigerant then enters a storage vessel before finally passing through an expansion valve to reduce its pressure prior to beginning the cycle again in the evaporator.

This cycle is repeated over and over again until the temperature reaches the desired level (about 1–2°C in the food chamber of a domestic refrigerator and minus 15–18°C in the deep-freeze compartment). The compressor is then switched off, and on again later, by a thermostat to maintain a steady temperature. In order to transfer heat from the cold interior to the warm surroundings without contravening the second law of thermodynamics, energy has to be supplied to the cycle by the electric current that drives the compressor.

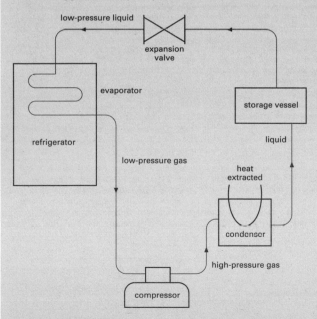

Vapour-compression cycle

## The vapour-absorption cycle

In this cycle there are no moving parts and energy is supplied as heat either by an electric heater or a gas burner. The refrigerant is usually ammonia, which is liberated from a water solution and moved through the evaporator by a stream of hydrogen gas under pressure. Heat is applied to the generator, raising the ammonia and water vapour to the separator; the ammonia separates from the water and passes to the condenser, where it cools and liquefies, giving off its latent heat to the surroundings. The liquid ammonia is then mixed with hydrogen gas, which carries it through the evaporator and helps in the process of evaporation. Subsequently the hydrogen and ammonia vapour enter the absorber, where the water returned from the separator dissolves the ammonia before returning to the generator.

## Small-scale refrigerator

These two systems account for all domestic and industrial refrigerators including those on in ships, trains, and refrigerated lorries. However, small-scale refrigeration is sometimes achieved by means of the *Peltier effect at junctions for $n$-type and $p$-type semiconductors.

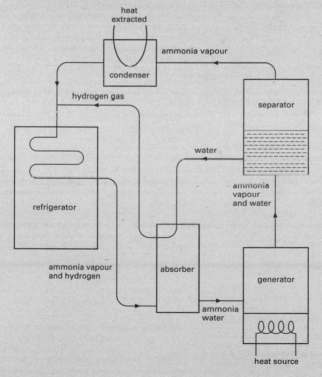

Vapour-absorption cycle

ways. Much of development in quite different organisms, such as mammals and insects, is controlled by genes that are structurally very similar, thought to be descended from genes in ancient common ancestors. Prime examples are the *homeotic genes.

**Reines, Frederick** *See* PAULI, WOLFGANG ERNST.

**reinforcement** (in animal behaviour) Increasing (or decreasing) the frequency of a particular behaviour through *conditioning, by arranging for some biologically important event (the **reinforcer**) always to follow another event. In instrumental conditioning an **appetitive reinforcer**, or **reward** (e.g. food), given after a response made by the animal, increases that response; an **aversive reinforcer**, or **punishment** (e.g. an electric shock) decreases the response.

**relational theory** A type of theory in which *absolute space and *absolute time do not occur, with the theory depending on relationships between events. Newtonian mechanics is not a relational theory but the general theory of relativity is nearer to being a relational theory. A theory that fully incorporated *Mach's principle would be a relational theory.

**relative aperture** *See* APERTURE.

**relative atomic mass (atomic weight; r.a.m.)** Symbol $A_r$. The ratio of the average mass per atom of the naturally occurring form of an element to 1/12 of the mass of a carbon–12 atom.

**(()) SEE WEB LINKS**

• Values of r.a.m. at the NIST website

**relative density (r.d.)** The ratio of the *density of a substance to the density of some reference substance. For liquids or solids it is the ratio of the density (usually at 20°C) to the density of water (at its maximum density). This quantity was formerly called **specific gravity**. Sometimes relative densities of gases are used; for example, relative to dry air, both gases being at s.t.p.

**relative humidity** *See* HUMIDITY.

**relative molecular mass (molecular weight)** Symbol $M_r$. The ratio of the average mass per molecule of the naturally occurring form of an element or compound to 1/12 of the mass of a carbon–12 atom. It is equal to

the sum of the relative atomic masses of all the atoms that comprise a molecule.

**relative permeability** *See* PERMEABILITY.

**relative permittivity** *See* PERMITTIVITY.

**relativistic electronics** A branch of *electronics that involves the special theory of relativity. The *free-electron laser is an example of an application in relativistic electronics.

**relativistic jets** Jets of matter or electromagnetic radiation that move away from some *active galaxies at very large speeds. It is thought that the energy for these jets is supplied by the activity around black holes, either through a *Blandford–Znajek process or through a *Penrose process.

**relativistic mass** The mass of a moving body as measured by an observer in the same frame of reference as the body. According to the special theory of *relativity the mass $m$ of a body moving at speed $v$ is given by $m = m_0/\sqrt{(1 - v^2/c^2)}$, where $m_0$ is its *rest mass and $c$ is the speed of light. The relativistic mass therefore only differs significantly from the rest mass if its speed is a substantial fraction of the speed of light. If $v = c/2$, for example, the relativistic mass is 15% greater than the rest mass.

**relativistic mechanics** An extension of Newtonian mechanics that takes into account the theory of *relativity.

**relativistic quantum theory** *See* QUANTUM THEORY; QUANTUM FIELD THEORY.

**relativistic runaway theory** A theory of the origin of lightning in which it is proposed that lightning is initiated by a cosmic ray hitting a molecule in a cloud with an electric field. This releases an electron into the cloud, where it is accelerated in the field. The fast-moving electron then ionizes other atoms and molecules and there is a chain reaction producing many rapidly moving electrons. This theory explains how lightning can occur in spite of the electric field of the cloud being insufficient to cause breakdown ionization in air.

**relativistic speed** A speed that is sufficiently large to make the mass of a body significantly greater than its *rest mass. It is usually expressed as a proportion of the speed of light. *See* RELATIVITY; RELATIVISTIC MASS.

**relativity** Two widely accepted theories

proposed by Albert *Einstein to account for departures from Newtonian mechanics. The special theory, of 1905, refers to non-accelerated frames of reference, while the general theory, of 1915, extends to accelerated systems.

**The special theory.** For Galileo and Newton, all uniformly moving frames of reference (Galilean frames) are equivalent for describing the dynamics of moving bodies. There is no experiment in dynamics that can distinguish between a stationary laboratory and a laboratory that is moving at uniform velocity. Einstein's special theory of relativity takes this notion of equivalent frames one step further: he required all physical phenomena, not only those of dynamics, to be independent of the uniform motion of the laboratory.

When Einstein published his special theory, he realized that it could explain the apparent lack of experimental evidence for an ether, which was supposed to be the medium required for the propagation of electromagnetic waves (*see* Michelson–Morley experiment). Einstein recognized that the existence of an ether would render invalid any equivalent relativity principle for electromagnetic phenomena; i.e. uniform movement of the laboratory through the ether would lead to measurable differences in the propagation of electromagnetic waves *in vacuo*. Since no experimental evidence of the ether was forthcoming, Einstein was encouraged to continue his search for a relativity principle that encompassed all physical phenomena. Since light is a physical phenomenon, its propagation *in vacuo* could not be used to distinguish between uniformly moving frames of reference. Therefore in all such frames the measured speed of light *in vacuo* must be the same.

This conclusion has some important consequences for the nature of space and time. In his popular exposition of 1916, Einstein illustrated these consequences with thought experiments. In one such experiment, he invites the reader to imagine a very long train travelling along an embankment with a constant velocity v in a given direction (see diagram).

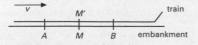

**Relativity.** Einstein's thought experiment.

Observers on the train use it as a rigid reference body, regarding all events with reference to the train. Einstein posed a simple question: Are two events, which are simultaneous relative to the railway embankment, also simultaneous relative to the observer on the train? For example, two lightning strokes strike the embankment at points A and B simultaneously with respect to the embankment, so that an observer at M (the mid-point of the line AB) will record no time lapse between them. However, the events A and B also correspond to positions A and B on the train. M′ is the mid-point of the distance AB on the moving train. When the flashes occur, from the point of view of the embankment, M′ coincides with M. However, M′ moves with speed v towards the right and therefore hastens towards the beam of light coming from B, while moving on ahead of the beam from A. The observer at M′ would not agree with the observer at M on the simultaneity of the events A and B because the beam of light from B will be seen to be emitted before the beam of light at A.

At first sight there may seem to be a problem here. If the observer at M′ is 'hastening towards the beam of light from B', is this not equivalent to saying that the beam of light is travelling towards M′ at a combined speed of v + c, where c is the speed of light *in vacuo*?

The resolution of this problem is the basis of special relativity. According to Einstein, the moving observer at M′ must measure the speed of light *in vacuo* to be c, since there can be no experiment that distinguishes the train's moving frame from any other Galilean frame. It is therefore the concept of time measurement that requires revision; that is, the time required for a particular event to occur with respect to the train cannot have the same duration as the same event when judged from the embankment.

This remarkable result also has implications for the measurement of spatial intervals. The measurement of a spatial interval requires the time coincidence of two points along a measuring rod. The relativity of simultaneity means that one cannot contend that an observer who traverses a distance x m per second in the train, traverses the same distance x m also with respect to the embankment in each second. In trying to include the law of propagation of light into a relativity principle, Einstein questioned the way in which measurements of space and time in different Galilean frames are compared. Place and time measurements in two

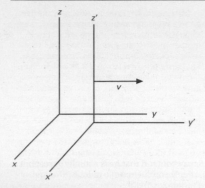

The frame $(x', y', z')$ is moving uniformly with respect to $(x, y, z)$ at a velocity $v$ in the $y$ direction.

Coordinates in the two frames are related as follows by the Lorentz transformations

$$y' = \frac{y - vt}{\sqrt{1 - \frac{v^2}{c^2}}}$$

$$t' = \frac{t - vy/c^2}{\sqrt{1 - \frac{v^2}{c^2}}}$$

$$x' = x$$
$$z' = z$$

**Relativity.** Lorentz transformations.

different Galilean frames must be related by a transformation preserving the relativity principle that every ray of light has a velocity of transmission $c$ relative to observers in both frames.

Transformations that preserve the relativity principle are called *Lorentz transformations. The form of these looks complicated at first (see diagram). However, they arise from the simple requirement that there can be no experiment in dynamics or electromagnetism that will distinguish between two different Galilean frames of reference.

These transformations suggest that observers in the two frames will not agree on measurements of length made in the $y$-direction. Indeed, the duration of intervals of time cannot be agreed upon in the two frames. This is exactly what was suggested in Einstein's thought experiment on the train. Simple manipulations lead to the following formulae, which relate lengths and time intervals in the $x', y', z'$ frame to their equivalent quantities in the $x, y, z$ frame:

$$\Delta l = \Delta l' \, (1 - v^2/c^2)^{\frac{1}{2}}$$

and

$$\Delta t = \Delta t' \, (1 - v^2/c^2)^{-\frac{1}{2}},$$

where $\Delta l$ and $\Delta t$ are respectively intervals in space and time. Motion therefore leads to a length contraction of the $x', y', z'$ lengths with respect to the $x, y, z$ lengths. Similarly, the equation relating the time intervals in the two frames leads to a time dilation in the $x', y', z'$ system compared to the $x, y, z$ system. From these expressions it is clear that the velocity $c$ plays the part of a limiting velocity, which can neither be reached nor exceeded by any material body.

The velocity $c$ is often said to be the limiting velocity for the transfer of information in the universe. Faster-than-light signals violate *causality when taken to their logical conclusions. The universe, therefore, according to the special theory, of relativity, updates itself at the maximum speed of light $c$: any local changes in the properties within a region of space are not communicated to the rest of the universe instantaneously. Rather, the universe is updated through a wave of reality, which emanates at speed $c$ from the region in which the change took place.

**The general theory.** In his special theory, Einstein updated the Galilean principle of relativity by including electromagnetic phenomena. Galileo, and later Newton, were well aware that no experiment in the dynamics of moving bodies could distinguish between frames of reference moving relative to each other at constant velocity (Galilean frames). If two Galilean frames move with respect to each other at uniform velocity, no experiment could determine which frame was in absolute motion and which frame was at absolute rest.

This is the basic principle of Einstein's special theory of relativity. However, Einstein was not content with the apparent absolute status conferred to accelerating frames by the behaviour of bodies within them. Einstein sought a general principle of relativity that would require all frames of reference, whatever their relative state of motion, to be equivalent for the formulation of the general laws of nature. In his popular exposition of 1916, Einstein explains this by describing the experiences of an observer within a railway carriage that is decelerating. In his own words: "At all events it is clear that the Galilean law does not hold with respect to the non-uniformly moving carriage. Because of this, we feel compelled at the present juncture to grant a kind of absolute physical reality to non-uniform motion, in opposition to the general principle of relativity."

Once again it was one of Galileo's observations that provided the starting point for the formulation of Einstein's ideas. Galileo observed that bodies moving under the sole influence of a *gravitational field acquire an acceleration that does not depend upon the material or the physical state of the body. Einstein realized that this property of gravitational fields implied equivalence between gravity and accelerating frames of reference. This equivalence, which became the basis of his general theory of relativity, is well illustrated by one of Einstein's thought experiments. Imagine an elevator, so far removed from stars and other large masses that there is no appreciable gravitational field. An observer inside the elevator is equipped with the appropriate apparatus and uses the elevator as a reference frame. Initially, if the elevator is a Galilean frame, the observer would feel weightless with only inferences from decorations inside the elevator to distinguish between 'up' and 'down'. However, if a rope attached to the top of the elevator were to be pulled with a constant acceleration of 9.81 m s$^{-2}$, the observer would detect this acceleration as a force reaction on the floor of the elevator. The experiences of the observer in the elevator are equivalent to the experiences of an observer in an elevator in the earth's gravitational field of strength 9.81 N kg$^{-1}$. Moreover, the force reaction at the feet of the observer in the accelerating frame is due to the observer's inertial *mass (the mass that represents the reluctance of the observer's body to accelerate under the influence of a force).

An observer in the earth-bound elevator would feel the same force reaction at the floor of the elevator, but for this observer the force is due to the influence of the earth's gravitational field on the observer's gravitational mass. Guided by this example, Einstein realized that his extension of the principle of relativity to include accelerations implies the equality of inertial and gravitational mass, which had been established experimentally by Lóránd Eötvös (1848–1919) in 1888.

These considerations have significant implications for the nature of space and time under the influence of a gravitational field. Another of Einstein's thought experiments illustrates these implications. Imagine a Galilean frame of reference K from which an observer A takes measurements of a non-Galilean frame K', which is a rotating disc inhabited by an observer B. A notes that B is in circular motion and experiences a centrifugal acceleration. This acceleration is produced by a force, which may be interpreted as an effect of B's inertial mass. However, on the basis of the general principle of relativity, B may contend that he is actually at rest but under the influence of a radially directed gravitational field.

A comparison of time-measuring devices placed at the centre and edge of the rotating disc would show a remarkable result. For although the devices would both be at rest with respect to K', the motion of the disc with respect to K would lead to a *time dilation at the edge with respect to measured time at the centre. It follows that the clock at the disc's periphery runs at a permanently slower rate than that at the centre, i.e. as observed from K. The same effect would be noted by an observer who is sitting next to the clock at the centre of the disc. Thus, on the disc, or indeed in any gravitational field, a timing device will run at different rates depending on where it is situated.

The measurement of spatial intervals on the rotating disc will also incur a similar lack of definition. Standard measuring rods placed tangentially around the circumference C of the disc will all be contracted in length due to relativistic length contraction with respect to K. However, measuring rods will not experience shortening in length, as judged from K, if they are applied across a diameter D. Dividing the circumference by the diameter would produce a surprising result from K's point of view. Normally such a quotient would have the value $\pi = 3.14159...$, but in this situation the quotient is larger. Euclidean geometry does not seem to hold in an accelerating frame, or indeed by the principle of relativity, within a gravitational field. Spaces in which the propositions of Euclid are not valid are sometimes called curved spaces. For example, the sum of the internal angles of a triangle drawn on a flat sheet of paper will be 180°; however, a triangle drawn on the curved surface of a sphere will not follow this Euclidean rule.

Einstein fully expected to see this effect in

gravitational fields, such was his belief in the general principle of relativity. In fact, it was the effect gravitational fields have on the propagation of light that was heralded as the major verification of his general relativity. Einstein realized that rays of light would be perceived as curving in an accelerating frame. This led him to conclude that, in general, rays of light are propagated curvilinearly in gravitational fields. By means of photographs taken of stars during the solar eclipse of 29 May 1919, the existence of the deflection of starlight around the sun's mass was confirmed.

The mathematics required to describe the curvature of space in the presence of gravitational fields existed before Einstein had need for it, but it was essentially rediscovered by Einstein to solve his general relativistic problems. In general relativity, material bodies follow lines of shortest distance, called **geodesics**. The line formed by stretching an elastic band over a curved surface would be a geodesic on the curved surface. Light follows geodesics called **null-geodesics**. The motions of material bodies are therefore determined by the curvature of the space in the region through which they pass. However, it is the mass of the bodies that causes the curvature of the space in the first place, which demonstrates the elegant self-consistency of Einstein's general theory.

**relaxation oscillator** *See* OSCILLATOR.

**relay** An electrical or electronic device in which a variation in the current in one circuit controls the current in a second circuit. These devices are used in an enormous number of different applications in which electrical control is required. The simplest is the electromechanical relay in which the first circuit energizes an electromagnet, which operates a switch in a second circuit. The *thyratron gas-filled relay found many uses in the past but has now been largely replaced by the *thyristor solid-state relay.

**releaser** *See* SIGN STIMULUS.

**releasing hormone (releasing factor)** A hormone that is produced by the hypothalamus and stimulates the release of a hormone from the anterior *pituitary gland into the bloodstream. Each hormone has a specific releasing hormone; for example, thyrotrophin-releasing hormone stimulates the release of *thyroid-stimulating hormone.

**reluctance** Symbol $R$. The ratio of the

magnetomotive force to the total magnetic flux in a magnetic circuit or component. It is measured in henries.

**reluctivity** The reciprocal of magnetic *permeability.

**rem** *See* RADIATION UNITS.

**remanence (retentivity)** The magnetic flux density remaining in a ferromagnetic substance when the saturating field is reduced to zero. *See* HYSTERESIS.

**remote sensing** The gathering and recording of information concerning the earth's surface by techniques that do not involve actual contact with the object or area under study. These techniques include photography (e.g. aerial photography), multispectral imagery, infrared imagery, and radar. Remote sensing is generally carried out from aircraft and, increasingly, satellites. The techniques are used, for example, in cartography (map making).

**renal** Of or relating to the *kidney. For example, the **renal artery** and **renal vein** convey blood towards and away from the kidney, respectively.

**renal capsule** *See* BOWMAN'S CAPSULE.

**renewable energy sources** Sources of energy that do not use up the earth's finite mineral resources. **Nonrenewable energy sources** are *fossil fuels and fission fuels (*see* NUCLEAR FISSION). Various renewable energy sources are being used or investigated. *See* GEOTHERMAL ENERGY; HYDROELECTRIC POWER; NUCLEAR FUSION; SOLAR ENERGY; TIDES; WIND POWER; WAVE POWER.

**renin** A proteolytic enzyme (*see* PROTEASE) that is involved in the formation of the hormone *angiotensin, which raises blood pressure. Renin is secreted into the blood by cells of the kidney glomeruli under the control of the sympathetic nervous system; its release also occurs in response to a fall in blood-sodium levels and to falling blood pressure. Renin catalyses cleavage of circulating $\alpha_2$-globulin to produce angiotensin I, precursor of the active hormone.

**rennin (chymosin)** An enzyme secreted by cells lining the stomach in mammals that is responsible for clotting milk. It acts on a soluble milk protein (**caseinogen**), which it converts to the insoluble form *casein. This ensures that milk remains in the stomach long enough to be acted on by protein-digesting enzymes.

**renormalization** A procedure used in relativistic *quantum field theory to deal with the fact that in *perturbation theory calculations give rise to infinities beyond the first term. Renormalization was first used in *quantum electrodynamics, where the infinities were removed by taking the observed mass and charge of the electron as 'renormalized' parameters rather than the 'bare' mass and charge.

Theories for which finite results for all perturbation-theory calculations exist, by taking a finite number of parameters from experiment and using renormalization, are called **renormalizable**. Only certain types of quantum field theories are renormalizable. Theories that need an infinite number of parameters are said to be **nonrenormalizable** and are regarded as incomplete physical theories. The *gauge theories that describe the strong, weak, and electromagnetic interactions are renormalizable. The quantum theory of gravitational interactions is a nonrenormalizable theory, which perhaps indicates that gravity needs to be unified with other fundamental interactions before one can have a consistent quantum theory of gravity. Renormalization theory has been expressed in terms of noncommutative geometry.

**renormalization group** A technique used to understand systems in which many length-scales are involved. Such systems include phase transitions, turbulence, polymers, many-electron systems, and the localization of electrons in disordered systems. The renormalization group has its origin in *quantum field theory, in which it is used to calculate how *coupling constants change with energy. The way in which this change with energy takes place involves a *group and the procedure of *renormalization.

**repetitive DNA** DNA whose base sequence is repeated many times throughout the genome of an organism. It is common in eukaryotes, accounting for about half of the total DNA in mammals, for example, and can be divided into various types. Some serves a useful purpose, but a significant proportion is of uncertain function, and may be 'junk', or *selfish DNA. One important type consists of multiple copies of particular genes or gene sequences, which may be duplicates of

genes encoding histones or ribosomal RNAs. **Satellite DNA** – repeats of short DNA sequences (typically less than 10 bp) flanking the centromeres of each chromosome and stretching for hundreds of kilobases along either arm of the chromosome – and telomeric DNA (*see* TELOMERE) are important for maintaining chromosome structure. Other distinct types of repetitive DNA dispersed throughout the genome, both in noncoding introns within genes and between genes, include **minisatellite DNA** (variable number tandem repeats) – sequences of 15–100 bp repeated hundreds or thousands of times – and **microsatellite DNA**, repeats of shorter sequences (2–6 bp). Both these types are of value as *molecular markers in DNA fingerprinting.

**replacing bone** *See* CARTILAGE BONE.

**replicon** A DNA sequence that is replicated as a unit from a single initiation site (origin of replication). The genome of a bacterium or a virus comprises a single replicon; eukaryotes contain a number of replicons on each chromosome.

**repolarization** The restoration of the *resting potential in neurons or muscle fibres following the passage of a nerve impulse. Repolarization is brought about by diffusion of potassium ions out of the neuron and by active elimination of sodium ions (*see* SODIUM PUMP).

**Reppe processes** A set of related industrial reactions of acetylene to produce vinyl compounds, such as

$$HC{\equiv}CH + ROH \rightarrow H_2C{=}CHR$$

$$HC{\equiv}CH + RCN \rightarrow H_2C{=}CR(CN)$$

They take place at high pressure using metal acetylide catalysts. The processes are named after the German chemist Walter Reppe (1892–1969), who pioneered techniques for the safe industrial use of acetylene.

**repressor** A protein that can prevent the expression of a gene. *See* OPERON.

**reproduction** The production of new individuals more or less similar in form to the parent organisms. This may be achieved by a number of means (*see* SEXUAL REPRODUCTION; ASEXUAL REPRODUCTION) and serves to perpetuate or increase a species.

**reproductive system** The organs that are involved in the process of *sexual reproduction in an organism. The reproductive

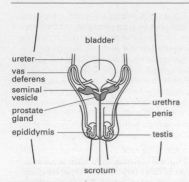
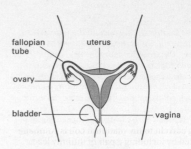

**Reproductive system.** The human male (left) and female (right) systems.

system of a flowering plant is found in the *flower and consists of the stamens (male organs) and carpels (female organs). In mammals the reproductive system consists of the testes, epididymis, sperm duct, and penis in the male and the ovaries, fallopian tubes, and uterus in the female.

**Reptilia** The class that contains the first entirely terrestrial vertebrates, which can live in dry terrestrial habitats as their skin is covered by a layer of horny scales, preventing water loss. They breathe atmospheric oxygen by means of lungs assisted by respiratory movements principally involving the ribs (there is no diaphragm). Reptiles are cold-blooded (*see* POIKILOTHERMY) but behavioural patterns make it possible for them to maintain a fairly even body temperature throughout the day. Fertilization is internal and the majority of reptiles lay eggs on land. These eggs have a porous shell to provide protection from desiccation and allow gas exchange. In some reptiles the eggs are retained within the body of the mother until the young are ready to hatch, thereby greatly reducing juvenile mortality (*see* OVOVIVIPARITY).

The class includes the modern crocodiles, lizards and snakes (*see* SQUAMATA), and tortoises and turtles, as well as many extinct forms, such as the *dinosaurs and *Pterosauria.

(⊕) **SEE WEB LINKS**

• This Amniota page from the Tree of Life Project shows the phylogenetic relationships of reptile groups, birds, and mammals

**residual volume** The amount of air remaining in the lungs after maximum expiration, which cannot be expelled from the lungs voluntarily. An average human has a residual volume of about 1 litre. *See also* VITAL CAPACITY.

**resin** A synthetic or naturally occurring *polymer. Synthetic resins are used in making *plastics. Natural resins are acidic chemicals secreted by many trees (especially conifers) into ducts or canals. They are found either as brittle glassy substances or dissolved in essential oils. Their functions are probably similar to those of gums and mucilages, i.e. protective.

**resistance** **1.** (in physics) Symbol $R$. The ratio of the potential difference across an electrical component to the current passing through it. It is thus a measure of the component's opposition to the flow of electric charge. In general, the resistance of a metallic conductor increases with temperature, whereas the resistance of a *semiconductor decreases with temperature. **2.** (in microbiology) The degree to which pathogenic microorganisms remain unaffected by antibiotics and other drugs. Genes for antibiotic resistance are often carried on *plasmids or *transposons, which can spread across species barriers. **3.** (in ecology) **a.** The degree to which a *pest can withstand the effects of a pesticide. It depends on the selection and spread within a pest population of genes that confer the ability to destroy, or minimize the effects of, a pesticide. **b.** *See* ENVIRONMENTAL RESISTANCE. **4.** (in immunology) The degree of *immunity to infection that an animal possesses.

**resistance thermometer (resistance pyrometer)** A *thermometer that relies on the increase of electrical resistance of a metal wire with rising temperature, according to the approximate relationship

$$R = R_0(1 + aT + bT^2),$$

where $R$ is the resistance of the wire at temperature $T$ and $R_0$ is the resistance of the wire at a reference temperature, usually 0°C; $a$ and $b$ are constants characteristic of the metal of the wire. The metal most frequently used is platinum and the platinum resistance coil is usually incorporated into one arm of a *Wheatstone bridge. The effect of the temperature change on the leads carrying current to the platinum coil is compensated by including a pair of dummy leads within the casing carrying the coil. *See also* THERMISTOR.

**resistivity** Symbol ρ. A measure of a material's ability to oppose the flow of an electric current. It is given by $RA/l$, where $R$ is the resistance of a uniform specimen of the material, having a length $l$ and a cross-sectional area $A$. It is usually given at 0°C or 20°C and is measured in ohm metres. It was formerly known as **specific resistance**.

**(( SEE WEB LINKS**

• Values of resistivity for a range of materials at the NPL website

**resistor** A component in an electrical or electronic circuit that is present because of its electrical resistance. For electronic purposes many resistors are either wire-wound or consist of carbon particles in a ceramic binder. The ceramic coating carries a number or colour code indicating the value of the resistance. Some resistors can be varied manually by means of a sliding contact; others are markedly dependent on temperature or illumination.

**resolution** 1. (in chemistry) The process of separating a racemic mixture into its optically active constituents. In some cases the crystals of the two forms have a different appearance, and the separation can be done by hand. In general, however, physical methods (distillation, crystallization, etc.) cannot be used because the optical isomers have identical physical properties. The most common technique is to react the mixture with a compound that is itself optically active, and then separate the two. For instance, a racemic mixture of (–)-A and (+)-A reacted with (–)-B, gives two compounds AB that are not optical isomers but diastereoisomers and can be separated and reconverted into the pure

(–)-A and (+)-A. Biological techniques using bacteria that convert one form but not the other can also be used. **2.** (in mathematics) The separation of a vector quantity into two components, which are usually at right angles to each other. Thus, a force $F$ acting on a body in a vertical plane at an angle θ to the horizontal can be resolved into a horizontal component $F\cosθ$ and a vertical component $F\sinθ$, both in the same plane as the original force. **3.** (in optics) *See* RESOLVING POWER.

**resolving power** A measure of the ability of an optical instrument to form separable images of close objects or to separate close wavelengths of radiation. The **chromatic resolving power** for any spectroscopic instrument is equal to $λ/δλ$, where $δλ$ is the difference in wavelength of two equally strong spectral lines that can barely be separated by the instrument and $λ$ is the average wavelength of these two lines. For a telescope forming images of stars the **angular resolving power** is the smallest angular separation of the images; the **linear resolving power** is the linear separation of the images in the focal plane. In a telescope forming images of two stars, as a result of diffraction by the lens aperture each image consists of a bright central blob surrounded by light and dark rings. According to the **Rayleigh criterion** for resolution, the central ring of one image should fall on the first dark ring of the other. The angular resolving power in radians is then $1.22λ/d$, where $d$ is the diameter of the objective lens in centimetres and $λ$ is the wavelength of the light (usually taken as 560 nanometres). For microscopes, the resolving power is usually taken as the minimum distance between two points that can be separated. In both cases, the smaller the resolving power, the better the resolution; to avoid this apparent paradox the resolving power is now sometimes taken as the reciprocals of the quantities stated above.

**resonance** 1. (in physics) An oscillation of a system at its natural frequency of vibration, as determined by the physical parameters of the system. It has the characteristic that large amplitude vibrations will ultimately result from low-power driving of the system. Resonance can occur in atoms and molecules, mechanical systems, and electrical circuits (*see* RESONANT CIRCUIT; RESONANT CAVITY). **2.** (in particle physics) A very short-lived *elementary particle that can be regarded as an excited state of a more stable particle. Resonances decay by the strong interaction (*see* FUNDAMENTAL INTERACTIONS) in $10^{-24}$ second. **3.** (in chemistry) The representation of the structure of a molecule by two or more conventional formulae. For example, the formula of methanal can be represented by a covalent structure $H_2C=O$, in which there is a double bond in the carbonyl group. It is known that in such compounds the oxygen has some negative charge and the carbon some positive charge. The true bonding in the molecule is somewhere between $H_2C=O$ and the ionic compound $H_2C^+O^-$. It is said to be a **resonance hybrid** of the two, indicated by

$$H_2C=O \leftrightarrow H_2C^+O^-$$

The two possible structures are called **canonical forms**, and they need not contribute equally to the actual form. The double-headed arrow does not imply that the two forms are in equilibrium.

**resonant cavity (cavity resonator)** A closed space within a conductor in which an electromagnetic field can be made to oscillate at frequencies above those at which a *resonant circuit will operate. The resonant frequency of the oscillation will depend on the dimensions and the shape of the cavity. The device is used to produce microwaves (*see* KLYSTRON; MAGNETRON).

**resonant circuit** A reactive circuit (*see* REACTANCE) so arranged that it is capable of *resonance. In a **series resonant circuit** a resistor, inductor, and capacitor are arranged in series. Resonance occurs when the *impedance ($Z$) is a minimum and the current amplitude therefore a maximum. In a **parallel resonant circuit** the inductance and capacitance are in parallel and resonance (with minimal current amplitude) occurs at maximum impedance. The frequency at which resonance occurs is called the **resonant frequency**. In a series resonant circuit

$$Z = R + i[\omega L - 1/\omega C],$$

where $\omega = 2\pi f$ and $f$ is the frequency, $R$ is the resistance, $L$ is the inductance, and $C$ is the capacitance. At resonance, $Z$ is a minimum and $\omega L = 1/\omega C$, i.e. the circuit behaves as if it is purely resistive. In the parallel circuit, resonance occurs when $R^2 + \omega^2 L^2 = L/C$, which in most cases also approximates to $\omega L = 1/\omega C$. Resonant circuits are widely used in *radio to select one signal frequency in preference to others.

**resonating valence bond theory** An extension of *valence bond theory to solids in which there is quantum-mechanical *resonance between different structures throughout the solid. Attempts have been made to describe high-temperature *superconductivity in terms of resonating valence bond theory.

**respiration** The metabolic process in animals and plants in which organic substances are broken down to simpler products with the release of energy, which is incorporated into special energy-carrying molecules (*see* ATP) and subsequently used for other metabolic processes. In most plants and animals respiration requires oxygen, and carbon dioxide is an end product. The exchange of oxygen and carbon dioxide between the body tissues and the environment is called **external respiration** (*see* VENTILATION). In many animals the exchange of gases takes place at *respiratory organs (e.g. *lungs in air-breathing vertebrates) and is assisted by *respiratory movements (e.g. breathing). In plants oxygen enters through pores on the plant surface and diffuses through the tissues via intercellular spaces or dissolved in tissue fluids.

Respiration at the cellular level is known as **internal** (or **tissue**) **respiration** and can be divided into two stages. In the first, *glycolysis, glucose is broken down to pyruvate. This does not require oxygen and is a form of *anaerobic respiration. In the second stage, the *Krebs cycle, pyruvate is broken down by a cyclic series of reactions to carbon dioxide and water. This is the main energy-yielding stage and requires oxygen. The processes of glycolysis and the Krebs cycle are common to all plants and animals that respire aerobically (*see* AEROBIC RESPIRATION).

**((⊕)) SEE WEB LINKS**
• Overview of cellular respiration, with animation and illustrations; from About.com.

**respiratory chain** *See* ELECTRON TRANSPORT CHAIN.

**respiratory movement** The muscular movement that enables the passage of air to and from the lungs or other *respiratory organs of an animal. The mechanism of the movement varies with the species. In insects abdominal muscles relax and contract rhythmically to encourage the flow of air through the *tracheae. In amphibians air is drawn into the lungs by a pumping action of the muscles in the floor of the mouth. **Breathing** in mammals involves the muscle of the *diaphragm and the *intercostal muscles between the ribs. Contraction of these muscles lowers the diaphragm and raises the ribs, so that the lungs expand and air is drawn in (*see* INSPIRATION). Relaxation has the opposite effect and forces air out during *expiration.

**respiratory organ** Any animal organ across which exchange of carbon dioxide and oxygen takes place. The surface membranes of such organs are always moist, thin, and well supplied with blood. Examples are the *lungs of air-breathing vertebrates, the *gills of fish, and the *tracheae of insects.

**respiratory pigment** A coloured compound that is capable of reversibly binding with oxygen at high oxygen concentrations and releasing it at low oxygen concentrations. Such pigments are present in the blood, transporting oxygen within the circulatory system from the *respiratory organs to the tissues of the body. In vertebrates the respiratory pigment is *haemoglobin, contained in the erythrocytes (red blood cells). *See also* HAEMOCYANIN.

**respiratory quotient (RQ)** The ratio of the volume of carbon dioxide produced by an organism during respiration to the volume of oxygen consumed. The RQ is usually about 0.8.

**respirometer** Any device that measures an organism's oxygen uptake. Simple respirometers consist of a chamber (in which the organism is placed) connected to a *manometer. Carbon dioxide is chemically removed from the chamber so that only oxygen uptake is measured. Human oxygen consumption is generally measured by a device known as a **spirometer**, which can also be used to measure depth and frequency of breathing.

**response** The physiological, muscular, or behavioural activity that can be elicited by a *stimulus.

**rest energy** The *rest mass of a body expressed in energy terms according to the relationship $E_0 = m_0c^2$, where $m_0$ is the rest mass of the body and $c$ is the speed of light.

**resting potential** The difference in electrical potential that exists across the membrane of a nerve cell that is not in the process of transmitting a nerve impulse. The resting potential is maintained by means of the *sodium pump. *Compare* ACTION POTENTIAL.

**restitution coefficient** Symbol $e$. A measure of the elasticity of colliding bodies. For two spheres moving in the same straight line,

$$e = (v_2 - v_1)/(u_1 - u_2),$$

where $u_1$ and $u_2$ are the velocities of bodies 1 and 2 before collision ($u_1 > u_2$) and $v_1$ and $v_2$ are the velocities of 1 and 2 after impact ($v_2 > v_1$). If the collision is perfectly elastic $e = 1$ and the kinetic energy is conserved; for an inelastic collision $e < 1$.

**rest mass** The mass of a body at rest when measured by an observer who is at rest in the same frame of reference. *Compare* RELATIVISTIC MASS.

**restriction enzyme (restriction endonuclease)** A type of enzyme that can cleave molecules of foreign DNA at a particular site. Restriction enzymes are produced by many bacteria and protect the cell by cleaving (and therefore destroying) the DNA of invading viruses. The bacterial cell is protected from attack by its own restriction enzymes by modifying the bases of its DNA during replication. Restriction enzymes are widely used in the techniques of genetic engineering (*see* DNA PROFILING; DNA LIBRARY; DNA SEQUENCING; GENE CLONING; RESTRICTION MAPPING).

**restriction fragment length polymorphism (RFLP)** The occurrence of different cleavage sites for *restriction enzymes in the DNA of different individuals of the same species. Cleavage of DNA from different individuals with restriction enzymes thus produces differing sets of restriction fragments. The deletion of existing restriction sites or the creation of new ones is the result of random base changes in the noncoding stretches of DNA (*introns) between genes. RFLPs have provided geneticists with a powerful set of genetic markers for mapping the genome (*see* RESTRICTION MAPPING) and for identifying particular genes (*see* GENE TRACKING).

**restriction mapping** A technique for determining the sites at which a length of DNA (e.g. from a chromosome) is cleaved by *restriction enzymes. By cleaving the DNA with various such enzymes, both individually and in combination, and analysing the resultant number and size of fragments by electrophoresis, a **restriction map**, indicating the order of restriction sites in the original DNA, can be deduced (*see also* LINKAGE MAP). Gene deletions or rearrangements that alter the restriction sites can be detected as changes in the pattern of fragments obtained. This may be used, for instance, to diagnose certain genetic abnormalities in the fetus. The fragments are separated by gel electrophoresis and identified using specific *gene probes, as in the *Southern blotting technique. The absence of a certain fragment in a fetal DNA digest can be diagnostic of a pathological change in the fetal gene containing the corresponding restriction site.

**resultant** A *vector quantity that has the same effect as two or more other vector quantities of the same kind. *See* PARALLELO-GRAM OF VECTORS.

**retardation (deceleration)** The rate of reduction of speed, velocity, or rate of change.

**retardation plate** A transparent plate of a birefringent material, such as quartz, cut parallel to the optic axis. Light falling on the plate at 90° to the optic axis is split into an ordinary ray and an extraordinary ray (*see* DOUBLE REFRACTION), which travel through the plate at different speeds. By cutting the plate to different thicknesses a specific phase difference can be introduced between the transmitted rays. In the **half-wave plate** a phase difference of $\pi$ radians, equivalent to a path difference of half a wavelength, is introduced. In the **quarter-wave plate** the waves are out of step by one quarter of a wavelength.

**reticular formation** *See* BRAINSTEM.

**reticulum** The first of four chambers that form the stomach of ruminants. *See* RUMI-NANTIA.

**retina** The light-sensitive membrane that lines the interior of the eye. The retina consists of two layers. The inner layer contains nerve cells, blood vessels, and two types of light-sensitive cells (*rods and *cones). The outer layer is pigmented, which prevents the back reflection of light and consequent decrease in visual acuity. Light passing through the lens stimulates individual rods and cones, which generates nerve impulses that are transmitted through the optic nerve to the brain, where the visual image is formed.

**retinal** *See* RHODOPSIN; VITAMIN A.

**retinol** *See* VITAMIN A.

**retort** 1. A laboratory apparatus consisting of a glass bulb with a long neck. 2. A vessel

used for reaction or distillation in industrial chemical processes.

**retrograde motion** 1. The apparent motion of a planet from east to west as seen from the earth against the background of the stars. 2. The clockwise rotation of a planet, as seen from its north pole. *Compare* DIRECT MOTION.

**retrorocket** A small rocket motor that produces thrust in the opposite direction to a *rocket's main motor or motors in order to decelerate it.

**retrotransposon** A type of *transposon found in the DNA of various organisms, including yeast, *Drosophila*, and mammals, that forms copies of itself using a mechanism similar to that of retroviruses. It undergoes transcription to RNA, then creates a DNA copy of the transcript with the aid of the enzyme *reverse transcriptase. This DNA copy can then reintegrate into the cell's genome.

**retrovirus** An RNA-containing virus that converts its RNA into DNA by means of the enzyme *reverse transcriptase; this enables it to become integrated into its host's DNA. Some retroviruses can cause cancer in animals: they contain *oncogenes, which are activated when the virus enters its host cell and starts to replicate. Retroviruses are useful as *vectors for inserting genetic material into eukaryotic cells. The best-known retrovirus is *HIV, responsible for AIDS in humans. *See also* PROVIRUS.

**reverberation time** The time taken for the energy density of a sound to fall to the threshold of audibility from a value $10^6$ times as great; i.e. a fall of 60 decibels. It is an important characteristic of an auditorium. The optimum value is proportional to the linear dimensions of the auditorium.

**reverberatory furnace** A metallurgical furnace in which the charge to be heated is kept separate from the fuel. It consists of a shallow hearth on which the charge is heated by flames that pass over it and by radiation reflected onto it from a low roof.

**reverse genetics** Any approach to genetic investigation that aims to find the function for some known protein or gene. It contrasts with the more traditional *forward genetics approach. For example, analysis of gene sequences reveals open reading frames, which are the hallmarks of functional genes (*see* READING FRAME). Reverse genetics methods can be used to discover the function of such genes, which can be cloned, subjected to mutation, and then reinserted into the organism (e.g. a bacterium or yeast cell) to see what effect the mutations have on function. A similar approach can be taken starting with a protein of unknown function. The amino-acid sequence can be back-translated into genetic code, a DNA probe constructed for part of the DNA sequence, and the relevant gene selected from a *DNA library of the organism.

**reverse osmosis** A method of obtaining pure water from water containing a salt, as in *desalination. Pure water and the salt water are separated by a semipermeable membrane and the pressure of the salt water is raised above the osmotic pressure, causing water from the brine to pass through the membrane into the pure water. This process requires a pressure of some 25 atmospheres, which makes it difficult to apply on a large scale. The process is used for the purification of drinking water.

**reverse transcriptase** An enzyme, occurring in *retroviruses, that catalyses the formation of double-stranded DNA using the single RNA strand of the viral genome as template. This enables the viral genome to be inserted into the host's DNA and replicated by the host. Reverse transcriptase is thus an RNA-directed DNA *polymerase. The enzyme is used in genetic engineering for producing *complementary DNA from messenger RNA.

**reversible process** Any process in which the variables that define the state of the system can be made to change in such a way that they pass through the same values in the reverse order when the process is reversed. It is also a condition of a reversible process that any exchanges of energy, work, or matter with the surroundings should be reversed in direction and order when the process is reversed. Any process that does not comply with these conditions when it is reversed is said to be an **irreversible process**. All natural processes are irreversible, although some processes can be made to approach closely to a reversible process.

**Reynolds number** Symbol *Re*. A dimensionless number used in fluid dynamics to determine the type of flow of a fluid through a pipe, to design prototypes from small-scale

models, etc. It is the ratio $\nu \rho l/\eta$, where $\nu$ is the flow velocity, $\rho$ is the fluid density, $l$ is a characteristic linear dimension, such as the diameter of a pipe, and $\eta$ is the fluid viscosity. In a smooth straight uniform pipe, laminar flow usually occurs if $Re < 2000$ and turbulent flow is established if $Re > 3000$. It is named after Osborne Reynolds (1842–1912).

**RFLP** *See* RESTRICTION FRAGMENT LENGTH POLYMORPHISM.

**$R_F$ value** (in chromatography) The distance travelled by a given component divided by the distance travelled by the solvent front. For a given system at a known temperature, it is a characteristic of the component and can be used to identify components.

**Rh** *See* RHESUS FACTOR.

**rhachis** *See* RACHIS.

**rhe** A unit of fluidity equal to the reciprocal of the *poise.

**rhenium** Symbol Re. A silvery-white metallic *transition element; a.n. 75; r.a.m. 186.2; r.d. 20.53; m.p. 3180°C; b.p. 5627 (estimated)°C. The element is obtained as a by-product in refining molybdenum, and is used in certain alloys (e.g. rhenium–molybdenum alloys are superconducting). It forms a number of complexes with oxidation states in the range 1–7. It was discovered by Walter Noddack (1893–1960) and Ida Tacke in 1925.

((⊕)) SEE WEB LINKS
• Information from the WebElements site

**rheology** The study of the deformation and flow of matter.

**rheopexy** The process by which certain thixotropic substances set more rapidly when they are stirred, shaken, or tapped. Gypsum in water is such a **rheopectic substance**.

**rheostat** A variable *resistor, the value of which can be changed without interrupting the current flow. In the common wire-wound rheostat, a sliding contact moves along the length of the coil of wire.

**rhesus factor** (Rh factor) An *antigen whose presence or absence on the surface of red blood cells forms the basis of the rhesus *blood group system. (The factor was first recognized in rhesus monkeys.) Most people possess the Rh factor, i.e. they are rhesus positive (Rh+). People who lack the factor are Rh–. If Rh+ blood is given to an Rh– patient, the latter develops anti-Rh antibodies. Subsequent transfusion of Rh+ blood results in *agglutination, with serious consequences. Similarly, an Rh– pregnant woman carrying an Rh+ fetus may develop anti-Rh antibodies in her blood; these will react with the blood of a subsequent Rh+ fetus, causing anaemia in the newborn baby.

**rhizoid** One of a group of delicate and often colourless hairlike outgrowths found in certain algae and the gametophyte generation of bryophytes and ferns. They anchor the plant to the substrate and absorb water and mineral salts.

**rhizome** A horizontal underground stem. It enables the plant to survive from one growing season to the next and in some species it also serves to propagate the plant vegetatively. It may be thin and wiry, as in couch grass, or fleshy and swollen, as in *Iris*. Compact upright underground stems, as in rhubarb, strawberry, and primrose, are often called **rootstocks**.

**rhizosphere** The zone immediately surrounding the actively growing region of a plant root. Typically 1–2 mm thick, it consists of a *biofilm of water and soluble substances derived from the plant, soil constituents, and a community of fungi, bacteria, and other microorganisms that interact with each other and with the plant. The rhizosphere has a profound influence on the growth and survival of the plant, notably in helping it to absorb nutrients from the soil and inhibiting root pathogens, and on soil composition and structure.

**rhodium** Symbol Rh. A silvery-white metallic *transition element; a.n. 45; r.a.m. 102.9; r.d. 12.4; m.p. 1966°C; b.p. 3727°C. It occurs with platinum and is used in certain platinum alloys (e.g. for thermocouples) and in plating jewellery and optical reflectors. Chemically, it is not attacked by acids (dissolves only slowly in aqua regia) and reacts with nonmetals (e.g. oxygen and chlorine) at red heat. Its main oxidation state is +3 although it also forms complexes in the +4 state. The element was discovered in 1803 by William Wollaston (1766–1828).

**(((⊕))) SEE WEB LINKS**
• Information from the WebElements site

**Rhodophyta (red algae)** A phylum of *algae that are often pink or red in colour due to the presence of the pigments phycocyanin and phycoerythrin. Members of the Rhodophyta may be unicellular or multicellular; the latter form branched flattened thalli or filaments. They are commonly found along the coasts of tropical areas. Red algae are now regarded as members of the assemblage Archaeplastida (*see* PLANT).

**rhodopsin (visual purple)** The light-sensitive pigment found in the *rods of the vertebrate retina. It consists of a protein component, **opsin**, linked to a nonprotein part, **retinal** (a derivative of *vitamin A). Light falling on the rod is absorbed by the retinal, which changes its form and separates from the opsin component; this initiates the transmission of a nerve impulse to the brain. The great sensitivity of rhodopsin allows vision in dim light (night vision).

**rhombencephalon** *See* HINDBRAIN.

**rhombus** A parallelogram in which all the sides are of equal length.

**rhumb line (loxodrome)** (in navigation) A line of constant compass direction that cuts across all lines of longitude at the same angle. The rhumb line is not the shortest distance between two points unless the two points are on the same meridian or on the equator. On the Mercator map projection a rhumb line is represented by a straight line.

**rhyolite** An igneous rock, the volcanic equivalent of granite. It is usually glassy or cryptocrystalline and consists of quartz, feldspars, and mica or amphibole. It may contain larger crystals (phenocrysts) set in a much finer-grained matrix.

**rhytidome** *See* BARK.

**rib** One of a series of slender curved bones that form a cage to enclose, support, and protect the heart and lungs (*see* THORAX). Ribs occur in pairs, articulating with the *thoracic vertebrae of the spinal column at the back and (in reptiles, birds, and mammals) with the *sternum (breastbone) in front. Movements of the rib cage, controlled by **intercostal muscles** between the ribs, are important in breathing (*see* RESPIRATORY MOVEMENT).

**riboflavin** *See* VITAMIN B COMPLEX.

**ribonuclease** *See* RNASE.

**ribonucleic acid** *See* RNA.

**ribonucleoprotein (RNP)** Any complex of protein and RNA that forms during the synthesis of RNA in eukaryotes; the protein is involved in the packaging and condensation of the RNA. Certain RNPs are restricted to the nucleus whereas others are found in both the nucleus and the cytoplasm. The most common RNP occurring in the nucleus is **heterogeneous nuclear RNP (hnRNP)**, which consists of protein bound to the primary transcript of DNA (*see* TRANSCRIPTION). It may be associated with **small nuclear RNP (snRNP)**, which is involved in the removal of intron sequences from the primary transcript to form messenger RNA, which eventually leaves the nucleus (*see* GENE SPLICING).

**ribose**  A *monosaccharide, $C_5H_{10}O_5$, rarely occurring free in nature but important as a component of *RNA (ribonucleic acid). Its derivative **deoxyribose**, $C_5H_{10}O_4$, is equally important as a constituent of *DNA (deoxyribonucleic acid), which carries the genetic code in chromosomes.

**ribosomal RNA**  *See* RIBOSOME; RNA.

**ribosome**  A small spherical body within a living cell that is the site of *protein synthesis. Ribosomes consist of two subunits, one large and one small, each of which comprises a type of RNA (called **ribosomal RNA**) and protein. Usually there are many ribosomes in a cell, either attached to the *endoplasmic reticulum or free in the cytoplasm. During protein synthesis they are associated with messenger RNA as *polyribosomes in the process of *translation.

**ribozyme (catalytic RNA)**  Any RNA molecule that can catalyse changes to its own molecular structure. Self-splicing introns (*see* GENE SPLICING) are examples of ribozymes.

**ribulose**  A ketopentose sugar (*see* MONOSACCHARIDE), $C_5H_{11}O_5$, that is involved in carbon dioxide fixation in photosynthesis as a component of *ribulose bisphosphate.

**ribulose bisphosphate (RuBP)**  A five-carbon sugar that is combined with carbon dioxide to form two three-carbon intermediates in the first stage of the light-independent reactions of *photosynthesis (*see* CALVIN CYCLE). The enzyme that mediates the carboxylation of ribulose bisphosphate is **ribulose bisphosphate carboxylase/oxygenase (rubisco)**.

**Richardson equation (Richardson–Dushman equation)**  *See* THERMIONIC EMISSION.

**Richter scale**  A logarithmic scale devised

in 1935 by Charles Richter (1900–85) to compare the magnitude of earthquakes. The scale ranges from 0 to 10 and the Richter scale value is related to the logarithm of the amplitude of the ground motion divided by the period of the dominant wave, subject to certain corrections. On this scale a value of 2 can just be felt as a tremor and damage to buildings occurs for values in excess of 6. The largest shock recorded had a magnitude of 9.5.

**ricin**  A highly toxic protein present in the castor oil plant (*Ricinus communis*), in particular in the seeds of the plant (castor beans). It is probably the most poisonous substance present in plants – the fatal dose is about 1 milligram per kilogram of body weight.

**rickets**  A childhood condition caused by decalcification of bone, resulting in deformed bones. Rickets is associated with chronic deficiency of *vitamin D or calcium and with disorders that cause poor phosphate reabsorption from the kidney *nephrons.

**rickettsia**  A very small spherical or rod-shaped Gram-negative bacterium belonging to the phylum Proteobacteria. Most rickettsias are obligate parasites, being unable to reproduce outside the cells of their hosts. Rickettsias can infect such arthropods as ticks, fleas, lice, and mites, through which they can be transmitted to vertebrates, including humans. The group includes the causal agents of trench fever, Rocky Mountain spotted fever, and forms of typhus.

**rift valley**  A steep-sided depression that occurs in regions in which there is *plate tectonic activity. Upwelling of magma causes a part of a plate to dome up and stretch. Subsequently a pair of long normal faults form, and the valley collapses between them; it is in effect a large *graben. The valleys may be extremely large, e.g. the East African Rift Valley extends for more than 4000 km, with cliff-like edges 2–3 km tall.

**right ascension**  A coordinate used with *declination for locating an object on the *celestial sphere; it is equivalent to longitude in the earth's *latitude and longitude system, except that its zero point is not a prime meridian. Right ascension is measured along the celestial equator. The 360° around the equator are divided into 24 hours (1 hour = 15°) with subdivisions into minutes and seconds; the intersection of the ecliptic and the equator at the vernal equinox is taken as 0 hours. A celestial object's right ascension is defined as the angular distance from the vernal equinox eastward along the celestial equator to the **hour circle**, the perpendicular great circle passing through the object.

**rigidity modulus** *See* ELASTIC MODULUS.

**ring** A closed chain of atoms in a molecule. In compounds, such as naphthalene, in which two rings share a common side, the rings are **fused rings**. **Ring closures** are chemical reactions in which one part of a chain reacts with another to form a ring, as in the formation of *lactams and *lactones.

**(())) SEE WEB LINKS**

• Information about IUPAC nomenclature of fused rings

**Ringer's solution** *See* PHYSIOLOGICAL SALINE.

**R-isomer** *See* ABSOLUTE CONFIGURATION.

**ritualization** An evolutionary process in which the form or context of an action is altered because it comes to play a role in social communication. For example, many *courtship and greeting ceremonies in animals include ritual food presentation (though the quantities of food may be negligible), derived from the action of feeding the young.

**RME** *See* BIOFUEL.

**RMS value** *See* ROOT-MEAN-SQUARE VALUE.

**RNA (ribonucleic acid)** A complex organic compound (a nucleic acid) in living cells that is concerned with *protein synthesis. In some viruses, RNA is also the hereditary material. Most RNA is synthesized in the nucleus and then distributed to various parts of the cytoplasm. An RNA molecule consists of a long chain of *nucleotides in which the sugar is *ribose and the bases are adenine, cytosine, guanine, and uracil (see illustration; *compare* DNA). **Messenger RNA (mRNA)** is responsible for carrying the *genetic code transcribed from DNA to specialized sites within the cell (known as *ribosomes), where the information is translated into protein composition (*see* TRANSCRIPTION; TRANSLATION). **Ribosomal RNA (rRNA)** is present in ribosomes; it is single-stranded but helical regions are formed by *base pairing within the strand. **Transfer RNA (tRNA, soluble RNA, sRNA)** is involved

in the assembly of amino acids in a protein chain being synthesized at a ribosome. Each tRNA is specific for an amino acid and bears a triplet of bases complementary with a triplet on mRNA (*see* CODON). *See also* ANTISENSE RNA; RIBONUCLEOPROTEIN.

**(())) SEE WEB LINKS**

• Animated account of the roles of RNA in protein synthesis; part of the 'DNA from the Beginning' website

**RNAase** *See* RNASE.

**RNA interference (RNAi)** The ability of double-stranded RNA to interfere with or suppress (silence) the expression of a gene with a corresponding base sequence. It involves a ribonuclease called **Dicer** cutting double-stranded RNA into fragments (21–22 nucleotides), one strand of which (the antisense strand) possesses a base sequence complementary to that of the target gene's mRNA; this strand is then incorporated into an assemblage of proteins, the **RNA-induced silencing complex** (**RISC**), which binds to the target mRNA and causes **gene silencing** by degrading the mRNA or preventing its translation into protein (*see* MICRORNA). RNAi is a regulating mechanism for an estimated 30% of all protein-coding genes in mammals. It also helps protect cells against certain viruses by targeting viral RNA, and it helps to silence potentially disruptive *transposons by destroying RNA copies arising from transposon replication.

**RNA polymerase** *See* POLYMERASE.

**RNase (ribonuclease; RNAase)** Any enzyme that catalyses the cleavage of nucleotides in RNA. Each RNase has a specificity for a different cleavage site. For example, RNase A is a digestive enzyme secreted by the pancreas that hydrolyses phosphodiester bonds in the nucleotide chain. Other RNases are active at the cellular level, for instance in modifying transfer RNA and ribosomal RNA after transcription.

**roasting** The heating of a finely ground ore, especially a sulphide, in air prior to *smelting. The roasting process expels moisture, chemically combined water, and volatile matter; in the case of sulphides, the sulphur is expelled as sulphur dioxide and the ore is converted into an oxide. Part of the heat may be provided by the combustion of the sulphur.

**Roche limit** The minimum distance from

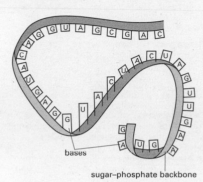

Detail of molecular structure of sugar–phosphate backbone. Each ribose unit is attached to a phosphate group and a base, forming a nucleotide.

Single-stranded structure of RNA

The four bases of RNA

**RNA.** Its molecular structure.

the centre of a celestial body at which a satellite orbiting the body can safely remain in equilibrium under the influence of its own gravitation and that of its *primary.

**Rochelle salt** Potassium sodium tartrate tetrahydrate, $KNaC_4H_4O_6.4H_2O$. A colourless crystalline salt used for its piezoelectric properties. It was first prepared at La Rocelle in France.

**Roche lobe** One of the two pear-shaped regions that surround each of a pair of *bi-

nary stars. It is the region in which a particular star's gravitational field is predominant. It is named after Edouard Roche (1820–83).

**Rochon prism** An optical device consisting of two quartz prisms; the first, cut parallel to the optic axis, receives the light; the second, with the optic axis at right angles, transmits the ordinary ray without deviation but the extraordinary ray is deflected and can be absorbed by a screen. The device can be used to produce plane-polarized light and it can also be used with ultraviolet radiation.

It was invented by the French astronomer Alexis-Marie de Rochon (1741–1817).

**rock** An aggregate of mineral particles that makes up part of the earth's crust. It may be consolidated or unconsolidated (e.g. sand, gravel, mud, shells, coral, and clay).

**rock crystal** See QUARTZ.

**rocket** A space vehicle or projectile that is forced through space or the atmosphere by *jet propulsion and that carries its own propellants and oxidizers. It is therefore independent of the earth's atmosphere for lift, thrust, or oxygen and is the only known vehicle for travel outside the earth's atmosphere. Rocket motors (or rocket engines) are currently driven by solid or liquid chemical propellants, which burn in an oxidizer carried within the rocket. Typical liquid bipropellant combinations include liquid hydrogen with liquid oxygen for main engines and hydrazine with dinitrogen tetroxide oxidizer for smaller positioning rockets. Experimental rocket motors have also been tested using ionized gases and plasmas to provide thrust (*see also* ION ENGINE). The measure of a rocket motor's performance is its *specific impulse.

**rocking-chair cell** See INTERCALATION CELL.

**rock salt** See HALITE.

**rock salt structure (sodium chloride structure)** A type of ionic crystal structure in which the cations have a face-centred cubic arrangement, with anions occupying all the octahedral holes. It can equally be described as an fcc array of anions with cations in the octahedral holes. Each type of ion has a coordination number of 6. Examples of compounds that have the structure are NaCl, KBr, AgCl, AgBr, HgO, CaO, FeO, NiO, and SnAs.

**(⊕) SEE WEB LINKS**
• An interactive version of the structure

**rod** A type of light-sensitive receptor cell present in the retinas of vertebrates. Rods contain the pigment *rhodopsin and are essential for vision in dim light. They are not evenly distributed on the retina, being absent in the *fovea and occupying all of the retinal margin. *Compare* CONE.

**Rodentia** An order of mammals characterized by a single pair of long curved incisors in each jaw. These teeth are specialized for gnawing: they continue growing throughout

life and have enamel only on the front so that they wear to a chisel-shaped cutting edge. Rodents often breed throughout the year and produce large numbers of quickly maturing young. The order includes the squirrels, beavers, rats, mice, and porcupines.

**roentgen** The former unit of dose equivalent (*see* RADIATION UNITS). It is named after W. K. Roentgen.

**Roentgen, William Konrad** (1845–1923) German physicist, who made many contributions to physics, the best known being his discovery of X-rays in 1895. For this work he was awarded the first Nobel Prize for physics in 1901.

**roentgenium** Symbol Rg. A radioactive transactinide; a.n. 111. It was made by fusion of $^{209}$Bi with $^{64}$Ni. Only a few atoms have been detected.

**(⊕) SEE WEB LINKS**
• Information from the WebElements site

**Rohypnol** See FLUNITRAZEPAM.

**rolling friction** *Friction between a rolling wheel and the plane surface on which it is rotating. As a result of any small distortions of the two surfaces, there is a frictional force with a component, $F_1$, that opposes the motion. If $N$ is the normal force, $F_r = N\mu_r$, where $\mu_r$ is called the **coefficient of rolling friction**.

**ROM** Read-only memory. A form of computer memory, fabricated from *integrated circuits, whose contents are permanently recorded at the time of manufacture. It is thus used to store data that never require modification. (The contents of **programmable ROM** (or **PROM**) are recorded in a separate process after manufacture.) Like *RAM it consists of an array of 'cells' to which there is direct and extremely rapid access.

**root** 1. (in botany) The part of a vascular plant that grows beneath the soil surface in response to gravity and water. It anchors the plant in the soil and absorbs water and mineral salts. Unlike the stem, it never produces leaves, buds, or flowers and never contains chlorophyll. The *radicle (embryonic root) may give rise either to a **tap root system** with a single main **tap root** from which lateral roots develop, or a **fibrous root system**, with many roots of equal size. The *apical meristem at the root tip gives rise to a protective sheath, the *root cap, and to the primary tis-

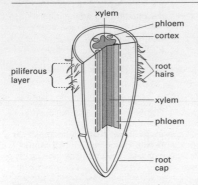

**Root.** Section through the tip of a plant root.

sues of the root. The vascular tissues usually form a central core (see illustration). This distinguishes roots from stems, in which the vascular tissue often forms a ring. A short distance behind the root tip **root hairs** develop from the epidermis and greatly increase the surface area for absorption of water and minerals. Beyond this, lateral roots develop.

Roots may be modified in various ways. Some are swollen with food to survive the winter, as in the carrot. Certain plants, such as orchids, have absorptive aerial roots; others, such as ivy, have short clasping roots for climbing. The roots of leguminous plants, such as beans and peas, contain *root nodules, which have an important role in nitrogen fixation. Other modifications include *prop roots, stilt roots, and buttress roots, which support the plant.

**2.** (in dentistry) The portion of a *tooth that is not covered with enamel and is embedded in a socket in the jawbone. Incisors, canines, and premolars have single roots; molars normally have several roots.

**3.** (in anatomy) The point of origin of a nerve in the central nervous system. There are two roots for every *spinal nerve (*see* DORSAL ROOT; VENTRAL ROOT).

**4.** (in mathematics) **a.** One of the equal factors of a number or quantity, e.g. the cube root of 8 is 2. In general, the $n$th root of a number or quantity $a$ is a number or quantity $x$ that satisfies the equation $x^n = a$. **b.** The value or values of an independent variable in an equation that satisfies that equation.

**root cap (calyptra)** A cone-shaped structure that covers the root tip and develops as a result of cell division by a meristem at the

root apex (*see* CALYPTROGEN). It protects the root tip as it grows between the soil particles. The cells are constantly worn away by friction and are replaced by the meristem.

**root hair** *See* ROOT.

**root-mean-square value (RMS value)**
**1.** (in statistics) A typical value of a number ($n$) of values of a quantity ($x_1, x_2, x_3, \ldots$) equal to the square root of the sum of the squares of the values divided by $n$, i.e.

$$\text{RMS value} = \sqrt{[(x_1{}^2 + x_2{}^2 + x_3{}^2 \ldots)/n]}$$

**2.** (in physics) A typical value of a continuously varying quantity, such as an alternating electric current, obtained similarly from many samples taken at regular time intervals during a cycle. Theoretically this can be shown to be the **effective value**, i.e. the value of the equivalent direct current that would produce the same power dissipation in a given resistor. For a sinusoidal current this is equal to $I_m/\sqrt{2}$, where $I_m$ is the maximum value of the current.

**root nodule** A swelling on the roots of certain plants, especially those of the family Fabaceae (Leguminosae), that contains bacteria (notably *Rhizobium*) capable of fixing atmospheric nitrogen into ammonia, which is subsequently converted to nitrates and amino acids (*see* NITROGEN FIXATION). Plants that possess root nodules increase soil fertility by increasing the nitrate content of the soil. The practice of *crop rotation will normally include the cultivation of a leguminous species.

**root pressure** The pressure that forces water, absorbed from the soil, to move through the roots and up the stem of a plant. This pressure can be demonstrated by cutting a stem, from which water will exude. A *manometer can be attached to a plant stem to measure the root pressure. Root pressure is believed to be due to both the osmosis of water, from the soil into the root cells, and the active pumping of salts into the *xylem tissue, which maintains a concentration gradient along which the water will move. *See also* TRANSPIRATION.

**rootstock** *See* RHIZOME.

**Rose's metal** An alloy of low melting point (about 100°C) consisting of 50% bismuth, 25–28% lead, and 22–25% tin.

**rot** (in mathematics) *See* CURL.

**rotary converter** A device for converting

direct current to alternating current or one d.c. voltage to another. It consists of an electric motor coupled to a generator.

**rotational motion** The laws relating to the rotation of a body about an axis are analogous to those describing linear motion. The **angular displacement** ($\theta$) of a body is the angle in radians through which a point or line has been rotated in a specified sense about a specified axis. The **angular velocity** ($\omega$) is the rate of change of angular displacement with time, i.e. $\omega = d\theta/dt$, and the **angular acceleration** ($\alpha$) is the rate of change of angular velocity, i.e. $\alpha = d\omega/dt = d^2\theta/dt^2$.

The equations of linear motion have analogous rotational equivalents, e.g.:

$$\omega_2 = \omega_1 + \alpha t$$
$$\theta = \omega_1 t + \alpha t^2/2$$
$$\omega_2^2 = \omega_1^2 + 2\theta\alpha$$

The counterpart of Newton's second law of motion is $T = I\alpha$, where $T$ is the *torque causing the angular acceleration and $I$ is the *moment of inertia of the rotating body.

**rotaxane** A type of compound that has a dumbbell-shaped molecule with a cyclic molecule around its axis. The dumbbell has a chain with large groups at each end, these being large enough to trap the ring. There is no formal chemical bonding between the dumbbell and the ring. Rotaxanes are examples of compounds with *mechanical bonding. A number of natural peptide rotaxanes have been identified. Synthesis of new rotaxanes is a matter of interest because of their possible use as 'molecular machines' in nanotechnology (e.g. as molecular switches or information storage units).

**rotor** The rotating part of an electric motor, electric generator, turbine, etc. *Compare* STATOR.

**roughage** *See* (DIETARY) FIBRE.

**rounding error** The difference between the exact value of a number and its approximate value that results from considering only a fixed number of decimal places or *significant figures (by rounding up or down). Because only a finite number of *bytes are available to stand for a number in a computer system, computer calculations are subject to rounding errors.

**round window (fenestra rotunda)** A membrane-covered opening between the middle ear and the inner ear (*see* EAR), situated below the *oval window. Pressure

waves transmitted through the perilymph in the *cochlea are released into the middle ear through the round window.

**roundworms** *See* NEMATODA.

**r-process** *See* ORIGIN OF THE ELEMENTS.

**RQ** *See* RESPIRATORY QUOTIENT.

**RR Lyrae variable star** A type of short-period pulsating *variable star that occurs in globular clusters and in the galactic nucleus. RR Lyrae variables are old population II stars (*see* POPULATION TYPE), typically giant stars whose brightness varies widely every few hours.

**rRNA** *See* RNA.

**R–S convention** *See* ABSOLUTE CONFIGURATION.

**r selection** A type of selection that favours organisms with a high rate of reproduction (*r* **value**). Organisms that are *r* selected (*r* **strategists**) are able to colonize a habitat rapidly, utilizing the food and other resources before other organisms are established and begin to compete. They tend to be relatively small with short life spans (e.g. bacteria) and often live in unstable environments; characteristically their survival depends on their ability to produce large numbers of offspring rather than on their ability to compete. *Compare* K SELECTION.

**rubber** A polymeric substance obtained from the sap of the tree *Hevea brasiliensis*. Crude natural rubber is obtained by coagulating and drying the sap (latex), and is then modified by *vulcanization and compounding with fillers. It is a polymer of *isoprene containing the unit $-CH_2C(CH_3){:}CHCH_2-$. Various synthetic rubbers can also be made. *See* NEOPRENE; NITRILE RUBBER; SILICONES.

**rubidium** Symbol Rb. A soft silvery-white metallic element belonging to *group 1 (formerly IA) of the periodic table; a.n. 37; r.a.m. 85.47; r.d. 1.53; m.p. 38.89°C; b.p. 688°C. It is found in a number of minerals (e.g. lepidolite) and in certain brines. The metal is obtained by electrolysis of molten rubidium chloride. The naturally occurring isotope rubidium–87 is radioactive (*see* RUBIDIUM–STRONTIUM DATING). The metal is highly reactive, igniting spontaneously in air. It was discovered spectroscopically by Robert Bunsen and Gustav Kirchhoff in 1861.

(((●))) **SEE WEB LINKS**

• Information from the WebElements site

**rubidium–strontium dating** A method of dating geological specimens based on the decay of the radioisotope rubidium–87 into the stable isotope strontium–87. Natural rubidium contains 27.85% of rubidium–87, which has a half-life of $4.7 \times 10^{10}$ years. The ratio $^{87}Rb/^{87}Sr$ in a specimen gives an estimate of its age (up to several thousand million years).

**rubisco** *See* RIBULOSE BISPHOSPHATE.

**ruby** The transparent red variety of the mineral *corundum, the colour being due to the presence of traces of chromium. It is a valuable gemstone, more precious than diamonds. The finest rubies are obtained from Mogok in Burma, where they occur in metamorphic limestones; Sri Lanka and Thailand are the only other important sources. Rubies have been produced synthetically by the Verneuil flame-fusion process. Industrial rubies are used in lasers, watches, and other precision instruments.

**rumen** The second of four chambers that form the stomach of ruminants. *See* RUMINANTIA.

**Rumford, Count** (Benjamin Thompson; 1753–1814) American-born British physicist, who acted as an English spy during the American Revolution. As a result he was forced to flee in 1775, first to England and then to Munich. There he observed the boring of cannon barrels, which led him to his best-known proposition, that *friction produces heat. While in Munich he was made a count of the Holy Roman Empire. Returning to England in 1795, he helped to demolish the *caloric theory.

**Ruminantia** A suborder of hooved mammals (*see* ARTIODACTYLA) comprising the sheep, cattle, goats, deer, and antelopes. They are characterized by a four-chambered stomach (see illustration). Swallowed food passes from the first chamber, the **reticulum**, to the **rumen**, where food is digested by *cellulase and other enzymes secreted by symbiotic anaerobic microorganisms that live in the rumen. Some products of digestion are absorbed in the rumen; the remaining partly digested food is regurgitated and chewed to a pulp – the process known as 'chewing the cud'. This food mass is then swallowed and passes from the reticulum to the third chamber, the **omasum**, where water and some nutrients

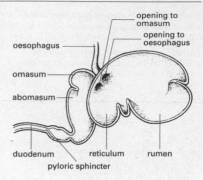

**Ruminantia.** Section of the stomach of a ruminant.

are absorbed; and finally to the **abomasum**, in which further digestion takes place.

**runner** A stem that grows horizontally along the soil surface and gives rise to new plants from axillary or terminal buds. Runners are seen in the creeping buttercup and the strawberry. **Offsets**, e.g. those of the houseleek, are short runners.

**Russell–Saunders coupling (L–S coupling)** A type of coupling in systems involving many *fermions. These systems include electrons in atoms and nucleons in nuclei, in which the energies associated with electrostatic repulsion are much greater than the energies associated with *spin–orbit coupling. *Multiplets of many-electron atoms with a low *atomic number are characterized by Russell–Saunders coupling. It is named after the US physicists Henry Norris Russell (1877–1957) and F. A. Saunders, who postulated this type of coupling to explain the spectra of many-electron atoms with low atomic number in 1925. The multiplets of heavy atoms and nuclei are better described by *j-j coupling or **intermediate coupling**, i.e. a coupling in which the energies of electrostatic repulsion and spin–orbit coupling are similar in size.

**rusting** Corrosion of iron (or steel) to form a hydrated iron(III) oxide $Fe_2O_3.xH_2O$. Rusting occurs only in the presence of both water and oxygen. It is an electrochemical process in which different parts of the iron surface act as electrodes in a cell reaction. At the anode, iron atoms dissolve as $Fe^{2+}$ ions:

$$Fe(s) \rightarrow Fe^{2+}(aq) + 2e$$

At the cathode, hydroxide ions are formed:

$$O_2(aq) + 2H_2O(l) + 4e \rightarrow 4OH^-(aq)$$

The $Fe(OH)_2$ in solution is oxidized to $Fe_2O_3$. Rusting is accelerated by impurities in the iron and by the presence of acids or other electrolytes in the water.

**rusts** A group of parasitic fungi of the phylum *Basidiomycota. Many of these species attack the leaves and stems of cereal crops: characteristic rust-coloured streaks of spores appear on infected plants. The life cycles of some rusts may be complex; many form a number of different types of spore and some require two different host plants. *Compare* SMUTS.

**ruthenium** Symbol Ru. A hard white metallic *transition element; a.n. 44; r.a.m. 101.07; r.d. 12.3; m.p. 2310°C; b.p. 3900°C. It is found associated with platinum and is used as a catalyst and in certain platinum alloys. Chemically, it dissolves in fused alkalis but is not attacked by acids. It reacts with oxygen and halogens at high temperatures. It also forms complexes with a range of oxidation states. The element was isolated by K. K. Klaus in 1844.

**((⊕)) SEE WEB LINKS**
• Information from the WebElements site

**Rutherford, Ernest, Baron** (1871–1937) New Zealand-born British physicist, who worked under Sir J. J. *Thomson at Cambridge University (1895–98). He then took up a professorship at McGill University, Canada, and collaborated with Frederick *Soddy in studying radioactivity. In 1899 he discovered *alpha particles and beta particles, followed by the discovery of *gamma radiation the following year. In 1905, with Soddy, he announced that radioactive *decay involves a series of transformations. In 1907 he moved to Manchester University, where he directed the *Rutherford scattering experiments that led to the discovery of the atomic *nucleus. After moving to Cambridge in 1919 he achieved the artificial splitting of light atoms. In 1908 he was awarded the Nobel Prize for chemistry.

**Rutherford backscattering spectrometry (RBS)** A technique for analysing samples of material by irradiation with a beam of alpha particles and measurement of the energies of the alpha particles after they have been scattered by the sample. This enables the elements present and their amounts to be determined because the energy of a scat-

tered alpha particle depends on the mass of the nucleus with which it collides. RBS is used extensively in medicine and industry.

**rutherfordium** Symbol Rf. A radioactive *transactinide element; a.n. 104. It was first reported in 1964 at Dubna, near Moscow, and in 1969 it was detected by A. Ghiorso and a team at Berkeley, California. It can be made by bombarding californium–249 nuclei with carbon–12 nuclei.

**((⊕)) SEE WEB LINKS**
• Information from the WebElements site

**Rutherford scattering** The scattering of *alpha particles by thin films of heavy metals, notably gold. The experiments, performed in 1909 by Geiger and Marsden under Rutherford's direction, provided evidence for the existence of an atomic *nucleus. A narrow beam of alpha particles from a radon source was directed onto a thin metal foil. A glass screen coated with zinc sulphide (which scintillates on absorbing alpha particles) was placed at the end of a travelling microscope and was used to detect scattered alpha particles. The travelling microscope could be rotated about the metal foil; by counting the number of scintillations produced in various positions during equal intervals, the angular dependence of the scattering was determined. Since the range of alpha particles in air is limited, the central chamber of the apparatus was evacuated. Most alpha particles suffered only small angles of deflection. However, a very small number, about 1 in 8000, were deviated by more than an angle $\theta = 90°$.

Rutherford concluded that alpha particles deflected by angles greater than 90° had encountered a small intense positive charge of high inertia. In 1911 he proposed that an atom has a positively charged nucleus, which contains most of the mass of the atom and is surrounded by orbiting electrons (*see* BOHR THEORY). Since very few alpha particles were scattered through large angles, it follows that the probability of a head-on collision with the nucleus is small. The nucleus therefore occupies a very small part of the atomic volume. It is of the order of $10^{-15}$ m across, whereas the atomic radius is of the order of $10^{-10}$ m.

**rutile** A mineral form of titanium(IV) oxide, $TiO_2$.

**rutile structure** A type of ionic crystal structure in which the anions have a hexagonal close packed arrangement with cations in half the octahedral holes. The coordination number of the anions is 6 and the coordination number of the cations is 3. Compounds with this structure include $TiO_2$, $MnO_2$, $SnO_2$, $MgF_2$, and $NiF_2$.

**(⊕) SEE WEB LINKS**

• An interactive version of the structure

**Ruybal test** *See* SCOTT'S TEST.

**Rydberg constant** Symbol $R$. A constant that occurs in the formulae for atomic spectra and is related to the binding energy between an electron and a nucleon. It is connected to other constants by the relationship $R = \mu_0^2 me^4 c^3/8h^3$, where $\mu_0$ is the magnetic constant (*see* PERMEABILITY), $m$ and $e$ are the mass and charge of an electron, $c$ is the speed of light, and $h$ is the *Planck constant. It has the value $1.097 \times 10^7$ m$^{-1}$. It is named after the Swedish physicist Johannes Robert Rydberg (1854–1919), who developed a formula for the spectrum of hydrogen.

**Rydberg spectrum** An absorption spectrum of a gas in the ultraviolet region, consisting of a series of lines that become closer together towards shorter wavelengths, merging into a continuous absorption region. The absorption lines correspond to electronic transitions to successively higher energy levels. The onset of the continuum corresponds to photoionization of the atom or molecule, and can thus be used to determine the ionization potential.

**Ryle, Sir Martin** (1918–84) British radio astronomer, who became professor of radio astronomy at Cambridge University in 1959. He organized three surveys of celestial radio sources and developed the technique of aperture synthesis (*see* RADIO TELESCOPE). In 1974 he shared the Nobel Prize for physics with Antony Hewish (1924–  ), who led the team that discovered pulsars.

**Sabatier–Senderens process** A method of organic synthesis employing hydrogenation and a heated nickel catalyst. It is employed commercially for hydrogenating unsaturated vegetable oils to make margarine. It is named after Paul Sabatier (1854–1941) and Jean-Baptiste Senderens (1856–1937).

**saccharide** *See* SUGAR.

**saccharin** A white crystalline solid, $C_7H_5NO_3S$, m.p. 224°C. It is made from a compound of toluene, derived from petroleum or coal tar. It is a well-known artificial sweetener, being some 500 times as sweet as sugar (sucrose), and is usually marketed as its sodium salt. Because of an association with cancer in laboratory animals, its use is restricted in some countries.

***Saccharomyces*** An industrially important genus of yeasts. *S. cerevisiae*, of which there are at least 1000 strains, is used in baking (*see* BAKER'S YEAST), brewing, and wine making; it is also used in the production of various proteins and other compounds in biotechnology, including industrial alcohol. Other yeasts used in the production of beer include *S. uvarum* (or *carlsbergensis*); it is distinguished from *S. cerevisiae* by its ability to ferment the disaccharide melibose using α-galactosidase, an enzyme not produced by *S. cerevisiae*.

**saccharose** *See* SUCROSE.

**sacculus (saccule)** A chamber of the *inner ear from which the *cochlea arises in reptiles, birds, and mammals. It bears patches of sensory epithelium concerned with balance (*see* MACULA).

**Sachse reaction** A reaction of methane at high temperature to produce ethyne:

$$2CH_4 \rightarrow C_2H_2 + 3H_2$$

The reaction occurs at about 1500°C, the high temperature being obtained by burning part of the methane in air.

**Sachs–Wolfe effect** A phenomenon, predicted in 1967 by Rainer Kurt Sachs (1932– ) and Arthur Michael Wolfe (1932– ), in which density fluctuations associated with quantum-mechanical effects in the early universe caused 'ripples' in the cosmic microwave background radiation. It was first observed by the *COBE satellite and analysed in detail by *WMAP.

**Sackur–Tetrode equation** An equation for the entropy of a perfect monatomic gas. The entropy $S$ is given by:

$$S = nR \ln(e^{5/2} \, V/nN_A\Lambda^3),$$

where $\Lambda = h/(2\pi mkT)^{1/2}$, where $n$ is the amount of the gas, $R$ is the gas constant, e is the base of natural logarithms, $V$ is the volume of the system, $N_A$ is the Avogadro constant, $h$ is the Planck constant, $m$ is the mass of each atom, $k$ is the Boltzmann constant, and $T$ is the thermodynamic temperature. To calculate the **molar entropy** of the gas both sides are divided by $n$. The Sackur–Tetrode equation can be used to show that the entropy change $\Delta S$, when a perfect gas expands isothermally from $V_i$ to $V_f$, is given by:

$$\Delta S = nR \ln(aV_f) - nR \ln(aV_i) = nR \ln(V_f/V_i),$$

where $aV$ is the quantity inside the logarithm bracket in the Sackur–Tetrode equation. The equation was produced independently in 1912 by Hugo Tetrode (1895–1931) and Otto Sackur (1880–1914).

**sacral vertebrae** The vertebrae that lie between the lumbar and the caudal vertebrae in the *vertebral column. The function of the sacral vertebrae is to articulate securely with the *pelvic girdle, and they are usually fused to form a single bone (the **sacrum**) to provide a firm support. The number of sacral vertebrae varies from animal to animal. Amphibians have a single sacral vertebra, reptiles have two, and mammals have three or more.

**sacrificial protection (cathodic protection)** The protection of iron or steel against corrosion (*see* RUSTING) by using a more reactive metal. A common form is galvanizing (*see* GALVANIZED IRON), in which the iron surface is coated with a layer of zinc. Even if the zinc layer is scratched, the iron does not rust because zinc ions are formed in solution in preference to iron ions. Pieces of magnesium alloy are similarly used in protecting pipelines, etc.

**saddle point** For a three-dimensional surface, a point that is a minimum in one planar cross-section and a maximum in another plane. At this point, the surface is saddle-shaped.

**safranin** A stain used in optical microscopy that colours lignified tissues, cutinized tissues, and nuclei red and chloroplasts pink. It is used mainly for plant tissues, in conjunction with a green or blue counterstain.

**sal ammoniac** *See* AMMONIUM CHLORIDE.

**salicylic acid (1-hydroxybenzoic acid)** A naturally occurring carboxylic acid, $HOC_6H_4COOH$, found in certain plants; r.d. 1.44; m.p. 159°C; sublimes at 211°C. It is used in making *aspirin and in the foodstuffs and dyestuffs industries.

**saline** Describing a chemical compound that is a salt, or a solution containing a salt. *See also* PHYSIOLOGICAL SALINE.

**salinometer** An instrument for measuring the salinity of a solution. There are two main types: one is a type of *hydrometer to measure density; the other is an apparatus for measuring the electrical conductivity of the solution.

**saliva** A watery fluid secreted by the *salivary glands in the mouth. Production of saliva is stimulated by the presence of food in the mouth and also by the smell or thought of food. Saliva contains mucin, which lubricates food and eases its passage into the oesophagus, and in some animals salivary *amylase (or ptyalin), which begins the digestion of starch. The saliva of insects is rich in digestive enzymes, and that of bloodsucking animals contains an anticoagulant.

**salivary glands** Glands in many terrestrial animals that secrete *saliva into the mouth. In humans there are three pairs: the sublingual, submandibular, and the submaxillary glands. The salivary gland cells of some insect larvae produce giant (**polytene**) chromosomes, which are widely used in the study of genetics and protein synthesis.

***Salmonella*** A genus of rod-shaped Gram-negative bacteria that inhabit the intestine and cause disease (**salmonellosis**) in humans and animals. They are aerobic or facultatively anaerobic, and most are motile.

Salmonellae can exist for long periods outside their host, and may be found, for example, in sewage and surface water. Humans may become infected by consuming contaminated water or food, especially animal products, such as eggs, meat, and milk, or vegetables that have been fertilized with contaminated manure. The bacteria can also be transmitted from human or animal carriers by unhygienic food preparation. Various species of *Salmonella* cause gastroenteritis and septicaemia; typhoid fever and paratyphoid fever are caused by *S. typhi* and *S. paratyphi*, respectively.

**sal soda** Anhydrous *sodium carbonate, $Na_2CO_3$.

**salt** A compound formed by reaction of an acid with a base, in which the hydrogen of the acid has been replaced by metal or other positive ions. Typically, salts are crystalline ionic compounds such as $Na^+Cl^-$ and $NH_4^+NO_3^-$. Covalent metal compounds, such as $TiCl_4$, are also often regarded as salts.

**saltation** A mechanism by which the moving water in a river carries sedimentary material. The particles bound along the river bed in a series of small leaps. A similar process occurs on land, when wind acts on grains of material.

**salt bridge** An electrical connection made between two half cells. It usually consists of a glass U-tube filled with agar jelly containing a salt, such as potassium chloride. A strip of filter paper soaked in the salt solution can also be used.

**salt cake** Industrial *sodium sulphate.

**salting in** *See* SALTING OUT.

**salting out** The effect in which the solubility of a substance in a certain solvent is reduced by the presence of a second solute dissolved in the solvent. For example, certain substances dissolved in water can be precipitated (or evolved as a gas) by addition of an ionic salt. The substance is more soluble in pure water than in the salt solution. The opposite effect involving an increase in solubility may occur. This is known as **salting in**.

**saltpetre** *See* NITRE.

**samara** A dry single-seeded indehiscent fruit in which the fruit wall hardens and extends to form a long membranous winglike structure that aids dispersal. Examples are ash and elm fruits. The sycamore fruit is a double samara and technically a *schizo-carp. *See also* ACHENE.

**samarium** Symbol Sm. A soft silvery metallic element belonging to the *lanthanoids; a.n. 62; r.a.m. 150.35; r.d. 7.52 (20°C); m.p. 1077°C; b.p. 1791°C. It occurs in monazite and bastnaitie. There are seven naturally occurring isotopes, all of which are stable except samarium–147, which is weakly radioactive (half-life $2.5 \times 10^{11}$ years). The metal is used in special alloys for making nuclear-reactor parts as it is a neutron absorber. Samarium oxide ($Sm_2O_3$) is used in small quantities in special optical glasses. The largest use of the element is in the ferromagnetic alloy $SmCo_5$, which produces permanent magnets five times stronger than any other material. The element was discovered by François Lecoq de Boisbaudran in 1879.

(🌐) SEE WEB LINKS

• Information from the WebElements site

**sampling** The selection of small groups of entities to represent a large number of entities in *statistics. In **random sampling** each individual of a population has an equal chance of being selected as part of the sample. In **stratified random sampling**, the population is divided into strata, each of which is randomly sampled and the samples from the different strata are pooled. In **systematic sampling**, individuals are chosen at fixed intervals; for example, every tenth animal in a population. In **sampling with replacement**, each individual chosen is replaced before the next selection is made.

**sand** Particles of rock with diameters in the range 0.06–2.00 mm. Most sands are composed chiefly of particles of quartz, which are derived from the weathering of quartz-bearing rocks.

**Sandmeyer reaction** A reaction of diazonium salts used to prepare chloro- or bromo-substituted aromatic compounds. The method is to diazotize an aromatic amide at low temperature and add an equimolar solution of the halogen acid and copper(I) halide. A complex of the diazonium salt and copper halide forms, which decomposes when the temperature is raised. The copper halide acts as a catalyst in the reaction of the halide ions from the acid, for example

$$C_6H_5N_2^+(aq) + Cl^-(aq) + CuCl(aq) \rightarrow$$
$$C_6H_5Cl(l) + N_2(g) + CuCl(aq)$$

The reaction was discovered in 1884 by the German chemist Traugott Sandmeyer (1854–1922). *See also* GATTERMANN REACTION.

**sandstone** A common sedimentary rock composed of grains of sand. The sand accumulated originally underwater in shallow seas or lakes, or on the ground along shorelines or in desert regions. The rounded quartz grains are 0.06–2 mm across. They may be consolidated by pressure, but more often they are cemented together by calcite (calcareous sandstone), clay, or iron oxide (ferruginous sandstone), which determines the colour of the rock.

**sandwich compound** A transition-metal complex in which a metal atom or ion is 'sandwiched' between two rings of atoms. *Ferrocene was the first such compound to be prepared, having two parallel cyclopentadienyl rings with an iron ion between them. In such compounds (also known as **metallocenes**) the metal coordinates to the pi electron system of the ring, rather than to individual atoms. A wide variety of these compounds are known, having five-, six-, seven-, or eight-membered rings and involving such metals as Cr, Mn, Co, Ni, and Fe. Other similar compounds are known. A **multidecker sandwich** has three or more parallel rings with metal atoms between them. In a **bent sandwich**, the rings are not parallel. A **half sandwich** (or **piano stool**) has one ring, with single ligands on the other side of the metal.

**Sanger's reagent** 2,4-dinitrofluorobenzene, $C_6H_3F(NO_2)_2$, used to identify the end *amino acid in a protein chain. It is named after Frederick Sanger (1918–   ).

**sap** **1.** The sugary fluid that is found in the phloem tissue of plants. Sap is the medium in which carbohydrates, produced in photosynthesis, and other organic molecules are transported and stored in plants. **2. (cell sap)** The fluid that is contained in the *vacuoles of plant cells. It is a solution of organic and inorganic compounds, including sugars, amino acids, salts, pigments, and waste products.

**saponification** The reaction of esters with alkalis to give alcohols and salts of carboxylic acids:

$$RCOOR' + OH^- \rightarrow RCOO^- + R'OH$$

*See* ESTERIFICATION; SOAP.

**sapphire** Any of the gem varieties of

*corundum except ruby, especially the blue variety, but other colours of sapphire include yellow, brown, green, pink, orange, and purple. Sapphires are obtained from igneous and metamorphic rocks and from alluvial deposits. The chief sources are Sri Lanka, Kashmir, Burma, Thailand, East Africa, the USA, and Australia. Sapphires are used as gemstones and in record-player styluses and some types of laser. They are synthesized by the Verneuil flame-fusion process.

**saprotroph (saprobe; saprobiont)** Any organism that feeds by absorbing dead organic matter. Most saprotrophs are bacteria and fungi. Saprotrophs are important in *food chains as they bring about decay and release nutrients for plant growth. Compare PARASITISM.

**sapwood (alburnum)** The outer wood of a tree trunk or branch. It consists of living *xylem cells, which both conduct water and provide structural support. Compare HEARTWOOD.

**sarcoma** See CANCER.

**sarcomere** Any of the functional units that make up the myofibrils of *voluntary muscle. Each sarcomere is bounded by two membranes (**Z lines**), which provide the points of attachment of *actin filaments; another membrane (the **M band** or **line**) is the point of attachment of the *myosin filaments. The sarcomere is divided into various bands reflecting the arrangement of the filaments (see illustration). During muscle contraction

**A sarcomere.**

the actin and myosin filaments slide over each other and the length of the sarcomere shortens: the Z lines are drawn closer together and the I and H bands become narrower.

**sarin** A highly toxic colourless liquid, $C_4H_{10}FO_2P$; r.d. 1.09; m.p. $-56°C$; b.p. $158°C$. It is an organophosphorus compound, O-isopropyl methylphosphonofluoridate. Sarin was discovered in 1938 and belongs to the G-series of *nerve agents (GB). It was used by Iraq in the Iran–Iraq war (1980–88). In 1988, sarin was also used by Iraq in a poison gas attack on the Kurd city of Halabja in the north of Iraq. About 5000 people died. In 1994, a Japanese religious sect released sarin in Matsumoto and in 1995 released it in the Tokyo subway.

**saros** A cycle of 6586.32 days (almost exactly 18 years) that governs the order and recurrence of solar and lunar eclipses. At the end of each saros the earth, moon, and sun return to approximately the same positions relative to each other, and the eclipses are repeated in the same sequence and separated by approximately the same intervals.

**satellite 1. (natural satellite)** A relatively small natural body that orbits a planet. For example, the earth's only natural satellite is the moon. **2. (artificial satellite)** A manmade spacecraft that orbits the earth, moon, sun, or a planet. Artificial satellites are used for a variety of purposes. **Communication satellites** are used for relaying telephone, radio, and television signals round the curved surface of the earth (see SYNCHRONOUS ORBIT). They are of two types: **passive satellites** reflect signals from one point on the earth's surface to another; **active satellites** are able to amplify and retransmit the signals that they pick up. **Astronomical satellites** are equipped to gather and transmit to earth astronomical information from space, including conditions in the earth's atmosphere, which is of great value in weather forecasting.

**satellite DNA** See REPETITIVE DNA.

**saturated 1. (of a solution)** Containing the maximum equilibrium amount of solute at a given temperature. In a saturated solution the dissolved substance is in equilibrium with undissolved substance; i.e. the rate at which solute particles leave the solution is exactly balanced by the rate at which they dissolve. A solution containing less than the equilibrium amount is said to be **unsaturated**. One containing more than the equilibrium amount is **supersaturated**. Supersaturated solutions can be made by slowly cooling a saturated solution. Such solutions are metastable; if a small crystal seed is added the excess solute crystallizes out of solution. **2.** (of a vapour) *See* VAPOUR PRESSURE. **3.** (of a ferromagnetic material) Unable to be magnetized more strongly as all the domains are orientated in the direction of the field. **4.** (of a compound) Consisting of molecules that have only single bonds (i.e. no double or triple bonds). Saturated compounds can undergo substitution reactions but not addition reactions. *Compare* UNSATURATED.

**saturation** 1. *See* COLOUR. **2.** *See* SUPER-SATURATION.

**Saturn** The second largest planet in the *solar system and the sixth in order from the sun (1433.45 × 10⁶ km distant). Its equatorial diameter is 120 536 km and its *sidereal period is 29.46 years. Saturn is a *gas giant. Its mass is $5.685 \times 10^{26}$ kg or 95 times that of the earth. Although it is the second largest planet, its mean density is lower than any other and is less than that of water, its relative density being 0.7. It has at least 61 satellites, of which the largest, *Titan, is the only planetary satellite to have a dense atmosphere. Like Jupiter, Saturn is believed to consist of a dense rocky core, surrounded by hydrogen compressed to such an extent that it behaves like a metal; this layer merges with an atmosphere of mostly hydrogen. Saturn is best known for the spectacular and complex system of rings that surrounds it in its equatorial plane; the rings have an overall diameter of about 273 000 km. They are believed to consist of millions of particles of ice (possibly with a rock core) with diameters between <1 mm and 10 m. Saturn was given a preliminary examination by the spacecraft Pioneer II in 1979, more detailed studies by Voyager I in 1980 and Voyager II in 1981, and a comprehensive survey (including its rings and satellites) by the Cassini-Huygens space probe in 2004.

**(())) SEE WEB LINKS**
• NASA's profile to the planet and its moons

**savanna** *See* GRASSLAND.

**sawtooth waveform** A waveform in which the variable increases uniformly with time for a fixed period, drops sharply to its

initial value, and then repeats the sequence periodically. The illustration shows the ideal waveform and the waveform generated by practical electrical circuits. Sawtooth generators are frequently used to provide a time base for electronic circuits, as in the *cathode-ray oscilloscope.

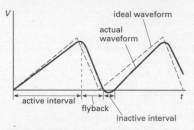

Sawtooth waveform.

**SAX** Simple API for XML: a specification for an event-driven XML parser. SAX parsers read XML files and raise appropriate events for each type of XML object encountered (start of element, end of element, character-data block, etc.). Unlike DOM parsers, SAX parsers do not create a representation of the document in RAM, which makes them faster and less memory-hungry. It is the application's responsibility to use the SAX events to extract and maintain whatever information it requires. The current version is 2.0 (**SAX2**), with Java implementation generally regarded as normative in the absence of a formal specification.

**(())) SEE WEB LINKS**
• The SAX home page

**s-block elements** The elements of the first two groups of the *periodic table; i.e. groups 1 (Li, Na, K, Rb, Cs, Fr) and 2 (Be, Mg, Ca, Sr, Ba, Ra). The outer electronic configurations of these elements all have inert-gas structures plus outer $ns^1$ (group 1) or $ns^2$ (group 2) electrons. The term thus excludes elements with incomplete inner $d$-levels (transition metals) or with incomplete inner $f$-levels (lanthanoids and actinoids) even though these often have outer $ns^2$ or occasionally $ns^1$ configurations. Typically, the $s$-block elements are reactive metals forming stable ionic compounds containing $M^+$ or $M^{2+}$ ions. *See* ALKALI METALS; ALKALINE-EARTH METALS.

**scalar product (dot product)** The product of two vectors $U$ and $V$, with components

$U_1$, $U_2$, $U_3$ and $V_1$, $V_2$, $V_3$, respectively, given by:

$$U.V = U_1V_1 + U_2V_2 + U_3V_3.$$

It can also be written as $UV\cos\theta$, where $U$ and $V$ are the lengths of $U$ and $V$, respectively, and $\theta$ is the angle between them. *Compare* VECTOR PRODUCT.

**scalar quantity** A quantity in which direction is either not applicable (as in temperature) or not specified (as in speed). *Compare* VECTOR.

**scalar triple product** *See* TRIPLE PRODUCT.

**scalene** Denoting a triangle having three unequal sides.

**scaler (scaling circuit)** An electronic counting circuit that provides an output when it has been activated by a prescribed number of input pulses. A **decade scaler** produces an output pulse when it has received ten or a multiple of ten input pulses; a **binary scaler** produces its output after two input pulses.

**scales** The small bony or horny plates forming the body covering of fish and reptiles. The wings of some insects, notably the Lepidoptera (butterflies and moths), are covered with tiny scales that are modified cuticular hairs.

In fish there are three types of scales. **Placoid scales (denticles)**, characteristic of cartilaginous fish, are small and toothlike, with a projecting spine and a flattened base embedded in the skin. They are made of *dentine, have a pulp cavity, and the spine is covered with a layer of enamel. Teeth are probably modified placoid scales. **Cosmoid scales**, characteristic of lungfish and coelacanths, have an outer layer of hard **cosmin** (similar to dentine) covered by modified enamel (**ganoine**) and inner layers of bone. The scale grows by adding to the inner layer only. In modern lungfish the scales are reduced to large bony plates. **Ganoid scales** are characteristic of primitive ray-finned fishes, such as sturgeons. They are similar to cosmoid scales but have a much thicker layer of ganoine and grow by the addition of material all round. The scales of modern teleost fish are reduced to thin bony plates.

In reptiles there are two types of scales: horny epidermal **corneoscutes** sometimes fused with underlying bony dermal **osteoscutes**.

**scandium** Symbol Sc. A rare soft silvery metallic element belonging to group 3 (formerly IIIA) of the periodic table; a.n. 21; r.a.m. 44.956; r.d. 2.989 (alpha form), 3.19 (beta form); m.p. 1541°C; b.p. 2831°C. Scandium often occurs in *lanthanoid ores, from which it can be separated on account of the greater solubility of its thiocyanate in ether. The only natural isotope, which is not radioactive, is scandium–45, and there are nine radioactive isotopes, all with relatively short half-lives. Because of the metal's high reactivity and high cost no substantial uses have been found for either the metal or its compounds. Predicted in 1869 by Dmitri Mendeleev, and then called **ekaboron**, the oxide (called **scandia**) was isolated by Lars Nilson (1840–99) in 1879.

**SEE WEB LINKS**
• Information from the WebElements site

**scanning** The process of repeatedly crossing a surface or volume with a beam, aerial, or moving detector in order to bring about some change to the surface or volume, to measure some activity, or to detect some object. The fluorescent screen of a television picture tube is scanned by an electron beam in order to produce the picture; an area of the sky may be scanned by the movable dish aerial of a radio telescope in order to detect celestial bodies, etc.

**scanning electron microscope** *See* ELECTRON MICROSCOPE.

**scanning probe microscopy (SPM)** Any of several microscopic techniques that are based on measuring the interaction between a very sharp-tipped probe and the surface of the sample. The probe scans the sample surface and records the surface topography, which can be processed by a computer to produce images. Widely used in chemistry, such techniques are now increasingly used in biology to study biomolecules and cell surfaces at the nanometre scale. *See* ATOMIC FORCE MICROSCOPY; SCANNING TUNNELLING MICROSCOPY.

**SEE WEB LINKS**
• Overview of scanning probe microscopy techniques produced by Nanoscience Instruments Inc

**scanning tunnelling microscopy (STM)** A variation of *scanning probe microscopy in which electrons tunnel between the sample and the probe, producing an electrical signal. The probe is slowly moved across the surface and raised and lowered so as to keep the signal constant. A profile of the surface is produced, and a computer-generated contour map of the surface is produced. Although most suitable for imaging substances that are electrical conductors, STM is now used for studying biological materials at the nanometre scale.

**⊕ SEE WEB LINKS**

- An account of the technique produced by Nanoscience Instruments Inc

**scapula (shoulder blade)** The largest of the bones that make up each half of the *pectoral (shoulder) girdle. It is a flat triangular bone, providing anchorage for the muscles of the forelimb and an articulation for the *humerus at the *glenoid cavity. It is joined to the *clavicle (collar bone) in front.

**scattered disc** A remote region of the outer *solar system that is home to a sparse population of *trans-Neptunian objects following extremely erratic orbits around the sun. The scattered disc appears to overlap the *Kuiper belt but extends far beyond its outer limits. Scattered disc objects (SDOs) are thought by some astronomers to be former members of the Kuiper belt that were ejected into their present orbits by the gravitational influence of the gas giants. The scattered disc may also be the present home of short-period *comets and possibly the original home of the *centaurs.

**scattering of electromagnetic radiation** The process in which electromagnetic radiation is deflected by particles in the matter through which it passes. In **elastic scattering** the photons of the radiation are reflected; i.e. they bounce off the atoms and molecules without any change of energy. In this type of scattering, known as **Rayleigh scattering** (after Lord Rayleigh; 1842–1919), there is a change of phase but no frequency change. In **inelastic scattering** and **superelastic scattering**, there is interchange of energy between the photons and the particles. Consequently, the scattered photons have a different wavelength as well as a different phase. Examples include the *Raman effect and the *Compton effect. *See also* TYNDALL EFFECT.

**scavenger** An animal that feeds on dead organic matter. Scavengers (such as hyenas) may feed on animals killed by predators or they may be *detritivores.

**Scheele, Karl Wilhelm** (1742–86) Swedish chemist, who became an apothecary and in 1775 set up his own pharmacy at Köping. He made many chemical discoveries. In 1772 he prepared oxygen (*see also* LAVOISIER, ANTOINE LAURENT; PRIESTLEY, JOSEPH) and in 1774 he isolated chlorine. He also discovered manganese, glycerol, hydrocyanic (prussic) acid, citric acid, and many other substances.

**scheelite** A mineral form of calcium tungstate, $CaWO_4$, used as an ore of tungsten. It occurs in contact metamorphosed deposits and vein deposits as colourless or white tetragonal crystals.

**Schiaparelli, Giovanni Virginio** (1835–1910) Italian astronomer, who became director of the Milan Observatory in 1860. There he studied asteroids, meteors, and planets. In 1877 he described *canali* among the surface features of *Mars. The Italian word (which means 'channels') was mistranslated as 'canals', establishing a long controversy about the possibility of intelligent life on Mars.

**Schiff base** A compound formed by a condensation reaction between an aromatic amine and an aldehyde or ketone, for example

$$RNH_2 + R'CHO \rightarrow RN{:}CHR' + H_2O$$

The compounds are often crystalline and are used in organic chemistry for characterizing aromatic amines (by preparing the Schiff base and measuring the melting point). They are named after the German chemist Hugo Schiff (1834–1915).

**Schiff reagent** A reagent used for testing for aldehydes and ketones; it consists of a solution of fuchsin dye that has been decolorized by sulphur dioxide. Aliphatic aldehydes restore the pink immediately, whereas aromatic ketones have no effect on the reagent. Aromatic aldehydes and aliphatic ketones restore the colour slowly.

**schist** A group of coarse-grained metamorphic rocks characterized by the presence of platy minerals (e.g. micas, chlorite, talc, hornblende, and graphite) that show parallel alignment at right angles to the direction of stress. The original bedding of the rock is absent. Schists readily split into layers along schistosity planes, parallel to the alignment of the minerals.

**schizocarp** A dry indehiscent fruit formed from carpels that develop into separate one-seeded fragments called **mericarps**, which may be dehiscent, as in the *regma, or indehiscent, as in the *cremocarp and *carcerulus.

**schizogeny** The localized separation of plant cells to form a cavity (surrounded by the intact cells) in which secretions accumulate. Examples are the resin canals of some conifers and the oil ducts of caraway and aniseed fruits. *Compare* LYSIGENY.

**Schleiden, Matthias Jakob** *See* CELL THEORY; SCHWANN, THEODOR.

**schlieren photography** A technique that enables density differences in a moving fluid to be photographed. In the turbulent flow of a fluid, for example, short-lived localized differences in density create differences of refractive index, which show up on photographs taken by short flashes of light as streaks (German: *Schliere*). Schlieren photography is used in wind-tunnel studies to show the density gradients created by turbulence and the shock waves around a stationary model.

**Schmidt camera** *See* TELESCOPE.

**Schoenflies system** A system for categorizing symmetries of molecules. $C_n$ groups contain only an $n$-fold rotation axis. $C_{nv}$ groups, in addition to the $n$-fold rotation axis, have a mirror plane that contains the axis of rotation (and mirror planes associated with the existence of the $n$-fold axis). $C_{nh}$ groups, in addition to the $n$-fold rotation axis, have a mirror plane perpendicular to the axis. $S_n$ groups have an $n$-fold rotation–reflection axis. $D_n$ groups have an $n$-fold rotation axis and a two-fold axis perpendicular to the $n$-fold axis (and two-fold axes associated with the existence of the $n$-fold axis). $D_{nh}$ groups have all the symmetry operations of $D_n$ and also a mirror plane perpendicular to the $n$-fold axis. $D_{nd}$ groups contain all the symmetry operations of $D_n$ and also mirror planes that contain the $n$-fold axis and bisect the angles between the two-fold axes. In the Schoenflies notation $C$ stands for 'cyclic', $S$ stands for 'spiegel' (mirror), and $D$ stands for 'dihedral'. The subscripts $h$, $v$, and $d$ stand for horizontal, vertical, and diagonal respectively, where these words refer to the position of the mirror planes with respect to the $n$-fold axis (considered to be vertical). In addition to the noncubic groups referred to so far, there are cubic groups, which have several rotation axes with the same value of $n$. These are the tetrahedral groups $T$, $T_h$, and $T_d$, the octahedral groups $O$ and $O_h$, and the icosahedral group $I$. The Schoenflies system is commonly used for isolated molecules, while the *Hermann–Mauguin system is commonly used in crystallography.

**schönite** A mineral form of potassium sulphate, $K_2SO_4$.

**Schottky defect** *See* CRYSTAL DEFECT.

**Schottky effect** A reduction in the *work function of a substance when an external accelerating electric field is applied to its surface in a vacuum. The field reduces the potential energy of electrons outside the substance, distorting the potential barrier at the surface and causing *field emission. A similar effect occurs when a metal surface is in contact with a *semiconductor rather than a vacuum, when it is known as a **Schottky barrier**. The effect was discovered by the German physicist Walter Schottky (1886–1976).

**Schrieffer, John** *See* BARDEEN, JOHN.

**Schrödinger, Erwin** (1887–1961) Austrian physicist, who became professor of physics at Berlin University in 1927. He left for Oxford to escape the Nazis in 1933, returned to Graz in Austria in 1936, and then left again in 1938 for Dublin's Institute of Advanced Studies. He finally returned to Austria in 1956. He is best known for the development of *wave mechanics and the *Schrödinger equation, work that earned him a share of the 1933 Nobel Prize for physics with Paul Dirac (1902–84).

**Schrödinger equation** An equation used in wave mechanics (*see* QUANTUM MECHANICS) for the wave function of a particle. The time-independent Schrödinger equation is:

$$\nabla^2\psi + 8\pi^2 m(E-U)\psi/h^2 = 0$$

where $\psi$ is the wave function, $\nabla^2$ the Laplace operator (*see* LAPLACE EQUATION), $h$ the Planck constant, $m$ the particle's mass, $E$ its total energy, and $U$ its potential energy. The equation can be solved exactly for simple systems, such as the harmonic oscillator and the hydrogen atom. It was devised by Erwin Schrödinger, who was mainly responsible for wave mechanics.

**Schrödinger's cat** A thought experiment introduced by Erwin Schrödinger in 1935 to illustrate the paradox in *quantum mechanics regarding the probability of finding, say, a subatomic particle at a specific point in space. According to Niels Bohr, the position of such a particle remains indeterminate until it has been observed. Schrödinger postulated a sealed vessel containing a live cat and a device triggered by a quantum event, such as the radioactive decay of a nucleus. If the quantum event occurs, cyanide is released and the cat dies; if the event does not occur the cat lives. Schrödinger argued that Bohr's interpretation of events in quantum mechanics means that the cat could only be said to be alive or dead when the vessel has been opened and the situation inside it had been observed.

**Wigner's friend** is a variation of the Schrödinger's cat paradox in which a friend of the physicist Eugene Wigner is the first to look inside the vessel. The friend will either find a live or dead cat. However, if Professor Wigner has both the vessel with the cat and the friend in a closed room, the state of mind of the friend (happy if there is a live cat but sad if there is a dead cat) cannot be determined in Bohr's interpretation of quantum mechanics until the professor has looked into the room although the friend has looked at the cat. This paradox has been extensively discussed since its introduction with many proposals made to resolve it. It is thought that the explanation depends on *decoherence.

**Schwann, Theodor** (1810–82) German physiologist, who trained in medicine. After working in Berlin, he moved to Belgium. In 1838 the German botanist Matthias Schleiden (1804–81) had stated that plant tissues were composed of cells. Schwann demonstrated the same fact for animal tissues, and in 1839 concluded that all tissues are made up of cells: this laid the foundations for the *cell theory. Schwann also worked on fermentation and discovered the enzyme *pepsin. *Schwann cells are named after him.

**Schwann cell** A cell that forms the *myelin sheath of nerve fibres (axons). Each cell is responsible for a given length of a particular axon (called an **internode**); adjacent internodes are separated by small gaps (**nodes of Ranvier**) where the axon is bare. During its development the cell wraps itself around the fibre, so the sheath consists of concentric layers of Schwann cell membrane. These cells are named after Theodor Schwann.

**Schwarzschild radius** A critical radius of a body of given mass that must be exceeded if light is to escape from that body. It equals $2GM/c^2$, where $G$ is the gravitational constant, $c$ is the speed of light, and $M$ is the mass of the body. If the body collapses to such an extent that its radius is less than the Schwarzschild radius the escape velocity becomes equal to the speed of light and the object becomes a *black hole. The Schwarzschild radius is then the radius of the hole's event horizon. It is named after Karl Schwarzschild (1873–1916).

**Schweizer's reagent** A solution made by dissolving copper(II) hydroxide in concentrated ammonia solution. It has a deep blue colour and is used as a solvent for cellulose in the cuprammonium process for making rayon. When the cellulose solution is forced through spinnarets into an acid bath, fibres of cellulose are reformed.

**scientific notation** *See* STANDARD FORM.

**scintillation (twinkling)** The rapid intermittent changes in brightness in the light from a celestial object, especially a star, caused by the earth's atmosphere.

**scintillation counter** A type of particle or radiation counter that makes use of the flash of light (scintillation) emitted by an excited atom falling back to its ground state after having been excited by a passing photon or particle. The scintillating medium is usually either solid or liquid and is used in connection with a *photomultiplier, which produces a pulse of current for each scintillation. The pulses are counted with a *scaler. In certain cases, a pulse-height analyser can be used to give an energy spectrum of the incident radiation.

**scintillator** *See* PHOSPHOR.

**scion** *See* GRAFT.

**sclera** *See* SCLEROTIC.

**sclerenchyma** A plant tissue whose cell walls have become impregnated with lignin. Due to the added strength that this confers, sclerenchyma plays an important role in support; it is found in the stems and also in the midribs of leaves. Mature sclerenchyma cells are dead, since the lignin makes the cell wall impermeable to water and gases. The presence of *plasmodesmata prevents lignin being deposited in areas of the cell wall called **pits**; these form shallow depressions enabling the exchange of substances between adjacent cells. *Compare* COLLENCHYMA (*see* GROUND TISSUES); PARENCHYMA.

**sclerometer** A device for measuring the hardness of a material by determining the pressure on a standard point that is required to scratch it or by determining the height to which a standard ball will rebound from it when dropped from a fixed height. The rebound type is sometimes called a **scleroscope**.

**scleroprotein** Any of a group of proteins found in the exoskeletons of some invertebrates, notably insects. Scleroproteins are formed by conversion of the relatively soft elastic larval protein by a natural tanning process (**sclerotization**) involving orthoquinones. These are secreted and form cross linkages between polypeptides of the proteins, producing a hard rigid covering.

**sclerotic (sclera)** The tough external layer of the vertebrate eye. At the front of the eye, the sclera is modified to form the *cornea.

**scorpions** *See* ARACHNIDA.

**scotopic vision** The type of vision that occurs when the *rods in the eye are the principal receptors, i.e. when the level of illumination is low. With scotopic vision colours cannot be identified. *Compare* PHOTOPIC VISION.

**Scott's test (cobalt thiocyanate test; Ruybal test)** A *presumptive test for cocaine. **Scott's reagent** has an initial solution of 2% cobalt thiocyanate ($Co(SCN)_2$) in glycerine and water. This is followed by concentrated hydrochloric acid, and then chloroform. A positive test is indicated by a blue colour in the chloroform layer.

**SCP** *See* SINGLE-CELL PROTEIN.

**screen grid** A wire grid in a tetrode or pentode *thermionic valve, placed between the anode and the *control grid to reduce the grid–anode capacitance. *See also* SUPPRESSOR GRID.

**screw** **1.** A simple *machine effectively consisting of an inclined plane wrapped around a cylinder. The mechanical advantage of a screw is $2\pi r/p$, where $r$ is the radius of the thread and $p$ is the **pitch**, i.e. the distance between adjacent threads of the screw measured parallel to its axis. **2.** A symmetry element in a crystal lattice that consists of a combination of a rotation and a translation. *See also* GLIDE.

**scrotum** The sac of skin and tissue that contains and supports the *testes in most mammals. It is situated outside the body cavity and allows sperm to develop at the optimum temperature, which is slightly lower than body temperature.

**scurvy** A disease caused by deficiency of *vitamin C, which results in poor collagen formation. Symptoms include anaemia, skin discoloration, and tooth loss. Scurvy was a common disease among sailors in the 16th–18th centuries, when no fresh food was available on long sea voyages.

**S-drop** *See* STRANGE MATTER.

**seaborgium** Symbol Sg. A radioactive *transactinide element; a.n. 106. It was first detected in 1974 by Albert Ghiorso and a team in California. It can be produced by bombarding californium–249 nuclei with oxygen–18 nuclei. It is named after the US physicist Glenn Seaborg (1912–99).

(((●))) **SEE WEB LINKS**
• Information from the WebElements site

**seamount** An isolated steep-sided hill up to 1000 m tall on the sea floor. Most are conical in shape and volcanic in origin, with summits 1000–2000 m below the sea surface. *See also* GUYOT.

**search coil** A small coil in which a current can be induced to detect and measure a magnetic field. It is used in conjunction with a *fluxmeter.

**Searle's bar** An apparatus for determining the thermal conductivity of a bar of material. The bar is lagged and one end is heated while the other end is cooled, by steam and cold water respectively. At two points $d$ apart along the length of the bar the temperature is measured using a thermometer or thermocouple. The conductivity can then be calculated from the measured temperature gradient.

**seaweeds** Large multicellular *algae living in the sea or in the intertidal zone. They are commonly species of the *Chlorophyta, *Phaeophyta, and *Rhodophyta.

**sebaceous gland** A small gland occurring in mammalian *skin. Its duct opens into a hair follicle, through which it discharges *sebum onto the skin surface.

**sebacic acid** *See* DECANEDIOIC ACID.

**sebum** The substance secreted by *sebaceous glands onto the surface of the *skin. It is a fatty mildly antiseptic material that protects, lubricates, and waterproofs the skin and hair and helps prevent desiccation.

**secant** 1. *See* TRIGONOMETRIC FUNCTIONS. 2. A line that cuts a circle or other curve.

**sech** *See* HYPERBOLIC FUNCTIONS.

**second** 1. Symbol s. The SI unit of time equal to the duration of 9 192 631 770 periods of the radiation corresponding to the transition between two hyperfine levels of the ground state of the caesium–133 atom. 2. Symbol ″. A unit of angle equal to 1/3600 of a degree or 1/60 of a minute.

**secondary alcohol** *See* ALCOHOLS.

**secondary amine** *See* AMINES.

**secondary cell** A *voltaic cell in which the chemical reaction producing the e.m.f. is reversible and the cell can therefore be charged by passing a current through it. *See* ACCUMULATOR; INTERCALATION CELL. *Compare* PRIMARY CELL.

**secondary colour** Any colour that can be obtained by mixing two *primary colours. For example, if beams of red light and green light are made to overlap, the secondary colour, yellow, will be formed. Secondary colours of light are sometimes referred to as the pigmentary primary colours. For example, if transparent yellow and magenta pigments are overlapped in white light, red will be observed. In this case the red is a pigmentary secondary although it is a primary colour of light.

**secondary consumer** *See* CONSUMER.

**secondary emission** The emission of electrons from a surface as a result of the impact of other charged particles, especially as a result of bombardment with (primary) electrons. As the number of secondary electrons can exceed the number of primary electrons, the process is important in *photomultipliers. *See also* AUGER EFFECT.

**secondary growth (secondary thickening)** The increase in thickness of plant shoots and roots through the activities of the vascular *cambium and *cork cambium. It is seen in most dicotyledons and gymnosperms but not in monocotyledons. The tissues produced by secondary growth are called **secondary tissues** and the resultant plant or plant part is the **secondary plant body**. *Compare* PRIMARY GROWTH.

**secondary sexual characteristics** External features of a sexually mature animal that, although not directly involved in copulation, are significant in reproductive behaviour. The development of such features is controlled by sex hormones (androgens or oestrogens); they may be seasonal (e.g. the antlers of male deer or the body colour of male sticklebacks) or permanent (e.g. breasts in women or facial hair in men). In humans they develop during *adolescence.

**secondary structure** *See* PROTEIN.

**secondary thickening** *See* SECONDARY GROWTH.

**secondary winding** The winding on the output side of a *transformer or *induction coil. *Compare* PRIMARY WINDING.

**second convoluted tubule** *See* DISTAL CONVOLUTED TUBULE.

**second messenger** A chemical within a cell that is responsible for initiating the response to a signal from a chemical messenger (such as a hormone, neurotransmitter, or growth factor) that cannot enter the target cell itself, for example because it is not lipid-soluble and is therefore unable to cross the plasma membrane. A common second messenger is *cyclic AMP; the signal for its formation within the cell by *adenylate cyclase is transmitted from hormone receptors on the cell surface by a *G protein.

**second-order reaction** *See* ORDER.

**secretin** A hormone produced by the anterior part of the small intestine (the *duodenum and *jejunum) in response to the presence of hydrochloric acid from the stomach. It causes the pancreas to secrete alkaline pancreatic juice and stimulates bile production in the liver. Secretin, whose function was first demonstrated in 1902, was the first substance to be described as a hormone.

**secretion** 1. The manufacture and discharge of specific substances into the external medium by cells in living organisms. (The substance secreted is also called the secretion.) Secretory cells are often specialized and organized in groups to form *glands. The substances produced may be released directly into the blood (**endocrine secretion**; *see* ENDOCRINE GLAND) or through a duct (**exocrine secretion**; *see* EXOCRINE GLAND). Secretions can be classified according to the manner of their discharge. **Merocrine (ec-**

crine) **secretion** occurs without the secretory cells sustaining any permanent change; in **apocrine secretion** the cells release a secretory vesicle incorporating part of the secretory cell membrane; and **holocrine secretion** involves the disruption of the entire cell to release its accumulated secretory vesicles. **2.** The process by which a substance is pumped out of a cell against a concentration gradient. Secretion has an important role in adjusting the composition of urine as it passes through the *nephrons of the kidney.

**secular magnetic variation** *See* GEO-MAGNETISM.

**sedimentary rocks** A group of rocks formed as a result of the accumulation and consolidation of sediments. Sedimentary rocks are one of the three major rock groups forming the earth's crust (*see also* IGNEOUS ROCKS; METAMORPHIC ROCKS). Most sedimentary rocks are formed from pre-existing rocks that have been broken down through mechanical processes into small particles, which have then been transported and redeposited; these form the clastic sedimentary rocks. The clastic rocks are subdivided according to the size of the dominant constituent particles: arenaceous being composed largely of sand-grade particles (e.g. sandstone, grit), argillaceous of silt- or clay-grade (e.g. mud, mudstone, clay, shale), and rudaceous of gravel-grade or larger particles (e.g. conglomerate). Chemical sedimentary rocks form the second large division of sedimentary rocks and originate as chemical precipitates at the site of deposition. These include the evaporites and sedimentary iron ores, and the organic sedimentary rocks, which are derived largely from the remains of plants and animals and include coal and many limestones.

**sedimentation** The settling of the solid particles through a liquid either to produce a concentrated slurry from a dilute suspension or to clarify a liquid containing solid particles. Usually this relies on the force of gravity, but if the particles are too small or the difference in density between the solid and liquid phases is too small, a *centrifuge may be used. In the simplest case the rate of sedimentation is determined by *Stokes' law, but in practice the predicted rate is rarely reached. Measurement of the rate of sedimentation in an *ultracentrifuge can be used to estimate the size of macromolecules.

**Seebeck effect (thermoelectric effect)** The generation of an e.m.f. in a circuit containing two different metals or semiconductors, when the junctions between the two are maintained at different temperatures. The magnitude of the e.m.f. depends on the nature of the metals and the difference in temperature. The Seebeck effect is the basis of the *thermocouple. It was named after Thomas Seebeck (1770–1831), who actually found that a magnetic field surrounded a circuit consisting of two metal conductors only if the junctions between the metals were maintained at different temperatures. He wrongly assumed that the conductors were magnetized directly by the temperature difference. *Compare* PELTIER EFFECT.

**seed 1.** (in botany) The structure in angiosperms and gymnosperms that develops from the ovule after fertilization. Occasionally seeds may develop without fertilization taking place (*see* APOMIXIS). The seed contains the *embryo and nutritive tissue, either as *endosperm or food stored in the *cotyledons. Angiosperm seeds are contained within a *fruit that develops from the ovary wall. Gymnosperm seeds lack an enclosing fruit and are thus termed **naked**. The seed is covered by a protective layer, the *testa. During development of the testa the seed dries out and enters a resting phase (dormancy) until conditions are suitable for germination.

Annual plants survive the winter or dry season as seeds. The evolution of the seed habit enabled plants to colonize the land, since seed plants do not depend on water for fertilization (unlike the lower plants). **2.** (in chemistry) A crystal used to induce other crystals to form from a gas, liquid, or solution.

**seed coat** *See* TESTA.

**seed ferns** *See* CYCADOFILICALES.

**seed leaf** *See* COTYLEDON.

**seed plant** Any plant that produces seeds. Most seed plants belong to the phyla *Anthophyta (flowering plants) or *Coniferophyta (conifers).

**Seger cones (pyrometric cones)** A series of cones used to indicate the temperature inside a furnace or kiln. The cones are made from different mixtures of clay, limestone, feldspars, etc., and each one softens at a different temperature. The drooping of the vertex is an indication that the known softening temperature has been reached and thus the furnace temperature can be estimated.

**segmentation 1.** *See* METAMERIC SEGMENTATION. **2.** *See* CLEAVAGE.

**Segrè chart** A diagram in which the number of protons in a nucleus is plotted against the number of neutrons. This chart, named after the Italian–American physicist Emilio Segrè (1905–89), enables the narrow range of stable nuclei to be seen clearly and also illustrates that for more than 20 protons more neutrons than protons are required for stable nuclei.

**segregation** The separation of pairs of *alleles during the formation of reproductive cells so that they contain one allele only of each pair. Segregation is the result of the separation of *homologous chromosomes during *meiosis. *See also* MENDEL'S LAWS.

**Segrè plot** A plot of *neutron number ($N$) against *atomic number ($Z$) for all stable nuclides. The stability of nuclei can be understood qualitatively on the basis of the nature of the strong interaction (*see* FUNDAMENTAL INTERACTIONS) and the competition between this attractive force and the repulsive electrical force. The strong interaction is independent of electric charge, i.e. at any given separation the strong force between two neutrons is the same as that between two protons or between a proton or a neutron. Therefore, in the absence of the electrical repulsion between protons, the most stable nuclei would be those having equal numbers of neutrons and protons. The electrical repulsion shifts the balance to favour a greater number of neutrons, but a nucleus with too many neutrons is unstable, because not enough of them are paired with protons.

As the number of nucleons increases, the total energy of the electrical interaction increases faster than that of the nuclear interaction. The (positive, repulsive) electric potential energy of the nucleus increases approximately as $Z^2$, while the (negative, attractive) nuclear potential energy increases approximately as $N + Z$ with corrections for pairing effects. For large $N + Z$ values, the electrical potential energy per nucleon grows much faster than the nuclear potential energy per nucleon, until a point is reached where the formation of a stable nucleus is impossible. The competition between the electric and nuclear forces therefore accounts for the increase with $Z$ of the neutron–proton ratio in stable nuclei as well as the existence of maximum values for $N + Z$ for stability. The plot is named after Emilio Segrè (1905–89). *See* BINDING ENERGY; LIQUID-DROP MODEL; DECAY.

**seif dune** *See* DUNE.

**seismic waves** A vibration propagated within the earth or along its surface as a result of an *earthquake or explosion. Earthquakes generate two types of body waves that travel within the earth and two types of surface wave. The body waves consist of primary (or longitudinal) waves that impart a back-and-forth motion to rock particles along their path. They travel at speeds between 6 km per second in surface rock and 10.4 km per second near the earth's core. Secondary (or transverse or shear) waves cause rock particles to move back and forth perpendicularly to their direction of propagation. They travel at between 3.4 km per second in surface rock and 7.2 km per second near the core.

The surface waves consist of Rayleigh waves (after Lord Rayleigh, who predicted them) and Love waves (after A. E. Love). The Love waves displace particles perpendicularly to the direction of propagation and have no longitudinal or vertical components. They travel in the surface layer above a solid layer of rock with different elastic characteristics. Rayleigh waves travel over the surface of an elastic solid giving an elliptical motion to rock particles. It is these Rayleigh waves that have the strongest effect on distant seismographs.

**seismograph** An instrument that records ground oscillations, e.g. those caused by earthquakes, volcanic activity, and explosions. Most modern seismographs are based on the inertia of a delicately suspended mass and depend on the measurement of the displacement between the mass and a point fixed to the earth. Others measure the relative displacement between two points on earth. The record made by a seismograph is known as a **seismogram**.

**seismology** The branch of geology concerned with the study of earthquakes.

(((ο))) SEE WEB LINKS

• The British Geological Survey seismology home page

**Selachii** The major subclass of the Chondrichthyes (cartilaginous fishes), containing the sharks, rays, skates, and similar but extinct forms. Their sharp teeth develop from the toothlike placoid *scales (denticles) and are rapidly replaced as they wear out.

**selection** (in biology) The process by which one or more factors acting on a population produce differential mortality and favour the transmission of specific characteristics to subsequent generations. *See* ARTIFICIAL SELECTION; NATURAL SELECTION; SEXUAL SELECTION.

**selection pressure** The extent to which organisms possessing a particular characteristic are either eliminated or favoured by environmental demands. It indicates the degree of intensity of *natural selection.

**selection rules** Rules that determine which transitions between different energy levels are possible in a system, such as an elementary particle, nucleus, atom, molecule, or crystal, described by quantum mechanics. Transitions cannot take place between any two energy levels. *Group theory, associated with the *symmetry of the system, determines which transitions, called **allowed transitions**, can take place and which transitions, called *forbidden transitions, cannot take place. Selection rules determined in this way are very useful in analysing the *spectra of quantum-mechanical systems.

**selective breeding** *See* BREEDING.

**selectron** *See* SUPERSYMMETRY.

**selenides** Binary compounds of selenium with other more electropositive elements. Selenides of nonmetals are covalent (e.g. $H_2Se$). Most metal selenides can be prepared by direct combination of the elements. Some are well-defined ionic compounds (containing $Se^{2-}$), while others are non-stoichiometric interstitial compounds (e.g. $Pd_4Se$, $PdSe_2$).

**selenium** Symbol Se. A metalloid element belonging to group 16 (formerly VIB) of the periodic table; a.n. 34; r.a.m. 78.96; r.d. 4.81 (grey); m.p. 217°C (grey); b.p. 684.9°C. There are a number of allotropic forms, including grey, red, and black selenium. It occurs in sulphide ores of other metals and is obtained as a by-product (e.g. from the anode sludge in electrolytic refining). The element is a semiconductor; the grey allotrope is light-sensitive and is used in photocells, xerography, and similar applications. Chemically, it resembles sulphur, and forms compounds with selenium in the +2, +4, and +6 oxidation states. Selenium was discovered in 1817 by Jöns Berzelius.

 **SEE WEB LINKS**

• Information from the WebElements site

**selenium cell** Either of two types of *photoelectric cell; one type relies on the photoconductive effect, the other on the photovoltaic effect (*see* PHOTOELECTRIC EFFECT). In the photoconductive selenium cell an external e.m.f. must be applied; as the selenium changes its resistance on exposure to light, the current produced is a measure of the light energy falling on the selenium. In the photovoltaic selenium cell, the e.m.f. is generated within the cell. In this type of cell, a thin film of vitreous or metallic selenium is applied to a metal surface, a transparent film of another metal, usually gold or platinum, being placed over the selenium. Both types of cell are used as light meters in photography.

**selenology** The branch of astronomy concerned with the scientific study of the *moon.

**self-assembly** *See* SELF-ORGANIZATION.

**self-exciting generator** A type of electrical generator in which the field electromagnets are excited by current from the generator output.

**self-fertilization** *See* FERTILIZATION.

**self-inductance** *See* INDUCTANCE.

**selfish DNA** Regions of DNA that apparently have no function (they are also known as 'junk' DNA) and exist between those regions of DNA that represent the genes. *Transposons are good examples. Certain types of *repetitive DNA also have 'selfish' characteristics. This DNA is said to be 'selfish' as it seemingly exists only to pass copies of itself from one generation to another; it does so by using the organism in which it is contained as a survival machine. This is known as the **selfish DNA theory**. The greatest amounts of selfish DNA are found in vertebrates and higher plants. The presence of selfish DNA may be due to an unrecognizable function that it performs or because the cell has no way of halting its increase in the genome.

**self-organization** The spontaneous order arising in a system when certain parameters of the system reach critical values. Self-organization occurs in many systems in physics, chemistry, and biology. An example in physics is the *Bénard cell. In chemistry, **self-assembly** is one of the features of *supramolecular chemistry. Self-organization can occur when a system is driven far from thermal *equilibrium. Since a self-organizing system is open to its environment, the second law of *thermodynamics is not violated by the formation of an ordered phase, as entropy can be transferred to the environment. Self-organization is related to the concepts of *broken symmetry, *complexity, nonlinearity, and *nonequilibrium statistical mechanics. Many systems that undergo transitions to self-organization can also undergo transitions to *chaos.

**self-pollination**  See POLLINATION.

**self-sterility**  The condition found in many hermaphrodite organisms in which male and female reproductive cells produced by the same individual will not fuse to form a zygote, or if they do, the zygote is unable to develop into an embryo. In plants this is usually termed **self-incompatibility** (see INCOMPATIBILITY).

**Seliwanoff's test**  A biochemical test to identify the presence of ketonic sugars, such as fructose, in solution. It was devised by the Russian chemist F. F. Seliwanoff. A few drops of the reagent, consisting of resorcinol crystals dissolved in equal amounts of water and hydrochloric acid, are heated with the test solution and the formation of a red precipitate indicates a positive result.

**semaphorin**  One of a class of proteins that act as guidance molecules during the development of nerve cells, immune cells, blood vessels, bone, and other tissues. There are several classes, some occurring in invertebrates and others in vertebrates; some semaphorins are secreted by cells, whereas others remain bound to the plasma membrane. However, all semaphorin molecules have a characteristic region called a sema domain. Class 3 semaphorins act as short-range cues to guide the *growth cone at the tip of elongating nerve fibres along appropriate pathways through tissue. They may either attract the growth cone or repel it, depending on the particular semaphorin and the type of nerve cell. The action of semaphorins is effected by binding to any of various cell surface receptors.

**semen**  A slightly alkaline fluid (pH 7.2–7.6) containing sperm and various secretions that is produced by a male mammal during copulation and is introduced into the body of the female by *ejaculation. Spermatozoa are produced by the *testes and the secretions by the *prostate gland, *seminal vesicles, and *Cowper's glands. Semen also contains enzymes that activate the sperm after ejaculation.

**semicarbazones**  Organic compounds containing the unsaturated group $=C:N.NH.CO.NH_2$. They are formed when aldehydes or ketones react with a semicarbazide ($H_2N.NH.CO.NH_2$). Semicarbazones are crystalline compounds with relatively high melting points. They are used to identify aldehydes and ketones in quantitative analysis: the semicarbazone derivative is made and identified by its melting point. Semicarbazones are also used in separating ketones from reaction mixtures: the derivative is crystallized out and hydrolysed to give the ketone.

 **SEE WEB LINKS**
• Information about IUPAC nomenclature

**semicircular canals**  The sense organ in vertebrates that is concerned with the maintenance of physical equilibrium (sense of balance). It occurs in the *inner ear and consists of three looped canals set at right angles to each other and attached to the *utriculus. The canals contain a fluid (**endolymph**) that flows in response to movements of the head and body. A swelling (*ampulla) at one attachment point of each canal contains sensory cells that respond to movement of the endolymph in any of the three planes. These sensory cells initiate nervous impulses to the brain.

**semiclassical approximation**  An approximation technique used to calculate quantities in quantum mechanics. This technique is called the semiclassical approximation because the *wave function is written as an asymptotic series with ascending powers of the Planck constant $h$, with the first term being purely classical. The semiclassical approximation is also known as the **Wentzel–Kramers–Brillouin (WKB) approximation**, named after its inventors Gregor Wentzel (1898–1978), Hendrik Anton Kramers (1894– 1952), and Léon Brillouin (1889–1969), who invented it independently in 1926. The semiclassical approximation is particularly successful for calculations involving the *tunnel effect, such as *field emission and radioactive decay producing *alpha particles.

**semiconductor** A crystalline solid with an electrical conductivity (typically $10^5$–$10^{-7}$ siemens per metre) intermediate between that of a conductor (up to $10^9$ S m$^{-1}$) and an insulator (as low as $10^{-15}$ S m$^{-1}$). Semiconducting properties are a feature of *metalloid elements, such as silicon and germanium. As the atoms in a crystalline solid are close together, the orbitals of their electrons overlap and their individual *energy levels are spread out into *energy bands. Conduction occurs in semiconductors as the result of a net movement, under the influence of an electric field, of electrons in the conduction band and empty states, called **holes**, in the valence band. A hole behaves as if it was an electron with a positive charge. Electrons and holes are known as the **charge carriers** in a semiconductor. The type of charge carrier that predominates in a particular region or material is called the **majority carrier** and that with the lower concentration is the **minority carrier**. An **intrinsic semiconductor** is one in which the concentration of charge carriers is a characteristic of the material itself; electrons jump to the conduction band from the valence band as a result of thermal excitation, each electron that makes the jump leaving behind a hole in the valence band. Therefore, in an intrinsic semiconductor the charge carriers are equally divided between electrons and holes. In **extrinsic semiconductors** the type of conduction that predominates depends on the number and valence of the impurity atoms present. Germanium and silicon atoms have a valence of four. If impurity atoms with a valence of five, such as arsenic, antimony, or phosphorus, are added to the lattice, there will be an extra electron per atom available for conduction, i.e. one that is not required to pair with the four valence electrons of the germanium or silicon. Thus extrinsic semiconductors doped with atoms of valence five give rise to crystals with electrons as majority carriers, the so-called **n-type conductors**. Similarly, if the impurity atoms have a valence of three, such as boron, aluminium, indium, or gallium, one hole per atom is created by an unsatisfied bond. The majority carriers are therefore holes, i.e. **p-type conductors**.

Semiconductor devices have virtually replaced thermionic devices, because they are several orders of magnitude smaller, cheaper in energy consumption, and more reliable. The basic structure for electronic semiconductor devices is the **semiconductor diode** (*see also* TRANSISTOR). This consists of a silicon crystal doped in such a way that half is *p*-type and half is *n*-type. At the junction between the two halves there is a depletion layer in which electrons from the *n*-type have filled holes from the *p*-type. This sets up a potential barrier, which tends to keep the remaining electrons in the *n*-region and the remaining holes in the *p*-region. However, if the *p*-region is biased with a positive potential, the height of the barrier is reduced; the diode is said to be forward biased, because the majority holes in the *p*-region can then flow to the *n*-region and majority electrons in the *n*-region flow to the *p*-region. When forward biased there is a good current flow across the barrier. On the other hand if the *p*-region is negatively biased, the height of the potential barrier is increased and there is only a small leakage current of minority electrons from the *p*-region able to flow to the *n*-region. Thus the *p–n* junction acts as an efficient rectifier, for which purpose it is widely used.

**semiconductor diode** *See* DIODE; SEMICONDUCTOR.

**semiconservative replication** The generally accepted method of *DNA replication, in which the two strands of the DNA helix separate and free nucleotides pair with the exposed bases on the single chains to form two new DNA molecules, each containing one original and one newly synthesized strand of DNA.

**semimetal** *See* METALLOID.

**seminal receptacle semimetal** *See* SPERMATHECA.

**seminal vesicle** 1. A pouch or sac in many male invertebrates and lower vertebrates that is used for storing sperm. 2. One of a pair of glands in male mammals that secrete a liquid component of *semen into the vas deferens. This secretion is alkaline, which neutralizes the acidic conditions in the female genital tract, and contains fructose, used by the sperm as a source of energy.

**seminiferous tubules** *See* TESTIS.

**semipermeable membrane** A membrane that is permeable to molecules of the solvent but not the solute in *osmosis. Semipermeable membranes can be made by supporting a film of material (e.g. cellulose) on a wire gauze or porous pot.

**semipolar bond** See CHEMICAL BOND.

**Semtex** A nitrogen-based stable odourless plastic *explosive.

**senescence** The changes that occur in an organism (or a part of an organism) between maturity and death, i.e. ageing. Characteristically there is a deterioration in functioning as the cells become less efficient in maintaining and replacing vital cell components. In animals this results in a decline in physical ability and, in humans, there is also often a reduction in mental ability. Not all the parts of the body necessarily become senescent at the same time or age at the same rate. For example, in deciduous trees the shedding of senescent leaves in the autumn is a normal physiological process.

**sense organ** A part of the body of an animal that contains or consists of a concentration of *receptors that are sensitive to specific stimuli (e.g. sound, light, pressure, heat). Stimulation of these receptors initiates the transmission of nervous impulses to the brain, where sensory information is analysed and interpreted. Examples of sense organs are the *eye, *ear, *nose, and *taste bud.

**senses** The faculties that enable animals to perceive information about their external environment or about the state of their bodies in relation to this environment (see SENSE ORGAN; VISION; HEARING; BALANCE; OLFACTION (SMELL); TASTE; TOUCH). Specific *receptors are sensitive to pain, temperature, chemicals, etc.

**sensitivity (irritability)** One of the fundamental properties of all organisms: the capacity to detect, interpret, and respond to changes in the environment (e.g. the stimuli of light, touch, chemicals, etc.). Multicellular animals have specialized *sense organs and *effector organs for this purpose; in unicellular organisms, which lack a nervous system, the reception of, and response to, a stimulus occur in the same cell.

**sensitization** 1. (of a cell) The alteration of the integrity of a plasma membrane resulting from the reaction of specific *antibodies with *antigens on the surface of the cell. In the presence of *complement, the cell ruptures. 2. (of an individual) Initial exposure to a specific antigen such that re-

exposure to the same antigen causes a severe immune response (see ANAPHYLAXIS).

**sensory cell** See RECEPTOR.

**sensory neuron** A nerve cell (see NEURON) that transmits information about changes in the internal and external environment to the central nervous system. Sensory neurons are of two types. **Somatic sensory neurons** occur in peripheral nerves in the skin, skeletal muscle, joints, and bones. **Visceral sensory neurons** are located in sympathetic and parasympathetic nerves in the heart, lungs, and other organs.

**sepal** One of the parts of a flower making up the *calyx. Sepals are considered to be modified leaves with a simpler structure. They are usually green and often hairy but in some plants, e.g. monk's hood, they may be brightly coloured.

**septivalent (heptavalent)** Having a valency of seven.

**septum** Any dividing wall in a plant or animal. Examples are the septa that separate the chambers of the heart.

**sequence analysis** The process of characterizing sequences of biomolecules, particularly the nucleotides of nucleic acids (DNA or RNA) or the amino acids of proteins. Once the order of nucleotides of, say, a genome fragment has been established by *DNA sequencing, the sequence data can be analysed using computer software. The unknown sequence is compared with existing sequence data held on any of numerous databases. Likely homology is revealed by its degree of *alignment with other DNA sequences, which will provide clues about its evolutionary relationships with other biomolecules (see PHYLOGENOMICS). See also BIOINFORMATICS.

**sequence database** A database containing the sequences of biomolecules, which may be the sequences of nucleotides for nucleic acids (DNA or RNA) or of amino acids for proteins. See INTERNATIONAL NUCLEOTIDE SEQUENCE DATABASE COLLABORATION.

**sequestration** The process of forming coordination complexes of an ion in solution. Sequestration often involves the formation of chelate complexes, and is used to prevent the chemical effect of an ion without removing it from the solution (e.g. the sequestration of $Ca^{2+}$ ions in water softening). It is also a way of supplying ions in a protected form, e.g. sequestered iron solutions for plants in regions having alkaline soil.

**sere** A complete *succession of plant communities, which results in the climax community. A sere is composed of a series of different plant communities that change with time. These communities are known as **seral stages** or **seral communities**.

**series** A sequence of terms each of which can be written in a form that is an algebraic function of its position in the series. For example, the *exponential series $1 + x + x^2/2! + x^3/3!$ has an $n$th term $x^n/n!$. The sum of all the terms from $n = 0$ to $n = \infty$ is written:

$$\sum_{n=0}^{\infty} x^n/n!$$

This series has an infinite number of terms and is therefore called an **infinite series**. A **finite series** has a fixed number of terms. *See also* ASYMPTOTIC SERIES; CONVERGENT SERIES.

**series circuits** Circuits in which the circuit elements are arranged in sequence so that the same current flows through each of them in turn. For resistances in series, the total resistance is the sum of the individual resistances. For capacitors in series, the total capacitance, $C$, is given by $1/C = 1/C_1 + 1/C_2 + 1/C_3 \ldots$.

**series-wound machine** *See* SHUNT.

**serine** *See* AMINO ACID.

**seroconversion** The stage in an immune response when antibodies to the infecting agent are first detected in the bloodstream. For example, people infected with HIV typically seroconvert about 4–6 weeks following the initial infection, when antibodies against viral proteins are first produced.

**serology** The laboratory study of blood serum and its constituents, particularly *antibodies and *complement, which play a part in the *immune response.

**serotonin (5-hydroxytryptamine; 5-HT)** A compound, synthesized from the amino acid tryptophan, that (among other actions) affects the diameter of blood vessels and also functions as a *neurotransmitter. Serotonin plays a key role in arousal, mood, aggression, and the sleep–wake cycle; reduced levels are associated with various disorders, including depression and migraine. Several types of antidepressant drugs act by enhancing serotonin levels in the brain, including the **SSRIs** (selective serotonin reuptake inhibitors),

which slow the rate of uptake of serotonin by neurons.

**serous membrane (serosa)** A tissue consisting of a layer of *mesothelium attached to a surface by a thin layer of connective tissue. Serous membrane lines body cavities that do not open to the exterior; the *peritoneum, *pleura, and serous *pericardium are examples.

**serpentine** Any of a group of hydrous magnesium silicate minerals with the general composition $Mg_3Si_2O_5(OH)_4$. Serpentine is monoclinic and occurs in two main forms: **chrysotile**, which is fibrous and the chief source of *asbestos; and **antigorite**, which occurs as platy masses. It is generally green or white with a mottled appearance, sometimes resembling a snakeskin – hence the name. It is formed through the metamorphic alteration of ultrabasic rocks rich in olivine, pyroxene, and amphibole. **Serpentinite** is a rock consisting mainly of serpentine; it is used as an ornamental stone.

**Sertoli cells (sustentacular cells)** Cells that line the seminiferous tubules in the *testis, named after the Italian histologist Enrico Sertoli (1842–1910). Sertoli cells protect the spermatids (developing germ cells) and convey nutrients to both the developing and mature spermatozoa. They also produce a hormone, **inhibin**, which can inhibit *follicle-stimulating hormone and thereby regulate production of spermatozoa.

**serum** *See* BLOOD SERUM.

**sesqui-** Prefix indicating a ratio of 2:3 in a chemical compound. For example, a sesquioxide has the formula $M_2O_3$.

**sesquiterpene** *See* TERPENES.

**sessile** 1. Describing animals that live permanently attached to a surface, i.e. sedentary animals. Many marine animals, e.g. sea anemones and limpets, are sessile. 2. Describing any organ that does not possess a stalk where one might be expected. For example, the leaves of the oak (*Quercus robur*) are attached directly to the twigs.

**seta** 1. A bristle or hair in many invertebrates. Setae are produced by the epidermis and consist either of a hollow projection of cuticle containing all or part of an epidermal cell (as in insects) or are composed of chitin (as in the *chaetae of annelid worms). 2. *See* SPOROGONIUM.

**sets** Collections of objects or elements that have at least one characteristic in common. For example, the set $X$ may consist of all the elements $x_1$, $x_2$, $x_3$, etc. This is written $\{x_1, x_2, x_3, \ldots\} = X$. A specific element in a set is characterized by $x_1 \in X$, meaning $x_1$ is a member of set $X$. A **subset** of set $X$, say $M$, would be written $M \subset X$, i.e. $M$ is contained in $X$. If $x_3$ is a member of both subsets $M$ and $N$, then $x_3 \in (M \cap N)$, i.e. $x_3$ belongs to the **intersection** of $M$ and $N$. $M \cup N$ means the **union** of $M$ and $N$. For example, if $M$ consists of $\{1, 4, 5, 8\}$ and $N$ consists of $\{2, 3, 4, 5\}$ then $M \cap N = \{4, 5\}$ and $M \cup N = \{1, 2, 3, 4, 5, 8\}$. In the diagram, the rectangle represents the universal set E, circles represent sets or subsets. These diagrams are called **Venn diagrams**, after John Venn (1834–1923), who invented them.

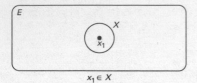

$$x_1 \in X$$

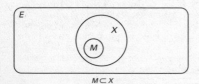

$$M \subset X$$

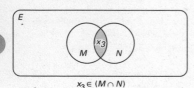

$$x_3 \in (M \cap N)$$

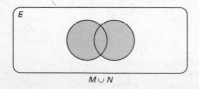

$$M \cup N$$

**Sets.**

**sewage** Waste matter from industrial and domestic sources that is dissolved or suspended in water. Raw (untreated) sewage is a pollutant. It has a high content of organic matter (notably faeces and nitrogenous waste) and therefore provides a rich source of food for many decomposers (bacteria, fungi) and *detritivores, some of which are pathogenic to humans. The release of raw (untreated) sewage into a river causes eutrophication (*see* EUTROPHIC); there is a sudden increase in the *biochemical oxygen demand (BOD), as the organisms that feed on sewage proliferate and use up the available dissolved oxygen in the river. Oxygen-sensitive organisms, such as fish, will die. Certain organisms can proliferate in particular concentrations of sewage, depending on their tolerance, and can be used as markers of the extent to which a river is polluted by sewage. For example, *Tubifex* worms are able to tolerate high concentrations of sewage.

Sewage can be treated before release. This involves a number of stages, including filtration, sedimentation, and microbial degradation (notably by *methanogens). When most of the solid waste has been removed, the remaining liquid (**effluent**) is discharged into rivers, etc. During sedimentation, particulate organic matter accumulates at the bottom of large tanks. This material, known as **sludge**, is periodically removed, further decomposed by microorganisms, and then sold as fertilizer or dumped.

**sex chromosome** A chromosome that operates in the sex-determining mechanism of a species. In many animals there are two kinds of sex chromosomes; for example, in mammals there is a large **X chromosome** and a much smaller **Y chromosome**. A combination of two X chromosomes occurs in a female while one X and one Y chromosome is found in males. Sex chromosomes carry genes governing the development of sex organs and secondary sexual characteristics (*see* TESTIS-DETERMINING FACTOR). They also carry other genes unrelated to sex (*see* SEX LINKAGE). *See also* SEX DETERMINATION.

**sex determination** The method by which the distinction between males and females is established in a species. It is usually under genetic control. Equal numbers of males and females are produced when sex is determined by *sex chromosomes or by a contrasting pair of alleles. In some species (e.g. bees) females develop from fertilized eggs and males from unfertilized eggs. This does not produce equal numbers. Environmental factors (e.g. temperature) can also play a role in determining the sex of developing individuals.

**sex hormones** Steroid hormones that control sexual development. The most important are the *androgens and *oestrogens.

**sexivalent (hexavalent)** Having a valency of six.

**sex linkage** The tendency for certain inherited characteristics to occur far more frequently in one sex than the other. For example, red–green colour blindness and *haemophilia affect men more often than women. This is because the genes governing normal colour vision and blood clotting occur on the X *sex chromosome. Women have two X chromosomes. If one carries an abnormal allele it is likely that its effects will be masked by a normal allele on the other X chromosome. However, men only have one X chromosome and any abnormal alleles therefore will not be masked. *See also* CARRIER.

**sex ratio** The ratio of the number of females to the number of males in a *population. Because the mortality rates in the two sexes may be different, the sex ratios in different age classes may differ.

**sextant** An instrument used in navigation to measure the altitude of a celestial body. Originally it had an arc of 60° (one sixth of a circle, hence its name) but modern instruments have various angles. The sextant uses two mirrors: the horizon glass, in which only the lower half is silvered, and the index mirror, which can be rotated about an axis perpendicular to the plane of the instrument (see illustration). An arm attached to the index glass sweeps round the calibrated arc, from which angles are read. The instrument is aimed at the horizon and the index mirror rotated until the celestial object can also be seen through the telescope. After careful adjustment to make the image of the celestial body just touch the horizon, the angle is read off the graduated scale.

**sexual intercourse (coitus; copulation; mating)** The process by which spermatozoa from a male are deposited in the body of a female during *sexual reproduction. In mammals the penis of the male becomes erect and stiff as its tissues become filled with blood, enabling it to be inserted into the vagina of the female. Thrusting movements of the penis result in *ejaculation, in which *semen, containing spermatozoa, is deposited in the vagina.

**sexually transmitted disease (STD)** Any disease that is passed from one individual to another during sexual intercourse or other types of sexual activity. These diseases have been traditionally referred to as **venereal diseases**. They include gonorrhoea, caused by the bacterium *Neisseria gonorrhoeae*; syphilis, due to infection by the bacterium *Treponema pallidum*; genital herpes, which is caused by a herpesvirus; and *AIDS, resulting from infection with *HIV, a retrovirus. The transmission of sexually transmitted diseases can be reduced by limiting the number of sexual partners and by the use of condoms (*see* BIRTH CONTROL), which reduces the risk of contact with body fluids that harbour the microorganisms that cause these diseases.

**sexual reproduction** A form of reproduction that involves the fusion of two reproductive cells (*gametes) in the process of *fertilization. Normally, especially in animals, it requires two parents, one male and the other female. However, most plants bear both male and female reproductive organs and self-fertilization may occur, as it does in hermaphrodite animals. Gametes are formed by *meiosis, a special kind of cell division in the parent reproductive organs that both reassorts the genetic material and halves the chromosome number. Meiosis thus ensures genetic variability in the gametes and therefore in the offspring result-

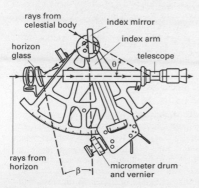

rays from celestial body

index mirror

horizon glass

index arm

telescope

θ

rays from horizon

β

micrometer drum and vernier

**Sextant.**

S

ing from their subsequent fusion. Sexual reproduction, unlike *asexual reproduction, therefore generates variability within a species. However, it depends on there being reliable means of bringing together male and female gametes, and many elaborate mechanisms have evolved to ensure this.

**sexual selection** The means by which it is assumed that certain *secondary sexual characteristics, particularly of male animals, have evolved. Females presumably choose to mate with the male that gives the best courtship display and therefore has the brightest coloration, etc.: these features would be inherited by its male offspring and would thus tend to become exaggerated down the generations.

**Seyfert galaxy** A type of galaxy, typically a spiral or barred spiral, with an active *galactic nucleus. The nucleus is brighter than the spiral arms, too bright to derive its radiation only from stars. It is thought that there is probably a low-power *quasar-like object at the centre or possibly an extremely massive black hole. It is named after the American astronomer Carl Keenan Seyfert (1911–60), who identified this category in 1943.

**SGML** Standard generalized mark-up language: an international standard metalanguage (ISO 8859) used for defining the syntax of textual mark-up languages. This enables both sender and receiver of the text to identify its structure (e.g. title, author, header, paragraph, etc.). *See also* XML, DTD, HTML.

**shadow** An area of darkness formed on a surface when an object intercepts the light falling on the surface from a source. In the case of a point source the shadow has a sharply defined outline. If the source has an

appreciable size the shadow has two distinct regions; one of full-shadow, called the **umbra**, the other of half-shadow, called the **penumbra** (see illustration).

**shale** A form of *clay that occurs in thin layers. Shales are very common *sedimentary rocks. *See also* OIL SHALE.

**shearing force** A force that acts parallel to a plane rather than perpendicularly, as with a tensile or compressive force. A **shear stress** requires a combination of four forces acting over (most simply) four sides of a plane and produces two equal and opposite couples. It is measured as the ratio of one shearing force to the area over which it acts, $F/(ab)$ in the diagram. The shear strain is the angular deformation, $\theta$, in circular measure. The **shear modulus** is the ratio of the shear stress to the shear strain (*see also* ELASTIC MODULUS).

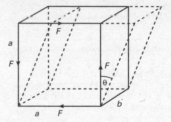

**Shearing force.**

**shear modulus** *See* ELASTIC MODULUS; SHEARING FORCE.

**shell** *See* ATOM.

**shell model** 1. *See* ATOM. 2. A model of the atomic nucleus in which nucleons are assumed to move under the influence of a central field in shells that are analogous to atomic electron shells. The model provides a good explanation of the stability of nuclei that have *magic numbers and is successful in predicting other properties of many nuclei.

**sherardizing** The process of coating iron or steel with a zinc corrosion-resistant layer by heating the iron or steel in contact with zinc dust to a temperature slightly below the melting point of zinc. At a temperature of about 371°C the two metals amalgamate to form internal layers of zinc–iron alloys and an external layer of pure zinc. The process

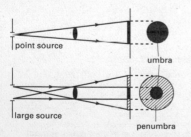

**Shadow.**

was invented by Sherard Cowper-Coles (1867–1935).

**Sherrington, Sir Charles Scott**
(1857–1952) British physiologist, who became professor of physiology at Oxford University in 1913. After early work on antitoxins he began studying human reflex reactions. Like *Pavlov in Russia, he discovered conditioned reflexes. His subsequent research concerned the functioning of neurons, for which he shared the 1932 Nobel Prize for physiology or medicine with Sir Edgar *Adrian.

**shielding 1.** A barrier surrounding a region to exclude it from the influence of an energy field. For example, to protect a region from an electric field an earthed barrier is required; to protect it from a magnetic field a shield of high magnetic permeability is needed. **2.** A barrier used to surround a source of harmful or unwanted radiations. For example, the core of a *nuclear reactor is surrounded by a cement or lead shield to absorb neutrons and other dangerous radiation. **3.** (in atoms) the barrier provided by inner electron shells to the influence of nuclear charge on outer electrons. Shielding has an effect on ionic radius, as in the *lanthanoids.

**shivering** *See* THERMOGENESIS.

**SHM** *See* SIMPLE HARMONIC MOTION.

**Shockley, William** *See* BARDEEN, JOHN.

**shock wave** A very narrow region of high pressure and temperature formed in a fluid when the fluid flows supersonically over a stationary object or a projectile flying supersonically passes through a stationary fluid. A shock wave may also be generated by violent disturbances in a fluid, such as a lightning stroke or a bomb blast.

**shoot** The aerial part of a vascular plant. It develops from the *plumule and consists of a stem supporting leaves, buds, and flowers.

**Shor's algorithm** An algorithm in quantum computing that enables large numbers to be factorized into prime numbers in a way which is much quicker than using traditional computers. This algorithm, which was proposed by the American computer scientist Peter Shor (1959–  ) in 1994, has major implications for the security of Internet information transfer.

**short-day plant** A plant in which flowering can be induced or enhanced by short days, usually of less than 12 hours of daylight. Examples are strawberry and chrysanthemum. *See* PHOTOPERIODISM. *Compare* DAY-NEUTRAL PLANT; LONG-DAY PLANT.

**short period** *See* PERIODIC TABLE.

**short-sightedness** *See* MYOPIA.

**shoulder girdle** *See* PECTORAL GIRDLE.

**shower** *See* COSMIC RADIATION.

**shunt** An electrical resistor or other element connected in parallel with some other circuit or device, to take part of the current passing through it. For example, a shunt is used across the terminals of a galvanometer to increase the current that can pass through the system. A **shunt-wound** electric generator or motor is one in which the field winding is in parallel with the armature circuit. In a **series-wound** electrical machine the field coils and the armature circuit are in series.

**sial** The rocks that form the earth's continental crust. These are granite rock types rich in *si*lica ($SiO_2$) and *al*uminium (Al), hence the name. *Compare* SIMA.

**siblings** Individuals that have both parents in common.

**sickle-cell disease** *See* POLYMORPHISM.

**sideband** The band of frequencies above or below the frequency of the carrier wave in a telecommunications system within which the frequency components of the wave produced by *modulation fall. For example, if a carrier wave of frequency $f$ is modulated by a signal of frequency $x$, the **upper sideband** will have a frequency $f + x$ and the **lower sideband** a frequency $f - x$.

**side chain** *See* CHAIN.

**side reaction** A chemical reaction that occurs at the same time as a main reaction but to a lesser extent, thus leading to other products mixed with the main products.

**sidereal day** *See* DAY.

**sidereal period** The time taken for a planet or satellite to complete one revolution of its orbit measured with reference to the background of the stars. *See also* DAY; SYNODIC PERIOD; YEAR.

**siderite** A brown or grey-green mineral form of iron(II) carbonate, $FeCO_3$, often with magnesium and manganese substituting for the iron. It occurs in sedimentary deposits or in hydrothermal veins and is an important iron ore. It is found in England, Greenland, Spain, N Africa, and the USA.

**siemens** Symbol S. The SI unit of electrical conductance equal to the conductance of a circuit or element that has a resistance of 1 ohm. 1 S = $10^{-1}$ Ω. The unit was formerly called the mho or reciprocal ohm. It is named after Ernst Werner von Siemens (1816–92).

**sieve element** A type of plant cell occurring within the *phloem. Sieve elements combine to form a series of tubes (**sieve tubes**) connecting the leaves, shoots, and roots in a fine network. Food materials are transported from one element to another via perforations termed **sieve areas** or **sieve plates**. Sieve elements contain little cytoplasm and no nucleus. Their metabolic activities are supplemented by *companion cells in angiosperms and by albuminous cells in gymnosperms.

**sievert** The SI unit of dose equivalent (*see* RADIATION UNITS). It is named after the Swedish physicist Rolf Sievert (1896–1966).

**sieve tube** A tube within the *phloem tissue of a plant, composed of joined *sieve elements.

**sigma bond** *See* ORBITAL.

**sigma electron** An electron in a sigma orbital. *See* ORBITAL.

**sigma particle** A type of spin ½ *baryon. There are three types of sigma particles, denoted $\Sigma^-$, $\Sigma^0$, $\Sigma^+$, for the negatively charged, electrically neutral, and positively charged forms, respectively. The quark content of the sigma particles are $\Sigma^-$ (dds), $\Sigma^0$ (dus), $\Sigma^+$ (uus), where d, u, and s denote down, up, and strange, respectively. The masses of the sigma particles are 1189.36 MeV ($\Sigma^+$), 1192.46 MeV ($\Sigma^0$), 1197.34 MeV ($\Sigma^-$); their average lifetimes are $0.8 \times 10^{-10}$ s ($\Sigma^+$), $5.8 \times 10^{-20}$ s ($\Sigma^0$), and $1.5 \times 10^{-10}$ s ($\Sigma^-$).

**sigmatropic reaction** A type of rearrangement in which a sigma bond is formed between two nonlinked atoms at the same time as an existing sigma bond is broken.

**signal** The variable parameter that contains information and by which information is transmitted in an electronic system or circuit. The signal is often a voltage source in which the amplitude, frequency, and waveform can be varied.

**signal molecule** A molecule of a hormone, neurotransmitter, growth factor, or other chemical that binds specifically with a cell-surface *receptor and thereby initiates a sequence of activities that triggers a response inside the cell. *See also* SECOND MESSENGER; G PROTEIN.

**signature of space–time** The division of space–time into space dimensions and time dimensions. It has been claimed that the combination of duality and *supersymmetry can explain why there only appears to be one time dimension, with there being duality relations between theories with one time dimension and any other possible theories that can have more than one time dimension.

**sign convention** A set of rules determined by convention for giving plus or minus signs to distances in the formulae involving lenses and mirrors. The *real-is-positive is the convention now usually adopted. The **New Cartesian convention** is now not widely used. In this convention distances to the right of the pole are treated as positive, those to the left as negative. This system has the advantage of conforming to the sign convention used with Cartesian coordinates in mathematics and is therefore preferred by some for the more complicated calculations.

**significant figures** The number of digits used in a number to specify its accuracy. The number 6.532 is a value taken to be accurate to four significant figures. The number $7.3 \times 10^3$ is accurate only to three significant figures. Similarly 0.0732 is also only accurate to three significant figures. In these cases the zeros only indicate the order of magnitude of the number, whereas 7.065 is accurate to four significant figures as the zero in this case is significant in expressing the value of the number.

**sign stimulus (releaser)** The essential feature of a stimulus, which is necessary to elicit a response. For example, a red belly (characteristic of courting male sticklebacks) is the sign stimulus necessary to provoke an attack from a rival male; even a very crude model fish is attacked if it has a red undersurface.

**silane (silicane) 1.** A colourless gas, $SiH_4$,

which is insoluble in water; d. 1.44 g dm$^{-3}$; r.d. 0.68 (liquid); m.p. −185°C; b.p. −112°C. Silane is produced by reduction of silicon with lithium tetrahydridoaluminate(III). It is also formed by the reaction of magnesium silicide (Mg$_2$Si) with acids, although other silicon hydrides are also produced at the same time. Silane itself is stable in the absence of air but is spontaneously flammable, even at low temperatures. It is a reducing agent and has been used for the removal of corrosion in inaccessible plants (e.g. pipes in nuclear reactors). **2. (silicon hydride)** Any of a class of compounds of silicon and hydrogen. They have the general formula Si$_n$H$_{2n+2}$. The first three in the series are silane itself (SiH$_4$), **disilane** (Si$_2$H$_6$), and **trisilane** (Si$_3$H$_8$). The compounds are analogous to the alkanes but are much less stable and only the lower members of the series can be prepared in any quantity (up to Si$_6$H$_{14}$). No silicon hydrides containing double or triple bonds exist (i.e. there are no analogues of the alkenes and alkynes).

**silica** *See* SILICON(IV) OXIDE.

**silica gel** A rigid gel made by coagulating a sol of sodium silicate and heating to drive off water. It is used as a support for catalysts and also as a drying agent because it readily absorbs moisture from the air. The gel itself is colourless but, when used in desiccators, etc., a blue cobalt salt is added. As moisture is taken up, the salt turns pink, indicating that the gel needs to be regenerated (by heating).

**silicane** *See* SILANE.

**silicate** Any of a group of substances containing negative ions composed of silicon and oxygen. The silicates are a very extensive group and natural silicates form the major component of most rocks (*see* SILICATE MINERALS). The basic structural unit is the tetrahedral SiO$_4$ group. This may occur as a simple discrete SiO$_4^{4-}$ anion as in the **orthosilicates**, e.g. phenacite (Be$_2$SiO$_4$) and **willemite** (Zn$_2$SiO$_4$). Many larger silicate species are also found (see illustration). These are composed of SiO$_4$ tetrahedra linked by sharing oxygen atoms as in the **pyrosilicates**, Si$_2$O$_7^{6-}$, e.g. Sc$_2$Si$_2$O$_7$. The linking can extend to such forms as bentonite, BaTiSi$_3$O$_9$, or alternatively infinite chain anions, which are single strand (*pyroxenes*) or double strand (*amphiboles*). Spodumene, LiAl(SiO$_3$)$_2$, is a pyroxene and the asbestos minerals are amphiboles. Large two-

dimensional sheets are also possible, as in the various *micas* (see illustration overleaf), and the linking can extend to full three-dimensional framework structures, often with substituted trivalent atoms in the lattice. The *zeolites* are examples of this.

SiO$_4^{4-}$ as in Be$_2$SiO$_4$ (phenacite)

Si$_2$O$_5^{2-}$ as in Sc$_2$Si$_2$O$_7$ (thortveitite)

Si$_3$O$_9^{6-}$ as in BaTiSi$_3$O$_9$ (bentonite)

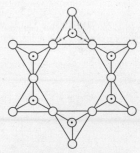

Si$_6$O$_{18}^{12-}$ as in Be$_3$Al$_2$Si$_6$O$_{18}$ (beryl)

**Silicate.** Structure of some discrete silicate ions.

**silicate minerals** A group of rock-forming minerals that make up the bulk of the earth's outer crust (about 90%) and constitute one-third of all minerals. All silicate minerals are based on a fundamental structural unit – the SiO₄ tetrahedron (*see* SILICATE). They consist of a metal (e.g. calcium, magnesium, aluminium) combined with silicon and oxygen. The silicate minerals are classified on a structural basis according to how the tetrahedra are linked together. The six groups are: nesosilicates (e.g. olivine and *garnet); sorosilicates (e.g. hemimorphite); cyclosilicates (e.g. axinite, *beryl, and *tourmaline); inosilicates (e.g. *amphiboles and *pyroxenes); phyllosilicates (e.g. *micas, *clay minerals, and *talc); and tektosilicates (e.g. *feldspars and *feldspathoids). Many silicate minerals are of economic importance.

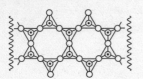

single chain : pyroxenes

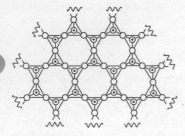

double chain : amphiboles

sheet : micas

**Silicate.** Structure of some polymeric silicate ions.

**silicide** A compound of silicon with a more electropositive element. The silicides are structurally similar to the interstitial carbides but the range encountered is more diverse. They react with mineral acids to form a range of *silanes.

**silicon** Symbol Si. A metalloid element belonging to *group 14 (formerly IVB) of the periodic table; a.n. 14; r.a.m. 28.086; r.d. 2.33; m.p. 1410°C; b.p. 2355°C. Silicon is the second most abundant element in the earth's crust (25.7% by weight) occurring in various forms of silicon(IV) oxide (e.g. *quartz) and in *silicate minerals. The element is extracted by reducing the oxide with carbon in an electric furnace and is used extensively for its semiconductor properties. It has a diamond-like crystal structure; an amorphous form also exists. Chemically, silicon is less reactive than carbon. The element combines with oxygen at red heat and is also dissolved by molten alkali. There is a large number of organosilicon compounds (e.g. *siloxanes) although silicon does not form the range of silicon–hydrogen compounds and derivatives that carbon does (*see* SILANE). The element was identified by Antoine Lavoisier in 1787 and first isolated in 1823 by Jöns Berzelius.

**(((⊕))) SEE WEB LINKS**

• Information from the WebElements site

**silicon carbide (carborundum)** A black solid compound, SiC, insoluble in water and soluble in molten alkali; r.d. 3.217; m.p. *c.* 2700°C. Silicon carbide is made by heating silicon(IV) oxide with carbon in an electric furnace (depending on the grade required sand and coke may be used). It is extremely hard and is widely used as an abrasive. The solid exists in both zinc blende and wurtzite structures. There is a clear form known as moissanite, used as a diamond substitute.

**silicon chip** A single crystal of a semiconducting silicon material, typically having millimetre dimensions, fabricated in such a way that it can perform a large number of independent electronic functions (*see* INTEGRATED CIRCUIT).

**silicon dioxide** *See* SILICON(IV) OXIDE.

**silicones** Polymeric compounds containing chains of silicon atoms alternating with oxygen atoms, with the silicon atoms linked to organic groups. A variety of silicone materials exist, including oils, waxes, and rubbers. They tend to be more resistant to temperature and chemical attack than their carbon analogues.

**silicon hydride** *See* SILANE.

**silicon(IV) oxide (silicon dioxide; silica)**

S

A colourless or white vitreous solid, $SiO_2$, insoluble in water and soluble (by reaction) in hydrofluoric acid and in strong alkali; m.p. 1713°C; b.p. 2230°C. The following forms occur naturally: **cristobalite** (cubic or tetragonal crystals; r.d. 2.32); **tridymite** (rhombic; r.d. 2.26); *quartz (hexagonal; r.d. 2.63–2.66); **lechatelierite** (r.d. 2.19). Quartz has two modifications: α-quartz below 575°C and β-quartz above 575°C; above 870°C β-quartz is slowly transformed to tridymite and above 1470°C this is slowly converted to cristobalite. Various forms of silicon(IV) oxide occur widely in the earth's crust; yellow sand for example is quartz with iron(III) oxide impurities and flint is essentially amorphous silica. The gemstones amethyst, opal, and rock crystal are also forms of quartz.

Silica is an important commercial material in the form of **silica brick**, a highly refractive furnace lining, which is also resistant to abrasion and to corrosion. Silicon(IV) oxide is also the basis of both clear and opaque silica glass, which is used on account of its transparency to ultraviolet radiation and its resistance to both thermal and mechanical shock. A certain proportion of silicon(IV) oxide is also used in ordinary glass and in some glazes and enamels. It also finds many applications as a drying agent in the form of *silica gel.

**silicula** A type of *capsule formed from a bicarpellary ovary. It is longitudinally flattened and divided lengthwise into two cavities (**loculi**). It is broader than a *siliqua. Examples include the fruits of *Alyssum* and candytuft.

**siliqua** A type of *capsule formed from a bicarpellary ovary. It resembles a *silicula but is longer than it is broad; an example is the fruit of the wallflower. *See also* LOMENTUM.

**siloxanes** A group of compounds containing silicon atoms bound to oxygen atoms, with organic groups linked to the silicon atoms, e.g. $R_3SiOSiR_3$, where R is an organic group. *Silicones are polymers of siloxanes.

**Silurian** A geological period of the Palaeozoic era following the Ordovician period and extending until the beginning of the Devonian period. It began about 444 million years ago and lasted for about 28 million years. The Silurian was named by Roderick Murchison (1792–1871) after an ancient British tribe that inhabited South Wales,

where he observed rocks of this period. The majority of Silurian life was marine but during the later part of the period primitive plants began to make their appearance on land. Trilobites and graptolites became less common, brachiopods were numerous and varied, crinoids became common for the first time, and corals also increased. The only known vertebrates during the Silurian were primitive fish; the first jawed fish appeared later in the period. The Caledonian orogeny (mountain-building period) reached its peak towards the end of the Silurian.

**silver** Symbol Ag. A white lustrous soft metallic *transition element; a.n. 47; r.a.m. 107.87; r.d. 10.5; m.p. 961.93°C; b.p. 2212°C. It occurs as the element and as the minerals argentite ($Ag_2S$) and horn silver (AgCl). It is also present in ores of lead and copper, and is extracted as a by-product of smelting and refining these metals. The element is used in jewellery, tableware, etc., and silver compounds are used in photography. Chemically, silver is less reactive than copper. A dark silver sulphide forms when silver tarnishes in air because of the presence of sulphur compounds. Silver(I) ionic salts exist (e.g. $AgNO_3$, AgCl) and there are a number of silver(II) complexes.

(((●))) SEE WEB LINKS
• Information from the WebElements site

**silver(I) bromide** A yellowish solid compound, AgBr; r.d. 6.5; m.p. 432°C. It can be precipitated from silver(I) nitrate solution by adding a solution containing bromide ions. It dissolves in concentrated ammonia solutions (but, unlike the chloride, does not dissolve in dilute ammonia). The compound is used in photographic emulsions.

**silver(I) chloride** A white solid compound, AgCl; r.d. 5.6; m.p. 455°C; b.p. 1550°C. It can be precipitated from silver(I) nitrate solution by adding a solution of chloride ions. It dissolves in ammonia solution (due to formation of the complex ion $[Ag(NH_3)_2]^+$). The compound is used in photographic emulsions.

**silver(I) iodide** A yellow solid compound, AgI; r.d. 6.01; m.p. 558°C; b.p. 1506°C. It can be precipitated from silver(I) nitrate solutions by adding a solution of iodide ions. Unlike the chloride and bromide, it does not dissolve in ammonia solutions.

**silver-mirror test** *See* TOLLENS REAGENT.

**silver(I) nitrate** A colourless solid, $AgNO_3$; r.d. 4.3; m.p. 212°C. It is an important silver salt because it is water-soluble. It is used in photography. In the laboratory, it is used as a test for chloride, bromide, and iodide ions and in volumetric analysis of chlorides using an absorption indicator (*see* ADSORPTION INDICATOR).

**silver(I) oxide** A brown slightly water-soluble amorphous powder, $Ag_2O$; r.d. 7.14. It can be made by adding sodium hydroxide solution to silver(I) nitrate solution. Silver(I) oxide is strongly basic and is also an oxidizing agent. It is used in certain reactions in preparative organic chemistry; for example, moist silver(I) oxide converts haloalkanes into alcohols; dry silver oxide converts haloalkanes into ethers. The compound decomposes to the elements at 300°C and can be reduced by hydrogen to silver. With ozone it gives the oxide AgO (which is diamagnetic and probably $Ag^IAg^{III}O_2$).

**sima** The rocks that form the earth's oceanic crust and underlie the upper crust. These are basaltic rock types rich in *si*lica ($SiO_2$) and *ma*gnesium (Mg), hence the name. The sima is denser and more plastic than the *sial that forms the continental crust.

**Simmons–Smith reaction** A reaction in which a cyclopropane ring is produced from an alkene. It uses the **Simmons–Smith reagent**, which was originally diiodomethane ($CH_2I_2$) with a Zn/Cu couple. Usually, diethyl zinc is used rather than Zn/Cu. The mechanism involves the formation of $H_2C(I)(ZnI)$ and *carbene transfer from the zinc to the double bond of the alkene.

**simple harmonic motion (SHM)** A form of periodic motion in which a point or body oscillates along a line about a central point in such a way that it ranges an equal distance on either side of the central point and that its acceleration towards the central point is always proportional to its distance from it. One way of visualizing SHM is to imagine a point rotating around a circle of radius $r$ at a constant angular velocity $\omega$. If the distance from the centre of the circle to the projection of this point on a vertical diameter is $y$ at time $t$, this projection of the point will move about the centre of the circle with simple harmonic motion. A graph of $y$ against $t$ will be a sine wave, whose equation is $y = r\sin\omega t$ (see diagram). *See also* PENDULUM.

**simulation** *See* MONTE CARLO SIMULATION.

**simultaneous equations** A set of equations that jointly specify the values of the variables they contain. If the number of variables is equal to the number of equations, each variable has a unique value, i.e. the equations can be solved.

**sine** *See* TRIGONOMETRIC FUNCTIONS.

**sine wave (sinusoidal wave)** Any waveform that has an equation in which one variable is proportional to the sine of the other. Such a waveform can be generated by an oscillator that executes *simple harmonic motion.

**single bond** *See* CHEMICAL BOND.

**single-cell protein (SCP)** Protein produced by microorganisms, such as bacteria, yeasts, and unicellular algae, that is extracted for use as a component of human and animal foods.

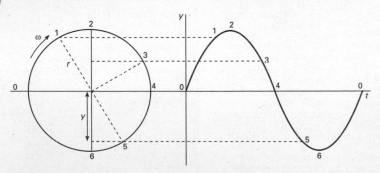

**Simple harmonic motion.**

**single circulation** The type of circulatory system that occurs in fishes, in which the blood passes only once through the heart in each complete circuit of the body *Compare* DOUBLE CIRCULATION.

**single nucleotide polymorphism (SNP)** A variation in the base sequence occurring at any given single position in the genome (for example A instead of C) that is found in more than 1% of the population. It thus differs from a *point mutation only in its greater frequency. SNPs can be found in all parts of the genome, including structural genes, regulatory regions, and noncoding 'junk' DNA. In the human genome overall, about 10 million SNPs are thought to occur, making them exceptionally useful as *molecular markers. Some are known to be linked to disease-causing alleles.

**singularity** An infinitely dense point located at the centre of a *black hole, where the laws of physics no longer apply. It is surrounded by the event horizon and so cannot be seen (because no light can escape from it).

**sinh** *See* HYPERBOLIC FUNCTIONS.

**sink hole (pot-hole)** A saucer-shaped or funnel-shaped depression in a limestone landscape formed when the rock is dissolved away by standing water. If a stream or river flows into the depression and disappears underground, it is called a swallow hole. Some sink holes are formed when the roof of a cave collapses.

**sinoatrial node** *See* PACEMAKER.

**sintered glass** Porous glass made by sintering powdered glass, used for filtration of precipitates in gravimetric analysis.

**sintering** The process of heating and compacting a powdered material at a temperature below its melting point in order to weld the particles together into a single rigid shape. Materials commonly sintered include metals and alloys, glass, and ceramic oxides. Sintered magnetic materials, cooled in a magnetic field, make especially retentive permanent magnets.

**sinus** A saclike cavity or organ in an animal, e.g. the **sinus venosus** in the heart of lower vertebrates.

**sinusoid** A tiny blood vessel or blood-filled space in an organ. Sinusoids replace capillaries in certain organs, notably the liver;

they allow more direct contact between the blood and the tissue it is supplying.

**sinusoidal oscillator** *See* OSCILLATOR.

**sinusoidal wave** *See* SINE WAVE.

**siphon** An inverted U-tube with one limb longer than the other. Liquid will be transferred from a reservoir at the base of the shorter limb to the end of the longer limb, provided that the U-tube is filled with liquid (see illustration). The device is useful for emptying an inaccessible container, such as a car's petrol tank.

The pressure ($p_1$) on the liquid at the base of the short limb (length $h_1$) is $p - h_1 k$, where $p$ is the atmospheric pressure and $k$ is a constant equal to the product of the density of the liquid and the acceleration of free fall. The pressure ($p_2$) on the liquid at the base of the long limb (length $h_2$) is $p - h_2 k$. Thus for fluid flow to occur through the tube, from short limb to long limb, $p_1 > p_2$, and for this to occur $h_2 > h_1$. Thus if the limbs are of equal length no flow will occur; it will only occur if the limb dipping into the reservoir is shorter than the delivering limb.

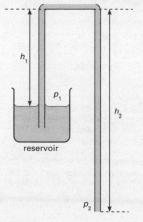

**Siphon.**

**Siphonaptera** An order of wingless insects comprising the fleas. The body of a flea is laterally compressed and bears numerous backward-directed spines. Fleas live as blood-sucking ectoparasites of mammals and birds, having mouthparts adapted to piercing their host, injecting saliva to prevent clotting, and sucking up the blood. The

S

long bristly legs can transmit energy stored in the elastic body wall to leap relatively long distances (over 300 mm horizontally). Apart from causing irritation, fleas can transmit disease organisms, most notably bubonic plague bacteria, which can be carried from rats to humans by the rat flea (*Xenopsylla cheopsis*). The whitish wormlike legless larvae feed on organic matter. After two moults the larva spins a cocoon and undergoes metamorphosis into the adult.

**Siphunculata (Anoplura)** An order of wingless insects comprising the sucking lice: blood-sucking ectoparasites of mammals, with piercing and sucking mouthparts forming a snoutlike proboscis. They constitute an irritating pest to humans and domestic animals and can transmit diseases, including typhoid. The human louse (*Pediculus humanus*) exists in two forms: the head louse (*P. humanus capitis*) and the body louse (*P. humanus corporis*).

**s-isomer** *See* ABSOLUTE CONFIGURATION.

**Sisyphus effect** *See* LASER COOLING.

**Site of Special Scientific Interest** *See* SSSI.

**SI units** Système International d'Unités: the international system of units now recommended for all scientific purposes. A coherent and rationalized system of units derived from the *m.k.s. units, SI units have now replaced *c.g.s. units and *Imperial units. The system has seven **base units** and two **dimensionless** (formerly called **supplementary**) **units** (see Appendix), all other units being derived from these nine units. There are 18 derived units with special names. Each unit has an agreed symbol (a capital letter or an initial capital letter if it is named after a scientist, otherwise the symbol consists of one or two lower-case letters). Decimal multiples of the units are indicated by a set of prefixes; whenever possible a prefix representing 10 raised to a power that is a multiple of three should be used.

**Sivapithecus** A genus of extinct primates that lived about 12–18 million years ago. Fossil remains of sivapithecines have been found in India and Pakistan, the Near East, and East Africa. Early discoveries of jaw fragments suggested that they chewed from side to side and had fairly short muzzles, both of which are humanoid features. However, subsequent finds, including a complete jaw, were not hominoid, and sivapithecines are

now regarded as ancestral to the Asian great apes (e.g. orang-utans), not the hominids. *See also* DRYOPITHECUS; AUSTRALOPITHECUS.

**skeletal electrons** *See* WADE'S RULES.

**skeletal muscle** *See* VOLUNTARY MUSCLE.

**skeleton** The structure in an animal that provides mechanical support for the body, protection for internal organs, and a framework for anchoring the muscles. The skeleton may be external (*see* EXOSKELETON) or internal (*see* ENDOSKELETON). Both types require *joints to allow locomotion. The skeleton of higher vertebrates consists of a system of *bones (*see* APPENDICULAR SKELETON; AXIAL SKELETON). Soft-bodied animals have a *hydrostatic skeleton.

(((●))) SEE WEB LINKS

- The e-Skeletons Project site explores the comparative skeletal anatomy of humans and other primates

**skin** The outer layer of the body of a vertebrate (see illustration). It is composed of two layers, the *epidermis and *dermis, with a complex nervous and blood supply. The skin may bear a variety of specialized structures, including *hair, *scales, and *feathers. This skin has an important role in protecting the body from mechanical injury, water loss, and the entry of harmful agents (e.g. disease-causing bacteria). It is also a sense organ, containing receptors sensitive to pain, temperature, and pressure. In warm-blooded animals it helps regulate body temperature by means of hair, fur, or feathers and *sweat glands.

**skip distance** The minimum distance from the transmitter of a radio wave at which reception is possible by means of a sky wave (*see* RADIO TRANSMISSION). If a radio wave strikes the ionosphere at a small angle of incidence the wave passes through it and is not reflected. There is therefore a minimum angle of incidence at which reflection occurs for a given frequency. This leads to a region around a transmitter in which sky waves cannot be received. As the frequency of the transmission increases the minimum angle of incidence at which ionospheric reflection occurs becomes greater. Above about 4 megahertz there may be a region of several hundred kilometres around a transmitter, which is within the skip distance and in which ground waves are too attenuated to be effectively received. In this region no reception is possible.

**skull** The skeleton of the head. In mammals it consists of a \*cranium enclosing the brain and the bones of the face and jaw. All the joints between the individual bones of the skull are immovable (*see* SUTURE) except for the joint between the mandible (lower jaw) and the rest of the skull. There is a large opening (**foramen magnum**) at the base of the skull through which the spinal cord passes from the brain.

**skyrmion** A representation of a baryon in which the baryon is regarded as a soliton for a meson field theory. This picture of baryons can be justified using \*quantum chromodynamics and gives very useful insights into both baryons and nuclear structure. It is named after the British physicist Tony Skyrme (1922–87), who proposed this theory of baryons in the 1960s.

**sky wave** *See* RADIO TRANSMISSION.

**slag** Material produced during the \*smelting or refining of metals by reaction of the flux with impurities (e.g. calcium silicate formed by reaction of calcium oxide flux with silicon dioxide impurities). The liquid slag can be separated from the liquid metal because it floats on the surface. *See also* BASIC SLAG.

**slaked lime** *See* CALCIUM HYDROXIDE.

**slate** A blue to grey fine-grained metamorphic rock characterized by the ease with which it cleaves into large thin sheets. It is formed mainly by the metamorphosis of mudstone or shale, in which platy minerals become aligned in parallel planes. Slate is traditionally used as a roofing material.

**sleep** A readily reversible state of reduced awareness and metabolic activity that occurs periodically in many animals. Usually accompanied by physical relaxation, the onset of sleep in humans and other mammals is marked by a change in the electrical activity of the brain, which is recorded by an \*electroencephalogram as waves of low frequency and high amplitude (**slow-wave sleep**). This is interspersed by short bouts of high-

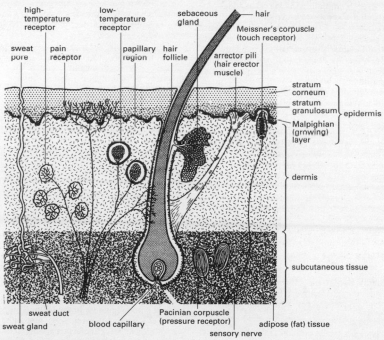

**Skin.** Structure of mammalian skin.

frequency low-amplitude waves (similar to wave patterns produced when awake) associated with restlessness, dreaming, and rapid eye movement (REM); this is called **REM** (or **paradoxical**) **sleep**. Several regions of the brain are involved in sleep, especially the reticular formation of the *brainstem.

**sleep movements** *See* NYCTINASTY.

**slepton** *See* SUPERSYMMETRY.

**slime moulds** Any of various small simple eukaryotic organisms that live in damp terrestrial habitats and superficially resemble fungi, to which they are unrelated. They are often seen as slimy masses on rotting wood and show *amoeboid movement, feeding by ingesting small particles of food. They exist either as free cells (**myxamoebas**) or as multinucleate aggregates of cells depending on the stage of the life cycle. When conditions become unfavourable, slime moulds form fruiting bodies (sporangia), from which spores are released. These disperse and subsequently germinate into small amoebas, thereby completing the life cycle. **Plasmodial slime moulds** live as 'giant cells' (plasmodia), formed by the fusion of individual flagellated cells and containing many nuclei. **Cellular slime moulds** live mainly as separate amoeboid cells, but aggregate to form a cellular swarm called a **pseudoplasmodium**, in which the plasma membranes of individual cells are retained. Another unrelated group, the Labyrinthomycota, consists of the **slime nets**, protists that secrete filaments along which the cells glide.

**slow neutron** A neutron with a kinetic energy of less than $10^2$ eV ($10^{-17}$ joule). *See also* FAST NEUTRON; THERMALIZATION.

**sludge** *See* SEWAGE.

**slug** 1. (in physics) An f.p.s. unit of mass equal to the mass that will acquire an acceleration of 1 ft sec$^{-2}$ when acted on by a force of one pound-force. 2. (in zoology) *See* GASTROPODA.

**slurry** A paste consisting of a suspension of a solid in a liquid.

**small intestine** The portion of the *alimentary canal between the stomach and the large intestine. It is subdivided into the *duodenum, *jejunum, and *ileum. It plays an essential role in the final digestion and absorption of food.

**small solar system body (SSSB)** Any

object in orbit around the sun that is not classified as a *planet or *dwarf planet.

**smectic** *See* LIQUID CRYSTAL.

**smell** *See* OLFACTION.

**smelting** The process of separating a metal from its ore by heating the ore to a high temperature in a suitable furnace in the presence of a reducing agent, such as carbon, and a fluxing agent, such as limestone. Iron ore is smelted in this way so that the metal melts and, being denser than the molten *slag, sinks below the slag, enabling it to be removed from the furnace separately.

**smoke** A fine suspension of solid particles in a gas.

**smoker** An active hydrothermal vent on the sea floor that emits mineral-containing fluids at high pressure. Minerals precipitating out of solution as they rise in the water give the appearance of smoke rising in the air. Dark sulphur compounds released from mid-ocean ridges form **black smokers**; the light-coloured emissions containing barytes or silica are **white smokers**. Sometimes deposits build up to form a tube-shaped chimney round the vent.

**smooth muscle** *See* INVOLUNTARY MUSCLE.

**smuts** A group of parasitic fungi of the phylum *Basidiomycota. Many of these species attack the ears of cereal crops, replacing the grain by a mass of dark spores. *Compare* RUSTS.

**S$_N$1 reaction** *See* NUCLEOPHILIC SUBSTITUTION.

**S$_N$2 reaction** *See* NUCLEOPHILIC SUBSTITUTION.

**snakes** *See* SQUAMATA.

**Snell's law** *See* REFRACTION.

**SNG** Substitute (or synthetic) natural gas; a mixture of gaseous hydrocarbons produced from coal, petroleum, etc., and suitable for use as a fuel. Before the discovery of natural gas *coal gas was widely used as a domestic and industrial fuel. This gave way to natural gas in the early part of this century in the US and other countries where natural gas was plentiful. The replacement of coal gas occurred somewhat later in the UK and other parts of Europe. More recently, interest has developed in ways of manufacturing hydrocarbon gas fuels. The main sources are coal and the naphtha fraction of petroleum. In the case of coal three methods have been used: (1) pyrolysis – i.e. more efficient forms of destructive distillation, often with further hydrogenation of the hydrocarbon products; (2) heating the coal with hydrogen and catalysts to give hydrocarbons – a process known as **hydroliquefaction** (*see also* BERGIUS PROCESS); (3) producing carbon monoxide and hydrogen and obtaining hydrocarbons by the *Fischer–Tropsch process. SNG from naptha is made by steam *reforming.

**SNP** *See* SINGLE NUCLEOTIDE POLYMORPHISM.

**snRNP** *See* RIBONUCLEOPROTEIN.

**soap** A substance made by boiling animal fats with sodium hydroxide. The reaction involves the hydrolysis of *glyceride esters of fatty acids to glycerol and sodium salts of the acids present (mainly the stearate, oleate, and palmitate), giving a soft semisolid with *detergent action. Potassium hydroxide gives a more liquid product (**soft soap**). By extension, other metal salts of long-chain fatty acids are also called soaps. *See also* SAPONIFICATION.

**SOAP** A protocol for exchanging structured and typed information between networked computers, especially in web services over the Internet. SOAP messages are formatted in *XML and generally use *HTTP as their transport protocol; the most common type is the remote procedure call.

(((●))) **SEE WEB LINKS**
• The W3C's XML Protocol Working Group page

**social behaviour** Any behaviour exhibited by a group of animals that interact with each other. Social behaviour ranges from moving as a herd in order to minimize the effects of predators to performing designated roles in highly organized societies. For example, within a colony of bees specific tasks, including tending the larvae, foraging for food, and controlling the temperature within the colony by wing fanning, are performed by different individuals. The application of evolutionary theory to social behaviour is called **sociobiology**, which is concerned primarily with genetically determined aspects of behaviour and their adaptive significance.

**soda** Any of a number of sodium compounds, such as caustic soda (NaOH) or, especially, washing soda ($Na_2CO_3.10H_2O$).

**soda ash** Anhydrous *sodium carbonate, $Na_2CO_3$.

**soda lime** A mixed hydroxide of sodium and calcium made by slaking lime with caustic soda solution (to give NaOH + $Ca(OH)_2$) and recovering greyish white granules by evaporation. The material is produced largely for industrial adsorption of carbon dioxide and water, but also finds some applications in pollution and effluent control. It is also used as a laboratory drying agent.

**sodamide** *See* SODIUM AMIDE.

**Soddy, Frederick** (1877–1956) British chemist, who worked with Ernest *Rutherford in Canada and William *Ramsay in London before finally settling in Oxford in 1919. His announcement in 1913 of the existence of *isotopes won him the 1921 Nobel Prize for physics.

**sodium** Symbol Na. A soft silvery reactive element belonging to group 1 (formerly IA) of the periodic table (*see* ALKALI METALS); a.n. 11; r.a.m. 22.9898; r.d. 0.97; m.p. 97.8°C; b.p. 882–889°C. Sodium occurs as the chloride in sea water and in the mineral halite. It is extracted by electrolysis in a *Downs cell. The metal is used as a reducing agent in certain reactions and liquid sodium is also a coolant in nuclear reactors. Chemically, it is highly reactive, oxidizing in air and reacting violently with water (it is kept under oil). It dissolves in liquid ammonia to form blue solutions containing solvated electrons. Sodium is a major *essential element required by animals. It is important in maintaining the *acid–base balance and in controlling the volume of extracellular fluid and functions in the transmission of nerve impulses (*see* SODIUM PUMP). The element was first isolated by Humphry Davy in 1807.

(((●))) **SEE WEB LINKS**
• Information from the WebElements site

**sodium acetate** *See* SODIUM ETHANOATE.

**sodium aluminate** A white solid, $NaAlO_2$ or $Na_2Al_2O_4$, which is insoluble in ethanol and soluble in water giving strongly alkaline solutions; m.p. 1800°C. It is manufactured by heating bauxite with sodium carbonate and extracting the residue with water, or it may be prepared in the laboratory by adding excess aluminium to hot concentrated sodium hydroxide. In solution the ion $Al(OH)_4^-$ predominates. Sodium aluminate is used as a mordant, in the production of zeolites, in

S

effluent treatment, in glass manufacture, and in cleansing compounds.

**sodium amide (sodamide)** A white crystalline powder, $NaNH_2$, which decomposes in water and in warm ethanol, and has an odour of ammonia; m.p. 210°C; b.p. 400°C. It is produced by passing dry ammonia over metallic sodium at 350°C. It reacts with red-hot carbon to give sodium cyanide and with nitrogen(I) oxide to give sodium azide.

**sodium azide** A white or colourless crystalline solid, $NaN_3$, soluble in water and slightly soluble in alcohol; hexagonal; r.d. 1.846; decomposes on heating. It is made by the action of nitrogen(I) oxide on hot sodamide ($NaNH_2$) and is used as an organic reagent and in the manufacture of detonators.

**sodium benzenecarboxylate (sodium benzoate)** An either colourless crystalline or white amorphous powder, $C_6H_5COONa$, soluble in water and slightly soluble in ethanol. It is made by the reaction of sodium hydroxide with benzoic acid and is used in the dyestuffs industry and as a food preservative. It was formerly used as an antiseptic.

**sodium benzoate** *See* SODIUM BENZENECARBOXYLATE.

**sodium bicarbonate** *See* SODIUM HYDROGENCARBONATE.

**sodium bisulphate** *See* SODIUM HYDROGENSULPHATE.

**sodium bisulphite** *See* SODIUM HYDROGENSULPHITE.

**sodium bromide** A white crystalline solid, $NaBr$, known chiefly as the dihydrate (monoclinic; r.d. 2.17), and as the anhydrous salt (cubic; r.d. 3.20; m.p. 747°C; b.p. 1390°C). The dihydrate loses water at about 52°C and is very slightly soluble in alcohol. Sodium bromide is prepared by the reaction of bromine on hot sodium hydroxide solution or of hydrogen bromide on sodium carbonate solution. It is used in photographic processing and in analytical chemistry.

**sodium carbonate** Anhydrous sodium carbonate (**soda ash**, **sal soda**) is a white powder, which cakes and aggregates on exposure to air due to the formation of hydrates. The monohydrate, $Na_2CO_3.H_2O$, is a white crystalline material, which is soluble in water and insoluble in alcohol; r.d. 2.532; loses water at 109°C; m.p. 851°C.

The decahydrate, $Na_2CO_3.10H_2O$ (**washing soda**), is a translucent efflorescent crystalline solid; r.d. 1.44; loses water at 32–34°C to give the monohydrate; m.p. 851°C. Sodium carbonate may be manufactured by the *Solvay process or by suitable crystallization procedures from any one of a number of natural deposits, such as:

**trona** ($Na_2CO_3.NaHCO_3.2H_2O$),
**natron** ($Na_2CO_3.10H_2O$),
**ranksite** ($2Na_2CO_3.9Na_2SO_4.KCl$),
**pirssonite** ($Na_2CO_3.CaCO_3.2H_2O$),
**gaylussite** ($Na_2CO_3.CaCO_3.5H_2O$).

The method of extraction is very sensitive to the relative energy costs and transport costs in the region involved. Sodium carbonate is used in photography, in cleaning, in pH control of water, in textile treatment, glasses and glazes, and as a food additive and volumetric reagent. *See also* SODIUM SESQUICARBONATE.

**sodium chlorate(V)** A white crystalline solid, $NaClO_3$; cubic; r.d. 2.49; m.p. 250°C. It decomposes above its melting point to give oxygen and sodium chloride. The compound is soluble in water and in ethanol and is prepared by the reaction of chlorine on hot concentrated sodium hydroxide. Sodium chlorate is a powerful oxidizing agent and is used in the manufacture of matches and soft explosives, in calico printing, and as a garden weedkiller.

**sodium chloride (common salt)** A colourless crystalline solid, $NaCl$, soluble in water and very slightly soluble in ethanol; cubic; r.d. 2.17; m.p. 801°C; b.p. 1413°C. It occurs as the mineral *halite (rock salt) and in natural brines and sea water. It has the interesting property of a solubility in water that changes very little with temperature. It is used industrially as the starting point for a range of sodium-based products (e.g. Solvay process for $Na_2CO_3$, Castner–Kellner process for $NaOH$), and is known universally as a preservative and seasoner of foods. Sodium chloride has a key role in biological systems in maintaining electrolyte balances.

**sodium chloride structure** *See* ROCK SALT STRUCTURE.

**sodium cyanide** A white or colourless crystalline solid, $NaCN$, deliquescent, soluble in water and in liquid ammonia, and slightly soluble in ethanol; cubic; m.p. 564°C; b.p. 1496°C. Sodium cyanide is now made by absorbing hydrogen cyanide in sodium hydroxide or sodium carbonate solution. The compound is extremely poisonous because it reacts with the iron in haemoglobin in the blood, so preventing oxygen reaching the tissues of the body. It is used in the extraction of precious metals and in electroplating industries. Aqueous solutions are alkaline due to salt hydrolysis.

**sodium dichromate** A red crystalline solid, $Na_2Cr_2O_7.2H_2O$, soluble in water and insoluble in ethanol. It is usually known as the dihydrate (r.d. 2.52), which starts to lose water above 100°C; the compound decomposes above 400°C. It is made by melting chrome iron ore with lime and soda ash and acidification of the chromate thus formed. Sodium dichromate is cheaper than the corresponding potassium compound but has the disadvantage of being hygroscopic. It is used as a mordant in dyeing, as an oxidizing agent in organic chemistry, and in analytical chemistry.

**sodium dihydrogenorthophosphate** *See* SODIUM DIHYDROGENPHOSPHATE(V).

**sodium dihydrogenphosphate(V) (sodium dihydrogenorthophosphate)** A colourless crystalline solid, $NaH_2PO_4$, which is soluble in water and insoluble in alcohol, known as the monohydrate (r.d. 2.04) and the dihydrate (r.d. 1.91). The dihydrate loses one water molecule at 60°C and the second molecule of water at 100°C, followed by decomposition at 204°C. The compound may be prepared by treating sodium carbonate with an equimolar quantity of phosphoric acid or by neutralizing phosphoric acid with sodium hydroxide. It is used in the preparation of sodium phosphate ($Na_3PO_4$), in baking powders, as a food additive, and as a constituent of buffering systems. Both sodium dihydrogenphosphate and trisodium phosphate enriched in $^{32}P$ have been used to study phosphate participation in metabolic processes.

**sodium dioxide** *See* SODIUM SUPEROXIDE.

**sodium ethanoate (sodium acetate)** A colourless crystalline compound, $CH_3COONa$, which is known as the anhydrous salt (r.d. 1.52; m.p. 324°C) or the trihydrate (r.d. 1.45; loses water at 58°C). Both forms are soluble in water and in ethoxyethane, and slightly soluble in ethanol. The compound may be prepared by the reaction of ethanoic acid (acetic acid) with sodium carbonate or with sodium hydroxide. Because it is a salt of a strong base and a weak acid, sodium ethanoate is used in buffers for pH control in many laboratory applications, in foodstuffs, and in electroplating. It is also used in dyeing, soaps, pharmaceuticals, and in photography.

**sodium fluoride** A crystalline compound, NaF, soluble in water and very slightly soluble in ethanol; cubic; r.d. 2.56; m.p. 993°C; b.p. 1695°C. It occurs naturally as villiaumite and may be prepared by the reaction of sodium hydroxide or of sodium carbonate with hydrogen fluoride. The reaction of sodium fluoride with concentrated sulphuric acid may be used as a source of hydrogen fluoride. The compound is used in ceramic enamels and as a preservative agent for fermentation. It is highly toxic but in very dilute solution (less than 1 part per million) it is used in the fluoridation of water for the prevention of tooth decay on account of its ability to replace OH groups with F atoms in the material of dental enamel.

**sodium formate** *See* SODIUM METHANOATE.

**sodium hexafluoraluminate** A colourless monoclinic solid, $Na_3AlF_6$, very slightly soluble in water; r.d. 2.9; m.p. 1000°C. It changes to a cubic form at 580°C. The compound occurs naturally as the mineral *cryolite but a considerable amount is manufactured by the reaction of aluminium fluoride wth alumina and sodium hydroxide or directly with sodium aluminate. Its most important use is in the manufacture of aluminium in the *Hall–Heroult cell. It is also used in the manufacture of enamels, opaque glasses, and ceramic glazes.

**sodium hydride** A white crystalline solid, NaH; cubic; r.d. 0.92; decomposes above 300°C (slow); completely decomposed at 800°C. Sodium hydride is prepared by the reaction of pure dry hydrogen with sodium at 350°C. Electrolysis of sodium hydride in molten LiCl/KCl leads to the evolution of hydrogen; this is taken as evidence for the ionic nature of NaH and the presence of the hydride ion ($H^-$). It reacts violently with water to give sodium hydroxide and hydrogen, with halogens to give the halide and appropriate hydrogen halide, and ignites spontaneously with oxygen at 230°C. It is a powerful reducing agent with several laboratory applications.

**sodium hydrogencarbonate (bicarbonate of soda; sodium bicarbonate)** A white

**S**

crystalline solid, $NaHCO_3$, soluble in water and slightly soluble in ethanol; monoclinic; r.d. 2.159; loses carbon dioxide above 270°C. It is manufactured in the *Solvay process and may be prepared in the laboratory by passing carbon dioxide through sodium carbonate or sodium hydroxide solution. Sodium hydrogencarbonate reacts with acids to give carbon dioxide and, as it does not have strongly corrosive or strongly basic properties itself, it is employed in bulk for the treatment of acid spillage and in medicinal applications as an antacid. Sodium hydrogencarbonate is also used in baking powders (and is known as **baking soda**), dry-powder fire extinguishers, and in the textiles, tanning, paper, and ceramics industries. The hydrogencarbonate ion has an important biological role as an intermediate between atmospheric $CO_2/H_2CO_3$ and the carbonate ion $CO_3^{2-}$. For water-living organisms this is the most important and in some cases the only source of carbon.

**sodium hydrogensulphate (sodium bisulphate)** A colourless solid, $NaHSO_4$, known in anhydrous and monohydrate forms. The anhydrous solid is triclinic (r.d. 2.435; m.p. >315°C). The monohydrate is monoclinic and deliquescent (r.d. 2.103; m.p. 59°C). Both forms are soluble in water and slightly soluble in alcohol. Sodium hydrogensulphate was originally made by the reaction between sodium nitrate and sulphuric acid, hence its old name of **nitre cake**. It may be manufactured by the reaction of sodium hydroxide with sulphuric acid, or by heating equimolar proportions of sodium chloride and concentrated sulphuric acid. Solutions of sodium hydrogensulphate are acidic. On heating the compound decomposes (via $Na_2S_2O_7$) to give sulphur trioxide. It is used in paper making, glass making, and textile finishing.

**sodium hydrogensulphite (sodium bisulphite)** A white solid, $NaHSO_3$, which is very soluble in water (yellow in solution) and slightly soluble in ethanol; monoclinic; r.d. 1.48. It decomposes on heating to give sodium sulphate, sulphur dioxide, and sulphur. It is formed by saturating a solution of sodium carbonate with sulphur dioxide. The compound is used in the brewing industry and in the sterilization of wine casks. It is a general antiseptic and bleaching agent. *See also* ALDEHYDES.

**sodium hydroxide (caustic soda)** A white transluscent deliquescent solid, NaOH, soluble in water and ethanol but insoluble in ether; r.d. 2.13; m.p. 318°C; b.p. 1390°C. Hydrates containing 7, 5, 4, 3.5, 3, 2, and 1 molecule of water are known.

Sodium hydroxide was formerly made by the treatment of sodium carbonate with lime but its main source today is from the electrolysis of brine using mercury cells or any of a variety of diaphragm cells. The principal product demanded from these cells is chlorine (for use in plastics) and sodium hydroxide is almost reduced to the status of a by-product. It is strongly alkaline and finds many applications in the chemical industry, particularly in the production of soaps and paper. It is also used to adsorb acidic gases, such as carbon dioxide and sulphur dioxide, and is used in the treatment of effluent for the removal of heavy metals (as hydroxides) and of acidity. Sodium hydroxide solutions are extremely corrosive to body tissue and are particularly hazardous to the eyes.

**sodium iodide** A white crystalline solid, NaI, very soluble in water and soluble in both ethanol and ethanoic acid. It is known in both the anhydrous form (cubic; r.d. 3.67; m.p. 661°C; b.p. 1304°C) and as the dihydrate (monoclinic; r.d. 2.45). It is prepared by the reaction of hydrogen iodide with sodium carbonate or sodium hydroxide in solution. Like potassium iodide, sodium iodide in aqueous solution dissolves iodine to form a brown solution containing the $I_3^-$ ion. It finds applications in photography and is also used in medicine as an expectorant and in the administration of radioactive iodine for studies of thyroid function and for treatment of diseases of the thyroid.

**sodium methanoate (sodium formate)** A colourless deliquescent solid, HCOONa, soluble in water and slightly soluble in ethanol; monoclinic; r.d. 1.92; m.p. 253°C; decomposes on further heating. The monohydrate is also known. The compound may be produced by the reaction of carbon monoxide with solid sodium hydroxide at 200°C and 10 atmospheres pressure; in the laboratory it can be conveniently prepared by the reaction of methanoic acid and sodium hydroxide. Its uses are in the production of oxalic acid (ethanedioic acid) and methanoic acid and in the laboratory it is a convenient source of carbon monoxide.

**sodium monoxide** A whitish-grey deliquescent solid, $Na_2O$; r.d. 2.27; sublimes at

1275°C. It is manufactured by oxidation of the metal in a limited supply of oxygen and purified by sublimation. Reaction with water produces sodium hydroxide. Its commercial applications are similar to those of sodium hydroxide.

**sodium nitrate (Chile saltpetre)** A white solid, $NaNO_3$, soluble in water and in ethanol; trigonal; r.d. 2.261; m.p. 306°C; decomposes at 380°C. A rhombohedral form is also known. It is obtained from deposits of caliche or may be prepared by the reaction of nitric acid with sodium hydroxide or sodium carbonate. It was previously used for the manufacture of nitric acid by heating with concentrated sulphuric acid. Its main use is in nitrate fertilizers.

**sodium nitrite** A yellow hygroscopic crystalline compound, $NaNO_2$, soluble in water, slightly soluble in ether and in ethanol; rhombohedral; r.d. 2.17; m.p. 271°C; decomposes above 320°C. It is formed by the thermal decomposition of sodium nitrate and is used in the preparation of nitrous acid (reaction with cold dilute hydrochloric acid). Sodium nitrite is used in organic *diazotization and as a corrosion inhibitor.

**sodium orthophosphate** *See* TRISODIUM PHOSPHATE(V).

**sodium peroxide** A whitish solid (yellow when hot), $Na_2O_2$, soluble in ice-water and decomposed in warm water or alcohol; r.d. 2.80; decomposes at 460°C. A crystalline octahydrate (hexagonal) is obtained by crystallization from ice water. The compound is formed by the combustion of sodium metal in excess oxygen. At normal temperatures it reacts with water to give sodium hydroxide and hydrogen peroxide. It is a powerful oxidizing agent reacting with iodine vapour to give the iodate and periodate, with carbon at 300°C to give the carbonate, and with nitrogen(II) oxide to give the nitrate. It is used as a bleaching agent in wool and yarn processing, in the refining of oils and fats, and in the production of wood pulp.

**sodium pump** A mechanism by which sodium ions are transported out of a eukaryotic cell across the plasma membrane. The process requires energy in the form of ATP, being a form of *active transport. It maintains the differential concentrations of sodium and potassium ions on either side of the plasma membrane, which is necessary,

for example, for establishing the *resting potential of a neuron.

**sodium sesquicarbonate** A white crystalline hydrated double salt, $Na_2CO_3$. $NaHCO_3.2H_2O$, soluble in water but less alkaline than sodium carbonate; r.d. 2.12; decomposes on heating. It may be prepared by crystallizing equimolar quantities of the constituent materials; it also occurs naturally as **trona** and in Searles Lake brines. It is widely used as a detergent and soap builder and, because of its mild alkaline properties, as a water-softening agent and bath-salt base. *See also* SODIUM CARBONATE.

**sodium sulphate** A white crystalline compound, $Na_2SO_4$, usually known as the anhydrous compound (orthorhombic; r.d. 2.67; m.p. 888°C) or the decahydrate (monoclinic; r.d. 1.46; which loses water at 100°C). The decahydrate is known as **Glauber's salt**. A metastable heptahydrate ($Na_2SO_4.7H_2O$) also exists. All forms are soluble in water, dissolving to give a neutral solution. The compound occurs naturally as
  **mirabilite** ($Na_2SO_4.10H_2O$),
  **thenardite** ($Na_2SO_4$), and
  **glauberite** ($Na_2SO_4.CaSO_4$).
  Sodium sulphate may be produced industrially by the reaction of magnesium sulphate with sodium chloride in solution followed by crystallization, or by the reaction of concentrated sulphuric acid with solid sodium chloride. The latter method was used in the *Leblanc process for the production of alkali and has given the name **salt cake** to impure industrial sodium sulphate. Sodium sulphate is used in the manufacture of glass and soft glazes and in dyeing to promote an even finish. It also finds medicinal application as a purgative and in commercial aperient salts.

**sodium sulphide** A yellow-red solid, $Na_2S$, formed by the reduction of sodium sulphate with carbon (coke) at elevated temperatures. It is a corrosive and readily oxidized material of variable composition and usually contains polysulphides of the type $Na_2S_2$, $Na_2S_3$, and $Na_2S_4$, which cause the variety of colours. It is known in an anhydrous form (r.d. 1.85; m.p. 1180°C) and as a nonahydrate, $Na_2S.9H_2O$ (r.d. 1.43; decomposes at 920°C). Other hydrates of sodium sulphide have been reported. The compound is deliquescent, soluble in water with extensive hydrolysis, and slightly soluble in alcohol. It is used in wood pulping, dyestuffs manufac-

S

ture, and metallurgy on account of its reducing properties. It has also been used for the production of sodium thiosulphate (for the photographic industry) and as a depilatory agent in leather preparation. It is a strong skin irritant.

**sodium sulphite** A white solid, $Na_2SO_3$, existing in an anhydrous form (r.d. 2.63) and as a heptahydrate (r.d. 1.59). Sodium sulphite is soluble in water and because it is readily oxidized it is widely used as a convenient reducing agent. It is prepared by reacting sulphur dioxide with either sodium carbonate or sodium hydroxide. Dilute mineral acids reverse this process and release sulphur dioxide. Sodium sulphite is used as a bleaching agent in textiles and in paper manufacture. Its use as an antioxidant in some canned foodstuffs gives rise to a slightly sulphurous smell immediately on opening, but its use is prohibited in meats or foods that contain vitamin $B_1$. Sodium sulphite solutions are occasionally used as biological preservatives.

**sodium–sulphur cell** A type of *secondary cell that has molten electrodes of sodium and sulphur separated by a solid electrolyte consisting of beta alumina (a crystalline form of aluminium oxide). When the cell is producing current, sodium ions flow through the alumina to the sulphur, where they form sodium polysulphide. Electrons from the sodium flow in the external circuit. The opposite process takes place during charging of the cell. Sodium–sulphur batteries have been considered for use in electric vehicles because of their high peak power levels and relatively low weight. However, some of the output has to be used to maintain the operating temperature (about 370°C) and the cost of sodium is high.

**sodium superoxide (sodium dioxide)** A whitish-yellow solid, $NaO_2$, formed by the reaction of sodium peroxide with excess oxygen at elevated temperatures and pressures. It reacts with water to form hydrogen peroxide and oxygen.

**sodium thiosulphate (hypo)** A colourless efflorescent solid, $Na_2S_2O_3$, soluble in water but insoluble in ethanol, commonly encountered as the pentahydrate (monoclinic; r.d. 1.73; m.p. 42°C), which loses water at 100°C to give the anhydrous form (r.d. 1.66). It is prepared by the reaction of sulphur dioxide with a suspension of sulphur in boiling sodium hydroxide solution. Aqueous

solutions of sodium thiosulphate are readily oxidized in the presence of air to sodium tetrathionate and sodium sulphate. The reaction with dilute acids gives sulphur and sulphur dioxide. It is used in the photographic industry and in analytical chemistry.

**sodium-vapour lamp** A form of *electric lighting that gives a yellow light as a result of the luminous discharge obtained by the passage of a stream of electrons between tungsten electrodes in a tube containing sodium vapour. To facilitate starting, the tube also contains some neon; for this reason, until the lamp is warm the neon emits a characteristic pink glow. As the sodium vaporizes, the yellow light predominates. Sodium-vapour lamps are widely used as street lights because of their high luminous efficiency and because the yellow light is less absorbed than white light by fog and mist. Low-pressure sodium lamps emit a characteristic yellow light; in high-pressure lamps the atoms are sufficiently close to each other to interact and broaden the spectral lines into the orange and green regions.

**soft acid** *See* HSAB PRINCIPLE.

**soft base** *See* HSAB PRINCIPLE.

**soft iron** A form of iron that contains little carbon, has high relative permeability, is easily magnetized and demagnetized, and has a small hysteresis loss. Soft iron and other **soft ferromagnetic materials**, such as silicon steel, are used in making parts exposed to rapid changes of magnetic flux, such as the cores of electromagnets, motors, generators, and transformers.

By comparison, **hard ferromagnetic materials**, such as cobalt steel and various alloys of nickel, aluminium, and cobalt, have low relative permeability, are difficult to magnetize, and have a high hysteresis loss. They are used in making permanent magnets.

**soft matter** A general name given to non-crystalline condensed matter. This includes liquids, disordered solids (including glasses), liquid crystals, and random networks of polymers.

**soft radiation** Ionizing radiation of low penetrating power, usually used with reference to X-rays of long wavelength. *Compare* HARD RADIATION.

**soft soap** *See* SOAP.

**software** *See* COMPUTER.

**soft water** *See* HARDNESS OF WATER.

**softwood** *See* WOOD.

**soil** The layer of unconsolidated particles derived from weathered rock, organic material (*humus), water, and air that forms the upper surface over much of the earth and supports plant growth. The formation of soil depends on the parent material (i.e. the original material from which the soil is derived), the climate and topography of the area, the organisms present in the soil, and the time over which the soil has been developing. Soils are often classified in terms of their structure and texture. The structure of a soil is the way in which the individual soil particles are bound together to form aggregates or peds. The structure types include platy, blocky, granular, and crumbs. The texture of a soil denotes the proportion of the various particle sizes that it contains. The four main texture classes are sand, silt, clay, and *loam, of which loams are generally the best agricultural soils as they contain a mixture of all particle sizes. A number of distinct horizontal layers can often be distinguished in a vertical section (profile) of soil – these are known as **soil horizons**. Four basic horizons are common to most soils: an uppermost A horizon (or **topsoil**) containing the organic matter; an underlying B horizon (or **subsoil**), which contains little organic material and is strongly leached; a C horizon consisting of weathered rock; and a D horizon comprising the bedrock. *See also* BROWN EARTH; CHERNOZEM; PODZOL.

**soil erosion** The removal and thinning of the soil layer due to climatic and physical processes, such as high rainfall, which is greatly accelerated by certain human activities, such as *deforestation. Soil erosion can lead to a loss of agricultural land and if unchecked, eventually results in *desertification.

**sol** A *colloid in which small solid particles are dispersed in a liquid continuous phase.

**solar cell** An electric cell that uses the sun's radiation to produce usable electric current. Most solar cells consist of a single-crystal silicon *p–n* junction. When photons of light energy from the sun fall on or near the *semiconductor junction the electron–hole pairs created are forced by the electric field at the junction to separate so that the holes pass to the *p*-region and the electrons pass to the *n*-region. This displacement of free charge creates an electric current when a load is connected across the terminals of the device (see illustration). Individual silicon solar cells cannot be made with a surface area much in excess of 4000 mm² and the maximum power delivered by such a cell is approximately 0.6 W at about 0.5 V in full sun. The efficiency of such devices is about 15%. For practical use, therefore, solar cells have to be assembled in arrays. Panels of solar cells have been the exclusive source of power for satellites and space capsules. Their use on earth has been largely limited by their high cost, a reduction in the cost by a factor of 10 being required to make them competitive with other energy sources at present.

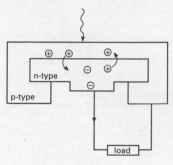

**Solar cell.** A silicon p-n junction.

**solar constant** The rate at which solar energy is received per unit area at the outer limit of the earth's atmosphere at the mean distance between the earth and the sun. The value is 1.366 kW m⁻².

**solar cycle (sunspot cycle)** The 11-year period over which the sun's activity varies. The principal variable is in the number and disposition of sunspots. At the solar minimum, at the beginning of the cycle, there are no sunspots. Several weeks later they begin to appear at middle latitudes, then gradually drift towards the equator. This behaviour continues for about five years, accompanied by a general increase in the number of sunspots. The number decreases over the next six years until they disappear altogether. All activity within and on the sun, including the generation of solar flares and large prominences, can be associated with the cycle.

**solar day** *See* DAY.

**solar energy** The electromagnetic energy radiated from the sun. The tiny proportion (about $5 \times 10^{-10}$ of the total) that falls on the earth is indicated by the *solar constant. The total quantity of solar energy falling on the earth in one year is about $4 \times 10^{18}$ J, whereas the total annual energy consumption of the earth's inhabitants is only some $3 \times 10^{14}$ J. The sun, therefore, could provide all the energy needed. The direct ways of making use of solar energy can be divided into thermal methods (*see* SOLAR HEATING) and nonthermal methods (*see* SOLAR CELL).

**solar flare** A sudden explosive release of particles and energy from the sun's chromosphere. It characteristically takes a few minutes to reach maximum brightness, and then fades during the next hour. When charged particles reach the earth, they cause radio interference, magnetic storms, and aurorae.

**solar heating** A form of domestic or industrial heating that relies on the direct use of solar energy. The basic form of **solar heater** is a thermal device in which a fluid is heated by the sun's rays in a collector (see illustrations) and pumped or allowed to flow round a circuit that provides some form of heat storage and some form of auxiliary heat source for use when the sun is not shining. More complicated systems are combined heating-and-cooling devices, providing heat in the winter and air-conditioning in the summer. The simplest form of solar collector is the flat-plate collector, in which a blackened receiving surface is covered by one or more glass plates that acts like a greenhouse (*see* GREENHOUSE EFFECT) and traps the maximum amount of solar energy. Tubes attached to the receiving surface carry air, water, or some other fluid to which the absorbed heat is transferred. The whole panel is insulated at the back and can thus form part of the roof of a building. More sophisticated collectors focus the sun's rays using reflectors. *See also* SOLAR CELL.

**solar parallax** The angle subtended by the earth's equatorial radius at the centre of the sun at the mean distance between the earth and the sun (i.e. at 1 astronomical unit). It has the value 8.794 148 arc seconds.

**solar prominence** *See* PROMINENCE.

**solar system** The sun, the eight major planets (Mercury, Venus, the earth, Mars, Jupiter, Saturn, Uranus, and Neptune) and their natural satellites, the five dwarf planets (Ceres, Pluto, Haumea, Makemake, and Eris) and their satellites, all *trans-Neptunian objects, the asteroids, the centaurs, the comets, and meteoroids. See Feature overleaf.

**((( ))) SEE WEB LINKS**
• Data for the solar system from NASA/JPL

**solar units** Units based on the physical properties of the sun used to describe other celestial objects. Examples include solar luminosity (symbol L$\odot$), solar mass (M$\odot$), and solar radius (R$\odot$).

**solar wind** A continuous outward flow of charged particles, mostly protons and electrons, from the sun's *corona into interplanetary space. The particles are controlled by the sun's magnetic field and are able to escape from the sun's gravitational field because of their high thermal energy. The average velocity of the particles in the vicinity of the earth is about 450 km s$^{-1}$ and their density at this range is about $8 \times 10^6$ protons per cubic metre.

**solar year** *See* YEAR.

**solder** An alloy used to join metal surfaces.

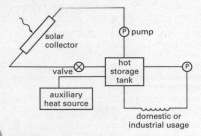

Typical solar heating system

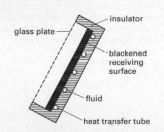

Flat-plate solar collector

**Solar heating.**

A **soft solder** melts at a temperature in the range 200–300°C and consists of a tin–lead alloy. The tin content varies between 80% for the lower end of the melting range and 31% for the higher end. **Hard solders** contain substantial quantities of silver in the alloy. **Brazing solders** are usually alloys of copper and zinc, which melt at over 800°C.

**solenoid** A coil of wire wound on a cylindrical former in which the length of the former is greater than its diameter. When a current is passed through the coil a magnetic field is produced inside the coil parallel to its axis. This field can be made to operate a plunger inside the former so that the solenoid can be used to operate a circuit breaker, valve, or other electromechanical device.

**solid** A state of matter in which there is a three-dimensional regularity of structure, resulting from the proximity of the component atoms, ions, or molecules and the strength of the forces between them. True solids are crystalline (*see also* AMORPHOUS). If a crystalline solid is heated, the kinetic energy of the components increases. At a specific temperature, called the **melting point**, the forces between the components become unable to contain them within the crystal structure. At this temperature, the lattice breaks down and the solid becomes a liquid.

**solid angle** Symbol $\Omega$. The three-dimensional 'angle' formed by the vertex of a cone. When this vertex is the centre of a sphere of radius $r$ and the base of the cone cuts out an area $s$ on the surface of the sphere, the solid angle in *steradians is defined as $s/r^2$.

**solid electrolyte** *See* FAST-ION CONDUCTOR.

**solid solution** A crystalline material that is a mixture of two or more components, with ions, atoms, or molecules of one component replacing some of the ions, atoms, or molecules of the other component in its normal crystal lattice. Solid solutions are found in certain alloys. For example, gold and copper form solid solutions in which some of the copper atoms in the lattice are replaced by gold atoms. In general, the gold atoms are distributed at random, and a range of gold–copper compositions is possible. At a certain composition, the gold and copper atoms can each form regular individual lattices (referred to as **superlattices**). Mixed crystals of double salts (such as alums) are also examples of solid solutions. Compounds can form solid solutions if they are isomorphous (*see* ISOMORPHISM).

**solid-state detector** *See* JUNCTION DETECTOR.

**solid-state physics** The study of the physical properties of solids, with special emphasis on the electrical properties of semiconducting materials in relation to their electronic structure. **Solid-state devices** are electronic components consisting entirely of solids (e.g. semiconductors, transistors, etc.) without heating elements, as in thermionic valves.

Recently the term **condensed-matter physics** has been introduced to include the study of crystalline solids, amorphous solids, and liquids.

**solid superfluid (supersolid)** A substance that has a lattice structure but can flow through a solid with no friction. A solid superfluid is a *Bose–Einstein condensate of atoms. The existence of supersolids was predicted theoretically in 1970. There is some experimental evidence for their existence but it cannot be regarded as conclusive.

**solidus** A line on a phase diagram below which a substance is solid.

**soliton** A stable particle-like solitary wave state that is a solution of certain equations for propagation. Solitons are thought to occur in many areas of physics and applied mathematics, such as plasmas, fluid mechanics, lasers, optics, solid-state physics, and elementary-particle physics.

**solstice** **1.** Either of the two points on the *ecliptic midway between the *equinoxes, at which the sun is at its greatest angular distance north (**summer solstice**) or south (**winter solstice**) of the celestial equator. **2.** The time at which the sun reaches either of these points. The summer solstice occurs on June 21 and the winter solstice on December 21 in the northern hemisphere; the dates are reversed in the southern hemisphere.

**solubility** The quantity of solute that dissolves in a given quantity of solvent to form a saturated solution. Solubility is measured in kilograms per metre cubed, moles per kilogram of solvent, etc. The solubility of a substance in a given solvent depends on the temperature. Generally, for a solid in a liquid, solubility increases with temperature;

## SOLAR SYSTEM

In general, a solar system is an astronomical feature consisting of a central star together with the planets and other celestial bodies held in orbit around it by gravitational attraction. The term most commonly refers to the particular solar system to which our earth belongs. This is dominated by our own central star, the *sun, and contains numerous different objects, including planets and SSSBs (small solar system bodies) – dwarf planets, asteroids, comets, etc. Many objects are under secondary gravitational influence, i.e. satellites (such as our moon) held in orbit around the larger bodies and travelling with them around the sun.

The solar system itself is part of a larger system, the Galaxy, which is itself rotating. Our solar system revolves around the centre of the Galaxy once every 220 million years.

### Origin of the solar system

The current theory of the origin of the solar system is the **solar nebular disc model.** It holds that the system originated within a giant molecular cloud, a vast nebula composed of hydrogen, in which clumps of denser material formed, possibly produced by shock waves from supernova explosions. One such clump, rotating and collapsing under its own gravitation, formed a flattened spinning disc, the solar nebula. The sun formed at the hot dense centre of the solar nebula, while in its cooler outer regions the planets grew by accretion. By about 4.6 billion years ago the earth had formed.

### Regions of the solar system

Many astronomers divide the solar system into three regions:

- the inner solar system, containing the planets Mercury, Venus, earth, and Mars, the earth's moon and the two moons of Mars, and the main asteroid belt.
- the outer solar system (beyond 5 AU from the sun), containing the planets Jupiter, Saturn, Uranus, and Neptune, with their satellites and other objects.
- the trans-Neptunian region (beyond 30 AU), a largely unmapped region lying outside the orbit of Neptune.

Some writers define the region containing the main planets as the 'inner solar system' and the region beyond Neptune as the 'outer solar system'. The trans-Neptunian region may be partitioned further into three other regions:

- the Kuiper belt, ranging from 30–50 AU, the existence of which was confirmed in 1992 by David C. Jewett and Jane X. Luu. It was named after the Dutch-American astronomer Gerard P. Kuiper (1905–73), one of several experts who had theorized about the outer reaches of the solar system. It contains thousands of small icy bodies, called **Kuiper belt objects** or **KBOs**. It includes a number of significant objects; for example, Pluto, once regarded as a planet, is a Kuiper belt object (now also classified as a dwarf planet).
- the **scattered disc** lying beyond the Kuiper belt and perhaps an outer part of it. It is home to the dwarf planet Eris and probably to most periodic comets and centaurs.
- the **Oort cloud**, a hypothetical region way outside the Kuiper belt and scattered disc, about 1 light year from the sun, thought to be the home of long-period comets. Originally proposed in 1932 by the Estonian astronomer Ernst Öpik (1893–1985), it is named after the Dutch astronomer Jan Hendrik Oort (1900–92), who revived the idea in 1950.

### The planets

There are eight bodies now recognized as *planets. They are divided into two groups. The **inner** or **terrestrial planets** – Mercury, Venus, earth, and Mars – are comparatively small. They are composed of rock and metal, with the metal part forming a dense central core. Mercury and Venus have no natural satellites. The earth has one satellite (the moon) and Mars has two (Phobos and Deimos).

The **outer** or **giant planets** – Jupiter. Saturn, Uranus, and Neptune – are massive low-density bodies with a rocky core surrounded by deep layers consisting mainly of solid, liquid, and gaseous hydrogen and helium. They are much further from the sun and therefore much cooler. All have large numbers of satellites: Jupiter has at least 63; Saturn at least 61; Uranus 27; and Neptune 13. The outer planets also have ring systems composed of smaller bodies, rocks, dust, and ice particles.

The four largest satellites of Jupiter – Ganymede, Callisto, Io, and Europa – are known as the **Galilean satellites** after their discovery by Galileo in 1610. Ganymede is the largest natural satellite in the solar system and is in fact larger than Mercury; it is followed by Titan (Saturn), Callisto (Jupiter), Io (Jupiter), the moon (earth), Europa (Jupiter), and Triton (Neptune).

## Dwarf planets

Some smaller bodies orbiting the sun are classed as *dwarf planets. Like planets, they have reached hydrostatic equilibrium but have not cleared their neighbourhoods of planetesimals. Five objects are presently classified as dwarf planets:

- **Pluto** – a Kuiper belt object ((KBO), considered to be a planet from its discovery in 1930 until 2006. It is named after the Greek god of the underworld.
- **Makemake** – another KBO, discovered in 2005 and about 75% of the size of Pluto. Its name is that of the creator of humanity in the mythology of Easter Island.
- **Ceres** – the largest member of the asteroid belt, named after the Roman goddess of agriculture.
- **Eris** – the largest dwarf planet, about 27% more massive than Pluto, found in the scattered disc. It is named after the Greek goddess of strife.
- **Haumea** – a KBO about one third the mass of Pluto. It is named after the Hawaiian goddess of fertility.

Pluto has three satellites, Haumea two, and Eris one.

Many bodies are candidates for dwarf planet status. They include the trans-Neptunian objects Orcus, Ixion, Huya, Varuna, Quaoar, and Sedna.

## Other objects

The space between the orbits of Mars and Jupiter is occupied by thousands of asteroids. Most are small, often irregularly shaped chunks of rock, with perhaps only 150 of them more than 100 km across. Even smaller are the innumerable meteoroids, some no larger than grains of dust originating in comet tails. Those that enter the earth's atmosphere and burn up as trails of light (shooting stars) are termed *meteors. The largest ones that reach the ground are called meteorites.

The remaining solar system members include *comets and *centaurs. Comets consist of a nucleus of dust and ice a kilometre or two across surrounded by a gaseous coma and with a long tail that appears when the comet nears the sun. Centaurs are more like asteroids in size, but some of them develop comet-like comas. Their orbits are unstable because of the gravitational influence of the giant outer planets.

for a gas, solubility decreases. *See also* CONCENTRATION.

**solubility product** Symbol $K_s$. The product of the concentrations of ions in a saturated solution. For instance, if a compound $A_xB_y$ is in equilibrium with its solution

$$A_xB_y(s) \rightleftharpoons xA^+(aq) + yB^-(aq)$$

the equilibrium constant is

$$K_c = [A^+]^x[B^-]^y / [A_xB_y]$$

Since the concentration of the undissolved solid can be put equal to 1, the solubility product is given by

$$K_s = [A^+]^x[B^-]^y$$

The expression is only true for sparingly soluble salts. If the product of ionic concentrations in a solution exceeds the solubility product, then precipitation occurs.

**solute** The substance dissolved in a solvent in forming a *solution.

**solution** A homogeneous mixture of a liquid (the *solvent) with a gas or solid (the **solute**). In a solution, the molecules of the solute are discrete and mixed with the molecules of solvent. There is usually some interaction between the solvent and solute molecules (*see* SOLVATION). Two liquids that can mix on the molecular level are said to be **miscible**. In this case, the solvent is the major component and the solute the minor component. *See also* SOLID SOLUTION.

**solvation** The interaction of ions of a solute with the molecules of solvent. For instance, when sodium chloride is dissolved in water the sodium ions attract polar water molecules, with the negative oxygen atoms pointing towards the positive $Na^+$ ion. Solvation of transition-metal ions can also occur by formation of coordinate bonds, as in the hexaquocopper(II) ion $[Cu(H_2O)_6]^{2+}$. Solvation is the process that causes ionic solids to dissolve, because the energy released compensates for the energy necessary to break down the crystal lattice. It occurs only with polar solvents. Solvation in which the solvent is water is called **hydration**.

**Solvay process (ammonia–soda process)** An industrial method of making sodium carbonate from calcium carbonate and sodium chloride. The calcium carbonate is first heated to give calcium oxide and carbon dioxide, which is bubbled into a solution of sodium chloride in ammonia. Sodium hydrogencarbonate is precipitated:

$$H_2O + CO_2(g) + NaCl(aq) + NH_3(aq) \rightarrow$$
$$NaHCO_3(s) + NH_4Cl(aq)$$

The sodium hydrogencarbonate is heated to give sodium carbonate and carbon dioxide. The ammonium chloride is heated with calcium oxide (from the first stage) to regenerate the ammonia. The process was patented in 1861 by the Belgian chemist Ernest Solvay (1838–1922).

**solvent** A liquid that dissolves another substance or substances to form a *solution. **Polar solvents** are compounds such as water and liquid ammonia, which have dipole moments and consequently high dielectric constants. These solvents are capable of dissolving ionic compounds or covalent compounds that ionize (*see* SOLVATION). **Nonpolar solvents** are compounds such as ethoxyethane and benzene, which do not have permanent dipole moments. These do not dissolve ionic compounds but will dissolve nonpolar covalent compounds. Solvents can be further categorized according to their proton-donating and accepting properties. **Amphiprotic solvents** self-ionize and can therefore act both as proton donators and acceptors. A typical example is water:

$$2H_2O \rightleftharpoons H_3O^+ + OH^-$$

**Aprotic solvents** neither accept nor donate protons; tetrachloromethane (carbon tetrachloride) is an example.

**solvent extraction** The process of separating one constituent from a mixture by dissolving it in a solvent in which it is soluble but in which the other constituents of the mixture are not. The process is usually carried out in the liquid phase, in which case it is also known as **liquid–liquid extraction**. In liquid–liquid extraction, the solution containing the desired constituent must be immiscible with the rest of the mixture. The process is widely used in extracting oil from oil-bearing materials.

**solvolysis** A reaction between a compound and its solvent. *See* HYDROLYSIS.

**soman** A highly colourless volatile liquid, $C_7H_{16}FO_2P$; r.d. 1.02; m.p. –42°C; b.p. 198°C. It is an organophosphorus compound, O-pinacolyl methylphosphonofluoridate. Soman was discovered in 1944 and belongs to the G-series of *nerve agents (GD).

**somatic** **1.** Relating to all the cells of an animal or plant other than the reproductive cells. Thus a somatic *mutation is one that is not heritable. **2.** Relating to organs and tissues of the body other than the gut and its associated structures. The term is applied especially to voluntary muscles, the sense organs, and the nervous system. *Compare* VISCERAL.

**somatostatin (growth hormone inhibiting hormone; GHIH)** A hormone, secreted by the hypothalamus, that inhibits the release of *growth hormone from the anterior pituitary gland. The secretion of somatostatin is stimulated by various factors, including very high blood glucose levels, which result from the effect that growth hormone has on glucose metabolism. It is also produced by the D cells of the *islets of Langerhans in the pancreas and can inhibit the release of glucagon and insulin from the islets.

**somatotrophin** *See* GROWTH HORMONE.

**sonar** *See* ECHO.

**sonic boom** A strong *shock wave generated by an aircraft when it is flying in the earth's atmosphere at supersonic speeds. This shock wave is radiated from the aircraft and where it intercepts the surface of the earth a loud booming sound is heard. The loudness depends on the speed and altitude of the aircraft and is lower in level flight than when the aircraft is undertaking a manoeuvre. The maximum increase of pressure in the shock wave during a transoceanic flight of a commercial supersonic transport (SST) is 120 Pa, equivalent to 136 decibels.

**Sonic hedgehog (SHH)** A protein that has a crucial role in patterning and development of tissues in vertebrates, particularly of the nervous and skeletal systems. It is homologous to **Hedgehog protein**, which performs a similar function in the fruit fly *Drosophila* – Hedgehog was so named because mutation of its gene in *Drosophila* produced flies bearing spiky denticles resembling the spines of a hedgehog. Sonic hedgehog was named after a video game character.

**sonochemistry** The study of chemical reactions in liquids subjected to high-intensity sound or ultrasound. This causes the formation, growth, and collapse of tiny bubbles within the liquid, generating localized centres of very high temperature and pressure, with extremely rapid cooling rates. Such conditions are suitable for studying novel reactions, decomposing polymers, and producing amorphous materials.

**sonometer** A device consisting essentially of a hollow sounding box with two bridges attached to its top. The string, fixed to the box at one end, is stretched between the two bridges so that the free end can run over a pulley and support a measured load. When the string is plucked the frequency of the note can be matched with that of another sound source, such as a tuning fork. It can be used to verify that the frequency ($f$) of a stretched string is given by $f = (1/2l)\sqrt{(T/m)}$, where $l$ is the length of the string, $m$ is its mass per unit length, and $T$ is its tension.

Originally called the **monochord**, the sonometer was widely used as a tuning aid, but is now used only in teaching laboratories.

**sorosis** A type of *composite fruit formed from an entire inflorescence spike. Mulberry and pineapple fruits are examples.

**sorption** *Absorption of a gas by a solid.

**sorption pump** A type of vacuum pump in which gas is removed from a system by absorption on a solid (e.g. activated charcoal or a zeolite) at low temperature.

**sorus** **1.** Any of the spore-producing structures on the undersurface of a fern frond, visible as rows of small brown dots. **2.** A reproductive area on the thallus of some algae, e.g. *Laminaria*. **3.** Any of various spore-producing structures in certain fungi.

**sound** A vibration in an elastic medium at a frequency and intensity that is capable of being heard by the human ear. The frequency of sounds lie in the range 20–20 000 Hz, but the ability to hear sounds in the upper part of the frequency range declines with age (*see also* PITCH). Vibrations that have a lower frequency than sound are called **infrasounds** and those with a higher frequency are called **ultrasounds**.

Sound is propagated through an elastic fluid as a longitudinal **sound wave**, in which a region of high pressure travels through the fluid at the *speed of sound in that medium. At a frequency of about 10 kilohertz the maximum excess pressure of a sound wave in air lies between $10^{-4}$ Pa and $10^3$ Pa. Sound travels through solids as either longitudinal or transverse waves.

**source** **1.** The electrode in a field-effect *transistor from which electrons or holes enter the interelectrode space. **2.** *See* MASS FLOW.

**Southern blotting** A chromatographic

technique for isolating and identifying specific fragments of DNA, such as the fragments formed as a result of DNA cleavage by *restriction enzymes. The mixture of fragments is subjected to electrophoresis through an agarose gel, followed by denaturation to form single-stranded fragments. These are transferred, or 'blotted', onto a nitrocellulose filter where they are immobilized in their relative positions. Specific *gene probes labelled with a radioisotope or fluorescent marker are then added. These hybridize with any complementary fragments on the filter, which are subsequently revealed by autoradiography or a fluorescence detector. The technique was devised by E. M. Southern. A similar technique for detecting RNA fragments is called **Northern blotting**, by analogy. *See also* WESTERN BLOTTING.

**Soxhlet apparatus** An apparatus for extracting components from a solid (e.g. extracting natural products from plant material). The material used is placed in a thimble made of thick filter paper and this is held in a specially designed reflex condenser with a suitable solvent. The chamber holding the thimble fills with warm solvent and this is led back to the source via a side arm. The apparatus can be operated for long periods, with components concentrating in the source vessel. It is named after Franz Soxhlet, who devised it in 1879.

**space** 1. A property of the universe that enables physical phenomena to be extended into three mutually perpendicular directions. In Newtonian physics, space, time, and matter are treated as quite separate entities. In Einsteinian physics, space and time are combined into a four-dimensional continuum (*see* SPACE–TIME) and in the general theory of *relativity matter is regarded as having an effect on space, causing it to curve. 2. **(outer space)** The part of the universe that lies outside the earth's atmosphere.

**space group** A *group formed by the set of all symmetry operations of a crystal lattice. This set consists of translations, rotations, and reflections and their combinations, such as *glide and *screw. It was discovered in the late 19th century that there are 230 possible space groups for a lattice in three dimensions. Space groups are used in the quantum theory of solids and in structure analysis in crystallography.

**space probe** A crewless spacecraft that investigates features within the solar system. A **planetary probe** examines the conditions on or in the vicinity of one or more planets and a **lunar probe** is designed to obtain information about the moon. **Interplanetary probes**, such as the two Voyager probes launched in 1977, have toured the solar system and are now on the way out of it. Probes are propelled by rocket motors and once out of the earth's gravitational field use their propulsion systems for course changes. Many use panels of *solar cells, for powering internal computer operations, on-board sensors, and radio communications.

**space-reflection symmetry** *See* PARITY.

**space–time (space–time continuum)** A geometry that includes the three dimensions and a **fourth dimension** of time. In Newtonian physics, space and time are considered as separate entities and whether or not events are simultaneous is a matter that is regarded as obvious to any competent observer. In Einstein's concept of the physical universe, based on a system of geometry devised by Hermann Minkowski (1864–1909), space and time are regarded as entwined, so that two observers in relative motion could disagree regarding the simultaneity of distant events. In Minkowski's geometry, an event is identified by a **world point** in a four-dimensional continuum.

**spadix** A flowering shoot (a type of *spike) with a large fleshy floral axis bearing small, usually unisexual, flowers. It is protected by a large petal-like bract, the **spathe**, and is characteristic of plants of the family Araceae (e.g. calla lily).

**spallation** A type of nuclear reaction in which the interacting nuclei disintegrate into a large number of protons, neutrons, and other light particles, rather than exchanging nucleons between them. It is thought that most of the nuclei of light elements, such as boron, are made in this way. Spallation reactions of this type are thought to occur in interstellar space when a high-energy particle, such as a proton, hits a nucleus.

**spark** *See* ELECTRIC SPARK.

**spark chamber** A device for detecting charged particles. It consists of a chamber, filled with helium and neon at atmospheric pressure, in which a stack of 20 to 100 plates are placed; the plates are connected alternately to the positive and negative terminals of a source of high potential (10 000 V or more). An incoming particle creates ion pairs in its track, causing the gas to become conducting and sparks to jump between the plates. The light from the sparks is focused to obtain stereoscopic photographs of the particles' tracks. It can also function as a counter (called a **spark counter**) when connected to suitable counting circuits. Some versions use crossed sets of parallel wires rather than plates; simple patterns may have a single wire near a plate, in the open atmosphere.

**spark counter** *See* SPARK CHAMBER.

**spathe** *See* SPADIX.

**special creation** The belief, in accordance with the Book of Genesis, that every species was individually created by God in the form in which it exists today and is not capable of undergoing any change. It was the generally accepted explanation of the origin of life until the advent of *Darwinism. The idea has recently enjoyed a revival, especially among members of the fundamentalist movement in the USA, partly because there still remain problems that cannot be explained entirely by Darwinian theory. However, special creation is contradicted by fossil evidence and genetic studies, and the pseudoscientific arguments of **creation science** cannot stand up to logical examination.

**specialization** 1. Increasing *adaptation of an organism to a particular environment. 2. *See* PHYSIOLOGICAL SPECIALIZATION.

**special theory of relativity** *See* RELATIVITY.

**speciation** The development of one or more species from an existing species. It occurs when *sympatric or *allopatric populations diverge so much from the parent population that interbreeding can no longer occur between them.

**species** 1. (in biology) A category used in the *classification of organisms. Similar species are grouped into a genus and a single species may be subdivided into *subspecies or *races (*see also* BINOMIAL NOMENCLATURE). According to the biological species concept, a species comprises a group of individuals that can usually breed among themselves and produce fertile offspring. Typically, a species consists of numerous local populations distributed over a geographical range.

Within a species, groups of individuals become reproductively isolated because of geographical or behavioural factors (*see* ISOLATING MECHANISM), and over time may evolve different characteristics and form new and distinct species. 2. (in chemistry) A chemical entity, such as a particular atom, ion, or molecule.

**species diversity** *See* BIODIVERSITY.

**specific** 1. (in physics) **a.** Denoting that an extensive physical quantity so described is expressed per unit mass. For example, the **specific latent heat** of a body is its latent heat per unit mass. When the extensive physical quantity is denoted by a capital letter (e.g. $L$ for latent heat), the specific quantity is denoted by the corresponding lower-case letter (e.g. $l$ for specific latent heat). **b.** In some older physical quantities the adjective 'specific' was added for other reasons (e.g. specific gravity, specific resistance). These names are now no longer used. 2. (in biology) Relating to a species.

**specific activity** *See* ACTIVITY.

**specific charge** The ratio of the charge of an *elementary particle or other charged body to its mass.

**specific gravity** *See* RELATIVE DENSITY; SPECIFIC.

**specific heat capacity** *See* HEAT CAPACITY.

**specific humidity** *See* HUMIDITY.

**specific impulse** A measure of the thrust available from a rocket propellant. It is the ratio of the thrust produced to the fuel consumption.

**specific intensity** *See* PLANCK'S RADIATION LAW.

**specific latent heat** *See* LATENT HEAT.

**specific resistance** *See* RESISTIVITY; SPECIFIC.

**specific surface** The surface area of a particular substance per unit mass, expressed in $m^2 kg^{-1}$. It provides a measure of the surface area available for a process, such as adsorption, for a given mass of a powder or porous substance.

**specific volume** The volume of a substance per unit mass. The reciprocal of density, it has the units $m^3 kg^{-1}$.

**speckle interferometer** An instrument

that improves the resolving power of an astronomical telescope by reducing the distortion produced by atmospheric turbulence (which greatly mars long-exposure photographs). Many short-exposure photographs are taken, one after the other, effectively 'freezing' the turbulence effects and producing a series of point images free from distortion. The overall image is obtained by combining these images.

**spectral class (spectral type)** Any of a set of categories to which stars can be assigned, based on the characteristics of their spectra. The **Harvard classification**, introduced in 1890 and modified in the 1920s, is based on the seven star types known as O, B, A, F, G, K, M:

O hottest blue stars; ionized helium lines dominant

B hot blue stars; neutral helium lines dominant, no ionized helium

A blue blue-white stars; hydrogen lines dominant

F white stars; metallic lines strengthen, hydrogen lines weaken

G yellow stars; ionized calcium lines dominant

K orange-red stars; neutral metallic lines dominant, some molecular bands

M coolest red stars; molecular bands dominant

Each class is divided into 10 subclasses denoted by the digits 0 to 9. Thus the sun is described as a yellow dwarf of spectral class G2. Further categorization will assign a star to a luminosity class, indicate the presence of emission lines, etc. Other classification letters are also used, e.g. DA for white dwarf stars, or W for Wolf-Rayet stars.

**spectrochemical series** A series of ligands arranged in the order in which they cause splitting of the energy levels of d orbitals in metal complexes (*see* CRYSTAL-FIELD THEORY). The series for some common ligands has the form:

$$CN^->NO_2^->NH_3>C_5H_5N>H_2O>OH^->F^->Cl^->Br^->I^-$$

**spectrograph** *See* SPECTROSCOPE.

**spectroheliogram** A photograph of the sun's chromosphere taken at a particular wavelength of light with the aid of a high-dispersion *spectroscope. Various levels of the chromosphere can be studied by varying the wavelength slightly.

**spectrometer** Any of various instruments for producing a spectrum and measuring the wavelengths, energies, etc., involved. A simple type, for visible radiation, is a spectroscope equipped with a calibrated scale allowing wavelengths to be read off or calculated. In the X-ray to infrared region of the electromagnetic spectrum, the spectrum is produced by dispersing the radiation with a prism or diffraction grating (or crystal, in the case of hard X-rays). Some form of photoelectric detector is used, and the spectrum can be obtained as a graphical plot, which shows how the intensity of the radiation varies with wavelength. Such instruments are also called **spectrophotometers**. Spectrometers also exist for investigating the gamma-ray region and the microwave and radio-wave regions of the spectrum (*see* ELECTRON PARAMAGNETIC RESONANCE; NUCLEAR MAGNETIC RESONANCE). Instruments for obtaining spectra of particle beams are also called spectrometers (*see* SPECTRUM; PHOTOELECTRON SPECTROSCOPY).

**spectrophotometer** *See* SPECTROMETER.

**spectroscope** An optical instrument that produces a *spectrum for visual observation. The first such instrument was made by R. W. Bunsen; in its simplest form it consists of a hollow tube with a slit at one end by which the light enters and a collimating lens at the other end to produce a parallel beam, a prism to disperse the light, and a telescope for viewing the spectrum (see illustration). In the **spectrograph**, the spectroscope is provided with a camera to record the spectrum. For a broad range of spectroscopic work, from the ultraviolet to the infrared, a diffraction grating is used instead of a prism. *See also* SPECTROMETER.

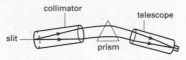

**Spectroscope.**

**spectroscopic binary** *See* BINARY STARS.

**spectroscopy** The study of methods of producing and analysing *spectra using *spectroscopes, *spectrometers, spectrographs, and spectrophotometers. The interpretations of the spectra so produced can be used for chemical analysis, examining atomic and molecular energy levels and molecular structures, and for determining the composition and motions of celestial bodies (*see* REDSHIFT).

**SEE WEB LINKS**
- The NIST atomic spectra database
- NIST Handbook of Basic Atomic Spectroscopic Data
- A history of spectroscopy from the Massachusetts Institute of Technology

**spectrum** (*pl.* **spectra**) **1.** A distribution of entities or properties arrayed in order of increasing or decreasing magnitude. For example, a beam of ions passed through a mass spectrograph, in which they are deflected according to their charge-to-mass ratios, will have a range of masses called a **mass spectrum**. A **sound spectrum** is the distribution of energy over a range of frequencies of a particular source. **2.** A range of electromagnetic energies arrayed in order of increasing or decreasing wavelength or frequency (*see* ELECTROMAGNETIC SPECTRUM). The **emission spectrum** of a body or substance is the characteristic range of radiations it emits when it is heated, bombarded by electron or ions, or absorbs photons. The **absorption spectrum** of a substance is produced by examining, through the substance and through a spectroscope, a continuous spectrum of radiation. The energies removed from the continuous spectrum by the absorbing medium show up as black lines or bands. With a substance capable of emitting a spectrum, these are in exactly the same positions in the spectrum as some of the lines and bands in the emission spectrum.

Emission and absorption spectra may show a **continuous spectrum**, a **line spectrum**, or a **band spectrum**. A continuous spectrum contains an unbroken sequence of frequencies over a relatively wide range; it is produced by incandescent solids, liquids, and compressed gases. Line spectra are discontinuous lines produced by excited atoms and ions as they fall back to a lower energy level. Band spectra (closely grouped bands of lines) are characteristic of molecular gases or chemical compounds. *See also* ACTION SPECTRUM; SPECTROSCOPY.

**speculum** An alloy of copper and tin formerly used in reflecting telescopes to make the main mirror as it could be cast, ground, and polished to make a highly reflective surface. It has now been largely replaced by silvered glass for this purpose.

**speed** The ratio of a distance covered by a body to the time taken. Speed is a *scalar quantity, i.e. no direction is given. Velocity is a *vector quantity, i.e. both the rate of travel and the direction are specified.

**speed of light** Symbol *c*. The speed at which electromagnetic radiation travels. The speed of light in a vacuum is $2.997\,924\,58 \times 10^8$ m s$^{-1}$. When light passes through any material medium its speed is reduced (*see* REFRACTIVE INDEX). The speed of light in a vacuum is the highest speed attainable in the universe (*see* RELATIVITY; CERENKOV RADIATION). It is a universal constant and is independent of the speed of the observer. Since October 1983 it has formed the basis of the definition of the *metre.

**speed of sound** Symbol *c* or $c_s$. The speed at which sound waves are propagated through a material medium. In air at 20°C sound travels at 344 m s$^{-1}$, in water at 20°C it travels at 1461 m s$^{-1}$, and in steel at 20°C at 5000 m s$^{-1}$. The speed of sound in a medium depends on the medium's modulus of elasticity ($E$) and its density (ρ) according to the relationship $c = \sqrt{(E/\rho)}$. For longitudinal waves in a narrow solid specimen, $E$ is the Young modulus; for a liquid $E$ is the bulk modulus (*see* ELASTIC MODULUS); and for a gas $E = \gamma p$, where γ is the ratio of the principal specific *heat capacities and $p$ is the pressure of the gas. For an ideal gas the relationship takes the form $c = \sqrt{(\gamma r T)}$, where $r$ is the gas constant per unit mass and $T$ is the thermodynamic temperature. This equation shows how the speed of sound in a gas is related to its temperature. This relationship can be written $c = c_0\sqrt{(1 + t/273)}$, where $c_0$ is the speed of sound in a particular gas at 0°C and $t$ is the temperature in °C.

**sperm** **1.** A single *spermatozoon. **2.** Spermatozoa, collectively.

**spermatheca (seminal receptacle)** A sac or receptacle in some female or hermaphrodite animals (e.g. earthworms) in which sperm from the mate is stored until the eggs are ready to be fertilized.

**spermatid** A nonmotile cell, produced during *spermatogenesis, that subsequently differentiates into a mature spermatozoon. Four spermatids are formed after two meiotic divisions of a primary spermatocyte and therefore contain the *haploid number of chromosomes.

**spermatogenesis** The series of cell divisions in the testis that results in the production of spermatozoa. Within the seminiferous tubules of the testis germ cells grow and divide by mitosis to produce **spermatogonia**. These divide by mitosis to produce **spermatocytes**, which divide by meiosis to produce *spermatids. The spermatids, which thus have half the number of chromosomes of the original germ cells, then develop into spermatozoa.

**Spermatophyta** In traditional classifications, a division of the plant kingdom containing plants that reproduce by means of *seeds. In modern systems seed plants are grouped into separate phyla, the most important of which are the *Anthophyta and *Coniferophyta.

**spermatozoid** *See* ANTHEROZOID.

**spermatozoon (sperm)** The mature mobile reproductive cell (*see* GAMETE) of male animals, which is produced by the testis (*see* SPERMATOGENESIS). It consists of a head section containing a *haploid nucleus and an **acrosome**, a membranous sac that releases enzymes allowing the sperm to penetrate the egg at fertilization; a middle section containing *mitochondria to provide the energy for movement; and a tail section, which lashes to drive the sperm forward.

**sperm competition** Competition between sperm from different males to reach and fertilize the egg cell of a single female. Sperm competition can occur among rodents in which a male mates a number of times with the same female, with a rest period between successive matings during which the sperm journeys towards the egg. If a second male mates with the female during a rest period its own sperm may disrupt the movement of sperm from the first male and succeed in fertilizing the egg cell. Certain animals in which sperm competition is possible have evolved features to minimize this interference. For example, in moths and butterflies the male cements the opening of the female genitalia after mating, thereby preventing further matings with other males. An ingenious mechanism operates in the fly *Johannseniella nitida*, in which the female eats a copulating male except for his genitalia, which remain in the body of the female and prevent further mating.

**sphalerite (zinc blende)** A mineral form of zinc sulphide, ZnS, crystallizing in the cubic system; the principal ore of zinc. It is usually yellow-brown to brownish-black in colour and occurs, often with galena, in metasomatic deposits and also in hydrothermal veins and replacement deposits. Sphalerite is mined on every continent, the chief sources including the USA, Canada, Mexico, Australia, Peru, and Poland.

**sphalerite structure (zinc-blende structure)** A type of ionic crystal structure in which the anions have an expanded face-centred cubic arrangement with the cations occupying one type of tetrahedral hole. The coordination number of each type of ion is 4. Examples of compounds with the sphalerite structure are ZnS, CuCl, CdS, and InAs.

 **SEE WEB LINKS**

• An interactive version of the structure

**Sphenophyta (Arthrophyta)** A phylum of *tracheophyte plants, the only living members of which are the horsetails (*Equisetum*). Horsetails have a perennial creeping rhizome supporting erect jointed stems bearing whorls of thin leaves. Spores are produced by terminal conelike structures. The group has a fossil record extending back to the Palaeozoic with its greatest development in the Carboniferous period, when giant tree forms were the dominant vegetation with the *Lycophyta.

**sphere** The figure generated when a circle is rotated about a diameter. The volume of a sphere is $4\pi r^3/3$ and its surface area is $4\pi r^2$, where $r$ is its radius. In Cartesian coordinates the equation of a sphere centred at the origin is $x^2 + y^2 + z^2 = r^2$.

**spherical aberration** *See* ABERRATION.

**spherical mirror** *See* MIRROR.

**spherical polar coordinates** *See* POLAR COORDINATES.

**spherometer** An instrument for measuring the curvature of a surface. The usual instrument for this purpose consists of a tripod, the pointed legs of which rest on the spherical surface at the corners of an equilateral triangle. In the centre of this triangle is a fourth point, the height of which is adjusted by means of a micrometer screw (see illustration). If the distance between each leg and the axis through the micrometer screw is $l$, and the height of the micrometer point above (or below) a flat surface is $x$, the radius ($r$) of the sphere is given by $r = (l^2 + x^2)/2x$.

**sphincter** A specialized muscle encircling an opening or orifice. Contraction of the sphincter tends to close the orifice. Examples are the **anal sphincter** (round the opening of the anus) and the **pyloric sphincter** (at the lower opening of the *stomach).

**sphingolipid** *See* PHOSPHOLIPID.

**spiders** *See* ARACHNIDA.

**spiegel (spiegeleisen)** A form of *pig iron containing 15–30% of manganese and 4–5% of carbon. It is added to steel in a Bessemer converter as a deoxidizing agent and to raise the manganese content of steel.

**spike** A type of *racemose inflorescence in which stalkless flowers arise from an undivided floral axis, as in plantain and *Orchis*. In the family Gramineae (Poaceae; sedges and grasses) the flowers are grouped in clusters called **spikelets**, which may be arranged to form a compound spike (as in wheat).

**spin (intrinsic angular momentum)** Symbol $s$. The part of the total angular momentum of a particle, atom, nucleus, etc., that can continue to exist even when the particle is apparently at rest, i.e. when its translational motion is zero and therefore its orbital angular momentum is zero. A molecule, atom, or nucleus in a specified energy level, or a particular elementary particle, has a particular spin, just as it has a particular charge or mass. According to *quantum theory, this is quantized and is restricted to multiples of $h/2\pi$, where $h$ is the *Planck constant. Spin is characterized by a quantum number $s$ (or $m_s$). For example, for an electron $s = \pm\frac{1}{2}$, implying a spin of $+ h/4\pi$ when it is spinning in one direction and $-h/4\pi$ when it is spinning in the other. Because of their spin, particles also have their own intrinsic *magnetic moments and in a magnetic field the spin of the particles lines up at an angle to the direction of the field, precessing around this direction. *See also* ELECTRON-SPIN RESONANCE; NUCLEAR MAGNETIC RESONANCE.

**spinal column** *See* VERTEBRAL COLUMN.

**spinal cord** The part of the vertebrate central nervous system that is posterior to the brain and enclosed within the *vertebral column. It consists of a hollow core of *grey matter (H-shaped in cross section) surrounded by an outer layer of *white matter; the central cavity contains *cerebrospinal fluid. The white matter contains numerous longitudinal nerve fibres organized into distinct tracts: **ascending tracts** consist of sensory neurons, conducting impulses towards the brain; **descending tracts** consist of motor neurons, transmitting impulses from the brain. Paired *spinal nerves arise from the spinal cord.

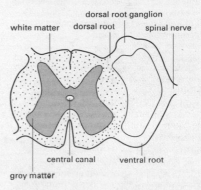

**Spinal cord.** Transverse section of the spinal cord.

**spinal nerves** Pairs of nerves that arise from the *spinal cord (*compare* CRANIAL NERVES). In humans there are 31 pairs (one from each of the vertebrae). Each nerve arises from a *dorsal root and a *ventral root and contains both motor and sensory fibres (i.e. they are mixed nerves). The spinal nerves form an important part of the *peripheral nervous system.

**spinal reflex** *See* REFLEX.

**spindle** 1. A structure formed from *microtubules in the cytoplasm during cell division that moves chromatids (*see* MITOSIS) or chromosomes (*see* MEIOSIS) diametrically apart and gathers them in two clusters at opposite

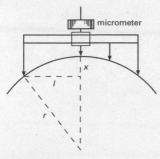

**Spherometer.**

ends (poles) of the cell. Broadest in the middle (the **spindle equator**) and narrowing to a point at either pole, its construction is directed by the *centrosome. The spindle becomes fully formed by metaphase, when the chromatids are attached to spindle fibres via their *centromeres and lie at the spindle equator. During anaphase this set of fibres shortens and hauls the attached chromatids towards the corresponding pole of the cell. Also, the overlapping fibres at the equator actively engage and slide past each other to elongate the entire spindle. **2.** See MUSCLE SPINDLE.

**spindle attachment** See CENTROMERE.

**spin drag** The resistance to the transport of spins in a quantum-mechanical system.

**spine 1.** See VERTEBRAL COLUMN. **2.** A hard pointed protective structure on a plant that is formed through modification of a leaf, part of a leaf, or a stipule. The edge of the holly leaf is drawn out into spines, but in cacti the whole leaf is modified as a spine. Compare PRICKLE; THORN.

**spinel** A group of oxide minerals with the general formula $F^{2+}R_2^{3+}O_4$, where $F^{2+}$ = Mg, Fe, Zn, Mn, or Ni and $R^{3+}$ = Al, Fe, or Cr, crystallizing in the cubic system. The spinels are divided into three series: spinel ($MgAl_2O_4$), *magnetite, and *chromite. They occur in high-temperature igneous or metamorphic rocks.

**spinel structure** An ionic crystal structure shown by compounds of the type $AB_2O_4$. There is a face-centred cubic arrangement of $O^{2-}$ ions. The A cations occupy one eighth of the tetrahedral holes and the B cations occupy the octahedral holes. Examples of the spinel structure are $MgAl_2O_4$, $Fe_3O_4$, and $Mn_3O_4$.

**SEE WEB LINKS**

• An interactive version of the structure

**spin glass** An alloy of a small amount of a magnetic metal (0.1–10%) with a nonmagnetic metal, in which the atoms of the magnetic element are randomly distributed through the crystal lattice of the nonmagnetic element. Examples are AuFe and CuMn. Theories of the magnetic and other properties of spin glasses are complicated by the random distribution of the magnetic atoms.

**spin Hall effect** An analogue of the *Hall effect for spin, i.e. there is an accumulation

of particles with opposite spins on the opposite surfaces of a material in which a current is flowing. However, the analogy is not complete since an external magnetic field destroys the effect, resulting in the spins precessing around the magnetic field. The spin Hall effect is due to *spin–orbit coupling.

**spin label** A molecule or group that contains an unpaired electron and can be attached to another molecule. The spin of the unpaired electron can be detected by electron paramagnetic resonance. The technique of spin labelling is used to investigate proteins and biological systems.

**spinneret** A small tubular appendage from which silk is produced in spiders and some insects. Spiders have four to six spinnerets on the hind part of the abdomen, into which numerous silk glands open. The silk is secreted as a fluid and hardens on contact with the air. It is composed of α-keratin crystals embedded in a matrix of amino-acid chains, giving the material its flexibility and strength. Various types of silk are produced depending on its use (e.g. for webs, egg cocoons, etc.). The spinnerets that produce the cocoons of insects are not homologous with those of spiders. For example, the spinneret of the silkworm is in the pharynx and the silk is produced by modified salivary glands.

**spinor** A mathematical entity similar to a vector but having the property that it changes sign on each rotation through 360°. The wave function of a spin-½ particle, such as an electron, in relativistic quantum mechanics is an example of a spinor. Spinors have also been used extensively in general relativity theory. There have also been many attempts to base the theory of elementary particles on spinors.

**spin–orbit coupling** An interaction between the orbital angular momentum and the spin angular momentum of an individual particle, such as an electron. For light atoms, spin–orbit coupling is small so that *multiplets of many-electron atoms are described by *Russell–Saunders coupling. For heavy atoms, spin–orbit coupling is large so that multiplets of many-electron atoms are described by *j-j coupling. For medium-sized atoms the sizes of the energies associated with spin–orbit coupling are comparable with the sizes of energies associated with electrostatic repulsion between the electrons, the multiplets in this case being described as having **intermediate coupling**. Spin–orbit coupling is large in many nuclei, particularly heavy nuclei.

**spin–statistics theorem** A fundamental theorem of relativistic *quantum field theory that states that half-integer *spins can only be quantized consistently if they obey Fermi–Dirac statistics and even-integer spins can only be quantized consistently if they obey Bose–Einstein statistics (*see* QUANTUM STATISTICS). This theorem enables one to understand the result of quantum statistics that wave functions for bosons are symmetric and wave functions for fermions are antisymmetric. It also provides the foundation for the *Pauli exclusion principle. It was first proved by Wolfgang Pauli in 1940.

**spin transport** The transport of spin in a quantum mechanical system. Spin transport is of key importance in spintronics, the branch of technology that makes use of the spin of electrons in electronic devices. The physical origin of spin transport is *spin–orbit coupling. Spin transport can be used to enhance the technological effectiveness of systems based on magnetism.

**spin wave (magnon)** A *collective excitation associated with magnetic systems. Spin waves occur in both ferromagnetic and antiferromagnetic systems (*see* MAGNETISM).

**spiracle** 1. A small paired opening that occurs on each side of the head in cartilaginous fish. It is the reduced first *gill slit, its small size resulting from adaptations of the skeleton for the firm attachment of the jaws. In modern teleosts (bony fish) the spiracle is closed up. In tetrapods the first gill slit develops into the middle ear cavity. 2. Any of the external openings of the *tracheae along the side of the body of an insect.

**spiral galaxy** *See* GALAXY.

**spirillum** Any rigid spiral-shaped bacterium. Generally, spirilla are Gram-negative (*see* GRAM'S STAIN), aerobic, and highly motile, bearing flagella either singly or in tufts. They occur in soil and water, feeding on organic matter.

**spirits of salt** A name formerly given to hydrogen chloride because this compound can be made by adding sulphuric acid to common salt (sodium chloride).

**spirochaete** Any nonrigid corkscrew-shaped bacterium that moves by means of flexions of the cell. Most spirochaetes are

Gram-negative (*see* GRAM'S STAIN), anaerobic, and feed on dead organic matter. They are very common in sewage-polluted waters. Some, however, can cause disease; *Treponema*, the agent of syphilis, is an example.

**spirometer** *See* RESPIROMETER.

**spleen** A vertebrate organ, lying behind the stomach, that is basically a collection of *lymphoid tissue. Its functions include producing lymphocytes and destroying foreign particles. It acts as a reservoir for erythrocytes and can regulate the number in circulation. It is also the site for the breakdown of worn-out erythrocytes and it stores the iron they contain.

**splicing** *See* GENE SPLICING.

**SPM** *See* SCANNING PROBE MICROSCOPY.

**sponges** *See* PORIFERA.

**spongy bone** *See* BONE.

**spongy mesophyll** *See* MESOPHYLL.

**spontaneous combustion** Combustion in which a substance produces sufficient heat within itself, usually by a slow oxidation process, for ignition to take place without the need for an external high-temperature energy source.

**spontaneous emission** The emission of a photon by an atom as it makes a transition from an excited state to the ground state. Spontaneous emission occurs independently of any external electromagnetic radiation; the transition is caused by interactions between atoms and vacuum fluctuations (*see* VACUUM STATE) of the quantized electromagnetic field. The process of spontaneous emission, which cannot be described by non-relativistic *quantum mechanics, as given by formulations such as the *Schrödinger equation, is responsible for the limited lifetime of an excited state of an atom before it emits a photon. *See also* EINSTEIN COEFFICIENTS; INDUCED EMISSION; LASER; QUANTUM THEORY OF RADIATION.

**spontaneous generation** The discredited belief that living organisms can somehow be produced by nonliving matter. For example, it was once thought that microorganisms arose by the process of decay and even that vermin spontaneously developed from household rubbish. Controlled experiments using sterilized media by Pasteur and others finally disproved these notions. *Compare* BIOGENESIS. *See also* BIOPOIESIS.

S

**sporangium** A reproductive structure in plants that produces asexual spores. *See* SPOROPHYLL.

**spore** A reproductive cell that can develop into an individual without first fusing with another reproductive cell (*compare* GAMETE). Spores are produced by plants, fungi, bacteria, and some protozoa. A spore may develop into an organism resembling the parent or into another stage in the life cycle, either immediately or after a period of dormancy. In plants showing *alternation of generations, spores are formed by the *sporophyte generation and give rise to the *gametophyte generation. In ferns, the rows of brown reproductive structures on the undersurface of the fronds are spore-producing bodies.

**spore mother cell (sporocyte)** A diploid cell that gives rise to four haploid spores by meiosis.

**sporocyte** *See* SPORE MOTHER CELL.

**sporogonium** The *sporophyte generation in mosses and liverworts. It is made up of an absorptive **foot**, a stalk (**seta**), and a spore-producing **capsule**. It may be completely or partially dependent on the *gametophyte.

**sporophore (fructification)** The aerial spore-producing part of certain fungi; for example, the stalk and cap of a mushroom.

**sporophyll** A leaf that bears **sporangia** (spore-producing structures). In ferns the sporophylls are the normal foliage leaves, but in other plants the sporophylls are modified and arise in specialized structures such as the strobilus (cone) of clubmosses, horsetails, and gymnosperms and the flower of angiosperms. Most plants produce spores of two different sizes (small **microspores** and large **megaspores**). The sporophylls bearing these are called **microsporophylls** and **megasporophylls** respectively.

**sporophyte** The generation in the life cycle of a plant that produces spores. The sporophyte is *diploid but its spores are *haploid. It is either completely or partially dependent on the *gametophyte generation in mosses and liverworts but is the dominant plant in the life cycle of clubmosses, horsetails, ferns, and seed plants. *See also* ALTERNATION OF GENERATIONS.

**spot test** A simple test for a given substance using a reagent that changes colour when mixed with the substance. In forensic science spot tests are often called *presumptive tests.

**spreadsheet** A *computer program that enables a user to make mathematical calculations on rows and columns of figures. If figures are interrelated, changing (updating) one figure also automatically changes all the others that are dependent on it. The user can also enter formulae for manipulating selected figures or groups of figures in a particular way.

**spring balance** A simple form of *balance in which a force is measured by the extension it produces in a helical spring. The extension, which is read off a scale, is directly proportional to the force, provided that the spring is not overstretched. The device is often used to measure the weight of a body approximately.

**s-process** *See* ORIGIN OF THE ELEMENTS.

**SPS** Super Proton Synchrotron. *See* CERN.

**sputtering** The process by which some of the atoms of an electrode (usually a cathode) are ejected as a result of bombardment by heavy positive ions. Although the process is generally unwanted, it can be used to produce a clean surface or to deposit a uniform film of a metal on an object in an evacuated enclosure.

**Squamata** An order of reptiles comprising the lizards and snakes. They appeared at the end of the Triassic period, about 170 million years ago, and have invaded a wide variety of habitats. Most lizards have four legs and a long tail, eardrums, and movable eyelids. Snakes are limbless reptiles that lack eardrums; the eyes are covered by transparent immovable eyelids and the articulation of the jaws is very loose, enabling a wide gape to facilitate swallowing prey whole.

**square-planar** Describing a coordination compound in which four ligands positioned at the corners of a square coordinate to a metal ion at the centre of the square. *See* COMPLEX.

**square wave** A train of rectangular voltage pulses that alternate between two fixed values for equal lengths of time. The time of transition between each fixed value is negligible compared to the duration of the fixed value. See diagram.

**squark** *See* SUPERSYMMETRY.

**squeezed state** A quantum state in a system in which the product in the Heisenberg *uncertainty principle takes the lowest possible value, i.e. the product $\delta x \, \delta p$, where $\delta x$ is the uncertainty in position and $\delta p$ is the uncertainty in momentum, is equal to $h/(4\pi)$, where $h$ is the *Planck constant, rather than greater than $h/(4\pi)$, with analogous results for the uncertainty principle for other pairs of variables. The concept of a squeezed state is used extensively in quantum optics and in precision measurements.

**SSSB** See SMALL SOLAR SYSTEM BODY.

**SSSI (Site of Special Scientific Interest)** The legal designation for an area of land in England, Scotland, or Wales that has been identified by Natural England, Scottish Natural Heritage, or the Countryside Council for Wales as being of special interest because of its flora, fauna, or geological or physiographical features. Such sites are protected from development activities and funds are available for their conservation and management. There are over 6000 SSSIs in Britain; similar sites in Northern Ireland are designated **Areas of Special Scientific Interest (ASSIs)**.

**stability of matter** The conclusion that matter consisting of a very large number of protons and electrons described by non-relativistic *quantum mechanics is stable. An essential element in the proof of this conclusion, which was established by several authors in the 1960s, is the *Pauli exclusion principle.

**stabilization energy** The amount by which the energy of a delocalized chemical structure is less than the theoretical energy of a structure with localized bonds. It is obtained by subtracting the experimental heat of formation of the compound (in kJ mol$^{-1}$) from that calculated on the basis of a classical structure with localized bonds.

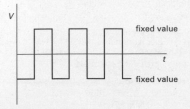

**Square wave.**

**stabilizer 1.** A substance used to inhibit a chemical reaction, i.e. a negative catalyst. **2.** A substance used to prevent a colloid from coagulating.

**stable equilibrium** See EQUILIBRIUM.

**staggered conformation** See CONFORMATION.

**staining** A technique in which cells or thin sections of biological tissue that are normally transparent are immersed in one or more coloured dyes (**stains**) to make them more clearly visible through a microscope. Staining heightens the contrast between the various cell or tissue components. Stains are usually organic salts with a positive and negative ion. If the colour comes from the negative ion (organic anion), the stain is described as **acidic**, e.g. *eosin. If the colour comes from the positive ion (organic cation), the stain is described as **basic**, e.g. *haematoxylin. **Neutral stains** have a coloured cation and a coloured anion; an example is *Leishman's stain. Cell constituents are described as being **acidophilic** if they are stained with acidic dyes, **basophilic** if receptive to basic dyes, and **neutrophilic** if receptive to neutral dyes. **Vital stains** are used to colour the constituents of living cells without harming them (*see* VITAL STAINING); **nonvital stains** are used for dead tissue.

Counterstaining involves the use of two or more stains in succession, each of which colours different cell or tissue constituents. **Temporary staining** is used for immediate microscopical observation of material, but the colour soon fades and the tissue is subsequently damaged. **Permanent staining** does not distort the cells and is used for tissue that is to be preserved for a considerable period of time.

Electron stains, used in the preparation of material for electron microscopy, are described as **electron-dense** as they interfere with the transmission of electrons. Examples are lead citrate, phosphotungstic acid (PTA), and uranyl acetate (UA).

**stainless steel** A form of *steel containing at least 11–12% of chromium, a low percentage of carbon, and often some other elements, notably nickel and molybdenum. Stainless steel does not rust or stain and therefore has a wide variety of uses in industrial, chemical, and domestic environments. A particularly successful alloy is the steel known as 18–8, which contains 18% Cr, 8% Ni, and 0.08% C.

**stalactites and stalagmites** Accretions of calcium carbonate in limestone caves. Stalactites are tapering cones or pendants that hang down from the roofs of caves; stalagmites are upward projections from the cave floor and tend to be broader at their bases than stalactites. Both are formed from drips of water containing calcium hydrogencarbonate in solution and may take thousands of years to grow.

**stamen** One of the male reproductive parts of a flower. It consists of an upper fertile part (the *anther) on a thin sterile stalk (the **filament**).

**staminode** A sterile stamen.

**standard cell** A *voltaic cell, such as a *Clark cell, or *Weston cell, used as a standard of e.m.f.

**standard deviation** A measure of the dispersion of data in statistics. For a set of values $a_1, a_2, a_3, ..., a_n$, the mean $m$ is given by $(a_1 + a_2 + ... + a_n)/n$. The **deviation** of each value is the absolute value of the difference from the mean: $|m - a_1|$, etc. The standard deviation is the square root of the mean of the squares of these values, i.e.

$$\sqrt{[(|m - a_1|^2 + ... |m - a_n|^2)/n]}$$

When the data is continuous the sum is replaced by an integral.

**standard electrode** An electrode (a half cell) used in measuring electrode potential. *See* HYDROGEN HALF CELL.

**standard electrode potential** *See* ELECTRODE POTENTIAL.

**standard form (scientific notation)** A way of writing a number, especially a large or small number, in which only one integer appears before the decimal point, the value being adjusted by multiplying by the appropriate power of 10. For example, 236,214 would be written in the standard form as $2.362\ 14 \times 10^5$; likewise 0.006821047 would be written $6.821\ 047 \times 10^{-3}$. Note that in the standard form, commas are not used, the digits are grouped into threes and a space is left between groups.

**standard model** *See* ELEMENTARY PARTICLES.

**standard solution** A solution of known concentration for use in volumetric analysis.

**standard state** A state of a system used as a reference value in thermodynamic measurements. Standard states involve a reference value of pressure (usually one atmosphere, 101.325 kPa) or concentration (usually 1 M). Thermodynamic functions are designated as 'standard' when they refer to changes in which reactants and products are all in their standard and their normal physical state. For example, the standard molar enthalpy of formation of water at 298 K is the enthalpy change for the reaction

$$H_2(g) + \tfrac{1}{2}O_2(g) \rightarrow H_2O(l)$$

$\Delta H^{\ominus}_{298} = -285.83$ kJ mol$^{-1}$.

Note that the superscript $^{\ominus}$ is used to denote standard state and the temperature should be indicated.

**standard temperature and pressure** *See* S.T.P.

**standing biomass** *See* STANDING CROP.

**standing crop** The total amount of living material in a specified population at a particular time, expressed as *biomass (**standing biomass**) or its equivalent in terms of energy. The standing crop may vary at different times of the year; for example, in a population of deciduous trees between summer and winter.

**standing wave** *See* STATIONARY WAVE.

**stannane** *See* TIN(IV) HYDRIDE.

**stannate** A compound formed by reaction of tin oxides (or hydroxides) with alkali. Tin oxides are amphoteric (weakly acidic) and react to give stannate ions. Tin(IV) oxide with molten alkali gives the stannate(IV) ion:

$$SnO_2 + 2OH^- \rightarrow SnO_3^{2-} + H_2O$$

In fact, there are various ions present in which the tin is bound to hydroxide groups, the main one being the hexahydroxostannate(IV) ion, $Sn(OH)_6^{2-}$. This is the negative ion present in crystalline 'trihydrates' of the type $K_2Sn_2O_3.3H_2O$.

Tin(II) oxide gives the trihydroxostannate(II) ion in alkaline solutions

$$SnO(s) + OH^-(aq) + H_2O(l) \rightarrow Sn(OH)_3^-(aq)$$

Stannate(IV) compounds were formerly referred to as **orthostannates** ($SnO_4^{4-}$) or **metastannates** ($SnO_3^{2-}$). Stannate(II) compounds were called **stannites**.

**stannic compounds** Compounds of tin in its higher (+4) oxidation state; e.g. stannic chloride is tin(IV) chloride.

**stannite** *See* STANNATE.

**stannous compounds** Compounds of tin in its lower (+2) oxidation state; e.g. stannous chloride is tin(II) chloride.

**stapes (stirrup)** The third of the three *ear ossicles of the mammalian middle ear.

***Staphylococcus*** A genus of spherical nonmotile Gram-positive bacteria that occur widely as saprotrophs or parasites. The cells occur in grapelike clusters. Many species inhabit the skin and mucous membranes, and some cause disease in humans and animals. *S. aureus* infection can lead to boils and abscesses in humans; this species also produces *toxins that irritate the gastrointestinal tract and result in staphylococcal food poisoning. Certain strains are resistant to antibiotics, and infection with these is very difficult to treat. For example, some strains of methicillin-resistant *S. aureus* (**MRSA**) are now resistant to nearly all antibiotics and pose a grave threat, both to patients in hospitals and to individuals in the wider community.

**star** A self-luminous celestial body, such as the *sun, that generates nuclear energy within its core. Stars are not distributed uniformly throughout the universe, but are collected together in *galaxies. The age and lifetime of a star are related to its mass (*see* STELLAR EVOLUTION; HERTZSPRUNG–RUSSELL DIAGRAM).

**starch** A *polysaccharide consisting of various proportions of two glucose polymers, *amylose and *amylopectin. It occurs widely in plants, especially in roots, tubers, seeds, and fruits, as a carbohydrate energy store. Starch is therefore a major energy source for animals. When digested it ultimately yields glucose. Starch granules are insoluble in cold water but disrupt if heated to form a gelatinous solution. This gives an intense blue colour with iodine solutions and starch is used as an *indicator in certain titrations.

**star cluster** A group of stars that are sufficiently close to each other for them to be physically associated. Stars belonging to the cluster are formed together from the same cloud of interstellar gas and have approximately the same age and initial chemical composition. Because of this, and since the stars in a given cluster are at roughly the same distance from earth, observations of star clusters are of great importance in studies of stellar evolution.

There are two types of star cluster. **Open** (or **galactic**) **clusters** are fairly loose systems of between a few hundred and a few thousand members. The stars in open clusters are quite young by astronomical standards (some as young as a few million years) and have relatively high abundances of heavy elements. **Globular clusters** are approximately spherical collections of between ten thousand and a million stars. These are very old (of order $10^{10}$ years) and have low heavy-element abundances.

**Stark effect** The splitting of lines in the *spectra of atoms due to the presence of a strong electric field. It is named after the German physicist Johannes Stark (1874–1957), who discovered it in 1913. Like the normal *Zeeman effect, the Stark effect can be understood in terms of the classical electron theory of Lorentz. The Stark effect for hydrogen atoms was also described by the *Bohr theory of the atom. In terms of *quantum mechanics, the Stark effect is described by regarding the electric field as a *perturbation on the quantum states and energy levels of an atom in the absence of an electric field. This application of perturbation theory was its first use in quantum mechanics.

**starquake** A sudden change in the crust of a neutron star. In the case of neutron stars which are pulsars, starquakes are associated with 'glitches' in the regularity of the production of electromagnetic radiation from the pulsar. A starquake is the analogue of an earthquake for a neutron star. Ordinary stars do not have starquakes because they do not have crusts. It is possible to obtain information about the thickness and rigidity of the crust from analysing the patterns of glitches. The starquakes are associated with discontinuous slowing down of the rotating neutron star.

**start codon (initiation codon)** The triplet of nucleotides on a messenger *RNA molecule (*see* CODON) at which the process of *translation is initiated. In eukaryotes the start codon is AUG (*see* GENETIC CODE), which codes for the amino acid methionine; in bacteria the start codon can be either AUG, coding for *N*-formyl methionine, or GUG, coding for valine. *Compare* STOP CODON.

**stat-** A prefix attached to the name of a practical electrical unit to provide a name for a unit in the electrostatic system of units, e.g. statcoulomb, statvolt. *Compare* AB-. In modern practice both absolute and electrostatic units have been replaced by *SI units.

**state of matter** One of the three physical states in which matter can exist, i.e. *solid, *liquid, or *gas. *Plasma is sometimes regarded as the fourth state of matter.

**static electricity** The effects produced by electric charges at rest, including the forces between charged bodies (*see* COULOMB'S LAW) and the field they produce (*see* ELECTRIC FIELD).

**statics** The branch of mechanics concerned with bodies that are acted upon by balanced forces and couples so that they remain at rest or in unaccelerated motion. *Compare* DYNAMICS.

**stationary orbit** *See* SYNCHRONOUS ORBIT.

**stationary phase** 1. *See* CHROMATOGRAPHY. 2. *See* BACTERIAL GROWTH CURVE.

**stationary state** A state of a system when it has an energy level permitted by *quantum mechanics. Transitions from one stationary state to another can occur by the emission or absorption of an appropriate quanta of energy (e.g. in the form of photons).

**stationary wave (standing wave)** A form of *wave in which the profile of the wave does not move through the medium but remains stationary. This is in contrast to a **travelling** (or **progressive**) **wave**, in which the profile moves through the medium at the speed of the wave. A stationary wave results when a travelling wave is reflected back along its own path. In a stationary wave there are points at which the displacement is zero; these are called **nodes**. Points of maximum displacement are called **antinodes**. The distance between a node and its neighbouring antinode is one quarter of a wavelength. In a stationary wave all the points along the wave have different amplitudes and the points between successive nodes are in phase; in a travelling wave every point vibrates with the same amplitude and the phase of vibration changes for different points along its path.

**statistical mechanics** The branch of physics in which statistical methods are applied to the microscopic constituents of a system in order to predict its macroscopic properties. The earliest application of this method was Boltzmann's attempt to explain the thermodynamic properties of gases on the basis of the statistical properties of large assemblies of molecules.

In classical statistical mechanics, each particle is regarded as occupying a point in *phase space, i.e. to have an exact position and momentum at any particular instant. The probability that this point will occupy any small volume of the phase space is taken to be proportional to the volume. The Maxwell–Boltzmann law gives the most probable distribution of the particles in phase space.

With the advent of quantum theory, the exactness of these premises was disturbed (by the Heisenberg uncertainty principle). In the *quantum statistics that evolved as a result, the phase space is divided into cells, each having a volume $h^f$, where $h$ is the Planck constant and $f$ is the number of degrees of freedom of the particles. This new concept led to Bose–Einstein statistics, and for particles obeying the Pauli exclusion principle, to Fermi–Dirac statistics.

**statistics** The branch of mathematics concerned with the inferences that can be drawn from numerical data on the basis of probability. A **statistical inference** is a conclusion drawn about a population as a result of an analysis of a representative sample. *See* SAMPLING.

**statocyst (otocyst)** A balancing organ found in many invertebrates. It consists of a fluid-filled sac lined with sensory hairs and contains granules of calcium carbonate, sand, etc. (**statoliths**). As the animal moves the statoliths stimulate different hairs, giving a sense of the position of the body or part of it. The *semicircular canals in the ears of vertebrates act on the same principle and have a similar function.

**stator** The stationary electromagnetic structure of an electric motor or electric generator. *Compare* ROTOR.

**steady-state theory** The cosmological theory that the universe has always existed in a steady state, that it had no beginning, will have no end, and has a constant mean density. To compensate for the observed *expansion of the universe this theory postulates that matter is created throughout the universe at a rate of about $10^{-10}$ nucleon per metre cubed per year as a property of space. Because it has failed to account for the *microwave background radiation or the evidence of evolution in the universe it has lost favour to the *big-bang theory. It was first proposed by Hermann Bondi (1919–2005), Thomas Gold (1920–2004), and Fred Hoyle in 1948.

**steam distillation** A method of distilling liquids that are immiscible with water by bubbling steam through them. It depends on the fact that the vapour pressure (and hence the boiling point) of a mixture of two immiscible liquids is lower than the vapour pressure of either pure liquid.

**steam engine** A *heat engine in which the thermal energy of steam is converted into mechanical energy. It consists of a cylinder fitted with a piston and valve gear to enable the high-pressure steam to be admitted to the cylinder when the piston is near the top of its stroke. The steam forces the piston to the bottom of its stroke and is then exhausted from the cylinder usually into a condenser. The reciprocating motion of the piston is converted to rotary motion of the flywheel by means of a connecting rod, crosshead, and crank. The steam engine reached its zenith at the end of the 19th century, since when it has been replaced by the steam turbine and the internal-combustion engine. *See also* RANKINE CYCLE.

**steam point** The temperature at which the maximum vapour pressure of water is equal to the standard atmospheric pressure (101 325 Pa). On the Celsius scale it has the value 100°C.

**stearate (octadecanoate)** A salt or ester of stearic acid.

**stearic acid (octadecanoic acid)** A solid saturated *fatty acid, $CH_3(CH_2)_{16}COOH$; r.d. 0.94; m.p. 71.5–72°C; b.p. 360°C (with decomposition). It occurs widely (as *glycerides) in animal and vegetable fats.

**steel** Any of a number of alloys consisting predominantly of iron with varying proportions of carbon (up to 1.7%) and, in some cases, small quantities of other elements (**alloy steels**), such as manganese, silicon, chromium, molybdenum, and nickel. Steels containing over 11–12% of chromium are known as *stainless steels.

Carbon steels exist in three stable crystalline phases: **ferrite** has a body-centred cubic crystal, **austenite** has a face-centred cubic crystal, and **cementite** has an orthorhombic crystal. **Pearlite** is a mixture of ferrite and cementite arranged in parallel plates. The phase diagram shows how the phases form at different temperatures and compositions.

Steels are manufactured by the *basic-oxygen process (L–D process), which has largely replaced the *Bessemer process and the *open-hearth process, or in electrical furnaces.

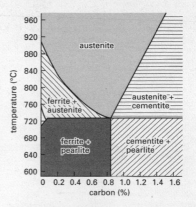

**Steel.** Phase diagram for steel.

**Stefan's law (Stefan–Boltzmann law)** The total energy radiated per unit surface area of a *black body in unit time is proportional to the fourth power of its thermodynamic temperature. The constant of proportionality, the **Stefan constant** (or **Stefan–Boltzmann constant**) has the value $5.6697 \times 10^{-8}$ J s$^{-1}$ m$^{-2}$ K$^{-4}$. The law was discovered by Joseph Stefan (1853–93) and theoretically derived by Ludwig Boltzmann.

**stele** The vascular tissue (i.e. *xylem and *phloem) of *tracheophyte plants, together with the endodermis and pericycle (when present). The arrangement of stelar tissues is very variable. In roots the stele often forms a solid core, which better enables the root to withstand tension and compression. In stems it is often a hollow cylinder separating the cortex and pith. This arrangement makes the stem more resistant to bending stresses. Monocotyledons and dicotyledons can usually be distinguished by the pattern of their stelar tissue. In monocotyledons the vascular bundles are scattered throughout the stem whereas in dicotyledons (and gymnosperms) they are arranged in a circle around the pith.

**stellar association** A very loose grouping of 10–100 young stars that share a common origin, usually having been born together in a *giant molecular cloud, and although no longer bound to each other by gravitational forces, are moving through space together.

**stellar evolution** The changes that occur to a *star during its lifetime, from birth to final extinction. A star is believed to form from a condensation of interstellar matter, which collects either by chance or for unexplained reasons, and grows by attracting other matter towards itself as a result of its gravitational field. This initial cloud of cold contracting matter, called a **protostar**, builds up an internal pressure as a result of its gravitational contraction. The pressure raises the temperature until it reaches $5–10 \times 10^6$ K, at which temperature the thermonuclear conversion of hydrogen to helium begins. In our *sun, a typical star, hydrogen is converted at a rate of some $10^{11}$ kg s$^{-1}$ with the evolution of some $6 \times 10^{25}$ J s$^{-1}$ of energy. It is estimated that the sun contains sufficient hydrogen to burn at this rate for $10^{10}$ years and that it still has half its life to live as a main-sequence star (*see* HERTZSPRUNG–RUSSELL DIAGRAM). Eventually, however, this period of stability comes to an end, because the thermonuclear energy generated in the interior is no longer sufficient to counterbalance the gravitational contraction. The core, which is now mostly helium, collapses until a sufficiently high temperature is reached in a shell of unburnt hydrogen round the core to start a new phase of thermonuclear reaction. This burning of the shell causes the star's outer envelope to expand and cool, the temperature drop changes the colour from white to red and the star becomes a **red giant** or a **supergiant** if the original star was very large. The core now contracts, reaching a temperature of $10^8$ K, and the helium in the core acts as the thermonuclear energy source. This reaction produces carbon, but a star of low mass relatively soon runs out of helium and the core collapses into a *white dwarf, while the outer regions drift away into space, possibly forming a **planetary nebula**. Larger stars (several times larger than the sun) have sufficient helium for the process to continue so that heavier elements, up to iron, are formed. But iron is the heaviest element that can be formed with the production of energy and when the helium has all been consumed there is a catastrophic collapse of the core, resulting in a *supernova explosion, blowing the outer layers away. The current theory suggests that thereafter the collapsed core becomes a *neutron star or a *black hole depending on its mass.

**stem** The part of a plant that usually grows vertically upwards towards the light and supports the leaves, buds, and reproductive structures (see illustration). The leaves develop at the *nodes and side or branch stems develop from buds at the nodes. The stems of certain species are modified as bulbs, corms, rhizomes, and tubers. Some species have twining stems; others have horizontal stems, such as *runners. Another modification is the *cladode. Erect stems may be cylindrical or angular; they may be covered with hairs, prickles, or spines and many exhibit secondary growth and become woody (*see* GROWTH RING). In addition to its supportive function, the stem contains *vascular tissue that conducts food, water, and mineral salts between the roots and leaves. It may also contain chloroplasts and carry out photosynthesis.

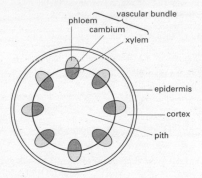

**Stem.** Transverse section through a herbaceous stem.

**stem cell** A cell that is not differentiated itself but can undergo unlimited division to form other cells, which either remain as stem cells or differentiate to form specialized cells. For example, **haemopoietic stem cells** in the bone marrow divide to produce daughter cells that differentiate into various types of blood cell (*see* HAEMOPOIETIC TISSUE). **Embryonic stem cells**, such as those taken from an early human embryo, are capable of differentiating into many or all of the various tissue cells found in a fully developed individual – they are described as **pluripotent**. Cultures of such cells have the potential to provide replacement tissues and organs for medical use, including transplantation. However, ethical concerns have led to tight controls on research using human embryonic stem cells in many countries, including the USA and UK.

**(⊕) SEE WEB LINKS**

• Website of the International Stem Cell Forum, which promotes good practice and progress in stem cell research

**step** A single stage in a chemical reaction. For example, the addition of hydrogen chloride to ethene involves three steps:

$$HCl \rightarrow H^+ + Cl^-$$

$$H^+ + C_2H_4 \rightarrow CH_3CH_2^+$$

$$CH_3CH_2^+ + Cl^- \rightarrow CH_3CH_2Cl$$

**steradian** Symbol sr. The dimensionless *SI unit of solid angle equal to the solid angle that encloses a surface on a sphere equal to the square of the radius of the sphere.

**stere** A unit of volume equal to 1 m³. It is not now used for scientific purposes.

**stereochemistry** The branch of chemistry concerned with the structure of molecules and the way the arrangement of atoms and groups affects the chemical properties.

**stereographic projection** A type of azimuthal projection used for making maps (*see* MAP PROJECTIONS) and for specifying the structures of crystals (*see* CRYSTAL). A point *p* on a sphere (called a pole of the projection) is projected onto a plane that is a tangent to the sphere at a point diametrically opposite *p*. The sizes and sense of angles between lines or curves are preserved in this projection.

**stereoisomerism** *See* ISOMERISM.

**stereoregular** Describing a *polymer that has a regular pattern of side groups along its chain.

**stereospecific** Describing chemical reac-

tions that give products with a particular arrangement of atoms in space. An example of a stereospecific reaction is the *Ziegler process for making polyethene.

**steric effect** An effect in which the rate or path of a chemical reaction depends on the size or arrangement of groups in a molecule.

**steric hindrance** An effect in which a chemical reaction is slowed down or prevented because large groups on a reactant molecule hinder the approach of another reactant molecule.

**sterile** 1. (of living organisms) Unable to produce offspring. *See also* HYBRID; INCOMPATIBILITY; SELF-STERILITY; STERILIZATION. 2. Free from contaminating microorganisms. *See* STERILIZATION.

**sterilization** 1. The process of destroying microorganisms that contaminate food, wounds, surgical instruments, etc. Common methods of sterilization include heat treatment (*see* AUTOCLAVE; PASTEURIZATION) and the use of *disinfectants and *antiseptics. 2. The operation of making an animal or human incapable of producing offspring. Men are usually sterilized by tying and then cutting the *vas deferens (**vasectomy**); in women the operation often involves permanently blocking the fallopian tubes by means of clips (**tubal occlusion**). *See also* BIRTH CONTROL.

**sternum (breastbone)** 1. A shield-shaped or rod-shaped bone in terrestrial vertebrates, on the ventral side of the thorax, that articulates with the *clavicle (collar bone) of the pectoral girdle and with most of the ribs. It is absent in fish, and in birds it bears a *keel. 2. The ventral portion of each segment of the exoskeleton of arthropods.

**steroid** Any of a group of lipids derived from a saturated compound called cyclopentanoperhydrophenanthrene, which has a nucleus of four rings (see formulae).

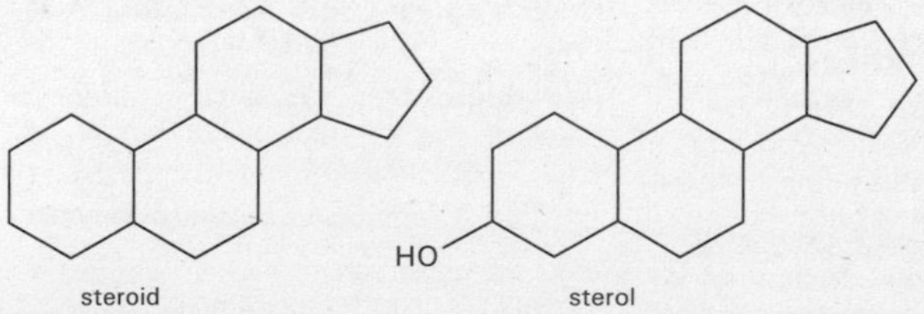

steroid                                sterol

**Steroid.**

Some of the most important steroid derivatives are the steroid alcohols, or *sterols. Other steroids include the *bile acids, which aid digestion of fats in the intestine; the sex hormones (*androgens and *oestrogens); and the *corticosteroid hormones, produced by the adrenal cortex. *Vitamin D is also based on the steroid structure.

**SEE WEB LINKS**
• Information about IUPAC nomenclature

**sterol** Any of a group of *steroid-based alcohols having a hydrocarbon side-chain of 8–10 carbon atoms. Sterols exist either as free sterols or as esters of fatty acids. Animal sterols (**zoosterols**) include *cholesterol and lanosterol. The major plant sterol (**phytosterol**) is beta-sitosterol, while fungal sterols (**mycosterols**) include *ergosterol.

**stibnite** A mineral form of antimony(III) sulphide, $Sb_2S_3$, the chief ore of antimony. It is a steel-grey crystalline solid, often containing extractable amounts of lead, mercury, and silver in addition to the antimony.

**sticky end** A single unpaired strand of nucleotides protruding from the end of a double-stranded DNA molecule. It is able to join with a complementary single strand, e.g. the sticky end of another DNA molecule, thus forming a single large double-stranded molecule. Sticky ends provide a means of annealing segments of DNA in genetic engineering, e.g. in the packaging of *vectors.

**stigma** 1. The glandular sticky surface at the tip of a carpel of a flower, which receives the pollen. In insect-pollinated plants the stigmas are held within the flower, whereas in wind-pollinated species they hang outside it. 2. See EYESPOT.

**stilt root** See PROP ROOT.

**stimulated emission** See INDUCED EMISSION; LASER.

**stimulus** Any change in the external or internal environment of an organism that provokes a physiological or behavioural response in the organism. In an animal specific *receptors are sensitive to stimuli.

**stipe** 1. The stalk that forms the lower portion of the fruiting body of certain fungi, such as mushrooms, and supports the umbrella-shaped cap. 2. The stalk between the holdfast and blade (**lamina**) of certain brown algae, notably kelps.

**stipule** An outgrowth from the petiole or leaf base of certain plants. Those of the garden pea are leaflike photosynthetic organs. The stipules of the lime tree are scalelike and protect the winter buds, whereas those of the false acacia (*Robinia*) are modified as spines.

**Stirling engine** A heat engine consisting of a hot cylinder and a cold cylinder separated by a regenerator acting as a heat exchanger. The cylinders enclose oscillating pistons. Heat is applied externally to the hot cylinder, causing the working fluid within it to expand and drive the piston. The fluid is cooled in the regenerator before entering the cold cylinder, where it is compressed by the piston and driven back to be heated in the regenerator before entering the hot cylinder again. Stirling engines, which were invented by Robert Stirling (1790–1878) in 1816, are silent and efficient but costly to produce. They have found limited use; interest in them revived in the 1960s.

**STM** See SCANNING TUNNELLING MICROSCOPY.

**stochastic process** Any process in which there is a random variable.

**stock** See GRAFT.

**Stockholm Convention** See PERSISTENT ORGANIC POLLUTANT.

**stoichiometric** Describing chemical reactions in which the reactants combine in simple whole-number ratios.

**stoichiometric coefficient** See CHEMICAL EQUATION.

**stoichiometric compound** A compound in which atoms are combined in exact whole-number ratios. *Compare* NONSTOICHIOMETRIC COMPOUND.

**stoichiometric mixture** A mixture of substances that can react to give products with no excess reactant.

**stoichiometric sum** See CHEMICAL EQUATION.

**stoichiometry** The relative proportions in which elements form compounds or in which substances react.

**stokes** Symbol St. A c.g.s. unit of kinematic viscosity equal to the ratio of the viscosity of a fluid in poises to its density in grams per cubic centimetre. 1 stokes = $10^{-4}$ m$^2$ s$^{-1}$. It is named after Sir George Stokes.

**Stokes, Sir George Gabriel** (1819–1903) British physicist and mathematician, born in Ireland, who worked at Cambridge University all his life. He is best known for *Stokes' law, concerning the movement of objects in a fluid. The *stokes is named after him.

**Stokes' law** A law that predicts the frictional force $F$ on a spherical ball moving through a viscous medium. According to this law $F = 6\pi r \eta v$, where $r$ is the radius of the ball, $v$ is its velocity, and $\eta$ is the viscosity of the medium. The sphere accelerates until it reaches a steady terminal speed. For a falling ball, $F$ is equal to the gravitational force on the sphere, less any upthrust. The law was discovered by Sir George Stokes.

**Stokes theorem** A theorem that is the analogue of the *divergence theorem for the *curl of a vector. Stokes theorem states that if a surface $S$, which is smooth and simply connected (i.e. any closed curve on the surface can be contracted continuously into a point without leaving the surface), is bounded by a line $L$ the vector $F$ defined in $S$ satisfies

$$\int_S \text{curl} F \cdot dS = \int_L F \cdot dl$$

where $l$ is distance. Stokes theorem was stated by Sir George Stokes as a Cambridge examination question, having been raised by Lord Kelvin in a letter to Stokes in 1850.

**stolon** A long aerial side stem that gives rise to a new daughter plant when the bud at its apex touches the soil. Plants that multiply in this way include blackberry and currant bushes. Gardeners often pin down stolons to the soil to aid the propagation of such plants. This process is termed **layering**.

**stoma** (*pl.* **stomata**) A pore, large numbers of which are present in the epidermis of leaves (especially on the undersurface) and young shoots. Stomata function in gas exchange between the plant and the atmosphere. Each stoma is bordered by two semicircular **guard cells** (specialized epidermal cells), whose movements (due to changes in water content) control the size of the aperture. The term stoma is also used to mean both the pore and its associated guard cells.

**stomach** The portion of the vertebrate *alimentary canal between the oesophagus and the small intestine. It is a muscular organ, capable of dramatic changes in size and shape, in which ingested food is stored and undergoes preliminary digestion. Cells lining the stomach produce *gastric juice, which is thoroughly mixed with the food by muscular contractions of the stomach. The resultant acidic partly digested food mass (*chyme) is discharged into the *duodenum through the pyloric *sphincter for final digestion and absorption. Some herbivorous animals (the Ruminantia) have multichambered stomachs from which food is regurgitated, rechewed, and swallowed again.

**stomium** A region of thin-walled cells in certain spore-producing structures that ruptures to release the spores. For example, in the sporangium of the fern *Dryopteris* the stomium ruptures when the annulus dries out.

**stop** A circular aperture that limits the effective size of a lens in an optical system. It may be adjustable, as the iris diaphragm in a camera, or have a fixed diameter, as the disk used in some telescopes.

**stop codon** The triplet of nucleotides on a messenger *RNA molecule (*see* CODON) at which the process of *translation ends. It is recognized by proteins called release factors, which attach to the binding site for an aminoacyl tRNA molecule. This effectively stops the formation of a polypeptide chain at that point. The three stop codons are UGA, UAA, and UAG (*see* GENETIC CODE). *Compare* START CODON.

**stopping power** A measure of the ability of matter to reduce the kinetic energy of a particle passing through it. The **linear stopping power**, $-dE/dx$, is energy loss of a particle per unit distance. The **mass stopping power**, $(1/\rho)dE/dx$, is the linear stopping power divided by the density ($\rho$) of the substance. The **atomic stopping power**, $(1/n)dE/dx = (A/\rho N)dE/dx$, is the energy loss per atom per unit area perpendicular to the particle's motion, i.e. $n$ is the number of atoms in unit volume of the substance, $N$ is the Avogadro number, and $A$ is the relative atomic mass of the substance. The relative stopping power is the ratio of the stopping power of a substance to that of a standard substance, usually aluminium, oxygen, or air.

**storage compound** *See* FOOD RESERVES.

**storage ring** A large evacuated toroidal ring forming a part of some particle accelerators. The rings are designed like *synchrotrons, except that they do not accelerate the particles circling within them but supply just sufficient energy to make up for losses (mainly *synchrotron radiation). The storage rings are usually built tangentially to the associated accelerator so that particles can be transferred accurately between them. At *CERN in Geneva, two interlaced storage rings are used, containing protons rotating in opposite directions. At the intersections very high collision energies (up to 1700 GeV) can be achieved.

**s.t.p.** Standard temperature and pressure, formerly known as **N.T.P.** (normal temperature and pressure). The standard conditions used as a basis for calculations involving quantities that vary with temperature and pressure. These conditions are used when comparing the properties of gases. They are 273.15 K (or 0°C) and 101 325 Pa (or 760.0 mmHg).

**straight chain** See CHAIN.

**strain** A measure of the extent to which a body is deformed when it is subjected to a *stress. The **linear strain** or **tensile strain** is the ratio of the change in length to the original length. The **bulk strain** or **volume strain** is the ratio of the change in volume to the original volume. The **shear strain** is the angular distortion in radians of a body subjected to a *shearing force. See also ELASTICITY; ELASTIC MODULUS.

**strain gauge** A device used to measure a small mechanical deformation in a body (see STRAIN). The most widely used devices are metal wires or foils or semiconductor materials, such as a single silicon crystal, which are attached to structural members; when the members are stretched under tensile *stress the resistance of the metal or semiconductor element increases. By making the resistance a component in a *Wheatstone-bridge circuit an estimate of the strain can be made by noting the change in resistance. Other types of strain gauge rely on changes of capacitance or the magnetic induction between two coils, one of which is attached to the stressed member.

**strain hardening (work hardening)** An increase in the resistance to the further plastic deformation of a body as a result of a rearrangement of its internal structure when it is strained, particularly by repeated stress. See also ELASTICITY.

**strange attractor** See ATTRACTOR.

**strange matter** Matter composed of up, down, and strange quarks (rather than the up and down quarks found in normal nucleons). It has been suggested that strange matter may have been formed in the *early universe, and that pieces of this matter (called **S-drops**) may still exist.

**strangeness** Symbol *s*. A property of certain elementary particles called hadrons (K-mesons and hyperons) that decay more slowly than would have been expected from the large amount of energy released in the process. These particles were assigned the quantum number *s* to account for this behaviour. For nucleons and other nonstrange particles $s = 0$; for strange particles *s* does not equal zero but has an integral value. In quark theory (see ELEMENTARY PARTICLES) hadrons with the property of strangeness contain a strange quark or its antiquark.

**stratification** 1. The arrangement of the components of an entity in layers (**strata**). Stratification is a feature of sedimentary rocks and *soils. Thermal stratification can occur in some lakes (see THERMOCLINE). 2. The practice of placing certain seeds between layers of peat or sand and then exposing them to low temperatures for a period, which is required before they will germinate. See VERNALIZATION.

**stratigraphy** The branch of geology concerned with the origin, composition, sequence, and correlation of rock strata. It forms the basis of historical geology and has also found practical application in mineral exploration, especially that of petroleum.

**stratosphere** See EARTH'S ATMOSPHERE.

**stratum corneum** The layer of dead keratinized cells that forms the outermost layer of mammalian *epidermis. It provides a water-resistant barrier between the external environment and the living cells of the *skin.

**streamline flow** A type of fluid flow in which no *turbulence occurs and the particles of the fluid follow continuous paths, either at constant velocity or at a velocity that alters in a predictable and regular way (see also LAMINAR FLOW).

**Streptococcus** A genus of spherical Gram-positive bacteria occurring widely in nature, typically as chains or pairs of cells. Many are saprotrophic and exist as usually harmless commensals inhabiting the skin, mucous membranes, and intestine of humans and animals. Others are parasites, some of which cause diseases, including scarlet fever (S.

*pyogenes*, group A streptococci), endocarditis (*S. viridans*), and pneumonia and meningitis (*S. pneumoniae*).

**streptomycin** See ACTINOBACTERIA; ANTIBIOTICS.

**stress** The force per unit area on a body that tends to cause it to deform (*see* STRAIN). It is a measure of the internal forces in a body between particles of the material of which it consists as they resist separation, compression, or sliding in response to externally applied forces. **Tensile stress** and **compressive stress** are axial forces per unit area applied to a body that tend either to extend it or compress it linearly. **Shear stress** is a tangential force per unit area that tends to shear a body. *See also* ELASTICITY; ELASTIC MODULUS.

**stretch reflex (myotatic reflex)** The *reflex initiated when a muscle is stretched; an example is the **knee-jerk reflex**. Stretching of a muscle causes impulses to be generated in the *muscle spindles. These impulses are transmitted by sensory neurons to the spinal cord, where the sensory neurons synapse with motor neurons; these initiate contraction of the same muscle so that it returns to its original length. Since the reflex action involves the transmission of impulses across only one set of synapses, the response is rapid and described as **monosynaptic**.

**striated muscle** See VOLUNTARY MUSCLE.

**stridulation** The production of sounds by insects rubbing one part of the body against another. The parts of the body involved vary from species to species. Stridulation is typical of the Orthoptera (grasshoppers, crickets, cicadas), in which the purpose of the sounds is usually to bring the sexes together, although they are also used in territorial behaviour, warning, etc.

**string** A one-dimensional object used in theories of elementary particles and in cosmology (**cosmic string**). **String theory** replaces the idea of a pointlike elementary particle (used in quantum field theory) by a line or loop (a closed string). States of a particle may be produced by standing waves along this string. The combination of string theory with supersymmetry leads to *superstring theory.

**string landscape** The very large number (about $10^{500}$) of possible states in string theory. It is postulated that the universe we inhabit, including the small but non-zero value of the *cosmological constant, is therefore one of $10^{500}$ possible universes. The string landscape is a very controversial idea, which is far from being universally accepted.

**stripping reaction** A nuclear or chemical reaction in which part of an incident particle combines with the target while the rest of the incident particle proceeds almost unchanged. A standard example of a stripping reaction is a (d, p) reaction in which a *deuteron is the incident particle and a nucleus is the target particle with the neutron from the deuteron combining with the target nucleus and the proton from the deuteron proceeding almost without disruption.

**strobilus** 1. A type of *composite fruit that is formed from a complete inflorescence. It produces *achenes enclosed in bracts and when mature becomes cone-shaped. The hop fruit is an example. 2. *See* CONE.

**stroboscope** A device for making a moving body intermittently visible in order to make it appear stationary. It may consist of a lamp flashing at regular intervals or a shutter that enables it to be seen intermittently. **Stroboscopic photography** is the taking of very short-exposure pictures of moving objects using an electronically controlled flash lamp.

**stroma** Tissue that forms the framework of an organ; for example, the tissue of the ovary that surrounds the reproductive cells, or the gel-like matrix of *chloroplasts that surrounds the grana.

**stromatolite** A rocky cushion-like mass formed by the unchecked growth of millions of lime-secreting cyanobacteria. Stromatolites are found only in areas where other organisms that would normally keep down the bacterial numbers cannot survive, such as extremely salty bays. Such bacteria were abundant during the *Proterozoic and *Archaean eons, from as early as 3500 million years ago. The white rings of fossilized microorganisms found in rocks of this age are the remains of stromatolites.

**strong acid** An *acid that is completely dissociated in aqueous solution.

**strong electrolyte** An *electrolyte that is completely dissociated into its component ions in aqueous solution (as opposed to a weak electrolyte, which is incompletely dissociated).

S

**strong interaction** *See* FUNDAMENTAL IN-
TERACTIONS.

**strontia** *See* STRONTIUM OXIDE.

**strontianite** A mineral form of *strontium
carbonate, $SrCO_3$.

**strontium** Symbol Sr. A soft yellowish
metallic element belonging to group 2 (formerly IIA) of the periodic table (*see* ALKALINE-
EARTH METALS); a.n. 38; r.a.m. 87.62; r.d. 2.6;
m.p. 769°C; b.p. 1384°C. The element is
found in the minerals strontianite ($SrCO_3$)
and celestine ($SrSO_4$). It can be obtained by
roasting the ore to give the oxide, followed
by reduction with aluminium (i.e. the *Goldschmidt process). The element, which is
highly reactive, is used in certain alloys and
as a vacuum getter. The isotope strontium–90 is present in radioactive fallout
(half-life 28 years), and can be metabolized
with calcium so that it collects in bone.
Strontium was discovered by Martin
Klaproth (1743–1817) and Thomas Hope
(1766–1844) in 1798 and isolated by
Humphry Davy in 1808.

**strontium bicarbonate** *See* STRONTIUM
HYDROGENCARBONATE.

**strontium carbonate** A white solid,
$SrCO_3$; orthorhombic; r.d. 3.7; decomposes
at 1340°C. It occurs naturally as the mineral
**strontianite** and is prepared industrially by
boiling celestine (strontium sulphate) with
ammonium carbonate. It can also be prepared by passing carbon dioxide over strontium oxide or hydroxide or by passing the
gas through a solution of strontium salt. It is
a phosphor, used to coat the glass of cathode-ray screens, and is also used in the refining of sugar, as a slagging agent in certain
metal furnaces, and to provide a red flame in
fireworks.

**strontium chloride** A white compound,
$SrCl_2$. The anhydrous salt (cubic; r.d. 3.05;
m.p. 872°C; b.p. 1250°C) can be prepared by
passing chlorine over heated strontium. It is
deliquescent and readily forms the hexahydrate, $SrCl_2.6H_2O$ (r.d. 2.67). This can be
made by neutralizing hydrochloric acid with
strontium carbonate, oxide, or hydroxide.
Strontium chloride is used for military flares.

**strontium hydrogencarbonate (stron-
tium bicarbonate)** A compound, $Sr(HCO_3)_2$,
which is stable only in solution. It is formed

by the action of carbon dioxide on a suspension of strontium carbonate in water. On
heating, this process is reversed.

**strontium oxide (strontia)** A white compound, SrO; r.d. 4.7; m.p. 2430°C; b.p. 3000°C.
It can be prepared by the decomposition of
heated strontium carbonate, hydroxide, or
nitrate, and is used in the manufacture of
other strontium salts, in pigments, soaps
and greases, and as a drying agent.

**strontium sulphate** A white solid, $SrSO_4$;
r.d. 3.96; m.p. 1605°C. It can be made by dissolving strontium oxide, hydroxide, or carbonate in sulphuric acid. It is used as a
pigment in paints and ceramic glazes and to
provide a red colour in fireworks.

**structural formula** *See* FORMULA.

**structural gene** *See* OPERON.

**structural isomerism** *See* ISOMERISM.

**structure formation** The process by
which *large-scale structure, such as galaxies, forms in the universe. It is thought that
quantum fluctuations in the early universe
and the *Jeans instability in an expanding
universe are the key factors in understanding
structure formation. A complete quantitative
theory of structure formation does not exist
at present.

**Strukturbericht** A system used to describe crystal structure. It consists of a capital letter followed by a number. The capital
letter indicates the atomic nature of the material. A stands for monoatomic, B for diatomic with equal numbers of atoms of each
type, C for diatomic with a ratio of 2:1 for
atoms of each type, D is as C but with a ratio
of 3:1, E ... K are more complicated types of
compounds, L are alloys, O are organic compounds, and S are silicates. The number
gives the order of the discovery of that particular structure. The term is German for
'structure report'.

**strychnine** A colourless poisonous crystalline alkaloid found in certain plants.

**style** The stalk of a carpel, between the
stigma and the ovary. In many plants it is
elongated to aid pollination.

**styrene** *See* PHENYLETHENE.

**subarachnoid space** The space between
the *arachnoid membrane and the *pia
mater, two of the membranes (*meninges)

that surround the brain and spinal cord. It is filled with *cerebrospinal fluid.

**subatomic particle**  *See* ELEMENTARY PARTICLES.

**subclavian artery**  A paired artery that passes beneath the collar bone (clavicle) and branches to supply blood to the arm. The left subclavian artery arises from the aorta; the right from the innominate artery. ·

**subcritical**  *See* CRITICAL MASS; CRITICAL REACTION; MULTIPLICATION FACTOR.

**subcutaneous tissue**  The tissue that lies immediately beneath the *dermis (*see* SKIN). It is made up of loose fibrous *connective tissue, muscle, and fat (*see* ADIPOSE TISSUE), which in some animals (e.g. whales and hibernating mammals) forms an insulating layer or an important food store.

**suberin**  A mixture of waxy substances, similar to *cutin, present in the thickened cell walls of many trees and shrubs, particularly in corky tissues. The deposition of suberin (**suberization**) provides a protective water-impermeable layer.

**sublimate**  A solid formed by sublimation.

**sublimation**  A direct change of state from solid to gas.

**sublittoral**  **1.** Designating or occurring in the shallow-water zone of a sea, over the continental shelf and below the low tide mark.  **2.** Designating or occurring in the zone of a lake below the littoral zone, to a depth of 6–10 metres.

**submillimetre astronomy**  A combination of radio and infrared astronomy techniques that focus on collecting radiation in the wavelength range 0.3–1.0 mm. This is the part of the spectrum in which a large number of molecular emission lines are present. *Giant molecular clouds are a primary source of such emissions.

**subnet mask**  In computing, a specification of which part of an IP address (*see* TCP/IP) represents the network identity. It is a 32-bit bitmap where '1' means that bit position is part of the network identity and '0' that it is part of the host address. This bitmap always take the form of a single block of 1's (the network identity part) followed by a single block of 0's (the host address part). It is important that routers can distinguish efficiently between those packets originating on its network whose destination is on the same

network and those that that require routing over an inter-network. This operation requires isolating the network identity part of the IPA address, and an AND operation with the subnet mask is a very efficient way to do this. For convenience, subnet masks are usually represented either as the decimal form of four 8-bit numbers or as an integer representing the network identity. Thus, 255.255.255.0 and /24 both indicate a 24-bit network identity and an 8-bit host address.

**subshell**  *See* ATOM.

**subsoil**  *See* SOIL.

**subsonic speed**  A speed that is less than *Mach 1.

**subspecies**  A group of individuals within a *species that breed more freely among themselves than with other members of the species and resemble each other in more characteristics. Reproductive isolation of a subspecies may become so extreme that a new species is formed (*see* SPECIATION). Subspecies are sometimes given a third Latin name, e.g. the mountain gorilla, *Gorilla gorilla beringei* (*see also* BINOMIAL NOMENCLATURE).

**substantive dye**  *See* DYES.

**substantivity**  The affinity of a dye for its substrate.

**substituent**  **1.** An atom or group that replaces another in a substitution reaction.  **2.** An atom or group regarded as having replaced a hydrogen atom in a chemical derivative. For example, dibromobenzene ($C_6H_4Br_2$) is a derivative of benzene with bromine substituents.

**substitute natural gas**  *See* SNG.

**substitution**  (in genetics) A *point mutation in which one base pair in the DNA sequence is replaced by another. Substitutions may or may not cause the incorporation of an incorrect amino acid in a protein chain; when an incorrect amino acid is inserted, it may or may not affect the functioning of that protein. Sickle-cell anaemia is an example of a substitution mutation in which thymine is replaced by adenine in the triplet coding for the sixth amino acid in the β-chain of haemoglobin.

**substitution reaction** (**displacement reaction**)  A reaction in which one atom or molecule is replaced by another atom or

S

molecule. *See* ELECTROPHILIC SUBSTITUTION; NUCLEOPHILIC SUBSTITUTION.

**substrate 1.** The substance that is affected by the action of a catalyst; for example, the substance upon which an *enzyme acts in a biochemical reaction. **2.** The substance on which some other substance is adsorbed or in which it is absorbed. Examples include the material to which a dye is attached, the porous solid absorbing a gas, and the *matrix trapping isolated atoms, radicals, etc. **3.** (in biology) The material on which a sedentary organism (such as a barnacle or a plant) lives or grows. The substrate may provide nutrients for the organism or it may simply act as a support.

**subtractive process** *See* COLOUR.

**succession** (in ecology) The sequence of communities that develops in an area from the initial stages of colonization until a stable mature **climax community** is achieved. Many factors, including climate and changes brought about by the colonizing organisms, influence the nature of a succession; for example, after many years shrubs produce soil deep enough to support trees, which then shade out the shrubs. *See also* SERE.

**succinate** A salt of succinic acid (*butanedioic acid), a four-carbon fatty acid. Succinate occurs in living organisms as an intermediate in metabolism, especially in the *Krebs cycle.

**succulent** A plant that conserves water by storing it in fleshy leaves or stems. Succulents are found either in dry regions or in areas where there is sufficient water but it is not easily obtained, as in salt marshes. Such plants are often modified to reduce water loss by transpiration. For example, the leaves of cacti are reduced to spines.

**succus entericus** *See* INTESTINAL JUICE.

**sucker (turion)** A shoot that arises from an underground root or stem and grows at the expense of the parent plant. Suckers can be dug up with a portion of root attached and used to propagate a plant. If, however, a plant is grafted onto a different rootstock, as many roses are, any suckers will be of the wild rootstock, rather than the ornamental scion, and must be removed.

**sucrase** A carbohydrate-digesting enzyme, produced in the brush border of the small intestine, that breaks down the disaccharide

sucrose into the monosaccharides glucose and fructose.

**sucrose (cane sugar; beet sugar; saccharose)** A sugar comprising one molecule of glucose linked to a fructose molecule. It occurs widely in plants and is particularly abundant in sugar cane and sugar beet (15–20%), from which it is extracted and refined for table sugar. If heated to 200°C, sucrose becomes caramel.

**sugar (saccharide)** Any of a group of water-soluble *carbohydrates of relatively low molecular weight and typically having a sweet taste. The simple sugars are called *monosaccharides. More complex sugars comprise between two and ten monosaccharides linked together: *disaccharides contain two, trisaccharides three, and so on. The name is often used to refer specifically to *sucrose (cane or beet sugar).

(((●))) **SEE WEB LINKS**
• Information about IUPAC nomenclature

**sugar of lead** *See* LEAD(II) ETHANOATE.

**sulpha drugs** *See* SULPHONAMIDES.

**sulphamic acid** A colourless crystalline solid, $NH_2SO_2OH$, which is extremely soluble in water and normally exists as the *zwitterion $H_3N^+.SO_3^-$. It is a strong acid, readily forming sulphamate salts. It is used in electroplating, hard-water scale removers, herbicides, and artificial sweeteners.

**sulphanes** Compounds of hydrogen and sulphur containing chains of sulphur atoms. They have the general formula $H_2S_n$. The simplest is hydrogen sulphide, $H_2S$; other members of the series are $H_2S_2$, $H_2S_3$, $H_2S_4$, etc. *See* SULPHIDES.

(((●))) **SEE WEB LINKS**
• Information about IUPAC nomenclature

**sulphanilic acid (4-aminobenzene sulphonic acid)** A colourless crystalline solid, $H_2NC_6H_4SO_2OH$, made by prolonged heating of *phenylamine (aniline) sulphate. It readily forms *diazo compounds and is used to make dyes and sulpha drugs.

**sulphate** A salt or ester of sulphuric(VI) acid. Organic sulphates have the formula $R_2SO_4$, where R is an organic group. Sulphate salts contain the ion $SO_4^{2-}$.

**sulphides 1.** Inorganic compounds of sulphur with more electropositive elements. Compounds of sulphur with nonmetals are covalent compounds, e.g. hydrogen sulphide ($H_2S$). Metals form ionic sulphides containing the $S^{2-}$ ion; these are salts of hydrogen sulphide. **Polysulphides** can also be produced containing the polymeric ion $S_x^{2-}$.

**2. (thio ethers)** Organic compounds that contain the group –S– linked to two hydrocarbon groups. Organic sulphides are named from the linking groups, e.g. dimethyl sulphide ($CH_3SCH_3$), ethyl methyl sulphide ($C_2H_5SCH_3$). They are analogues of ethers in which the oxygen is replaced by sulphur (hence the alternative name) but are generally more reactive than ethers. Thus they react with halogen compounds to form *sulphonium compounds and can be oxidized to *sulphoxides.

**sulphinate (dithionite; hyposulphite)** A salt that contains the negative ion $S_2O_4^{2-}$, usually formed by the reduction of sulphites with excess $SO_2$. Solutions are not very stable and decompose to give thiosulphate and hydrogensulphite ions. The structure is $^-O_2S-SO_2^-$.

**sulphinic acid (dithionous acid; hyposulphurous acid)** An unstable acid, $H_2S_2O_4$, known in the form of its salts (sulphinates). *See also* SULPHURIC ACID.

**sulphite** A salt or ester derived from sulphurous acid. The salts contain the trioxosulphate(IV) ion $SO_3^{2-}$. The sulphites generally have reducing properties.

**sulphonamides** Organic compounds containing the group –$SO_2.NH_2$. The sulphonamides are amides of sulphonic acids. Many have antibacterial action and are also known as **sulpha drugs**, including sulphadiazine, $NH_2C_6H_4SO_2NHC_4H_3N_2$, and several others. They act by preventing bacteria from reproducing and are used to treat a variety of bacterial infections, especially of the gut and urinary system.

**sulphonate** A salt or ester of a sulphonic acid.

**sulphonation** A type of chemical reaction in which a –$SO_3H$ group is substituted on a benzene ring to form a *sulphonic acid. The reaction is carried out by refluxing with concentrated sulphuric(VI) acid for a long period. It can also occur with cold disulphuric(VI) acid ($H_2S_2O_7$). Sulphonation is an example of electrophilic substitution in which the electrophile is a sulphur trioxide molecule, $SO_3$.

**sulphonic acids** Organic compounds containing the –$SO_2.OH$ group. Sulphonic acids are formed by reaction of aromatic hydrocarbons with concentrated sulphuric acid. They are strong acids, ionizing completely in solution to form the sulphonate ion, –$SO_2.O^-$.

**((⊕)) SEE WEB LINKS**
• Information about IUPAC nomenclature

**sulphonium compounds** Compounds containing the ion $R_3S^+$ (sulphonium ion), where R is any organic group. Sulphonium compounds can be formed by reaction of organic sulphides with halogen compounds. For example, diethyl sulphide, $C_2H_5SC_2H_5$, reacts with chloromethane, $CH_3Cl$, to give diethylmethylsulphonium chloride, $(C_2H_5)_2.CH_3.S^+Cl^-$.

R — S — R          sulphide (thio ether)

R\$\overset{+}{\underset{|\ R}{S}}$/R          sulphonium ion

R — S — H          thiol (mercaptan)

R\$\underset{R}{S}$= O          sulphoxide

$\overset{O}{\underset{O}{\overset{\|}{R-S-OH}}}$          sulphonic acid

$\overset{O}{\underset{O}{\overset{\|}{R-S-O^-}}}$          sulphonate ion

$\overset{O}{\underset{O}{\overset{\|}{R-S-NH_2}}}$          sulphonamide

**Sulphonium compounds.** Examples of organic sulphur compounds.

**sulphoxides** Organic compounds containing the group =S=O (**sulphoxide group**) linked to two other groups, e.g. dimethyl sulphoxide, $(CH_3)_2SO$.

**((⊕)) SEE WEB LINKS**
• Information about IUPAC nomenclature

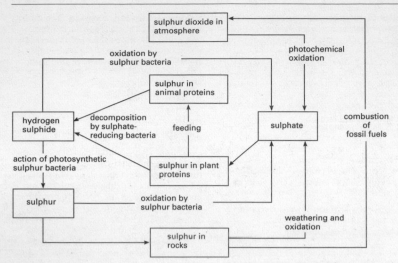

**Sulphur cycle.**

**sulphur**  Symbol S. A yellow nonmetallic element belonging to *group 16 (formerly VIB) of the periodic table; a.n. 16; r.a.m. 32.06; r.d. 2.07 (rhombic); m.p. 112.8°C; b.p. 444.674°C. The element occurs in many sulphide and sulphate minerals and native sulphur is also found in Sicily and the USA (obtained by the *Frasch process). It can also be obtained from hydrogen sulphide by the *Claus process.

Sulphur has various allotropic forms. Below 95.6°C the stable crystal form is rhombic; above this temperature the element transforms into a triclinic form. These crystalline forms both contain cyclic $S_8$ molecules. At temperatures just above its melting point, molten sulphur is a yellow liquid containing $S_8$ rings (as in the solid form). At about 160°C, the sulphur atoms form chains and the liquid becomes more viscous and dark brown. If the molten sulphur is cooled quickly from this temperature (e.g. by pouring into cold water) a reddish-brown solid known as **plastic sulphur** is obtained. Above 200°C the viscosity decreases. Sulphur vapour contains a mixture of $S_2$, $S_4$, $S_6$, and $S_8$ molecules. **Flowers of sulphur** is a yellow powder obtained by subliming the vapour. It is used as a plant fungicide. The element is also used to produce sulphuric acid and other sulphur compounds.

Sulphur is an *essential element in living organisms, occurring in the amino acids cysteine and methionine and therefore in many proteins. It is also a constituent of various cell metabolites, e.g. coenzyme A. Sulphur is absorbed by plants from the soil as the sulphate ion ($SO_4^{2-}$). *See* SULPHUR CYCLE.

(((●))) SEE WEB LINKS
• Information from the WebElements site

**sulphur bridge**  *See* DISULPHIDE BRIDGE.

**sulphur cycle**  The cycling of sulphur between the biotic (living) and abiotic (nonliving) components of the environment (*see* BIOGEOCHEMICAL CYCLE). Most of the sulphur in the abiotic environment is found in rocks, although a small amount is present in the atmosphere as sulphur dioxide ($SO_2$), produced by combustion of fossil fuels. Sulphate ($SO_4^{2-}$), derived from the weathering and oxidation of rocks, is taken up by plants and incorporated into sulphur-containing proteins. In this form sulphur is passed along food chains to animals. Decomposition of dead organic matter and faeces by anaerobic sulphate-reducing bacteria returns sulphur to the abiotic environment in the form of hydrogen sulphide ($H_2S$). Hydrogen sulphide can be converted back to sulphate or to elemental sulphur by the action of photosynthetic and sulphide-oxidizing bacteria. Elemental sulphur becomes incorporated into rocks.

**sulphur dichloride** *See* DISULPHUR DICHLORIDE.

**sulphur dichloride dioxide (sulphuryl chloride)** A colourless liquid, $SO_2Cl_2$; r.d. 1.67; m.p. $-54.1°C$; b.p. $69°C$. It decomposes in water but is soluble in benzene. The compound is formed by the action of chlorine on sulphur dioxide in the presence of an iron(III) chloride catalyst or sunlight. It is used as a chlorinating agent and a source of the related fluoride, $SO_2F_2$.

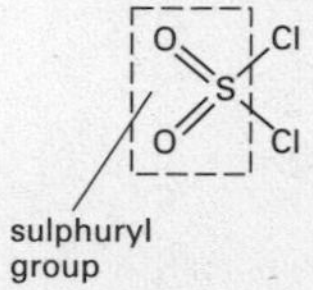

sulphuryl
group

**Sulphur dichloride dioxide.**

**sulphur dichloride oxide (thionyl chloride)** A colourless fuming liquid, $SOCl_2$; m.p. $-105°C$; b.p. $78.8°C$. It hydrolyses rapidly in water but is soluble in benzene. It may be prepared by the direct action of sulphur on chlorine monoxide or, more commonly, by the reaction of phosphorus(V) chloride with sulphur dioxide. It is used as a chlorinating agent in synthetic organic chemistry (replacing $-OH$ groups with Cl).

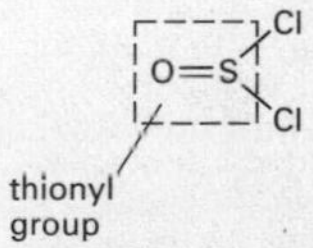

thionyl
group

**Sulphur dichloride oxide.**

**sulphur dioxide (sulphur(IV) oxide)** A colourless liquid or pungent gas, $SO_2$, formed by sulphur burning in air; r.d. 1.43 (liquid); m.p. $-72.7°C$; b.p. $-10°C$. It can be made by heating iron sulphide (pyrites) in air. The compound is a reducing agent and is used in bleaching and as a fumigant and food preservative. Large quantities are also used in the *contact process for manufacturing sulphuric acid. It dissolves in water to give a mixture of sulphuric and sulphurous acids. *See also* ACID RAIN.

**sulphuretted hydrogen** *See* HYDROGEN SULPHIDE.

**sulphuric acid (oil of vitriol)** A colourless oily liquid, $H_2SO_4$; r.d. 1.84; m.p. $10.36°C$; b.p. $338°C$. The pure acid is rarely used; it is commonly available as a 96–98% solution (m.p. $3.0°C$). The compound also forms a range of hydrates:
$H_2SO_4.H_2O$ (m.p. $8.62°C$);
$H_2SO_4.2H_2O$ (m.p. $-38/39°C$);
$H_2SO_4.6H_2O$ (m.p. $-54°C$);
$H_2SO_4.8H_2O$ (m.p. $-62°C$).
Its full systematic name is **tetraoxosulphuric(VI) acid**.

Until the 1930s, sulphuric acid was manufactured by the *lead-chamber process, but this has now been replaced by the *contact process (catalytic oxidation of sulphur dioxide). It is extensively used in industry, the main applications being fertilizers (32%), chemicals (16%), paints and pigments (15%), detergents (11%), and fibres (9%).

In concentrated sulphuric acid there is extensive hydrogen bonding and several competing equilibria, to give species such as

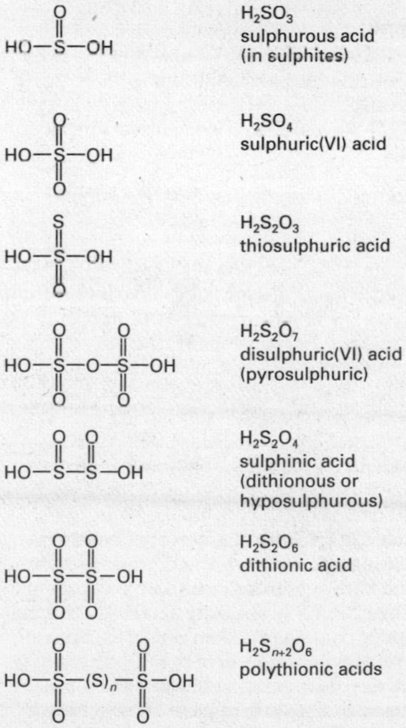

**Sulphuric acid.** Structures of some oxo acids of sulphur.

$H_3O^+$, $HSO_4^-$, $H_3SO_4^+$, and $H_2S_2O_7$. Apart from being a powerful protonating agent (it protonates chlorides and nitrates producing hydrogen chloride and nitric acid), the compound is a moderately strong oxidizing agent. Thus, it will dissolve copper:

$$Cu(s) + H_2SO_4(l) \rightarrow CuO(s) + H_2O(l) + SO_2(g)$$

$$CuO(s) + H_2SO_4(l) \rightarrow CuSO_4(aq) + H_2O(l)$$

It is also a powerful dehydrating agent, capable of removing $H_2O$ from many organic compounds (as in the production of acid *anhydrides). In dilute solution it is a strong dibasic acid forming two series of salts, the sulphates and the hydrogensulphates.

**sulphuric(IV) acid** *See* SULPHUROUS ACID.

**sulphur monochloride** *See* DISULPHUR DICHLORIDE.

**sulphur mustard** A chemical warfare agent, $C_4H_8Cl_2S$; r.d. 1.27; m.p. 14.4°C; b.p. 217°C. It is a potent blistering agent, first used in 1915 by Germany in World War I. It is often known simply as **mustard gas**, although it is an oily liquid, which can be dispersed as an aerosol. It was one of the early chemicals used in chemotherapy treatment of cancer. The systematic name is bis(2-chlorethyl) sulphide. *See also* NITROGEN MUSTARD.

**sulphurous acid (sulphuric(IV) acid)** A weak dibasic acid, $H_2SO_3$, known in the form of its salts: the sulphites and hydrogensulphites. It is considered to be formed (along with sulphuric acid) when sulphur dioxide is dissolved in water. It is probable, however, that the molecule $H_2SO_3$ is not present and that the solution contains hydrated $SO_2$. It is a reducing agent. The systematic name is **trioxosulphuric(IV) acid**. *See also* SULPHURIC ACID.

**sulphur(IV) oxide** *See* SULPHUR DIOXIDE.

**sulphur(VI) oxide** *See* SULPHUR TRIOXIDE.

**sulphur trioxide (sulphur(VI) oxide)** A colourless fuming solid, $SO_3$, which has three crystalline modifications. In decreasing order of stability these are: α, r.d. 1.97; m.p. 16.83°C; b.p. 44.8°C; β, m.p. 16.24°C; sublimes at 50°C; r.d. 2.29; γ, m.p. 16.8°C; b.p. 44.8°C. All are polymeric, with linked $SO_4$ tetrahedra: the γ-form has an icelike structure and is obtained by rapid quenching of the vapour; the β-form has infinite helical chains; and the α-form has infinite chains

with some cross-linking of the $SO_4$ tetrahedra. Even in the vapour, there are polymeric species, and not discrete sulphur trioxide molecules (hence the compound is more correctly called by its systematic name **sulphur(VI) oxide**).

Sulphur trioxide is prepared by the oxidation of sulphur dioxide with oxygen in the presence of a vanadium(V) oxide catalyst. It may be prepared in the laboratory by distilling a mixture of concentrated sulphuric acid and phosphorus(V) oxide. It reacts violently with water to give sulphuric(VI) acid and is an important intermediate in the preparation of sulphuric acid and oleum.

**sulphuryl chloride** *See* SULPHUR DICHLORIDE DIOXIDE.

**sulphuryl group** The group $=SO_2$, as in *sulphur dichloride oxide.

**sulphydryl group** *See* THIOLS.

**summation** 1. (in neurophysiology) The combined effect of the changes in electric potential elicited in one or more postsynaptic membranes by the transmission of impulses at *synapses that is sufficient to trigger an action potential in the postsynaptic neuron. Summation occurs when one or a few postsynaptic potentials alone are insufficient to elicit a response in the postsynaptic neuron; it may consist of the effect of two or more potentials evoked simultaneously at different synapses on the same neuron (**spatial summation**) or in rapid succession at the same synapse (**temporal summation**). 2. *See* SYNERGISM.

**sun** The *star at the centre of the *solar system of which the planet earth is a member. A typical main-sequence dwarf star (*see* HERTZSPRUNG–RUSSELL DIAGRAM; STELLAR EVOLUTION), the sun is some 149 600 000 km from earth. It has a diameter of about 1 392 000 km and a mass of $1.9 \times 10^{30}$ kg. Hydrogen and helium are the primary constituents (about 75% hydrogen, 25% helium), with less than 1% of heavier elements. In the central core, some 400 000 km in diameter, hydrogen is converted into helium by thermonuclear reactions, which generate vast quantities of energy. This energy is radiated into space and provides the earth with all the light and heat necessary to have created and maintained life on earth (*see* SOLAR CONSTANT). The surface of the sun, the *photosphere, forms the boundary between its opaque interior and its transparent atmosphere. It is here that *sunspots occur. Above the photosphere is the *chromosphere and above this the *corona, which extends tenuously into interplanetary space. *See also* SOLAR WIND.

**sunspot** A dark patch in the sun's *photosphere resulting from a localized fall in temperature to about 4000 K. Most spots have a central very dark umbra surrounded by a lighter penumbra. Sunspots tend to occur in clusters and to last about two weeks. The number of sunspots visible fluctuates over an an 11-year cycle – often called the **sunspot cycle** (*see* SOLAR CYCLE). The cause of sunspots is thought to be the presence of intense localized magnetic fields, which suppress the convection currents that bring hot gases to the photosphere.

**superatom** *See* BOSE–EINSTEIN CONDENSATION.

**supercluster** *See* GALAXY CLUSTER.

**supercomputer** An extremely high-power computer that has a large amount of main *memory and very fast processors, capable of several billion operations per second. Often the processors run in parallel (*see* PARALLEL PROCESSING). Examples include the Cray computers, which are used for weather forecasting and other applications that need rapid real-time processing of large amounts of data.

**superconductivity** The absence of measurable electrical resistance in certain substances at temperatures close to 0 K. First discovered in 1911 in mercury, superconductivity is now known to occur in some 26 metallic elements and many compounds and alloys. The temperature below which a substance becomes superconducting is called the **transition temperature** (or **critical temperature**). Compounds are now known that show superconductivity at liquid-nitrogen temperatures.

The theoretical explanation of the phenomenon was given by John Bardeen, Leon Cooper (1930–  ), and John Schrieffer (1931–  ) in 1957 and is known as the **BCS theory**. According to this theory an electron moving through an elastic crystal lattice creates a slight distortion of the lattice as a result of Coulomb forces between the positively charged lattice and the negatively charged electron. If this distortion persists for a finite time it can affect a second passing electron. In 1956 Cooper showed that the ef-

fect of this phenomenon is for the current to be carried in superconductors not by individual electrons but by bound pairs of electrons, the **Cooper pairs**. The BCS theory is based on a *wave function in which all the electrons are paired. Because the total momentum of a Cooper pair is unchanged by the interaction between one of its electrons and the lattice, the flow of electrons continues indefinitely.

Superconducting coils in which large currents can circulate indefinitely can be used to create powerful magnetic fields and are used for this purpose in some particle accelerators and in other devices.

Superconductivity can also occur by a slightly more complicated mechanism than BCS theory in *heavy-fermion systems. In 1986, Georg Bednorz (1950–  ) and Karl Müller (1927–  ) found an apparently completely different type of superconductivity. This is called **high-temperature superconductivity**, since the critical temperature is very much higher than for BCS superconductors; some high-temperature superconductors have critical temperatures greater than 100 K. A typical high-temperature superconductor is $YBa_2Cu_3O_{1-7}$.

At the present time a theory of high-temperature superconductivity has not yet been established.

🌐 **SEE WEB LINKS**

• Properties of superconducting elements at the NPL website

**supercooling 1.** The cooling of a liquid to below its freezing point without a change from the liquid to solid state taking place. In this metastable state the particles of the liquid lose energy but do not fall into place in the lattice of the solid crystal. If the liquid is seeded with a small crystal, crystallization usually takes place and the temperature returns to the freezing point. Crystallization can also be induced by the presence of particles of dust, by mechanical vibration, or by rough surfaces. This is a common occurrence in the atmosphere where water droplets frequently remain unfrozen at temperatures well below 0°C until disturbed, following which they rapidly freeze. The supercooled droplets, for example, rapidly freeze on passing aircraft forming 'icing', which can be a hazard. **2.** The analogous cooling of a vapour to make it supersaturated until a disturbance causes condensation to occur, as in the Wilson *cloud chamber.

**supercritical** *See* CRITICAL MASS; CRITICAL REACTION; MULTIPLICATION FACTOR.

**superexchange** *See* MAGNETISM.

**superficial expansivity** *See* EXPANSIVITY.

**superfluidity** The property of liquid helium at very low temperatures that enables it to flow without friction. Both helium isotopes possess this property, but $^4$He becomes superfluid at 2.172 K, whereas $^3$He does not become superfluid until a temperature of 0.00093 K is reached. There is a basic connection between superfluidity and *superconductivity, so that sometimes a superconductor is called a charged superfluid. The temperature at which superfluidity occurs is called the **lambda point**.

**supergiant** The largest and most luminous type of star. They are formed from the most massive stars and are therefore very rare. They lie above the giants on the *Hertzsprung–Russell diagram. *See also* STELLAR EVOLUTION.

**supergravity** A *unified-field theory for all the known fundamental interactions that involves *supersymmetry. Supergravity is most naturally formulated as a *Kaluza–Klein theory in eleven dimensions. The theory contains particles of spin 2, spin 3/2, spin 1, spin 1/2, and spin 0. Although supersymmetry means that the infinities in the calculations are less severe than in other attempts to construct a quantum theory of gravity, it is not clear whether perturbation theory in supergravity gives finite answers due to the great complexity of the calculations. It is thought by many physicists that to obtain a consistent quantum theory of gravity one has to abandon *quantum field theories, since they deal with point objects, and move to theories based on extended objects, such as *superstrings and *supermembranes, and therefore that supergravity is not a complete theory of the fundamental interactions. However, it may well be a key ingredient in such a theory since it is related to *superstring theory by duality.

**superheating** The heating of a liquid to above its normal boiling point by increasing the pressure.

**superheterodyne receiver** A widely used type of radio receiver in which the incoming radio-frequency signal is mixed with an internally generated signal from a local oscillator. The output of the mixer has a carrier frequency equal to the difference between the transmitted frequency and the locally generated frequency, still retains the transmitted modulation, and is called the **intermediate frequency** (IF). The IF signal is amplified and demodulated before being passed to the audio-frequency amplifier. This system enables the IF signal to be amplified with less distortion, greater gain, better selectivity, and easier elimination of noise than can be achieved by amplifying the radio-frequency signal.

**super high frequency (SHF)** A radio frequency in the range 3–30 gigahertz.

**superionic conductor** An ionic solid in which the electrical conductivity due to the motion of ions is similar to that of a molten salt, i.e. a much higher conductivity than is usually observed in ionic solids.

**superior** Describing a structure that is positioned above or higher than another structure in the body. For example, in flowering plants the ovary is described as superior when located above the other organs of the flower. *Compare* INFERIOR.

**superlattice** *See* SOLID SOLUTION.

**supermembrane theory** A unified theory of the *fundamental interactions involving *supersymmetry, in which the basic entities are two-dimensional extended objects (**supermembranes**). They are thought to have about the same length scale as *superstrings, i.e. $10^{-35}$ m. At the present time there is no experimental evidence for supermembranes.

**supernatant liquid** The clear liquid remaining when a precipitate has settled.

**supernova** The explosive death of a star in which the energy radiated by the star suddenly increases by a factor of $10^{10}$. It takes several weeks or months to fade and while it lasts it dominates the whole galaxy in which it is observed. It is estimated that there could be a supernova explosion in a galaxy as big as the Milky Way every 30 years, although only six have actually been observed in our Galaxy in the last 1000 years. A supernova explosion occurs when a star has burnt up all its available nuclear fuel and the core collapses catastrophically (*see* STELLAR EVOLUTION). Normal main-sequence stars like the sun can expect to end their lives as *white dwarfs but more massive stars explode catastrophically, producing in some cases a *neutron star, in others a *black hole. *Compare* NOVA.

**supernova remnant** The expanding gas shell left behind by a supernova, at the heart of which lies a fast-rotating *neutron star or *pulsar.

**superoxides** A group of inorganic compounds that contain the $O_2^-$ ion. They are formed in significant quantities only for sodium, potassium, rubidium, and caesium. They are very powerful oxidizing agents and react vigorously with water to give oxygen gas and $OH^-$ ions. The superoxide ion has an unpaired electron and is paramagnetic and coloured (orange).

**superphosphate** A commercial phosphate mixture consisting mainly of monocalcium phosphate. It is made by treating phosphate rock with sulphuric acid; the product contains 16–20% 'available' $P_2O_5$:

$$Ca_{10}(PO_4)_6F_2 + 7H_2SO_4 \rightarrow 3Ca(H_2PO_4)_2 + 7CaSO_4 + 2HF$$

Triple-superphosphate is made by using phosphoric(V) acid in place of sulphuric acid; the product contains 45–50% 'available' $P_2O_5$:

$$Ca_{10}(PO_4)_6F_2 + 14H_3PO_4 \rightarrow 10Ca(H_2PO_4)_2 + 2HF$$

**superplasticity** The ability of some metals and alloys to stretch uniformly by several thousand percent at high temperatures, unlike normal alloys, which fail after being stretched 100% or less. Since 1962, when this property was discovered in an alloy of zinc and aluminium (22%), many alloys and ceramics have been shown to possess this property. For superplasticity to occur, the metal grain must be small and rounded and the alloy must have a slow rate of deformation.

**Super Proton Synchrotron** *See* CERN.

**supersaturated solution** *See* SATURATED.

**supersaturation 1.** The state of the atmosphere in which the relative humidity is over 100%. This occurs in pure air where no condensation nuclei are available. Supersaturation is usually prevented in the atmosphere by the abundance of condensation nuclei (e.g. dust, sea salt, and smoke particles). **2.** The state of any vapour whose pressure exceeds that at which condensation usually occurs (at the prevailing temperature).

**supersonic** *See* MACH NUMBER.

**superstring theory** A unified theory of the *fundamental interactions involving supersymmetry, in which the basic objects are one-dimensional objects (**superstrings**). Superstrings are thought to have a length scale of about $10^{-35}$ m and, since very short distances are associated with very high energies, they should have energy scales of about $10^{19}$ GeV, which is far beyond the energy of any accelerator that can be envisaged.

Strings associated with bosons are only consistent as quantum theories in a 26-dimensional *space–time; those associated with fermions are only consistent as quantum theories in 10-dimensional space–time. It is thought that four macroscopic dimensions arise by a *Kaluza–Klein theory mechanism, with the remaining dimensions being 'curled up' to become very small, although other possibilities for the higher dimensions have been put forward.

One of the most attractive features of the theory of superstrings is that it leads to spin 2 particles, which are identified as *gravitons. Thus, a superstring theory automatically contains a quantum theory of the gravitational interaction. It is thought that superstrings are free of the infinities that cannot be removed by *renormalization, which plague attempts to construct a quantum field theory incorporating gravity. There is some evidence that superstring theory is free of infinities but not a complete proof yet.

**supersymmetry** A *symmetry that can be applied to elementary particles so as to include both bosons and fermions. In the simplest supersymmetry theories, every boson has a corresponding fermion partner and every fermion has a corresponding boson partner. The boson partners of existing fermions have names formed by adding 's' to the beginning of the name of the fermion, e.g. **selectron**, **squark**, and **slepton**. The fermion partners of existing bosons have names formed by replacing '-on' at the end of the boson's name by '-ino' or by adding '-ino', e.g. **gluino**, **photino**, **wino**, and **zino**.

It is thought that particles associated with supersymmetry may be one of the ingredients of dark matter (*see* MISSING MASS).

If supersymmetry is relevant to observed elementary particles then it must be a *broken symmetry, although there is no convincing evidence at present to show at what energy it would be broken. There is, in fact, no experimental evidence for the theory, although it is thought that the idea of strings with supersymmetry may be the best approach to unifying the four fundamental interactions (*see* SUPERSTRING THEORY).

**supervolcano** *See* VOLCANO.

**supination** Rotation of the lower forearm so that the hand faces forwards or upwards with the radius and ulna parallel. *Compare* PRONATION.

**supplementary units** *See* SI UNITS.

**suppressed-carrier transmission** *See* TRANSMITTER.

**suppressor grid** A wire grid in a pentode *thermionic valve placed between the *screen grid and the anode to prevent electrons produced by *secondary emission from the anode from reaching the screen grid.

**supramolecular chemistry** A field of chemical research concerned with the formation and properties of large assemblies of molecules held together by intramolecular forces (hydrogen bonds, van der Waals' forces, etc.). One feature of supramolecular chemistry is that of **self-assembly** (*see* SELF-ORGANIZATION), in which the structure forms spontaneously as a consequence of the nature of the molecules. The molecular units are sometimes known as **synthons**. Another aspect is the study of very large molecules able to be used in complex chemical reactions in a fashion similar to, for example, the actions of the naturally occurring haemoglobin and nucleic acid molecules. Typical examples are the *helicate and *texaphyrin molecules and *dendrimers. Such molecules have great potential in such areas as medicine, electronics, and optics. The field also includes **host–guest chemistry**, which is concerned with molecules specifically designed to accept other molecules. Examples include *crown ethers, *cryptands, and *calixarenes.

**suprarenal glands** *See* ADRENAL GLANDS.

**surd** A quantity that cannot be expressed as a *rational number. It consists of the root of an arithmetic member (e.g. $\sqrt{3}$), which cannot be exactly determined, or the sum or difference of such roots.

**surface tension** Symbol $\gamma$. The property of a liquid that makes it behave as if its surface is enclosed in an elastic skin. The property results from intermolecular forces: a molecule in the interior of a liquid experiences a force of attraction from other molecules equally from all sides, whereas a molecule at the surface is only attracted by molecules below it in the liquid. The surface tension is defined as the force acting over the surface per unit length of surface perpendicular to the force. It is measured in newtons per metre. It can equally be defined as the energy required to increase the surface area by one square metre, i.e. it can be measured in joules per metre squared (which is equivalent to $N\ m^{-1}$).

The surface tension of water is very strong, due to the intermolecular hydrogen bonding, and is responsible for the formation of drops, bubbles, and meniscuses, as well as the rise of water in a capillary tube (**capillarity**), the absorption of liquids by porous substances, and the ability of liquids to wet a surface. Capillarity is very important in plants as it is largely responsible for the transport of water, against gravity, within the plant.

**((⊕)) SEE WEB LINKS**

• Values of surface tension at the NPL website

**surfactant (surface active agent)** A substance, such as a *detergent, added to a liquid to increase its wetting properties by reducing its *surface tension.

**surveying** The practice of accurately measuring and recording the relative altitudes, angles, and distances of features on, above, or below the land surface from which maps and plans can be plotted. Surveying is necessary to locate and measure property lines; to lay out buildings, bridges, roads, dams, and other constructions; and to obtain topographic information for mapping and charting. A number of methods are used depending on the degree of precision that is required. The chief methods include triangulation, trilateration, levelling (*see* LEVEL), plane tabling, and traversing.

**susceptance** Symbol $B$. The reciprocal of the *reactance of a circuit and thus the imaginary part of its *admittance. It is measured in siemens.

**susceptibility** **1. (magnetic susceptibility)** Symbol $\chi_m$. The dimensionless quantity describing the contribution made by a substance when subjected to a magnetic field to the total magnetic flux density present. It is equal to $\mu_r - 1$, where $\mu_r$ is the relative *permeability of the material. Diamagnetic materials have a low negative susceptibility, paramagnetic materials have a low positive susceptibility, and ferromagnetic materials have a high positive value. **2. (electric susceptibility)** Symbol $\chi_e$. The dimensionless quantity referring to a *dielectric equal to $P/\varepsilon_0 E$, where $P$ is the electric polarization, $E$ is the electric intensity producing it, and $\varepsilon_0$ is the electric constant. The electric susceptibility is also equal to $\varepsilon_r - 1$, where $\varepsilon_r$ is the relative *permittivity of the dielectric.

**suspension** A mixture in which small solid or liquid particles are suspended in a liquid or gas.

**suture** The line marking the junction of two body structures. Examples are the immovable joints between the bones of the skull and, in plants, the seam along the edge of a pea or bean pod.

**swallowing** *See* DEGLUTITION.

**swash** A surge of turbulent seawater that rushes up the shore after a wave breaks; it runs back down the slope as a backwash. The swash can carry materials, such as driftwood, seashells, and seaweed, which are often left on the beach as a line marking the extent of high tide. On a falling tide the backwash may form a series of channels.

**sweat** The salty fluid secreted by the *sweat glands onto the surface of the skin. Excess body heat is used to evaporate sweat, thereby resulting in cooling of the skin surface. Small amounts of urea are excreted in sweat.

**sweat gland** A small gland in mammalian skin that secretes *sweat. The distribution of sweat glands on the body surface varies between species: they occur over most of the body surface in humans and higher primates but have a more restricted distribution in other mammals.

**swim bladder (air bladder)** An air-filled sac lying above the alimentary canal in bony fish that regulates the buoyancy of the animal. Air enters or leaves the bladder either via a pneumatic duct opening into the oesophagus or stomach or via capillary blood vessels, so that the specific gravity of the fish always matches the depth at which it is swimming. This makes the fish weightless,

so less energy is required for locomotion. In lungfish it also has a respiratory function. The lungs of tetrapods are homologous with the swim bladder, which has developed its hydrostatic function by specialization.

**syconus** A type of *composite fruit formed from a hollow fleshy inflorescence stalk inside which tiny flowers develop. Small *drupes, the 'pips', are produced by the female flowers. An example is the fig.

**sylvite (sylvine)** A mineral form of *potassium chloride, KCl.

**symbiont** An organism that is a partner in a symbiotic relationship (*see* SYMBIOSIS).

**symbiosis** An interaction between individuals of different species (**symbionts**). The term symbiosis is usually restricted to interactions in which both species benefit (*see* MUTUALISM), but it may be used for other close associations, such as *commensalism. Many symbioses are obligatory (i.e. the participants cannot survive without the interaction); for example, a lichen is an obligatory symbiotic relationship between an alga or a cyanobacterium and a fungus.

**symmetry 1.** (in physics) The set of invariances of a system. Upon application of a symmetry operation on a system, the system is unchanged. Symmetry is studied mathematically using *group theory. Some of the symmetries are directly physical. Examples include reflections and rotation for molecules and translation in crystal lattices. Symmetries can be **discrete** (i.e. have a finite number), such as the set of rotations for an octahedral molecule, or **continuous** (i.e. do not have a finite number), such as the set of rotations for atoms or nuclei. More general and abstract symmetries can occur, as in the symmetries associated with *gauge theories. *See also* BROKEN SYMMETRY; SUPERSYMMETRY. **2.** (in biology) *See* BILATERAL SYMMETRY; RADIAL SYMMETRY.

**sympathetic nervous system** Part of the *autononomic nervous system. Its postganglionic neurons release mainly noradrenaline; preganglionic neurons release acetylcholine. Its actions tend to antagonize those of the *parasympathetic nervous system, thus achieving a balance in the organs they serve. For example, the sympathetic nervous system decreases salivary gland secretion, increases heart rate, and constricts blood vessels, while the parasympathetic nervous system has opposite effects.

**sympatric** Describing groups of similar organisms that, although in close proximity and theoretically capable of interbreeding, do not interbreed because of differences in behaviour, flowering time, etc. *See* ISOLATING MECHANISM. *Compare* ALLOPATRIC.

**symphysis** A *joint that is only slightly movable; examples are the joints between the vertebrae of the vertebral column and that between the two pubic bones in the pelvic girdle. The bones at a symphysis articulate by means of smooth layers of cartilage and strong fibres.

**symplast** The system of *protoplasts in plants, which are interconnected by *plasmodesmata. This forms a continuous system of cytoplasm bounded by the plasma membranes of the cells. The movement of water through the symplast is known as the **symplast pathway**. It is the only means by which water crosses the *endodermis. *Compare* APOPLAST.

**sympodium** The composite primary axis of growth in such plants as lime and horse chestnut. After each season's growth the shoot tip of the main stem stops growing (sometimes terminating in a flower spike); growth is continued by the tip of one or more of the lateral buds. *Compare* MONOPODIUM.

**synapse** The junction between two adjacent neurons (nerve cells), i.e. between the axon ending of one (the **presynaptic neuron**) and the dendrites of the next (the **postsynaptic neuron**). The swollen tip of the axon of the presynaptic neuron, called the **synaptic knob** (or **bouton**), contains vesicles of *neurotransmitter substance. At a synapse, the membranes of the two cells (the **pre-** and **postsynaptic membranes**) are in close contact, with only a minute gap (the **synaptic cleft**) between them. A nerve *impulse is transmitted across the synapse by the release from the presynaptic membrane of neurotransmitter, which diffuses across the synaptic cleft to the postsynaptic membrane. This triggers the propagation of the impulse from the dendrite along the length of the postsynaptic neuron. Most neurons have more than one synapse. *See also* EXCITATORY POSTSYNAPTIC POTENTIAL.

**synapsis** (in genetics) *See* PAIRING.

**syncarpy** The condition in which the female reproductive organs (*carpels) of a flower are joined to each other. It occurs, for example, in the primrose. *Compare* APOCARPY.

**synchrocyclotron** A form of *cyclotron in which the frequency of the accelerating potential is synchronized with the increasing period of revolution of a group of the accelerated particles, resulting from their rela-

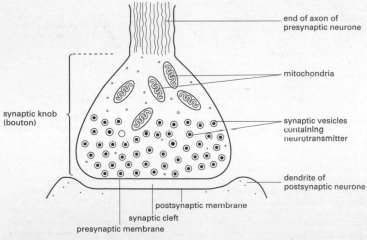

Synapse.

tivistic increase in mass as they reach \*relativistic speeds. The accelerator is used with protons, deuterons, and alpha particles.

**synchronous motor**  *See* ELECTRIC MOTOR.

**synchronous orbit (geosynchronous orbit)**  An orbit of the earth made by an artificial \*satellite with a period exactly equal to the earth's period of rotation on its axis, i.e. 23 hours 56 minutes 4.1 seconds. If the orbit is inclined to the equatorial plane the satellite will appear from the earth to trace out a figure-of-eight track once every 24 hours. If the orbit lies in the equatorial plane and is circular, the satellite will appear to be stationary. This is called a **stationary orbit** (or **geostationary orbit**) and it occurs at an altitude of 35 900 km. Most communication satellites are in stationary orbits, with three or more spaced round the orbit to give worldwide coverage.

**synchronous rotation**  The rotation of a natural satellite in which the period of rotation is equal to its orbital period. The moon, for example, is in synchronous rotation about the earth and therefore completes one axial rotation in 27.322 days, the same time as it takes to complete one orbit of the earth. The result is that the moon always presents the same face to the earth.

**synchrotron**  A particle accelerator used to impart energy to electrons and protons in order to carry out experiments in particle physics and in some cases to make use of the \*synchrotron radiation produced. The particles are accelerated in closed orbits (often circular) by radio-frequency fields. Magnets are spaced round the orbit to bend the trajectory of the particles and separate focusing magnets are used to keep the particles in a narrow beam. The radio-frequency accelerating cavities are interspersed between the magnets. The motion of the particles is automatically synchronized with the rising magnetic field, as the field strength has to increase as the particle energy increases; the frequency of the accelerating field also has to increase synchronously.

**synchrotron radiation (magnetobremsstrahlung)**  Electromagnetic radiation that is emitted by charged particles moving at relativistic speeds in circular orbits in a magnetic field. The rate of emission is inversely proportional to the product of the radius of curvature of the orbit and the fourth power of the mass of the particles. For this reason, synchrotron radiation is not a problem in the design of proton \*synchrotrons but it is significant in electron synchrotrons. The greater the circumference of a synchrotron, the less important is the loss of energy by synchrotron radiation. In \*storage rings, synchrotron radiation is the principal cause of energy loss.

However, since the 1950s it has been realized that synchrotron radiation is itself a very useful tool and many accelerator laboratories have research projects making use of the radiation on a secondary basis to the main high-energy research. The radiation used for these purposes is primarily in the ultraviolet and X-ray frequencies.

Much of the microwave radiation from celestial radio sources outside the Galaxy is believed to originate from electrons moving in curved paths in celestial magnetic fields; it is also called synchrotron radiation as it is analogous to the radiation occurring in a synchrotron. Synchrotron radiation is also predicted to exist for gravitational radiation.

( (()) SEE WEB LINKS )

- Website of the Advanced Light Source facility at Berkeley
- A history of synchrotron radiation

**syncline**  *See* FOLD.

**syncytium**  A group of animal cells in which cytoplasmic continuity is maintained. For example, the cells of striated muscle form a syncytium. In some syncytia the cells remain discrete but are joined together by cytoplasmic bridges.

**syndiotactic**  *See* POLYMER.

**synecology**  The study of ecology at the level of the \*community. A synecological study aims to investigate the relationships between different species that form a community and their interactions with the surrounding environment. Synecology involves both \*biotic and \*abiotic factors. *Compare* AUTECOLOGY.

**synergism (summation)  1.** The phenomenon in which the combined action of two substances (e.g. drugs or hormones) produces a greater effect than would be expected from adding the individual effects of each substance. **2.** The combined action of one muscle (the **synergist**) with another (the **agonist**) in producing movement. *Compare* ANTAGONISM.

S

**syngamy** *See* FERTILIZATION.

**synodic month (lunation)** The interval between two successive new *moons. It is equal to 29 days, 12 hours, and 44 minutes.

**synodic period** The mean time taken by any object in the solar system to move between successive returns to the same position, relative to the sun as seen from the earth. Since a planet is best observed at opposition the synodic period of a planet, $S$, is easier to measure than its *sidereal period, $P$. For inferior planets $1/S = 1/P - 1/E$; for superior planets $1/S = 1/E - 1/P$, where $E$ is the sidereal period of the earth.

**synoptic chart** *See* SYNOPTIC METEOROLOGY.

**synoptic meteorology** The branch of meteorology concerned with the study and analysis of weather information that is obtained simultaneously over a large area. It is based on the analysis of the **synoptic chart**, which is built up from simultaneous observations at weather stations of such elements as wind speed and direction, air temperature, cloud cover, and air pressure over an area at a particular time. Synoptic meteorology is applied chiefly to weather forecasting.

**synovial membrane** The membrane that lines the ligament surrounding a freely movable joint (such as that at the hip or elbow). It secretes a fluid (**synovial fluid**) that lubricates the layers of cartilage forming the articulating surfaces of the joint.

**synthesis** The formation of chemical compounds from more simple compounds.

**synthesis gas** A mixture of two parts hydrogen and one part carbon monoxide made from methane and steam heated under pressure. It is used in the manufacture of various organic chemicals, including hydrocarbons, methanol, and other alcohols. *See also* HABER PROCESS.

**synthetic** Describing a substance that has been made artificially; i.e. one that does not come from a natural source.

**synthetic metal** A substance that is not a metal but has free electrons that can contribute to electrical conductivity. Conducting polymers are examples of synthetic metals.

**synthon** *See* SUPRAMOLECULAR CHEMISTRY.

**syrinx** The sound-producing organ of a bird, situated at the lower end of the trachea where it splits into the bronchi. It has a complex structure with a number of vibrating membranes.

**systematics** The study of the diversity of organisms and their natural relationships. It is sometimes used as a synonym for *taxonomy. The term **biosystematics** describes the experimental study of diversity, especially at the species level. Biosystematic methods include breeding experiments, biochemical work (known as **chemosystematics**), and cytotaxonomy. *See also* MOLECULAR SYSTEMATICS.

**Système International d'Unités** *See* SI UNITS.

**systemic circulation** The part of the circulatory system of birds and mammals that transports oxygenated blood from the left ventricle of the heart to the tissues in the body and returns deoxygenated blood from the tissues to the right atrium of the heart. *Compare* PULMONARY CIRCULATION. *See* DOUBLE CIRCULATION.

**systems biology** An approach that seeks to study organisms as complete systems – networks of interacting genes, biomolecules, and biochemical reactions. It thus attempts to integrate all relevant structural and functional information, rather than focusing on, say, just one particular gene or protein at a time. This involves amassing and organizing data obtained from genomics, proteomics, and other areas of bioinformatics and managing and analysing the data to identify patterns, formulate hypotheses, and ultimately create computer models that will enable accurate predictions of cellular and organismal responses. Such models will have a radical impact on medicine and biology in the future.

**systems software** *See* APPLICATIONS SOFTWARE.

**systole** The phase of the heart beat during which the ventricles of the heart contract to force blood into the arteries. *Compare* DIASTOLE. *See* BLOOD PRESSURE.

**syzygy** The situation that occurs when the sun, the moon (or a planet), and the earth are in a straight line. This occurs when the moon (or planet) is at *conjunction or *opposition.

S

**2,4,5-T** 2,4,5-trichlorophenoxyacetic acid (2,4,5-trichlorophenoxyethanoic acid): a synthetic *auxin formerly widely used as a herbicide and defoliant. It is now banned in many countries as it tends to become contaminated with the toxic chemical *dioxin.

**tabun** A highly toxic colourless or brown liquid, $C_5H_{11}N_2O_2P$; r.d. 1.09; m.p. –50°C; b.p. 247.5°C. It is an organophosphorus compound, ethyl N,N-dimethylphosphoramidocyanidate. Tabun was discovered in 1936 and belongs to the G-series of *nerve agents (GA). It was used by Iraq in the Iran–Iraq war (1980–88).

**tachometer** An instrument for measuring angular speed, especially the number of revolutions made by a rotating shaft in unit time. Various types of instrument are used, including mechanical, electrical, and electronic devices. The widely used electrical-generator tachometer consists of a small generator in which the output voltage is a measure of the rate of rotation of the shaft that drives it.

**tachyon** A hypothetical particle that has a speed in excess of the *speed of light. According to electromagnetic theory, a charged particle travelling through a medium at a speed in excess of the speed of light in that medium emits *Cerenkov radiation. A charged tachyon would emit Cerenkov radiation even in a vacuum. No such particle has yet been detected. According to the special theory of *relativity, it is impossible to accelerate a particle up to the speed of light because its energy $E$, given by $E = mc^2/\sqrt{(1 - v^2/c^2)}$, would have to become infinite. The theory, however, does not forbid the existence of particles with $v > c$ (where $c$ is the speed of light). In such cases the expression in the brackets becomes negative and the energy would be imaginary.

**tactic movement** *See* TAXIS.

**tactic polymer** *See* POLYMER.

**taiga** A terrestrial *biome consisting mainly of evergreen coniferous forests (mainly pine, fir, and spruce), which occurs across subarctic North America and Eurasia. In certain parts, such as northeastern Siberia, deciduous conifers and broadleaved trees, such as larch and birch, are dominant. Over most of the taiga the ground is permanently frozen within about one metre of the surface, which prevents water from filtering down to deeper levels in the soil. This means that bogs may form in depressions. For at least six months of the year temperatures are below freezing but there is a short growing season lasting 3–5 months. The soil in taiga areas is acidic and infertile. *Compare* TUNDRA.

**Takayama test** *See* HAEMOCHROMOGEN TEST.

**talc** A white or pale-green mineral form of magnesium silicate, $Mg_3Si_4O_{10}(OH)_2$, crystallizing in the triclinic system. It forms as a secondary mineral by alteration of magnesium-rich olivines, pyroxenes, and amphiboles of ultrabasic rocks. It is soaplike to touch and very soft, having a hardness of 1 on the Mohs' scale. Massive fine-grained talc is known as **soapstone** or **steatite**. Talc in powdered form is used as a lubricant, as a filler in paper, paints, and rubber, and in cosmetics, ceramics, and French chalk. It occurs chiefly in the USA, Russia, France, and Japan.

**tandem generator** A type of particle generator, essentially consisting of a *Van de Graaff generator that maintains one electrode at a high positive potential; this electrode is placed between two earthed electrodes. Negative ions are accelerated from earth potential to the positively charged electrode, where surplus electrons are stripped from the ions to produce positive ions, which are accelerated again from the positive electrode back to earth. Thus the ions are accelerated twice over by a single high potential. This tandem arrangement enables energies up to 30 MeV to be achieved.

**tangent** 1. A line that touches a curve or a plane that touches a surface. 2. *See* TRIGONOMETRIC FUNCTIONS.

**tangent galvanometer** A type of galvanometer, now rarely used, in which a small magnetic needle is pivoted horizontally at the centre of a vertical coil that is adjusted to be parallel to the horizontal component of the earth's magnetic field. When a current $I$ is passed through the coil, the needle is deflected so that it makes an angle $\theta$ with its equilibrium position parallel to the earth's field. The value of $I$ is given by $I = (2Hr\tan\theta)/n$, where $H$ is the strength of the earth's horizontal component of magnetizing force, $r$ is the radius of the coil, and $n$ is the number of turns in the coil. Although not now used for measuring current, the instrument provides a means of measuring the earth's magnetizing force.

**tanh** *See* HYPERBOLIC FUNCTIONS.

**tannic acid** A yellowish complex organic compound present in certain plants. It is used in dyeing as a mordant.

**tannin** One of a group of complex organic chemicals commonly found in leaves, unripe fruits, and the bark of trees. Their function is uncertain though the unpleasant taste may discourage grazing animals. Some tannins have commercial uses, notably in the production of leather and ink.

**tantalum** Symbol Ta. A heavy blue-grey metallic *transition element; a.n. 73; r.a.m. 180.948; r.d. 16.63; m.p. 2996°C; b.p. 5427°C. It is found with niobium in the ore columbite–tantalite $(Fe,Mn)(Ta,Nb)_2O_6$. It is extracted by dissolving in hydrofluoric acid, separating the tantalum and niobium fluorides to give $K_2TaF_7$, and reduction of this with sodium. The element contains the stable isotope tantalum–181 and the long-lived radioactive isotope tantalum–180 (0.012%; half-life >$10^7$ years). There are several other short-lived isotopes. The element is used in certain alloys and in electronic components. Tantalum parts are also used in surgery because of the unreactive nature of the metal (e.g. in pins to join bones). Chemically, the metal forms a passive oxide layer in air. It forms complexes in the +2, +3, +4, and +5 oxidation states. Tantalum was identified in 1802 by Anders Ekeberg (1767–1813) and first isolated in 1820 by Jöns Berzelius.

(((🌐))) **SEE WEB LINKS**
• Information from the WebElements site

**tapetum** A reflecting layer, containing crystals of guanine, in the *choroid of the eye of many nocturnal vertebrates. It reflects light back onto the retina, thus improving vision and causing the eyes to shine in the dark.

**tapeworms** *See* CESTODA.

**tap root** *See* ROOT.

**tar** Any of various black semisolid mixtures of hydrocarbons and free carbon, produced by destructive distillation of *coal or by *petroleum refining.

**tarsal (tarsal bone)** One of the bones that form the ankle (*see* TARSUS) in terrestrial vertebrates.

**tar sand** *See* OIL SAND.

**tarsus** The ankle (or corresponding part of the hindlimb) in terrestrial vertebrates, consisting of a number of small bones (**tarsals**). The number of tarsal bones varies with the species: humans, for example, have seven.

**tartaric acid** A white crystalline naturally occurring carboxylic acid, $(CHOH)_2$-$(COOH)_2$; r.d. 1.8; m.p. 171–174°C. It can be obtained from tartar (potassium hydrogen tartrate) deposits from wine vats, and is used in baking powders and as a foodstuffs additive. The compound is optically active (*see* OPTICAL ACTIVITY). The systematic name is **2,3-dihydroxybutanedioic acid**.

**tartrate** A salt or ester of *tartaric acid.

**taste** **1.** The sense that enables the flavour of different substances to be distinguished (*see* TASTE BUD). **2.** The flavour of a substance.

(((🌐))) **SEE WEB LINKS**
• A tutorial on the sense of taste, compiled by Tim Jacob of Cardiff University

**taste bud** A small sense organ in most vertebrates, specialized for the detection of taste. In terrestrial animals taste buds are concentrated on the upper surface of the *tongue. They are sensitive to four types of taste: sweet, salt, bitter, or sour. The taste bud transmits information about a particular type of taste to the brain via nerve fibres. The four types of taste bud show distinct distribution patterns on the surface of the human tongue.

In fishes, taste buds are distributed over the entire surface of the body and provide information about the surrounding water.

**TATA box (Hogness box)** A sequence of nucleotides that serves as the main recognition site for the attachment of RNA polymerase in the *promoter of eukaryotic genes. Located at around 25 nucleotides before the start of transcription, it consists of the seven-base *consensus sequence TATAAAA, and is analogous to the *Pribnow box in prokaryotic promoters.

**TATP** *See* TRIACETONE TRIPEROXIDE.

**Tatum, Edward** *See* BEADLE, GEORGE WELLS.

**tau particle** *See* ELEMENTARY PARTICLES; LEPTON.

**tautomerism** A type of *isomerism in which the two isomers (**tautomers**) are in equilibrium. *See* KETO–ENOL TAUTOMERISM.

**taxis (taxic response; tactic movement)** The movement of a cell (e.g. a gamete) or a microorganism in response to an external stimulus. Certain microorganisms have a light-sensitive region that enables them to move towards or away from high light intensities (positive and negative *phototaxis respectively). Many bacteria move in response to chemical stimuli (**chemotaxis**); a specific example is **aerotaxis**, in which atmospheric oxygen is the stimulus. Taxic responses are restricted to cells that possess cilia, flagella, or some other means of locomotion. The term is usually not applied to the movements of higher animals. *Compare* KINESIS; TROPISM.

**taxon** (*pl.* **taxa**) Any named taxonomic group of any *rank in the hierarchical *classification of organisms. Thus the taxa Papilionidae, Lepidoptera, Hexapoda, and Uniramia are named examples of a family, order, class, and phylum, respectively.

**taxonomy** The study of the theory, practice, and rules of *classification of living and extinct organisms. The naming, description, and classification of a given organism draws on evidence from a number of fields. **Classical taxonomy** is based on morphology and anatomy. **Cytotaxonomy** compares the size, shape, and number of chromosomes of different organisms. **Numerical taxonomy** uses mathematical procedures to assess similarities and differences and establish taxonomic groups. *See also* SYSTEMATICS.

**Taylor series** The infinite power series of derivatives into which a function f($x$) can be expanded, for a fixed value of the variable $x = a$:

$$f(x) = f(a) + f'(a)(x - a) + f''(a)(x - a)^2/2! + \dots$$

When $a = 0$, the series formed is known as **Maclaurin's series**:

$$f(x) = f(0) + f'(0)x + f''(0)x^2/2! + \dots$$

The series was discovered by Brook Taylor (1685–1731) and the special case was named after Colin Maclaurin (1698–1746).

**TCA cycle** *See* KREBS CYCLE.

**T cell (T lymphocyte)** Any of a population of *lymphocytes that are the principal agents of cell-mediated *immunity. T cells are derived from the bone marrow but migrate to the thymus to mature (hence *T* cell). After leaving the thymus, and when presented with antigen, a T cell becomes 'armed' to act as an **effector T cell** when it subsequently encounters its specific antigen. Subpopulations of T cells play different roles in the immune response and can be characterized by their surface antigens (*see* CD). **Helper T cells** carry the CD4 antigen and recognize foreign antigens provided these are presented by cells (such as macrophages and B lymphocytes) bearing class II *histocompatibility antigens. The helper T cell binds to its target cell by means of **T-cell receptors**. Helper T cells are essential in the majority of infections for stimulating B cells to proliferate and differentiate into clones of antibody-producing plasma cells.

**Cytotoxic T cells**, which carry the CD8 antigen, recognize foreign antigen on the surface of virus-infected cells and destroy the cell by releasing cytolytic proteins. **Suppressor T cells** (or **regulatory T cells**) are important in regulating the activity of other lymphocytes and are crucial in maintaining tolerance to self tissues.

**TCP/IP** *Trademark.* Transmission Control Protocol/Internet Protocol: the obligatory standard to be used by any system connecting to the *Internet. The two protocols were originally developed on the DARPA net. They were devised to optimize the performance of networks that are based on unreliable data-transmission systems operating at relatively low data rates.

The Internet Protocol, IP, is the lower of the two protocols. It provides a connectionless datagram service, and a managed address structure for data transmission. In IP version 4 (**IPv4**), the dominant version on the Internet, an **IP address** is a 32-bit number. The interpretation of these bits was formerly rigid and divided IP addresses into four classes, A to D. This system has been superseded since 1993 by Classless Inter-Domain Routing.

The explosive growth of the Internet has resulted in the IPv4 32-bit address space becoming restrictive. IP version 6 (**IPv6**) seeks to remedy this by using 128-bit IP addresses, with 64 bits being used for both the network identity and the host address. IPv6 is gradually being adopted on the Internet. IP allows a long datagram to be fragmented into numbered packets, which can then be transmitted and reassembled in their correct sequence at the destination system. It is intended to be used in conjunction with the Transmission Control Protocol, TCP.

TCP provides error-free delivery of arbitrarily long messages, known as **segments**, with the data being released to the host system in the same order as the original transmission. It achieves this by a 'sliding window' mechanism. As data are transmitted, they are accompanied by a checksum; at the receiving end the checksum is verified and an acknowledgment is returned to the transmitter, which indicates the position of the last data to be successfully received. The transmitter will not send data beyond a certain point, determined by the size of the window, i.e. the gap between the last data to be sent and the last data for which an acknowledgment has been received. If the checksum fails at any point, the transmitter will retransmit data from the point immediately following the latest acknowledgment of correct receipt.

**(⊕) SEE WEB LINKS**
• The TCP (version 4) specification
• The IP (version 4) specification
• The IP (version 6) specification

***t*-distribution** In *statistics, a probability distribution made up of the *means of random samples from a collection of samples that have a *normal distribution of unknown *variance. *See also* NORMAL DISTRIBUTION; POISSON DISTRIBUTION.

**tebi-** *See* BINARY PREFIXES.

**technetium** Symbol Tc. A radioactive metallic *transition element; a.n. 43; m.p. 2172°C; b.p. 4877°C. The element can be detected in certain stars and is present in the fission products of uranium. It was first made by Carlo Perrier and Emilio Segrè (1905–89) by bombarding molybdenum with deuterons to give technetium–97. The most stable isotope is technetium–99 (half-life 2.6 × 10⁶ years); this is used to some extent in la-

belling for medical diagnosis. There are sixteen known isotopes. Chemically, the metal has properties intermediate between manganese and rhenium.

**(⊕) SEE WEB LINKS**
• Information from the WebElements site

**technicolour theory** *See* HIGGS FIELD.

**teeth** *See* DECIDUOUS TEETH; PERMANENT TEETH; DENTITION; TOOTH.

**Teflon** Trade name for a form of *polytetrafluoroethene.

**Teichmann test** *See* HAEMATIN TEST.

**tektite** A small black, greenish, or yellowish glassy object found in groups on the earth's surface and consisting of a silicaceous material unrelated to the geological formations in which it is found. Tektites are believed to have formed on earth as a result of the impact of meteorites.

**telecommunications** The study and application of means of transmitting information, either by wires or by electromagnetic radiation.

**Teleostei** The major superorder of the *Osteichthyes (bony fish), containing about 20 000 species. Teleosts have colonized an extensive variety of habitats and show great diversity of form. The group includes the eel, seahorse, plaice, and salmon. They have been the dominant fish since the Cretaceous period (about 70 million years ago).

**telescope** An instrument that collects electromagnetic radiation from a distant object in order to produce an image of it or enable the radiation to be analysed. See Feature.

**television** The transmission and reception of moving images by means of radio waves or cable. The scene to be transmitted is focused onto a photoelectric screen in the television *camera. This screen is scanned by an electron beam. The camera produces an electric current, the instantaneous magnitude of which is proportional to the brightness of the portion of the screen being scanned. In Europe the screen is scanned by 625 lines and 25 such frames are produced every second. In the USA 525 lines and 30 frames per second are used. The picture signal so produced is used to modulate a VHF or UHF carrier wave and is transmitted with an independent sound signal, but with colour information (if any) incorporated into

## OPTICAL ASTRONOMICAL TELESCOPES

Optical astronomical telescopes fall into two main classes: **refracting telescopes** (or **refractors**), which use lenses to form the primary image, and **reflecting telescopes** (or **reflectors**), which use mirrors.

### Refracting telescopes

The refracting telescopes use a converging lens to collect the light and the resulting image is magnified by the eyepiece. This type of instrument was first constructed in 1608 by Hans Lippershey (1587–1619) in Holland and developed in the following year as an astronomical instrument by Galileo, who used a diverging lens as eyepiece. The **Galilean telescope** was later improved by Johannes Kepler (1571–1630), who substituted a converging eyepiece lens. This form is still in use for small astronomical telescopes (the **Keplerian telescope**).

### Reflecting telescopes

The first reflecting telescope was produced by Newton in 1668. This used a concave mirror to collect and focus the light and a small secondary mirror at an angle of 45° to the main beam to reflect the light into the magnifying eyepiece. This design is known as the **Newtonian telescope**. The **Gregorian telescope**, designed by James Gregory (1638–75), and the **Cassegrainian telescope**, designed by N. Cassegrain (*fl.* 1670s), use different secondary optical systems. The **coudé telescope** (French: angled) is sometimes used with larger instruments as it increases their focal lengths.

### Catadioptic telescopes

**Catadioptic telescopes** use both lenses and mirrors. The most widely used astronomical instruments in this class are the **Maksutov telescope** and the **Schmidt camera**.

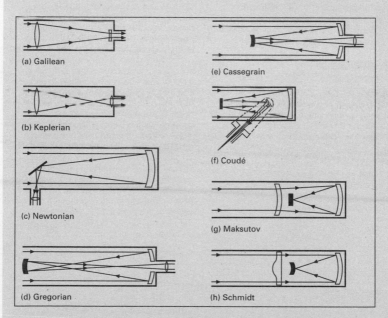

(a) Galilean

(b) Keplerian

(c) Newtonian

(d) Gregorian

(e) Cassegrain

(f) Coudé

(g) Maksutov

(h) Schmidt

the brief gaps between the picture lines. The signals received by the receiving aerial are demodulated in the receiver; the demodulated picture signal controls the electron beam in a cathode-ray tube, on the screen of which the picture is reconstructed. *See also* COLOUR TELEVISION.

**television tube** *See* CATHODE RAYS.

**tellurides** Binary compounds of tellurium with other more electropositive elements. Compounds of tellurium with nonmetals are covalent (e.g. $H_2Te$). Metal tellurides can be made by direct combination of the elements and are ionic (containing $Te^{2-}$) or nonstoichiometric interstitial compounds (e.g. $Pd_4Te$, $PdTe_2$).

**tellurium** Symbol Te. A silvery metalloid element belonging to *group 16 (formerly VIB) of the periodic table; a.n. 52; r.a.m. 127.60; r.d. 6.24 (crystalline); m.p. 449.5°C; b.p. 989.8°C. It occurs mainly as *tellurides in ores of gold, silver, copper, and nickel and it is obtained as a by-product in copper refining. There are eight natural isotopes and nine radioactive isotopes. The element is used in semiconductors and small amounts are added to certain steels. Tellurium is also added in small quantities to lead. Its chemistry is similar to that of sulphur. It was discovered by Franz Müller (1740–1825) in 1782.

**(⊕) SEE WEB LINKS**
• Information from the WebElements site

**telomere** The end of a chromosome, which consists of tandemly repeated short sequences of DNA that perform the function of ensuring that each cycle of *DNA replication has been completed.

**telophase** A stage of cell division. In *mitosis the chromatids that separated from each other at *anaphase collect at the poles of the spindle. A nuclear membrane forms around each group, producing two daughter nuclei with the same number and kind of chromosomes as the original cell nucleus. In the first telophase of *meiosis, complete chromosomes from the pairs that separated at first anaphase form the daughter nuclei. The number of chromosomes in these nuclei is therefore half the number in the original one. In the second telophase, daughter nuclei are formed from chromatids (as in mitosis).

**temperament** The way in which the intervals between notes on keyboard instruments are distributed throughout the scale to ensure that music in all keys sounds in tune. The problem can be illustrated by a piano keyboard. Taking a low C and a high C seven octaves above, the interval should be $2^7 = 128$. However, in passing through the cycle of 12 keys, each using as its fundamental the fifth of its predecessor, the interval between Cs becomes $(3/2)^{12} = 129.75$. The difference between 129.75 and 128 is known as the **comma of Pythagoras**. The **equal-temperament scale**, which has been in use since the time of J. S. Bach, distributes the comma of Pythagoras equally between the 12 intervals of the scale over seven octaves. Thus each fifth becomes $(128)^{1/12} = 1.4983$. All forms of temperament involve a measure of compromise; this system is now regarded as the best.

**temperature** The property of a body or region of space that determines whether or not there will be a net flow of heat into it or out of it from a neighbouring body or region and in which direction (if any) the heat will flow. If there is no heat flow the bodies or regions are said to be in **thermodynamic equilibrium** and at the same temperature. If there is a flow of heat, the direction of the flow is from the body or region of higher temperature. Broadly, there are two methods of quantifying this property. The empirical method is to take two or more reproducible temperature-dependent events and assign **fixed points** on a scale of values to these events. For example, the Celsius temperature scale uses the freezing point and boiling point of water as the two fixed points, assigns the values 0 and 100 to them, respectively, and divides the scale between them into 100 degrees. This method is serviceable for many practical purposes (*see* TEMPERATURE SCALES), but lacking a theoretical basis it is awkward to use in many scientific contexts. In the 19th century, Lord Kelvin proposed a thermodynamic method to specify temperature, based on the measurement of the quantity of heat flowing between bodies at different temperatures. This concept relies on an absolute scale of temperature with an *absolute zero of temperature, at which no body can give up heat. He also used Sadi Carnot's concept of an ideal frictionless perfectly efficient heat engine (*see* CARNOT CYCLE). This Carnot engine takes in a quantity of heat $q_1$ at a temperature $T_1$, and exhausts heat $q_2$ at $T_2$, so that $T_1/T_2 = q_1/q_2$. If

$T_2$ has a value fixed by definition, a Carnot engine can be run between this fixed temperature and any unknown temperature $T_1$, enabling $T_1$ to be calculated by measuring the values of $q_1$ and $q_2$. This concept remains the basis for defining **thermodynamic temperature**, quite independently of the nature of the working substance. The unit in which thermodynamic temperature is now expressed is the *kelvin. In practice thermodynamic temperatures cannot be measured directly; they are usually inferred from measurements with a gas thermometer containing a nearly ideal gas. This is possible because another aspect of thermodynamic temperature is its relationship to the *internal energy of a given amount of substance. This can be shown most simply in the case of an ideal monatomic gas, in which the internal energy per mole ($U$) is equal to the total kinetic energy of translation of the atoms in one mole of the gas (a monatomic gas has no rotational or vibrational energy). According to *kinetic theory, the thermodynamic temperature of such a gas is given by $T = 2U/3R$, where $R$ is the universal *gas constant.

**temperature coefficient** A coefficient that determines the rate of change of some physical property with change in temperature. For example, the dependence of the resistance ($R$) of a material on the Celsius temperature $t$, is given by $R = R_0 + \alpha t + \beta t^2$, where $R_0$ is the resistance at 0°C and $\alpha$ and $\beta$ are constants. If $\beta$ is negligible, then $\alpha$ is the **temperature coefficient of resistance**.

**temperature inversion** An abnormal increase in air temperature that occurs in the troposphere, the lowest level of the earth's atmosphere. This can lead to pollutants becoming trapped in the troposphere (*see* AIR POLLUTION).

**temperature scales** A number of empirical scales of *temperature have been in use: the *Celsius scale is widely used for many purposes and in certain countries the *Fahrenheit scale is still used. These scales both rely on the use of **fixed points**, such as the freezing point and the boiling point of water, and the division of the **fundamental interval** between these two points into units of temperature (100 degrees in the case of the Celsius scale and 180 degrees in the Fahrenheit scale).

However, for scientific purposes the scale in use is the **International Practical Temperature Scale (IPTS)**, which is designed to conform as closely as possible to thermodynamic temperature and is expressed in the unit of thermodynamic temperature, the *kelvin. The 1968 version of the table (known as IPTS-68) had 11 fixed points defined by both Celsius and thermodynamic temperatures. The most recent version (IPTS-90), introduced in 1990, has 16 fixed points with temperatures expressed in kelvins:

Triple point of hydrogen: 13.8033
Boiling point of hydrogen (33 321.3 Pa): 17.035
Boiling point of hydrogen (101 292 Pa): 20.27
Triple point of neon: 24.5561
Triple point of oxygen: 54.3584
Triple point of argon: 83.8058
Triple point of mercury: 234.3156
Triple point of water: 273.16 (0.01°C)
Melting point of gallium: 302.9146
Freezing point of indium: 429.7485
Freezing point of tin: 505.078
Freezing point of zinc: 692.677
Freezing point of aluminium: 933.473
Freezing point of silver: 1234.93
Freezing point of gold: 1337.33
Freezing point of copper: 1357.77

Methods for measuring intermediate temperatures between these fixed points are specified; for example, at low temperatures (0–5 K) they are measured by means of vapour-pressure determinations of $^3$He and $^4$He; at high temperatures (above 1234.93 K) a radiation pyrometer is used.

**tempering** The process of increasing the toughness of an alloy, such as steel, by heating it to a predetermined temperature, maintaining it at this temperature for a predetermined time, and cooling it to room temperature at a predetermined rate. In steel, the purpose of the process is to heat the alloy to a temperature that will enable the excess carbide to precipitate out of the supersaturated solid solution of *martensite and then to cool the saturated solution fast enough to prevent further precipitation or grain growth. For this reason steel is quenched rapidly by dipping into cold water.

**template** Any molecule that acts as a pattern for the synthesis of a new molecule. For example, the two nucleotide chains of a DNA molecule can separate and each acts as a template for the synthesis of the missing chain (*see* DNA REPLICATION).

**temporary hardness** *See* HARDNESS OF WATER.

**temporary magnetism** Magnetism in a

body that is present when the body is in a magnetic field but that largely disappears when it is removed from the field.

**tendon** A thick strand or sheet of tissue that attaches a muscle to a bone. Tendons consist of *collagen fibres and are therefore inelastic: they ensure that the force exerted by muscular contraction is transmitted to the relevant part of the body to be moved.

**tendril** A slender branched or unbranched structure found in many climbing plants. It may be a modified stem, leaf, leaflet, or petiole. Tendrils respond to contact with solid objects by twining around them (*see* THIGMOTROPISM). The cells that touch the object lose water and decrease in volume in comparison to the outer cells, thus causing the tendril to curve.

**tensile strength** A measure of the resistance that a material offers to tensile *stress. It is defined as the stress, expressed as the force per unit cross-sectional area, required to break it.

**tensimeter** A form of differential manometer with two sealed bulbs attached to the limbs. It is used to measure the difference in vapour pressure between liquids sealed into the bulbs. If one liquid has a known vapour pressure (often water is used) that of the other can be determined.

**tensiometer** Any apparatus for measuring *surface tension.

**tentacle** Any of the soft flexible appendages in aquatic invertebrate animals that are used principally for feeding. Water flows over the tentacles, which are able to capture food and direct it to the oral aperture. Tentacles are possessed by many cnidarians, some echinoderms (including sea cucumbers), and by cephalopod molluscs, in which the tentacles bear rows of suckers.

**tera-** Symbol T. A prefix used in the metric system to denote one million million times. For example, $10^{12}$ volts = 1 teravolt (TV).

**teratogen** Any environmental factor that acts on a fetus to cause congenital abnormality. Examples include ionizing radiation (e.g. X-rays), nutritional deficiencies, drugs (e.g. thalidomide), toxic chemicals, and virus infections (e.g. rubella).

**terbium** Symbol Tb. A silvery metallic element belonging to the *lanthanoids; a.n. 65;

r.a.m. 158.92; r.d. 8.23 (20°C); m.p. 1356°C; b.p. 3123°C. It occurs in apatite and xenotime, from which it is obtained by an ion-exchange process. There is only one natural isotope, terbium–159, which is stable. Seventeen artificial isotopes have been identified. It is used as a dopant in semiconducting devices. It was discovered by Carl Mosander (1797–1858) in 1843.

**(())) SEE WEB LINKS**
• Information from the WebElements site

**terephthalic acid (1,4-benzenedicarboxylic acid)** A colourless crystalline solid, $C_6H_4(COOH)_2$, m.p. 300°C. It is made by oxidizing p-xylene (1,4-dimethylbenzene) and used for making polyesters, such as Terylene.

**terminal** 1. The point at which electrical connection is made to a device or system. 2. A device at which data is put into a *computer or taken from it.

**terminal speed** The constant speed finally attained by a body moving through a fluid under gravity when there is a zero resultant force acting on it. *See* STOKES' LAW.

**terminator** The boundary, on the surface of the moon or a planet, between the sunlit area and the dark area.

**ternary compound** A chemical compound containing three different elements.

**terpenes** A group of unsaturated hydrocarbons present in plants (*see* ESSENTIAL OIL). Terpenes consist of isoprene units, $CH_2:C(CH_3)CH:CH_2$. Monoterpenes have two units, $C_{10}H_{16}$, sesquiterpenes three units, $C_{15}H_{24}$, diterpenes four units, $C_{20}H_{32}$, etc. **Terpenoids**, which are derivatives of terpenes, include abscisic acid and gibberellin (*plant hormones) and the *carotenoid and *chlorophyll pigments.

**(())) SEE WEB LINKS**
• Information about IUPAC nomenclature

**terrestrial magnetism** *See* GEOMAGNETISM.

**territory** A fixed area that an animal or group of animals defends against intrusion from others of its species by various types of **territorial behaviour**. Outside the territory (which may contain food sources, hiding places, and nesting sites) others are not threatened. Many mammals indicate their territory boundaries with scent markings, while birds sing territorial songs that repel would-be intruders. Animals in neighbouring territories normally respect each other's boundaries, which reduces overt *aggression. Some animals are territorial only at certain times of the year, usually the breeding season (*see* COURTSHIP).

**Tertiary** Formerly, the older geological period of the Cenozoic era (*compare* QUATERNARY). It began about 65 million years ago, following the Cretaceous period, and extended to the beginning of the Quaternary, about 2 million years ago. The Tertiary period was characterized by the rise of the modern mammals and the development of shrubs, grasses, and other flowering plants. It has been replaced by the *Palaeogene and *Neogene periods.

**tertiary alcohol** *See* ALCOHOLS.

**tertiary amine** *See* AMINES.

**tertiary colour** A colour obtained by mixing two *secondary colours.

**tertiary consumer** *See* CONSUMER.

**tertiary structure** *See* PROTEIN.

**tervalent (trivalent)** Having a valency of three.

**Terylene** Trade name for a type of *polyester used in synthetic fibres.

**tesla** Symbol T. The SI unit of magnetic flux density equal to one weber of magnetic flux per square metre, i.e. $1 T = 1 Wb m^{-2}$. It is named after Nikola Tesla (1870–1943), Croatian-born US electrical engineer.

**Tesla coil** A device for producing a high-frequency high-voltage current. It consists of a *transformer with a high turns ratio, the primary circuit of which includes a spark gap and a fixed capacitor; the secondary circuit is tuned by means of a variable capacitor to resonate with the primary. It was devised by Nikola Tesla. Tesla coils are commonly used to excite luminous discharges in glass vacuum apparatus, in order to detect leaks.

**testa (seed coat)** The lignified or fibrous protective covering of a seed that develops from the integuments of the ovule after fertilization. *See also* HILUM; MICROPYLE.

**test cross** A mating (cross) made to identify hidden *recessive alleles in an individual of unknown genotype. This individual is crossed with one that is *homozygous for the allele being investigated (i.e. a homozygous recessive). The homozygous recessive indi-

vidual may be the parent of the individual being investigated (*see* BACK CROSS).

**testis (testicle)** The reproductive organ in male animals in which spermatozoa are produced. In vertebrates there are two testes; as well as sperm, they produce steroid hormones (*see* ANDROGEN). In most animals the testes are within the body cavity but in mammals, although they develop within the body near the kidneys, they come to hang outside the body cavity in a *scrotum. Most of the vertebrate testis is made up of a mass of **seminiferous tubules**, in which the sperms develop. It is connected to the outside by means of the *vas deferens. *See* REPRODUCTIVE SYSTEM.

**testis-determining factor (TDF; SRY protein)** A protein that plays a crucial role in sex determination in mammals. It is encoded by the **SRY** (sex reversal on Y) gene on the Y chromosome, and switches embryonic development from the default female pathway to the male pathway, by driving testis formation. Male development is then consolidated by secretion of the male sex hormone testosterone by the testes.

**testosterone** The principal male sex hormone. *See* ANDROGEN.

**tetrachloroethene** A colourless nonflammable volatile liquid, $CCl_2:CCl_2$; r.d. 1.6; m.p. –22°C; b.p. 121°C. It is used as a solvent.

**tetrachloromethane (carbon tetrachloride)** A colourless volatile liquid with a characteristic odour, virtually insoluble in water but miscible with many organic liquids, such as ethanol and benzene; r.d. 1.586; m.p. –23°C; b.p. 76.54°C. It is made by the chlorination of methane (previously by chlorination of carbon disulphide). The compound is a good solvent for waxes, lacquers, and rubbers and the main industrial use is as a solvent, but safer substances (e.g. 1,1,1-trichloroethane) are increasingly being used. Moist carbon tetrachloride is partly decomposed to phosgene and hydrogen chloride and this provides a further restriction on its use.

**tetrad** A group of four *haploid cells formed at the end of the second division of *meiosis.

**tetradecanoic acid (myristic acid)** A saturated carboxylic acid, $CH_3(CH_2)_{12}COOH$; r.d. 0.86; m.p. 58.8°C; b.p. 250.5°C (100 mmHg). Its glycerides are found in nutmeg, palm oil, and butter fat. Compounds are used in cosmetics and skin-care preparations. The traditional name comes from the Latin name for nutmeg, *Myristica fragrans*.

**tetraethyl lead** *See* LEAD(IV) TETRAETHYL.

**tetragonal** *See* CRYSTAL SYSTEM.

**tetrahedral angle** **1.** (in geometry) The solid angle bounded by three faces of a tetrahedron. **2.** (in chemistry) The angle between the bonds in a *tetrahedral compound (approximately 109° for a regular tetrahedron).

**tetrahedral compound** A compound in which four atoms or groups situated at the corners of a tetrahedron are linked (by covalent or coordinate bonds) to an atom at the centre of the tetrahedron. *See also* COMPLEX.

**tetrahedron** A polyhedron with four triangular faces. In a **regular tetrahedron** all four triangles are congruent equilateral triangles. It constitutes a regular triangular *pyramid.

**tetrahydrate** A crystalline hydrate containing four moles of water per mole of compound.

**tetrahydrofuran (THF)** A colourless volatile liquid, $C_4H_8O$; r.d. 0.89; m.p. –65°C; b.p. 67°C. It is made by the acid hydrolysis of polysaccharides in oat husks, and is widely used as a solvent.

**tetrahydroxomonoxodiboric(III) acid** *See* BORIC ACID.

**tetraoxophosphoric(V) acid** *See* PHOS-PHORIC(V) ACID.

**tetraoxosulphuric(VI) acid** *See* SUL-PHURIC ACID.

**tetraploid** Describing a nucleus, cell, or organism that has four times ($4n$) the haploid number ($n$) of chromosomes. *See also* POLYPLOID.

**Tetrapoda** In some classifications, a superclass of jawed chordates (*Gnathostomata) comprising all vertebrate animals with four limbs, i.e. the amphibians, reptiles, birds, and mammals. The skeleton of the limbs of all tetrapods is based on the same five-digit pattern (*see* PENTADACTYL LIMB).

**tetravalent (quadrivalent)** Having a valency of four.

**tetrode** A *thermionic valve with a **screen grid** placed between the anode and the control grid of a *triode to reduce the capaci-

tance between these two electrodes and so improve the valve's performance as an amplifier or oscillator at high frequencies. The screen grid is maintained at a fixed potential.

**texaphyrin** A synthetic molecule similar to a porphyrin but containing five central nitrogen atoms rather than four, thus increasing the size of the central 'hole' and enabling larger cations, such as cadmium, to be bound stably. *See also* SUPRAMOLECULAR CHEMISTRY.

**thalamus** **1.** (in anatomy) Part of the vertebrate *forebrain that lies above the hypothalamus. It relays sensory information to the cerebral cortex and is also concerned with the translation of impulses into conscious sensations. **2.** (in botany) *See* RECEPTACLE.

**thallium** Symbol Tl. A greyish metallic element belonging to *group 13 (formerly IIIB) of the periodic table; a.n. 81; r.a.m. 204.39; r.d. 11.85 (20°C); m.p. 303.5°C; b.p. 1457±10°C. It occurs in zinc blende and some iron ores and is recovered in small quantities from lead and zinc concentrates. The naturally occurring isotopes are thallium–203 and thallium–205; eleven radioisotopes have been identified. It has few uses – experimental alloys for special purposes and some minor uses in electronics. The sulphate has been used as a rodenticide. Thallium(I) compounds resemble those of the alkali metals. Thallium(III) compounds are easily reduced to the thallium(I) state and are therefore strong oxidizing agents. The element was discovered by Sir William Crookes in 1861.

( SEE WEB LINKS )

• Information from the WebElements site

**thallus** A relatively undifferentiated vegetative body with no true roots, stems, leaves, or vascular system. It is found in the algae, fungi, mosses, and liverworts and in the gametophyte generation of clubmosses, horsetails, and ferns.

**theca** *See* CAPSULE.

**theobromine (3,7-dimethylxanthine)** An alkaloid, $C_7H_8N_4O_2$, structurally related to caffeine. It is found in small quantities in tea and larger quantities in the cacao tree, hence its presence in cocoa and chocolate. Like theophylline it is used in treating certain respiratory conditions. Note that the compound contains no bromine. The name comes from the name of the cacao tree, *Theobroma cacao. See also* METHYL-XANTHINES.

**theodolite** An optical surveying instrument for measuring horizontal and vertical angles. It consists of a sighting telescope, with crosshairs in the eyepiece for focusing on the target, which can be rotated in both the horizontal and vertical planes. It is mounted on a tripod and a spirit level is used to indicate when the instrument is horizontal. The angles are read off graduated circles seen through a second eyepiece in the instrument.

**theophylline (1,3-dimethylxanthine)** An alkaloid, $C_7 H_8 N_4 O_2$, structurally related to caffein. It is found in small quantities in tea and is used medicinally for certain respiratory conditions. *See also* METHYLXANTHINES.

**theoretical physics** The study of physics by formulating and analysing theories that describe natural processes. Theoretical physics is complementary to the study of physics by experiment, which is called **experimental physics**. A large part of theoretical physics consists of analysing the results of experiments to see whether or not they obey particular theories. The branch of theoretical physics concerned with the mathematical aspects of theories in physics is called **mathematical physics**.

**theory** *See* LAWS, THEORIES, AND HYPOTHESES.

**theory of everything (TOE)** A theory that provides a unified description of all known types of elementary particles, all known forces in the universe, and the evolution of the universe. Some believe that *superstring theory is potentially a theory of everything. Others believe that it is impossible to formulate a theory of everything and that any such theory could only claim to be a theory for all forces, particles, and observations concerning the evolution of the universe known at the time. It is thought that a *quantum field theory or a *unified-field theory cannot be a theory of everything. A theory of everything should explain the number of dimensions in the universe and why the number of observable dimensions is four.

**therapeutic half-life** (in pharmacology) The time taken for half the dose of a drug to be excreted: used to calculate the most effective and nontoxic dosing intervals. It can be determined by administering a therapeutic dose of the drug labelled with a radioisotope (*see* LABELLING) and measuring the time for half of it to be excreted in the urine.

**therm** A practical unit of energy defined as $10^5$ British thermal units. 1 therm is equal to $1.055 \times 10^8$ joules.

**thermal analysis** A technique for chemical analysis and the investigation of the products formed by heating a substance. In **differential thermal analysis (DTA)** a sample is heated, usually in an inert atmosphere, and a plot of weight against temperature made. In **differential scanning calorimetry (DSC)** heat is added to or removed from a sample electrically as the temperature is increased, thus allowing the enthalpy changes due to thermal decomposition to be studied.

**thermal capacity** *See* HEAT CAPACITY.

**thermal conductivity** *See* CONDUCTIVITY.

**thermal diffusion** The diffusion that takes place in a fluid as a result of a temperature gradient. If a column of gas is maintained so that the lower end is cooler than the upper end, the heavier molecules in the gas will tend to remain at the lower-temperature end and the lighter molecules will diffuse to the higher-temperature end. This property has been used in the separation of gaseous isotopes (*see* CLUSIUS COLUMN).

**thermal equilibrium** *See* EQUILIBRIUM.

**thermal expansion** *See* EXPANSIVITY.

**thermalization** The reduction of the kinetic energy of neutrons in a thermal *nuclear reactor by means of a *moderator; the process of producing thermal neutrons.

**thermal neutrons** *See* MODERATOR; NUCLEAR REACTOR; THERMALIZATION.

**thermal reactor** *See* NUCLEAR REACTOR.

**thermionic emission** The emission of electrons, usually into a vacuum, from a heated conductor. The emitted current density, $J$, is given by the **Richardson** (or **Richardson–Dushman**) **equation**, i.e. $J = AT^2 \exp(-W/kT)$, where $T$ is the thermodynamic temperature of the emitter, $W$ is its *work function, $k$ is the Boltzmann constant, and $A$ is a constant. Thermionic emission is the basis of the *thermionic valve and the *electron gun in cathode-ray tubes.

**thermionics** The branch of electronics concerned with the study and design of devices based on the emission of electrons from metal or metal-oxide surfaces as a result of high temperatures. The primary concern of thermionics is the design of *thermionic valves and the electron guns of cathode-ray tubes and other devices.

**thermionic valve** An electronic valve based on *thermionic emission. In such valves the cathode is either directly heated by passing a current through it or indirectly heated by placing it close to a heated filament. Directly heated cathodes are usually made of tungsten wire, whereas indirectly heated cathodes are usually coated with barium and strontium oxides. Most electronic valves are thermionic vacuum devices, although a few have cold cathodes and some are gas-filled (*see* THYRATRON). *See* DIODE; TRIODE; TETRODE; PENTODE.

**thermistor** An electronic device the resistance of which decreases as its temperature increases. It consists of a bead, rod, or disk of various oxides of manganese, nickel, cobalt, copper, iron, or other metals. Thermistors are used as thermometers, often forming one element in a resistance bridge. They are used for this purpose in such applications as bearings, cylinder heads, and transformer cores. They are also used to compensate for the increased resistance of ordinary resistors when hot, and in vacuum gauges, time-delay switches, and voltage regulators.

**thermite** A stoichiometric powdered mixture of iron(III) oxide and aluminium for the reaction:

$$2Al + Fe_2O_3 \rightarrow Al_2O_3 + 2Fe$$

The reaction is highly exothermic and the increase in temperature is sufficient to melt the iron produced. It has been used for localized welding of steel objects (e.g. railway lines) in the **Thermit process**. Thermite is also used in incendiary bombs.

**thermochemistry** The branch of physical chemistry concerned with heats of chemical reaction, heats of formation of chemical compounds, etc.

**thermocline** A steep temperature gradient that exists in the middle zone (the **metalimnion**) of a lake and gives rise to thermally induced vertical stratification of the water. The metalimnion lies between the relatively warm **epilimnion** above and the cold **hypolimnion** below. The thermocline may repre-

sent a temperature change of 1°C for every incremental depth of 1 metre of water. It may be short-lived, especially in shallow lakes where wind action can mix the water from different levels. However, it can exist for most of the summer period in temperate lakes and sometimes nearly all year in tropical lakes. A thermocline can speed up the process of eutrophication by preventing the diffusion of oxygen from the epilimnion to the hypolimnion (*see* EUTROPHIC).

**thermocouple** A device consisting of two dissimilar metal wires or semiconducting rods welded together at their ends. A thermoelectric e.m.f. is generated in the device when the ends are maintained at different temperatures, the magnitude of the e.m.f. being related to the temperature difference. This enables a thermocouple to be used as a thermometer over a limited temperature range. One of the two junctions, called the **hot** or **measuring junction**, is exposed to the temperature to be measured. The other, the **cold** or **reference junction**, is maintained at a known reference temperature. The e.m.f. generated is measured by a suitable millivoltmeter or potentiometer incorporated into the circuit. *See* SEEBECK EFFECT; THERMOPILE.

**thermodynamics** The study of the laws that govern the conversion of energy from one form to another, the direction in which heat will flow, and the availability of energy to do work. It is based on the concept that in an isolated system anywhere in the universe there is a measurable quantity of energy called the *internal energy ($U$) of the system. This is the total kinetic and potential energy of the atoms and molecules of the system of all kinds that can be transferred directly as heat; it therefore excludes chemical and nuclear energy. The value of $U$ can only be changed if the system ceases to be isolated. In these circumstances $U$ can change by the transfer of mass to or from the system, the transfer of heat ($Q$) to or from the system, or by the work ($W$) being done on or by the system. For an adiabatic ($Q = 0$) system of constant mass, $\Delta U = W$. By convention, $W$ is taken to be positive if work is done on the system and negative if work is done by the system. For nonadiabatic systems of constant mass, $\Delta U = Q + W$. This statement, which is equivalent to the law of conservation of energy, is known as the **first law of thermodynamics**.

All natural processes conform to this law,

but not all processes conforming to it can occur in nature. Most natural processes are irreversible, i.e. they will only proceed in one direction (*see* REVERSIBLE PROCESS). The direction that a natural process can take is the subject of the **second law of thermodynamics**, which can be stated in a variety of ways. R. Clausius (1822–88) stated the law in two ways: "heat cannot be transferred from one body to a second body at a higher temperature without producing some other effect" and "the entropy of a closed system increases with time". These statements introduce the thermodynamic concepts of *temperature ($T$) and *entropy ($S$), both of which are parameters determining the direction in which an irreversible process can go. The temperature of a body or system determines whether heat will flow into it or out of it; its entropy is a measure of the unavailability of its energy to do work. Thus $T$ and $S$ determine the relationship between $Q$ and $W$ in the statement of the first law. This is usually presented by stating the second law in the form $\Delta U = T\Delta S - W$.

The second law is concerned with changes in entropy ($\Delta S$). The **third law of thermodynamics** provides an absolute scale of values for entropy by stating that for changes involving only perfect crystalline solids at *absolute zero, the change of the total entropy is zero. This law enables absolute values to be stated for entropies.

One other law is used in thermodynamics. Because it is fundamental to, and assumed by, the other laws of thermodynamics it is usually known as the **zeroth law of thermodynamics**. This states that if two bodies are each in thermal equilibrium with a third body, then all three bodies are in thermal equilibrium with each other. *See also* ENTHALPY; FREE ENERGY.

(🌐) SEE WEB LINKS
• A tutorial from the Division of Chemical Education, Purdue University, Indiana
• Thermodynamic properties from the Committee on Data for Science and Technology

**thermodynamic temperature** *See* TEMPERATURE.

**thermoelectricity** An electric current generated by temperature difference. *See* SEEBECK EFFECT. The converse effects, the *Peltier effect and the *Thomson effect, are also sometimes known as thermoelectric effects.

**thermogenesis** The production of heat

within tissues to raise body temperature or as an adaptive response (**adaptive thermogenesis**) to 'burn off' excess food energy intake. Heat production occurs especially in birds and mammals, animals that maintain their temperature within a narrow range (i.e. *endotherms), but is also found in some 'cold-blooded' vertebrates and invertebrates. There are two types of thermogenesis. **Shivering** involves repeated rapid contractions of antagonistic sets of skeletal muscles, which produce little net movement so that most of the chemical energy (in the form of ATP) is converted to heat rather than mechanical work. **Nonshivering thermogenesis** takes place in *fat cells (adipose tissue) and involves the breakdown of stored fat to generate heat in situ instead of its being transported to the liver for conversion to ATP. This process is activated by the sympathetic nervous system. Certain mammals have deposits of a special adipose tissue called *brown fat that is adapted to provide the body with bursts of intense heat. Stimulation of brown fat oxidation enables rapid warming during the arousal of hibernating animals, for example. Brown fat deposits are also present in human babies and other neonate mammals, to help protect them against hypothermia.

**thermograph** **1.** A recording thermometer used in meteorology to obtain a continuous record of temperature changes over a period on a graph. **2.** A record so obtained. **3. (thermogram)** A record obtained by the technique of *thermography.

**thermography** A medical technique that makes use of the infrared radiation from the human skin to detect an area of elevated skin temperature that could be associated with an underlying cancer. The heat radiated from the body varies according to the local blood flow, thus an area of poor circulation produces less radiation. A tumour, on the other hand, has an abnormally increased blood supply and is revealed on the **thermogram** (or **thermograph**) as a 'hot spot'. Thermography was formerly used in mammography before the advent of more sensitive techniques.

**thermoluminescence** *Luminescence produced in a solid when its temperature is raised. It arises when free electrons and *holes, trapped in a solid as a result of exposure to ionizing radiation, unite and emit photons of light. The process is made use of in **thermoluminescent dating**, which assumes that the number of electrons and holes trapped in a sample of pottery is related to the length of time that has elapsed since the pottery was fired. By comparing the luminescence produced by heating a piece of pottery of unknown age with the luminescence produced by heating similar materials of known age, a fairly accurate estimate of the age of an object can be made.

**thermoluminescent dating** *See* THERMOLUMINESCENCE.

**thermolysis (pyrolysis)** The chemical decomposition of a substance by heat. It is an important process in chemical manufacture, such as the thermal *cracking of hydrocarbons in the petroleum industry.

**thermometer** An instrument used for measuring the *temperature of a substance. A number of techniques and forms are used in thermometers depending on such factors as the degree of accuracy required and the range of temperatures to be measured, but they all measure temperature by making use of some property of a substance that varies with temperature. For example, **liquid-in-glass thermometers** depend on the expansion of a liquid, usually mercury or alcohol coloured with dye. These consist of a liquid-filled glass bulb attached to a partially filled capillary tube. In the **bimetallic thermometer** the unequal expansion of two dissimilar metals that have been bonded together into a narrow strip and coiled is used to move a pointer round a dial. The **gas thermometer**, which is more accurate than the liquid-in-glass thermometer, measures the variation in the pressure of a gas kept at constant volume. The **resistance thermometer** is based on the change in resistance of conductors or semiconductors with temperature change. Platinum, nickel, and copper are the metals most commonly used in resistance thermometers. *See also* PYROMETRY; THERMISTOR; THERMOCOUPLE.

**thermonuclear reaction** *See* NUCLEAR FUSION; THERMONUCLEAR REACTOR.

**thermonuclear reactor (fusion reactor)** A reactor in which *nuclear fusion takes place with the controlled release of energy. Although thermonuclear reactors do not yet exist, intense research in many parts of the world is being carried out with a view to achieving such a machine. There are two central problems in the creation of a self-

sustaining thermonuclear reactor: heating the reacting nuclides to the enormous *ignition temperature (about $40 \times 10^6$ K for a deuterium–tritium reaction) and containing the reacting nuclides for long enough for the fusion energy released to exceed the energy required to achieve the ignition temperature (*see* LAWSON CRITERION). The two methods being explored are **magnetic containment** and **pellet fusion**.

In the closed magnetic-containment device the fusion *plasma is contained in a toroidal-shaped reactor, called a **tokamak**, in which strong magnetic fields guide the charged plasma particles round the toroid without allowing them to contact the container walls. In open-ended magnetic systems the plasma is trapped between magnetic mirrors (strong magnetic fields) at the two ends of a straight containment vessel.

In pellet fusion the objective is to heat and compress a tiny pellet of the nuclear fuels, by means of a laser or an electron beam, so rapidly that fusion is achieved before the pellet flies apart. Results with this type of equipment have been comparable to those achieved by magnetic confinement.

**thermonuclear weapon** *See* NUCLEAR WEAPONS.

**thermophilic** Describing an organism that lives and grows optimally at extremely high temperatures, typically over 40°C. The majority are prokaryotes, such as the archaebacteria found in hot springs and in undersea hydrothermal vents. *See* EXTREMOPHILE.

**thermopile** A device used to detect and measure the intensity of radiant energy. It consists of a number of *thermocouples connected together in series to achieve greater sensitivity. The hot junctions of the thermocouples are blackened and exposed to the radiation to be detected or measured, while the cold junctions are shielded from the radiation. The thermoelectric e.m.f. generated enables the hot junction excess temperature to be calculated and the radiant intensity to be deduced. They are used in various applications, from a safety device that ceases to produce an electric current if a pilot light blows out to an instrument to measure the heat radiation received from the sun.

**thermoplastic** *See* PLASTICS.

**thermoregulation** Regulation of body temperature by any means, whether physiological or behavioural. Some animals, particularly mammals and birds, can maintain a fairly constant internal body temperature (*see* HOMOIOTHERMY), whereas in others the body temperature varies with the temperature of the environment (*see* POIKILOTHERMY). *See also* THERMOGENESIS.

**thermosetting** *See* PLASTICS.

**thermostat** A device that controls the heating or cooling of a substance in order to maintain it at a constant temperature. It consists of a temperature-sensing instrument connected to a switching device. When the temperature reaches a predetermined level the sensor switches the heating or cooling source on or off according to a predetermined program. The sensing thermometer is often a *bimetallic strip that triggers a simple electrical switch. Thermostats are used for space-heating controls, in water heaters and refrigerators, and to maintain the environment of a scientific experiment at a constant temperature.

**THF** *See* TETRAHYDROFURAN.

**thiamine** *See* VITAMIN B COMPLEX.

**thigmotropism (haptotropism)** The growth of an aerial plant organ in response to localized physical contact. For example, when a tendril of sweet pea touches a supporting structure, it curves in the direction of the support and coils around it. *See* TROPISM.

**thin-layer chromatography** A technique for the analysis of liquid mixtures using *chromatography. The stationary phase is a thin layer of an absorbing solid (e.g. alumina) prepared by spreading a slurry of the solid on a plate (usually glass) and drying it in an oven. A spot of the mixture to be analysed is placed near one edge and the plate is stood upright in a solvent. The solvent rises through the layer by capillary action carrying the components up the plate at different rates (depending on the extent to which they are absorbed by the solid). After a given time, the plate is dried and the location of spots noted. It is possible to identify constituents of the mixture by the distance moved in a given time. The technique needs careful control of the thickness of the layer and of the temperature. *See also* $R_F$ VALUE.

**thiocyanate** A salt or ester of thiocyanic acid.

**thiocyanic acid** An unstable gas, HSCN.

**thio ethers** *See* SULPHIDES.

**thiol group** *See* THIOLS.

**thiols (mercaptans; thio alcohols)** Organic compounds that contain the group –SH (called the **thiol group**, **mercapto group**, or **sulphydryl group**). Thiols are analogues of alcohols in which the oxygen atom is replaced by a sulphur atom. They are named according to the parent hydrocarbon; e.g. ethane thiol ($C_2H_5SH$). A characteristic property is their strong disagreeable odour. For example the odour of garlic is produced by ethane thiol. Unlike alcohols they are acidic, reacting with alkalis and certain metals to form saltlike compounds. The older name, mercaptan, comes from their ability to react with ('seize') mercury.

**thionyl chloride** *See* SULPHUR DICHLORIDE OXIDE.

**thionyl group** The group =SO, as in *sulphur dichloride oxide.

**thiophene** A colourless liquid, $C_4H_4S$, b.p. 84°C. It is a cyclic aromatic compound with a ring containing four carbon atoms and one sulphur atom. It is made from butane and sulphur and occurs as an impurity in commercial benzene. Thienyl compounds are formed by substituting various groups in the ring. Thiophene has a smell resembling benzene and is used as a solvent.

**thiosulphate** A salt containing the ion $S_2O_3^{2-}$ formally derived from thiosulphuric acid. Thiosulphates readily decompose in acid solution to give elemental sulphur and hydrogensulphite ($HSO_3^-$) ions.

**thiosulphuric acid** An unstable acid, $H_2S_2O_3$, formed by the reaction of sulphur trioxide with hydrogen sulphide. *See also* SULPHURIC ACID.

**thiourea** A white crystalline solid, $(NH_2)_2CS$; r.d. 1.4; m.p. 182°C. It is used as a fixer in photography.

**thixotropy** *See* NEWTONIAN FLUID.

**Thompson, Benjamin** *See* RUMFORD, COUNT.

**Thomson, Sir Joseph John** (1856–1940) British physicist, who became a professor at Cambridge University in 1884. He is best known for his work on *cathode rays, which led to his discovery of the *electron in 1897. He went on to study the conduction of electricity through gases, and it is for this work that he was awarded the Nobel Prize for physics in 1906.

t

**Thomson, William** See KELVIN, BARON.

**Thomson effect (Kelvin effect)** When an electric current flows through a conductor, the ends of which are maintained at different temperatures, heat is evolved at a rate approximately proportional to the product of the current and the temperature gradient. If either the current or the temperature gradient is reversed heat is absorbed rather than being evolved. It is named after Sir William Thomson (later Lord Kelvin).

**Thomson scattering** The scattering of electromagnetic radiation by free charged particles, especially electrons, when the photon energy is small compared with the energy equivalent to the *rest mass of the charged particles. The energy lost by the radiation is accounted for by classical theory as a result of the radiation emitted by the charged particles when they are accelerated in the transverse electric field of the radiation. It is named after Sir J. J. Thomson.

**thoracic cavity** The space within the *thorax, which in vertebrates contains the heart, lungs, and rib cage.

**thoracic duct** The main collecting vessel of the *lymphatic system, running longitudinally in front of the backbone. The thoracic duct drains its lymph into the superior vena cava.

**thoracic vertebrae** The *vertebrae of the upper back, which articulate with the *ribs. They lie between the *cervical vertebrae and the *lumbar vertebrae and are distinguished by a number of articulating facets for attachment of the ribs. In humans there are 12 thoracic vertebrae.

**thorax** The anterior region of the body trunk of animals. In vertebrates it contains the heart and lungs within the rib cage. It is particularly well-defined in mammals, being separated from the *abdomen by the *diaphragm. In insects the thorax is divided into an anterior **prothorax**, a middle **mesothorax**, and a posterior **metathorax**, each of which bears a pair of legs; the hindmost two segments also both carry a pair of wings. In other arthropods, especially crustaceans and arachnids, the thorax is fused with the head to form a **cephalothorax**.

**thoria** See THORIUM.

**thorium** Symbol Th. A grey radioactive metallic element belonging to the *actinoids; a.n. 90; r.a.m. 232.038; r.d. 11.5–11.9 (17°C);

m.p. 1740–1760°C; b.p. 4780–4800°C. It occurs in monazite sand in Brazil, India, and USA. The isotopes of thorium have mass numbers from 223 to 234 inclusive; the most stable isotope, thorium–232, has a half-life of $1.39 \times 10^{10}$ years. It has an oxidation state of (+4) and its chemistry resembles that of the other actinoids. It can be used as a nuclear fuel for breeder reactors as thorium–232 captures slow neutrons to breed uranium–233. Thorium dioxide (**thoria**, $ThO_2$) is used on gas mantles and in special refractories. The element was discovered by J. J. Berzelius in 1829.

**(((●))) SEE WEB LINKS**
- Information from the WebElements site

**thorium series** See RADIOACTIVE SERIES.

**thorn** A hard side stem with a sharp point at the tip, replacing the growing point. In some plants the development of thorns and subsequent suppression of the growing points may be a response to dry conditions. Examples are the thorns of gorse and hawthorn. Compare PRICKLE; SPINE.

**thread cell (nematoblast; cnidoblast)** A specialized cell found only in the ectoderm of the *Cnidaria. It contains a **nematocyst**, a fluid-filled sac within which lies a long hollow coiled thread. When a small sensory projection (**cnidocil**) on the surface of the thread cell is touched, e.g. by prey, the thread is shot out and adheres to the prey, coils round it, or injects poison into it. Numerous thread cells on the tentacles of jellyfish produce their sting.

**three-body problem** See MANY-BODY PROBLEM.

**threnardite** A mineral form of *sodium sulphate, $Na_2SO_4$.

**threonine** See AMINO ACID.

**threshold** 1. (in physics) The minimum value of a parameter or variable that will produce a specified effect. 2. (in physiology) The minimum intensity of a stimulus that is necessary to initiate a response.

**threshold frequency** See PHOTOELECTRIC EFFECT.

**thrombin** An enzyme that catalyses the conversion of fibrinogen to fibrin. See BLOOD CLOTTING; PROTHROMBIN.

**thrombocyte** See PLATELET.

**thromboplastin** A glycoprotein, released

from damaged tissues at the site of a wound, that initiates the cascade of reactions leading to the formation of a blood clot. *See* BLOOD CLOTTING.

**thrombosis** The obstruction of a blood vessel by a mass of blood cells and fibrin (**thrombus**), which can result from excessive *blood clotting.

**thromboxane A₂** *See* PROSTAGLANDIN.

**thrust** The propelling force generated by an aircraft engine or rocket. It is usually calculated as the product of the rate of mass discharge and the velocity of the exhaust gases relative to the vehicle.

**thulium** Symbol Tm. A soft grey metallic element belonging to the *lanthanoids; a.n. 69; r.a.m. 168.934; r.d. 9.321 (20°C); m.p. 1545°C; b.p. 1947°C. It occurs in apatite and xenotime. There is one natural isotope, thulium–169, and seventeen artificial isotopes have been produced. There are no uses for the element, which was discovered by Per Cleve (1840–1905) in 1879.

(((●))) **SEE WEB LINKS**
• Information from the WebElements site

**thumb drive** *See* USB DRIVE.

**thunderstorm** A convective storm accompanied by *lightning and thunder and a

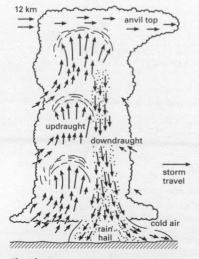

**Thunderstorm.** Cross section through a thunderstorm cell.

variety of weather conditions, especially heavy rain or hail, high winds, and sudden temperature changes. Thunderstorms originate when intense heating causes a parcel of moist air to rise, leading to instability and the development of cumulonimbus cloud – a towering cloud with a characteristic anvil-shaped top (see illustration). The exact mechanisms of thunderstorms are not fully understood. They occur most frequently in the tropics but are also common in the mid-latitudes.

**thymidine** A nucleoside consisting of one thymine molecule linked to a D-deoxyribose sugar molecule.

**thymine** A *pyrimidine derivative and one of the major component bases of *nucleotides and the nucleic acid *DNA.

**thymus** An organ, present only in vertebrates, that is concerned with development of *lymphoid tissue, particularly the white blood cells involved in cell-mediated *immune responses (*see* T CELL). In mammals it is a bilobed organ in the region of the lower neck, above and in front of the heart. In humans it undergoes progressive shrinkage throughout life, starting after the first 12 months of age.

**thyratron** A thermionic valve (usually a triode) that functions as a gas-filled relay. A positive pulse fed to a correctly biased thyratron causes a discharge to start and to continue until the anode voltage has been reduced. It has now been replaced by its solid-state counterpart, the silicon-controlled rectifier.

**thyristor** A silicon-controlled rectifier whose anode–cathode current is controlled by a signal applied to a third electrode (the gate) in much the same way as in a thyratron valve. It consists usually of a four-layer chip comprising three $p$–$n$ junctions.

**thyrocalcitonin** *See* CALCITONIN.

**thyroglobulin (TGB)** A glycoprotein, made in the thyroid gland, that consists of about 5000 amino acids, some of which are tyrosine residues. TGB is the precursor of the thyroid hormones, thyroxine and triiodothyronine. Iodine binds to the tyrosine residues in thyroglobulin, which is then hydrolysed into **iodotyrosines** that combine to form triiodothyronine ($T_3$) or thyroxine (tetraiodothyronine or $T_4$).

**thyroid gland** A bilobed endocrine gland

in vertebrates, situated in the base of the neck. It secretes two iodine-containing **thyroid hormones, thyroxine ($T_4$)** and **triiodothyronine ($T_3$)**, which are formed in the gland from *thyroglobulin; they control the rate of all metabolic processes in the body and influence physical development and activity of the nervous system. Growth and activity of the thyroid is controlled by *thyroid-stimulating hormone, secreted by the anterior *pituitary gland.

**thyroid-stimulating hormone (TSH; thyrotrophin)** A hormone, secreted by the anterior pituitary gland, that controls the synthesis and secretion of the two thyroid hormones, thyroxine and triiodothyronine, in the *thyroid gland. The secretion of thyroid-stimulating hormone is controlled by **thyrotrophin-releasing hormone (TRH)** from the hypothalamus. The release of TRH depends on many factors, including the levels of TSH, glucose, and thyroxine in the blood and the rate of metabolism in the body.

**thyrotrophin-releasing hormone (TRH)** *See* THYROID-STIMULATING HORMONE.

**thyroxine ($T_4$)** The principal hormone secreted by the *thyroid gland. *See also* THYROGLOBULIN.

**tibia** **1.** The larger of the two bones of the lower hindlimb of terrestrial vertebrates (*compare* FIBULA). It articulates with the *femur at the knee and with the *tarsus at the ankle. The tibia is the major load-bearing bone of the lower leg. **2.** The fourth segment of an insect's leg, which is attached to the femur.

**tidal energy** *See* TIDES.

**tidal volume** The volume of air taken in or expelled by an animal breathing normally at rest during each cycle of *ventilation. The average human has a tidal volume of approximately 500 cm$^3$.

**tides** The regular rise and fall of the water level in the earth's oceans as a result of the gravitational forces between the earth, moon, and sun. The forces involved are complex, but the moon is approximately twice as effective as the sun in causing tides. In illustration (a) the resultant gravitational forces between the moon and various points on the earth (solid lines) are shown as the vector sums of the tide-generating forces (broken lines) and a constant force (dotted

lines) that is the same at all points on the earth and is equal to the moon's attraction on the earth's centre. The resultant force when the moon is in zenith (Z in the illustration) is greater than that at nadir (N) because Z is closer to the moon than N and the force is inversely proportional to the square of the distance according to *Newton's law of gravitation. Illustration (b) shows how at full and new moon the sun and moon act together to produce the high-range **spring tides**, while at quarter moon the forces are at right angles to each other causing the low-range **neap tides**.

The use of **tidal energy**, estimated at some $4 \times 10^{18}$ J per annum at known tidal sites, dates back to medieval tidal mills. Modern tidal power stations use specially designed turbines, operated by tidal waters, to drive generators.

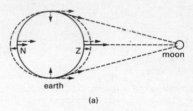

(a)

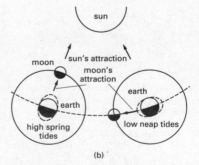

(b)

**Tides.**

**timbre** *See* QUALITY OF SOUND.

**time** A dimension that enables two otherwise identical events that occur at the same point in space to be distinguished (*see* SPACE–TIME). The interval between two such events forms the basis of time measurement. For general purposes, the earth's rotation on its axis provides the units of the clock (*see*

DAY) and the earth's orbit round the sun (*see* YEAR) provides the units of the calendar. For scientific purposes, intervals of time are now defined in terms of the frequency of a specified electromagnetic radiation (*see* SECOND). *See also* TIME DILATION; TIME REVERSAL.

**timebase** A voltage applied to the electron beam in a *cathode-ray tube at regular intervals so that the luminous spot on the screen is deflected in a predetermined manner. The timebase is usually designed to make the beam sweep the screen horizontally, the period during which the spot returns to its starting point (the **flyback**) being suppressed in some contexts (e.g. in television) although in a *cathode-ray oscilloscope the timebase can be put to more complicated uses.

**time dependence of fundamental constants** The possible variation with time in the values of the fundamental constants of nature, such as the speed of light. Such a variation has been proposed in various theories but there is no conclusive evidence at present for any such variation.

**time dilation (time dilatation)** The principle, predicted by Einstein's special theory of *relativity, that intervals of time are not absolute but are relative to the motion of the observers. If two identical clocks are synchronized and placed side by side in an inertial frame of reference they will read the same time for as long as they both remain side by side. However, if one of the clocks has a velocity relative to the other, which remains beside a stationary observer, the travelling clock will show, to that observer, that less time has elapsed than the stationary clock. In general, the travelling clock goes more slowly by a factor $\sqrt{(1 - v^2/c^2)}$, when measured in a frame of reference travelling at a velocity $v$ relative to another frame of reference; $c$ is the speed of light. The principle has been verified in a number of ways; for example, by comparing the lifetimes of fast muons, which increase with the speed of the particles to an extent predicted by this factor.

**time-lapse photography** A form of photography used to record a slow process, such as plant growth. A series of single exposures of the object is made on film at predetermined regular intervals. The film produced can then be projected at normal cine speeds and the process appears to be taking place at an extremely high rate.

**time-of-flight mass spectrometer** *See* MASS SPECTROMETRY.

**time reversal** Symbol T. The operation of replacing time $t$ by time $-t$. The *symmetry of time reversal is known as **T invariance**. As with CP violation, **T violation** occurs in certain weak interactions, notably kaon decay, and in processes involving mesons with bottom quarks. *See also* CP INVARIANCE; CPT THEOREM.

**time sharing** A means by which several jobs – data plus the programs to manipulate the data – share the processing time and other resources of a computer. A brief period is allocated to each job by the computer's operating system, and the computer switches rapidly between jobs. A *multiaccess system relies on time sharing.

**tin** Symbol Sn. A silvery malleable metallic element belonging to *group 14 (formerly IVB) of the periodic table; a.n. 50; r.a.m. 118.69; r.d. 7.28; m.p. 231.88°C; b.p. 2260°C. It is found as tin(IV) oxide in ores, such as cassiterite, and is extracted by reduction with carbon. The metal (called **white tin**) has a powdery nonmetallic allotrope **grey tin**, into which it changes below 18°C. The formation of this allotrope is called **tin plague**; it can be reversed by heating to 100°C. The natural element has 21 isotopes (the largest number of any element); five radioactive isotopes are also known. The metal is used as a thin protective coating for steel plate and is a constituent of a number of alloys (e.g. phosphor bronze, gun metal, solder, Babbitt metal, and pewter) Chemically it is reactive. It combines directly with chlorine and oxygen and displaces hydrogen from dilute acids. It also dissolves in alkalis to form *stannates. There are two series of compounds with tin in the +2 and +4 oxidation states.

**(()) SEE WEB LINKS**
• Information from the WebElements site

**Tinbergen, Niko(laas)** (1907–88) Dutch-born British zoologist and ethologist. Working first at Leiden University, he moved to Oxford in 1947, becoming professor of animal behaviour in 1966. With Konrad *Lorenz he was a pioneer of ethology, working with animals in their natural setting. In later years Tinbergen attempted to apply ethological principles to human beings, especially autistic children. He shared the Nobel Prize for physiology or medicine with Lorenz and Karl von Frisch (1886–1982).

**tin(II) chloride** A white solid, $SnCl_2$, soluble in water and ethanol. It exists in the anhydrous form (rhombic; r.d. 3.95; m.p. 246°C; b.p. 652°C) and as a dihydrate, $SnCl_2.2H_2O$ (monoclinic; r.d. 2.71; m.p. 37.7°C). The compound is made by dissolving metallic tin in hydrochloric acid and is partially hydrolysed in solution.

$$Sn^{2+} + H_2O \rightleftharpoons SnOH^+ + H^+$$

Excess acid must be present to prevent the precipitation of basic salts. In the presence of additional chloride ions the pyramidal ion $[SnCl_3]^-$ is formed; in the gas phase the $SnCl_2$ molecule is bent. It is a reducing agent in acid solutions and oxidizes slowly in air:

$$Sn^{2+} \rightarrow Sn^{4+} + 2e$$

**tin(IV) chloride** A colourless fuming liquid, $SnCl_4$, hydrolysed in cold water, decomposed by hot water, and soluble in ethers; r.d. 2.226; m.p. 33°C; b.p. 114°C. Tin(IV) chloride is a covalent compound, which may be prepared directly from the elements. It dissolves sulphur, phosphorus, bromine, and iodine, and there is evidence for the presence of species such as $SnCl_2I_2$. In hydrochloric acid and in chloride solutions the coordination is extended from four to six by the formation of the $SnCl_6^{2-}$ ion.

**tincture** A solution with alcohol as the solvent (e.g. tincture of iodine).

**tin(IV) hydride (stannane)** A highly reactive and volatile gas (b.p. –53°), $SnH_4$, which decomposes on moderate heating (150°C). It is prepared by the reduction of tin chlorides using lithium tetrahydridoaluminate(III) and is used in the synthesis of some organo-tin compounds. The compound has reducing properties.

**tin(IV) oxide (tin dioxide)** A white solid, $SnO_2$, insoluble in water; tetrahedral; r.d. 6.95; m.p. 1127°C; sublimes between 1800°C and 1900°C. Tin(IV) oxide is trimorphic: the common form, which occurs naturally as the ore *cassiterite, has a rutile lattice but hexagonal and rhombic forms are also known. There are also two so-called dihydrates, $SnO_2.2H_2O$, known as α- and β-stannic acid. These are essentially tin hydroxides. Tin(IV) oxide is amphoteric, dissolving in molten alkalis to form *stannates; in the presence of sulphur, thiostannates are produced.

**tin plague** *See* TIN.

**tin(II) sulphide** A grey-black cubic or monoclinic solid, SnS, virtually insoluble in water; r.d. 5.22; m.p. 882°C; b.p. 1230°C. It has a layer structure similar to that of black phosphorus. Its heat of formation is low and it can be made by heating the elements together. Above 265°C it slowly decomposes (disproportionates) to tin(IV) sulphide and tin metal. The compound reacts with hydrochloric acid to give tin(II) chloride and hydrogen sulphide.

**tin(IV) sulphide (mosaic gold)** A bronze or golden yellow crystalline compound, $SnS_2$, insoluble in water and in ethanol; hexagonal; r.d. 4.5; decomposes at 600°C. It is prepared by the reaction of hydrogen sulphide with a soluble tin(IV) salt or by the action of heat on thiostannic acid, $H_2SnS_3$. The golden-yellow form used for producing a gilded effect on wood is prepared by heating tin, sulphur, and ammonium chloride.

**T invariance** *See* TIME REVERSAL.

**tissue** A collection of similar cells organized to carry out one or more particular functions. For example, in animals nervous tissue is specialized to perceive and transmit stimuli. An organ, such as a lung or kidney, contains many different types of tissues.

**tissue culture** The growth of the tissues of living organisms outside the body in a suitable culture medium. Culture (or nutrient) media contain a mixture of nutrients either in solid form (e.g. in *agar) or in liquid form (e.g. in *physiological saline). Tissue culture has proved to be invaluable for gaining information about factors that control the growth and differentiation of cells. Culture of plant tissues has resulted in the regeneration of complete plants, enabling commercial propagation (e.g. of orchids) and – through culture of meristem tissues – the production of virus-free crop plants. *See also* EXPLANTATION; TISSUE ENGINEERING.

**tissue engineering** The creation of synthetic or semisynthetic tissue that can be used instead of human tissue in surgery. Different kinds of tissue have been developed or are currently being researched, including skin, bone, cartilage, cornea, and spinal tissue. For example, the first such product to gain approval for clinical use was a form of artificial skin consisting of a thin sheet of collagen gel infiltrated with two layers of cultured human cells – keratinocytes on the outer surface to form the 'epidermis', and fibroblasts on the inner surface to form the 'dermis'. More rigid tissues, such as synthetic bone and cartilage, are typically based on a biopolymer scaffold, which is treated with growth factors and seeded with cultured bone cells or cartilage cells to secrete the natural tissue material.

**tissue fluid** The fluid, consisting of water, ions, and dissolved gases and food substances, that is formed when blood is ultrafiltered (*see* ULTRAFILTRATION) from the capillaries into the intercellular spaces. The pressure in the arterial capillaries causes most components of the blood to pass across the capillary walls; blood cells and most of the plasma proteins are retained in the capillaries. The tissue fluid surrounds the body cells, facilitating the exchange of nutrients and waste materials. At the venous end of the capillaries, the tissue fluid is drawn into the capillaries by *osmosis.

**Titan** The largest satellite of *Saturn and the only one to possess a thick planet-like atmosphere. With a mean diameter of 5152 km and a mass of $1.345 \times 10^{23}$ kg, it is the second largest satellite in the solar system after Ganymede (*see* GALILEAN SATELLITES). Titan orbits Saturn in synchronous rotation at a distance of 1 221 900 km once every 15.95 days. Titan's disc appears reddish-orange, due to the photochemically produced fog that obscures its surface. The atmosphere is composed almost entirely of nitrogen. It exerts a pressure at the surface that is equal to about 1.6 times that of the earth's atmosphere. In January 2005 the space probe Huygens landed on Titan. It found a world littered with ice boulders, where liquid methane falls like rain and forms lakes and channels of standing liquid hydrocarbons.

**titania** *See* TITANIUM(IV) OXIDE.

**titanium** Symbol Ti. A white metallic *transition element; a.n. 22; r.a.m. 47.9; r.d. 4.5; m.p. 1660±10°C; b.p. 3287°C. The main sources are rutile ($TiO_2$) and, to a lesser extent, ilmenite ($FeTiO_3$). The element also occurs in numerous other minerals. It is obtained by heating the oxide with carbon and chlorine to give $TiCl_4$, which is reduced by the *Kroll process. The main use is in a large number of strong light corrosion-resistant alloys for aircraft, ships, chemical plant, etc. The element forms a passive oxide coating in air. At higher temperatures it reacts with oxygen, nitrogen, chlorine, and other nonmetals. It dissolves in dilute acids.

The main compounds are titanium(IV) salts and complexes; titanium(II) and titanium(III) compounds are also known. The element was first discovered by William Gregor (1761–1817) in 1789.

(((🌐))) **SEE WEB LINKS**

• Information from the WebElements site

**titanium dioxide** *See* TITANIUM(IV) OXIDE.

**titanium(IV) oxide** (titania; titanium dioxide) A white oxide, $TiO_2$, occurring naturally in various forms, particularly the mineral rutile. It is used as a white pigment and as a filler for plastics, rubber, etc.

**Titius–Bode law** *See* BODE, JOHANN ELERT.

**titration** A method of volumetric analysis in which a volume of one reagent (the **titrant**) is added to a known volume of another reagent slowly from a burette until an end point is reached (*see* INDICATOR). The volume added before the end point is reached is noted. If one of the solutions has a known concentration, that of the other can be calculated.

**titre** 1. The number of infectious virus particles present in a suspension. 2. A measure of the amount of *antibody present in a sample of serum, given by the highest dilution of the sample that results in the formation of visible clumps with the appropriate antigen (*see* AGGLUTINATION). 3. The concentration of a solution as measured by *titration. 4. The minimum quantity of solution required to complete a reaction in a titration.

**T lymphocyte** *See* T CELL.

**TNT** *See* TRINITROTOLUENE.

**toads** *See* AMPHIBIA.

**tobacco mosaic virus (TMV)** A rigid rod-shaped RNA-containing virus that causes distortion and blistering of leaves in a wide range of plants, especially the tobacco plant. It is transmitted by insects when they feed on plant tissue. TMV was the first virus to be discovered.

**tocopherol** *See* VITAMIN E.

**TOE** *See* THEORY OF EVERYTHING.

**tokamak** *See* THERMONUCLEAR REACTOR.

**Tollens reagent** A reagent used in testing for aldehydes. It is made by adding sodium hydroxide to silver nitrate to give silver(I) oxide, which is dissolved in aqueous ammonia (giving the complex ion $[Ag(NH_3)_2]^+$).

The sample is warmed with the reagent in a test tube. Aldehydes reduce the complex Ag⁺ ion to metallic silver, forming a bright silver mirror on the inside of the tube (hence the name **silver-mirror test**). Ketones give a negative result. It is named after Bernhard Tollens (1841–1918).

**toluene** *See* METHYLBENZENE.

**tomography** The use of X-rays to photograph a selected plane of a human body with other planes eliminated. The **CT** (**computerized tomography**) **scanner** is a ring-shaped X-ray machine that rotates through 180° around the horizontal patient, making numerous X-ray measurements every few degrees. The vast amount of information acquired is built into a three-dimensional image of the tissues under examination by the scanner's own computer. The patient is exposed to a dose of X-rays only some 20% of that used in a normal diagnostic X-ray.

**Tomonaga–Luttinger liquid** *See* LUTTINGER LIQUID.

**tone (tonus)** The state of sustained tension in muscles that is necessary for the maintenance of posture. In a tonic muscle contraction, only a certain proportion of the muscle fibres are contracting at any given time; the rest are relaxed and recovering for subsequent contractions. The fibres involved in tone contract more slowly than the fast fibres used for rapid responses by the same muscle. The proportions of slow and fast fibres depend on the function of the muscle.

**tongue** A muscular organ of vertebrates that in most species is attached to the floor of the mouth. It plays an important role in manipulating food during chewing and swallowing and in terrestrial species it bears numerous *taste buds on its upper surface. In some advanced vertebrates the tongue is used in the articulation of sounds, particularly in human speech.

**tonoplast (vacuole membrane)** The single membrane that bounds the *vacuole of plant cells.

**tonsil** A mass of *lymphoid tissue, several of which are situated at the back of the mouth and throat in higher vertebrates. In humans there are the **palatine tonsils** at the back of the mouth, **lingual tonsils** below the tongue, and **pharyngeal tonsils** (or **adenoids**) in the pharynx. They are concerned

with the production of *lymphocytes and therefore with defence against infection.

**tonus** *See* TONE.

**tooth** Any of the hard structures in vertebrates that are used principally for biting and chewing food but also for attack, grooming, and other functions. In fish and amphibians the teeth occur all over the palate, but in higher vertebrates they are concentrated on the jaws. They evolved in cartilaginous fish as modified placoid *scales, and this is reflected in their structure: a body of bony *dentine with a central *pulp cavity and an outer covering of *enamel on the exposed surface (**crown**). The portion of the tooth embedded in the jawbone is the **root** (see illustration).

In mammals there are four different types of teeth, specialized for different functions (*see* CANINE TOOTH; INCISOR; MOLAR; PREMOLAR). Their number varies with the species (*see* DENTAL FORMULA). *See also* DECIDUOUS TEETH; PERMANENT TEETH.

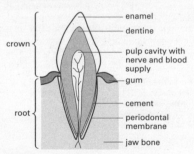

**Tooth.** Section through an incisor tooth.

**topaz** A variably coloured aluminium silicate mineral, $Al_2(SiO_4)(OH,F)_2$, that forms orthorhombic crystals. It occurs chiefly in acid igneous rocks, such as granites and pegmatites. Topaz is valued as a gemstone because of its transparency, variety of colours (the wine-yellow variety being most highly prized), and great hardness (8 on the Mohs' scale). When heated, yellow or brownish topaz often becomes a rose-pink colour. The main sources of topaz are Brazil, Russia, and the USA.

**top carnivore** *See* CONSUMER.

**topography** The relief and position of the natural and cultural features of the earth's surface.

**topological defect** A stable configuration of a system that results from topologically distinct solutions to differential equations describing the system. A number of different phenomena are identifiable as topological defects, including screw dislocations in crystals, *solitons, and *skyrmions. In some cosmological theories, topological defects are thought to have formed in the early universe.

**topological insulator** An insulator in which there is an excitation gap due to *spin–orbit coupling. It is so named because such an insulator is associated with topology.

**topology** The branch of geometry concerned with the properties of geometrical objects that are unchanged by continuous deformations, such as twisting or stretching. Mathematical approaches employing topology are of great importance in modern theories of the *fundamental interactions.

**topsoil** *See* SOIL.

**tornado** A violently rotating column of air, 10–100 m in diameter, usually made visible by a funnel cloud, that extends below cloud, usually cumulonimbus, and is in contact with the ground surface. Wind speeds of up to 100 metres per second may occur, but the damage to meteorological instruments caused by the passage of tornadoes makes exact measurement difficult. Tornadoes frequently occur in groups and are most common in the central USA and Australia, where they cause considerable destruction.

**torque (moment of a force or couple)** The product of a force and its perpendicular distance from a point about which it causes rotation or *torsion. The unit of torque is the newton metre, a vector product, unlike the joule, also equal to a newton metre, which is a scalar product. A turbine produces a torque on its central rotating shaft. *See also* COUPLE.

**torr** A unit of pressure, used in high-vacuum technology, defined as 1 mmHg. 1 torr is equal to 133.322 pascals. The unit is named after Evangelista Torricelli (1609–47).

**Torricellian vacuum** The vacuum formed when a long tube, closed at one end and filled with mercury, is inverted into a mercury reservoir so that the open end of the tube is below the surface of the mercury. The pressure inside the Torricellian vacuum is

the vapour pressure of mercury, about $10^{-3}$ torr.

**torsion** A twisting deformation produced by a *torque or *couple. A **torsion bar** is a form of spring in which one end of a bar is fixed and a torque is applied to the other end. Torsion bars are used in the suspension systems of motor vehicles.

**torsion balance** An instrument for measuring very weak forces. It consists of a horizontal rod fixed to the end of a vertical wire or fibre or to the centre of a taut horizontal wire. The forces to be measured are applied to the end or ends of the rod. The turning of the rod may be measured by the displacement of a beam of light reflected from a plane mirror attached to it. The best-known form is that used by Henry Cavendish (1731–1810) and later by Sir Charles Boys (1855–1944) to determine the *gravitational constant; in this form the balance is calibrated by determining the torsional coefficient of the suspension by treating the device as a torsional pendulum.

**torus** 1. (in mathematics) A solid generated by rotating a circle about an external line in its plane, also called an **anchor ring**. It has the shape of the inner tube of a tyre. If $r$ is the radius of the rotating circle and $R$ the distance between the centre of the circle and the axis of rotation, the volume of the torus is $2\pi^2 R r^2$ and the surface area is $4\pi^2 R r$. In Cartesian coordinates, if the $z$-axis is the axis of rotation, the equation of the torus is $[\sqrt{(x^2 + y^2)} - R]^2 + z^2 = r^2$. 2. (in botany) *See* RECEPTACLE.

**total internal reflection** The total reflection of a beam of light at the interface of one medium and another medium of lower refractive index, when the angle of incidence to the second medium exceeds a specific **critical angle**.

If a beam of light passing through a medium $A$ (say glass) strikes the boundary to a medium $B$ of lower refractive index (say air) with a small angle of incidence $i$, part will be refracted, with an angle of refraction $r$, and part will be reflected (see illustration a overleaf). If $i$ is increased it will reach a critical angle $c$, at which $r = 90°$ (see illustration b overleaf). If $i$ is now increased further, no refraction can occur and all the light energy is reflected at the interface (see illustration c overleaf). This total internal reflection occurs when $c$ (given by $n \sin c = 1$) is exceeded ($n$ is the refractive index of $A$ relative to $B$). The

**Total internal reflection.**

critical angle of optical glass is usually about 40° and total internal reflection is made use of by incorporating prisms in some optical instruments instead of mirrors.

**totality** The period during a total *eclipse of the sun in which the view of the sun's surface from a point on the earth is totally obscured by the moon. The maximum duration of totality is 7.67 minutes, but it is usually less.

**total-radiation pyrometer** *See* PYROMETRY.

**touch** The sense that enables the texture of objects and substances to be perceived. Touch receptors occur in the *skin, being concentrated in the tips of the finger in humans.

**tourmaline** A group of minerals composed of complex cyclosilicates containing boron with the general formula $NaR_3^{2+}Al_6B_3Si_6O_{27}(H,F)_4$, where $R = Fe^{2+}$, Mg, or (Al + Li). The crystals are trigonal, elongated, and variably coloured, the two ends of the crystals often having different colours. Tourmaline is used as a gemstone and because of its double refraction and piezoelectric properties is also used in polarizers and some pressure gauges.

**Townes, Charles** *See* BASOV, NIKOLAI GENNEDIYEVITCH.

**toxicogenomics** The study of the toxic effects of drugs or other substances on patterns of gene expression in particular cells or tissues. It is thus a synthesis of toxicology and genomics, and is aimed at identifying potential new drugs more efficiently and revealing the genetic basis of toxic reactions to compounds.

**toxicology** The science of the study of poisons. Originally developed by Paracelsus (1493–1541), toxicology is concerned with

the investigation of the deleterious effects of all foreign substances (*xenobiotics) on living organisms.

**toxin** A poison produced by a living organism, especially a bacterium. An **endotoxin** is released only when the bacterial cell dies or disintegrates. An **exotoxin** is secreted by a bacterial cell into the surrounding medium. In the body a toxin acts as an *antigen, producing an *immune response.

**trace element** *See* ESSENTIAL ELEMENT.

**trace fossil** *See* FOSSIL.

**trachea 1. (windpipe)** The tube in air-breathing vertebrates that conducts air from the throat to the *bronchi. It is strengthened with incomplete rings of cartilage. **2.** An air channel in insects and most other terrestrial arthropods. Tracheae occur as ingrowths of the body wall. They open to the exterior by **spiracles** and branch into finer channels (**tracheoles**) that terminate in the tissues (*see also* AIR SAC). Pumping movements of the abdominal muscles cause air to be drawn into and out of the tracheae.

**tracheid** A type of cell occurring within the *xylem of conifers, ferns, and related plants. Tracheids are elongated and their walls are usually extensively thickened by deposits of lignin. Water flows from one tracheid to another through unthickened regions (pits) in the cell walls. *Compare* VESSEL ELEMENT.

**tracheophyte** Any plant that has elaborate tissues, including *vascular tissue; a conspicuous *sporophyte generation; and complex leaves with waterproof cuticles. Tracheophytes include plants of the phyla *Lycophyta, *Sphenophyta, *Filicinophyta, *Coniferophyta, and *Anthophyta. In traditional classification systems these were regarded as classes of the division Tracheophyta.

**tracing (radioactive tracing)** *See* LA-BELLING.

**trajectory** *See* PHASE SPACE.

**transactinide elements** Elements with an atomic number greater than 103, i.e. elements above lawrencium in the *periodic table. So far, elements up to 112 have been detected. Because of the highly radioactive and transient nature of these elements, there has been much dispute about priority of discovery and, consequently, naming of the elements. The International Union of Pure and Applied Chemistry (IUPAC) introduced a set of systematic temporary names based on affixes, as shown in the table.

| Affix | Number | Symbol |
| --- | --- | --- |
| nil | 0 | n |
| un | 1 | u |
| bi | 2 | b |
| tri | 3 | t |
| quad | 4 | q |
| pent | 5 | p |
| hex | 6 | h |
| sept | 7 | s |
| oct | 8 | o |
| enn | 9 | e |

**Transactinide elements.**

All these element names end in -ium. So, for example, element 109 in this system is called un + nil + enn + ium, i.e. unnilennium, and given the symbol u+n+e, i.e. Une.

One long-standing dispute was about the element 104 (rutherfordium), which has also been called kurchatovium (Ku). There have also been disputes between IUPAC and the American Chemical Union about element names.

In 1994 IUPAC suggested the following list:
mendelevium (Md, 101)
nobelium (No, 102)
lawrencium (Lr, 103)
dubnium (Db, 104)
joliotium (Jl, 105)
rutherfordium (Rf, 106)
bohrium (Bh, 107)
hahnium (Hn, 108)
meitnerium (Mt, 109)

The ACU favoured a different set of names:
mendelevium (Md, 101)
nobelium (No, 102)
lawrencium (Lr, 103)

rutherfordium (Rf, 104)
hahnium (Ha, 105)
seaborgium (Sg, 106)
nielsbohrium (Ns, 107)
hassium (Hs, 108)
meitnerium (Mt, 109)

A compromise list was adopted by IUPAC in 1997 and is generally accepted:
mendelevium (Md, 101)
nobelium (No, 102)
lawrencium (Lr, 103)
rutherfordium (Rf, 104)
dubnium (Db, 105)
seaborgium (Sg, 106)
bohrium (Bh, 107)
hassium (Hs, 108)
meitnerium (Mt, 109)

Element 110 was named as darmstadtium in 2003 and element 111 was named roentgenium in 2004. So far elements 112 (ununbium, Uub), 113 (ununtrium, Uut), 114 (ununquadium, Uuq), 115 (ununpentium, Uup), and 116 (ununhexium, Uuh) are not officially named. All these elements are unstable and have very short half-lives.

**transamination** A biochemical reaction in amino acid metabolism in which an amine group is transferred from an amino acid to a keto acid to form a new amino acid and keto acid. The coenzyme required for this reaction is pyridoxal phosphate.

**transcendental number** A number that is not algebraic, such as $\pi$ or $e$. A **transcendental function** is also nonalgebraic, such as $a^x$, $\sin x$, or $\log x$.

**transcriptase** *See* REVERSE TRANSCRIPTASE.

**transcription** The process in living cells in which the genetic information of *DNA is transferred to a molecule of messenger *RNA (mRNA) as the first step in *protein synthesis (*see also* GENETIC CODE). Transcription takes place in the cell nucleus or nuclear region. It involves the action of RNA *polymerase enzymes in assembling the nucleotides necessary to form a complementary strand of mRNA from the DNA template, and (in eukaryote cells) the subsequent removal of the noncoding sequences from this primary transcript (*see* GENE SPLICING) to form a functional mRNA molecule. *See also* REVERSE TRANSCRIPTASE. *Compare* TRANSLATION.

**transcription factor** Any of a group of proteins that work synergistically to regulate gene activity by increasing or decreasing the binding of RNA *polymerases to the DNA molecule during the process of *transcription. This is achieved by the ability of the transcription factors to bind to the DNA molecule.

**transcriptome** The full complement of RNA transcripts of the genes of a cell or organism. The types and relative abundance of different transcripts, i.e. the messenger RNAs (mRNAs), can be obtained by analysing cell contents using oligonucleotide *DNA microarrays. Such an analysis provides a 'snapshot' of the expression pattern of the cell's genes. *See* TRANSCRIPTOMICS.

**transcriptomics** The study of the RNA transcripts of a cell, tissue, or organism (i.e. the *transcriptome). Transcriptomics is concerned with determining how the transcriptome, and hence pattern of gene expression, changes with respect to various factors, such as type of tissue, stage of development, hormones, drugs, or disease. It complements and overlaps with *proteomics.

**transducer** A device for converting a non-electrical signal, such as sound, light, heat, etc., into an electrical signal, or vice versa. Thus microphones and loudspeakers are **electroacoustic transducers**. An **active transducer** is one that can itself introduce a power gain and has its own power source. A **passive transducer** has no power source other than the actuating signal and cannot introduce gain.

**transduction** 1. The transfer of genetic material from one bacterial cell to another by means of a *bacteriophage. 2. The conversion of stimuli detected in sensory *receptor cells into electric impulses, which are transmitted to the brain by the nervous system. 3. **(signal transduction)** (in cell biology) Any mechanism by which binding of a *signal molecule to a cell-surface receptor triggers a response inside the cell. It often involves *second messengers.

**transect** A straight line across an expanse of ground along which ecological measurements are taken, continuously or at regular intervals. Thus an ecologist wishing to study the numbers and types of organisms at different distances above the low-tide line might sample at five-metre intervals along a number of transects perpendicular to the shore.

**trans effect** An effect in the substitution of inorganic square-planar complexes, in which certain ligands in the original complex are able to direct the incoming ligand into the trans position. The order of ligands in decreasing trans-directing power is: $CN^- > NO_2^- > I^- > Br^- > Cl^- > NH_3 > H_2O$.

**transferase** Any of a class of enzymes that catalyse the transfer of a group of atoms from one molecule to another.

**transfer RNA** *See* RNA.

**transformer** A device for transferring electrical energy from one alternating-current circuit to another with a change of voltage, current, phase, or impedance. It consists of a primary winding of $N_p$ turns magnetically linked by a ferromagnetic core or by proximity to the secondary winding of $N_s$ turns. The **turns ratio** ($N_s/N_p$) is approximately equal to $V_s/V_p$ and to $I_p/I_s$, where $V_p$ and $I_p$ are the voltage and current fed to the primary winding and $V_s$ and $I_s$ are the voltage and current induced in the secondary winding, assuming that there are no power losses in the core. In practice, however, there are *eddy-current and *hysteresis losses in the core, incomplete magnetic linkage between the coils, and heating losses in the coils themselves. By the use of a *laminated core and careful design, transformers with 98% efficiency can be achieved.

**transgene** A gene that is taken from one organism and inserted into the germ line of another organism so that it is replicated as part of the genome and present in all the recipient's cells. *See* TRANSGENIC.

**transgenic** Describing an organism whose genome incorporates and expresses genes from another species (*transgenes). *See* GENETICALLY MODIFIED ORGANISMS (Feature).

**transient** A brief disturbance or oscillation in a circuit caused by a sudden rise in the current or e.m.f.

**trans-isomer** *See* ISOMERISM.

**transistor** A *semiconductor device capable of amplification in addition to rectification. It is the basic unit in radio, television, and computer circuits, having almost completely replaced the *thermionic valve. The **point-contact transistor**, which is now obsolete, was invented in 1948. It consists of a small germanium crystal with two rectifying point contacts attached to it; a third contact, called the **base**, makes a low-resistance non-rectifying (ohmic) connection with the crystal. Current flowing through the device between the point contacts is modulated by the signal fed to the base. This type of transistor was replaced by the **junction transistor**, which was developed in 1949–50. The **field-effect transistor** (FET) was a later invention. **Bipolar transistors**, such as the junction transistor, depend on the flow of both majority and minority carriers, whereas in **unipolar transistors**, such as the FET, the current is carried by majority carriers only.

In the bipolar junction transistor, two *p*-type semiconductor regions are separated by a thin *n*-type region, making a *p–n–p* structure. Alternatively, an *n–p–n* structure can also be used. In both cases the thin central region is called the base and one outer region of the sandwich is called the **emitter**, the other the **collector**. The emitter–base junction is forward-biased and the collector–base junction is reverse-biased. In the *p–n–p* transistor, the forward bias causes holes in the emitter region to flow across the junction into the base; as the base is thin, the majority of holes are swept right across it (helped by the reverse bias), into the collector. The minority of holes that do not flow from the base to the collector combine with electrons in the *n*-type base. This recombi-

nation is balanced by a small electron flow in the base circuit. The diagram illustrates the (conventional) current flow using the **common-base** type of connection. If the emitter, base, and collector currents are $I_e$, $I_b$, and $I_c$, respectively, then $I_e = I_b + I_c$ and the current gain is $I_c/I_b$.

Field-effect transistors are of two kinds, the **junction FET** (**JFET** or **JUGFET**) and the **insulated-gate FET** (**IGFET**; also known as a **MOSFET**, i.e. metal-oxide-semiconductor FET). Both are unipolar devices and in both the current flows through a narrow **channel** between two electrodes (the **gate**) from one region, called the **source**, to another, called the **drain**. The modulating signal is applied to the gate. In the JFET, the channel consists of a semiconductor material of relatively low conductivity sandwiched between two regions of high conductivity of the opposite polarity. When the junctions between these regions are reverse-biased, *depletion layers form, which narrow the channel. At high bias the depletion layers meet and pinch-off the channel completely. Thus the voltage applied to the two gates controls the thickness of the channel and thus its conductivity.

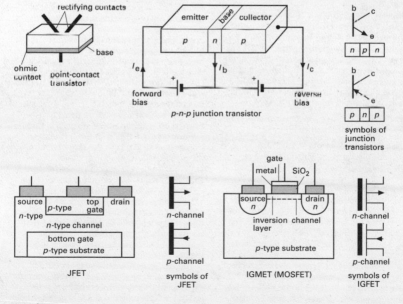

**Transistors.**

JFETs are made with both *n*-type and *p*-type channels.

In the IGFET, a wafer of semiconductor material has two highly doped regions of opposite polarity diffused into it, to form the source and drain regions. An insulating layer of silicon dioxide is formed on the surface between these regions and a metal conductor is evaporated on to the top of this layer to form the gate. When a positive voltage is applied to the gate, electrons move along the surface of the *p*-type substrate below the gate, producing a thin surface of *n*-type material, which forms the channel between the source and drain. This surface layer is called an **inversion layer**, as it has opposite conductivity to that of the substrate. The number of induced electrons is directly proportional to the gate voltage, thus the conductivity of the channel increases with gate voltage. IGFETs are also made with both *p*-type and *n*-type channels. Because MOS devices cannot be formed on gallium arsenide (there are no stable native oxides of GaAs), metal semiconductor FETs (MESFET) devices are used. This makes use of Schottky barrier (*see* SCHOTTKY EFFECT) as the gate electrode rather than a semiconductor junction.

**transit** (in astronomy) **1.** The passage of a planet across the face of a star, especially the passage of Mercury or Venus across the face of the sun as viewed from earth. **2.** The passage of a planetary satellite and its shadow across the disc of the parent planet. **3.** The passage of a celestial object across an observer's meridian as near as possible to the observer's zenith.

**transition** A change of a system from one quantum state to another.

**transition elements** A set of elements in the *periodic table in which filling of electrons in an inner *d*- or *f*-level occurs. With increasing proton number, electrons fill atomic levels up to argon, which has the electron configuration $1s^2 2s^2 2p^6 3s^2 3p^6$. In this shell, there are 5 *d*-orbitals, which can each contain 2 electrons. However, at this point the subshell of lowest energy is not the 3*d* but the 4*s*. The next two elements, potassium and calcium, have the configurations [Ar]$4s^1$ and [Ar]$4s^2$ respectively. For the next element, scandium, the 3*d* level is of lower energy than the 4*p* level, and scandium has the configuration [Ar]$3d^1 4s^2$. This filling of the inner *d*-level continues up to zinc

[Ar]$3d^{10} 4s^2$, giving the first transition series. There is a further series of this type in the next period of the table: between yttrium ([Kr]$4d5s^2$) and cadmium ([Kr]$4d^{10} 5s^2$). This is the second transition series. In the next period of the table the situation is rather more complicated. Lanthanum has the configuration [Xe]$5d^1 6s^2$. The level of lowest energy then becomes the 4*f* level and the next element, cerium, has the configuration [Xe]$4f^1 5d^1 6s^2$. There are 7 of these *f*-orbitals, each of which can contain 2 electrons, and filling of the *f*-levels continues up to lutetium ([Xe]$4f^{14} 5d^1 6s^2$). Then the filling of the 5*d* levels continues from hafnium to mercury. The series of 14 elements from cerium to lutetium is a 'series within a series' called an **inner transition series**. This one is the *lanthanoid series. In the next period there is a similar inner transition series, the *actinoid series, from thorium to lawrencium. Then filling of the *d*-level continues from element 104 onwards.

In fact, the classification of chemical elements is valuable only in so far as it illustrates chemical behaviour, and it is conventional to use the term 'transition elements' in a more restricted sense. The elements in the inner transition series from cerium (58) to lutetium (71) are called the lanthanoids; those in the series from thorium (90) to lawrencium (103) are the actinoids. These two series together make up the *f*-block in the periodic table. It is also common to include scandium, yttrium, and lanthanum with the lanthanoids (because of chemical similarity) and to include actinium with the actinoids. Of the remaining transition elements, it is usual to speak of three **main transition series**: from titanium to copper; from zirconium to silver; and from hafnium to gold. All these elements have similar chemical properties that result from the presence of unfilled *d*-orbitals in the element or (in the case of copper, silver, and gold) in the ions. The elements from 104 to 109 and the undiscovered elements 110 and 111 make up a fourth transition series. The elements zinc, cadmium, and mercury have filled *d*-orbitals both in the elements and in compounds, and are usually regarded as nontransition elements forming group 12 of the periodic table.

The elements of the three main transition series are all typical metals (in the nonchemical sense), i.e. most are strong hard materials that are good conductors of heat and electricity and have high melting and boiling points. Chemically, their behaviour depends on the existence of unfilled *d*-orbitals. They exhibit variable valency, have coloured compounds, and form *coordination compounds. Many of their compounds are paramagnetic as a result of the presence of unpaired electrons. Many of them are good catalysts. They are less reactive than the *s*- and *p*-block metals.

**transition point (transition temperature)** **1.** The temperature at which one crystalline form of a substance changes to another form. **2.** The temperature at which a substance changes phase. **3.** The temperature at which a substance becomes superconducting (*see* SUPERCONDUCTIVITY). **4.** The temperature at which some other change, such as a change of magnetic properties (*see also* CURIE POINT), takes place.

**transition state (activated complex)** The association of atoms of highest energy formed during a chemical reaction. The

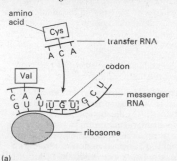

amino acid
Cys
transfer RNA
A C A
codon
Val
C A A
G U U U G U G C U
messenger RNA
ribosome

(a)

transition state can be regarded as a short-lived intermediate that breaks down to give the products. For example, in a $S_N2$ substitution reaction, one atom or group approaches the molecule as the other leaves. The transition state is an intermediate state in which both attacking and leaving groups are partly bound to the molecule, e.g.

$$B + RA \rightarrow B\text{---}R\text{---}A \rightarrow BR + A$$

In the theory of reaction rates, the reactants are assumed to be in equilibrium with this activated complex, which decomposes to the products.

**transition zone** *See* HYPOCOTYL.

**translation** **1.** (in biochemistry) The process in living cells in which the genetic information encoded in messenger *RNA (mRNA) in the form of a sequence of nucleotide triplets (*codons) is translated into a sequence of amino acids in a polypeptide chain during *protein synthesis (see illustration). Translation takes place on *ribosomes in the cell cytoplasm. The ribosomes move along the mRNA 'reading' each codon in turn. Molecules of transfer RNA (tRNA), each bearing a particular amino acid, are brought to their correct positions along the mRNA molecule: base pairing occurs between the bases of the codons and the complementary base triplets of tRNA (*see* ANTICODON). In this way amino acids are assembled in the correct sequence to form the polypeptide chain. **2.** (in physics) Motion of a body in which all the points in the body follow parallel paths.

**translocation** **1.** (in botany) The movement of minerals and chemical compounds within a plant. There are two main processes. The first is the uptake of soluble

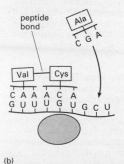

peptide bond
Ala
C G A
Val   Cys
C A A A C A
G U U U G U G C U

(b)

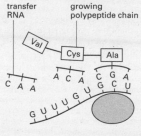

transfer RNA
growing polypeptide chain
Val
Cys   Ala
A C A   C G A
C A A   G C U
G U U U G U

(c)

**Translation.** The stages of translation in protein synthesis.

minerals from the soil and their passage upwards from the roots to various organs by means of the water-conducting vessels (*xylem). The second is the transfer of organic compounds, synthesized by the leaves, both upwards and downwards to various organs, particularly the growing points. This movement occurs within the *phloem tubes. *See also* MASS FLOW. **2.** (in genetics) A type of *chromosome mutation in which a section of one chromosome is broken off and becomes attached to another chromosome, resulting in a loss of genetic information from the first chromosome.

**translucent** Permitting the passage of radiation but not without some scattering or diffusion. For example, frosted glass allows light to pass through it but an object cannot be seen clearly through it because the light rays are scattered by it. *Compare* TRANSPARENT.

**transmission** **1.** (in neurophysiology) The one-way transfer of a nerve *impulse from one neuron to another across a *synapse. *See also* NEUROTRANSMITTER. *Compare* PROPAGATION. **2.** (in medicine) The spread of an *infection from person to person. This can occur in various ways, such as close contact with an infected person, including sexual contact (*see* SEXUALLY TRANSMITTED DISEASE); contact with a *vector or a *carrier of the disease; consuming food or drink contaminated with the infecting microorganism; and breathing in contaminated droplets of moisture, produced by coughing and sneezing. **3.** (in radio) *See* TRANSMITTER.

**transmission coefficient** *See* TRANSMITTANCE.

**transmission electron microscope** *See* ELECTRON MICROSCOPE.

**transmittance (transmission coefficient)** The ratio of the energy of some form of radiation transmitted through a surface to the energy falling on it. The reciprocal of the transmittance is the opacity.

**transmitter** **1.** The equipment used to generate and broadcast radio-frequency electromagnetic waves for communication purposes. In **transmitted-carrier transmission** it consists of a carrier-wave generator, a device for modulating the carrier wave in accordance with the information to be broadcast, amplifiers, and an aerial system. In **suppressed-carrier transmission**, the carrier component of the carrier wave is

not transmitted; one *sideband (**single-sideband transmission**) or both sidebands (**double-sideband transmission**) are transmitted and a local oscillator in the receiver regenerates the carrier frequency and mixes it with the received signal to detect the modulating wave. **2.** The part of a telephone system that converts sound into electrical signals. **3.** *See* NEUROTRANSMITTER.

**transmutation** The transformation of one element into another by bombardment of nuclei with particles. For example, plutonium is obtained by the neutron bombardment of uranium.

**trans-Neptunian object (TNO)** Any of the large number of objects orbiting the sun out beyond the planet *Neptune. They include the objects of the *Kuiper belt and the *scattered disc.

**transparent** Permitting the passage of radiation without significant deviation or absorption. *Compare* TRANSLUCENT. A substance may be transparent to radiation of one wavelength but not to radiation of another wavelength. For example, some forms of glass are transparent to light but not to ultraviolet radiation, while other forms of glass may be transparent to all visible radiation except red light. *See also* RADIOTRANSPARENT.

**transpiration** The loss of water vapour by plants to the atmosphere. It occurs mainly from the leaves through pores (stomata) whose primary function is gas exchange. The water is replaced by a continuous column of water (and dissolved nutrients) moving upwards from the roots within the *xylem vessels. The flow of this column of water is known as the **transpiration stream**, which is maintained by *root pressure and a combination of cohesive and adhesive forces in the xylem vessels according to the **cohesion–tension theory** (*see* COHESION). *See also* POTOMETER.

**transplantation** *See* GRAFT.

**transport coefficients** Quantities that characterize transport in a system. Examples of transport coefficients include electrical and thermal *conductivity. One of the main purposes of non-equilibrium *statistical mechanics is to calculate such coefficients from first principles. It is difficult to calculate transport coefficients exactly for non-interacting systems and it is therefore necessary to use *approximation techniques and/or *model systems. A transport coefficient gives a measure for flow in a system. An **inverse transport coefficient** gives a measure of resistance to flow in a system. An example of an inverse transport coefficient is *resistivity.

**transport number** Symbol *t*. The fraction of the total charge carried by a particular type of ion in the conduction of electricity through electrolytes.

**transport protein** A protein that penetrates or spans a cell membrane to permit the passage of a substance through the membrane. Some transport proteins form pores, or *channels, through which particular ions or molecules can pass. Other types of transport protein bind the substance on one face of the membrane, then change shape so that the substance is carried by the protein through the membrane to be released at the other face. Transport proteins often require energy to drive the transport process; this is provided by hydrolysis of ATP or by an existing concentration gradient. *See* ACTIVE TRANSPORT.

**transport theory** The theory of phenomena involving the transfer of matter or heat. The calculation of *transport coefficients and inverse transport coefficients, such as *conductivity and *viscosity, is an aim of transport theory. Calculations from first principles in transport theory start from *non-equilibrium statistical mechanics. Because of the difficulties involved in calculations in non-equilibrium statistical mechanics, transport theory uses approximate methods, including the *kinetic theory of gases and *kinetic equations, such as the *Boltzmann equation.

**transposon (transposable genetic element)** A mobile genetic element, known informally as a 'jumping gene', that can become integrated at many different sites along a chromosome. The simplest types of transposon are known as **insertion sequences**, typically found in bacteria and consisting of some 700–2500 base pairs and with numerous short repeated nucleotide sequences at either end. Larger and more complex are the **composite transposons**, which consist of a central portion, possibly containing functional genes, flanked by insertion sequences at either end. Transposons were first discovered by Barbara McClintock in maize in the 1940s and have since been found in other eukaryotes and in bacteria. They can disrupt gene expression

or cause deletions and inversions, and hence affect both the genotype and phenotype of the organisms concerned. Most eukaryotic transposons are *retrotransposons. Transposons account for a sizable proportion of the *repetitive DNA in eukaryotes.

**transuranic elements** Elements with an atomic number greater than 92, i.e. elements above uranium in the *periodic table. Most of these elements are unstable and have short half-lives. *See also* TRANSACTINIDE ELEMENTS.

**transverse wave** *See* WAVE.

**travelling wave** *See* WAVE.

**travelling-wave accelerator** *See* LINEAR ACCELERATOR.

**Travers, Morris** *See* RAMSAY, SIR WILLIAM.

**Trematoda** A class of parasitic flatworms (*see* PLATYHELMINTHES) comprising the flukes, such as *Fasciola* (liver fluke). Flukes have suckers and hooks to anchor themselves to the host and their body surface is covered by a protective cuticle. The whole life cycle may either occur within one host or require one or more intermediate hosts to transmit the infective eggs or larvae. *Fasciola hepatica*, for example, undergoes larval development in a land snail (the intermediate host) and infects sheep (the primary host) when contaminated grass containing the larvae is swallowed.

**triacetone triperoxide (TATP)** An organic peroxide, $C_9H_{18}O_6$, derived from acetone; r.d. 1.18, m.p. 91°C. It is a highly explosive white crystalline substance, very sensitive to heat and shock. It is favoured by some terrorist groups because it can easily be made from commonly available compounds. For example, it is probable that it was used in the London bombings of 7 July 2005.

**triacylglycerol** *See* TRIGLYCERIDE.

**trial-and-error learning** *See* LEARNING.

**triangle of vectors** A triangle constructed so that each of its sides represents one of three coplanar *vectors acting at a point with no resultant. If the triangle is completed, with the sides representing the vectors in both magnitude and direction, so that there are no gaps between the sides, then the vectors are in equilibrium. If the three vectors are forces, the figure is called a

**triangle of forces**; if they are velocities, it is a **triangle of velocities**.

**Triassic** The earliest period of the Mesozoic era. It began about 251 million years ago, following the Permian, the last period of the Palaeozoic era, and extended until about 200 million years ago when it was succeeded by the Jurassic. It was named, by F. von Alberti in 1834, after the sequence of three divisions of strata that he studied in central Germany – Bunter, Muschelkalk, and Keuper. The Triassic rocks are frequently difficult to distinguish from the underlying Permian strata and the term **New Red Sandstone** is often applied to rocks of the Permo-Triassic. During the period marine animals diversified: molluscs were the dominant invertebrates – ammonites were abundant and bivalves replaced the declining brachiopods. Reptiles were the dominant vertebrates and included turtles, phytosaurs, dinosaurs, and the marine ichthyosaurs.

**triatomic molecule** A molecule formed from three atoms (e.g. $H_2O$ or $CO_2$).

**triazine** See AZINE.

**tribe** A category used in the *classification of plants and animals that consists of several similar or closely related genera within a family. For example the Bambuseae, Oryzeae, Paniceae, and Aveneae are tribes of grasses.

**triboelectricity** *Static electricity produced as a result of friction.

**tribology** The study of friction, lubrication, and lubricants.

**triboluminescence** *Luminescence caused by friction; for example, some crystalline substances emit light when they are crushed as a result of static electric charges generated by the friction.

**tribromomethane (bromoform)** A colourless liquid *haloform, $CHBr_3$; r.d. 2.9; m.p. 8°C; b.p. 150°C.

**tricarbon dioxide (carbon suboxide)** A colourless gas, $C_3O_2$, with an unpleasant odour; r.d. 1.114 (liquid at 0°C); m.p. –111.3°C; b.p. 7°C. It is the acid anhydride of malonic acid, from which it can be prepared by dehydration using phosphorus(V) oxide. The molecule is linear (O:C:C:C:O).

**tricarboxylic acid cycle** See KREBS CYCLE.

**trichloroethanal (chloral)** A liquid aldehyde, $CCl_3CHO$; r.d. 1.51; m.p. –57.5°C; b.p. 97.8°C. It is made by chlorinating ethanal and used in making DDT. *See also* 2,2,2-TRICHLOROETHANEDIOL.

**2,2,2-trichloroethanediol (chloral hydrate)** A colourless crystalline solid, $CCl_3CH(OH)_2$; r.d. 1.91; m.p. 57°C; b.p. 96.3°C. It is made by the hydrolysis of trichloroethanal and is unusual in having two –OH groups on the same carbon atom. Gem diols of this type are usually unstable; in this case the compound is stabilized by the presence of the three Cl atoms. It is used as a sedative.

**trichloroethene (trichlorethylene)** A colourless liquid, $CCl_2=CHCl$, b.p. 87°C. It is toxic and nonflammable, with a smell resembling that of chloroform (trichloromethane). It is widely used as a solvent in dry cleaning and degreasing. It is also used to extract oils from nuts and fruit, as an anaesthetic, and as a fire extinguisher.

**trichloromethane (chloroform)** A colourless volatile sweet-smelling liquid *haloform, $CHCl_3$; r.d. 1.48; m.p. –63.5°C; b.p. 61.7°C. It can be made by chlorination of methane (followed by separation of the mixture of products) or by the haloform reaction. It is an effective anaesthetic but can cause liver damage and it has now been replaced by other halogenated hydrocarbons. Chloroform is used as a solvent and raw material for making other compounds.

**trichome** A hairlike projection from a plant epidermal cell. Examples include root hairs and the stinging hairs of nettle leaves.

**triclinic** See CRYSTAL SYSTEM.

**tricuspid valve** A valve, consisting of three flaps, situated between the right atrium and the right ventricle of the mammalian heart. When the right ventricle contracts, forcing blood into the pulmonary artery, the tricuspid valve closes the aperture to the atrium, thereby preventing any backflow of blood. The valve reopens to allow blood to flow from the atrium into the ventricle. *Compare* BICUSPID VALVE.

**tridymite** A mineral form of *silicon(IV) oxide, $SiO_2$.

**triglyceride (triacylglycerol)** An ester of glycerol (propane-1,2,3-triol) in which all three hydroxyl groups are esterified with a fatty acid. Triglycerides are the major constituent of fats and oils and provide a concentrated food energy store in living organisms as well as cooking fats and oils, margarines, etc. Their physical and chemical properties depend on the nature of their constituent fatty acids. In **simple triglycerides** all three fatty acids are identical; in **mixed triglycerides** two or three different fatty acids are present.

**trigonal bipyramid** *See illustration at* COMPLEX.

**trigonometric functions** Functions defined in terms of a right-angled triangle (see diagram) and widely used in the solution of many mathematical problems. They are defined as:

  tangent of angle $A$, written $\tan A = a/b$
  sine of angle $A$, written $\sin A = a/c$
  cosine of angle $A$, written $\cos A = b/c$,
where $a$ is the length of the side opposite the angle $A$, $b$ is the length of the side opposite the angle $B$, and $c$ is the hypotenuse of the triangle.

  The reciprocal functions are:
  cotangent of angle $A$, written $\cot A = 1/\tan A$
    $= b/a$
  secant of angle $A$, written $\sec A = 1/\cos A =$
    $c/b$
  cosecant of angle $A$, written $\operatorname{cosec} A =$
    $1/\sin A = c/a$.

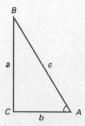

**Trigonometric functions.**

**trihydrate** A crystalline hydrate that contains three moles of water per mole of compound.

**trihydric alcohol** *See* TRIOL.

**triiodomethane (iodoform)** A yellow volatile solid sweet-smelling *haloform, $CHI_3$; r.d. 4.1; m.p. 115°C. It is made by the haloform reaction.

**triiodothyronine (T₃)** A hormone secreted by the *thyroid gland. *See also* THYROGLOBULIN.

**triiron tetroxide (ferrosoferric oxide)** A black magnetic oxide, $Fe_3O_4$; r.d. 5.2. It is formed when iron is heated in steam and also occurs naturally as the mineral *magnetite. The oxide dissolves in acids to give a mixture of iron(II) and iron(III) salts.

**trilobite** An extinct marine arthropod belonging to the class Trilobita (some 4000 species), fossils of which are found in deposits dating from the Precambrian to the Permian period (590–280 million years ago). Trilobites were typically small (1–7 cm long); the oval flattened body comprised a head (covered by a semicircular dorsal shield) and a thorax and abdomen, which were protected by overlapping dorsal plates with a raised central part and flattened lateral portions, presenting a three-lobed appearance. The head bore a pair of antenna-like appendages and a pair of compound eyes; nearly all body segments bore a pair of Y-shaped (biramous) appendages – one branch for locomotion and the other fringed for respiratory exchange. Trilobites were bottom-dwelling scavengers.

**trimethylaluminium (aluminium trimethyl)** A colourless liquid, $Al(CH_3)_3$, which ignites in air and reacts with water to give aluminium hydroxide and methane, usually with extreme vigour; r.d. 0.752; m.p. 0°C; b.p. 130°C. Like other aluminium alkyls it may be prepared by reacting a Grignard reagent with aluminium trichloride. Aluminium alkyls are used in the *Ziegler process for the manufacture of high-density polyethene (polythene).

**2,4,6-trinitrophenol** *See* PICRIC ACID.

**trinitrotoluene (TNT)** A yellow highly explosive crystalline solid, $CH_3C_6H_2(NO_2)_3$; r.d. 1.65; m.p. 82°C. It is made by nitrating toluene (methylbenzene), the systematic name being 1-methyl-2,4,6-trinitrobenzene.

**triode** A *thermionic valve with three electrodes. Electrons produced by the heated cathode flow to the anode after passing through the negatively biased *control grid. Small voltage fluctuations superimposed on the grid bias cause large fluctuations in the anode current. The triode was thus the first electronic device capable of amplification. Its role has now been taken over by the transistor, except where high power (radio-frequency transmitters producing more than 1 kW in power) is required.

**triol (trihydric alcohol)** An *alcohol containing three hydroxyl groups per molecule.

**triose** A sugar molecule that contains three carbon atoms. *See* MONOSACCHARIDE.

**trioxoboric(III) acid** *See* BORIC ACID.

**trioxosulphuric(IV) acid** *See* SULPHUROUS ACID.

**trioxygen** *See* OZONE.

**triple bond** *See* CHEMICAL BOND.

**triple point** The temperature and pressure at which the vapour, liquid, and solid phases of a substance are in equilibrium. For water the triple point occurs at 273.16 K and 611.2 Pa. This value forms the basis of the definition of the *kelvin and the thermodynamic *temperature scale.

**triple product** Either a *scalar product or a *vector product each having three components. A **scalar triple product** is obtained by multiplying three *vectors *a*, *h* and *c* in the manner *a*. (*b* × *c*); the result is a scalar. If the three vectors represent the positions of three points with respect to the origin, the magnitude of the scalar triple product is the volume of the parallelepiped with corners at the three points and the origin. A **vector triple product** is obtained by multiplying three vectors *a*, *b* and *c* in the manner *a* × (*b* × *c*); the result is a vector. It also equals (*a.c*)*b* − (*a.b*)*c* (but note it does not equal (*a* × *b*) × *c*).

**triplet code** *See* CODON; GENETIC CODE.

**triploblastic** Describing an animal having a body composed of three embryonic cell layers: the *ectoderm, *mesoderm, and *endoderm. Most multicellular animals are triploblastic; the coelenterates, which are *diploblastic, are an exception.

**triploid** Describing a nucleus, cell, or organism that has three times (3*n*) the haploid number (*n*) of chromosomes (*see also* POLYPLOID). Triploid organisms are normally sterile as their lack of *homologous chromosomes prevents pairing during meiosis. This can be useful to plant breeders, for example in banana cultivation: sterile triploid bananas can be propagated asexually and will not contain any seeds.

**trisilane** *See* SILANE.

**trisodium phosphate(V) (sodium orthophosphate)** A colourless crystalline compound, $Na_3PO_4$, soluble in water and insoluble in ethanol. It is known both as the decahydrate (octagonal; r.d. 2.54) and the dodecahydrate (trigonal; r.d. 1.62) The dodecahydrate loses water at about 76°C and the decahydrate melts at 100°C. Trisodium phosphate may be prepared by boiling sodium carbonate with the stoichiometric amount of phosphoric acid and subsequently adding sodium hydroxide to the disodium salt thus formed. It is useful as an additive for high-pressure boiler feed water (for removal of calcium and magnesium as phosphates), in emulsifiers, as a water-softening agent, and as a component in detergents and cleaning agents. Sodium phosphate labelled with the radioactive isotope $^{32}P$ is used in the study of the role of phosphate in biological processes and is also used (intravenously) in the treatment of polycythaemia.

**trisomy** The condition of a nucleus, cell, or organism in which one of the pairs of homologous chromosomes has gained an additional chromosome, resulting in a chromosome number of $2n + 1$. Trisomy is the cause of a number of human genetic abnormalities, including *Down's syndrome; Patau's syndrome, in which there is an extra chromosome 13 (**trisomy 13**); and Edwards' syndrome, in which there is an extra chromosome 18 (**trisomy 18**).

**tritiated compound** *See* LABELLING.

**tritium** Symbol T. An isotope of hydrogen with mass number 3; i.e. the nucleus contains 2 neutrons and 1 proton. It is radioactive (half-life 12.3 years), undergoing beta decay to helium–3. Tritium is used in *labelling.

**triton** The nucleus of a tritium atom.

**trivalent (tervalent)** Having a valency of three.

**tRNA** *See* RNA.

**trochanter** 1. Any of several bony knobs on the femur of vertebrates to which muscles are attached. 2. The second segment of an insect's leg, between the *coxa and the *femur.

**trona** A mineral form of sodium sesquicarbonate, consisting of a mixed hydrated sodium carbonate and sodium hydrogencarbonate, $Na_2CO_3.NaHCO_3.2H_2O$.

**trophic level** The position that an organism occupies in a *food chain. For example, green plants (which obtain their energy directly from sunlight) are the primary *producers; herbivores are primary *consumers (and secondary producers). A carnivore that eats only herbivores is a secondary consumer and a tertiary producer. Many animals feed at several different trophic levels.

**tropical cyclone** A *cyclone that develops over tropical or subtropical waters, in which sea temperatures are above 27°C and at least 5° of latitude away from the equator. The term encompasses tropical depression, with wind speeds of 33 knots (62 km/h) or less, and tropical storm, with wind speeds of 34–63 knots (63–117 km/h). A tropical cyclone with wind speeds of over 64 knots (117 km/h) is known as a *hurricane where it occurs in the North Atlantic Ocean, Caribbean Sea, Gulf of Mexico, and the east and central North Pacific Ocean (east of the dateline); as a typhoon where it occurs in the North Pacific Ocean; and as a severe tropical cyclone where it occurs in the southwest Pacific Ocean (west of 160°E) and southeast Indian Ocean (east of 90°E).

**tropical year** *See* YEAR.

**tropism** The directional growth of a plant organ in response to an external stimulus, such as light, touch, or gravity. Growth towards the stimulus is a **positive tropism**; growth away from the stimulus is a **negative tropism**. *See also* GEOTROPISM; HYDROTROPISM; ORTHOTROPISM; PHOTOTROPISM; PLAGIOTROPISM; THIGMOTROPISM. *Compare* NASTIC MOVEMENTS; TAXIS.

**tropomyosin** A protein found in the *actin filaments in muscles. The molecule consists of two elongated strands that run along the length of the filament. When the muscle is at rest, the tropomyosin molecule covers the region of the actin molecule where interaction with myosin occurs. On contraction of the muscle, the tropomyosin is displaced by another protein, *troponin, allowing the interaction of actin with myosin.

**troponin** A complex of three polypeptide chains that are found at regular intervals along the length of an *actin filament. During muscle contraction, troponin binds to calcium ions, displacing *tropomyosin and exposing the active site on the actin filament. This allows the interaction of actin and myosin to occur.

**troposphere** *See* EARTH'S ATMOSPHERE.

**tropylium ion** The positive ion $C_7H_7^+$, having a ring of seven carbon atoms. The ion is symmetrical and has characteristic properties of *aromatic compounds.

**truth table** A table that summarizes all possible outcomes of a logical operation. For example, for an AND *gate with inputs A and B and output C, the truth table is

| A | B | C |
|---|---|---|
| 0 | 0 | 0 |
| 0 | 1 | 0 |
| 1 | 0 | 0 |
| 1 | 1 | 1 |

This indicates that the output will be 0 unless both inputs are 1.

**trypsin** An enzyme that digests proteins (*see* ENDOPEPTIDASE; PROTEASE). It is secreted in an inactive form (**trypsinogen**) by the pancreas into the duodenum. There, trypsinogen is acted on by an enzyme (**enterokinase**) produced in the duodenum to yield trypsin. The active enzyme plays an important role in the digestion of proteins in the anterior portion of the small intestine. It also activates other proteases in the pancreatic juice (*see* CARBOXYPEPTIDASE; CHYMOTRYPSIN).

**trypsinogen** *See* TRYPSIN.

**tryptamine** A naturally occurring alkaloid, $C_{10}H_{12}N_2$, having an indole ring system with a $-CH_2-CH_2-NH_2$ side chain in the 3-position of the nitrogen-containing ring. Derivatives of tryptamine have hallucinogenic effects. An example is *psilocybin.

**tryptophan** *See* AMINO ACID.

**TSH** *See* THYROID-STIMULATING HORMONE.

**tsunami** A large sea wave usually generated by a submarine earthquake or volcanic eruption. It may also be caused by a mass underwater mudslide. The waves, which can be over 10 m high, spread in concentric circles from the focus of the earthquake, often travelling hundreds of kilometres and reaching speeds of 700 km/h. A tsunami can be extremely destructive when it breaks on the shore. A devasting tsunami followed a powerful earthquake of magnitude 9.0 (*see* RICHTER SCALE) off the west coast of N Sumatra, Indonesia, on 26 December 2004. It swept over land causing widespread destruction in Indonesia, Sri Lanka, Thailand, Malaysia, Myanmar, S India, and the Maldives, with lesser effects as faraway as E Africa, and resulted in over 283 100 deaths.

**T Tauri star** An unstable young variable star in its pre-main sequence phase (*see* HERTZSPRUNG–RUSSELL DIAGRAM). The instability, brought about by the beginning of nuclear fusion in the core of the star, causes pulsations and stellar winds, possibly with *bipolar outflows. Groups of such stars, often associated with *Herbig–Haro objects, are called T Tauri associations.

**tuber** A swollen underground stem or root in certain plants. It enables the plant to survive the winter or dry season and is also a means of propagation. A **stem tuber**, such as the potato, forms at the end of an underground stem. Each tuber represents several nodes and internodes. The following season several new plants develop from the terminal and axillary buds (eyes). **Root tubers**, such as those of the dahlia, are modified food-storing adventitious roots and may also give rise to new plants.

**tubulin** A protein of which the *microtubules of cells are formed.

**Tullgren funnel** A device used to remove and collect small animals, such as insects, from a sample of soil or leaf litter. The sample is placed on a coarse sieve fixed across the wide end of a funnel and a 100-watt light bulb, in a metal reflector, is placed about 25 cm above the funnel. The heat from the bulb dries and warms the sample, causing the animals to move downwards and fall through the sieve into the funnel, which directs them into a collecting dish or tube below. The dish can contain water or alcohol to prevent the animals from escaping.

**tumour** *See* NEOPLASM.

**tundra** A terrestrial *biome characterized by a lack of trees and a permanently frozen subsoil. Tundra lies to the north of the *taiga in North America and Eurasia; the vegetation is dominated by grasses, sedges, lichens, mosses, heathers, and low shrubs. The growing season, which occurs during the warmest part of the year when the average daily mean temperature is about 10°C, lasts only 2–4 months, during which the topsoil thaws to a depth of 30 cm, allowing roots to penetrate it. However, below this level the soil is permanently frozen (**permafrost**); water cannot filter through the soil and may lie in surface depressions during the growing season.

Global warming is now affecting the ecology and economy of tundra regions dramatically. By the mid-21st century, the area of permafrost is predicted to decline by around 20–35%. *Compare* TAIGA.

**tuneable laser** *See* DYE LASER.

**tungsten** Symbol W. A white or grey metallic *transition element (formerly called **wolfram**); a.n. 74; r.a.m. 183.85; r.d. 19.3; m.p. 3410°C; b.p. 5660°C. It is found in a number of ores, including the oxides wolframite, $(Fe,Mn)WO_4$, and scheelite, $CaWO_4$. The ore is heated with concentrated sodium hydroxide solution to form a soluble **tungstate**. The oxide $WO_3$ is precipitated from this by adding acid, and is reduced to the metal using hydrogen. It is used in various alloys, especially high-speed steels (for cutting tools) and in lamp filaments. Tungsten forms a protective oxide in air and can be oxidized at high temperature. It does not dissolve in dilute acids. It forms compounds in which the oxidation state ranges from +2 to +6. The metal was first isolated by Juan and Fausto d'Elhuyer in 1783.

**((()) SEE WEB LINKS**
• Information from the WebElements site

**tungsten carbide** A black powder, WC, made by heating powdered tungsten metal with lamp black at 1600°C. It is extremely hard (9.5 on Mohs' scale) and is used in dies and cutting tools. A ditungsten carbide, $W_2C$, also exists.

**tuning fork** A metal two-pronged fork that when struck produces an almost pure tone of a predetermined frequency. It is used for tuning musical instruments and in experiments in acoustics.

**tunnel diode (Esaki diode)** A semiconductor diode, discovered in 1957 by L. Esaki (1925–   ), based on the *tunnel effect. It consists of a highly doped $p$–$n$ semiconductor junction, which short circuits with negative bias and has negative resistance over part of its range when forward biased. Its fast speed of operation makes it a useful device in many electronic fields.

**tunnel effect** An effect in which electrons are able to tunnel through a narrow *potential barrier that would constitute a forbidden region if the electrons were treated as classical particles. That there is a finite probability of an electron tunnelling from one classically allowed region to another arises as a consequence of *quantum mechanics. The effect is made use of in the *tunnel diode. Alpha decay (*see* ALPHA PARTICLE) is an example of a tunnelling process.

**Turbellaria** A class of free-living flatworms (*see* PLATYHELMINTHES) comprising the planarians, which occur in wet soils, fresh water, and marine environments. Their undersurface is covered with cilia, used for gliding over stones and weeds. Planarians can also swim by means of undulations of the body.

**turbine** A machine in which a fluid is used to produce rotational motion. The most widely used turbines are the **steam turbines** and **water turbines** that provide some 95% of the world's electric power (in the form of *turbogenerators) and the **gas turbines** that power all the world's jet-propelled aircraft. In the **impulse turbine** a high-pressure low-velocity fluid is expanded through stationary nozzles, producing low-pressure high-velocity jets, which are directed onto the blades of a rotor. The rotor blades reduce the speed of the jets and thus convert some of the fluid's kinetic energy into rotational kinetic energy of the rotor shaft. In the **reaction turbine** the discharge nozzles are themselves attached to the rotor. The acceleration of the fluid leaving the nozzles produces a force of reaction on the pipes, causing the rotor to move in the opposite direction to that of the fluid. (See illustrations.) Many turbines work on a combination of the impulse and reaction principles.

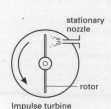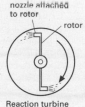

Impulse turbine        Reaction turbine

stationary nozzle    rotor

nozzle attached to rotor    rotor

**Turbine.**

**turbogenerator** A steam turbine driving an electric generator. This is the normal method of generating electricity in power stations. In a conventional power station the steam is raised by burning a fossil fuel (coal, oil, or natural gas); in a nuclear power station the steam is raised by heat transfer from a nuclear reactor.

**turbojet** *See* JET PROPULSION.

**turbulence** A form of fluid flow in which the particles of the fluid move in a disordered manner in irregular paths, resulting in an exchange of momentum from one portion of a fluid to another. Turbulent flow takes over from *laminar flow when high values of the *Reynolds number are reached.

**turgor** The condition in a plant cell when its *vacuole is distended with water, pushing the protoplast against the cell wall. In this condition the force causing water to enter the cell by *osmosis is balanced by the hydrostatic pressure of the protoplast against the cell wall (*see also* WATER POTENTIAL). Turgidity assists in maintaining the rigidity of plants; a decrease in turgidity leads to *wilting. *Compare* PLASMOLYSIS.

**Turing, Alan Mathison** (1912–54) British mathematician, who after studying at Cambridge University went to Princeton, where in 1937 he published his most important work on computable numbers, which contained a description of the hypothetical *Turing machine. He returned to Britain at the outbreak of World War II and worked on cracking German codes. This led to his involvement in the development of computers. He committed suicide after being convicted of indecency (as a homosexual).

**Turing machine** A hypothetical machine that determines whether or not a problem is computable. It has an infinite memory represented by an infinitely long ribbon of paper tape passing through the machine, which can be in several discrete internal states. The tape is divided into cells that can each hold one of a given number of symbols. The machine can move left or right along the tape, acting on one cell at a time. It is programmed by a set of instructions that make it change symbols, change state, and move one cell left or right (or remain at the same cell). If an operation can be performed by using an algorithm (i.e. if it is computable), a Turing machine can do it. It was devised by Alan Turing.

**turion** 1. A winter bud, covered with scale leaves and mucilage, that is produced by certain aquatic plants, such as frogbit. Turions become detached and remain dormant on the pond or lake bottom during the winter before developing into new plants the following season. 2. *See* SUCKER.

**Turner's syndrome** A genetic disorder of

women caused by the absence of the second *sex chromosome (such women are XO, rather than the normal XX). It is characterized by a lack of ovaries and menstrual cycle. Affected women are sterile and lack secondary sexual characteristics, although the external genitalia are present. The syndrome is named after the US endocrinologist Henry Turner (1892–1970), who first described it.

**turns ratio** *See* TRANSFORMER.

**turpentine** An oily liquid extracted from pine resin. It contains pinene, $C_{10}H_{16}$, and other terpenes and is mainly used as a solvent.

**turquoise** A mineral consisting of a hydrated phosphate of aluminium and copper, $CuAl_6(PO_4)_4(OH)_8.4H_2O$, that is prized as a semiprecious stone. It crystallizes in the triclinic system and is generally blue in colour, the 'robin's egg' blue variety being the most sought after. It usually occurs in veinlets and as masses and is formed by the action of surface waters on aluminium-rich rocks. The finest specimens are obtained from Iran.

**T violation** *See* TIME REVERSAL.

**tweeter** A small loudspeaker capable of reproducing sounds of relatively high frequency, i.e. 5 kilohertz upwards. In high-fidelity equipment a tweeter is used in conjunction with a *woofer.

**twinning** A process in which two crystals of the same material form with orientations such that the two crystals are related to each other by a symmetry operation. This may be either reflection in a plane (the **twinning plane**) or rotation about an axis (the **twinning axis**). The plane or axis is common to the two crystals.

**twins** Two individuals born to the same mother at the same time. Twins can develop from the same egg (*see* IDENTICAL TWINS) or from two separately fertilized eggs (*see* FRATERNAL TWINS).

**twistor** A generalization of the concept of a *spinor that involves the use of complex numbers in an essential way. Twistors were invented by Roger *Penrose in the hope that they would be a key concept in unifying the general theory of relativity and quantum mechanics. It has not done so yet, although some very intriguing relations have been established between twistors and noncommutative geometry, *supergravity theory, *superstring theory, and supersymmetry.

Twistors have also proved useful for finding *soliton solutions to gauge theories.

**tympanic cavity** *See* MIDDLE EAR.

**tympanum (tympanic membrane; eardrum)** The membrane that separates the *outer ear from the *middle ear. It vibrates in response to sound waves and transmits these vibrations via the *ear ossicles of the middle ear to the site of hearing (the *cochlea of the *inner ear). In amphibians and some reptiles there is no external ear and the tympanum is exposed at the skin surface.

**Tyndall effect** The scattering of light as it passes through a medium containing small particles. If a polychromatic beam of light is passed through a medium containing particles with diameters less than about one-twentieth of the wavelength of the light, the scattered light appears blue. This accounts for the blue appearance of tobacco smoke. At higher particle diameters, the scattered light remains polychromatic. It is named after John Tyndall (1820–93). *See also* SCATTERING OF ELECTROMAGNETIC RADIATION.

**type A and B metals** A classification of metal ions according to the stability of their complexes for a given ligand. **Type A metal** cations include the ions of group 1 ($Li^+$ to $Cs^+$), the ions of group 2 ($Be^{2+}$ to $Ba^{2+}$), and ions of lighter transition metals in high oxidation states (e.g. $Co^{3+}$, $Ti^{4+}$, $Fe^{3+}$). The **type B metal** cations are those of heavier transition metals in lower oxidation states (e.g. $Ag^+$, $Cu^+$, $Ni^{2+}$, $Pd^{2+}$, $Pt^{2+}$). Certain ligands tend to form more stable complexes with type A metals; others form more stable complexes with type B. For example, the tendency of halide anions to complex with type A metals is in the sequence

$$F^- > Cl^- > Br^- > I^-.$$

Their tendency to complex with type B metals is the opposite sequence. This led to a classification of ligands into type A ligands (c.g. $I^-$), which tend to complex with type A metals, and type B ligands (e.g. $I^-$), which tend to complex with type B metals. The classification was introduced by Ahrland in 1958. *See also* HSAB PRINCIPLE.

**type A metal** *See* TYPE A AND B METALS.

**type B metal** *See* TYPE A AND B METALS.

**type specimen** The specimen used for naming and describing a *species or subspecies. If this is the original specimen collected by the author who named the species it is termed a **holotype**. The type specimen is not necessarily the most characteristic representative of the species. The term **type** is also applied to any taxon selected as being representative of the rank to which it belongs. For example, the genus *Solanum* (potato) is said to be the type genus of the family Solanaceae.

**typhoon** A *tropical cyclone or *hurricane that occurs in the W and N Pacific Ocean and the South China Sea.

**tyrosine** *See* AMINO ACID.

**tyrosine kinase** Any of a large family of proteins that catalyse the phosphorylation of a tyrosine residue of a protein by ATP (*see* PROTEIN KINASE). They are components of numerous signalling pathways inside cells, notably ones regulating cell growth and differentiation.

**ubiquinone (coenzyme Q)** Any of a group of related quinone-derived compounds that serve as electron carriers in the *electron transport chain reactions of cellular respiration. Ubiquinone molecules have side chains of different lengths in different types of organisms but function in similar ways.

**ubiquitous computing (ambient computing; pervasive computing)** A model for the development of computing in the early 21st century. It envisages a movement away from computers as distinct specialized devices; rather, many objects used in everyday life will contain embedded computing devices that can recognize and interact in useful ways with each other and with their environment. Commonly cited examples are the refrigerator that is aware of its contents via embedded tags and is thus able to suggest shopping lists, plan menus, warn of approaching expiry dates, etc.; a home environment (heating, lighting, etc.) that continuously adjusts itself according to data from biometric monitors incorporated into clothing; and automatic recognition of a returning home owner so that security alarms are switched off and appropriate in-house systems activated. Such a model becomes possible as the size of computing devices decreases and their power increases, with wireless networking making the ad-hoc self-configuring networks necessary for ubiquitous computing viable (*see* BLUETOOTH). A key issue is how humans interact with such a system. The ideal is that such interaction should be unobtrusive and natural for the user, emphasizing speech recognition and *artificial intelligence. Another issue is privacy: a truly useful ubiquitous computing system will inevitably acquire a large amount of personal data about its users. Ubiquitous computing is currently a subject of both academic and commercial research.

**UHV** Ultrahigh vacuum. *See* VACUUM.

**ulna** The larger of the two bones in the forearm of vertebrates (*compare* RADIUS). It articulates with the outer carpals at the wrist and with the humerus at the elbow.

**ultracentrifuge** A high-speed centrifuge used to measure the rate of sedimentation of colloidal particles or to separate macromolecules, such as proteins or nucleic acids, from solutions. Ultracentrifuges are electrically driven and capable of speeds up to 60 000 rpm.

**ultradian rhythm** *See* BIORHYTHM.

**ultrafiltration** The process in which hydrostatic pressure causes water and small dissolved molecules and ions to move across a membrane against a *concentration gradient. Ultrafiltration is responsible for the formation of *tissue fluid and *glomerular filtrate from blood. In both these processes the ultrafiltered fluid has the same composition as the plasma except that it does not contain blood cells or large protein molecules.

**ultrahigh frequency (UHF)** A radio frequency in the range $3 \times 10^9$– $0.3 \times 10^9$ Hz; i.e. having a wavelength in the range 10 cm to 1 m.

**ultramicroscope** A form of microscope that uses the *Tyndall effect to reveal the presence of particles that cannot be seen with a normal optical microscope. Colloidal particles, smoke particles, etc., are suspended in a liquid or gas in a cell with a black background and illuminated by an intense cone of light that enters the cell from the side and has its apex in the field of view. The particles then produce diffraction-ring systems, appearing as bright specks on the dark background.

**ultramicrotome** *See* MICROTOME.

**ultrasonics** The study and use of pressure waves that have a frequency in excess of 20 000 Hz and are therefore inaudible to the human ear. **Ultrasonic generators** make use of the *piezoelectric effect, *ferroelectric materials, or *magnetostriction to act as transducers in converting electrical energy into mechanical energy. Ultrasonics are used in medicine for diagnosis, particularly in conditions such as pregnancy, in which X-rays could have a harmful effect, and for treatment, the vibratory effect being used to break up kidney stones, etc. Ultrasonic techniques are also used industrially to test for flaws in metals, to clean surfaces, to test the thickness of parts, and to form colloids.

**ultrastructure** The submicroscopic, almost molecular, structure of living cells, which is revealed by the use of an electron microscope.

**ultraviolet microscope** A *microscope that has quartz lenses and slides and uses *ultraviolet radiation as the illumination. The use of shorter wavelengths than the visible range enables the instrument to resolve smaller objects and to provide greater magnification than the normal optical microscope. The final image is either photographed or made visible by means of an *image converter.

**ultraviolet radiation (UV)** Electromagnetic radiation having wavelengths between that of violet light and long X-rays, i.e. between 400 nanometres and 4 nm. In the range 400–300 nm the radiation is known as the **near ultraviolet**. In the range 300–200 nm it is known as the **far ultraviolet**. Below 200 nm it is known as the **extreme ultraviolet** or the **vacuum ultraviolet**, as absorption by the oxygen in the air makes the use of evacuated apparatus essential. The sun is a strong emitter of UV radiation but only the near UV reaches the surface of the earth as the *ozone layer of the atmosphere absorbs all wavelengths below 290 nm. Ultraviolet radiation is classified in three ranges according to its effect on the skin. The ranges are:
UV-A (320–400 nm);
UV-B (290–320 nm);
UV-C (230–290 nm).
The longest-wavelength range, UV-A, is not harmful in normal doses and is used clinically in the treatment of certain skin complaints, such as psoriasis. It is also used to induce *vitamin D formation in patients that are allergic to vitamin D preparations. UV-B causes reddening of the skin followed by pigmentation (tanning). Excessive exposure can cause severe blistering. UV-C, with the shortest wavelengths, is particularly damaging. It is thought that short-wavelength ultraviolet radiation causes skin cancer and that the risk of contracting this has been increased by the depletion of the ozone layer.

Most UV radiation for practical use is produced by various types of *mercury-vapour lamps. Ordinary glass absorbs UV radiation

and therefore lenses and prisms for use in the UV are made from quartz.

**ultraviolet–visible spectroscopy (UV–visible spectroscopy)** A technique for chemical analysis and the determination of structure. It is based on the principle that electronic transitions in molecules occur in the visible and ultraviolet regions of the electromagnetic spectrum, and that a given transition occurs at a characteristic wavelength. The spectrometer has two sources, one of ultraviolet and the other of white visible light, which together cover the whole wavelength range of the instrument. If the whole wavelength range is used, the source is changed over at the appropriate point. The radiation from the source is split into two beams of equal intensity. One beam is passed through a dilute solution of the sample while the other is passed through the pure solvent and is used as a reference against which the first is compared after transmittance. The cells used for the sample and reference solutions are usually made of silica and are matched.

**umbel** A type of *racemose inflorescence in which stalked flowers arise from the same point on the flower axis, resembling the spokes of an umbrella. An involucre (cluster) of bracts may occur at the point where the stalks emerge. This arrangement is characteristic of the family Umbelliferae (Apiaceae; e.g. carrot, hogweed, parsley, parsnip), in which the inflorescence is usually a compound umbel.

**umbilical cord** The cord that connects the embryo to the *placenta in mammals. It contains a vein and two arteries that carry blood between the embryo and placenta. It is severed after birth to free the newly born animal from the placenta, and shrivels to leave a scar, the navel, on the animal.

**umbra** See SHADOW.

**uncertainty principle (Heisenberg uncertainty principle; principle of indeterminism)** The principle that it is not possible to know with unlimited accuracy both the position and momentum of a particle. This principle, discovered in 1927 by Werner Heisenberg, is usually stated in the form: $\Delta x \Delta p_x \geq h/4\pi$, where $\Delta x$ is the uncertainty in the $x$-coordinate of the particle, $\Delta p_x$ is the uncertainty in the $x$-component of the particle's momentum, and $h$ is the *Planck constant. An explanation of the uncertainty is that in order to locate a particle exactly, an observer must be able to bounce off it a photon of radiation; this act of location itself alters the position of the particle in an unpredictable way. To locate the position accurately, photons of short wavelength would have to be used. These would have associated large momenta and cause a large effect on the position. On the other hand, using long-wavelength photons would have less effect on the particle's position, but would be less accurate because of the longer wavelength. The principle has had a profound effect on scientific thought as it appears to upset the classical relationship between cause and effect at the atomic level.

**underdamped** *See* DAMPING.

**undernourishment** *See* MALNUTRITION.

**undulipodium** (*pl.* **undulipodia**) A slender flexible outgrowth of a eukaryote cell used for locomotion or propelling fluids over the surface of the cell. The term 'undulipodium' is used to designate a eukaryotic 'flagellum' or a *cilium (which have the same structure), to emphasize the distinction between these structures and the *flagellum of a bacterium. Many protists and sperm cells swim by means of undulipodia, and various organisms use them to establish feeding currents, or to clear debris from epithelial surfaces. All undulipodia have a shaft, about 0.25 μm in diameter, consisting of a longitudinal array of *microtubules, the **axoneme**, which is surrounded by an extension of the cell's plasma membrane. The axoneme has two single microtubules running down the middle surrounded by nine pairs of microtubules. At its base the axoneme connects with a **basal body** (or **kinetosome**), which organizes assembly of the axoneme microtubules. Cilia are shorter than flagella and move by a whiplike power stroke followed by a recovery stroke in the opposite direction. Flagella generate successive waves that pass from the base to the tip. In both cases, flexing of the shaft is produced by a sliding motion of the microtubule pairs relative to each other. This involves the successive formation and breakage of molecular bridges between adjacent pairs. The bridges are composed of a protein, dynein, and their formation requires energy in the form of ATP.

**UNFO** Urea nitrate–fuel oil. An explosive based on the fertilizer urea nitrate mixed with fuel oil, similar in its action to ammonium nitrate–fuel oil (ANFO). It has been used in a number of terrorist attacks, most notably a car bombing in the basement of the World Trade Centre, New York, on 26 February 1993.

**ungulate** A herbivorous mammal with hoofed feet (*see* UNGULIGRADE). Ungulates are grouped into two orders: *Artiodactyla and *Perissodactyla.

**unguligrade** Describing the gait of ungulates (e.g. horses and cows), in which only the tips of the digits (i.e. the hooves) are on the ground and the rest of the foot is off the ground. *Compare* DIGITIGRADE; PLANTIGRADE.

**uniaxial crystal** A double-refracting crystal (*see* DOUBLE REFRACTION) having only one *optic axis.

**unicellular** Describing tissues, organs, or organisms consisting of a single cell. For example, the reproductive organs of some algae and fungi are unicellular. Unicellular organisms include bacteria, protozoans, and certain fungi and algae. *Compare* ACELLULAR; MULTICELLULAR.

**Unicode** A standard for storing, manipulating, and displaying textual data. The Unicode character set currently (2008) allows for 1 114 112 **codepoints**, of which over 90 000 have been assigned characters. A codepoint is an unsigned integer denoting a position in a character encoding system. Unicode codepoints are generally represented as U+ followed by a four-byte hexadecimal number: for example, U+0041 (decimal 65, the letter capital A). Unicode also specifies various normative classifications for each character (upper-case letter, lower-case letter, decimal number, etc.), rules (e.g. how to decompose a composite character, such as an accented letter, into its component characters), and algorithms (e.g. for collation) as well as reference charts showing the visual form of each character. For backward compatibility the characters assigned to codepoints 0 to 127 are the same as *ASCII character set; and those assigned to 0 to 255 Unicode are the same as ISO-8859-1, a superset of Latin alphabet no. 1.

(((●))) **SEE WEB LINKS**
• The Unicode Consortium home page

**unified-field theory** A comprehensive theory that would relate the electromagnetic, gravitational, strong, and weak interactions (*see* FUNDAMENTAL INTERACTIONS) in one set of equations. In its original context the expression referred only to the unification of general *relativity and classical electromagnetic theory. No such theory has yet been found but some progress has been made in the unification of the electromagnetic and weak interactions (*see* ELECTROWEAK THEORY).

Einstein attempted to derive *quantum mechanics from unified-field theory, but it is now thought that any unified-field theory has to start with quantum mechanics. Attempts to construct unified-field theories, such as *supergravity and *Kaluza–Klein theory, have run into great difficulties. At the present time it is not clear whether the framework of relativistic *quantum field theory is adequate to give a unified theory for all the known fundamental interactions and elementary particles, or whether one has to go to extended objects, such as superstrings or supermembranes. Unified-field theories and other fundamental theories, such as *superstring theory and *supermembrane theory, are of great importance in understanding cosmology, particularly the *early universe. In turn cosmology puts constraints on unified-field theories. *See also* GRAND UNIFIED THEORY.

**unimolecular reaction** A chemical reaction or step involving only one molecule. An example is the decomposition of dinitrogen tetroxide:

$$N_2O_4 \rightarrow 2NO_2$$

Molecules colliding with other molecules acquire sufficient activation energy to react, and the activated complex only involves the atoms of a single molecule.

**union** *See* SETS.

**unisexual** Describing animals or plants with either male or female reproductive organs but not both. Most of the more advanced animals are unisexual but plants are often *hermaphrodite. Flowers that contain either stamens or carpels but not both are also described as unisexual. *See also* MONOECIOUS; DIOECIOUS.

**unit** A specified measure of a physical quantity, such as length, mass, time, etc., specified multiples of which are used to express magnitudes of that physical quantity. For many scientific purposes previous systems of units have now been replaced by *SI units.

**unit cell** The group of particles (atoms, ions, or molecules) in a crystal that is repeated in three dimensions in the *crystal lattice. *See also* CRYSTAL SYSTEM.

**unit magnetic pole** *See* MAGNETIC POLES.

**unit vector** A vector that has the magnitude 1. If $a$ is any non-zero vector the unit vector in the direction of $a$ is given by $a/|a|$ and is denoted $\hat{a}$.

**univalent (monovalent)** Having a valency of one.

**universal constants** *See* FUNDAMENTAL CONSTANTS.

**universal indicator** A mixture of acid–base *indicators that changes colour (e.g. red-yellow-orange-green-blue) over a range of pH.

**universality** *See* PHASE TRANSITION.

**universal motor** *See* ELECTRIC MOTOR.

**universe** All the matter, energy, and space that exists. *See* COSMOLOGY; EARLY UNIVERSE; HEAT DEATH OF THE UNIVERSE.

**UNIX** A general-purpose *computer operating system that allows several users, at different terminals, to use the machine at the same time. It was developed in 1969 and became generally available in 1971.

**(()) SEE WEB LINKS**
• The Open Group's UNIX page

**unnil-** *See* TRANSACTINIDE ELEMENTS.

**Unruh effect** The phenomenon, predicted in 1976 by the Canadian physicist William Unruh, that an accelerating body would seem to be surrounded by particles at a non-zero temperature, which is proportional to the acceleration. The vacuum state of a non-accelerating observer is different to that of an accelerating observer because of distortion of the zero-point fluctuations. The effect itself is very small and has not been verified experimentally. There is **Unruh radiation** associated with this effect. The effect itself is very small and has not been verified experimentally.

**unsaturated** 1. (of a compound) Having double or triple bonds in its molecules. Unsaturated compounds can undergo addition reactions as well as substitution. *Compare* SATURATED. 2. (of a solution) *See* SATURATED.

**unstable equilibrium** *See* EQUILIBRIUM.

**upper atmosphere** The upper part of the *earth's atmosphere above about 30 km.

u

This is the part of the atmosphere that cannot be reached by balloons.

**upthrust** *See* ARCHIMEDES' PRINCIPLE.

**UPVC** Unplasticized PVC: a tough hard-wearing form of PVC used for window frames and similar applications.

**upwelling** In the oceans and some inland seas, the process by which colder water, often rich in nutrients, is brought up from a lower depth to the surface layers. Coastal upwelling occurs where persistent surface winds blow parallel to the coastline, with the coast to the left of the wind in the northern hemisphere and to the right in the southern hemisphere. The warmer surface water is deflected away from the coast and colder water rises to replace it. Regions of coastal upwelling often support important fisheries and birdlife, e.g. off the coasts of California, Peru, and Ghana. Equatorial upwelling occurs in the Atlantic and Pacific Oceans along the equator as a result of the effects of the trade winds.

**uracil** A *pyrimidine derivative and one of the major component bases of *nucleotides and the nucleic acid *RNA.

**uraninite** A mineral form of uranium(IV) oxide, containing minute amounts of radium, thorium, polonium, lead, and helium. When uraninite occurs in a massive form with a pitchy lustre it is known as **pitchblende**, the chief ore of uranium. Uraninite occurs in Saxony (east central Germany), Romania, Norway, the UK (Cornwall), E Africa (Congo), USA, and Canada (Great Bear Lake).

**uranium** Symbol U. A white radioactive metallic element belonging to the *actinoids; a.n. 92; r.a.m. 238.03; r.d. 19.05 (20°C); m.p. 1132±1°C; b.p. 3818°C. It occurs as *uraninite, from which the metal is extracted by an ion-exchange process. Three isotopes are found in nature: uranium–238 (99.28%), uranium–235 (0.71%), and uranium–234 (0.006%). As uranium–235 undergoes *nuclear fission with slow neutrons it is the fuel used in *nuclear reactors and *nuclear weapons; uranium has therefore assumed enormous technical and political importance since their invention. It was discovered by Martin Klaproth (1743–1817) in 1789.

(((●))) SEE WEB LINKS
• Information from the WebElements site

**uranium(VI) fluoride (uranium hexafluoride)** A volatile white solid, $UF_6$; r.d.

4.68; m.p. 64.5°C. It is used in the separation of uranium isotopes by gas diffusion.

**uranium hexafluoride** *See* URANIUM(VI) FLUORIDE.

**uranium–lead dating** A group of *dating techniques for certain rocks that depends on the decay of the radioisotopes uranium–238 to lead–206 (half-life $4.5 \times 10^9$ years) or the decay of uranium–235 to lead–207 (half-life $7.1 \times 10^8$ years). One form of uranium–lead dating depends on measuring the ratio of the amount of helium trapped in the rock to the amount of uranium present (since the decay $^{238}U \rightarrow {}^{206}Pb$ releases eight alpha particles). Another method of calculating the age of the rocks is to measure the ratio of radiogenic lead ($^{206}Pb$, $^{207}Pb$, and $^{208}Pb$) present to non-radiogenic lead ($^{204}Pb$). These methods give reliable results for ages of the order $10^7$–$10^9$ years.

**uranium(IV) oxide** A black solid, $UO_2$; r.d. 10.96; m.p. 2500°C. It occurs naturally as *uraninite and is used in nuclear reactors.

**uranium series** *See* RADIOACTIVE SERIES.

**Uranus** The third largest *planet in the *solar system and the seventh in order from the *sun. Its mean distance from the sun is $2876.679 \times 10^6$ km, its mass is $8.6810 \times 10^{25}$ kg (14.5 times that of earth), and its mean equatorial diameter is 51 119 km; it has a *sidereal period of 84.32 years. Its average period of axial rotation is 17h 14.4m. The equator of Uranus is tilted at 98° with respect to its orbit, so that each pole is facing the sun continuously during half of each orbit, giving each pole 42 years of sunshine followed by 42 years of darkness. The equatorial region experiences a rapid alternation of day and night. The temperature of Uranus is very low, about 50 K, and its atmosphere contains molecular hydrogen and helium, with 2.3% methane. Methane forms an icy cloud layer in the upper atmosphere, and its absorption of red and infrared light gives the planet its aquamarine or cyan colour. The planet itself is believed to have an ice mantle some 8000 km thick surrounding a rocky core. The planet has 27 known satellites; the largest, Triton, is thought to be a captured *Kuiper belt object. Uranus also has a system of about 20 rings, nine of which were discovered in 1977 with the rest being photographed in 1986 by the US Voyager II probe.

(((●))) SEE WEB LINKS
• NASA's introduction to Uranus and its satellites

**urea (carbamide)** A white crystalline solid, $CO(NH_2)_2$; r.d. 1.3; m.p. 135°C. It is soluble in water but insoluble in certain organic solvents. Urea is the major end product of nitrogen excretion in mammals, being synthesized by the *urea cycle. Urea is synthesized industrially from ammonia and carbon dioxide for use in *urea–formaldehyde resins and pharmaceuticals, as a source of nonprotein nitrogen for ruminant livestock, and as a nitrogen fertilizer.

**urea cycle (ornithine cycle)** The series of biochemical reactions that converts ammonia, which is highly toxic, and carbon dioxide to the much less toxic *urea during the excretion of metabolic nitrogen. These reactions take place in the liver in mammals and, to a lesser extent, in some other animals. The urea is ultimately excreted in solution in *urine.

**urea–formaldehyde resins** Synthetic resins made by copolymerizing urea with formaldehyde (methanal). They are used as adhesives or thermosetting plastics.

**urea nitrate** *See* UNFO.

**ureter** The duct in vertebrates that conveys urine from the *kidney to the *bladder.

**urethane resins (polyurethanes)** Synthetic resins containing the repeating group –NH–CO–O–. There are numerous types made by copolymerizing isocyanate esters with polyhydric alcohols. They have a variety of uses in plastics, paints, and solid foams.

**urethra** The duct in mammals that conveys urine from the *bladder to be discharged to the outside of the body. In males the urethra passes through the penis and is joined by the *vas deferens; it therefore also serves as a channel for sperm.

**Urey, Harold Clayton** (1894–1981) US physical chemist, who became a professor at the University of California in 1958. His best-known work was the discovery of *deuterium (heavy hydrogen) in 1932, for which he was awarded the 1939 Nobel Prize for physics.

**URI** Uniform resource indicator: an identifier that uses the syntax defined in RFC 3986. URIs are divided into two parts by a colon. The first part is a label indicating the **scheme** of the URI, and the second part identifies the resource using the syntax appropriate to that scheme. For example, the URI 'tel:+44-020-7...' uses the 'tel' scheme, which identifies telephone numbers; there-

fore, the second part uses the standard syntax for telephone numbers (in this case specifying a number in London, England). Internet URLs (*see* WORLD WIDE WEB) are a subset of URIs, with the schemes http, ftp, smtp, etc.

(((●))) SEE WEB LINKS
• The URI specification

**uric acid** The end product of purine breakdown in most mammals, birds, terrestrial reptiles, and insects and also (except in mammals; *see* UREA) the major form in which metabolic nitrogen is excreted. Being fairly insoluble, uric acid can be expelled in solid form, which conserves valuable water in arid environments. The accumulation of uric acid in the synovial fluid of joints causes gout.

**uridine** A nucleoside consisting of one uracil molecule linked to a D-ribose sugar molecule. The derived nucleotide uridine diphosphate (UDP) is important in carbohydrate metabolism.

**urinary system** The collection of organs and tissues that perform *osmoregulation and *excretion. The mammalian urinary system consists of two *kidneys each linked to the bladder by a ureter.

**urine** The aqueous fluid formed by the excretory organs of animals for the removal of metabolic waste products. In higher animals, urine is produced by the *kidneys, stored in the *bladder, and excreted through the *urethra or *cloaca. Apart from water, the major constituents of urine are one or more of the end products of nitrogen metabolism – ammonia, urea, uric acid, and creatinine. It may also contain various inorganic ions, the pigments urochrome and urobilin, amino acids, and purines. Precise composition depends on many factors, especially the habitat of a particular species: aquatic animals produce copious volumes; terrestrial animals need to conserve water and produce much less (about 1.0–1.5 litres per day in humans).

**uriniferous tubule** *See* NEPHRON.

**URL** *See* WORLD WIDE WEB.

**USB drive** In general, any storage device that can be attached to a computer through a special type of connection (universal serial bus connection). The term is particularly used for small portable storage devices typically sealed in plastic. Their physical size is reflected in the various names for this type of

**u**

device – **thumb drive**, **pen drive**, **keyring drive**. They have capacities as high as 1 gigabyte and on most modern personal computers they can be recognized as an additional drive without the need for a special driver. Storage devices of this type are increasingly used for backup, data transfer, storage of photographs or MP3 files, etc.

**uterus (womb)** The organ of female mammals in which the embryo develops. Paired in most mammals but single in humans, it is situated between the bladder and rectum and is connected to the *fallopian tubes and to the *vagina. The lining (*see* ENDOMETRIUM) shows cyclical changes (*see* MENSTRUAL CYCLE; OESTROUS CYCLE) associated with egg production and provides a thick spongy layer in which the fertilized egg becomes embedded. The outer wall of the uterus is thick and muscular; by contracting, it forces the fully grown fetus through the vagina to the outside.

**UTF-8** 8-bit Unicode Transformation Format. A method of encoding Unicode codepoints using one-byte unsigned integers; from one to four such integers are used depending on the codepoint. UTF-8 is the most widely used Unicode encoding scheme and is the default encoding for *XML documents.

**(((⊕))) SEE WEB LINKS**

• Details of UTF-8 encoding from the Unicode (version 5) standard

**utriculus (utricle)** A chamber of the *inner ear from which the *semicircular canals arise. It bears patches of sensory epithelium concerned with detecting changes in the direction and speed of movement (*see* MACULA).

**UV** *See* ULTRAVIOLET RADIATION.

**UV–visible spectroscopy** *See* ULTRAVIOLET–VISIBLE SPECTROSCOPY.

u

**vacancy** *See* CRYSTAL DEFECT.

**vaccination** *See* IMMUNIZATION.

**vaccine** A liquid preparation of treated disease-producing microorganisms or their products used to stimulate an *immune response in the body and so confer resistance to the disease (*see* IMMUNIZATION). Vaccines are administered orally or by injection (**inoculation**). They take the form of dead viruses or bacteria that can still act as antigens, live but weakened microorganisms (*see* ATTENUATION), specially treated *toxins, or antigenic extracts of the microorganism.

**vacuole** A space within the cytoplasm of a living *cell that is filled with air, water or other liquid, sap, or food particles. In plant cells there is usually one large vacuole bounded by a single-layered membrane (**tonoplast** or **vacuole membrane**); animal cells usually have several small vacuoles. *See also* CONTRACTILE VACUOLE.

**vacuum** A space in which there is a low pressure of gas, i.e. relatively few atoms or molecules. A **perfect vacuum** would contain no atoms or molecules, but this is unobtainable as all the materials that surround such a space have a finite *vapour pressure. In a **soft** (or **low**) **vacuum** the pressure is reduced to about $10^{-2}$ pascal, whereas a **hard** (or **high**) vacuum has a pressure of $10^{-2}$–$10^{-7}$ pascal. Below $10^{-7}$ pascal is known as an **ultrahigh vacuum** (**UHV**). *See also* VACUUM PUMP.

**vacuum distillation** Distillation under reduced pressure. The depression in the boiling point of the substance distilled means that the temperature is lower, which may prevent the substance from decomposing.

**vacuum pump** A pump used to reduce the gas pressure in a container. The normal laboratory rotary oil-seal pump can maintain a pressure of $10^{-1}$ Pa. For pressures down to $10^{-7}$ Pa a *diffusion pump is required. *Ion pumps can achieve a pressure of $10^{-9}$ Pa and a *cryogenic pump combined with a diffusion pump can reach $10^{-13}$ Pa.

**vacuum state** The ground state in a relativistic *quantum field theory. A vacuum state does not mean a state of nothing. Because one is dealing with *quantum mechanics, the vacuum state has a *zero-point energy, which gives rise to **vacuum fluctuations**. The existence of vacuum fluctuations has observable consequences in *quantum electrodynamics.

**vacuum tube** *See* THERMIONIC VALVE.

**vagina** The tube leading from the uterus to the outside. Sperm are deposited in the vagina during copulation and the fully developed fetus is born through it. In a number of mammals the vagina may be sealed when the animal is not sexually receptive and only open during oestrus. Its lining produces mucus, which prevents friction and the entry of infective organisms.

**vagus nerve** The tenth *cranial nerve: a paired nerve that supplies branches to many major internal organs. It carries motor nerve fibres to the heart, lungs, and viscera and sensory fibres from the viscera.

**valence** *See* VALENCY.

**valence band** *See* ENERGY BANDS.

**valence bond theory** A theory of the electronic structure of molecules and solids in which the quantum state of the system corresponds to a mixture of molecular structures involving chemical bonds formed by pairs of electrons. Valence bond theory is a complementary starting point to *molecular orbital theory for understanding the electronic structure of molecules and solids.

**valence electron** An electron in one of the outer shells of an atom that takes part in forming chemical bonds.

**valency (valence)** The combining power of an atom or radical, equal to the number of hydrogen atoms that the atom could combine with or displace in a chemical compound (hydrogen has a valency of 1). It is equal to the ionic charge in ionic compounds; for example, in $Na_2S$, sodium has a valency of 1 ($Na^+$) and sulphur a valency of 2

($S^{2-}$). In covalent compounds it is equal to the number of bonds formed; in $CO_2$ oxygen has a valency of 2 and carbon has a valency of 4.

**valine** *See* AMINO ACID.

**Valium** *See* DIAZEPAM.

**valley of stability** The bottom region of the surface defined by plotting the numbers of protons and neutrons of a nucleus horizontally and the energy per nucleon of the nucleus vertically. Stable nuclei form a 'valley' with the most stable nuclei, such as iron and nickel, being at the bottom of the valley. The concept of the valley of stability enables several important topics in nuclear physics, such as *nuclear fission and *nuclear fusion, to be understood in a clear physical way. There are occasional 'pits' in the valley of stability due to nuclei with *magic numbers being more stable than nuclei with similar numbers of protons and neutrons.

**valve** 1. (in anatomy) Any of various structures for restricting the flow of a fluid through an aperture or along a tube to one direction. Valves in the heart (*see* BICUSPID VALVE; TRICUSPID VALVE), veins, and lymphatic vessels consist of two or three flaps of tissue (**cusps**) fastened to the walls. The cusps are flattened to the walls to allow the normal passage of blood or lymph, but a reverse flow causes them to block the vessel or aperture, so preventing further backflow. 2. (in biology) **a.** Any of the parts that make up a capsule or other dry fruit that sheds its seeds. **b.** One of the two halves of the cell wall of a diatom. **c.** Either of the two hinged portions of the shell of a bivalve mollusc. 3. (in electronics) *See* THERMIONIC VALVE.

**vanadium** Symbol V. A silvery-white metallic *transition element; a.n. 23; r.a.m. 50.94; r.d. 5.96; m.p. 1890°C; b.p. 3380°C. It occurs in a number of complex ores, including vanadinite ($Pb_5Cl(VO_4)_3$) and carnotite ($K_2(ClO_2)_2(VO_4)_2$). The pure metal can be obtained by reducing the oxide with calcium. The element is used in a large number of alloy steels. Chemically, it reacts with non-metals at high temperatures but is not affected by hydrochloric acid or alkalis. It forms a range of complexes with oxidation states from +2 to +5. Vanadium was discovered in 1801 by Andrés del Rio (1764–1849), who allowed himself to be persuaded that what he had discovered was an impure form of chromium. The element was rediscovered

and named by Nils Sefström (1787–1854) in 1830.

 **SEE WEB LINKS**
• Information from the WebElements site

**vanadium(V) oxide (vanadium pentoxide)** A crystalline compound, $V_2O_5$, used extensively as a catalyst in industrial gas-phase oxidation processes.

**vanadium pentoxide** *See* VANADIUM(V) OXIDE.

**Van Allen belts (radiation belts)** Belts that are sources of intense radiation surrounding the earth, consisting of high-energy charged particles trapped in the earth's magnetic field within which they follow roughly helical paths. They were discovered in 1958 by James Van Allen (1914–2006) as a result of radiation detectors carried by Explorer satellites. The lower belt, extending from 1000 to 5000 km above the equator, contains electrons and protons, while the upper belt, 15 000–25 000 km above the equator, contains mainly electrons (see illustration).

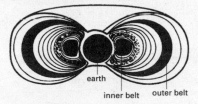

**Van Allen belts.**

**Van de Graaff accelerator** *See* LINEAR ACCELERATOR.

**Van de Graaff generator** An electrostatic generator used to produce a high voltage, usually in the megavolt range. It consists of a large metal dome-shaped terminal mounted on a hollow insulating support. An endless insulating belt runs through the support from the base to a pulley within the spherical terminal. In the original type, charge is sprayed by point discharge from metal needles, held at a potential of about 10 kV, on to the bottom of the belt. A row of needles near the upper belt pulley removes the charge from the belt and passes it to the outer surface of the spherical terminal. The voltage achieved by the device is proportional to the radius of the spherical terminal. A typical device with a terminal having a radius of 1 m will produce about 1 MV. However, terminals can be made smaller, for a given voltage, by enclosing the apparatus in nitrogen at a pressure of 10–20 atmospheres (1–2 MPa) to reduce sparking. Generators having a positive-ion source are fitted with an evacuated tube through which the particles can be accelerated for research purposes (see LINEAR ACCELERATOR). Machines having an electron source are used for various medical and industrial purposes. The generator was invented by Robert Van de Graaff (1901–67).

Modern patterns of the generator have a chainlike belt of alternate links of metal and insulator. The metal links are charged by contact with a metal pulley, and discharge to the dome in the same way. This permits much higher current drain that the point discharge.

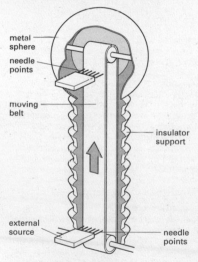

metal sphere

needle points

moving belt

insulator support

external source

needle points

**Van de Graaff generator.**

**van der Waals' equation** See EQUATION OF STATE.

**van der Waals' force** An attractive force between atoms or molecules, named after Johannes van der Waals (1837–1923). The force accounts for the term $a/V^2$ in the van der Waals equation (see EQUATION OF STATE). These forces are much weaker than those arising from valence bonds and are inversely proportional to the seventh power of the distance between the atoms or molecules. They are the forces responsible for nonideal behaviour of gases and for the lattice energy of molecular crystals. There are three factors causing such forces: (1) dipole–dipole interaction, i.e. electrostatic attractions between two molecules with permanent dipole moments; (2) dipole-induced dipole interactions, in which the dipole of one molecule polarizes a neighbouring molecule; (3) dispersion forces arising because of small instantaneous dipoles in atoms.

**vanillin (4-hydroxy-3-methoxybenzaldehyde)** A white crystalline solid, $C_8H_8O_3$, m.p. 82°C. It has the characteristic taste and smell of vanilla, in which it occurs; it is also made from the by-products of wood pulp manufacture. It is widely used in flavourings and perfumes, and in some pharmaceutical products.

**van't Hoff factor** Symbol $i$. A factor appearing in equations for *colligative properties, equal to the ratio of the number of actual particles present to the number of undissociated particles. It was first suggested by Jacobus van't Hoff (1852–1911).

**van't Hoff's isochore** An equation for the variation of equilibrium constant with temperature:

$$(\mathrm{d}\log_e K)/\mathrm{d}T = \Delta H/RT^2,$$

where $K$ is the equilibrium constant, $R$ is the gas constant, $T$ is the thermodynamic temperature, and $\Delta H$ the enthalpy of the reaction.

**vapour density** The density of a gas or vapour relative to hydrogen, oxygen, or air. Taking hydrogen as the reference substance, the vapour density is the ratio of the mass of a particular volume of a gas to the mass of an equal volume of hydrogen under identical conditions of pressure and temperature. Taking the density of hydrogen as 1, this ratio is equal to half the relative molecular mass of the gas.

**vapour pressure** The pressure exerted by a vapour. All solids and liquids give off vapours, consisting of atoms or molecules of the substances that have evaporated from the condensed forms. These atoms or molecules exert a vapour pressure. If the substance is in an enclosed space, the vapour pressure will reach an equilibrium value that depends only on the nature of the substance and the temperature. This equilibrium value occurs when there is a dynamic equilibrium between the atoms or molecules escaping from the liquid or solid and those that strike the surface of the liquid or solid and return to it. The vapour is then said to be a **saturated vapour** and the pressure it exerts is the **saturated vapour pressure**.

**variable star** Any star that varies in brightness. There are several types, depending on the cause of the variation. Some, such as eclipsing *binary stars, are actually a pair of stars whose combined light output varies as one star passes in front (or behind) the other. **Intrinsic variables** are caused by some internal phenomenon. The brightness of *Cepheid variables, *Mira-type variables, and *RR Lyrae variables alternates regularly (with timescales from hours to years); *flare stars and *T Tauri stars vary irregularly. Other irregular variable stars, sometimes called cataclysmic variables, include *novae and *supernovae.

**variance** In *statistics, the square of the *standard deviation, usually written $\sigma^2$. It represents the dispersion of a random variable's distribution, equal to $E[(X − E(X))^2]$, where $X$ is the random variable and $E$ is the expected value of the square of the difference between the variable and its *mean. *See also* COVARIANCE.

**variation** 1. (in biology) The differences between individuals of a plant or animal species. Variation may be the result of environmental conditions; for example, water supply and light intensity affect the height and leaf size of a plant. Differences of this kind, acquired during the lifetime of an individual, are not transmitted to succeeding generations since the genes are not affected. **Genetic variation**, due to differences in genetic constitution, is inherited (*see* CONTINUOUS VARIATION; DISCONTINUOUS VARIATION). The most important sources of genetic variation are *mutation and *recombination (*see also* CROSSING OVER). It is also increased by *outbreeding. Wide genetic variation improves the ability of a species to survive in a changing environment, since the chances that some individuals will tolerate a particular change are increased. Such individuals will survive and transmit the advantageous genes to their offspring. 2. (in magnetism) *See* GEOMAGNETISM.

**variegation** The occurrence of differently coloured patches, spots, or streaks in plant leaves, petals, or other parts, due to absence of pigment or different combinations of pigment in the affected area of the part. Variegation may be brought about by infection, for example *tobacco mosaic virus infection, or by genetic differences between the cells of the variegated part.

**variety** A category used in the *classification of plants and animals below the *species level. A variety consists of a group of individuals that differ distinctly from but can interbreed with other varieties of the same species. The characteristics of a variety are genetically inherited. Examples of varieties include breeds of domestic animals and the human *races. *See also* CULTIVAR. *Compare* SUBSPECIES.

**variometer** 1. A variable inductor consisting of two coils connected in series and able to move relative to each other. It is used to measure inductance as part of an a.c. bridge. 2. Any of several devices for detecting and measuring changes in the geomagnetic elements (*see* GEOMAGNETISM).

**varve dating (geochronology)** An absolute *dating technique using thin sedimentary layers of clays called **varves**. The varves, which are particularly common in Scandinavia, have alternate light and dark bands corresponding to winter and summer deposition. Most of them are found in the Pleistocene series, where the edges of varve deposits can be correlated with the annual retreat of the ice sheet, although some varve formation is taking place in the present day. By counting varves it is possible to establish an absolute time scale for fossils up to about 20 000 years ago.

**vascular bundle (fascicle)** A long continuous strand of conducting (vascular) tissue in tracheophyte plants that extends from the roots through the stem and into the leaves. It consists of *xylem and *phloem, which are separated by a *cambium in plants that undergo secondary thickening. *See* VASCULAR TISSUE.

**vascular cambium** *See* CAMBIUM.

**vascular plants** All plants possessing organized *vascular tissue. *See* TRACHEOPHYTE.

**vascular system** 1. A specialized network of vessels for the circulation of fluids throughout the body tissues of an animal. All animals, apart from simple invertebrate groups, possess a **blood vascular system**, which enables the passage of respiratory gases, nutrients, excretory products, and other metabolites into and out of the cells. In vertebrates it consists of a muscular *heart, which pumps blood through major blood vessels (*arteries) into increasingly finer branches until in the *capillaries it is in intimate contact with tissues. It then returns to the heart via another network of vessels (the *veins). This *circulation also enables a stable *internal environment for tissue function (*see* HOMEOSTASIS), the transmission of chemical messengers (*hormones) around the body, and a means of defending the body against pathogens and damage via the immune system. A **water vascular system** is characteristic of the *Echinodermata. **2.** The system of *vascular tissue in plants.

**vascular tissue (vascular system)** The tissue that conducts water and nutrients through the plant body in higher plants (*tracheophytes). It consists of *xylem and *phloem. Since the xylem and phloem tissues are always in close proximity to each other, distinct regions of vascular tissue can be identified (*see* VASCULAR BUNDLE). The possession of vascular tissue has enabled the higher plants to attain a considerable size and dominate most terrestrial habitats.

**vas deferens** One of a pair of ducts carrying sperm from the testis (or *epididymis) to the outside, in mammals through the *urethra.

**vas efferens** Any of various small ducts carrying sperm. In reptiles, birds, and mammals they convey sperm from the seminiferous tubules of the testis to the *epididymis; in invertebrates they carry sperm from the testis to the vas deferens.

**Vaseline** *See* PETROLEUM JELLY.

**vasoactive intestinal peptide** *See* VIP.

**vasoconstriction** The reduction in the internal diameter of blood vessels, especially arterioles or capillaries. The constriction of arterioles is mediated by the action of nerves on the smooth muscle fibres of the arteriole walls and results in an increase in blood pressure.

**vasodilation (vasodilatation)** The increase in the internal diameter of blood vessels, especially arterioles or capillaries. The vasodilation of arterioles is mediated by the action of nerves on the smooth muscle fibres of the arteriole walls and results in a decrease in blood pressure.

**vasomotor nerves** The nerves of the *autonomic nervous system that control the diameter of blood vessels. **Vasoconstrictor nerves** decrease the diameter (*see* VASOCONSTRICTION); **vasodilator nerves** increase it (*see* VASODILATION).

**vasopressin** *See* ANTIDIURETIC HORMONE.

**vector 1.** (in mathematics) A quantity in which both the magnitude and the direction must be stated (*compare* SCALAR QUANTITY). Force, velocity, and field strength are examples of vector quantities. Note that distance and speed are scalar quantities, whereas displacement and velocity are vector quantities. Vector quantities must be treated by **vector algebra**, for example, the resultant of two vectors may be found by a *parallelogram of vectors. A (three-dimensional) vector $V$ may be written in terms of components $V_1$, $V_2$, and $V_3$ along the $x$, $y$, and $z$ axes (say) as $V_1\boldsymbol{i} + V_2\boldsymbol{j} + V_3\boldsymbol{k}$, where $\boldsymbol{i}$, $\boldsymbol{j}$, and $\boldsymbol{k}$ are **unit vectors** (i.e. vectors of unit length) along the $x$, $y$, and $z$ axes. *See also* TRIANGLE OF VECTORS. **2.** (in medicine) An animal, usually an insect, that passively transmits disease-causing microorganisms from one animal or plant to another or from an animal to a human. *Compare* CARRIER. **3. (cloning vector)** (in genetics) A vehicle used in *gene cloning to insert a foreign DNA fragment into the genome of a host cell. For bacterial hosts various different types of vector are used: *bacteriophages, *artificial chromosomes, *plasmids, and their hybrid derivatives, *cosmids. The foreign DNA is spliced into the vector using specific *restriction enzymes and *ligases to cleave the vector DNA and join the foreign DNA to the two ends created (**insertional vectors**). In some phage vectors, part of the viral genome is enzymically removed and replaced with the foreign DNA (**replacement vectors**). *Retroviruses can be effective vectors for introducing recombinant DNA into mammalian cells. In plants, derivatives of the tumour-inducing (Ti) plasmid of the crown gall bacterium, *Agrobacterium tumefaciens*, are used as vectors.

**vector product (cross product)** The product of two *vectors $U$ and $V$, with components $U_1$, $U_2$, $U_3$ and $V_1$, $V_2$, $V_3$, respectively, given by:

$$U \times V = (U_2V_3 - U_3V_2)\boldsymbol{i} + (U_3V_1 - U_1V_3)\boldsymbol{j} + (U_1V_2 - U_2V_1)\boldsymbol{k}.$$

It is itself a vector, perpendicular to both $U$ and $V$, and of length $UV\sin\theta$, where $U$ and $V$ are the lengths of $U$ and $V$, respectively, and θ is the angle between them. *Compare* SCALAR PRODUCT.

**vector space** A set of *vectors for which an operation of addition is defined so that if $v_1$ and $v_2$ are vectors, the sum $v_1 + v_2$ is also a vector; an operation of *scalar multiplication is defined so that if $v$ is a vector and $c$ is a scalar, the product $cv$ is also a vector. *See also* HILBERT SPACE.

**vector triple product** *See* TRIPLE PRODUCT.

**vegetative propagation (vegetative reproduction) 1.** A form of *asexual reproduction in plants whereby new individuals develop from specialized multicellular structures (e.g. *tubers, *bulbs) that become detached from the parent plant. Examples are the production of strawberry plants from *runners and of gladioli from daughter *corms. Artificial methods of vegetative propagation include grafting (*see* GRAFT), *budding, and making *cuttings. **2.** Asexual reproduction in animals, e.g. budding in *Hydra*.

**vein 1.** A blood vessel that carries blood towards the heart. Most veins carry deoxygenated blood (the *pulmonary vein is an exception). The largest veins are fed by smaller ones, which are formed by the merger of *venules. Veins have thin walls and a relatively large internal diameter. *Valves within the veins ensure that the flow of blood is always towards the heart. *Compare* ARTERY. **2.** A vascular bundle in a leaf (*see* VENATION). **3.** Any of the tubes of chitin that strengthen an insect's wing.

**velamen** A whitish spongy sheath of dead empty cells that surrounds the aerial roots of epiphytic plants, such as certain orchids. It absorbs any surface water on the roots.

**velocity** Symbol $v$. The rate of displacement of a body. It is the *speed of a body in a specified direction. Velocity is thus a *vector quantity, whereas speed is a scalar quantity.

**velocity modulation** *See* KLYSTRON.

**velocity ratio (distance ratio)** The ratio of the distance moved by the point of application effort in a simple *machine to the distance moved by the point of application load in the same time.

**velum** *See* ANNULUS.

**vena cava** Either of the two large veins that carry deoxygenated blood into the right atrium of the heart. The **precaval vein (anterior** or **superior vena cava)** receives blood from the head and forelimbs; the **postcaval vein (posterior** or **inferior vena cava)** drains blood from the trunk and hindlimbs.

**venation 1.** The arrangement of veins (vascular bundles) in a leaf. The leaves of dicotyledons have a central main vein (midrib) with side branches that themselves further subdivide to form a network (**net** or **reticulate venation**). The leaves of monocotyledons have parallel veins (**parallel venation**). **2.** The arrangement of the veins in an insect's wing, which is often important in classification.

**Venn diagram** *See* SETS.

**venter** (in botany) The swollen base of an *archegonium, in which the egg cell (oosphere) develops.

**ventilation** The process by which a continuous exchange of gases is maintained across respiratory surfaces. Often called external *respiration, this is achieved by *respiratory movements; in air-breathing vertebrates it is movement of air into and out of the lungs (*see also* AIR SAC; EXPIRATION; INSPIRATION; TRACHEA). The **ventilation rate** (or **respiration rate**) of an animal is the volume of air breathed per minute, i.e. *tidal volume × number of breaths per minute. It can be measured with the aid of a *respirometer.

**ventilation centre** The group of neurons in the *medulla oblongata of the brain that controls the process of *ventilation. The partial pressure of carbon dioxide in the blood and the pH of the blood are monitored by chemoreceptors in the arteries. These include the *carotid bodies in the carotid arteries and the **aortic bodies** in the wall of the aorta close to the heart. The ventilation centre responds to an increase in the amount of carbon dioxide in the blood by increasing the rate of breathing. Within the ventilation centre are subcentres that control inspiration (**inspiratory centre**) and expiration (**expiratory centre**).

**ventral** Describing the surface of a plant or animal that is nearest or next to the ground or other support, i.e. the lower surface. In bipedal animals, such as humans, it is the forward-directed (*anterior) surface. *Compare* DORSAL.

**ventral root** The part of a *spinal nerve

that leaves the spinal cord on the ventral side and contains motor fibres. *Compare* DORSAL ROOT. *See* SPINAL CORD.

**ventricle 1.** A chamber of the *heart that receives blood from an *atrium and pumps it into the arterial system. Amphibians and fish have a single ventricle, but mammals, birds, and reptiles have two, pumping deoxygenated blood to the lungs and oxygenated blood to the rest of the body, respectively. **2.** Any of the four linked fluid-filled cavities in the brain of vertebrates. One of these cavities is in the *medulla oblongata, two are in the cerebral hemispheres (*see* CEREBRUM), and the fourth is in the posterior part of the *forebrain. The ventricles contain cerebrospinal fluid filtered from the blood by the *choroid plexus.

**Venturi tube** A device for mixing a fine spray of liquid with a gas or measuring a flow rate of a gas. It consists of two tapered sections of pipe joined by a narrow throat. The fluid velocity in the throat is increased and the pressure is therefore reduced. By attaching manometers to the three sections of the tube, the pressure drop can be measured and the flow rate through the throat can be calculated. In a carburettor, the petrol from the float chamber is made into a fine spray by being drawn through a jet into the low pressure in the throat of a Venturi tube, where it mixes with the air being drawn into the engine. The device was invented by the Italian physicist Giovanni Venturi (1746–1822).

**venule** A small blood vessel that receives blood from the capillaries and transports it to a vein.

**Venus** The sixth largest *planet in the *solar system and the second in order from the *sun. Its mean distance from the sun is $108.21 \times 10^6$ km, its mass is $4.868\,5 \times 10^{24}$ kg (about 81% that of earth), and its equatorial diameter is 12 104 km; it has a *sidereal period of 224.7 days. Because Venus is so close to the earth, it appears as the brightest object in the sky after the sun and moon. Venus is never more than 47° from the sun and usually appears as a morning or evening 'star'. Venus is shrouded in a thick atmosphere topped with a dense layer of cloud about 20 km thick, and consisting largely of concentrated sulphuric acid, that completely obscures the surface from external view. Therefore, most of our knowledge about it comes from a series of planetary probes

launched between 1961 and 2005, including the Soviet Venera 7, which in 1970 became the first spacecraft to send back data from the surface of another planet.

The atmosphere is 98% carbon dioxide and most of the remainder is nitrogen. The atmospheric pressure on the surface of Venus is about 90 times that on earth. This has meant that landers have only been able to function for an hour or less before being crushed and destroyed. Images taken by the dim sunlight reaching the surface reveal an arid landscape strewn with rocks rounded by windblown sand; radar has revealed very high mountains and plateaux, impact craters, and volcanic lava plains. The temperature on the surface of Venus soars to 750 K. This is the result of the *greenhouse effect, not proximity to the sun. Venus is unusual among the planets of the solar system because it rotates in a retrograde sense. In addition, its axial tilt is 177.3° and its period of axial rotation is the slowest of any planet so far known – 243 .01 days relative to the stars, but 116.8 days relative to the sun.

(((●))) **SEE WEB LINKS**
- NASA's profile of Venus
- Maps of the surface of Venus

**verdigris** A green patina of basic copper salts formed on copper. The composition of verdigris varies depending on the atmospheric conditions, but includes the basic carbonate $CuCO_3.Cu(OH)_2$, the basic sulphate $CuSO_4.Cu(OH)_2.H_2O$, and in some cases the basic chloride $CuCl_2.Cu(OH)_2$.

**vermiculite** *See* CLAY MINERALS.

**vermiform appendix** *See* APPENDIX.

**vernalization** The promotion of flowering by exposure of a plant to low temperatures. For example, winter cereals will not flower unless subjected to a period of chilling early in their development. Winter cereals are therefore sown in the autumn for flowering the following year. However, if germinating seeds are artificially vernalized they can be sown in the spring for flowering the same year.

**vernier** A short auxiliary scale placed beside the main scale on a measuring instrument to enable subdivisions of the main scale to be read accurately. The vernier scale is usually calibrated so that each of its divisions is 0.9 of the main scale divisions. The zero on the vernier scale is set to the observed measurement on the main scale and

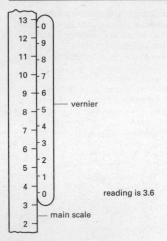

**Vernier.**

by noting which division on the vernier scale is exactly in line with a main scale division, the second decimal place of the measurement is obtained (see illustration). The device was invented by Pierre Vernier (1580–1637) in about 1630.

**vertebra** Any of the bones that make up the *vertebral column. In mammals each vertebra typically consists of a main body, or **centrum**, from which arises a **neural arch** through which the spinal cord passes, and **transverse processes** projecting from the side. In all vertebrates there are five groups of vertebrae, specialized for various functions and varying in number with the species. In humans, for example, there are 7 *cervical vertebrae, 12 *thoracic vertebrae, 5 *lumbar vertebrae, 5 fused *sacral vertebrae, and 5 fused *caudal vertebrae (forming the *coccyx).

**vertebral column (backbone; spinal column; spine)** A flexible bony column in vertebrates that extends down the long axis of the body and provides the main skeletal support. It also encloses and protects the *spinal cord and provides attachment for the muscles of the back. The vertebral column consists of a series of bones (*see* VERTEBRA) separated by discs of cartilage (*intervertebral discs). It articulates with the skull by means of the *atlas vertebra, with the ribs at the *thoracic vertebrae, and with the pelvic girdle at the sacrum (*see* SACRAL VERTEBRAE).

**vertebrate** Any one of a large group of animals comprising all those members of the subphylum *Craniata that have backbones (*see* VERTEBRAL COLUMN). Vertebrates include the fishes, amphibians, reptiles, birds, and mammals.

**very high frequency (VHF)** A radio frequency in the range $3 \times 10^8 - 0.3 \times 10^8$ Hz, i.e. having a wavelength in the range 1–10 m.

**very long baseline interferometry (VLBI)** A technique used in *radio astronomy in which radio telescopes at widely separated locations on the earth combine their observations of the same object to emulate a single radio telescope with a dish whose diameter equals the greatest separation between the participating telescopes.

**very low frequency (VLF)** A radio frequency in the range $3 \times 10^4 - 0.3 \times 10^4$ Hz, i.e. having a wavelength in the range 10–100 km.

**Vesalius, Andreas** (1514–64) Belgian physician and anatomist, who was a professor at Padua for six years before becoming a physician to the Habsburg court. He is remembered for producing in 1538–43 definitive text and anatomical drawings of the human body, which were made from actual dissections.

**vesicle** A small, usually fluid-filled, membrane-bound sac within the cytoplasm of a living cell. Vesicles occur, for example, as part of the *Golgi apparatus.

**vessel** **1.** (in botany) A tube within the *xylem composed of joined *vessel elements. Vessels facilitate the efficient movement of water from the roots to the shoots and leaves of a plant. **2.** (in zoology) Any of various tubular structures through which substances are transported, especially a blood vessel or a lymphatic vessel.

**vessel element** A type of cell occurring within the *xylem of flowering plants, many of which, end to end, form water-conducting **vessels**. Vessel elements are frequently very broad and have side walls thickened by deposits of lignin over most of the surface area. However, the end walls are broken down to provide connections with the cells both above and below them. *Compare* TRACHEID.

**vestibular apparatus** The part of the inner ear that is responsible for balance. The vestibular apparatus is continuous with the cochlea. It consists of the three *semicircular canals, which detect movements of the head (*see* AMPULLA), and the *utriculus and *sacculus, which detect the position of the head (*see* MACULA). *See* EAR.

**vestigial organ** Any part of an organism that has diminished in size during its evolution because the function it served decreased in importance or became totally unnecessary. Examples are the human appendix and the wings of the ostrich.

**vibrio** Any comma-shaped bacterium. Generally, vibrios are Gram-negative (*see* GRAM'S STAIN), motile, and aerobic. They are widely distributed in soil and water and while most feed on dead organic matter some are parasitic, e.g. *Vibrio cholerae*, the causal agent of cholera.

**vicinal (vic)** Designating a molecule in which two atoms or groups are linked to adjacent atoms. For example, 1,2-dichloroethane ($CH_2ClCH_2Cl$) is a vicinal (or vic) dihalide and can be named *vic*-dichloroethane.

**Victor Meyer's method** A method of measuring vapour density, devised by Victor Meyer (1848–97). A weighed sample in a small tube is dropped into a heated bulb with a long neck. The sample vaporizes and displaces air, which is collected over water and the volume measured. The vapour density can then be calculated.

**video camera (camcorder)** A hand-held *camera used to record moving pictures on video-tape cassettes for playing back on a television set. In a video camera the standard optical lenses focus the scene to be recorded onto an electronic camera tube, as in a television camera. The electronic data so obtained are recorded on video tape within the camcorder. At the same time, sound is picked up by a microphone attached to the camcorder, and recorded as a sound track along the edge of the video tape. The contents of the tape can usually be viewed on a miniscreen in the camcorder or played back through a television screen. Modern camcorders store their data in digital rather than analogue form – the same principle as used in digital photography. The latest camcorders use DVDs or electronic memory instead of tape.

**video recording** The recording of films, television programmes, etc., on magnetic tape. Because the demodulated video (vision) signal can have frequencies in the megahertz range, a video tape cannot be run like a sound tape. Sound has a maximum frequency of 15–20 kHz, which means that using the same system as sound, a video tape would need to run 1000 times faster than a sound tape. As this would be impractical the signal is recorded diagonally on the tape (each diagonal line representing one line of the picture) and the tape is run slowly over a drum on which the recording and reading heads rotate at high speeds. Devices using this mechanism are available for use with domestic television sets and video cameras (camcorders).

**villiaumite** A mineral form of sodium fluoride, NaF.

**villus** A microscopic outgrowth from the surface of some tissues and organs, which serves to increase the surface area of the organ. Numerous villi line the interior of the small intestine. Their shape may vary from finger-like (in the *duodenum) to spadelike (in the *ileum). Intestinal villi are specialized for the absorption of soluble food material: each contains blood vessels and a lymph vessel (*see* LACTEAL).

**Chorionic villi** occur on the chorion of the mammalian placenta, where they increase the surface area for the exchange of materials between the fetal and maternal blood.

**vinyl acetate** *See* ETHENYL ETHANOATE.

**vinyl chloride** *See* CHLOROETHENE.

**vinyl group** The organic group $CH_2{:}CH–$.

**VIP (vasoactive intestinal peptide)** A widely distributed *neuropeptide that acts as a neurotransmitter in the central and peripheral nervous systems and many other organs and tissues and also as a peptide hormone, being secreted by endocrine cells of the upper part of the small intestine in response to the entry of partially digested food from the stomach. VIP, along with *secretin, stimulates the pancreas to produce a thin watery secretion containing bicarbonate. This raises the pH in the intestine in preparation for secretion of pancreatic enzymes. VIP also inhibits gastric secretion, causes widening of blood vessals and airways, increases cardiac output, and relaxes smooth muscle.

**virial equation** A gas law that attempts to account for the behaviour of real gases, as opposed to an ideal gas. It takes the form

$$pV = RT + Bp + Cp^2 + Dp^3 + \ldots,$$

where $B$, $C$, and $D$ are known as **virial coefficients**.

**virion** *See* VIRUS.

**viroid** Any of various small naked single-stranded RNA molecules that infect plant cells and cause disease. Smaller than viruses, viroids are not enclosed in a protein coat of any kind: they generally consist of less than 400 nucleotides and do not contain any genes. The circular RNA strand undergoes extensive base pairing within itself, forming a double-stranded structure that mimics DNA and is replicated by the host cell's enzymes. This behaviour is similar to that of certain *introns, prompting the suggestion that viroids are escaped introns. Viroids include many commercially important disease agents, such as coconut cadang-cadang, citrus exocortis, and potato spindle tuber viroid.

**virology** The scientific study of *viruses. *See* MICROBIOLOGY.

( SEE WEB LINKS )

• Website for All the Virology on the WWW, a major portal for virology resources via the internet

**virtual image** *See* IMAGE.

**virtual reality** A form of computer simulation in which the user has the impression of being in an artificial environment. Typically, the user wears a visor into which are built two small screens, one for each eye, giving a three-dimensional view of a computer-generated environment. Sensors in the visor detect head movements and cause the perspective of the scene to change. It is also possible to wear special gloves (known as 'datagloves') containing sensors. These allow the user to move objects in the environment by making hand movements. Virtual-reality systems are used for training purposes as well as for entertainment.

**virtual state** The state of the **virtual particles** that are exchanged between two interacting charged particles. These particles, called *photons, are not in the real state, i.e. directly observable; they are constructs to enable the phenomenon to be explained in terms of *quantum mechanics.

**virtual work** The imaginary work done when a system is subjected to infinitesimal hypothetical displacements. According to the **principle of virtual work**, the total work done by all the forces acting on a system in

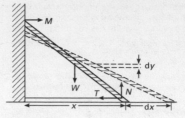

**Virtual work.**

equilibrium is zero. This principle can be used to determine the forces acting on a system in equilibrium. For example, the illustration shows a ladder leaning against a wall, with the bottom of the ladder attached to the wall by a horizontal weightless string. The tension, $T$, in the string can be calculated by assuming that infinitesimal movement d$x$ and d$y$ take place as shown. Then by applying the principle of virtual work, $Tdx + Wdy = 0$. As d$x$ and d$y$ can be calculated from the geometry, $T$ can be found.

**virulence** The disease-producing ability of a microorganism. *See also* PATHOGEN.

**virus** 1. (in microbiology) A particle that is too small to be seen with a light microscope or to be trapped by filters but is capable of independent metabolism and reproduction within a living cell. Outside its host cell a virus is completely inert. A mature virus (a **virion**) ranges in size from 20 to 400 nm in diameter. It consists of a core of nucleic acid (DNA or RNA) surrounded by a protein coat (**capsid**). Some bear an outer envelope (**enveloped viruses**). Inside its host cell the virus initiates the synthesis of viral proteins and undergoes replication. The new virions are released when the host cell disintegrates. Viruses are parasites of animals, plants, and some bacteria (*see* BACTERIOPHAGE). Viral diseases of animals include the common cold, influenza, smallpox, AIDS, herpes, hepatitis, polio, and rabies (*see* ADENOVIRUS; ARBOVIRUS; HERPESVIRUS; HIV; MYXOVIRUS; PAPOVAVIRUS; PICORNAVIRUS; POXVIRUS); some viruses are also implicated in the development of cancer (*see* RETROVIRUS). Plant viral diseases include various forms of yellowing and blistering of leaves and stems (*see* TOBACCO MOSAIC VIRUS). *Antiviral drugs are effective against certain viral diseases and *vaccines (if available) provide protection against others. 2. (in computing) A computer program that can replicate itself and be transferred from one computer to another without the user being aware of it. Viruses are often designed to destroy or damage the data on the user's computer. They generally spread by exchange of floppy disks or by telephone links.

**visceral** Relating to the internal organs (the **viscera**) that lie in the coelomic cavities of animals, i.e. in the thoracic and abdominal cavities of mammals. *Compare* SOMATIC.

**viscoelasticity** The property of certain materials of exhibiting both viscous behaviour when subject to shear flow and elastic properties when deformed. Some complex polymers exhibit this behaviour.

**viscometer** An instrument for measuring the viscosity of a fluid. In the **Ostwald viscometer**, used for liquids, a bulb in a capillary tube is filled with the liquid and the time taken for the meniscus to reach a mark on the capillary, below the bulb, is a measure of the viscosity. The **falling-sphere viscometer**, based on *Stokes' law, enables the speed of fall of a ball falling through a sample of the fluid to be measured. Various other devices are used to measure viscosity.

**viscose** Cellulose xanthate, the sticky brown liquid formed by adding carbon disulphide and sodium hydroxide to cellulose from wood pulp. It is forced through spinnarets into an acid bath, which reforms cellulose fibres as rayon (artificial silk).

**viscose process** *See* RAYON.

**viscosity** A measure of the resistance to flow that a fluid offers when it is subjected to shear stress. For a *Newtonian fluid, the force, $F$, needed to maintain a velocity gradient, $dv/dx$, between adjacent planes of a fluid of area $A$ is given by: $F = \eta A(dv/dx)$, where $\eta$ is a constant, the coefficient of viscosity. In *SI units it has the unit pascal second (in the c.g.s. system it is measured in *poise). Non-Newtonian fluids, such as clays, do not conform to this simple model. *See also* KINEMATIC VISCOSITY.

**(((⊕))) SEE WEB LINKS**

• Values of viscosities of liquids at the NPL website

**visible spectrum** The *spectrum of electromagnetic radiations to which the human eye is sensitive. *See* COLOUR.

**vision** The sense that enables perception of objects in the environment by means of the *eyes.

**visual acuity** Sharpness of vision: the ability of the eye to distinguish between objects that lie close together. This hinges on the ability of the eye to focus incoming light to form a sharp image on the retina. Visual acuity depends on the *cone cells, which are most densely packed in the *fovea, close to the centre of the retina, and are therefore in the optimum position to receive focused light. In addition, each cone cell synapses with a single nerve cell and is thus able to send a separate signal, via the optic nerve fibres, to the brain.

**visual binary** *See* BINARY STARS.

**visual-display unit (VDU)** The part of a *computer system or word processor on which text or diagrams are displayed. It consists of a *cathode-ray tube and usually has its own input keyboard attached.

**visual purple** *See* RHODOPSIN.

**vital capacity** The total amount of air that can be exhaled after maximum inspiration. The vital capacity of an average human is about 4.5 litres; in trained male athletes it can be 6 litres or more. However, some air always remains in the lungs (*see* RESIDUAL VOLUME).

**vital staining** A technique in which a harmless dye is used to stain living tissue for microscopical observation. The stain may be injected into a living animal and the stained tissue removed and examined (**intravital staining**) or the living tissue may be removed directly and subsequently stained (**supravital staining**). Microscopic organisms, such as protozoa, may be completely immersed in the dye solution. Vital stains include trypan blue, vital red, and Janus green, the latter being especially suitable for observing mitochondria.

**vitamin** One of a number of organic compounds required by living organisms in relatively small amounts to maintain normal health. There are some 14 generally recognized major vitamins: the water-soluble *vitamin B complex (containing 9) and *vitamin C and the fat-soluble *vitamin A, *vitamin D, *vitamin E, and *vitamin K. Most B vitamins and vitamin C occur in plants, animals, and microorganisms; they function typically as *coenzymes. Vitamins A, D, E, and K occur only in animals, especially vertebrates, and perform a variety of metabolic roles. Animals are unable to manufacture many vitamins themselves and must have adequate amounts in the diet. Foods may contain vitamin precursors (called **provitamins**) that are chemically changed to the actual vitamin on entering the body. Many vitamins are destroyed by light and heat, e.g. during cooking. See Chronology.

**((())) SEE WEB LINKS**
• Information and advice on vitamins and minerals from the UK Food Standards Agency

**vitamin A (retinol)** A fat-soluble vitamin that cannot be synthesized by mammals and other vertebrates and must be provided in the diet. Green plants contain precursors of the vitamin, notably carotenes, that are converted to vitamin A in the intestinal wall and liver. The aldehyde derivative of vitamin A, **retinal**, is a constituent of the visual pigment *rhodopsin. Deficiency affects the eyes, causing night blindness, xerophthalmia, and eventually total blindness. Vitamin A is also important in maintaining the integrity of epidermal and mucosal surfaces, which act as barriers against infection.

**((())) SEE WEB LINKS**
• Information about IUPAC nomenclature

**vitamin B complex** A group of water-soluble vitamins that characteristically serve as components of *coenzymes. Plants and many microorganisms can manufacture B vitamins but dietary sources are essential for most animals. Heat and light tend to destroy B vitamins.

**Vitamin B$_1$ (thiamine)** is a precursor of the coenzyme thiamine pyrophosphate, which functions in the Krebs cycle and carbohydrate metabolism. Deficiency leads to *beriberi in humans and to polyneuritis in birds. Good sources include wholegrain or fortified cereals, beans, peas, and nuts.

**Vitamin B$_2$ (riboflavin)** occurs in green vegetables, yeast, liver, and milk. It is a constituent of the coenzymes *FAD and FMN, which have an important role in the metabolism of all major nutrients as well as in the oxidative phosphorylation reactions of the *electron transport chain. Deficiency of B$_2$ causes inflammation of the tongue and lips, mouth sores, and conjunctivitis.

**Vitamin B$_6$ (pyridoxine)** is widely distributed in cereal grains, yeast, liver, milk, etc. It is a constituent of a coenzyme (pyridoxal phosphate) involved in amino acid metabolism. Deficiency can cause anaemia, dermatitis, and fatigue.

**Vitamin B$_{12}$ (cyanocobalamin or cobalamin)** is manufactured only by microorganisms and natural sources are entirely of animal origin. Liver is especially rich in it. One form of B$_{12}$ functions as a coenzyme in a number of reactions, including the oxidation of fatty acids and the synthesis of DNA. It also works in conjunction with *folic acid (another B vitamin) in the synthesis of the amino acid methionine and it is required for normal production of red blood cells. Vitamin B$_{12}$ can only be absorbed from the gut in the presence of a glycoprotein called **intrinsic factor**; lack of this factor or deficiency of B$_{12}$ results in pernicious anaemia.

Other vitamins in the B complex include *nicotinic acid, *pantothenic acid, *biotin, and *lipoic acid. *See also* CHOLINE; INOSITOL.

**((())) SEE WEB LINKS**
• Information about IUPAC nomenclature

**vitamin C (ascorbic acid)** A colourless crystalline water-soluble vitamin found especially in citrus fruits and green vegetables. Most organisms synthesize it from glucose, but humans and other primates and various other species must obtain it from their diet. It functions as a scavenger of free radicals; deficiency leads to *scurvy. Vitamin C is readily destroyed by heat and light.

**((())) SEE WEB LINKS**
• Information about IUPAC nomenclature

**vitamin D** A fat-soluble vitamin occurring in the form of two steroid derivatives: **vitamin D$_2$ (ergocalciferol**, or **calciferol**), found in yeast; and **vitamin D$_3$ (cholecalciferol**), which occurs in animals. Vitamin D$_2$ is formed from a steroid by the action of ultraviolet light and D$_3$ is produced by the action of sunlight on a cholesterol derivative in the skin. Fish-liver oils are the major dietary source. The active form of vitamin D (**calcitriol**) is manufactured in the kidneys in response to the secretion of *parathyroid hormone, which occurs when blood calcium levels are low. It causes increased uptake of calcium from the gut, which increases the supply of calcium for bone synthesis. Vitamin D deficiency causes *rickets in growing animals and osteomalacia in mature animals. Both conditions are characterized by weak deformed bones.

**((())) SEE WEB LINKS**
• Information about IUPAC nomenclature

**vitamin E (tocopherol)** A fat-soluble vitamin, consisting of several closely related compounds, that is the main *antioxidant of cell membranes and other lipid-rich tissue

## VITAMINS

1897 Dutch physician Christiaan Eijkman (1858–1930) cures beriberi in chickens with diet of whole rice.

1906–07 British biochemist Sir Frederick Hopkins demonstrates existence of accessory dietary elements essential for growth.

1912 Polish-born US biochemist Casimir Funk (1884–1967) extracts antiberiberi factor (an amine) from rice husks and coins the term 'vitamine' (vital amine; later changed to 'vitamin').

1913 US biochemist Elmer McCollum (1879–1967) discovers and names vitamin A (retinol) and names antiberiberi factor vitamin B.

1920 McCollum names antirachitic factor vitamin D.

1922 US embryologist Herbert Evans (1882–1971) discovers vitamin E (tocopherol).

1926 German chemist Adolf Windaus (1876–1959) discovers that ergosterol is converted to vitamin D in the presence of sunlight.

1931 German chemist Paul Karrer (1889–1971) determines the structure of (and synthesizes) vitamin A.

1932 Hungarian-born US biochemist Albert Szent-Györgyi (1893–1986) and US biochemist Charles King (1896–1986) independently isolate vitamin C (ascorbic acid).

1933 Polish-born Swiss chemist Tadeus Reichstein (1897–1996) and British chemist Walter Haworth (1883–1950) independently synthesize vitamin C.
US chemist Roger Williams (1893–1988) discovers the B vitamin pantothenic acid.

1934 Danish biochemist Carl Dam (1895–1976) discovers vitamin K.

1935 Karrer and Austrian-born German chemist Richard Kuhn (1900–67) independently synthesize vitamin $B_2$ (riboflavin).

1937 US chemist Robert Williams (1886–1965) synthesizes vitamin $B_1$ (thiamine).

1938 Karrer synthesizes vitamin E.
Kuhn isolates and synthesizes vitamin $B_6$ (pyridoxine).

1939 Dam and Karrer isolate vitamin K.

1940 Szent-Györgyi and US biochemist Vincent Du Vigneaud (1901–78) discover 'vitamin H' (the B vitamin biotin).
Roger Williams determines the structure of pantothenic acid.
US biochemist Edward Doisey synthesizes vitamin K.

1948 US biochemist Karl Folkers (1906–97) isolates vitamin $B_{12}$ (cyanocobalamin).

1956 British chemist Dorothy Hodgkin (1910–94) determines the structure of vitamin $B_{12}$.

1971 US chemist Robert Woodward (1917–79) and Swiss chemist Albert Eschenmoser (1925– ) synthesize vitamin $B_{12}$.

components. Deficiency leads to a range of disorders in different species, including muscular dystrophy, liver damage, and infertility. Good sources are cereal grains and green vegetables.

(🌐) **SEE WEB LINKS**
• Information about IUPAC nomenclature

**vitamin K** Any of several related fat-soluble compounds, including vitamins $K_1$ (**phylloquinone**) and $K_2$ (**menaquinone**), that act as coenzymes in the synthesis of several proteins (including prothrombin) necessary for blood clotting. Deficiency of vitamin K, which leads to extensive bleeding, is rare because a form of the vitamin is manufactured by intestinal bacteria. Green vegetables and egg yolk are good sources.

**vitelline membrane** *See* EGG MEMBRANE.

**vitreous** Having a glasslike appearance or structure.

**vitreous humour** The colourless jelly that fills the space between the lens and the retina of the vertebrate eye.

**vitriol (oil of vitriol)** An old name for sulphuric acid. As a result, hydrated iron(II) (ferrous) sulphate was known as **green vitriol**, and hydrated copper(II) (cupric) sulphate as **blue vitriol**.

**viviparity** 1. (in zoology) A form of reproduction in animals in which the developing embryo obtains its nourishment directly from the mother via a *placenta or by other means. Viviparity occurs in some insects and other arthropods, in certain fishes, amphibians, and reptiles, and in the majority of mammals. *Compare* OVIPARITY; OVOVIVIPARITY. 2. (in botany). **a.** A form of *asexual reproduction in certain plants, such as the onion, in which the flower develops into a budlike structure that forms a new plant when detached from the parent. **b.** The development of young plants on the inflorescence of the parent plant, as seen in certain grasses and the spider plant.

**vocal cords** A pair of elastic membranes that project into the *larynx in air-breathing vertebrates. Vocal sounds are produced when expelled air passing through the larynx vibrates the cords. The pitch of the sound produced depends on the tension of the cords, which is controlled by muscles and cartilages in the larynx.

**Voice over Internet Protocol** *See* VoIP.

**void** A large region of space at the centre of a collection of galaxy superclusters; there are hardly any galaxies in the void itself, indeed there is little evidence of any matter at all. Empty 'corridors' connect voids, rather like the holes in a piece of sponge, making them part of the large-scale structure of the universe.

**VoIP** Voice over Internet Protocol: the use of an IP network, in particular the Internet, to carry verbal conversations. This has been a major growth area in recent years, with several companies offering services that not only allow conversations between their members but also connect to the global telephone network. Indeed, many traditional telephone providers now use VoIP internally. Implementing conversations in real time imposes special constraints because they are particularly sensitive to any delay or lost data. For example, the scope for using buffering to assemble an incoming message from data packets that might arrive out of order, be delayed, be lost and require retransmission, etc., is severely limited. There are currently (2008) two main standards used to implement VoIP: the ITU's **H323** and the IETF's **SIP**.

(🌐) **SEE WEB LINKS**
• A comparison of H323 and SIP
• The SIP specification

**volcano** A fissure or vent in the earth's surface, connected to a magma source in the earth's interior by a conduit or series of fractures, from which solid, molten, and gaseous material is ejected. The resultant geological structure, also called volcano, can take a number of forms. **Central volcanoes**, which have a circular vent or number of vents, may be composite volcanoes or stratovolcanoes, comprising alternate layers of tephra (fragmental material including ash) and lava (e.g. Vesuvius, Italy), or (where only solid material is ejected) steep-sided cinder cones. **Shield volcanoes** are large structures with gentle slopes (e.g. the islands of Hawaii). **Fissure volcanoes** are linear fractures in the earth's surface where fluid material is emitted and, on land, spreads over large areas (e.g. volcanoes in Iceland). A number of types of volcanic eruption are recognized: Hawaiian, which are generally quiet with fluid lava erupted freely from fissures or pits, and – with increasing viscosity of magma – Strombolian, Vulcanian, Vesuvian, Plinian, and Peléean, which are more explosive. An exceptionally large form is the **supervolcano**, in which magma rises to form a vast reservoir in the earth's crust and builds in pressure over time before erupting in devastating explosions. The last supervolcano to erupt was Toba Caldera in Sumatra, 74 000 years ago. The *caldera of Yellowstone National Park is one of the largest in the world. Volcanoes occur principally along constructive or destructive plate margins (see PLATE TECTONICS) but some volcanic activity occurs away from the margins.

**volcanology (vulcanology)** The scientific study of volcanism, i.e. the processes by which magma and associated gases rise from the earth's interior and are emitted from the surface, and the resultant structures (see VOLCANO).

**volt** Symbol V. The SI unit of electric potential, potential difference, or e.m.f. defined as the difference of potential between two points on a conductor carrying a constant current of one ampere when the power dissipated between the points is one watt. It is named after Alessandro Volta.

**Volta, Alessandro Giuseppe Antonio Anastasio** (1745–1827) Italian physicist. In 1774 he began teaching in Como and in that year invented the *electrophorus. He moved to Pavia University in 1778. In 1800 he made the *voltaic cell, thus providing the first practical source of electric current (see also GALVANI, LUIGI). The SI unit of voltage is named after him.

**voltage** Symbol *V*. An e.m.f. or potential difference expressed in volts.

**voltage divider (potential divider; potentiometer)** A resistor or a chain of resistors connected in series that can be tapped at one or more points to obtain a known fraction of the total voltage across the whole resistor or chain. In the illustration, *V* is the total voltage across the divider and *v* is required voltage, then

$$v/V = R_2/(R_1 + R_2).$$

**voltaic cell (galvanic cell)** A device that produces an e.m.f. as a result of chemical reactions that take place within it. These reactions occur at the surfaces of two electrodes, each of which dips into an electrolyte. The first voltaic cell, devised by Alessandro Volta (1745–1827), had electrodes of two different metals dipping into brine. See PRIMARY CELL; SECONDARY CELL.

**voltaic pile** An early form of battery, devised by Alessandro Volta, consisting of a number of flat *voltaic cells joined in series. The liquid electrolyte was absorbed into paper or leather discs.

**voltameter (coulometer) 1.** An electrolytic cell formerly used to measure quantity of electric charge. The increase in mass ($m$) of the cathode of the cell as a result of the deposition on it of a metal from a solution of its salt enables the charge ($Q$) to be determined from the relationship $Q = m/z$, where $z$ is the electrochemical equivalent of the metal. **2.** Any other type of electrolytic cell used for measurement.

**voltmeter** An instrument used to measure voltage. *Moving-coil instruments are widely used for this purpose; generally a galvanometer is used in series with a resistor of high values (sometimes called a **multiplier**). To measure an alternating potential difference a rectifier must be included in the circuit. A moving-iron instrument can be used for either d.c. or a.c. without a rectifier. *Cathode-ray oscilloscopes are also used as voltmeters. The electronic **digital voltmeter** displays the value of the voltage in digits. The input is repeatedly sampled by the voltmeter and the instantaneous values are displayed.

**volume** Symbol *V*. The space occupied by a body or mass of fluid.

**volumetric analysis** A method of quantitative analysis using measurement of volumes. For gases, the main technique is in reacting or absorbing gases in graduated containers over mercury, and measuring the

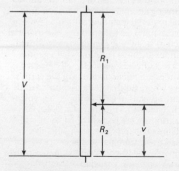

**Voltage divider.**

volume changes. For liquids, it involves *titrations.

**voluntary** (in biology) Controlled by conscious thought. *See* VOLUNTARY MUSCLE. *Compare* INVOLUNTARY.

**voluntary muscle (skeletal, striped,** *or* **striated muscle)** Muscle that is under the control of the will and is generally attached to the skeleton. An individual muscle consists of bundles of long **muscle fibres**, each containing many nuclei, the whole muscle being covered with a strong connective tissue sheath (**epimysium**) and attached at each end to a bone by inextensible *tendons. Each fibre contains smaller fibres (**myofibrils**) having alternate light and dark bands, which contain protein filaments responsible for the muscle's contractile ability and give the muscle its typical striped appearance under the microscope. The functional unit of a myofibril is the *sarcomere. See illustration.

The end of the muscle that is attached to a nonmoving bone is called the **origin** of the muscle; the end attached to a moving bone is the **insertion**. As a muscle contracts it becomes shorter and fatter, moving one bone closer to the other. Since a muscle cannot expand, another muscle (the **extensor**) is required to move the bone in the opposite direction and stretch the first muscle (known as the **flexor**). The flexor and extensor are described as **antagonistic muscles**. See illustration.

**von Baeyer nomenclature** A system of naming polycyclic compounds originally introduced by Adolf von Baeyer in 1900 and since extended.

**(⊕) SEE WEB LINKS**
• Information about the IUPAC system

**von Laue, Max** *See* LAUE, MAX THEODOR FELIX VON.

**vortex dynamics** The branch of fluid mechanics that is concerned with the dynamics of **vortices**, i.e. fluid which is rotating. There are many applications of vortex dynamics in engineering, atmospheric dynamics, and oceanography. Familiar examples of vortex dynamics include hurricanes and tornadoes. Vortex dynamics is also important in quantum liquids, such as superfluids.

**vortex pinning** The fixing in space of a vortex so that it does not 'slip'. In some prac-

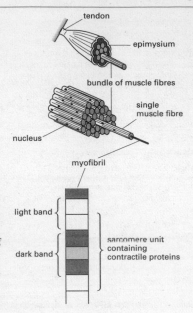

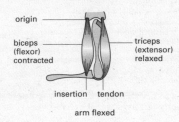

arm flexed

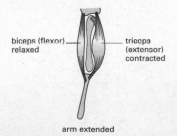

arm extended

**Voluntary muscle.** Structure and action of a voluntary muscle.

tical applications of *superconductivity vortex pinning is desirable.

**V-series** *See* NERVE AGENTS.

**vulcanite (ebonite)** A hard black insulating material made by the vulcanization of rubber with a high proportion of sulphur (up to 30%).

**vulcanization** A process for hardening rubber by heating it with sulphur or sulphur compounds.

**vulcanology** *See* VOLCANOLOGY.

**vulva** The female external genitalia, comprising in women two pairs of fleshy folds of tissue, the labia (*see* LABIUM); the *clitoris; and the vaginal opening.

**VX** A highly toxic colourless oily liquid, $C_{11}H_{26}NO_2PS$; r.d. 1.01; m.p. $-50°C$; b.p. $298°C$. It is an organophosphorus compound and is probably one of the most toxic of all *nerve agents. VX was discovered in 1952 at the Porton Down research station in Wiltshire. VX is the best known of the V-series of nerve agents (the V denotes very high persistence in the environment). Other, lesser known, members of the series include VG and VM, which are both organophosphorus compounds related to VX.

V

**Wacker process** A process for the manufacture of ethanal by the air oxidation of ethene. A mixture of air and ethene is bubbled through a solution containing palladium(II) chloride and copper(II) chloride. The $Pd^{2+}$ ions form a complex with the ethene in which the ion is bound to the pi electrons in the C=C bond. This decreases the electron density in the bond, making it susceptible to nucleophilic attack by water molecules. The complex formed breaks down to ethanal and palladium metal. The $Cu^{2+}$ ions oxidize the palladium back to $Pd^{2+}$, being reduced to $Cu^+$ ions in the process. The air present oxidizes $Cu^+$ back to $Cu^{2+}$. Thus the copper(II) and palladium(II) ions effectively act as catalysts in the process, which is now the main source of ethanal and, by further oxidation, ethanoic acid. It can also be applied to other alkenes. It is named after Alexander von Wacker (1846–1922).

**Wade's rules** A set of rules for predicting the structure of a cluster compound based on the number of electrons in the framework counted in a particular way. The electrons counted are known as **skeletal electrons**. The rules apply to polyhedra that have triangular faces (known as **deltahedra**). They were originally applied to the boron hydrides. Electron pairs in bonding between two boron atoms are counted as skeletal electrons but the pairs in B–H units are ignored. However, if a boron atom is connected to two hydrogens ($BH_2$), the second bond is counted with the skeletal electrons. According to the rules, if the formula is

$$[B_nH_n]^{2-}$$

and there are $n+1$ skeletal electron pairs, then the structure is **closo**. If the formula is

$$B_nH_{n+4}$$

and these are $n+2$ skeletal electron pairs, the structure is **nido**. If the formula is

$$B_nH_{n+6}$$

and these are $n+3$ skeletal electron pairs, the structure is **arachno**. The rules are named after the British chemist Kenneth Wade, who first formulated them in the early 1970s.

**Wagenaar test** *See* ACETONE–CHLOR–HAEMIN TEST.

**Wagner-Meerwein rearrangement** A rearrangement in which an alkyl group moves from a carbon atom to an adjacent carbon atom during a reaction. It often occurs to stabilize a carbocation formed as an intermediate during the reaction. Rearrangements of this type have been extensively studied in terpene chemistry.

**Wallace, Alfred Russel** (1823–1913) British naturalist, who in 1848 went on an expedition to the Amazon, and in 1854 travelled to the Malay Archipelago. There he noticed the differences between the animals of Asia and Australasia and devised *Wallace's line, which separates them. This led him to develop a theory of *evolution through *natural selection, which coincided with the views of Charles *Darwin; their theories were presented jointly to the Linnaean Society in 1858.

**Wallace's line** An imaginary line that runs between the Indonesian islands of Bali and Lombok and represents the separation of the Australian and Oriental faunas. It was proposed by Alfred Russel Wallace, who had noted that the mammals in SE Asia are different from and more advanced than their Australian counterparts. He suggested this was because the Australian continent had split away from Asia before the better adapted placental mammals evolved in Asia. Hence the isolated Australian marsupials and monotremes were able to thrive while those in Asia were driven to extinction by competition from placental mammals. *See also* ZOOGEOGRAPHY.

**wall effect** Any effect resulting from the nature or presence of the inside wall of a container on the system it encloses.

**Walton, Ernest** *See* COCKCROFT, SIR JOHN DOUGLAS.

**warfarin** 3-(alpha-acetonylbenzyl)-4-hydroxycoumarin: a synthetic *anticoagulant used both therapeutically in clinical medicine and, in lethal doses, as a rodenticide (*see* PESTICIDE).

**warm-blooded animal** *See* ENDOTHERM.

**warm front** *See* FRONT.

**warning coloration (aposematic coloration)** The conspicuous markings of an animal that make it easily recognizable and warn would-be predators that it is a poisonous, foul-tasting, or dangerous species. For example, the yellow-and-black striped abdomen of the wasp warns of its sting. *See also* MIMICRY.

**washing soda** *Sodium carbonate decahydrate, $Na_2CO_3.10H_2O$.

**waste product** 1. Any product of metabolism that is not required for further metabolic processes and is therefore excreted from the body. Common waste products include nitrogenous compounds (such as *urea and ammonia), carbon dioxide, and *bile. **2.** *See* RADIOACTIVE WASTE.

**water** A colourless liquid, $H_2O$; r.d. 1.000 (4°C); m.p. 0.000°C; b.p. 100.000°C. In the gas phase water consists of single $H_2O$ molecules in which the H–O–H angle is 105°. The structure of liquid water is still controversial; hydrogen bonding of the type $H_2O...H–O–H$ imposes a high degree of structure and current models supported by X-ray scattering studies have short-range ordered regions, which are constantly disintegrating and reforming. This ordering of the liquid state is sufficient to make the density of water at about 0°C higher than that of the relatively open structured ice; the maximum density occurs at 3.98°C. This accounts for the well-known phenomenon of ice floating on water and the contraction of water below ice, a fact of enormous biological significance for all aquatic organisms.

Ice has nine distinct structural modifications of which ordinary ice, or ice I, has an open structure built of puckered six-membered rings in which each $H_2O$ unit is tetrahedrally surrounded by four other $H_2O$ units.

Because of its angular shape the water molecule has a permanent dipole moment and in addition it is strongly hydrogen bonded and has a high dielectric constant. These properties combine to make water a powerful solvent for both polar and ionic compounds. Species in solution are frequently strongly hydrated and in fact ions frequently written as, for example, $Cu^{2+}$ are

essentially $[Cu(H_2O)_6]^{2+}$. Crystalline *hydrates are also common for inorganic substances; polar organic compounds, particularly those with O–H and N–H bonds, also form hydrates.

Pure liquid water is very weakly dissociated into $H_3O^+$ and $OH^-$ ions by self ionization:

$$H_2O \rightleftharpoons H^+ + OH^-$$

(*see* IONIC PRODUCT) and consequently any species that increases the concentration of the positive species, $H_3O^+$, is acidic and species increasing the concentration of the negative species, $OH^-$, are basic (*see* ACID). The phenomena of ion transport in water and the division of materials into **hydrophilic** (water loving) and **hydrophobic** (water hating) substances are central features of almost all biological chemistry. A further property of water that is of fundamental importance to the whole planet is its strong absorption in the infrared range of the spectrum and its transparency to visible and near ultraviolet radiation. This allows solar radiation to reach the earth during hours of daylight but restricts rapid heat loss at night. Thus atmospheric water prevents violent diurnal oscillations in the earth's ambient temperature. *See also* GREENHOUSE EFFECT.

**water cycle** *See* HYDROLOGICAL CYCLE.

**water gas** A mixture of carbon monoxide and hydrogen produced by passing steam over hot carbon (coke):

$$H_2O(g) + C(s) \rightarrow CO(g) + H_2(g)$$

The reaction is strongly endothermic but the reaction can be used in conjunction with that for *producer gas for making fuel gas. The main use of water gas before World War II was in producing hydrogen for the *Haber process. Here the above reaction was combined with the **water-gas shift reaction** to increase the amount of hydrogen:

$$CO + H_2O \rightleftharpoons CO_2 + H_2$$

Most hydrogen for the Haber process is now made from natural gas by steam *reforming.

**water glass** A viscous colloidal solution of sodium silicates in water, used to make silica gel and as a size and preservative.

**water of crystallization** Water present in crystalline compounds in definite proportions. Many crystalline salts form hydrates containing 1, 2, 3, or more moles of water per mole of compound, and the water may be held in the crystal in various ways. Thus, the water molecules may simply occupy lattice positions in the crystal, or they may form bonds with the anions or the cations present. In the pentahydrate of copper sulphate ($CuSO_4.5H_2O$), for instance, each copper ion is coordinated to four water molecules through the lone pairs on the oxygen to form the *complex $[Cu(H_2O)_4]^{2+}$. Each sulphate ion has one water molecule held by hydrogen bonding. The difference between the two types of bonding is demonstrated by the fact that the pentahydrate converts to the monohydrate at 100°C and only becomes anhydrous above 250°C. **Water of constitution** is an obsolete term for water combined in a compound (as in a metal hydroxide $M(OH)_2$ regarded as a hydrated oxide $MO.H_2O$).

**water potential** Symbol Ψ. The difference between the chemical potential of the water in a biological system and the chemical potential of pure water at the same temperature and pressure. It is manifested as a force acting on water molecules in a solution separated from pure water by a membrane that is permeable to water molecules only. Water potential is measured in kilopascals (kPa). The water potential of pure water is zero; aqueous solutions of increasing concentration have increasingly negative values. Water tends to move from areas of high (less negative) water potential to areas of low (more negative) water potential. *Osmosis in plants is now described in terms of water potential.

**water softening** *See* HARDNESS OF WATER.

**Watson, James Dewey** (1928–  ) US biochemist, who moved to the Cavendish Laboratory, Cambridge, in 1951 to study the structure of *DNA. In 1953 he and Francis *Crick announced the now accepted two-stranded helical structure for the DNA molecule. In 1962 they shared the Nobel Prize for physiology or medicine with Maurice Wilkins (1916–2004), who with Rosalind Franklin (1920–58) had made X-ray diffraction studies of DNA.

**Watson–Crick model** The double-stranded twisted ladder-like molecular structure of *DNA as determined by James Watson and Francis Crick at Cambridge, England, in 1953. It is commonly known as the **double helix**.

**watt** Symbol W. The SI unit of power,

defined as a power of one joule per second. In electrical contexts it is equal to the rate of energy transformation by an electric current of one ampere flowing through a conductor the ends of which are maintained at a potential difference of one volt. The unit is named after James Watt (1736–1819).

**wattmeter** An instrument for measuring the power in watts in an alternating-current electric circuit. In a direct-current circuit, power is usually determined by separate measurements of the voltage and the current.

The **electrodynamic wattmeter** consists of two coils, one fixed (current) coil and one movable (potential) coil. The fixed coil carries the load current, and the movable coil carries a current proportional to the voltage applied to the measured circuit. The deflection of the needle attached to the movable coil indicates the power.

**wave** A periodic disturbance in a medium or in space. In a **travelling wave** (or **progressive wave**) energy is transferred from one place to another by the vibrations (*see also* STATIONARY WAVE). In a wave passing over the surface of water, for example, the water rises and falls as the wave passes but the particles of water on average do not move forward with the wave. This is called a **transverse wave** because the disturbances are at right angles to the direction of propagation. The water surface moves up and down while the waves travel across the surface of the water. Electromagnetic waves (see diagram) are also of this kind, with electric and magnetic fields varying in a periodic way at right angles to each other and to the direction of propagation. In sound waves, the air is alternately compressed and rarefied by displacements in the direction of propagation. Such waves are called **longitudinal waves**.

The chief characteristics of a wave are its

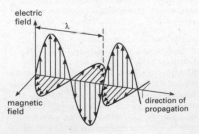

**Wave.** Electromagnetic waves.

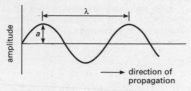

**Wave.** Sine wave.

**speed of propagation**, its **frequency**, its **wavelength**, and its **amplitude**. The speed of propagation is the distance covered by the wave in unit time. The frequency is the number of complete disturbances (cycles) in unit time, usually expressed in *hertz. The wavelength is the distance in metres between successive points of equal phase in a wave. The amplitude is the maximum difference of the disturbed quantity from its mean value.

Generally, the amplitude ($a$) is half the peak-to-peak value. There is a simple relationship between the wavelength ($\lambda$) and the frequency ($f$), i.e. $\lambda = c/f$, where $c$ is the speed of propagation. The energy transferred by a progressive *sine wave (see diagram) is proportional to $a^2 f^2$. *See also* SIMPLE HARMONIC MOTION.

**wave equation** A partial differential equation of the form:

$$\nabla^2 u = (1/c^2)\partial^2 u/\partial t^2$$

where

$$\nabla^2 = \partial^2/\partial x^2 + \partial^2/\partial y^2 + \partial^2/\partial z^2$$

is the Laplace operator (*see* LAPLACE EQUATION). It represents the propagation of a wave, where $u$ is the displacement and $c$ the speed of propagation. *See also* SCHRÖDINGER EQUATION.

**wave form** The shape of a wave or the pattern representing a vibration. It can be illustrated by drawing a graph of the periodically varying quantity against distance for one complete wavelength. *See also* SINE WAVE.

**wavefront** A line or surface within a two- or three-dimensional medium through which waves are passing, being the locus of all adjacent points at which the disturbances are in phase. At large distances from a small source in a uniform medium, the fronts are small parts of a sphere of very large radius and they can be considered as plane. For example, sunlight reaches the earth with plane wavefronts.

**wave function** A function $\psi(x,y,z)$ appearing in the *Schrödinger equation in *quantum mechanics. The wave function is a mathematical expression involving the coordinates of a particle in space. If the Schrödinger equation can be solved for a particle in a given system (e.g. an electron in an atom) then, depending on the boundary conditions, the solution is a set of allowed wave functions (**eigenfunctions**) of the particle, each corresponding to an allowed energy level (**eigenvalue**). The physical significance of the wave function is that the square of its absolute value, $|\psi|^2$, at a point is proportional to the probability of finding the particle in a small element of volume, $dxdydz$, at that point. For an electron in an atom, this gives rise to the idea of atomic and molecular *orbitals.

**wave guide** A hollow tube through which microwave electromagnetic radiation can be transmitted with relatively little attenuation. They often have a rectangular cross section, but some have a circular cross section. In transverse electric (TE) modes the electric vector of the field has no component in the direction of propagation. In transverse magnetic (TM) modes, the magnetic vector has no such component.

**wavelength** *See* WAVE.

**wave mechanics** A formulation of *quantum mechanics in which the dual wave–particle nature (*see* COMPLEMENTARITY) of such entities as electrons is described by the *Schrödinger equation. Schrödinger put forward this formulation of quantum mechanics in 1926 and in the same year showed that it was equivalent to *matrix mechanics. Taking into account the *de Broglie wavelength, Schrödinger postulated a wave mechanics that bears the same relation to *Newtonian mechanics as physical optics does to geometrical optics (*see* OPTICS).

**wavemeter** A device for measuring the wavelength of electromagnetic radiation. For frequencies up to about 100 MHz a wavemeter consists of a tuned circuit with a suitable indicator to establish when resonance occurs. Usually the tuned circuit includes a variable capacitor calibrated to read wavelengths and resonance is indicated by a current-detecting instrument. At higher frequencies a cavity-resonator in a waveguide is often used. The cavity resonator is fitted with a piston, the position of which determines the resonant frequency of the cavity.

**wave number** Symbol $k$. The number of cycles of a wave in unit length. It is the reciprocal of the wavelength (*see* WAVE).

**wave–particle duality** The concept that waves carrying energy may have a corpuscular aspect and that particles may have a wave aspect; which of the two models is the more appropriate will depend on the properties the model is seeking to explain. For example, waves of electromagnetic radiation need to be visualized as particles, called *photons, to explain the *photoelectric effect while electrons need to be thought of as de Broglie waves in *electron diffraction. *See also* COMPLEMENTARITY; DE BROGLIE WAVELENGTH; LIGHT.

**wave power** The use of wave motion in the sea to generate energy. The technique used is to anchor a series of bobbing floats offshore; the energy of the motion of the floats is used to turn a generator. It has been estimated that there are enough suitable sites to generate over 100 GW of electricity in the UK.

**wave theory** *See* LIGHT.

**wave-vector** A quantity that simultaneously defines the magnitude of a *wave and its direction. The magnitude is equal to $2\pi/\lambda$, where $\lambda$ is the wavelength, or $2\pi k$, where $k$ is the *wave number.

**wax** Any of various solid or semisolid substances. There are two main types. Mineral waxes are mixtures of hydrocarbons with high molecular weights. Paraffin wax, obtained from *petroleum, is an example. Waxes secreted by plants or animals are mainly esters of fatty acids and usually have a protective function. Examples are the beeswax forming part of a honeycomb and the wax coating on some leaves, fruits, and seed coats, which acts as a protective water-impermeable layer supplementing the functions of the cuticle. The seeds of a few plants contain wax as a food reserve.

**W boson (W particle)** Either of a pair of elementary particles (W⁺ or W⁻), classified as **intermediate vector bosons**, that are believed to transmit the weak interaction (*see* FUNDAMENTAL INTERACTIONS) in much the same way as photons transmit the electromagnetic interaction. They are not, however, massless like photons, and are believed to

have a rest mass of the order of $10^{-25}$ kg (80.4 GeV). W bosons were discovered at CERN in 1983 with the expected mass. *See also* Z BOSON.

**weak acid** An *acid that is only partially dissociated in aqueous solution.

**weak interaction** *See* FUNDAMENTAL INTERACTIONS.

**weakly interacting massive particle (WIMP)** *See* MISSING MASS.

**weather** The state of atmospheric conditions, including humidity, precipitation (e.g. rain, snow, hail), temperature, pressure, cloud cover, visibility, and wind, at any one place and time. A **weather forecast** is a prediction of the weather conditions to be expected at a particular place over a given period. These may be short-range (1–2 days), medium-range (5–7 days), or long-range (1 month or a season). The two chief methods employed in obtaining short-range and medium-range forecasts are **synoptic forecasting** and **numerical forecasting**. Synoptic forecasting involves the simultaneous observation of weather elements at a series of weather stations, the collection of data, and the plotting of the information obtained on synoptic charts (weather maps), from which forecasts can be made. Since the early 1960s data collected by satellites, such as pictures of cloud cover and infrared measurements, have been increasingly used in forecasting. Numerical forecasting involves the numerical solution of equations governing the motions and changes of atmospheric conditions. Computers are used to carry out the vast number of calculations. Both synoptic and numerical methods are unsuitable for long-range forecasts and instead statistical and analogue methods are used.

**weathering** The process of breakdown and alteration of rocks on the earth's surface by mechanical or chemical processes. Mechanical (physical) weathering includes the splitting of rocks through the action of frost and extreme temperature changes. Chemical weathering includes solution (the dissolving of solid materials by water); carbonation (the dissolving of soluble rocks and minerals by a weak carbonic acid formed by the combination of water with atmospheric carbon dioxide); oxidation (the combination of atmospheric oxygen with rock materials); and hydration (the chemical combination of rock materials with water). Organic weathering, which may involve both chemical and mechanical processes, is caused by plants and animals. For example, burrowing animals and plant roots may physically break up rocks; lichens, which can exist on bare rock surfaces, cause decomposition through the removal of nutrients.

**Web** *See* WORLD WIDE WEB.

**weber** Symbol Wb. The SI unit of magnetic flux equal to the flux that, linking a circuit of one turn, produces in it an e.m.f. of one volt as it is reduced to zero at a uniform rate in one second. It is named after Wilhelm Weber.

**Weber, Wilhelm Eduard** (1804–91) German physicist, who became a professor at Göttingen. In 1833 he and Karl *Gauss built an electric telegraph between their laboratories. In 1843 Weber moved to Leipzig, where his main work was to develop a system of self-consistent elecrical units (as Gauss had already done for magnetism). Both systems were adopted in 1881. The SI unit of magnetic flux is named after him.

**web log** *See* BLOG.

**weed** *See* PEST.

**Wegener, Alfred Lothar** (1880–1930) German geologist and meteorologist, who became a professor at the University of Graz in 1924. He is best known for his theory of *continental drift, which he formulated in 1915.

**weight** The force by which a body is attracted to the earth. *See also* MASS.

**weightlessness** A condition of a body when it is an infinite distance from any other body. In practice the appearance of weightlessness occurs in space when the gravitational attraction of the earth on a body in space is equal to the centripetal force required by its orbital motion so that the body is effectively in free fall. Weightlessness can also be simulated for short periods in an aircraft flying a parabolic flight path, so that its occupants are again in free fall.

**Weinberg–Salam model (WS model)** *See* ELECTROWEAK THEORY.

**Weismannism** The theory of the **continuity of the germ plasm** published by August Weismann (1834–1914) in 1886. It proposes that the contents of the reproductive cells (sperms and ova) are passed on unchanged from one generation to the next, unaffected by any changes undergone by the rest of the body. It thus rules out any possibility of the inheritance of acquired characteristics, and has become fundamental to neo-Darwinian theory.

**Western blotting (protein blotting)** An *immunoassay for determining very small amounts of a particular protein in tissue samples or cells. The sample is subjected to electrophoresis on SDS-polyacrylamide gel to separate constituent proteins. The resultant protein bands are then 'blotted' onto a polymer sheet. A radiolabelled or fluorescently labelled antibody specific for the target protein is added; this binds to the protein, which can then be detected by autoradiography or a fluorescence detector. A variation of this technique is used to screen bacterial colonies containing cDNA clones in order to isolate those colonies expressing a particular protein. The name is derived by analogy to that of *Southern blotting.

**Weston cell (cadmium cell)** A type of primary *voltaic cell, which is used as a standard; it produces a constant e.m.f. of 1.0186 volts at 20°C. The cell is usually made in an H-shaped glass vessel with a mercury anode covered with a paste of cadmium sulphate and mercury(I) sulphate in one leg and a cadmium amalgam cathode covered with cadmium sulphate in the other leg. The electrolyte, which connects the two electrodes by means of the bar of the H, is a saturated solution of cadmium sulphate. In some cells sulphuric acid is added to prevent the hydrolysis of mercury sulphate. It is named after Edward Weston (1850–1936).

**wet-and-dry bulb hygrometer** *See* HYGROMETER.

**whalebone (baleen)** Transverse horny plates hanging down from the upper jaw on each side of the mouth of the toothless whales (*see* CETACEA), forming a sieve. Water, containing plankton on which the whale feeds, enters the open mouth and is then expelled with the mouth slightly closed, so that food is retained on the baleen plates.

**whales** *See* CETACEA.

**Wheatstone, Sir Charles** (1802–75) British physicist, who set up as a musical instrument-maker in London. He studied acoustics and optics, inventing a stereoscope in 1838. His most important work, done with William Cooke (1806–79), was the development of an electric telegraph, which they achieved in 1837. He gave his name to the *Wheatstone bridge, although he did not invent it.

**Wheatstone bridge** An electrical circuit for measuring the value of a resistance. In the illustration, $R_1$ is a resistance of unknown value, $R_2$ is a fixed resistance of known value, $R_3$ and $R_4$ are variable resistances with known values. When no current flows between A and B the bridge is said to be balanced, the galvanometer registers no deflection, and $R_1/R_2 = R_3/R_4$. $R_1$ can therefore be calculated. The Wheatstone bridge is used in various forms. In the **metre bridge**, a wire 1 metre long of uniform resistance is attached to the top of a board alongside a metre rule. A sliding contact is run along the wire, which corresponds to $R_3$ and $R_4$, until the galvanometer registers zero. Most practical forms use one or more rotary rheostats to provide the variation. The device was popularized though not invented by Sir Charles Wheatstone.

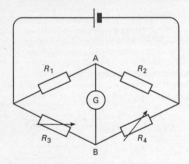

**Wheatstone bridge.**

**whey** *See* CURD.

**white arsenic** *See* ARSENIC(III) OXIDE.

**white blood cell** *See* LEUCOCYTE.

**white dwarf** A compact stellar object that is supported against collapse under self-gravity by the *degeneracy pressure of electrons. White dwarfs are formed as the end products of the evolution of stars of relatively low mass (about that of the sun); high-mass stars may end up as *neutron stars or *black holes (*see* STELLAR EVOLUTION). White dwarfs consist of helium nuclei (and carbon and oxygen nuclei in the more massive cases) and a *degenerate gas of electrons. A typical white-dwarf density is $10^9$ kg m$^{-3}$;

white dwarf masses and radii are in the region of 0.7 solar masses and $10^3$ km respectively. There is a maximum mass for white dwarfs, above which they are unstable to gravitational collapse – this is known as the *Chandrasekhar limit and is about 1.4 solar masses.

**white hole** *See* WORMHOLE.

**white matter** Part of the tissue that makes up the central nervous system of vertebrates. It consists chiefly of nerve fibres enclosed in whitish *myelin sheaths. *Compare* GREY MATTER.

**white mica** *See* MUSCOVITE.

**white spirit** A liquid mixture of petroleum hydrocarbons used as a solvent for paint ('turpentine substitute').

**Wiedemann–Franz law** The ratio of the thermal conductivity of any pure metal to its electrical conductivity is approximately constant at a given temperature. The law is fairly well obeyed, except at low temperatures. The law is named after Gustav Wiedemann and Rudolph Franz, who discovered it empirically in 1853.

**Wien formula** *See* PLANCK'S RADIATION LAW.

**Wien's displacement law** For a *black body, $\lambda_m T = $ constant, where $\lambda_m$ is the wavelength corresponding to the maximum radiation of energy and $T$ is the thermodynamic temperature of the body. Thus as the temperature rises the maximum of the spectral energy distribution curve is displaced towards the short-wavelength end of the spectrum. The law was stated by Wilhelm Wien (1864–1928).

**Wigner energy** Energy stored in a crystalline substance as a result of irradiation. This phenomenon is known as the **Wigner effect**. For example, some of the energy lost by neutrons in a *nuclear reactor is stored by the graphite moderator. As a result, the crystal lattice is changed and there is a consequent change in the physical dimensions of the moderator. It is named after Eugene Wigner (1902–95).

**Wigner nuclides** Pairs of isobars with odd nucleon numbers in which the atomic number and the neutron number differ by one. $^3$H and $^3$He are examples.

**Wigner's friend** *See* SCHRÖDINGER'S CAT.

**wiki** A web site whose content can be changed and extended by users via their web browsers. The term is also used for web applications that facilitate this. Wikis are a useful technique for rapidly assembling and disseminating 'collective wisdom' ('wiki wiki' is Hawaiian for 'fast' or 'quick'). They have become popular on many levels, ranging from small private wikis through internal departmental or corporate wikis to such popular web sites as Wikipedia. Their major problem, for which they have been criticized, is the possibility of inappropriate or malicious changes. All wikis maintain lists of recent changes to and often a complete history of each web page, to aid the 'user community' in discovering and reverting unwanted changes. Often each web page has an associated discussion page where the merits of existing and potential changes can be debated. However, such self-policing has been found inadequate for highly exposed web sites with their large but unknown user base. These impose additional security measures: finely graded levels of trust that control the extent to which a user can make alterations; programmatic techniques that seek to identify suspicious changes; etc.

**wild type** Describing the form of an *allele possessed by most members of a population in their natural environment. Wild-type alleles are usually *dominant.

**Wilkins, Maurice** *See* CRICK, FRANCIS HARRY COMPTON.

**Williamson's synthesis** Either of two methods of producing ethers, both named after the British chemist Alexander Williamson (1824–1904). **1.** The dehydration of alcohols using concentrated sulphuric acid. The overall reaction can be written

$$2ROH \rightarrow H_2O + ROR$$

The method is used for making ethoxyethane ($C_2H_5OC_2H_5$) from ethanol by heating at 140°C with excess of alcohol (excess acid at 170°C gives ethene). Although the steps in the reaction are all reversible, the ether is distilled off so the reaction can proceed to completion. This is **Williamson's continuous process**. In general, there are two possible mechanisms for this synthesis. In the first (favoured by primary alcohols), an alkylhydrogen sulphate is formed

$$ROH + H_2SO_4 \rightleftharpoons ROSO_3H + H_2O$$

This reacts with another alcohol molecule to give an oxonium ion

$$ROH + ROSO_3H \rightarrow ROHR^+$$

This loses a proton to give ROR.

The second mechanism (favoured by tertiary alcohols) is formation of a carbonium ion

$$ROH + H^+ \rightarrow H_2O + R^+$$

This is attacked by the lone pair on the other alcohol molecule

$$R^+ + ROH \rightarrow ROHR^+$$

and the oxonium ion formed again gives the product by loss of a proton.

The method can be used for making symmetric ethers (i.e. having both R groups the same). It can successfully be used for mixed ethers only when one alcohol is primary and the other tertiary (otherwise a mixture of the three possible products results).
**2.** A method of preparing ethers by reacting a haloalkane with an alkoxide. The reaction, discovered in 1850, is a nucleophilic substitution in which the negative alkoxide ion displaces a halide ion; for example:

$$RI + {}^-OR' \rightarrow ROR' + I^-$$

A mixture of the reagents is refluxed in ethanol. The method is particularly useful for preparing mixed ethers, although a possible side reaction under some conditions is an elimination to give an alcohol and an alkene.

**Wilson, Charles Thomson Rees** (1869–1959) British physicist, born in Scotland, who studied physics with J. J. *Thomson in Cambridge. His best-known achievement was the development of the *cloud chamber in 1911, for which he was awarded the 1927 Nobel Prize for physics.

**Wilson cloud chamber** *See* CLOUD CHAMBER.

**wilting** The condition that arises in plants when more water is lost by evaporation than is absorbed from the soil. This causes the cells to lose their *turgor and the plant structure droops. Plants can normally recover from wilting if water is added to the soil, but permanent wilting and possible death can result if the plant does not have access to water for a long period of time. In certain plants wilting is important as a mechanism to avoid overheating: when the leaves droop they are taken out of direct contact with the sun's rays. When the sun sets the plant can begin to transpire at the normal rate and the cells of the leaves regain their turgor.

**WIMP**  Weakly interacting massive particle. *See* MISSING MASS.

**Wimshurst machine**  A laboratory electrostatic generator. It consists of two insulating discs to which radial strips of metal foil are attached. After a few strips have been charged individually, the discs are rotated in opposite directions and the charge produced on the strips by induction is collected by metal combs or brushes. It was invented by James Wimshurst (1836–1903).

**wind**  The motion (usually horizontal) of air relative to the earth's surface. The **general circulation of the atmosphere** – the large-scale patterns of wind and pressure that persist throughout the year or recur seasonally – results largely from differences in the net radiation received at the earth's surface. However, it is modified by the rotation of the earth, the presence of mountain barriers, the relative distribution of land and sea, and the positions of the ocean currents. Although the resulting wind circulation patterns are complex, they can be simplified into a series of belts (see illustration). Surface heating, which is at its greatest along the equator, creates a belt of low pressure – the intertropical convergence zone (ITCZ) – towards which the airstreams of the northern and southern hemispheres converge. On either side of the ITCZ lie the trade wind belts: the northeast trade winds in the northern hemisphere and the southeast trade winds of the southern hemisphere. The trade winds are separated from the westerlies – the predominant winds in the mid-latitudes of both hemispheres – by the subtropical high-pressure belt (the horse latitudes), which lies between about 30° and 35° latitude. Near to the poles are the polar easterlies, separated from the westerlies by the sub-polar low-pressure troughs.

**window**  **1.** A band of electromagnetic wavelengths that is able to pass through a particular medium with little reflection or absorption. For example, there is a **radio window** in the atmosphere allowing radio waves of wavelengths 5 mm to 30 m to pass through. This radio window enables *radio telescopes to be used on the surface of the earth. **2.** A period of time during which an event may occur in order to achieve a desired result. For example a **launch window** is the period during which a space vehicle must be launched to achieve a planned encounter.

**windpipe**  *See* TRACHEA.

**wind pollination**  *See* ANEMOPHILY.

**wind power**  The use of winds in the

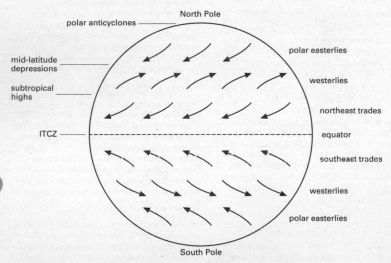

**Wind.** Global surface wind and pressure systems.

earth's atmosphere to drive machinery, especially to drive an electrical generator. Practical land-based **wind generators (aerogenerators)** are probably capable of providing some $10^{20}$ J ($10^{14}$ kW h) of energy per year throughout the world and interest in this form of renewable energy is increasing. The power, $P$, available to drive a wind generator is given by $P = kd^2v^3$, where $k$ is the air density, $d$ is the diameter of the blades, and $v$ is the average wind speed. Wind farms now exist in many parts of the world; California, for example, has the capacity to produce over 1200 MW from wind energy.

**wing** See FLIGHT.

**wino** See SUPERSYMMETRY.

**wireless access point** See ACCESS POINT.

**withdrawal reflex** See REFLEX.

**witherite** A mineral form of *barium carbonate, $BaCO_3$.

**WMAP (Wilkinson Microwave Anisotropy Probe)** A satellite launched in 2001 to investigate the *microwave background radiation. In 2003 a full-sky picture was obtained of the early universe (380 000 years after the big bang) showing temperature fluctuations at high resolution. The results supported the big-bang and inflation theories, and gave an age for the universe of $13.7 \times 10^9$ years. WMAP is named in honour of the US cosmologist and WMAP team member David Todd Wilkinson (1935–2002).

((()) SEE WEB LINKS
• The NASA site for WMAP

**Wöhler, Friedrich** (1800–82) German physician and chemist, who became a professor of chemistry at Göttingen. In 1828 he made his best-known discovery, the synthesis of urea (an organic compound) from ammonium cyanate (an inorganic salt): see WÖHLER'S SYNTHESIS. This finally disproved the assertion that organic substances can be formed only in living things. Wöhler also isolated *aluminium (1827), *beryllium (1828), and *yttrium (1828).

**Wöhler's synthesis** A synthesis of urea performed by Friedrich Wöhler in 1828. He discovered that urea ($CO(NH_2)_2$) was formed when a solution of ammonium isocyanate ($NH_4NCO$) was evaporated. At the time it was believed that organic substances such as urea could be made only by living organisms, and its production from an inorganic

compound was a notable discovery. It is sometimes (erroneously) cited as ending the belief in vitalism.

**wolfram** See TUNGSTEN.

**wolframite (iron manganese tungsten)** A mineral consisting of a mixed iron–manganese tungstate, $(FeMn)WO_4$, crystallizing in the monoclinic system; the principal ore of tungsten. It commonly occurs as blackish or brownish tabular crystal groups. It is found chiefly in quartz veins associated with granitic rocks. China is the major producer of wolframite.

**Wolf–Rayet star** An extremely hot bright type of star, often found inside a developing planetary nebula. Strong stellar winds give rise to a large loss of mass, which takes the form of a cloud of gas surrounding the star. As a result, its spectrum contains emission lines caused because the starlight has to pass through this gas. They were discovered in 1867 by C. J. E. Wolf (1827–1918) and G. A. P. Rayet (1839–1906).

**Wollaston prism** A type of quartz prism for producing plane-polarized light. It deviates the ordinary and extraordinary rays in opposite directions by approximately the same amount. The Wollaston prism, like the *Rochon prism, can be used with ultraviolet radiation. It is named after the inventor William Wollaston (1766–1828).

**womb** See UTERUS.

**wood** The hard structural and water-conducting tissue that is found in many perennial plants and forms the bulk of trees and shrubs. It is composed of secondary *xylem and associated cells, such as fibres. The wood of angiosperms is termed **hardwood**, e.g. oak and mahogany, and that of gymnosperms **softwood**, e.g. pine and fir. New wood is added to the outside of the old wood each growing season by divisions of the vascular cambium (see GROWTH RING). Only the outermost new wood (*sapwood) functions in water conduction; the inner wood (*heartwood) provides only structural support.

**wood alcohol** See METHANOL.

**Wood's metal** A low-melting (71°C) alloy of bismuth (50%), lead (25%), tin (12.5%), and cadmium (12.5%). It is used for fusible links in automatic sprinkler systems. The melting point can be changed by varying the composition. It is named after William Wood (1671–1730).

**Woodward–Hoffmann rules** Rules governing the formation of products during certain types of organic concerted reactions. The theory of such reactions was put forward in 1969 by the American chemists Robert Burns Woodward (1917–79) and Roald Hoffmann (1937– ), and is concerned with the way that orbitals of the reactants change continuously into orbitals of the products during reaction and with conservation of orbital symmetry during this process. *See also* FRONTIER-ORBITAL THEORY.

**woofer** A large loudspeaker designed to reproduce sounds of relatively low frequency, in conjunction with a *tweeter and often a mid-range speaker, in a high-fidelity sound reproducing system.

**word** A number of *bits, often 32, 48, or 64, processed by a computer as a single unit.

**work** The work done by a force acting on a body is the product of the force and the distance moved by its point of application in the direction of the force. If a force $F$ acts in such a way that the displacement $s$ is in a direction that makes an angle $\theta$ with the direction of the force, the work done is given by: $W = Fs\cos\theta$. Work is the scalar product of the force and displacement vectors. It is measured in joules.

**work function** A quantity that determines the extent to which *thermionic or photoelectric emission will occur according to the Richardson equation or Einstein's photoelectric equation (*see* EINSTEIN EQUATION). It is sometimes expressed as a potential difference (symbol $\phi$) in volts and sometimes as the energy required to remove an electron (symbol $W$) in electronvolts or joules. The former has been called the **work function potential** and the latter the **work function energy**.

**(((●))) SEE WEB LINKS**

• Values given on Carl Nave's HyperPhysics site at Georgia State University

**work hardening** An increase in the hardness of metals as a result of working them cold. It causes a permanent distortion of the crystal structure and is particularly apparent with iron, copper, aluminium, etc., whereas with lead and zinc it does not occur as these metals are capable of recrystallizing at room temperature.

**World Wide Web (WWW; Web)** A computer-based information service developed at CERN in the early 1990s. It is a hypermedia system (*see* HYPERTEXT) distributed over a large number of computer sites that allows users to view and retrieve information from 'documents' contining 'links'. It is accessed by a computer connected to the *Internet that is running a suitable program.

Web documents may consist of textual material or a number of other forms, such as graphics, still or moving images, or audio clips. Within a document there will be material to be displayed and usually one or more links, which in a text document appear as highlighted words or phrases, or as icons. The links 'point' to other documents located elsewhere on the Web by means of a **URL** (universal resource locator), which contains information specifying, for example, the network address of the device holding the document and the local index entry for that document. Activating a link will result in the display of the requested document.

**wormhole** A theoretical structure in space–time bridging two universes. Early calculations on the physics of black holes indicated that such structures might exist as a result of extreme distortion of space–time. Matter falling into a black hole in one universe could then appear pouring out of a **white hole** in another universe. It is now thought that such links do not occur and that white holes, which spontaneously generate matter, do not exist.

**W particle** *See* W BOSON.

**wrought iron** A highly refined form of iron containing 1–3% of slag (mostly iron silicate), which is evenly distributed throughout the material in threads and fibres so that the product has a fibrous structure quite dissimilar to that of crystalline cast iron. Wrought iron rusts less readily than other forms of metallic iron and it welds and works more easily. It is used for chains, hooks, tubes, etc.

**WS model** Weinberg–Salam model. *See* ELECTROWEAK THEORY.

**wurtzite structure** A type of ionic crystal structure in which the anions have a hexagonal close packed arrangement with the cations occupying one type of tetrahedral hole. Each type of ion has a coordination number of 4. Examples of this structure are found in ZnS, ZnO, AlN, SiC, and $NH_4F$.

• An interactive version of the structure

**Wurtz reaction** A reaction to prepare alkanes by reacting a haloalkane with sodium:

$$2RX + 2Na \rightarrow 2NaX + RR$$

The haloalkane is refluxed with sodium in dry ether. The method is named after the French chemist Charles-Adolphe Wurtz (1817–84). The analogous reaction using a haloalkane and a haloarene, for example:

$$C_6H_5Cl + CH_3Cl + 2Na \rightarrow 2NaCl + C_6H_5CH_3$$

is called the **Fittig reaction** after the German chemist Rudolph Fittig (1835–1910).

**WWW** *See* WORLD WIDE WEB.

**xanthates** Salts or esters containing the group –SCS(OR), where R is an organic group. Cellulose xanthate is an intermediate in the manufacture of *rayon by the viscose process.

**xanthene** A heterocyclic compound having three fused rings with one oxygen atom, $C_{13}H_{10}O$; m.p. 101–102°C b.p. 310–312°C. The ring structure is present in a class of **xanthene dyes**.

**xanthine** A purine base, $C_5H_4N_4O_2$, found in many organisms. *See* METHYLXANTHINES.

**xanthophyll** A member of a class of oxygen-containing *carotenoid pigments, which provide the characteristic yellow and brown colours of autumn leaves.

**X chromosome** *See* SEX CHROMOSOME.

**xenobiotic** Any substance foreign to living systems. Xenobiotics include drugs, pesticides, and carcinogens. *Detoxification of such substances occurs mainly in the liver.

**xenolith** A piece of pre-existing rock that occurs as an inclusion within an igneous *intrusion. Often it is a fragment of the country rock surrounding the inclusion, although it may have been modified to a hybrid rock by the intense heat of the intruding magma.

**xenon** Symbol Xe. A colourless odourless gas belonging to group 18 of the periodic table (*see* NOBLE GASES); a.n. 54; r.a.m. 131.30; d. 5.887 g dm$^{-3}$; m.p. –111.9°C; b.p. –107.1°C. It is present in the atmosphere (0.00087%) from which it is extracted by distillation of liquid air. There are nine natural isotopes with mass numbers 124, 126, 128–132, 134, and 136. Seven radioactive isotopes are also known. The element is used in fluorescent lamps and bubble chambers. Liquid xenon in a supercritical state at high temperatures is used as a solvent for infrared spectroscopy and for chemical reactions. The compound $Xe^+PtF_6^-$ was the first noble-gas compound to be synthesized. Several other compounds of xenon are known, including $XeF_2$, $XeF_4$, $XeSiF_6$, $XeO_2F_2$, and $XeO_3$. Recently, compounds have been isolated that contain xenon–carbon bonds, such as $[C_6H_5Xe]$

$[B(C_6H_5)_3F]$ (pentafluorophenylxenon fluoroborate), which is stable under normal conditions. The element was discovered in 1898 by Ramsey and Travers.

**⊕ SEE WEB LINKS**

• Information from the WebElements site

**xeric** Denoting conditions characterized by an inadequate supply of water. Xeric conditions exist in arid habitats, extremely cold habitats, and in salt marshes. Certain plants are adapted to live in such conditions. *See* HALOPHYTE; XEROPHYTE.

**xeromorphic** Describing the structural modifications of certain plants (*xerophytes) that enable them to reduce water loss, particularly from their leaves and stems.

**xerophyte** A plant that is adapted to live in conditions in which there is either a scarcity of water in the soil, or the atmosphere is dry enough to provoke excessive transpiration, or both. Xerophytes have special structural (**xeromorphic**) and functional modifications, including swollen water-storing stems or leaves (*see* SUCCULENT) and specialized leaves that may be hairy, rolled, or reduced to spines or have a thick cuticle to lower the rate of transpiration. Examples of xerophytes are desert cacti and many species that grow on sand dunes and exposed moorlands. Some *halophytes have xeromorphic features. *Compare* MESOPHYTE; HYDROPHYTE.

**XML** Extensible mark-up language: a metalanguage used to define the syntax of textual mark-up languages. Closely based on *SGML (Standard Generalized Mark-up Language), it removes much of SGML's complexity while extending its facilities in immediately useful ways. XML was released in 1998 and has spread rapidly in the computing world.

All languages defined by XML share a common mark-up system: their content is held as human-readable text using the *Unicode character set; and their logical structure is indicated by named **tags** embedded in the data. Such tags are always paired, marking the beginning and end of the content they apply to; and they always occur in the form <tag_name>&e</tag_name>. An opening tag, its paired closing tag, and the content they enclose is called an **element**. For example, in this dictionary's data cross-references are indicated by xr elements, and so part of the first sentence of this paragraph is held like this: &etext using the <xr>Unicode</xr> character set &e Data relating to an element may also be stored as key–value pairs (**attributes**) within the element's opening tag. Elements may be nested inside other elements, and the element structure of a document may be validated against a normative structure specified in a *DTD or *XML schema.

This seemingly simple scheme offers great power and flexibility while maintaining firm underlying standards. As well as storing dictionary data and other structured text, applications of XML are used for such different purposes as computer-to-computer protocols (e.g. *SOAP) and graphics-file formats. Yet the data in all these document types can be manipulated according to standard models, such as the *Document Object Model and *XSL, often using publicly available utilities. Such standardization and ease of manipulation is increasingly important, especially on the Internet, and XML's rise looks set to continue.

( (()) SEE WEB LINKS
• The W3C's XML page

**XML schema** A formalized description of the structure of an XML document. XML schemas serve the same purpose as *DTDs but are much more powerful and flexible. For example, a single document may use namespaces to take element definitions from more than one schema, with each element being validated against the appropriate schema. Schemas also allow element content to be restricted to specified data types (string, boolean, decimal number, etc.). Although the World Wide Web Consortium has issued a specification for XML schemas, other schema languages are also used.

( (()) SEE WEB LINKS
• The W3C's XML schema page

**XPath** A language for addressing parts of an *XML document, selecting nodes (elements, attributes, character data, etc.) by their name, their value, and/or their location relative to specified other nodes in a tree view of the document's structure. XPath is used within other XML standards, for example XQuery and *XSL; indeed, XPath version 2, released in 2007, is a subset of XQuery.

( (()) SEE WEB LINKS
• The XPath (version 2.0) specification
• The XPath (version 1.0) specification (still very widely used)

**X-ray astronomy** The study of *X-ray sources by rockets and balloons in the earth's atmosphere and by satellites beyond it. The first nonsolar X-ray source was detected during a rocket flight in 1962, and this observation heralded an entirely new branch of astronomy which developed rapidly with the availability of satellites from 1970 onward. The latest in a long line of increasingly sophisticated satellites have included NASA's Chandra X-ray Observatory and the European Space Agency's XMM-Newton Observatory, both launched in 1999.

**X-ray crystallography** The use of *X-ray diffraction to determine the structure of crystals or molecules, such as nucleic acids. The technique involves directing a beam of X-rays at a crystalline sample and recording the diffracted X-rays on a photographic plate. The diffraction pattern consists of a pattern of spots on the plate, and the crystal structure can be worked out from the positions and intensities of the diffraction spots. X-rays are diffracted by the electrons in the molecules and if molecular crystals of a compound are used, the electron density distribution in the molecule can be determined. *See also* NEUTRON DIFFRACTION.

( (()) SEE WEB LINKS
• An illustrated introduction to the principles of X-ray crystallography, compiled by Randy Read, University of Cambridge

**X-ray diffraction** The diffraction of X-rays by a crystal. The wavelengths of X-rays are comparable in size to the distances between atoms in most crystals, and the repeated pattern of the crystal lattice acts like a diffraction grating for X-rays. Thus, a crystal of suitable type can be used to disperse X-rays in a spectrometer. X-ray diffraction is also the basis of X-ray crystallography. *See also* BRAGG'S LAW.

**X-ray fluorescence** The emission of *X-rays from excited atoms produced by the impact of high-energy electrons, other particles, or a primary beam of other X-rays. The wavelengths of the fluorescent X-rays can be measured by an X-ray spectrometer as a means of chemical analysis. X-ray fluores-

X

cence is used in such techniques as *electron probe microanalysis.

**X-rays** Electromagnetic radiation of shorter wavelength than ultraviolet radiation produced by bombardment of atoms by high-quantum-energy particles. The range of wavelengths is $10^{-11}$ m to $10^{-9}$ m. Atoms of all the elements emit a characteristic **X-ray spectrum** when they are bombarded by electrons. The X-ray photons are emitted when the incident electrons knock an inner orbital electron out of an atom. When this happens an outer electron falls into the inner shell to replace it, losing potential energy ($\Delta E$) in doing so. The wavelength $\lambda$ of the emitted photon will then be given by $\lambda = ch/\Delta E$, where $c$ is the speed of light and $h$ is the Planck constant. *See also* BREMSSTRAHLUNG.

X-rays can pass through many forms of matter and they are therefore used medically and industrially to examine internal structures. X-rays are produced for these purposes by an *X-ray tube.

**X-ray sources** Sources of X-radiation from outside the solar system. Some 100 sources within the Galaxy have been observed as objects that emit most of their energy in the X-ray region of the electromagnetic spectrum and only a relatively small proportion of their energy in the visible spectrum. Many of these X-ray sources appear to be members of a binary system, consisting of one optically visible star and one very compact object; it is thought that the latter is either a *neutron star or (if very massive) a *black hole. Owing to the absorption of X-rays by the earth's atmosphere these X-ray sources are only visible by **X-ray telescopes** carried by space probes and satellites, although some high-energy X-rays can penetrate the upper atmosphere and are detectable by X-ray telescopes mounted on balloons.

**X-ray spectrum** *See* X-RAYS.

**X-ray tube** A device for generating *X-rays by accelerating electrons to a high energy by an electrostatic field and making them strike a metal target either in a tube containing a low-pressure gas or, as in modern tubes, in a high vacuum. The target is made from a heavy metal, usually tungsten, and is backed by a massive metal anode to conduct the heat away (see illustration showing a liquid-cooled copper anode). The electron beam is produced by heating the cathode by means of a white-hot tungsten filament. A transformer supplies the high voltage, often

100 kV, the tube acting as its own rectifier. On the half-cycles when the target is negative nothing happens. When the target becomes positive, the electrons bombarding it generate X-rays.

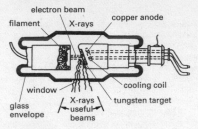

**X-ray tube.**

**XSL** Extensible stylesheet language: two related languages for restructuring and formatting *XML data. **XSL** (**XSL Transformations**) allow rules to be specified that restructure XML data, while **XSL-FO** (XSL Formatting Objects) specify formatting details for presentation in print or on screen. Both languages are themselves applications of XML, and both use *XPath as a sublanguage for the identification of those nodes (elements, attributes, character data, etc.) subject to each rule. Although they were intended to be used together, in practice XSLT is far more widely used than XSL-FO. Its abilities exactly fulfilled a widespread need for a standard and convenient method of transforming XML data, whereas formatting on web browsers was already adequately addressed by the less sophisticated but widely established CSS.

**(((⊕))) SEE WEB LINKS**

- The XSLT (version 2.0) specification
- The XSLT (version 1.0) specification (still very widely used)
- The XSL-FO (version 1.1) specification

**XSLT** *See* XSL.

**xylem** A tissue that transports water and dissolved mineral nutrients in vascular plants. In flowering plants it consists of hollow **vessels** that are formed from cells (*vessel elements) joined end to end. The end walls of the vessel elements are perforated to allow the passage of water. In less advanced vascular plants, such as conifers and ferns, the constituent cells of the xylem are called *tracheids. In young plants and at the shoot and root tips of older plants the xylem is formed by the apical meristems. In plants showing secondary growth this xylem is replaced in most of the plant by secondary xylem, formed by the vascular *cambium. The walls of the xylem cells are thickened with lignin, the extent of this thickening being greatest in secondary xylem. Xylem contributes greatly to the mechanical strength of the plant: *wood is mostly made up of secondary xylem. *See also* FIBRE. *Compare* PHLOEM.

**xylenes** *See* DIMETHYLBENZENES.

**YAC** *See* ARTIFICIAL CHROMOSOME.

**Yagi aerial** A directional aerial array widely used for television and *radio telescopes. It consists of one or two dipoles, a parallel reflector, and a series of closely spaced directors (0.15–0.25 wavelength apart) in front of the dipole. When used for reception this arrangement focuses the incoming signal on the dipole. For transmission, the output of the dipole is reinforced by the directors. It is named after Hidetsuga Yagi (1886–1976).

**Yang–Mills theory** *See* GAUGE THEORY.

**yard** The former Imperial standard unit of length. In 1963 the yard was redefined as 0.9144 metre exactly.

**Y chromosome** *See* SEX CHROMOSOME.

**year** The measure of time on which the calendar is based. It is the time taken for the earth to complete one orbit of the sun. The **calendar year** consists of an average of 365.25 mean solar days – three successive years of 365 days followed by one (leap) year of 366 days. The **solar year** (or **astronomical year**) is the average interval between two successive returns of the sun to the first point of Aries; it is 365.242 19 mean solar days. This is effectively the same as the **tropical year**, the interval between two successive passages of the sun through the vernal equinox. The **sidereal year** is the average period of revolution of the earth with respect to the fixed stars; it is 365.256 mean solar days. The **anomalistic year** is the average interval between successive perihelions of the earth; it is 365.259 mean solar days. *See also* EPHEMERIS TIME.

**yeast artificial chromosome (YAC)** *See* ARTIFICIAL CHROMOSOME.

**yeasts** A group of unicellular fungi within the class Hemiascomycetae of the phylum *Ascomycota. They occur as single cells or as groups or chains of cells; yeasts reproduce asexually by *budding and sexually by producing ascospores. Yeasts of the genus *Saccharomyces* ferment sugars and are used in the baking and brewing industries (*see* BAKER'S YEAST).

**yellow body** *See* CORPUS LUTEUM.

**yield point** *See* ELASTICITY.

**yocto-** Symbol y. A prefix used in the metric system to indicate $10^{-24}$. For example, $10^{-24}$ second = 1 yoctosecond (ys).

**yolk** The food stored in an egg for the use of the embryo. It can consist mainly of protein (**protein yolk**) or of phospholipids and fats (**fatty yolk**). The eggs of oviparous animals (e.g. birds) contain a relatively large yolk.

**yolk sac** One of the protective membranes surrounding the embryos of birds, reptiles, and mammals (*see* EXTRAEMBRYONIC MEMBRANES). The embryo derives nourishment from the yolk sac via a system of blood vessels. In birds and reptiles the yolk sac encloses the yolk; in most mammals a fluid replaces the yolk.

**yotta-** Symbol Y. A prefix used in the metric system to indicate $10^{24}$. For example, $10^{24}$ metres = 1 yottametre (Ym).

**Young, Thomas** (1773–1829) British physician and physicist, who was a child prodigy and could speak 14 languages before he was 19. His early researches concerned the eye and vision, but he is best known for establishing the wave theory of *light (1800–1804) and explaining the phenomenon of *interference (1807).

**Young modulus of elasticity** *See* ELASTIC MODULUS.

**Young's slits** *See* INTERFERENCE.

**ytterbium** Symbol Yb. A silvery metallic element belonging to the *lanthanoids; a.n. 70; r.a.m. 173.04; r.d. 6.965 (20°C); m.p. 819°C; b.p. 1194°C. It occurs in gadolinite, monazite, and xenotime. There are seven natural isotopes and ten artificial isotopes are known. It is used in certain steels. The element was discovered by Jean de Marignac (1817–94) in 1878.

 **SEE WEB LINKS**

• Information from the WebElements site

**yttrium** Symbol Y. A silvery-grey metallic element belonging to group 3 (formerly IIIA) of the periodic table; a.n. 39; r.a.m. 88.905; r.d. 4.469 (20°C); m.p. 1522°C; b.p. 3338°C. It occurs in uranium ores and in *lanthanoid ores, from which it can be extracted by an ion exchange process. The natural isotope is yttrium–89, and there are 14 known artificial isotopes. The metal is used in superconducting alloys and in alloys for strong permanent magnets (in both cases, with cobalt). The oxide ($Y_2O_3$) is used in colour-television phosphors, neodymium-doped lasers, and microwave components. Chemically it resembles the lanthanoids, forming ionic compounds containing $Y^{3+}$ ions. The metal is stable in air below 400°C. It was discovered in 1828 by Friedrich Wöhler.

(((⊕))) **SEE WEB LINKS**

• Information from the WebElements site

**Z boson** An electrically neutral elementary particle, $Z^0$, which – like *W bosons – is thought to mediate the weak interactions in the *electroweak theory. The $Z^0$ boson was discovered at CERN in 1983 and has a mass of about 90 GeV as had been predicted from theory.

**Zeeman effect** The splitting of the lines in a spectrum when the source of the spectrum is exposed to a magnetic field. It was discovered in 1896 by Pieter Zeeman (1865–1943). In the **normal Zeeman effect** a single line is split into three if the field is perpendicular to the light path or two lines if the field is parallel to the light path. This effect can be explained by classical electromagnetic principles in terms of the speeding up and slowing down of orbital electrons in the source as a result of the applied field. The **anomalous Zeeman effect** is a complicated splitting of the lines into several closely spaced lines, so called because it does not agree with classical predictions. This effect is explained by quantum mechanics in terms of electron spin.

**(⊕) SEE WEB LINKS**
• Pieter Zeeman's Nobel Lecture (1903)

**Zeilinger's principle** The principle that any elementary system carries just one bit of information. This principle was put forward by the Austrian physicist Anton Zeilinger in 1999 and subsequently developed by him to derive several aspects of quantum mechanics.

**Zeisel reaction** A method of determining the number of methoxy ($-OCH_3$) groups in an organic compound. The compound is heated wih excess hydriodic acid, forming an alcohol and iodomethane

$$R-O-CH_3 + HI \rightarrow ROH + CH_3I$$

The iodomethane is distilled off and led into an alcoholic solution of silver nitrate, where it precipitates silver iodide. This is filtered and weighed, and the number of iodine atoms and hence methoxy groups can be calculated. The method was developed by S. Zeisel in 1886.

**Zeise's salt** A complex of platinum and ethene, $PtCl_3(CH_2CH_2)$, in which the Pt coordinates to the pi bond of the ethene. It was the first example of an enyl complex, synthesized by W. C. Zeise in 1827.

**Zener diode** A type of semiconductor diode, consisting of a $p$-$n$ junction with high doping concentrations on either side of the junction. It acts as a rectifier until the applied reverse voltage reaches a certain value, the **Zener breakdown voltage**, when the device becomes conducting. This effect occurs as a result of electrons being excited directly from the valence band into the conduction band (*see* ENERGY BANDS). Zener diodes are used in voltage-limiting circuits; they are named after C. M. Zener (1905–93).

**zenith** The point on the *celestial sphere that lies directly above an observer. *Compare* NADIR.

**zeolite** A natural or synthetic hydrated aluminosilicate with an open three-dimensional crystal structure, in which water molecules are held in cavities in the lattice. The water can be driven off by heating and the zeolite can then absorb other molecules of suitable size. Zeolites are used for separating mixtures by selective absorption – for this reason they are often called **molecular sieves**. They are also used in sorption pumps for vacuum systems and certain types (e.g. **Permutit**) are used in ion-exchange (e.g. water-softening).

**zepto-** Symbol z. A prefix used in the metric system to indicate $10^{-21}$. For example, $10^{-21}$ second = 1 zeptosecond (zs).

**zero order** *See* ORDER.

**zero-point energy** The energy remaining in a substance at the *absolute zero of temperature (0 K). This is in accordance with quantum theory, in which a particle oscillating with simple harmonic motion does not have a stationary state of zero kinetic energy. Moreover, the *uncertainty principle does not allow such a particle to be at rest at exactly the centrepoint of its oscillations.

**zeroth law of thermodynamics** *See* THERMODYNAMICS.

**zetta-** Symbol Z. A prefix used in the metric system to indicate $10^{21}$. For example, $10^{21}$ metres = 1 zettametre (Zm).

**Ziegler process** An industrial process for manufacturing high-density polyethene using catalysts of titanium(IV) chloride ($TiCl_4$) and aluminium alkyls (e.g. triethylaluminium, $Al(C_2H_5)_3$). The process was introduced in 1953 by the German chemist Karl Ziegler (1898–1973). It allowed the manufacture of polythene at lower temperatures (about 60°C) and pressures (about 1 atm.) than used in the original process. Moreover, the polyethene produced had more straight-chain molecules, giving the product more rigidity and a higher melting point than the earlier low-density polyethene. The reaction involves the formation of a titanium alkyl in which the titanium can coordinate directly to the pi bond in ethene.

In 1954 the process was developed further by the Italian chemist Giulio Natta (1903–79), who extended the use of Ziegler's catalysts (and similar catalysts) to other alkenes. In particular he showed how to produce stereospecific polymers of propene.

**zinc** Symbol Zn. A blue-white metallic element; a.n. 30; r.a.m. 65.38; r.d. 7.1; m.p. 419.88°C; b.p. 907°C. It occurs in sphalerite (or zinc blende, ZnS), which is found associated with the lead sulphide, and in smithsonite ($ZnCO_3$). Ores are roasted to give the oxide and this is reduced with carbon (coke) at high temperature, the zinc vapour being condensed. Alternatively, the oxide is dissolved in sulphuric acid and the zinc obtained by electrolysis. There are five stable isotopes (mass numbers 64, 66, 67, 68, and 70) and six radioactive isotopes are known. The metal is used in galvanizing and in a number of alloys (brass, bronze, etc.). Chemically it is a reactive metal, combining with oxygen and other nonmetals and reacting with dilute acids to release hydrogen. It also dissolves in alkalis to give *zincates. Most of its compounds contain the $Zn^{2+}$ ion.

(**)) SEE WEB LINKS

• Information from the WebElements site

**zincate** A salt formed in solution by dissolving zinc or zinc oxide in alkali. The formula is often written $ZnO_2^{2-}$ although in aqueous solution the ions present are probably complex ions in which the $Zn^{2+}$ is coordi-

nated to $OH^-$ ions. $ZnO_2^{2-}$ ions may exist in molten sodium zincate, but most solid 'zincates' are mixed oxides.

**zinc blende** A mineral form of *zinc sulphide, ZnS, the principal ore of zinc (*see* SPHALERITE). The **zinc-blende structure** is the crystal structure of this compound (and of other compounds). It has zinc atoms surrounded by four sulphur atoms at the corners of a tetrahedron. Each sulphur is similarly surrounded by four zinc atoms. The crystals belong to the cubic system.

**zinc-blende structure** *See* SPHALERITE STRUCTURE.

**zinc chloride** A white crystalline compound, $ZnCl_2$. The anhydrous salt, which is deliquescent, can be made by the action of hydrogen chloride gas on hot zinc; r.d. 2.9; m.p. 283°C; b.p. 732°C. It has a relatively low melting point and sublimes easily, indicating that it is a molecular compound rather than ionic. Various hydrates also exist. Zinc chloride is used as a catalyst, dehydrating agent, and flux for hard solder. It was once known as **butter of zinc**.

**zinc chloride cell** *See* DRY CELL.

**zinc group** The group of elements in the periodic table forming group 12 and consisting of zinc (Zn), cadmium (Cd), and mercury (Hg). They were formerly classified in group IIB of the table (*see* GROUP 2 ELEMENTS).

**zincite** A mineral form of *zinc oxide, ZnO.

**zinc oxide** A powder, white when cold and yellow when hot, ZnO; r.d. 5.606; m.p. 1975°C. It occurs naturally as a reddish orange ore **zincite**, and can also be made by oxidizing hot zinc in air. It is amphoteric, forming *zincates with bases. It is used as a pigment (**Chinese white**) and a mild antiseptic in zinc ointments. An archaic name is **philosopher's wool**.

**zinc sulphate** A white crystalline water-soluble compound made by heating zinc sulphide ore in air and dissolving out and recrystallizing the sulphate. The common form is the heptahydrate, $ZnSO_4.7H_2O$; r.d. 1.9. This loses water above 30°C to give the hexahydrate and more water is lost above 70°C to form the monohydrate. The anhydrous salt forms at 280°C and this decomposes above 500°C. The compound, which was formerly called **white vitriol**, is used as a mordant and as a styptic (to check bleeding).

Z

**zinc sulphide** A yellow-white water-soluble solid, ZnS. It occurs naturally as *sphalerite (*see also* ZINC BLENDE) and wurtzite. The compound sublimes at 1180°C. It is used as a pigment and phosphor.

**zino** *See* SUPERSYMMETRY.

**zircon** A naturally occurring silicate of zirconium, $ZrSiO_4$, used as a gemstone. The colour depends in small amounts of other metals and may be red, brown, yellow, or green. Red gem-quality zircon is sometimes called **jacinth**; gem-quality zircons with other colours are called **jargoons**. There is also a naturally occurring colourless variety. Zircon gems can be given other colours, or made colourless, by heat treatment. The colourless varieties (either natural or treated) are sometimes called **Matura diamonds** (after Matura in Sri Lanka). The name 'zircon' is often erroneously applied to a synthetic form of the oxide *cubic zircona, which is used as a diamond substitute.

**zirconia** *See* ZIRCONIUM.

**zirconium** Symbol Zr. A grey-white metallic *transition element; a.n. 40; r.a.m. 91.22; r.d. 6.49; m.p. 1852°C; b.p. 4377°C. It is found in zircon ($ZrSiO_4$; the main source) and in baddeleyite ($ZnO_2$). Extraction is by chlorination to give $ZrCl_4$ which is purified by solvent extraction and reduced with magnesium (Kroll process). There are five natural isotopes (mass numbers 90, 91, 92, 94, and 96) and six radioactive isotopes are known. The element is used in nuclear reactors (it is an effective neutron absorber) and in certain alloys. The metal forms a passive layer of oxide in air and burns at 500°C. Most of its compounds are complexes of zirconium(IV). **Zirconium(IV) oxide** (**zirconia**) is used as an electrolyte in fuel cells. The element was identified in 1789 by Klaproth and was first isolated by Berzelius in 1824.

( ( ) ) SEE WEB LINKS

• Information from the WebElements site

**zirconium(IV) oxide** *See* ZIRCONIUM.

**zodiac** A band that passes round the *celestial sphere, extending 9° on either side of the *ecliptic. It includes the apparent paths of the sun, the moon, and the eight major planets of the solar system. The band is divided into the twelve **signs of the zodiac**, each 30° wide. These signs indicate the sun's position each month in the year and were named by the ancient Greeks after the **zodia-cal constellations** that occupied the signs some 2000 years ago. However, as a result of the *precession of the equinoxes the constellations have since moved eastwards by over 30° and no longer coincide with the signs.

**zodiacal light** A faint luminous glow in the sky that can be observed on a moonless night on the western horizon after sunset or on the eastern horizon before sunrise. It is caused by the scattering of sunlight by dust particles in interplanetary space.

**zonation** The distribution of the different species of a community into separate zones, which are created by variations in the environment. A clear example of zonation occurs on a rocky shore, where different species of seaweed (*Fucus*) occupy different zones, according to their ability to withstand desiccation. For example, the species found in the splash zone, which is never completely submerged in water, is better adapted to exposure than those found in zones lower down the shore, where they are submerged for longer periods. Animals, particularly stationary species, such as barnacles, also exhibit zonation on a rocky shore; as with the seaweeds, this may depend on the ability of different species to withstand desiccation. Competition between species may also contribute to zonation.

**zone fossil** *See* INDEX FOSSIL.

**zone refining** A technique used to reduce the level of impurities in certain metals, alloys, semiconductors, and other materials. It is based on the observation that the solubility of an impurity may be different in the liquid and solid phases of a material. To take advantage of this observation, a narrow molten zone is moved along the length of a specimen of the material, with the result that the impurities are segregated at one end of the bar and the pure material at the other. In general, if the impurities lower the melting point of the material they are moved in the same direction as the molten zone moves, and vice versa.

**zoogeography** The study of the geographical distributions of animals. The earth can be divided into several zoogeographical regions separated by natural barriers, such as oceans, deserts, and mountain ranges. The characteristics of the fauna of each region are believed to depend particularly on the process of *continental drift and the stage of evolution reached when the various land masses became isolated. For example Australia, which has been isolated since Cretaceous times, has the most primitive native mammalian fauna, consisting solely of marsupials and monotremes. *See also* WALLACE'S LINE.

**zoology** The scientific study of animals, including their anatomy, physiology, biochemistry, genetics, ecology, evolution, and behaviour.

**zooplankton** The animal component of *plankton. All major animal phyla are represented in zooplankton, as adults, larvae, or eggs; some are just visible to the naked eye but most cannot be seen without magnification. Near the surface of the sea there may be many thousands of such animals per cubic metre.

**zoospore** A spore that possesses one or more flagella and is therefore motile. Released from a sporangium (called a **zoosporangium**), zoospores are produced by many algae and certain protists, such as the potato blight (*Phytophthora infestans*).

**zwitterion (ampholyte ion)** An ion that has a positive and negative charge on the same group of atoms. Zwitterions can be formed from compounds that contain both acid groups and basic groups in their molecules. For example, aminoethanoic acid (the amino acid glycine) has the formula $H_2N.CH_2.COOH$. However, under neutral conditions, it exists in the form of the zwitterion $^+H_3N.CH_2.COO^-$, which can be regarded as having been produced by an internal neutralization reaction (transfer of a proton from the carboxyl group to the amino group). Aminoethanoic acid, as a consequence, has some properties characteristic of ionic compounds; e.g. a high melting point and solubility in water. In acid solutions, the positive ion $^+H_3NCH_2COOH$ is formed. In basic solutions, the negative ion $H_2NCH_2COO^-$ predominates. The name comes from the German *zwei*, two.

**zygomorphy** *See* BILATERAL SYMMETRY.

**Zygomycota** A phylum of saprotrophic or parasitic fungi that includes the bread mould (*Mucor*). Their hyphae lack cross walls and they can reproduce asexually by sporangiospores formed within a *sporangium or sexually by means of *zygospores.

**zygospore** A zygote with a thick resistant wall, formed by some algae and fungi (*see* ZYGOMYCOTA). It results from the fusion of two gametes, neither of which is retained by the parent in any specialized sex organ (such as an oogonium). It enters a resting phase before germination. *Compare* OOSPORE.

**zygote** A fertilized female *gamete: the product of the fusion of the nucleus of the ovum or ovule with the nucleus of the sperm or pollen grain. *See* FERTILIZATION.

**zygotene** The second phase of the first *prophase of meiosis, in which *pairing (synapsis) of homologous chromosomes takes place. Intimate contact is made between identical regions of homologues, in a process involving proteins and DNA organized to form a **synaptonemal complex**.

**zymogen** Any inactive enzyme precursor that, following secretion, is chemically altered to the active form of the enzyme. For example, the protein-digesting enzyme *trypsin is secreted by the pancreas as the zymogen trypsinogen. This is changed in the small intestine by the action of another enzyme, enterokinase, to the active form.

Z

# Appendices

# SI units

## Base and dimensionless SI units

| Physical quantity | Name | Symbol |
|---|---|---|
| length | metre | m |
| mass | kilogram | kg |
| time | second | s |
| electric current | ampere | A |
| thermodynamic temperature | kelvin | K |
| luminous intensity | candela | cd |
| amount of substance | mole | mol |
| *plane angle | radian | rad |
| *solid angle | steradian | sr |

*dimensionless units

## Derived SI units with special names

| Physical quantity | Name of SI unit | Symbol of SI unit |
|---|---|---|
| frequency | hertz | Hz |
| energy | joule | J |
| force | newton | N |
| power | watt | W |
| pressure | pascal | Pa |
| electric charge | coulomb | C |
| electric potential difference | volt | V |
| electric resistance | ohm | Ω |
| electric conductance | siemens | S |
| electric capacitance | farad | F |
| magnetic flux | weber | Wb |
| inductance | henry | H |
| magnetic flux density (magnetic induction) | tesla | T |
| luminous flux | lumen | lm |
| illuminance | lux | lx |
| absorbed dose | gray | Gy |
| activity | becquerel | Bq |
| dose equivalent | sievert | Sv |

# Decimal multiples and submultiples to be used with SI units

| Submultiple | Prefix | Symbol | Multiple | Prefix | Symbol |
|---|---|---|---|---|---|
| $10^{-1}$ | deci | d | 10 | deca | da |
| $10^{-2}$ | centi | c | $10^2$ | hecto | h |
| $10^{-3}$ | milli | m | $10^3$ | kilo | k |
| $10^{-6}$ | micro | $\mu$ | $10^6$ | mega | M |
| $10^{-9}$ | nano | n | $10^9$ | giga | G |
| $10^{-12}$ | pico | p | $10^{12}$ | tera | T |
| $10^{-15}$ | femto | f | $10^{15}$ | peta | P |
| $10^{-18}$ | atto | a | $10^{18}$ | exa | E |
| $10^{-21}$ | zepto | z | $10^{21}$ | zetta | Z |
| $10^{-24}$ | yocto | y | $10^{24}$ | yotta | Y |

# Conversion of units to SI units

| From | To | Multiply by |
|---|---|---|
| in | m | $2.54 \times 10^{-2}$ |
| ft | m | 0.3048 |
| sq. in | $m^2$ | $6.4516 \times 10^{-4}$ |
| sq. ft | $m^2$ | $9.2903 \times 10^{-2}$ |
| cu. in | $m^3$ | $1.638\,71 \times 10^{-5}$ |
| cu. ft | $m^3$ | $2.831\,68 \times 10^{-2}$ |
| l(itre) | $m^3$ | $10^{-3}$ |
| gal(lon) | l(itre) | 4.546 09 |
| miles/hr | m s$^{-1}$ | 0.477 04 |
| km/hr | m s$^{-1}$ | 0.277 78 |
| lb | kg | 0.453 592 |
| g cm$^{-3}$ | kg m$^{-3}$ | $10^3$ |
| lb/in$^3$ | kg m$^{-3}$ | $2.767\,99 \times 10^4$ |
| dyne | N | $10^{-5}$ |
| poundal | N | 0.138 255 |
| lbf | N | 4.448 22 |
| mmHg | Pa | 133.322 |
| atmosphere | Pa | $1.013\,25 \times 10^5$ |
| hp | W | 745.7 |
| erg | J | $10^{-7}$ |
| eV | J | $1.602\,10 \times 10^{-19}$ |
| kW h | J | $3.6 \times 10^6$ |
| cal | J | 4.1868 |

# Fundamental constants

| Constant | Symbol | Value in SI units |
|---|---|---|
| acceleration of free fall | $g$ | $9.806\,65$ m s$^{-2}$ |
| Avogadro constant | $L, N_A$ | $6.022\,141\,79(30) \times 10^{23}$ mol$^{-1}$ |
| Boltzmann constant | $k = R/N_A$ | $1.380\,6504(24) \times 10^{-23}$ J K$^{-1}$ |
| electric constant | $\varepsilon_0$ | $8.854\,187\,817 \times 10^{-12}$ F m$^{-1}$ |
| electronic charge | $e$ | $1.602\,176\,487(40) \times 10^{-19}$ C |
| electronic rest mass | $m_e$ | $9.109\,382\,15(45) \times 10^{-31}$ kg |
| Faraday constant | $F$ | $9.648\,3399(24) \times 10^4$ C mol$^{-1}$ |
| gas constant | $R$ | $8.314\,472(15)$ J K$^{-1}$ mol$^{-1}$ |
| gravitational constant | $G$ | $6.674\,28(67) \times 10^{-11}$ m$^3$ kg$^{-1}$ s$^{-2}$ |
| Loschmidt's constant | $N_L$ | $2.686\,7774(47) \times 10^{25}$ m$^{-3}$ |
| magnetic constant | $\mu_0$ | $4\pi \times 10^{-7}$ H m$^{-1}$ |
| neutron rest mass | $m_n$ | $1.674\,927\,211(84) \times 10^{-27}$ kg |
| Planck constant | $h$ | $6.626\,068\,96(33) \times 10^{-34}$ J s |
| proton rest mass | $m_p$ | $1.672\,621\,637(83) \times 10^{-27}$ kg |
| speed of light | $c$ | $2.997\,924\,58 \times 10^8$ m s$^{-1}$ |
| Stefan–Boltzmann constant | $\sigma$ | $5.670\,400(40) \times 10^{-8}$ W m$^{-2}$ K$^{-4}$ |

# The solar system

| Planet | Equatorial diameter (km) | Mean distance from sun ($10^6$ km) | Sidereal period | |
|---|---|---|---|---|
| Mercury | 4879.4 | 57.91 | 87.97 | days |
| Venus | 12 104 | 108.21 | 224.7 | days |
| Earth | 12 756.3 | 149.6 | 0.999 | years |
| Mars | 6795 | 227.94 | 686.98 | days |
| Jupiter | 142 985 | 778.54 | 11.86 | years |
| Saturn | 120 536 | 1433.45 | 29.46 | years |
| Uranus | 51 119 | 2876.679 | 84.32 | years |
| Neptune | 49 528 | 4452.94 | 164.79 | years |
| Pluto | 2300 | 5906.376 | 248.02 | years |

# Geological time scale

| millions of years ago | Eon | Era | Period | Epoch | | millions of years ago |
|---|---|---|---|---|---|---|
| | Phanerozoic | Cenozoic | Neogene | Holocene | | |
| | | | | Pleistocene | | |
| | | | | Pliocene | | |
| | | | | Miocene | 23 | |
| | | | Palaeogene | Oligocene | | |
| | | | | Eocene | | |
| | | | | Palaeocene | 65 | |
| | | Mesozoic | Cretaceous | | 145 | |
| | | | Jurassic | | 200 | |
| | | | Triassic | | 251 | |
| | | Palaeozoic | Permian | | 299 | |
| | | | Carboniferous | | 359 | |
| | | | Devonian | | 416 | |
| | | | Silurian | | 444 | |
| | | | Ordovician | | 488 | |
| | | | Cambrian | | | |
| 542 | | | | | 542 | |
| | Proterozoic | | Precambrian time | | | |
| 2500 | | | | | | |
| | Archaean | | | | | |
| 3600 | | | | | | |
| | Hadean | | | | | |
| 4500 | | | | | 4500 | |

# Simplified classification of land plants

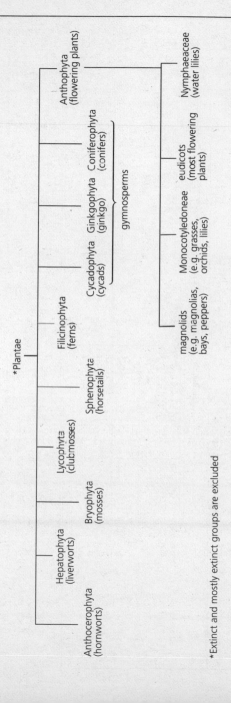

*Extinct and mostly extinct groups are excluded

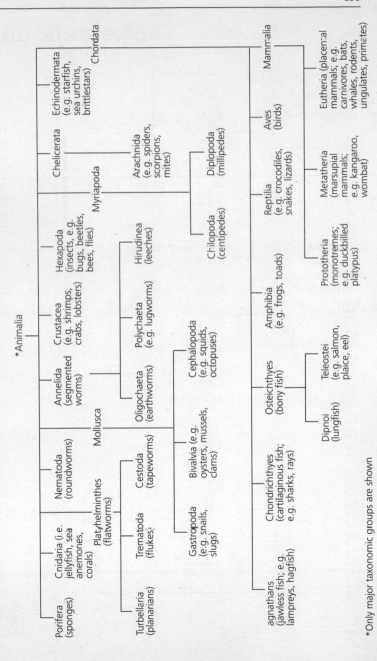

# Simplified classification of the animal kingdom

*Animalia

Porifera (sponges)

Cnidaria (i.e. jellyfish, sea anemones, corals)

Platyhelminthes (flatworms)

Turbellaria (planarians)

Trematoda (flukes)

Cestoda (tapeworms)

Nematoda (roundworms)

Mollusca

Gastropoda (e.g. snails, slugs)

Bivalvia (e.g. oysters, mussels, clams)

Cephalopoda (e.g. squids, octopuses)

Annelida (segmented worms)

Oligochaeta (earthworms)

Polychaeta (e.g. lugworms)

Hirudinea (leeches)

Crustacea (e.g. shrimps, crabs, lobsters)

Hexapoda (insects, e.g. bugs, beetles, bees, flies)

Myriapoda

Chilopoda (centipedes)

Diplopoda (millipedes)

Chelicerata

Arachnida (e.g. spiders, scorpions, mites)

Echinodermata (e.g. starfish, sea urchins, brittlestars)

Chordata

agnathans (jawless fish; e.g. lampreys, hagfish)

Chondrichthyes (cartilaginous fish; e.g. sharks, rays)

Osteichthyes (bony fish)

Dipnoi (lungfish)

Teleostei (e.g. salmon, plaice, eel)

Amphibia (e.g. frogs, toads)

Reptilia (e.g. crocodiles, snakes, lizards)

Aves (birds)

Mammalia

Prototheria (monotremes; e.g. duckbilled platypus)

Metatheria (marsupial mammals; e.g. kangaroo, wombat)

Eutheria (placental mammals; e.g. carnivores, bats, whales, rodents, ungulates, primates)

*Only major taxonomic groups are shown

898

# The periodic table

Period

| Group | 1 | 2 | 3 | 4 | 5 | 6 | 7 | 8 | 9 | 10 | 11 | 12 | 13 | 14 | 15 | 16 | 17 | 18 | n |
|---|---|---|---|---|---|---|---|---|---|---|---|---|---|---|---|---|---|---|---|
| | 1 H | | | | | | | | | | | | | | | | | 2 He | 1 |
| | 3 Li | 4 Be | | | | | | | | | | | 5 B | 6 C | 7 N | 8 O | 9 F | 10 Ne | 2 |
| | 11 Na | 12 Mg | | | | | | | | | | | 13 Al | 14 Si | 15 P | 16 S | 17 Cl | 18 Ar | 3 |
| | 19 K | 20 Ca | 21 Sc | 22 Ti | 23 V | 24 Cr | 25 Mn | 26 Fe | 27 Co | 28 Ni | 29 Cu | 30 Zn | 31 Gs | 32 Ge | 33 As | 34 Se | 35 Br | 36 Kr | 4 |
| | 37 Rb | 38 Sr | 39 Y | 40 Zr | 41 Nb | 42 Mo | 43 Tc | 44 Ru | 45 Rh | 46 Pd | 47 Ag | 48 Cd | 49 In | 50 Sn | 51 Sb | 52 Te | 53 I | 54 Xe | 5 |
| | 55 Cs | 56 Ba | 57–71 La–Lu | 72 Hf | 73 Ta | 74 W | 75 Re | 76 Os | 77 Ir | 78 Pt | 79 Au | 80 Hg | 81 Tl | 82 Pb | 83 Bi | 84 Po | 85 At | 86 Rn | 6 |
| | 87 Fr | 88 Ra | 89–103 Ac–Lr | 104 Rf | 105 Db | 106 Sg | 107 Bh | 108 Hs | 109 Mt | 110 Ds | 111 Rg | 112 Uub | 113 Uut | 114 Uuq | 115 Uup | 116 Uuh | | | 7 |

|  | 3 | 4 | 5 | 6 | 7 | 8 | 9 | 10 | 11 | 12 | 13 | 14 | 15 | 16 | 17 | n |
|---|---|---|---|---|---|---|---|---|---|---|---|---|---|---|---|---|
| Lanthanoids | 57 La | 58 Ce | 59 Pr | 60 Nd | 61 Pm | 62 Sm | 63 Eu | 64 Gd | 65 Tb | 66 Dy | 67 Ho | 68 Er | 69 Tm | 70 Yb | 71 Lu | 6 |
| Actinoids | 89 Ac | 90 Th | 91 Pa | 92 U | 93 Np | 94 Pu | 95 Am | 96 Cm | 97 Bk | 98 Cf | 99 Fs | 100 Fm | 101 Md | 102 No | 103 Lr | 7 |

Correspondance of recommended group designations to other designations in recent use

| | 1 | 2 | 3 | 4 | 5 | 6 | 7 | 8 | 9 | 10 | 11 | 12 | 13 | 14 | 15 | 16 | 17 | 18 |
|---|---|---|---|---|---|---|---|---|---|---|---|---|---|---|---|---|---|---|
| IUPAC Recommendations 1990 | 1 | 2 | 3 | 4 | 5 | 6 | 7 | 8 | 9 | 10 | 11 | 12 | 13 | 14 | 15 | 16 | 17 | 18 |
| Usual European Convention | IA | IIA | IIIA | IVA | VA | VIA | VIIA | VIII (or VIIIA) | | | IB | IIB | IIIB | IVB | VB | VIB | VIIB | 0 (or VIIIB) |
| Usual US Convention | IA | IIA | IIIB | IVB | VB | VIB | VIIB | VIII | | | IB | IIB | IIIA | IVA | VA | VIA | VIIA | VIIIA (or 0) |

# Useful websites

 SEE WEB LINKS

To access any of these websites, go to the dictionary's web page at
http://www.oup.com/uk/reference/resources/science, click on **Web links** in the Resources
section, go to **Appendix web links**, and then click through to the relevant site.

**Chemdex**  An online directory of chemistry on the web established in 1993 and containing over 5000 links to further resources.

**Human Proteome Organization**  A starting point for insights into the pioneering
and rapidly developing world of proteomics. HUPO is the international body responsible
for consolidating and coordinating the work of national and regional proteomics groups.

**Nature Magazine Online**  An online weekly journal that offers news articles and features, complete reference works online, and information on the latest science research.

**NASA**  Contains a large amount of information on the solar system and space science.

**National Institute of Standards and Technology**  A large amount of information about units and constants.

**New Scientist**  A popular news and archive site for all branches of science.

**Nobel Foundation**  Information about Nobel prize winners and their work.

**Physics Central**  A site run by the American Physical Society. It contains a searchable
collection of articles on all branches of physics.

**Royal Botanic Gardens, Kew**  Besides an introduction and special features about
the UK's premier plant collection, there is a wealth of information about horticulture,
plant conservation, and other plant sciences.

**Royal Society of Chemistry**  An exploration of key events in the history of science
with a particular emphasis on chemistry.

**Scientific American**  A popular science news site containing selected recent articles.

**Society for Conservation Biology**  An information site from the international professional organization dedicated to promoting the scientific study of the phenomena that
affect the maintenance, loss, and restoration of biological diversity.

**ZSL (Zoological Society of London)**  This site contains links to London Zoo, Whip-
snade Wild Animal Park, and the ZSL's research division, the Institute of Zoology.

# Oxford Paperback Reference

**A Dictionary of Chemistry**

Over 4,200 entries covering all aspects of chemistry, including physical chemistry and biochemistry.

'It should be in every classroom and library ... the reader is drawn inevitably from one entry to the next merely to satisfy curiosity.'
*School Science Review*

**A Dictionary of Physics**

Ranging from crystal defects to the solar system, 3,500 clear and concise entries cover all commonly encountered terms and concepts of physics.

**A Dictionary of Biology**

The perfect guide for those studying biology – with over 4,700 entries on key terms from biology, biochemistry, medicine, and palaeontology.

'lives up to its expectations; the entries are concise, but explanatory'
*Biologist*

'ideally suited to students of biology, at either secondary or university level, or as a general reference source for anyone with an interest in the life sciences'
*Journal of Anatomy*

# Oxford Paperback Reference

**Concise Medical Dictionary**

Over 10,000 clear entries covering all the major medical and surgical specialities make this one of our best-selling dictionaries.

'"No home should be without one" certainly applies to this splendid medical dictionary'

*Journal of the Institute of Health Education*

'An extraordinary bargain'

*New Scientist*

'Excellent layout and jargon-free style'

*Nursing Times*

**A Dictionary of Nursing**

Comprehensive coverage of the ever-expanding vocabulary of the nursing professions. Features over 10,000 entries written by medical and nursing specialists.

**An A-Z of Medicinal Drugs**

Over 4,000 entries cover the full range of over-the-counter and prescription medicines available today. An ideal reference source for both the patient and the medical professional.

# Oxford Paperback Reference

### A Dictionary of Psychology
Andrew M. Colman

Over 10,500 authoritative entries make up the most wide-ranging dictionary of psychology available.

'impressive ... certainly to be recommended'
*Times Higher Educational Supplement*

'Comprehensive, sound, readable, and up-to-date, this is probably the best single-volume dictionary of its kind.'
*Library Journal*

### A Dictionary of Economics
John Black

Fully up-to-date and jargon-free coverage of economics. Over 2,500 terms on all aspects of economic theory and practice.

### A Dictionary of Law

An ideal source of legal terminology for systems based on English law. Over 4,000 clear and concise entries.

'The entries are clearly drafted and succinctly written ... Precision for the professional is combined with a layman's enlightenment.'
*Times Literary Supplement*

# Oxford Paperback Reference

**The Concise Oxford Dictionary of Art & Artists**
Ian Chilvers

Based on the highly praised *Oxford Dictionary of Art*, over 2,500 up-to-date entries on painting, sculpture, and the graphic arts.

'the best and most inclusive single volume available, immensely useful and very well written'

Marina Vaizey, *Sunday Times*

**The Concise Oxford Dictionary of Art Terms**
Michael Clarke

Written by the Director of the National Gallery of Scotland, over 1,800 entries cover periods, styles, materials, techniques, and foreign terms.

**A Dictionary of Architecture**
James Stevens Curl

Over 5,000 entries and 250 illustrations cover all periods of Western architectural history.

'splendid ... you can't have a more concise, entertaining, and informative guide to the words of architecture'

*Architectural Review*

'excellent, and amazing value for money ... by far the best thing of its kind'

Professor David Walker

# OXFORD